AF669730

Physics of Galaxies

Observations and Investigation of Galactic Disks

Physics of Galaxies

Observations and Investigation of Galactic Disks

A.M. Fridman
A.V. Khoperskov

New Delhi | Mumbai | Chennai | Kolkata | Bengaluru | Hyderabad | Kochi | Guwahati

First Indian Edition 2014

VIVA BOOKS PRIVATE LIMITED

- 4737/23 Ansari Road, Daryaganj, New Delhi 110 002
  E-mail: vivadelhi@vivagroupindia.net, Tel. 42242200
- 76, Service Industries, Shirvane, Sector 1, Nerul, Navi Mumbai 400 706
  E-mail: vivamumbai@vivagroupindia.net, Tel. 27721273, 27721274
- Megh Towers, Old No. 307, New No. 165, Poonamallee High Road, Maduravoyal
  Chennai - 600 095 E-mail: vivachennai@vivagroupindia.net, Tel. 23780991, 23780992
- B-103, Jindal Towers, 21/1A/3 Darga Road, Kolkata 700 017
  E-mail: vivakolkata@vivagroupindia.net, Tel. 22836381, 22816713
- 7, Sovereign Park Apartments, 56-58, K. R. Road, Basavanagudi, Bengaluru 560 004
  E-mail: vivabangalore@vivagroupindia.net, Tel. 26607409, 26607410
- 101-102 Mughal Marc Apartments, 3-4-637 to 641, Narayanguda, Hyderabad 500 029
  E-mail: vivahyderabad@vivagroupindia.net, Tel. 27564481, 27564482
- Beevi Towers, First Floor, SRM Road, Kaloor, Kochi 682 018, Kerala
  E-mail: vivakochi@vivagroupindia.net, Tel: 0484-2403055, 2403056
- 232, GNB Road, Beside UCO Bank, Silpukhuri, Guwahati 781 003
  E-mail: vivaguwahati@vivagroupindia.net, Tel: 0361-2666386

www.vivagroupindia.com

Published by arrangement with
Cambridge International Science Publishing
7 Meadow Walk, Great Abington,
Cambridge CB21 6AZ, UK

ISBN : 978-81-309-2675-9

Published by Vinod Vasishtha for Viva Books Private Limited, 4737/23, Ansari Road, Daryaganj, New Delhi - 110 002. Printed and bound by Saku Printers Pvt. Ltd., New Delhi.

The authors

Alexey Maksimovich Fridman was a Russian physicist specializing in astrophysics, physics of gravitating systems and plasma physics. He discovered new types of instabilities in gravitating media, created the theory of planetary rings and predicted the existence of small Uranus satellites that were later discovered. He also developed the hydrodynamic theory of the spiral structure in galaxies. A. M. Fridman predicted the existence of the Alfven's solutions, and worked out the stability theory of thermonuclear plasma at finite high pressure. He discovered new types of non-Jeans instabilities in gravitating media, some of which develop at timescales less than Jeans' timescale, while others exist in systems that are stable according to Jeans.

Alexander Valentinovich Khoperskov is the Head of the Department of Information Systems and Computering Modelling at the Volgograd State University, Russia. His interests include mathematical modelling, gas dynamics, ecology, astrophysics

Contents

Preface

As a result of the development of new observational methods in the last 1015 years it has become possible to determine the distribution of the kinematic parameters (the velocity and dispersion of velocity of the stars and gas) in the disk for hundreds of galaxies, and these data are related to the regions far away from the centre. This plays a key role in the problems of constructing realistic models of the galaxies taking into account dark matter in the halo. It should be mentioned that up to the middle of the 90s of the 20th century, the number of the galaxies with the known radial profiles of the dispersion of velocity along the visual ray outside the limits of the bulge was less than 10. This information is of considerable importance for determination of the mass of the disk and the halo, the mass–luminosity ratio and for understanding the phenomenon of many large-scale structures (spiral pattern, bars, rings, etc).

The approach to studying the galaxies, based on the construction of the dynamic models of stellar, gas and stellar–gas gravitating systems has become generally accepted. As regards the capabilities and accuracy, these models become closer and closer to the full-size experiments. In the so-called models of *N*-bodies, the number of particles rapidly approaches the number of stars in the galaxies and is already greater than the number of the stars in large spherical clusters and dwarf of galaxies ($N \sim 10^6$–10^7 or greater).

Special attention in the monograph is paid to the advanced numerical models of describing the dynamics of galaxies with both stellar and gas components. In particular, this is the subject of chapter 4 in which the algorithms of constructing numerical models are described in considerable detail and a number of the most important results are presented. In addition to this, the material in other chapters is based to a large extent on dynamic modelling. In particular, we should mention the chapters concerned with our Galaxy and the problem of dark matter. Unfortunately, the last fundamental monograph in the Russian language dealing with the numerical models for modelling of the galaxies (*N*-body) was the translated book R.W. Hockney and J.W. Eastwood, Computer Simulation Using Particles (the original book was published in 1981).

The book includes a number of new data obtained in recent years, in particular the material relating to the recently discovered new structures in the spiral galaxies: giant anticyclones, cyclones and slow bars. New methods for the detection these structures are described. The role of vertical motion in the disk systems and the turbulence theory are discussed in detail.

Chapter 1 presents the main information of the observational nature, providing information on different systems containing disks. The extensive and exhausting presentation of this material is not the task here. The main task is to provide information on the main facts which, on the one hand, can be used to construct theoretical models and, on the other hand, stimulate the desire to understand and clarify the observations. We do not pretend that the literature cited here is complete. The task of the first chapter is to provide information on the orders of the physical quantities and the main structural special features of the actual objects so that in constructing the theoretical models we do not deviate greatly from the actual situation.

The main object of investigations in chapter 2 is the stellar disk which is in most cases the brightest subsystem of spiral galaxies with impressive structural features. Section 2.1 describes the relatively simple self-consistent model of a heterogeneous differentially rotating hot stellar disk of a finite thickness. In section 2.2 we derive an equation describing, within the framework of the constructed model, the dispersion properties of the non-axisymmetric perturbations with a small amplitude in the plane of the stellar disk and investigate branches of its oscillations. The effect of different factors on the stability of the gravitating disks is investigated on the qualitative level in section 2.3. The following section determined the condition of stability of the stellar disk taking into account the poor stabilisation of non-axisymmetric perturbations in comparison with axially symmetric ones. Finally, section 2.2 deals with the investigation of the stability of the stellar disk in relation to bending perturbations and formulation of the appropriate stability conditions. In this chapter, the results of theoretical predictions are compared with the observational data. The same chapter in section 2.5 discusses the methods of dynamics of collisionless systems based on the application of the action–angle variables and the possibility of development of loss cone instability in a gravitating disc (this section was kindly written by V.L. Polyachenko, E.V. Polyachenko and I.G. Shukhman). It should also be mentioned that the questions of the origin of the spiral pattern of the galaxies are not discussed in this chapter because they are considered in other chapters.

The models, described in chapter 3, are used mostly for gas disks of flat galaxies. However, in a number of cases the area of application of the investigated models is outside the framework of the physics of galaxies and can be used efficiently for protostellar discs, gas subsystems in close binary stars and active galactic nuclei, protoplanetary disks and circumplanetary rings. These objects can be described by the gas dynamics equations and, from this viewpoint, if we do not consider the differences in the spatial and time scales, we are faced with very similar problems with similar solutions. Because of this, we tried, where possible, to obtain universal results relating to any gas disk. Naturally, this is not always possible. In particular, this chapter includes the results relating to the large-scale processes in the gas disks of the flat galaxies.

In addition to this, chapter 3 is concerned with the collective processes, whose characteristic timescales are of the order of or greater than the revolution period of the disk, and the characteristic spatial scales in its symmetry plane are considerably greater than the thickness of the gas subsystems. We discuss the conditions of equilibrium of the gas disk and explain the conditions of applicability of the model of the thin disk for describing collective processes in the gas subsystems of the galaxies and accretion disks. Both the dynamics of non-axisymmetric perturbations in the gas disk and, in particular, perturbations of the type of Rossby waves, and the gravitational stability of the gas disk are discussed. The properties of gradient–entropy instability are described and the restrictions for the values of the parameters of the gas subsystems of the flat galaxies in accretion disks, determined by these instabilities, are determined. The dissipative effects and, in particular, the fast dissipative instability of the gas disk and the fluctuation parameters in it are investigated in section 3.9. Special attention is given to the properties of the main instabilities, determined by the differential nature of the rotation of the gas disks; the class of the rotation curves, permitting buildup of these instabilities is outlined, and the structure of their eigenfunctions is determined. In addition to this, the large-scale shockwaves are considered. This chapter describes in detail the derivation of the system of exact two-dimensional equations from initial three-dimensional equations.

The main results obtained by numerical experiments are discussed in chapter 4. The method for constructing the numerical dynamic models of the gas and collisionless stellar disks is described in detail. We examine the problems of formation of the stellar bars (including binary ones), gravitational stability of the stellar disk, the formation of lopsided galaxies, and the mechanisms of bending instabilities.

Analysis the possibilities of development of hydrodynamic instabilities in a rotating gas and their astrophysical applications are the subject of chapter 5.

The chapter 6 describes the results of laboratory studies in Spiral equipment in application to galactic disks. The gravitational–hydrodynamic concept of the generation of the spiral structure is discussed in detail.

Chapter 6 describes briefly three main results of full-size experiments in 'shallow water' carried out at the Kurchatov Institute, Russian Scientific Centre [163, 183, 184, 241, 588, 966, 967]:

1. The possibility of generation in galactic disks of a spiral structure in the non-linear stage of development of centrifugal instability;

2. Formation of large-scale vortices (cyclones and anticyclones) which are a compulsory element of the united spiral-vortex structure in the galaxies;

3. The instability of super-reflection in Spiral equipment is demonstrated. It is studied up to the formation of a non-linear quasi-stationary structure.

The large-scale wave galactic structures are investigated in chapter 7. Recent discoveries of new galactic structures are described here. Special attention is given to the detection of large-scale vortex structures in the galaxies, slow bars, mass detection of double-hump rotation curves, vertical movement in gas galactic disks. This has been possible as a result of the development of the methods of restoration of the three-dimensional field of velocities of the gas using the observational data. For these discoveries, one of the authors (A.M. Fridman) received the State Prize of the Russian Federation in the area of science and technology in 2003 and 2008. Section 7.8 compares the differences between the floccular galaxies and grand design spirals on the basis of the results of Fourier analysis of brightness charts and the velocity field along the visual ray for the floccular and regular spiral galaxies.

Chapter 8 deals with several problems of the dynamics of disks taking into account non-linear factors. In particular, the chapter describes the results of discovery of the structure of oscillations of the type of the envelope soliton in the spiral arm of the NGC1365 galaxies. Special features of shockwaves in the galaxies are also considered.

Chapter 9 deals exclusively with our Galaxy (Milky Way). The numerical dynamic models are an important tool in these investigations. The dynamic role of different components: bulge, bar, halo, spirals, is discussed. Restrictions were obtained for one of the most important parameters of the Galaxy – the surface density of the disk subsystem in the solar neighbourhood $\sigma_{\odot}$. The mass of the dark matter of the halo is estimated. Section 9.1 describes the reason why the mini-spiral in the centre of the Galaxy is the result of instability of super-reflection. Section 9.5 gives the review of the main problems of chemical evolution and was written by B.M. Shustov and D.Z. Vibe.

According to the cosmological data, the fraction of dark matter (dark matter, dark mass, hidden mass) is approximately 30% of the total energy/mass which in turn is almost 10 times greater than the contribution of the conventional visible (baryonic) matter which is concentrated mostly in the galaxies where it is in the form of stars of different types and gas, and also in the inter-galactic gas. We should mention that the 'Problem of dark matter (dark matter) and its detection has been included in V.L. Ginzburg's list of 'Very important and interesting problems' in recent years.

The nature of dark matter is not known, evidently, it consists of non-relativistic weakly interacting particles whose mass is hundreds of times greater than the proton mass. It is important to note that the dark matter exerts a gravitational effect on the surrounding visible matter (stars, galaxies, galaxy clusters), its morphology and kinematics so that we can evaluate not only the mass but also the nature of distribution of dark matter on the scales of tens of kiloparsec and even in the range 1–10 kpc.

Chapter 10 discusses a number of observational data for the galaxies which can be used to obtain the estimates of the densities of dark matter directly inside the galaxy (in the range of the optical radius $r > R_{max}$) and in its immediate vicinity $r \leqslant (2–3)R_{max}$. The results of numerical modelling of the dynamics of the gas and stellar disks, immersed in the triaxial massive dark halo, are presented.

The final chapter is concerned with the dynamics of accretion disks. In particular, detailed analysis of the large-scale processes in the accretion disks of the system of binary stars has made it possible to obtain for the first time the coefficient of turbulent viscosity from the first principles and the value of the parameter $\alpha = 0.035$, which is in agreement with the currently available observational and theoretical estimates.

The authors are very grateful to all co-authors, especially Prof. A.V. Zasov.

1

Observed properties of disk galaxies

1.1. Advances in the investigation of galaxies

1.1.1. The main stages of investigation of the galaxies

The first reports on the nebulosity of the Andromeda relate to the 10th century and are associated with the name Al Sufi who stressed that this object differs from the stars because of its blurred form. In 1612, S. Marius observed the nebulosity of the Andromeda in a telescope. Thomas Wright in his study 'An Original Theory of the Universe' (1750), explaining the observed Milky Way, assumed that the stars are distributed in a thick disk – this was the first step in understanding that the Milky Way is a galaxy. Immanuel Kant, impressed by this concept, said that the Milky Way, rotating in accordance with the Newton's law, is only one of the many 'Island Universes', scattered in the infinite Universe.

An important stage is associated with the compilation of the catalogue by Charles Messier in 1784 in which, as explained at a considerably later date, there were 39 galaxies developed amongst 108 objects. In 1864, John Gerschel (the son of William Kershaw) published the General Catalogue of Nebulae and Clusters of Stars which became the basis of the New General Catalogue.

However, the modern extragalactic astronoimy was born in the 20's of the 20th century by the confirmation of the extragalactic nature of the nebulosity of the Andromeda and, subsequently, of other spiral nebulae in studies of B. Lindblad, K. Lundmark, P. Doig, H. Shapley, E. Epik, E. Hubble, J. Oort, and others. In these years, the concept of the general structure of our galaxy was also developed. Evaluations of the dimensions of the systems were obtained, the centre of the Milky Way was determined, the spherical and flat subsystems were discovered, and the rotation of the Galaxy was also confirmed.

This was accompanied by the discovery of the general expansion of the Universe reflected in the recession (red shift) of the galaxies. This marks the beginning of the scientific stage of extragalactic astronomy together with the start of observatory cosmology. The Hubble law for the galactic recession

$$V_{sys} = H \cdot r \tag{1.1}$$

links the systematic velocity of the galaxy V_{sys} and the distance from it r, being the most important method of evaluating the extragalactic distances. The determination of the Hubble constant H, km/(s·Mpc) started with the initial value of $H = 500$ km/(s·Mpc) (1929) to the current estimates of $H = 70$–75 km/(s·Mpc).

Already in the period of the establishment of extragalactic astronomy, the number of the galaxies which could be observed was estimated at 120 000. The NGC catalogue (New General Catalogue, 1888–908), together with the additions and IC (Index Catalogue) includes more than 13 000 objects (the catalogue NGC 2000.0 is the current variant of the classic catalogues NGC and the Index Catalogues (IC)). The UGC catalogue (Uppsala General Catalogue, 1973) is the almost complete catalogue of the galaxies with the size not smaller than 1′ and/or the visible star value not lower than 14.5^m at the Palomar Observatory, and contains approximately 13 000 objects. The MCG catalogue (Morphological Catalogue of Galaxies, 1962–1974), includes 32 000 galaxies brighter than 15.1^m. In the group of the latest large catalogues, it is important to mention the Catalogue of Bright Galaxies (RC3), 1991, Southern Galaxy Catalogue). Even larger are the catalogues and databases IRAS, PGC, 2MASS. The 90s of the 20nd century mark the start of the era of electronic catalogues, for example, the HyperLeda database contains different information on more than 2 million galaxies.

The current estimate of the total number of galaxies is $\sim 10^{12}$. Note that the Hubble Space Telescope in the area of 1 square degree is able to resolve up to ~400 000 galaxies. Modern sky surveys (DSS, 2MASS, 2dF, SDSS, etc.) provide information on millions of galaxies [216].

In the 40's of the XXth century a new period of rapid development of astronomy started, inextricably linked to scientific and technological revolution. There has been an invasion of new methods of astronomy research. The radio band became available. By the beginning of the 70's astronomical observations covered the entire wave range. The diameter of the collecting surface D of modern optical telescopes on the Earth is greater than about 10 m, and that of the radiotelescopes – tens and hundreds of meters. Radio interferometers have been constructed, uniting dozens of radio telescopes. Astronomical instruments were launched into space, employing infrared, ultraviolet, X-ray and gamma-ray observatories. Thus, the main progress of extragalactic astronomy is related to the complex study of galaxies with virtually the entire electromagnetic spectrum. Research into neutrino and gravitational-wave astronomy started.

The achievements of the past 15 years have been largely based on the work of the Hubble Space Telescope (optical range and adjacent ranges of IR and UV), the orbital space telescope Galex (UV range 135–280 nm) and the infrared space telescope, Spitzer (3–180 μm) of the Chandra Space X-ray Observatory (0.07–10 keV). New opportunities have emerged with the beginning of operation of large ground-based telescopes in the optical and submillimetre bands (Keck I, II, $D = 10$ m; VLT, 4 telescopes with $D = 8.2$ m, the Gemini observatory with two 8-meter telescopes; Subaru, $D = 8.3$ m; LBT – binocular with 2×8.4 m; HET, spherical multi-

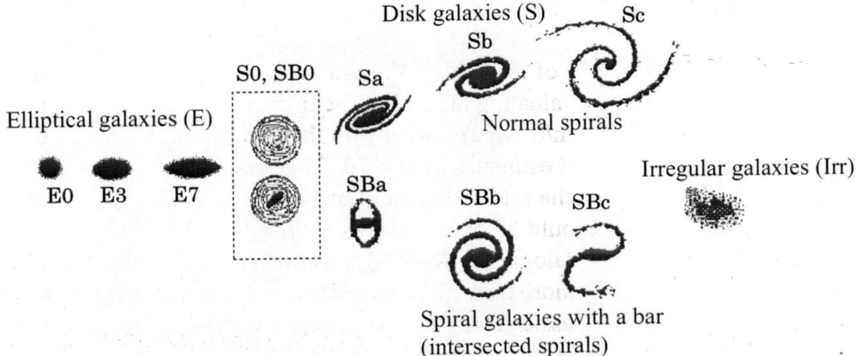

Fig. 1.1. Classification of galaxies based on the Hubble tuning fork diagram. S0 galaxie have a flat component with no spiral branches.

segment mirror with $D = 11$ m; SALT, $D = 10$ m; MMT, $D = 6.5$ m; the Observatory Magellan, two telescopes with $D = 6.5$ m; GTC, $D = 10.4$ m). The use of active optics significantly enhances the possibility of astronomical observations. The largest optical tool in Europe is the telescope at the Special Astrophysical Observatory of the Russian Academy of Sciences ($D = 6$ m).

The radio range is important for extragalactic astronomy. In particular, it serves as a source of information on the gas component of the galaxies. The use of radio interferometers with very long baselines has significantly improved the resolution, surpassing the optical telescopes. The largest radio telescopes include: VLA (New Mexico), RATAN-600 (Zelenchukskaya), DKR-1000 (Pushchino), GBT (Virginia)MERLIN (UK), UTR-2 (Khar'kov), Westerbork (Holland), ANTF (Australia), etc. The progress of observational astronomy is due to a large extent to qualitative changes in the technologies of information processing associated with extensive use of computers and numerical methods in all phases of research. There has been a full transition to digital representation of images utilize not only the capacities of computer technology, but also increase the access to the raw oservational data for subsequent treatment. The complex nature of studies of galaxies is based on the joint application of the photometric and spectroscopic methods in different regions of the electromagnetic spectrum.

1.1.2. Morphological types

Hubble's classification of 1925 (amended in 1936 by introduction of a new class of lenticular galaxies, S0) is constructed on the morphological princiuple [695], based directly on the observed shape of the object (Fig. 1.1). Numerous attempts to produce own classification schemes, or modify Hubble's chart in the middle of the XX century have not been generally accepted.

We mention de Vaucouleurs' classification, in which it is attempted to describe all forms of spiral galaxies, highlighting the three classes (E, S0-S, I), three families of spirals (SA – no bars, SB – with a bar, SAB – intermediate structures with

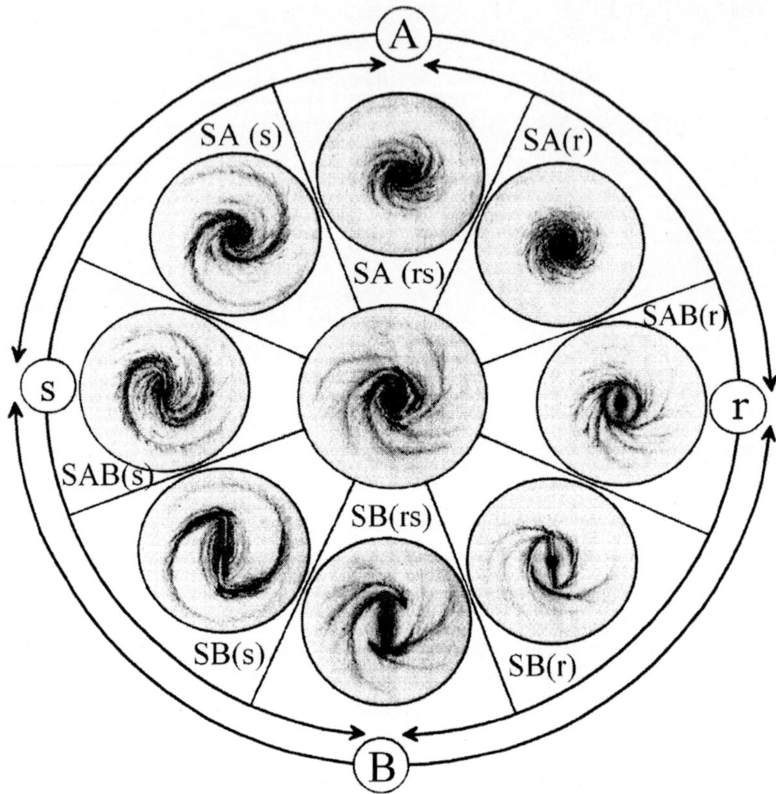

Fig. 1.2. de Vaucouleurs' classification of galaxies [433].

indistinctive oval structures) and three kinds: (r) – a galaxy with an inner ring, (s) – spirals, starting from the nucleus or the ends of the bar, (rs) – intermediate. The symbol r indicates the presence of the outer ring. The classification is complemented by later galaxies than Sc, for which the types Sd (SBd) and Sm (SB (s), m) were introduced. The latter characterize the galaxies from spiral to irregular, so-called Magellanic type – prototype is the Large Magellanic Cloud. Finally, the galaxies in which no structure can be allocated, are denoted by de Vaucouleurs by IAm, IBm, Im. Additional subclasses have also been introduced for lenticular galaxies: $S0^-$, $S0^0$, $S0^+$, S0/a. Compact objects (elliptical and irregular) can be characterized by symbol c: cE and cI, respectively. The proposed classification is three-dimensional (Fig. 1.2).

B.A. Vorontsov-Velyaminov, in describing the objects in MGC gave up trying to use the classification, used the explicit description of the most characteristic structural elements (nucleus, lens, bar, shape distortion, dust bandes, the number of branches, rings, gamma-shape, tails and jumpers in clearly interacting systems, the nature of spin and the width of the branches, etc.).

Table 1.1. Relationship between the morphological type of a galaxy and the value of parameter t which in dependence on the morphological properties of the galaxies can have values $-6 \leqslant t \leqslant 11$

Type	t	Type	t
cE	–6	Sb	3
E	–5	Sbc	4
E^{+}	–4	Sc	5
$S0^{-}$	–3	Scd	6
$S0^{0}$	–2	Sd	7
$S0^{+}$	–1	Sdm	8
S0/a	0	Sm	9
Sa	1	Im	10
Sab	2	cI	11

However, the Hubble diagram is currently the working tool in the physics of galaxies, not only because of its simplicity and visibility. Spiral galaxies are divided into subtypes on the basis of the characteristics of the spiral arms (structure, degree of twist) and the ratio of luminosities of the stellar disk and central star thickening (the bulge). However, the change of the external form has deep physical roots, because the most important physical characteristics change systematically along the Hubble tuning fork chart.

Here are characteristic average changes in the properties of galaxies during movement along the morphological sequence (left to right in Fig. 1.1).

1. Increasing the degree of oblateness a/b. The role of the spheroidal subsystem (lens, bulge) decreases.
2. The degree of twisting of the spirals decreases (spirals become more open)
3. The relative fraction of the mass attributable to the gas increases.
4. The colour indices change from red to blue.
5. The rate of star formation per unit of luminosity increases.
6. The dust content increases.
7. The maximum speed of rotation decreases from S0-Sa to Irr.
8. The mass decreases during the transition from spiral to irregular galaxies.

It should be emphasized that all trends are of the statistical character.

In conclusion, let us discuss the evolutionary status of the charts such as the Hubble or de Vaucouleurs. E. Hubble originally proposed an evolutionary scheme of transition of each galaxy from the E-type to irregular; this is embedded in the terminology: the early and later morphological types of galaxies. Later, in the middle of the XX century, the classification lost its evolutionary status; it was thought that the morphological properties of galaxies do not change qualitatively. In recent decades, attitudes have changed again. Although the full return to evolutionary thinking of the Hubble diagram has not happened, we can point to the possibility of changing the type of galaxies in the process of evolution. In particular, we note the following.

- Lenticular galaxies (S0, SB0) may be the result of development of S-galaxies associated with the depletion or loss of gas. When the density is below a certain level ($\lesssim$1%), star formation is almost stopped and there are no conditions for maintaining strong spiral density waves. Without the gas the amplitude of the spirals in the old star disk of lenticular galaxies usually does not exceed a few percent. Causes of depletion of gas reserves in this case are not important. This may be due to a higher rate of star formation (flares of star formation) or the sweeping of gas from the galaxy due to gravitational interaction with other galaxies (tidal interaction), while maintaining the stellar disk on the whole. Note that even the presence of a bar in lenticular galaxies (SB) did not lead to any substantial spiral structure in the stellar disk.

- Collision of spiral galaxies of comparable masses can be accompanied by the formation of elliptical galaxies (S + S → E and S + E → E). The common point of view is that the formation of giant elliptical galaxies (cD) is associated with this phenomenon of mutual absorption of galaxies in which the gas is often lost, partially passes into stars. The total angular momentum of the new object is reduced, and the temperature of the system associated with the chaotic component of the stellar velocity increases. The largest E-galaxies found in the centres of rich clusters (e.e. NGC 4881 in Coma or M87 in Virgo). There are observations of systems that are in various stages of such mutual absorption (e.g. NGC 5128 contains a gas-dust band and an atypical content of young blue stars; NGC 1700 is also at an early stage of formation of elliptical galaxies by the scheme S + E). With a large probability, after of a few billion years our Galaxy the Andromeda Nebula (M31) will interact in accordance with the scenario S + S → E. Star–gas disks are observed inside some elliptical galaxies.
- Galactic bars 'appear before our eyes'. The evolutionary status of stellar bars is indicate by the increased frequency of occurrence of SB-galaxies in the case of strong interactions, for example, in rich clusters. We should observe an increase in the relative proportion of galaxies with a bar with the aging of the Universe (with decreasing z). Therefore, the transitions S → SB are apparently common. Reverse transformation (SB → S) is also possible, but it requires the destruction of the bar which is quite a stable dynamic entity. But the mechanisms of its dissipation, for example, in the presence of a massive central massive black hole or a concentrated bulge, are known.
- Late acquisitions are such important morphological features as polar and inner rings.
- The basis of changes in our views on the evolution of galaxies is the realization that stellar systems are open and strong gravitational interactions between galaxies, accretion of intergalactic gas, the absorption of dwarf galaxies (small merging) are the most important factors of evolution, substantially affecting their morphological characteristics.

In conclusion, it should be mentioned that the classic classification of galaxies described here (extended Hubble sequence) is not complete. It does not include the giant cD-galaxies, a miniature 'zoo' of dwarf galaxies (there are several types of them), LSB-galaxies (galaxies with low luminosity), and several others.

1.2. Stellar disks of flat galaxies

Using Hubble's classification, most of the observed galaxies could be classified as flat[1]. These galaxies have a bright and a rather thin rotating disk with certain structural features (spirals, jumpers, rings, etc.) and a spheroidal subsystem. The disk subsystem usually consists of a rather massive stellar disk and much less massive gas. Here we give an overview of the main characteristics of the subsystems of flat galaxies necessary for further discussion. More detailed observations of flat galaxies can be found in books by B.A. Vorontsov-Vel'yaminov [37], V.G. Gorbatskii [41], I.D. Karachentsev [85], A.V. Zasov [61], A.V. Zasov and K.A. Postnov [72], Yu.N. Efremov [52–54], P. Hodge [251], V.P. Reshetnikov [215] (see also the review by A.V. Zasov [58]), and for the Galaxy – the book by L.S. Marochnikand A.A. Suchkov [131]. Note also [30, 43, 61, 94, 203, 366, 367, 372, 590, 1207, 1300].

1.2.1. Distribution of surface brightness in disk galaxies

The brightness distribution of $\mu(r)$ along the radial coordinate averaged over the azimuthal direction in the stellar disks of galaxies is generally well described by an exponential law

$$\mu(r) / \mu(0) \simeq \exp(-r / L_\sigma), \tag{1.2}$$

where the characteristic radial scale L_σ in each galaxy is similar in different spectral ranges [558]. It is usually assumed that the distribution of the surface density in the stellar disks $\sigma_*(r)$ is also exponential ($\sigma \propto \mu$):

$$\sigma_*(r) = \sigma_0 \exp(-r / L_\sigma), \tag{1.3}$$

where $\sigma_0 = \sigma_*(0)$ and $L_\sigma \simeq 6$ kpc [559, 733, 800]. In particular, in the Galaxy, according to different authors, $L_\sigma \simeq 2$ and 4 kpc and in the solar neighborhood the value of the surface density of gravitating mass (star + gas + dark matter inside $|z| \lesssim 1$ kpc, the stellar component dominates) is ~$40 \div 100\ M_\odot/\text{pc}^2$ [2] [260, 449, 495, 508, 561, 563, 604, 646, 763, 792, 2074, 1102, 1259] the values of $L_\sigma = 3$ kpc, $\sigma_\odot \simeq 50 \div 60\ M_\odot/\text{pc}^2$, are more likely see chapter 9).

According to the observations in the photometric bands B and K the radial scale of the disks L_σ is weakly dependent on the morphological type of S-galaxies [733]. But among the galaxies of more early types ($t = 2 \div 5$) more common objects have the scale of $L_\sigma > 6$ kpc. More significantly, the very value of L_σ is systematically dependent on the spectral range. In the transition from B to K or I the radial scale decreases: $\langle L_\sigma(B) / L_\sigma(K) \rangle = 1.2 \pm 0.2$ for galaxies with a small inclination angle [732], $\langle L_\sigma(B) / L_\sigma(K) \rangle = 1.56 \pm 0.46$, $\langle L_\sigma(B) / L_\sigma(I) \rangle = 1.32 \pm 0.24$ for galaxies with large inclination angle $i \simeq 90^\circ$ [637]. Moreover, there is a systematic dependence of $\langle L_\sigma \rangle(B) / L_\sigma(K, I)$ on the morphological type of galaxy t: this ratio increases with increasing t. These results are related to the effects of absorption, depending on the

[1]We will use the term 'flat' as a synonym of 'disk', regardless of the orientation of the galaxy.

[2]The sign $\odot$ corresponds to the vicinity of the Sun in the Galaxy.

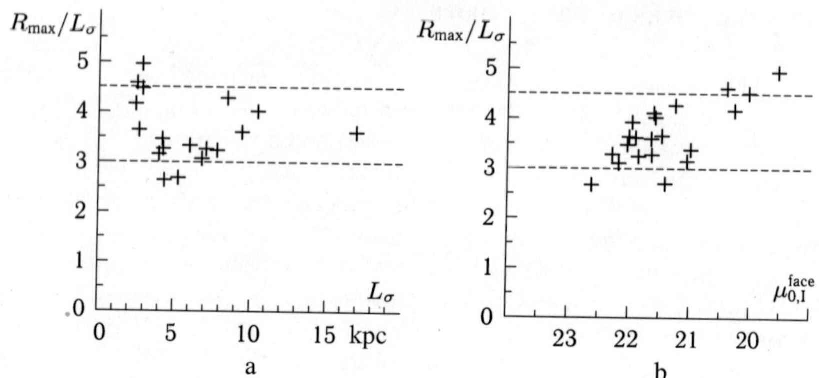

Fig. 1.3. Dependence of the ratio R_{max}/L_σ on the radial scale of the disk (a) and central surface brightness in the band *I*, corrected, taking into account the slope (b)) (+ – [798], for this sample $R_{max}/L_\sigma = 4 \pm 1.1$). The position of galaxies (bullet), for which the models are constructed in this book.

content of the dust component in galaxies of different types and changes of the relative share of brightness attributable to the young stars in various distances.

The exponential nature of disks is easily found in, for example, the model of galaxy formation from a uniformly rotating spherical system with its compression with the conservation of angular momentum [800, 907]. There are other scenarios of the evolution of galaxies, taking into account the viscosity of the gas and star formation, leading to the law (1.3) [664, 843].

Deviations from the law (1.2) usually occur in both the central parts of galaxy size $r \lesssim L_\sigma$ [558] and their periphery. The first are due, apparently, to the presence of either a bright bulge or a lens in the central regions of flat galaxies [431, 558] or a reduction in the density in the central part of the stellar disk [17, 57] [3].

According to the photometric data for the galaxies (most distinctive for the galaxies visible edge on), the second deviation from the law (1.2) is that in the range $r \gtrsim (3\text{–}5)\, L_\sigma$ the brightness of the disk decreases at a considerably higher rate than that indicated by the law (1.2). At the same time, the stellar disks may have in fact a sufficiently sharp boundary R_{max} in the radial direction [798, 805]. The ratio R_{max}/L_σ depends on the central surface brightness of the disk (Fig. 1.3). There is a tendency in the galaxies with low surface brightness for breaks in the disks at small distances from the centre in the units of the scale of the disk.

It is justified to assumed that the outer boundary of the stellar disk of the Galaxy is positioned at a distance of 12–15 kpc from the centre [1074] and, at the same time, $R_{max}/L_\sigma \simeq 4\text{–}5$.

However, the rate of the break of the disk is not high. Outside the limits of the radius R_{max} the profile of brightness also decreases in accordance with exponential law but on a smaller radial scale. Thus, the disk of the galaxies consists of two

[3]The lack of matter in the central regions of the stellar disks in comparison with the law (1.3) (a hole or depression of density) is incorporated, for example, in models of the Galaxy [441, 526] and a number of others [17].

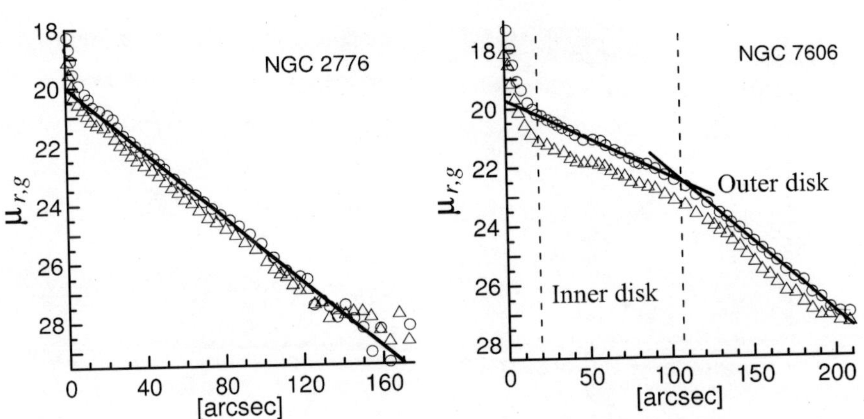

Fig. 1.4. Azimuth-averaged radial profiles of surface brightness in different regions ($\Delta - g'$, $o - r'$) [456]. In the NGC 7606 at the radius r = 106″ there are large changes in the brightness profile.

exponential disks with equal exponential scales L_σ (Fig. 1.4). The outer disk may extend to $2R_{max}$, as in the case of NGC 7606, and up to 27 stellar values the faster decrease of brightness in comparison with the inner disk remains unchanged. The reasons for the formation of these disks with two exponential profiles have not as yet been explained.

It should be mentioned that in the zone of the variation of the law of the radial profile of brightness (in Figs. 1.4, 1.5 this corresponds to the transition from one exponential law to another) the colour changes. At the same time, the value of the surface density/brightness ratio, σ/I (or mass/luminosity, $M/\mathcal{L}$) differs for regions of the disk with different exponential scales. If this effect is taken into account, then in transition from the distribution of brightness to the profile of surface density we obtain 'smoother' radial dependences $\sigma(r)$ and the distribution of mass follows more accurately the exponential law with a single radial scale in contrast to brightness.

However, in some galaxies it has not been possible to detect deviations from the exponential law with a single scale up to 29 stellar value/square second, as in the case of NGC 2776 (Fig. 1.4). The problem of explaining the special features of the distribution of density of the stellar disk in the outer regions is still open because in many cases the formation of stellar disks cannot be regarded as completed.

The masses of the stellar disks of the flat galaxies are in the range $10^9 \div 10^{12}\ M_\odot$ (this value for the Galaxy does not exceed $M_d \sim 10^{11}\ M_\odot$, section 9.2).

1.2.2. Vertical structure of stellar disks

The volume density of matter (stars and gas) $\rho_d(r, z)$ in the disks of the galaxies is maximum in their symmetry plane ($z = 0$). With increase of the distance from this plane, the reduction of the density of the stellar component of the disk $\rho_*(r, z)$, according to the distribution of brightness along the z-coordinate [636, 805], can often be described satisfactorily by the law:

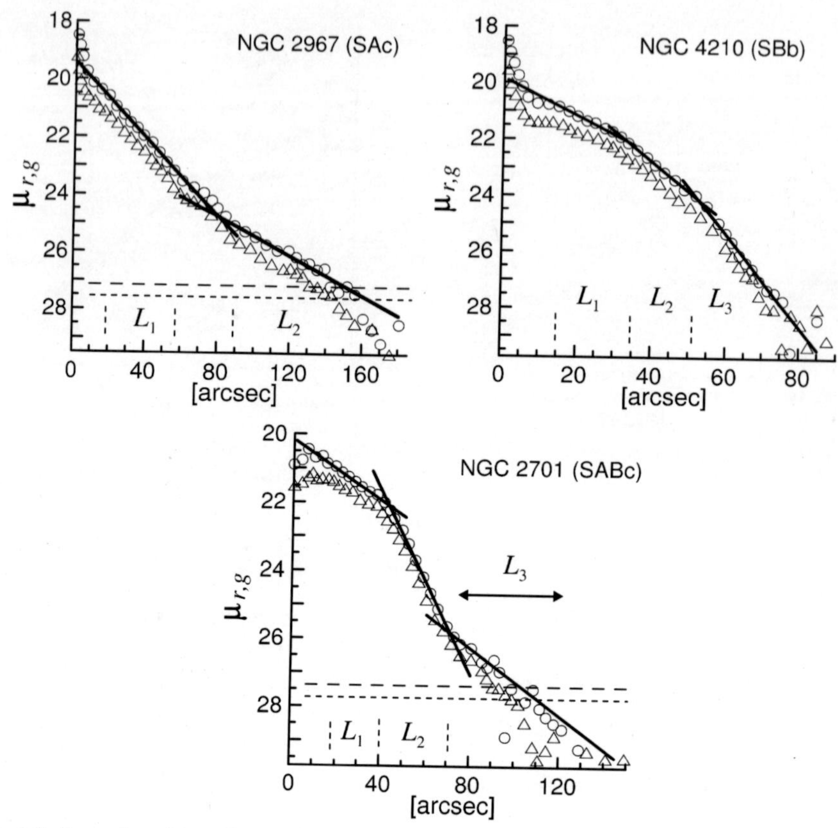

Fig. 1.5. Examples of less frequently observed profiles of surface density (in different ranges: $\Delta - g'$, $\circ - r'$): NGC 2967, NGC 4210, NGC 2701 [1033]. The regions of the disk characterised by the exponential profiles with different scales $L_{1,2,3}$ are clearly visible.

$$\rho_*(r,z) \propto \mathrm{ch}^{-2}(z/\Delta_*), \tag{1.4}$$

here Δ_* is the vertical scale (effective half thickness) of the stellar disk[4]. It should be mentioned that at this dependence $\rho_*(r, z)$ the surface density of the stellar disk $\sigma^*(r) = 2\Delta_*\rho_*(r, z = 0)$. Profile (1.4) is obtained for the self-gravitating layer with constant temperature along z (section 2.1.4).

Approximating distributions of the volume density along the z-coordinate, different from (1.4), are also used. In some galaxies in the vicinity of the plane $z = 0$ there is a surplus of brightness in comparison with the distribution (1.4) and for the approximation of the observed profiles we use the exponential law [636, 638, 1275]:

---

[4]Here we are discussing the mean value of Δ^* for all types of stars. The difference in the nature of distribution of the stars of different spectral classes along the disk plane was measured a long time ago by Bartaya [18].

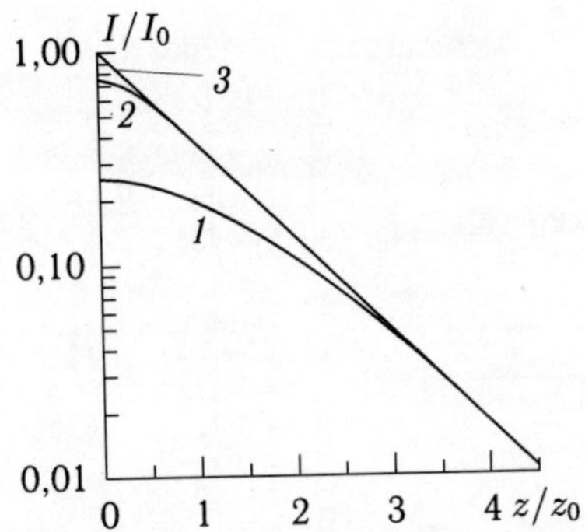

Fig. 1.6. Vertical profiles (1.6) at: 1) $n = 1$; 2) $n = 5$; 3) the exponential law.

$$I(z) = I_0 \cdot \exp(-|z|/h_z), \tag{1.5}$$

where h_z is the exponential vertical scale. The distributions (1.4) and (1.5) yielded the identical profile starting with $|z| \gtrsim \Delta_*$ at $\Delta_* = 2h_z$. The exponential law can be regarded as a typical distribution at higher altitudes, starting at $|z| \gtrsim \Delta_* = 2h_z$.

The generalisation of (1.4) and (1.5) was proposed for a more adequate description of the nature of brightness in the range $|z| \gtrsim \Delta_*$ [801]:

$$I(z) = \frac{I_0}{2^{2/n}} \operatorname{sech}^{2/n} \left\{ \frac{nz}{2z_0} \right\}, \tag{1.6}$$

where the case with $n = 1$ corresponds to (1.4), and for (1.5) it should be excepted that $n \to \infty$ (Fig. 1.6)

The vertical structure of the disks can be examined conveniently on the galaxies visible edge-on (see the catalogue of flat galaxies [745]). The value Δ_* in the majority of flat galaxies is evidently in the range 0.1 kpc $\lesssim \Delta_* \lesssim$1.0 kpc. In particular, the approximation of different data for $\rho_*(z)/\rho_*(0)$ in the vicinity of the Sun in the Galaxy by the function (1.4) leads to the estimate $\Delta_{*\odot} \simeq 0.2 \div 0.6$ kpc. The results obtained from the data in the reviews COBE/DIRBE and 2MASS lead to the vertical scale of the disk of the Galaxy in the vicinity of the Sun of approximately 0.28 kpc for the old stellar disk. A thick stellar disk with a large vertical scale is found but it provides a small contribution to the mass of the disk component. Similar results for the values of the vertical scale of the old stellar disk were obtained in the review RAVE: 225 $\pm$ 10 kpc [1263], and using these results for the thick disk we have 1048 $\pm$ 36 kpc.

In the galaxies of later morphological types, the vertical scale does not depend strongly on the radial component [639, 805, 1062]. In the galaxies $t \simeq 2$ in the presence of a massive bulge or lens the vertical scale often increases with the radius at the periphery [636, 638, 639] (Fig. 1.7). Evidently, this effect is also characteristic of our Galaxy [292].

There is also a thick disk characterised by the vertical scale ~1–5 kpc. Its volume density in the plane $z = 0$ is several percent of the density of the disk system (for example, in the solar neighbourhood it is approximately 2% [562]). It should

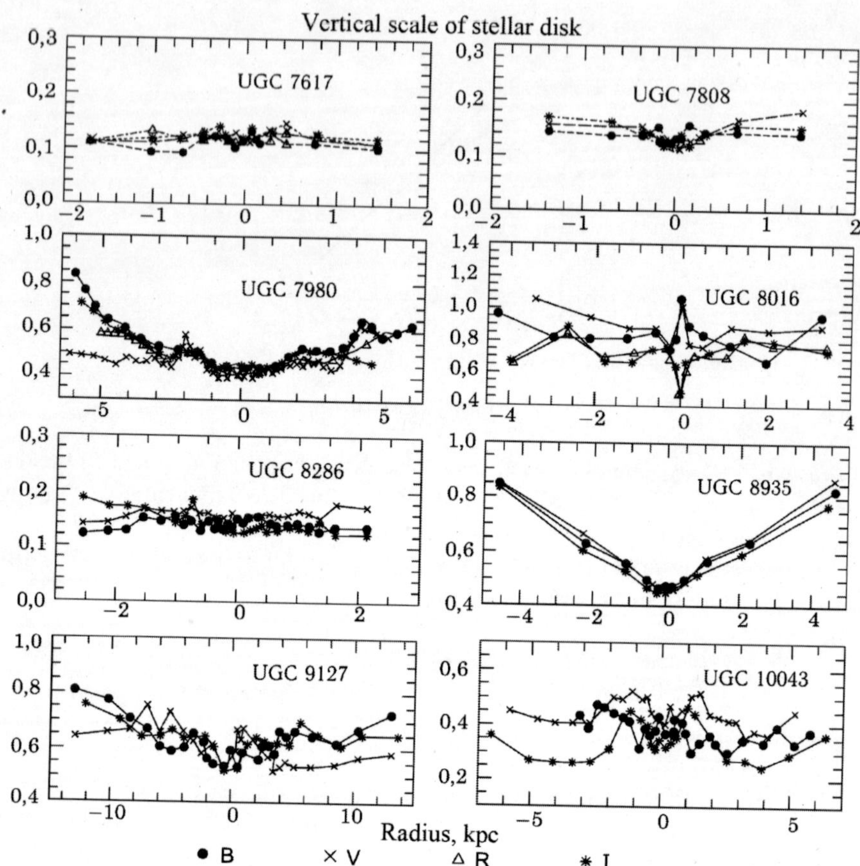

Fig. 1.7. Examples of the dependences of the vertical scale of the stellar disk on the distance to the centre in different bands for different galaxies indicated by the symbols •, ×, Δ,*[639]. There are two typical profiles of the disk: with a constant value of the scale and with thickening at the periphery.

be mentioned that the disk system is usually divided to a old disk (the stars of later classes – they provide the major contribution to the mass) and a flat subsystem which is the thinnest and consists of gas, dust and young stars.

The mean values of the ratio of the radial scale of the disk with the vertical scale are in the range 5–7 [636, 638, 639]. However, there are also relatively thick disks in which the vertical scale is 2–3 times smaller than the radial scale (as in UGC 8935). There is also a reversed situation in which this ratio of the scales is >10, and such galaxies are usually referred to as super thin (for example, UGC 7321).

Comparison of the characteristic scales of the stellar disks ($\Delta_* \ll L_\sigma$) shows clearly that the equilibrium gravitational potential $\Phi_0(r, z)$ changes far less frequently in the z-direction than in the radial direction. Therefore, in the vicinity of the plane $z = 0$ it can be naturally assumed that $|\partial^2\Phi_0 / \partial z^2| \gg |\partial(r\partial\Phi_0 / \partial r) / r\partial r|$ and, consequently, the Poisson equation can be written in the form $\partial^2\Phi_0 / \partial z^2 \simeq 4\pi G\rho_D$. From this, in the simplest approximation of the disk with a homogeneous thickness, we obtain the estimate of the frequency of oscillations of the star (or another object) across the disk plane

$$\omega_z \simeq \sqrt{4\pi G\rho_D(z=0)} = \sqrt{2\pi G\sigma_D / \Delta_*}. \tag{1.7}$$

1.2.3. Kinematics of the disks

The main movement of the objects, both stellar and gas disks of the galaxies, is their rotation. The chaotic motion of the stars and the gas clouds is usually characterised by considerably lower velocities and, in the first approximation, they can be ignored. In accordance with this, the centrifugal force per unit mass at the given distance from the disk centre which compensates the gradient of the general gravitational potential is similar both for the stars and gas clouds. However, there is also a small but systematically and principally important difference between the velocity of the stars V_* and the velocity of the gas V_{gas}, since the contribution of thermal motion to the radial balance of the forces for the dynamically hotter stellar component is greater than for the colder gas component.

The rotation curve of the gas, in any case at the periphery of the Galaxy ($r \gtrsim 2L_\sigma$), characterises more accurately the distribution of the gravitating mass. From this viewpoint, it is preferred to use the velocity of rotation $V_{rot}(r)$ with respect to the interstellar gas.

Relatively complete reviews of the rotation curves of a large number of spiral galaxies have already been published (see, for example [972, 1090, 1092, 1093, 1183, 1186]). There is a large number of data for the rotation curve of the Galaxy [19, 40, 48, 193, 214, 250, 297, 300, 340, 395, 1176). These results show that the rotation curves of the flat galaxies can in the first approximation be divided into two types which can then be applied equally to both the stellar and gas disks.

The first of these is characterised by the almost solid-state rotation of the central part of the disk ($V_{rot} \propto r$) with increase of the distance from the centre of the disk, transferring to the rotation curve of the 'plateau' type [$V_{rot}(r) \simeq$ const]. At the far periphery of such a disk the value V_{rot} may increase slowly, decrease or remain approximately constant (M33, NGC 157, 300, 681, 1055, 6503, etc). These rotation curves can be commercially referred to as 'single-hump'.

The second type of rotation curves (referred to as 'double-hump', Fig. 1.8) is characterised by the fact that in the central part of the disk the value $V_{rot}(r)$ rapidly increases in the range $0 < r \lesssim 0.3 \div 1$ kpc, and this is then followed by the range of decrease of V_{rot} and after this $V_{rot}(r)$ again (although not always) increases and reaches a plateau, as in the case of the galaxies with the single-hump rotation curves (Fig. 1.9). The entire internal 'hump' of the rotation curves of this type appears to be superposed on the section of the almost solid-state rotation of the galaxies with the single-hump curves (Figs. 1.8, 1.10). The maximum value V_{rot} on the inner hump usually differs only slightly from the level of V_{rot} on the plateau. This type probably also includes the rotation curves of the Galaxy ([250, 1176], section 9.1.1), galaxies M31 [1091], M81 [626], and many others.

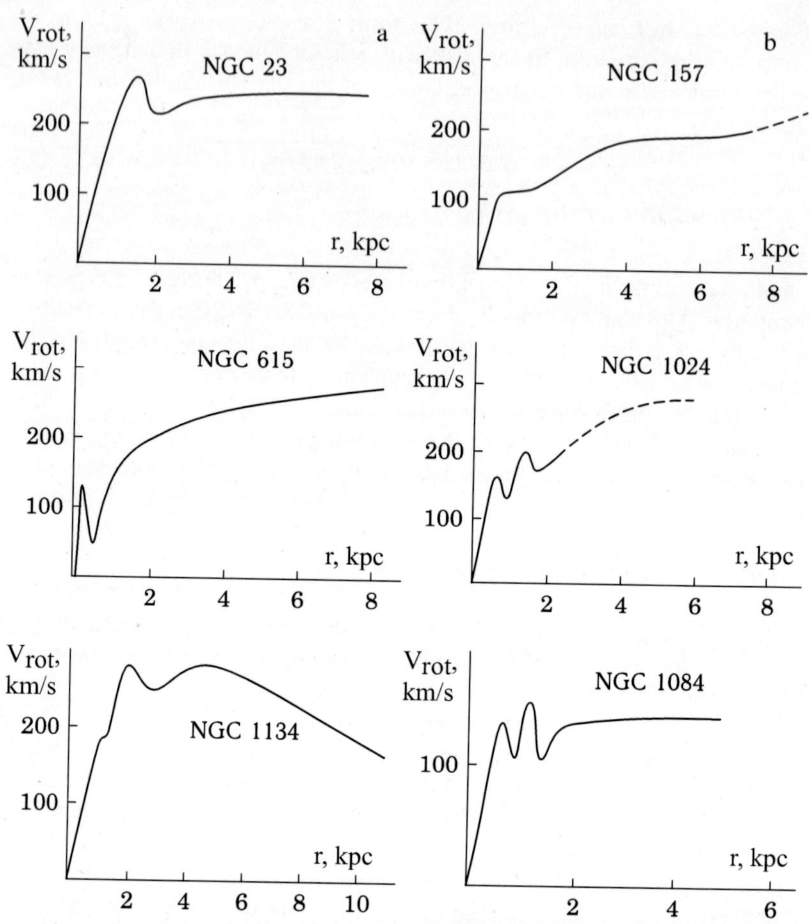

Fig. 1.8. Typical examples of the rotation curves obtained in the 6-m telescope (Special Astrophysical Observatory (SAO), Russian Academy of Sciences) in the 80s of the previous century.

Up to the middle of the 80s, almost all galaxies were regarded as having the single-hump rotation curve. However, with increase of the resolution power of the observations, the double-hump form of the rotation curves was detected in many cases, for example, in the galaxies NGC 1566, 2590, 2608, 2708, 3200, 4321, 1553, and others [191, 793, 1093, 1094]. In fact, it can be clearly seen that the interval between the disk centre and the inner hump of the rotation curve (0.5÷1 kpc) already at a distance greater than 10÷20 Mpc, is visible under the angle smaller than 10 angular seconds. For this reason, the detection of the inner hump on the rotation curves of not too close galaxies up to a certain time was a relatively complicated observational programme. In 1986–1992, using a 6-m telescope (SAO) by the initiative one of the authors (A.M. Fridman) experiments were carried out with an observational programme to investigate the rotation of the inner regions of the spiral galaxies [9–12, 73]. In the group of the investigated objects a large part was represented by the double-hump type with the amplitude of $\gtrsim$10–15%.

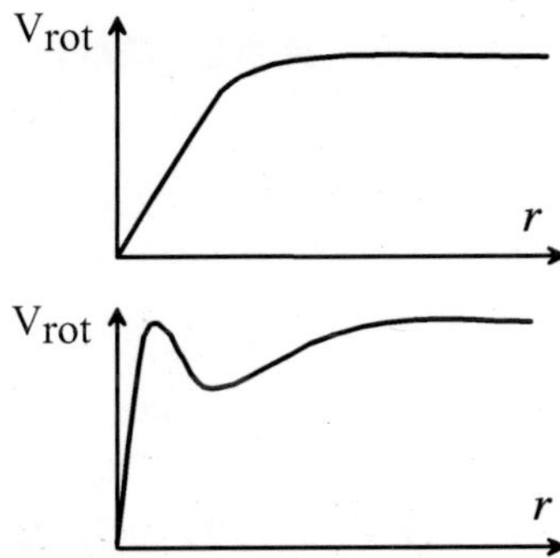

Fig. 1.9. Typical form of the 'single-hump' and 'double-hump' rotation curves.

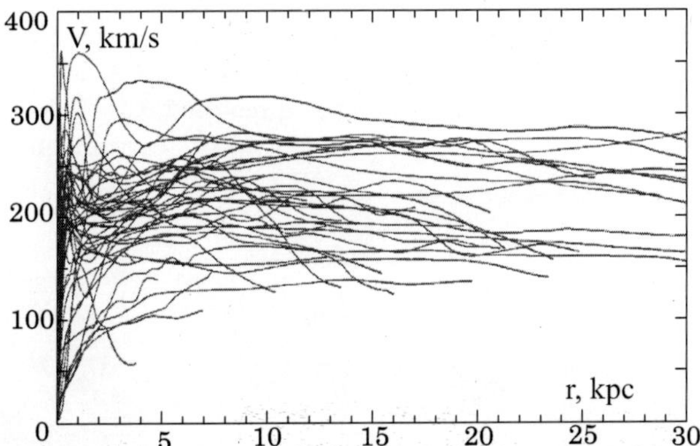

Fig. 1.10. Examples of rotation curves for a large number of galaxies [1186].

The reason for the formation of the double-hump rotation curves of the flat galaxies has not as yet been completely explained. However, we shall indicate two variants of explaining this phenomenon. Firstly, the concentration and, at the same time, massive central formation (bulge) with the parameters typical of the observed galaxies in not a very dense disk create in its inner part a gravitational potential gradient in which the value $V_{rot}(r)$ for a large part of the galaxies should decrease in the transition zone from the bulge to the disk [62]. Also, if such a bulge is flat (lens-like), V_{rot} may decrease at a rate higher than that indicated by the Kepler law ($V_{rot} \propto r^{-1/2}$). This possibility was mentioned for the first time in [168] and was discussed in detailed in the monograph by Polyachenko and Fridman [590]. The second variant was proposed by Zasov [57] (see also [17]) and may be described as follows – the dense stellar disk with a 'hole' in this central part, observed in certain galaxies, often permits the two-hump rotation curve to be explained even at a relatively moderate concentration of the bulge density in the centre.

It should be stressed that the central region may be characterised by relatively significant non-circular movements, for example, with the bar or a triaxial bulge leading to the formation of a hump not associated unambiguously with the distribution of the gravitating masses. Dynamic processes, associated with the activity of the nucleus, may also contribute to this.

The typical values of the maximum velocity of rotation of the galaxies are in the range 100÷300 km/s. For the Galaxy, this value is 220 ÷260 km/s (chapter 9).

The most important special feature of the large majority of the galaxies is that the rotation curve does not disappear at the remote periphery ($r \gg 2L_\sigma$), where the density of the disk matter decreases by an order of magnitude or more, for example, in the Galaxy there is no large reduction of the value $V_{rot}(r)$ up to 16 ÷ 18 kpc ([40, 131, 1080], section 9.1.1), and possibly up to ~60 kpc [297]. This observational fact may indicate the presence of a massive halo outside the limits of the optical radius of the Galaxy.

We now return to the problem of the deviations of motion of the objects of the disk from the purely circular trajectory, characterised by the orbit radius $r = r_0 =$ const and angular velocity $\Omega(r) = V_{rot}(r)/r$. Let the equilibrium gravitational potential, produced by the distribution of the entire matter of the Galaxy in its plane, be $\Phi_0(r)$, the orbital momentum of the object of the unit mass $\mathcal{L} = r^2\Omega(r)$. Therefore, the movement of such an object in the radial direction takes place in the effective gravitational potential $\Phi_{eff} = \Phi_0(\mathrm{r}) + \mathcal{L}^2/2r^2$ [113]. The corresponding equation of motion at $\mathcal{L} =$ const has the form

$$\ddot{r} = -\frac{\partial \Phi_{\text{eff}}}{\partial r} = -\frac{\partial \Phi_0}{\partial r} + \frac{\mathcal{L}^2}{r^3}.$$

However, at the minimum point $\Phi_{eff}(r)$ (circular orbit with $r = r_0$)

$$\left.\frac{\partial \Phi_0}{\partial r}\right|_{r=r_0} \equiv \frac{\mathcal{L}_0^2}{r_0^3} = r_0\,\Omega^2(r_0).$$

Therefore, at $|r - r_0| \ll r_0$, the equation of motion has the following form

$$\ddot{r} = -\left[4\Omega^2 + 2r\Omega\frac{d\Omega}{dr}\right]\cdot(r - r_0),$$

This shows that the movement of the investigated object in the radial direction at low $|r - r_0|$ is harmonic and characterised by the frequency

$$\varkappa = 2\Omega\sqrt{1 + \frac{r\,d\Omega}{2\Omega\,dr}}, \tag{1.8}$$

referred to as the epicyclic frequency. It should be mentioned that in the case of solid-state rotation $\varkappa = 2\Omega$, for the Kepler rotation the rotation frequency and the epicyclic frequency coincide, and in the sections of the rotation curves of the 'plateau' type $\varkappa \simeq \sqrt{2}\,\Omega$.

It is interesting to evaluate, at least in the first approximation, the ratio of the frequencies of oscillations of the star in the disk plane (1.8) and across its plane (1.7). Evidently, in the central part of the disk, where the distribution of matter with the bulge taken into account is close to spherical, these frequencies are comparable.

However, at the periphery of the galaxy this is not so. Actually, inside a sphere with a radius $R \simeq (4 \div 5) L_\sigma$ the mass of the exponential $[\sigma_*(r) = \sigma_0 \exp(-r/L_\sigma)]$ disk is equal to $M_d \simeq 2\pi \sigma_0 L_\sigma^2$ and, consequently, as regards the order of magnitude

$$\varkappa \sim \Omega \sim (M_d G / R^3)^{1/2} \simeq (2\pi G \sigma_0 L_\sigma^2 / R^3)^{1/2}$$

(in these estimates, it is assumed that the mass of the spheroidal subsystem is equal to or smaller than the disk mass). Consequently

$$\frac{\varkappa}{\omega_z} \simeq \left(\frac{\Delta_*}{L_\sigma}\right)^{1/2} \left(\frac{L_\sigma}{R}\right)^{3/2} \exp\left\{\frac{R}{2L_\sigma}\right\}\Bigg|_{R=2\div 5L_\sigma} \simeq \left(\frac{\Delta_*}{L_\sigma}\right)^{1/2} \ll 1.$$

In particular, according to the observational data in the solar neighbourhood in the Galaxy $\varkappa / \omega_z \sim 0.3$.

1.2.4. 'Temperature' of the stellar disk

The stars in the disks of the flat galaxies move with deviation from the circular orbits. In the galaxies without the bar[5] or way from it, these deviations are not large and maybe characterised by the presence in the star of the components v_r, v_φ, v_z of the peculiar (less rotational $V_{\text{rot}}(r)$) velocity. The distributions of the stars with respect to the residual velocities are described satisfactorily by the Schwarzschild distribution function (Maxwell distributions with respect to each component of the velocities with different 'temperatures'):

$$f(v_r, v_\varphi, v_z) \propto \exp\left\{-\frac{v_r^2}{2c_r^2} - \frac{v_\varphi^2}{2c_\varphi^2} - \frac{v_z^2}{2c_z^2}\right\}, \tag{1.9}$$

in which the dispersions of the velocities c_r, c_φ, c_z are functions of the coordinate r. If the collisionless stellar disk is sufficiently cold, i.e. $c_r \ll$ V, we have the following theoretical relationship between the dispersion of the radial velocities and the dispersion of the azimuthal velocities of the stars:

$$\frac{c_r}{c_\varphi} = \frac{2\Omega(r)}{\varkappa(r)}, \tag{1.10}$$

which is determined only by the rotation curve $V(r) = r\Omega$.

In the solar vicinity of the Galaxy for the oldest stellar population of the disk $(c_r, c_\varphi, c_z) = (38, 24, 20)$ km/s [507]. We have the ratio of the values $c_r{:}c_\varphi{:}c_z$ = 1:0.63:0.33. Earlier investigations by Wielen [1301, 1302], Ogorodnikov and Osipkov [188] resulted in higher values of the dispersion of the velocities: $(c_r, c_\varphi, c_z) = (48, 29, 25)$ km/s. The observed relationship between c_r and c_φ (see also [89, 229]) is close to that resulting from the equilibrium condition of the disk: $c_\varphi/c_r = \varkappa / 2$ (see chapter 2). The value of the ratio c_z/c_r depends on the age of the

[5]The bar is the elongated oval-like formation in the centre of some (SB) galaxies (section 1.4).

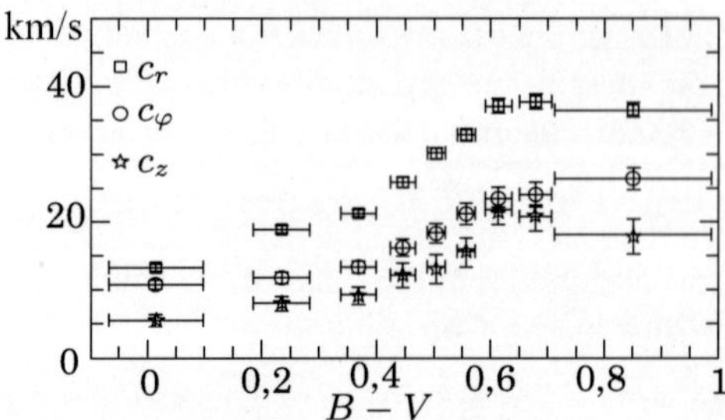

Fig. 1.11. Dependence of the component of the dispersions of the velocities of the stars in the vicinity of the Sun according to the HIPPARCS data on the colour index *B–V* which on average is greater for older stars [507].

stars [229]. For the oldest stars (with the age not shorter than $9{\cdot}10^9$ years) of the disk of the Galaxy $c_z/c_r \simeq 0.4$ [1301]. The observational data for the young stars give $c_z/c_r \simeq 0.7 \div 0.85$ [90–92]. There is a relatively thick stellar disk characterised by high values of the dispersion of the velocities of the stars.

The HIPPARCS data show convincingly the increase of all three components of the dispersion of the velocities (c_r, c_φ, c_z) with increase of the colour index *B–V* [373, 507], indicating the evolution of the random component of the velocities in the stellar disk (Fig. 1.11). Heating of the stellar component is observed up to the values $(B\text{–}V) \simeq 0.6$ (Fig. 1.12). In the stars with the higher value of (*B–V*) (older stars) no further heating was detected. A small reduction of the dispersion of the velocities of the stars for $(B\text{–}V) > 0.6$ was associated with the special features of the chemical composition. Figure 1.13 shows the heating effect of the stellar disk of the Galaxy with time.

The ratios c_z/c_r and c_φ/c_r characterise the anisotropic nature of the dispersion of the velocities (the temperature of the stellar disk) and are the most important characteristics for constructing the realistic models of the galaxies (Fig. 1.14). The value c_z/c_r increases with evolution of the stellar disk. The fact that c_φ/c_r is close to unity for the youngest objects ($B - V \lesssim 0.1$) may indicate the initial conditions of formation of the stars in the gas clouds where the anisotropy of the dispersion of the velocities is insignificant. In less than a period of rotation of the disk in the young stars the ratio c_φ/c_r decreases as a result of the collisionless nature to the value determined by (1.10).

The anisotropy of the distribution function (1.9) stresses the collisionless nature of the stellar disks of the galaxies [266] and, consequently, dictates the method of describing their dynamics using the collisionless kinetic equation.

As mentioned previously, the velocity of rotation of the matter of the disk is considerably higher than the residual velocities of the stars characterised by the dispersions c_r, c_φ, c_z (for example, in the solar neighbourhood $V_{\text{rot}} \simeq 200$–$250$ km/s). As a result of this fact, we can use in subsequent considerations the epicyclic approximation (mainly with respect to the small parameter $\zeta = c_r/V_{\text{rot}}$).

In the last 10–15 years it has been possible to determine the radial profiles of the dispersion of the velocities of the stars for a large number of galaxies (see [77, 78,

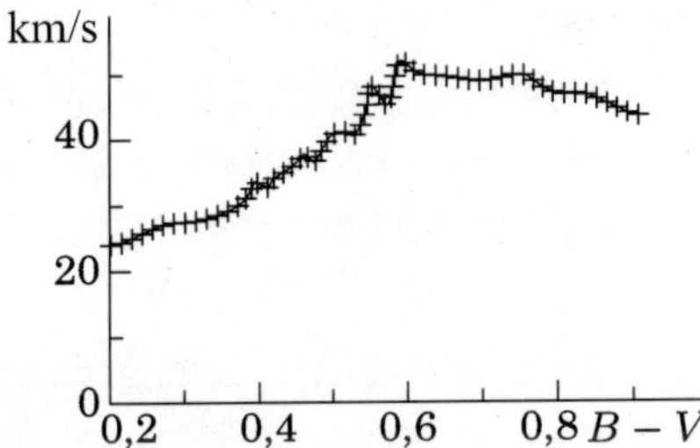

Fig. 1.12. Dependence of the total dispersion of the velocities of the stars in the vicinity of the Sun according to the HIPPARCS data on the colour index $B–V$ indicates the increase to 50 km/s after $B–V \gtrsim 0.6$ with subsequent tendency for a small reduction [373].

1327] and references therein, project SAURON [1021], see chapter 4, 10). The dispersion of the velocities of the stars c_{obs}, observed along the visual ray, is determined in the general case by all three quantities c_r, c_φ, c_z and geometrical factors (the angle of inclination of the Galaxy i). Only for the galaxy with $i = 0°$ we have $c_z = c_{obs}$. In the general case, the following equation is fulfilled

$$c_{obs} = \sqrt{c_z^2 \cos^2(i) + c_\varphi^2 \sin^2(i)\cos^2(\alpha) + c_r^2 \sin^2(i)\sin^2(\alpha)}, \qquad (1.11)$$

where α is the angle between the direction of the slit and the major axis measured in the projection on the plane of the Galaxy. In the case of galaxies visible under a high angle of inclination ($85° \lesssim i \leqslant 90°$), it is necessary to take into account the integral contribution to the radiation at different distances along the visual ray, and equation (1.11) cannot be used (see section 2.7).

On the whole, the value of the dispersion c_r decreases with increase of the distance from the centre and in almost all cases the reduction of dispersion with increasing r is slower than for the surface density of the stellar disk [137, 258, 413, 1021] (SAURON project). With increase from the centre the ratio $\zeta = c_r / V_{rot}$ usually decreases. For the spiral galaxies in the range $r \gtrsim L_\sigma$ the typical values of the parameter equal 0.1÷0.3 [59, 77, 78, 409, 411, 793, 794].

To conclude this section, it should be mentioned that we can find weak disks in the equatorial plane of some E-galaxies. The luminosity of these disks is several percent of the total luminosity, and they are relatively thick [453, 467, 1085, 1130, 1171].

1.2.5. Chemical composition of the stellar component (D.Z. Vibe, B.M. Shustov)

The elemental composition of galactic stellar populations is a reflection of its chemical and dynamical evolution. Pregalactic matter composed of hydrogen and helium with small admixture of lithium. Then, in the process of stellar evolution synthesized heavier elements, are ejected into the interstellar evolved star environment and become raw material for star formation future generations.

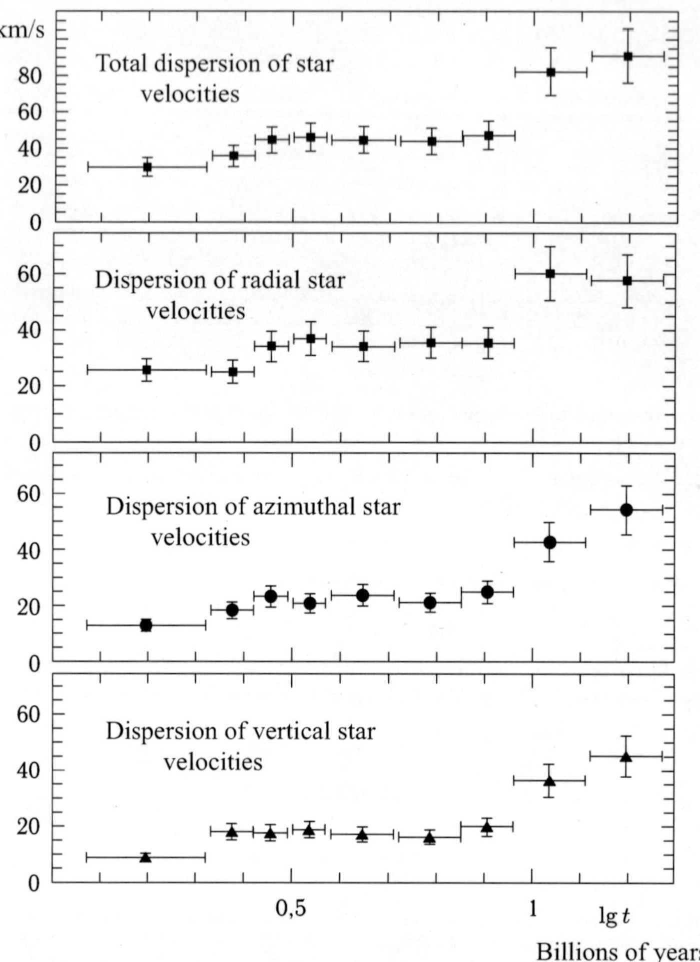

Fig. 1.13. Different components of the dispersion of the velocities of the star in the vicinity of the Sun in our Galaxy for different ages of stars according to the HIPPARCS data [1048]. Heating of the disk (increase of the velocity dispersion) is shown for the young stars with the age of $\lesssim$ 3 billion years. The dispersion of the velocities of the older objects remains almost completely constant over $(3\div9)\cdot 10^9$ years. These stars form an old stellar disk and provide the main contribution to the mass of the disk component. The oldest objects with the age of more than 10 billion years have the dispersion of the velocity approximately twice as large but they belong to the thick stellar disk.

The chemical composition of stars and the interstellar medium is usually characterized by the term 'metallicity', which implies the general content of elements heavier than helium, i.e. products of fusion and nuclear reactions that take place mainly in the stars. The chemical composition of stars is determined by spectral observations. Since this method cannot determine the content of all the elements, the numerical characteristics of metallicity is often represented separately by the content of iron, oxygen, etc. Quantitatively, the value of metallicity is expressed

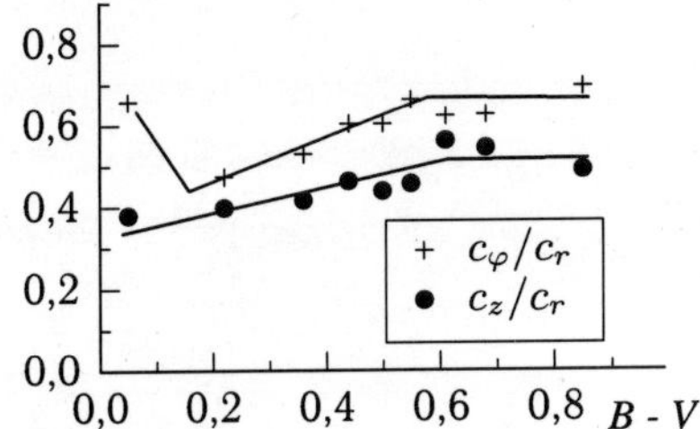

Fig. 1.14. Observed values of c_φ/c_r (+), c_z/c_r (•) for the stars with different colour indices B–V [507]. The lines show the approximation of the data.

$$[\mathrm{M}/\mathrm{H}] = \lg(N_{\mathrm{M}}/N_{\mathrm{H}}) - \lg(N_{\mathrm{M}}/N_{\mathrm{H}})_{\odot}, \tag{1.12}$$

where N_{M} is the number of atoms of element M, and N_{H} is the number of hydrogen atoms. As usually, index $\odot$ marks the relevant characteristics of the Sun. It should, of course, be considered that the methods of spectroscopy determine the chemical composition of the stellar atmosphere, which may not correspond to the chemical the composition of stars in general.

During the evolution of the Galaxy th heavy elements should gradually accumulate in it, so one of the important areas of study of the chemical evolution of the Milky Way is the search for the 'age–metallicity' relationships for the objects of the Galaxy. Observations indicate that the 'smooth' dependence of the chemical composition on time does not exist in the Galaxy. Rather, it is possible to speak of the difference between the average metallicity for three different kinematic galactic populations – a thin disk, a thick disk and a halo [714]. The average metallicity of halo stars and globular star clusters (GSC) is about –1.5 (in units of [Fe/H]). In other words, the stars of the halo contain on average 30 times less heavy elements than the Sun. Thin disk stars (in the solar neighborhood) have a mean metallicity of the order of –0.2, i.e. the metallicity of the Sun is slightly higher than the local average. Finally, the metallicity of the thick disk is approximately –0.6.

Because of the above components of the Galaxy the halo is the oldest and the thin disk the youngest, we can say that on the whole in the Galaxy the metallicity does depend on the age. However, in each of these subsystems presence of the 'internal' relation of age and metallicity is in doubt. However, the population of globular clusters is divided into two subsystems [406]: the old clusters with the age of the order of the age of the Galaxy (or rather, the age of the Galaxy is the age of globular clusters!) and low metallicity, reaching –3, and younger ones (by about 3 billion years) and metallic clusters may be related to the thick disk. Formation of the bulk of the old clusters occurred probably during a single episode of star formation [1066], so that simply there were no preconditions for the emergence of the relationship between age and metallicity. The chemical composition of field stars may have a more complex distribution associated with the history of absorption of galaxies–satellites by the Milky Way (see, e.g. [451]), but in this case, the scatter of the metallicity values is not associated with age, but with different properties of the

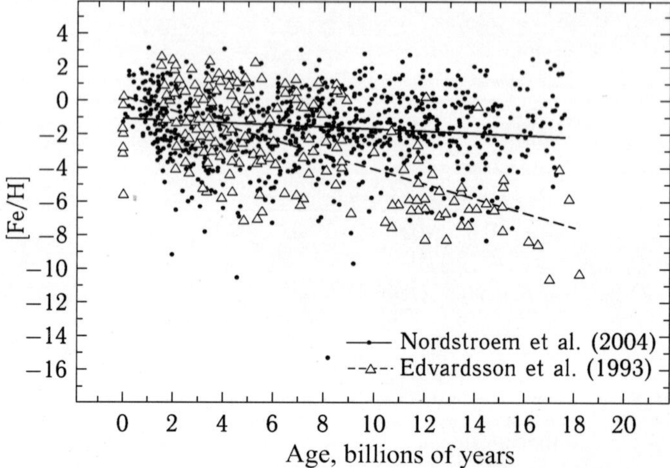

Fig. 1.15. The 'age–metallicity' relationship according to the of data from two different directories. Triangles – Edvardsoon's catalogue [524]; points – Nordstroem et al catalogue [976]. The straight lines show the results of linear regression for each of the directories.

absorbed stellar systems. It is interesting to note that the dependenceof the metallicity on the age can be observed in the most massive globular clusters [1204].

In the thick and thin disks the 'age–metallicity' ratio can theoretically exist, but there is as yet no unambiguous observational evidence in favour of its reality (cf. [452] and [686]). In general, it can be argued that a drop in metallicity with age seems to occur, but its value is negligible and is accompanied by a noticeable spread in the metallicity for stars of a certain age. A stronger dependence of metallicity on age, according to some studies, may be largely due to the effects of selection [686] (Fig. 1.15).

An exception from the general trend is another important component of the Galaxy with a specific chemical composition – the bulge. At the age probably close to the age of the thick disk, some bulge stars have the metallicity three times higher than solar [313].

In addition to the difference between the metallicity of subsystems of different ages, the chemical composition of the Galaxy has another important feature of – a radial gradient of the negative content of various elements. In the halo, the gradient, determined by the metallicity of the globular clusters is not particularly pronounced and may be associated with the division of GSC into the cluster of the halo and the thick disk [1013]. However, in a thin (and a thick) disk the reality of the gradient is not in doubt, despite the many open questions.

In the disk of our Galaxy the gradient is observed for almost all the most abundant elements (C, N, O, Ne, S, Fe, Ar, Al). Its value for the objects with the age of 10 to 10^4 million years(scattered clusters, regions of HII, planetary nebulae, B-stars) within a galactocentric distance of $R \sim 5$–15 kpc is approximately the same, namely –0.05÷0.08 dex/kpc [1097]. However, the uncertainty, even for a single element and one type of object is very high. For example, estimates of the gradient of oxygen in the HII regions ranged from –0.13 [1020] to –0.04 [504]. According to [750], in B-stars, the relative oxygen content in the same range of galactocentric distances practically does not change (the gradient is equal to 0).

Nevertheless, we can consider it established that the content of the most of the heavy elements in the central region of the Galaxy is 2–5times higher than their content on its periphery [302]. The proximity of the gradients determined from the young and old objects, indicates that the significant dependence of the gradient of age does not exist, at least for most of the life of the Galaxy. There may be a vertical gradient of metallicity [1193], but it may partly be a reflection of the mentioned differences in the chemical composition of the various subsystems of the Galaxy with different height scales. Finally, another important feature of the chemical composition of the Galaxy – the difference in the relative contents of elements between the halo and thick and thin disks (see Figure 1.16).

Most notable among these differences is the relative overabundance in the halo and bulge stars of the so-called α-elements, characterised by their relative content relative to iron, i.e. the value

$$[\alpha / \mathrm{Fe}] = \lg(N_\alpha / N_{\mathrm{Fe}}) - \lg(N_\alpha / N_{\mathrm{Fe}})_\odot . \tag{1.13}$$

The α-elements are elements the nuclei of which contain an equal even number of protons and neutrons (i.e. an integer number of α-particles). It is these elements that are the main products of stellar fusion. In the old stars, their relative content is about 3 times higher than that of the Sun [1295]. There are also differences in the relative contents of various elements between the stars of the thin and thick disks [133, 361].

Recently, much attention has been paid to the so-called stars with the extremely low metallicity ([Fe/H] < –3), which should be the oldest objects in the Galaxy. They are characterised by unusual values of the relative content of most elements and the large scatter of these contents. It is expected that here we are dealing with

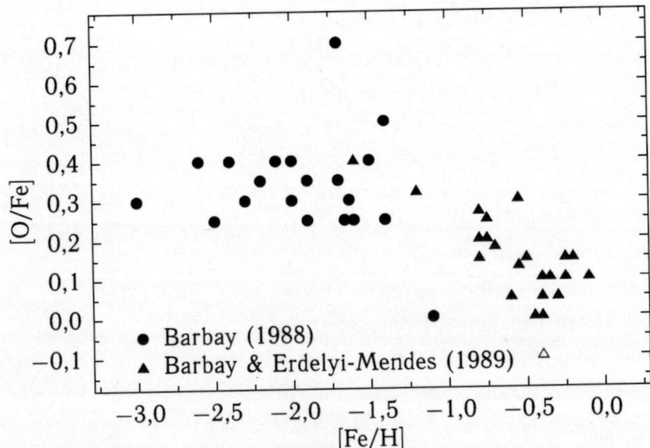

Fig. 1.16. The dependence of the relative content of oxygen to the iron content in the stars of the halo and the disk according to [333] and [334].

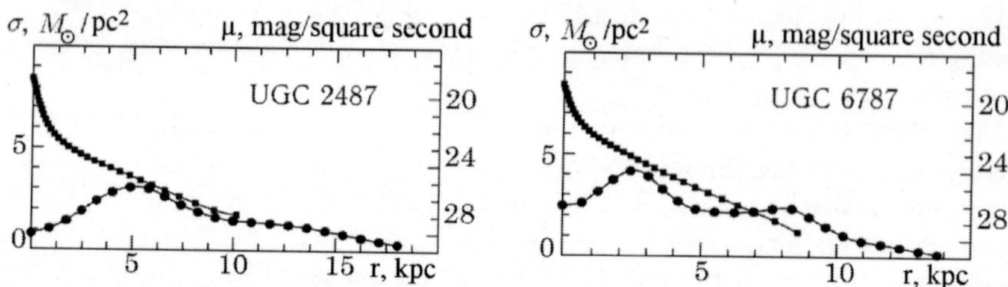

Fig. 1.17. Radial profiles of the stellar and gas components for the galaxies UGC 2487 and UGC 6787 [975]: μ(■) is the surface brightness at the band B, σ(•) is the surface density of the gas.

objects that, due to insufficient mixing in the early stages of the evolution of the Galaxy, are enriched by single supernova explosion products [898].

1.3. Gas components of disk galaxies

1.3.1. Distribution of the surface density of the gas

Gas in flat galaxies is observed mainly in the form of a thin disk, the plane of symmetry ofn which coincides with the plane of symmetry of the stellar disk[6]. The main components of the mass of the gas disk are atomic and molecular hydrogen and helium. Until recently, due to difficulties in assessing the overall mass of the disk the total fraction of gas in the galaxies was systematically underestimated. Thus, in the 60s it was assumed that the contribution of the surface density of gas in the galactic disk does not exceed 2÷5%, but then this estimate (with H_2 and helium taken into account and with the reduced estimates of the density of stars) increased to 10÷15% [1112]. The relative fraction of gas increases when going from galaxies of early morphological types (S0, Sa) to late-type objects (Sd).

The distribution of the surface density of gas in the disk $\sigma_{gas}(r)$ averaged in the azimuthal direction is usually quite different from the distribution of the surface density of stellar disks, especially in the central regions of galaxies (see reviews [58, 61] and studies cited therein). Figure 1.17 shows the typical distribution of the brightness of the gaseous and stellar components. In contrast to the stars, the profile for gas is almost always non-monotonic. There is often a situation where there is a pronounced maximum at a certain distance from the centre. A more complex patter, as in UGC 6787, is also found. But the central reduction of the density of the gaseous component is a common phenomenon (see the region $r \lesssim 2$ kpc in Fig. 1.17), particularly in galaxies with a bright bulge.

As an example, we describe in greater detail the distribution $\sigma_{gas}(r)$ in the Galaxy [1109, 1111, 1112]. In the central part of the disk of the Galaxy ($r \lesssim 1$ kpc), the surface density of the gas disk (consisting of molecular hydrogen) reaches several hundreds of $M_\odot/\text{pc}^2$ and rapidly decreases in the direction to the range $r \simeq 2 \div 4$ kpc, where $\sigma_g \simeq 5 \div 8\ M_\odot/\text{pc}^2$ (Fig. 1.18). This is followed by a wide and relatively dense

[6]At the periphery of flat galaxies outside their the spiral pattern the gas layer is often distorted, deviating from the plane of symmetry of the stellar disk at a distance of several kiloparsecs.

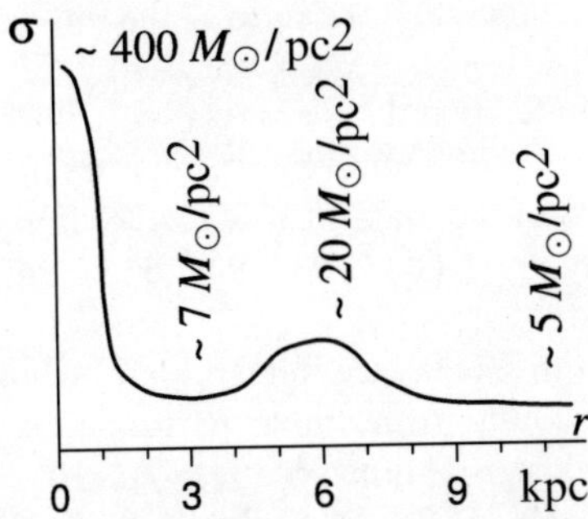

Fig. 1.18. Schematic radial profile of the density of the gas disk of our Galaxy ($M_\odot$ is the mass of the Sun).

($\sigma_{gas} \simeq 20 \div 30\ M_\odot/pc^2$) molecular ring in the range $r \simeq 5 \div 8$ kpc, and outside this range (actually, starting at $r \simeq 6$ kpc) $\sigma_{gas} \propto \exp(-r/L_g)$, where $L_g \simeq 4$ kpc ($\simeq L_\sigma$). In the range 12 kpc $\lesssim r \lesssim$ 15 kpc, the value $\sigma_{gas}(r)$ remains almost completely constant ($\simeq 5 \div 6\ M_\odot/pc^2$). The identical distribution $\sigma_{gas}(r)$ takes place in M 81 [1086]. It should be mentioned that the identical molecular rings are also detected in other spiral galaxies at a distance of 3÷8 kpc from the centre [499, 532].

Giant molecular clouds

It is also important to mention the fact that a large fraction of the mass of the gas disk of the Galaxy (according to different estimates 20÷90%) is present in the clouds [1188]. The mass of the individual clouds may reach the values of the order $(2 \div 5) \cdot 10^6\ M_\odot$, and their diameters up to 80÷100 pc. The concentration of molecular hydrogen is $n \simeq 10^3$ cm^{-3}, and the gas temperature only $T \sim 10 \div 100$ K. The mean values of the diameter and mass are equal to respectively $d \simeq 30 \div 40$ pc and $M \simeq 5 \cdot 10^5\ M_\odot$, and their total amount in the Galaxy is estimated at 4000. These objects are referred to as giant molecular clouds (GMC). It is not justified to assume that our Galaxy is unique in this respect; for example, M31 is known to contain a GMC with the size of 160 pc and a mass of $1.3 \times 10^6\ M_\odot$ [457].

The total mass of the GMC in the ring $2 < r < 10$ kpc is, according to the data in [1187] (see also references therein to other studies) $(0.7 \div 3.9) \cdot 10^9\ M_\odot$. The giant molecular clouds are concentrated preferentially within the ring 4 kpc $\lesssim r \lesssim$ 8 kpc. However, the problem of the total amount of the molecular gas in the Galaxy has not as yet been completely solved because there is a number of difficulties in interpreting the observations. Since the clouds are cold (10÷100 K), the hydrogen molecules can not be detected directly. The transitions between the adjacent rotation levels of energy in the hydrogen molecule are forbidden, and for the molecule to receive the energy corresponding to the difference between two rotation levels, the gas is not sufficiently hot. Therefore, the main source of information are the molecules of carbon oxide for which the transitions between the adjacent rotation levels are permitted ($\lambda_{CO} = 1.3$; 2.6 mm).

Evidently, it can be assumed that the ratio of the mass of the gas in the Galaxy to the mass of the stellar disk is ~10–15%. Considering the observations of other flat galaxies, it may be concluded that there are relatively different situations: according to the data in [1023] the mass of neutral hydrogen is approximately 20% of the mass of NGC 289 and only 2% of the mass of NGC 7496. It is justified to assume that in certain galaxies of later types (NGC 7793, 6946, 5457, 2403), the fraction of the gas may reach ~30% of the disk mass [317].

The outer region of the Galaxy contains clouds whose size and mass are not inferior to those of the clouds in the inner part of the Galaxy. The total mass of molecular gas at $r > 11$ kpc is estimated at $(1 \div 7) \cdot 10^8 M_\odot$ [514]. The GMCs are found in the nearest galaxies: M 31, M 33, M 64, M 81, M 83, NGC 1569, Magellanic Clouds, etc.

1.3.2. Multicomponent nature of the gas medium

In addition to the GMC, the gas forms small molecular clouds ($d \sim 10$ pc), diffusion clouds of neutral hydrogen HI ($n \simeq 40$ cm^{-3}, $T \simeq 50 \div 100$ K), inter-cloud gas HI $(T \simeq 7 \div 9) \cdot 10^3$ K, the zones of ionised hydrogen (zones HII) and the regions of the rarefied hot gas (coronal gas, $T \sim 10^6$ K).

Since a large part of the mass of the gas is in the clouds, it is natural to regard them as unique macromolecules and determine the effective 'temperature' of the gas disk by the dispersion of the velocities of the gas clouds. The latter, according to the observations in the Galaxy, equals 5÷7 km/s [321, 600, 1112, 1205]. This result characterises the uniform dispersion of the velocities $c_g^{(1)}$. At the same time, the general distribution of the gas clouds with respect to the velocity is practically isotropic – correspondingly, the three-dimensional dispersion of the velocities $c_g^{(3)} \simeq 9 \div 12$ km/s. The latter circumstance indicates the collision nature of the gas disk consisting of the 'macromolecules' – the gas clouds and, consequently, the need to describe its dynamics by a system of gas dynamic equations. It is important to stress the collision nature of the movement of the atoms of the interstellar gas, even if the gas is not included in the gas clouds: at the mean concentration of the particles of the order of 0.1–1 cm^{-3}, the free path is considerably smaller than 1 pc.

For the speed of sound it should be accepted that $c_s = c_g^{(1)}$ for the adiabatic parameter $\gamma = 1$.

It is important to mention two further aspects of the discussed problem. Firstly, the dispersion of the velocities of the gas clouds, in contrast to the value σ_g, usually depends only slightly on the radial coordinate and remains close to the constant value in a large part of the disk, with the exception of the regions of high-rate star formation. For example, in the Galaxy, according to the data obtained by Barton [42], the characteristic spatial scale of reduction of $c_g^{(1)}$ to the disk periphery in approximation by the exponential law $c_g^{(1)} \exp \propto (-r/L_{cg})$ equals $L_{cg} \simeq 16 \div 17$ kpc (compare with $L_g \simeq 4$ kpc [1109]). Secondly, the dispersion of the velocities of the gas clouds depends only slightly on the type and mass of the Galaxy (see, for example, the data obtained by Unwin M 31 [1244]] and discussion of this problem by van der Kruit [799], and for the estimates it can evidently be assumed that in the flat galaxies $c_g^{(3)} \simeq 8\div 12$ km/s.

In the vicinity of $r \simeq 0.4$ kpc, the Galaxy contains a region of rapid changes of the value $c_g^{(3)}$ from 20 km/s to 8 km/s in the section with the length of $\simeq$0.3 kpc. This special feature coincides with the region in which the density of the gas decreases and where V_{rot} reaches the internal maximum, although the region of the large variation of the velocity is considerably wider $r = 0.2\div 1.2$ K parsec (see the references in [566], chapter 9).

The controlling process for the evolution of the disk is star formation. The stars form from molecular gas and, consequently, it is important to determine the conditions in which the atomic gas changes to the molecular phase. Observation show that the fraction of the molecular gas increases in the direction to the centre of the galaxies. This is determined by the effect of different factors, but primarily depends on the gas pressure and its chemical composition. Actually, as shown by the calculations, the gas pressure in the plane of the Galactic disk (lg (P/k) ~3–5 cm^{-3} K) correlates satisfactorily with the fraction of the gas in the molecular phase, with the exception of the regions inside massive bulges where, evidently, it is essential to take into account the pressure of the hot gas, filling the bulge, on the gas in the disk [93].

1.3.3 Vertical structure of gas disks

In most cases, the distribution of the volume density of the gas (molecules – clouds) across the disk plane is described satisfactorily by the Gauss law [492, 1079, 1112]:

$$\rho_g(z) \propto \exp\left\{-\left[z/(1.2h_g)\right]^2\right\}, \tag{1.14}$$

where h_g is the characteristic half thickness of the gas disk ($\rho_g(h_g)/\rho_g(0) = 1/2$). However, each gas component is characterised by its vertical distribution. In the solar vicinity of the Galaxy we can accept for different components [492]:

$$\rho_g(z) = \rho_c \cdot \begin{cases} \exp[-(z/h_g^{(\mathrm{mol})})^2] & \text{for molecular hydrogen, } h_g^{(\mathrm{mol})} = 81\,\text{pc}, \\ 0.7\cdot\exp[-(z/h_g^{(\mathrm{HI})})^2] & \text{for cold HI – component, } h_g^{(\mathrm{HI})} = 127\,\text{pc}, \\ 0.18\cdot\exp[-(z/h_g^{(\mathrm{HIa})})^2] & \text{for warm HIa – component, } h_g^{(\mathrm{HIa})} = 318\,\text{pc}, \\ 0.11\cdot\exp[-|z|/h_g^{(\mathrm{HIb})}] & \text{for warm HIb – components, } h_g^{(\mathrm{HIb})} = 403\,\text{pc}, \\ 0.015\cdot\exp[-|z|/h_g^{(\mathrm{HII})}] & \text{for HI regions, } h_g^{(\mathrm{HIb})} = 70\,\text{pc}, \\ 0.025\cdot\exp[-|z|/h_g^{(\mathrm{dHII})}] & \text{for diffusion HII component, } h_g^{(\mathrm{dHII})} = 1000\,\text{pc}, \end{cases} \tag{1.15}$$

where $\rho_c = 0.57$ hydrogen atoms/cm^3.

Since the dispersion of the velocities of the gas clouds is considerably smaller than the dispersion of the velocities of the stars across the disk plane (with the exception of far away peripheral regions) (for example, $(c_g^1 / c_z)_\odot \simeq 0.2 \div 0.3$), the thickness of the gas disk is also small in comparison with the thickness of the stellar disk. For example, the value h_g in the Galaxy is 40÷100 pc in a relatively wide region of the disk 3 kpc $\lesssim r \lesssim$ 10 kpc. At the same time, at the periphery of the S-galaxies ($r \gtrsim 3\div 5\ L_\sigma$), where the density of the stellar disk is $\sigma_* \sim \sigma_g$, the thickness of the gas disk may become quite large ($h_g \sim \Delta^*$).

1.3.4. Gas kinematics

The general form of the rotation curve of the gas component is similar to the stellar form (see section 1.2.3). It is important to mention several characteristic differences:

- the gas rotates on the whole at a higher velocity than the stellar disk, usually by 5–15%. The reversed situation is found only seldom. There are also a small number of cases in which the stellar and gas components rotate in the opposite directions;
- the mean value $(V_{\rm gas} - V_*)/V_{\rm gas}$ increases in transition from the galaxies of late morphological types to earlier ones;
- the rotation curve of the gas disk shows more fine-scale inhomogeneities than that of the stellar disk;
- the rotational curve for the gas is influenced more extensively by non-circular motion, associated in particular with the presence of the spiral structure. Consequently, the rotation curve in the region of the plateau has fund the form of a 'wave' (Fig. 1.19);
- the rotation curves for the gas components can often be seen only far away outside the limits of the stellar disk;
- the distribution of the speed of sound c_s in the gas disk changes with the radius at a lower rate than the dispersion of the velocities of the stars $c_r(r)$ in the stellar disk. Often we can accept for the molecular gas that $c_s \simeq$ const ~10 km/s.

Because of the collisions, the velocity dispersion of gas clouds (speed of sound) c_s is isotropic in contrast to the stellar component (section 1.2.4). Different components of the gas (hot and cold) significantly different in the values of c_s and the nature of the distribution in the disk. Note that the giant molecular clouds are collisionless like the stars [585].

1.3.5. Dust component

Dust forms only a small fraction of the total mass of the interstellar medium and can be regarded as the collisionless ensemble of particles. Consequently, dust has no direct strong effect on the large-scale dynamics. However, the photometric parameters strongly depend on the content and nature of the distribution of the dust

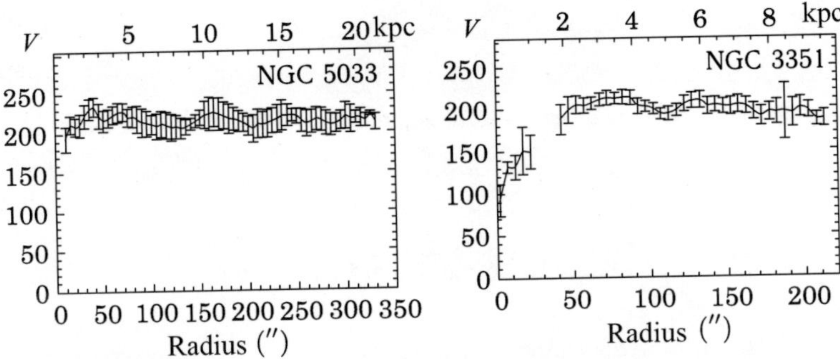

Fig. 1.19. Examples of the rotation curves, demonstrating the 'wave character' of the rotation curve determined evidently by non-circular motions in the disk [513].

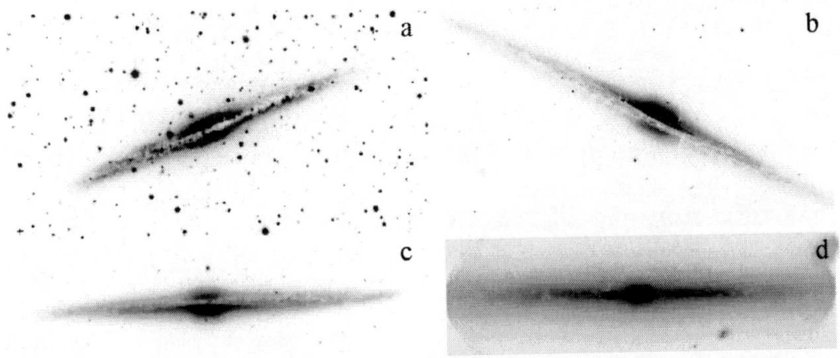

Fig. 1.20. The images of spiral galaxies with the disk seen edge-on, showing the presence of a thin gas–dust band by the absorption of dust in the optical range: NGC 891 [1063] (a), NGC 4594 (SHT) (b) NGC 4565 [1312] (c) our Galaxy (MW) according to 2MASS data (d).

component. This is most evident for the galaxies visible edge-on where the radiation from the stars from the band in the vicinity of the plane $z = 0$ does not reach us (Fig. 1.20).

The nature of the distribution of dust in the disk is complex and on a small scale more heterogeneous than that of the gas, which is associated with the origin of the dust: in slow-flowing atmospheres of red dwarfs, resulting in explosive processes in stars and gas emission from galactic nuclei; in the planetary and protostellar nebulae, stellar atmospheres. Dust attenuates the radiation from the star, scattering and absorbing it. In the optical wavelength range of the value of attenuation is inversely proportional to the wavelength, which leads to reddening of the colour of the stars. Dust is a powerful cooling agent for the interstellar environment. Since the energy of the absorbed photon changes to the chaotic motion of atoms of the dust particles, the particle radiation occurs in the infrared range of 50–200 μm in the form the continuous spectrum. The contribution of dust in the infrared range can be 10–40% of the total luminosity of stars in spiral galaxies.

The most important manifestation of the dust component are dust veins that extend along the spiral arms and inside the central star jumpers. Note the global

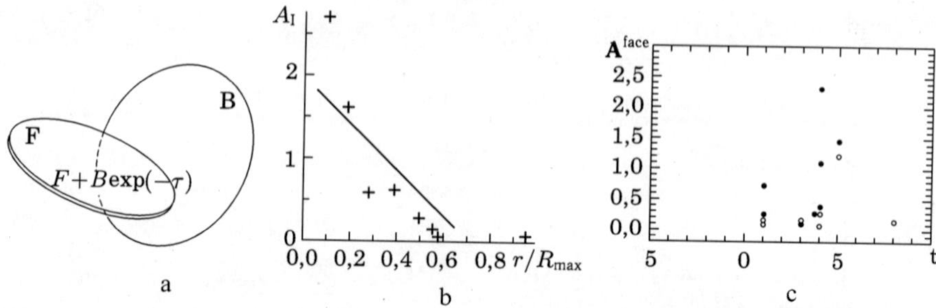

Fig. 1.21. a) if the galaxies do not interact but the disks of the galaxies partially overlap, such systems enable the most direct method of examining the internal absorption in the disks [1296]; b) the absorption in the stellar units in the band *I* in in the disks of the Sa–Sbc galaxies at different distances from the centre [215, 753]; c) the absorption in the stellar units for galaxies of different morphological types after reduction to the 'broadside' position (● – the regions of spiral arms, ○ – the regions between the spirals) [516].

distribution of dust on the example of NGC 891 [1314] for which the radial scale of the dust components is 8.1 kpc (5.7 kpc for the stars), the vertical scale is 0.26 kpc (0.39 kpc for the stars).

The vertical distribution of dust is strongly affected by the dynamic processes in the gas component, the processes of star formation and evolution of stars in the final phase, when the interstellar medium is enriched with heavy elements. Absorption along the visual ray (coordinate s) is defined by the expression:

$$I \propto \int \varrho_*(s) \exp\{-\tau(s)\}\, ds, \quad \tau \propto \int \varrho_d(s')\, ds', \tag{1.16}$$

which depends on the laws of the density distribution of stars $\varrho_*(r, \varphi, z)$ and dust $\varrho_d(r, \varphi, z)$ in space. It should be noted that in order to explain the actual distribution of dust one must consider spiral waves and bars that violate the axial symmetry, and warping of the disks – as a result there is no symmetry with respect to the plane $z = 0$. These effects play a more significant role for galaxies with an angle of inclination of $i \gtrsim 80^\circ$.

From observations we know that the absorption is greatest in spiral arms and the ring structures, and here it very poorly depends on the distance to the centre [753, 12 960 (Fig. 1.21). Between the spiral arms the absorption is less marked, and the function $I(r)$ decreases with increasing r.

In [1246], based on analysis of data on the surface brightness of 16 000 galaxies it was concluded that the optical thickness of the inner regions of spiral galaxies is large. This question was investigated in a number of works, and it should be assume that the S-galaxies are not optically thin [439, 501, 1063, 1315]. In the centre the absorption is maximum (equal to several units), but it decreases rapidly with distance from the centre of the galaxies, varying approximately proportionately to the amount of gas in the HI on the visual ray. External fields of the galaxies are almost completely transparent. The absorption and optical thickness τ depend to a large extent on the vertical and radial distributions of dust and stars in the disk and the bulge, the nature of the scattering of stellar radiation by dust, and the geometric parameters of the galaxy.

1.4. Large-scale structures in S-galaxies

1.4.1. Spiral galactic pattern

The Hubble classification already divided the flat galaxies into subsets (Fig. 1.1): S0-galaxies without any distinctive and contrasting patterns and, typically, with little or no gas; S – natural spiral galaxies, and SB – the spiral galaxies with jumpers in which the two spiral arms diverge almost at right angles from the jumper (bar) rotating in the centre of the galaxy.

The geometry of the spiral structures is striking for its diversity. In normal spiral galaxies (type S) there are mostly two global spiral arms – as, for example, in M 51, M 74, M 81, M 100, M 106, NGC 157, NGC 2997. However, there are also galaxies with numerous spiral segments (for example, M 33, NGC 5055, NGC 7793 – see the book by Vorontsov-Vel'yaminov [37]), NGC 4414, stacked with different numbers of spiral arms in different regions of the disk radius (NGC 1232), galaxies with branching of the spiral arms (NGC 151, 1288, 2997, 4725, 4736, 5033, 6946, 7412 et al.), the existence of a dense gas ring superimposed on the spiral structure (NGC 1024, 2223, 3124, 3344, 7329, et al.), with the presence of a spiral structure in the central area of the bar (NGC 1512), etc. In NGC 4622 there is a rare two-tiered structure, in which the twisting of the external and internal branches is not the same.

As regards appearance, it is most natural to divide the galactic spiral structures into two main types. The first includes symmetrical (in relation to the centre) long arms, forming a global spiral pattern, extending to several radial disk scales[7]. The second group combines fragmentary spiral patterns, which are ragged, irregular geometric fibre formations – short pieces of arms that do not create a regular spiral pattern. They are also called flocculent spirals. According to the observations, the number of fragmentary galaxies is substantially larger than that of the galaxies with a regular structure [52]. However, a clear distinction between them is not easy, and there is a continuous series from grand design to flocculent and irregular galaxies, in which it is difficult to identify any extended segments of spirals. Moreover, the branches of the grand design and flocculent spirals are completely compatible in the same galaxy.

Spiral galaxies are differentially rotating with the angular rate $\Omega(r) = V(r) / r$, so that any gravitationally unlinked structural formation should be rapidly eroded by differential rotation of the galaxy.

According to modern concepts, in most cases the mechanisms of formation of global and fragmentary spiral patterns can behave differently. The first is due to the global spiral density wave propagating through the galactic disk. In the second case, the main reason is the stretching of areas of star formation due to differential galactic rotation and breaking of the spiral arms. An important factor is also the possible spreading of star formation into the neighbouring areas. Flocculent spirals

[7]Such a spiral pattern is called grand design.

often disappear in the near IR range, so the density perturbation of the old stellar disk does not occur.

Thus, the spiral structure of galaxies is extremely diverse and it is difficult to expect a single physical mechanism of formation of spirals. The individual spirals may have a wave nature, some may be caused by the tidal interaction. Regular galaxies very strongly differ in their characteristics: the presence of gas, the presence of satellites and/or bar, the rotation curve, active nuclei, etc. Even in the same galaxy there can operate simultaneously several factors that lead to the real observed spiral structure.

In almost all spiral galaxies, spiral arms are lagging – they rotate with their ends backwards [192] (see Fig. 6.10). Perhaps the leading spiral branches may form in close pairs during movement of the satellites against the rotation of the galaxy [1226]; however, such spirals are rare. Different geometries of the spiral pattern (e.g. logarithmic or Archimedes helix [88, 498]) are found.

We discuss the connection between star formation and spiral pattern. In the arms there is an increased concentration of young stars, implying a higher star formation rate (*SFR*) than in the rest of the disk. The question arises about the effectiveness of star formation ($SFE = SFR/M_{gas}$) in the arms and out of them in the light of a higher content of gas in spirals. The presence of a galactic shock wave (SW) in the gas in the arms with higher stellar–gas density can increase the rate of star formation, providing the trigger mechanism for the formation of star birth regions due to compression of gas at the shock wave front in grand design galaxies. The effectiveness of star formation is complicated by a number of factors: it is difficult to directly assess the rate and efficiency of star formation, to address the issue of the lifetime of the spirals – whether they are quasi-stationary structures or short-lived transient entities; the definition of the characteristic times of the various stages of star formation also requires additional research.

These observations indicate that the region between the arms provides a significant contribution to star formation, which in some cases is only 30% less than that of the spiral arms, even in grand design galaxies. In this case, the spirals contain a larger fraction of molecular gas in comparison with the area between the arms. But the efficiency of star formation in the arms is not much greater than the SFE between the spirals (~10–20%). However, such research so far has been based on small samples of galaxies.

The spiral pattern in galaxies that are visible almost flat is easy to see. With decreasing angle of inclination of the plane of the galaxy to the visual ray the determination of the parameters of the spiral structure becomes more difficult. The main difficulty is to identify the geometric parameters of the spiral pattern of the Galaxy. It was generally assumed that it is a spiral galaxy without the jumper. Since the early 90s there have been grounds for doubts about the validity of this point of view. The analysis of primarily kinematics of the gas in the central region indicates the presence of a bar in the Galaxy [396]. There is now reliable photometric evidence of the existence of the bar.

Figure 1.22 shows a diagram of the spiral structure of our Galaxy, refined using the data from the survey in the Spitzer telescope. These data allowed us to separate

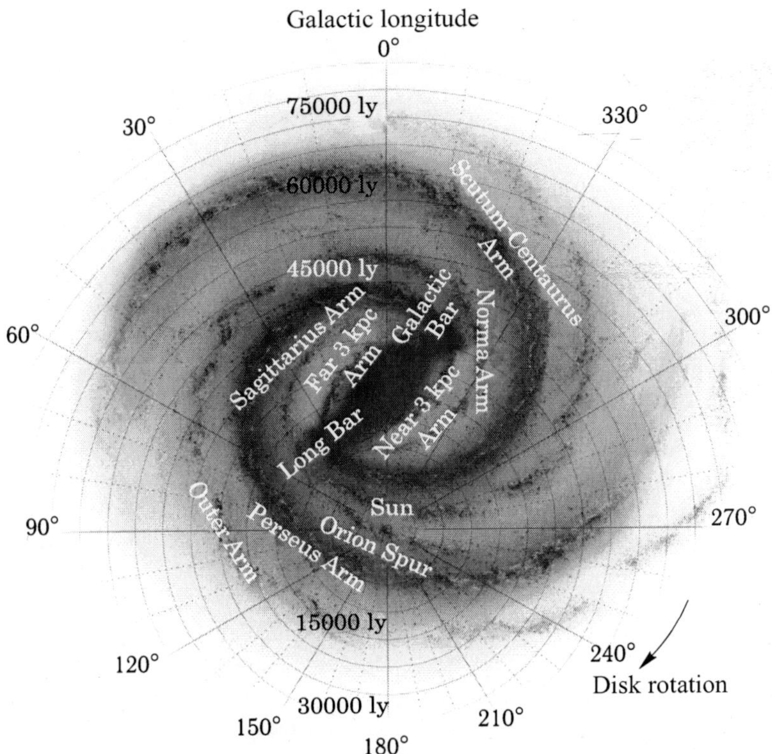

Fig. 1.22. The spiral structure of the Galaxy refined on the basis of the results of data of the review using the Spitzer telescope. In general, there is quite a classic picture of the two-arm pattern with a central bar and the presence of smaller fragments of additional spirals. The arms sleeves–Centaur' Shield and Perseus – start at the ends of the bar. There are secondary arms in the central region (Archer, Norma) and the periphery (spur of the Orion = Orion arm = Orion trace, where is the Sun and the outer arm are situated outside the the main Perseus arm).

the arms of the wave nature (where there is a concentration of the density of old stars)from the non-wave arms. The question of the number of arms and their geometry in our system is actively being discussed. In the inner region of the Galaxy ($r \lesssim$ 7 kpc), we have two long arms from the ends of the bar and two shorter and weaker structures (the so-called 3-kpc arms) near the jumper. The outer spiral pattern is more complicated. A reliable marker is the distribution of neutral hydrogen (Fig. 1.23). These observations point to the possible branching of the main spirals in the area of 7–8 kpc, leading to the four-arm spiral structure in the outer region. Note the lack of symmetry of the outer arm that can be due to the long-known bends of the disk at the periphery. The observed pattern may be due to the superposition of density waves with different azimuthal numbers $m = 2, 3, 4$ [533].

Branching spiral structures (within a certain radius there are two arms, and on the periphery four arms) are often observed in galaxies. Examples are NGC 3184 (SABc), NGC 3992 (SB (rs) bc), NGC 6946, which on the basis of the spiral struc-

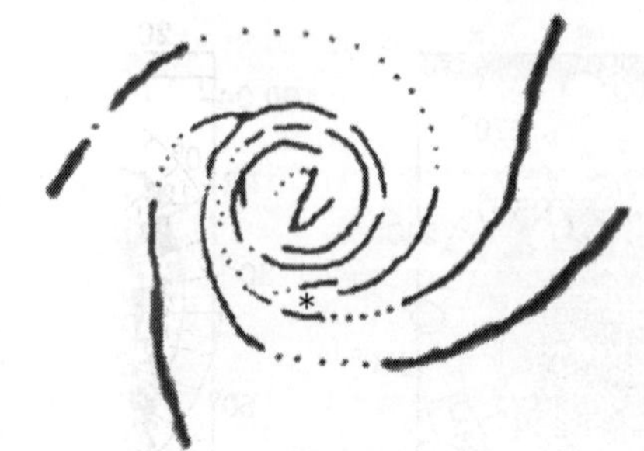

Fig. 1.23. The spiral structure in the Galaxy according to the distribution of molecular gas CO [533]. The distribution of gas and arms behind the area of the galactic centre can be seen. The asterisk below the centre indicates the position of the Sun.

ture can be regarded as 'twins' for the Galaxy. Moreover, in NGC 3184 the old stars show the classic two-arm structure.

Typically, the contrast of brightness (and therefore density) of the spiral pattern of galaxies in the old disk stars (in the red region of the spectrum) is small – ranging from a few percent 30÷40% of the axisymmetric background[8]. But the gas density difference between the sleeve and the sleeve may differ by several times (near the front of the compression wave (shock wave) density jump may be even higher). Young stars of early spectral classes are concentrated to the spiral branches and, most striking, in fact highlight the spiral pattern on the background of the disk. In general, these facts indicate the important role of gaseous subsystems and processes of star formation in the creation of the spiral pattern of galaxies.

Contrast, brightness, and the length of the spiral pattern substantially depend on the speed of rotation of the disk components. In the case of slow rotation, $V \lesssim 100$ km/s, spirals are either not observed or are very weak (chaotic). In these objects there are either no conditions for the excitation of density waves or they have a small amplitude.

The observed parameters of the spirals are highly dependent on the spectral range: in the ultraviolet (U band) range the spirals are brighter and contrasting. In the V band and near-infrared the spirals are smoother and less bright. In the middle and far infrared ranges the spirals again become clearly visible, because their radiation is associated with the dust heated by young stars. The spiral arms are characterized by high density of relativistic particles and magnetic fields.

The problem of the geometry and mechanisms of amplification of the magnetic field in the spiral galaxies is actively studied (for example [348, 593, 1332]. The magnetic disks in the galactic disks consists of two components: ordered and diordered with respect to the geometry of the lines of force of the field. The ordered component often dominates between the spirals, and the spirals appear to be shifted in relation to each other as regards polarized radio radiation and star radiation (Fig.

[8]In some rare types of early galaxies with an extremely low content of gas there are visible only low contrast spiral arms of the old stars [1304]. In galaxies with a bar there may be a very powerful (up to 50%) density wave in the old disk population [52].

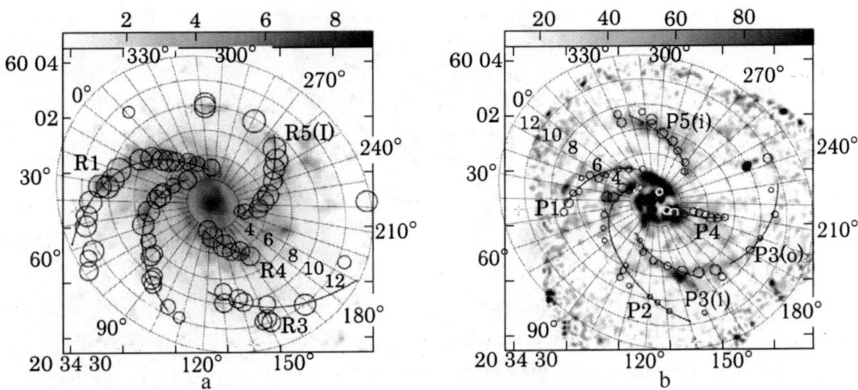

Fig. 1.24. The galaxy NGC 6946: a) – the brightness distribution in the R band; b) – the intensity of the polarized component at a wavelength λ= 3.5 cm [593]. The circles show the structure of the spirals, which is determined from the maxima of the wavelets W. 'Magnetic arms' (polarized radio emission) are localized almost precisely between the optical arms.

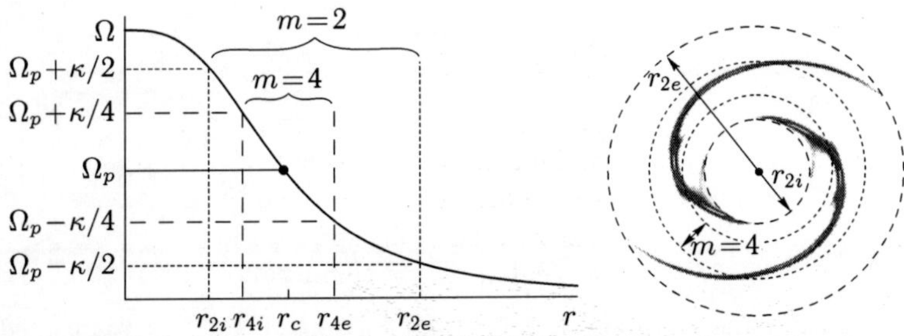

Fig. 1.25. The position of the corotation radius and Lindblad resonances. The scheme of the positions of the spirals (m – the number of arms). Two-arm spiral is located between r_{2i} and r_{2e}. As the number of arms m increases the zone of the existence of a stationary wave becomes narrower. The arrows show the corresponding region for the mode m = 4.

1.24). On the whole, the strength of the magnetic field in the spiral arms is greater than between them as a result of gas compression.

The long-lived spiral pattern rotates with constant angular velocity Ω_p. Due to the heterogeneity of the average angular rotational speed of the disk $\Omega(r)$ there are characteristic radii (Fig. 1.25, m – the number of arms):

- corrotation radius r_c from the condition $\Omega_p = \Omega(r_c)$;
- inner Lindblad resonance (ILR) on r_{mi} corresponds to

$$\Omega_p - \Omega(r_{mi}) = -\varkappa / m.$$

- outer Lindblad resonance (OLR) on r_{mo}:

$$\Omega_p - \Omega(r_{mo}) = \varkappa / m.$$

The linear density wave in the stellar disk does not penetrate into the central zone $r < r_{mi}$ and is located between the Lindblad resonances r_{mi} and r_{mo}. In accordance with this simplified picture only the two-arm wave can be extended, and in the case of $m \geqslant 3$ the structures are localized close to corotation radiu r_c. That is why the observed spiral pattern, as a rule, does not begin from the centre and ends before reaching the outer boundary of the disk.

The spiral structure is manifested not only photometrically but also kinematically. In the spiral region the velocities of stars and of gas differ from the circular velocities, and in addition to the azimuthal velocity perturbations the radial and vertical velocity components also appear (see [568, 584, 587], chapter 7).

1.4.2. Stellar groups and spiral patterns

In galaxies of later morphological types (Sb–Sd) spiral arms are well traced by the chains of the individual gas complexes with enhanced star formation [52, 527].

The dimensions of young stellar systems (SS) reach 1 kpc, and they combine star clusters of up to 20 pc or more sparse formations – associations. Star complexes have the age up to 100 million years, after which they decay. Therefore, within a single SS there occur multiple acts of birth and collapse of stellar associations. Originally discovered in the Galaxy, the majority of SSs are observed in nearby galaxies (M 31, LMC, M 109). In large galaxies, the number of SSs is 200.

There are also superassociations – young stellar complexes in the phase of rapid star formation throughout the area. This indicates the synchronization of the process of star formation across the area of the complex in contrast to the usual stellar systems, where star formation has a focal character. In general, here we should talk about the star–gas complexes.

The physical nature of star complexes and superassociations is associated with the development of a large-scale gravitational instability, accompanied by powerful turbulent motions. Explosions of supernova stars condense the gas, initiating the process of spreading of the areas star formation. There is also a negative feedback when the large number of young massive stars interrupts bursts of star formation due to gas heating. Thermal and magnetic-gravitational instabilities can also play a certain role [823].

The complexes in the wave spiral arms at their formation must be gravitationally bound, and their preferred size is associated with the parameters of gravitational instability along the arm (Fig. 1.26). The characteristic observed nodal structure of the spiral arms reflects the location of such arms star–gas complexes.

In the absence of a global wave structure the stellar complexes can give rise to a flocculent spiral structure – gravitationally unbound circular areas with the size of the order of the thickness of the disk which are stretched by the differential galactic rotation. Thus, the complexes give rise to flocculent arms but, on the other hand, are a product of the spiral arms in grand design galaxies.

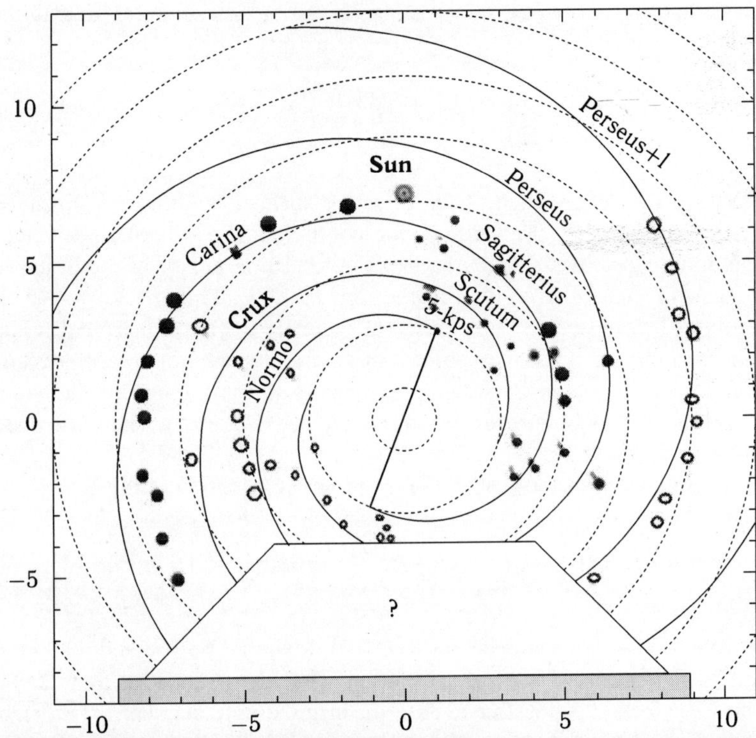

Fig. 1.26. The positions of the gas complexes HI (filled circles) in our Galaxy on the scheme of the spiral structure [54]. Gray circles are constructed by rotating the positions of the complexes HI by 180° around the galactic centre. Distances are given in kpc. The selected area (indicated by '?') is not available for observation of HI because of the opacity of the gaseous medium.

1.4.3 The bulge – central spheroidal subsystem

In disk galaxies of early morphological types there is distinctive spherical or flattened thickening in the centre (bulge), the characteristic dimensions of which are generally small in compared with the radius of the disk. The bulge consists primarily of old stars and is clearly visible in the images of galaxies, especially those visible edge-on. The total mass of the bulge M_b is negligible compared with the total mass of the disk M_d in Sb-, Sc-galaxies, and can be very tangible in some Sa- and S0-galaxies. So, for example, according to [552], in NGC 4450 $M_b / M_d \simeq 2.7$. The form of the bulge is satisfactorily represented by an oblate ellipsoid of rotation. According to the degree of flattening there are both almost spherical and greatly flattened bulges – lenses (with the semiaxis ratio to $\simeq$1/6).

There are several widely used empirical formulas for describing the surface brightness distributions in elliptical galaxies and bulges of spiral galaxies, which are generally similar [215]. Note de Vaucouleurs law ('the $r^{1/4}$ law')

$$\lg\left(\frac{I(r)}{I_e}\right) = -\beta\left[\left(\frac{r}{r_e}\right)^{1/4} - 1\right], \tag{1.17}$$

characterising the dependence of the effective surface brightness I_e on the appropriate effective radius r_e, the value of the coefficient $\beta > 0$ is often selected from the condition according to which half of the total luminosity of the bulge (or the elliptic galaxy) should be emitted in the range $r \leqslant r_e$. This gives $\beta = 3.331$. The de Vaucoulers law describes with sufficient accuracy the real spheroidal stellar systems, but from the physical viewpoint it has a significant shortcoming because the corresponding distribution of the volume density in the vicinity of the centre has the form $\varrho(R) \propto R^{-3/4}$, which leads to $\varrho(0) = \infty$ $\left(R = \sqrt{x^2 + y^2 + z^2}\right)$. The mass inside the radius R remains finite.

The generalisation of the law (1.17) is the Sersic equation [458]:

$$I(r) = I_0 \cdot \exp\left\{-\nu_n \left(\frac{r}{r_e}\right)^{1/n}\right\} \text{ or } \frac{I(r)}{I_e} = \exp\left\{-\nu_n\left[\left(\frac{r}{r_e}\right)^{1/n} - 1\right]\right\}, \tag{1.18}$$

which at $n = 4$, $\nu_4 = 7.669$ transforms to (1.17). It should be stressed that the Sersic law (and its partial cases) is purely empirical. The advantage of using (1.18) is that using the same form it is possible to describe bright elliptical galaxies ($n \simeq 4$), bulges and E-galaxies with low luminosity require the values $n \simeq 1 \div 4$, and for the disks of the spiral galaxies $n = 1$ in accordance with (1.2).

The volume density is often described by the King law which in the simplest form can be written as follows

$$\varrho(R) = \begin{cases} \dfrac{\varrho_0}{\left[1 + (R/b)^2\right]^{3/2}}, & R < R_{\max}, \\ 0, & R > R_{\max}^{(b)}, \end{cases} \tag{1.19}$$

which is in agreement with the de Vaucouleurs law in the range $R = \sqrt{r^2 + z^2} > b$ at $b \ll R^{(b)}_{\max}$, but (1.19) requires truncation at some radius $R^{(b)}_{\max}$. However, in contrast to (1.17), equation (1.19) gives the finite volume density in the centre. The distribution (1.19) is written for marked truncation after the radius $R^{(b)}_{\max}$. If $b \sim R^{(b)}_{\max}$, then it is necessary to carry out continuous truncation in (1.19) after the radius $R^{(b)}_{\max}$, for example, in accordance with the linear or exponential law.

Coefficient n in equation (1.18) correlates with the ratio of the luminosities of the bulge to the disk, the morphological type of the spiral galaxy, the parameters of the bulge (its luminosity, effective radius r_e and central surface brightness) [642, 772]. For the bulges of the galaxies of later morphological types $n \simeq 1$, and in the case of the galaxies of early types the luminosity of the bulge on average and its effec-

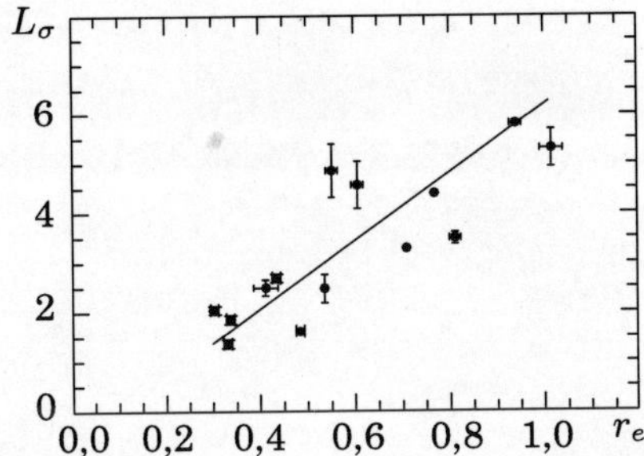

Fig. 1.27. Relationship between the radial scale of the disk L_σ and the effective bulge radius r_e.

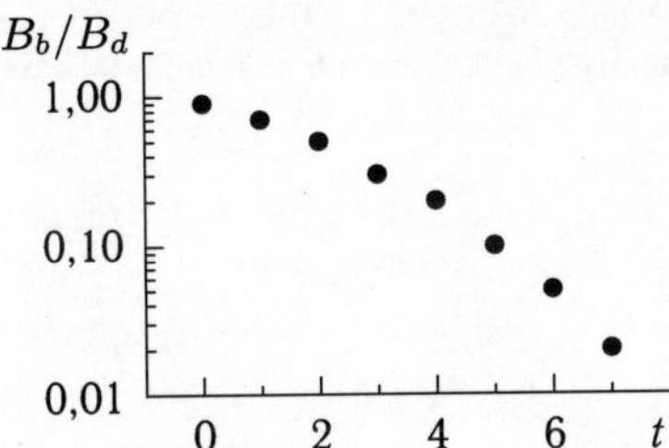

Fig. 1.28. The ratio of the luminosity of the bulge B_b to that of the disk B_d along the morphological sequence for S-galaxies.

tive surface brightness are higher and $n \simeq 4$. On average, as the exponential scale of the disk L_σ in (1.2) increases, the effective radius of the bulge becomes greater (Fig. 1.27) [289].

There is a strong dependence of the ratio of luminosity of the bulge to the luminosity of the disk on the morphological type [733], which is one of the criteria for constructing the morphological sequence of the disk galaxies from S0 to Sd (Fig. 1.28).

1.4.4. Galactic bars

In approximately half of the spiral galaxies the central region containts a bright, strongly elongated oval-like solid-state rotating jumper (the bar). These stellar systems are referred to as the intersected (SB) spiral galaxies. The length of the major semi-axis of the bars is ~1÷5 kpc. In the galaxies of later types, the bars contain a large amount of gas and dust which take part in the relatively complicated non-circular motion.

It is interesting to note that our views regarding the relative number of the galaxies with the bar are that the number has been increasing in the last couple of decades. At the boundary of the 70s and 80s of the 20th century, the fraction of the SB-galaxies was estimated at 10%. The current data give the values of 50–70%. The fraction of the SB-galaxies amongst the isolated spiral galaxies is close to 60%, whereas in the group of the Seyfert galaxies the fraction of the objects with the bar reaches 80% [784]. Starting with the morphological types $t > -1$, the frequency of

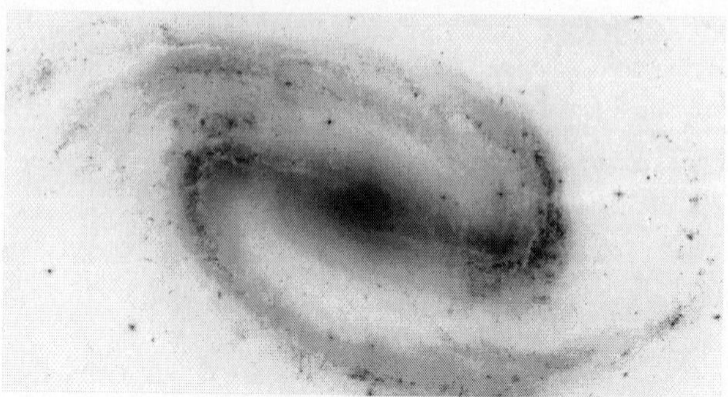

Fig. 1.29. One of the close galaxies NGC 1300 is a classic example of the spiral galaxy with the bar (HST image). The clearly visible elongated dust bands are linked with the position of the shock waves.

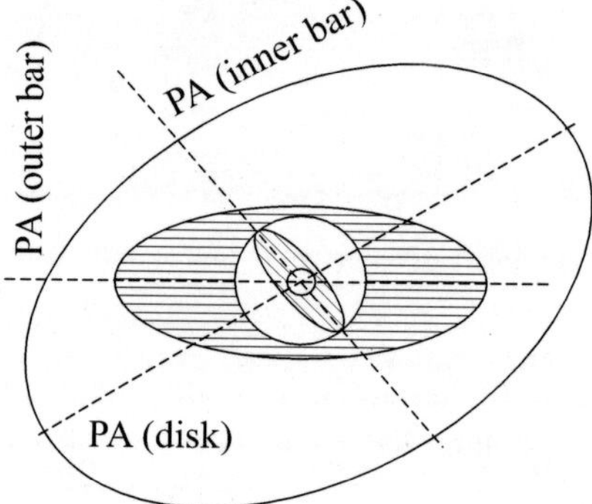

Fig. 1.30. Diagram of a system with a binary bar. The dashed lines indicate the major axes (PA) of the galaxy, the main and inner bar. The ring corresponds to the inner Lindblad resonance for the main bar.

occurrence of the system with the bar shows no systematic dependence on the type of galaxy.

The bar is characterised by solid-state rotation with the angular velocity Ω_b. The stars of the bar move along complicated elongated orbits whose major axis also forms the axis of the bar. The movement of gas in the region of the bar is far more complicated as a result of the non-axisymmetric gravitational effect of the stellar bar (Fig. 1.29). It is important to note to characteristic features of the bars. Firstly, the presence in the bars of narrow zones with strong compression of the interstellar medium where as a result of the increase of the concentration of dust and the increase of the opacity of the medium observations showed long thin veins. The second special feature of the bars is the formation in the bars of radial flows of gas along the bar to the centre of the galaxy. This is accompanied by the buildup of gas in the central region of the galaxy.

In the central region of the bar different complicated structures were detected in a number of galaxies: the inner bars, inner polar rings, nuclear rings (nuclear rings

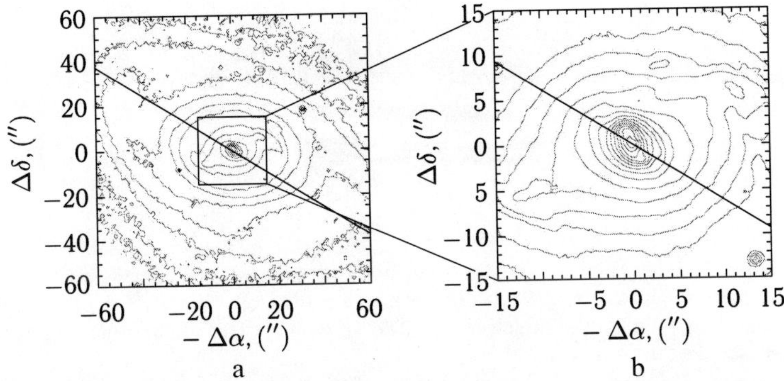

Fig. 1.31. a) isophots NGC 20 to 73 in the band R [147]; b is the structure of the central zone in the region of the binary bar. The thick line indicates the major axis of the Galaxy.

are also found in the galaxies without the bar [1169]), inclined disks, minispirals (for example [146, 147, 536, 856, 930, 931, 1168, 1169]). More than double-barred 100 galaxies are known to exist (Fig. 1.30).

1.4.5. Galactic halo and corona

As already mentioned, the flat galaxies are also characterised by a spherical subsystem whose volume density decreases with increase of the distance from the centre of the galaxy (the inner bulge, the stellar halo and the dark halo are usually specified here). The dispersion of the velocities of the stars is comparable with their circular velocity [561]. Usually, this feature is often used to separate them from the general disk of the stars in the solar vicinity of the Galaxy, relating the stars with lower radial velocities to the population of the disk. According to the data published in [940], the dispersion of the stars of the halo in the Galaxy in the solar neighbourhood $(c_r, c_\theta, c_\varphi) = (133\pm8, 98\pm13, 94\pm6)$ km/s and the velocity of rotation $V = 25\pm15$ km/s. Similar results were also obtained for other stellar samples, e.g., $(c_r, c_\theta, c_\varphi) = (142\pm2, 81\pm2, 77\pm2)$ km/s [1182].

The direct determination of the volume density of the stars of the spheroidal subsystem of the Galaxy and, consequently, of its mass is associated with difficulties due to the fact that many of the stars of the subsystem are characterised by low luminosity. For example, according to Schmidt's estimates [1122], the mass of the stars of the spheroidal subsystem[9] M_h in the sphere with the radius of the solar orbit does not exceed 2–6% of the mass of the stellar disk in the same region. On the other hand, in [937] it was concluded that in the sphere with the radius of the solar orbit the fraction of the disk represents no more half the total mass of the Galaxy: $M_h/M_d\,(r < R_\odot) \gtrsim 1$. In the dynamic model, constructed in [260], it is assumed that in

[9]The mass of the halo $M_h(r)$ increases with a radius. In this book, if not mentioning a specific radius, it will be always assumed that $M_h = M_h(R_{\max})$ is the mass of the halo inside the optical radius of the stellar disk and it is also usually assumed that $R_{\max} = 4L_\sigma$ (if no other value is given).

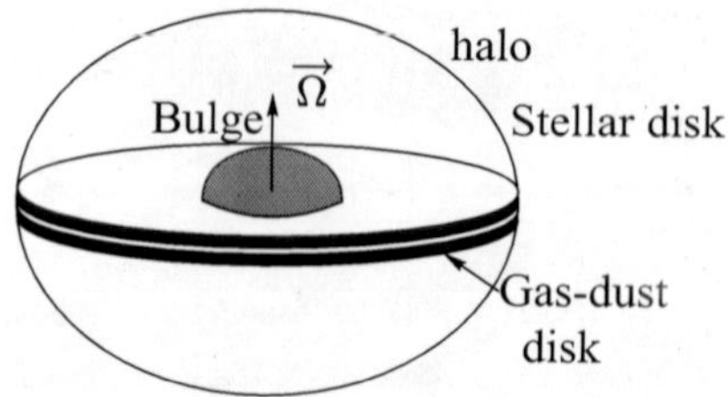

Fig. 1.32. Main structural elements of the S-galaxy. Inside the stellar disk there is a gas disk which is almost always extended along the radius further than the stellar radius. The centre contains a bulge – an almost spherical stellar subsystem. The disk subsystems rotate more rapidly than the bulge. The halo does not rotate at all.

the 12 kpc range the mass of the spheroidal components (bulge + halo) is 1.7 times greater than the mass of the flat subsystem (Fig. 1.32).

The important observed components of the stellar halo are the spherical stellar clusters with the mass of up to $10^6 M_\odot$. The number of spherical clusters in our Galaxy is approximately 250, and the total mass of these objects is not large: $\lesssim 10^8 M_\odot$, which is typical of large spiral galaxies.

At the beginning of the 80s of the 20th century, a thick disk was found in the Galaxy. The disk consists of stars with a lower metal content than that of the stars of the disk which is approximately 2 kpc thick [620]. At the altitudes of $z \simeq 1 \div 2$ kpc, the thick disk is characterised by high values of the dispersion of velocities $c_r \sim 60$ km/s, $c_z \sim 40$ km/s, which is considerably greater than for the so-called old stellar disk (section 1.2.4) in which the main mass of the stellar component is concentrated. A slightly thicker subsystem was found in lenticular galaxies visible edge-on.

A stellar halo has been found in a large number of galaxies, in particular those edge-on galaxies (NGC 891, M 81, NGC 300) [1223]. The thick disks extends the distances of the order of 6–10 kpc and then changes to halo stars. The observations indicate the sharp boundary between these subsystems (Fig. 1.33). On the basis of these results is may also be concluded that the halo may be oblated from the poles.

An additional argument in favour of the relatively massive spheroidal subsystems of flat galaxies is also the form of their rotation curves at the disk periphery. If the masses of the spheroidal subsystems were negligibly small in comparison with the mass of the disk within the limits of the optical radius, then the value $V_{\rm rot}(r)$ at the periphery of these galaxies would greatly decrease, as in the case of NGC 157 [1098]. The number of these galaxies is however very small [58]. In a considerably larger number of flat galaxies at the periphery of their disks (at $r \gtrsim 4 \div 5\ L_\sigma$, where the stellar disk actually disappears) $V_{\rm rot}(r) \simeq$ const or even slightly increases with increase of the distance from the centre of the Galaxy [193, 1090, 1093, 1094]. For example, in the spiral galaxy with low luminosity NGC 5585 within the limits of the measured rotation curve ($r \lesssim 7L_\sigma \simeq 10$ kpc) the mass of the dark halo should equal 90% of the total mass of the system [486]. Models of 48 specific spiral galaxies were constructed in [317]. Of these only 15 may not have any halo. In 23 cases, the presence of a low-mass halo ($0.2 < M_h / M_d < 1$) is assumed, and 10 models include massive spheroidal subsystems ($M_h / M_d \geqslant 1$). If we take into account additional fac-

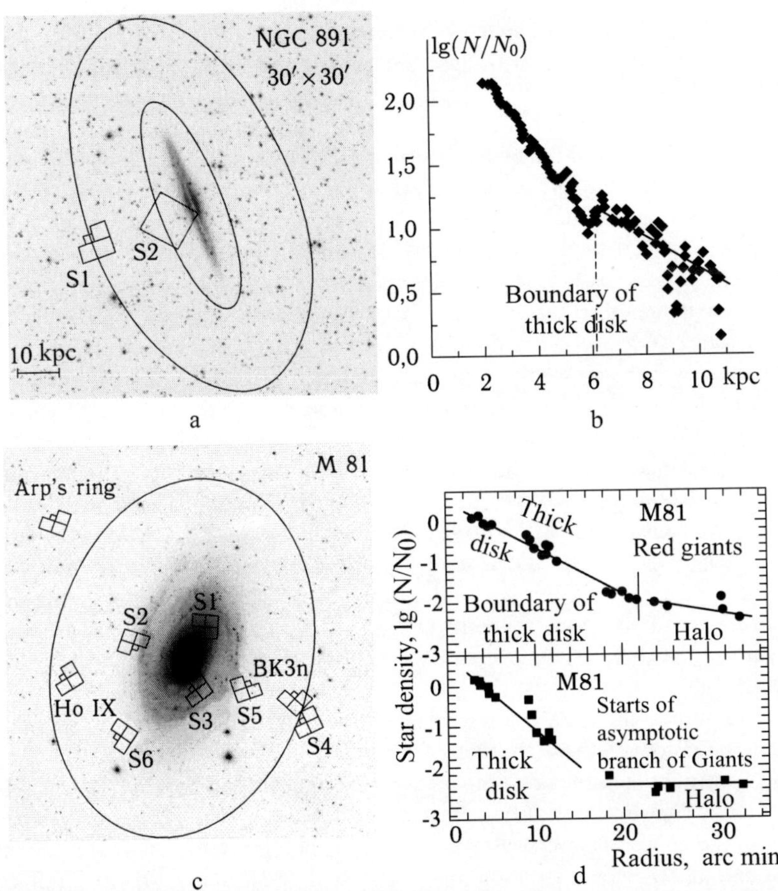

Fig. 1.33. Images of the spiral galaxies NGC 891 (a) and M 81 (c) [1223]. The ellipse indicates the boundaries of the thick disk according to the observations. The dependences of the variation of the logarithm of the density of the stars lg (N/N_0) along the radius of the Galaxy are shown (b, d). The position of the inflection corresponds to the boundary between the thick disk and the halo.

tors (chapters 4, 10), then to explain the observational data for the very large majority of the S-galaxies it is necessary to take into account the presence of relatively massive halo is with $M_h(R_{\max}) \simeq (0.5 \div 3) \cdot M_d$.

There is also a number of other features indicating the existence of massive spherical subsystems around the flat galaxies which can easily explain the dynamic behaviour of the disks. For example, in the absence of a halo at a low dispersion of the velocities of the stars of the disk a bar-like structure develops in the centre but not all galaxies with the dynamically cold disk belong to the SB type; the ratio of the dispersion of the stars to the circular velocity $c_r/V_{\rm rot}$ is inversely proportional to the ratio M_h/M_d, and to explain the observed values of $c_r/V_{\rm rot}$, it is necessary to assume that $M_h \sim M_d$. The condition for long-term existence of the region with reduced density (depression, see section 1.2.1) in the central part of the disk may also be represented by the massive spheroidal subsystem – the bulge + halo [66]. The thin galaxies observed edge-on with the ratio of the axes $b/a \leqslant 0.1$ have a geometrical thickness which can be explained only in the presence of a dark massive halo [63]. These effects are discussed in greater detail in chapters 4 and 10.

The x-ray coronas of the galaxies (mainly for the elliptical galaxies but there are also some in spiral galaxies), detected in the 80s, are considerably greater than the visible dimensions of the systems and are associated with the hot gas ($T \sim 10^7$ K). For gravitational sustainment of the gas, accumulated in the process of evolution of the system, it is necessary to provide an additional massive component (dark matter) with no detected radiation.

Very surprising are the galaxies with the stellar–gas ring at the distance of approximately 2–3 radii of the stellar disk situated in the plane perpendicular to the disk, with the ring rotating in its plane of symmetry with the velocity of rotation of the periphery of the stellar disk. This phenomenon of the polar rings not only indicates the presence of a dark subsystem but also makes it possible to determine the form of the dark halo. The velocities of rotation of the polar rings and disks for three SO-galaxies NGC 2685, A0136-0801, NGC 4650A are similar to each other: $V_{\rm ring}/V_{\rm disk} = 0.97 \pm 0.08$ [1298]. The kinematics of these systems indicates the existence of a relatively massive spheroidal halo. In the NGC 2787 galaxy the ring is situated at a distance of two optical dimensions of the Galaxy (20 kpc). The observational data shows that this system is characterised by the ratio $M_h / M_d \simeq 4$ in the range $\lesssim 20$ kpc.

The nature of the dark matter forming the halo of the galaxies has not been explained. Of course, there are many suggestions. However, they still only hypotheses. A very popular concept is the one which suggests that the dark matter consists of elementary particles weakly interacting with matter. The properties of the dark matter in the galaxies are discussed in detail in chapters 4 and 10. It should only be mentioned that the conclusion on the presence of the dark matter, comparable with or exceeding the total mass of the gas and the stars, is unavoidable at least for the majority of galaxies. The main mass of the dark matter should be located outside the disk. The estimates of the ratio of the dark matter to baryonic matter inside the optical radius using different methods and different observational material are in agreement with the accuracy not better than a factor of 2 at the moment.

1.4.6. Circular velocities of different components

The circular velocity V_c characterises the distribution of gravitating matter:

$$\frac{V_c^2}{r} = \frac{\partial \Phi}{\partial r}, \tag{1.20}$$

where Φ is the appropriate potential in the plane $z = 0$. For a multicomponent system (disk + halo + bulge + nucleus +...) the total circular velocity V_c is determined by the circular velocities of the components $V_c^{(j)}$:

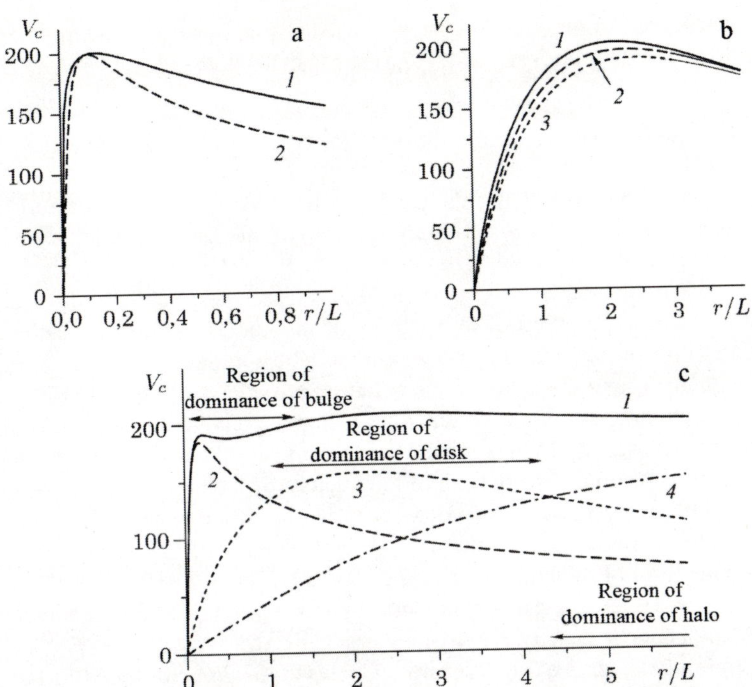

Fig. 1.34. Radial dependences of the circular velocities $V_c(r)$, km/s: a – for a bulge in the de Vaucouleurs model (1.17) (1) and in the King model (1.19); b – for the disk: 1 – infinitely thin disk, 2 – the ratio of the vertical scale radial scale $\Delta^*/L_\sigma = 0.2$, 3 $-\Delta^*/L_\sigma = 0.4$; c – for the three-component model of the galaxy V_c according to (1.21), bulge (2), the disk (3) and the halo (4).

$$V_c = \sqrt{\left(V_c^{\mathrm{halo}}\right)^2 + \left(V_c^{\mathrm{disc}}\right)^2 + \left(V_c^{\mathrm{bulge}}\right)^2 + \ldots} \quad (1.21)$$

In the case of the axisymmetric distribution of matter, the test particle rotates along a circular orbit (in the absence of the radial component of velocity) with the azimuthal velocity V_c. The test particle may be represented by cold gas whose speed of sound c_s is small in comparison with V_c. In this case, the contribution of the pressure gradient to the radial balance of the forces can be ignored and this yields (1.20).

The circular velocity for the spherical mass M outside its limits is equal to Kepler velocity: $V_c = \sqrt{GM/r}$. In real galaxies, the function $V_c(r)$ strongly depends on the nature of distribution of the mass in the system. We examine the circular velocities of different galactic components. The bulge dominates in the very central part of the galaxy (Fig. 1.34a) The disk component with the radial scale L for the infinitely thin exponential disk gives the maximum circumferential velocity at the radius $r \simeq 2.2\,L$ (Fig. 1.34b). For the fixed central density of the disk an increase of the ratio Δ_*/L increases the maximum velocity and the corresponding radius decreases. The circular velocity at the periphery of the galaxy is determined usually by the halo (Fig. 1.34c). The dimensions of the regions in which specific components dominate depend strongly on the Galaxy parameters.

The distribution of mass in the galaxies is usually described by the multicomponent models including both stellar and gas subsystems [228].

1.5. Active galaxies

In the group of the observed galaxies there is a small group characterised by extremely high activity (non-stationary nature) of the galactic nuclei – central regions of the galaxies with the size not exceeding 1÷10 pc. The galaxies with the non-stationary nuclei are usually subdivided into four types: Seyfert galaxies, radio galaxes, lacertides and quasars. The activity of the nuclei is evident mostly in the generation of powerful radiation in the volume whose radius is negligible in comparison with the size of the galaxy. This radiation provides a significant contribution to the general luminosity of the system or even completely determines the luminosity in the case of quasars. In most cases, radiation is non-thermal and in many cases the dominant factor in the irradiation of the nucleus of the objects is not the visible light but radio-, infrared, ultraviolet or x-ray radiation. Another special feature is the short duration of variability of radiation (typical values $\tau \sim 10^5 \div 10^6$ s) so that we can determine the limit of the linear size of the source $l \lesssim c\tau \simeq 3 \cdot 10^{15} \div 3 \cdot 10^{16}$ cm (c is the speed of light). Thus, the active nuclei are compact objects and are characterised by a very high energy density.

The Seyfert galaxies are usually spiral galaxies of the type S0–Sb, often having a bar (~70%). The spectrum of the Seyfert galaxies consists of a bright non-continuous component and also wide lines. The nuclei of the Seyfert galaxies belong to the most powerful ($10^{39} \div 10^{45}$ erg/s) sources of non-thermal radiation with a continuous spectrum in the range from 10^{12} to 10^{22} Hz. In the x-ray range, the luminosity may reach $10^{42} \div 10^{45}$ erg/s. As regards the normal spiral galaxies, the frequency of observation of the safer the galaxies is ~1%; the latter can be used to estimate the duration of the active phase of the nucleus at ~10^8 years.

Radiation varies rapidly. In different objects, the characteristic time of variation of the luminosity is $10^3 \div 10^5$ s, and Seyfert galaxies in which the variation of the x-ray flux takes place only within 100 seconds are also known (for example, NGC 6814). The observed wide emission lines indicate the movement of gas at high velocity.

It should be mentioned that more than 50% of the Seyfert galaxies are found in multiple systems [438]. There is a number of studies in which the results of comparison of the active and non-active galaxies are used to conclude that the environment has only a slight effect on the activity of the nuclei [1123]. However, it is quite possible that the role of interactions with small satellites and their subsequent absorption is underestimated [1181]. The comparative analysis of the central regions of the Seyfert and normal galaxies gives on average a bluer colour for the active nuclei. This is associated with the increase in intensity of star formation in the inner regions of the Seyfert galaxies [71]. The observations of some normal spiral galaxies (including our Galaxy, see below) indicate that they are characterised by weak nuclear activity. Evidently, there is a smooth transition between the 'normal' and Seyfert galaxies.

Radio galaxies

The power of radio radiation of the normal galaxies is lower than the power of their optical radiation. The radio galaxies are usually objects in which the radio luminosity is several orders of magnitude greater than in the normal galaxies. The power of radio radiation in them is $\sim 10^{42} \div 10^{44}$ erg/s, and they are usually classified as giant elliptical galaxies (type E).

The charts of distribution of radio brightness indicate that the radio galaxies (with the exception of those found in rich clusters) usually have two radiating clouds (radio 'ears'), connected with the nucleus by radio jets. Usually, these clouds are situated at distances of $10 \div 500$ kpc from the galaxy far outside the limits of its stellar component. The radio sources form as a result of generation of energy in the nucleus of the galaxy which is accompanied by the ejection of plasma jets of relativistic particles in two opposite directions. Evidently, an important role is played here by the bipolar nature of the magnetic field of the nucleus of the galaxy. This is indicated by jets detected on the basis of synchrotron radiation in some radio galaxies.

Lacertides (according to the object BL Lacertac) represent a small group of galaxies (no more than 100 have been found). Their distinguishing feature is: there is a point source, characterised by optical variability with a high amplitude (by up to 100 times), variable radio radiation and marked polarisation of radiation. The characteristic duration of the variability of radiation, equalling weeks and months, can be used to determine the size of the radiating region of the given objects at $\sim 10^{16} \div 10^{17}$ cm.

Quasars

The characteristic feature of the quasars with extremely high luminosities $\sim 10^{46} \div 10^{47}$ erg/s is their rapid variability in different ranges of the spectrum. The estimates of their size gives the values $\geqslant 3 \cdot 10^{15}$ cm. The quasars represent a larger scale form of the Seyfert galaxies where the galaxy itself is simply not visible on the background of the active nucleus, which determines the energy parameters of the entire objects. In some cases, traces of the appropriate galaxy can be detected around quasars situated at not too large distances (for example [398, 408]).

Supermassive black holes (SBH)

Of the most widely used models at the present time, which explains the activity of the nuclei of the galaxies in the quasars, is the disk accretion of matter on supermassive black holes with the mass from 10^5 to several $10^9\ M_\odot$ (for example [28, 36, 490, 1057]). The observed luminosity (up to 10^{47} erg/s) and compactness are explained if it is assumed that $\sim 10\ M_\odot$ of matter arrives in the accretion disk per annum and this matter then accretes (falls) onto the massive black hole. Some spectral special features of the radiation of the active nuclei actually indicate the disk accretion on the massive black hole. They include the surplus of soft x-ray radiation in the range $\lesssim 1$ keV which is associated with the shortwave tail of ultraviolet radiation of the accretion disk [1040]. The high dispersion of the velocities of the stars $c(r)$ in the range $r \lesssim 1 \div 10$ pc is a direct confirmation of the existence of the supermassive black holes at the galaxies [268]. The dependence $c\ (r)$ gives the estimates of the mass

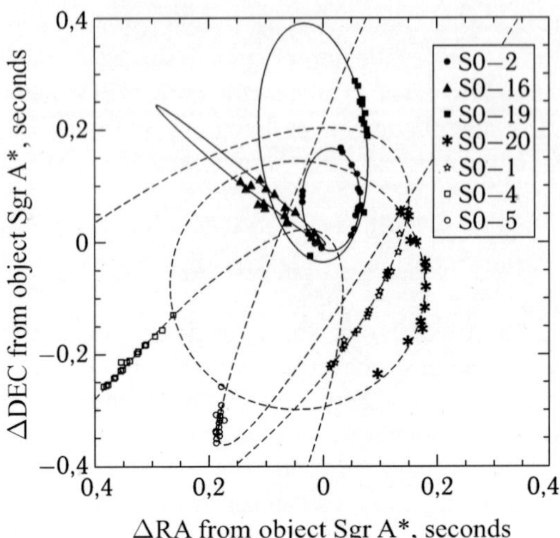

Fig. 1.35. Orbits of seven stars around the dynamic centre of the Galaxy (SBH) constructed on the basis of observations in the period 1995–2003 [617].

of the SBH for different galaxies in the range $M_{BH} = 10^6 \div 6 \cdot 10^9 M_\odot$ [873, 1236]. A more accurate method of determining the mass of the SBH is the method of echo mapping, based on the measurement of the delay of changes in the brightness of the emission lines in relation to the changes of the luminosity of the active nucleus in the continuous spectrum. It is quite possible that the centre of the majority of galaxies contains relatively massive black holes. Formation of the SBH, their role in the evolution of the parent galaxy, and the relationship with the integral parameters and special features of the kinematics are the subject of active studies and, therefore, the observational data on the kinematics of the gas and stars in the central part of the disk are very important [464].

The centre of our Galaxy

The nucleus of our Galaxy also shows signs of activity but the luminosity of the central region (1 kpc) does not evidently exceed 10^{42} erg/s. This region contains the rotating gas disk with a radius of $\simeq$600 pc and the mass of $\simeq 2 \cdot 10^7 M_\odot$. In the internal region ($r \lesssim 150$ pc) hydrogen is ionised and there discrete sources of thermal radiation with the size of ~10 pc and the mass of $\sim 10^3 \div 104\ M_\odot$. The rate of star formation is very high. The dynamics of gas in the very centre of the Galaxy can be explained by the presence of a massive black hole ($4 \cdot 10^6\ M_\odot$). The presence of the SBH in our Galaxy may be regarded as proven, taking into account direct observations of the movement of stars in the very centre on the Kepler orbits (Fig. 1.35).

The obvious activity (although very weak at the present time) of the centre of the Galaxy indicates that, evidently, our Galaxy passed approximately 10^7 years ago through the Seyfert stage in its evolution.

It can be seen that the activity of the galactic nuclei is determined to a large extent by the presence of gas. A mechanism for supplying the gas into the centre of

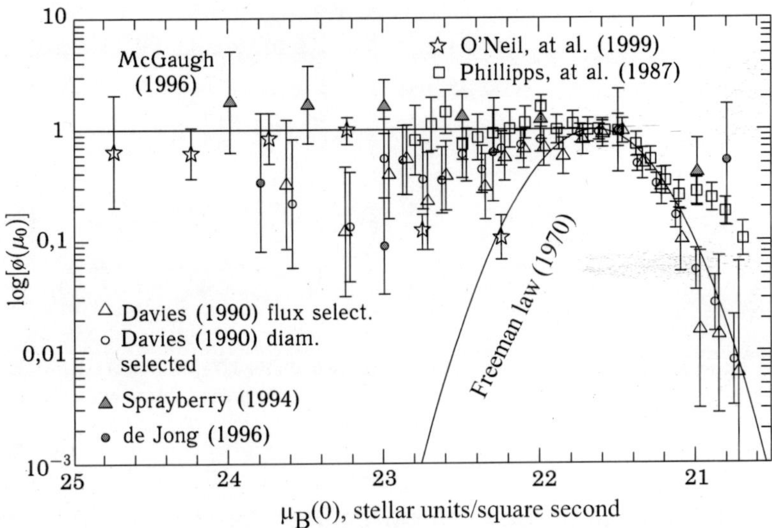

Fig. 1.36. Normalised distribution of the values of the central surface brightness of the galactic disks in the band B from study [992]. The symbols show the data of reviews by different authors (see the names in the figure) corrected with respect to observational selection. The continuous curve indicates the 'Freeman law'.

the galaxy is essential (see more details in section 8.4.4). The radial movement of matter may be associated with the non-axisymmetric potential (bar), gravitational instability of the gas in the nucleus. The destruction of stars may also be a significant source of gas.

1.6. Physical dependences for the parameters of the galaxies

1.6.1. The Distribution of the central surface brightness of Galactic disks

In [558] it was shown for a sample of spiral and S0 galaxies that the central surface brightness of the exponential disks in the band B is situated in a relatively narrow range $\langle \mu_0 \rangle = 21.65 \pm 0.30$ (σ) (the Freeman law). Investigations of the last couple of decades show that this hypothesis is not accurate. Figure 1.36 shows the distribution of the central brightness of the disk demonstrating the flat form in the range $\mu_0 = 21.5 \div 25$. Only objects brighter than $\mu_0 < 21.5$ are found far less frequently.

There is a systematic increase of the central surface brightness in the interacting galaxies for which $\langle \mu_0 \rangle \simeq 20.5$ [540] which is manifested by the effects of the gravitational and hydrodynamic interaction and a higher rate of star formation.

According to the data in [73], the central surface brightness in the band K depends only slightly on the morphological type (Fig. 1.37)

It should also be noted that the central surface brightness of the spiral galaxies increases with increase of the total luminosity (and, consequently, mass) of the galaxy [609].

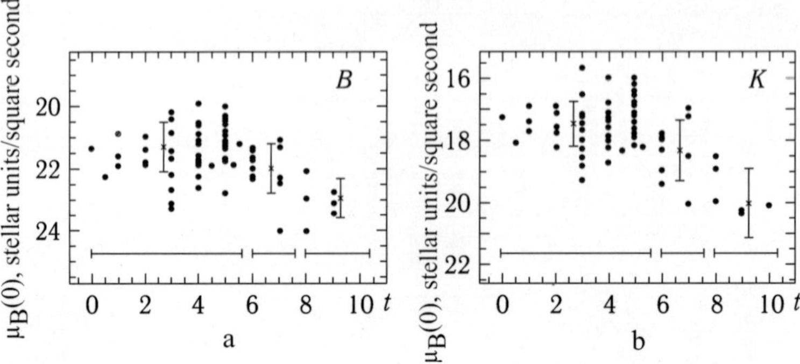

Fig. 1.37. Dependence of the central brightness of the disk on the morphological type of the galaxy t in the filters B (a) and K (b) [733]. The crosses indicate the mean values within the range of the values indicated along the horizontal axis of the parameter t.

1.6.2. Fundamental plane for the S-galaxies

A number of photometric and kinematic parameters of the galaxies are linked by statistical dependences. Usually, three parameters are defined:

- the energy characteristic which may be represented by the luminosity of the galaxy L, the absolute stellar value, for example M_I in the band I, the central surface brightness I_0;
- the spatial scale of the galaxy: the radius of the galaxy R (the radius of different isophots, for example $R_{23.5}$ or R_{25}), the radial exponential scale of the disk L_σ;
- the kinematic characteristics: the maximum velocity of rotation V, the width of the line W_{20} in respect of HI.

Combining any three parameters from the given set, any galaxy can be represented in the form of a point in a three-dimensional space (L, R, V) or (M_I, $R_{23.5}$, W_{20}) (Fig. 1.38). It appears that the galaxies are positioned mostly on some plane which is referred to as the fundamental plane [786]. Other scale relationships (Tully–Fisher, Freeman, the radius–velocity dependence) between M_I, $R_{23.5}$, W_{20} are projections on the appropriate planes. Evidently, the existence of the relationships of the type

$$L \propto V^m \cdot R^\ell \tag{1.22}$$

is determined by the dominance of the defined physical factors in the formation of spiral galaxies. A number of relationships of the type (1.22) have been proposed: $L \propto (VR)^{1/3}$ [786], $L_\sigma \propto V^{1.47} \cdot I_0^{-0.61}$ (in band H).

1.6.3. The Tully–Fisher dependence

We examine in greater detail the fundamental planes (L, V) and (V, R) which are referred to as the Tully–Fisher dependence [1240].

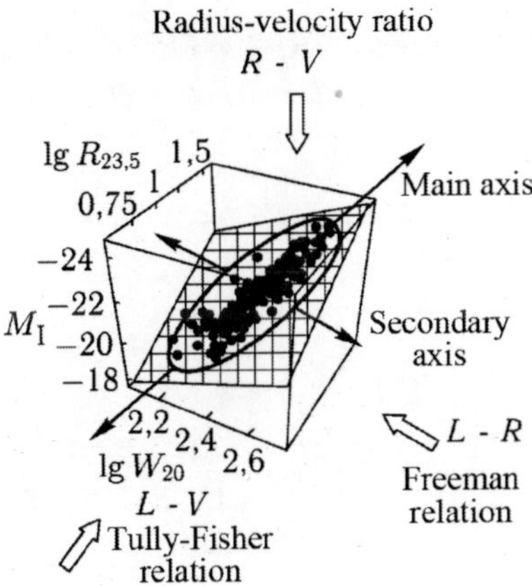

Fig. 1.38.

R, Tully and G. Fisher [1240] discovered the existence of a close correlation between the integral luminosity of the spiral galaxies and the amplitude of internal motion in them, determined on the basis of the width of the emission line of neutral hydrogen 21 cm. Since then, this relationship has become an effective tool for the mass determination of the distances to the spiral galaxies, regardless of their beam velocities. Identical relationships between the global parameters were determined for elliptical and lenticular galaxies. The application of these dependences makes it possible to evaluate the peculiar velocity of the galaxies and, the same time, investigate the field of non-Hubble motions of the galaxies. The Tully–Fisher dependence was constructed for greatly different groups of the galaxies, for example, visible edge-on [748], for the galaxies with extremely small mass [355].

The Tully–Fisher dependence shows that the galaxies with high velocities of internal motion are characterised by high luminosity. Or: as the luminosity of the galaxy increases, the velocity of the stars and the gas of the galaxy also increases. For the disk systems: as the spiral galaxy becomes brighter its velocity of rotation also increases. The maximum velocity of rotation V is linked with the luminosity (or mass) of the disk and, therefore, the Tully–Fisher dependence estimates the distances to the galaxies. This is especially important for relatively close objects. The Tully–Fisher dependence for spiral galaxies has the form:

$$L_B \propto V^{2,7}, \quad L_K \propto V^4 . \tag{1.23}$$

1.6.4. Surface brightness, bulge parameters

We will discuss several statistical dependences for the surface brightness of spiral galaxies [624, 625].

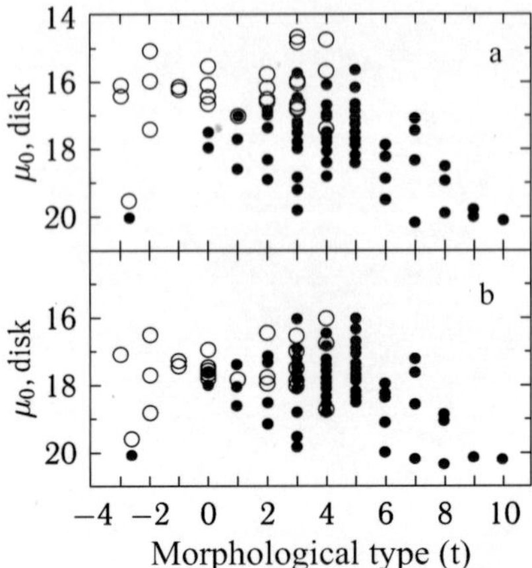

Fig. 1.39. Central surface brightness of the disk without taking into account the inclination (a) and corrected taking inclination into account (b) for the galaxies of different morphological types. The samples of the galaxies with the angle of inclination: $i > 60°$ (○) and $i \leqslant 50°$ (●) [625].

- In bright galaxies of the morphological types $t < 5$, the central surface brightness of the disk μ_{disk} does not correlate with the type (Fig. 1.39). μ_{disk} starts to decrease from $t > 5$.
- There is a correlation between the central surface brightness of the disk and the radial scale of the disk (Fig. 1.40). It should be mentioned that a weak dependence of the ratio of the luminosities of the bulge to the disk on the radial scale has been detected. The relative luminosity of the bulge is lower in the galaxies with the small radial scale of the disk.

The ratio of the bulge to the disk B/D is a key factor in the determination of the morphological type. We will discuss the correlation relationships between some other parameters of the bulge and the disk.

- As the effective radius of the bulge increases, the radial scale of the disk becomes greater [289]. For the galaxies of early morphological types we have $\langle r_e/L_e\rangle \simeq 0.2$ (Fig. 1.41).

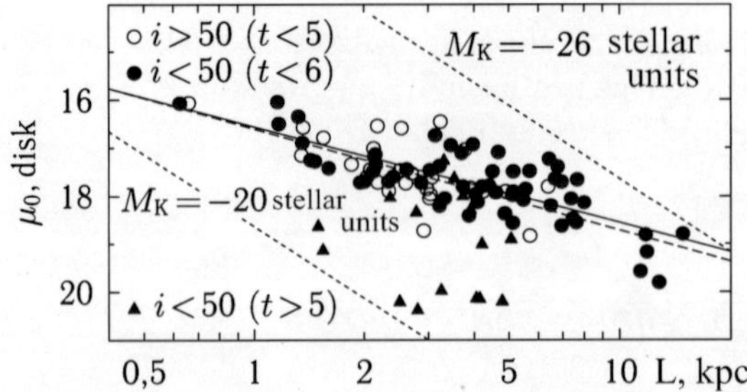

Fig. 1.40. The central surface brightness of the disk in the band K and the radial scale of the disk for different galaxies (t – morphological type, i – the angle of inclination), corrected taking the inclination of the disk into account [625].

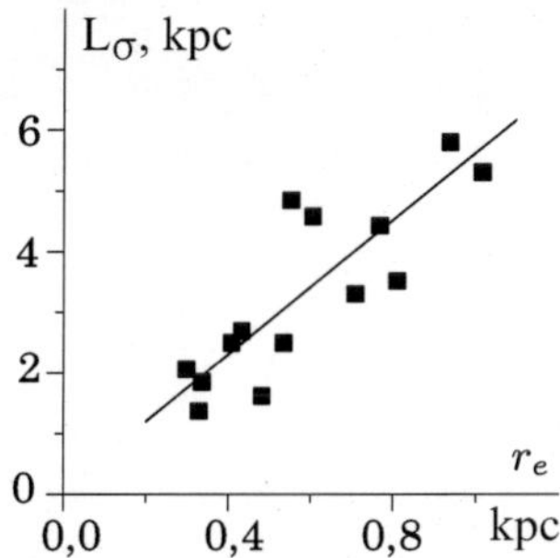

Fig. 1.41. The position of the galaxies on the 'radial scale of the disk L – effective bulge radius r_e' plane.

- The logarithm of the effective radius of the bulge lg (r_e) depends in a linear manner on the logarithm of the central dispersion of the velocities of the stars in the bulge c_0:

$$\lg(r_e) = 1.12(\pm 0.10)\lg(c_0) + 0.33(\pm 0.05)\mu_e - 8.68(\pm 0.08), \qquad (1.24)$$

  which describes the fundamental plane for the elliptical galaxies. Figure 1.42 shows the relationship (1.24) for the sample of E- and S0-galaxies [289], indicating agreement between the E-galaxies and the bulges in the disk galaxies;
- the properties of the bulges depend on the presence of the bar (Fig. 1.43). In particular, the relative luminosity of the bulge in the absence of the bar is on average higher than in the SB-galaxies. The distribution of brightness in the bulges of the galaxies without the bar is closer to the de Vacouleurs law than in the SB-galaxies in which the profiles of the brightness of the bulges is closer to exponential.

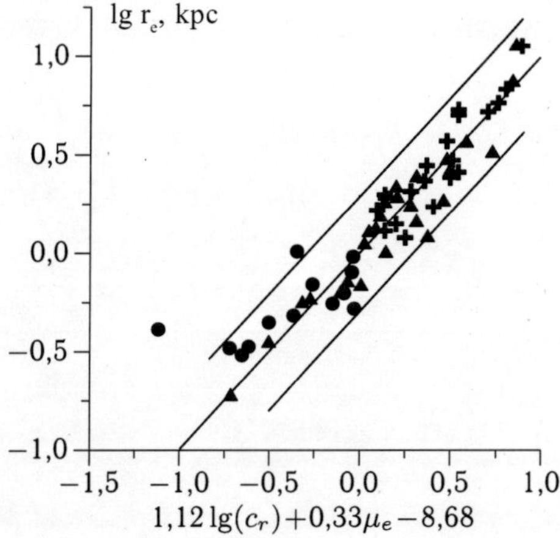

Fig. 1.42. Different symbols indicate the position of elliptical galaxies (triangles), bulges of lenticular galaxies with the bar (circle) and without the bar (crosses). Along the horizontal axis – the value determined by equation (1.24) (c_r is the dispersion of the velocities of the stars; μ_e is the surface brightness at the effective radius r_e).

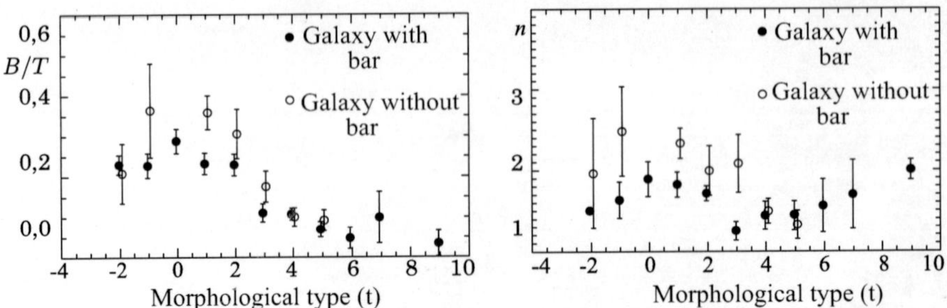

Fig. 1.43. Relative luminosity of the bulge B/T in the infrared band K and the parameter n (see (1.18)) for galaxies of different morphological types (t) [821]. The systems with and without the bulge are shown separately.

2

Dynamics of stellar disks

2.1. Equilibrium of the axisymmetric disk

2.1.1. Kinetic equation with self-gravitation taken into account

The characteristic time during which the parameters of motion of a separate star may greatly differ as a result of the paired gravitational interaction with other stars in the disks of flat galaxies is considerably longer than the duration of existence of these systems [266]. Therefore, the structure and dynamics of the stellar disks of the galaxies can be described with sufficient accuracy using the collisionless kinetic equation. In the Cartesian coordinates, this equation has the form

$$\frac{\partial f}{\partial t} + \mathbf{V}\frac{\partial f}{\partial \vec{r}} - \nabla\Phi\frac{\partial f}{\partial \mathbf{V}} = 0, \tag{2.1}$$

where $f\,(t, \mathbf{r}, \mathbf{V})$ is the function of distribution of the stars in the six-dimensional phase (coordinate–velocity) space, $\Phi(\mathbf{r}, t)$ is the gravitational potential. It should be stressed that the vector form has been used exclusively to shorten the equations, and in transition to another coordinate system the type of equation (2.1) changes[1].

The stellar disks of the galaxies are characterised by axial symmetry which is a natural consequence of their rotation. For this reason, to solve many problems of the dynamics of the star disk, it is convenient to use the equation (2.1) written in the cylindrical coordinates r, φ, z [203]:

$$\begin{gathered}\frac{\partial f}{\partial t} + V_r\frac{\partial f}{\partial r} + \frac{V_\varphi}{r}\frac{\partial f}{\partial \varphi} + V_z\frac{\partial f}{\partial z} + \left(\frac{V_\varphi^2}{r} - \frac{\partial \Phi}{\partial r}\right)\frac{\partial f}{\partial V_r} - \\ -\left(\frac{V_r V_\varphi}{r} + \frac{\partial \Phi}{r\partial \varphi}\right)\frac{\partial f}{\partial V_\varphi} - \frac{\partial \Phi}{\partial z}\frac{\partial f}{\partial V_z} = 0.\end{gathered} \tag{2.2}$$

[1]Details of the derivation of the kinetic equation in an arbitrary coordinate system can be found in the monograph by Polyachenko and Fridman [203].

Equation (2.2) could describe completely the dynamics of the star disk if the gravitational potential $\Phi\,(t, r, \varphi, z)$ was given. The value of the potential is determined by the distribution of the spatial densities of all subsystems of the galaxy, mainly the stars of the disk $\rho_*(t, \mathbf{r})$, spheroidal $\rho_h\,(t, \mathbf{r})$ and flat gas $\rho_g\,(t, \mathbf{r})$ components. In the volume occupied by the stellar disk, the quantity ρ_h is small in comparison with ρ_*, even if the mass of the disk is comparable with the mass of the halo (see section 1.4.5). Being interested in the dynamics of the stellar disk, the potential of the halo Φ_h can be regarded in the first approximation as external and independent of time.

In systems without the gas, the gravitational potential, which determines the dynamics of the stellar disk and can be regarded as almost completely self-consistent and, at the same time, the quantities Φ and f can be linked together through the volume density of the stellar disk ρ_* using the Poisson equation

$$\Delta\Phi \equiv \frac{1}{r}\frac{\partial}{\partial r}\left(r\frac{\partial\Phi}{\partial r}\right)+\frac{\partial^2\Phi}{r^2\partial\varphi^2}+\frac{\partial^2\Phi}{\partial z^2}=4\pi G\rho_*, \tag{2.3}$$

where G is the gravitational constant, and ρ_* and f are linked by the relationship

$$\rho_*(\mathbf{r},t)=\int f(\mathbf{r},\mathbf{V},t)d\mathbf{V}. \tag{2.4}$$

The relative smallness of the amplitudes of the non-axisymmetric perturbations of the density of the star disk has already been mentioned. In addition, if these special features are not taken into account, the distribution of different quantities in the disks remains almost completely constant during the time intervals of the order of several revolutions of the disk. Therefore, the distribution function of the stars of the disk can be represented in the form of the sum of stationary axisymmetric $f_0\ (r,\ z,\ V_r,\ V_\varphi,\ V_z)$ and 'perturbed' $f_1\ (t,\ r,\ \varphi,\ z,\ V_r,\ V_\varphi,\ V_z)$ components, where $|f_1/f_0| \ll 1$.

As already mentioned in the review of the observational data (section 1.2), the main type of motion of the stars in the disk is the rotation of the stars around the axis of symmetry of the flat galaxy. Consequently, it is convenient to separate the angular velocity of rotation of the disk

$$\Omega(r,z)=\int V_\varphi f_0 d\mathbf{V} \Big/ r\int f_0 d\mathbf{V} \tag{2.5}$$

and introduce the residual velocity $\mathbf{v}$ in the form of the relationships

$$v_r = V_r,\quad v_\varphi = V_\varphi - r\Omega,\quad v_z = V_z. \tag{2.6}$$

This transformation reduces the initial kinetic equation (2.2) to the form

$$\begin{aligned}&\frac{\partial f}{\partial t}+v_r\frac{\partial f}{\partial r}+\left(\Omega+\frac{v_\varphi}{r}\right)\frac{\partial f}{\partial\varphi}+v_z\frac{\partial f}{\partial z}+\left(\frac{v_\varphi^2}{r}+2\Omega v_\varphi+\Omega^2 r-\frac{\partial\Phi}{\partial r}\right)\frac{\partial f}{\partial v_r}-\\&-\left(\frac{v_r v_\varphi}{r}+\frac{\varkappa^2}{2\Omega}v_r+v_z r\frac{\partial\Omega}{\partial z}+\frac{\partial\Phi}{r\partial\varphi}\right)\frac{\partial f}{\partial v_\varphi}-\frac{\partial\Phi}{\partial z}\frac{\partial f}{\partial v_z}=0,\end{aligned} \tag{2.7}$$

where $\varkappa = 2\Omega\sqrt{1 + rd\Omega / 2\Omega dr}$ is the epicyclic frequency (the frequency of small oscillations of the stars of the disk in the radial direction – section 1.2).

2.1.2. Equilibrium function of distribution of stars

The equilibrium (stationary and axisymmetric) components of the distribution function of the stars of the disk f_0 according to (2.7) should satisfy the equation

$$v_r \frac{\partial f_0}{\partial r} + v_z \frac{\partial f_0}{\partial z} + \left(\frac{v_\varphi^2}{r} + 2\Omega v_\varphi + \Omega^2 r - \frac{\partial \Phi_0}{\partial r} \right) \frac{\partial f_0}{\partial v_r} - \\ - \left(\frac{v_r v_\varphi}{r} + \frac{\varkappa^2}{2\Omega} v_r + v_z r \frac{\partial \Omega}{\partial z} \right) \frac{\partial f_0}{\partial v_\varphi} - \frac{\partial \Phi_0}{\partial z} \frac{\partial f_0}{\partial v_z} = 0, \tag{2.8}$$

Here $\Phi_0 = \Phi_0\,(r, z)$ is the stationary axisymmetric part of the gravitational potential, described by the Poisson equation

$$\frac{1}{r}\frac{\partial}{\partial r}\left(r \frac{\partial \Phi_0}{\partial r} \right) + \frac{\partial^2 \Phi_0}{\partial z^2} = 4\pi G \rho_{*0} = 4\pi G \int f_0 d\mathbf{v}. \tag{2.9}$$

The system of equations (2.8), (2.9) is self-consistent and, in accordance with the previous considerations, it can be used to determine the functions f_0 and Φ_0. The system is relatively complicated, and for further advances in the solution of the given problem it is necessary to use the approximations based on the oscillation data.

These data show in particular that the characteristic half thickness of the start disk is considerably smaller than the disk radius R. This means that in the equation (2.9) $\left|\partial^2\Phi_0 / \partial z^2\right| \gg \left|\partial(r\partial\Phi_0 / \partial r) / r\partial r\right|$ and in the simplest approximation of the disk with the uniform thickness (2.9) gives the estimate of the frequency of oscillations of the stars across the plane of the disk (see equation (1.7)):

$$\omega_z \simeq \sqrt{4\pi G \rho_{*0}(z = 0)} = \sqrt{2\pi G \sigma_{*0} / \Delta_*}, \tag{2.10}$$

where $\sigma_{*0}(r) = 2\,\Delta_*\rho_*0\;(r, z = 0)$ is the equilibrium surface density of the star disk. Evidently, for the given value of σ_{*0} the quantity $\Omega_z \to \infty$ at $\Delta_* \to 0$. In the same limit, the frequency of the oscillations of the stars in the disk plane remains a finite value [1253]. Therefore, in the theory which investigates the structure of relatively thin stellar disks, there is a small parameter (section 1.2.2)

$$\varepsilon = \varkappa / \omega_z \ll 1. \tag{2.11}$$

The part of the kinetic equation (2.8) linked with the z-motion of the stars has the form

$$\hat{\mathcal{L}} f_0 \equiv v_z \frac{\partial f_0}{\partial z} - \frac{\partial \Phi_0}{\partial z}\frac{\partial f_0}{\partial v_z} \sim \omega_z f_0 = \varepsilon^{-1} \varkappa f_0. \tag{2.12}$$

The observational data also show (section 1.2) that the characteristic dispersion c_r, c_φ, c_z of the components of the residual velocities of the stars (equation (2.6)) is small in comparison with the speed of rotation $V = r\Omega$ of the stars of the disk around the centre of the mass of the galaxy:

$$c / V \ll 1, \tag{2.13}$$

where $c = \max\{c_i\}$, $i = r, \varphi, z$. The small value of this parameter indicates that $v_\varphi^2 / r \ll 2\Omega | v_\varphi |$; $| v_r v_\varphi / r | \ll (\varkappa^2 / 2\Omega) | v_r |$ and, consequently, the terms $v_\varphi^2 / \partial f_0 / r\partial v_r \lesssim c f_0 / r$ and $v_r v_\varphi \partial f_0 / r\partial v_\varphi \lesssim c f_0 / r$ in equation (2.8) can be ignored. The approximation in which these terms are ignored is referred to as the epicyclic approximation. The term $v_r \partial f_0 / \partial r \sim c\, f_0 / L$ is not ignored because the characteristic scales L of the radial inhomogeneity of the parameters of the disks outside the limits of the central part of the disk are small in comparison with the distance to the disk centre [558].

The approximation, determined by the inequality (2.13), shows that the equilibrium of the disk in the radial direction is determined by the balance of the centrifugal and equilibrium of the gravitational forces:

$$\Omega^2 r - \frac{\partial \Phi_0}{\partial r} = 0 \tag{2.14}$$

with the accuracy to the terms $\lesssim c^2 / L \ll \Omega^2 r$.

The previously described estimate $\partial\Phi_0/\partial z \propto \varepsilon^0$ shows that the z-dependent part $\delta\Phi_0$ of the equilibrium gravitational potential $\delta\Phi_0 \propto \varepsilon^2$. This quantity also determines the order with respect to the parameter ε of the component of the angular velocity of rotation of the disk which depends on the z-coordinate. Actually, since the complete $\Omega(r, z) = \Omega_0(r) + \delta\Omega(r, z)$, where $\delta\Omega\,(r, z = 0) = 0$, from the equilibrium condition of the disk (2.14) one obtains

$$\delta\Omega \simeq \frac{\partial \delta\Phi_0}{2r\Omega_0 \partial r} \propto \varepsilon^2, \tag{2.15}$$

and, consequently,

$$v_r r \frac{\partial \Omega}{\partial z} \frac{\partial f_0}{\partial v_\varphi} \propto \varepsilon f_0. \tag{2.16}$$

Deducting these contribution and $\hat{\mathcal{L}} f_0$ (see (2.12)), the remaining set of the terms in (2.8) is

$$\hat{D}_0 f_0 \equiv v_r \frac{\partial f_0}{\partial r} + 2\Omega v_\varphi \frac{\partial f_0}{\partial v_r} - \frac{\varkappa^2}{2\Omega} v_r \frac{\partial f_0}{\partial v_\varphi} \sim \varkappa f_0 \propto \varepsilon f_0. \tag{2.17}$$

The results, estimating in accordance with (2.12), (2.16) and (2.17) the orders of the terms of the kinetic equation with respect to the small parameter ε, enabled the transition to solving the given problem determination of f_0 and Φ_0 by the method of successive approximations. To utilise this possibility, it should be mentioned that from the elementary estimate $\partial^2\Phi_0/\partial z^2 \propto \varepsilon^{-2}$ we obtain $\rho_{*0} \propto \varepsilon^{-2}$ and, therefore, $f_0 \propto \varepsilon^{-3}$. Therefore, the expansion of the required functions f_0 and Φ_0 into series in respect of the powers of the parameter ε should have the following form

$$f_0 = f_{0,-3} + f_{0,-2} + f_{0,-1} + ..., \quad \Phi_0 = \Phi_{0,0}(r) + \Phi_{0,2}(r,z) + \Phi_{0,4}(r,z) + ..., \tag{2.18}$$

where $f_{0,n} \propto \varepsilon^n$; $\Phi_{0,n} \propto \varepsilon^n$. The expansion $\Phi_0(r, z)$ with regard to the even powers is determined by the fact that its dependence on z is determined only by the parameters of the disk thickness $\Delta_* \propto \varepsilon^2$. This claim is also valid for ρ_{*0} (r, z) but should also be fulfilled for f_0 $(r, z, \mathbf{v})$.

The expansions (2.18) are substituted in the initial system of the equations (2.8), (2.9). Consequently, in the two main orders with respect to parameter ε we obtain

$$\hat{\mathcal{L}}_0 f_{0,-3} = 0, \tag{2.19}$$

$$\hat{\mathcal{L}}_0 f_{0,-2} + \hat{D}_0 f_{0,-3} = 0, \tag{2.20}$$

$$\frac{\partial^2 \Phi_{0,2}}{\partial z^2} = 4\pi G \rho_{*0,-2}, \tag{2.21}$$

where

$$\hat{\mathcal{L}}_0 = v_z \frac{\partial}{\partial z} - \frac{\partial \Phi_{0,2}}{\partial z} \frac{\partial}{\partial z}. \tag{2.22}$$

If the projection of the trajectory of the star on the plane $z = 0$ would have the form of a circle, then as a result of the stationary form and axisymmetry of the equilibrium state of the disk the value

$$E_z = \frac{v_z^2}{2} + \Phi_0(r,z) - \Phi_{0,0}(r) \tag{2.23}$$

would be the integral of motion. In reality, the radial coordinate of the star is a slowly changing parameter (owing to the fact that $\varepsilon = \varkappa / \Omega_z \ll 1$). Therefore, the adiabatic invariant $J = \text{const}\ E_z/\Omega_z$ can be the approximately retained [113], and to solve the system of equations described above it would be convenient to transfer from the variables z, v_z to the variables of the action–angle type, i.e. J, w [1253]:

$$J = \oint v_z\, dz, \quad w = \int_0^z dz' / v_z. \tag{2.24}$$

In these variables, the system (2.19), (2.20) has the following form:

$$\hat{\mathcal{L}}_0 f_{0,-3} \equiv \frac{\partial f_{0,-3}}{\partial w} = 0, \tag{2.25}$$

$$\frac{\partial f_{0,-2}}{\partial w} + \hat{D}_0 f_{0,-3} = 0. \tag{2.26}$$

This shows that $f_{0,-3}$ is independent of the variable w. Since the definition of $f_{0,-3}$ is also our task, the equation (2.25) will be integrated with respect to the total period of variation of w. Consequently, we obtain

$$v_r \frac{\partial f_{0,-3}}{\partial r} + 2\Omega v_\varphi \frac{\partial f_{0,-3}}{\partial v_r} - \frac{\varkappa^2}{2\Omega} v_r \frac{\partial f_{0,-3}}{\partial v_\varphi} = 0. \tag{2.27}$$

We now transfer to the variables r_0, $v_\perp$, linked with the values r, v_r, v_φ by the relationships:

$$v_r = v_\perp \cos\alpha; \quad v_\varphi = v_\perp \frac{\varkappa}{2\Omega} \sin\alpha; \quad r = r_0 - \frac{v_\perp}{\varkappa} \sin\alpha, \tag{2.28}$$

$$v_\perp = \sqrt{v_r^2 + \left(\frac{2\Omega v_\varphi}{\varkappa}\right)^2}; \; r_0 = r + \frac{2\Omega}{\varkappa^2} v_\varphi; \; \alpha = \mathrm{arctg}\left(\frac{2\Omega v_\varphi}{\varkappa v_r}\right). \tag{2.29}$$

As indicated in the following section, the value $v_\perp$ has the meaning of the amplitudes of the radial velocity of the star, r_0 is the radius of the circular orbit of the centre of the epicycle of the star, and the quantity α is the phase of movement of the star along the epicycle. Since in the framework of the epicyclic approximation used here $|(r - r_0)/r_0| \sim c/V \ll 1$, then in (2.28), (2.29) it should be assumed that $\Omega = \Omega(r_0)$, $\varkappa = \varkappa(r_0)$. It can easily be seen that in the new variables (2.29) the equation (2.25) has the form

$$-\varkappa \frac{\partial f_{0,-3}}{\partial \alpha} = 0. \tag{2.30}$$

Taking into account equation (2.25), the required equilibrium distribution function of the stars is

$$f_{0,-3} = f_0(r_0, v_\perp, J). \tag{2.31}.$$

This shows that in the framework of the approximations used here (the epicyclic and main order with respect to parameter ε), the quantities r_0, $v_\perp$, J are the integrals of motion.

The specific form of this function should be determined on the basis of additional data. The distribution function f_0 can be constructed directly on the basis of the observations at the present time only in the solar vicinity of the Galaxy. The inequality of the dispersions of the velocities of the galactic stars in three mutually perpendicular directions is the obvious experimental fact (section 1.2.4). There is a number of analytical approximations of the observed distributions of the velocity of the stars which are in agreement with the observations only in limited velocity ranges (see the literature review [274]). The best known distribution is the Schwarzschild distribution which is an anisotropic Maxwellian distribution:

$$f(\mathbf{v}) = A \cdot \exp\left\{-\frac{v_r^2}{2c_r^2} - \frac{v_\varphi^2}{2c_\varphi^2} - \frac{v_z^2}{2c_z^2}\right\}. \tag{2.32}$$

Considerable effort has been made on the one side to justify this law theoretically [862, 1159] and, on the other side, to eliminate contradictions with the observations by increasing the number and accuracy of observations. In the final analysis, it may be concluded that the satisfactory agreement of the law (2.32) with the observations can be achieved only for low and medium velocities of the stars. This problem was studied in detail in [46]. The differences for higher velocities are significant [274]. However, regardless of certain difficulties, mainly of the theoretical aspects, the Schwarzschild law is most popular in star astronomy. We shall use the distribution in the form (232)[2]. In accordance with this, it is assumed that

$$f_{0,-3} = A(r_0)\ \exp\left\{-v_\perp^2 / 2c_r^2(r_0) - J / \tau(r_0)c_z^2(r_0)\right\}, \tag{2.33}$$

where $\tau = 2\pi/_{\Omega z}$, the parameters $c_r(r_0)$, $c_z(r_0)$ have the obvious meaning of the dispersions of the velocities of the stars in the radial and perpendicular directions in relation to the disk plane, respectively, and equation (2.29) shows clearly that the role of the dispersions of the velocities of the stars in the azimuthal direction c_φ is played by the quantity

$$c_\varphi(r_0) = \frac{\varkappa(r_0)}{2\Omega(r_0)} c_r(r_0). \tag{2.34}$$

This relationship between c_r and c_φ corresponds with sufficient accuracy to the currently available data obtained in astronomical observations [89, 229, 507, 1048].

The equilibrium distribution of the volume density in the disk will be calculated. For this purpose, the equation (2.33) is integrated over the space of the velocities

$$\begin{aligned} \rho_{*0}(r,z) &= \int f_{0,-3} dv_r dv_\varphi dv_z = \frac{\varkappa}{2\Omega} \int f_{0,-3} v_\perp dv_\perp d\alpha dv_z = \\ &= \rho_{*0}(r_0, z=0) \cdot \exp\{-[\Phi_0(r_0,z) - \Phi_{0,0}(r_0)] / c_z^2(r_0)\}, \end{aligned} \tag{2.35}$$

where

$$\rho_{*0}(r_0, z=0) = (2\pi)^{3/2} c_r^2 c_z A \frac{\varkappa}{2\Omega} = (2\pi)^{3/2} c_r c_\varphi c_z A, \tag{2.36}$$

A is the dimensional constant [see (2.33)]. It is now assumed that

$$\Phi_0(r_0,z) - \Phi_{0,0}(r_0) \simeq \Phi_{0,2}(r_0,z) = -c_z^2 \ln\{F(r_0,z)\}. \tag{2.37}$$

Consequently, the Poisson equation (2.21) in the main order with respect to the small parameter ε changes to the equation which determines the function $F(r_0, z)$:

$$\frac{d^2F}{dz^2} - \frac{1}{F}\left(\frac{dF}{dz}\right)^2 + \frac{4\pi G \rho_{*0}(r_0, z=0)}{c_z^2} F^2 = 0. \tag{2.38}$$

It may be seen that the solution of this equation decreasing at $|z| \to \infty$ has the form

$$F(r_0,z) = \mathrm{ch}^{-2}\{z / \Delta_*(r_0)\}, \tag{2.39}$$

[2]Analysis of the effect of the possible deviation of the function of the residual velocity distribution of the stars from anisotropic Maxwell's distribution on the gravitational stability of the stellar disk is described in section 2.3.4.

where the parameter

$$\Delta_*(r_0) = \sqrt{c_z^2(r_0)/2\pi G\rho_{*0}(r_0, z=0)} = \frac{c_z^2(r_0)}{\pi G\sigma_{*0}(r_0)} \tag{2.40}$$

has the obvious meaning of the characteristic half thickness of the star disk, and the quantity

$$\sigma_{*0}(r_0) = \int_{-\infty}^{\infty} \rho_{*0}(r_0, z)dz = 2\Delta_*(r_0)\rho_{*0}(r_0, z=0) \tag{2.41}$$

is the equilibrium surface density of the star disk.

The results will be summed up. The equilibrium distribution function of the stars in the main order with respect to the small parameter $\varepsilon = \varkappa/\Omega_z \ll 1$ and in the epicyclic approximation, determined by the inequality (2.13), has the form

$$f_0 = \frac{\rho_{*0}(r_0, z)}{(2\pi)^{3/2} c_r(r_0)c_\varphi(r_0)c_z(r_0)} \exp\left\{-\frac{v_r^2}{2c_r^2(r_0)} - \frac{v_\varphi^2}{2c_\varphi^2(r_0)} - \frac{v_z^2}{2c_z^2(r_0)}\right\}, \tag{2.42}$$

where

$$\rho_{*0}(r_0, z) = \frac{\sigma_{0*}(r_0)}{2\Delta_*(r_0)} \mathrm{ch}^{-2}\left(\frac{z}{\Delta_*(r_0)}\right), \tag{2.43}$$

and the distribution of the equilibrium gravitational potential across the disk plane is described by the relationship

$$\Phi_0(r_0, z) = \Phi_0(r_0, z=0) + 2\pi G\sigma_{*0}(r_0)\Delta_*(r_0)\ln\left\{\mathrm{ch}\left(\frac{z}{\Delta_*(r_0)}\right)\right\}. \tag{2.44}$$

As indicated by (2.42)–(2.44), the equilibrium of the disk is determined by the distributions of the following parameters: $\Omega(r_0), \varkappa(r_0), c_r(r_0), c_\varphi(r_0), c_z(r_0), \sigma_{*0}(r_0), \Delta_*(r_0)$. However, not all these quantities are independent. For example, $\varkappa(r_0)$ is unambiguously determined through the angular velocity $\Omega(r_0)$ by the relationship (1.8), and the parameters c_r, c_φ, and also c_z, σ_{*0}, Δ_* are connected by the equations (2.34) and (2.40), respectively. Another relationship between the equilibrium parameters of the star disk could be provided by the equation (2.14), if the mass of the disk in the flat galaxies would be considerably greater than the mass of the spheroidal subsystems. In this case, we could link $\Phi_{0,0}(r_0, 0)$ [and through the Poisson equation also $\sigma_{*0}(r_0)$] with the angular velocity of rotation of the disk $\Omega(r_0)$. Unfortunately, this condition is not fulfilled and, consequently, in the framework of the previously described theory there are four independent parameters describing the star disk. As indicated by further considerations, further two relationships between the equilibrium parameters of the star disk can in principle be obtained from the conditions of its marginal[3] stability in relation to the bending perturbations and perturbations in its plane. Consequently, only two parameters remain dependent and, therefore,

[3]i.e. At the stability boundary.

requiring definition from the observations. One of these parameters is usually the observed angular velocity of rotation of the disk Ω. The second parameter can be either one of the dispersions of the velocities of the stars (c_r, c_φ, c_z) in the disk or its surface density σ_{*0}. The value of the latter parameter is not determined convincingly from the observations, but the data obtained in the multicolour photometry of the stellar disks shows that $\sigma_{*0} \propto \exp\{-r/L_\sigma\}$ beyond the limits of the central ($r \gtrsim L_\sigma$) part of the disk [558, 800].

In many cases, the Miyamoto–Nagai model is used for estimation of the disk potential in the form

$$\Phi(r,z) = -\frac{GM_d}{\sqrt{r^2 + \left(b + \sqrt{z^2 + c^2}\right)^2}}, \tag{2.45}$$

where b, c determine the radial and vertical scale, respectively.

In this section we have constructed, following van der Voort [1253], a relatively simple model of the star disk. More complicated models of the structure of the disk across explain can be found in, for example, the already mentioned study by van der Voort and Bahcall [323]. The latter takes into account in particular the contribution of the volume density of the stars of the spheroidal subsystem (see also study [174]). However, we will not discuss this. Firstly, this is due to the fact that the relatively thin components of the models are difficult to compare with the results of observations. Secondly, the problem of construction of the theory of stability of these models of stellar disks in relation to both the bending perturbations and the perturbations in the plane of the disks has not been solved and consequently the dynamics of the star disk is not completely understood within the framework of these models.

2.1.3. The Jeans equations

Investigation of the consequences of the Boltzmann collisionless equation will be continued. Equations will be written for the moments of the distribution function. These equations, which link the macroscopic parameters (density, mean velocity of motion, dispersion of the velocities) are referred to as the Jeans equations. Since the collision analogue of these equations in accordance with the derivation and structure is represented by the system of equations of the solid medium (hydrodynamics equations), the term 'collisionless hydrodynamics' is often used in these cases.

We confine ourselves to the stationary and axisymmetric cases. Integration of the Boltzmann equation leaves to the continuity equation:

$$\frac{\partial(r\varrho\langle u\rangle)}{r\partial r} + \frac{(\varrho\langle w\rangle)}{\partial z} = 0, \tag{2.46}$$

which is identical with the law of conservation of mass in hydrodynamics with the accuracy to the notations ($\langle ... \rangle$ are the mean values of the appropriate quantities).

After multiplication of the Boltzmann equation by V_r and V_z and integration, we obtain the Jeans equations:

$$\frac{\partial(\varrho\langle u^2\rangle)}{\partial r}+\frac{\partial(\varrho\langle uw\rangle)}{\partial z}+\varrho\frac{\langle u^2\rangle-\langle v^2\rangle}{r}=-\varrho\frac{\partial\Phi}{\partial r}, \tag{2.47}$$

$$\frac{\partial(\varrho\langle w^2\rangle)}{\partial z}+\frac{\partial(\varrho\langle uw\rangle)}{\partial r}+\frac{\varrho\langle uw\rangle}{r}=-\varrho\frac{\partial\Phi}{\partial z}. \tag{2.48}$$

In a general case, the motions in the radial and vertical directions prove to be dependent because of the terms containing $\langle uw\rangle$. The Jeans equations are discussed in greater detail in, for example [94, 131, 203, 266, 1249].

The circular velocity V_c is determined from the equation

$$\frac{V_c^2}{r}=\frac{\partial\Phi}{\partial r}, \tag{2.49}$$

and the residual velocity of the star $\tilde{v}$ after deducting the mean velocity $V_{\varphi 0}$ is:

$$\langle v^2\rangle = V_{\varphi 0}^2+\langle\tilde{v}^2\rangle. \tag{2.50}$$

Taking into account (2.47), (2.48), we obtain the equation of the asymmetric drift

$$V_{\varphi c}^2-V_{\varphi 0}^2=-\langle u^2\rangle\left[\frac{\partial\ln(\varrho\langle u^2\rangle)}{\partial\ln(r)}+\left(1-\frac{\langle\tilde{v}^2\rangle}{\langle u^2\rangle}\right)+\frac{r}{\varrho\langle u^2\rangle}\frac{\partial(\varrho\langle uw\rangle)}{\partial z}\right], \tag{2.51}$$

which is valid for the arbitrary distribution function. The mean velocity of the stars is lower than the circular velocity in typical situations.

We derive an expression for the mean tangential velocity of the stars in a disk in terms more suitable for applications, introducing the dispersions of the velocities c_r, c_φ, c_z and the degree of anisotropy $\alpha_{rz}=\langle uw\rangle$ which is determined by the random component of the radial velocity u and vertical velocity w. From the Jeans equations, assuming the absence of systematic radial motion, axisymmetr and symmetry in relation to the plane $z = 0$, we obtain

$$V^2=(<v>)^2=V_c^2+c_r^2\left\{1-\frac{c_\varphi^2}{c_r^2}+\frac{r}{\varrho c_r^2}\frac{\partial(\varrho c_r^2)}{\partial r}+\frac{r}{c_r^2}\frac{\partial\alpha_{rz}}{\partial z}\right\}, \tag{2.52}$$

where the circular velocity V_c in the plane $z = 0$ is determined by the distribution of matter in three main components: disk, bulge and halo. Consequently

$$V_c=\sqrt{(V_c^{\text{disk}})^2+(V_c^{\text{bulge}})^2+(V_c^{\text{halo}})^2}, \tag{2.53}$$

where V_c^{disc}, V_c^{bulge} and V_c^{halo} is the corresponding contribution of the individual components to the circular velocity.

We make an important comment associated with the selection of the distribution function. The traditionally used Schwarzschild distribution (anisotropic Maxwellian distribution) cannot be realised in the actual star systems. In particular, this problem has been examined for the solar vicinity of our Galaxy where the direct construction of the velocity distribution function is possible – evidently, there are systematic deviations of the observed distribution from the Schwarzschild distribution.

2.1.4. The vertical structure of the star disk

We return to the problem of the vertical balance of the forces in the disk. The vertical structure of the disk is determined by the equations (section 2.1.3)

$$\frac{\partial(\varrho c_z^2)}{\partial z}+\frac{\partial(r\varrho\alpha_{rz})}{r\partial r}=-\frac{\partial\Phi}{\partial z}\varrho,$$
$$\frac{\partial}{r\partial r}\left(r\frac{\partial\Phi}{\partial r}\right)+\frac{\partial^2\Phi}{\partial z^2}=4\pi G(\varrho+\varrho_s), \tag{2.54}$$

where ϱ and ϱ_s is the volume density of matter in the disk and spheroidal subsystems, respectively; c_z is the dispersion of the vertical velocities, $\alpha_{rz}=\langle uw\rangle$. Excluding the potential Φ in the equations (2.54), and introducing the circular velocity

$$V_c(r)\equiv\sqrt{r\left(\frac{\partial\Phi}{\partial r}\right)_{|z=0}}, \tag{2.55}$$

for the given values c_z, α_{rz} we can transfer from the system of the differential equations in the partial derivatives to the approximate equation for the volume density of the disk $\varrho(z)$ in the form of the conventional differential equation (it is assumed that all the derivatives with respect to r are known):

$$\varrho\frac{d}{dz}\left(c_z^2\frac{d\varrho}{dz}\right)-c_z^2\left(\frac{d\varrho}{dz}\right)^2+\varrho^2\frac{d^2c_z^2}{dz^2}+4\pi G\varrho^2\Big(\varrho+E+\varrho_s(z)\Big)+\varrho^2\frac{d}{dz}\frac{E_\alpha}{\varrho}=0, \tag{2.56}$$
$$E=-\frac{1}{4\pi Gr}\frac{\partial V_c^2}{\partial r},\quad E_\alpha=\frac{\partial(r\varrho\alpha_{rz})}{r\partial r},$$

which is compatible with the conditions:

$$\varrho(z=0)=\varrho_0,\ d\varrho(0)/dz=0,\ \int_{-\infty}^{\infty}\varrho(z;r)dz=\sigma(r)$$

and for the given distribution of the surface density σ determines the vertical structure of the disk at the fixed radius r. For $c_z=\text{const}$ and $\alpha_{rz}=0$, the equation coincides with that investigated in [323]. The third term in (2.56) in the case of weak vertical inhomogeneity c_z is small in comparison with the first two terms.

Equation (2.56) is approximate because the Vc, introduced by the relation (2.55), does not depend on the z-coordinate. In the initial equation (2.54), the term

$\frac{1}{r}\frac{\partial}{\partial r}\left(r\frac{\partial\Phi}{\partial r}\right)$ depends in a general case on the z-coordinate. Therefore, in transition

to V_c, the dependence on the vertical coordinate disappears and it is then assumed that E (see (2.56) does not depend on z (but it depends parametrically on r). This approximation is the 'payment' for the transition from the differential equations in the partial derivatives (2.54) to the conventional differential equation (2.56). It may easily be seen that the relative error has the order $(h/r)^2$ (h is the vertical scale of the disk). In the case in which $\partial c_z/\partial_z = 0$, $\varrho_s = 0$, $V_c = \text{const}$, $E_\alpha = 0$, the solution of equation (2.56) is

$$\varrho(z) = \frac{\sigma}{2z_0}\mathrm{ch}^{-2}(z/z_0) \tag{2.57}$$

with the characteristic vertical scale $z_0 = \sqrt{c_z^2/2\pi G\varrho(0)} = c_z^2/\pi G\sigma$. The law (2.57) together with the exponential law

$$\varrho(z) = \varrho(0)\cdot\exp(-z/h_{\exp}), \tag{2.58}$$

is often used for the approximation of the observed profiles of brightness for the edge-on galaxies[4]. At large distances from the plane $z = 0$, these laws have a single asymptotics and in this case $z_0 \simeq 2h_{\exp} \simeq 2h_{\mathrm{ch}}$.

2.2. The dynamics of perturbations in the disk plane

In the previous section, we constructed a model of the stationary equilibrium stellar disk. The deviations of the disk parameters (distribution function, potential, etc.) from the equilibrium values will be referred to as perturbations of a specific quantity. The assumption of the low value of the amplitude of deflections from the equilibrium state is very attractive for theoretical investigations because in this case one of the main problems in solving the kinetic equation – its non-linearity – disappears. The investigations of possible special structures on the background of the axisymmetric stationary state are very popular. At the same time, in the case of real systems, the assumption of the low value of the amplitudes of the perturbations is often violated. Only small advances have been made in the construction of non-linear models (in this case, we are not talking about numerical models, see Chapter 3). However, the linear approach solves quite successfully the classic problem of the stability of the equilibrium system, answering the question of the reasons for the increase with time of specific perturbations with some small initial amplitudes. The analysis of linear stability is based on the dispersion equations which determine the temporal dynamics of the small perturbations in relation to their spatial structure. Because of the stationary nature of the initial disk, the perturbations are proportional to $\exp\{-i\Omega t\}$, and the future of the system is determined by the presence and sign of the imaginary part of frequency Ω (increment). The increase of the amplitude with time in the case $\mathrm{Im}(\Omega) > 0$ indicates that the system is unstable.

In this section we derive the dispersion equation which takes into account the inhomogeneity of the distribution of the equilibrium parameters of the stellar disk.

[4]The dependence $\varrho \propto \mathrm{ch}^{-1}(z/h_{ch})$ is used for a number of galaxies [636, 639, 805, 806].

2.2.1. Formulation of the problem

Following van der Voort [1254] we obtain the kinetic equation which describes the dynamics of the perturbations with a small amplitude in the plane of a thin stellar disk. For this purpose, the total functions of distribution of the stars and the gravitational potential are presented in the form of a sum of equilibrium (f_0, Φ_0) and perturbed (f_1, Φ_1) quantities:

$$f = f_0 + f_1,\ \ |f_1| \ll |f_0|, \qquad \Phi = \Phi_0 + \Phi_1,\ \ |\Phi_1| \ll |\Phi_0|. \qquad (2.59)$$

The kinetic equation (2.7) linearised with respect to the perturbed values f_1, Φ_1 has the following form in the epicyclic approximation

$$\frac{\partial f_1}{\partial t} + v_r \frac{\partial f_1}{\partial r} + \left(\Omega + \frac{v_\varphi}{r}\right)\frac{\partial f_1}{\partial \varphi} + 2\Omega v_\varphi \frac{\partial f_1}{\partial v_r} - \left(\frac{\varkappa^2}{2\Omega} v_r + v_z r \frac{\partial \Omega}{\partial z}\right)\frac{\partial f_1}{\partial v_\varphi} +$$
$$+ v_z \frac{\partial f_1}{\partial z} - \frac{\partial \Phi_0}{\partial z}\frac{\partial f_1}{\partial v_z} = \frac{\partial \Phi_1}{\partial r}\frac{\partial f_0}{\partial v_r} + \frac{\partial \Phi_1}{r\partial \varphi}\frac{\partial f_0}{\partial v_\varphi} + \frac{\partial \Phi_1}{\partial z}\frac{\partial f_0}{\partial v_z}. \qquad (2.60)$$

To calculate the perturbed quantities f_1, Φ_1, as in the determination of f_0, Φ_0, we shall use the approximation of a thin disk ($\varkappa \ll \Omega_z$). It is evident that, like equilibrium σ_{*0}, the value of the perturbed surface density is $\sigma_{*1} \propto \varepsilon^0$. Therefore, the perturbed volume density $\rho_{*1} \sim \sigma_{*1}/\Delta_* \propto \varepsilon^{-2}$, $f_1 \propto \varepsilon^{-3}$ and the part $\delta\Phi_1$ of the perturbed gravitational potential, which depends on the z-coordinate, is proportional to ε^2. In accordance with this f_1, Φ_1 is expanded into a series with respect to the powers of the small parameter ε (compare with (2.18)):

$$f_1 = f_{1,-3} + f_{1,-2} + \dots, \qquad \Phi_1 = \Phi_{1,0}(r,\varphi) + \Phi_{1,2}(r,\varphi) + \dots, \qquad (2.61)$$

where $f_{1,n} \propto \varepsilon^n$ and $\Phi_{1,n} \propto \varepsilon^n$. Consequently, the equation (2.60) can be split into a system of two equations within the framework of the two main orders with respect to parameter ε:

$$\hat{\mathcal{L}}_0 f_{1,-3} = \frac{\partial \Phi_{1,2}}{\partial z}\frac{\partial f_{0,-3}}{\partial v_z} \equiv \delta\hat{\mathcal{L}} f_{0,-3}, \qquad (2.62)$$

$$\frac{\partial f_{1,-3}}{\partial t} + \left(\Omega + \frac{v_\varphi}{r}\right)\frac{\partial f_{1,-3}}{\partial \varphi} + \hat{\mathcal{D}}_0 f_{1,-3} + \hat{\mathcal{L}}_0 f_{1,-2} =$$
$$= \frac{\partial \Phi_{1,0}}{\partial r}\frac{\partial f_{0,-3}}{\partial v_r} + \frac{\partial \Phi_{1,0}}{r\partial \varphi}\frac{\partial f_{0,-3}}{\partial v_\varphi} + \delta\hat{\mathcal{L}}_0 f_{0,-2}, \qquad (2.63)$$

Here the operators $\hat{\mathcal{L}}_0$ and $\hat{\mathcal{D}}_0$ are determined by the relationships (2.22) and (2.17), respectively.

As in section 2.1.2, to solve this system, we transfer to the action–angle J, w variables (2.24). Since $J = \tau E_z$, where $\tau = \oint dz/v_z$, then

$$\delta\hat{\mathcal{L}} = \tau \frac{\partial \Phi_{1,2}}{\partial w} \frac{\partial}{\partial J}, \tag{2.64}$$

Equation (2.62) has the following form

$$\frac{\partial f_{1,-3}}{\partial w} = \tau \frac{\partial f_{0,-3}}{\partial J} \frac{\partial \Phi_{1,2}}{\partial w}. \tag{2.65}$$

The general solution of this equation in the case in which the equilibrium distribution function $f_{0,-3}$ is determined by the relationship (2.42), can be written in the following form

$$f_{1,-3} = \left(\langle \Phi_{1,2} \rangle - \Phi_{1,2}\right) \frac{f_{0,-3}}{c_z^2} + g_1(t, r, \varphi, v_r, v_\varphi, J), \tag{2.66}$$

where the symbol $\langle ... \rangle$ denotes the z-motion is averaged out with respect to the phase:

$$\langle Q \rangle = \frac{1}{\tau} \oint Q \frac{dz}{v_z}. \tag{2.67}$$

To determine $g_1 = \langle f_{1,-3} \rangle$ the equation (2.63) will be averaged out with respect to the phase of z-motion in accordance with the rule (2.67). As a result, we obtain

$$\begin{aligned} \frac{dg_1}{dt} \equiv \frac{\partial g_1}{\partial t} + v_r \frac{\partial g_1}{\partial r} + \left(\Omega + \frac{v_\varphi}{r}\right) \frac{\partial g_1}{\partial \varphi} + 2\Omega v_\varphi \frac{\partial g_1}{\partial v_r} - \frac{\varkappa^2}{2\Omega} v_r \frac{\partial g_1}{\partial v_\varphi} = \\ = \frac{\partial \langle \Phi_{1,0} \rangle}{\partial r} \frac{\partial f_{0,-3}}{\partial v_r} + \frac{\partial \langle \Phi_{1,0} \rangle}{r \partial \varphi} \frac{\partial f_{0,-3}}{\partial v_\varphi}. \end{aligned} \tag{2.68}$$

The structure of this equation does not describe the effects associated with the movement of the stars along the z-coordinate. At the same time, it is possible to formulate one of the stages of solution of the problem as the task of determination of the perturbed distribution function g_1 in the model of the thin ($\Delta_* \to 0$) stellar disk. The calculation of g_1 and, subsequently, of the perturbed volume density of the disk of a finite thickness is described in the following section.

2.2.2. *Perturbed density of the star disk*

Equation (2.68) is a linear differential equation in the partial derivatives. Its characteristics, which determine the unperturbed trajectories of the stars in the disk plane, are described by the equation

$$\frac{dr}{dt} = v_r, \ \frac{d\varphi}{dt} = \Omega(r) + \frac{v_\varphi}{r}; \ \frac{dv_r}{dt} = 2\Omega(r) v_\varphi, \ \frac{dv_\varphi}{dt} = -\frac{\varkappa^2(r)}{2\Omega(r)} v_r. \tag{2.69}$$

In the framework of the epicyclic approximation, the solution of these equations has the form

$$v_r = v_\perp \cos(\alpha), \qquad v_\varphi = v_\perp \frac{\varkappa}{2\Omega}\sin(\alpha),$$
$$r = r_0 - \frac{v_\perp}{\varkappa}\sin(\alpha), \qquad \varphi = \Omega t + \frac{v_\perp}{r}\frac{2\Omega}{\varkappa^2}\cos(\alpha), \tag{2.70}$$

where r_0, $v_\perp$ are the integration constants, $\Omega = \Omega(r_0)$, $\varkappa = \varkappa(r_0)$, $\alpha = \alpha_0 - \varkappa t$ is the phase of movement of the star along the epicyclic trajectory (α_0 is the initial phase). Equation (2.70) shows clearly that the unperturbed movement of the star in the epicyclic approximation is the movement along an ellipse (epicycle), where one of the half-axes, oriented to the centre of the disk, is equal to $v_\perp / \varkappa$, and the other half-axis, oriented in the azimuthal direction, is equal to $2\,\Omega v_\perp / \varkappa^2$. The centre of this ellipse (epicycle) moves around the centre of the disk along a circular orbit with a radius r_0 with the angular velocity $\Omega(r_0)$.

To calculate the perturbed distribution function g_1, equation (2.68) is written in the form suitable for application of the method of integration over the trajectories [138]:

$$g_1(t) = \int_{-\infty}^{t} dt' \left\{ \frac{2\Omega}{\varkappa^2}\frac{\partial f_{0,-3}}{\partial I_2}\frac{\partial\langle\Phi_{1,0}(t')\rangle}{r_0\,\partial\varphi(t')} + \frac{\partial f_{0,-3}}{\partial I_1} \times \left[v_r(t')\frac{\partial\langle\Phi_{1,0}(t')\rangle}{\partial r(t')} + \frac{4\Omega^2}{\varkappa^2} v_\varphi(t')\frac{\partial\langle\Phi_{1,0}(t')\rangle}{r_0\,\partial\varphi(t')} \right] \right\}, \tag{2.71}$$

where $I_1 = v_\perp^2/2$; $I_2 = r_0$ are the integrals of motion of the star in the plane of the disk in the epicyclic approximation, and it is taken into account [see (2.29)] that

$$\frac{\partial f_{0,-3}}{\partial v_r} = v_r \frac{\partial f_{0,-3}}{\partial I_1}, \tag{2.72}$$

$$\frac{\partial f_{0,-3}}{\partial v_\varphi} = \frac{4\Omega^2}{\varkappa^2} v_\varphi \frac{\partial f_{0,-3}}{\partial I_1} + \frac{2\Omega}{\varkappa^2}\frac{\partial f_{0,-3}}{\partial I_2}. \tag{2.73}$$

The presentation of $g_1(t)$ in the form (2.71) assumes that the perturbation takes place at the moment of time $t = -\infty$ with the amplitude negligible in comparison with its level at the time t.

The coefficients of equation (2.62) do not depend explicitly on time and the azimuthal coordinate because of the stationary nature and the axisymmetric equilibrium state of the disk. Consequently, the dependence of the perturbed quantities f_1, Φ_1 on time t and angle φ can be represented in the exponential form $f_1, \Phi_1 \exp \propto \{-i\,\Omega\, t + im\varphi\}$, where Ω is the frequency of the perturbations, $m = 0, 1, 2, \ldots$ is the number of the mode in respect of the azimuth, $k_\varphi = m/r$ is the azimuthal number. In particular, the representation indicates that the increase of the amplitude of perturbations with time, assumed previously in transition from (2.68) to (2.71), is equivalent to the presence in the frequency Ω of a small positive imaginary part – increment.

In the radial direction, the equilibrium star disk is inhomogeneous. However, we can confine ourselves to investigating shortwave perturbations in this direction – the perturbations whose characteristic scale of variation is small in comparison with the minimum scale of the radial inhomogeneity of the disk. The role of the disk in the galaxies outside the limits of their central areas is usually played by the quantity $L_\sigma = | d \ln(\sigma_{*0})/dr |^{-1}$. The properties of these perturbations are investigated by the WKB-approximation in which the radial dependence of the perturbed quantities is assumed to be $\propto$ exp $(ik_r r)$, where k_r^{-1} is the previously mentioned characteristic scale of the variation of the perturbed quantities in the radial direction (k_r is the radial wavenumber). Thus, the previously formulated condition of applicability of the WKB-approximation can be written in the form[5]

$$k_r L_\sigma \gg 1. \tag{2.74}$$

This approach can be used to transform of the differential operators in the subintegrand expression in (2.71) into an algebraic expression and, consequently, link the values of Ω, m, k_r by the algebraic equation

$$F(\omega, k_r, m) = 0, \tag{2.75}$$

which is usually referred to as the dispersion equation. The results of solution of the equation can be interpreted as follows. The perturbations, characterised by specific values of m, k_r, evolve in accordance with the law exp $\{-\Omega it\}$, where $\Omega = \Omega(m, k_r)$ is, generally speaking, the complex quantity. If the roots Ω of the dispersion equation (2.75) at some values of m, k_r are such that Im $(\Omega) > 0$, then the amplitude of these perturbations increases exponentially with time (generally speaking, not necvessarily at every point). It should also be mentioned that, knowing the solutions of (2.75), we can investigate in a general case also the evolution of arbitrary perturbations, if the Fourier spectrum of such perturbation in the m–k_r-space at the initial moment of time is known.

For further considerations it is important to note that we do not link the solution of the investigated problem of studying the dynamics of small perturbations in the stellar disk with the investigations of the behaviour of some global structural peculiarities of the disk (for example, a spiral pattern). Therefore, in contrast to the approach used by Lin and Shu [844, 845], the perturbed azimuthal forces in (2.68) and (2.71) will not be ignored. This approach enables us to investigate the dispersion properties of the non-axisymmetric perturbations in an inhomogeneous

[5]In a number of studies, using different models of the gravitating disk, it has been shown that the WKB-approximation gives the results very close to the exact ones, also for the large-scale modes. For example, according to Kalnajs [737], the WKB solution of the Poisson equation for the mode $m = 2$, $k_r L_\sigma \simeq 1$ differs from the exact solution by no more than 2%, and in [167] the exact values of the frequency of oscillations of the disk coincide almost completely with the values calculated in the shortwave approximation up to the very largest scale modes, inclusively. Therefore, the application of the WKB-approximation can lead to the results with satisfactory accuracy up to $k_r L_\sigma \simeq 1$. However, this does not apply to the vicinity of the dynamic resonances of such modes where $(\text{Re}\Omega - m\Omega)^2 \simeq n^2 \varkappa^2$, $n = \pm 0, 1, 2, \ldots$ – see (2.80).

star disk and obtain the condition of instability in relation to the non-axisymmetric perturbations in its plane [151, 153].

The perturbation phase $m\varphi + k_r r$ and will be calculated. Here, in the two-dimensional space of the wave vectors (k_r, k_φ) we transfer to the quantities

$$\tilde{k} = \sqrt{k_r^2 + \left(\frac{2\Omega}{\varkappa} k_\varphi\right)^2}, \quad \beta = \operatorname{arctg}\left\{\frac{2\Omega k_\varphi}{\varkappa k_r}\right\}; \tag{2.76}$$

where $k_\varphi = m/r_0$. Consequently, using equations (2.70), which describe the unperturbed trajectories of the stars, we easily obtain

$$k_r r(t) + m\varphi(t) = m\Omega t + \tilde{k}\left[r_0 \cos(\beta) - \frac{v_\perp}{\varkappa}\sin(\alpha - \beta)\right]. \tag{2.77}$$

Using this result, equation (2.71) is transformed to the form

$$\begin{aligned} g_1 = i\tilde{k}\langle\Phi_{1,0}\rangle \int_{-\infty}^{t} dt' \left\{ v_\perp \cos(\alpha' - \beta) \frac{\partial f_{0,-3}}{\partial I_1} + \frac{\sin(\beta)}{\varkappa} \frac{\partial f_{0,-3}}{\partial I_2} \right\} \times \\ \times \exp\{i\hat{\omega}(t - t') + i\xi[\sin(\alpha - \beta) - \sin(\alpha' - \beta)]\}, \end{aligned} \tag{2.78}$$

where $\hat{\omega} = \omega - m\Omega(r)$; $\xi = \tilde{k}v_\perp / \varkappa$; $\alpha' = \alpha(t') = \alpha + \varkappa(t - t')$. Subsequently, as a result of expansion of the derivative function [45]

$$\exp\{i\xi\sin(\delta)\} = \sum_{n=-\infty}^{\infty} J_n(\xi)\ \exp\{in\delta\} \tag{2.79}$$

into a series with respect to the Bessell functions of the first order $J_n(\xi)$ we integrate (2.78) with respect to time t'. Consequently, we obtain

$$\begin{aligned} g_1 = -\langle\Phi_{1,0}\rangle \sum_{n=-\infty}^{\infty} \frac{J_n(\xi)}{\hat{\omega} - n\varkappa} \exp[i\xi\sin(\alpha - \beta) - in(\alpha - \beta)] \times \\ \times \left\{ n\varkappa \frac{\partial f_{0,-3}}{\partial I_1} + \frac{\tilde{k}\sin(\beta)}{\varkappa} \frac{\partial f_{0,-3}}{\partial I_2} \right\}. \end{aligned} \tag{2.80}$$

The volume density of the disk, generated by the perturbed distribution function (2.66) with the quantity $g_1(t)$, determined from (2.80), can be determined by integration of the latter quantity over the space of velocities. In the case of an equilibrium disk [see (2.42)], this operation, with the expansion (2.79) taken into account, leads to the following equation:

$$\rho_1 = \int f_{1,-3}\, dv_r dv_\varphi dv_z = \frac{\rho_{*0}(r_0,z)}{\sqrt{2\pi}c_z}\int\limits_{-\infty}^{\infty}(\langle\Phi_{1,0}\rangle - \Phi_{1,0})\times$$

$$\times\exp\{-v_z^2/2c_z^2\}dv_z - \frac{2\pi\varkappa}{2\Omega}\sum_{n=-\infty}^{\infty}\int\limits_{-\infty}^{\infty}dv_z\int\limits_0^{\infty}dI_1\frac{J_n^2(\xi)}{\hat{\omega}-n\varkappa}\langle\Phi_{1,0}\rangle\times \qquad (2.81)$$

$$\times\left\{n\varkappa\frac{\partial f_{0,-3}}{\partial I_1} + \frac{\tilde{k}\sin(\beta)}{\varkappa}\frac{\partial f_{0,-3}}{\partial I_2}\right\}.$$

Polyachenko and Fridman [203] show that the perturbations in the plane of the disk (the properties of these perturbations are the subject of the present section) within the framework of the linear theory in the model of the thin disk split off from the bending (membrane) oscillations of the disk. In the latter, the local surface density is not perturbed, and the perturbations of volume density are of the 'dipolar' form. Therefore, when discussing the properties of the perturbations in the plane of the disk in the expression (2.81) it is necessary to reject the terms which do not provide any contribution to the perturbed surface density:

$$\sigma_1 = \int_{-\infty}^{\infty}\rho_1 dz. \qquad (2.82)$$

It may easily be seen that in particular the first term in (2.81) does not provide any con tribution to σ_1. In fact, integrating the quantity $\rho_{*0}(r_0,z)\int_{-\infty}^{\infty}\Phi_{1,0}\exp\{-v_z^2/2c_z^2\}dv_z$ in respect of the z-coordinate with (2.43), taken into account, we obtain

$$\int_{-\infty}^{\infty}\rho_{*0}(r_0,z)dz\int_{-\infty}^{\infty}\Phi_{1,0}\exp\{-v_z^2/2c_z^2\}dv_z = \rho_{*0}(r_0,0)\times$$

$$\times\int_0^{\infty}dE_z\oint dw\Phi_{1,0}\exp(-E_z/c_z^2) \equiv \rho_{*0}(r_0,0)\int_0^{\infty}dE_z\langle\Phi_{1,0}\rangle\times \qquad (2.83)$$

$$\times\oint dw\exp\left(-\frac{E_z}{c_z^2}\right) = \int_{-\infty}^{\infty}\rho_{*0}(r_0,z)dz\int_{-\infty}^{\infty}\langle\Phi_{1,0}\rangle\exp\left(-\frac{v_z^2}{2c_z^2}\right)dv_z.$$

The perturbed volume density, determined by the second term in (2.81), is calculated using the specific form of the equilibrium distribution function of the stars $f_{0,-3}$ (2.42). In addition, in accordance with the observational data (section 1.2), showing that the characteristic thicknesses of the stellar disks of the galaxies change only slightly along the radial coordinate, it is assumed that $\Delta_*(r_0)$ = const. In this case, with equation (2.40) taken into account, we have three independent parameters of the star disk represented by σ_{*0}, Ω, c_r. Consequently, from (2.33) we obtain:

$$\frac{\partial f_{0,-3}}{\partial I_1} = -\frac{f_{0,-3}}{c_r^2}, \qquad (2.84)$$

$$\frac{\partial f_{0,-3}}{\partial I_2} = f_{0,-3}\frac{\partial \ln(2\Omega/\varkappa)}{\partial r_0} + \frac{\partial f_{0,-3}}{\partial \sigma_{*0}}\frac{d\sigma_{*0}}{dr} + \frac{\partial f_{0,-3}}{\partial c_r^2}\frac{dc_r^2}{dr_0}. \tag{2.85}$$

Integrating subsequently (2.82) over I_1, v_z and rejecting the terms which do not provide any contribution to the perturbed the surface density of the disk, as described previously, we obtain

$$\rho_1 = -\frac{\sigma_{*0}\Phi_{1,0}}{2\Delta_* c_r^2 \mathrm{ch}^2(z/\Delta_*)}\left\{1 - \sum_{n=-\infty}^{\infty}\frac{\hat{\omega}}{\hat{\omega} - n\varkappa}\times \right.$$
$$\left.\times\left[1 - \frac{\omega_*}{\hat{\omega}}\left(1 + \xi + 2\eta\tilde{z}\frac{\partial}{\partial\tilde{z}}\right)\right] I_n(\tilde{z})\ \exp\{-\tilde{z}\}\right\}, \tag{2.86}$$

where $\tilde{z} = \tilde{k}^2 c_r^2/\varkappa^2; I_n(\tilde{z})$ is the modified Bessell function of the first kind

$$\omega_* = \frac{\tilde{k}\sin(\beta)}{\varkappa}c_r^2\frac{\partial \ln(\sigma_{*0})}{\partial r_0}, \tag{2.87}$$

$$\xi = \frac{\partial \ln(2\Omega/\varkappa)}{\partial \ln(\sigma_{*0})}, \qquad \eta = \frac{\partial \ln(c_r)}{\partial \ln(\sigma_{*0})} \tag{2.88}$$

and it is taken into account that [45]: $\sum_{n=-\infty}^{\infty} I_n(x)e^{-x} = 1$.

2.2.3. The dispersion equation

The perturbations of the disk density ρ_1 result in perturbations of the gravitational potential Φ_1, and the relationship between these quantities is determined by the Poisson equation [see (2.3)]:

$$\frac{\partial}{r\partial r}\left(r\frac{\partial\Phi_1}{\partial r}\right) + \frac{\partial^2\Phi_1}{r^2\partial\varphi^2} + \frac{\partial^2\Phi_1}{\partial z^2} = 4\pi G\rho_1. \tag{2.89}$$

In the WKB-approximation discussed here, this equation for the harmonics of the perturbed potential, characterised by the wavenumber $k = \sqrt{k_r^2 + m^2/r^2}$, has the form

$$\frac{\partial^2\Phi_{1,0}}{\partial z^2} - k^2\Phi_{1,0} = -\frac{\hat{\mu}\Phi_{1,0}}{\Delta_*^2\mathrm{ch}^2(z/\Delta_*)}, \tag{2.19}$$

where the quantity $\hat{\mu}$ is determined from (2.56) and the identity

$$\rho_1 \equiv -\frac{\hat{\mu}\Phi_{1,0}}{4\pi G\Delta_*^2\mathrm{ch}^2(z/\Delta_*)}. \tag{2.91}$$

Equation (2.90) is similar to the Schrödinger equation $U = -U_0/\mathrm{ch}^2(z/\Delta_*)$ which describes the movement of the particle in the potential well along the z coordinate [109]. However, in the case of (2.90), the problem is formulated in a slightly different manner: for the fixed value of 'energy' ($-k^2$) it is necessary to find the 'depth of the potential well' ($\hat{\mu}/\Delta_*^2$) in which the given 'energy level' ($-k^2$) can exist. It may easily be seen that the minimum depth of the well is determined by the eigenfunction

$$\Phi_{1,0} = \text{const} \cdot \text{ch}^{-k\Delta_*}(z/\Delta_*), \tag{2.92}$$

which is 'nodeless' in the z-direction

$$\Phi_{1,0} = \text{const} \cdot \text{ch}^{-k\Delta_*}(z/\Delta_*), \tag{2.92}$$

and the value of the 'depth of the well', corresponding to this function, is determined by the relation[6]:

$$\frac{\hat{\mu}}{\Delta_*^2} = (1 + k\Delta_*)\frac{k}{\Delta_*}. \tag{2.93}$$

Using this result and the definition of $\hat{\mu}$ by (2.86), (2.91), we obtain the required dispersion equation describing the dynamics of the perturbations in the star disk with the wave vector situated in the disk plane:

$$\frac{kc_r^2(1+k\Delta_*)}{2\pi G\sigma_{*0}} = 1 - \left[1 - \frac{\omega_*}{\hat{\omega}}\left(1 + \xi + 2\eta\tilde{z}\frac{\partial}{\partial \tilde{z}}\right)\right] \times \\ \times \left(I_0(\tilde{z})\exp(-\tilde{z}) + 2\sum_{n=1}^{\infty}\frac{\hat{\omega}^2 I_n(\tilde{z})\exp(-\tilde{z})}{\hat{\omega}^2 - n^2\varkappa^2}\right). \tag{2.94}$$

It may easily be seen that this dispersion equation in the limit of the axisymmetric perturbations ($m = 0$ and, consequently, $\Omega^* \equiv 0$) in the model of the thin disk coincides with the dispersion equation derived by Lin and Shu [845] (see also monographs by Fridman and Polyachenko [203 and 590]).

2.2.4. *Dynamics of weakly non-axisymmetric perturbations*

Attention will be given to the dispersion properties of the perturbations, described by equation (2.94), whose frequency in the reference system, rotating together with the matter of the disk, is lower than epicyclic ($|\Omega\hat{}| < \varkappa$). In approximate calculations of these frequencies the terms of the series with $n \geqslant 2$ in equation (2.94) can be omitted. The dispersion equation simplified in this manner has the form

$$kk_T\rho^2(1+k\Delta_*) = 1 - \left[1 - \frac{\omega_*}{\hat{\omega}}\left(1 + \xi + 2\eta\tilde{z}\frac{\partial}{\partial \tilde{z}}\right)\right]\left[a(\tilde{z}) + 2\frac{\hat{\omega}^2}{\hat{\omega}^2 - \varkappa^2}b(\tilde{z})\right], \tag{2.95}$$

where $a(\tilde{z}) = I_0(\tilde{z})e^{-\tilde{z}}$; $b(\tilde{z}) = I_1(\tilde{z})e^{-\tilde{z}}$; $\rho = c_r/\varkappa$; $k_T = \varkappa^2/2\pi G\sigma_{*0}$.

If only the axisymmetric perturbations ($m = 0$ and, consequently, $\Omega^* = 0$) are investigated, then (2.95) describes two gravitational (Jeans) branches of the

[6]Addition of 'deep wells' for eigenfunctions $\Phi_{1,0}$ even with respect to the z-coordinate according to [1254] reduces the value $\mu/k\Delta_*$ in the range $0 \leqslant k\Delta_* \leqslant 1$ by no more than 2%.

oscillations of the stellar disk, with the frequency of these oscillations differing in the sign

$$\hat{\omega}^2 = \varkappa^2 \frac{1-a(k^2\rho^2)-kk_T\rho^2(1+k\Delta_*)}{1-a(k^2\rho^2)-2b(k^2\rho^2)-kk_T\rho^2(1+k\Delta_*)}. \tag{2.96}$$

In the approximation of the thin Toomre disk ($\Delta_*= 0$) it was shown that these perturbations are stable [Im ($\Omega\hat{}$) = 0 for (2.96)] if the following conditions is fulfilled

$$c_r \geqslant c_T = \frac{3.36\ G\sigma_{*0}}{\varkappa}. \tag{2.97}$$

In a thin disk, characterised by the dispersion of the radial velocities $c_r \equiv c_T$, the radial scale of the marginally stable axisymmetric perturbations can be determined from the relationship

$$k_r = k_0 = 0.974\varkappa / c_T. \tag{2.98}$$

If the stabilising effect of the finite thickness of the stellar disk is taken into account (this effect was analysed by Toomre in [1229], it can be seen that in the model (2.42) this disk is stable if the condition [1254] is fulfilled

$$c_r \geqslant c_{TV} \simeq c_T(1+0.974\Delta_*\varkappa / c_T)^{-1}. \tag{2.99}$$

We transfer to investigating the spectrum of non-axisymmetric perturbations. It is noted that $|\omega_* / \varkappa| \sim k_\varphi \rho^2 / L_\sigma \lesssim (k\rho)^2 / kL_\sigma$. Taking into account the fact that for the perturbations in the marginally stable disk (according to Toomre and van der Voort) closest to the instability threshold $k\rho \sim 1$, and also the conditions (2.74), this means that for these perturbations

$$|\omega_* / \varkappa| \ll 1. \tag{2.100}$$

In the longwave part of the spectrum ($k < k_T$) in the marginally stable (according to Toomre) disk, the condition (2.100) is also fulfilled and, consequently, the effect of the inhomogeneity of the disk is small. In this limit, the laws of dispersion of the two gravitational branches of the oscillations of the stellar disk according to (2.94) have the form

$$\hat{\omega}_{1,2} = \pm\varkappa[1-\tilde{k}^2 / kk_T(1+k\Delta_*)]^{1/2} + \frac{\omega_*\tilde{k}^2[1+\xi+2\eta-(1+\xi)/kk_T\rho^2(1+k\Delta_*)]}{2kk_T(1+k\Delta_*-\tilde{k}^2/kk_T)}. \tag{2.101}$$

In addition to these branches, the dispersion equation (2.95) predicts the existence of another – gradient branch[7] – of non-axisymmetric perturbations. The dispersion law of this branch

[7]The term 'gradient' is used due to the fact that in the disk homogeneous with respect to c_r and σ^* this branch of the oscillations degenerates ($\Omega\hat{}_3 = 0$).

$$\hat{\omega}_3 \simeq \omega_* \frac{[1+\xi-\tilde{k}^2\rho^2(1+\xi+2\eta)]}{kk_T\rho^2(1+k\Delta_*-\tilde{k}^2/kk_T)}. \qquad (2.102)$$

In the longwave range ($k < k_T$) according to the order of magnitude $|\hat{\omega}_{1,2}| \sim \varkappa$; $|\hat{\omega}_3| \lesssim \varkappa / k_T L \ll \varkappa$.

However, in the shortwave part of the spectrum ($k \gg k_T$), using the asymptotics of the modified Bessel functions, (2.95) gives

$$\hat{\omega}_{1,2} \simeq \pm\varkappa\left\{1-\frac{1}{\sqrt{2\pi\tilde{z}}}\frac{2}{[kk_T\rho^2(1+k\Delta_*)-1]+3}\right\}^{1/2}+\frac{\hat{\omega}_3\sqrt{2\pi\tilde{z}}[kk_T\rho^2(1+k\Delta_*)-1]}{\sqrt{2\pi\tilde{z}}[kk_T\rho^2(1+k\Delta_*)-1]+3}, \qquad (2.103)$$

$$\hat{\omega}_3 \simeq \frac{\omega_*(1+\xi-\eta)}{\sqrt{2\pi\tilde{z}}[kk_T\rho^2(1+k\Delta_*)-1]+1}. \qquad (2.104)$$

In this range of the wavelength $|\hat{\omega}_{1,2}| \sim \varkappa$; $|\hat{\omega}_3| \lesssim \varkappa/k_T Lk\rho \ll \varkappa$.

Thus, the gradient branch $\hat{\Omega}_3$ in both the longwave and shortwave parts of the spectrum is of the low-frequency type and can be clearly separated from the gravitational branches. All three branches in these parts of the spectrum are stable in the marginally stable disk in relation to the axisymmetric perturbations.

A different result is obtained in the intermediate wavelength range $\tilde{z}\sim 1$. To show this, we examine the simple model of the thin ($\Delta_* = 0$) solid-state rotating ($\varkappa \equiv 2\Omega$) inhomogeneous disk in which $c_r \equiv c_T$ and the same time, the scales of the inhomogeneities σ_{*0} and c_r are identical and, consequently, $\eta = \partial \ln c_r/\partial \ln \sigma_{*0} = 1$). In the immediate vicinity $q = (k - k_0)$ of the marginally stable (according to Toomre) perturbations with $k = k_0$ [see (2.98)], the dispersion equation (2.95) has the form

$$x^3+1.4\nu x^2(1-1.111q-0.473q^2)-0.121q^2x+ \\ +0.097\nu(1+9.65q-24.4q^2)=0, \qquad (2.105)$$

where $x = \hat{\omega}/\varkappa; \nu = (c_T/\varkappa L_\sigma)\sin(\Theta)$; $\sin(\Theta) = m/kr$, and quantity σ_{*0} in accordance with the observational data is assumed to be decreasing towards the periphery of the disk. In (2.105) we ever restricted ourselves to expansion of the terms of equation (2.95) into series in respect of the powers q to the second one inclusive, taking into account also the type of instability in addition to the calculation of the frequencies of the oscillations of the disk.

It is assumed that the disk is slightly inhomogeneous $\nu \ll 1$. Consequently, in the main order with respect to from (2.105) for the frequencies of the oscillations of the disk we obtain:

$$\hat{\omega}_{1,3} \simeq 0.23\nu^{1/3}\varkappa(1+3.22q-18.3q^2)(1+O(\nu^{2/3})\pm i\sqrt{3}), \qquad (2.106)$$

$$\hat{\omega}_2 \simeq -0.46\nu^{1/3}\varkappa(1+3.22q-18.3q^2)+O(\nu^{2/3}). \qquad (2.107)$$

The instability of the investigated model in the range of the Jeans wavelength ($k = k_0$) is obvious. What is the reason for this?

It is well-known that in the marginally stable (according to Toomre) disk in the vicinity of the point $k = k_0$ the frequency of both gravitational branches of the axisymmetric perturbations are proportional to $(k - k_0)^2$. It is clear that the frequency of the gradient perturbations in some close vicinity (because of $|\Omega^*| \ll \varkappa$) $k = k_0$ the frequency of the gradient perturbations becomes comparable with the frequency of one of the gravitational branches of the non-axisymmetric perturbations. Consequently, the mutual effect of these branches, distorting the perturbations spectra, results in the instability of the non-axisymmetric perturbations. Thus, the main reason for gravitational–gradient instability (2.106) is the inhomogeneity of the disk. It is quite evident that this instability is also of the gravitational type

The gravitational-gradient instability belongs to the type of 'absolute' instabilities, i.e. instabilities where the amplitude of perturbations at excitation of these instabilities increases at every point of the space, moving together with the matter of the disk. Actually, the instability is 'absolute' and not 'convective' (in this case, the unstable perturbation is removed by the flow so rapidly that the perturbation at every point of the space does not manage to grow), if the condition [82] is fulfilled

$$U_{\text{gr}}^2 \equiv \left(\frac{\partial \text{Re}(\hat{\omega})}{\partial k}\right)^2 < U_{\text{cr}}^2 = 2\text{Im}(\hat{\omega})\frac{\hat{\alpha}^2 + \hat{\beta}^2}{\hat{\alpha}}, \tag{2.108}$$

where $\hat{\alpha} = -\partial^2 \text{Im}(\hat{\omega}) / \partial k^2; \hat{\beta} = \partial^2 \text{Re}(\hat{\omega}) / \partial k^2$. It may easily be seen that for the perturbations with $k = k_0 (q \equiv 0) U_{\text{gr}} \simeq 0.75 \nu^{1/3} c_T$ and $U_{\text{cr}} \simeq 4\nu^{1/3} c_T$. At the same time, the condition (2.108) for the gravitational–gradient instability is fulfilled.

The gradient branch was determined for the first time by Hunter [697] in the model of a cold ($c_r \equiv 0$) gravitating disk. The results are described here relate to a relatively hot (cr cT) collisionless disk. Nevertheless, the results obtained by Hunter follow from the dispersion equation (2.94) with the chain of the inequalities k $k_T \rho \ll k\rho \ll 1$ (in the actual systems $kT\rho \sim 1$) fulfilled.

2.3. Physics of gravitational instability. Qualitative investigation

The main 'disturber of peace' in the disks examined here is gravitation. The gravitational interaction between different parts of the system (in many cases this is referred to as 'self-gravitation') compresses the matter. This process is referred to as gravitational or Jeans instability. The process leads to the redistribution of mass – in one area the density increases whereas in the other area it decreases if necessary. A number of factors counteract self-gravitation, whereas others support it. Prior to transferring to strict explanation in the following sections, we will discuss the physics of gravitational instability and attempt to understand qualitatively how the different properties of the system influence the tendency of self-gravitation to 'compress matter into a point'. Jeans instability has complicated features in the stellar and gas disks and, therefore, we will also consider the gas system (for more details see chapter 3) taking into account fact that it is easier to analyse. In this case, the differences between them become more evident and it is easier to explain the effect of the gas subsystem on the gravitational stability of the stellar disk. At the present time, the entire difference of the gas from stars is based on the collisions of the gas so that it is described by the gas dynamics equations.

The equations in this section will be derived in the following part and, therefore, here we only mention the main approximations under which they were obtained. It is the epicyclic approximation (equation (2.13)), the disk is assumed to be thin (equation (2.11)), the perturbations are situated in the disk plane and are of the shortwave type (2.74)). It should be mentioned that the relationships presented here for the stellar disk result from the dispersion equation (2.94) and for the gas disk from (3.330). The indexes '*', 'g' indicate the affiliation of the quantity to the stellar or gas disk. To ensure the detailed understanding of the mechanism of gravitational instability it is recommended to referred to the monographs published by Polyachenko and Fridman [203, 590], Rohlfs [1079], and Saslaw [218, 372].

2.3.1. *The role of self-gravitation*

We ignore the effect all factors, with the exception of self-gravitation, i.e. we investigate a flat cold[8] infinitely thin gravitating layer[9]. In this model, compression is not accompanied by the formation of the force acting in the opposite direction. The matter is accelerated and moves into the region with higher density continuously increasing the density and, consequently, the attraction force. The gravitational instability (collapse) develops, and the frequency of perturbations of this instability is purely imaginary

$$\omega_*^2 = \omega_g^2 = -2\pi G\sigma_0 k, \tag{2.109}$$

and we have Re $(\Omega) \equiv 0$, i.e. as expected, there is no oscillatory process because of the absence of the returning force. In the case of cold systems there is no difference between the collisionless and gas-dynamic media. Instability forms at any initial perturbation but the highest rate of increase is recorded for the fine-scale (higher k) perturbations and in this the gravitation layer differs from the density uniform in all directions of the medium ρ_0, for which $\Omega^2 = -4\pi G\rho_0$ [81]. This equation can be easily explained by the following example. If the distance between two identical stationary (in relation to each other) gravitationally interacting solids is equal to a at the initial moment of time, then after the time $t_0 = \pi/2\ (a^3/2mG)^{1/2}$ the particles collide (this is easily explained by the third Kepler law). Accepting for estimation the mean density of the system $\rho_0 = 2m/\left(\frac{4}{3}\pi a^3\right)$ we obtain $t_0 \approx 1/\sqrt{G\rho_0}$. The imaginary part of frequency (increment) is inversely proportional to the characteristic time of increase of perturbations and as regards the order of magnitude $\mathrm{Im}(\omega) \sim 1/t_0 \sim \sqrt{G\rho_0}$.

---

[8]'Temperature' of the stellar system is characterised by the squares of dispersion of the velocity of the stars (c_r, $c\varphi$, c_z), and of the gas – by the square of dispersion of the speed of sound c_s.
[9]Strictly speaking, the system is non-stationary in this case.

Below, we will gradually consider the random movement of the boundaries, rotation of the disk, various inhomogeneities of equilibrium quantities, etc. Because of some of the factors, the disk is more unstable (they increase the increment), and they are therefore referred to as destabilising. Others reduce the increment to the stabilisation of gravitational instability.

2.3.2. *Effect of the temperature of the medium*

It is well-known that if the two previously investigated gravitationally interacting particles have the momentum of the amount of motion (initially, they do not move along the same straight line), then this relative 'random' movement can prevent collisions. In transition to a system with a large number of particles, the role of these random movements is fulfilled by thermal motion ('temperature'), and its motion works against gravitational crowding. If a region with higher density with the size of $\lambda \sim 1/k$ forms, then the stars due to random movement can leave the 'dangerous' zone, reduce the density in the zone and, at the same time, arrest the fall of the surrounding matter. The stability condition is that the typical velocity of the star c_r becomes higher than the characteristic velocity of gravitational for $\sim \lambda / t_0 \sim \sqrt{G\sigma_0 / k}$, leading to the requirement $k \gtrsim G\sigma_0 / c_r^2$. Naturally, a small perturbation is stabilised more easily by random movement. In the case of gas, the identical estimate can be obtained from the condition of equality of the characteristic time of gravitational fall and the duration of passage of the soundwave in the gas through the region with the size $\lambda \sim 1/k$. A sound wave starts to propagate when compression stars take place. If the characteristic time of gravitational increase t_0 exceeds the oscillation period $\tau = 2\pi/\Omega = 2\pi/kc_s$, the perturbations are stable. Taking this estimate into account, (2.109) can be generalised to the case of finite values of cc:

$$\omega_g^2 = c_s^2 k^2 - 2\pi G\sigma_{g0} k. \tag{2.110}$$

It can be seen that the random movement of the particles can stabilise short waves (from (2.110) for the gas $k > 2\pi G\sigma_{g0} / c_s^2$) but is ineffective against long waves.

2.3.3. *The effect of rotation*

Taking rotation into account transforms the flat layer into a disk and makes the long-wave axisymmetric ($k = k_r$) perturbations are stable. This can be easily understood from the following considerations. If a region with the size $\lambda \sim 1/k$ with solid-state rotation with the angular velocity Ω of the homogeneous disk is compressed by $\Delta r \ll R$, then because of the law of conservation of the momentum the matter at the radius $(R - \Delta r)$ will rotate at $\Omega' = \Omega R^2/(R - \Delta r)^2$. This results in the formation of a rotating centrifugal force $(R - \Delta r)\Omega'^2 - (R - \Delta r)\,\Omega^2 \simeq 4\Omega^2\,\Delta r$ with the accuracy to a small $\Delta r/\lambda$. If we compare it with the additional gravitational attraction forces, associated with compression of the disk $\pi G\sigma_0\lambda^2\,(1/(\lambda - \Delta r)^2 - 1/\lambda^2) \simeq -2\pi\,G\,\sigma_0\,\Delta r / \lambda^2$, we can see that only the large-scale perturbations $\lambda \sim 1/k \gtrsim \pi\,G\,\sigma_0/\Omega^2$ will be stable. The dispersion relationship for sound waves in the rotating medium is $\Omega^2 = 4\Omega^2 + k^2c_s^2$ (first term describes epicyclic oscillations), and the equations (2.109), (2.110) can be generalised:

$$\omega_g^2 = \varkappa^2 - 2\pi G\sigma_{g0}k + k^2c_s^2 \qquad (2.111)$$

($\varkappa = 2\Omega$ in the case of solid-state rotation). If we ignore the random motion ($c_s = 0$), the stability condition $\Omega^2 \geqslant 0$ leads to the previously obtained restriction on the wavenumber. Equation (2.111) can be written conveniently in the dimensionless form: $v_g^2 \equiv \omega_g^2 / \varkappa^2 = 1 - 2K + Q_g^2K^2 (K = k / k_0$, $Q_g = c_s / c_0, k_0 = \varkappa^2 / \pi G\sigma_{g0}, c_0 = \pi G\sigma_{g0} / \varkappa)$. As explained previously, the long waves are stabilised by rotation, and the short waves by the random motion of the particles.

The conditions

$$v_g^2 = 0, \qquad \frac{dv_g^2}{dK} = 0 \qquad (2.112)$$

determine the stability boundary. The solution of the system (2.112) does not cause any difficulties, and $K = 1$, $Q_g = 1$. Figure 2.1a shows the dispersion curves for relatively hot ($Q_g = 1.2$; 1) and, consequently, stable disks. Two Jeans branches are symmetric in relation to the abscissa. At $Q_g = 1$ the disk is at the stability boundary. If the value of this parameter is reduced ($Q_g < 1$), the perturbations with $k \simeq k_0$ become unstable (Fig. 2.1b). The minimum of the function v_g^2 (K) determines the most unstable wavenumbers, for which $k = k_0/Q_g^2$. It can be seen that the value of the unique parameter Q_g fully determines the stability of the model. The condition $Q_g = 1$ sets the boundary between the gravitationally stable and unstable systems.

Although the dispersion equations for the collisionless stellar disk are more complicated, the stability condition differs only slightly from the case of the gas disk. Figure 2.1c, d shows two Jeans branches in the region $\Omega < \varkappa$.[10] They are identical with the gravitational branches of the gas disk (Fig. 2.1a, b) but their behaviour differs in the range of small wavelength values. The difference is associated with special features of the random motion in the collisional and collisionless systems. In the former, sonic waves form, in the second the random motion for fine-scale perturbations does not lead to any wave processes, and the dispersion law is determined by the rotation $\Omega \simeq \pm\varkappa$. As we know, in the reference system, rotating with the angular velocity Ω, the trajectory of movement of the star in the model of the solid-state rotating disk has the form of a circle with the characteristic epicyclic radius $\rho = c_r/\varkappa$ (see section 1.2.3). Therefore, only the perturbations with $k \simeq 1/\rho$ are subjected to a strong effect of the random motion of the stars.

In the case of the collisionless disk, the role of the parameter c_0 is played by the Toomre parameter $c_T = 3.36\, G\, \sigma_{*0}/\varkappa$ [1229], and to ensure stability it should be

[10] In the collisionless disk there are also high-frequency oscillations with $\Omega \simeq \pm n\, \varkappa$, $n = 2, 3, \ldots$ (see (2.94)). At least in the homogeneous disk they have no effect on gravitational stability.

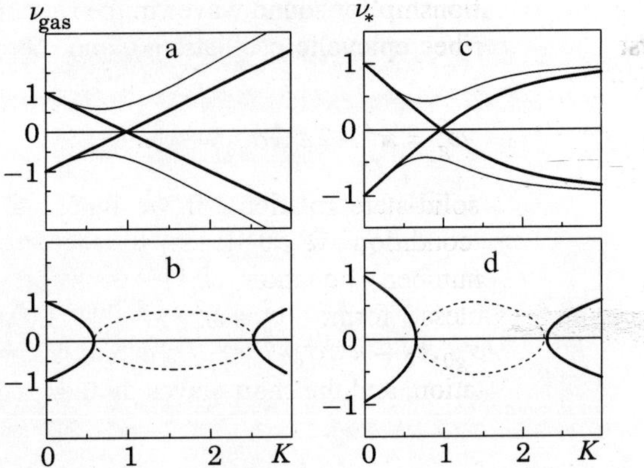

Fig. 2.1. Dependence of the frequency of the Jeans oscillations $\nu = \Omega/\varkappa$ on the dimensionless wavenumber K. For a gas disk: a) $Q_g = 1.2$ (thin lines), 1 (thick lines); b) $Q_g = 0.8$, Re (Ω) (solid line), Im (Ω) (dotted line); c and d) identical dependences for the stellar disk.

$Q_* = c_r/cT \geqslant 1$. The wavenumber can be conveniently normalised for the value $1/\rho_T$ ($\rho_T = c_T/\varkappa$ is the epicyclic radius at $c_r = c_T$). Consequently, at the stability boundary ($\Omega_* = 0$, $Q_* = 1$) there are waves with $K \equiv k\rho_T = 0.974$ (see Fig. 2.1c). At $Q_* < 1$, as in the case of the gas disk, the perturbations with $K \simeq 1$ are absolutely unstable because for them Re $(\Omega) \equiv 0$ (Fig. 2.1d). It should be mentioned that since in the galaxies $\sigma_{g0}/\sigma_{*0} \ll 1$, then $k_0\rho_T \gg 1$.

2.3.4. Velocity distribution function of the stars. Multicomponent systems

The stability of the stellar and gas disks is determined by the values of the parameters $c_0 = \pi G\sigma_{g0}/\varkappa$ and $c_T = 3.36G\sigma_{*0}/\varkappa$, respectively, which formally coincides with the accuracy to the numerical coefficient. Replacement of π by 3.36 is caused by the fact that the system is collisionless and by the specific selection of the velocity distribution function of the stars $f(\mathbf{v})$ (2.32). As already mentioned, the function of the type (2.32) describes the actual distribution of the star velocities approximately [39, 131, 274, 275]. This is associated with star formation which becomes more intensive during the passage of the spiral density wave, the collisionless nature of the system in the sense of the stellar–stellar interaction (the problem of relaxation and the initial conditions) and, consequently, the discrete nature of the stellar population with respect to kinematics. The construction of the distribution function $f(\mathbf{v})$ directly from the observations in the solar vicinity of the Galaxy gives a systematic deviation from (2.32) [274, 275]. As previously, the disk is assumed to be infinitely thin and homogeneous. According to definition, the distribution function satisfies the conditions

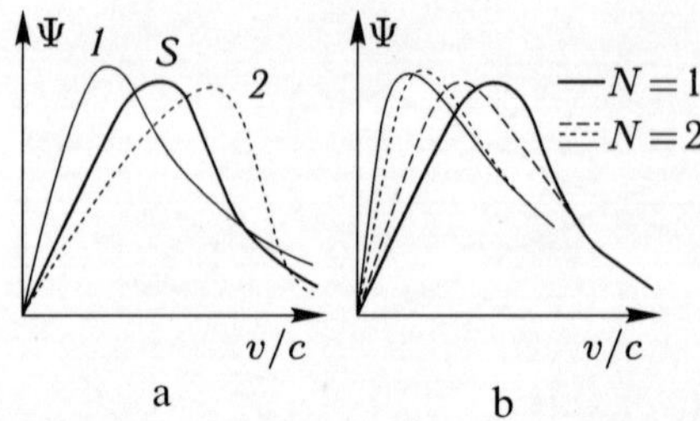

Fig. 2.2. Distribution function of the length of the velocity vector $\Psi = v \cdot f(v)$, a) Schwarzschild (S) with a surplus of slow stars (1), with a surplus of hot stars (2); b) for a multicomponent disk at different $\alpha = \sigma_2/\sigma_1$ and $\beta = c_2/c_1$.

$$\int_{-\infty}^{\infty}\int_{-\infty}^{\infty} f(\vec{v})dv_\varphi dv_r = \sigma_{*0}, \quad \int_{-\infty}^{\infty}\int_{-\infty}^{\infty} v_r^2 f(\vec{v})dv_\varphi dv_r = \sigma_{*0}c_r^2. \tag{2.113}$$

The same dispersion of the velocities can be produced by different distributions of the velocities. We will discuss the effect of the possible deviations of the distribution function from the Schwarzschild distribution f_s (2.32). Figure 2.2a shows two qualitatively different cases differing in the surplus of the slow stars f_1 and in their shortage f_2 in comparison with function f_s. In all three cases, the quantity c_r in (2.113) is the same. For the gravitational stability of the model with f_1, in order to stabilise the subsystem of the slow particles, it is necessary on the whole to heat the system at a higher rate, i.e. the minimum dispersion of the velocities required for stabilisation c_{cr} should be greater than c_T. In the case with f_2 to ensure stability it is sufficient to have $c_{\text{cr}} < c_T$.

The stellar population of the disks of the flat galaxies can be divided into N subsystems each of which is characterised by its surface density σ_j and the dispersion c_j, since because of (2.113)

$$\sum_{j=1}^{N} \sigma_j = \sigma_{*0}, \quad \sum_{j=1}^{N} \sigma_j c_j^2 = \sigma_{*0}c_r^2. \tag{2.114}$$

It is assumed that each subsystem is described by the Schwarzschild distribution $f_j = \dfrac{\sigma_j}{2\pi c_j^2}\exp(-v^2/2c_j^2)$. Figure 2.2b shows the distribution function of the length of the velocity vector $\Psi = v f(\mathbf{v})$ for $N = 1$, $N = 2$, at different ratios σ_1/σ_2 and c_1/c_2 The cold and low-mass subsystem ($\sigma_2 < \sigma_1$ and $c_2 < c_1$) can be represented by young stars.

For gravitational stability of such a model it is necessary on the whole to heat the system more rapidly $Q^* = c_r/c_T > 1$. Evidently, at $N > 1$ any ratio of the sets σ_j, c_j (naturally, satisfying the conditions (2.114)) produces a disk that is more unstable in comparison with $N = 1$ because in these models there is always a surplus of cold stars. Calculations with the observed velocity distribution function of the stars in

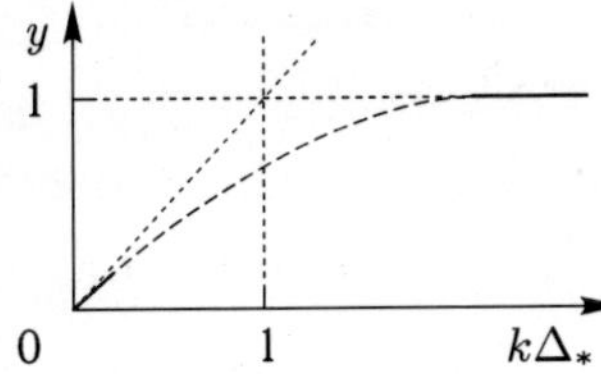

Fig. 2.3. The effect of disk thickness on the gravitational stability of the system. Here $y = -\Omega^2/4\,\pi\, G\rho_0$.

the solar neighbourhood indicate that this correction is not large and not greater than 15%.

In the approach considered here, we will investigate a stellar–gas disk. The characteristic velocity of the gas clouds c_s is considerably smaller than the dispersion of the velocity of the stars c_r. In addition to this, $\sigma_{g0} < \sigma_{*0}$. From this viewpoint, consideration of the gas subsystem is equivalent to the presence of young stars which, as will be shown later, is a destabilising factor (see for more detail section 8.5.1).

2.3.5. *Effect of the finite thickness of the disk*

In order to answer the principal question as to whether the finite thickness of the disk $2\Delta_*$ is a stabilising or destabilising factor, we turn to equation (2.109) and will write it in the form $\Omega_*^2 = -4\pi G\rho_0 k\Delta_*$, where the mean density $\rho_0 = \sigma_{*0}/\Delta_*$. Naturally, this equation holds only in the limit $k\Delta_* \to 0$. It was shown in section 2.3.1 that in the reversed limit $k\,\Delta_* \gg 1$ we have $\Omega_*^2 = -4\pi G\rho_0$. As indicated by Fig. 2.3, we should expect that the thickness of the disk will have a stabilising effect on its gravitational stability. Actually, the value Δ_* and the surface density σ_{*0} are included in the dispersion equation (2.94) only in the combination $\sigma_{*0}/(1 + k\Delta_*)$ and this reduces the increment at $k\Delta_* > 0$. Since the most unstable perturbations have $k = 0.974\,/\rho_T$, we can generalise the stability criterion of the stellar disk taking into account Δ_* and we can write

$$Q_* = \frac{c_r}{c_T} \geq \frac{1}{1 + 0{,}974\Delta_*\varkappa / c_T}. \tag{2.115}$$

The identical considerations also apply to a gas disk of finite thickness $2h$: $Q_g \geqslant 1/(1 + hk_0)$.

2.3.6. *Differentiality of rotation of the disk components*

We have considered the axisymmetric oscillations $k = k_r$ which is valid for the homogeneous solid-state rotating disk because consideration of the non-axisymmetric perturbations [$\propto \exp\{im\varphi + ik_r r - i\Omega w\}$] leads only to the Doppler frequency shift $\Omega \to \Omega\hat{} \equiv \Omega - m\Omega$ and has no effect on stability. In a differentially rotating disk with $\Omega(r)$, the axi-symmetric perturbations are described by the dispersion equation (2. 111) for $\varkappa^2 = 4\Omega^2 + r\Omega^2/dr$. Another aspect is more important. As a result of the dependence of the angular velocity on r the non-axisymmetric ('oblique')

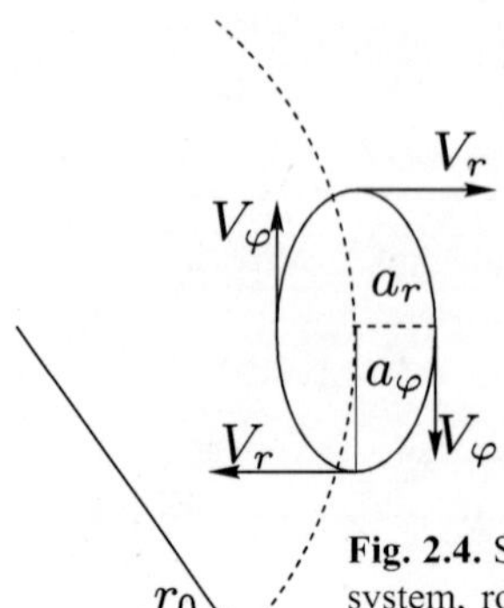

Fig. 2.4. Stability of the differentially rotating stellar disk. In the coordinate system, rotating with the angular velocity $\Omega(r_0)$, the star travels along the ellipse whose minor axis is directed along the radius.

perturbations ($k_\varphi = m/r \neq 0$) are more unstable. The physics of this effect will be investigated separately for the stellar and gas disks.

When constructing the equilibrium model of the stellar disk in section 2.1.2 we obtained $c_\varphi/c_r = \varkappa/2\Omega = (1 + r\,d\Omega/2\Omega dr)^{1/2}$, and for the angular velocity of rotation Ω, which decreases with the radius, we have $c_\varphi < c_r$. Taking into account the differentiality of rotation, the trajectory of motion of the star becomes elliptical with the characteristic scales of the axes $a_r = c_r/$ and $a_\varphi = 2\Omega a_r/\varkappa > a_r$ (Fig. 2.4).

In movement along the elllipse the mean velocity of the star in the azimuthal direction is lower than in the radial direction, and the non-axisymmetric perturbations ('spoke-shaped' in the limit) are more difficult to suppress in comparison with the axisymmetric ones. The dispersion of the azimuthal velocities determines the elasticity of the medium for oblique perturbations, and to suppress the gravitational instability, the disk should be heated at a rate $2\Omega/\varkappa$ times greater. Thus, the stability is determined by the lower of the values c_r and c_φ:

$$Q_* \geq \frac{2\Omega}{\varkappa} = \sqrt{\frac{2}{2-n}}, \tag{2.116}$$

where $\Omega \propto r^{-n}$.

We consider a gas disk. As a result of collisions, the 'macromolecules' of the gas do not move along the epicyclic path, and the pressure is isotropic in the radial and azimuthal directions. The destabilising effect of the differentiality of rotation is associated with the effect of two factors. Firstly, the law of dispersion of the non-axisymmetric sound waves (2.111) changes. This effect is associated with the effect of the Coriolis force. The oblique perturbations produce the radial component of the Coriolis force, and in the first order with respect to the differentiality of rotation, this component does not depend on the sign of k_φ; therefore, the dispersion equation should include k_φ^2. Calculations give

$$\hat{\omega}_g^2 = \varkappa^2 + \left(k^2 c_s^2 - 2\pi G\sigma_{g0} k\right)\left(1 - 2s^2 \frac{r}{\varkappa^2}\frac{d\Omega^2}{dr}\right), \tag{2.117}$$

where $s = k_\varphi/k$, $k^2 = k_r^2 + k_\varphi^2$. The instability increment increases with s^2. However, strictly speaking, according to the variation conditions the equation (2.117) cannot be used for the perturbations with $s^2 = 1$. When used for actual galaxies, the formal substitution $s^2 = 1$ results in the indeterminacy $\leqslant$25–30%. If we accept for estimates $s \to 1$, then taking into account (2.117), the stability condition has the form

$$Q_g = \frac{c_s}{c_0} = \sqrt{\frac{2+n}{2-n}}. \qquad (2.118)$$

For a flat rotation curve ($n = 1$) we have $Q_g = \sqrt{3}$ (identical result was obtained in [201]). Both conditions (2.116) and (2.118) at $n \to 2$ give the infinitely higher values of the parameter Q. This is associated with the instability of the circular orbits for the law with $n \geqslant 2$. This instability is determined by the type of external potential, is not associated with self-gravitation and cannot be suppressed by temperature. It should be mentioned that for the system with $n \geqslant 1$ as the result of violation of the conditions used as a basis for deciding these criteria, the error of these equations may be quite high.[11]

There is another destabilising factor associated with the differentiality of rotation but this factor will be examined in the following section.

2.3.7. The role of the heterogeneity of the radial distribution of parameters

The heterogeneity of the disk is the dependence of the surface density and/or dispersion of the velocities (the speed of sound for the gas disk) on the radial coordinate. In a homogeneous system there are two Jeans branches of oscillations (Fig. 2.1). The heterogeneity of the disk (or differentiality of rotation) leads to the formation of another branch of oscillations referred to as gradient. The wave vector **k** for this type of oscillations should be directed under an angle to the gradient of the equilibrium quantities. In most cases, as the intensity of the oblique perturbations increases, the frequency of oscillations also increases. The gradient waves are determined by the additional 'elasticity' of the heterogeneous medium. There is a large number of examples of manifestation of such oscillations in greatly different areas of physics.

The surface gravitational waves (SGW) on the surface of the interface of two media are the best-known types of waves associated with the heterogeneity of the system caused by the vertical gravitational force g. For incompressible liquids with the density ρ_{01} and ρ_{02} the dispersion law has the following form $\Omega^2_{\rm SGW} = k_\perp g(\rho_{02} - \rho_{01})/(\rho_{02} + \rho_{01})$.

The inner gravitational waves (IGW) can propagate in an ocean or the Earth atmosphere as a result of the heterogeneity in the vertical direction of the volume density of matter $\rho_0(z)$. For them

[11]Unfortunately, these difficulties are of principal nature. The situation can be solved by comparison with the reference numerical experiments (see chapter 4) in which the heterogeneous systems are shifted to the stability boundary and, if necessary, the stability criteria are determined more accurately.

$$\omega_{\mathrm{IGW}}^2 = -g\frac{d\ln(\rho_0)}{dz}\frac{k_\perp^2}{k_\perp^2 + k_z^2}.$$

Both the SGW and IGW are transverse (or shear). Shear elasticity of the medium forms as a result of the heterogeneity of the Archimedes force in the vertical direction, and such waves cannot propagate along the z-coordinate.

The gravitational–gyroscopic waves (GGW) are a suitable example of the large-scale perturbations in the ocean of constant depth h taking into account the rotation of the planet with the angular velocity $\mathbf{\Omega}$. The dispersion equation for them has the form $\Omega_{\mathrm{GGW}}^2 = k_\perp^2 c^2 + 4\Omega_z^2$, and here $c^2 = gh$. From the equilibrium condition $g = -\,dp/\rho dz$.

The Rossby waves are associated with the variation $\Omega_z = \Omega \cdot \sin\Theta$ along the geographical latitude (Fig. 2.5): $\omega_R = -\dfrac{2k_x\Omega_y}{R(k_\perp^2 + 1/r_R^2)}$, where $r_R = c/2\Omega_z$, $\Omega_y = d\Omega_z/d\Theta$. The returning force is the Coriolis force.

All these parameters are well-known and easily detected in the nature. A very large number of waves of the gradient type was obtained in the investigation of plasma physics (they are often referred to as drift waves). Thus, the presence of the gradient branches of the oscillations in the heterogeneous astrophysical disks is regarded as natural.

The investigation of the mechanism of the destabilising effect of the heterogeneity for the stellar and gas disks gives qualitatively similar patterns. On the basis of the previously mentioned examples it can be seen that the frequency of the shortwave gradient oscillations is approximately equal to c/L (c is the characteristic velocity of the random motion, L is the characteristic scale of the heterogeneity), i.e. they are low-frequency oscillations $|\hat{\Omega}| \ll \varkappa$. If in some region of the wavelength two branches of the oscillations have similar real frequencies, a 'weak link' can form between them with the formation of the condition $\mathrm{Im}\,(\Omega) > 0$, i.e. instabilities (see, for example [122]). In the case of the gravitationally stable disk with $Q \gg 1$ the Jeans and gradient branches are separated from each other (Fig. 2.6a). With the reduction of the parameter Q the frequency of the gravitational branches decreases and at some Q in the case $d\sigma/dr < 0$ the negative Jeans branch starts to interact with

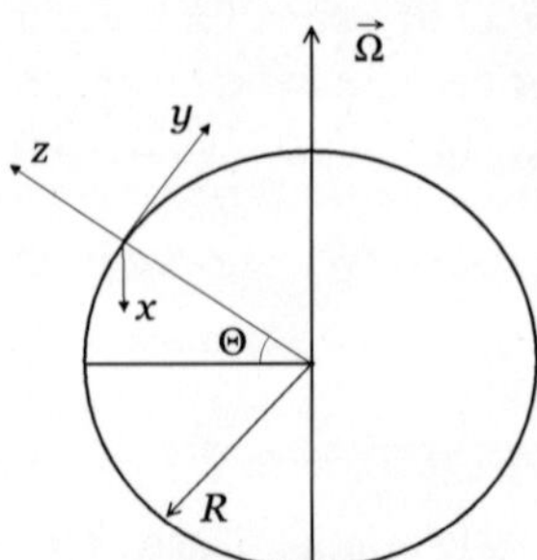

Fig. 2.5. Rossby waves on a planet, rotating with the angular velocity $\mathbf{\Omega}$. The x-axis is directed along the latitude, the y-axis along the longitude in the northern direction.

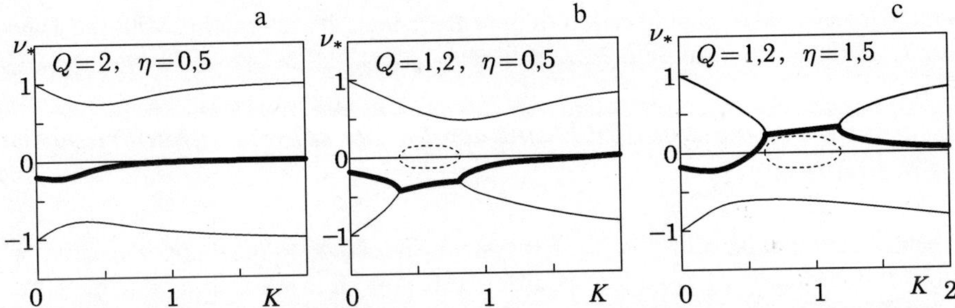

Fig. 2.6. Dispersion curves for the non-axisymmetric perturbations ($s = 0.5$) in a stellar disk. The thin lines show the actual frequencies of Jeans oscillations, the thick lines – gradient, the dotted line – imaginary parts of the frequencies.

the gradient branch and, consequently, two complexly conjugate branches (Fig. 2.6b) appear instead of two actual frequencies. The solution with Im ($\hat{\Omega}$) > 0 corresponds to the gravitational–gradient instability. The gradient branch can, under certain conditions, interact not only with the negative but also positive Jeans brand (Fig. 2.6c). A more complicated case is when the gradient branch of the oscillations interacts with the negative and positive Jeans branches. Consequently, we have two regions of instability (this situation will be investigated in section 2.4.6).

The gradient oscillations can be determined not only by the heterogeneity but also by the differential rotation. This is clearly visible by considering the previously mentioned example of the Rossby waves for the gas.

For the gas disk under the typical conditions, the destabilisation of the system as a result of the effects of interaction of the gradient and Jeans oscillations is not significant. The condition (2.118) in this case has the form:

$$Q_g \geqslant \sqrt{\frac{2+n}{2-n}}\left\{1+0.6\left|2\sqrt{\frac{2}{2-n}}\left(\frac{n}{k_0 r}+\frac{d\ln(\sigma_{g0})}{k_0\,dr}\right)\right|^{2/3}\right\}, \tag{2.119}$$

and at typical values of the parameters the correction does not exceed 10%.

For a stellar disk with all the previously mentioned factors taken into account we can write the identical condition:

$$Q_* \geqslant \frac{N}{1+0.974\delta / N^2}\left\{1+1.07\left|N\frac{d\ln(\sigma_{*0})}{k_T dr}(1+\xi-1.086\eta)\right|^{2/3}\right\} \tag{2.120}$$

where

$$N = 2\Omega / \varkappa = \sqrt{2/(2-n)},\ \eta = L_\sigma / L_c = d\ln(c_r) / d\ln(\sigma_{*0}),$$
$$\delta = \Delta_* / \rho_T,\ \xi = d\ln(2\Omega / \varkappa) / d\ln(\sigma_{*0})\cdot$$

The heterogeneity in the stellar disk has a stronger effect than in the gas disk.

In conclusion, it should be mentioned that another branch of oscillations – entropy – can exist in the gas disk. In the model with the equilibrium entropy $s_0\,(p_0, \sigma_0)$ = const, the frequency of these oscillations is $\hat{\Omega} = 0$. If $ds_0/dr \neq 0$, this leads to a new entropy branch of the oscillations. Naturally, this branch relates to the gradient type and is characterised by low frequency. Its interaction with the Jeans branches results in an additional destabilising effect.

Evaluation of the stability boundary for the solar neighbourhood of the Galaxy

The gravitational instability of the flat systems in relation to the fine-scale perturbations is determined by the effect of different factors. It is noteworthy that the criterion of local stability can be written in the form of a sufficiently simple condition $Q \geqslant Q_{cr}$ (see (2.119), (2.120)). The value of the parameter Q_{cr} is very important for the investigated problem. The combined effect of self-gravitation, rotation and random motion of the particles requires $Q_{cr} = 1$. The finite thickness of the disk reduces its value, and the dependences of the equilibrium quantities on the radial coordinate lead on the whole to $Q_{cr} > 1$. As an example, we examine the solar neighbourhood of the stellar disk of the Galaxy. For estimates we accept: $\sigma_{*0} = 60\ M_{\odot}/\text{pc}^2$, $\Omega =$ 25 km/(s · kpc), $\Omega \propto r^{-1}$, $\Delta_* = 300$ pc, $L_\sigma = 3$ kpc, $L_c = 7$ kpc and, consequently, $N = \sqrt{2}$, $\eta = 0.4$, $\rho_T/L_\sigma = 0.23$, $\xi = 0$, $cT_{\odot} = 22$ km/s. Substituting the values of these parameters into (2.120) we obtain $Q^* \simeq 1.6$.

2.4. Condition of gravitational stability of the disk

2.4.1. Is the Toomre condition sufficient for the stability of the actual stellar disk?

The results discussed previously show that for the stability of the solar disk, the dispersion of the radial velocities of the stars should exceed some critical value. The minimum required value $(c_r)_{\min}$ (see (2.97)) for the stability of a thin ($\Delta_* = 0$) disk in relation to the axisymmetric perturbations was calculated by Toomre [1229]. Later, van der Voort [1254] corrected this value by taking into account the finite thickness of the disk $(c_r)_{\min} \simeq c_{TV}$ (2.99). The observational data [188, 1301] show that in the solar neighbourhood $c_r > c_{TV}$ and, consequently, it should be expected that the stability condition (2.99) is essential.

However, the numerical experiments [154, 155, 528, 681, 682, 914, 915, 918, 920, 996, 1142], started in the 70s, show that the stellar disks with the initial value $c_r \simeq c_{TV} < c_T$ in the period of time equal to only one or two rotations of the disk are heated to the state with $c_r \geqslant (1.5 \div 2.5)\ c_T$. This results confirm the necessity of the stability condition of the stellar disk (2.99) but indicate that it is not completely sufficient. Therefore, the 'heating' of the disk, detected in the numerical experiments, together with the LeChatelier principle (see the book [110]) can be naturally interpreted as the result of the evolution of the unstable system to the stable state. Consequently, it is important to understand the nature of the instability leading to such heating.

To explain this question, it is necessary initially to mention that the equilibrium collisionless disk is anisotropic – its 'elasticity' in the azimuthal direction, characterised by the value c_φ, is smaller than the radial 'elasticity': $c_\varphi = (\varkappa/2\Omega)c_r < c_r$, and this is the destabilising factor. Therefore, if we would investigate, in contrast to Toomre [1229] and van der Voort [1254], who investigated the dynamics of only axisymmetric perturbations, the behaviour of the non-symmetric perturbations, it should be expected that the value c_φ for their stabilisation should be close to c_{TV}. Taking into account the relationship (2.34) this means that the value $(c_r)_{\min}$ should be close to $(2\Omega/\varkappa)\, c_{TV}$ [699].

The nature of the second factor, destabilising the stellar disk, is determined evidently by the gravitational–gradient instability whose intensity increases with an increase of the strength of non-axisymmetric perturbations. This instability, as shown later, may also be suppressed by increasing the dispersion of the radial velocities of the stars.

Thus, it should be hoped that the necessary and sufficient condition for the stability of the heterogeneous differentially rotating stellar disk can be obtained as the condition of absence of the complex roots in the dispersion equation (2.94) for the non-axisymmetric perturbations. This condition in the limit of the homogeneous solid-state rotating disk should naturally transform to the condition (2.99).

It should also be mentioned that qualitatively the effect of insufficient stabilisation of the non-axisymmetric perturbations was known already earlier from the investigations of the stability of the models of the stellar disks with the quadratic potential [168, 203, 590, 738]. However, in these models, the gradients of the density and dispersion of the velocities of the star are rigidly connected with each other and with other disk parameters, and the rotation is of the solid-state type. The artificial nature of such models is also indicated by the fact that even the non-rotating disk, i.e. the disk whose equilibrium is maintained only by the 'pressure' gradient (density and dispersion of the velocities of the stars) is gravitationally unstable in relation to the axisymmetric perturbations. For this reason, the applicability of the models of the disks with the quadratic potential to the actual galaxies is associated with many problems. At the same time, the numerical experiments mentioned at the start of the section and also the very fact of existence of flat galaxies indicate that the rotating stellar disk can be stable and that the condition $c_r \simeq 1.5\text{–}2.5\ c_T$ should be fulfilled everywhere for the disk (with the exception of the central part of the disk). The model of the stellar disk (2.100)–(2.102) is in good agreement with the observational data and we shall therefore attempt to determine the condition of its stability from the dispersion equation (2.94) corresponding to this model.

2.4.2. Destabilising effect of non-axisymmetric perturbations in the differentially rotating disk of finite thickness

According to the previously mentioned estimate of the frequency of the gradient perturbations, the stability boundary of the stellar disk should be located in the frequency range $|\hat{\Omega}|^2 \ll \varkappa^2$. Therefore, it is convenient to use the simplified dispersion equation (2.95). Initially, we examine the model of a homogeneous ($\partial\sigma_{*0}/\partial r = 0$, $\partial c_r/\partial r = 0$) disk, rotating with the angular velocity which depends in the exponential manner on the radial coordinate: $\Omega \propto r^{n}$, $n = \text{const} > 0$. In this case, $\xi = 0$, $\Omega_* = 0$. At the same time, the gradient branch and the effects associated with it are not considered. Equation (2.95) in this model has the following form

$$\frac{\hat{\omega}^2}{\varkappa^2} = \frac{1 - \sqrt{z}\, k_T \rho (1 + \sqrt{z}\, \Delta_* / \rho) - a(\tilde{z})}{1 - \sqrt{z}\, k_T \rho (1 + \sqrt{z}\, \Delta_* / \rho) - a(\tilde{z}) - b(\tilde{z})}, \tag{2.121}$$

where $z = k^2 \rho^2$; $\tilde{z} = \tilde{k}^2 \rho^2 = z(1+h)$; $h = \left(\frac{4\Omega^2}{\varkappa^2} - 1\right) \cdot \sin^2 \Theta = \frac{n \sin^2 \Theta}{2-n}$; $\rho = c_r / \varkappa$.

The stability boundary of the perturbations with the given $\sin \Theta = m/kr$ is determined from (2.121) by the relationships $\hat{\omega}^2 = 0$; $d\hat{\omega}^2 / dk = 0$ [$\hat{\omega}^2 = 0$ at the minimum of the dispersion curve $\hat{\omega}^2 = \hat{\omega}^2(k)$]; this is equivalent to the system of equations

$$1 - \sqrt{z} k_T \rho (1 + \sqrt{z} \Delta_* / \rho) - a(\tilde{z}) = 0, \tag{2.122}$$

$$2\sqrt{z}[b(\tilde{z}) - a(\tilde{z})](1+h) + k_T \rho (1 + 2\sqrt{z} \Delta_* / \rho) = 0. \tag{2.123}$$

Initially, we solve this system of equations in the model of the thin disk ($\Delta_* = 0$). It may easily be seen that for the stability of the perturbations with the given $\sin \Theta = m/kr$ it is necessary to fulfill the condition (compare with (2.97))

$$c_r \geq c_{r\min 1} = c_T (1+h)^{1/2}, \tag{2.124}$$

and in the case $c_r = c_{r\,\min}$ the wavelength of the marginally stable perturbations is determined by the relation (compare with (2.98)]

$$k = k_1 = 0.974 \left(\frac{\varkappa}{c_{r\min 1}} \right) (1+h)^{-1/2} = k_0 (1+h)^{-1}. \tag{2.125}$$

As expected, the non-axisymmetric perturbations in the differentially rotating disk are stable than the axisymmetric ones. In addition to this, equation (2.125) shows that the boundary of marginal stability is shifted into the longwave region with the increase of the strength of non-axisymmetric perturbations.

We now transfer to a finite thickness disk. According to the observational data (see section 1.2.4), (2.40), (2.97), the relationship $(\Delta_*/\rho) \sim (c_z/c_r)^2 \ll 1$. Therefore, we find the correction for (2.124), (2.125), linked with the finiteness of the ratio Δ_*/ρ in the first order with respect to this value. As a result, from (2.122), (2.123) we obtain [compared with (2.99)]

$$\begin{aligned} c_r \geqslant c_{r_{\min}} &= c_T (1+h)^{1/2} \left[1 - \frac{0.974 \Delta_* / \rho}{(1+h)^{1/2}} \right] \simeq \\ &\simeq c_T (1+h)^{1/2} \left[1 + 0.974 \Delta_* \varkappa / c_T (1+h) \right]^{-1}, \end{aligned} \tag{2.126}$$

and at $c_r = c_{r\,\min}$, according to (2.126), the wavelength of the marginally stable perturbations is characterised by the value

$$k \simeq \frac{k_0}{(1+h)}\left[1+0.2\Delta_* \varkappa / c_T(1+h)\right]. \tag{2.127}$$

it can be seen that the non-axisymmetric perturbations, like the axisymmetric ones, are stabilised by the finite thickness of the disk, also the degree of this stabilisation is lower than in the case of the axisymmetric perturbations.

2.4.3. Gravitational stability of the heterogeneous disk

We now determine the effect of the heterogeneity of surface density on the value of the minimum required dispersion of the radial velocities of the stars for the stability of the non-axisymmetric perturbations with the given m/kr and on the wavelength of the marginally stable perturbations (consideration of $\eta \neq 0$ in section 2.4.5). As the initial approximation, we use the model of a thin differentially rotating disk, see (2.124), (2.125), assuming that the effect of its small thickness is already determined by the multiplicative form factors in (2.126), (2.127). In this case, in the vicinity of the minimum of the dispersion curve of the gravitational perturbations ($k \simeq k_0$) the effect of the gradient branch is significant (see paragraph 2.2.4). Therefore, the condition of stability of such a disk should follow from the conditions of absence of the complex roots in the cubic (with respect to $\hat{\Omega}$) dispersion equation (2.95). We write this equation in the form $\hat{\Omega}^3 + s(k)\,\hat{\Omega}^2 + p(k)\,\hat{\Omega} + q(k) = 0$ and by linear transformation $y = \hat{\Omega} + s(k)/3$ it is reduced to the following form: $y^3 + \hat{p}(k)\,y + \hat{q}(k) = 0$. Consequently, the disk will be stable in relation to such perturbations for which $D(k) = (\hat{p}/3)^3 + (\hat{q}/2)^2 \leqslant 0$. At $c_r \simeq c_T$ in both the longwave ($k \ll k_T$) and shortwave ($k \gg k_T$) parts of the spectrum we have $D(k) < 0$ (see (2.101)–(2.104)). However, in the intermediate range we may also have $D(k) > 0$. Thus, D (k) will be the convex upward function and will reach its maximum value somewhere in the vicinity of $k\rho \sim 1$. It is therefore clear that in (2.95) the complex roots disappear at any $k\rho$, as soon as the condition $D(k) = 0$ is fulfilled for the perturbations for which $dD(k)/dk = 0$. From the system of these two equations we can calculate $(c_r)_{\min}$ for the given type of perturbation and the wavelength of the marginally stable perturbations.

Calculations are carried out mostly with respect to the small parameter

$$\nu = \frac{2\Omega}{\varkappa k_T} \sin\Theta \frac{\partial \ln(\sigma_{*0})}{\partial r}. \tag{2.128}$$

consequently, from the equations $D(k) = 0$; $dD(k)/dk = 0$ we obtain the stability criterion

$$c_r \geqslant c_{r_{\min}} \simeq c_T(1+h)^{1/2}(1+1.07\nu^{2/3}) \tag{2.129}$$

and the value of k for the marginally stable perturbations

$$k = k_0(1-0.72\nu^{2/3})/(1+h). \tag{2.130}$$

Using these results, from equation (2.95) we easily calculate also the frequency of marginally stable perturbations

$$\hat{\omega} \approx 0.673 \varkappa \nu^{1/3} + O(\nu). \tag{2.131}$$

2.4.4. The conditions of applicability of the stability criterion

We will discuss the conditions of applicability of the stability criterion of collisionless stellar disks, resulting from the dispersion equation (2.94).

The value of the minimum dispersion of the radial velocities of the stars required for stability increases with increase of |sin| [(sin $\Theta = m/kr$, see (2.129)]. We calculate the maximum value equal to $c_{r\,\min}$ (sin $\Theta = 1) = c_*$, and assume that the stability of the disk in relation to the arbitrary perturbations forms at $c_r \geqslant c_*$. However, in this case it is important to take into account the effect of two factors. Firstly, the dispersion equation (2.94), obtained in the framework of the WKB-approximation. Even if the boundary of applicability of WKB-approximation is reached (see the reference in section 2.2.2), then $(\sin\Theta)_{\max} \leq m/\sqrt{m^2 + (r/L_\sigma)^2} < 1$. The stability boundary of the disk is situated in the region $k \sim 1/\rho$ and, therefore, for $(m/r) \gg k_r$, the value $(m/r) < k \sim 1/\rho$ and at the same time $(\sin\Theta)_{\max} \leqslant 1/\sqrt{1+(\rho/L_\sigma)^2} < 1$ (in the solar neighbourhood $\rho/L_\sigma \sim 1/3$). Secondly, there is a more important restriction determined by the differentiallity of rotation of the disk. Actually, the value $\hat{\Omega} = \Omega - m(r)$ changes along the radial coordinate. Naturally, it is necessary to assume that the variation $|\hat{\Omega}|$ on the scale, characterising the variation of the perturbation along the radial coordinate, should be small in comparison with $|\hat{\Omega}|$:

$$\left|\frac{d\hat{\omega}}{k_r dr}\right| \ll |\hat{\omega}|. \tag{2.132}$$

We consider the marginally stable perturbations, using for estimates (2.131). We assume the law of rotation $\Omega \sim r^{-n}$ and, consequently, the condition (2.132) can be written in the following form

$$\frac{nm\Omega}{k_r r} \ll 0.673 \varkappa \left|\frac{2\Omega}{\varkappa}\frac{m}{kr}\frac{1}{k_T L_\sigma}\right|^{1/3}. \tag{2.133}$$

Equation (2.133) will be transformed as a condition for the value n which determines the degree of differentiality of rotation of the disk:

$$n \ll n_{\max} = 1.35\left(\frac{\pi G \sigma_{*0}}{2\Omega^2 L_\sigma}\right)^{1/3} \frac{(1-\sin^2\Theta)^{1/2}}{(\sin\Theta)^{2/3}}. \tag{2.134}$$

Equation (2.134) shows that for the perturbations with sin $(\Theta) = 1 - \varepsilon$ at $\varepsilon \ll 1$

$$n_{\max} \simeq 1.9\left(\frac{\pi G\sigma_{*0}}{2\Omega^2 L_\sigma}\right)^{1/3} \varepsilon^{1/2} \sim \varepsilon^{1/2}. \qquad (2.135)$$

We now assume that we use the criterion of stability of the disk at the limit sin (Θ). We evaluate the relative variation of the value $c_{r\,\min}$ in substitution s = sin (Θ) = 1 by $s = 1 - \varepsilon$. Since (2.131) and (2.134), resulting from it, were derived mostly with respect to the parameter ν, this value should be evaluated using the criterion which do not take into account the heterogeneities of the disk (2.124). Consequently, the error for the relative variation of the theoretically calculated value $c_{r\,\min}$ is:

$$\Pi_{\text{th}} = \frac{c_{r_{\min}}(s=1) - c_{r_{\min}}(s=1-\varepsilon)}{c_{r_{\min}}(s=1) + c_{r_{\min}}(s=1-\varepsilon)} = \frac{1-\sqrt{1-\frac{n}{2}(1-s^2)}}{1+\sqrt{1-\frac{n}{2}(1-s^2)}} \simeq \frac{n\varepsilon}{4}, \qquad (2.136)$$

which, taking into account (2.135) gives $\Pi_{\text{th}} \sim \varepsilon^{3/2}$ or $n_{\max} \sim (\Pi_{\text{th}})^{1/3}$. It may be seen that at low Π_{th} the value n_{max} may prove to be relatively high.

On the other hand, the dispersion of the radial velocities of the stars is determined from the observations already with some error

$$\Pi = \left(\frac{c_{r_{\max}} - c_{r_{\min}}}{c_{r_{\max}} + c_{r_{\min}}}\right)_{\text{obs}} \ll 1. \qquad (2.137)$$

Evidently, it is irrational to require from the theory that the error of calculations of the values in this theory does not exceed the error of determination of these values from the observations. Therefore, the value determined from (2.136) should be

$$\Pi_{\text{th}} = \Pi. \qquad (2.138)$$

Expressing from this relationship sin (Θ) as a function of Π and substituting into (2.134), we obtain the equation for $n_{\max}$:

$$n_{\max} = 1.54\left(\frac{\pi G\sigma_{*0}}{2\Omega^2 L_\sigma}\right)^{2/9} \frac{(4\Pi)^{1/3}/(1+\Pi)^{2/3}}{\left\{1-2\left[1-\left\{\frac{1-\Pi}{1+\Pi}\right\}^2\right]/n_{\max}\right\}^{2/9}}. \qquad (2.139)$$

As an example, we present the solutions of (2.139) for the typical values of the parameters $\sigma_{*0} = 80\ M_\odot/\text{pc}^2$, $\Omega = 25$ km/s/kpc, $L_\sigma = 4$ kpc:

$$n_{\max}(\Pi = 0.05) \simeq 0.73;\ n_{\max}(\Pi = 0.1) \simeq 1;\ n_{\max}(\Pi = 0.15) \simeq 1.2.$$

These solutions are quite typical because for the disks rotating with $V_{\text{rot}} \simeq$ const $\simeq$ ($n \simeq 1$), the value $\pi G_{\sigma *0}/(2\Omega^2 L_\sigma) \simeq c_T/(\varkappa L_\sigma) = \rho_T/L_\sigma$ changes, evidently in not too wide a range: $\rho_T/L_\sigma \simeq 0.15 \div 0.5$ (in the solar vicinity $\rho_T/L_\sigma \simeq 0.2$) and the dependence of n_{max} on ρ_T/L_σ is relatively weak. At the same time, the error, characterised by the scatter of the observational data, for the value c_r only in the solar neighbourhood is less than 10%, and for other galaxies it may exceed 20÷25% (see, for example [77, 409-411, section 4.4). Therefore, for the majority of stellar disks of the flat galaxies ($n \leqslant 1$) the stability criterion of the non-axisymmetric perturbations in the limit $\sin\Theta = 1$ on the basis of equation (2.94) can be used for evaluating the dispersion of the radial velocities of the stars of the stellar disk required for its stability.

2.4.5. The stability criteria of the stellar disk

We now determine the general condition of the stability of the stellar disk taking into account the gradient of the dispersion of the velocities of the stars ($\eta \neq 0$) and the possible deviation of the differential rotation of the disk from the power law ($\xi \neq 0$). In this case, the calculations are identical with those in section 2.4.3. Taking into account also the effect of the finite thickness of the disk, described by the relation (2.126), we obtain the following result:

$$c_r \geqslant \frac{c_T(1+h)^{1/2}}{1+0.974\dfrac{\Delta_* \varkappa}{c_T(1+h)}} \cdot \left\{1+1.07\nu^{2/3}\left|1+\xi-1.086\eta\right|^{2/3}\right\}. \qquad (2.140)$$

If now in accordance with the above considerations we set in (2.140) $\sin(\Theta) = 1$, then to determine the upper boundary of the dispersion of the radial velocities of the stars, required for the stability of the stellar disk, we obtain the following estimate [153]:

$$\sup\left\{c_{r_{min}}\right\} \simeq c_* = c_T\left(\frac{2\Omega}{\varkappa}\right)\left[1+0.974\frac{\Delta_*\varkappa}{c_T}\left(\frac{\varkappa}{2\Omega}\right)^2\right]^{-1} \times$$
$$\times\left\{1+1.07\left|\frac{2\Omega}{\varkappa k_T}\frac{d\ln\sigma_{*0}}{dr}(1+\xi-1.086\,\eta)\right|^{2/3}\right\}. \qquad (2.141)$$

This result, like the results obtained by Toomre (2.97) and van der Voort (2.99), is local and applicable only in the regions of the disk in which the initial approximations are fulfilled – epicyclic and the approximation of the smallness of the parameter $\varkappa/\Omega_z$ (evidently, the estimate (2.141) cannot be fulfilled in the central regions of the stellar disks). It should also be mentioned that the calculation of c_* should, if necessary, be iterative, because the value c_r is also included in the right hand part of the equality (through $\eta = d\ln c_r/d\ln\sigma_{*0}$).

The results of verification of the estimate (2.141) in the numerical experiments and for several constructed models of the Galaxy are discussed in chapter 4.9.

2.4.6. *Relationship between characteristic scales of heterogeneities of surface density and dispersion of radial velocities of the stellar disks*

When discussing the Toomre–van der Voort stability criterion (section 2.3.1), we already mentioned the numerical experiments with the models of the stellar disks. In particular, these experiments (see also chapter 4) show that the process of evolution to the stationary state in these models is accompanied by the redistribution of the equilibrium surface density σ_{*0} and the velocity of rotation $V_{\text{rot}}(r)$ and 'heating' of the disks to the state in which the experimental value c_r is close to $c_*(r)$ calculated from (2.141). At the same time, the value $c_*(r)$ depends not only on the local values of $\sigma_{*0}(r)$, $\Omega(r)$, but also on their gradients as well as on the gradient $c_r(r)$. Therefore, it may be expected that in the process of evolution to the marginally stable state in the stellar disk, the distribution of the parameters is such that the value $c_*(r)$ will be close to the minimum possible value. We will analyse from this viewpoint the dispersion properties of the branches of oscillations of the stellar disk and the condition of its gravitational stability.

As explained previously, the destabilising effect of the radial heterogeneity of the disk parameters is determined by two factors. Firstly, in the stationary stellar disk for the dispersion of the azimuthal velocities of the stars we have $c_\varphi = [\varkappa/2\Omega]\, c_r < c_r$ [203] and, consequently, for the stabilisation of the limiting non-axisymmetric perturbations as a result of the lower azimuthal 'elasticity of the disk in comparison with the radial elasticity, the value c_r should be $(2\Omega/\varkappa) > 1$ times greater than c_T. Secondly, the dispersion equation (2.94) in the frequency range $|\hat{\omega}| \leqslant \varkappa$ describes three branches of the oscillations in the disk plane: two gravitational and one gradient. The gradient branch is determined by the presence of the heterogeneity of the dispersion of the radial velocities of the stars or the heterogeneities of the surface density of the disk, or by both. The additional destabilisation of the perturbations in the heterogeneous disk is associated with the formation of 'interaction' between the gradient and gravitational branches in the case of the insufficiently hot ($c_r \simeq c_T$) stellar disk. Below, we will assume that the law of rotation is exponential $\Omega \sim r^{-n}$ with $n = \text{const}$ ($\xi \equiv 0$). The dispersion curves in the frequency range $|\hat{\omega}| \leq \varkappa$ in the disk with $c_r \simeq c_T$ are shown in Figs. 2.6 and 2.7.

At the same time, an increase of 'temperature' of the disk (increase of parameter c_r) increases the absolute value of the frequency of gravitational perturbations in the range $k \leqslant 2k_T$, and at some $c_r = c_* > c_T$, the 'weak' relationship between the gradient and gravitational branches disappears. This leads to the stabilisation of the gravitational–gradient instability of the disk (Fig. 2.7b). As indicated by Fig. 2.7a, in the sufficiently hot stellar disk ($c_r < c_*$) in the space of the wavenumbers there can be two regions of instability. This effect is determined by the following circumstance. The law of dispersion of the gradient branch of the oscillations of the disk, close to the stability boundary, in the wavelength range $k \leqslant k_T$ is written in the form

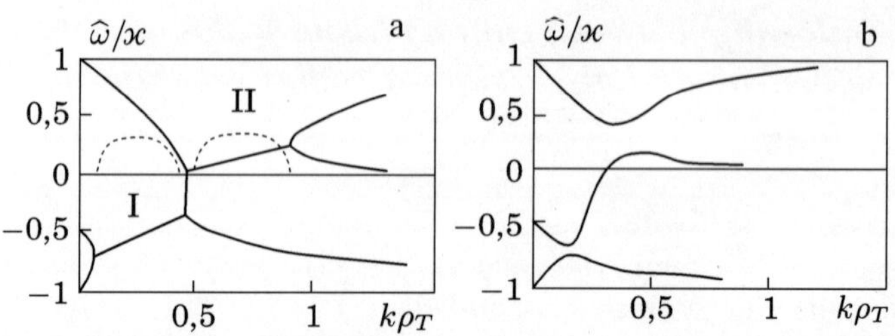

Fig. 2.7. Branches of non-axisymmetric perturbations in the model of the stellar disk, characterised by the parameters $\eta = 1.1$; $\Delta_*/\rho_T = 0.3$; $\rho_T \sin\Theta/L_\sigma = 0.18$: a) at $c_r = 1.2\, c_T$ we have an unstable disk, as indicated by the presence of Im $\hat{\omega} > 0$; b) at $c_r = 2.6\, c_T$ the disk is stabilized. The solid line shows $\mathrm{Re}(\hat{\omega}/\varkappa)$, the dotted line $\mathrm{Im}(\hat{\omega}/\varkappa)$.

$$\hat{\omega}_3 \simeq -\frac{2k_\varphi \Omega(1-\tilde{z}(1+2\eta))}{kk_T \mid L_\sigma \mid (1+k\Delta_* - \tilde{z}/kk_T\rho^2 + 3\tilde{z}^2/4kk_T\rho^2)}. \tag{2.142}$$

This shows that in the longwave limit ($k \ll k_T$) the frequency of the gradient branch is negative, and in the wavelength range $\tilde{z} > 1/(1+2\eta)$ it is positive. Thus, the gradient perturbations may interact with both negative and positive Jeans branches of the oscillations of the stellar disk (Fig. 2.7).

If the parameters of the disk are such that $\eta = L_\sigma/L_c = 0$ (c_r = const), then there is another region of instability in the k-space. At $\eta \neq 0$ there maybe two regions of instability in the k-space. In the dependence on the parameter η but increasing c_r one of them this years at lower values of c_r, and the other one – at higher values. In the disks with c $0 < = L_\sigma/L_c \leq 1$, the instability in the region II is suppressed at lower values of c_r than the instability in the region I. However, if in the stellar disk $\eta \geqslant 1$, then the instability in the region I is suppressed at lower values of c_r. In the disks with $\eta = \eta_{cr} \simeq 1$ the two regions of instability disappear almost at the same value c_r (Fig. 2.7b). The accurate value of η_{cr}, calculated from the total dispersion equation (2.94), depends only slightly on the disk parameters [175].

Thus, the function c_* (L_σ/L_c) at $|\xi| \ll 1$ reaches its minimum at $\eta = d\ln c_r/d\ln \sigma_{*0} = L_\sigma/L_c \simeq 1$. It would appear that it would be advantageous for the stellar disk to have at every point r similar values of L_σ and L_c. However, this conclusion is based on the local analysis at the fixed value r and, consequently, simultaneous fulfillment of two conditions – the disk is marginally stable and parameter η = const $\simeq 1$ – in a general case this cannot be realised for a relatively long region.

We will explain what the requirement for the entire disk (with the exception of the central regions) to have the minimum dispersion of the radial velocities of the stars for stabilility would mean [137]. Confining ourselves to qualitative examination, we use the stabilility criterion (2.141), written in the form

$$c_r \geqslant c_* = NDc_T\left\{1+1.07\left|1.87N\frac{c_T}{\varkappa}\left(\frac{d\ln\sigma_{*0}}{dr}-1.09\frac{d\ln c_r}{dr}\right)\right|^{2/3}\right\}, \tag{2.143}$$

for analysis. Here $N = 2\Omega/\varkappa$, $D = (1+0.974\,\varkappa\Delta_*/(N^2c_T))^{-1}$. We treat equation (2.143) as the differential equation for the function $c_*(r) = c_r(r)$, assuming the dependences $\sigma_{*0}(r)$, $\Omega(r)$ are known. In accordance with this, equation (2.143) can be written as follows

$$\frac{dc_*}{dr} = \frac{c_*}{1.09}\left\{\frac{d\sigma_{*0}}{\sigma_{*0}dr} + \gamma\right\}, \quad \gamma = \pm\frac{\varkappa}{1.87Nc_T}\left(\frac{c_*/NDc_T - 1}{1.07}\right)^{3/2}. \qquad (2.144)$$

As indicated clearly by equation (2.143), the value c_* reaches is minimum at $\eta = \eta_{cr} = 0.91$ and in this case if necessary to have $\gamma = 0$. Good approximation for real flat galaxies is the dependence $\sigma_{*0} \sim \exp(-r/|L_\sigma|)$ with $L_\sigma = \text{const}$. Therefore, the condition $\eta = \eta_{cr} \simeq 1$ requires $c_r \sim \exp(-r/|L_c|)$ with $L_c \simeq L_\sigma$. However, in this case, for arbitrary r the condition $\gamma = 0$ cannot be fulfilled. At $\gamma \neq 0$, depending on the sign + or – in (2.144), the value η becomes correspondingly higher or lower than η_{cr}.

The numerical integration of the equation (2.144) leads to the following result: restricted solutions of equation (2.144) are possible only at the – sign, i.e. $\gamma < 0$. Thus, although the parameter η can depend on the radial coordinate, but $\eta < \eta_{cr} \simeq 1$.

Observations

The values $\Omega(r)$ and L_σ (but not the density itself) are determined from the observations quite accurately. For a number of galaxies, the dispersions of the radial velocities have been determined at several points (to determine the value L_c it is sufficient to have two points) with respect to the radial coordinate. Therefore, it is interesting to verify for these objects that the condition $\gamma = 0$ is fulfilled. This condition is equivalent to [137]

$$\eta < \eta_{cr} = 0.9. \qquad (2.145)$$

The values of the parameters L_σ, L_c for different galaxies are distributed in a wide range. However, the condition (2.145) is fulfilled for the galaxies with the known distribution of the dispersion of the velocities of the stars, including our Galaxy [77, 78, 137, 258, 260, 1327].

In numerical experiments with modelling of the collisionless stellar disk we can determine the radial dependences of the equilibrium parameters $c_r(r)$, $\sigma_{*0}(r)$, $\Omega(r)$, $\Delta_*(r)$ and, therefore, calculate the characteristic scales of the heterogeneities of these values (chapter 4). The results of these studies also confirm that outside the central regions, the characteristic scale of the heterogeneity of the surface density does not exceed in value the characteristic scale of the heterogeneity of the dispersion of the radial velocities of the stars ($\eta = L_\sigma/L_c \leqslant 1$). As mentioned previously, this result can be explained on the basis of the requirement that the entire disk is situated at the boundary of gravitational stability.

It is also interesting to note that in the first models of the Galaxy constructed by Rohlfs and Kreitsehmann [1081], according to the calculation results of $c_*(r)$ the value $\eta_\odot \simeq 0.4$, and in the Caldwell and Ostriker model [441] $\eta_\odot \simeq 0.9$.

2.5. The models based on action–angle variables (V.L. Polyachenko, E.V. Polyachenko and I.G. Shukhman)

In this section, we derive the dispersion equation for the thin disk using the action–angle variables. This equation is then used to examine the conditions of formation of an instability [209], identical to the loss-cone instability in plasma physics, in the vicinity of massive black holes.

2.5.1. Natural oscillations of integrated systems

In canonic action–angle variables, the collisionless kinetic equation for the arbitrary integrated system has the following form:

$$\frac{\partial f}{\partial t} = [H, f] \equiv \sum_i \frac{\partial H}{\partial w_i}\frac{\partial f}{\partial I_i} - \sum_i \frac{\partial f}{\partial w_i}\frac{\partial H}{\partial I_i}, \qquad (2.146)$$

where [..., ...] is the Poisson bracket, H is the total Hamiltonian of the star, consisting of the integrated unperturbed part

$$H_0 = \frac{\mathbf{v}^2}{2} + \Phi_0(\mathbf{r}) = H_0(\mathbf{I}) \qquad (2.147)$$

and the perturbed part Φ which is a function of both variables of the action and angular variables.

The total distribution function f will be presented in the form of the unperturbed and perturbed parts, $f = \mathcal{F}_0 + \mathcal{F}$. These parts correspond to the unperturbed and perturbed potential of the simple layer:

$$\Phi_0(\mathbf{r}) = -G\int d\mathbf{r}'d\mathbf{v}' \frac{\mathcal{F}_0(\mathbf{r}',\mathbf{v}')}{|\mathbf{r}-\mathbf{r}'|}, \qquad \Phi(\mathbf{r}) = -G\int d\mathbf{r}'d\bar{\mathbf{v}}' \frac{\mathcal{F}(\mathbf{r}',\mathbf{v}')}{|\mathbf{r}-\mathbf{r}'|}. \qquad (2.148)$$

The perturbed functions will be expanded into a Fourier series with respect to the angular variables of the unperturbed orbits:

$$\mathcal{F}(\mathrm{I},\mathrm{w},t) = \sum_{\mathrm{n}} \mathcal{F}_{\mathrm{n}}(\mathrm{I}) e^{i\mathbf{n}\cdot\mathbf{w} - i\omega t}, \quad \Phi(\mathbf{I},\mathrm{w},t) = \sum_{\mathbf{n}} \Phi_{\mathbf{n}}(\mathrm{I}) e^{i\mathbf{n}\cdot\mathbf{w} - i\omega t}, \qquad (2.149)$$

where N is the dimension of space, $\mathbf{n} = (n_1, \dots n_N)$ is the N-dimensional integer vector, ω is the perturbation frequency. Subsequently, these values are substituted in the linearised kinetic equation (2.146) and the equations for the potential of the simple layer (2.148):

$$[-\omega + \mathbf{n}\cdot\mathbf{\Omega}(\mathbf{I})]\mathcal{F}_{\mathbf{n}}(\mathbf{I}) = \left(\mathbf{n}\cdot\frac{\partial \mathcal{F}_0}{\partial \mathbf{I}}\right)\Phi_{\mathbf{n}}(\mathbf{I}) \qquad (2.150)$$

and

$$\sum_{\mathbf{n}} \Phi_{\mathbf{n}}(\mathbf{I}) e^{i\mathbf{n}\cdot\mathbf{w}} = -G\int d\mathbf{I}'d\mathbf{w}' \frac{1}{|\mathbf{r}-\mathbf{r}'|} \sum_{\mathbf{n}'} \mathcal{F}_{\mathbf{n}'}(\mathbf{I}') e^{i\mathbf{n}'\cdot\mathbf{w}'}. \qquad (2.151)$$

from the equations (2.150) and (2.151) we obtain the required general equation for natural modes:

$$\left[\omega - \mathbf{n}\cdot\mathbf{\Omega}(\mathbf{I})\right]\mathcal{F}_{\mathbf{n}}(\mathbf{I}) = G\left(\mathbf{n}\cdot\frac{\partial\mathcal{F}_0}{\partial\mathbf{I}}\right)\int d\mathbf{I}'\sum_{\mathbf{n}'}\Pi_{\mathbf{n},\mathbf{n}'}(\mathbf{I},\mathbf{I}')\mathcal{F}_{\mathbf{n}'}(\mathbf{I}'), \qquad (2.152)$$

where the kernel of the integral equation $\Pi_{n,n'}$ (**I**, **I'**) is determined by the expression

$$\Pi_{\mathbf{n},\mathbf{n}'}(\mathbf{I},\mathbf{I}') = \frac{1}{(2\pi)^N}\int d\mathbf{w}d\mathbf{w}'\frac{e^{i(\mathbf{n}'\cdot\mathbf{w}'-\mathbf{n}\cdot\mathbf{w})}}{|\mathbf{r}-\mathbf{r}'|}. \qquad (2.153)$$

If we are interested only in unstable modes, Im $\omega > 0$, in the expression (2.152) integration can be carried out with respect to the real variables $\mathbf{I}'$. However, in the case of neutral and attenuating modes it is necessary to change the integration contour so that in the plane ω it passes below the required natural frequencies.

In the operator form, this equation is converted to the classic problem of eigenvalues

$$\omega\mathbf{F} = \left[\mathbf{n}\cdot\Omega(\mathbf{I})\mathcal{E} + G\left(\mathbf{n}\cdot\frac{\partial\mathcal{F}_0}{\partial\mathbf{I}}\right)\hat{\Pi}\right]\mathbf{F}. \qquad (2.154)$$

Here $\mathcal{E}$ is the unit operator, with the notation $\mathbf{F} = \{\mathcal{F}_{\mathbf{n}}\}$, and the eigenvalue its nothing else but the frequency of the natural oscillations of the system. The program (2.154) can be solved numerically. Using the quadrature formula

$$\int g_{\mathbf{n}}(\mathbf{I})d\mathbf{I} \to \sum_{\mathbf{I}} g_{\mathbf{n}}(\mathbf{I})w_{\mathbf{n}}(\mathbf{I}) \qquad (2.155)$$

the problem is reduced to the determination of the eigenvalues of the matrix [1043]:

$$\det\left|G\left(\mathbf{n}\cdot\frac{\partial\mathcal{F}_0(\mathbf{I})}{\partial\mathbf{I}}\right)\Pi_{\mathbf{n},\mathbf{n}'}(\mathbf{I},\mathbf{I}')w_{\mathbf{n}'}(\mathbf{I}') + \delta_{\mathbf{n},\mathbf{n}'}(\mathbf{I},\mathbf{I}')(\mathbf{n}\cdot\Omega(\mathbf{I})-\omega)\right| = 0, \qquad (2.156)$$

where $\delta_{\mathbf{n},\mathbf{n}'}$ (**I**, **I'**) denotes the Kronecker symbol.

We now examine a partial case of the integrated systems – thin stellar disks and write the appropriate problem for the eigenmodes.

2.5.2. Natural oscillations of thin stellar disks

We write the Hamiltonian of the unperturbed thin stellar disk in the form

$$H_0 = \frac{\mathbf{v}^2}{2} + \Phi_0(r). \qquad (2.157)$$

The unperturbed potential Φ_0 (r) includes both the potential of the stellar disk and the contribution from the halo of dark matter which is assumed to be constant and axisymmetric. In this system there are two integrals of motion: the angular momentum L and energy E. Consequently, the system is integrable. Instead of energy E it is more efficient to use the variable of the radial action I_r

$$I_r = \int\limits_{r_{\min}}^{r_{\max}} dr' v_r, \qquad v_r = (2[E - \Phi_0(r')] - L^2 / r'^2)^{1/2}. \tag{2.158}$$

In the action–angle variables $\mathbf{I} = (I_r, L)$ and $\mathbf{w} = (w_1, w_2)$, the unperturbed Hamiltonian of the system depends only on the action variables $H_0 = H_0(I_r, L)$. The motion on the unperturbed trajectories is described by the angular variables $w_i = \Omega_i(\mathbf{I})t + w_i^0$, where

$$\Omega_1 = \frac{\partial H_0(\mathbf{I})}{\partial I_r}, \qquad \Omega_2 = \frac{\partial H_0(\mathbf{I})}{\partial L}. \tag{2.159}$$

There are only several types of potentials Φ_0 permitting canonic transformation to the action variables I_r and L in the explicit form. For arbitrary Φ_0 the transformation is carried out by means of the function $r = r(\mathbf{I}, \overline{w}_1)$ which is obtained from the equation:

$$w_1(r, \mathbf{I}) = \Omega_1(\mathbf{I}) \int\limits_{r_{\min}(\mathbf{I})}^{r} \frac{dr'}{v_r} \tag{2.160}$$

(it is assumed that $0 \leqslant w_1 \leqslant \pi$), and the polar angle θ, satisfying the equation

$$w_2 = \theta + \varphi(\mathbf{I}, \overline{w}_1), \tag{2.161}$$

where

$$\varphi(\mathbf{I}, w_1) = \Omega_2(\mathbf{I}) \int\limits_{r_{\min}(\mathbf{I})}^{r} \frac{dr'}{v_r} - L \int\limits_{r_{\min}(\mathbf{I})}^{r} \frac{dr'}{r'^2 v_r}. \tag{2.162}$$

Frequencies Ω_i are defined by the following integrals:

$$\frac{\pi}{\Omega_1(\mathbf{I})} = \int\limits_{r_{\min}(\mathbf{I})}^{r_{\max}(\mathbf{I})} \frac{dr'}{v_r}, \quad \Omega_2(\mathbf{I}) = \frac{\Omega_1(\mathbf{I})}{\pi} \int\limits_{r_{\min}(\mathbf{I})}^{r_{\max}(\mathbf{I})} dr' \frac{L}{r'^2 v_r}. \tag{2.163}$$

Since the modes of the harmonics with different azimuthal numbers are separated, the expansion (2.149) contains a series for only one variable w_1:

$$\mathcal{F} = e^{imw_2 - i\omega t} \sum_l \mathcal{F}_l(\mathbf{I}) e^{ilw_1}, \qquad \Phi = e^{imw_2 - i\omega t} \sum_l \Phi_l(\mathbf{I}) e^{ilw_1}. \tag{2.164}$$

These equations describe the perturbations with the azimuthal number m. Then, equation (2.153) has the following form:

$$\Pi_{l,l'}(\mathbf{I}, \mathbf{I}') = \frac{1}{2\pi} \int w_1 dw_1' \psi(r, r') e^{il'w_1' - ilw_1} e^{im\delta\varphi}. \tag{2.165}$$

where $\delta\varphi = \varphi(w'_1) - \varphi(w_1)$ and

$$\psi(r,r') = \int_0^{2\pi} d\theta \frac{\cos m\theta}{[r^2 + r'^2 - 2rr'\cos\theta]^{1/2}} \quad (2.166)$$

(in this equation r and r' are functions of the corresponding action–angle variables: $r = r(\mathbf{I}, w_1)$, $r'=r'(\mathbf{I}', w'_1)$). The function $\psi(r, r')$ has a weak singularity $\propto \ln |r-r'|$ and, therefore, equation (2.165) of the kernel $\Pi_{l,l'}$ includes calculation of an improper integral. Substituting (2.165) into (2.153), we obtain equations for the eigenvalues for unstable modes of the thin stellar disk:

$$\mathcal{F}_l(\mathbf{I})(\omega - l\Omega_1 - m\Omega_2) = G\mathcal{F}'_{0,l}(\mathbf{I}) \int d\mathbf{I}' \sum_{l'} \Pi_{l,l'}(\mathbf{I},\mathbf{I}')\mathcal{F}_{l'}(\mathbf{I}'). \quad (2.167)$$

Here $\mathcal{F}_{0,l}'(\mathbf{I})$ denotes the derivatives with respect to action variables

$$\mathcal{F}'_{0,l}(\mathbf{I}) \equiv= l\frac{\partial \mathcal{F}_0(\mathbf{I})}{\partial I_1} + m\frac{\partial \mathcal{F}_0(\mathbf{I})}{\partial L}. \quad (2.168)$$

The system of equations (2.167) is used to find unstable modes in a thin stellar disk with an arbitrary rotation curve and an arbitrary degree of elongation of the orbits. Thus, this system is an alternative to the well-known Kalnajs matrix approach [739]. However, by comparison with this approach, our approach has several advantages: 1) it is simpler, because it consistently uses the formalism of the action–angle variables without going to the variables of the configuration space, and 2) does not need biorthonormal system of the function for the expansion of the potential and density, and finally, 3) leads to an linear eigenvalue problem, which allows one to define all unstable modes simultaneously. Note that in the Kalnajs matrix method the frequencies of natural oscillations are in the equation in the non-linear form, so that to determine the frequency we must know their approximate values.

2.5.3. Approximation of slow modes

The application of our integral equation is particularly fruitful when studying systems in which the orbits change over time slowly. This situation occurs when a thin self-gravitating disk is in the gravitational field of a heavy point mass (stellar disk near the centre of the galaxy), or in the field of a heavy halo which produces the approximately harmonic potential. As we know, in the case of an exact Kepler pr exact quadratic potential any orbit is closed. In the retreat from the exact law the precession of the orbit appears and its frequency is small compared with the frequency of rotation of the star in its orbit. Unstable processes that can occur when the effects of self-gravitation are taken into account, must be of the order of frequency of the slow precession frequency. Thus, for fixed m the system of equations (2.167) is characterised by the dominance of a single term for which the combination of frequencies $l\Omega_1 + m\Omega_2$ is small. For symmetric orbits in the harmonic potential $\Omega_1 = 2\Omega_2$, so the simplest resonance occurs for $m = 2$, $l = -1$. For Keplerian orbits, for which $\Omega_1 = \Omega_2$, the combination of frequencies converts to zero when $m = 1$, $l = -1$. We consider this case in more detail.

If the point mass M_c is centred in the focus, then not far from the centre this ratio will be close to 1:1. The deviation from exact resonance (and the associated slow precession of the orbits) is due in this case to the distribution of stars surrounding the central point mass. In this case, we assume that

$$\epsilon = M_G / M_c \ll 1, \tag{2.169}$$

where M_G is the mass of a cluster of stars surrounding a large central mass.

For simplicity, we consider the description of the models of stellar systems in which stars have the same energy $E = E_0$ (monoenergetic models). Then, the distribution function can be written as

$$F(E,L) = A\delta(E - E_0)f(L). \tag{2.170}$$

Such a dependence on energy, on the one hand, allows one to do most of the research analytically and on the qualitative level. On the other hand, it should not be essential for the establishment of stability or instability of each individual investigated distribution.

The model is constructed by defining the specific function f, which depends only on a single variable. Such a variable can be conveniently represented by the dimensionless angular momentum $\alpha = L/L_{circ}(E_0)$, where L_{circ} is the angular momentum of the star with energy E_0 on the circular orbit $L_{circ}(E_0) = GM_c/(2|E_0|)^{1/2} = \sqrt{\frac{1}{2}GM_cR}$, M_c is the central point mass, G is the gravitational constant. The frequency of the radial oscillations of the star $\Omega_1(E_0) = (2|E_0|)^{3/2}/GM_c = \sqrt{8GM_c/R^3}$ does not depend on the angular momentum. Here R is the radius of the maximum apocentric distance, corresponding to the orbit with the zero momentum, $R(E_0) = GM_c/|E_0|$. In order to reduce the number of notations, we will omit the argument E_0 and accept $R = 1$.

The permissible range of the variation of variable α is $-1 \leqslant \alpha \leqslant 1$. The stars with the positive angular momentum $\alpha > 0$ are directly rotating stars. However, the stellar disks may also contain retrograde stars with the opposite direction of rotation, $\alpha < 0$.

The normalisation constant A in (2.170) is selected to ensure that the mass of the spherical system, surrounding the central point mass, is equal to M_G:

$$M_G = \int f d\Gamma = (2\pi)^2 \int \frac{dE}{\Omega_1(E)} \int dL\, f(E,L).$$

Normalising the dimensionless distribution function by the condition $\int_{-1}^{1} d\alpha f(\alpha) = 1$, we obtain

$$A = \frac{M_G \Omega_1(E_0)}{4\pi^2 L_{circ}(E_0)}. \tag{2.171}$$

For the slow processes considered here, characterised by the characteristic frequencies of the order of the typical angular precession velocities, $\Omega \sim \Omega_{pr} \sim \varepsilon\Omega_1$, it is convenient to transfer to the dimensionless frequencies, measuring them in the natural 'slow units'

$$\bar{\omega} = \frac{\Omega_p}{\epsilon\Omega_1}, \quad \nu(\alpha) = \frac{\Omega_{\mathrm{pr}}(\alpha)}{\epsilon\Omega_1}. \tag{2.172}$$

The formulation of the distribution function (2.170) in the equation (2.167) leads to the following equation:

$$\phi(\alpha) = \frac{2C_m}{\pi^3}\int_0^1 \frac{d\alpha'\nu(\alpha')f'(\alpha')}{\bar{\omega}^2 - \nu^2(\alpha')}\mathcal{K}_m(\alpha,\alpha')\phi(\alpha'), \tag{2.173}$$

where the kernel of the integral operator is determined by the equation

$$\mathcal{K}_m(\alpha,\alpha') = \frac{1}{C_m}\int_0^\pi d\tau r(\alpha;\tau)\cos m\zeta(\alpha;\tau)\times$$
$$\times\int_0^\pi d\tau' r(\alpha';\tau')\cos m\zeta(\alpha';\tau')\mathcal{F}_m[r(\alpha;\tau),r(\alpha';\tau')]. \tag{2.174}$$

Equation (2.173) describes the slow eigenmodes of the thin stellar disk in the Kepler dominating potential. In the expression for the kernel r and ζ are the radial coordinate and the true anomaly of the star on the Kepler orbit, which can be conveniently expressed by the eccentric anomaly r:

$$r = \frac{1}{2}(1 - e\cos\tau), \qquad \cos\zeta = \frac{\cos\tau - e}{1 - e\cos\tau}, \tag{2.175}$$

where $e = (1 - \alpha^2)^{1/2}$ is the eccentricity of the orbit.

The kernel is normalised to ensure that on the purely radial orbits it is equal to unity: $\mathcal{K}_m(0, 0) = 1$. Coefficient C_m is calculated from the equation

$$C_m = \int_0^1 dx x^{1/2}(1-x)^{-1/2}\int_0^1 dy y^{1/2}(1-y)^{-1/2}\mathcal{F}_m(x,y).$$

where

$$\mathcal{F}_m(x,y) = \int_{-\pi}^{\pi}\frac{\cos m\theta\, d\theta}{\sqrt{x^2 + y^2 - 2xy\cos\theta}}. \tag{2.176}$$

In particular, for $m = 1$ and $m = 2$ we have $C_1 \approx 10.884$, $C_2 \approx 7.45$. For $m = 1$ and $m = 2$ we have the explicit form of the functions $\mathcal{F}_m(x, y)$. For the case $m = 1$ (see, for example, [1235]):

$$\mathcal{F}_1(x,y) = \frac{4}{\max(x,y)}\frac{\mathbf{K}(q) - \mathbf{E}(q)}{q}, \qquad q = \frac{\min(x,y)}{\max(x,y)}, \tag{2.177}$$

and for $m = 2$:

$$\mathcal{F}_2(x,y) = \frac{4}{3\max(x,y)}\left[\left(\frac{2}{q^2}+1\right)\mathbf{K}(q) - 2\left(\frac{1}{q^2}+1\right)\mathbf{E}(q)\right]. \tag{2.178}$$

Here $\mathbf{K}$ (q) and $\mathbf{E}$ (q) are the elliptical integrals of the first and second kind, respectively. It should be mentioned that in the accepted approach it is assumed that the slow precession of the orbits is caused by the small difference of the potential of the system from the Kepler potential as a result of taking into account the self-consistent gravitational potential of the stellar system. Therefore, in the main order with respect to the appropriate small parameter, the kernel of the equation can be calculated assuming that the orbit is closed, but the precession velocity ν (α) should be taken into account in the denominators of the equations in the right-hand part of the system (2.173).

2.5.4. Gravitational loss-cone instability

From the viewpoint of application of the theory to the central regions of the galaxies, of special interest are the systems with the dominance of radial orbits which could form as a result of the collapse of the non-rotating gas cloud, accompanied by star formation. It is well-known that these systems may be characterised by the development of the instability of radial orbits, initially detected theoretically in a study by Polyachenko and Shukhmann [206] and subsequently confirmed by numerical experiments [196, 336]. The quantitative description of the physical mechanism of this instability was proposed by Lynden–Bell [864] who examined the slow precessing orbits in the symmetric potential of the halo. According to this theory, the main role in the formation of the instability is played by the orbits in which the precession direction coincides with the direction of motion of the star on the orbit because only they are capable of being carried away by the weak random perturbation of the potential and, at the same time, cause the potential to increase. The instability of the radial orbits develops in the scatter of the precession velocities is not too large; otherwise, the instability is suppressed.

If the precession of the orbits is directed in the direction opposite to the direction of rotation of the star on the orbit (retrograde precession), as is the case in, for example, systems with the dominant Kepler potential in the vicinity of the massive black holes, another instability can develop. For this purpose it is necessary that in the range of small angular moments the distribution function has a shortage of stars or, in other words, that the cone of the losses is empty or almost empty as a result of tidal failure or direct absorption of the stars by a black hole. Because of the obvious analogy with the simplest plasma traps, this instability is referred to as the gravitational loss-cone instability. It should be mentioned that as a result of departure of the plasma particles with the small transverse velocities $v_\perp$ (in relation to the axis of the system) to the 'cone of the losses' a specific 'beam' distribution of the particles forms with respect to these transverse velocities so that at a relatively low $v_\perp$, $\partial f_0/\partial v > 0$ (f_0 is the distribution function of plasma). This distribution is unstable [138, 1087]. In this section, we show the presence of the gravitational loss-cone instability on the example of the simplest model of the stellar disk:

$$f(\alpha)=\frac{N_d}{\alpha_T}x^2 e^{-x^2}, \qquad x=\frac{\alpha}{\alpha_T}, \qquad \int_{-1}^{1} f(\alpha)d\alpha=1, \tag{2.179}$$

in which constant N_d is determined from the normalisation conditions.

For this purpose it is necessary to calculate the kernel of the integral equation (2.174) and the velocity of recession of the orbits, included in equation (2.173). The expression for the precession velocity of the orbits in the almost-Kepler potential was presented for the first time in [1237] (see equation (42)). In the variables we use here, it is reduced to the form

$$\nu(\alpha)=\frac{2\alpha}{\pi^3 e}\int_0^{\pi}\Psi_G\left[\frac{1-e\cos\tau}{2}\right](\cos\tau-e)d\tau, \tag{2.180}$$

where Ψ_G is the dimensionless derivative of the gravitational potential,

$$\Psi_G(r)=\frac{\pi^2R^2}{8GM_G}\frac{d\Phi_G}{dr}=-\frac{1}{2r}\int_0^R r'dr'\Sigma_0(r')\left[\frac{\mathbf{E}(\varkappa)}{r'-r}-\frac{\mathbf{K}(\varkappa)}{r+r'}\right], \quad \varkappa=\frac{2\sqrt{rr'}}{r+r'}, \tag{2.181}$$

$\Sigma_0(r)$ is the dimensionless surface density

$$\Sigma_0(r)=\int_{-\alpha_{\max}(r)}^{\alpha_{\max}(r)}\frac{f(\alpha)d\alpha}{\sqrt{\alpha_{\max}^2(r)-\alpha^2}}, \quad \alpha_{\max}^2(r)=4(r/R)(1-r/R). \tag{2.182}$$

It should be mentioned that the integral with a singularity at $r'=r$ in the subintegrand expression in (2.181) is understood to be the main value. The precession velocity $\nu(\alpha)$ can be represented by a single potential with respect to the distribution function $f(\alpha)$:

$$\nu(\alpha)=\alpha\int_{-1}^{1}\mathcal{Q}(\alpha,\alpha')f(\alpha')\alpha'd\alpha', \tag{2.183}$$

where $\mathcal{Q}(\alpha,\alpha')$ is a universal function which does not depend on the model. Actually, substituting (2.182) into (2.181) and then into (2.180), and rearranging the integration variables, we obtain

$$\mathcal{Q}(\alpha,\alpha')=\frac{1}{\pi^3 e}\int_0^{\pi}d\tau\frac{e-\cos\tau}{1-e\cos\tau}\int_0^{\pi}d\tau'(1-e'\cos\tau')\left[\frac{\mathbf{E}(\varkappa)}{r'-r}-\frac{\mathbf{K}(\varkappa)}{r'+r}\right]. \tag{2.184}$$

The derived expression for the function $\mathcal{Q}(\alpha,\alpha')$ is two complicated for direct calculations. However, $\mathcal{Q}(\alpha,\alpha')$ can be expressed by the derivative of the continuous function

$$\mathcal{Q}(\alpha,\alpha')=\frac{1}{\alpha'}\frac{\partial\mathcal{P}(\alpha,\alpha')}{\partial\alpha'}, \tag{2.185}$$

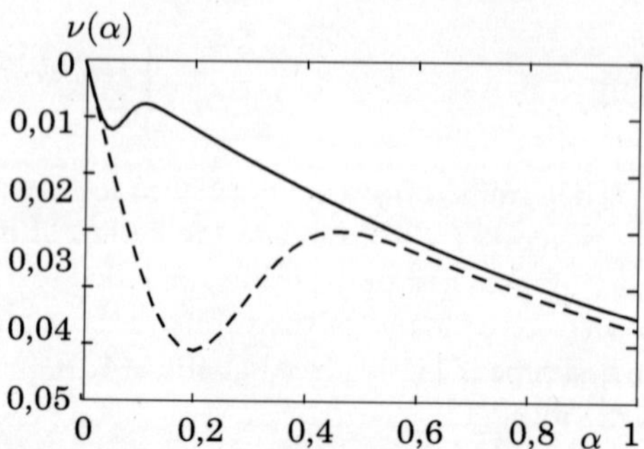

Fig. 2.8. Dependence of the precession velocity $\nu\,(\alpha)$ for different values of the thermal scatter of the distribution function α_T of the model of the stellar disk (2.179), situated in the dominant potential for $\alpha_T = 0.05$ (solid curve) and $\alpha_T = 0.2$ (the dashed curve).

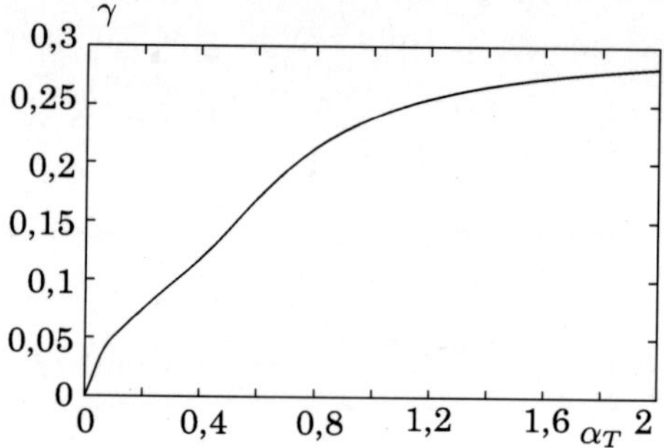

Fig. 2.9. Dependence of the increment of the gravitational loss-cone instability $\gamma = \text{Im}\,\omega$ on the thermal scatter of the distribution function α_T for the model of the stellar disk (2.179).

which does not contain the integral in the meaning of the main value. Direct verification shows that the following function satisfies equation (2.185):

$$\mathcal{P}(\alpha,\alpha') = -\frac{4e'}{\pi^3 e}\int_0^{\pi} d\tau \frac{e-\cos\tau}{1-e\cos\tau}\int_0^{\pi} d\tau' r' \frac{K(q)}{r'+r}(\cos\tau' - 2e'\cos^2\tau' + e'). \quad (2.186)$$

The characteristic functions $\nu(\alpha)$ for different α_T are shown in Fig. 2.8. We see that the precession of the orbit is retrograde, i.e. directed in the opposite direction of rotation of the star in its orbit. A characteristic feature of disk systems is the existence of non-monotonic changes of the angular momentum. This feature, as well as the analytical properties of the nucleus, lead to the fact that in order to study the stability of disk systems with highly eccentric orbits we can not use the simplified description of the orbits in the form of elongated orbits – radial spokes [210]. Note, however, that such a description is possible in the case of star clusters of spherical geometry.

Figure 2.9 shows the results of calculations of the growth rate of unstable modes $m = 1$, depending on the dispersion of α_T with respect to the angular momentum of the simulation distribution function. There is a branch of unstable oscillations caused by the development of gravitational loss-cone instability. In the low angular moment range the increment is directly proportional to the dispersion of α_T and increases to approximately $\alpha_T \approx 1$ at which the distribution function is monotonically increasing. After this the increment practically ceases to grow and strives to a finite limit.

In conclusion, we note once more that the emergence of gravitational loss-cone instability is associated with the presence of the loss cone at low angular moments and retrograde precession of the orbits. If one of these conditions is not satisfeed, this instability does not develop.

2.6. Stability of stellar disks in relation to bending perturbations

2.6.1. Dynamics of bending perturbations

In the previous sections we examined the dynamics of perturbations in the plane of the stellar disk and the restrictions on the disk parameters resulting from the requirement of the stability of such perturbations. In this section we focus on another type of perturbation – perturbations bending the stellar disk plane – and, accordingly, we find the restrictions on the disk parameters and the system as a whole, which follow from the stability condition such disturbances.

For the first time, apparently, the dynamics of bending (membrane) oscillations in models of cold thin disks was considered by Hunter and Toomre [700]. This study was designed to explain the large-scale warping of the periphery of the disks in a number of isolated galaxies. Another problem, which can be solved by studying the dynamics of bending perturbations, is associated with the task of explaining significant differences in the thickness and z-dispersion of the velocity of objects of stellar and gas disks of flat galaxies.

However, the models of the cold thin disks studied by Hunter and Toomre [700] were stable with respect to bending perturbations (for a detailed exposition of the theory of this issue see the Polyachenko and Fridman [590]). In [496] it was shown that for solutions of the above-mentioned problems it is necessary to study the dynamics of bending perturbations in the disk models, hot in their plane.

It is obvious that such local parameters such as the z-dispersion of the velocity of the stars and the disk thickness should be determined from the condition of stability of the perturbations which are small in scale compared to its thickness. The corresponding dispersion WKB equation of the bending perturbations in the simplest model of a homogeneous thin non-rotating stellar layer can be written as [877]

$$\omega^2 = k^2 c_{\|}^2 - 2\pi G\sigma_{*0} k, \tag{2.187}$$

where ω is the frequency of perturbations; k is their wavenumber; c_0 is the dispersion of the velocities of the stars in the layer plane. Evidently, the instability forms for the perturbations with the wavelength $\lambda < c_{\|}^2 / G\sigma_{*0}$ ($k > 2\pi G\sigma_{*0} / c_{\|}^2$). Using the relationship (2.40) of the half thickness of the disk Δ_* with the values σ_{*0} and c_z, it is easy to write the stabilility condition in the form

$$\left(c_z / c_{\|}\right) > \sqrt{k\Delta_* / 2}. \tag{2.188}$$

On the other hand, the bending perturbations which are small in scale compared with the thickness of the disk must also be stable [496]. This means that there is the minimum values of the ratio $c_z / c_{\|}$ necessary for the stability of the disk. However, since the dispersion of radial velocities of stars is bounded from below by the condition of gravitational stability of the disk, the value of the z-dispersion of the velocity of the stars and the associated thickness of the disk can not be arbitrarily small. The definition of the maximum value of anisotropy $\alpha = c_{\|} / c_z$ in the distribution of the velocities of stars due to the requirement of stability of the disk with respect to bending perturbations is discussed in section 2.6.1.

The dynamics of large-scale bending perturbations, covering the entire disk, must obviously be studied taking into account the rotation and structure of the disk as a whole. Such a study for a number of models of thin disks [195] showed that the most large-scale modes can be unstable if the balance of the disks in the radial direction is provided mainly by pressure rather than rotation. With a decrease in the contribution of the pressure to the equilibrium condition the most large-scale modes are stabilised, and those with the shorter wavelength remain unstable. This result, together with the dispersion equation (2.187), shows that the instability of the bending perturbations of the disk is in some sense an addition to the gravitational instability of the perturbations in the disk plane. Indeed, gravity destabilizes the perturbations in the disk plane and stabilizes the bending perturbations; pressure turns out to be a stabilising factor for the perturbations in the disk plane and a destabilising factor for the bending perturbations. Therefore, one would expect that most large-scale bending modes are, in contrast to the perturbations in the disk plane (see chapter 3), destabilised by a sufficiently massive spheroidal subsystem. This issue is discussed in section 2.6.3.

2.6.2. *What should the value of z-dispersion of the velocity of stars be?*

Following Polyachenko and Shukhman [207], we define the maximum anisotropy $\alpha = c_{\|} / c_z$ of the velocities in the stellar disk. Since rotation is important only for perturbations whose scope is comparable with the disk radius, we proceed from the simple model of a collisionless non-rotating plane layer of finite thickness [4] described by the distribution function

$$f(z,\mathbf{v}) = \frac{\rho_0}{\pi \omega_z \Delta} \mathbf{F}(v_x, v_y) \left[1 - \frac{z^2}{\Delta^2} - \frac{v_z^2}{\omega_z^2 \Delta^2}\right]^{-1/2}, \tag{2.189}$$

where ρ_0 is the density, Δ is the half thickness of the layer, $\Omega_z = (4\pi G\rho_0)^{1/2}$, and $\mathbf{F}\,(v_x, v_y)$ can be the arbitrary function of its arguments.

The perturbations are regarded as longwave in the sense that the characteristic scale of the perturbation in the layer plane is larger in comparison with its half thickness ($k\Delta \ll 1$). However, the parameter $kc_{\|} / \Omega_z = k\,\Delta\,\alpha$ is not assumed to be small. The axis x is oriented along the direction of the wave vector. Consequently, the dependence of $\mathbf{F}(v_x, v_y)$ on v_y in the non-rotating layer is assumed to be insignificant and the problem is reduced to a two-dimensional problem.

The distribution function with respect to v_x will be presented in the form of superposition of the flow

$$\mathbf{F}(v_x) = \int \mathbf{F}(v_0)\delta(v_x - v_0)dv_0 \tag{2.190}$$

and we initially examine a single flow with the velocity v_0 and density $\delta\rho_0 = \rho_0 \mathbf{F}(v_0)\, dv_0$. The perturbed distribution function for the flow is presented in the form

$$\begin{aligned} \tilde{f} = \frac{\rho_0}{\pi\omega_z\Delta}\left(1 - \frac{z^2}{\Delta^2} - \frac{v_z^2}{\omega_z^2\Delta^2} - \chi\right)^{-1/2} \delta(v_x - v_0) + \\ + \frac{\rho_0}{\pi\omega_z\Delta}\left(1 - \frac{z^2}{\Delta^2} - \frac{v_z^2}{\omega_z^2\Delta^2}\right)^{-1/2} \mathrm{B}\delta'(v_x - v_0). \end{aligned} \tag{2.191}$$

The substitution of this equation into the linearised kinetic equation results in the following equations for the functions χ, B:

$$\frac{d\chi}{dt} = 2ik\mathrm{B}[1 - z^2/\Delta^2 - v_z^2/(\omega_z^2\Delta^2)] + \frac{2v_z}{\omega_z^2\Delta^2}\frac{\partial\Phi_1}{\partial z}, \tag{2.192}$$

$$\frac{d\mathrm{B}}{dt} = ik\Phi_1, \tag{2.193}$$

where $\Phi_1 \propto \exp\{-i\omega t + ikx\}$ is the perturbed gravitational potential and

$$\frac{d}{dt} = \frac{\partial}{\partial t} + v_0\frac{\partial}{\partial x} + v_z\frac{\partial}{\partial z} - \omega_z^2 z\frac{\partial}{\partial v_z}. \tag{2.194}$$

Estimating the ratio of the values B and Φ_1 from (2.193), it can easily be seen that the first term in the right-hand part of (2.192) in the limit $k\Delta \ll 1$ is small in comparison with the second one (their ratio is of the order of $k^2\Delta^2 \ll 1$) and, consequently, can be ignored. It may also be easily verified that in this limit for the bending perturbations the local density is not perturbed ($\rho_1 = 0$) and it should therefore be assumed that $\Phi_1 = \beta z$. Consequently, from (2.192) we obtain

$$\chi = \frac{2\beta z}{\Delta^2(\omega_z^2 - \hat{\omega}^2)}, \tag{2.195}$$

where $\hat{\omega} = \omega - kv_0$. In equation (2.195), the term proportional to v_z, is omitted because it does not provide any contribution to the displacement of the boundary of the layer nor to its volume density.

The displacement of the boundary of the layer ξ is determined from the condition

$$\left(1 - \frac{z^2}{\Delta^2} - \frac{v_z^2}{\omega_z^2\Delta^2} - \chi\right)\Bigg|_{z=\Delta, v_z=0} = 0.$$

Consequently

$$\xi\Big|_{z=\Delta} = -\frac{1}{2}\chi\Delta = -\frac{\beta}{\omega_z^2 - \hat{\omega}^2}. \tag{2.196}$$

We now calculate the appropriate (2.196) perturbation of the gravitational potential. For this purpose, we connect the solutions of the equation

$$\frac{d^2\Phi_1}{dz^2} - k^2\Phi_1 = 0$$

at $z = \Delta$ in accordance with the boundary conditions

$$\Phi_1\Big|_{z=\Delta+0} - \Phi_1\Big|_{z=\Delta-0} = 0, \tag{2.197}$$

$$\frac{\partial\Phi_1}{\partial z}\Big|_{z=\Delta+0} - \frac{\partial\Phi_1}{\partial z}\Big|_{z=\Delta-0} = \omega_z^2\xi\Big|_{z=\Delta}. \tag{2.198}$$

Consequently, with the accuracy to the terms of the order of $k\Delta \ll 1$ inclusive we obtain the dispersion equation for the investigated flow:

$$1 + k\Delta = \frac{\omega_z^2}{\omega_z^2 - \hat{\omega}^2}. \tag{2.199}$$

Summing up these equations with respect to all flows, we obtain the required dispersion equation

$$1 + k\Delta + \omega_z^2 \int \frac{\mathbf{F}(v_x)dv_x}{(\omega - kv_x)^2 - \omega_z^2} = 0. \tag{2.200}$$

For specific calculations we start with the Schwarzschild distribution function $\mathbf{F}(v_x) = (2\pi c_x^2)^{-1/2}\exp\{-v_x^2/2c_x^2\}$. In this case, the dispersion equation (2.200) leads to the following equation

$$\frac{i\sqrt{\pi}\omega_z}{kc_x}\left\{\mathbf{W}\left(\frac{\omega - \omega_z}{\sqrt{2}\,kc_x}\right) - \mathbf{W}\left(\frac{\omega + \omega_z}{\sqrt{2}\,kc_x}\right)\right\} = 2\sqrt{2}\,(1 + k\Delta), \tag{2.201}$$

where $\mathbf{W}(x)$ is the Kramp function [235]. The dispersion of the velocities in the z-direction is determined as follows. Pressure with respect to z is equal to $p = \rho_0\left\langle v_z^2\right\rangle = \rho_0\omega_z^2\Delta^2(1 - z^2/\Delta^2)/2$, since temperature $T_z = \omega_z^2\Delta^2(1 - z^2/\Delta^2)/2$. Averaging this value over the thickness of the layer, we obtain $c_z = \sqrt{\langle T_z\rangle} = \omega_z\Delta/\sqrt{3}$. Therefore, for the anisotropy in the velocity distribution of the stars we have

$$\alpha = \frac{c_{\|}}{c_z} = \frac{c_{\|}\sqrt{3}}{\omega_z\Delta}. \tag{2.202}$$

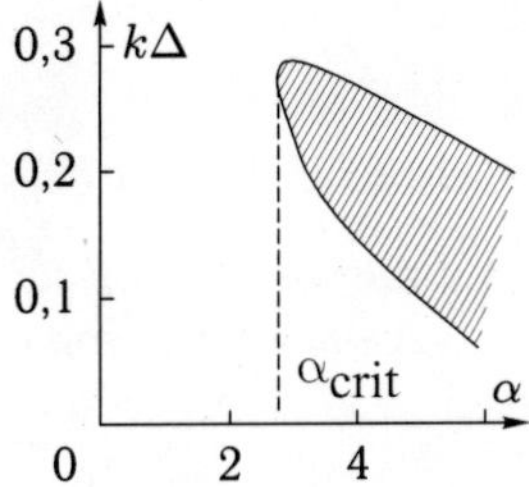

Fig. 2.10. The region of hose-pipe instability (crosshatched) on the plane of the parameters $\alpha = c_{\parallel}/c_z$ and $k\Delta$.

At $\min(|\omega-\omega_z|,|\omega+\omega_z|) \gg k_{c\parallel}$ from equation (2.201) we can obtain the simplified dispersion equation (2.187). This equation shows that the hose-pipe[12] (anisotropic) instability of the bending perturbations of the layer takes place at $k\Delta>3/\alpha^2$. At the same time, the perturbations whose characteristic scale is small in comparison with the layer thickness, should be stable and, consequently, the regional instability should be restricted on the side of 'high' $k\Delta$. For the exact determination of the boundaries of this region we assume in (2.201) $\omega = 0$. The resultant equation can be presented in the form

$$1+k\Delta = 2AF_D(A), \tag{2.203}$$

where $A=\sqrt{3/2}\,/\,k\Delta\alpha; F_D(A)=R_0^A\exp(t^2-A^2)dt$ is the Doson integral. It may easily be seen that at $\alpha \gg 1$ equation (2.203) has two solutions, $k_1\Delta \simeq 3/\alpha^2$ and $k_2\Delta \simeq 1.4/\alpha$. Thus, the region of the unstable wavelength values in the limit $\alpha \gg 1$ is restricted by the band $k_1 \lesssim k \lesssim k_2$. With the reduction of the parameter α the width of this region decreases and, as shown by the numerical solution of (2.203), disappears at $\alpha < \alpha_{cr} \simeq 2.70$ (Fig. 2.10). In application to the galaxies it is evidently necessary to assume that $\alpha = c/c_z \equiv c_r/c_z$, since $c_r > c_\varphi$. This means that the stellar disks of the flat galaxies should be stable at

$$c_z \gtrsim c_{z_{cr}} \simeq 0.37c_r. \tag{2.204}$$

The observational data in the near-solar neighbourhood of the Galaxy do not contradict the estimate (2.204) – according to [89, 229, 301] for not too young stars $c_z/c_r \simeq 0.6$. Thus, in the solar neighbourhood of the Galaxy the stellar disk is stable (according to (2.204) with a margin) in relation to the bending perturbations of the intermediate scales $1/R \ll k \ll 1/\Delta$ (R is the radius of the galactic disk). Evidently, the same can be said of the stellar disks of other flat galaxies [409, 410, 803]. The results of numerical experiments are discussed in section 4.8. The results of the calculations $((c_z/c_r)_{cr} \simeq 0.29)$ with the vertical heterogeneity of the disk taken into account, were published by S. Araki and are discussed in [906].

[12]The term often used in plasma physics and hydrodynamics.

2.6.3. Large-scale bending perturbations of the disk in a massive halo

To investigate the dynamics of large-scale bending modes affecting the entire disk of the flat galaxy, it is evidently necessary to take into account the structure of the disk and the spherical system surrounding the disk as a whole. The analytical solution of this problem is possible only on relatively simple models, and in this section we describe the initial results obtained in the investigations of this type [208].

The model of the galaxy is represented in the form of a biaxial solid-state rotating stellar ellipsoid with homogeneous density with the half axes *a, c* ($a > c$), immersed in an elongated homogeneous spherical halo. The equilibrium gravitational potential inside the ellipsoid (the model of the disk) is

$$\Phi_0 = \frac{\Omega^2}{2}(x^2 + y^2) + \frac{\omega_0^2}{2}z^2, \tag{2.205}$$

where $\Omega^2 = GM_h / a^3 + \Omega_d^2; \omega_0^2 = \omega_h^2 + \omega_d^2$; M_h is the mass of the halo in the region $r < \alpha$, and the values ω_h, Ω_d, ω_d depend on the density, the ratio of the half axes of the ellipsoid and the ratio of the masses of the subsystems (explicit expressions for them can be found in a monograph by Polyachenko and Fridman [590]). We confine ourselves to the investigation of the dynamics of the largest-scale modes:

a) $m = 0$ ('cupola')

$$\Phi_1 = z(x + iy)\left[Az^2 + B(x^2 + y^2)\right]; \tag{2.206}$$

b) $m = 1$ ('sombrero')

$$\Phi_1 = z(x + iy)\left[Az^2 + B(x^2 + y^2)\right]; \tag{2.207}$$

c) $m = 2$ ('saddle')

$$\Phi_1 = Az(x + iy)^2, \tag{2.208}$$

where Φ_1 is the perturbation of the gravitational potential, and the coefficients *A, B, D* depend only on time (the names 'cupola', 'sombrero' and 'saddle' are determined by the form of the perturbation of the symmetry plane of the ellipsoid).

These modes are interesting mostly as having the largest scale and, therefore, being most 'dangerous' from the viewpoint of the loss of stability. In addition to this, it should be expected that the behaviour of similar modes, affecting the entire system, is determined by a small number of its 'global' parameters. This means that the investigation of the stability of the examined model with respect to the perturbation of the previously mentioned modes does not require specifications of the distribution function – it is necessary to know all the several main moments of the function. For the investigated model of the homogeneous ellipsoid, the equations for the linear and quadratic (with respect to velocities) moments is determined unambiguously, if we only assume the isotropy in the distribution of the velocities in the rotation plane. These moments have the form

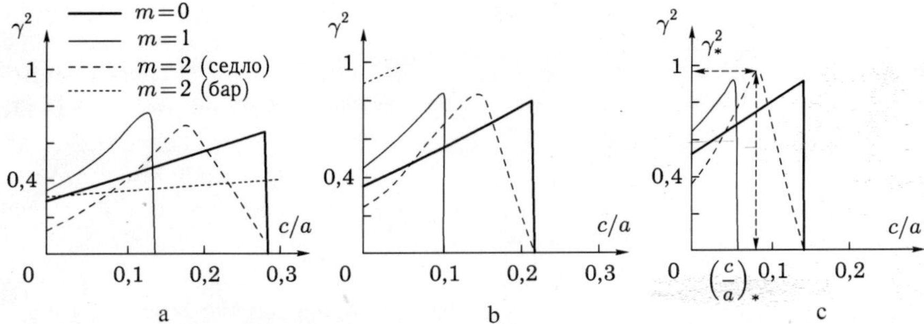

Fig. 2.11. Regions of instability of the bending modes (2.206)–(2.208) for homogeneous ellipsoids, immersed in the homogeneous halo: a) $M_h = 0$; b) $M_h = M_d$; c) $M_h = 3M_d$. For the bending modes the regions of instability are below the appropriate curves; for the bar mode (Fig. a, b) they are above the appropriate curves. Different modes are shown: the thick line $m = 6$, the thin solid line $m = 1$, the dashed line $m = 2$ (saddle), the dotted line $m = 2$ (bar).

$$\langle v_r \rangle = \langle v_z \rangle = 0, \quad \langle v_\varphi \rangle = \gamma \Omega r, \quad \langle v_z^2 \rangle = \frac{\omega_0^2 c^2}{2}(1 - r^2 / a^2 - z^2 / c^2),$$

$$\langle v_r^2 \rangle = \langle (v_\varphi - \langle v_\varphi \rangle)^2 \rangle = \frac{a^2 \Omega^2}{2}(1 - \gamma^2)(1 - r^2 / a^2 - z^2 / c^2). \tag{2.209}$$

These equations show clearly that the model investigated here is described by three 'global' parameters: the ratio of the half axes of the ellipsoid c/a; the ratio of the masses of the halo and the disk (ellipsoid) $\mu = M_h / M_d$ and the parameter γ, equal to the ratio of the angular velocity of rotation of the hot ellipsoid to the angular velocity of rotation of the cold ellipsoid (thus, the value γ^2 determines the contribution of rotation to the radial equilibrium of the ellipsoid, and $(1 - \gamma^2)$ – contribution of 'pressure' $(0 \leqslant \gamma \leqslant 1)$).

The method of investigating the stability is standard [4, 203, 590]. Initially, using the equations for Lagrangian displacements x, y, z, these values are expressed through the perturbed potential Φ_1. This is followed by calculating the perturbations of density in the normal displacement of the ellipsoid boundary. Solving subsequently the Poisson equation and comparing the resultant potential with the initial one (2.206)–(2.208), we obtain the system of linear algebraic equations in relation to the coefficients included in the equation for the perturbed potential. Equating, finally, the determinant of this system to 0, we obtain the required dispersion equation.

The dispersion equations (2.206)–(2.208) for the previously mentioned modes were solved numerically (they are quite cumbersome and are not presented here). The results of solving these equations are presented in Fig. 2.11 *a–c*, corresponding to different values of the parameter $\mu = M_d / M_d$. These figures show that at the disk limit ($c/a \to 0$) each of the modes is unstable at a relatively high dispersion of the velocities of the stars (the regions below the curves). With increase of the 'thickness' of the ellipsoid, the region of instability with respect to the parameter γ^2 initially increases (i.e. more and more 'colder' systems become unstable) and then disappears quite rapidly.

It should be mentioned that a rapid decrease of the curves in Fig. 2.11 – the boundaries of the instability regions – takes place at the oblateness of the ellipsoid c/a at which the investigated homogeneous model has resonances between the frequencies of the oscillations of the stars across the plane of the ellipsoid and in the plane of its rotation. For the modes $m = 0$ and $m = 2$ there is a resonance $\omega_0 = 2\Omega$, and for the mode $m = 1$ resonance $\omega_0 = 3\Omega$. Evidently, the strong effect of these resonances is associated with the fact that the model is regarded as ideal. For actual differentially rotating systems these curves are 'smoother'.

As indicated by Fig. 2.11, the mode $m = 2$ ('saddle') is only slightly affected by resonance. However, it is important that this mode has the largest region of instability with respect to γ^2 at $M_h \gtrsim M_d$. It is also important to note that with the increase of mass the region of instability with respect to the parameter of the oblateness of the ellipsoid c/a becomes smaller (only the highly oblate systems remain unstable). However, for the mode $m = 2$ even at very high values of the halo mass ($M_h \simeq 13\ M_d$) the ellipsoid is unstable up to $c/a \simeq 1/15$, i.e. the values characteristic of the flat subsystems of the spiral galaxies. At the same time, we can conclude that the saddle-shaped mode stands out amongst other bending type modes. It should be mentioned that the identical conclusion also applies to the perturbations not bending the disk plane – here the bar-like mode $m = 2$ is also distinctive.

On the whole, an important result is the detection of the dependence of the position of the boundaries of the regions of instability on the halo mass: at a relatively large halo mass only the systems with the small dispersion of the velocities in the ellipsoid plane [with low $(1 - \gamma^2)$] can become unstable. This describes the destabilising role of the halo.

At the same time, with increasing halo mass the dispersion of the velocities of the stars in the plane of the disk, required for the gravitational stability of the disk, decreases: $c_r/V_{sp} \simeq (1 + \mu)/2$ (see chapter 4 and Fig. 4.30). Therefore, it may be that the destabilising (in relation to the large-scale bending perturbations) role of the massive halo will not become evident. This problem is investigated in greater detail in [64, 65, 911].

It is assumed that the stellar disk has a margin of gravitational stability Q_* which we determine as the ratio of the observed value c_r to the value c_{re} determined in the numerical experiments (it is assumed that c_{re} corresponds to the boundary of gravitational stability). Consequently, everywhere outside the limits of the central part of the disk

$$\frac{c_r}{V_{\text{rot}}} \simeq \frac{Q_*}{2(1+\mu)}. \tag{2.210}$$

On the other hand, taking into account the investigations by Polyachenko and Shukhman [208] of the model in the plane $z = 0$ (see (2.209)], we obtain

$$\frac{c_r}{V_{\text{rot}}} \simeq \frac{1}{\sqrt{2}\gamma x}\left[(1-\gamma^2)(1-x^2)\right]^{1/2}, \tag{2.211}$$

where $x = r/a$. Comparing these expressions, we obtain

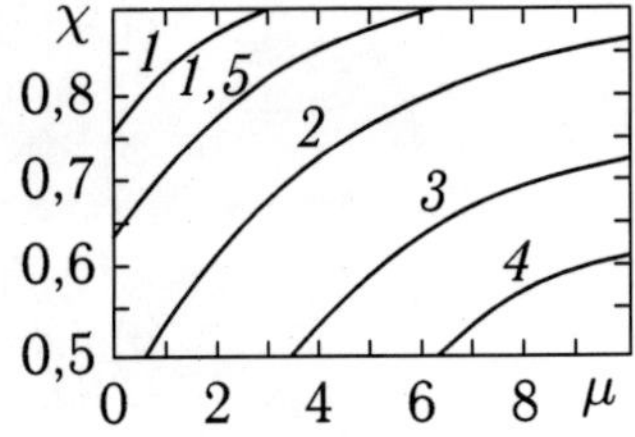

Fig. 2.12. The isolines of the parameter $\hat{Q}_*(\mu, x)$ in the plane $x = r/a$; $\mu = M_h/M_d$. In the region where $\hat{Q}_* > 1$ the most 'dangerous' mode $m = 2$ ('saddle') is stable.

$$Q_* \simeq \left\{2(1+\mu)^2\left(\frac{1-x^2}{x^2}\right)\left(\frac{1-\gamma^2}{\gamma^2}\right)\right\}^{1/2} . \tag{2.212}$$

The state with $Q_* = 1$ is physically distinctive – it corresponds to the boundary of gravitational stability. We now assume that the disk is in the state corresponding to the highest point of the stability boundary in relation to the most 'dangerous' (at $\mu \gtrsim 1$) bending mode $m = 2$. This means that $\gamma^2 = \gamma^2_*(\mu)$ (see Fig. 2.11), and then according to (2.212) $Q_*(\mu, x, \gamma) = \hat{Q}_*(\mu, x)$. The results of the calculation of this value can be both higher or lower than unity. Let it be that, e.g., $\hat{Q}_*(\gamma^2_*, x) > 1$. To understand what this means, note that the value of $Q_*(\gamma, \mu, x)$, determined from (2.212), is a monotonically decreasing function of the parameter γ: $(\partial Q_*/\partial\gamma)_{\mu,\, x} < 0$. Consequently, in the galaxies with $Q_* \simeq 1$ it will be evidently $\gamma^2 > \gamma^2_*$. Conversely, if $\hat{Q}_* < 1$, then in the galactic disks with $Q_* \simeq 1$ we will have $\gamma^2 > \gamma^2_*$. In the former case, the disk will have a certain margin of stability in relation to the most 'dangerous' bending mode $m = 2$, in the second case it will be unstable in relation to this mode. Thus, if as a result of the calculations we have $\hat{Q}_*(\gamma^2_*(\mu), x) > 1$, the instability of the bending modes of the marginally gravitationally stable disk will be suppressed.

The results of calculations of $\hat{Q}_*(\gamma^2_*(\mu), x)$ are presented in Fig. 2.12 (it should be mentioned that in this model, the parameter $\gamma^2_*(\mu)$ (Fig. 2.11) reaches at $\mu \gg 1$ the asymptotic value $\gamma^2_* = 0.96$). It may be seen that only the peripheral regions of the systems with small μ are unstable. In the systems with $\mu \gg 1$ the large-scale bending modes are stabilised. The regions close to the periphery of the ellipsoid ($x \gtrsim 0.9$) were not studied because in this model, in contradiction to the observational data, they are too cold: $c_r = 0$ at $x \to 1$. Thus, it should be expected that in the galaxies with a relatively massive halo the instability of the large-scale bending modes will not be observed.

2.7. Problems of investigating the kinematics of edge-on galaxies

We will examine some special features of the kinematics of the edge-on galaxies, taking into account the effect of projection, internal absorption and the dispersion of the velocities of the gas or stars on the measured rotation curve of the galaxies.

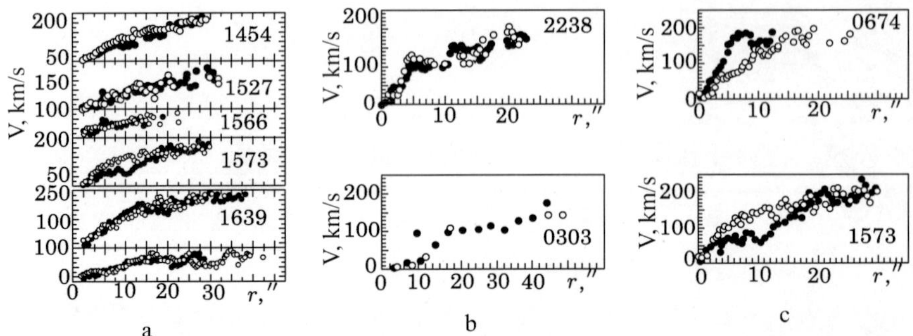

Fig. 2.13. Examples of the observed rotation curves for the edge-on galaxies: a) typical form of the rotation curve; b) examples of the galaxies with the 'step' on the rotation curve; c) examples of galaxies with greatly differing distributions of the measured velocities in different sides from the centre. The black circles corresponds to the side of the galaxy moving away, open circles – to the incoming side of the galaxy (according to the studies [87, 128, 129]). The number of the galaxy in the FGC catalogue is given [745].

2.7.1. The form of the rotation curves according to observations

For the spiral galaxies with edge-on disks the problem of the construction of the rotation curve is non-trivial because of two complicating factors: the projection effects and the internal absorption effect. Both effects depend both on the shape of the rotation curve and on the distribution of the radiation sources and the absorbing medium in the galaxy. The distinguishing feature of the rotation curves and, more accurately, one-dimensional distributions of the beam velocity in the disk is the long region of 'solid-state' rotation (monotonic increase of velocity). The almost linear increase of the beam velocity to large distances from the centre was already detected in the initial studies concerned with the determination of the kinematics of galaxies with thin disks [627]. Evidently, these authors were the first to pay attention to the fact that the solid-state part of the rotation curves in the edge-on galaxies, may be an artefact and that these effects of the heterogeneous distribution of dust required drawing an envelope line restricting at the top the position of the points on the 'radius–measured velocity of rotation' diagram.

The comparison of the shape of the rotation curve in the inner region of the galaxies with the degree of inclination of the disk to the visual ray, carried out at a later stage, showed that the gradient of the measured velocity of rotation in the inner region of the galaxy is actually lower in the greatly inclined disk, and this effect is stronger for the galaxies with high luminosity, characterised on average by stronger inner absorption [622].

More than 300 rotation curves for the edge-on galaxies (from the catalogue of flat galaxies FGC [745]) were obtained in the 6 m telescope of the Special Astrophysical Observatory of the Russian Academy of Sciences (see [870] and references therein). The measurement results confirm that the large majority of these galaxies have a very long 'solid-state' section which sometimes extends to the outer boundary of the measured curve, although there are exceptions [128–130]. As an example, Fig. 2.13a shows several quite typical rotation curves for the galaxies published in [129]. Figure 2.13 bc illustrate less frequently observed special features of the curves which are discussed below (in section 2.7.4). As regards the interpretation of the rotation curves it is necessary to answer the following question: what is the effect of different factors on the measured velocities of rotation of such galaxies and is it possible to obtain directly from the observations the actual rotation curve of the galaxy and estimate the maximum velocity of rotation of the disks? For the axisymmetric models of the galaxies with the given rotation curves we will calculate the distribution of the beam velocity $V_s(x)$ along the major axis of the edge-on disk [76].

2.7.2. Radial profiles of the velocity of rotation and dispersion of the velocities without taking absorption into account

The case with slight internal absorption is encountered in the measurement of the velocity of rotation in the radial range either in the galaxies with a very low content of the interstellar medium or in the measurements of the velocity of the outer zone with small optical thickness τ.

In the absence of inner absorption the maximum values of the Doppler component of the velocity of rotation are obtained in the region situated at the diameter of the galaxy normal to the visual ray (conventionally, it will be referred to as the major axis of the galaxy). Therefore, the estimates of the maximum velocity of the sources at the given distance from the centre should in fact represent (with the accuracy to the value of the dispersion of the velocities) the actual velocity of circular rotation. However, this value of the velocity in a general case does not necessarily correspond to the centre of gravity (baricentre) of the line profile because its position depends on the distribution of not only velocity but also volume luminosity along the visual ray.

It will be assumed that the velocity of rotation of the disk, like its brightness in the spectral line, used for measuring the velocities, is distributed axisymmetrically. The volume luminosity in the line will be denoted by $S(r)$. Let it be that the axis x corresponds to the major axis of the galaxy, and the axis y is directed along the visual ray (Fig. 2.14a). Consequently, the mean velocity, weighted on the basis of the specific luminosity along the visual ray, is determined by the expression

$$V_s(x) = \frac{\int_{-y_0}^{y_0} S(r) V_\varphi(r) \frac{x}{\sqrt{x^2+y^2}} dy}{\int_{-y_0}^{y_0} S(r) dy}, \tag{2.213}$$

here R is the disk radius, $r = \sqrt{x^2+y^2}$, and the integration limit is $y_0 = \sqrt{R^2-x^2}$. Because of the axial symmetry of the models, the distance from the centre of the disk in the x projection will always assumed to be positive.

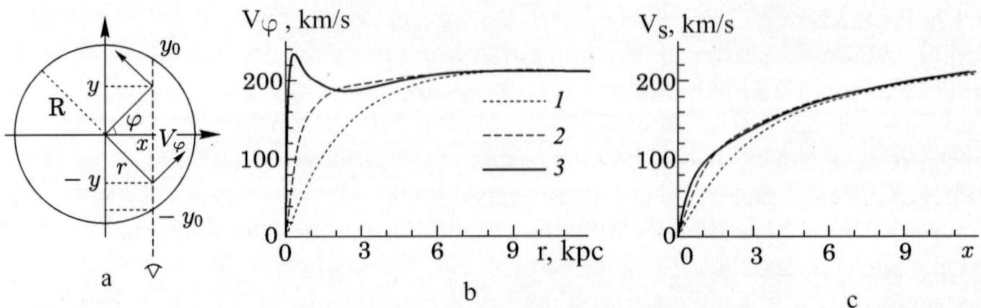

Fig. 2.14. a) problem of determination of the velocity of revolution of the edge-on disk with radius R. The velocity of rotation of the disk at the point (x,y) is V_φ. Integration is carried out from $-y_0$ to y_0; b) three types of investigated rotation curves V_φ; c) the distribution of the beam velocity in relation to the centre of the galaxy along its major axis $V_s(x)$ for rotation curves in Fig. 2.14b, without considering inner absorption.

Equations for the dispersion of the velocities, detected along the visual ray, will also be derived. At a high angle of inclination of the galaxy i ($i \to 90°$), in addition to the effects associated with integration along the visual ray, which gives

$$c_\ell = \frac{\int_{-y_0}^{y_0} \sqrt{c_r^2 \sin^2\varphi + c_\varphi^2 \cos^2\varphi}\; \varrho(r)dy}{\int_{-y_0}^{y_0} \varrho(r)dy}, \tag{2.214}$$

there is also a second factor associated with the effect of the spatial heterogeneity of the velocity of rotation:

$$c_V^2 = \frac{\int_{-y_0}^{y_0} (V_s(r) - V_y)^2\, \varrho(r)dy}{\int_{-y_0}^{y_0} \varrho(r)dy}, \tag{2.215}$$

where V_y is the projection of the velocity on the axis y at the point (x, y) (see Fig. 2.14a).

The effect of the actual form of the rotation curve $V_\varphi(r)$ and of special features of the distribution of matter in the disk and the dependence $V_s(x)$ (the observed rotation curve) will be determined.

Let the volume luminosity of the disk in the line $S(r)$ for which the velocity is evaluated decreases with r in accordance with exponential law $S(r) = S_0 \cdot \exp(-r/L)$, where L is the radial scale of variation of the brightness of the disk in the emission line, and in the case of determination of the velocities on the basis of the absorption lines it is the radial scale of the stellar disk. For determinacy $R = 4L$. At distances $r > R$ the contribution of the disk to the investigated line profile will be assumed to be equal to 0. For illustration, we restrict ourselves to examining three types of the rotation curve whose form differsin the inner part of the disk ($r \lesssim 2L$) (Fig. 2.14b):

1. The rotation curve $V_{\varphi 1}$, characteristic of the galaxies without the massive bulge (the dotted line in Fig. 2.14b);
2. The second type curve $V_{\varphi 2}$ obtained in the case of a low-mass bulge, responsible for the steeper gradient of the velocity in the inner region of the galaxy (the dashed line in Fig. 2.14b). The region of 'solid-state' rotation is small;
3. The curve $V_{\varphi 3}$ with the circumnuclear maximum of the velocity of rotation (solid line in Fig. 2.14b) which reflects the existence of a massive concentrated bulge.

In all cases, the rotation curve reaches a plateau at high r. Figure 2.14c shows the radial dependences of the velocity of the disk along the visual ray $V_s(x)$, calculated from equation (2.213), for the three investigated types of the rotation curve in the absence of absorption. It may be seen that in all cases the entire disk is characterised by underestimation of the rotation velocity, especially in the central region. The difference between the second and third type of rotation curve disappears almost completely, and the underestimation of the velocity of rotation is the largest for the rotation curve $V_{\varphi 3}$. The approximate equality $V_s \simeq V_\varphi$ is observed only at the very edge of the disk. Since the curves of variation of velocity $V_s(x)$ depend only slightly on the actual form of the rotation curve $V_\varphi(r)$, they do not permit the restoration of the distribution of mass in the galaxy [76].

2.7.3. The models with absorption of light taken into account

In passage of radiation through the disk matter the contribution from faraway regions of the galaxy is smaller than that from the regions closer to the observer. For the light, propagating in the disk under a small angle to the disk plane, the galaxies, containing dust, are almost completely opaque. Therefore, it may be assumed that the radiation from the zone furthest away does not reach the observer. The radiation of the central region characterised by the maximum values of the projection of the velocity of rotation on the visual ray, is greatly weakened, and the strength of the effect of absorption decreases with a reduction of the distance to the centre of the system. As a result, the main contribution to radiation in the central region of the galaxy $x \lesssim R/2$ comes from the nearest zone in which the velocity along the visual ray is not high. Taking absorption into account, expression (2.213) has the form

$$V_s(x) = \frac{\int_{-y_0}^{y_0} S(r)\,\exp(-\tau(x,y))V_\varphi(r)\frac{x}{\sqrt{x^2+y^2}}dy}{\int_{-y_0}^{y_0} S(r)\,\exp(-\tau(x,y))dy}, \tag{2.216}$$

where $\tau(x, y)$, characterising the optical thickness and calculations for the section of unit length in the disk plane, is determined by the distribution function of dust $f_{\text{dust}}(x, y)$:

$$\tau = \int_{y_0}^{y} f_{\text{dust}}(x,\xi)d\xi, \tag{2.217}$$

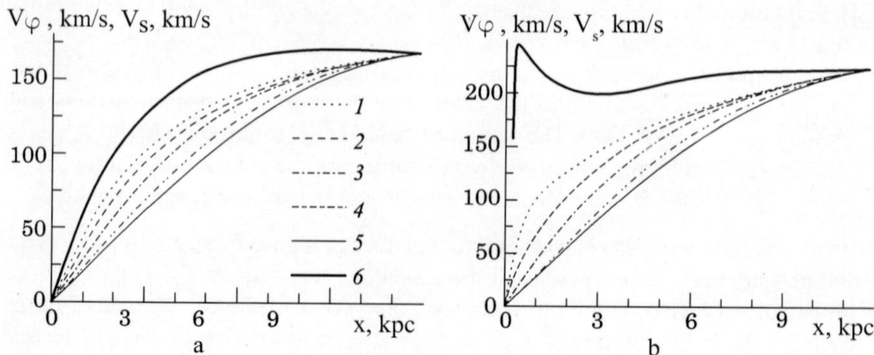

Fig. 2.15. Calculation of the distribution of the beam velocity $V_s(x)$ taking into account absorption for the radial scale of distribution of dust $L = L_d = 3$ kpc: a) dependences $V_\varphi(r)$ (6, the model without the bulge) and $V_s(x)$ at different values of optical thickness at $r = 3$ kpc: $\tau_0 = 0$ (1), 0.37 kpc^{-1} (2), 0.73 kpc^{-1} (3), 1.46 kpc^{-1} (4), 2.2 kpc^{-1} (5); b) the same for the rotation curve of the third type (the thick solid line V_φ, the model with a massive concentrated bulge, see Fig. 2.14b).

where ξ is the coordinate in the y direction.

We restrict our considerations to the axisymmetric exponential dust distribution

$$f_{\text{dust}}(x,\xi) = \alpha_d \cdot \exp\left(-\frac{\sqrt{x^2+\xi^2}}{L_d}\right), \tag{2.218}$$

where L_d is the radial distribution scale of the absorbing medium, α_d is the normalisation parameter. Following the density distribution of dust, the function τ decreases with increase of the distance from the centre and converts to 0 at $r = R$. Parameter $\tau_0(x)$ is determined as the optical thickness in calculation for unit length (1 kpc) at the major axis of the Galaxy at the distance $r = x$ from the centre. This value, being proportional to f_{dust}, reflects the density of the absorbing medium at the given distance x. The variation of parameter α_d in the range from 0 to 6 corresponds to the variation of the optical thickness τ_0 at the distance $x = 3$ kpc from 0 to 2.2 kpc^{-1}. The degree of opaqueness, expressed in the units of τ kpc^{-1}, approximately corresponds to the weakening in the stellar units by 1 kpc.

Figure 2.15 shows the results of calculations using equation (2.216), with (2/217) and (2.218) taken into account, for radial scales of the distribution of the brightness of emission and density of dust $L = 3$ kpc and $L_d = 3$ kpc respectively, for different values of parameter τ_0. In all cases, absorption greatly reduces the velocity $V_s(x)$ with the exception of the outer, relatively transparent region of the galaxy. Consequently, the observed rotation curve 'straightens' so that the almost linear increase of $V_s(t)$ in a highly dusty disk may be detected up to the outer boundary of the disk, regardless of the actual form of the rotation curve (Fig. 2.15). The presence of the 'ring' and the central 'hole' in the distribution of the gas intensifies even more appreciably this special features of the form of the curve $V_s(x)$ in the models with absorption [76].

2.7.4. Effect of random motions on the radial profile of the velocity of rotation of the disk

In the models discussed previously it was assumed that the dispersion of the velocities of the radiation source is equal to 0 and that there is only regular rotation. We now take into account the residual velocities which will be characterised by the values of the dispersion of the radial c_r and azimuthal c_φ velocities. If the rotation curve is evaluated on the basis of the stellar spectra then in accordance with the observations (see, for example [413]) it may be assumed that both c_r and c_φ decrease with increase of the distance from the centre. For c_r we accept a simple expression $c_r = c_{r0} \exp(-r/L_c)$. The radial scale of the dispersion of the velocities L_c is usually considerably greater than the scale of variation of density L [137]. The velocity distribution function is selected in the form (see equation (4.20))

$$F(v_r, v_\varphi) = F_0 \cdot \exp\left\{ -\frac{(V_\varphi - v_\varphi)^2}{2c_\varphi^2} - \frac{v_r^2}{2c_r^2} \right\}, \tag{2.219}$$

where c_r, c_φ are the dispersions of the velocities for the random component of velocity v_r and v_φ, respectively. The profile of the spectral line, reflecting the variation of the velocity along the visual ray, at the fixed value of the observed distance from the centre of the disk of the galaxy x will be determined by the expression

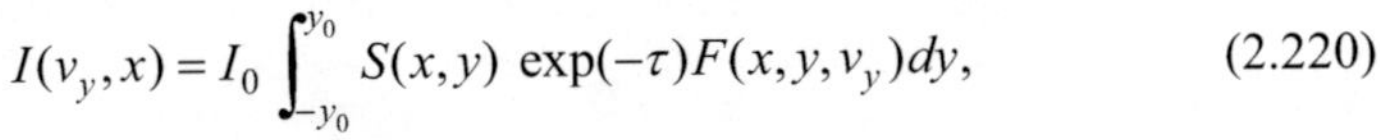

$$I(v_y, x) = I_0 \int_{-y_0}^{y_0} S(x,y)\, \exp(-\tau) F(x,y,v_y)\, dy, \tag{2.220}$$

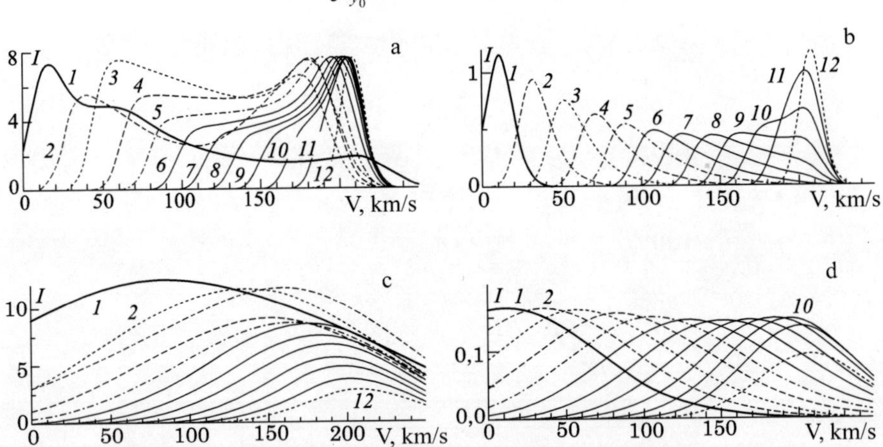

Fig. 2.16. a) the profiles of the spectral line $I(V)$ for different distances from the centre of the galaxy for the model of the gas disk with the dispersion of velocities $c_r(0)$ = 15 km/s and the radial scale L_c = 9 kpc (without taking absorption into account). Functions $I(V)$ are normalised arbitrarily. The disk is divided into 12 zones, the numbers indicate the number of the zone in the sequence of increasing distance from the disk centre; d) the same for the model of a highly dusted ring with τ_0 = 2.2 kpc^{-1}; c) the same, for the model of the stellar disk with $c_r(0)$ = 100 km/s, L_c = 3 kpc without absorption; d) the same, for the model of the strongly dusted stellar disk with τ_0 = 2.7 kpc^{-1}.

where I_0 is the normalisation constant, and τ is determined by the integral (2.217).

We examine separately the effect of the previously discussed factors on the observed rotation curves of the dynamically hot (stellar disk) and cold (gas disk) subsystems for the edge-on galaxies. For a gas disk $c_r/V_{max} << 1$ (V_{max} is the maximum velocity of rotation), and the velocity distribution is isotropic ($c_r = c_\varphi$). Figure 2.16 shows the velocity profiles (Doppler line profiles) $I(V) = I(v_y, x)$ at different distances z from the disk centre. In the absence of dust (Fig. 2.16a) the profiles greatly differ for different values of x. The characteristic special feature of the profiles is their asymmetry, and for the central region (curves 1–3 in Fig. 2.16a) – the presence of two maxima or a long wing with the 'step' instead of the second maxima (curve 4–7). The first maximum (at high velocities) is determined by the rapid rotation of matter in the region $|\varphi| <<1$ in the vicinity of the 'major' axis (Fig. 2.14a). The second maximum or 'step' (in the region of lower velocities) are associated with two factors: with the presence of residual velocities but, to a larger degree, with a large reduction of the projection V_φ on the visual ray $|\varphi| > \pi/4$.

In a strongly dusted disk the situation changes quantitatively (Fig. 2.16b). The profiles $I(V)$ in the central region have only one maximum (from two maxima observed in the former case, only the maximum corresponding to the lower velocity remains). The maximum at the higher velocity forms only when the absorption is sufficiently small in order to enable observation of the areas in the vicinity of the major axis at the given distance x.

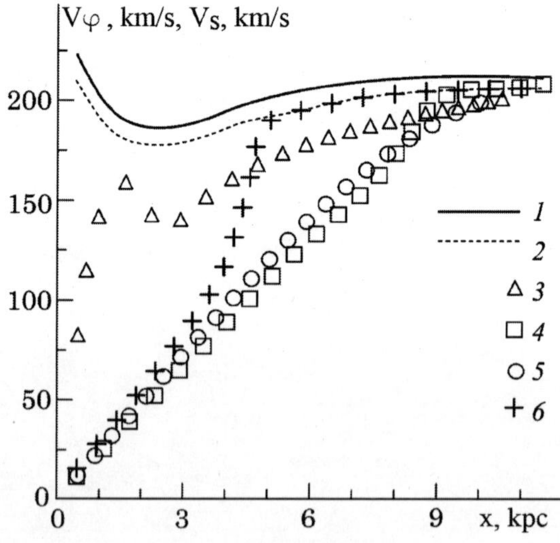

Fig. 2.17. The given rotation curve of the galaxy ($V_\varphi(r)$, curve 1) in comparison with the 'measured' velocities of rotation $V_s(x)$, constructed on the basis of the maximum of the function $I(V)$ (see the text) for the following models: 2) the model of the gas disk without absorption; 3) the model of the stellar disk without absorption; 4) the model of the gas disk with strong absorption $\tau_0 = 2.2$ kpc^{-1}); 5) the model of the stellar disk with strong absorption ($\tau_0 = 2.2$ kpc^{-1}); 6) model of the gas disk with weak absorption ($\tau_0 = 0.73$ kpc^{-1}). In all cases, the radial scale of distribution of the gas and dust was accepted equal to 3 kpc and the value τ_0 relates to 3 kpc.

The dispersion of the velocities for the stellar disk is greater than for the gas. The axisymmetric model is described by the relationship (2.34). In the case of the stellar disk with the dispersion of the radial velocities of the stars in the centre $c_r(0)/V_{max} = 0.5$ the distributions $I(V)$ (where $I(V)$ is taken in accordance with the modulus because in this case we are concerned with the absorption lines) are presented in Fig. 2.16c, d. In this case, as also in the case of the emission gas, in the presence of absorption the maximum values of $I(V)$ are displaced into the range of lower velocities.

The complicated form of the line profiles results in the differences in the estimates of the beam velocity on the basis of measurement of the wavelengths, corresponding to the intensity maximum, and the wavelength of the baricentre of the line. Figure 2.17 shows the dependences $V_\varphi(r)$ and $V_S(x)$, constructed on the basis of the position of the maxima of the profiles $I(V)$, shown in Fig. 2.16. In accordance with these considerations, in the presence of two maxima we select the one which is associated with the higher velocity. In the model for the gas disk without absorption (Fig. 2.16a) in this method of evaluation of the velocity, the difference between the actual velocity of rotation V_φ and $V_S(x)$ (curves 1 and 2 in Fig. 2.17)) is small, and for the investigated case does not exceed 7%.

For a stellar population with a high dispersion of the velocities $c_r(0)/V_{\max} = 0.5$ the difference $V_\varphi - V_s(x)$ is considerably greater, but in this case the presence of the internal maximum on the rotation curve can be clearly seen (curve 3), although its amplitude is considerably lower.

Thus, the approach developed on the basis of determination of the maximum $I(V)$, gives a more accurate result in the construction of the rotation curve than the method of the mean weighted value of the beam velocity, determined on the basis of the baricentre of the spectral line, described in section 2.7.3. Unfortunately, in the actual galaxies the line profiles can be distorted by the non-uniform distribution of the emission regions and the absorbing medium. However, the asymmetry of the profile, if it does exist, can be measured efficiently and taken into account in the evaluation of velocity.

In a highly dusted disk the dependences $V_s(x)$ for both the gas and stellar population, determined by the two methods, differ only slightly (curves 4 and 5 in Fig. 2.17). In both cases, the resultant curve $V_s(x)$ does not give accurate information on the form of the actual rotation curve in the inner regions of the disk.

The intermediate case is most interesting; here, the absorption in the central part of the galaxy is strong (line profile has only one maximum – at low velocities), and starting from some value x, a significant contribution to the radiation is provided by the regions in the vicinity of the major axis, responsible for the appearance of the maximum with a high beam velocity. In this case, the rotation curve, constructed on the basis of the position of the maxima in the line profiles, shows a distinctive 'jump' – from the velocities considerably lower than the velocity of rotation to the velocities close to the velocity of rotation (curve 6 in Fig. 2.17). As the degree of absorption in the galaxy decreases, the distance of this transition from the centre also decreases. Varying the value τ_0 it can be seen that the increase of the measured velocity of rotation takes place at a radius r at which the absorption decreases below

0.3–0.53 stellar units per kpc. It should be mentioned that a similar step on the rotation curve is actually detected in a number edge-on galaxies (see, for example, Fig. 2.13b).

It should be mentioned that in some edge-on galaxies, the curves of the variation of beam velocities $V_s(x)$, in contrast to the expected variation, do not show solid–state rotation. In these galaxies, the central regions is characterised by a high gradient of beam velocity (Fig. 2.13c). This behaviour indicates either a very low dust content of the galaxy or (which is more likely) a very large difference between the angle of inclination of the disk and $i = 90°$. The differences of several degrees can be sufficient for the projection effects and the effects of internal absorption become insignificant (exact evaluation of the angle at which this takes place depends on the disk thickness and the spatial distribution of both these sources and the absorbing medium). It should be mentioned that the influence of the projection effect depends not only on the angle of inclination of the disk i but also on the width of the slit of the spectrograph. For example, if the slit is oriented along the major axis of the galaxy and its width is comparable with the invisible thickness of the disk (more accurately – with the size of the minor axis of the ellipse, which restricts the measured region of the disk), the disk will be regarded as observable edge-on, even if its inclination angle differs from 90°.

The calculations described here relate to the rotation curve $V_{\varphi,3}$ (see Fig. 2.14b). For the cases $V_{\varphi,2}$ or $V_{\varphi,1}$ the main conclusions remain valid.

2.7.5. The effect of non-uniform distribution of dust

As shown by the observations, the distribution of stars in the galactic disks is often highly non-axisymmetric and non-uniform on the fine scale ($\ll L$). To illustrate this, we consider a model in which the absorbing regions are distributed in 'spots', scattered randomly throughout the disk. The dust concentration in the i-th cloud ($i = 1, \ldots, m$) with the coordinates of the centre $(x_d^{(i)}, y_d^{(i)})$ is characterised by the value $\tau^{(i)}$ which is governed by the law (2.218). The dimensions of the cloud $l_d^{(i)}$ in the investigated model were specified randomly in the range $0.1L \leqslant l_d^{(i)} \leqslant 0.5\,L$.

Figure 2.17 shows the radial dependence of the velocity $V_s(x)$ in the model with $m = 50$, $\tau_0 = 2.2$ (curves 3 and 4). It is evident that the scatter of the points or the uneven form of the curve $V_s(x)$, which is non-symmetric in relation to the galaxy centre, is the result of existence of randomly distributed corridors of transparency which enable the regions in the disk depth to be observed in certain directions. The amplitude of the variation of the velocity can be high, and to obtain the rotation curve it is necessary to produce the upper envelope of the points on the diagram $V_s(x)$. However, in this case the form of the curve can be determined only approximately [627].

If the mean density of the dust is relatively small so that we can detect at different distances x the areas situated in the vicinity of the major axis, this procedure actually can be used to produce the rotation curve. However, this applies especially to the outer regions of the Galaxy where τ_0 is small. In the investigated model, even the envelope of the points on the V_c–r diagram (the broken line in Fig. 2.17) is far away from the given form of the rotation curve.

The second effect, generated by the non-uniform distribution of the dust, can actually be expected for the galaxies with a developed spiral structure. A certain orientation of the spiral arms to which the interstellar medium concentrates, results in different transparency of the disk in different sides of the centre. In the areas where the spiral branch is situated on the side of the galaxy, facing the observer, the visual ray penetrates into the disk to a smaller depth and this reduces the estimate of the velocity of rotation. Consequently, the measured curves $V_s(x)$ are non-symmetric in relation to the centre. Similar cases have actually been detected (see Fig. 2.13c). The curve, constructed with the higher values of the measured velocities, can be regarded as approximating better the actual form of the rotation curve [76].

It should be mentioned that all the observed rotation curves (dependences $V_s(x)$) at high x reach a 'plateau' corresponding to the given velocity of rotation of the modelled galaxy. Consequently, the measured maximum velocities of rotation of the edge-on disks should differ only slightly from the actual velocities.

3

Main models of the dynamics of galactic gas disks

The title of the chapter indicates that the main subject of the chapter are the gas disks. Some aspects of the gas disks whose structure and observed manifestations are associated with accretion (fall of matter on a gravitational centre) were investigated in chapter 2.

Strictly speaking, the gas subsystems of the galactics (evidently, in contrast to the accretion disks) are not single-phase systems. If the gas in the disks of the galactics did not group in clouds, the results of observations of the distribution of the gas in the Galaxy [1112] would indicate that the free path length of the molecules of the gas ($l \sim 10^{-3} \div 10^{-2}$ pc) is considerably smaller than the characteristic thickness of the gas disk ($h \leqslant 10^2$ pc). Similar estimates as regards the order of magnitude of the quantity also apply to other flat galaxies. Evidently, in this case, the pressure of the gas should be regarded as isotropic, and the properties of the gas should be described by the gas-dynamics equations.

A large part of the gas in the disks of the galaxies is concentrated in low-temperature clouds ($\leqslant 10^2$ K) and a small part is in the form of an inter-cloud gas with a temperature of $\sim 10^4$ K [84]. The characteristic dispersion of the velocities of the gas clouds is, as regards the order of magnitude, equal to the thermal velocity of the particles of the inter-cloud gas ($\sim 5 \div 10$ km/s). The dispersion of the velocities of the clouds both across the plane of the disk and in its plane is the same. This indicates that the pressure in the gas subsystems of the galaxies its isotropic.

3.1. The model of the three-dimensional gas astrophysical disk

3.1.1. Complete system of equations for the gas disk

In the construction of the models of the gas disk, the initial equations were represented by three-dimensional gas-dynamics equations and the gravitational field (in a number of cases the dissipative factors (radiation and viscosity) were taken into account):

$$\frac{\partial \varrho}{\partial t} + \nabla(\varrho \mathbf{u}) = 0, \tag{3.1}$$

$$\frac{d(\varrho\mathbf{u})}{dt} = -\nabla\mathcal{P} + \nabla\cdot\hat{\Pi} - \varrho\nabla\Phi, \tag{3.2}$$

$$\varrho\frac{d}{dt}\left(\frac{\epsilon}{\varrho}\right) - \frac{\mathcal{P}}{\varrho}\frac{d\varrho}{dt} = q^{+} - q^{-}, \tag{3.3}$$

where ϱ is volume density, ∇ is the nabla operator, $\mathbf{u}$ is the gas velocity; $\frac{d}{dt} = \frac{\partial}{\partial t} + \mathbf{u}\nabla$ is the total derivative in respect of time, pressusre $\mathcal{P} = \mathcal{P}_{\text{gas}} + \mathcal{P}_{\text{rad}}$ takes into account the contributions from the gas and radiation, $\hat{\Pi}$ is the tensor of the viscous stresses; ϵ is the internal energy of the unit volume; q^{+} and q^{-} are respectively the energy generated in unit time as a result of low dissociation and the energy carried away by radiation.

Equations (3.1)–(3.3) can be re-written in the cylindrical coordinate system (r, φ, z) which is the most suitable system in the investigation of the rotating systems. The continuity equation (3.1) has the form

$$\frac{\partial\varrho}{\partial t} + \frac{\partial(r\varrho u)}{r\partial r} + \frac{\partial(\varrho v)}{r\partial\varphi} + \frac{\partial(\varrho w)}{\partial z} = 0. \tag{3.4}$$

The equations for the radial (u), azimuthal (v) and vertical (w) velocity components are presented in the form:

$$\frac{\partial u}{\partial t} + u\frac{\partial u}{\partial r} + \frac{v}{r}\frac{\partial u}{\partial\varphi} + w\frac{\partial u}{\partial z} - \frac{v^2}{r} = -\frac{\partial P}{\varrho\partial r} - \frac{\partial\Phi}{\partial r} + \frac{1}{\varrho}(\nabla\cdot\hat{\Pi})_r, \tag{3.5}$$

$$\frac{\partial v}{\partial t} + u\frac{\partial v}{\partial r} + \frac{v}{r}\frac{\partial v}{\partial\varphi} + w\frac{\partial v}{\partial z} + \frac{uv}{r} = -\frac{1}{r\varrho}\frac{\partial P}{\partial\varphi} - \frac{\partial\Phi}{r\partial\varphi} + \frac{1}{\varrho}(\nabla\cdot\hat{\Pi})_\varphi, \tag{3.6}$$

$$\frac{\partial w}{\partial t} + u\frac{\partial w}{\partial r} + \frac{v}{r}\frac{\partial w}{\partial\varphi} + w\frac{\partial w}{\partial z} = -\frac{\partial P}{\varrho\partial z} - \frac{\partial\Phi}{\partial z} + \frac{1}{\varrho}(\nabla\cdot\hat{\Pi})_z, \tag{3.7}$$

where $(\nabla\cdot\hat{\Pi})_i$, $i = r, \varphi, z$ are respectively the r-, φ- and z-components of the specific viscous force. The explicit equations for the components of the tensor of the viscous stresses can be written using expression (see [112]):

$$\Pi_{ik} = \eta\left(\frac{\partial v_i}{\partial x_k} + \frac{\partial v_k}{\partial x_i} - \frac{2}{3}\delta_{ik}\frac{\partial v_l}{\partial x_l}\right) + \zeta\delta_{ik}\frac{\partial v_l}{\partial x_l}, \tag{3.8}$$

and the quantities η and ζ are referred to as the coefficients of dynamic viscosity, but the double indexes indicate the summation.

For the self-gravitating disk, the system should be supplemented by the Poisson equation

$$\frac{\partial}{r\partial r}\left(r\frac{\partial\Phi}{\partial r}\right) + \frac{\partial^2\Phi}{r^2\partial\varphi^2} + \frac{\partial^2\Phi}{\partial z^2} = 4\pi G\varrho(r, \varphi, z, t). \tag{3.9}$$

3.1.2. Traditional two-dimensional equations

The model of the 'shallow water 'for a thin layer of an incompressible fluid with the free surface in a homogeneous gravity field [112] is well-known. The astrophysical gas disks are described by the model of the thin layer of the incompressible gas where the spatial scale of the investigated structures in the plane of the system λ is larger in comparison with the characteristic thickness of the disk $2h$ ($\lambda \gg 2h$). In the model of the thin disk, the volume density $\varrho(r, \varphi, z, t)$ is replaced by the surface density

$$\sigma(r,\varphi,t) = \int_{-\infty}^{\infty} \varrho\, dz, \tag{3.10}$$

and the pressure (r, φ, z, t) is replaced by the plane (surface) pressure

$$p = \int_{-\infty}^{\infty} P\, dz. \tag{3.11}$$

The reduction of the dimensionality of the problems associated with additional conditions for the symmetry of motion, the slow rate of propagation of perturbations in the plane of the layer in comparison with the time required for the establishment of equilibrium in the vertical z-direction. The mean value of the density and pressure in the vertical direction is denoted by $\overline{\rho}(r,\varphi,t)$ and $P(r, \varphi, t)$. Consequently, the half thickness of the disk $h(r, \varphi, t)$ is described by the relationships

$$\sigma = 2h\overline{\rho}, \quad p = 2hP. \tag{3.12}$$

The fact that the astrophysical disks are thin systems leads to the requirement to ignore the vertical movement and replace the volume density with the surface density and the pressure with the so-called 'flat pressure'. As a result of this form of substitution, the continuity equation (3.4) is replaced by the equation for :

$$\frac{\partial\sigma}{\partial t} + \frac{\partial(r\sigma u)}{r\partial r} + \frac{\partial(\sigma v)}{r\partial\varphi} = 0. \tag{3.13}$$

In the equations of motion (3.5), (the .6), substitution $\mathcal{P} \to p$, $\varrho \to \sigma$, $\gamma \to \gamma_s$ is carried out and, taking into account the symmetry of the vertical velocity, we obtain:

$$\frac{\partial u}{\partial t} + u\frac{\partial u}{\partial r} + \frac{v}{r}\frac{\partial u}{\partial\varphi} - \frac{v^2}{r} = -\frac{\partial\Psi}{\partial r} - \frac{1}{\sigma}\frac{\partial p}{\partial r}, \tag{3.14}$$

$$\frac{\partial v}{\partial t} + u\frac{\partial v}{\partial r} + \frac{v}{r}\frac{\partial v}{\partial\varphi} + \frac{uv}{r} = -\frac{1}{r}\frac{\partial\Psi}{\partial\varphi} - \frac{1}{\sigma r}\frac{\partial p}{\partial\varphi}, \tag{3.15}$$

$$\frac{\partial p}{\partial t} + \gamma_s p\left(\frac{\partial ru}{r\partial r} + \frac{\partial v}{r\partial\varphi}\right) + u\frac{\partial p}{\partial r} + v\frac{\partial p}{r\partial\varphi} = 0, \tag{3.16}$$

where $\Psi = \Phi(z = 0)$, γ_s is the so-called 'flat' adiabatic index. It is assumed in this case that the stationary condition of hydrostatic equilibrium exists at every point in the disk at every moment of time. The disk manages to rearrange itself in the z- direction for any movement in its plane.

The system of equations (3.13)–(3.16) is used additionally for describing the dynamics of astrophysical disks (other forms of description are also used in addition to (3.16)). For this reason, this system of equation will be referred to as the traditional two-dimensional (flat) model.

Equation (3.13) is correct, and transition to the equations (3.14)–(3.16) requires that certain conditions are fulfilled. If the radial component of the total Euler equation is integrated to ensure determinacy and the result is compared with the traditional equation (3.14), we obtain

$$\int_{-\infty}^{\infty} \varrho \frac{du}{dt} dz \neq \sigma \frac{du}{dt}, \tag{3.17}$$

$$\int_{-\infty}^{\infty} \varrho \frac{\partial \Phi}{\partial r} dz \neq \sigma \frac{\partial \Psi}{\partial r}. \tag{3.18}$$

In (3.17) it should be assumed that the velocity u does not depend on the z-coordinate. Equation (3.18) is more important because the potential always depends on the vertical coordinate.

3.1.3. Two-dimensional observational data – the basis for constructing three-dimensional theoretical models

The area of astronomy known as the dynamics of galactic disks has become, after many years of development, the classic discipline, but until recently was developed as part of theoretical astrophysics (for example [372, 590]). The observational data, representing the basis for dynamic investigations, were until recently mostly one-dimensional: the reconstruction of the distribution of mass in the disk was usually carried out on the basis of the radial profiles of surface brightness and velocity along the major galactic axis or the rotation curve. This approach assumed the strictly axisymmetric galactic disks. Evidently, in a general case, these disks are not axisymmetric. The most distinctive manifestation of non-axisymmetric nature are the bars and spiral structures. Thus, to carry out more reliable dynamic investigations, it is important to obtain two-dimensional (2D) data.

Two-dimensional photometric investigations were carried out as a result of the construction of a device with a charge coupling (CCD). In particular, Kent [759, 760] carried out two-dimensional decomposition of the images of the galaxies using the CCD and for a large sample of the galaxies determined the exponential scales of the disks. In a number of studies by Athanassoula et al [480, 607] the investigators used the azimuthal Fourier analysis of the images of the galactic disss to determine the properties of the spiral structure: the number of 'arms' and their twist angle. However, the two-dimensional fields of velocities were not often regarded as the basis of advanced dynamic investigations. In the majority of cases, the rotation curve was plotted assuming the axisymmetric form of the system. However, the two-dimensional fields of the velocities contain a large amount of important information.

The observed velocity of the gas along the visual ray in spiral galaxies contains a contribution not only from regular rotation but also from the perturbations of velocities of the spiral density waves. In principle, analysis of the field of velocities permits the separation of the individual components of movement of the gas, but this problem is non-trivial: the expected amplitude of the main harmonics is linked with the wave movement approximately an order of magnitude smaller than the maximum velocity of rotation of the galaxies. In addition, the presence of also non-flat oscillations of the gas along the axis of rotation and the local non-circular movement make the observed field of the velocities difficult to interpret. Although the perturbations in the movement of the gas, determined by the spiral pattern, were also observed in our and in many other galaxies, starting with the classic study [1086] for M81, the amplitude of the perturbed velocities and the angular velocity is not known even for the galaxies studied to the largest extent.

Various methods have been proposed for determining the kinematic parameters of the density waves on the basis of the observed fields of the velocity along the visual ray (see [220, 1124, 1285] and references therein). However, all these methods have one or two major disadvantages. Firstly, they are based on the assumption that the equilibrium velocity of rotation can be restored without analysis of the residual velocities. This possibility does exist if we consider the results of simulation and know the form of the gravitational potential [446]. However, if we analyse the field of velocities along the visual ray of an actual galaxy, the independent restoration of the rotation curve becomes impossible [127]. The rotation curve, determined within the framework of the model of the purely circular motion, contains systematic errors of the order of residual velocities and, consequently, the residual field, formed on the basis of this curve is not the actual field of perturbations of the velocity, caused by the density wave [127]. Secondly, all the previously mentioned approaches, are based on the assumption of the two-dimensional nature of galactic movement in the disk, whereas any actual galactic disk is a three-dimensional object and the regular movements, induced by the density wave, are in principle also three-dimensional [581, 586]. Thus, the only direct approach to analysis of the observed fields of velocities is to find the self-consistent solutions for the total vector field of the velocities. In other words, the speed of rotation and all three components of the perturbed velocity should be determined by the analysis of the observational data taking into account also the three-dimensional nature of the galactic disks.

Successful attempts for the restoration of the total velocity vector (of all three components) in the gas disk of galaxies with the regular spiral pattern, using the observational data on kinematics [127], indicate that the dynamics of galactic disks will be constructed in future on the basis of direct observations. The data on the total field of velocities will immediately provide: 1) the rotation curve, 2) all main resonances: Lindblad and corotation, 3) the residual field of the velocities contains information on the large-scale structures, such as gigantic anticyclones [581] and cyclones [586] which are 'compound parts' of the spiral density waves. Finally, the knowledge of the total field of velocities helps in investigating the collective processes – various types of instabilities which can be responsible for the spiral-vortex structure of the disk. Thus, it may be assumed without exaggeration that the observed field of velocities is the essential observational basis for constructing the dynamic portrait of the gWalaxy.

3.1.4. Possibilities of investigating the dynamics of astrophysical disks on the basis of two-dimensional approximation

Prior to investigating some highly three-dimensional non-linear effects in thin astrophysical disks, it is important to discuss the problem of two-dimensional approximation and explained why in a number of cases it is essential to take into account the finite (although small) thickness of the disk [43, Appendix I], [577, Application I], [579].

It could be expected that the dynamics of the thin disks can be described using exclusively two-dimensional functions which depend on the coordinates in the disk plane, the type of surface density, the two-dimensional filed of the velocities or the gravitational potential in the disk plane. In fact, the dynamics equation for these functions are used widely in the astrophysical literature.

It is natural to expect that this description is valid only if a number of conditions is fulfilled and that these equations should be derived from general three-dimensional gas dynamics equations simultaneously with the formulation of the given approximations. However, until now, this approach has not been used widely and the conditions of applicability of the two-dimensional equations are often violated.

The derivation of a closed system of two-dimensional equations for a thin disk for two limiting cases will be discussed: the disk is situated in the external gravitational field which dominates in comparison with self-gravitation, and at the reversed limit. In both cases, it is assumed that the matter in the disk satisfies the barotropic equation of state $P = P(\varrho)$. The problem of the possibility of writing the closed system of equations for the equation of state in the general form $P = P(\varrho, S)$ remains unsolved.

If the barotropic equation of state $P = P(\varrho)$ is used in the general form, the system of dynamic equations contains, in addition to the differential equations, an integro-differential equation. In a partial case of the polytropic equation of state $P = A\varrho\gamma$, the integral equation of state for the two-dimensional functions can be reduced to an algebraic equation and the system of equations has the form which is very similar to the traditional form.

However, there are considerable differences.

Firstly, the dynamics equation differs from the previously described equations by the presence of additional terms or, in the case of a self-gravitating disk, by the values of the constants in the equations.

Secondly, the conditions of applicability of the two-dimensional equations of this type require examination of not only the relatively non-wave movements for which the characteristic scale of the inhomogeneity in the plane of the disk should be large $L \gg h$ ($2h$ is the disk thickness); this is evident because in this case we are concerned with the approximation of the thin disk. A more important aspect is the requirement on the characteristic timescales of the processes taking place $T^2 \gg h^2/c^2$ (c is the speed of sound), and this is often ignored. In a general case, this results in a constraint on the frequency of the process in comparison with the frequency of rotation of the disk: $\omega^2 \ll \Omega^2$.

3.1.5. The system of initial equations

The initial dynamics equations (3.5–3.7) can be written in the following form:

$$\hat{L}_1 v_r - \frac{v_\varphi^2}{r} = -\frac{1}{\varrho}\frac{\partial P}{\partial r} - \frac{\partial \Phi}{\partial r}, \tag{3.19}$$

$$\hat{L}_1 v_\varphi + \frac{v_r v_\varphi}{r} = -\frac{1}{\varrho}\frac{\partial P}{r\partial \varphi} - \frac{1}{r}\frac{\partial \Phi}{\partial \varphi}, \tag{3.20}$$

$$\hat{L}_1 v_z = -\frac{1}{\varrho}\frac{\partial P}{\partial z} - \frac{\partial \Phi}{\partial z}, \tag{3.21}$$

$$\frac{\partial \varrho}{\partial t} + \frac{1}{r}\frac{\partial}{\partial r}(r\varrho v_r) + \frac{1}{r}\frac{\partial}{\partial \varphi}(\varrho v_\varphi) + \frac{\partial}{\partial z}(\varrho v_z) = 0, \tag{3.22}$$

here P this pressure. The operator $\hat{L}_1$ is determined by the expression:

$$\hat{L}_1 \equiv \frac{\partial}{\partial t} + v_r\frac{\partial}{\partial r} + \frac{v_\varphi}{r}\frac{\partial}{\partial \varphi} + v_z\frac{\partial}{\partial z}. \tag{3.23}$$

Potential Φ is a given function, if the disk is situated in the external gravity field, and its intrinsic gravity field is a negligible small value in comparison with the external field. In the opposite case, the gravitational potential is fully determined by the disk density and the Poisson equation:

$$\frac{1}{r}\frac{\partial}{\partial r}\left(r\frac{\partial \Phi}{\partial r}\right) + \frac{1}{r^2}\frac{\partial^2 \Phi}{\partial \varphi^2} + \frac{\partial^2 \Phi}{\partial z^2} = 4\pi G\varrho. \tag{3.24}$$

Thus, for five (or six) unknown functions

$$v_r,\ v_\varphi,\ v_z,\ \varrho,\ P,\ (\Phi) \tag{3.25}$$

we have four (or five) equations. The fifth (or sixth) equation is required to produce a closed system of equations. This equation is the equation of state whose form can differ depending on the selection of thermodynamic functions. If the density and temperature are regarded as a basis, the equation of state can be written in the following form

$$P = P(\varrho, T). \tag{3.26}$$

If, however, in addition to density measurements we also have the entropy function in the disk, then for the pressure

$$P = P(\varrho, S). \tag{3.27}$$

(3.26) and (3.27) can be regarded as two general forms of writing the equation of state.

The investigated processes should be relatively slow in the sense that the characteristic time is long in comparison with the heat exchange time. Therefore, it may be assumed that the process will be isothermal, i.e. at any point of space we have $T = T_0 = \text{const}$. Consequently, instead of (3.26) we have

$$P = P(\varrho, T_0) = P(\varrho). \tag{3.28}$$

$$P = P(\varrho, S_0) = P(\varrho). \tag{3.29}$$

Thus, in both limiting cases of 'slow' or 'fast' processes we can use the barotropic equation of state.

3.1.6. Analysis of equations in transition from three-dimensional to two-dimensional equations

The degradation of the equations in this section is identical with the approach used in deriving the 'shallow water' equations [1019].

The individual terms in the initial equations will be estimated on the basis of the order of magnitude. For the functions (3.25) and their daily lattice we introduced the characteristic 'orders of magnitude of the values':

$$|v_r| \sim U, |v_\varphi| \sim V, |v_z| \sim W, |\varrho| \sim \varrho, |P| \sim P, (|\Phi| \sim \Phi),$$
$$\left|\frac{\partial}{\partial r}\right| \sim L^{-1}, \left|\frac{1}{r}\frac{\partial}{\partial \varphi}\right| \sim \zeta^{-1}, \left|\frac{\partial}{\partial z}\right| \sim D^{-1}, \left|\frac{\partial}{\partial t}\right| \sim T^{-1}, \tag{3.30}$$

here we introduce three characteristic length scales: L is the scale along the radial coordinate r, ζ – along the azimuthal angle, D – along the vertical axis of rotation (characteristic disk thickness), and T is the characteristic time scale. After substituting (3.30) into the continuity equation, we obtain the following values in accordance with the order of magnitude:

$$W \le (DT^{-1}, \varepsilon_1 U, \varepsilon_2 V)_{\max}, \tag{3.31}$$

where it is necessary to use the maximum of the terms in the brackets and introduce dimensionless parameters:

$$\varepsilon_1 \equiv \frac{D}{L}, \ \varepsilon_2 \equiv \frac{D}{\zeta}. \tag{3.32}$$

Substituting (3.30) into (3.19) and (3.20) and omitting the terms containing the derivatives with respect to the z-coordinate because of the reasons which will be explained later, the following can be written on the basis of the order of magnitude

$$\frac{P}{\varrho} \sim \left(\frac{L}{T}U, U^2, \frac{L}{\zeta}UV, \frac{L}{R}V^2, \Phi, \frac{L}{R}\Phi, \frac{\zeta}{T}V, \frac{\zeta}{L}UV, V^2, \frac{\zeta}{R}UV \right)_{\max} . \tag{3.33}$$

The dimensionless parameters in the round brackets L/R, ζ/R, L/ζ are determined by the differences in the scales of variation of pressure, potential and velocity. Using (3.30), we determine the ratio P/ϱ from (3.21) on the basis of the order of magnitude

$$\frac{P}{\varrho} \sim \left(\frac{D}{T}W, \varepsilon_1 UW, \varepsilon_2 VW, W^2, \Phi \right)_{\max} . \tag{3.34}$$

We determine conditions under which all the terms in the round brackets, with the exception of the last term, are considerably smaller than the value of P/ϱ determined from (3.33). Substituting $W \sim D/T$, we obtain the estimate

$$\frac{P}{\varrho} \sim \left(\frac{D^2}{T^2}, \varepsilon_1 \frac{D}{T}U, \varepsilon_2 \frac{D}{T}V, \Phi \right)_{\max} . \tag{3.35}$$

Using the ratio of the second term in (3.35) to the first term in (3.33) and the first term in (3.35) of the seventh term in (3.33) gives

$$\frac{\varepsilon_1 D T^{-1} U}{L T^{-1} U} = \varepsilon_1^2 , \quad \frac{\varepsilon_2 D T^{-1} V}{\zeta T^{-1} V} = \varepsilon_2^2 . \tag{3.36}$$

If the conditions

$$\varepsilon_1^2 \ll 1, \quad \varepsilon_2^2 \ll 1 \tag{3.37}$$

are fulfilled, the second and third term in the left-handed part of the equation (3.21) can be ignored. The inequalities (3.37) determine the approximation of the thin disk.

The first and fourth terms are of the same order and can be ignored if $D^2/T^2 \ll P/\varrho$ is fulfilled. The quantity on the right as regards the order is equal to the speed of sound $P/\varrho \sim c^2$. Consequently, we obtain a constraint for the characteristic time of the problem:

$$T^2 \gg \frac{D^2}{c^2} . \tag{3.38}$$

Condition (3.37) shows that the disk thickness is small in comparison with the scales of perturbations in the plane, and the inequality (3.38) requires that the investigated processes are slow in comparison with the time for the establishment of pressure in the thickness of the disk.

Assuming that the conditions (3.37) and (3.38) are fulfilled, the equation (3.21) can be written in the following form

$$\frac{1}{\varrho}\frac{\partial P}{\partial z}=-\frac{\partial \Phi}{\partial z}, \tag{3.39}$$

which is the condition of hydrostatic equilibrium on the z-coordinate. It is clear that in the absence of rapid movement along the vertical coordinate, the force of gravitational attraction to the plane $z = 0$ can be compensated only by the gradient of movement along the z-coordinate.

Taking into account the second value W in (3.31), from (3.34) we obtain

$$\frac{P}{\varrho}\sim\left(\frac{D}{T}\varepsilon_1 U, \varepsilon_1^2 U^2, \varepsilon_2 U \varepsilon_1 V, \Phi\right)_{\max}. \tag{3.40}$$

Using the ratio of the second term in (3.40) to the first three terms in (3.33) respectively we obtain

$$\frac{\varepsilon_1 D T^{-1} U}{L T^{-1} U}=\varepsilon_1^2, \frac{\varepsilon_1^2 U^2}{U^2}=\varepsilon_1^2, \frac{\varepsilon_1 U \varepsilon_2 V}{L \zeta^{-1} U V}=\varepsilon_1 \varepsilon_2 \frac{\zeta}{L}. \tag{3.41}$$

Thus, if the conditions (3.37) are fulfilled together with

$$\varepsilon_1 \varepsilon_2 \frac{\zeta}{L} \ll 1,\ \varepsilon_1 \varepsilon_2 \frac{L}{\zeta} \ll 1,\ \varepsilon_1 \varepsilon_2 \frac{R}{\zeta} \ll 1, \tag{3.42}$$

then from (3.21) we obtain the hydrostatic balance condition (3.39).

Finally, if in the right-hand part of (3.34) we use the value $W \sim \varepsilon_2 V$ for W, we obtain

$$\frac{P}{\varrho}\sim\left(\frac{D}{T}\varepsilon_2 V, \varepsilon_1 U \varepsilon_2 V, \varepsilon_2^2 V^2, \Phi\right)_{\max}. \tag{3.43}$$

Using the ratio of the first three terms in the brackets of (3.43), respectively, to the seventh, third and fourth terms in the round brackets of (3.33), we obtain the equations

$$\frac{\varepsilon_2 D T^{-1} V}{\zeta T^{-1} V}=\varepsilon_2^2,\quad \frac{\varepsilon_1 U \varepsilon_2 V}{L \zeta^{-1} U V}=\varepsilon_1 \varepsilon_2^2 \frac{\zeta}{L},\quad \frac{\varepsilon_2^2 V^2}{L R^{-1} V^2}=\varepsilon_1^2 \frac{R}{L}. \tag{3.44}$$

Consequently, if the conditions (3.37) and the inequalities

$$\varepsilon_1 \varepsilon_2 \frac{\zeta}{L} \ll 1,\quad \varepsilon_1 \varepsilon_2 \frac{L}{\zeta} \ll 1,\quad \varepsilon_2^2 \frac{R}{L} \ll 1,\quad \varepsilon_1 \varepsilon_2 \frac{R}{\zeta} \ll 1 \tag{3.45}$$

are fulfilled, then instead of (3.21) we confine our considerations to (3.39).

It may easily be seen that all the inequalities (3.42) and (3.45) can be reduced to two conditions (3.37) and the third inequality $\varepsilon_2^2 R / L \ll 1$. Since in all cases $R/L > 1$, instead of the inequalities (3.37), (3.42) and (3.45) we can use the conditions

$$\varepsilon_1^2 \ll 1,\quad \varepsilon_2^2 \frac{R}{L} \ll 1. \tag{3.46}$$

Equation (3.21) is reduced to (3.39) when (3.46) and (3.38) are fulfilled.

3.1.7. Two limiting models of astrophysical disks

In this section, special attention is given to the two limiting variants of the thin astrophysical disks: the disk in the dominant external gravitational field, and the 'purely' self-gravitating disk. The first case is the case in which the force along the axis of rotation from the intrinsic gravitational field of the disk can be ignored in comparison with the pressure force. The second variant is when the vertical (along the axis of rotation) structure of the disk is determined by the balance of pressure forces and gravitation of the disk itself.

The latter case requires additional explanation. Let it be that there is no external gravitational field, i.e. the rotation curve of the disk is determined by its intrinsic gravitational field. For the linearised equations, the condition (3.37) indicates that $(kh)^2 \ll 1$, where h is the half thickness of the disk, $k = 2\pi/\lambda$, λ is the wavelength of the perturbations. To ensure that the condition can be satisfied at least for the perturbations with the longest wavelength of the order of the radius of the disk ($\lambda \sim R$), it is necessary to fulfill the inequality $(R/h)^2 \gg (2\pi)^2$. It will be shown that in this case the disk is highly unstable because the characteristic time of development of the perturbations is considerably shorter than a single rotation of the disk.

Actually, to simplify considerations, we consider a self-gravitating disk rotating with the constant velocity Ω_0 = const, with the characteristic half thickness h and the radius R. From the equilibrium condition of this disk it is quite easy to calculate the surface density as the function of the radius $\sigma_0(r) = \sigma_0(0) \cdot (1 - r^2/R^2)^2$ and the dependence of the angular velocity of rotation on the disk radius and the value of the surface density in its centre [236, 590]:

$$\Omega_0^2 = \frac{\pi^2 G\sigma_0(0)}{2R}.$$

The dispersion equation, describing the frequency of small oscillations of such a disk as the function of the radial wave number k, has the form

$$\omega^2 = 4\Omega_0^2 - 2\pi G\sigma_0 |k| + k^2c^2.$$

At the point of the minimum of the dispersion curve $k = k_0 = \pi G\sigma_0/c^2$, determined from the equality $\partial\omega^2/\partial k = 0$, it is reduced to

$$\omega^2 = 4\Omega_0^2 - \left(\frac{\pi G\sigma_0}{c}\right)^2.$$

Taking into account the equation for the half thickness of the disk $h \simeq \dfrac{c^2}{\pi G\sigma_0}$, the value of the square of frequency at the point k_0 is

$$\omega^2 = 4\Omega_0^2\left(1 - \frac{R}{2\pi h}\right),$$

and this transforms to the instability condition

$$\frac{R}{h} > 2\pi.$$

Thus, the condition of fulfilling the last inequality at least for the perturbations with the longest wavelength corresponds to the condition of strong instability of the disk, $R/h \gg 2\pi$. In this case, structures will form in the gas self-gravitating disk already during part of the rotation and, consequently, the disk will be 'heated' and its thickness will increase until the system is found at the stability boundary. Consequently, at the condition $R/h \gg 2\pi$, the concept of the initial stationary self-gravitating disk can be regarded as non-existing.

It may easily be seen that the previously derived condition of marginal stability for the value R/h depends only slightly on the equation of state of the disk, i.e. on the value γ_V. In fact, we obtain the instability condition in the opposite (in comparison with isothermal) limiting case $\gamma_V = 2$.

To use this argument for the claim which is of the procedural nature, it is necessary to move forward and use some of the equations derived below together with some definitions.

In the case of arbitrary γ_V from equation (3.39) we obtain

$$-A_V\gamma_V\varrho^{\gamma_V-2}\frac{\partial\varrho}{\partial z}=\sqrt{8\pi G A_V[\varrho_c^{\gamma_V}-\varrho^{\gamma_V}(z)]},$$

and, consequently, for the half thickness of the disk

$$h=\int_0^h dz=\int_0^{\varrho_c}\frac{A_V\gamma_V\varrho^{\gamma_V-2}d\varrho}{\sqrt{8\pi G A_V[\varrho_c^{\gamma_V}-\varrho^{\gamma_V}(z)]}}=$$

$$=-\frac{A_V\gamma_V\varrho_c^{\gamma_V-1}}{\sqrt{8\pi G A_V\varrho_c^{\gamma_V}}}\int_0^1\frac{x^{\gamma_V-2}dx}{\sqrt{1-x^{\gamma_V}}}=\frac{c_c^2 I(\gamma_V)}{\sqrt{8\pi G P_c}}=\sqrt{\frac{\gamma_V}{8\pi G\varrho_c}}\,c_c I(\gamma_V). \tag{3.47}$$

Here we use the definition of P and the definition $c_c^2=(\partial P/\partial\varrho)_c=\gamma_V A_V\varrho_c^{\gamma_V-1}$, and also

$$I(\gamma_V)\equiv\int_0^1\frac{x^{\gamma_V-2}dx}{\sqrt{1-x^{\gamma_V}}},\quad x\equiv\frac{\varrho}{\varrho_c}.$$

For a partial case $\gamma_V = 2$ we have $I(2) = \pi/2$ and, correspondingly

$$h=\frac{\pi}{2}c_c\sqrt{\frac{1}{4\pi G\varrho_c}}=\sqrt{\frac{\pi c_c^2}{16G\varrho_c}}=\sqrt{\frac{\pi A_V}{8G}}.$$

To determine a relationship between ϱ_c and σ, we use the profile of density along the axis of rotation of the solid-state rotating disk with $\gamma_V = 2$, presented in [628]:

$$\varrho_0(z)=\varrho_c\cos\left(\sqrt{\frac{2\pi G}{A_V}}z\right),$$

from which

$$\sigma_0 = \int_{-h}^{h} \varrho_0(z)dz = \frac{4}{\pi}\varrho_c h.$$

Thus, the relationship between h and σ_0 for the investigated case is as follows:

$$h = \frac{c_c^2}{4G\sigma_0},$$

from which

$$\omega^2 = 4\Omega_0^2\left(1 - \frac{R}{8h}\right).$$

Finally, the instability condition for the disk with $\gamma_V = 2$ is obtained in the form

$$\frac{R}{h} > 8.$$

This instability condition differs only slightly from the same condition for the isothermal disk. Thus, all the considerations regarding the absence of the initial stationart state for the self-gravitating disk apply regardless of the dependence on γ_V.

Thus, the 'purely' self-gravitating gas disk is highly unstable, and the highest instability of the disk is observed for the wavelengths corresponding to $kh \sim 1$ which do not satisfy the condition of longwave approximation.

Consequently, the dynamics of the self-gravitating gas disk can be described in principle within the framework of the traditional two-dimensional approximation despite the fact that this is carried out usually in the literature. Naturally, the question arises as to the use of the external field which, stabilising the instability of the disk, would leave the disk self-gravitating along the axis of rotation, i.e. the structure of the disk in this direction would be determined only by its intrinsic field. If we require from the external field stabilisation of the instability of the disk, the presence of such a field should move the disk not into the stable region but at least to the boundary of this region. The only method of achieving this is to increase greatly the angular velocity of rotation of the disk. In other words, the velocity of revolution of the stable disk should be determined mainly by the external field.

Let us assume that the external gravitational field is spherically-symmetric. For example, this is so in the case of a spherical stellar halo for a galactic disks, or in the case of a large central mass for the accretion circumstellar or circumplanetary disks. In this case, it is quite easy to show that the derivatives along the axis of rotation from the external gravitational potential are of the same order as the derivatives from the intrinsic potential of the disk. Actually, for determinacy, we use the external potential from the central mass M

$$\Phi_{\text{ext}}(r,z) = \frac{GM}{\sqrt{r^2 + z^2}}$$

and calculate its derivatives with respect to z:

$$\frac{\partial \Phi_{\text{ext}}}{\partial z} = -\frac{G M_z}{(r^2 + z^2)^{3/2}} \approx \Omega^2 z, \quad \frac{\partial^2 \Phi_{\text{ext}}}{\partial z^2} \approx -\frac{G M}{(r^2 + z^2)^{3/2}} \approx \Omega^2.$$

On the other hand, the derivatives with respect to z from the potential of the thin self-gravitating disk can be estimated using the Poisson equation:

$$\frac{\partial^2 \Phi_d}{\partial z^2} \approx 4\pi G \varrho, \quad \frac{\partial \Phi_d}{\partial z} \approx 4\pi G \int_0^z \varrho dz.$$

Here ϱ is the disk density. Since the disk density decreases with altitude, the upper estimate of the ratio of the derivatives of the disk potential and the derivatives of the external potential (both first and second) can be represented by its value in the central plane of the disk (at $z = 0$). For an isothermal disk (and as regards the order of magnitude for any disk) we have $\varrho(z = 0) = \sigma/2h$. Using this relationship and the previous relationships we obtain

$$\frac{\partial^2 \Phi_d / \partial z^2}{\partial^2 \Phi_{\text{ext}} / \partial z^2} \leq \frac{2\pi G \sigma}{h \Omega^2} \approx \frac{1.7}{Q^2}.$$

Here Q is the Toomre coefficient and we use the expression for the thickness of the isothermal disk described previously. It may be seen that the ratio of the derivatives is proportional to Q^{-2}. The condition of disk stability indicates that $Q^{-2} \leqslant 1$, i.e. $\partial^2\Phi_d/\partial z^2 \leqslant \partial^2\Phi_{\text{ext}}/\partial z^2$ or $\partial^2\Phi_d/\partial z^2 \leqslant \partial^2\Phi_{\text{ext}}/\Omega^2$. Integrating with respect to z, we obtain the identical relationship for the first derivatives: $\partial\Phi_d/\partial z \leqslant \partial\Phi_{\text{ext}}/\partial z$. Thus, movement of the stability boundary should also lead to the situation in which the structure of the disk along the axis of rotation is determined by both its intrinsic and external gravitational fields – the disk ceases to be self-gravitating.

It should be mentioned that since the rotation curve of a stable disk is determined by the external field then in all cases

$$\frac{\partial^2 \Phi_{\text{ext}}}{\partial r^2} \approx \Omega^2.$$

The above derivation is also valid for the non-spherical external field, if its anisotropy is not very strong:

$$\frac{\partial^2 \Phi_{\text{ext}}}{\partial r^2} \simeq \frac{\partial^2 \Phi_{\text{ext}}}{\partial z^2}.$$

An exception is the case in which the external field has a cylindrical geometry with the generating line along the axis of rotation of the disk. The angular velocity of rotation of the disk is determined by the radial component of the gravitational force of the cylinder and may be sufficiently high to ensure stability of the disk. At the same time, the structure of the disk along the axis of rotation is determined exclusively by its self-gravitation. This model will be taken into account when discussing the self-gravitating disk. The model represents the opposite limiting case in comparison with the case of the dominant external potential. After all, this case is artificial because if these 'cylindrical' fields are found in the real systems (needle galaxies, jumpers), they do not contain inserted disks. However, we believe that the investigation of this self-gravitating disk inserted into the cylinder is of considerable procedural interest. Firstly, in this case we can specify the effects to which self-gravitation leads in the 'pure' form and, consequently, determine the maximum differences in comparison with the case of the dominant external field. Secondly, this model is also investigated in a large number of classic studies of astrophysical disks (for example [628, 698, 1229]).

3.1.8. Restrictions on the characteristic time of the processes in the two-dimensional disk model

Now, we return to the explanation of the restrictions imposed by the condition (3.38) on the characteristic time (frequency) of the processes which can be investigated in the two-dimensional formulation.

It should be mentioned that if the disk matter is incompressible, i.e. $c^2 \to \infty$, the right-hand part of the inequality (3.38) tends to 0 and this condition is automatically fulfilled for all processes. In particular, this situation is observed in the case of 'shallow water', and consequently the only condition of applicability of the two-dimensional equations are the 'special' restrictions (3.46).

At first sight, the condition (3.38) should be easily satisfied because of the small disk thickness D. To understand that this is not so, it is necessary to remember that the small disk thickness is the result of the low pressure force (essentially c^2) in comparison with the gravitational and inertia forces. Consequently, the quantity in the right-hand part of (3.38) is the ratio of two small values and requires special attention.

We consider separately each of these cases: a self-gravitating disk (in the previously mentioned sense) and a disk in the dominant external field. In the latter case $c^2 / D^2 \simeq \Omega_0^2$ [590], which results in the restriction of the characteristic frequencies of the processes by the frequency of rotation of the disk:

$$T^{-2} \simeq \omega^2 \ll \Omega_0^2 . \tag{3.48}$$

This condition requires additional comments. In fact, generally speaking, the characteristic frequency of the process changes if the process is studied in a specific rotating reference system, i.e. it is necessary to determine what the characteristic frequency is. Since the investigated restriction is determined by the finiteness of the time to establishment of vertical equilibrium, it is clear from the physical viewpoint that the time to establishment of equilibrium in a specific area of the disk should be compared with the characteristic time of the variation of the parameters in the same area of the disk, i.e. the characteristic process time should be estimated in the accompanying local coordinate system. The more accurate (and, consequently, more cumbersome) derivation of the condition (3.48) confirms this.

In the opposite case of the self-gravitating disk, its thickness can be calculated from the previously presented relationships. For the isothermal disk

$$h \approx \frac{c^2}{\pi G \sigma_0}$$

(it should be mentioned that the result depends only slightly on the equation of state). Consequently, equation (3.38) can be presented in the following form

$$T^{-2} \simeq \omega^2 \ll \frac{c^2}{h^2} \approx \left(\frac{\pi G \sigma_0}{c}\right)^2 \leq \varkappa^2. \tag{3.49}$$

The last inequality is the condition for the epicyclic frequency of the marginally stable self-gravitating disk [590]: $\varkappa \geqslant G\sigma_0/c$.

Thus, for the self-gravitating disk, as for the disk in the dominant external field, condition (3.38) leads to the restriction of the characteristic frequencies of the processes, which can be investigated in the two-dimensional formulation, by the frequency of rotation of the disk.

An exception is the case in which the external field, acting on the gas disk, is formed by a thicker and larger stellar disk. Let the stellar disk have the thickness h_* and the dispersion of velocities along the axis of rotation c_{z*}, consequently, according to [590] the thickness of the gas disk is

$$h_g = h_* \frac{c_g}{c_{z*}} = \frac{c_g}{\omega_{z*}} = \frac{c_g}{\varkappa_*}\sqrt{\frac{h_*}{R_*}},$$

were ω_{z*} is the frequency of oscillations of the stars along the axis of rotation; R_* and $\varkappa_*$ is the radius and epicyclic frequency of the stellar disk. From the last equation we have

$$\frac{c_g}{h_g} = \varkappa_* \sqrt{\frac{R_*}{h_*}}.$$

Since according to the order of magnitude $\varkappa_* \simeq \varkappa_g \simeq \varkappa$, then

$$\frac{c}{D} \simeq \varkappa \sqrt{\frac{R_*}{h_*}} \gg \varkappa.$$

However, here it should be mentioned that the previously used inequality $\omega_{z*} = \varkappa_*(R_*/h_*)^{1/2} \gg \varkappa_*$ holes only in the case of the purely self-gravitating stellar disk without the external halo. As already mentioned, this disk is highly unstable. If the external spherical halo, stabilising the stellar disk already exists, we again return to the condition (3.48).

Thus, the condition (3.38) for an arbitrary disk is highly 'rigid' and must be taken into account together with the conditions (3.46) in the analysis of the applicability of the two-dimensional equations for specific problems.

3.2. Correct two-dimensional model of the gas disk

3.2.1. Closed system of integro-differential equations for a barotropic disk

Here, the investigated barotropic processes whose partial cases have already been mentioned previously, see equations (3.28) and (3.29): $\varrho = \varrho(P)$. We introduce the pressure function[1]:

$$\mathcal{P}(P)=\int\frac{dP}{\varrho(P)}. \tag{3.50}$$

The gradient of this function is equal to

$$\nabla\mathcal{P}=\frac{d\mathcal{P}}{dP}\nabla P=\frac{1}{\varrho}\nabla P. \tag{3.51}$$

Consequently, $\varrho^{-1}\,\partial P/\partial z = \partial\mathcal{P}/\partial z$ and instead of (3.39) we have

$$\frac{\partial}{\partial z}(\mathcal{P}+\Phi)=0. \tag{3.52}$$

The general solution of the last equation is an arbitrary function independent of z:

$$\mathcal{P}(r, \varphi, z, t) + \Phi(r, \varphi, z, t) = \chi(r, \varphi, z, t). \tag{3.53}$$

Using (3.51) and (3.53), the right-hand part of the equations (3.19) and (3.20) can be written respectively in the form of functions $\partial\chi/\partial z$ and $r^{-1} = \partial\chi/\partial\varphi$, which are also independent of z. Consequently, the left-hand part of the equations (3.19) and (3.20) should also be independent of z, i.e. the functions

$$v_r = v_r(r, \varphi, t),\ v_\varphi = v_\varphi(r, \varphi, t). \tag{3.54}$$

Since $\partial v_r/\partial z = \partial v_\varphi/\partial z = 0$, we can see why in deriving (3.33) the terms containing the derivatives with respect to z from v_r in (3.19) and from v_φ in (3.20) were not taken into account. Using (3.51) and substituting (3.53), (3.54) into (3.19) and (3.20) we obtain the following dynamic equations for the 'planar' functions:

$$\hat{L}_2 v_r - \frac{v_\varphi^2}{r} = -\frac{\partial\chi}{\partial r}, \tag{3.55}$$

$$\hat{L}_2 v_\varphi + \frac{v_r v_\varphi}{r} = -\frac{1}{r}\frac{\partial\chi}{\partial\varphi}, \tag{3.56}$$

were

$$\hat{L}_2 \equiv \frac{\partial}{\partial t} + v_r\frac{\partial}{\partial r} + \frac{v_\varphi}{r}\frac{\partial}{\partial\varphi}.$$

The continuity equation (3.22) will be integrated with respect to z from $-\infty$ to $+\infty$:

---

[1]In this section, the letter $\mathcal{P}$ denotes three-dimensional pressure.

$$\frac{\partial}{\partial t}\left(\int_{-\infty}^{+\infty}\varrho\,dz\right)+\frac{1}{r}\frac{\partial}{\partial r}\left[rv_r\left(\int_{-\infty}^{+\infty}\varrho\,dz\right)\right]+$$

$$+\frac{1}{r}\frac{\partial}{\partial\varphi}\left[v_\varphi\left(\int_{-\infty}^{+\infty}\varrho dz\right)\right]+\int_{-\infty}^{+\infty}\left[\frac{\partial}{\partial z}(\varrho v_z)\right]dz=0. \tag{3.57}$$

This form can be used since v_r and v_φ are independent of z. The last integral is equal to 0 because there is no matter at infinity. Introducing the definition of surface density

$$\sigma(r,\varphi,t)\equiv\int_{-\infty}^{+\infty}\varrho(r,\varphi,z,t)\,dz, \tag{3.58}$$

we obtain the 'two-dimensional equation of continuity of the thin disk:

$$\frac{\partial\sigma}{\partial t}+\frac{1}{r}\frac{\partial}{\partial r}(r\sigma v_r)+\frac{1}{r}\frac{\partial}{\partial\varphi}(\sigma v_\varphi)=0. \tag{3.59}$$

The system of the equations disclosed using the barotropic equation of state

$$\varrho=\varrho(P)=\varrho(\mathcal{P})=\varrho(\chi-\Phi), \tag{3.60}$$

which is substituted into (3.58):

$$\sigma(\varrho,\varphi,t)\equiv\int_{+\infty}^{-\infty}\varrho\left[\chi(r,\varphi,t)-\Phi(r,\varphi,z,t)\right]dz, \tag{3.61}$$

where the expression in the square brackets is the argument of functions $\varrho(\mathcal{P})$.

For the five (six) unknown functions

$$\varrho,\ v_r,\ v_\varphi,\ \sigma,\ \chi,\ (\Phi)$$

we write five (six) the equations: (3.55), (3.56), (3.59)–(3.61) (and Poisson equation (3.24)). It can be seen that for the integral type of equation for surface density (3.61) the closed system of equations is integro-differential. For specific calculations, the barotropic equation of state is still relatively general. The class of the equations of state will be narrowed to the polytropic state.

3.2.2. A polytropic disk in the external gravitational field

As indicated in this and subsequent section, to obtain the closed system of the differential equations for the 'planar' functions, in addition to the previously mentioned conditions (3.46) and (3.48) (or (3.59) for the next section) it is sufficient to consider the assumption on the polytropic relation between P_V and ϱ:

$$P_V=A_V\,\varrho^\gamma, \tag{3.62}$$

where A_V, γ are constants, γ is the 'volume' exponent of the polytrope. Consequently, according to (3.50), the pressure function is

$$P=\int_0^P \frac{dP_V}{\varrho}=A_V\frac{\gamma}{\gamma-1}\varrho^{\gamma-1}. \tag{3.63}$$

Here, the lower integration limit is selected at the point situated at an infinite distance along z where pressure (and density) is equal to 0.

Substituting (3.63) into (3.53) gives g:

$$\varrho=\left[\frac{\gamma-1}{\gamma A_V}(\chi-\Phi)\right]^{1/(\gamma-1)}. \tag{3.64}$$

Subsequently, using (3.61), we obtain the following expression for ϱ:

$$\sigma(r,\varphi,t)=B_V\int_{z_1}^{z_2}\left[\chi(r,\varphi,t)-\Phi(r,\varphi,z,t)\right]^{1/(\gamma-1)}dz, \tag{3.65}$$

where the function at $z_{1,2}$, converts to 0

$$B_V\equiv\left[\frac{\gamma-1}{A_V\gamma}\right]^{1/(\gamma-1)}. \tag{3.66}$$

If the subintegrand expression in (3.61) denotes some function ϱ of the difference of two other functions χ and Φ, the subintegrand expression in (3.65) describes the fully determined function of z, since Φ is the defined external gravitational field. After integration with respect to z we obtain the specific relationship between the two 'planar' functions σ and χ, which closes the system of equations (3.55), (3.56) and (3.59) for the 'planar' functions: v_r, v_φ, σ and χ. The condition of smallness of the disk thickness enables us to write the relationship between σ and χ for the arbitrary function Φ using its expansion in the vicinity of the plane $z = 0$:

$$\Phi(r,\varphi,z,t)=\Phi_0(r,\varphi,t)+\Phi_0'(r,\varphi,t)\cdot z+\frac{1}{2}\Phi_0''(r,\varphi,t)\cdot z^2.$$

Here index 0 denotes that the appropriate quantity is taken at the point $z = 0$, and the prime denotes the differentiation with respect to z.

Substituting the expansion of the function into equation (3.65) gives

$$\sigma(r,\varphi,t)=B_V\left(\frac{\Phi_0''}{2}\right)^{1/(\gamma-1)}\int_{z_1}^{z_2}\left\{\frac{2(\chi-\Phi_0)}{\Phi_0''}+\left(\frac{\Phi_0'}{\Phi_0''}\right)^2-\right.$$

$$\left.-\left[z^2+2\frac{\Phi_0'}{\Phi_0''}z+\left(\frac{\Phi_0'}{\Phi_0''}\right)^2\right]\right\}^{1/(\gamma-1)}dz=$$

$$=B_V\left(\frac{\Phi_0''}{2}\right)^{1/(\gamma-1)}\left[\left(\frac{\Phi_0'}{\Phi_0''}\right)^2+\frac{2(\chi-\Phi_0)}{\Phi_0''}\right]^{1/(\gamma-1)+\frac{1}{2}}\int_{-1}^{1}(1-x^2)^{1/(\gamma-1)}dx, \tag{3.67}$$

where

$$x \equiv \left(z + \frac{\Phi_0'}{\Phi_0''} \right) \Big/ \sqrt{\left(\frac{\Phi_0'}{\Phi_0''} \right)^2 + \frac{2(\chi - \Phi_0)}{\Phi_0''}}. \tag{3.68}$$

The integration limits are determined from the condition $x = \pm 1$ which gives

$$z_{1,2} = \pm \sqrt{\left(\frac{\Phi_0'}{\Phi_0''} \right)^2 + \frac{2(\chi - \Phi_0)}{\Phi_0''}} - \frac{\Phi_0'}{\Phi_0''}. \tag{3.69}$$

It may be seen that when the external gravitational potential is symmetric in relation to the plane $z = 0$, i.e. $\Phi' = 0$, the integration limits with respect to z are naturally also symmetric.

The integral in (3.67) is converted to the form

$$\int_{-1}^{1} (1 - x^2)^{1/(\gamma-1)} dx = 2 \int_{0}^{1} (1 - x^2)^{\mu - 1} \equiv 2I, \tag{3.70}$$

where

$$\mu = \frac{1}{\gamma - 1} + 1 = \frac{\gamma}{\gamma - 1}.$$

In turn, using the handbook [45], we obtain (B is the beta function)

$$I = \frac{1}{2} B(1/2, \mu), \quad \mu > 0. \tag{3.71}$$

This equation is valid only at positive μ which in this case is always fulfilled, since $\gamma > 1$. We can write [45]:

$$B(1/2, \mu) = \frac{\Gamma(1/2)\Gamma(\mu)}{\Gamma(\mu + 1/2)}, \tag{3.72}$$

where γ is the gamma function. Substituting (3.72) into (3.71) and the result into (3.70), we obtain:

$$\int_{-1}^{1} (1 - x^2)^{1/(\gamma-1)} dx = \frac{\Gamma(1/2)\Gamma(\mu)}{\Gamma(\mu + 1/2)} = \frac{\sqrt{\pi}\, \Gamma\left(\frac{\gamma}{\gamma - 1} \right)}{\Gamma\left(\frac{\gamma}{\gamma - 1} + \frac{1}{2} \right)}. \tag{3.73}$$

Substituting (3.66) into (3.67) gives

$$\sigma(r,\varphi,t,)=\sqrt{\pi}\left[\frac{\gamma-1}{A_V\gamma}\right]^{1/(\gamma-1)}\frac{\Gamma\left(\frac{\gamma}{\gamma-1}\right)}{\Gamma\left(\frac{\gamma}{\gamma-1}+\frac{1}{2}\right)}\times$$

$$\times\left(\frac{\Phi_0''}{2}\right)^{1/(\gamma-1)}\left[\left(\frac{\Phi_0'}{\Phi_0''}\right)^2+\frac{2(\chi-\Phi_0)}{\Phi_0''}\right]^{(\gamma+1)/2(\gamma-1)}, \tag{3.74}$$

from which

$$\chi=C\sigma^{2\lambda}+\Phi_0-\frac{(\Phi_0')^2}{2\Phi_0''}, \tag{3.75}$$

where

$$C\equiv\left[\frac{\Phi_0''}{2\pi}\frac{\Gamma^2\left(\frac{\gamma}{\gamma-1}+\frac{1}{2}\right)}{\Gamma^2\left(\frac{\gamma}{\gamma-1}\right)}\right]^{\lambda}\left(\frac{A_V\gamma}{\gamma-1}\right)^{2/(\gamma+1)}, \tag{3.76}$$

$$\lambda\equiv\frac{\gamma-1}{\gamma+1}. \tag{3.77}$$

It should be noted that in the general case $\Phi_0 = \Phi_0\ (r, \varphi, t)$ and, therefore, C, generally speaking, is not constant and is a function of the coordinates and time: $C = C\ (r, \varphi, t)$. Φ is the given function and, therefore, the relationship between σ and χ, defined by the equation (3.75), together with the equations (3.55), (3.56) and (3.59) is a closed system of equations for 'flat' functions.

Substitution of (3.75) into (3.55) and (3.56) gives

$$\hat{L}_2 v_r-\frac{v_\varphi^2}{r}=-2\lambda C\sigma^{2\lambda-1}\frac{\partial\sigma}{\partial r}-\frac{\partial\Phi_0}{\partial r}-\sigma^{2\lambda}\frac{\partial C}{\partial r}+\frac{\partial}{\partial r}\left[\frac{(\Phi_0')^2}{2\Phi_0''}\right], \tag{3.78}$$

$$\hat{L}_2 v_\varphi+\frac{v_r v_\varphi}{r}=-\frac{2\lambda C\sigma^{2\lambda-1}}{r}\frac{\partial\sigma}{\partial\varphi}-\frac{1}{r}\frac{\partial\Phi_0}{\partial\varphi}-\frac{\sigma^{2\lambda}}{r}\frac{\partial C}{\partial\varphi}+\frac{\partial}{\partial\varphi}\left[\frac{(\Phi_0')^2}{2r\Phi_0''}\right]. \tag{3.79}$$

It should be mentioned that the equations of motion derived here greatly differ from the conventional equations of motion of the gas disk in an external gravitational field (3.14), (3.15).

3.2.3. *A disk in a stationary axisymmetric potential*

We investigated a partial case in which the external potential $\Phi_0\ (r, \varphi, t)$ satisfies the following three additional conditions:

1. is stationary;
2. is axisymmetric;
3. is symmetric in relation to the central plane of the disk.

Under these conditions, equation (3.75) has the following form

$$\chi(r, \varphi, t) = C(r)\,\sigma^{2\lambda}(r, \varphi, t) + \Phi_0(r). \tag{3.80}$$

The substitution of (3.80) into the equations of motion (3.55) and (3.56) gives

$$\hat{L}_2 v_r - \frac{v_\varphi^2}{r} = -2\lambda\, C(r)\sigma^{2\lambda-1}\frac{\partial\sigma}{\partial r} - \frac{\partial\Phi_0}{\partial r} - \sigma^{2\lambda}\frac{\partial C}{\partial r}, \tag{3.81}$$

$$\hat{L}_2 v_\varphi + \frac{v_r v_\varphi}{r} = -\frac{2\lambda C\sigma^{2\lambda-1}}{r}\frac{\partial\sigma}{\partial\varphi}. \tag{3.32}$$

The equations (3.31), (2.32) and (3.59) form a closed system of equations at the previously mentioned constraints.

We now write the equations of motion for the 'flat' functions in the 'traditional' form:

$$\hat{L}_2 v_r - \frac{v_\varphi^2}{r} = -\frac{1}{\sigma}\frac{\partial P_S}{\partial r} - \frac{\partial\Psi}{\partial r}, \tag{3.83}$$

$$\hat{L}_2 v_\varphi + \frac{v_r v_\varphi}{r} = -\frac{1}{\sigma r}\frac{\partial P_S}{\partial\varphi} - \frac{1}{r}\frac{\partial\Psi}{\partial\varphi}, \tag{3.84}$$

where P_S is the 'flat' pressure; Ψ is the value of the gravitational potential in the plane $z = 0$. The latter value coincides with Φ_0. As regards the first terms, they coincide only in the completely uniform case where the coefficient $C(r) = \text{const}$.

Actually, following the conventional method [271, 698], we introduce the 'flat' polytrope

$$P_S = A_S\sigma^{\gamma_S}, \tag{3.85}$$

where A_S and γ_S are constants, γ_S is the flat polytrope exponent. Substituting (3.85) into the first terms in the right-hand part of (3.83) and (3.84) gives

$$\frac{\nabla P}{\sigma} = A_S\,\gamma_S\,\sigma^{\gamma_S-2}\nabla\sigma. \tag{3.86}$$

Comparing (3.86) with the first terms in the right-hand part of (3.81) and (3.82) at $C = \text{const}$ we obtain

$$A_S = \frac{2\lambda}{\gamma_S}C, \quad \gamma_S = 2\lambda + 1,$$

or, using (3.77) we finally obtain

$$A_S = \frac{2(\gamma-1)}{3\gamma-1}C, \tag{3.87}$$

$$\gamma_S = 3 - \frac{4}{\gamma+1} = 1 + 2\frac{\gamma-1}{\gamma+1}. \tag{3.88}$$

These conditions were obtained for the first time by Churilov and Shukhman [271].

Thus, in the case $C\ (r)$ = const and if the inequalities (3.87) and (3.88) are fulfilled, the 'traditional' equations (3.83) and (3.84) coincide with the correctly determined equations (3.81) and (3.82). However, if $C = C(r) \neq$const, the equations (3.81) and (3.82) differ from (3.83) and (3.84) by the additional terms, proportional to $C'(r)$. The difference becomes even greater in the case of the external gravitational potential of the general type $\Phi_0 = \Phi'_0\ (r, \varphi, t)$ where the additional terms appear not only in the expressions describing the gravitational potential (see (3.75)).

3.2.4. The meaning of the function C and applicability of C = const

The fact that the function C is independent of the angle φ means that the external gravitational field is axially symmetric. This is fulfilled almost always in galactic disks (if the regions around the bars are not considered), in accretion disks, the protoplanet cloud, and rings of the planets.

As regards the radial dependence of C, we do not know of any examples where, strictly speaking, the coefficient C could be regarded as independent of r. However, in this case it is important that function C changes slightly at distances of the order of L (it should be noted that L is the characteristic radial scale of the structure).

When discussing the large-scale spiral structure (grand design) in a gas disk of our Galaxy, the distance along the radius between the main helical arms (L) in the solar neighbourhood is no less than 4 kpc [2]. On approach to the centre this distance decreases but no more than by 50%, i.e. remains greater than 2 kpc. These values are comparable with the radial scale of the exponential variation of the density of a stellar component $\simeq$3 kpc.

The size of the 'mini spiral' in the centre of our Galaxy also coincides with the characteristic scale of the radial variation of the density of the central stellar cluster [396].

It was shown in [378] that the scale of the spiral structures in the accretion disk is comparable with the size of the disk.

Thus, in all these cases it is important to take into account the additional terms which consider the dependence $C(r)$. In the ring-shaped structures of the protoplanet cloud and, in particular, in the disks in the vicinity of the planets, the characteristic dimensions of the structure L is considerably smaller than the scale of variation $C(r)$ which remains at least several times greater than the disk thickness. To show that in these cases the previously used 'traditional' two-dimensional equations are correct, it is necessary to fulfill the conditions (3.48) which can be realised only for a dissipative branch (when in the absence of dissociation there is the indiferent equilibrium $\omega \approx 0$). Further investigations are essential in other cases.

The terms, containing the function $C(r, \varphi, t)$ in the equations (3.78), (3.79) will be written in a slightly different form and we confine ourselves to considering only the external potential $\Phi(r,z) = \Psi(r) + \frac{z^2}{2}\Omega_z^2$, where $\Omega_z(r)$ is the frequency of oscillations of a test particle in the vertical direction. In the case of the Kepler problem $\Omega_z = \Omega_K = \sqrt{\frac{GM}{r^3}}$.

In the equation of motion (see (3.5), (3.6)) we integrate the terms with the gradient of the potential along the z-coordinate

$$\int_{-h}^{h} \varrho \frac{\partial}{\partial r}\left(\Psi(r) + \frac{1}{2}\Omega_z^2 z^2\right) dz = \sigma\frac{\partial \Psi}{\partial r} + \frac{1}{2}\frac{\partial \Omega_z^2}{\partial r}\int_{-h}^{h} \varrho z^2\, dz = \sigma\frac{\partial \Psi}{\partial r} + D_z \frac{p}{\sigma}\frac{\partial}{\partial r}\ln(\Omega_z), \tag{3.89}$$

where for the quantity

$$D_z \equiv \Omega_z^2 \frac{\int_{-h}^{h} \varrho z^2\, dz}{\int_{-h}^{h} P\, dz}$$

in the case of adiabatic motion the condition $D_z = 1$ is fulfilled. Here $\mathcal{P}$ is volume pressure, p is flat pressure (see section 3.6.1). Comparing with (3.78), we obtain the additional relationship between C and z:

$$2\lambda C \sigma^{2\lambda-1}\frac{\partial \sigma}{\partial r} + \sigma^{2\lambda}\frac{\partial C}{\partial r} = \frac{\partial p}{\partial r} + \frac{p}{\sigma}\frac{\partial}{\partial r}\ln \Omega_z. \tag{3.90}$$

3.2.5. A closed system of differential equations for the polytropic self-gravitating disk

Taking into account the inequalities (3.46), the Poisson equation can be written in the form

$$\frac{\partial^2 \Phi}{\partial z^2} = 4\pi G \varrho, \tag{3.91}$$

where $-O(D) < z < O(D)$ and D, as previously, is the characteristic half thickness of the disk. From (3.52) and (3.63) we have

$$\frac{\partial \Phi}{\partial z} = -A_V \frac{\gamma}{\gamma - 1}\frac{\partial}{\partial z}\left(\varrho^{\gamma-1}\right) = -A_V \gamma \varrho^{\gamma-2}\frac{\partial \varrho}{\partial z}. \tag{3.92}$$

Multiplying the left and right parts of the equations (3.91) and (3.92) we obtain

$$[(\Phi')^2] = 2\Phi'\Phi'' = -8\pi G A_v \gamma \varrho^{\gamma-1} \varrho' = -8\pi G A_v (\varrho^\gamma)'.$$

Here and in further equations the 'prime' denotes the derivative with respect to z. Integrating the resultant equation from 0 to z, we obtain

$$(\Phi')^2 - (\Phi_c')^2 = -8\pi G A_V\left[\varrho^\gamma - \varrho_c^\gamma\right], \tag{3.93}$$

where the index c corresponds to the value of the function at the point $z = 0$.

At the moment we are not interested in the membrane oscillations of the disk plane, for example, bending instability [590], and attention will be given to the perturbations in the disk plane. For this purpose, it is sufficient to make an assumption of the symmetry of the distribution of the density at any moment of time in relation to the plane $z = 0$. We make this assumption, in accordance with Hunter [698], to ensure that it is more efficient to compare our results with the results of this classic study.

This assumption is made to ensure that the z-th component of the force of gravitational attraction does not exist at any point of the plane z = 0, i.e.

$$\Phi'_c = 0. \tag{3.94}$$

Consequently, from (3.93) we obtain

$$\Phi'(z) = \sqrt{8\pi G A_V \left(\varrho_c^\gamma - \varrho^\gamma(z)\right)}. \tag{3.95}$$

It is now possible to calculate the surface density of the self-gravitating disk

$$\sigma = 2\int_0^D \varrho dz = \frac{1}{2\pi G}\int_0^D \Phi'' dz = \frac{\Phi'(D)}{2\pi G}.$$

Using the equation (3.95), we finally obtained

$$\sigma = \sqrt{\frac{2A_V}{\pi G}}\, \varrho_c^{\gamma/2}. \tag{3.96}$$

Here we use the condition $\varrho(D) = 0$. It should be noted that the expression (3.96), linking the surface density of the disk with its volume density in the plane $z = 0$, was derived for the first time by Hunter [698].

The equality (3.53) is valid at any point z and, in particular, point $z = 0$:

$$\chi = \Phi_c + \mathcal{P}_c = \Phi_c + A_V \frac{\gamma}{\gamma - 1} \varrho_c^{\gamma-1}. \tag{3.97}$$

From (3.96) we obtain

$$\varrho_c = \left(\frac{\pi G}{2 A_V}\right)^{1/\gamma} \sigma^{2/\gamma}.$$

From the last two equalities obtain the value of the pressure function in the plane $z = 0$:

$$\mathcal{P}_c = A_V \frac{\gamma}{\gamma - 1}\left(\frac{\pi G}{2A_V}\right)^{(\gamma-1)/\gamma} \sigma^{2(\gamma-1)/\gamma}. \tag{3.98}$$

The right part of the equation of motion, according to (3.55) and (3.56) is

$$\cdots = -\nabla_\chi = -\nabla\Phi_c - \nabla\mathcal{P}_c. \tag{3.99}$$

According to Hunter [698], the right-hand part of the equation of motion can be written in the following form

$$\cdots = -\nabla\Phi_c - \frac{1}{\sigma}\nabla P_S, \tag{3.100}$$

where

$$P_S = A_S \sigma^{\gamma_S}. \tag{3.101}$$

Using (3.98) we obtain

$$\nabla \mathcal{P}_c = 2A_V \left(\frac{\pi G}{2A_V}\right)^{(\gamma-1)/\gamma} \sigma^{(\gamma-2)/\gamma} \nabla\sigma. \tag{3.102}$$

On the other hand, using (3.101), we have

$$\frac{\nabla P_S}{\sigma} = A_S\, \gamma_S \sigma^{\gamma_S - 2} \nabla\sigma. \tag{3.103}$$

Assuming that the left-hand parts of the equations (3.102) and (3.103) are equal, we determine A_S and γ_S as functions of A_V and γ:

$$\gamma_S - 1 = \frac{2(\gamma-1)}{\gamma}, \tag{3.104}$$

$$\gamma_S A_S = 2\, A_V \left(\frac{\pi G}{2\, A_V}\right)^{(\gamma-1)/\gamma}. \tag{3.105}$$

From (3.104) we finally obtain

$$\gamma_S = 3 - \frac{2}{\gamma}. \tag{3.106}$$

Substituting this expression into (3.105) we obtain

$$A_S = \frac{2}{3\gamma - 2} A_V \left(\frac{\pi G}{2\, A_V}\right)^{(\gamma-1)/\gamma}. \tag{3.107}$$

Comparing this equation with the identical coefficient A_S derived by Hunter [698]:

$$A_S = \frac{\pi^{(3\gamma-2)/2\gamma}\, \Gamma\left(\frac{2\gamma-1}{\gamma}\right)}{2^{(2\gamma-1)/\gamma}\, \Gamma\left(\frac{5\gamma-2}{2\gamma}\right)} A_V \left(\frac{G}{A_V}\right)^{(\gamma-1)/\gamma}, \tag{3.108}$$

it may be seen that these equations differ appreciably[2]. The reason for this is simple: the right-hand part of the equation of motion contains the gradient of the pressure

[2]Bisnovatyi-Kogan and Blinnikov were the first [25] who paid attention to the error of calculations in Hunter's study in the equations of dynamic equilibrium of the self-gravitating disk.

function but it does not contain the gradient of the 'flat' pressure, i.e. volume pressure integrated with respect to z. It should be mentioned that the last quantity, in contrast to the surface ('flat') density, does not have the physical meaning of the force.

3.2.6. Why the flat pressure gradient is not a force?

This can be illustrated in a simple manner and convincingly by examining the flow of a liquid in a conical section pipe. Evidently, if the pressure of the liquid is constant everywhere, there is no flow. At the same time, analysis from the viewpoint of 'flat' pressure gives a completely different result. For this case, the 'flat' pressure is equal to

$$P_S = S\,P_V,$$

where S is the cross-sectional area. Thus, there is a flat pressure gradient and, according to all studies, using P_S, this should cause a flow. On the other hand, if the volume pressure changes in inverse proportion to the cross-section, the liquid will flow in the direction of increasing cross-section of the pipe, regardless of the absence of the flat pressure gradient ($P_S = SP_V = \text{const}$). Accurate analysis of the presence or absence of the flow is possible only if in addition to the flat pressure we also take into account the longitudinal component of the reaction force on the side of the pipe walls. The latter can be determined from the condition of the transverse equilibrium of the liquid in the pipe.

The situation with the self-gravitating disk is identical with the considered one with the only difference being that the role of the walls is played by the gravitational potential of the disk. Flat pressure can be considered only in the case of the potential homogeneous in the plane of the disk ('constant cross-section pipe'). However, in the case of the self-gravitating disk this is equivalent to the case of an infinite disk with a constant surface density. The flat pressure gradient in this case does not form. However, if there is a gravitational potential, the resultant force of pressure in the disk plane can be calculated only if we take into account efficiently the appropriate component of the gravitational force (the contribution of the walls), which can be determined from the condition of equilibrium along the axis of rotation. In particular, this analysis was carried out previously and resulted in equation (3.107) for constant A_S, describing the flat polytrope. In this case, it is incorrect to use the flat pressure and, consequently, equation (3.107).

Substitution of equation (3.93) into equation (3.92) gives the equation for the volume density which can be integrated. In turn, knowing ϱ, we can integrate the Poisson equation and determine the function Φ_0. Thus, the resultant system of equations is close although its solution obviously is associated with certain mathematical difficulties.

3.3. Linear waves in a three-dimensional disk

Several solutions of the total initial three-dimensional system of hydrodynamic equations will be obtained. This is important in order to be able to carry out comparative analysis of different models and approximations.

The model of the thin disk imposes a constraint on the wavelength $\lambda \gg h$ and the frequency of perturbations [43]. Therefore, for the correct investigation of the dynamics of perturbations with $\lambda \lesssim h$ it is necessary to investigate the vertical motions and the z-structure of the disk. The model of the thin disk can be used to investigate only pinch-oscillations (S-mode). For these oscillations, the perturbed pressure is a symmetric function of the z-coordinate, the vertical displacement of the gas does not shift the centre of mass in the disk in relation to the symmetry plane $z = 0$. Thus, the bending oscillations (AS-mode) are not considered. In addition, the two-dimensional model is cannot be used to study high-frequency (deflecting) harmonics with $\omega \gg \Omega$. In this section, we consider the dynamics of acoustic perturbations taking into account the inhomogeneous vertical structure of the gas. In addition to this, the problem of the range of applicability of the thin disk model will also be considered.

3.3.1. The problem of determination of eigenfrequencies for the acoustic waves with self-gravitation not taken into account

We investigate the so-called dissipative instabilities, caused by the combined effect of two factors: differential nature of rotation $\left(n = -\dfrac{d\ln\Omega}{d\ln r} > 0 \right)$ and dynamic viscosity perturbation ($\tilde{\eta} \neq 0$) when the coefficient of dynamic viscosity depends on density and/or temperature [261, 262, 515, 765, 1148, 1311]. Thermal, viscous and acoustic oscillations modes can be unstable. Here, we consider only the sound waves and the Kepler disk.

Our aim is to compare the dispersion dependences for the sound waves, obtained in the two-dimensional formulation, with the results of calculations of the frequency of oscillations for three-dimensional perturbations. Taking into account the dissipative factors, we can investigate the perturbations with the complex eigenfrequency, including unstable ones.

Dynamics of longwave perturbations

The characteristic feature of the viscous models of the accretion disks is the dependence of the turbulent dissipative coefficients on the disk parameters [1147, 1148]. Using the simplest model, we consider the effect of the dependence of dynamic viscosity on density ($\eta = \sigma\, \nu \propto^{1+\delta}$) on the nature of sound oscillations in the disk plane.

Taking into account the perturbations of dynamic viscosity $\eta_1/\eta_0 = (1+\delta)\ \sigma_1/\sigma_0$ for a Kepler disk, the system of equations for relatively longwave perturbations (3.353)–(3.355) has the form

$$-i\omega\sigma_1 + i\sigma_0 k u_1 = 0, \tag{3.109}$$

$$-i\omega u_1 - 2\Omega v_1 = -ikc_T^2 p_1 / p_0 - \lambda_0 k^2 u_1, \tag{3.110}$$

$$-i\omega v_1 + \frac{1}{2}\Omega u_1 = -k^2 v_0 v_1 - ik\frac{3}{2}\Omega v_0 \eta_1 / \eta_0. \tag{3.111}$$

To a first approximation, it is assumed that $p_1/p_0 = \gamma_s \sigma_1/\sigma_0$ (in [261, 262] dissipative factors were taken into account in the thermal balance equation). Consequently, we obtain the dispersion equation

$$\omega^3 + ik^2(\lambda_0 + v_0)\omega^2 - (\Omega^2 + k^2 c_s^2 + k^4 v_0 \lambda_0)\omega - ik^2 v_0 \Omega^2 [3(1+\delta) - c_s^2 k^2 / \Omega^2] = 0. \tag{3.112}$$

We derive an expression for the frequency of the longwave sound oscillations in the form $\omega/\Omega = 1 + \varepsilon k^2$ ($|\varepsilon k^2| \ll 1$). Thus, for the imaginary part of the frequency we have

$$\mathrm{Im}(\omega) \simeq \frac{k^2 v_0}{2}\left(2 + 3\delta - \lambda_0 / v_0\right). \tag{3.113}$$

If the condition $(2 + 3\delta)v_0 > \lambda_0$ is fulfilled, acoustic oscillations become unstable. It should be mentioned that for the α-model in the region of dominance of the gas pressure and opacity k_T we have $\delta = 2/3$ (see section 11.2). The increase of the intensity of perturbations is determined exclusively by the dependence of dynamic viscosity on the disk parameters. Taking into account the thermal processes, radiation, radiation pressure, temperature dependence of viscosity, the vertical structure of the disk and vertical oscillations changes the values of the increment, and the condition Im $(\omega) > 0$ remains valid, i.e. instability remains unchanged [261, 262, 765].

The dependence of viscosity and opacity $\varkappa$ on the surface density and disk thickness ($\nu \propto \sigma^{\delta_\sigma} h^{\delta_h}$, $\bar{\varkappa} \propto \sigma^{\Delta_h} h^{\Delta_h}$) has a controlling effect on the dissipative instabilities of the acoustic and also viscous and thermal (see section 11.3.1) branches of the oscillations. The currently available numerous models of accretion disks (AD) were constructed for both the stable and unstable regions with respect to the parameters Δ_h, Δ_σ, δ_h, δ_σ for different branches of the oscillations [262].

Instability of the shortwave acoustic waves

The model of the thin disk imposes constraints on the wavelength $\lambda \gg h$ and the frequency of perturbations [43]. Therefore, for efficient investigation of the dynamics of perturbations with $\lambda \lesssim h$ it is necessary to consider the vertical motion and the z-structure of the accretion disk. The thin disk model can be used to examine only the pinch-oscillations (S-mode). For these oscillations, the perturbed pressure is the symmetric function of the z-coordinate, the vertical displacement of the gas does not shift the centre of the mass in the disk in relation to the plane of symmetry $z = 0$. Thus, the bending oscillations (*AS*-mode) are not considered. In addition to this, the two-dimensional models do not make it possible to investigate the high-frequency (reflective) harmonics with $\omega \gg \Omega$. The dynamics of acoustic perturbations will be investigated taking vertical movement into account. Thus, we investigate the problem of the range of applicability of the thin disk model.

Determination of eigenfrequency

We investigate a mathematical model which can be used to determine the eigenfrequency for different unstable modes in the z-inhomogeneous viscous asymmetric disk [765].

It is assumed that the equilibrium velocity in the disk has only r- and φ-components: $\mathbf{u}_0(r, z) = (u_0, v_0, 0)$. For the components of the tensor of viscous stresses we accept [1147]:

$$\Pi_{ij} = -\alpha_{ij}\,\mathcal{P}, \tag{3.114}$$

where $i, j = (r, \varphi, z)$, here and further $\mathcal{P}$ denotes the volume pressure. Parameters α_{rz}, $\alpha_{\varphi z}$ and $\alpha_{r\varphi}$ determine the level of turbulence in the disk. The values of α_{rz} and $\alpha_{\varphi z}$ are determined by the shear mechanism of the flow in the z-direction, and $\alpha_{r\varphi}$ is associated with the differential nature of rotation in the disk plane and coincides with the α-parameter of the standard theory of disk accretion [272]. Taking into account the estimates $\alpha_{r\varphi}:\alpha_{rz}:\alpha_{\varphi z}:\alpha_{rr}:\alpha_{\varphi\varphi} \sim \left|\frac{r\,\partial\Omega}{\partial r}\right|:\left|\frac{\partial u_0}{\partial z}\right|:\left|\frac{r\,\partial\Omega}{\partial z}\right|:\left|\frac{\partial u_0}{\partial r}\right|:\left|\frac{u_0}{r}\right|$, $h \ll r$, we restrict ourselves in the equations of motion only to the component $\Pi_{r\varphi}$. It is assumed that the equilibrium balance of the forces is determined by the system of equations

$$\frac{v_0^2}{r} = \frac{1}{\varrho_0}\frac{\partial P_0}{\partial r} + \frac{\partial \Phi}{\partial r}, \quad u_0\frac{\partial r v_0}{r\partial r} = -\frac{\alpha}{\varrho_0}\frac{\partial r^2 P_0}{r^2 \partial r}, \quad \frac{\partial P_0}{\partial z} = -\varrho_0\frac{\partial \Phi}{\partial z}. \tag{3.115}$$

Equilibrium functions are presented in the form $f_0(r, z) = f_{01}(r) \cdot f_{02}(z)$, restricting ourselves to the model $\partial(\mathcal{P}/\varrho^{\ell})/\partial z = 0$. We use the solutions in section 3.6.1. The relationship

$$c_s^2(r,z) = \frac{\gamma P_0}{\varrho_0} = c_s^2(r,0)\, f(z), \quad c_s^2(r,0) = \frac{\gamma}{a}\Omega_K^2 h^2, \tag{3.116}$$

determines the adiabatic speed of sound (γ is the adiabatic index, in the general case $\gamma \neq \ell$). For equilibrium velocities v_0 and u_0 from equations (3.115) with (3.116) taken into account we obtain

$$v_0(r,z) = r\,\Omega(r,z) = \sqrt{r^2\Omega_K^2 + \ell_P\, c_s^2/\gamma}, \tag{3.117}$$

$$u_0(r,z) = -\frac{\alpha(2+\ell_P)}{\gamma(1+\ell_v)}\frac{c_s^2}{v_0}, \tag{3.118}$$

where the parameters $\ell_P = \partial(\ln P_0)/\partial(\ln r)$, $\ell_\varrho = \partial(\ln \varrho_0)/\partial(\ln r)$, $\ell_v = \partial(\ln v_0)/\partial(\ln r) = 1 - n$ characterise radial inhomogeneities of the equilibrium quantities.

After the standard linearisation procedure, we obtain a system of equations in relation to the perturbed quantities:

$$\frac{\partial\tilde{\varrho}}{\partial t} + \frac{\partial}{r\partial r}\left[r(u_0\tilde{\varrho} + \varrho_0\tilde{u})\right] + \frac{\partial(\varrho_0\tilde{w})}{\partial z} = 0, \tag{3.119}$$

$$\frac{\partial\tilde{u}}{\partial t} + u_0\frac{\partial\tilde{u}}{\partial r} + \tilde{u}\frac{\partial u_0}{\partial r} + \tilde{w}\frac{\partial u_0}{\partial z} - \frac{2v_0}{r}\tilde{v} = -\frac{1}{\varrho_0}\left(\frac{\partial\tilde{P}}{\partial r} - \frac{\tilde{\varrho}}{\varrho_0}\frac{\partial P_0}{\partial r}\right), \tag{3.120}$$

$$\frac{\partial\tilde{v}}{\partial t} + u_0\frac{\partial(r\tilde{v})}{r\partial r} + \tilde{u}\frac{\partial(rv_0)}{r\partial r} + \tilde{w}\frac{\partial v_0}{\partial z} = -\frac{1}{\varrho_0}\left(\alpha\frac{\partial(r^2\tilde{P})}{r^2\partial r} - \frac{\alpha\tilde{\varrho}}{\varrho_0}\frac{\partial(r^2P_0)}{r^2\partial r}\right), \tag{3.121}$$

$$\frac{\partial\tilde{w}}{\partial t} + u_0\frac{\partial\tilde{w}}{\partial r} = -\frac{\partial\tilde{P}}{\varrho_0\partial z} - g\frac{\tilde{\varrho}}{\varrho_0}, \tag{3.122}$$

here $g \equiv \partial\Phi/\partial z$.

Since we intend to investigate the dynamics of only acoustic oscillations, the dissipation in the energy equation can be ignored. Consequently, the thermal equation in the linear approximation can be written in the form of the law of conservation of entropy:

$$\frac{\partial\tilde{s}}{\partial t} + u_0\frac{\partial\tilde{s}}{\partial r} + \tilde{u}\frac{\partial s_0}{\partial r} + \tilde{w}\frac{\partial s_0}{\partial z} = 0, \tag{3.123}$$

where s_0 is the specific entropy of the equilibrium gas. The perturbation of entropy will be excluded from (3.1 to 3) using the equation of state $s = s(\mathcal{P}, \varrho)$ which has the following form in the linear approximation:

$$\tilde{s} = \left(\frac{\partial s}{\partial P}\right)_\varrho \cdot\tilde{\mathcal{P}} + \left(\frac{\partial s}{\partial \varrho}\right)_P \cdot\tilde{\varrho} = c_V\frac{\tilde{\mathcal{P}}}{\mathcal{P}_0} - c_P\frac{\tilde{\varrho}}{\varrho_0}, \tag{3.124}$$

where $c_V = T(\partial s/\partial T)_\varrho$, $c_P = T(\partial s/\partial T)_P$ are the specific heat capacities at constant density and pressure. We introduce the function of displacement of matter in the vertical direction :

$$w = \frac{d\xi}{dt} = \frac{\partial\xi}{\partial t} + u\frac{\partial\xi}{\partial r} + v\frac{\partial\xi}{r\partial\varphi}. \tag{3.125}$$

Taking into account the WKB-approximation in the radial direction ($kr \gg 1$, k is the radial wave number), the solution is presented in the form

$$\tilde{f}(r,z,t) = \hat{f}(z)\cdot\exp\{-i\,\omega t + i\,kr\}. \tag{3.126}$$

Taking into account (3.124) and (3.126), the system (3.119)–(3.123) is reduced to two conventional differential equations with respect to the amplitude of the perturbed pressure $\hat{\mathcal{P}}(z)$ and vertical displacement $\hat{\xi}(z)$ from the plane $z = 0$:

$$\frac{d\hat{\xi}}{dz} = \frac{D_1(\hat{\omega},k,z)}{\hat{\omega}(\hat{\omega}^2 - \varkappa^2)c_s^2} \frac{\hat{\mathcal{P}}}{\varrho_0} + \frac{g}{c_s^2} \cdot \hat{\xi}, \tag{3.127}$$

$$\frac{d\hat{P}}{dz} = \varrho_0 \left\{ \hat{\omega}^2 - \frac{g}{\gamma} S_z \right\} \cdot \hat{\xi} - \frac{g}{c_s^2} \cdot \hat{P}, \tag{3.128}$$

here $\hat{\omega} = \omega - ku_0$, $\varkappa = \Omega\sqrt{2(2-n)}$ is epicyclic frequency,

$$D_1 = -(\hat{\omega}^3 - \hat{\omega}(\varkappa^2 + k^2c_s^2) - 2i\alpha\Omega k^2 c_s^2),\ S_z = \frac{1}{c_V}\frac{\partial s_0}{\partial z} = (\ln \mathcal{P}_0)' - \ell(\ln \varrho_0)'.$$

The prime indicates differentiation with respect to the z-coordinate ($f' \equiv \partial f/\partial z$), is the complex amplitude of the material z-displacement from the equilibrium state, so that taking into account (3.125) $\tilde{w} = d\tilde{\xi}/dt = -i\hat{\omega}\tilde{\xi}$ is fulfilled. In a more general form, taking into account the radial inhomogeneities, the system of equations with $\hat{\xi}$, $\hat{\mathcal{P}}$ was published in [765]. It should be mentioned that the WKB-approximation along r requires that the coefficients in the equations of the type (3.127), (3.12) in the dissipationless limit are real. In the non-viscous approximation without the magnetic field and without taking rotation into account, we obtain the limiting transition to the equations investigated in [173].

The system of equations (3.127), (3.128) is supplemented by the boundary conditions (Fig. 3.1). We consider two types of oscillations: 1) symmetric oscillations ($\hat{\xi}$(z) = – $\hat{\xi}$(– z) or $\hat{\mathcal{P}}$ (z) = – $\hat{\mathcal{P}}$ (– z)) at which

$$\hat{\xi}(0) = 0 \ \text{or} \ \frac{d\hat{\mathcal{P}}}{dz}\bigg|_{z=0} = 0, \tag{3.129}$$

such oscillations are referred to as the pinch-mode, or S-mode; 2) antisymmetric oscillations ($\hat{\xi}$ (z) = $\hat{\xi}$ (– z) or $\hat{\mathcal{P}}$ (z) = – $\hat{\mathcal{P}}$ (– z)), at which

$$\hat{\mathcal{P}}(0) = 0 \ \text{or} \ \frac{d\hat{\xi}}{dz}\bigg|_{z=0} = 0, \tag{3.130}$$

which corresponds to the bending oscillations of the disk or the AS-mode. The following condition should be fulfilled on the surface of the disk in the linear approximation

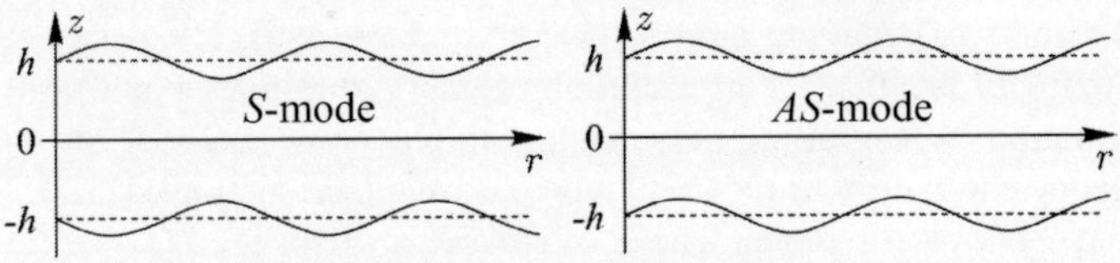

Fig. 3.1. Diagram of the structure of the S- and AS-modes.

$$\hat{\mathcal{P}}(h) + \frac{\partial \mathcal{P}_0}{\partial z}\bigg|_{z=h} \hat{\xi}(h) = 0. \tag{3.131}$$

In the case of the isothermal model, the density converts to 0 at infinity, and the boundary condition should be specified at $z \gtrsim 3 \div 4h$, which results in satisfactory accuracy.

Solving numerically the boundary problem for the system of equations (3.127), (3.128), with the boundary conditions (5.101) and (5.99) for the *S*-mode (or (5.100) for the *AS*-mode), we determine the eigenvalues of complex frequency ω for the given distributions of the equilibrium parameters of the disk along the z-coordinate. At the same time, we obtain the eigenfunctions $\hat{\xi}\,(z)$ and $\hat{\mathcal{P}}\,(z)$ which, taking into account (3.119)–(3.122), give the vertical dependences for all other perturbed quantities. The previously described scheme will be referred to as the '3-D model', and the thin disk model as the 2D-model. We shall use the dimensionless frequency $W = \omega/\Omega_K$ and the dimensionless wave number $K = kh$. If not specified otherwise, the parameters have the values: $\alpha = 0.2$; $\gamma = \ell = 5/3$; $h/r = 0.05$; $n = 3/2$.

3.3.2. Fundamental S- and AS-modes in 2D- and 3D-models

Initially, for the shortwave perturbations in the vertical direction we formally substitute $\hat{\mathcal{P}}$, $\hat{\xi} \propto \exp\,[ik_z z]$ into (3.127), (3.128) and, ignoring the inhomogeneity of the equilibrium quantities, we write the following dispersion equation:

$$\omega^4 - \omega^2\,[\varkappa^2 + c_s^2\,(k^2 + k_z^2)] - 2i\alpha\,\Omega c_s^2 k^2\,\omega + \varkappa^2 c_s^2\,k_z^2 = 0, \tag{3.132}$$

where k_z is the wave number in the z-direction. Naturally, equation (3.132) at $k_z = 0$ coincides with the result [262] for the 2D-model with the appropriate approximation in the case of replacement of γ by the flat adiabate index γ_s (3.88), (3.248), (11.97), and the square of the adiabatic speed of sound in the disk plane should be determined by the following expression:

$$c_s^2 = \gamma_s \frac{\int_0^h \mathcal{P}(z)\,dz}{\int_0^h \varrho(z)\,dz}. \tag{3.133}$$

Evidently, these oscillations correspond to the *S*-mode (5.99).

At $\alpha > 0$, equation (3.132) determines two unstable acoustic branches of the oscillations and two decaying modes (viscous and thermal). The decay of the latter is associated with the absence of radiation pressure, dissipation, and cooling in the thermal equation. When taking these factors into account, the viscous and thermal low-frequency oscillatory branches can be unstable (see section 11.3.1).

Figure 3.2 shows the dependences of $\partial W/\partial K$ on K in the 2D- and 3D-models, indicating that there is a small difference between them. The graph also gives the results of calculations with the flat adiabatic index

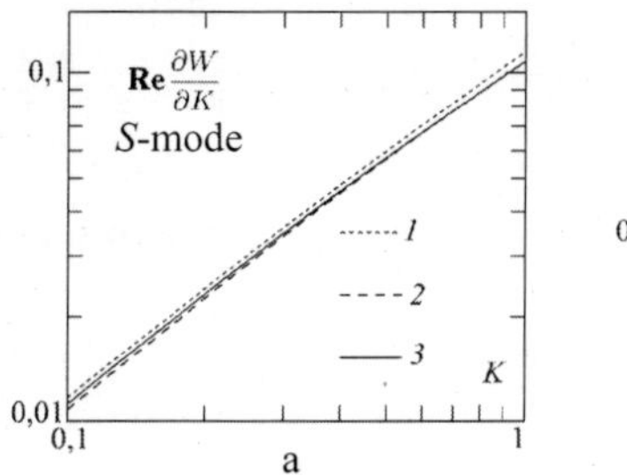

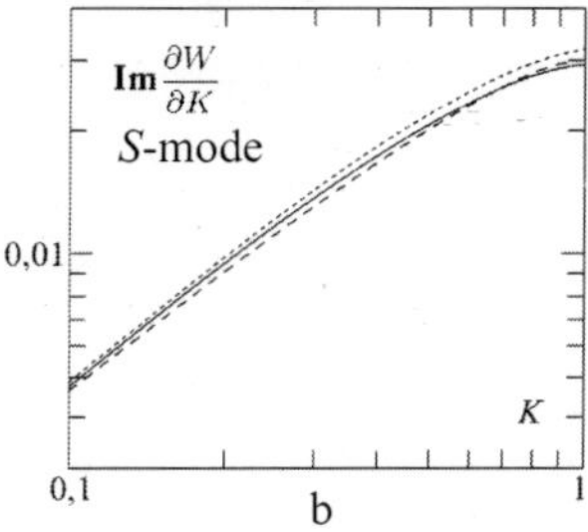

Fig. 3.2. Dependence of $\partial W/\partial$K on K at: 1) 2D-model with γ_s, calculated from (3.248); 2) 2D-model with $\gamma_s = \gamma_{s2}$; 3) 3D-model.

$$\gamma_{s2} = \frac{3\gamma - 1 - \gamma\,\delta}{1 + \gamma - \delta}(\delta = \varkappa^2 / \Omega_K^2 = 1)$$

obtained by more accurate consideration of the vertical movement using the hierarchic model (see [95], section 3.5.1). The linear dependence $\partial W/\partial K$ shows that $\omega \propto k^2$ in a wide range in accordance with the asymptotics (11.104).

Thus, direct comparison indicate satisfactory agreement between the model of the thin disk and the results of solving the 3D-problem. The eigenfrequencies of the oscillations, obtained from the equation (3.132) at $k_z = 0$ with (3.133) taken into account, and using the solution of the boundary problem (3.127)–(5.101), differ slightly in the range $kh \lesssim 3$ (Fig. 3.2). Large differences appear only at $kh \geqslant 3$. It should be mentioned that the formal condition of the applicability of the 2D-model ($kh \ll 4$) are knowingly violated in this region. In the case of small values of parameter γ differences appear at high kh since the characteristic scale of the inhomogeneity of the equilibrium parameters in the vertical direction increases with the reduction of γ [765]. In the shortwave range the increment and the phase velocity of the perturbations in the 3D-model are smaller than in the 2D-model. This is associated with the nonuniform z-distribution of the equilibrium quantities and the transverse gravitational force. The instability increment is linearly proportional to the α-parameter, and the value of the wave number k, at which differences appear between the models, depends only slightly on the value of the α-parameter.

The low-frequency mode discussed here $\left(\omega \simeq \sqrt{\Omega^2 + k^2 c_s^2}\right)$ depends only slightly on the z-structure, and it appears in both the 2D- and 3D-models. Consequently, it is referred to as the fundamental mode. In the case with $\alpha \ll 1$, equation (3.132) gives the following equation for the fundamental mode

$$\omega^2 = \varkappa^2 + k^2 c_s^2 + i\alpha \frac{\Omega c_s^2 k^2}{\varkappa^2 c_s^2 k^2}, \tag{3.134}$$

which is in agreement with the results of calculations carried out using the 3D-model.

Thus, the direct notification shows that 2D-model is suitable for investigating the dynamics of linear acoustic waves with the characteristic spatial scales of vertebrae shows in the disk plane in the range ~$(2h \div r)$.

3.3.3. *The special properties of high-frequency unstable modes*

The gas disk is a unique waveguide in which a discrete set of waves can propagate, with the waves differing in the vertical wavenumber, so that the integer or half-integer number of waves fits between $z = -h$ and $z = +h$. The vertical inhomogeneity of the equilibrium parameters results in a more complicated situation because the harmonic waves are not natural solutions of the boundary-value problem (3.127), (3.128).

In addition to the fundamental *S*- and *AS*-modes, the viscous accretion disks may always generate any number of unstable harmonics by the dissipative mechanism [765]. Figure 3.3 shows the dependences of the eigenfrequency W on the radial wave number K for the fundamental mode $j = 0$ and the first four deflecting harmonics $j = 1, 2, 3, 4$. Both the pinch-oscillations and bending modes are unstable. The increments of the investigated instabilities increase with the increase of k and reach the maximum value at some values of the wave number. The maximum Im (ω) is displaced into the region with shorter wavelength with the increase of the number of harmonics j, and the value of the maximum increases with the reduction of the characteristic scale of the perturbations in the z-direction. The increment for all the harmonics increases with increasing α-parameter, and Re (ω) is almost completely independent of α.

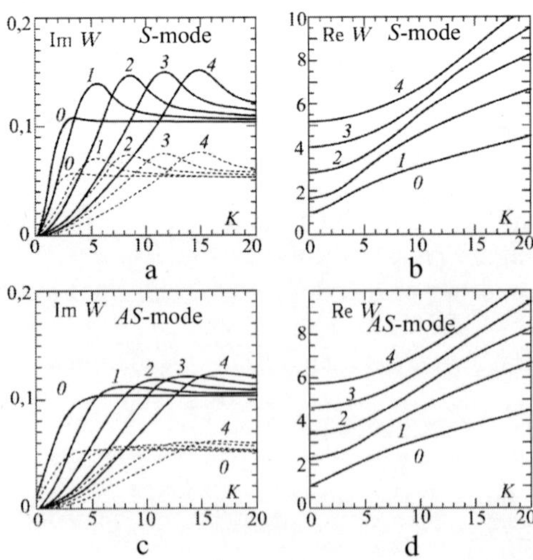

Fig. 3.3. Dependence of W on kh for *S*- (a, b) and *AS*-mode (c, d) at different values of α. The solid line corresponds to $\alpha = 0.2$, the dotted line to $\alpha = 0.1$. The numbers in the graph indicate the number of harmonics j. The solid and dotted lines in Figures b and d coincide.

For each harmonics j, the real part of the frequency of bending oscillations in the range of long waves is greater than the appropriate pinch mode, which is also indicated by the approximate dispersion equation $\omega^2 \simeq \varkappa^2 + (k^2 + k_z^2)\, c_s^2$ with $k_z h \simeq \pi j$ for the S-mode and $k_z h \simeq \pi(j + 1/2)$ for the AS-oscillations taken into account. On the whole, the pinch-perturbations are more unstable in comparison with the bending oscillations. At higher values of kh, the difference between the S- and AS-oscillations disappears. With increase of the number of harmonics j this takes place at high values of the radial wave number k. However, it should be mentioned that the very shortwave perturbations ($kh \geqslant 2 \div 10$) should be stable as a result of the presence of a radial gradient of the perturbed velocity in the viscous stress tensor which is not taken into account in the law (3.114). When this factor is taken into account, the maximum values of the increments decrease with increase of the number of harmonics j, and the waves with a very small scale (both with respect to r- and z-coordinate) should not be evident in the non-linear stage.

In this section we have demonstrated the principal possibility of the instability of the high-frequency acoustic waves in the differentially rotating gas disk for the running perturbations. They remain in the system for a limited period of time, and the presence of a positive increment does not yet mean that the perturbations will reach the non-linear stage of development. The perturbations propagating in the disk with the velocity $\partial\omega / \partial k \sim c_s kh / \sqrt{1 + k^2 h^2} < c_s$, and the characteristic holding time of the intensifying waves in the disk is equal to $\sim r/h \sim 10^2$ periods of disk rotation. If the estimate Im $(\omega) \sim 0.1\Omega$, is used, then at $kh > 1$ we obtain a large increase of the amplitude of the wave ($\exp\{0.1 \cdot r/h\}$). The non-axisymmetric oscillations may play a more significant role because they remain in the system for a longer period of time than the axisymmetric waves and, consequently, the non-linear stage of development may be safely reached (for more details see section 11.3.3).

The second and more important result is based on the direct comparison of the frequencies of oscillations, calculated using the thin disk model (TDM) with the exact solution which takes into account the vertical movement and the z-structure of the non-self-gravitating disk. Analysis shows that the TDM can be used to investigate not only the longwave acoustic perturbations but also study relatively short waves. Large quantitative differences appear only for the wave numbers $k \gtrsim 1/h$.

3.4. Integro-differential equations describing the dynamics of low amplitude waves symmetry in relation to the vertical plane

3.4.1. The linearised system of equations

The initial system of the equations is represented by the equations (3.90)–(3.24) from section 3.1, where the potential Φ is assumed to be the gravitational potential of the disk (and denoted below as Φ_d) associated with its density by the Poisson equation (3.24). Here we are interested in the waves generated in the region of the inner Lindblad resonance from a satellite and, therefore, the appropriate gravitational force of the satellite should be added to the right-hand part of the equations of motion.

One of the requirements essential for describing the dynamics of the disk in the two-dimensional approximation and described in the previous section, is the condition of slowness of the square of the eigenfrequency of the perturbations of the self-gravitating disk ω^2 in comparison with the square of the angular frequency of its rotation Ω_0^2, $\omega^2 \ll \Omega_0^2$. However, the gravitational–sound and acoustic waves are high-frequency waves and do not satisfy this condition. Consequently, these waves should be investigated using three-dimensional equations which is almost always the case. Linearising the initial system of the non-linear equations (3.19)–(3.22) from the previous section taking into account the action of the external satellite, we obtain

$$-i\hat{\omega}v_{r1}-2\Omega_0 v_{\varphi 1}=-(\mathcal{P}_1+\Phi_{d1}+\Phi_s)', \tag{3.135}$$

$$-i\hat{\omega}v_{\varphi 1}+\frac{\varkappa^2}{2\Omega_0}v_{r1}+\frac{\partial v_{0\varphi}}{\partial z}v_{z1}=-\frac{im}{r}(\mathcal{P}_1+\Phi_{d1}+\Phi_s), \tag{3.136}$$

$$-i\hat{\omega}v_{z1}=-\frac{\partial}{\partial z}(\mathcal{P}_1+\Phi_{d1}+\Phi_s), \tag{3.137}$$

$$-i\hat{\omega}\varrho_1+\frac{1}{r}(r\,\varrho_0 v_{r1})'+\frac{im}{r}\varrho_0\, v_{\varphi 1}+\frac{\partial}{\partial z}(\varrho_0\, v_{z1})=0. \tag{3.138}$$

Here the 'prime' denotes differentiation with respect to r, $\hat{\omega}\equiv\omega - m\Omega_0$. The stationary quantities are denoted by the index 0 and depend on both r and z[3]. The perturbed quantities are denoted by the index 1

$$A_1(r,\varphi,z,t)=\frac{1}{\sqrt{2\pi}}\int_{-\infty}^{+\infty}\sum_{m=-\infty}^{+\infty}A_{1m}(r,z,\omega)e^{i(m\varphi-\omega t)}d\omega. \tag{3.139}$$

The gravitational potential of the satellite Φ_s is determined as follows:

$$\Phi_s=-\frac{G\,M_s}{\sqrt{r^2+z^2+r_s^2-2rr_s\cos(\varphi-\Omega_s t)}}. \tag{3.140}$$

$$\Phi_s=\sum_{m=-\infty}^{+\infty}\Phi_{sm}\,e^{im\zeta}, \tag{3.141}$$

Denoting $\varphi-\Omega_S t\equiv\zeta$, the function $\Phi_s(r,\zeta,z)$ is expanded into a Fourier series remembering that the stationary quantities do not depend on the variable

$$\Phi_{sm}(r,z)=\frac{1}{2\pi}\int_{-\pi}^{\pi}\Phi_s(r,\zeta,z)e^{-im\zeta}d\zeta=-\frac{GM_s}{2\pi}\int_{-\pi}^{\pi}\frac{e^{-im\zeta}d\zeta}{\sqrt{r^2+z^2+r_s^2-2rr_s\cos\zeta}}. \tag{3.142}$$

---

[3]With the exception of Ω_0 which depends only on r (Poincaré).

$$\Phi_{sm}(r,z)=-\frac{GM_s}{\pi}\int_0^{\pi}\frac{\cos\zeta\, d\zeta}{\sqrt{r^2+z^2+r_s^2-2rr_s\cos\zeta}}. \tag{3.143}$$

Because of the linearity of the system (3.135)–(3.16), we select one of the harmonics ~ $\Phi_{sm}(r, z)\exp[im(\varphi-\omega_s t)]$, omitting the index m like at the amplitudes of other functions $Asm(r, z, \omega)$. Subsequently, when considering the wave, generated on the inner Lindblad resonance m: $(m-1)$, naturally, as the harmonics m we select the one which corresponds to this resonance.

It is assumed that all the potentials are functions of three spatial coordinates, $\mathcal{P}$ is the pressure function, determined from equation (3.50): $\mathcal{P}(P) = \int dP/\varrho$. If in accordance with equation (3.62) we introduce the assumption of the polytropic relationship between the volume pressure P_V and the density ϱ: $P_V = A_V\varrho^{\gamma V}$, where A_V, γ_V are constants, γ_V is the 'volume' exponent of the polytrope, then in accordance with (3.63) we obtain $\mathcal{P} = A_V\gamma_V(\gamma_V-1)^{-1}\varrho^{\gamma V-1}$.

Introducing the speed of sound using the standard relationship $c^2 \equiv (\partial P_V/\partial\varrho)_{\varrho=\varrho_0} = A_V\gamma_V\varrho_0^{\gamma_V-1}$, we obtain the perturbed pressure function:

$$\mathcal{P}_1=\mathcal{P}-\mathcal{P}_0=\left(\frac{\partial\mathcal{P}}{\partial\varrho}\right)_{\varrho=\varrho_0}(\varrho-\varrho_0)=A_V\gamma_V\varrho_0^{\gamma_V-2}\varrho_1, \tag{3.144}$$

or in the final form

$$\mathcal{P}_1=\frac{c^2\varrho_1}{\varrho_0}. \tag{3.145}$$

The quantity $\partial v_{0\varphi}/\partial z$ is equal to 0 according to the Poisson theorem (see, for example, the book The Theory of Rotating Stars, by Jean-Louis Tassoul).

Omitting in the equation (3.136) the small term with v_{z1} in the first two equations of motion we have two velocity constants v_{z1} and $v_{\varphi 1}$ which are expressed by the potentials

$$v_{r1}=\frac{-i\hat{\omega}(\mathcal{P}_1+\Psi_{d1}+\Psi_s)'+i2m\Omega_0(\mathcal{P}_1+\Psi_{d1}+\Psi_s)/r}{\hat{\omega}^2-\varkappa^2}, \tag{3.146}$$

$$v_{\varphi 1}=\frac{1}{\hat{\omega}^2-\varkappa^2}\left[-\frac{\varkappa^2}{2\Omega_0}(\mathcal{P}_1+\Psi_{d1}+\Psi_s)'+\frac{m\hat{\omega}}{r}(\mathcal{P}_1+\Psi_{d1}+\Psi_s)\right], \tag{3.147}$$

The last two equations can be written in a more compact form:

$$v_{r1}=\frac{1}{D}\left[-i\hat{\omega}\chi'+i\frac{2m\Omega_0}{r}\chi\right], \tag{3.148}$$

$$v_{\varphi 1}=\frac{1}{D}\left[-\frac{k^2}{2\Omega_0}\chi'+\frac{m\hat{\omega}}{r}\chi\right], \tag{3.149}$$

Or we can use the notations

$$\chi \equiv c^2 \frac{\varrho_1}{\varrho_0} + \Phi_{d1} + \Phi_s , \tag{3.150}$$

$$D \equiv \hat{\omega}^2 - k^2 . \tag{3.151}$$

Substituting the expressions for v_{r1} and $v_{\varphi 1}$ into the continuity equation (3.138), it can be seen that this equation, after several transformations, can be presented in the form

$$\hat{\omega}\left[\varrho_1 + \frac{1}{r}\left(\frac{r\varrho_0 \chi'}{D}\right)' - \frac{m^2 \varrho_0}{r^2 D}\chi\right] - \frac{2m}{r}\left(\frac{\Omega_0 \varrho_0}{D}\right)' \chi + i\frac{\partial}{\partial z}(\varrho v_{z1}) = 0 . \tag{3.152}$$

Integrating the last equation with respect to z:

$$I \equiv \int_0^z \left\{ \hat{\omega}\left[\varrho_1 + \frac{1}{r}\left(\frac{r\varrho_0 \chi'}{D}\right)' - \frac{m^2 \varrho_0}{r^2 D}\chi\right] - \frac{2m}{r}\left(\frac{\Omega_0 \varrho_0}{D}\right)' \chi \right\} dz = -i\varrho_0(z) v_{z1}(z). \tag{3.153}$$

Here we restrict ourselves to considering the symmetric ϱ_1, $v_{\varphi 1}$ and v_{R1} and non-symmetric v_{z1} perturbations with respect to the plane $z = 0$ which means in particular $v_{z1}(0) = 0$. In other words, here we are not interested in the bending oscillations for which $v_{z1}(0) \neq 0$. Multiplying the left and right parts of (3.153) by $\hat{\omega}$ and using the equation of motion (3.137), we obtain

$$-i\,\hat{\omega} v_{z1}(r,z) = \frac{I}{\varrho_0}\,\hat{\omega} = -\frac{\partial}{\partial z}\chi . \tag{3.154}$$

Integrating equation (3.154) with respect to z from 0 to z, and using the definition (3.150), we determine the perturbation of the volume density at any point z:

$$\varrho_1(r,z) = \frac{\varrho_0(r,z)}{c^2(r,z)}\left[-\Phi_{d1}(r,z) + \Phi_{d1}(r,0) + c^2(r,0)\frac{\varrho_1(r,0)}{\varrho_0(r,0)} - \int_0^z \frac{\hat{\omega} I}{\varrho_0(r,z)} dz \right] . \tag{3.155}$$

As indicated by (3.140), the difference $\Phi_s(z) - \Phi_s(0)$ is small:

$$\Phi_s(r,\varphi,z,t) - \Phi_s(r,\varphi,0,t) = -\frac{\Phi_s(r,\varphi,0,t)\, z^2}{2\,|\vec{r} - \vec{r}_s|^2} , \tag{3.156}$$

so this difference in (3.155) can be ignored. The linearised Poisson equation (3.24) and equation (3.155) represent a close system of integro-differential equations in relation to two functions $\varrho_1(z)$ and $\Phi_1(z)$.

3.4.2. The dispersion equation in the form of the boundary value problem for the integral equation

The identical problem arises in the investigation of the gravitational–acoustic waves in the stratified atmosphere [836]. The atmosphere may act as a one-sided waveguide in which the changes of pressure above the surface of the Earth are transferred most efficiently by low-frequency perturbations. However, the simple dispersion relationship, describing the strictly one-dimensional propagation of sound, cannot be used in this case because the speed of sound changes with altitude.

To solve a problem similar to this, the authors of [836] proposed the iteration method which will also be used here. As the first approximation, we use the solution obtained by ignoring z-motions, i.e. it is assumed that $v_{z1} = 0$. According to (3.154), both $I = 0$ and the independence of χ on variable z correspond to this approximation. In this case, from (3.155) we obtain

$$c^2(r,z)\frac{\varrho_1(r,z)}{\varrho_0(r,z)}+\Phi_{d1}(r,z)=c^2(r,0)\frac{\varrho_1(r,0)}{\varrho_0(r,0)}+\Phi_{d1}(r,0). \tag{3.157}$$

The solution of the Poisson equation taking into account the independence of χ on z, which also reflects the last inequality, is described in section 3.4.3.

Determining the function $\varrho_1(z)$ and $\Phi_{d1}(z)$ of the first approximation of the iteration and substituting them into the equation of motion (3.155), we obtain the functions $\varrho_1(z)$ and $\Phi_{d1}(z)$ of the second approximation, and so on.

If we are interested in the dispersion equation for the free wave or for a wave generated in the range of resonance from a satellite, this can be obtained by integrating equation (3.155) from one edge of the disk to another. The following zero boundary conditions [836] should be used in this case for z-flows

$$(\varrho_0 v_{z1})_{-h} = (\varrho_0 v_{z1})_h = 0. \tag{3.158}$$

The need for using (3.158) can be easily understood by considering the contradiction. Let the vertical streams at the disk boundaries be finite. If the density of the disk is equal to 0, this indicates the presence of infinite velocities v_{z1} which leads to an absurd result – infinite energy of vertical movement at the disk boundaries [836]: $(\varrho_0 v_{z1}^2/2)_{\pm h} \to \infty$

We integrate (3.152) with respect to z from $-h$ to h. Taking (3.158) into account we obtain

$$\int_{-h}^{+h}\left\{\hat{\omega}\left[D\varrho_1+\frac{D}{r}\left(\frac{r\varrho_0\chi'}{D}\right)'-\frac{m^2}{r^2}\varrho_0\chi\right]-\frac{2mD}{r}\left(\frac{\varrho_0\Omega_0}{D}\right)'\chi\right\}dz=0. \tag{3.159}$$

The resultant equation together with the boundary conditions with respect to r represents the boundary-value problem for the determination of the frequency of eigenoscillations $\hat{\omega}$. For the short wavelength perturbations, equation (3.159) is reduced directly to the dispersion equation. To indicate this, we analyse in greater detail the nature of the dependence on coordinate z along the axis of rotation of different terms, included in the subintegrand expression.

According to the Poisson theorem, quantity D does not depend on z. This was already taken into account because when deriving (3.159) we integrated equation (3.152) multiplied by D. It should also be mentioned that as a result of this we cannot use the resultant relationships at any of the points of the Lindblad resonance.

As shown previously, the gravitational potential of the satellite Φ_s is characterised by the same negligible dependence on z.

The density, pressure and speed of sound change in a general case on the scale of the order of half thickness of the disk. An exception is the case of specified equations of state – for example, the density is constant for an incompressible disk, and the speed of sound for an isothermal disk.

To determine the strength of the dependence of the perturbation of the gravitational potential of the disk Φ_{d1} on z, it is necessary to discuss is greater detail the solution of the Poisson equation for this case.

3.4.3. Solution of the Poisson equation for a finite thickness disk

A disk of finite half thickness h can be regarded as a set of infinitely thin layers situated at distances z_i from the central density of the disk $z = 0$ and having the perturbed surface density $d\sigma_i = \varrho(z_i)\,dz$. The perturbed surface density of the entire disk[4] is $\sigma = \sum d\sigma_i = \int_{-h}^{h} d\sigma$. Thus, because of the linearity of the Poisson equation, the problem of the potential of the disk of finite thickness is reduced to the problem of the potential of an infinitely thin layer.

Outside the i-th layer, the potential, generated by this layer, is described by the Laplace equation

$$\left[\frac{\partial^2}{\partial r^2}+\frac{1}{r}\frac{\partial}{\partial r}+\frac{1}{r^2}\frac{\partial^2}{\partial\varphi^2}+\frac{\partial^2}{\partial(z-z_i)^2}\right]\Phi_i=0. \tag{3.160}$$

The partial solution of the Laplace equation outside the disk (3.160) will be determined in the multiplicative form:

$$\Phi_i(r,\varphi,z)=R(r)F(\varphi)Z_i(z)=R(r)Z_i(z)e^{im\varphi}. \tag{3.161}$$

Substituting (3.161) into (3.160), we obtain

$$Z_i\left[\frac{\partial^2 R}{\partial r^2}+\frac{1}{r}\frac{\partial R}{\partial r}-\frac{m^2}{r^2}R\right]+R\frac{\partial^2 Z_i}{\partial(z-z_i)^2}=0. \tag{3.162}$$

The last equation will be solved by the method of separation of variables

$$\frac{1}{R}\left[\frac{\partial^2 R}{\partial r^2}+\frac{1}{r}\frac{\partial R}{\partial r}-\frac{m^2}{r^2}R\right]=-\frac{1}{Z_i}\frac{\partial^2 Z_i}{\partial(z-z_i)^2}=-k^2,\quad k^2>0. \tag{3.163}$$

The separation constant is selected negative to satisfy the homogeneous boundary conditions at $z\to\pm\infty$.

The last equation

[4]In contrast to the rest of the text, here the index 1 at the perturbed quantities will be omitted, and the unperturbed values will be denoted by index 0 as previously.

$$\frac{\partial^2 Z_i}{\partial(z-z_i)^2} - k^2 Z_i = 0 \tag{3.164}$$

as a general solution

$$Z_i(z) = A_1\, e^{-k(z-z_i)} + A_2 e^{k(z-z_i)}, \tag{3.165}$$

which, with the homogeneous boundary conditions taken into account, can be written in the form

$$Z_i(z) = A_1 e^{-k|z-z_i|},\ z \neq z_i. \tag{3.166}$$

From (3.160) we also obtain the equation for the cylindrical functions

$$\frac{\partial^2 R}{\partial x^2} + \frac{1}{x}\frac{\partial R}{\partial x} + \left[1 - \frac{m^2}{x^2}\right] R = 0, \tag{3.167}$$

whose general solution is

$$R = C_1 J_m(x) + C_2 Y_m(x), \tag{3.168}$$

where $x \equiv kr$; J_m, Y_m is the Bessel function of the first and second kind[5]. Since $Y_m(x) \to \infty$ at $x \to 0$, it is necessary to assume that $C_2 = 0$.

Thus, the partial solution of the Laplace equation outside the i-th layer is

$$\Phi_i(r,\varphi,z) = A_i\, J_m(kr)\, e^{im\varphi - k|z-z_i|},\ z \neq z_i. \tag{3.169}$$

The constant A_i is determined from the condition of cross-linking this potential with the potential in the layer, determined by the Poisson equation

$$\left[(\nabla^2)_\perp + \frac{\partial^2}{\partial(z-z_i)^2}\right]\Phi_i^{in} = 4\pi G\, d\sigma_i\, \delta(z-z_i). \tag{3.170}$$

Potential Φ_i should be a continuous function at any point z, in particular, at $z = z_i$, otherwise the z-th component of the force will be infinite. This means that at $z = z_i$ $\Phi_i = \Phi_i^{in}$. Substituting Φ_i^{in} in (3.172) by equation (3.169) we obtain

$$-2\delta(z-z_i)\,|k|\, A_i\, J_m(kr)\, e^{im\varphi - k|z-z_i|} = 4\pi G\, d\sigma_i(r,\varphi)\,\delta(z-z_i), \tag{3.171}$$

which gives the relationship between constant A_i and the perturbation of the density of the layer:

$$A_i\, J_m(kr)\, e^{im\varphi} = -\frac{2\pi G}{|k|} d\sigma_i(r,\varphi). \tag{3.172}$$

Substitution of (3.172) into (3.169) gives the final expression for the potential Φ_i:

$$\Phi_i(r,\varphi,z) = -\frac{2\pi G\, d\sigma_i}{|k|} e^{-k|z-z_i|}. \tag{3.173}$$

[5]The latter is also referred to the Webber function.

Adding up the contributions of all infinitely thin layers, we obtain the gravitational potential of the initial layer with finite thickness

$$\Phi(r,\varphi,z)=\sum_i \Phi_i(r,\varphi,z)=-\frac{2\pi G}{|k|}\int_{-h}^{h} e^{-k|z-z_i|}d\sigma_i =$$

$$=-\frac{2\pi G}{|k|}\int_{-h}^{h} e^{-k|z-z_i|}\varrho(z_i)\,dz_i=-\frac{2\pi G\sigma}{|k|}(1-\alpha(|z|)kh), \tag{3.174}$$

where α is the quantity of the order of unity:

$$\alpha(|z|)=\frac{1}{h\sigma}\int_{-h}^{h}|z-z_i|\varrho(z_i)\,dz_i\,. \tag{3.175}$$

Using the parity $\varrho(z)$ (it should be mentioned that here we consider only the perturbations symmetric in relation to $z = 0$), the last integral can be transformed to a more convenient form:

$$\int_{-h}^{h}|z-z_i|\,\varrho(z_i)dz_i = \int_{z}^{h}(z_i-z)\varrho(z_i)dz_i + \int_{-h}^{z}(z_i-z)\varrho(z_i)dz_i = \int_{0}^{h}(z_i-z)\varrho(z_i)dz_i +$$

$$+2\int_{0}^{z}(z_i-z)\varrho(z_i)dz_i + \int_{-h}^{0}(z_i-z)\varrho(z_i)dz_i = 2\int_{0}^{h}z_i\varrho(z_i)dz_i + \int_{0}^{z}\sigma_{zi}dz_i\,. \tag{3.176}$$

here $\sigma_{z_i} = \int_{-z_i}^{z_i}\varrho(z_i)\,dz_i$ is the surface density of the part of the disk restricted by the planes $z = \pm z_i$. The first term in the last equation is the value of the integral at $z = 0$, and the second one is evidently the even function of z. Finally, the expression for assumes the form:

$$\alpha(\tau)=1-\int_{\tau}^{1}f(x)\,dx, \tag{3.177}$$

where $\tau = |z|/h$ and we introduce the function $f(x) = \sigma_x/\sigma$, which changes from 0 to unity.

This equation gives the variation of the quantity α in the half thickness of the disk

$$\Delta\alpha=\int_{0}^{1}f(x)\,dx<1. \tag{3.178}$$

Correspondingly, the variation of the perturbed potential of the disk with variation of z from 0 to h

$$|\Delta\Phi| < 2\pi G|\sigma|h \approx kh|\Phi(z=0)| \ll |\Phi(z=0)|. \tag{3.179}$$

Thus, the change of the potential in the half thickness of the disk is very small.

Differentiation of the resultant expressions for the potential gives the following equation for the derivatives of the potential along the axis of rotation

$$\frac{\partial\Phi}{\partial z}=2\pi G\sigma_z=4\pi G\int_0^z \varrho dz, \tag{3.180}$$

$$\frac{\partial^2\Phi}{\partial z^2}=4\pi G\varrho. \tag{3.181}$$

Thus, according to the order of magnitude we have

$$\left|\frac{\partial\Phi}{\partial z}\right|\simeq k|\Phi(z=0)|\simeq\left|\frac{\partial\Phi}{\partial r}\right|, \tag{3.182}$$

$$\left|\frac{\partial^2\Phi}{\partial z^2}\right|\simeq\frac{k}{h}|\Phi(z=0)|\simeq\frac{1}{kh}\left|\frac{\partial\Phi}{\partial r}\right|. \tag{3.103}$$

3.4.4. Dynamics of waves in the disk plane. Transition to the traditional dispersion equation

We transfer to derivation, by equation (3.159), of the dispersion equation for the waves in the disk plane assuming for simplicity that the initial density of the disk is constant along the radius ($\varrho_0(r)$ = const). Ignoring the terms of the order $kd \ll 1$, instead of the exact values of the gravitational potentials we can use their values in the central plane of the disk

$$\Psi_1(r,\varphi,t)\equiv\Phi_{d1}(r,\varphi,0,t),\ \ \Psi_s(r,\varphi,t)\equiv\Phi_s(r,\varphi,0,t). \tag{3.184}$$

Taking this into account, equation (3.159) becomes

$$\int_{-h}^{h}\left\{\hat{\omega}\left[D\varrho_1+\frac{D}{r}\left(\frac{r(c^2\varrho_1+\varrho_0\Psi_1+\varrho_0\Psi_s}{D}\right)'-\frac{m^2}{r^2}\left(c^2\varrho_1+\varrho_0\Psi_1+\varrho_0\Psi_s\right)\right]-\right.$$
$$\left.-\frac{2mD}{r}\left(\frac{\Omega_0}{D}\right)\left(c^2\varrho_1+\varrho_0\Psi_1+\varrho_0\Psi_s\right)\right\}dz=0. \tag{3.185}$$

Integrating with respect to z gives

$$\hat{\omega}\left[D\sigma_1+\frac{D}{r}\left(\frac{r(\overline{c^2}\sigma_1+\sigma_0\Psi_1+\sigma_0\Psi_s)'}{D}\right)'-\frac{m^2}{r^2}(\overline{c^2}\sigma_1+\sigma_0\Psi_1+\sigma_0\Psi_s)\right]-$$
$$-\frac{2mD}{r}\left(\frac{\Omega_0}{D}\right)'(\overline{c^2}\sigma_1+\sigma_0\Psi_1+\sigma_0\Psi_s)dz=0. \tag{3.186}$$

with the following notations

$$\sigma_1 \equiv \int_{-h}^{h} \varrho_1 \, dz, \quad \overline{c^2} \equiv \frac{1}{\sigma} \int_{-h}^{h} c^2 \varrho_1 \, dz. \tag{3.187}$$

In the main order of the WKB-approximation, equation (3.186) is greatly simplified: we can ignore the derivatives from the non-perturbed quantities and use the simple local relationship between ψ_1 and σ_1 from (3.174):

$$\Psi_1 = -\frac{2\pi G\sigma}{|k|}. \tag{3.188}$$

In the absence of a satellite, we obtain the following dispersion equation:

$$\hat{\omega}\left[\hat{\omega}^2 - \varkappa^2 + 2\pi G\sigma_0 |k| - k^2 \overline{c^2}\right] = 0. \tag{3.189}$$

It may be seen that the equation obtains two branches of intrinsic perturbations. The first branch, $\hat{\omega} = 0$, describes the indifferent equilibrium. However, in the presence of dissipation in this branch there may be instabilities leading to the exponential growth of the perturbations, having a negative energy (see [43, 170, 577, 590], see chapter 8).

The second branch – gravitational–sound – is also well-known, with the exception of the fact that the square of the speed of sound is determined not through the flat pressure P_s: $c^2 = \partial P_s/\partial\sigma$, and is the weighted mean of the square of the conventional (three-dimensional) speed of sound in the thickness of the disk (see (3.187)). The weight function is represented by the profile of the perturbed volume density of the disk, i.e. for different perturbations $\overline{c^2}$ will, generally speaking, differ. An exception is the case of an isothermal disk where $\overline{c^2} = c^2 = \text{const}$ for all perturbations.

3.4.5. The role of perturbations along the axis of rotation

It was shown previously that the equations of dynamics of astrophysical disks cannot be investigated in the two-dimensional formulation. In this section, using the linearised equations, the problem of the role of perturbations along the axis of rotation can be investigated in greater detail. The analysis described below confirms the conditions determining the previous section for the spatial and time scales of the perturbations which can be investigated in the two-dimensional formulation. For small amplitude perturbations, these conditions have the form $(kh)^2 \ll 1$ and $\hat{\omega} \ll \Omega_0^2$. In addition, the analysis makes it possible to evaluate the possibility of investigating the viscous defects in the two-dimensional formulation.

In the WKB-approximation ($k^2r^2 \gg m^2$), the continuity equation (3.152) and the equation of motion (3.137) along the z axis have the form

$$\hat{\omega}\left[\varrho_1 + \frac{k^2\varrho_0}{D}\left(c^2\frac{\varrho_1}{\varrho_0} + \Phi_1\right)\right] + i\frac{\partial \varrho_0 v_{z1}}{\partial z} = 0, \tag{3.190}$$

$$i\hat{\omega}v_{z1} = \frac{\partial}{\partial z}\left(c^2\frac{\varrho_1}{\varrho_0} + \Phi_1\right). \tag{3.191}$$

For the problem discussed here, the presence of a satellite is not significant and, therefore, we accept that $\Phi_s = 0$. Together with the Poisson equation, these equations form the closed system of equations for three unknown functions ϱ_1, Φ_1, and v_{z1}. In particular, the fact that it is necessary to take into account the last of these variables, i.e. take into account the movement along the axis of rotation, makes the given equations virtually three-dimensional.

As indicated by the equations (3.190) and (3.191), the role of motion along the axis of rotation is dual. In the equation of motion (3.191), the term v_{z1} describes the impact of inertial effects associated with the z-th motions. It is clear that these effects are not important if the frequency of perturbations is low. This is the reason why the low-frequency condition $\hat{\omega}^2 \ll \Omega_0^2$ forms in the first section.

The term v_{z1} in the continuity equation (3.190) describes the transfer of mass along the z-axis, i.e. the redistribution of the mass, caused by movement along the axis of rotation. Since the remaining terms in the continuity equation are proportional to frequency, the contribution of the z-th motions can be ignored only for high-frequency processes.

Thus, any effect, determined by the motion along the axis, is always important. The perturbations in which the role of v_{z1} would be negligible strictly speaking do not exist, with the exception of possibly some specific cases.

We determine the conditions for the frequency of perturbations in which one of the factors of the effect of vertical motion is not important.

Conditions for ignoring mass transfer along the axis of rotation

Ignoring the mass transfer along the axis of rotation, from equation (3.189), taking into account the dispersion equation, we obtain the following equation for the density perturbations

$$\varrho_1 = \varrho_0\frac{k^2\Phi_1}{k^2\left(\overline{c^2} - c^2\right) - 2\pi G\sigma_0 |k|}, \tag{3.192}$$

and for the perturbed potential

$$\chi = c^2\frac{\varrho_1}{\varrho_0} + \Phi_1 = \frac{D}{k^2}\frac{\varrho_1}{\varrho_0}. \tag{3.193}$$

Substituting the value of the perturbed density from (3.192) into equation (3.193), we obtain

$$\chi = c^2\frac{\varrho_1}{\varrho_0} + \Phi_1 = \frac{D\Phi_1}{k^2\left(\overline{c^2} - c^2\right) - 2\pi G\sigma_0 |k|}. \tag{3.194}$$

When this expression is substituted into the equation of motion (3.191), it is important to take into account that the function in (3.194) which varies most markedly with respect to z is c^2. Consequently, we obtain the expression for the velocity of z-th motion:

$$v_{z1}=-\frac{i}{\hat{\omega}}\frac{k^2 D\Phi_1}{\left[k^2\left(\overline{c^2}-c^2\right)-2\pi G\sigma_0|k|\right]^2}\frac{\partial(c^2)}{\partial z}\sim$$

$$\sim\frac{k^2 D\Phi_1 c^2}{\hat{\omega}h\left[k^2\left(\overline{c^2}-c^2\right)-2\pi G\sigma_0|k|\right]^2}\sim\frac{Dc^2}{\hat{\omega}2\pi G\sigma_0|k|h}\frac{\varrho_1}{\varrho_0}. \tag{3.195}$$

Since $D\sim 2\pi G\sigma_0\,|k|$, the quantity v_{z1} is finally of the following order:

$$v_{z1}\sim\frac{c^2}{\hat{\omega}h}\frac{\varrho_1}{\varrho_0}. \tag{3.196}$$

We now estimate the order of the term describing the mass transfer along the axis which we have ignored

$$\frac{\partial}{\partial z}(\varrho_0 v_{z1})\sim\frac{\partial}{\partial z}\left[\frac{c^2}{\hat{\omega}h}\varrho_1\right]\sim\frac{c^2}{\hat{\omega}h^2}\varrho_1. \tag{3.197}$$

The condition of ignoring the mass transfer along the z-axis, as indicated by the continuity equation (3.190), is

$$\frac{\partial}{\partial z}(\varrho_0 v_{z1})\ll\hat{\omega}\varrho_1. \tag{3.198}$$

From this, using (3.197), we obtain

$$\frac{c^2}{h^2}\ll\hat{\omega}^2. \tag{3.199}$$

The resultant lower limit for frequency can be transformed using the equilibrium conditions of the disk. For a disk in the external field

$$h=\frac{c}{\Omega_0}, \tag{3.200}$$

and from (3.199) we obtain

$$\hat{\omega}^2\gg\Omega_0^2. \tag{3.201}$$

For a self-gravitating disk

$$\frac{c}{h}\sim\Omega_0\sqrt{\frac{R}{h}}, \tag{3.202}$$

and we have

$$\hat{\omega}^2\gg\Omega_0^2\frac{R}{h}. \tag{3.203}$$

The conditions (3.201), (3.203) are sufficiently rigid. An important exception is the case of an isothermal disk $c^2 = c^2 =$ const. We will discuss this approximation in a greater detail.

Isothermal disk

For this case, ignoring mass transfer along the axis of rotation, equation (3.194) for the potential χ is simplified as follows:

$$\chi = c^2 \frac{\varrho_1}{\varrho_0} + \Phi_1 = -\frac{D\Phi_1}{2\pi G\sigma_0 |k|}. \tag{3.204}$$

Consequently, from the equation of motion (3.191):

$$v_{z1} = \frac{iD}{2\pi G\sigma_0 |k| \hat{\omega}} \frac{\partial \Phi_1}{\partial z}. \tag{3.205}$$

Using (3.182) and (3.174), as regards the order of magnitude we have:

$$v_{z1} \sim \frac{D}{\hat{\omega}k} \frac{\sigma_1}{\sigma_0} \sim \frac{D}{\hat{\omega}k} \frac{\varrho_1}{\varrho_0}. \tag{3.206}$$

Now, instead of (3.197), we have the following expression for the isothermal disk

$$\frac{\partial}{\partial z}(\varrho_0 v_{z1}) \sim \frac{\partial}{\partial z}\left[\frac{D\varrho_1}{\hat{\omega}k}\right] \sim \frac{D\varrho_1}{\hat{\omega}hk}. \tag{3.207}$$

When obtaining the last estimate it was taken into account that the strongest dependence on z is shown by the function ϱ_1 – the function changes at half thickness h by the very value.

Substituting (3.207) into the condition for ignoring mass transfer along z (3.208) we obtain

$$D \ll \hat{\omega}^2 kh. \tag{3.208}$$

$$-2\pi G\sigma_0 |k| + k^2 c^2 \ll \hat{\omega}^2 kh. \tag{3.209}$$

In the case in which the contributions of the acoustic effects and of self-gravitation are comparable, $2\pi G\sigma_0 |k| \sim k^2 c^2$, we obtain

$$k^2 c^2 \ll \hat{\omega}^2 kh. \tag{3.210}$$

Two cases will be investigated:

1. A disk in an external field. Using the equilibrium condition (3.200), from (3.210) we obtain the condition for ignoring the mass flow in the disk in the external fields:

$$\Omega^2 kh \ll \hat{\omega}^2. \tag{3.211}$$

It can be seen that this condition can be fulfilled.

2. A self-gravitating disk. Using the equilibrium equation (3.202), from (3.210) we determine the conditions for ignoring the mass flow in the self-gravitating disk:

$$\Omega^2 k R \ll \hat{\omega}^2 . \tag{3.212}$$

Since $kR \gg 1$, this condition cannot be fulfilled for the long wavelength perturbations $kh \ll 1$.

Comparison of the condition for neglecting the mass flow in the general case of the self-gravitating disk (3.203) with the identical conditions for the partial case of the isothermal self-gravitating disk shows that in the latter case the conditions for neglecting the longwave perturbations are 'softer'. Nevertheless, as mentioned previously, these conditions cannot be fulfilled.

The conditions for ignoring the inertia term in the *z*-th equation of motion (the condition for ignoring the oscillations along the axis of rotation)

We now determine, more strictly than previously, the condition for ignoring the inertia term in the *z*-th equation of motion (3.191). It is evident that this is the approximation of low frequencies. Thus, ignoring the left-hand part of equation (3.188) in comparison with the right-hand part, we obtain

$$c^2 \frac{\varrho_1}{\varrho_0} + \Phi_1 = \chi(r, \varphi, t), \tag{3.213}$$

where the function χ is independent of z. The equation (3.190) can now be written in the following form

$$\frac{\partial}{\partial z}(\varrho_0 v_{z1}) = i\hat{\omega}\left[\varrho_1 - \frac{k^2}{D}\varrho_0\,\chi\right]. \tag{3.214}$$

As indicated by (3.213) is in the general case, the dependence of ϱ_1 on z should not coincide with the dependence ϱ_0. Actually, as indicated by (3.177), the value Φ_1 changes only slightly along the thickness of the disk. Consequently, the first term in the left-hand part of (3.230) is almost completely independent of z. The variation of c^2 in the thickness of the disk by the same value (in a general case) confirms the initial claim.

The disagreement in the general case of the dependences of the functions ϱ_1 and ϱ_0 on z shows that the difference in the round brackets in the right-hand part of 3.214) cannot be considerably smaller than both terms at all z. Consequently, the value of the left-hand part of the inequality (3.214) can be evaluated using, for example, the first term in the right-hand part of (3.214):

$$v_{z1} \sim \hat{\omega} h \frac{\varrho_1}{\varrho_0}. \tag{3.215}$$

Knowing v_z, it is now possible to determine the condition for ignoring the inertia force in (3.191) in comparison with the pressure force:

$$\hat{\omega}^2 \ll \frac{c^2}{h^2}. \tag{3.216}$$

The condition (3.216) corresponds to the resultant constraint (3.46) on the frequency (characteristic time) of the processes which can be described by the two-dimensional equations.

The only exception is the case of the self-gravitating disk with the polytropy exponent $\gamma = 2$ in a cylindrical external potential (the role of the latter has already been discussed). For this case, the profile of disk density in the direction of z has the form (see previously, [628])

$$\varrho_0(z) = \varrho_c \cos\left(\sqrt{\frac{2\pi G}{A_V}}\, z\right). \tag{3.217}$$

Let the profile of perturbed density have the same form:

$$\varrho_1(z) = \varrho_{1c} \cos\left(\sqrt{\frac{2\pi G}{A_V}}\, z\right). \tag{3.218}$$

For such profile ϱ_1 we have

$$\sigma_z = \int_{-z}^{z} \varrho_1(x)dx = 2\varrho_{c1}\sqrt{\frac{A_V}{2\pi G}} \sin\left(\sqrt{\frac{2\pi G}{A_V}}\, z\right). \tag{3.219}$$

From the equations (3.174) and (3.175) we obtain the following equation for the perturbed disk potential

$$\Phi_1(r,\varphi,z) = \text{const} - \varrho_c c^2 \frac{\varrho_1}{\varrho_0}. \tag{3.220}$$

Thus, for the investigated case, the conditions ϱ_1/ϱ_0 = const and γ = const are compatible and mass transfer along the axis of rotation is negligible.

Application of two-dimensional models of astrophysical disks

Thus, strictly speaking, complete ignorance of the contribution of the vertical motion is possible in only two specific cases: an isothermal disk in a dominant external field, and a self-gravitating disk with the entropy exponent $\gamma = 2$. In other cases, the perturbation of the disk leads to excitation of movement along the z-axis, and the effect of this motion on the disk dynamics requires separate investigation.

3.5. Quasi-two-dimensional hierarchic model of the disk

3.5.1. Equations for the vertical moments

The vertical motions can be taken into account within the framework of the so-called hierarchic quasi-two-dimensional model which was formulated in a study by Kovalenko and Lukin [95]. This approach is based on excluding the vertical component by averaging the three-dimensional equations with additional evolution equations taken into account. The velocity

$$\vec{\mathbf{u}}(r,\varphi,z,t)=\sum_{n=0}^{\infty}\vec{\mathbf{u}}^{(n)}(r,\varphi,t)\frac{z^n}{n!}. \qquad (3.221)$$

will be expanded into a series.

It is convenient to operate with the velocity component in the disk plane **v** and vertical velocity w separately:

$$\vec{\mathbf{v}}=\sum_{n=0}^{\infty}\vec{\mathbf{v}}^{(n)}(r,\varphi,t)\frac{z^n}{n!},\quad w=\sum_{n=0}^{\infty}w^{(n)}(r,\varphi,t)\frac{z^n}{n!}. \qquad (3.222)$$

We determine the integral moments of the density and pressure of the n-th order:

$$\sigma^{(n)}=\int_{-h_-}^{h_+}\varrho(r,\varphi,z,t)\frac{z^2}{n!}dz,\quad P^{(n)}=\int_{-h_-}^{h_+}P(r,\varphi,z,t)\frac{z^2}{n!}dz. \qquad (3.223)$$

For the the moments of the zero order we have the surface density $\sigma\equiv\sigma^{(0)}$ and flat pressure $p\equiv\mathcal{P}^{(0)}$. Additionally multiplying the total three-dimensional hydrodynamics equation by $z^n/n!$ at different n and integrating with respect to z, we obtain infinite chains of linked equations for the moments $\sigma^{(n)}$ (r,φ,t), $\mathcal{P}^{(n)}(r,\varphi,t)$ and the values $\mathbf{u}^{(n)}$ (r,φ,t). It is assumed that the following conditions are satisfied at the disk boundaries h_-, h_+:

$$\varrho(z=h_\pm)=0,\ \mathcal{P}(z=h_\pm)=0. \qquad (3.224)$$

At the same time, the effect of the intergalactic gas is not taken into account.

After integrating the continuity equation (3.1) with respect to z from h_- to h_+ we obtain

$$\frac{\partial\sigma}{\partial t}+\nabla_{\|}\left(\sigma\vec{\mathbf{v}}^{(0)}+\sigma^{(1)}\vec{\mathbf{v}}^{(1)}+\ldots\right)=0. \qquad (3.225)$$

This equation contains, strictly speaking, the infinite number of vertical momentsof density and velocity. If the equation (3 .1) is multiplied by z and integrated from h_- to h_+, we obtain the evolution equation for the first density moment:

$$\frac{\partial\sigma^{(1)}}{\partial t}+\nabla_{\|}\left(\sigma^{(1)}\vec{\mathbf{v}}^{(0)}+2\sigma^{(2)}\vec{\mathbf{v}}^{(1)}+\ldots\right)-\left(\sigma w^{(0)}+\sigma^{(1)}w^{(1)}+\ldots\right)=0, \qquad (3.226)$$

and also similarly for the second moment (2):

$$\frac{\partial\sigma^{(2)}}{\partial t}+\nabla_{\|}\left(\sigma^{(2)}\vec{\mathbf{v}}^{(0)}+3\sigma^{(3)}\vec{\mathbf{v}}^{(1)}+\ldots\right)-\left(\sigma^{(1)}w^{(0)}+2\sigma^{(2)}w^{(1)}+\ldots\right)=0, \qquad (3.227)$$

and subsequent values $\sigma^{(n)}$.

Additionally multiplying the equations (3.2) by z^n, after integration we again obtain an infinite chain of the equations for the vertical moments of velocities. The identical effects gives the chain of equations for the energy (3.3) at the dissipationless limit.

It was shown in [95] that if the system is interrupted at some n, this approach can be used to investigate the dynamics of fine scale, in the z-direction, perturbations, selecting n from the formulation of the problem and the required accuracy of the solution.

3.5.2. Taking into account vertical motion in the first approximation

In the approximation following the hydrostatic approximation which already takes into account the vertical movement, there are additional evolution equations for the vertical moments of density and pressure (3.223) and also the components of expansion of velocity (3.221). Within the framework of the formulated hierarchy approach we can write the evolution equations in the first approximation for the gas disk symmetric in relation to the plane $z = 0$ for the value σ, p, $u = u(0)$, $v = v(0)$, $\Sigma = \sigma^{(2)}$, $W = w(1)$:

$$\frac{\partial \sigma}{\partial t} + \nabla(\sigma \mathbf{u}) = 0, \tag{3.228}$$

$$\frac{\partial \Sigma}{\partial t} + \nabla(\Sigma \mathbf{u}) - 2\Sigma W = 0, \tag{3.229}$$

$$\sigma\left\{\frac{\partial \mathbf{u}}{\partial t} + (\mathbf{u}\nabla)\mathbf{u}\right\} = -\nabla p - \sigma \nabla \Phi_0 - \Sigma \nabla \Phi^{(2)}, \tag{3.230}$$

$$\Sigma\left\{\frac{\partial W}{\partial t} + (\mathbf{u}\nabla)W\right\} + 2\Sigma W^2 - \frac{p}{2} = -\Sigma \Phi^{(2)}, \tag{3.231}$$

$$\frac{\partial p}{\partial t} + \mathbf{u}\nabla p + \gamma p(\nabla)\mathbf{u}) + 2(\gamma - 1)\Sigma W \Phi^{(2)} = 0, \tag{3.232}$$

where the nabla operator in the polar coordinate system is equal to $\nabla = \left\{\frac{\partial}{\partial r}, \frac{\partial}{r\partial \varphi}\right\}$ and when writing (2.232) we use the equation of state of the ideal gas. This system of equations for the given potential is closed. Functions $\Sigma(r, \varphi, t)$ and $W(r, \varphi, t)$ are determined by the vertical movements in the disk.

Thus, remaining within the framework of the quasi-two-dimensional model, the hierarchic method can be used to describe the dynamics of thin gas disks taking into account the effect of vertical motion when the characteristic time to establishment of the vertical hydrostatic equilibrium is comparable or greater than the period of oscillations in the disk plane.

3.6. Dynamics of gas disks in the longwave approximation

3.6.1. The simplest model of the thin non-self-gravitating gas disk

In accordance with the considerations made in the sections 3.1 and 3.2 in formal replacement in the hydrodynamics equations of the volume quantities ϱ and $\mathcal{P}$ by σ and p, respectively, we lose important terms which are not generally small. Here, we construct the simplest model which will be used as an example for demonstrating the appearance of additional terms in the equations for velocity and energy.

We consider a thin gas disk is a gravitational potential

$$\Psi(\mathbf{r},z) = \Phi(\mathbf{r}) + \frac{1}{2}\Omega_z^2(\mathbf{r})z^2, \tag{3.233}$$

where $\mathbf{r}$ is the radius vector in the disk plane ($z = 0$). For a thin gas disk, situated in the field of a central object with mass M_1, the parameter Ω_z is the Kepler angular velocity $\Omega_z^2 = GM_1/r^3$. Equation (3.233) also holds for the gas disk in the gravitational field of a more massive and thick stellar galactic disk. We use the van der Voort model of the axisymmetric stellar disk in which the potential is determined by the equation (2.44). Since $\Omega_z^2 = \partial^2\Psi/\partial z^2|_{z=0}$, then without considering the halo, we obtain $\Omega_z = \sqrt{2\pi G\sigma_*/\Delta_*}$ (σ^* is the surface density of the stellar disk, Δ_* is its half thickness). The quantity Ω_z is equal to the characteristic frequency of oscillations of the stars across the plane of the stellar disk. The dispersion of the velocities of the start across the disk plane c_z and the half thickness of the stellar disk Δ_* are considerably greater than the speed of sound c_s and half thickness of the gas disk h, respectively (chapter 1). The difference in the surface densities of the stellar and gas disks can be even greater. Consequently, the thin gas disk is situated in the gravitational field of the almost uniform distribution of matter in the stellar disk in the region $z \ll \Delta_*$ (see equation (2.43)).

The model of the thin gas disks assumes the existence at every moment of time of hydrostatic equilibrium in the vertical direction[6]

$$\frac{\partial\mathcal{P}}{\partial z} = -\rho\frac{\partial\Psi}{\partial z}, \tag{3.234}$$

or, taking (3.233) into account

$$\frac{\partial\mathcal{P}}{\partial z} = -\Omega_z^2\,\rho z. \tag{3.235}$$

To integrate (3.235), it is necessary to take into account the structure of the disk in the z direction which is determined by the equation of state $\mathcal{P} = \mathcal{P}(\rho, T)$ and energy transfer.

---

[6]At every moment of time along the z-coordinate the disk should manage to arrange itself with respect to the changes taking place in the disk plane. However, as indicated by section 3.3, even for acoustic waves ($\omega \sim \Omega$) under specific conditions, the 2D-model may give sufficiently accurate results. In the points 7.8.3, 8.4.6 and 10.5.3 it is shown that the most characteristic large-scale peculiarities of the highly non-linear dynamics of the gas disk, including the shockwaves, coincide in the 2D- and 3D-models.

Estimates will be obtained on the basis of modelling considerations. For the polytropic law $\mathcal{P}(z) \propto [\rho(z)]^{\ell}$ equation (3.235) gives:

$$P(z) = P(0)\left[1 - A(\ell)\frac{z^2}{h^2}\right]^{\ell/(\ell-1)}, \quad \rho(z) = \rho(0)\left[1 - A(\ell)\frac{z^2}{h^2}\right]^{1/(\ell-1)}, \tag{3.236}$$

where

$$h^2 = B(\ell)P(0)/\rho(0)\Omega_z^2, \tag{3.237}$$

where A and B depend on parameter ℓ. Using the definition of the surface density σ and pressure p, equation (3.237) can be written in the form

$$p = 2hP = C_\ell^2\,\Omega_z^2\,\sigma h^2, \tag{3.238}$$

where if (3.236) for the dimensionless parameter C_ℓ^2 is fulfilled, we obtain

$$C_\ell^2 = \int_0^1 (1 - z^2)^{\ell/(\ell-1)}\,dz \Big/ B(\ell)\int_0^1 (1 - z^2)^{1/(\ell-1)}\,dz.$$

At $\ell \geqslant 3/2$, the density at the points $z = \pm h$ converts to 0, and if $\ell < 3/2$ the condition $\int_0^h \rho\,dz \gg \int_h^\infty \rho\,dz$ is fulfilled. The quantity h in this definition is naturally assumed to be equal to half thickness of the disk. At the limit $\ell \to \infty$ we obtain $\rho = \text{const}$ at $|z| \leqslant h$ and $C_2^\ell = 1/3$. In the other limiting case $\ell = 0$ the pressure and density are proportional to $\exp(-2z^2/h^2)$ and $C_2^\ell = 1/4$. The disruption of the polytropic law in the vertical direction (associated with the gradient and/or convective heat transfer, fine-scale magnetic fields and turbulence, ionisation of matter, etc.) may result in the relationship (3.238) in the dependence of parameter C_ℓ on the spatial coordinate r, φ. It should be mentioned that the equation (3.23) forms the basis of the α-model of accretion disks (chapter 11).

It will also be assumed that the velocity in the disk plane $\mathbf{u} = \{u, v\}$ does not depend on the z-coordinate. The law of conservation of mass has the form

$$\frac{\partial \sigma}{\partial t} + \nabla_\perp(\sigma \mathbf{u}) = 0, \tag{3.239}$$

where $\nabla_\perp$ is the nabla differential operator in the disk plane.

The Euler equation is integrated

$$\varrho\left(\frac{\partial \mathbf{u}}{\partial t} + (\mathbf{u}\nabla)\mathbf{u}\right) = -\nabla P - \varrho\nabla\Psi \tag{3.240}$$

with respect to the z-coordinate. Consequently, taking into account (3.233) and the condition $\mathcal{P}(z = \pm h) = 0$ we obtain the equation

$$\frac{\partial \mathbf{u}}{\partial t} + (\mathbf{u}\nabla_\perp)\mathbf{u} = -\frac{\nabla_\perp p}{\sigma} - \nabla_\perp\Phi - \frac{\nabla_\perp\Omega_z^2}{2\sigma}\int \varrho z^2\,dz. \tag{3.241}$$

For the given the distribution $\varrho(z)$ for the integral in the last term of equation (3.241) we can accept $\int \varrho z^2 dz \propto h^2\sigma$. Taking into account the equation (3.238), the equation (3.241) can be written in the form

$$\frac{\partial \mathbf{u}}{\partial t} + (\mathbf{u}\nabla_\perp)\mathbf{u} = -\frac{\nabla_\perp p}{\sigma} - \nabla_\perp \Phi - D\frac{p}{\sigma}\nabla_\perp \ln(\Omega_z). \tag{3.242}$$

The parameter $D = \Omega_z^2 \int_{-h_\pm}^{h_\pm} \varrho z^2\, dz \Big/ \int_{-h_\pm}^{h_\pm} P\, dz$ is determined by the dependence of the thermodynamic parameters on the z-coordinate. For example, in the case of (3.236) it can easily be shown that $D = 1$.

The equation of the radial equilibrium of the gas stationary axisymmetric disk has the form

$$\frac{V_{\mathrm{Bp}}^2}{r} - \frac{d\Phi_0}{dr} - \frac{dp_0}{\sigma_0 dr} - D\frac{p_0}{\sigma_0}\frac{d\ln\Omega_z}{dr} = 0, \tag{3.243}$$

where p_0, σ_0 is the equilibrium surface pressure and gas density. It should be mentioned that the third and fourth term give the correction $\propto (h/r)^2 \ll 1$, with the exception of possibly the regions of the rapid variation of the distribution of the density or angular velocity of rotation (see sections 1.2.3 and 1.3.2). Equation (3.242) differs from the traditional equation (3.14).

3.6.2. The energy conservation law

The system (3.239), (3.242) is supplemented by the energy conservation law. Taking into account that the variation of internal energy ε takes place as a result of the work of pressure forces, we obtain

$$\varrho\frac{d}{dt}\frac{\varepsilon}{\varrho} = \frac{\mathcal{P}}{\varrho}\frac{d\varrho}{dt}, \tag{3.244}$$

where $\frac{d}{dt} = \frac{\partial}{\partial t} + \mathbf{u}\nabla_\perp + w\frac{\partial}{\partial z}$ is the z-component of velocity. The equation (3.244) is integrated with respect to the z-coordinate in the range from $-h(\mathbf{r}, t)$ to $h(\mathbf{r}, t)$. As previously, it is assumed that the disk is symmetric in relation to the plane $z = 0$; ($\mathcal{P}(-z) = \mathcal{P}(z)$; $\varepsilon(-z) = \varepsilon(z)$; $w(-z) = -w(z)$); $\mathcal{P}(\pm h) = \varepsilon(\pm h) = 0$. Thus, we restrict ourselves to considering the movement of the interlayer where both boundaries are situated in the counter-phase, and the centre of the mass does not move in relation to the plane $z = 0$. Taking into account that

$$w(z = h) = \frac{dh}{dt} = \left(\frac{\partial}{\partial t} + \mathbf{u}\nabla_\perp\right)h,$$

we obtain the equation [261]

$$\frac{\partial (Eh)}{\partial t} + P\frac{dh}{dt} + \nabla_\perp(\mathbf{u}h(E+P)) - \mathbf{u}\nabla_\perp(Ph) = 0, \tag{3.245}$$

where E (**r**, t) is the mean density of thermal energy in the layer, $P = p/2h$ is the mean pressure in the layer. Here we restrict ourselves to analysis of the model of the ideal gas[7] which gives a simple relationship between energy and pressure:

$$E = \frac{P}{\gamma - 1}, \tag{3.246}$$

where γ is the 'volume' exponent of the adiabate.

Using the relationships (3.238), (3.246) and the continuity equation (3.239), it is quite easy to write equation (3.245) in relation to flat pressure

$$\frac{\partial p}{\partial t} + \gamma_s p \nabla_\perp \mathbf{u} + \mathbf{u}\nabla_\perp p = (\gamma_s - 1) p(\mathbf{u}\nabla_\perp)\ln(C_\ell \Omega_z), \tag{3.247}$$

where the quantity

$$\gamma_s = 1 + 2\,(\gamma - 1)/(\gamma + 1) \tag{3.248}$$

plays the role of the 'flat' exponent of the adiabate. It should be stressed that the combination (3.248) in equation (3.247) formed in a natural manner in deriving this equation (see also equation (3.88)). The right-hand part of (3.247) converts to 0 only at the isothermal limit ($\gamma = \gamma_s = 1$) or for $C_\ell \Omega_z$ = const.

The 'flat' exponent of the adiabate

When using the model of the thin disk it is important to consider the problem of the relationship of the 'volume' γ and 'surface' γ_s adiabate exponent. The problem can be formulated as follows. Let the polytropic equation of state $\mathcal{P} = A\varrho^\gamma$ be given. Consequently, what do the value s in the 'flat' polytropic equation of state $p = B\sigma^{\gamma_s}$.

For the gas disk which is not situated in the field of any other gravitating masses, this problem was solved by Hunter [698]:

$$\gamma_s = 3 - 2/\gamma. \tag{3.249}$$

Actually, assuming $p = \hat{C}\, A^\alpha G^\beta \sigma^{\gamma_s}$ where $\hat{C}$ and G are respectively the dimensionless and gravitational constants, from the dimensional considerations we obtain $\alpha = 1/\gamma$, $\beta = (\gamma - 1)/\gamma_s = 3 - 2/\gamma$. In the range $1 \leqslant \gamma \leqslant 2$, the quantity $\gamma_s \geqslant \gamma$. Assuming that for the 'macroatoms'– clouds system $\gamma = 5/3$, from (3.249) we obtain $\gamma_s = 9/5$.

In the second limiting case in which the gas disk is immersed into a considerably more massive stellar disk ($\varrho_* \gg \varrho_g$), it should be assumed that $p = \hat{C}\, A^\alpha (G\varrho_*)^\beta (\sigma_g)^{\gamma_s}$ [271]. According to the dimensional considerations, this equation leads to the previously derived relationship (3.248). In the case of an arbitrary relationship between the volume densities of the homogeneous spheroidal stellar halo and the gas disk, the relationship between γ and γ_s was determined by Abramyan [1].

Equation (3.247) shows that in a general case ($\gamma_s \neq 1$, $C_\ell \Omega_z \neq$ const) it can be assumed that the equality p/σ^{γ_s} = const is fulfilled. The presence of the right-hand part in equation (3.247) leads to the non-adiabatic state for the planar quantities p and σ which can be easily understood referring to the relationship (3.238) which according to its meaning is the equation of state for the flat layer because it links p, σ and h. The half thickness h plays the role of temperature. In the case in which

[7]Generalisation for the case of radiation pressure is described in [261], section 11.3.

$C_\ell\Omega_z(r,\varphi) \neq$ const we obtain the explicit dependence in the equation of state on the spatial coordinates and this also indicates the non-adiabatic state of the model. It should be mentioned that the discussed results can be easily obtained if we transfer in the equation for entropy $S \propto \ln(P/\bar{\rho}^{\gamma})$ to the surface values $p = 2hP$, $\sigma = 2h\bar{\rho}$, with equation (3.238) taken into account. We have $S \propto [\ln(p/\sigma^{\gamma_s}) - (\gamma_s - 1)\ln(C_\ell\Omega_z)]$ which also indicates the non-adiabatic nature of the thin disk for the case in which the quantity $C_\ell\Omega_z$ is inhomogeneous [253].

In conclusion, it would be useful to examine a simple approach which can be used to generalise the equation (3.248) for γ_s with additional factors taken into account. We obtain the relationship between γ and γ_s in the first approximation for the thin gas layer in the outer gravitational field with the dynamic (p_{gas}) and radiation (p_{rad}) pressures taken into account at the arbitrary value of the parameter $\beta \equiv p_{rad}/(p_{rad} + p_{gas})$ which is of interest for the physics of accretion flows in the vicinity of compact objects (chapter 11).

The mean density of the thermal energy in the layer $E(r, \varphi, t)$ is linked with the gas $P_g(r, \varphi, t)$ and radiation pressure $P_r(r, \varphi, t)$ by the relationship (compared with (3.246)):

$$E = \frac{P_{gas}}{\gamma - 1} + 3P_{rad} = \frac{1 + \beta(3\gamma - 4)}{\gamma - 1} P, \tag{3.250}$$

where $\beta = P_{rad}/P$ is the fraction of radiation pressure in the total pressure $P = P_{rad} + P_{gas}$. For the plasma, consisting of completely ionised hydrogen, in the case of local thermodynamic equilibrium $P_{gas} = \dfrac{2k_B}{m_p}\rho T$, $P_{rad} \simeq \dfrac{1}{3}aT^4$ (here k_B is the Boltzmann constant, m_p is the proton mass, a is the radiation constant). Equation (3.3) is integrated with respect to the z-coordinate and, as in the case of (3.245), we obtain

$$\frac{\partial(Eh)}{\partial t} + P\frac{dh}{dt} + \nabla\left[\mathbf{u}h(E + P)\right] - \mathbf{u}\nabla(Ph) = 0. \tag{3.251}$$

The law of conservation of energy (3.251) with the allowance made for (3.250) can be written in the following form in relation to the flat pressure:

$$\frac{dp_{gas}}{dt} + \left(1 + 2\frac{\gamma - 1}{\gamma + 1}\right)p_{gas}\nabla\mathbf{u} + 7\frac{\gamma - 1}{\gamma + 1}\left\{\frac{dp_{rad}}{dt} + \frac{9}{7}p_{rad}\nabla\mathbf{u}\right\} =$$

$$= 2\frac{\gamma - 1}{\gamma + 1} p(\mathbf{u}\nabla)\ln(C_\ell\Omega_z). \tag{3.252}$$

In the limiting case $p_{rad} \ll p_{gas}$, equation (3.252) results in the equation (3.248). In the inverse limit $p_{rad} \gg p_{gas}$, from the fact that the expression in the braces in (3.252) is equal to 0, it follows that $\gamma_s = 9/7$. Since for the radiation $\gamma = 4/3$, the equation (3.248) remains valid.

At an arbitrary value of the parameter $\beta = p_{\rm rad}/p$, the expression for the flat exponent of the adiabate can be derived for shortwave perturbations ($C_\ell \Omega_z = \text{const}$), considering the dynamics of linear waves. For the local system of the coordinates for the perturbations of the type exp $\{-i\omega t + ik_x x + ik_y y)$ we can write the linearised continuity equation, Euler equation and (3.252):

$$-\omega\tilde{\sigma} + \sigma_0 (\mathbf{k}\tilde{\mathbf{u}}) = 0, \tag{3.253}$$

$$-\omega\sigma_0 (\mathbf{k}\tilde{\mathbf{u}}) + k^2 \tilde{p} = 0, \tag{3.254}$$

$$(A_2 - 2A_1)\frac{\tilde{\sigma}}{\sigma_0} - A_2 \frac{\tilde{p}}{p_0} + 2\frac{A_3}{\omega}(\mathbf{k}\tilde{\mathbf{u}}) = 0, \tag{3.255}$$

here $\mathbf{k} = \{k_x, k_y\}$ is the wave vector, the symbol ~ denotes the amplitudes of the perturbed quantities, $A_1 = B_0 + 4(3\gamma - 4)\beta_0^2$, $A_2 = B_0(1 + \gamma) + (3\gamma - 4)\beta_0(9 - \beta_0)$, $A_3 = [\beta_0(3\gamma - 4) + \gamma]B_0$, $B_0 = 1 + 3\beta_0$.

The condition of existence of the non-trivial solutions of the system (3.253)-(3.255) results in the dispersion equation for the sound waves $\omega^2 = c_s^2 k^2$, where the velocity of propagation of the wave $c_s^2 = \gamma_s p_0/\sigma_0$, and the quantity

$$\gamma_s = 1 + 2\frac{A_3 - A_1}{A_2} \tag{3.256}$$

plays the role of the effective flat exponent of the adiabate and corresponds to the standard definition $\gamma_s = \left(\dfrac{\partial \ln p}{\partial \ln \sigma}\right)_s$.

At $\beta_0 = 1$ and any $\gamma > 1$, equation (3.256) gives $\gamma_s = 9/7$ (Fig. 3.4). For $\gamma > 4/3$ taking into account the radiation pressure ($\beta_0 > 0$) results in the monotonic decrease of γ_s. At $\gamma \leqslant 4/3$ the condition $\gamma_s \leqslant 9/7$ is fulfilled, and the dependence $\gamma_s(\beta_0)$ is not monotonic and has a minimum. This result is identical with the well-known effects: the speed of sound in a liquid containing gas bubbles can be lower than the speed of sound in the gas and the speed of sound in the liquid [185]. At $\gamma \to 1$, the presence of the range $\gamma_s < 0$ indicates the formation of unstable solutions when the surface pressure decreases with increasing surface density.

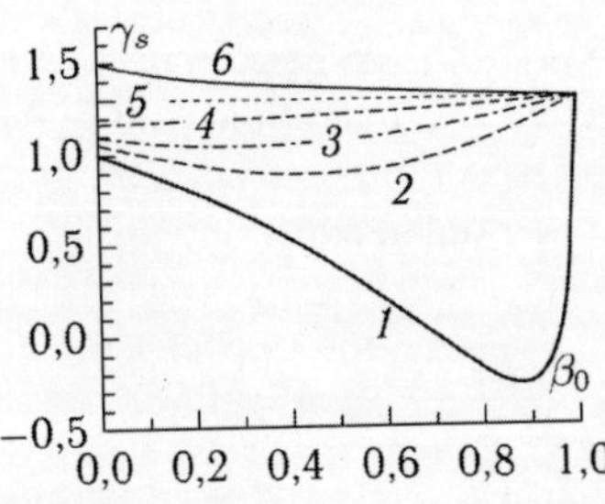

Fig. 3.4. Dependence $\gamma_s(\beta_0)$ at fixed values of γ: $\gamma = 1.001$ (1); 1.02 (2); 1.1 (3); 1.2 (4); 4.3 (5); 5.3 (6).

3.6.3. When can the gas disk be regarded as thin?

The determination of the stability of the real gas disks (gas subsystems of the galactics, accretion and protoplanetary disks, the ring-shaped systems of the planets, etc.) as the simplest investigation of the possible mechanisms of the evolution is unavoidably associated with the construction of relatively simple models. The simplest and, naturally, most popular is the model of an infinite thin disk, i.e. a disk whose half thickness h is small in comparison with the scale of the perturbations (instabilities) in which we are interested: $kh \ll 1$. It was already shown by Gildreich and Lynden–Bell [618] that in the models of the self-gravitating (i.e. the gas disk compressed across its plane only by the gravitational field generated by it) gas disk the wavelengths $kh \sim 1$ are most unstable from the viewpoint of gravity. This means that the model of the infinite thin disk for the self-gravitating gas systems cannot be used for the most unstable perturbations. Does this mean that we should use only the model of a finite thickness, and the investigation of the stability of such a model is a more time-consuming problem? [205, 270]. If the answer to this question can be negative, it is evident that this can be so only if certain conditions which were formulated in [43, 204] are fulfilled.

We examine the equilibrium of a system consisting of a gas disk immersed in a stellar disk. The volume densities of these subsystems are regarded as greatly differing:

$$\rho_{g0} \ll \rho_{*0}, \tag{3.257}$$

so that we can ignore in the first approximation the contribution of ρ_{g0} to the Poisson equation. It is assumed that the system is elongated in its plane to the extent that the following condition

$$|d^2\Phi_0 / dz^2| \gg |d(rd\Phi_0 / dr)/rdr|.$$

is fulfilled. Because of this, the Poisson equation assumes the form (2.21). The equilibrium condition of the stellar component along the z-axis and the equation of state $\mathcal{P}_{*z_0} = \mathcal{P}_{*z_0}(\rho_0)$ has the form

$$\frac{d\Phi_0}{dz} = -c_z^2 \frac{d\ln(\rho_{*0})}{dz}, \tag{3.258}$$

$c_z^2 = d\mathcal{P}_{*z_0} / d\rho_{*0}$ is the square of the dispersion of the velocities of the stars across the disk plane. Assuming that the condition $|d\ln(c_z^2)/d\ln(\rho_{*0})| \ll 1$ is fulfilled, differentiating (3.258) with respect to z, and comparing the results with the Poisson equation (2.21), we obtain the Emden equation

$$\frac{d^2u}{dz^2} + \frac{4\pi G}{c_z^2}\exp\{u\} = 0 \tag{3.259}$$

for the function $u(z) = \ln\{\rho_{*0}(z)/\rho_{*0}(0)\}$. The solution of this equation has the form

$$\rho_{*0}(z) = \rho_{*0}(z=0)/\mathrm{ch}^2(z/\Delta_*), \tag{3.260}$$

where

$$\Delta_* = c_z / \sqrt{2\pi G \rho_{*0}(z=0)} = c_z^2 / \pi G \sigma_{*0} \tag{3.261}$$

coincides with (2.40).

For the gas subsystem with the equation of state

$$P_0 = P_0(\rho_{g0}) \tag{3.262}$$

the equilibrium equation along the z-axis has the form

$$\frac{d\Phi_0}{dz} = -c_s^2 \frac{d\ln(\rho_{g0})}{dz}, \tag{3.263}$$

where $c_s^2 = dP_0 / d\rho_{g0}$. The left-hand parts of (3.263) and (3.258) are equal; equating the right-hand parts, we obtain

$$\rho_{g0}(z) = \rho_{g0}(0)[\mathrm{ch}^2(z/\Delta_*)]^{-c_z^2/c_s^2}. \tag{3.264}$$

In the galactic disks $c_z^2 / c_s^2 \gg 1$. At this limit, it may easily be seen that in the range $z \ll \Delta_*$, the distribution (3.264) is approximated quite well by the law

$$\rho_{g0}(z) \simeq \rho_{g0}(0)\exp\{-z^2/h^2\}, \tag{3.265}$$

where

$$h \simeq \Delta_* c_s / c_z \ll \Delta_*. \tag{3.266}$$

The possibility of using the model of the thin disk is evidently determined by the value of the parameter $k_g h$, where k_g corresponds to the most unstable (or close to the instability limit) mode. The value of k_g for the gravitational branch of the oscillations is estimated from the dispersion equation (see equation (3.312)) $\omega^2 = 4\Omega^2 - 2\pi G\sigma_{g0}k + k^2c_s^2$, which describes the properties of the shortwave perturbations in the simplest model of the homogeneous solid-state rotating gas disk. From the condition $d\omega^2/dk = 0$ we obtain

$$k_g = \pi G\sigma_{g0} / c_s^2. \tag{3.267}$$

In the system consisting only of the gas disk (a stellar disk or a compact massive object ensuring rotation of the gas disk, do not exist), the potential Φ_0 is determined only by the gas component and, therefore, $\rho_{g0}(z) = \rho_{g0}(0)\mathrm{ch}^{-2}\{z/h\}$, where $h = c_s / \sqrt{2\pi G\rho_{g0}(0)}$. From this it follows that $\sigma_{go} = 2h\rho_{g0}(0)$ and

$$k_g h = 1 \tag{3.268}$$

This relationship shows that the approximation of the infinitely thin disk for the investigation of the collective processes in isolated gas disks is not acceptable in the vicinity of the wave number $k = k_g$ which corresponds to the most unstable mode.

However, if we take into account the presence of a massive stellar disk, the situation changes. Actually, using the relationships (3.261), (3.266) and (3.267), it may easily be seen [205] that if the condition (3.257) is fulfilled then

$$k_g h \simeq \frac{\rho_{g0}}{\rho_{*0}} \ll 1. \tag{3.269}$$

Thus, if the density of the stellar component is considerably greater than the gas density, then to investigate the stability of the gas disk it is correct to assume that the disk is infinitely thin if some additional condition is fulfilled. The derivation of this condition will be described later.

This additional condition should evidently form from the condition of ignoring the contribution of the stellar disk to the perturbed gravitational potential. To estimate this contribution, we use the expressions for the perturbed surface density within the framework of the simplest homogeneous models of the gas and stellar disks

$$\sigma_{g1} = \sigma_{g0} k^2 \Phi_1 (\omega^2 - 4\Omega^2 - k^2 c_s^2)^{-1}, \tag{3.270}$$

$$\sigma_{*1} = \sigma_{*0} k^2 \Phi_1 (\omega^2 - 4\Omega^2 - k^2 c_r^2)^{-1} I_*, \tag{3.271}$$

where $I_* = (1 + k\Delta_*)^{-1}$ is the form factor which takes into account the finite thickness of the stellar disk (see section 2.2.3), and the gas disk in accordance with (3.269) is assumed to be thin.

It should be mentioned that although in Chapter 2 the form factor I_* was used at $k\Delta_* < 1$, direct calculations show that it provides the accurate asymptotics and in the limit $k\Delta_* \gg 1$ (this limit must be examined because the condition (3.269) can be applied at both $k_g\Delta_* \ll 1$, and at $k_g\Delta_* \gg 1$, see equation (3.266). Actually, the replacement of σ_{*0} at the limit $k\Delta_* \gg 1$ by $\sigma_{*0} I_* \simeq \sigma_{*0}/k\Delta_*$ in the dispersion equation $\omega^2 = 4\Omega^2 - 2\pi G\sigma_{*0}k + k^2 c_r^2$ taking into account that $\sigma_{*0} = 2\Delta_* \rho_{*0}(0)$ the results in the dispersion equation for the perturbations with $k_z = 0$ in the rotating gravitating cylinder $\omega^2 = 4\Omega^2 - 4\pi G\rho_{*0} + k^2 c_r^2$ [203, 590].

Initially, we consider the case $kh \ll k\Delta_* \ll 1$. At this limit, the condition of ignoring the contribution of the stellar disk to the perturbed gravitational potential

$$\Phi_1 = -2\pi G(\sigma_{*1} + \sigma_{g1})/k, \tag{3.272}$$

as indicated by (3.272), (3.279), has the form [879]

$$\frac{\sigma_{g0}}{c_s^2} \gg \frac{\sigma_{*0}}{c_r^2}. \tag{3.273}$$

Introducing the coefficient of anisotropy of the stellar disk

$$\alpha = c_r/c_z, \tag{3.274}$$

and using (3.266), we obtain

$$\frac{h}{\alpha^2 \Delta_*} \ll \frac{\rho_{g0}(0)}{\rho_{*0}(0)} \ll 1. \tag{3.275}$$

In the second case ($k\Delta_* \gg 1$; $I_* \simeq 1/k\Delta_*$), the condition $\sigma_{g1} \gg \sigma_{*1}$ has evidently the form

$$\frac{h^2}{\alpha^2 \Delta_*^2} \ll \frac{\rho_{g0}^2(0)}{\rho_{*0}^2(0)} \ll 1. \tag{3.276}$$

Since in the stellar galactic disks in most cases α >1, the inequalities (3.276) automatically follow from (3.275).

The following results are obtained from these estimates. The necessary (but in a general case insufficient) condition of the applicability of the approximation of the thin disk for investigating the gravitationally unstable perturbations in the gas subsystems of the galactics is the presence of the stellar component with the parameters satisfying the qualities (3.275), (3.276). As an example, we consider the typical situation: $\sigma_{g0} / \sigma_{*0} \simeq 0.1$; $c_z / c_s \simeq \Delta_* / h \simeq 3$; $c_r / c_z \simeq 2$ [441, 1112, 1205, 3002). From this $\rho_{g0}(0) / \rho_{*0}(0) \simeq 0.3$; $h / \alpha^2 \Delta_* \leqslant 0.1$ and, consequently, the conditions of applicability of the model of the thin disk in the equation (3.275) are not fulfilled.

Of course, when investigating the properties of the shortwave (k>>k_g) perturbations in the gas disk of a flat galaxy, it is important to take into account the structure of the latter across its plane of symmetry. These investigations [270] showed in particular that the law of dispersion of the Jeans oscillations of the disk in the range kh >>1 is similar to the dispersion dependence of the surface gravitational waves in deep water ($\omega^2 \simeq kg$, where $g = |\partial\Phi_0/\partial z|$ at $|z| = h$). However, in the intermediate range of the spectrum ($kh \leqslant 1$), consideration of the finite thickness of the gas disk requires a more accurate approach. Qualitative estimates can be obtained using the form factor $I_g = (1+ kh)^{-1}$, as in the case of a stellar disk.

3.7. Dynamics of perturbations in the disk plane

The analysis desribed above shows that the mathematical problem of determinatiom of the dispersion properties of the gravitating disks is reduced to solving a system of integro-differential equations even for axisymmetric waves in a radially homogeneous disk. For real systems, the factors of radial inhomogeneity of the equilibrium parameters and non-axisymmetric perturbations are not small. This complicates the task to an even greater extent. Therefore, we consider the simplest model of gravitational instability in order to determine the role of non-axisymmetric waves in the radial equilibrium gradients of the velocity of rotation, density and temperature on the qualitative level.

3.7.1. Formulation of the problem

The problem of investigating the dispersion properties of non-axisymmetric perturbations in the plane of a thin radially inhomogeneous gas disk will be formulated [114, 159]. We take into account the presence of the dependence $\Omega_z(r)$ only in equilibrium terms in the Euler equation and the thermal equation.

The initial equations of gas dynamics in this model with (3.239), (3.242), and (3.247), have the form

$$\frac{\partial\sigma}{\partial t} + \frac{\partial(r\sigma u)}{r\partial r} + \frac{\partial(\sigma v)}{r\partial\varphi} = 0, \qquad (3.277)$$

$$\frac{\partial u}{\partial t}+u\frac{\partial u}{\partial r}+\frac{v}{r}\frac{\partial u}{\partial \varphi}-\frac{v^2}{r}=-\frac{\partial \Phi}{\partial r}-\frac{1}{\sigma}\frac{\partial p}{\partial r}-D\frac{p}{\sigma}\frac{d\ln\Omega_z}{dr}, \tag{3.278}$$

$$\frac{\partial v}{\partial t}+u\frac{\partial v}{\partial r}+\frac{v}{r}\frac{\partial v}{\partial \varphi}+\frac{uv}{r}=-\frac{1}{r}\frac{\partial \Phi}{\partial \varphi}-\frac{1}{\sigma r}\frac{\partial p}{\partial \varphi}, \tag{3.279}$$

$$\frac{\partial p}{\partial t}+\gamma_s p\left(\frac{\partial ru}{r\partial r}+\frac{\partial v}{r\partial \varphi}\right)+u\frac{\partial p}{\partial r}+v\frac{\partial p}{r\partial \varphi}=(\gamma_s-1)\frac{d\ln(C_\ell\Omega_z)}{dr}, \tag{3.280}$$

where u, v are the radial and azimuthal components of gas velocity; σ, p are the surface density and pressure in the gas disk, and the dissipative terms are ignored (the dissipative effects are studied in section § 3.9).

To investigate the dynamics of small perturbations, the system (3.277)–(3.280) will be linearised. For this purpose, the variables included in the system are presented in the form of sums of equilibrium and perturbed quantities:

$$\sigma=\sigma_0(r)+\sigma_1(r,\varphi,t),\quad u=u_1(r,\varphi,t),\quad v=r\Omega(r)+v_1(r,\varphi,t),$$
$$p=p_0(r)+p_1(r,\varphi,t),\quad \Phi=\Phi_0(r)+\Phi_1(r,\varphi,t). \tag{3.281}$$

Consequently, we obtain

$$\frac{\partial \sigma_1}{\partial t}+\Omega\frac{\partial \sigma_1}{\partial \varphi}+\sigma_0\left(\frac{\partial u_1}{\partial r}+\frac{u_1}{r}+\frac{\partial v_1}{r\partial \varphi}\right)+u_1\sigma_0'=0, \tag{3.282}$$

$$\frac{\partial u_1}{\partial t}+\Omega\frac{\partial u_1}{\partial \varphi}-2\Omega v_1=-\frac{\partial \Phi_1}{\partial r}-\frac{1}{\sigma_0}\frac{\partial p_1}{\partial r}-g\frac{\sigma_1}{\sigma_0}-D(\ln\Omega_z)'\frac{p_1}{\sigma_0}, \tag{3.283}$$

$$\frac{\partial v_1}{\partial t}+\Omega\frac{\partial v_1}{\partial \varphi}+\frac{\varkappa^2}{2\Omega}u_1=-\frac{1}{r}\frac{\partial \Phi_1}{\partial \varphi}-\frac{1}{\sigma_0 r}\frac{\partial p_1}{\partial \varphi}, \tag{3.284}$$

$$\frac{\partial p_1}{\partial t}+\Omega\frac{\partial p_1}{\partial \varphi}+\gamma_s p_0\left(\frac{\partial u_1}{\partial r}+\frac{u_1}{r}+\frac{\partial v_1}{r\partial \varphi}\right)+u_1\left[p_{0'}-(\gamma_s-1)p_0(\ln C_\ell\Omega_z)'\right]=0, \tag{3.285}$$

where the prime indicates the derivatives along the radial component and, in accordance with the equilibrium equation (3.243),

$$g\equiv\frac{d\Phi_0}{dr}-r\Omega^2=-\frac{p_0'}{\sigma_0}-D\frac{p_0}{\sigma_0}(\ln\Omega_z)'. \tag{3.286}$$

The system (3.282)–(3.285) is supplemented by the linearised Poisson equation

$$\Delta\Phi_1=4\pi G\sigma_1\delta(z). \tag{3.287}$$

The dependence of the perturbed values

$$f_1=\{\sigma_1, p_1, u_1, \ldots\} \tag{3.288}$$

on the azimuthal component φ and the time t in accordance with stationarity and homogeneity of the investigated equilibrium model in the azimuthal direction is written in the form

$$f_1 = f(r)\exp\{-i\omega t + im\varphi\}. \tag{3.289}$$

Consequently, the system (3.282)÷(3.285), (3.287) is transformed to the following system (index '1' at the perturbed quantities is omitted and it is assumed that C_ℓ = const):

$$-i\hat{\omega}\Sigma / c_T^2 + (u' + u/r + imv/r) + u/L_\sigma = 0, \tag{3.290}$$

$$-i\hat{\omega}u - 2\Omega v = -R' - \Phi' - (1/L_\sigma + D/L_z)R + (1/L_p + D/L_z)\Sigma, \tag{3.291}$$

$$-i\hat{\omega}v + \varkappa^2 u / 2\Omega = -imR/r - im\Phi/r, \tag{3.292}$$

$$-i\hat{\omega}(R - \gamma_s\Sigma) + c_T^2 S_0' u = 0, \tag{3.293}$$

$$\Phi'' + \frac{\Phi'}{r} - \frac{m^2\Phi}{r^2} + \frac{\partial^2\Phi}{\partial z^2} = 4\pi G\sigma\delta(z), \tag{3.294}$$

where $\Sigma = c_T^2\sigma_1/\sigma_0$, $R = p_1/\sigma_0$, $\hat{\omega} = \omega - m\Omega$, $1/L_\sigma = \sigma_0'/\sigma_0$, $1/L_p = p_0'/p_0$, $1/L_z = \Omega_z'/\Omega_z$, $S_0 = \ln(p_0\Omega_z^{1-\gamma_s}/\sigma_0^{\gamma_s})$, $c_T = \sqrt{p_0/\sigma_0}$ is the isothermal speed of sound, $c_s = \sqrt{\gamma_s}c_T$ is the adiabatic speed of sound.

The perturbed velocities are omitted from this system. For this purpose, from (3.292) we obtain

$$v = -\frac{i\varkappa^2}{2\Omega}\frac{u}{\hat{\omega}} + \frac{m}{r\hat{\omega}}(R + \Phi) \tag{3.295}$$

and, substituting this expression into (3.291), we obtain

$$u = \frac{i}{\hat{\omega}^2 - \varkappa^2}\left\{\frac{2m\Omega}{r}(R + \Phi) + \hat{\omega}\left[(\ln p_0\Omega_z^D)'\Sigma - (\ln\sigma_0\Omega_z^D)'R - R' - \Phi'\right]\right\}. \tag{3.296}$$

Subsequently, differentiating (3.296) with respect to the radial component and substituting the result together with (3.295) and (3.296) into (3.290), we reduce (3.290) to the form

$$\begin{aligned}
&(R+\Phi)'' + \left(\ln\frac{\sigma_0 r}{\hat{\omega}^2 - \varkappa^2}\right)'(R+\Phi)' + \left(\ln\sigma_0\Omega_z^D\right)'R' - \left(\ln p_0\Omega_z^D\right)'\Sigma' - \\
&-(R+\Phi)\left\{\frac{m^2}{r^2} + \frac{1}{\hat{\omega}}\left(\frac{2m\Omega}{r}\right)' + \frac{2m\Omega}{r\hat{\omega}}\left(\ln\frac{\sigma_0 r}{\hat{\omega}^2 - \varkappa^2}\right)'\right\} + R\left\{\left(\ln\sigma_0\Omega_z^D\right)'' + \right. \\
&\left. + \left(\ln\frac{\sigma_0 r}{\hat{\omega}^2 - \varkappa^2}\right)'\left(\ln\sigma_0\Omega_z^D\right)' + \frac{2m\Omega}{r\hat{\omega}}\left(\ln\sigma_0\Omega_z^D\right)'\right\} + \Sigma\left\{\frac{\hat{\omega}^2 - \varkappa^2}{c_T^2} - \right. \\
&\left. -\left(\ln p_0\Omega_z^D\right)'' - \left(\ln\frac{\sigma_0 r}{\hat{\omega}^2 - \varkappa^2}\right)'\left(\ln p_0\Omega_z^D\right)' - \frac{2m\Omega}{r\hat{\omega}}\left(\ln p_0\Omega_z^D\right)'\right\} = 0.
\end{aligned} \tag{3.297}$$

Substituting also (3.296) into (3.293), we obtain the second equation, linking Σ, R, Φ:

$$S_{0'}\left\{\hat{\omega}\left[R'+\Phi'+(\ln\sigma_0\Omega_z^D)'R-(\ln p_0\Omega_z^D)'\Sigma\right]-\frac{2m\Omega}{r}(R+\Phi)\right\}+$$

$$+\hat{\omega}\frac{\hat{\omega}^2-\varkappa^2}{c_T^2}(R-\gamma_s\Sigma)=0. \quad (3.298)$$

The system of equations (3.297), (3.298), together with the Poisson equation, (3.294), is the initial system for further analysis of the dynamics of small perturbations in the model of the gas disk with the arbitrary distribution $\sigma_0(r)$, $p_0(r)$, $\Omega(r)$.

3.7.2. The dispersion equation in an isentropic disk

We investigate the isentropic model. In this model

$$S_0=\ln\frac{p_0}{\sigma_0^{\gamma_s}\Omega_z^{\gamma_s-1}}=\text{const}. \quad (3.299)$$

In this model, the relationship between L_c and L_σ is determined by the relationship

$$L_c=\left(\frac{d\ln c_s}{dr}\right)^{-1}=\frac{2L_\sigma}{(\gamma_s-1)(1+L_\sigma/L_z)}, \quad (3.300)$$

where γ_s is the 'flat' exponent of the polytrope (see section 3.6.1). If it is assumed that $\gamma \leqslant 5/3$, then, in accordance with (3.248) $\gamma_s \leqslant 3/2$ and, consequently, $L_c > L_\sigma$. It is clear from this that the isentropic model does not contradict the results of observation of a gas disk of a galaxy (Zeldovich and Pikel'ner showed that the gas disk of the Galaxy is isothermal). Therefore, we initially carry out further analysis within the framework of the isentropic model with $D = 1$ (a more general case will be examined later).

In the isentropic model from (3.298) it follows that

$$R=\gamma_s\Sigma. \quad (3.301)$$

This relationship brings the equation (3.297) to the form

$$(\Phi+\gamma_s\Sigma)''+\left(\ln\frac{\sigma_0 r}{\hat{\omega}^2-\varkappa^2}\right)'(\Phi+\gamma_s\Sigma)'+\frac{\hat{\omega}^2-\varkappa^2}{c_T^2}\Sigma-(\Phi+\gamma_s\Sigma)\times$$

$$\times\left\{\frac{m^2}{r^2}+\frac{1}{\hat{\omega}}\left(\frac{2m\Omega}{r}\right)'+\frac{2m\Omega}{r\hat{\omega}}\left(\ln\frac{\sigma_0 r}{\hat{\omega}^2-\varkappa^2}\right)'\right\}=0. \quad (3.302)$$

Significant advances in the understanding of the dynamics of the gravitating gas disk and, in particular, determination of the condition of its gravitational stability with respect to the axisymmetric perturbations were achieved by WKB analysis in the radial direction. This is caused by the fact that the length of the disk and the characteristic scales of its inhomogeneity in the radial direction are so large in comparison with its thickness that, in addition to fulfilling the condition (3.257), it is also possible to fulfill the condition of applicability of the WKB-approximation [$f(r) \propto \exp(ik_r r)$ – see (3.298)]:

$$k_r r \gg 1, \quad k_r L \gg 1, \tag{3.303}$$

where $L = \min\{|L_c|, |L_\sigma|, |L_z|, |\partial \ln\Omega/\partial r|^{-1}\}$.

However, it can easily be seen that the WKB-approximation can be used only for the non-axisymmetric perturbations described by the equation (3.302) for which the condition

$$\frac{m|\hat{\omega}|}{\varkappa} \ll kr = \sqrt{r^2 k_r^2 + m^2}, \tag{3.304}$$

is fulfilled which in a general case is not more stringent than the condition $|\hat{\omega}| \ll \varkappa$. Thus, we can investigate the properties of shortwave perturbations which greatly differ from the axisymmetric ones only in the low-frequency part of the spectrum with respect to $\hat{\omega}$. However, this restriction is not important for investigating the gravitational instability of the gas disk. Actually, the boundary of stability of the gas disk with respect to the axisymmetric perturbations is determined from the condition $\hat{\omega}^2 = 0$ in the minimum of the dispersion curve $\hat{\omega}^2 = \hat{\omega}^2(k)$. To solve the problem of the determination of the boundary of stability of the disk with respect to the non-axisymmetric perturbations, it is important to start from the fact that any inhomogeneous system should contain shortwave gradient perturbations whose maximum frequency is $\omega_g \sim \omega_0/kL \ll \omega_0$, where ω_0 is the frequency of natural oscillations of the system without taking its inhomogeneity into account. For example, in the Earth atmosphere the frequency of sound $\omega_0 = kc_s$, and the frequency of internal gravitational waves is $\omega_g \sim g/c_s \sim c_s/L_a \sim \omega_0/kL_a$ [112], where L_a is the vertical scale of inhomogeneity of the atmosphere, g is the gravitational acceleration. Similarly, in the stellar disk the frequency of Jeans perturbations is $\hat{\omega}_J \sim \varkappa$ everywhere, with the exception of the vicinity of the minimum of the dispersion curve $\hat{\omega}_J(k)$, and the maximum frequency of the gradient perturbations is $\hat{\omega}_g \sim \varkappa / kL \ll \varkappa$ (see section 2.2.4). It is evident that the same claim holds for the gas disk. In this case, the boundary of stability is determined in the range of 'contact' of the Jeans and gradient branches (the minimum of the dispersion curves $\hat{\omega}_J(k)$), where $\hat{\omega}_J \sim \hat{\omega}_g \sim \varkappa / kL \ll \varkappa$. In addition to this, the dispersion properties of the low-frequency gradient perturbations can also be of independent interest (see, for example, the sections concerned with the Rossby waves and gradient–entropy instability).

Therefore, we examine the perturbations whose frequency and azimuthal wave number satisfy the condition (3.304). If this condition is fulfilled, the terms with the first derivatives with respect to Φ and σ on the radial coordinate in (3.302) are negligibly small in comparison with the first term. Therefore, rejecting in (3.302) all small terms with respect to the conditions (3.303), (3.304), we obtain

$$(\Phi + \gamma_s \Sigma)\left\{k^2 + \frac{2m\Omega}{r\hat{\omega}}\frac{d}{dr}\ln\frac{2\Omega\sigma_0}{\varkappa^2} - \frac{2m^2}{r\varkappa^2}\frac{d\Omega^2}{dr}\right\} - \frac{\hat{\omega}^2 - \varkappa^2}{c_T^2}\Sigma = 0. \tag{3.305}$$

Using then the short wave solution of the Poisson equation for a thin disk (3.174) [203, 590]

$$\Phi = -\frac{2\pi G\sigma}{k}, \tag{3.306}$$

we obtain the sought dispersion equation for the isentropic gas disk

$$\hat{\omega}^3 - \hat{\omega}\left[\varkappa^2 + \left(k^2 c_s^2 - 2\pi G\sigma_0 k\right)\left(1 - 2s^2 \frac{r}{\varkappa^2}\frac{d\Omega^2}{dr}\right)\right] -$$
$$-2s\Omega(kc_s^2 - 2\pi G\sigma_0)\frac{d}{dr}\ln\left(\frac{2\Omega\sigma_0}{\varkappa^2}\right) = 0, \tag{3.307}$$

where $s = m/kr$ is a parameter characterising the degree of non-axisymmetry of perturbations.

In the [167] the authors analyse the stability of a solid-state rotating ($\Omega = \beta\,\Omega_0 =$ const) gas disk with a density

$$\sigma_0 = \sigma_0(0)\,(1 - r^2/R^2)^{1/2} \tag{3.308}$$

and the polytropic equation of state

$$p_0 = p_0(0)(1 - r^2/R^2)^{3/2} \propto \sigma_0^3, \tag{3.309}$$

where $\sigma_0(0) = 2R\Omega_0^2/\pi^2 G$; $p_0(0) = \sigma_0(0)\Omega_0^2 R^2(1-\beta^2)/3$. Without using the WKB approximation, the model yields the dispersion equation describing the properties of arbitrary perturbations in the disk plane:

$$\hat{\omega}(\hat{\omega}^2 - 4\Omega^2) + \left\{\hat{\omega}\left[l(l+1) - m^2\right] - 2m\Omega\right\}\left\{\Omega_0^2(1-\beta^2) - 4\Omega_0^2\gamma_l^m\right\} = 0, \tag{3.310}$$

where

$$\gamma_l^m = \frac{(l+m)!(l-m)!}{2^{2l+1}\left[\left(\frac{l+m}{2}\right)!\left(\frac{l-m}{2}\right)!\right]^2}. \tag{3.311}$$

Since the model (3.30 W.), (3.309) is also isentropic, it is interesting to compare the dispersion equation (3.310) in the shortwave limit ($l - m \gg 1$) with the dispersion equation (3.307) in the solid-state rotating limit ($d\Omega/dr \to 0$).

In the isentropic models $dp_0/dr = c_s^2\, d\sigma_0/dr$ and, consequently, it is concluded that in the model (3.309) $c_s^2 = \Omega_0^2(1-\beta^2)(R^2 - r^2)$. At the shortwave limit ($l - m \gg 1$), using the asymptotics of the gamma function, we obtain $\gamma_l^m \simeq 1/\pi\sqrt{l^2 - m^2}$ and, consequently

$$4\Omega_0^2\gamma_l^m = 2\pi G\sigma_0 / \sqrt{(l^2 - m^2)(R^2 - r^2)}.$$

To determine the analogue of the wavenumber $k = \sqrt{k_r^2 + m^2/r^2}$, it should be mentioned that [203, 590]

$$\Phi = -4\gamma_l^m R\Omega_0^2 \sigma \sqrt{R^2 - r^2} / \sigma_0(0) \simeq -2\pi G\sigma \sqrt{(R^2 - r^2)/(l^2 - m^2)}.$$

Comparing this expression with the WKB solution of the Poisson equation (3.306), it is concluded that $k \simeq \sqrt{(l^2 - m^2)/(R^2 - r^2)}$. Finally

$$d\ln(\sigma_0)/k^2 dr = -r/k^2(R^2 - r^2) = -r/(l^2 - m^2),$$

and, therefore, equation (3.10) can be written in the formula

$$\hat{\omega}^3 - \hat{\omega}[4\Omega^2 + k^2 c_s^2 - 2\pi G\sigma_0 k] - \frac{2m\Omega}{kr}(k c_s^2 - 2\pi G\sigma_0)\frac{d\ln(\sigma_0)}{dr} = 0, \quad (3.312)$$

which is virtually identical with the equation (3.307) at the limit $\Omega = \text{const}$. This fact is an additional argument in favour of the accuracy of approximation (3.304), used for the WKB analysis of the dispersion properties of the perturbations in a far more complicated (in comparison with (3.309)) model of the gas disk with the arbitrary distributions $\Omega(r)$ and $\sigma_0(r)$.

3.7.3. Rossby waves

The dispersion equation (3.307) describes three branches of oscillations of the gas disk. If the inhomogeneity of the disk and the differential form of its rotation are ignored, it can easily be seen that two of them are gravitational (Jeans) and their frequencies are determined from the condition of balance of the cubic and linear, with respect to $\hat{\omega}$, terms (the frequency of the third branch in this case is $\hat{\omega} = 0$). The appearance of the third type of perturbations js associated with the inhomogeneity of the disk and the differential form of its rotation (manifestation of shear elasticity of the inhomogeneous medium), and their frequency in the gravitationally stable disk (section 3.8) can be approximately determined from the condition of balance of linear (with respect to $\hat{\omega}$) and free in (3.307) terms [159]:

$$\hat{\omega}_g = -\frac{2m\Omega(k c_s^2 - 2\pi G\sigma_0) d\ln(2\Omega\sigma_0/\varkappa^2)/dr}{kr[\varkappa^2 + (k^2 c_s^2 - 2\pi G\sigma_0 k)(1 - 2m^2 d\Omega^2/rk^2\varkappa^2 dr)]}. \quad (3.313)$$

It may easily be seen that $\hat{\omega}_g$ according to (3.313) satisfies the condition (3.304) at any wavelength of the perturbations $(m\hat{\omega}_g/\varkappa \sim (s\Omega/kL)m/\varkappa \sim s^2 \ll kr)$. Identical branches of the oscillations, whose frequencies are proportional to the gradients of the non-perturbed quantities, occur in the atmosphere and oceans of the planets (internal gravitational waves and Rossby waves, see [31, 112]), plasma (drift waves – see [98]), stellar disks (Chapter 2), and other inhomogeneous media. The expression (3.313) describes the waves having the features of both the internal gravitational waves and Rossby waves [156]. To prove the second part of this claim, we transfer to the natural, for the planet atmospheres, limit of the homogeneous (along the surface of the planet) non-self gravitating medium (formally $G \to 0$, $\sigma_0' \to 0$). Consequently, assuming that the rotation of the disk is weakly differential ($n \ll 1$ for $\Omega \propto r^{-n}$), from (3.313) we obtain

$$\hat{\omega}_g = \hat{\omega}_{R\text{gal}} = \frac{2k_\varphi d\Omega / dr}{k_r^2 + k_\varphi^2 + 4\Omega^2 / c_s^2}, \tag{3.314}$$

where $k_\varphi = m/r$.

In the atmosphere of the planets, the law of dispersion of the shortwave barotropic Rossby perturbations [31] (here Ω_z is the z-component of the angular velocity of rotation) has the form

$$\hat{\omega}_R = -\frac{2k_\varphi (d\Omega_z / R d\theta)}{k_\theta^2 + k_\varphi^2 + 4\Omega_z^2 / c_s^2}. \tag{3.315}$$

The actual values of the parameters, characterising the dynamics and geometry of the previously mentioned vortex structures, identical with the planetary anticyclonic Rossby solitons in the gas disks of the galaxies should be evidently calculated in the non-linear theory. Therefore, it is very interesting to transfer directly the results of the theory of Rossby solitons obtained in 'shallow water' to the case of the gas disk studied by Korchagin and Petviashvili [102]. The soliton obtained by them has the characteristic radius of the order of or greater than the epicyclic radius ($\rho = c_s/\varkappa$) and, consequently, corresponds to the perturbations with $k\rho \leqslant 1$. Using the results obtained in section 3.6.2, it can easily be shown that in the gravitationally stable gas disk $k_g h \sim (\rho_g/\rho_*)_{|z=0} \sim \varkappa / \omega_z \sim h/\rho$ and, consequently $k_g \rho \sim 1$. Thus, the structure and dynamics of such a soliton are strongly affected by the perturbations of the gravitational potential determined by the perturbations of the surface density of the disk (see the difference of the dispersion properties of the planetary and galactic Rossby waves in the region of the spectrum $k \leqslant 1/\rho$).

This circumstance stresses the necessity for determining the sufficiently effective of excitation of Rossby waves in gravitating gas subsystems of the galaxies. One of these possible mechanisms will be described in section 3.8.5.

3.8. Instabilities of the gas gravitating disk

It is very convenient here to mention section 2.3, in which the mechanism of the gravitational instability of the gas disk is discussed on the qualitative level.

3.8.1. Condition of gravitational stability of the gas subsystem

The dynamics of small perturbations in the isentropic model is determined by the dispersion equation (3.307) which can be written in the following form

$$\hat{\omega}^3 + p(k)\hat{\omega} + q(k) = 0. \tag{3.316}$$

Therefore, the problem of the condition of gravitational stability of the gas disk with respect to the shortwave perturbations is reduced to the condition of absence of the complex roots of this equation

$$D(k) = (p/3)^3 + (q/2)^2 \leq 0. \tag{3.317}$$

To stabilise the gas disk with respect to the axisymmetric ($m = 0$) perturbations, it is sufficient to ensure that $c_s = c_0 = \pi G\sigma_0/\varkappa$ [618]. In this case, the perturbations with $k = k_0 = \pi G \sigma_0/c_0^2$ are marginally stable, and in this case

$$\omega^2 = \varkappa^2 + k_0^2 c_0^2 - 2\pi G\sigma_0 k_0 = 0.$$

Calculating the value D (k_0) at $c_s = c_0$ from (3.317), we obtain

$$D(k_0) = \left(-2s^2 r\frac{(\Omega^2)'}{3}\right)^3 + \left[s\Omega\varkappa^2\frac{\partial\ln(2\Omega\sigma_0/\varkappa^2)}{k_0\partial r}\right]^2 \geqslant 0 \tag{3.318}$$

for any functions $\Omega(r)$ not increasing with the increase of the radial coordinate. Since the observations do not indicate the existence of regions with $d\Omega/dr > 0$ in the flat galactic areaa (with the exception, possibly, of the central part of the SB-galaxies), equation (3.318) yields the important result: the value c_s required for stabilisation of non-axisymmetric ($s \neq 0$) perturbations in the gas disk is greater than c_0.

Initially, we calculate the value $c_s = c_1(s)$ required for stabilising the non-axisymmetric perturbations in the model of the disk with $\partial\ln(2\Omega\sigma_0/\varkappa^2)/\partial r = 0$. In this case, the free term in (3.307), converts to 0 and the dispersion equation becomes quadratic with respect to $\hat{\omega}$, and there is no term that is linear with respect to $\hat{\omega}$. The boundary of stability for the disk, described by such dispersion equation, is determined from the conditions $\hat{\omega}^2 = 0$, $\partial\hat{\omega}^2/\partial k = 0$ ($\hat{\omega}^2 = 0$ at the minimum of the dispersion curve). If these conditions are fulfilled, this means that for the stability of the perturbations with the fixed $s < 1$ the following condition should be satisfied

$$c_s \geqslant c_1(s) = c_0\sqrt{1 - 2s^2 r(\Omega^2)'/\varkappa^2}, \tag{3.319}$$

and the wavelength of the marginally stable ($c_s = c_1$ (c)) perturbations is determined by the relationship

$$k = k_1(s) = k_0\left[1 - 2s^2 r(\Omega^2)'/\varkappa^2\right]^{-1}. \tag{3.320}$$

This shows that for decreasing $\Omega(r)$ the stability boundary is displaced into the long wavelength region and the value $c_s = c_1$ (c), required for stabilising the perturbations, increases with increasing parameter s.

We now determine the condition of stability in the models without any specific relationship between $\sigma_0(r)$ and $\Omega(r)$. Evidently, in this case the stability boundary should be determined from the set of the conditions $D(k) = 0$, $dD(k)/dk = 0$ (see section 2.4.3). Calculations are carried out mainly with respect to the parameter

$$\mu = \left| s\frac{2\Omega}{\varkappa k_T}\frac{d\ln(2\Omega\sigma_0/\varkappa^2)}{dr}\right| \tag{3.321}$$

(where $k_T = \varkappa^2/2\pi G\sigma_0 = k_0/2$) using the solution (3.319) as the initial approximation. Consequently, from the condition $D = 0$ we obtain the criterion of stability of the perturbations for the given s [159]:

$$c_s \geqslant c_{s_{\min}}(s) \simeq c_0[1-2s^2 r(\Omega^2)'/\varkappa^2]^{1/2}(1+0.6\mu^{2/3}), \tag{3.322}$$

and from the condition $dD/dk = 0$ – the wavelengths of the marginally stable perturbation (at $c_s = c_2$ (s)):

$$k = k_2(s) \simeq k_0[1-2s^2 r(\Omega^2)'/\varkappa^2]^{-1}(1-0.6\mu^{2/3}). \tag{3.323}$$

The result (3.323) that the marginally stable perturbations in the gas disk become more and more longwave with the increase of the degree of their non-asymmetric state. According to (3.322), to stabilise these perturbations, it is necessary to ensure the high dispersion of the velocities of the objects (clouds) forming the gas disk in comparison with the dispersion required for stabilising the axisymmetric perturbations. Therefore, it is natural to formulate the question of practical application of the stability criterion (3.322).

If we confine ourselves only to axisymmetric ($s = 0$) perturbations, then to ensure the stability of the gas disk it is necessary to fulfil the inequality $c_s \geqslant c_0 = \pi G\sigma_0/\varkappa$. Estimating these values for the characteristic values $\sigma_0 \simeq 8\ M_\odot\ /\ \mathrm{pc}^2$, $\varkappa = 28$ km/s /kpc, we obtain $c_0 \simeq 4$ km/s.

With increase of the degree of non-asymmetric state of the perturbations, the dispersion of the velocities of the gas clouds required for suppressing the instability of the perturbations increases [see (3.82)]. Analysis, identical with that carried out in section 2.4.4 for a stellar disk, indicates that the relationship (3.332) at the limit $|s| \to 1$ can be used for estimating the upper boundary of the value c_s required for stability of the gas disk

$$\sup\{c_{s_{\min}}(s)\} = \hat{c}_s \simeq c_0\sqrt{\frac{2+n}{2-n}}\left\{1+0.6\left|\sqrt{\frac{2}{2-n}}\left(\frac{n}{k_T r}+\frac{d\ln\sigma_0}{k_T dr}\right)\right|^{2/3}\right\}, \tag{3.324}$$

where the approximation $\Omega \propto r^{-n}$ was applied to simplify calculations. The value $\hat{c}_s$ in the solar neighbourhood will be estimated. Assuming that $n = 1$ and $(d\ln\sigma_0/dr)^{-1} \simeq -4$ kpc [1109], we obtain $\hat{c}_s \simeq c_0\sqrt{3}(1+0.07) \simeq 7.5$ km/s. The one-dimensional dispersion of the velocities of the clouds $c_g^{(1)} \simeq 6$ km/s, corresponding to this estimate, fits in the observed range of this value. This results in the case that the gas disk of the Galaxy in the solar neighbourhood is marginally stable with the error not exceeding the observed error.

3.8.2. Can the condition of gravitational stability provide an estimate of the surface density of the gas subsystems of the galaxies?

As already mentioned, the gas disk of the galaxy marginally stable in relation to the axisymmetric perturbations would have approximately half the observed the dispersion of the velocities of the gas clouds. The same conclusion with respect to the gas subsystems of a number of other galaxies was made by Quirk [1050]. At the same time, the estimate of the upper boundary of the dispersion of the velocities of the gas clouds essential for the stability of the glass disk in relation to non-axisymmetric perturbations proved to be similar to that observed in the solar neighbourhood in the Galaxy. This result indicates that the gas disks of the galaxies may be marginally stable.

However, direct verification of this hypothesis is difficult because only atomic hydrogen was found with sufficiently reliability in the gas subsystems of the galaxies and the contribution to the surface density of the disk of molecular hydrogen and helium is difficult to determine. At the same time, the dispersion of the velocities of the gas clouds in the disks of the galaxies changes only slightly along the radial coordinate (with the exception of the internal regions) and is independent of the type and mass of the galaxy [432, 799]. It is therefore natural to try to verify the hypothesis of the marginal stability of the gas subsystems of the galaxies, paying special attention to the estimation of the value c_s, essential for the stability of the gas disk, in particular, the relationship (3.324) $c_s \geqslant \sup\{c_{s\min}(s)\} = \hat{c}_s(\sigma_0, n, L_\sigma) \propto \sigma_0$ should be written in the form $\sigma_0 \leqslant \hat{\sigma}_0(c_s^{\text{obs}}, n, L_\sigma)$. This approach (proposed in [70]) in the case of successful verification of the parts of the galaxies in which the fraction of molecular hydrogen is negligible, could provide a facility for estimating the upper boundary of the surface density (and, consequently, the general mass) of the gas subsystems of the flat galaxies.

For the simplest realisation of this concept, it should be mentioned that the correction $\propto \mu^{2/3}$ in (3.324) (determined by the term $\propto d\ln(2\Omega\sigma_0/\varkappa^2)/dr$ in the dispersion equation (3.307) is lower or similar to $(k_T L_\sigma)^{-1} \sim c_0/\varkappa L_\sigma \ll 1$ and, consequently, in the first approximation it can be ignored. Consequently, taking into account that $\varkappa = \sqrt{2(2-n)}\,\Omega$ at $\Omega \propto r^{-n}$ from (3.319) we obtain

$$\sigma_0 \leqslant \hat{\sigma}_0 \simeq \frac{\sqrt{2}\, c_s^{\text{obs}}\Omega}{\pi G} \frac{2-n}{\sqrt{2-n+2s^2 n}} \simeq \frac{\sqrt{2}\, c_s^{\text{obs}}\Omega}{\pi G} \frac{2-n}{\sqrt{2+n}}. \quad (3.325)$$

The initial estimates of $\hat{\sigma}_0$ in a number galaxies were obtained in [70] at a relatively large distance from the centre of the disk (equal to the photometric radius $D_0/2$), where the fraction of molecular hydrogen that is difficult to detect is usually small. In the estimates it was accepted that $c_s^{\text{obs}} \simeq 10$ km/s. The velocity of rotation of the disk was assumed to be constant ($n = 1$) with the exception of the case where this is not fulfilled: $n = 3/2$ (M81). The results of comparison of the observed σ_0 and the estimated $\hat{\sigma}_0$ are presented in Fig. 3.5. It may be seen that for the majority of the galaxies in accordance with the estimate (3.325) the observed gas density $\sigma_0 < \hat{\sigma}_0$ and the galaxies such as M31 and NGC 2403 are evidently characterised by a large reserve of the gravitational stability of the gas disk.

A more detailed comparative analysis of the observed distributions $\sigma_0(r)$ and estimates $\hat{\sigma}_0(r)$ for the observed values of $\Omega(r)$ for a sample of 19 galaxies (including SB systems) was carried out for the first time by Zasov and Simakov [74]. Within the framework of the investigated range of the values $c_s^{\text{obs}} \simeq 7 \div 10$ km/s [808, 829] at half the galaxies in the sample it was found that $\sigma_0(r) \simeq \hat{\sigma}_0(r)$. In the other half (including in all SB systems) $\sigma_0(r) < \hat{\sigma}_0(r)$. In all internal regions of the galaxies (in the range of several kiloparsec from the centre of the galaxies) in most cases $\sigma_0 < \hat{\sigma}_0$. This effect

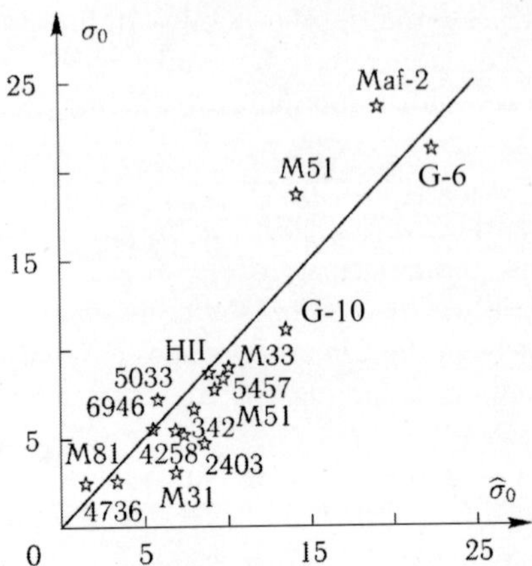

Fig. 3.5. Comparison of the observed σ_0 and estimated $\hat{\sigma}_0$ from (3.325) in the units $M_\odot$/pc^2. Two points correspond to galaxy M51: one (with H_2 not taken into account) relates to $r = D_0/2$, the second one to $r = 8$ kpc, where $\sigma_{H_2} \simeq 1.5\sigma_{HI}$ [1131]. Our Galaxy is represented by the points G-10 ($r = 10$ kpc) and G-6 ($r = 6$ kpc). In the latter case, the estimate $\hat{\sigma}_0$ represents the upper limit because in this region a large part of the mass is enclosed in the molecular clouds for which c_s^{obs} is slightly lower than the accepted value.

may be explained by additional stimulation of star formation in more intensive (in comparison with the outer part of the disk) helical density waves [74].

In [392, 1325, 1328] the density of the gas was evaluated on the basis of the assumption on the marginal stability of the gas for different samples of the galaxies.

We will mention another relatively unexpected application of the estimate (3.325) [60]. This estimate imposes restrictions on the density of the gas disk of the galaxy and, consequently, its mass, and the estimate of these parameters depends on the assumed distance to the investigated galaxy. At the same time, we may obtain information on the distance to the galaxy, bypassing the well-known problem of selection of the scale of distances. Actually, from (3.325) we obtain a restriction on the mass of the gas within the limits of the selected value $r \leqslant r_0$

$$M_{\text{gas}}(r_0) \leqslant 2\sqrt{2}\,\frac{c_s^{\text{obs}}}{G}\,\frac{V_0\, r_0}{\sqrt{2+n}}, \tag{3.326}$$

where $V_0 = V(r_0)$. Taking into account the fact that the gas subsystems of the galaxies contain approximately 30% of the elements heavier than hydrogen (mostly helium) and at the same time $M_{\text{H}} \simeq 0.7\, M_{\text{gas}}$, and transferring to the measurement units natural for applications, from (3.326) we obtain

$$M_{\text{H}} / M_\odot \leqslant 2.3 \cdot 10^6 r_0 V_0 \simeq 6.7 \cdot 10^5 V_0 r_0' \Delta,$$

where V_0 is measured in km/s; r_0 in kpc; r_0' in angular minutes; distance to the galaxies Δ in Mpc, and for further estimates it is assumed that $c_s^{\text{obs}} \simeq 10$ km/s; $n \simeq 1$. On the other hand [37]

$$M_{\text{H}} / M_\odot \simeq 2.36{\cdot}10^5 S_H \Delta^2,$$

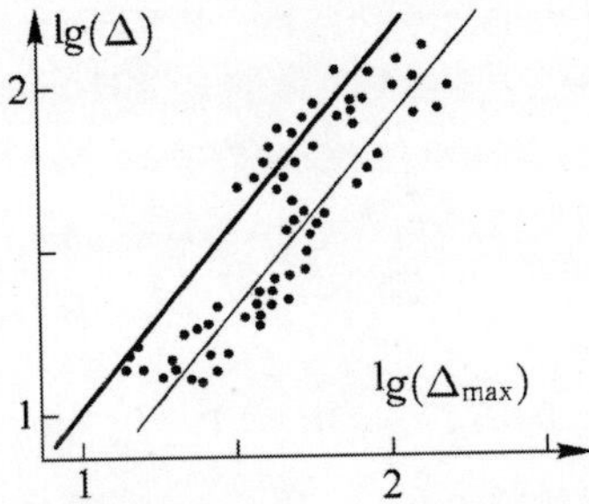

Fig. 3.6. Comparison of the distances to the galaxies observed on the basis of red shift (H_0 = 75 km/(s · Mpc) – thick line) with the estimate Δ_{max} according to (3.327). r_0 is represented by the value r_{70} at which 70% M_{HI} is found. The error in the estimation of Δ_{max}, determined by the error of observations of r_{70}, n, V_0 (r_{70}), is 8 ÷ 15% (see [60] and the studies cited there).

where $S_H = S_H(r_0)$ is the flow in the HI line, in the units n·km/s, investigated in the range r_0. Comparing the previously presented expressions for $M_H/M_\odot$, we obtain the estimate of the distance of the Galaxy:

$$\Delta \leqslant \Delta_{\text{max}} \simeq 2.83 V_0 r_0' / S_H. \tag{3.327}$$

The results of comparison of the limiting estimate Δ_{max} with the value of Δ determined on the basis of the red shift at the Hubble constant H_0 = 75 km/(s · Mpc) are presented in Fig. 3.6. It may be seen that the straight line $\Delta = \Delta_{\text{max}}$ actually illustrates the estimate Δ which indicates that the previously described approach is in fact valid. The reduction of H_0 to 50 km/(s · Mpc) (the thin line in Fig. 3.6) causes that a large part of the galaxies is transferred to the range $\Delta > \Delta_{\text{max}}$. Taking into account the hydrogen content H_2 and the frequently encountered central depression in the distribution of gas in the galaxies can only reduce the estimate Δ_{max} and aggravate further the contradiction with a lower value of H_0 (50 km/(s · Mpc)).

Thus, the method described in the above section can be used not only for restricting the estimates of the distances to the individual galaxies but also for making the scales of the extragalactic distances finer.

3.8.3. Condition of gravitational stability of the non-isentropic disk

In the sections (3.7.2, 3.8.1) we discussed a dispersion equation describing the properties of non-axisymmetric perturbations in the model of the isentropic gas disk and obtained the condition for its gravitational stability. To a first approximation, this model describes quite efficiently the state of the gas subsystem of the Galaxy in the range $r \geqslant 5 \div 6$ kpc but is obviously not universal. Therefore, in this section, we derive the dispersion equation describing the properties of the perturbations in the model of the gas disk with the arbitrary distributions $\sigma_0(r)$, $T_0(r)$, $\Omega(r)$, without restricting ourselves to the isentropy condition (3.299), and investigate the problem of gravitational stability of such a model [114].

The condition of applicability of the WKB-approximation to the non-axisymmetric perturbations in the gas disk (3.304), used in the section 3.7.2, reduced the system (3.297), (3.298) to the form

$$\left\{k^2+\frac{2m\Omega}{r\hat{\omega}}\frac{d\ln\left(2\Omega\sigma_0/\varkappa^2\right)}{dr}-\frac{2m^2}{r\varkappa^2}\frac{d\Omega^2}{dr}\right\}(\Phi+R)-$$

$$-\frac{\hat{\omega}^2-\varkappa^2}{c_T^2}\Sigma+\frac{2m\Omega}{r\hat{\omega}}\left\{\Sigma\frac{d\ln p_0\Omega_z}{dr}-R\frac{d\ln\sigma_0\Omega_z}{dr}\right\}=0, \qquad (3.328)$$

$$\frac{2m\Omega}{r\hat{\omega}}\frac{d}{dr}\left(\ln\frac{p_0}{\sigma_0^{\gamma_s}\Omega_z^{\gamma_s-1}}\right)(\Phi+R)-\frac{\hat{\omega}^2-\varkappa^2}{c_T^2}\left[R-\gamma_s\Sigma\right]=0, \qquad (3.329)$$

where $\Sigma=c_T^2\sigma/\sigma_0$, $R=p/\sigma_0$. Then, taking into account the relationship between Φ and Σ, resulting from (3.306), we obtain the required dispersion equation

$$\hat{\omega}^4-\hat{\omega}^2\left[\varkappa^2+\left(k^2c_s^2-2\pi G\sigma_0 k\right)\left(1-2s^2\frac{r\,d\Omega^2}{\varkappa^2 dr}\right)\right]-$$

$$-2s\Omega\hat{\omega}kc_T^2\left[\gamma_s\frac{d}{dr}\left(\ln\frac{2\Omega p_0^{2/\gamma_s}}{\varkappa^2\sigma_o}\Omega_z^{2(1-\gamma_s)/\gamma_s}\right)-\frac{2\pi G\sigma_0}{kc_T^2}\frac{d}{dr}\left(\ln\frac{2\Omega}{\varkappa^2}\sigma_0\right)\right]-$$

$$-\left(\frac{2s\Omega kc_T^2}{\varkappa}\right)^2\left[\frac{d\ln p_0\Omega_z}{dr}-\frac{2\pi G\sigma_0}{kc_T^2}\frac{d\ln\sigma_0\Omega_z}{dr}\right]\frac{d}{dr}\ln\frac{p_0}{\sigma_0^{\gamma_s}\Omega_z^{\gamma_s-1}}=0. \qquad (3.330)$$

It may easily be seen that in the limit of the isentropic disk (see (3.299)), equation (3.330) is transformed to the dispersion equation (3.307). For applications, it is useful to transfer from the value of pressure to temperature T_0 ($p_0\propto\sigma_0 T_0$) of the speed of sound ($c_s\propto\sqrt{T_0}$). In contrast to the previously mentioned limit of the isentropic disk, the state of the disk in a general case is described another parameter

$$\eta=\frac{d\ln c_T}{d\ln\sigma_0}=\frac{d\ln T_0}{2\,d\ln\sigma_0}, \qquad (3.331)$$

included in the dispersion equation (3.330) through the value

$$\frac{d}{dr}\ln\frac{p_0}{\sigma_0^{\gamma_s}}=2\frac{d\ln\sigma_0}{dr}(\eta-\eta_0), \qquad (3.332)$$

where $\eta_0=(\gamma_s-1)/2$) (at $\eta=\eta_0$ the disk is isentropic in the case $\Omega_z=\text{const}$).

It is clear that the estimate of the value $\hat{c}_s$ (see equation (3.324)), required for the gravitational stability of the gas disk, must depend on the parameter η. It is therefore naturally to ask the following question: at what value η will the value $\hat{c}_s$ be minimum? Evidently, this problem is identical with the problem of the value of the gradient of dispersion of the radial velocities of the stars in the disks

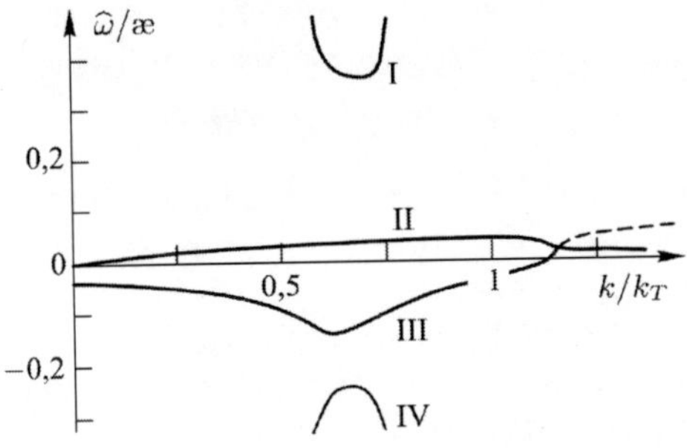

Fig. 3.7. Oscillation branches of a gravitating disk in a model with $d\ \ln(\Omega)/d\ \ln\ (r) = -1$; $\eta = d\ \ln\ (c_s)/d\ \ln(\sigma_0) = 0.3$; $c_s = 1.87$; $c_0 = 1.87\ \pi G\sigma_0/\varkappa$; $\gamma_s = 1.5$; $s/k_T L_\sigma = 0.05$; $s/k_T r = 0.02$ (here $k_T = \varkappa^2/2\ \pi\ G\sigma_0$; $L_\sigma = |d\ \ln\ \sigma_0\ dr|^{-1}$; $s = m/kr$. The dotted curve is the increment of gradient–entropy instability (see section 3.8.4).

of the galaxies, discussed in section 2.4.6. Also, if as a result of the solution of the given problem it appears that $\hat{c}_s$ is minimum at L_σ, which is considerably lower than L_c, then the previously mentioned observational fact – the very slow decrease of $c_s(r)$ in the direction to the disk periphery – becomes clear. At the same time, it should be mentioned that the energy of the gas, rotating in the given gravitational field of the galaxy, decreases with the decrease of the dispersion of the velocities of the gas clouds. If the observed value of η coincides with that at which the value $\hat{c}_s(\eta)$, required for the stability of the disk, is minimum, this will mean that the stationary distributions of the parameters of the gas subsystem are matched in such a manner that the energy of the gas disk has the minimum value compatible with the condition of its gravitational stability.

Equation (3.330) describes four branches of the oscillations of the gas disk [114]. Two of these branches are gravitational (see branches I, IV in Fig. 3.7); another branch, characterised by the properties of the Rossby waves (section 3.7.3) is gravitational (section III in Fig. 3.7). Finally, the last branch (branch II in Fig. 3.7) is also gradient but, in contrast of the branch III, its frequency $\hat{\omega} \equiv 0$ in the model of the isentropic disk. Therefore, branch II is referred to as the entropy branch. The gradient and entropy branches of the oscillations are of the low-frequency type ($|\hat{\omega}_{\text{II,III}}| \ll \varkappa$) at any wavelength of the perturbations. For the perturbations with the wavelengths determined by the relationship $k \simeq k_J = \pi G\sigma_0/c_S^2$, at $c_s \leqslant \hat{c}_s(\eta)$, the gravitation branches: $\hat{\omega}_{\text{I}} \div \hat{\omega}_{\text{IV}} \sim c_s / L_\sigma$, are also low-frequency branches. Naturally, this leads to the mutual effect of the gravitational and gradient branches in the given wavelength range and, consequently, to the gravitational–gradient instability of the gas disk at $c_s < \hat{c}_s$ (compare with the stellar disk, section 2.2.4).

For the estimates we restrict ourselves to the case Ω_z = const. Numerical calculation show that in the model with $\gamma_s = 3/2$ at $\eta = \eta_1 \simeq 1/2$, the relationship between the gravitational (IV) and the gradient branches is stronger than between the gravitational (I) and entropy branches, and the instability, determined by the relationship disappears at high values of c_s [114]. However, if $\eta > \eta_1$, then the instability, determined by the relationship with the entropy and gravitational (I) branches, disappears at high values of c_s. This transition, which determines the boundary of stability of the relationship from the pair of the branches (III–IV) at $\eta < \eta_1$ to the pair (I–II) at $\eta > \eta_1$ results in a minimum in the functional dependence of the value $\hat{c}_s$ (n) at $\eta = \eta_1 \simeq 1/2$ required for stability of the gas disk.

We now carry out approximate calculations of the values η_1 and $\hat{c}_s$ (η_1) (the latter – as the estimate of the upper boundary of the value c_s, required for the stability of the gas disk, in the sense indicated in section 3.8.1) directly from the dispersion equation (3.330) written in the form

$$\Delta(k,\hat{\omega}) = \hat{\omega}^4 + p(k)\,\hat{\omega}^2 + q(k)\,\hat{\omega} + d(k) = 0. \tag{3.333}$$

The above considerations show that at $\eta = \eta_1$ and $c_s = \hat{c}_s$ (η_1) there is contact between the branches III at some point A $(k_1,\ \hat{\omega}_1)$ and contact between the branches III and IV at another point B $(k_2,\ \hat{\omega}_2)$ in the plane $k,\ \hat{\omega}$. The points of contact A, B are the saddle points of the function Δ $(k,\ \hat{\omega})$ and, consequently, the following conditions should be fulfilled there

$$\left(\frac{\partial\Delta}{\partial k}\right)_{A,B} = 0, \quad \left(\frac{\partial\Delta}{\partial\hat{\omega}}\right)_{A,B} = 0, \quad \Delta(A) = \Delta(B) = 0. \tag{3.334}$$

The solution of this system are the required values of η_1, $\hat{c}_s$ (η_1) in the coordinates of the points A, B in the plane $(k,\ \hat{\omega})$. We will find the solution of the system (3.334) taking into account the fact that the coefficients q (k) and d (k) in (3.333) are small. Actually, $q \propto \lambda^{-1}$; $d \propto \lambda^{-2}$ $(\lambda = (\varkappa^2 / \pi G\sigma_0)\,|\,\partial \ln \sigma_0 / \partial r\,|^{-1} = 2k_T L_\sigma \gg 1)$ and in the limit of the homogeneous $(\lambda \to \infty)$ but differentially rotating $[\Omega\ (r) \propto r^{-n}]$ disk, the solution of the system (3.334) is

$$\hat{c}_s = c_0\sqrt{(2+n)/(2-n)}, \tag{3.335}$$

$$k_{1,2} = k_J = \pi\,G\sigma_0 \,/\, c_s^2, \tag{3.336}$$

$$\hat{\omega}_{1,2} = \pm\sqrt{-p(k_0)/2} = \hat{\omega}_{0_{1,2}}. \tag{3.337}$$

In the approximation of the weakly inhomogeneous disk $(\lambda \gg 1)$, the solution has the form $k_{1,2} = k_J + \delta_{1,2;}$ $\hat{\omega}_{1,2} = \omega_{0_{1,2}} + \epsilon_{1,2}$, where $\delta_{1,2}$, $\varepsilon_{1,2}$ are the small values. Consequently, from (3.334) we obtain

$$\epsilon_{1,2} = q(k_J)\,/\,4\,p(k_J), \tag{3.338}$$

$$\delta_{1,2} = \left(1/\hat{\omega}_{0_{1,2}}\right)\left(\frac{dq}{dk}\right)\left(\frac{d^2p}{dk^2}\right)^{-1}_{\Big|k=k_J}. \tag{3.339}$$

Substituting these results into (3.334) it can be seen that the following relationship should be fulfilled between the coefficients (3.333) at the points of contact of the branches

$$q(k_J)=0, \quad p^2(k_J)=4\,d(k_J)+2\,(dq/dk)^2/(d^2p/dk^2)\Big|_{k=k_J}, \tag{3.340}$$

using which we obtain the required results

$$\eta_1 \simeq \frac{\gamma_s}{4}(3-\Lambda)-\frac{1}{2}, \tag{3.341}$$

$$\hat{c}_s(\eta_1) \simeq \frac{\pi G\sigma_0}{\varkappa}\left\{\frac{\Lambda(2+n)/(2-n)}{\Lambda-\sqrt{2(1-\Lambda)(5-3\Lambda)/(2-n)}}\right\}^{1/2}, \tag{3.342}$$

where $\Lambda = nL_\sigma/r$.

We consider the results from the viewpoint of their applications. For example, in the solar neighbourhood of the galaxy $\lambda = 2k_T L_\sigma \simeq 35$; $\Lambda \simeq 0.4$ and, consequently, $\hat{c}_s(\eta_1)$ is almost identical with the value $\hat{c}_s(\eta_0)$ determined by the relationship (3.324) (in fact, $(\hat{c}_s(\eta_0)/\hat{c}_s(\eta_1)-1)<0.05$). In accordance with (3.341) for $\eta_1 = \eta_0 + \gamma_s(1-\Lambda)/4$ in the case $\gamma_s = 1.5$ and $\Lambda \simeq 0.4$ we have $\eta_1 \simeq 2\eta_0 \simeq 1/2$, which is not in agreement with the value observed in the Galaxy ($\eta^{obs} \simeq 1/4$). If we take into account $\Omega_z \neq \text{const}$, the difference becomes even greater. Thus, for the reasons not explained by the theory of gravitational stability of the gas disk, the relative gradient of the dispersion of the velocities of the gas clouds η^{obs}, observed in the Galaxy, is smaller than the one at which the value c_s, required for stability of these subsystem, is minimal. To determine the reasons for this difference, we will investigate separately the properties of the gradient and entropy branches of the oscillations of the gas disk.

3.8.4. Gravitational–entropy instability

As mentioned in the previous section, the entropy (II) and the gradient (III) branches of the oscillations of the gas disk at any wavelength of the perturbation are low-frequency ($|\hat{\omega}_{II,III}| \sim \varkappa / k L_\sigma \ll \varkappa$, see Fig. 3.7). At the same time, the gravitational perturbations (I, IV) outside the limits of the narrow zone of the wave numbers with $k \simeq k_J$ are characterised by the frequencies of the order of epicyclic frequency. Therefore, outside the limits of this zone, the gradient and entropy perturbations can be described by the simplified (quadratic with respect to $\hat{\omega}$) dispersion equation which can be derived from (3.330) by rejecting the small – the fourth degree with respect to $\hat{\omega}$ (and, consequently, the fourth degree with respect to the gradients of the unperturbed parameters of the disk) term. Since the results which will be obtained from such dispersion equation can also be interesting for understanding the nature of turbulent viscosity in the relatively thick accretion disks, this equation can be written taking into account the finite thickness of the disk. This fact will be taken into account by analogy with the stellar disk (see section 2.2.3), by replacing $G\sigma_0$ by $G\sigma_0/(1+kh)$, where h is the effective half thickness of the gas disk. Finally, we obtain [114]:

$$\hat{\omega}^2\left\{\varkappa^2+kc_s^2\left(k-\frac{2k_J}{1+kh}\right)\left(1-2s^2\frac{r\,d\Omega^2}{\varkappa^2 dr}\right)\right\}+2s\Omega\hat{\omega}\times$$

$$\times\left\{4kc_T^2(\eta-\eta_0)\frac{d\ln\sigma_0}{dr}+c_s^2\left(k-\frac{2k_J}{1+kh}\right)\frac{d}{dr}\ln\frac{2\Omega\sigma_0}{\varkappa^2}\right\}+8s^2\Omega^2\frac{kc_T^2}{\varkappa^2}\times$$

$$\times(\eta-\eta_0)\left(\frac{d\ln\sigma_0}{dr}\right)^2\left\{2kc_T^2(\eta-\eta_0)+c_s^2\left(k-\frac{2k_J}{1+kh}\right)\right\}=0. \tag{3.343}$$

It may easily be seen that the equation (3.343) predicts the development of instability if the following conditions is satisfied:

$$8\gamma_s\alpha(\eta-\eta_0)(\Lambda-\hat{z})-16\hat{z}(\eta-\eta_0)^2+\alpha^2\gamma_s^2(1+\Lambda)^2<0, \tag{3.344}$$

where $\Lambda = d\ln(2\Omega/\varkappa^2)/d\ln\sigma_0$;

$$\alpha = 1-2k_J/k(1+kh);\ \hat{z}=\alpha(kc_s/\varkappa)^2(1-2s^2 r\,d\Omega^2/\varkappa^2 dr).$$

The boundaries of the region of this instability, referred to as the gradient–entropy (GE) instability, are determined evidently by the relationship

$$\eta=\eta_0+\frac{\alpha\gamma_s}{4\hat{z}}\left\{(\Lambda-\hat{z})\pm\sqrt{(1+\hat{z})(\Lambda^2+\hat{z})}\right\}. \tag{3.345}$$

As an example, Fig. 3.8 shows the region of the GE-instability in the plane $(k;\eta)$. It may be seen that a large part of the region of GE-instability is situated in the range of the wavelength values $k > k_1$, where $k_1=\left(\sqrt{1+8k_J h}-1\right)/2h$ is the he positive root of the equation $k - 2k_J/(1+kh) = 0$ (at $k_J h \to 0$ we obtain $k_1 \to 2k_J$). The main distinguishing properties of GE-instability will be discussed. Firstly, according to (3.343), only the non-axisymmetric ($s = m/kr \neq 0$) perturbations can be unstable. Secondly, the GE-instability can also develop in a solid state rotating disk ($\Lambda = 0$) and disappears only if $\eta \equiv \eta_0$ (this case corresponds to an isentropic disk in which $c_s^2/(\sigma_0)^{\gamma_s-1}$ remains constant along the radial component). Finally, thirdly, equation (3.344) shows that the GE-instability can develop in any suitable 'hot' (in the limit in the non-self-gravitating) gas disk.

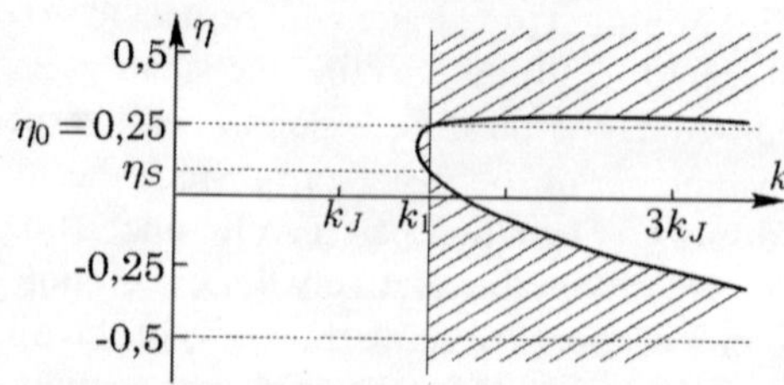

Fig. 3.8. The region of gradient–entropy instability (crosshatched) in the plane $(k;\eta)$ according to the parameters of the gas disk of the Galaxy in the solar neighbourhood assuming that $\gamma_s = 3/2$ ($\eta_0 = 1/4$). Only the model with $\eta = \eta_0$ is stable (the position of the point η_s depends on the degree of non-axisymmetry of the perturbations).

The latter circumstance indicates that the GE-instability may be also found in the non-self-gravitating systems and, consequently, its nature may be explained on the basis of the theory of collective processes in an inhomogeneous gas medium. Actually, in the limit of the non-self gravitating ($k_J \to 0$), non-rotating ($\Omega \to 0$) medium, from (3.343) we obtain

$$\omega = \pm i \frac{2 s c_s}{L_\sigma \gamma_s} \sqrt{(\eta - \eta_0)(\eta + 1/2)}. \tag{3.346}$$

Consequently, at the examined limit the instability is detected at

$$\eta > \eta_0 \tag{3.347}$$

and at

$$\eta < -1/2. \tag{3.348}$$

In order to understand the physical meaning of these restrictions, the radial gradient of the equilibrium entropy and pressure will be expressed by the radial density gradient:

$$\frac{dS_0}{dr} = \frac{c_T^2}{T_0 \eta_0} (\eta - \eta_0) \frac{d \ln \sigma_0}{dr}, \tag{3.349}$$

$$\frac{dp_0}{dr} = \sigma_0 c_T^2 (1 + 2\eta) \frac{d \ln \sigma_0}{dr}. \tag{3.350}$$

Equation (3.349) shows that the instability condition (3.347) corresponds accurately to the condition of formation of convection in an inhomogeneous medium – for this purpose, the gradients of equilibrium entropy and density of the medium should have the same sign [112]. On the other hand, the relationship (3.350) shows that the instability condition (3.348) is fulfilled if the gradients of the equilibrium pressure and density have different signs. It is well-known that in this case the Rayleigh-Taylor instability should develop in an inhomogeneous medium.

The above considerations explain the asymptotic behaviour of the boundaries of the region of GE-instability in Fig. 3.8 at the limit $k \to \infty$. Evidently, the distortions of the boundary of the region of instability in this graph for the long wavelength perturbations ($k \sim k_J$) are caused by the effect of self-gravitation and rotation.

In the isentropic model of the gas disk, the gradient and entropy perturbations are stable. It is therefore natural to assume that the local values of the dispersion of the velocities of the gas clouds and of the surface density of the gas subsystems of the galaxies (at least at their periphery) in accordance with section 3.8.1, are determined in most cases by the condition of gravitational stability of the disk, and the ratio of the gradients – by the condition of stability of the gradient and entropy perturbations. In this case, we should have $\eta = L_\sigma L_c \ll 1$. Evidently, because of this in discussing the observational data for the dispersion of the velocities of the gas clouds (and not only in the Galaxy) it is often assumed that $c_s \simeq$ const in the entire disk [70, 74, 392, 799, 1244].

3.8.5. Excitation of Rossby waves

Previously (section 3.7.3) we formulated the problem of determination of the mechanism of excitation of Rossby-type waves (3.314) in the gas subsystems of the galaxies. It is therefore natural to find a mechanism within the framework of gradient–entropy instability [115]. It may easily be seen that if the conditions of buildup of the GE-instability (3.344) in the non-isentropic ($\eta \neq \eta_0$) disk are fulfilled, the dispersion properties of the investigated type of perturbations slightly differ from those described by equation (3.314) which holds only in the isentropic case. However, this fact has no effect on the substance of the problem. The controlling factor is the following circumstance. The non-linear evolution of GE-instability may lead to the formation of a large-scale vortex structure in the disks. It is possible that structures of this type are also detected in the galaxies NGC157, 2814, 3788 as localised anomalies of the rotational curves in the azimuth (Fig. 3.9). These anomalies can be interpreted as a large-scale vortex formations at the periphery of the galaxies and their natural rotation may coincide (cyclone) or the opposite (anticyclone) to the rotation of the disk of the Galaxy. The centre of the vortex, situated at the distance r_c from the centre, drifts in the disk against the gas flow. The observed structures are characterised by the scales greatly exceeding the Jeans scale in their gas subsystems. The previously investigated GE-instability, at least with respect to the parameters of the gas disk of the Galaxy in the solar neighbourhood, can be excited only in the scales smaller than the Jeans scale (Figure 3.8). Therefore, it is necessary to find the range of the parameters of the gas disk in which the GE-instability would also occur the scales greater than the Jeans scale.

From this viewpoint, we examine the equations of the boundaries of the region of GE-instability with respect to the parameter η (3.345). Since the value $\alpha(k) = 1-2k_J/k\,(1+kh) < 0$ at $< k_1$ and $d\Omega/dr < 0$ in the disks of the galaxies, then $\hat{z}(k) = \alpha(k)(kc_s/\varkappa)^2(1-2s^2 r d\Omega^2/\varkappa^2 dr) < 0$ in the wavelength range in which we

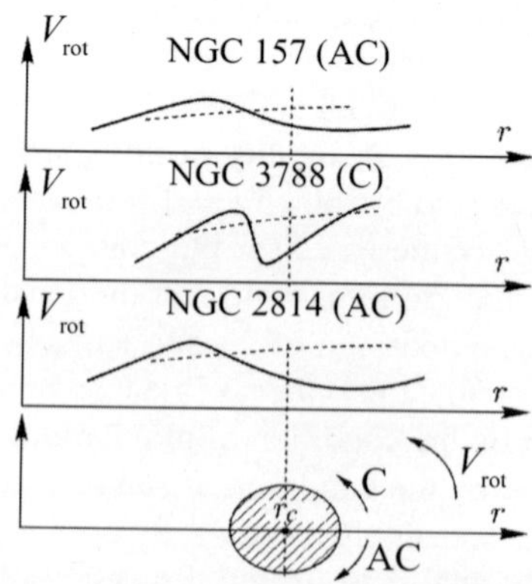

Fig. 3.9. Rotation curves of galaxies along the rays passing through the large-scale vortex condensations of the gas disk (broken lines – V in the direction opposite to the centre of the disk; C – cyclonic, AC – anticyclonic vortex; r_c is the distance from the centre of the disk to the centre of the vortex structure).

are interested ($k < k_1$). Since $z(0) = z(k_1) = 0$, it follows that $\hat{z}(k)$ has a minimum in the range $k < k_1$. The position of this minimum ($k = k_m$) can be determined from the condition $k(1+kh)^2 - k_J = 0$, and the value $\hat{z} = \hat{z}_m$ at this point

$$\hat{z}_m = \left(\frac{c_0}{c_s}\right)^2 \left(\frac{k_m}{k_J}\right)^{3/2} \left(\sqrt{k_m / k_J} - 2\right)\left(1 - 2s^2 r \frac{(\Omega^2)'}{\varkappa^2}\right). \tag{3.351}$$

Taking into account that $k_m < k_J$ in the disks with $h > 0$ and the condition of stability of the gravitational branches of the oscillations of the differentially rotating gas disks, equation (3.351) shows that $z_m > -1$. The latter means that in the gravitationally stable gas disks the band of the GE-instability with respect to the parameter η (see (3.345)) will exist in the entire range $k < k_1$ if the following condition is satisfied

$$\Lambda^2 > \Lambda_{\text{cr}}^2 = -\hat{z}_m. \tag{3.352}$$

If this condition is not fulfilled, the region of instability at $k < k_1$ breaks up into two subregions, and the Jeans scales ($k \sim k_J$) are stable. In particular, this case is also observed in the gas disk of the Galaxy in the solar neighbourhood ($|\Lambda|_\odot \simeq 0.4$; $\Lambda_{\text{cr}\odot} \gtrsim 0.6$) namely here we have the stability (see Fig. 3.8 in which the longwave sub-region ($k \ll k_J$) of GE-instability is not shown).

We return to the problem in which we are interested. The characteristic values $\Lambda_{\text{cr}} = \sqrt{-\hat{z}_m}$ can be estimated from the results shown in Fig. 3.10. It may be seen that for the gravitationally stable (but not excessively 'overheated') gas subsystems of the galaxies $\Lambda_{\text{cr}} \simeq 0.5 \div 0.8$. On the other hand, approximating $\Omega(r)$ by the exponential law $\Omega(r) \propto r^{-n}$ ($n > 0$), it may easily be seen that $|\Lambda| = nL_\sigma/r$, where $L_\sigma = |\partial \ln(\sigma_0)/\partial r|^{-1}$. Taking the previously given estimate Λ_{cr} it follows that the buildup of the Rossby-type waves of the Jeans and larger scales can take place if the distribution $\Omega(r)$ changes quite rapidly with the increase of the distance from the centre of the disk in comparison with the distribution $\sigma_0(r)$ ($L_\Omega \simeq r / n < L_\sigma / \Lambda_{\text{cr}}$).

Figure 3.11 shows the results of calculations of the increment of GE-instability (in the units $s\,\Omega / k_J L_\sigma$) ensuring buildup of the Rossby-type waves at the values of $\Lambda < 0$ ($\sigma_0(r)$ together with $\Omega(r)$ is assumed to be a function decreasing in the direction to the disk periphery). It may be seen that in the relevant range of the wavelengths ($k < k_J$), the instability increment reaches the maximum value at both $k \simeq (0.5 \div 1)\,k_J$ and $\max\{\text{Im}(\hat{\omega})\} \simeq \Omega / k_J L_\sigma \sim \Omega / k_J r$ (because of the condition $L_\sigma > \Lambda_{\text{cr}} L_\Omega$). Thus, the characteristic time of the increase of perturbations of the Rossby-type waves is at least several times greater than the duration of rotation of the disk.

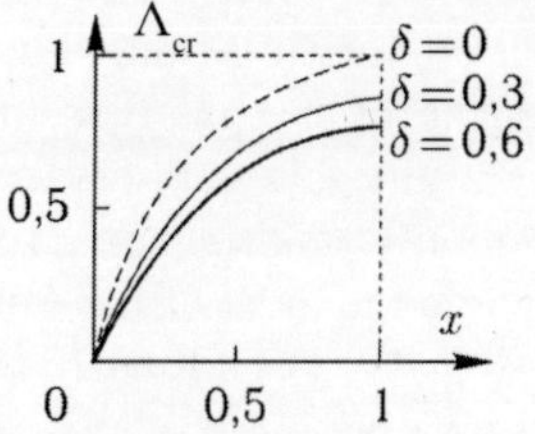

Fig. 3.10. Dependence of Λ_{cr} on the parameters $\delta = k_J h$ and $x = (c_0/c_s)^2(-2s^2\Omega\, r^2{}'/\varkappa^2)$.

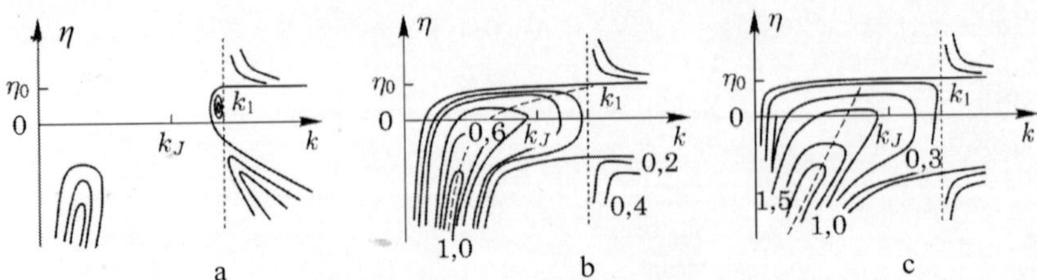

Fig. 3.11. Isolines of the increments (in the unit $s\Omega / k_J L_\sigma$) of the large-scale GE-instability in the models of the disks with $k_J h = 0.3$ and $\Lambda_{cr} = 0.633$: $\Lambda = -0.6$ (a); $= -0.75$ (b); $\Lambda = -1$ (c).

There are no satisfactory observational data for the parameters of the gas disks in the galaxies with the anticyclonic structures of the type of the Rossby solitons in which we are interested. Since we use the well-known parameters of the gas disk of the Galaxy (with the exception of the parameter Λ), and knowing its typical gas subsystem, we obtain (at least very rough) estimates of the duration of growth of the relative amplitude ($\tilde{\sigma} / \sigma_0$) of the perturbations of the Rossby-type waves to the value of the order of unity. For example, in the solar neighbourhood $k_J = \pi G \sigma_0 / c_s^2 \simeq 1.5$ kpc^{-1}. Consequently $\max\{\mathrm{Im}(\hat{\omega})\} \sim 10^{-1}$ and, therefore, the amplitude of perturbations increases *e* times during 10 revolutions of the disk. Therefore, at the characteristic amplitudes of the fluctuations of the Jeans scale in the gas galactic disks ($\tilde{\sigma}_{\text{fluct}} / \sigma_0 \sim (0.5 \div 1) \cdot 10^{-1}$ (see the estimate (3.385)) the required time for the growth of the perturbations of the Rossby waves to the amplitude ($\tilde{\sigma} / \sigma_0$) ~ 1 is of the order of 20÷30 revolutions of the disk. Thus, the examined mechanism of buildup of the Rossby waves in the gas disks of the galaxies can be quite effective for the excitation of the observed structures of the type of Rossby solitons.

3.9. The dissipative effects

In solving the problem of the gravitational stability of the gas disk and determination of the spectrum of oscillations in its plane it is not necessary to take into account the dissipative terms in the first approximation. This can be carried out using the following estimate. The value of the characteristic 'viscous' frequency $k_J^2 \sim 10^{-20}$ s^{-1} with respect to the parameters of the gas disk of the Galaxy in the solar neighbourhood (here ν is the molecular kinematic viscosity). At the same time, the epicyclic frequency $\varkappa \sim 10^{-15}$ s^{-1} $\gg \nu k_J^2$.

Nevertheless, the investigation of the effects associated with taking into account the dissipative terms may lead to important results. Firstly, this is so because some types of perturbations in the plane of the gas disk may have a negative energy [590] and, consequently, they are unstable with respect to dissipation even in the gravitationally stable disk. Secondly, taking into account the dissipative terms enables us to determine in principle the level of the equilibrium fluctuations, using the fluctuation–dissipative theory [110, 121]. Finally, the fine-scale turbulence can be taken into account in the dissipative model with effective (turbulent) viscosity (section 11.2.1).

3.9.1. *Effect of dissipation on gravitational and entropy perturbations*

We restrict ourselves to investigating the shortwave axisymmetric perturbations. For these perturbations, the Fourier harmonics of the linearised equations of gas dynamics with the dissipative terms taken into account have the form [156] (compared with (3.90)–(3.293))

$$-i\omega\sigma + i\sigma_0 ku = 0, \tag{3.353}$$

$$-i\omega u - 2\Omega v = -ik(\Phi + p/\sigma_0) - \lambda k^2 u, \tag{3.354}$$

$$-i\omega v + \frac{\varkappa^2}{2\Omega}u = -\nu k^2 v, \tag{3.355}$$

$$-i\omega T_0 S = 2\nu r\Omega' ikv - \chi k^2 T c_\sigma, \tag{3.356}$$

where $\lambda = 4\nu/3 + \mu$; ν, μ is the first and second kinematic viscosity; S is the entropy perturbation; χ is the thermal diffusivity coefficient; c_σ is the specific heat capacity at constant density.

The system (3.353)–(3.356) should be supplemented by the Poisson equation whose shortwave solution has the form $\Phi = -2G\sigma/k$, and by two thermodynamic relationships:

$$\frac{\sigma}{\sigma_0} = \beta\sigma_0 T + \frac{p}{\sigma_0 c_T^2}, \tag{3.357}$$

$$S = c_p \frac{T}{T_0} + \beta p, \tag{3.358}$$

where c_p is the specific heat capacity at constant pressure and

$$\beta^2 = \left(\frac{\partial S}{\partial p}\right)_T^2 = \frac{c_p - c_\sigma}{T_0 \sigma_0^2 c_T^2}. \tag{3.359}$$

Solving this system of algebraic equations, we obtain a dispersion equation describing the properties of the investigated perturbations [169]

$$\omega^4 + i(\nu + \lambda + \chi)k^2\omega^3 - \left\{(\nu\lambda + \nu\chi + \lambda\chi)k^4 + \omega_J^2\right\}\omega^2 -$$
$$-i\left\{\nu\lambda\chi k^6 + \chi k^2\omega_T^2 + \nu k^2(\omega_J^2 - \varkappa^2) + \nu k^2\varkappa^2(\gamma_s - 1)\frac{d\ln(\Omega)}{d\ln(r)}\right\}\omega +$$
$$+\nu\chi k^4(k^2 c_T^2 - 2\pi G\sigma_0 k) = 0, \tag{3.360}$$

where $\omega_J^2 = k^2 c_s^2 - 2\pi G\sigma_0 k + \varkappa^2$; $\omega_T^2 = \omega_J^2 - k^2(c_s^2 - c_T^2)$. In the dissipationless approximation, from (3.360) we obtain the dispersion equation of the axisymmetric gravitational perturbations $\omega^2 = \omega_J^2(k)$.

The effect of this equation on these perturbations will now be explained. In accordance with the previously given estimates it is assumed that $\check{\nu} k^2 \ll |\omega_J|$, where $\check{\nu} = \max\{\nu, \lambda, \chi\}$. Consequently, in the linear approximation with respect to the dissipative coefficients, we obtain ($\omega = \omega_J + \delta\omega_J; |\delta\omega_J| \ll |\omega_J|$)

$$\delta\omega_J = -\frac{ik^2}{2\omega_J^2}\left\{\nu\varkappa^2 + \lambda\omega_J^2 + \chi k^2(c_s^2 - c_T^2) - \nu\varkappa^2(\gamma_s - 1)\frac{d\ln\Omega}{d\ln r}\right\}. \quad (3.361)$$

Since $c_s^2 > c_T^2$, $\gamma_{sv} > 1$ and in the disks of the flat galaxies $d\ln\Omega/d\ln r < 0$, then in the gravitationally stable disk ($\omega_J^2 > 0$) the Jeans perturbations attenuate, and in the gravitationally unstable disk ($\omega_J^2 < 0$) they undergo additional destabilisation because of the dissipative effects. This result is natural because the Jeans perturbations $\omega^2 = \varkappa^2 - 2\pi G\sigma_0 k + k^2 c_s^2$ represent the sound ($k^2c_s^2$) perturbations with the allowance made for the gyroscopic effects ($\varkappa^2$) and self-consistent perturbations of the gravitational potential ($-2\pi G\sigma_0 k$). However, the sound perturbations decay[112]:

$$\omega \simeq k c_s - i\frac{k^2}{2c_s^2}\left\{\lambda c_s^2 + \chi(c_s^2 - c_T^2)\right\}, \quad (3.362)$$

and this result can be easily obtained from (3.361) in the limit of the shortwave ($k \gg k_J$) perturbations.

The general dispersion equation (3.360) is the equation of the fourth degree with respect to ω. In the dissipative approximation, this equation shows that in addition to the Jeans perturbations for which $\omega^2 = \omega_J^2(k)$ there are also two (entropy) types of perturbations with $\omega = 0$. Taking this into account, and assuming that $|\omega| \sim \check{\nu} k^2 \ll |\omega_J|$, we obtain the simplified (quadratic with respect to ω) dispersion equation for these perturbations from (3.360):

$$\omega^2 + i\omega\frac{k^2}{\omega_J^2}\left\{\chi\omega_T^2 + \nu(\omega_J^2 - \varkappa^2) + \nu\varkappa^2(\gamma_s - 1)\frac{d\ln\Omega}{d\ln r}\right\} - \frac{\nu\chi k^4}{\omega_J^2}(\omega_T^2 - \varkappa^2) = 0. \quad (3.363)$$

If we ignore heat conductivity ($\chi/\nu \to 0$) and assume that the disk is a solid-state rotating disk, the results of studies [145, 867] follow from (3.363):

$$\omega_D = i\nu k^2(\varkappa^2 - \omega_J^2)/\omega_J^2,$$

and this means that a dissipative instability can develop in the gravitationally stable ($\omega_J^2 > 0$) gas disk in the wavelength range $0 < k < 2k_J$.[8] Taking the finite heat conductivity into account expands the range of the dissipatively unstable wavelength [156] to

$$0 < k < 2k_J\gamma_s. \quad (3.364)$$

This result is in qualitative agreement with that obtained by Kumar [812] for the model of the gravitating cylinder.

[8] In fact, this interval can be slightly smaller because already at $k \to 0$ the WKB-approximation is violated.

3.9.2. Fast dissipative instability

Dispersion equation (3.363) shows that the increment of the dissipative instability with respect to the order of magnitude is equal to $\tilde{\nu} k^2$. It may also easily be seen that if the differential nature of rotation of the disk is taken in to account, the order of magnitude of this result does not change. However, this result can be regarded as correct only if the characteristic time of non-stationarity of the disk t_{ev} is considerably greater than the inverse increment. The characteristic time of dynamic non-stationarity is $t_{ev1} \sim r^2 / \nu$, the characteristic time of thermal non-stationarity is $t_{ev2} \sim (c_s/r\,\Omega)^2\, t_{ev1}$ (this estimate follows from the heat balance equation). In a general case,the increment of the dissipative perturbations is of the order νk^2 and $k \sim \varkappa / c_s$. It follows from this that $\nu k^2 t_{ev2} \sim 1$. Thus, the characteristic time of thermal non-stationarity of the disk and of the development of dissipative instability are of the same order.

Taking these considerations into account, we pay attention to the following circumstance[169]. The order of magnitude of the increment of dissipative instability (like decrement of attenuation of gravitational perturbations) is equal to $\tilde{\nu} k^2$ outside the limits of the relatively narrow adhesion of the wavenumbers situated in the vicinity $k \simeq k_J = \pi G \sigma_0 / c_s^2$. However, for the disk close to the boundary of gravitational stability, the increment of dissipative instability in the range of these wavenumbers is of the order $\tilde{\nu} k_J^2 \varkappa^2 / \omega_J^2 \gg \tilde{\nu} k_J^2$. It is clear that in the disk it may be $|\mathrm{Im}(\omega_D)| > |\omega_J(k_J)|$, or otherwise

$$|\,\omega_J(k_J)\,|^3 < \tilde{\nu} k_J^2 \varkappa^2, \tag{3.365}$$

and, consequently, the previously discussed results in the immediate vicinity of $k = k_J$ cannot be used. In this case, the term 'gravitational perturbations' loses its meaning, and the dissipative properties of all four types of perturbations are determined by the general dispersion equation (3.3.60) whose solution at $k = k_J$ and $|\,\omega_J^3(k_J)\,| \ll \tilde{\nu} k_J^2 \varkappa^2$ has the following form (without taking into account the differential nature of rotation of the disk)

$$\omega_1 \simeq i\,\omega_0;\ \ \omega_{2,3} \simeq \frac{\pm\sqrt{3} - i}{2}\,\omega_0;\ \ \omega_4 \simeq -i\,\chi \frac{k_J^2(2\gamma_s - 1)}{\gamma_s + \dfrac{\chi}{\nu}(\gamma_s - 1)}, \tag{3.366}$$

and

$$\omega_0 = \left\{ \nu\, k_J^2 \varkappa^2 \left[1 + \left(\frac{\gamma_s - 1}{\gamma_s} \right) \frac{\chi}{\nu} \right] \right\}^{1/3}. \tag{3.367}$$

In this spectrum of the solutions only the perturbations with $\omega = \omega_1$ are unstable and their increment is

$$\mathrm{Im}(\omega_1) \sim \tilde{\nu}^{1/3} \varkappa^{4/3} / c_s^{2/3} \gg \tilde{\nu} k_J^2,$$

and the remaining solutions correspond to the attenuating perturbations.

Taking into account the fact that as regards the order of magnitude of $k_J \sim 1/h$ and $k_J^2/\varkappa \sim l/h \ll 1$ (l is the free path of the particles), it may easily be seen that $\mathrm{Im}(\omega_1)t_{ev2} \sim (h/l)^{2/3} \gg 1$ .Therefore, the dissipative instability (3.366) is fast and for dynamic processes, determined by this instability, thermal non-stationarity of the disk is not important.

It is quite easy to generalise the results (3.366) for the case of the differentially rotating disk. At the limit $|\omega_J(k_J)|^3 \ll \tilde{\nu} k_J^2 \varkappa^2$ from the initial dispersion equation (3.360) we obtain

$$\omega_1 \simeq i\left\{\nu k_J^2 \varkappa^2 \left[1+\left(\frac{\gamma_s-1}{\gamma_s}\right)\frac{\chi}{\nu}-(\gamma_s-1)\frac{d\ln\Omega}{d\ln r}\right]\right\}^{1/3} . \qquad (3.368)$$

Since $\gamma_s > 1$ and for the astrophysical disks $d \ln \Omega / d \ln r < 0$, it may easily be seen that as regards the order of magnitude, the increment (3.368) is the same as in the case of solid-state rotation. Thus, if the differentiality of rotation of the disk is taken into account, this provides no significant contribution to the dynamics of dissipative instability.

At the shortest wavelength limit $k \gg k_J$ (see 3.364) the dissipative instability does not occur. Naturally, this indicates that the dissipative buildup of the perturbations in the gravitationally stable disk is determined by the effect of perturbations of the gravitational field which is important in the wavelength range $k \sim k_J$. As shown by Fridman and Polyachenko [590], this effect is reflected in the fact that the energy density of the dissipatively unstable perturbations is negative ($E_{\omega 1} < 0$). In this case, the energy dissipation of the perturbations ($dE_{\omega 1}/dt < 0$) is equivalent to the increase of its absolute value and, consequently, the increase of the perturbation amplitude. Actually, for the dissipatively unstable perturbations (3.366) with frequency ω_1 [169]

$$E_{\omega_1} \simeq -\frac{\sigma_0 c_s^2}{2}\left(\frac{\nu k_J^2}{\varkappa}\right)^{2/3} \frac{1-\chi(\gamma_s-1)/\nu\gamma_s}{[1+\chi(\gamma_s-1)/\nu\gamma_s]^{1/3}}\left|\frac{\sigma}{\sigma_0}\right|^2, \qquad (3.369)$$

where σ is the perturbed surface density of the disk.

Identical calculations of the energy density of the gravitational perturbations at $k \simeq k_J$ lead to the following result

$$E_J = \frac{\sigma_0 \omega_J^2(k_J)}{k_J^2}\left|\frac{\sigma}{\sigma_0}\right|^2, \qquad (3.370)$$

from which $\mathrm{sign}(E_J) = \mathrm{sign}[\omega_J^2(k_J)]$. Therefore, the Jeans perturbations, with the dissipative factors in the gravitationally stable disk taken into account, attenuate and are additionally unstable in the gravitationally unstable disk (see (3.361)).

3.9.3. Equilibrium fluctuations in a gas disk

To calculate the level of equilibrium noise in a gravitating gas disk, we use the fluctuation–dissipative theorem [110] and the hydrodynamic theory of fluctuations [121]. To simplify considerations, we restrict ourselves to the model of a homogeneous solid-state rotating ($\varkappa \equiv 2\Omega$) disk and, examining the shortwave perturbations, the x axis is oriented along the direction of the wave vector k. The Fourier harmonics of the gas-dynamic equations with the lateral forces taken into account and introduced for calculating the tensor of generalised susceptibility in the reference system rotating together with the disk, has the form

$$-i\omega u - \varkappa v = -ik(R + \Phi) - \lambda k^2 u + f_1^{\mathrm{CT}}, \tag{3.371}$$

$$-i\omega v + \varkappa u = -\nu k^2 v + f_2^{\mathrm{CT}}, \tag{3.372}$$

$$-i\omega \Sigma + iku = 0, \tag{3.373}$$

$$-i\omega S = -\chi k^2 c_\sigma \tau + f_0^{\mathrm{CT}}, \tag{3.374}$$

where $\Sigma = \sigma/\sigma_0$, $\tau = T/T_0$, $R = p/\sigma_0$. The system should be supplemented by the thermodynamic relationships (3.357), (3.358), and the Poisson equation ($\Phi = -2G\pi\sigma_0\Sigma/k$). The dissipative terms in (3.371)–(3.372) should be taken into account to bypass the poles in the complex ω-plane in inverse Fourier transform of the spectral density of noise. The final result – the level of noise – does not include the dissipative coefficients because $\nu k^2 \ll \varkappa$. Therefore, to simplify the calculations, it is assumed that $\nu = \chi$.

The dissipation of energy in the gas disk under the effect of the lateral forces is determined by the expression

$$\dot{E} = \sigma_0 \int\left\{u f_1 + v f_2 + T f_0\right\} dS = -\int \mathcal{X}_i \frac{\partial \mathcal{F}_i}{\partial t} dS, \tag{3.375}$$

where the integration is carried out over the surface of the disks, and $\mathcal{X}_i$, $\mathcal{F}_i$ are the generalised coordinates and the lateral forces, and the spectral relationship between them is determined by the equation [110]

$$\mathcal{X}_i(k,\omega) = \alpha_{ij}(k,\omega)\mathcal{F}_i(k,\omega), \tag{3.376}$$

where $\alpha_{ij}(k, \omega)$ is the tensor of generalised susceptibility.

The generalised coordinates are presented by $\mathcal{X}_1 = u$, $\mathcal{X}_2 = v$, $\mathcal{X}_0 = T$. Consequently $\mathcal{F}_i = -i\sigma_0 f_i \ / \omega (i = 0, 1, 2)$. Using these definitions and solving the system described above, we obtain the components of the tensor of generalised susceptibility. Using the main equation of the fluctuation–dissipative theory [110], the following equation can be written for the planar case

$$\langle \mathcal{X}_l \mathcal{X}_n \rangle_{k,\omega} = \frac{iT_0}{(2\pi)^3 \omega}(\alpha_{nl}^* - \alpha_{ln}), \tag{3.377}$$

and the spectral level of the surface density fluctuations is presented in the form [158]:

$$\left\langle \left(\frac{\sigma}{\sigma_0}\right)^2 \right\rangle_k = \frac{T_0(2+\gamma_s)}{4\pi\sigma_0 c_s^2} + \frac{T_0(2\pi G\sigma_0 k - \varkappa^2)}{2\pi^2 \sigma_0 c_s^2 \omega_J^2} + \frac{2GT_0}{4\pi c_T^2 (k c_T^2 - 2\pi G\sigma_0)}. \tag{3.378}$$

Here the first term which is independent of the wavenumber is the spectral density of the thermodynamic noise, and the second and third terms should be referred to as the spectral density of the 'gravitational–rotational' noise. Calculating the correlation function of the thermodynamic noise, we obtain

$$\left\langle \left(\frac{\sigma}{\sigma_0}\right)^2 \right\rangle_r^T = \frac{T_0(2+\gamma_s)}{4\pi^2\sigma_0 c_s^2}\int e^{i\mathbf{kr}}d\mathbf{k} = \frac{2+\gamma_s}{\sigma_0 c_s^2}T_0\delta(\mathbf{r}). \tag{3.379}$$

The absence of the spatial correlation of the thermodynamic noise should be regarded as the indication that the characteristic size of the correlation in this case is of the order of the minimum size possible in the hydrodynamics – the free path of the particles l. Thus, in (3.379) it should be assumed that $\delta(\mathbf{r}) \sim l^{-2}$.

The spectral density of the 'gravitational–rotational' noise, represented by the second term in (3.378), in the disk close to the boundary of gravitational stability $[|\omega_J(k_J)| \ll \varkappa]$, has a sharp maximum in the vicinity $k = k_J$, where $2\pi G\sigma_0 k - \varkappa^2 \simeq \varkappa^2$. Taking this circumstance into account, we obtain

$$\left\langle \left(\frac{\sigma}{\sigma_0}\right)^2 \right\rangle_r^{GR_1} \simeq \frac{T_0\varkappa^2 k_J J_0(k_J r)}{\sigma_0 c_s^2 \omega_J(k_J)}, \tag{3.380}$$

where $J_0(x)$ is the Bessel function of the first kind. Finally, integration of the last term in (3.378) gives

$$\left\langle \left(\frac{\sigma}{\sigma_0}\right)^2 \right\rangle_r^{GR_2} \simeq \frac{GT_0}{c_T^4}\left\{\frac{1}{r} - \frac{\pi}{2L_T}\left[Y_0(r/L_T) + H_0(r/L_T)\right]\right\}, \tag{3.381}$$

where $L_T = c_T^2/2\pi G\sigma_0$, $Y_0(x)$, $H_0(x)$ is the Neumann and Strood function, respectively.

We obtain the relationships for evaluating the noise level. Since $T_0 \sim mc_s^2$, where m is the characteristic mass of the 'particles' of the disk, we obtain the following equation for the thermodynamic noise (3.379), taking into account that $\delta(\mathbf{r}) \sim l^{-2}$

$$\alpha_T = \left\{ \left\langle \left(\frac{\sigma}{\sigma_0}\right)^2 \right\rangle^T \right\}^{1/2} \simeq \frac{1}{l}\left(\frac{m}{\sigma_0}\right)^{1/2}. \tag{3.382}$$

For the non-thermodynamic fluctuations (3.380), the characteristic spatial scale $L \sim k_J^{-1} \sim c_s/\varkappa = \rho$, and their intensity on the scales $\leqslant k_J^{-1}$ with the result (3.366) taken into account can be estimated as

$$\alpha_{GR_1} = \left\{ \left\langle \left(\frac{\sigma}{\sigma_0}\right)^2 \right\rangle^{GR_1} \right\}^{1/2} \simeq \left(\frac{m\varkappa}{\sigma_0\omega_J(k_J)\rho^2}\right)^{1/2} \simeq \frac{(m/\sigma_0)^{1/2}}{(\rho^5 l)^{1/6}}. \tag{3.383}$$

Finally, the characteristic scale of the non-thermodynamic noise (3.381) is of the order $L_T \sim \rho$, and their maximum intensity (at $r \sim l$) is

$$\alpha_{GR_2} = \left\{ \left\langle \left(\frac{\sigma}{\sigma_0} \right)^2 \right\rangle^{GR_2} \right\}^{1/2} \simeq \left(\frac{m}{\rho l \sigma_0} \right)^{1/2}. \qquad (3.384)$$

As an example of the application of this theory, we estimate the level of equilibrium noise in the solar vicinity of the gas subsystems of the Galaxy. It is assumed that the dispersion of the velocities of the gas clouds – 'macroatoms' of the disk is $c_{s\odot} \simeq 9$ km/s, their characteristic radius and the mass are equal to $r_m \simeq 4$ pc and $m \simeq 120\ M_\odot$, respectively. Using also the data on the parameters of the gas disk in the solar neighbourhood: $\varkappa \simeq 10^{-15}$ s^{-1}; $\sigma_0 \simeq 10\ M_\odot$/pc^2; $h \simeq 70 \div 80$ pc, we obtain $\rho = c_s/\varkappa \simeq 300$ pc, the number of clouds in the unit volume is $n \simeq \sigma_0/2hm \simeq 5 \cdot 10^{-4}$ pc^{-3}. Since the gas clouds are not solid spheres, it is assumed that the effective (collisional) radius of the mean cloud is slightly smaller than the eutectic radius (let it be that $r_{\text{eff}} \simeq 0.7\ r_m$). The interaction cross-section of the clouds is $\hat{\Sigma} \simeq \pi\,(2r_{\text{eff}})^2 \simeq 10^2$ pc^2, and the free path is $l \simeq 1/\sqrt{2}n\hat{\Sigma} \simeq 20\,\text{pc}$. As a result we obtain

$$\alpha_{T\odot} \sim 2{\cdot}10^{-1}; \qquad \alpha_{GR\odot} \sim 5{\cdot}10^{-2}. \qquad (3.385)$$

The high level of the noise, predicted by these estimates, indicates that the fluctuation structures in the gas subsystems of the Galaxy can in principle be detected. The characteristic dimensions of the most intensive of these structures (thermodynamic) are of the order of the free path of the clouds (several intercloud distances) and such complexes may contain up to several tens of clouds. The non-thermodynamic fluctuations with the slightly lower intensity have the characteristic dimensions $\sim 1 \div 3\ k_J^{-1}$ (according to the parameters of the gas subsystem in the solar neighbourhood it is $\sim 0.3 \div 1$ pc) and may evidently ensure the 'seeding' perturbations for special features of the spiral pattern (branching from the spirals, etc).

3.10. The effect of hot gas on disks of the galaxies

3.10.1. Hot gas of clusters

The static pressure of the hot gas, filling the clusters and some groups of the galaxies, may have a significant effect on the volume density and the thickness of the gas disks in the galaxies [1326].

The interstellar gas in the disks of the galaxies is inhomogeneous as regards the density, but since the regions with high density are usually also colder, the pressure fluctuations are considerably smaller than the density fluctuations, and we can talk about the characteristic equilibrium values of pressure P at the given distance r from the centre of the galaxy. In the solar neighbourhood, the pressure of the interstellar medium (normalised with respect to Boltzmann constant k), determined primarily by the turbulent movement of the gas, equals $\simeq 2{\cdot}10^4$ K$\cdot$cm^{-3} [492], i.e. $\log P/k \approx 4.3$. Gas pressure is determined primarily by its mean density and, consequently, the gas pressure decreases with increase of the distance from the centre of the Galaxy.

According to the calculations [93] of the equilibrium pressure of the gas disk in the plane of the stellar disk for the self-consistent models of a number of close spiral galaxies (including our Galaxy), the logarithm of gas pressure P in the selected unit system is 4÷5 at r = (0.2–0.3) R_{25} and 3.5÷4 at r = (0.7–0.8) R_{25}, where R_{25} is the photometric radius of the galaxy. At even greater distances from the centre, the gas disk rapidly expands and pressure decreases at a high rate. A very low gas pressure is recorded in low-brightness galaxies where the volume density of the gas is an order of magnitude lower than in the conventional spiral galaxies [1325].

If a galaxy is situated in a cluster, its interstellar medium is subjected to the effect of the hot inter-galactic gas. At rapid movement of the galaxy and the favourable orientation of the disk in relation to the vector of its velocity, the dynamic pressure of the gas (ram pressure) should become evident and proportional to $n_e V_c^2$, where n_e is the concentration of the electrons of the inter-galactic gas, and V_c is the velocity of the galaxy in relation to the gas. The ram pressure 'sweeps out' the rarefied gas (HI) from the outer regions of the galaxies, generates an asymmetry in the distribution of the HI and reduces the size of the region, occupied by the gas, in the galaxies with a 'shortage' of HI (see, for example [445, 1122] and references therein). It is far more difficult to 'sweep out' the gas from the internal regions of massive galaxies where the density of the stellar disk is higher, and the potential well, generated by the gravitational field of the galaxy, is considerably deeper.

However, the pressure of the surrounding gas at a relatively high temperature may prove to be very high even at a low velocity of movement of the galaxy in the medium. Since the hot gas fills the entire cluster, the speed of sound $c \approx (P/\rho)^{1/2}$ for the gas is close to the mean square velocity of movement of the galaxies and, therefore, the static pressure of the intergalactic gas $P = 2n_e kT$, determined by the electronic and ionic components, is close to the mean value of the dynamic pressure (with the accuracy of the factor ~1.5). However, in contrast to the latter, the static pressure acts independently of the velocity of the galaxy in relation to the medium or orientation of its disk, and manifests itself at all distances from the centre of the galaxy.

The estimates of the temperature and density of the intergalactic medium in the systems of the galaxies show that the static pressure P of the hot gas is usually comparable with the estimates of the pressure of the interstellar gas in the galactic disks and, therefore, may have a strong effect on the evolution of gas inside the galaxy. Actually, the characteristic values of the temperature of the hot intergalactic gas vary from several keV (more than 10^7 K) in small clusters to 10 keV (more than 10^8 K) in clusters such as Coma (see, for example [310, 555]). The gas concentration decreases with increase of the distance from the centre of of the cluster in accordance with the law which is usually approximated by the equation

$$n(R) = n_0 \left[1 + \left(\frac{R}{R_c} \right)^2 \right]^{-3\beta/2}, \tag{3.386}$$

where the parameter $\beta \simeq 0.4–0.7$ depends on the internal structure of the cluster, and the radius of the nucleus of the cluster R_c equals tens of kpc for small clusters and hundreds of kpc for the largest systems (see, for example [310]). At a distance of $(1–2)\cdot R_c$ the electron concentration is $n_e \simeq 10^{-2}–10^{-3}$ cm^{-3}. In the rich clusters, such as Coma, A7095, A3112, within the range of several hundreds of kpc from the centre $\log P/k \geqslant 5$. In the Virgo cluster at a distance of 100–200 kpc from the centre at temperature $T \approx 3$ keV the particle concentration is around 10^{-3} cm^{-3} which corresponds to $\log P/k \geqslant 4.5$. Even at the concentration of $n_e \sim 10^{-4}$, which is more characteristic of the cluster as a whole, for clusters such as Virgo one obtains $\log P/k \geqslant 3.5–4$, which is higher than the expected pressure of the interstellar gas in the external regions of the spiral galaxies. In small clusters, containing x-ray gas, the pressure is of the same order of magnitude as in the large clusters: $kT \approx 1.5$ keV, $n_e \approx 10^{-3}$ cm^{-3} (see references in [1326]).

The same considerations apply to the galaxies situated in groups, if the latter are filled with the x-ray gas. The slightly lower temperature of the hot gas (~1 keV) is compensated in this case by the higher particle concentration. In these groups, as in the clusters, there is a shortage of HI, and this shortage is greater than in the groups with low intensity x-ray radiation.

Thus, even at low velocities of movement of the galaxies in the systems of different scales, the pressure on the gas disk may exceed the internal pressure of the interstellar gas which would be obtained in the absence of the external effects. Consequently, the pressure of the interstellar gas in the galaxies, especially in the central regions of the clusters, should be on average higher, and the thickness of the gas disk $2h$ considerably smaller than in the galaxies not subjected to the external pressure. As the density of the gas in the Galaxy decreases, the effect of static pressure becomes stronger.

3.10.2. Pressure in the gas disk in the presence of an intergalactic medium

We investigate the effect of increase of the gas pressure in the plane of a galaxy quantitatively, within the framework of simple equilibrium models. The equation of hydrostatic equilibrium for the gas in the gravitational field of the stellar disk has the form

$$\frac{dP}{dz} = -\varrho(z)g(z) = -\frac{\mu}{R}\frac{P(z)}{T(z)}g(z). \tag{3.387}$$

Acceleration $g(z)$ can be expressed by the frequency of vertical oscillations Ω_z: $g(z) = z\,\Omega_z^2/(1 + |z|/\Delta)$, where Δ is the vertical scale of the stellar disk. This expression describes the linear growth of $g(z)$ at small z with transition to the constant at large altitudes $|z| \gg \Delta$. If we restrict ourselves to the approximation $g \propto z$, $T = T_0 =$ const, then equation (3.387) leads to the solution $P = P_0 \exp(-z^2/2h_0^2)$ where for the vertical scale of the isothermal gas disk we have $h_0^2 = \mathcal{R}T_0/\mu\Omega_z^2$.

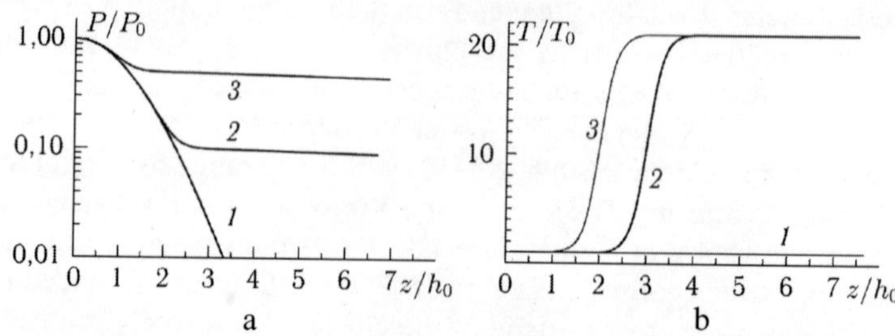

Fig. 3.12. Vertical distribution of pressure (P/P_0) and temperature (T/T_0) normalised with respect to the values in the plane $z = 0$, at the accepted ratio of the half thicknesses of the non-perturbed gas and the stellar disk $h_0/\Delta = 0.2$: 1) in the absence of atmosphere (T = const); 2 and 3) in the presence of an atmosphere. Curve 3 corresponds to the higher pressure of the atmosphere P_a.

We consider a hot gas surrounding a disk as an atmosphere with specific pressure P_a and temperature T_a. Let the temerature in the region of the disk be $T = T_0$ = const, and at large altitudes in the region of the atmosphere the temperature is also constant and $T(z \gg h_0) = T_a \gg T_0$, so that in some zone at $z > h_0$ the temperature changes from T_0 to T_a. Figure 3.12 shows the profiles of the distribution of pressure along z determined by numerical integration of (3.387) at different dependences of temperature $T(z)$ in the vertical coordinate [1326]. As indicated by the diagram, at a relatively large increase of temperature at the altitude $z \sim (2\text{–}3)\, h_0$ the pressure profile ceases to decrease with the altitude reaching the plateau at $P = P_a$.

The mean gas pressure in the galactic disk can be described by the equation

$$\langle P\rangle = P_a + \langle \varrho\rangle gh = P_a + \langle \varrho\rangle \Omega_z^2 h^2, \tag{3.388}$$

where h is the characteristic scale of the vertical inhomogeneity of the gas disk, and the 'vertical' acceleration in the range h, determined by the gravitation of the stellar disk, is accepted to be equal to $g = \Omega_z^2 h$, and the brackets $\langle\ldots\rangle$ denote averaging with respect to the z-coordinate. We introduce the notation $\langle \varrho\rangle = k_1\varrho_0$, $\langle P\rangle = k_2 P_0$, where index 0 here and later relates to the values in the plane of the disk $z = 0$, and the coefficients $k_{1,2}$ are determined by the nature of the vertical distribution of the density and pressure ($0 < k_{1,2} < 1$).

Let the volume density of the gas in the absence of the external atmosphere be equal to ϱ_1, and in the presence of the atmosphere to ϱ_2, and the appropriate values for the characteristic values of the half thickness of the gas disks are equal to H, h. Evidently, for the isothermal model with no atmosphere $H = h = h_0$, and in the presence of the atmosphere $h < H$.

It will be assumed that in the areas in which the gas is not 'swept out' from the disk by the ram pressure, the surface density of the gas in compression remains unchanged, and taking into account the conservation of mass in the disk, we can write the following equation

$$\int_0^{\infty} \varrho_1(z)dz = \int_0^{h} \varrho_2(z)dz, \tag{3.389}$$

where $H\varrho_{01} = h\varrho_{02}$.

Taking this equation into account and restricting ourselves to the polytropic law with the polytrope index n, we can write

$$P_0 = P_0\left(\frac{H}{h}\right)^n. \tag{3.390}$$

Substituting (3.389), (3.390) into (3.388) we obtain:

$$x^{n-1} + \frac{\delta}{k_1}x^n - \frac{k_2}{k_1} = 0, \tag{3.391}$$

where $x = h/H$, $\delta = P_a/P_0$. At $n = 2$ we obtain the following equation for the ratio of the vertical scales in the presence and absence of the atmosphere:

$$x = \frac{h}{H} = \frac{1}{2}\left(\sqrt{4\frac{k_2}{\delta} + \frac{k_1^2}{\delta^2}} - \frac{k_1}{\delta}\right). \tag{3.392}$$

In the case $\delta = 0$ (there is no hot atmosphere) to fulfil the condition $x = 1$, it is accepted that $k_1 = k_2$. In the second limiting case, at a large pressure difference, the asymptotics $x \propto 1/\sqrt{\delta}$ holds (Fig. 3.13a). The models with the low value of the exponent n are more realistic. As an example, Fig. 3.13b shows the dependence of the ratio H/h on the value δ at $n = 1.4$. It may be seen that the thickness of the disk rapidly decreases with increase of n.

Thus, with increasing external static pressure on the side of the intergalactic hot gas the gas disk thickness within the galaxy can significantly decrease and the coefficient of proportionality between H and h is defined by the equation of state of the gas and the nature of the vertical profile of density (and pressure) which in turn depends largely on the choice of thermodynamic model of the gas. For example, as follows from Fig. 3.13, at the external pressure equal to the unperturbed pressure in the disk plane, the disk thickness decreases and, therefore, the bulk density of the gas increases ~ 1.5–4 times. If the pressure of the atmosphere P_a

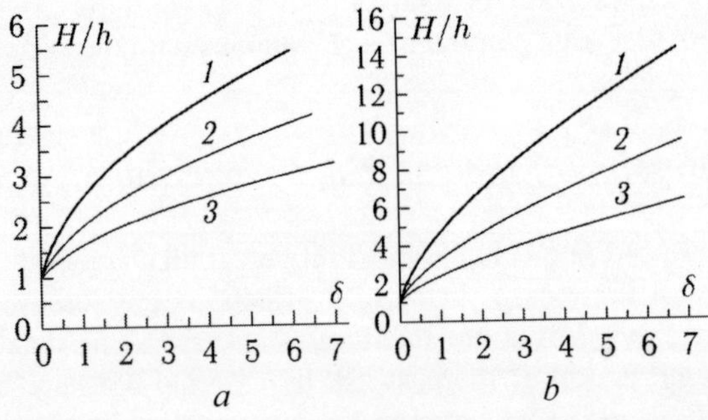

Fig. 3.13. The ratios of the characteristic values of the half thickness of the gas disk in the absence of the atmosphere (H) and in its presence (h) in relation to the ratio δ of the external pressure P_a to the non-perturbed pressure P_0 for different values of the parameters characterising the vertical distribution of gas density [1326]: $a - n = 2$, $b - n = 1.4$, 1) $k_1 = k_2 = 0.25$, 2 – $k_1 = k_2 = 0.5$, 3) $k_1 = k_2 = 1$. The latter case corresponds to the disk with the constant density in the region $|z| \leqslant h$.

exceeds P_0, say, four times, the thickness of the disk changes already 2.5–10 times, and lower values of these variables clearly do not relate to a realistic model of the homogeneous distribution of the density along the height of the disk. Note, however, that the presence of magnetic fields in the interstellar medium, whose pressure is comparable to the initial pressure of the interstellar gas, slightly reduces the degree of compression of the gas, because the preservation of the magnetic flux field pressure, counteracting compression, increases as $(H / h)^2$. In the limiting case when at strong compression the pressure of the interstellar gas will be small compared to the pressure of the magnetic field, the field strength $B = \sqrt{8\pi P_a}$. Note that the magnetic field can also dramatically reduce the thermal conductivity at the interface of two media, isolating the colder gas within the galaxy from the hot environment. This applies not only to the gas disks of galaxies in clusters, but also to the gas coronas of galaxies, which are compressed but 'survive', despite being surrounded by the hotter environment (see, for example, [1214]).

Thus, we can expect that the static pressure of the environment in clusters of galaxies and some groups of galaxies is able to increase several times the density of the interstellar medium, not 'swept' by the flow of the intergalactic gas. As a consequence, the gas disks should be in the mean thinner, and the bulk density of the gas higher.As the rate of star formation is proportional to the volume density to the degree $m = 1$–2 (Schmidt law), this means a significant increase in the rate of star formation per unit mass of gas and a more rapid exhaustion of the gas throughout the disk.

The authors of [1326] concluded that the content of the gas inside the galaxy, surrounded by hot gas, can be reduced not only by 'sweeping out' the gas, but also due to its more rapid transition to stars throughout the disk, since the static pressure, in contrast to dynamic, acts as on both the exterior and the interior of the disk (although with lower efficiency). In the process of evolution and in transition of the bulk of the gas into stars its surface density can be low enough so that the gas is almost completely 'swept out' from the disk by the ram pressure of the environment. Together with the mechanism of dynamic pressure and the process of merging of galaxies, the static pressure allows to explain the existence in the inner regions of clusters of a large number of disk galaxies with the extremely low content of the interstellar gas.

3.10.3. Compression of the gas disk by external pressure

The static pressure of the gas is the most universal mechanism of the influence of the environment on the galaxy, because it affects its gas disk in all cases where the galaxy is surrounded by a hot medium. There are some data that support the hypothesis of compression of the gas disk by external pressure in many galaxies [1326], especially if they are located in the inner regions of clusters, although the following arguments are indirect, so that in every case it is important to consider also the effect of other factors.

1. Many spiral galaxies of clusters, including those with the reduced content of HI, have indeed high rates of star formation per unit mass of gas, i.e. a short

timescale of gas depletion: active star formation continues, despite the 'deficit' of HI, indicating the high efficiency star formation [1129].

Note that the rates of star formation are statistically related with the gas density: SFR ~ ρ^n_{gas} where $n \approx 1$–2 (Schmidt law), so even the doubling of the gas density increases the rate of exhaustion of the gas in star formation 2–4 times. For spiral galaxies, the characteristic time of the exhaustion of gas $T_g = M_{gas}$ /SFR is usually a few billion years ([1324] and references therein). The same order of magnitude has the dynamic time during which the galaxy intersects the inner, most dense region of the cluster. Therefore, increasing the gas density and the corresponding decrease in T_g in compression of the gas layer can be a significant factor of the evolution of the gas content of the galaxies of the clusters. The decrease in gas density in the outer regions of the disk should facilitate 'sweeping' remnants of the interstellar medium by the flow of the intergalactic gas.

2. Galaxies with 'the deficit' of HI in clusters have a higher (average) molecular gas content in relation by the atomic gas [756], and this can not can always be explained by 'sweeping' of HI from the peripheral regions of galaxies at higher concentrations of H_2 to the centre. The example of spiral galaxies in the inner Virgo region shows that the unusually high proportion of the molecular gas is observed even in the inner disk region where HI remains, although it may be 'swept' from the periphery of the disk [950]. In the work cited it has been suggested that the external pressure may play a certain role increasing the amount of molecular gas. Compression of gas should really facilitate the transition of atomic gas to molecular, since, as shown by the analysis of available observational data, the relative content of the molecular gas is closely correlated with the pressure in the plane of the disk (see [93] and references therein).

3. The example of the galaxies of the Virgo cluster with a 'deficit' of HI shows that in some cases the decrease in the total number of atomic gas in a galaxy is not so much due to the reduction of the area occupied by this gas (which is naturally associated with the ram pressure) as with the decrease of the surface density of HI across the disk. The latter may be the result of gas compression and its faster 'exhausting'.

4. Amplification of the magnetic field, expected in compression of the gas disk in a galaxy, is consistent with the fact that, as noted by a number authors, a significant portion of the spiral galaxies in the clusters, is distinguished by a higher power of synchrotron radiation of the disk as compared to galaxies outside the clusters (see [1129] and references therein).

5. The role of the static pressure should be most significant for galaxies that are in the inner region of the cluster filled with X-ray gas, and thus have low velocities relative to the gas. This occurs if galaxies do not move far from the cluster centre. In this regard, special attention should be given to HI observations in the Pegasus I cluster. A small but clearly detectable 'deficit' of HI was found for galaxies in the central part of this cluster [828]. They have a very low velocity dispersion, so that the value of $n_e V_c^2$, which characterizes the dynamic pressure, is more than two orders of magnitude smaller than in Coma or Virgo, which does not allow the standard connection of the 'deficit' of HI with the

movement of galaxies in a gaseous environment. Electron density $\langle n_e \rangle$ in a cluster, calculated for the model of a homogeneous sphere filled with hot gas with $T = (0.6–3)\cdot 10^7$ K, is about $2\cdot 10^{-4}$ cm^{-3} [443]. It is almost coincides with the lower bound for the concentration of particles in the range 16 arcmin (360 kpc) from the central galaxy of the cluster NGC 7619, obtained from measurements of ROSAT. At $T \approx 1 \div 2$ keV the corresponding gas pressure is $P \approx (4 \div 8)\cdot 10^3$ K cm^{-3}. This static pressure, although not large, is still comparable to the pressure of the interstellar gas in the outer regions of galactic disks (see section 3.10.1) and is several times greater than the dynamic pressure experienced by galaxies in the central part of the cluster, which, based on assessment $n_e V^2 = 12$ (km/s)^2 cm^{-3} [828] is $\simeq 1.5\cdot 10^3$ K cm^{-3}. Therefore, in this case the static pressure of the gas may be more significant.

Note that the external static pressure on the gas disk of the galaxies may be exerted not only by the intergalactic gas, but also by the gas inside the galaxy, located in a bulge or inner corona, if it has a sufficiently high concentration of 10^{-2}–10^{-3} cm^{-3} at the virial temperature of a few million degrees. The existing estimates of the density and temperature of hot gas in the bulge and the halo of galaxies based on their radiation in the soft X-ray range are still scarce, but they show that at least in some part of the massive galaxies in systems such as M 104, NGC 4565, NGC 5746, NGC 4921, NGC 4911 the medium with the required values of density and temperature actually exists [1214, 1326]. In these cases, the pressure of hot gas on the disk manifest itself in the same way as the pressure the intergalactic medium: it compresses the gas in the inner region of the disk and increases its bulk density and the fraction of the mass falling on the molecular gas, which affects the star formation process. Indeed, as shown by observations, the proportion of the molecular gas in the interior of M 81, M 106, and, perhaps, M 31, dominated by the stellar bulge, has the surface density significantly higher than could be expected from the dependence of the relative weight of the molecular gas on the pressure if the latter is evaluated without consideration of the environment [93].

At an earlier stage of the evolution of disk galaxies their gas disk had a much higher surface density, which was then reduced by passing the bulk of gas mass to stars. For this reason, the efficiency of 'sweeping' the gas by ram pressure in movement in the gas medium was lower than at the present time. We can therefore assume that the static pressure, 'accelerating' the star formation process, has played an important role in the formation of the outer regions of stellar disks.

Thus, the pressure on the hot environment on the gas disk of the galaxy under some very real conditions may be an important factor of its evolution.

4

Numerical modelling of galactic disks

4.1. Review of the methods of dynamic modelling

At the present time, there are no methods of direct laboratory modelling of the collective processes in the continuous or discretely gravitating medium. It is also unlikely that they will be developed in the near future. However, the development of relatively powerful computers has opened unique possibilities of numerical modelling of such phenomena and, at the same time, created a unique experimental base for the physics of collective processes in gravitating systems. In the application to flat galaxies, this modelling may be carried out to determine the rate of collective relaxation processes, explain the spatial structure of quasi-stationary flat galaxies and individual sub-systems of these galaxies, and also can be used for examining the processes which could lead to the formation of various structures, such as spiral patterns, bars, ring-shaped structures, special features in interacting galaxies, etc.

The established term for denoting numerical modelling of gravitating N-bodies (and also gas-dynamic systems) is the term 'numerical (or computer) experiment'. The numerical experiment has become recently one of the main tools in the hands of theoreticians in the investigations of the dynamics of stellar systems. There is virtually 'gold rush', characterised by the exponential increase of the number of publications (Fig. 4.1). At the same time, since the number of unsolved problems is still greater than the number of the solved problems, the results are usually based on a small series of calculations of a small group of investigators. It appears that it is very important to repeat the previously obtained results so that they can be trusted, as has been carried out in experimental physics.

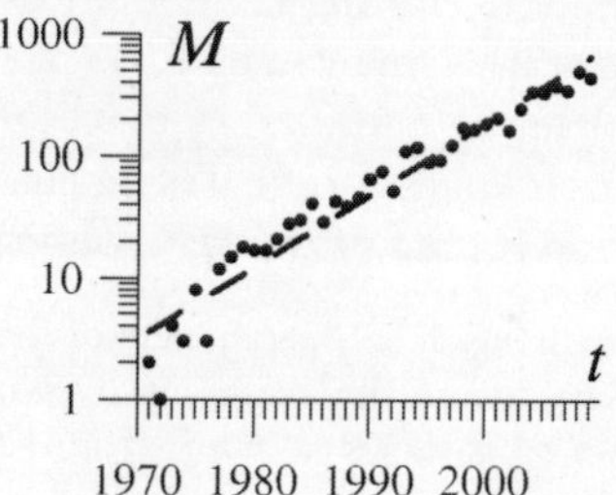

Fig. 4.1. The number of scientific publications in the physics of galaxies and cosmology in which the results are based mostly on the models of N-bodies. The dotted line corresponds to the law $\propto \exp(t/7 \text{ years})$.

The role of numerical modelling of the astrophysical phenomena is difficult to underevaluate, and this is clearly demonstrated by the specific increase of the number of studies based on numerical approaches. This is explained, on the one hand, by the colossal advances in the area of microprocessors and the availability of computing technology, in particular, computers with a massive-parallel architecture, and, on the other hand, by evident successes in the area of computing mathematics.

In this chapter, we describe the methods of constructing numerical models and use these models in subsequent chapters to investigate a number of problems of the physics of galaxies, using the numerical gas-dynamic models and the models of *N*-bodies. The results described in section 4.10 and in the section 7.7.1, 7.8.3, 8.4.6, 10.5.2, 10.5.3, 10.5, are new and have been obtained in cooperation with M.A. Butenko, M.A. Eremin, S.A. Khoperskov and S.S. Khrapov.

4.1.1. Models of N-bodies

In this section, we examine the main numerical methods of calculating the gravitational force (or potential) from the given distribution of the mass which form the basis of the dynamic models of the so-called models of *N*-bodies.

1) The direct method (*PP* = 'particle–particle') is based on taking into account the gravitational interaction of all particles with each other [252, 996]. This method is the simplest procedure for understanding and realisation and, most importantly, ensures, with other conditions being equal, the highest accuracy of calculation of the gravitational force. However, the method requires the largest machine resources because each particle interacts with other particles. Therefore, this approach can be used as a 'test model' and can be used to verify the most important principal results and verify other methods.

In the models of this type, the number of particles is usually not greater than the value $N \lesssim 10^4$ [1]. The number of arithmetic actions, required for the integration of a single time step within the framework of the PP model, is equal to $\sim 10 \times N^2 - N$. It may easily be seen that when using a computer in which 1 μs is required for a single operation, at $N = 10^3$ a single time step (and in most cases, several thousands of steps are required) requires 10 s, and at $N = 10^5$ it is more than 24 hours!

The advantage of the model of this type is associated with the fact that it is possible to determine the forces with the same accuracy as the arithmetic accuracy of the computer and, in particular, the experiments with these models yielded a number of important results [134, 154, 155, 258, 277, 989, 996, 1268, 1269], and others. Finally, the direct method can be used efficiently for testing faster but less accurate models (naturally, comparison must be made at the same value of *N*).

2) The group of models under the conventional term 'particles in a mesh' (PM = Particle–Mesh), including various modifications (Cloud-in-Cell, Triangular-Shaped-Cloud, P³M). They are characterised by the application of the fast Fourier

[1] In the PP models in the 'measurement' of macroscopic (including local) parameters of the disk which is in the state close to stationary, it is necessary to use the averaging over *n* time steps ($n \sim 100$). This effectively increases the number of particles taking part in the experiment).

transform for solving the Poisson equation. The algorithms differ in the method of calculating forces from near and remote particles [33, 252, 920].

The configuration space and, in a number of cases, the space of velocities are divided into cells in which the particles are also localised. The PM model does not take into account directly the interaction of the particles situated in the same cell, and the interaction of the particles, located in different cells, is modelled by the Newton gravitational forces calculated, in most cases, on the basis of the coordinates of the centres of the cells. This results in the complete 'switching off' of the gravitational interaction at the distance smaller than the size of the cell (the scale of 'truncation' of the potential is equal to the cell size). Since the number of the cells N_M is considerably smaller than N, the computing speed is increased. The models of this type have been described in detail in [376, 681, 682, 685, 914, 916, 920, 1049] and in a number of other studies.

The interaction between the centres of the spatial cells at a large number of particles in each cell is usually taken into account by the diskrete analogue of the convolution theorem and the algorithm of the fast Fourier transform.

Models have been proposed ('particle–particle–particle–mesh' = P^3M) which partly combine the high-speed of calculation by the PM method and take into account the short-range forces, as in the PP models. They are based on the splitting of the forces acting between the particles into a rapidly changing short-range part and a slowly changing long-range part.

3) A number of schemes have been proposed (PM^2 = Particle–Multiple–Mesh [611], MGPM = Nested Grid Particle–Mesh [1198]) which use several meshes with different steps, and the structure of the mesh changes dynamically with time taking into account the distribution of matter in the calculation domain [901]. This approach can be used to reduce the number of operations. These systems also include hierarchic schemes, in particular, different versions of TREEcode, for example TREECode (Top Down) [338, 339] taking into account improvements [897].

The system of the meshes is distributed in the order details from the finest in which the forces acting over the smallest distances are taken into account, with sufficient accuracy, to the meshes with the larger scale in which the long-range acting forces are taken into account. In other words, regularisation of the subsystems at different hierarchy levels is carried out. The algorithm requires $N \cdot \ln N$ calculations in each integration time step.

TREECode (Bottom Up) differs from the TREECode (Top Down) by the direction of passage of the graph-tree from the closest particles to larger blocks [724, 897]. This approach enables calculations to be carried out with different time steps Δt for different particles. Selecting for the particles situated in the region with larger field gradients the lower values of Δt (but multiples of the maximum value Δt^{max}) it is possible to increase the accuracy of calculations and save computing time. The TREEcode (Bottom Up) algorithm is most efficient in modelling the systems in which the role of paired convergence (including the close binary stars) is important. This algorithm is more difficult to apply but can be used to investigate the collisionless systems of the type of spherical accumulations.

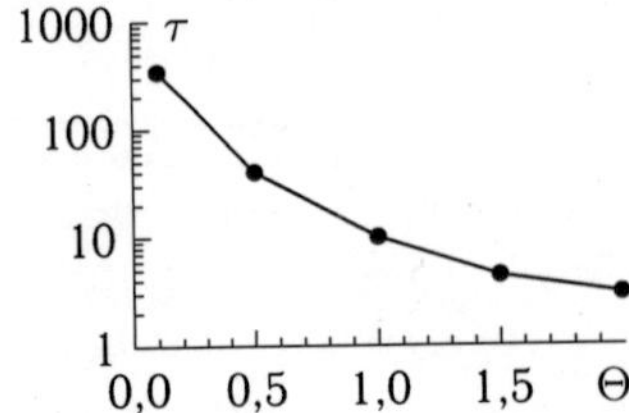

Fig. 4.2. Dependence of computing time of a single step in numerical integration τ (conventional units) on parameter Θ with the conditions being equal.

The only free parameter in practice which determines the accuracy of calculating forces in the TREECode method is the angle Θ. When the angle Θ decreases the accuracy increases, but the computing time for a single integration step with respect to time depends in the non-linear manner on the value of Θ (Fig. 4.2). The analysis carried out by the authors of the book confirms the conclusion [675] that the suitable value used for examining the galaxies is $\Theta \lesssim 0.6 \div 1$. In modelling large-scale structures, it may be accepted that $\Theta = 1$, and when examining highly inhomogeneous structures at small distribution scales it should be set $\Theta = 0.5$.

The method with slight modifications is described in sufficient detail in [724]. Comparison with the results showed a good agreement and also confirmed the advantages associated with the computing time. The TREE models can be used to investigate complicated systems, including three interacting components: the stellar and gas disks, 'live' halo [662, 1157]. The introduction to the practice of hierarchic schemes leads us to the natural limit in the construction of numerical models [674].

The further development of the TREECode is the FMM algorithm (Fast Multipole Method) which uses different multipolar expansions of the field for the far away and near regions [635, 897]. The potential at the given point is taken in the form of a series of multipolar expansion which is more economical in comparison with calculating the force with the required accuracy. Increasing the number of terms in the expansion makes it possible to obtain suitable exact solutions, of course at the expense of computing time.

The algorithms combining the TREECode and PM-schemes [521] have been proposed. This reduces the computing time at the expense of accuracy but these methods are not yet used widely.

4) The so-called SuperBox algorithm for the restoration of the potential on the basis of the given distribution of N gravitating particles [370, 546, 770], which belongs to the type Particle–Mesh, should be mentioned. A special feature of the method is the application of a fixed set of meshes, for example, three meshes with the different cell sizes can be used. The cell size increases in transition from the internal mesh to an intermediate one and then to an external one. A more complicated system of meshes, consisting of five or more meshes [370], can also be used. With other conditions being equal, this approach can be used to construct dynamic models of stellar systems with the maximum number of particles. The test calculations, carried out by the authors, have confirmed its efficiency up to $N = 10^9$ which is already comparable with the number of stars in small galaxies.

It is important to mention two shortcomings of this method. To improve the accuracy of calculating the gravitational force on the small scale, it is important to use very fine grids which require a large operating memory in modelling. Another special feature of these models is the presence of the zone of transition from one grid to another where non-physical effects may operate. It is therefore necessary to select the parameters of the models at which the perturbations in the system should be small in the vicinity of the boundary of the cells. If the latter condition is fulfilled, the SuperBox has important advantages associated primarily with high resolution as a result of the fact that calculations can be carried out with a very large number of particles N [770]. Finally, the SuperBox is a relatively fast algorithm [370, 770].

The SuperBox has been used to investigate the effects of interaction of the galaxies, movement of galactic satellites, and evolution of a supermassive black hole in a galactic nucleus [735, 780]. The SuperBox method has shown its efficiency in solving a difficult problem associated with the presence of a singularity in the rotational curves in the centre of the disk $V(0) > 0$ which is used in investigating the dynamics of an unstable stellar disk in the Jalali–Hunter analytical model [718]. Modelling using the SuperBox with the number of particles $N = (10 - 40) \cdot 10^6$ in study [770] enabled the unstable multimode perturbations to be described with high accuracy in complete agreement with the analytical theory.

Another direction of development is the formation of specialised computers for numerical solution of the gravitational problem of N-bodies (for example, GRAPE) in which the calculation of the interaction between the bodies is implemented using hardware [752, 1212]. Direct modelling in computers of this type indicates their considerable potential (see [871) and references therein).

A comparative review of individual methods of numerical modelling of N gravitating bodies can be found in, for example [252, 635, 1135].

The collisionless problem

The number of particles in the numerical models N is still considerably smaller than the number of stars in the galaxies N_*. With the exception of the central region, the stellar disks are collisionless, and this should be taken into account in the model of N-bodies. Another important problem of the models of N-bodies in application to the dynamics of galaxies is the effect of noise as a result of the small number of particles in comparison with the number of stars. The standard method of reducing the noise and ensuring the collisionless nature is the 'truncation' of the Newton potential at small distances [506, 1082, 1083]. To reduce the role of noise, work is being carried out on the development of wavelet methods which greatly reduce the intensity of this effect [1084].

In modelling a complete galaxy consisting of $N_* \simeq 10^{10} \div 10^{12}$ stars, it is necessary to form a collisionless system. The duration of a double collision in a three-dimensional system is [266]

$$T_c = \frac{v_*^3}{8\pi n G^2 m_*^2 H \ln(R v_*^2 / 2Gm_*)}, \tag{4.1}$$

where n is the volume concentration of the stars; m_* is the mass of the star; v_* is the relative velocity; R is the size of the system; $H \simeq 0.4$ is the geometrical factor.

Except for the very centre, the time T_c for the galaxies is greater than the age of the stellar systems. Since $m_* \propto N_*^{-1}$ and $n \propto N_*$ (N_* is the number of stars in the system), the proportionality $T_c \propto N_*$ is fulfilled. Since the number of particles in the numerical model is $N \ll N_*$, the particles in modelling of the galaxies are represented by 'macrostars' with the mass exceeding the mass of the actual stars N_*/N times[2]. To ensure the collisionless state of the system, it is necessary to 'truncate' the potential as previously mentioned.

4.1.2. Construction of experimental models of stellar disks

A relatively simple model of a disk galaxy will be described. The model is based on the numerical integration of the motion equation of N gravitationally interacting particles taking into account the external field determined by the given distribution of matter in the bulge and the halo. It is assumed that the parameters of the spheroidal subsystem are stationary – the density of matter in the bulge ϱ_b and the halo ϱ_h does not change in the process of evolution of the disk. This model will be referred to as the dynamic model.

The dynamics of gravitating N-bodies is described by the system of equations

$$\frac{d^2\mathbf{r}_i}{dt^2} = \sum_j^N \mathbf{f}_{ij} + \mathbf{F}_s \quad (i = 1,...,N), \tag{4.2}$$

where the radius-vector $\mathbf{r}_i\,(t)$ determines the position of the i-th particle; $\mathbf{f}_{ij}$ is the force of interaction between the i-th and j-th particles ($i \neq j$); the force $\mathbf{F}_s = \mathbf{F}_b + \mathbf{F}_h$ is determined by the spheroidal bulge/halo subsystem.

We use the spatial distribution of volume density in the halo in the form

$$\varrho_h(\xi) = \frac{\varrho_{h0}}{(1+\xi^2/a^2)^k}, \tag{4.3}$$

here $\xi = |\mathbf{r}| = \sqrt{r^2+z^2}$ is the radial spherical coordinate. Selection of $k = 1$ (the so-called quasi-isothermal halo – iso) ensures the flat circular velocity of rotation V_c = const in the region of dominance of the halo. Increase of k increases the relative concentration of the spherical component in the direction to the centre and this has an effect identical with that of the bulge.

One should note the model of the halo [960] which is produced in the cosmological models:

$$\varrho^{NWF}(\xi) = \frac{\varrho_0}{(\xi/r_s)(1+\xi/r_s)^2}, \tag{4.4}$$

where ϱ_0 is the constant, r_s is the halo scale. In the Burkert model [430]

$$\varrho_h(\xi) = \frac{\varrho_0}{(1+\xi/a)(1+(\xi/a)^2} \tag{4.5}$$

[2]The scattered stellar and spherical clusters contain respectively ~10^2 and $10^4 \div 10^6$ stars, and the PP method is used efficiently for modelling these systems.

there is no singularity in the centre of the halo, in contrast to (4.4).

The exponential profile [601]

$$\varrho_h = \frac{M_h}{8\pi e b^3} \exp\left\{-\frac{\xi}{b}\right\} \tag{4.6}$$

ensures the convergence of the integrals $M_h = \int_0^\infty \varrho_h 4\pi\xi^2 d\xi$. It is also important to note of the logarithmic potential of the halo

$$\Psi_h = v_h^2 \ln\left(\xi^2 + d_h^2\right). \tag{4.7}$$

The Einasto profile is used quite often:

$$\varrho_h = \varrho_s \exp\left\{-\frac{2}{\alpha}\left[(\xi / r_s)^\alpha - 1\right]\right\}, \tag{4.8}$$

where α is a free parameter.

The distribution of matter in the halo (4.3) at $k = 1$ with the central density $\varrho_{h0} = M_h / \{4\pi a^3 [R/a - \mathrm{arctg}(R/a)]\}$ gives the following equation for the force

$$\mathbf{F}_h(\mathbf{r}) = -\frac{4\pi G a^3 \varrho_{h0}}{\xi^2}\left\{\frac{\xi}{a} - \mathrm{arctg}\left(\frac{\xi}{a}\right)\right\}\frac{\mathbf{r}}{\xi} \tag{4.9}$$

and is determined by the spatial scale a and mass M_h inside the sphere $\xi < R$. The corresponding gravitational potential has the form

$$\Psi_h(\xi) = 4\pi G \varrho_{h0} a^2 \cdot \left\{\ln(\xi / a) + \frac{\mathrm{arctg}(\xi / a)}{\xi / a} + \frac{1}{2}\ln\frac{a^2 + \xi^2}{\xi^2}\right\}. \tag{4.10}$$

To describe the distribution of density in the spherical bulge, we can use the King model

$$\varrho_b = \begin{cases} \varrho_{b0} / \left[1 + (\xi / b)^2\right]^{3/2}, & \xi \le r_b^{\max} \\ 0, & \xi > r_b^{\max}, \end{cases} \tag{4.11}$$

for which the value

$$M_b = 4\pi b^3 \varrho_{b0}\left\{\ln\left[r_b^{\max} / b + \sqrt{1 + (r_b^{\max} / b)^2}\right] - \frac{r_b^{\max} / b}{\sqrt{1 + (r_b^{\max} / b)^2}}\right\} \tag{4.12}$$

is the bulge mass. Inside $\xi \le r_b^{\max}$ for the gravitational force we have

$$\mathbf{F}_b = -\frac{4\pi G b^3 \varrho_{b0}}{\xi^2}\left\{\ln\left(\frac{\xi}{b} + \sqrt{1 + \frac{\xi^2}{b^2}}\right) - \frac{\xi / b}{\sqrt{1 + \xi^2 / b^2}}\right\}\frac{\mathbf{r}}{\xi}. \tag{4.13}$$

Evidently, in the region $\xi > r_b^{\max}$ in the case (4.11) the condition $\mathbf{F}_b = -GM_h\mathbf{r} / \xi^2$ is fulfilled. For $\xi > r_b^{\max}$ attention was given to the generalisation of the law (4.11) by linear or exponential (we do not present here the cumbersome equations for this case) truncation with the characteristic scale L_b. Thus, the mass of the bulge M_b is finite, and the halo mass $M_h(r)$ in accordance with (4.3) increases with increasing radius. In the bulge, the subsystem in some models is represented by the nucleus in which the density ϱ_c (ξ) is also determined by the law (4.11) with the scale $b_n < r_n^{\max} \ll b$.

Taking into account the flattening of the bulge, more complicated equations are often used for the distribution of volume density, for example equation [508]:

$$\varrho(r,z) = \varrho_0 \left(\frac{m}{r_0}\right)^{-\gamma} \left(1+\frac{m}{r_0}\right)^{\gamma-\beta} \exp\left(-m^2 / r_t^2\right), \tag{4.14}$$

where $m = \sqrt{r^2 + z^2 / q^2}$ takes into account the non-spherical form of the bulge, the parameters γ, β, r_t, r_0, ϱ_0, q ensure the best agreement with the observational data.

The simplest model for the central spherical stellar component (the bulge) for the potential (Hernquist potential) [623] is

$$\Psi_b = -\frac{GM_b}{r_b + \xi}, \tag{4.15}$$

where M_b is the bulge mass, r_b is the scale of the bulge, $\xi = \sqrt{x^2 + y^2 + z^2}$.

In these equations one can often used the generalised expression for $\xi = a\sqrt{(x/a)^2 + (y/b)^2 + (z/c)^2}$. Taking into account the flattening of the halo or the bulge it is sufficient to assume $c < a = b$, and at $a \neq b$ we have the non-axisymmetric distribution. In a more general case $a \neq b \neq c$ we obtain triaxial bulges or halo (for more details see section 10.4).

The dynamic model should adequately describe the Newton interaction between the particles and ensure the collisionless state of the system. The latter is achieved by modification of the gravitational force at small distances, by introducing the truncation radius of the potential r_c between any two interacting particles i and j. The problem of the optimum selection of the radius of truncation and the number of particles has been discussed many times in the literature (see, for example, [318, 1082, 1083, 1189] and references therein). It should be stressed that the truncation of the potential is the only possibility of ensuring the collisionless state of the models of the galaxies in the framework of the problem of N-bodies. However, even if the value r_c is not present explicitly in the numerical scheme, as in the case of the PM-algorithm (it has been assumed that the method does not contain the truncation radius), it is present implicitly – its a role is played by the characteristic size of the cell.

The simple Plummer model for the potential

$$\Phi_{ij} = -G\frac{m_i m_j}{\sqrt{r_{ij}^2 + r_c^2}}, \tag{4.16}$$

where r_{ij} is the distance between the two particles, r_c is the truncation radius, solves the problem of ensuring the collisionless state. The application of the spline-nucleus for softening the gravitational force has an advantage in comparison with (4.16), because outside the given radius ($r > 2r_c$) the force in this case coincides accurately with the Newton force [830]. At a fixed number of particles N it is always possible to select the value r_c resulting in the collisionless state of the model. However, to reduce the error associated with the change of the interaction potential between the particles at short distances, the number of the particles should be relatively high.

The surface density of the disk is characterised by the scale L which determines the exponential law (1.3), and in this case for the mass of the disk we have $M_d = 2\pi L^2\sigma_0$, and approximately 95% of the disk is concentrated inside $r \leqslant 4L$. The relative mass of the halo is calculated inside the determined radius $\mu(r) = M_h(r)/M_d$.

If necessary, we can take into account the disk subsystem with the fixed distribution of matter. As the model we can select (2.43) or restrict our considerations to a simpler model of the disk component [623]

$$\Psi_d = -\frac{GM_d}{\sqrt{x^2 + y^2 + \left[d + (h^2 + z^2)^{1/2}\right]^2}}, \tag{4.17}$$

where M_d is the mass of the disk, d is the radial scale of the disk, h is the vertical scale.

In numerical calculations, it is convenient to select the system of units in which $G = 1$, $R = 4L = 1$, the disk mass $M_d = 1$. The mass of the halo M_h inside the radius $\xi \leqslant 4L$ is normalised with respect to the mass of the disk $\mu \equiv M_h / M_d$. In this system of units, a single period of rotation of the disk at the external edge is in the range $T_0 \sim 3 \div 6$ for the values $\mu = 1 \div 4$. The typical values of the velocity of rotation $V \sim 1 \div 2$, the dispersion of the radial velocities $c_r \lesssim 1$, at the periphery of the disk $c_r \ll 1$.

A simple procedure of defining the initial distribution of the density along the z-coordinate and the residual velocity of the equilibrium disk will be described. To define the coordinates and velocities of the particles at the initial moment of time, it is convenient to use the mesh in the cylindrical coordinates system. The coordinates of the particles inside the cell are determined using the triple application of the random number generator. Selecting the law of variation of the surface density $\sigma(r)$, the disk is divided into relatively narrow rings with the width Δr and from the given $\sigma(r)$ we calculate the number of particles n_i 'falling' into the k-th ring. For example, in the 'standard' Ostriker–Peebles model [996] $\sigma(r) \propto r^{-1}$ and $n_i = N\Delta r/R$, and in the model of the potential disk $\sigma \propto \exp(-r/L)$ the number of particles in the ring is equal to

$$n_i = \frac{N\exp(-i\Delta r / L)}{1-(1+1/L)\exp(-1/L)}\left\{\left(1+(i-1)\frac{\Delta r}{L}\right)\left(\exp\left(\frac{\Delta r}{L}\right)-1\right)-\frac{\Delta r}{L}\right\}, \tag{4.18}$$

where $r_i = (i - 1/2)\,\Delta r$ is the mean radius of the ring ($i = 1, 2, \ldots$), $R = 1$.

Each ring is divided to n_i equal parts in the azimuthal direction and a particle is placed in each cell produced in this manner. The coordinates of the particle inside the cell are naturally determined by the double application of the random number generator (the first time in the radial direction, the second time in the azimuthal direction), comparing the range of variation of the random numbers (usually [0, 1]) with the width of the ring Δr and the width of the sector $r_i\Delta\varphi = 2\pi r_i/n_i$. Subsequently, in each ring we calculate the mean (with respect to n_i particles) radial gravitational force F_{ri}, acting on each particle from the side of all other particles of the disk and the static halo (see below). All the particles of the investigated ring are given the azimuth velocity $V_i^{(1)} = (r_i\, Fr_i)^{1/2}$. It is clear that the disk prepares in this manner maybe equilibrium but is 'Cold' and, consequently, subjected to strong gravitational instability.

The vertical structure of the disk is determined by the equations (2.54). When defining the initial profile $\varrho(z)$ it is sufficient to use the equation (2.56) together with the conditions $\varrho(z=0)=\varrho_0$, $d\varrho(0)/dz=0$, $\int_{-\infty}^{\infty}\varrho(z;r)dz=\sigma(r)$.

This approach is used for any vertical distributions of the dispersion of velocities $c_z(z)$. It should be mentioned that in the very centre of the disk the term in (2.56) may result in a large error in the evaluation of density and in practice it should be assumed that $E\ (r \to 0) \to 0$.

For the given $\varrho_s(z, r)$, $c_z(z, r)$ and $\sigma(r)$, ϱ_0 and $\varrho(z)$ are determined from (2.56). After determining the law of distribution of density along the z-coordinate, the particles are distributed on the vertical axis in the given mesh $z_k = k\ \Delta z\ (k = -K, \ldots, K)$.

The particles in the proportion σ_k/σ, where $\sigma_k = \int_{z_{k-1}}^{z_k}\varrho(z)dz$, are placed in the k-th cell.

Strictly speaking, the disk constructed in this manner is not equilibrium. In investigating the instabilities, the absence of the exact equilibrium in the vertical direction plays the role of a small additional initial perturbation. In the case of the stable states, the presence of small initial perturbations does not change the initial stationary condition of the disk.

The half thickness of the disk in the numerical models is characterised by the value

$$\Delta^2 = \frac{\int_{-\infty}^{\infty}(z-\langle z\rangle)^2\varrho(z)dz}{\int_{-\infty}^{\infty}\varrho(z)dz}, \tag{4.19}$$

where $\langle\ldots\rangle$ is averaging with respect to z. Determination of the parameter Δ in accordance with (9.2) differs by only 10% from the vertical scale of the disk z_0 which is included in the law of vertical distribution of density (2.57). In [1191] it is proposed to determine the vertical scale on the basis of the median.

The initial velocity distribution function is of the Schwarzschild type and has the form of the anisotropy Maxwellian distribution

$$f(u,v,w) = A\cdot\exp\left\{-\frac{u^2}{2c_r^2}-\frac{(v-r\Omega)^2}{2c_\varphi^2}-\frac{w^2}{2c_z^2}\right\}, \tag{4.20}$$

where $\{u, v, w\}$ are the components of the velocity of the particles in the cylindrical coordinate system.

The mean tangential velocity of the points for the initial state of the model is determined from the Jeans equation (see section 2.1.3) assuming the absence of systematic radial movement, axisymmetricity and the symmetry with respect to the plane $z = 0$ [1249]:

$$V^2 = (<v>)^2 = V_c^2 + c_r^2 \left\{ 1 - \frac{c_\varphi^2}{c_r^2} + \frac{r}{\varrho c_r^2} \frac{\partial(\varrho c_r^2)}{\partial r} + \frac{r}{c_r^2} \frac{\partial \alpha_{rz}}{\partial z} \right\}, \tag{4.21}$$

where the circular velocity V_c in the plane $z = 0$ is determined by the distribution of matter in three main components: disk, bulge, and halo, $\alpha_{rz} = \langle uw \rangle$. Taking into account the multi-component form, it is convenient to write

$$V_c = \sqrt{(V_c^{\text{disc}})^2 + (V_c^{\text{bulge}})^2 + (V_c^{\text{halo}})^2}, \tag{4.22}$$

where V_c^{disc}, V_c^{bulge}, V_c^{halo} of the corresponding contribution to the circular velocity from the individual components. After specifying the initial state of the model, it may be assumed that $\langle u \rangle = 0$, $\langle w \rangle = 0$, and the speed of rotation $W \equiv \langle v \rangle$ can be defined in accordance with (4.21). In a hot disk at the periphery $c_r \lesssim W$ and the equation (4.21) may result in large errors. The applicability of (4.21) is approximately determined by the condition c_r $(r = 2L) < 0.5\ V^{\max}$. Thus, at the initial moment of time there is a balance of radial and vertical forces, and the disk stars evolution from the state close to equilibrium. At the initial moment of time the disk has finite 'temperature' which is determined by three functions $c_r(r)$, $c_\varphi(r)$, $c_z(r)$. The form of these dependencees is dictated by the formulation of the problem.

The particles of the investigated cell are assumed to have the azimuthal velocity equal to the circular velocity (2.55). Consequently, a cold disk is obtained. In the following search, on the basis of the given values c_r, c_φ, c_z, we generate the random velocities of the particles in accordance with the distribution (4.20). Finally, taking into account (4.21), we determine more accurately the velocity of rotation in relation to the final thickness and the thermal scatter of the velocities. In the final stage, the position of the particles and their velocities are corrected to ensure that the centre of the mass and the kinetic centre coincide with the centre of the coordinates[3]

A distinguishing special feature of the stellar collisionless disks is the anisotropy of the velocity dispersion of the stars; $c_r \neq c_\varphi \neq c_z$ [590], and $c_r > c_\varphi > c_z$. The dispersions of the components of the residual velocities of the stars $c_{r,\varphi,z}$ are small in comparison with the velocity of rotation of the stars of the disk V, with the exception of the central regions $r < Z$, where possibly $c_r > V^2$ [4]. Therefore, the epicyclic approximation holds for the axisymmetric distribution of the matter way the equilibrium of the disk in the radial direction is determined mainly by the balance of the centrifugal and gravitational forces with the accuracy to the terms $\sim (c_r/V)^2$ (see equation (4.21)). In the rotation of particles and the radius r, with the epicyclic motion taken into account, we obtain the relationship between c_φ and c_r:

$$c_r = \frac{2\Omega}{\varkappa} \cdot c_\varphi \quad \text{and} \quad Q_c \equiv \frac{\varkappa(r)}{2\Omega(r)} \frac{c_r(r)}{c_\varphi(r)} = 1, \tag{4.23}$$

where $\varkappa = 2\Omega\sqrt{1 + r\,d\Omega / 2\Omega dr}$ is the epicyclic frequency. It should be stressed that in the dynamic models the condition (4.23) is fulfilled for the angular velocity

[3]The latter procedure is important in the models with not too large numbers of particles $N \lesssim 10^4$, and its role decreases with increasing N).

[4]In the very centre ($r \leqslant L$) in all cases $c_r \gg V^2$.

of rotation of the particle $\Omega = V/r$, and not for the circular velocity (2.55). The fulfilment of $Q_c = 1$ is the essential condition for equilibrium of the collisionless dynamic models.

As a result of fulfilling of these procedures we obtain the model of the 'hot' stellar disk of a finite thickness in which it is necessary to correct the coordinates of the particles and their velocity in such a manner as to ensure that the centre of mass of the system of the particles remains at the origin of the coordinates during the experiments.

When formulating the experiments in which the disk of the flat galaxy is modelled together with its spherical subsystem (halo), it is necessary to take into account the fact that the dispersion of the velocities of the stars of the halo is considerably greater than the dispersion of the velocities of the stars of the disk ([561, 940], chapter 1). At the same time, the value of the perturbation of the density of any subsystem is inversely proportional to the square of its dispersion of the velocities ([845], section 2.2.2). Therefore, in the evolution of the disk, which does not change greatly the distribution of the mass in the disk, the halo may be regarded as non-perturbed and modelled by the potential $\Phi_h(r)$ acting on the stars of the disk.

Interaction of the particles in the disk

The interaction of particles with each other should evidently simulate the main properties of the stellar disk as a collisionless gravitating system. The experimental determination of the collisionless nature of the disk can be formulated as follows: the force, acting on the given particle, from the side of all other particles of the disk should be considerably greater than the force acting on the same particle from the side of the nearest neighbour of the particle. This determination leads naturally to the need for 'truncation' of the gravitational interaction of the particles over short distances. This is used in all numerical experiments. To evaluate the value of the appropriate parameter, the 'interaction radius' r_{int} is determined by the relationship

$$M_d(r)/r^2 = m/r_{\text{int}}^2, \tag{4.24}$$

where $m = 1/N$ is the mass of the particle; $M_d(r)$ is the mass of the disk in the sphere with radius r. For the exponential disk

$$\frac{r_{\text{int}}}{L} = \frac{(r/L)}{\sqrt{N}}\sqrt{\frac{1-(1+R/L)\exp\{-R/L\}}{1-(1+r/L)\exp\{-r/L\}}}. \tag{4.25}$$

According to the observational data, in the disks of the galaxies $L/R \simeq 0.2 \div 0.3$ [805], but in this case also the dependence of r_{int} on the radial coordinate differs slightly from the linear dependence. For example, at $L = 0.20\,5R$, the relationship (4.25) in

the range $r < R$ can be approximated by the function $r_{\text{int}} \simeq \dfrac{L}{\sqrt{N}}\left\{1.24 + 0.67\dfrac{r}{L}\right\}$.

It is now assumed that the 'truncation' of the gravitational interaction at short distances is performed by calculating the force acting on the i-th particle from the side of the j-th particle as follows [996]:

$$\vec{f}_{i,j} = \frac{1}{N^2} \frac{(\vec{r}_j - \vec{r}_i)}{\{(\vec{r}_j - \vec{r}_i)^2 + r_c^2\}^{3/2}}, \tag{4.26}$$

where r_c is the radius of 'truncation' of the interaction. Evidently, selection of r_c depends on the aim of the experiments. For example, when investigating the formation of the bar mode in the central part of the disk, Ostriker and Pebbles found the optimum value $r_c \simeq 0.05$ at a low value of $N = 150{\div}500$. When finer and finer scale structures are investigated, r_c should be lower retaining, however, the collisionless nature of the system. For this purpose, it is necessary to increase N. As regards the experiments, $r_c \simeq (1{\div}3)\ r_{\text{int}}$ is acceptable. It is important to note that it should not be assumed that the value r_c depends on the radial coordinate in the specific experiment because in this case the energy of the system is not the integral of motion.

A large amount of information in the investigation of the spiral and bar-like structures is provided by the Fourier analysis of the surface density in the coordinates φ and ln (r). The level of the non-axisymmetric state of the radially inhomogeneous distribution of matter in the disk at every moment of time can be conveniently characterised by the amplitude of the Fourier harmonics [1141]:

$$A(t;p,m) = \frac{1}{N}\sum_{j=1}^{N} \exp\left\{i\left[m\varphi_j(t) + p\ln(r_j(t))\right]\right\}, \tag{4.27}$$

here $i = \sqrt{-1}$, p and m characterise the structure of perturbations in the radial and azimuthal direction, respectively. The quantities $A\ (t;\ p,\ 2)$ determine the amplitude of the two-arm mode $m = 2$ and, in particular, of the bar for which the harmonics with $p = 0$ is maximum. Parameter p characterises the degree of twisting of the waves. We also introduce the integral amplitudes of the Fourier harmonics

$$\hat{A}(t;m) = \sqrt{\sum_p |A(p,m)|^2}, \tag{4.28}$$

which show which amplitude numbers of the modes characterise preferentially the global structure of the perturbations. The law of conservation of the mass of the disk gives the condition $A\ (T;\ 0,\ 0) = 1$, and the variation of $\hat{A}(t;\ 0)$ with time indicates the redistribution of matter in the radial direction.

4.1.3. The grid method for solving gas dynamics equations (M.A. Eremin)

In this section, we will discuss in some detail the currently available mesh methods used to construct the effective numerical solutions of the gas dynamics equations for describing the rotating gas disks.

The processes, taking place in the actual astrophysical systems are characterised by 'drops' in density and temperature, by the 'giant' range of spatial and time scales, and by the supersonic speeds of gas flow. All these factors create suitable conditions for the formation of strong impact waves which often play not only an important but also controlling role in the dynamic processes. In addition to this, it is justified to assume that the turbulence has a significant effect on the processes taking place, for example, in the interstellar medium and accreating systems.

The physical processes in the astrophysical systems are described by the equations in partial derivatives and are highly non-stationary and three-dimensional and this means that the solutions of a very large number of the problems interesting from the viewpoint of astrophysics can be obtained only numerically.

As a result of the historical reasons, numerical modelling in the astrophysics is carried out using either the methods of continuous counting on stationary Euler grids or using the Monte Carlo-type method in which attention is given to the movement of the Lagrangian particles within the framework of the hydrodynamics of the smoothed out particles (SPH, see section 4.1.4). The astrophysical flows, described by the gas dynamics equations, have two specific features which impose special requirements on the numerical methods. Firstly, the gas flow conditions are such that from the smoothed out initial data one obtains quite rapidly discontinuous solutions in the form of shockwaves. This means that the numerical methods should, on the one hand, retain the monotonic nature in the regions of sharp gradients and, on the other hand, should be characterised by a high order of approximation in the regions of he smooth flow. According to the Godunov theorem (1959), in the group of the linear difference schemes these two requirements cannot be satisfied simultaneously. The pass-through methods have been developed for solving this problem. These methods satisfythe condition of no increase of the total variation of the numerical solution or the total variation diminishing (TVD) condition [650]. At the same time, the quality of the numerical solutions has been greatly improved in comparison with the classic finite difference method, using the artificial viscosity of the von Neumann–Richtmayer type for suppressing the non-physical oscillations. Secondly, since it is important to take into account self-gravitation for the adequate modelling of a large number of astrophysical systems, clusterisation of matter takes place leading to the appearance of a wide range of spatial scales. The particle methods have been developed in most cases for solving this problem.

It should be mentioned that recently a large amount of work has been carried out on the development of the effective algorithms using adaptive moving grids, but they are not yet used widely in the investigation of astrophysical flows.

In this section, we discuss the minimum amount of information required for the realisation in practice of the finite-volume TVD scheme for the equations of gas dynamics in the three coordinate systems: Cartesian, cylindrical and spherical.

The gas dynamics equations in the orthogonal coordinate systems

The gas dynamics equations, describing the non-stationary flows of the ideal gas in the Euler coordinate system, in the divergent form can be written as

$$\frac{\partial \varrho}{\partial t} + \nabla \cdot (\varrho \mathbf{u}) = 0, \tag{4.29}$$

$$\frac{\partial}{\partial t}(\varrho \mathbf{u}) + \nabla \cdot (\varrho \mathbf{u}\mathbf{u}^{\mathrm{T}} + p\hat{\mathbf{I}}) = \mathbf{f}, \tag{4.30}$$

$$\frac{\partial E}{\partial t} + \nabla \cdot [(E + p)\mathbf{u}] = \mathbf{u} \cdot \mathbf{f}. \tag{4.31}$$

Here t is time, ϱ is the gas density, p is the pressure, $\mathbf{u} = [u_1, u_2, u_3]^T$ is the velocity vector,

$$E = \varrho\left(e + \frac{\mathbf{u}^2}{2}\right)$$

is the total energy of the unit volume, e is the specific internal energy. In the model of the ideal gas

$$e = \frac{p}{\varrho(\gamma - 1)},$$

where γ is the adiabatic index, $\hat{\mathbf{I}} = \mathrm{diag}[1,1,1]$ is the unit vector of the dimensions 3×3; the non-differential term $\mathrm{f} = [f_1, f_2, f_3]$, in the right-hand part of the equations (4.29)–(4.31) describes the action of external forces.

The gas dynamics equations describe the laws of conservation (variation) of the mass, momentum, and energy, respectively. This means that the variation of the quantities in the arbitrary volume can take place only as a result of the flow of the quantities through the boundaries of the volume. The flow of the gas can be described in terms of the vector of conservative $\mathbf{U} = [\varrho, \varrho\mu_1, \varrho\mu_2, \varrho\mu_3\, E]^T$ or simple $\mathbf{V} = [\varrho, \varrho\mu_1, \varrho\mu_2, \varrho\mu_3\, p]^T$ variables.

We examine a generalised system of curvilinear orthogonal coordinates (x^1, x^2, x^3) and determine the unit vectors $\mathbf{n}_l$ along the coordinate axes, $l = 1, 2, 3$. It is assumed that this system of vectors is ortho-normalised, i.e. $\mathbf{n}_l \cdot \mathbf{n}_m = \delta_{lm}$, where δ_{lm} is the Kronecker symbol. The geometrical properties of the curvilinear system of coordinates are described by the metric coefficient (h_1, h_2, h_3) which, in a general case, are the functions of the coordinates $h_i = h_i(x^1, x^2, x^3)$. The element of the volume in the curvilinear coordinates is expressed by the metric coefficient using the following equation:

$$\mathrm{d}V = h_1 h_2 h_3\, \mathrm{d}x^1 \mathrm{d}x^2 \mathrm{d}x^3.$$

In astrophysics, it is important to write the equations of hydrodynamics in three orthogonal coordinate systems: Cartesian, for which the coordinates of the point (x^1, x^2, x^3) and metric coefficients (h_1, h_2, h_3) are determined by the expressions

$$(x^1, x^2, x^3) = (x, y, z), \quad (h_1, h_2, h_3) = (1,1,1),$$

the cylindrical coordinate system, in which the coordinates of the points are presented in the polar radius r, the azimuthal angle φ and the applicate z:

$$(x^1, x^2, x^3) = (r, \varphi, z), \quad (h_1, h_2, h_3) = (1, r, 1),$$

and the spherical coordinate system where the position of the point is determined by the radius r, the zenith θ and azimuthal φ angles:

$$(x^1, x^2, x^3) = (r, \theta, \varphi), \quad (h_1, h_2, h_3) = (1, r, r\sin\theta).$$

The conservative form of the hydrodynamics equations differs in different geometries because the divergence operator $\nabla\cdot$ and the gradient operator ∇ differ in different coordinate systems. In a general case, the expression for the gradient of the scalar function Φ has the form

$$\nabla\Phi = \left(\frac{1}{h_1}\frac{\partial\Phi}{\partial x^1}, \frac{1}{h_2}\frac{\partial\Phi}{\partial x^2}, \frac{1}{h_3}\frac{\partial\Phi}{\partial x^3}\right),$$

the divergence of the vector $\mathbf{B} = [B_1, B_2, B_3]^{\mathrm{T}}$ is computed from the equation

$$\nabla\cdot\mathbf{B} = \frac{1}{h_1h_2h_3}\left[\frac{\partial}{\partial x^1}(h_2h_3B_1) + \frac{\partial}{\partial x^2}(h_1h_3B_2) + \frac{\partial}{\partial x^3}(h_1h_2B_3)\right].$$

After substituting these expressions into the equations (4.29)–(4.31) and regrouping the terms, we obtain the equations of hydrodynamics in the conservative form

$$\frac{\partial\mathbf{U}}{\partial t} + \frac{1}{h_1h_2h_3}\left\{\frac{\partial}{\partial x^1}(h_1h_2\mathbf{F}^1) + \frac{\partial}{\partial x^2}(h_1h_3\mathbf{F}^2) + \frac{\partial}{\partial x^3}(h_1h_2\mathbf{F}^3)\right\} = \mathbf{S}. \tag{4.32}$$

Here $\mathbf{F}^l$ are the vectors of the flows (l = 1, 2, 3), determined by the expressions

$$\mathbf{F}^l = \begin{bmatrix} \varrho u_l \\ \varrho u_1u_l + p\delta_{1l} \\ \varrho u_2u_l + p\delta_{2l} \\ \varrho u_3u_l + p\delta_{3l} \\ (E+p)u_l \end{bmatrix}, \tag{4.33}$$

and $\boldsymbol{S}$ is the vector of the sources targeted from the equation

$$\mathbf{S} = \begin{bmatrix} 0 \\ \varrho f_1 - \frac{\varrho u_1u_2}{h_1h_2}\frac{\partial h_1}{\partial x^2} - \frac{\varrho u_1u_3}{h_1h_3}\frac{\partial h_1}{\partial x^3} + \frac{\varrho u_2^2}{h_1h_2}\frac{\partial h_2}{\partial x^1} + \frac{\varrho u_3^2}{h_1h_3}\frac{\partial h_3}{\partial x^1} + \frac{p}{h_1h_2h_3}\frac{\partial(h_2h_3)}{\partial x^1} \\ \varrho f_2 - \frac{\varrho u_2u_1}{h_1h_2}\frac{\partial h_2}{\partial x^1} - \frac{\varrho u_2u_3}{h_2h_3}\frac{\partial h_2}{\partial x^3} + \frac{\varrho u_1^2}{h_1h_2}\frac{\partial h_1}{\partial x^2} + \frac{\varrho u_3^2}{h_2h_3}\frac{\partial h_3}{\partial x^2} + \frac{p}{h_1h_2h_3}\frac{\partial(h_1h_3)}{\partial x^2} \\ \varrho f_3 - \frac{\varrho u_3u_2}{h_3h_2}\frac{\partial h_3}{\partial x^2} - \frac{\varrho u_1u_3}{h_1h_3}\frac{\partial h_3}{\partial x^1} + \frac{\varrho u_1^2}{h_1h_3}\frac{\partial h_1}{\partial x^3} + \frac{\varrho u_2^2}{h_2h_3}\frac{\partial h_2}{\partial x^3} + \frac{p}{h_1h_2h_3}\frac{\partial(h_1h_2)}{\partial x^3} \\ \varrho\mathbf{u}\cdot\mathbf{f} \end{bmatrix}. \tag{4.34}$$

In the Cartesian coordinate system, the vector of the sources

$$\mathbf{S} = \left[0, \varrho f_1, \varrho f_2, \varrho f_3, \varrho\mathbf{u}\cdot\mathbf{f}\right]^{\mathrm{T}} \tag{4.35}$$

is of the physical nature. In the curvilinear coordinate systems, the equations of gas dynamics together with the terms which correspond to the physical factors include the terms determined by geometry. The above mentioned equation for the vector of the sources in the cylindrical and spherical coordinates is greatly simplified and written in the form of a cylindrical coordinate system:

$$\mathbf{S} = \begin{bmatrix} 0 \\ \varrho f_1 + \varrho \dfrac{u_2^2}{r} + \dfrac{p}{r} \\ \varrho f_2 - \varrho \dfrac{u_1 u_2}{r} \\ \varrho f_3 \\ \varrho \mathbf{u} \cdot \mathbf{f} \end{bmatrix} = \begin{bmatrix} 0 \\ \varrho f_r + \varrho \dfrac{u_\varphi^2}{r} + \dfrac{p}{r} \\ \varrho f_\varphi - \dfrac{\varrho u_r u_\varphi}{r} \\ \varrho f_z \\ \varrho (u_r f_r + u_\varphi f_\varphi + u_z f_z) \end{bmatrix}. \tag{4.36}$$

In the spherical coordinate system we obtain the following expression for the vector of sources

$$\mathbf{S} = \begin{bmatrix} 0 \\ \varrho f_1 + \varrho \dfrac{u_2^2 + u_3^2}{r} + \dfrac{2p}{r} \\ \varrho f_2 - \varrho \dfrac{u_2 u_1}{r} + \varrho \dfrac{u_3^2}{r} \operatorname{ctg} x^2 \\ \varrho f_3 - \dfrac{\varrho u_1 u_3}{r} - \varrho \dfrac{u_3 u_2}{r} \operatorname{ctg} x^2 \\ \varrho \mathbf{u} \cdot \mathbf{f} \end{bmatrix} = \begin{bmatrix} 0 \\ \varrho f_r + \varrho \dfrac{u_\theta^2 + u_\varphi^2}{r} + \dfrac{2p}{r} \\ \varrho f_\theta - \varrho \dfrac{u_\theta u_r}{r} + \varrho \dfrac{u_\varphi^2}{r} \operatorname{ctg} \theta \\ \varrho f_\varphi - \dfrac{\varrho u_r u_\varphi}{r} - \varrho \dfrac{u_\varphi u_\theta}{r} \operatorname{ctg} \theta \\ \varrho (u_r f_r + u_\theta f_\theta + u_\varphi f_\varphi) \end{bmatrix}. \tag{4.37}$$

The systems of the gas dynamics equations (4.29)-(4.31) is a system of the laws of conservation, written in the differential form. However, this formal writing assumes that the values to be determined **U** and the flows $\mathbf{F}^l = \mathbf{F}^l(\mathbf{U})$ are continuously differentiated functions. This holds only until the gas flow remains 'smooth'. It is well-known that the hyperbolic systems of the laws of conservation also permit discontinuous solutions, for example, solutions in the form of shock waves. In this case, the concept of the classic solution loses its meaning and to determine the generalised (weak) solution, it is necessary to turn to the integral form of the conservation law. Integrating the system of the equations (4.29)–(4.31) with respect to the arbitrarily defined volume V with the boundary ∂V, we obtain the integral equation, which holds for both smooth and discontinuous solutions

$$\frac{\mathrm{d}}{\mathrm{d}t} \int_V \mathbf{U} \mathrm{d}V + \int_{\partial V} \mathbf{F} \cdot \mathrm{d}\mathbf{A} = \int_V \mathbf{S} \mathrm{d}V, \tag{4.38}$$

where $d\mathbf{A}$ is the oriented element of the surface area, determined by the expression

$$\mathrm{d}\mathbf{A} = \mathbf{n}_1 \mathrm{d}A^1 + \mathbf{n}_2 \mathrm{d}A^2 + \mathbf{n}_3 \mathrm{d}A^3,$$

$\mathbf{n}_l$ are the components of the vector of the external normal to ∂V, dA^l is the area of the face calculated from the equation

$$\mathrm{d}A^l = \frac{\mathrm{d}V}{h_l \mathrm{d}x^l}, \tag{4.39}$$

the product $\mathbf{F} \cdot d\mathbf{A}$ is determined as follows: $\mathbf{F} \cdot \mathrm{d}\mathbf{A} = \sum_{l=1}^{3} \mathbf{n}^l \mathbf{F}^l \mathrm{d}A^l$.

Discretisation of the gas dynamics equations

There are various approaches to the construction of spatial discretisation of evolution equations. We shall mention the most important ones – the finite difference method (the grid methods), the finite volume method (integro-interpolation method), the finite element method, and the spectral method. In most cases, the hyperbolic equations are solved by the finite difference and finite volume methods. The difference between these methods may be described as follows. When using the finite difference method, the values to be determined are the values of the functions at some nodes of the discrete cells. In the finite volume method, the values of the grid functions are mean with respect to the volumes of the discrete cells. When using the finite volume methods for the equations, written in the divergent form, the numerical scheme ensures exact fulfilment of the laws of conservation on the discrete level with the accuracy to the approximation errors which from the viewpoint of physics is absolutely essential for the adequate modelling of the complex physical systems. Here, we describe the spatial discretisation of the gas dynamics equations in the curvilinear orthogonal coordinate systems in accordance with the finite element method.

The computation domain is divided into the following finite volume

$$\Delta V_{ijk} = \int\limits_{x^1_{i-\frac{1}{2}}}^{x^1_{i+\frac{1}{2}}} \int\limits_{x^2_{j-\frac{1}{2}}}^{x^2_{j+\frac{1}{2}}} \int\limits_{x^3_{k-\frac{1}{2}}}^{x^3_{k+\frac{1}{2}}} h_1 h_2 h_3 \, dx_1 dx_2 dx_3, \tag{4.40}$$

where the trio of the indexes (i, j, k) corresponds to the direction determined by the vectors $\mathbf{n}_1$, $\mathbf{n}_2$, $\mathbf{n}_3$. The integer indices denote the quantities related to the centre of the mass (i, j, k) of the cell. The half-integer indexes $(i \pm \frac{1}{2}, j, k)$, $(i, j \pm \frac{1}{2}, k)$, $(i, j, k \pm \frac{1}{2})$ correspond to the upper and lower boundaries between the cells (i, j, k) and $(i \pm 1, j, k)$, (i, j, k) and $(i, j \pm 1, k)$, (i, j, k) and $(i, j, k \pm 1)$, respectively. The relationship between the coordinate lines is described by the relationships

$$x^1_{i+\frac{1}{2}} - x^1_{i-\frac{1}{2}} = \Delta x^1_i, \quad x^1_{j+\frac{1}{2}} - x^2_{j-\frac{1}{2}} = \Delta x^1_j, \quad x^1_{k+\frac{1}{2}} - x^2_{k-\frac{1}{2}} = \Delta x^1_k, \tag{4.41}$$

where Δx^m_l is the width of the cell along the appropriate coordinate.

The areas of the faces of the cell is determined from equation (4.39))

$$A^1_{i\pm\frac{1}{2},j,k} = \int\limits_{x^2_{j-\frac{1}{2}}}^{x^2_{j+\frac{1}{2}}} \int\limits_{x^3_{k-\frac{1}{2}}}^{x^3_{k+\frac{1}{2}}} h_2(x^1_{i\pm\frac{1}{2}}, x^2, x^3) h_3(x^1_{i\pm\frac{1}{2}}, x^2, x^3) dx^2 dx^3, \tag{4.42}$$

$$A^2_{i,j\pm\frac{1}{2},k} = \int\limits_{x^1_{i-\frac{1}{2}}}^{x^1_{i+\frac{1}{2}}} \int\limits_{x^3_{k-\frac{1}{2}}}^{x^3_{k+\frac{1}{2}}} h_1(x^1, x^2_{j\pm\frac{1}{2}}, x^3) h_3(x^1, x^2_{j\pm\frac{1}{2}}, x^3) dx^1 dx^3, \tag{4.43}$$

$$A^3_{i,j,k\pm\frac{1}{2}} = \int_{x^1_{i-\frac{1}{2}}}^{x^1_{i+\frac{1}{2}}} \int_{x^2_{j-\frac{1}{2}}}^{x^2_{j+\frac{1}{2}}} h_1(x^1,x^2,x^3_{k\pm\frac{1}{2}})h_1(x^1,x^2,x^3_{k\pm\frac{1}{2}})\mathrm{d}x^1\mathrm{d}x^2. \tag{4.44}$$

The length of the arcs between the points, situated on the opposite boundaries of these cells $(i\pm\frac{1}{2}, j, k)$, $(i, j\pm\frac{1}{2}, k)$, $(i, j, k\pm\frac{1}{2})$, is calculated as follows

$$\Delta l^1_{ijk} = \int_{x^1_{i-\frac{1}{2}}}^{x^1_{i+\frac{1}{2}}} h_1(x^1,x^2_j,x^3_k)\mathrm{d}x^1, \tag{4.45}$$

$$\Delta l^2_{ijk} = \int_{x^2_{j-\frac{1}{2}}}^{x^2_{j+\frac{1}{2}}} h_2(x^1_i,x^2,x^3_k)\mathrm{d}x^2, \tag{4.46}$$

$$\Delta l^3_{ijk} = \int_{x^3_{k-\frac{1}{2}}}^{x^3_{k+\frac{1}{2}}} h_3(x^1_i,x^2_j,x^3)\mathrm{d}x^3. \tag{4.47}$$

The explicit expressions for the volume and areas of the faces of the cell in different coordinate systems will be presented. In the Cartesian coordinate system, the volume of the cell is $\mathrm{d}V_{ijk} = \Delta x_i \,\Delta y_j \,\Delta z_k$, and the areas of the six faces, defining the cell, are calculated as follows

$$A^x_{i+\frac{1}{2},j,k} = \Delta y_j \Delta z_k, \quad A^y_{i,j+\frac{1}{2},k} = \Delta x_i \Delta z_k, \quad A^z_{i,j,k+\frac{1}{2}} = \Delta x_i \Delta y_j. \tag{4.48}$$

In the cylindrical coordinate system, the volume of the cell is determined from the equation

$$\Delta V_{ijk} = \int_{r_{i-\frac{1}{2}}}^{r_{i+\frac{1}{2}}} \int_{\varphi_{j-\frac{1}{2}}}^{\varphi_{j+\frac{1}{2}}} \int_{z_{k-\frac{1}{2}}}^{z_{k+\frac{1}{2}}} r\mathrm{d}r\mathrm{d}\varphi dz = \frac{r^2_{i+\frac{1}{2}} - r^2_{i-\frac{1}{2}}}{2} \Delta\varphi_j \Delta z_k, \tag{4.49}$$

and the expressions for the areas of the faces have the following form

$$A^r_{i\pm\frac{1}{2},j,k} = r_{i\pm\frac{1}{2}} \Delta\varphi_j \Delta z_k, \quad A^\varphi_{i,j\pm\frac{1}{2},k} = \Delta r_i \Delta z_k, \quad A^z_{i,j,k\pm\frac{1}{2}} = \frac{r^2_{i+\frac{1}{2}} - r^2_{i-\frac{1}{2}}}{2} \Delta\varphi_j. \tag{4.50}$$

In the spherical coordinate system, the volume of the cell and the areas of the faces are calculated as follows

$$\Delta V_{ijk} = \int_{r_{i-\frac{1}{2}}}^{r_{i+\frac{1}{2}}} \int_{\theta_{j-\frac{1}{2}}}^{\theta_{j+\frac{1}{2}}} \int_{\varphi_{k-\frac{1}{2}}}^{\varphi_{k+\frac{1}{2}}} r^2 \mathrm{d}r\mathrm{d}\theta\mathrm{d}\varphi = \frac{r^3_{i+\frac{1}{2}} - r^3_{i-\frac{1}{2}}}{3}\left(\cos\theta_{j-\frac{1}{2}} - \cos\theta_{j+\frac{1}{2}}\right)\Delta\varphi_k, \quad (4.51)$$

$$A^r_{i\pm\frac{1}{2},j,k} = r_{i\pm\frac{1}{2}}\left(\cos\theta_{j-\frac{1}{2}} - \cos\theta_{j+\frac{1}{2}}\right)\Delta\varphi_k, \quad (4.52)$$

$$A^\theta_{i,j\pm\frac{1}{2},k} = \frac{r^2_{i+\frac{1}{2}} - r^2_{i-\frac{1}{2}}}{2}\sin\theta_{i\pm\frac{1}{2}}\Delta\varphi_k, \quad (4.53)$$

$$A^\varphi_{i,j,k\pm\frac{1}{2}} = \frac{r^2_{i+\frac{1}{2}} - r^2_{i-\frac{1}{2}}}{2}\Delta\theta_j. \quad (4.54)$$

Using the integral form of the equations (4.40) for the finite volumes ΔV_{ijk} and for the discrete time period $\Delta t^n = t^{n+1} - t^n$, we obtain the relationship linking the conservative variables averaged out over the volume and the surface integrals averaged out over time

$$\bar{\mathbf{U}}^{n+1}_{ijk} - \bar{\mathbf{U}}^n_{ijk} = -\frac{\Delta t^n}{\Delta V_{ijk}}\left[\oint_{ijk}\langle\mathbf{F}\rangle^n\cdot\mathrm{d}\mathbf{A} + \int_{ijk}\langle\mathbf{S}\rangle^n\mathrm{dV}\right]. \quad (4.55)$$

In these equations, the upper underscore indicates the averaging with respect to volume

$$\bar{\mathbf{U}}^n_{ijk} = \frac{1}{\Delta V_{ijk}}\int_{ijk}\mathbf{U}(x^1,x^2,x^3,t^n)\mathrm{d}V, \quad (4.56)$$

and averaging with respect to time is denoted by the angular brackets

$$\langle\mathbf{F}\rangle^n_{ijk} = \frac{1}{\Delta t^n}\int_{t^n}^{t^{n+1}}\mathbf{F}(x^1,x^2,x^3,t^n)\mathrm{d}t. \quad (4.57)$$

Further, to simplify considerations, the upper underscore and the angular brackets will be omitted and it is understood that the average values are included in the equations presented below.

On the basis of the relationships (4.55) we can write the numerical finite-volume explicit scheme of the first order with respect to time

$$\mathbf{U}^{n+1}_{ijk} = \mathbf{U}^n_{ijk} + \mathbf{L}^1\left(\mathbf{U}^n_{i+\frac{1}{2}}, \mathbf{U}^n_{i-\frac{1}{2}}\right) + + \mathbf{L}^2\left(\mathbf{U}^n_{j+\frac{1}{2}}, \mathbf{U}^n_{j-\frac{1}{2}}\right) + \\ + \mathbf{L}^3\left(\mathbf{U}^n_{\ 1}, \mathbf{U}^n_{\ 1}\right) + \Delta t\cdot\mathbf{S}^n_{ijk}. \quad (4.58)$$

Here L^m is the operator of the appropriate direction; as an example we present the explicit form of this operator along the direction $m = 1$:

$$\mathbf{L}^1\left(\mathbf{U}^n_{i+\frac{1}{2}}, \mathbf{U}^n_{i-\frac{1}{2}}\right) = -\frac{\Delta t^n}{\Delta V_{ijk}}\left[\mathbf{F}^1\left(\mathbf{U}^n_{i+\frac{1}{2}}\right)A^1_{i+\frac{1}{2}} - \mathbf{F}^1\left(\mathbf{U}^n_{i-\frac{1}{2}}\right)A^1_{i-\frac{1}{2}}\right], \quad (4.59)$$

where $\mathbf{F}^1\left(\mathbf{U}^n_{i\pm\frac{1}{2}}\right)$ is the numerical flow of the quantities through the face of the cell.

It may be shown that the averaging of the vector of sources over volume in the Cartesian cylindrical coordinate systems is actually reduced to calculating its components in the centre of the mass of the cells. Special attention should be given to the case of the spherical coordinate system. The following substitutions should be made in the equation (4.37)

$$\frac{1}{r_i} \rightarrow \frac{\frac{1}{2}r^2_{i+\frac{1}{2}} - \frac{1}{2}r^2_{i-\frac{1}{2}}}{\frac{1}{3}r^3_{i+\frac{1}{2}} - \frac{1}{3}r^3_{i-\frac{1}{2}}}, \quad \mathrm{ctg}\theta_i \rightarrow \frac{\sin\theta_{i+\frac{1}{2}} - \sin\theta_{i-\frac{1}{2}}}{\cos\theta_{i-\frac{1}{2}} - \cos\theta_{i+\frac{1}{2}}}. \quad (4.60)$$

Approximation of the vector of the sources by the values r and θ, related to the centre of the cells, results in a large error in the vicinity of the origin of the coordinates.

The cylindrical spherical system of the coordinates is characterised by the presence of a singularity at $r = 0$, in addition to this, the formulation of the boundary conditions at the origin of the coordinates must be carried out accurately. As an example, we consider the spherical coordinate system. The zenith θ and azimuthal φ angles vary in the ranges $\theta \in [0, \pi]$ and $\varphi \in [0, 2\pi]$. The boundary at any fixed radius has the form of a sphere. At $r = 0$, the sphere degenerates into a point. The values in the fictitious cells, corresponding to $r < 0$, in particular $(-r;\ \theta,\ \varphi)$ are determined by the values in the cells $(r;\ \pi - \theta,\ \varphi + \pi)$. Actually, in transition to the origin of the coordinates, the vectors $\mathbf{n}_r$ and $\mathbf{n}_\varphi$ change their direction to opposite, whereas vector $\mathbf{n}_\theta$ remains without changes. This means that the density, pressure and θ-component of the velocity in the fictitious cells with respect to the radius are equal to the values in the cells $(r;\ \pi - \theta,\ \varphi + \pi)$, and the radial and azimuthal components of the velocities should be equal to the negative values of the appropriate cells.

The boundary conditions at $\theta = 0$ and $\theta = \pi$ are determined by the same procedure. At any fixed value $\theta = \text{const}$ the boundary is represented by the surface of a cone. However, at $\theta = 0$ and $\theta = \pi$, the cone degenerates into a line. This means that at $\theta = 0$ the values in the fictitious cells $(r;\ -\theta,\ \varphi)$ are equal to the values in the cells $(r;\ \theta,\ \varphi + \pi)$. At $\theta = \pi$ the values in the fictitious cells $(r;\ \theta + \pi,\ \varphi)$ are determined by the values in the cells. At $\theta = \pi$, the values in the fictitious cells $(r;\ \theta + \pi,\ \varphi)$ are determined by the values in the cells $(r;\ \pi - \theta,\ \varphi + \pi)$.

To conclude this section, it should be stressed that to construct the numerical scheme it is sufficient to use the finite element method. Firstly, the application of the finite volume method to the systems of hyperbolic type equations, written in the integral form and expressing the laws of conservation, characterises the exact fulfilment of the laws of conversation on the discrete level, regardless of approximation errors. This means that the position of discontinuities and their magnitude are accurately reproduced by the scheme. Secondly, it should be stressed that the finite-difference approximation, in contrast to finite-volume one, in the cylindrical and spherical systems of coordinates results in areas of the order of $\mathcal{O}(1/r_i)$ in the vicinity of the origin of the coordinates $r = 0$. This may in turn cause the formation of cells with negative density and/or pressure in the calculations and result in failure of the programme.

Godunov-type numerical schemes

For the hyperbolic equations, which also include the gas dynamic equations, the problem of the arbitrary discontinuity – Riemann problem – is of special importance. We examine the one-dimensional system of homogeneous equations of gas dynamics:

$$\frac{\partial \mathbf{U}}{\partial t} + \frac{\partial \mathbf{F}}{\partial x} = 0, \tag{4.61}$$

where $t > 0$, $-\infty < x < \infty$. It is assumed that the initial data are of the piecewise-constant form (Fig. 4.3):

$$\mathbf{U}(x < 0, t) = \mathbf{U}_0 = \text{const},$$
$$\mathbf{U}(x \geq 0, t) = \mathbf{U}_1 = \text{const}.$$

The solution of the Riemann problem may be obtained completely in the analytical form and, from the mathematical viewpoint, is reduced to the numerical solution of the non-linear algebraic equation (see, for example, the monograph [106]). The discontinuity in the initial distribution may result in the appearance in the flow of shock waves, rarefaction waves, contact discontinuities, alternating with the allegiance with the constant values of the quantities. All these types of solution are referred to as the elementary solutions of the Riemann problem.

The concept of the Godunov method is the application of the exact solution of the Riemann problem for calculating the flows of the quantities through the boundaries of the cells $\mathbf{F}^l_{i\pm\frac{1}{2}} \equiv \mathbf{F}^l\left(\mathbf{U}^n_{i\pm\frac{1}{2}}\right)$. It is assumed that inside the disk with cells all the grid functions are piecewise-constant (Fig. 4.3). Thus, the values of all quantities to the left $\mathbf{U}^L_{i+\frac{1}{2}} = \mathbf{U}^n_i = \text{const}$ and to the right $\mathbf{U}^R_{i+\frac{1}{2}} = \mathbf{U}^n_{i+1} = \text{const}$ from the face of the cell are known. We use the determined values as the initial data for solving the Riemann problem of determination of the values $\mathbf{U}_{i+\frac{1}{2}}$ at the boundary of the cell (Fig. 4.4).

Investigation of the stability of the numerical Godunov scheme by the spectral analysis method gives the Courant–Friedrichs–Lewy conditions

$$\mathcal{C}_x = (|u_x| + c_s)\frac{\Delta t}{\Delta x} < 1, \tag{4.62}$$

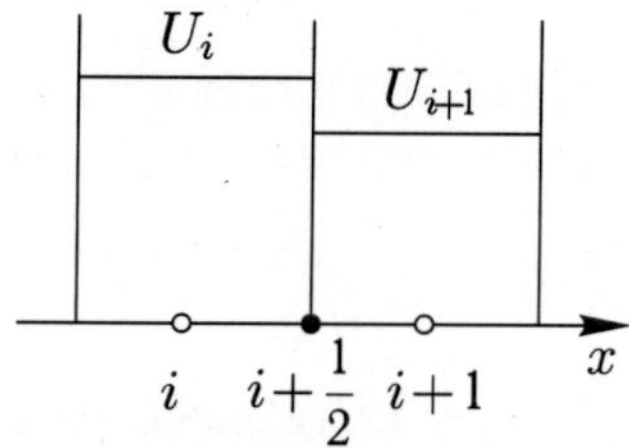

Fig. 4.3. Piecewise-constant distribution of U.

c_s is the adiabatic speed of sound in the gas. Thus, the integration step with respect to time Δt is selected from the condition

$$\Delta t = C_x \times \min_{ijk} \frac{\Delta x}{|u_x| + c_s}. \tag{4.63}$$

In the multidimensional schemes, using splitting with respect to space, the integration step is selected as follows:

$$\Delta t = C \times \min_{ijk} \left(\frac{\Delta l^1_{ijk}}{|u_1| + c_s}, \frac{\Delta l^2_{ijk}}{|u_2| + c_s}, \frac{\Delta l^3_{ijk}}{|u_3| + c_s} \right), \tag{4.64}$$

where $C < 1$, Δl^1_{ijk}, Δl^2_{ijk}, Δl^3_{ijk} and the lengths of the hardness of the discrete cells in the given direction. The stability condition for the multidimensional schemes without splitting with respect to the space has the form

$$C = \Delta t \times \min_{ijk} \left(\frac{|u_1| + c_s}{\Delta l^1_{ijk}} + \frac{|u_2| + c_s}{\Delta l^2_{ijk}} + \frac{|u_3| + c_s}{\Delta l^3_{ijk}} \right) < 1. \tag{4.65}$$

In the multidimensional schemes without splitting, the stability can be determined by, for example, condition (4.64), but in this case it should be assumed that $C < 1/n$, where n is the dimensionality of the problem.

To determine the numerical schemes of the second order of accuracy with respect to time and space, it is recommended to use the integration methods of the Runge–Kutta type combined with the piecewise-linear interpolation of the variables [824] (Fig. 4.5). The scheme, described below, is referred to as the Monotonic Upstream Scheme for Conservation Laws (MUSCL TVD).

It is assumed that the initial distribution of the values inside the cells has the form

$$\mathbf{U}(t^n, x) = \mathbf{U}_i^n + \mathbf{Q}_i^n (x - x_i), \quad x \in [x_{i-\frac{1}{2}}, x_{i+\frac{1}{2}}], \tag{4.66}$$

where $\mathbf{Q}_i^n$ is the vector of the angle of inclination of the distribution of the values of $\mathbf{U}$ inside the cell.

The flows $\mathbf{F}^n_{i+\frac{1}{2}}$ are determined by solving the Riemann problem with the following piecewise-constant initial data:

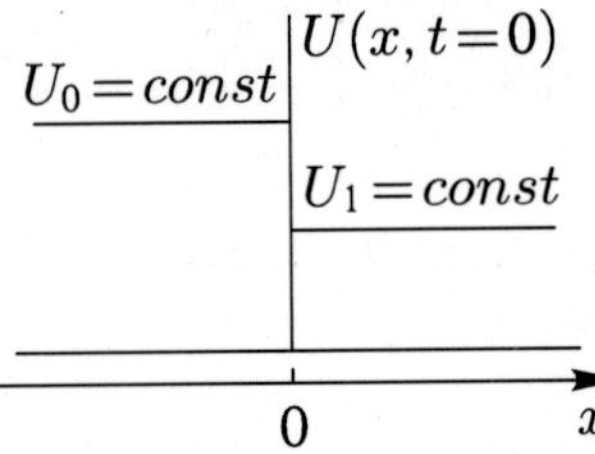

Fig. 4.4. Diagram of the Riemann problem.

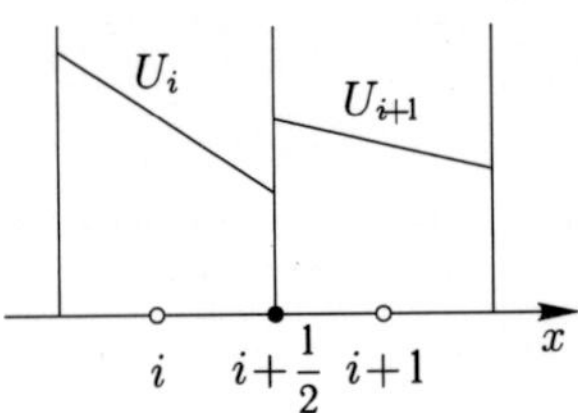

Fig. 4.5. Piecewise–linear distribution of U.

$$\mathbf{U}^L_{i+\frac{1}{2}} = \mathbf{U}^n_i + \frac{1}{2}\Delta x \mathbf{Q}^n_i, \quad x < x_{i+\frac{1}{2}}, \tag{4.67}$$

$$\mathbf{U}^R_{i+\frac{1}{2}} = \mathbf{U}^n_{i+1} - \frac{1}{2}\Delta x \mathbf{Q}^n_{i+1}, \quad x > x_{i+\frac{1}{2}}. \tag{4.68}$$

To obtain the numerical solution, monotonic in the sense of TVD [650, 1317], the angle of inclination of the quantities inside the cells should be restricted

$$\mathbf{Q}_i = \Psi\left(\frac{\mathbf{U}_{i+1} - \mathbf{U}_i}{\Delta x}, \frac{\mathbf{U}_i - \mathbf{U}_{i-1}}{\Delta x}\right), \tag{4.69}$$

where are Ψ is the function-constraint, for example

$$\Psi(a,b) = \text{minmod}(a,b) = \frac{1}{2}[\text{sign}(a) + \text{sign}(b)]\min(|a|,|b|).$$

The different procedures of reconstruction can be described by the two-parameter family [824]

$$\mathbf{u}^R_{i+\frac{1}{2}} = \mathbf{u}^n_{i+1} - \frac{1}{4}[(1-\eta)\bar{\Delta} + (1+\eta)\tilde{\Delta}]_{i+1}, \tag{4.70}$$

$$\mathbf{u}^L_{i+\frac{1}{2}} = \mathbf{u}^n_i + \frac{1}{4}[(1-\eta)\bar{\Delta} + (1+\eta)\tilde{\Delta}]_i, \tag{4.71}$$

where $\bar{\Delta}$ and $\tilde{\Delta}$ denote the calculation of the restricted differences

$$\bar{\Delta}_i = \text{minmod}[(\mathbf{u}^n_{i+1} - \mathbf{u}^n_i), \omega(\mathbf{u}^n_i - \mathbf{u}^n_{i-1})],$$

$$\tilde{\Delta}_i = \text{minmod}[(\mathbf{u}^n_i - \mathbf{u}^n_{i-1}), \omega(\mathbf{u}^n_{i+1} - \mathbf{u}^n_i)].$$

The reconstruction of the parameters of the flow at the boundary of the cell of the type (4.71)–(4.70) corresponds to the three-point template. Thus, the investigated scheme has a five-point template. The quantity η determines the type and order of the spatial approximation, in particular: if $\eta = -1$, we obtain the fully counterflow scheme of the second order (Fromm scheme); the case $\eta = 1$ corresponds to the three-point central scheme of the second order; at $\eta = 1/3$ we obtain when the counterflow-conditioned scheme of the third order. Parameter Ω is introduced for improving the resolution of the tangential discontinuities and is responsible for artificial compression:

$$1 \le \omega \le \frac{3-\eta}{1-\eta}. \tag{4.72}$$

The interpolated quantities $\mathbf{u}$ can be represented by the conservative $\mathbf{u} = \mathbf{U}$, simple $\mathbf{u} = \mathbf{V}$ or characteristic variables. Evidently, the interpolation of the simple variables is the most economical approach. The reconstruction of the characteristic variables, although it requires a longer computing time, gives the best results in the majority of the gas dynamics problems.

To discuss the questions associated with the increase of the order of approximation with respect to time, instead of (4.58) we consider the semi-discrete analogue

$$\frac{\mathrm{d}\mathbf{U}}{\mathrm{d}t} = \mathbf{L}\left(\mathbf{U}^n\right), \tag{4.73}$$

where $\mathbf{L}$ is the operator of the spatial discretisation of the flowes and sources (see 4.58). Thus, the time t is regarded as a continuous variable, and the remaining variables are regarded as discrete ones. This approach is referred to as the method of straight lines.

The equations (4.73) form a system of conventional differential equations which can be interpreted using various integration methods, for example, the methods of the Runge–Kutta type with high accuracy. However, it should be remembered that the increase of the order of approximation with respect to time may result in the loss of monotonicity by the numerical solution in the sense of TVD. It has been shown that the order of the Runge–Kutta TVD methods is not higher than the third order [1164}. We discuss the Runge–Kutta TVD method of the second order of accuracy

$$\mathbf{U}^1 = \mathbf{U}^n + \Delta t \cdot \mathbf{L}(\mathbf{U}^n) \tag{4.74}$$

$$\mathbf{U}^{n+1} = \frac{1}{2}\mathbf{U}^n + \frac{1}{2}\mathbf{U}^1 + \frac{1}{2}\Delta t \cdot \mathbf{L}(\mathbf{U}^1). \tag{4.75}$$

The resources-consuming Godunov method has stimulated the development of methods based on the approximate methods of solving the one-dimensional Riemann problem of the breakdown of an arbitrary discontinuity. These methods are referred to as the Godunov-type methods. We shall mention briefly some of them.

1. The Courant–Isakson–Ries (CIR) method is based on the approximate solution of the Riemann problem [488]. In this case, the numerical solution consists only of elementary solutions in the form of discontinuities separated from each other by regions with constant values of the quantities. This class of the methods makes it possible to obtain solutions for conservative and non-conservative forms of the hyperbolic systems of equations. The methods of this type include the Lax–Friedrichs TVD method characterised by exclusively high robustness and

efficiency. The numerical flow through the boundary of the cell is computed using the following equation:

$$\mathbf{F}^{n}_{i+\frac{1}{2}}=\frac{1}{2}\left[\mathbf{F}\left(\mathbf{U}^{R}_{i+\frac{1}{2}}\right)+\mathbf{F}\left(\mathbf{U}^{L}_{i+\frac{1}{2}}\right)-\frac{\Delta t}{\Delta x}\lambda^{max}_{i+\frac{1}{2}}\cdot\left(\mathbf{U}^{R}_{i+\frac{1}{2}}-\mathbf{U}^{L}_{i+\frac{1}{2}}\right)\right], \qquad (4.76)$$

where $\lambda^{\max} = |u_x| + c_e$ is the maximum eigenvalue of the Jacobi matrix, c_e is the adiabatic speed of sound in the gas. For example, it can be assumed that

$$\lambda^{max}_{i+\frac{1}{2}}=\max\left[\lambda^{max}\left(\mathbf{U}^{R}_{i+\frac{1}{2}}\right),\lambda^{max}\left(\mathbf{U}^{L}_{i+\frac{1}{2}}\right)\right].$$

2. In the Rowe method, the flows through the boundaries of the cells are calculated by the exact solution of the Riemann problem linearised in a special manner [1077, 1078]. The solution also consists of moving discontinuities separated from each other by regions with the constant values of the quantities. This method retains accurately the non-linear relationships at single discontinuities. The rarefaction waves in the solution of the Riemann problem are replaced by the rarefaction shock waves which may result in the appearance of non-physical rarefaction shock waves in the numerical solution. To eliminate the non-physical solutions, a procedure of entropy correction has been proposed in [650]. The method is suitable only for the systems of equations written in the conservative form, and requires the filing the specific form of the eigenvectors of the Jacobi matrix. Regardless of the popularity of the Rowe method, it should be mentioned that in application of this method to media with general equations of state there is a problem of the singularity of the Rowe equations.

3. In the Osher method, the solution of the problem of the breakdown of an arbitrary discontinuity is constructed using only the Riemann waves [106, 995]. It should be mentioned that the approximation of the shockwaves by the Riemann compression waves actually indicates that the ratio at the fronts of the shock waves is fulfilled only approximately.

4. In the Harten-Lax-van Leer (HLL) method, to calculate the numerical flow it is necessary to know only the maximum and minimum velocities of propagation of the waves inside the cell [651]. It has been shown that the method remains positive which is especially important for calculating hypersonic gas flows when the kinetic energy dominates over the thermal energy. There are various modifications of the HLL method. The HLLC method (see, for example, the monograph [1233]) takes into account not only discontinuities of the types of shock waves but also the discontinuities of the type of contact or tangential waves. The equation for the numerical flow, corresponding to the HLLC method has the following form:

$$\mathbf{F}_{i+\frac{1}{2}}^{\mathrm{HLLC}} = \begin{cases} \mathbf{F}_L, & \lambda_L > 0, \\ \mathbf{F}_L^* = \mathbf{F}_L + \lambda_L\left(\mathbf{U}_L^* - \mathbf{U}_L\right), & \lambda_L \le 0 \le \lambda_M, \\ \mathbf{F}_R^* = \mathbf{F}_R + \lambda_R\left(\mathbf{U}_R^* - \mathbf{U}_R\right), & \lambda_M \le 0 \le \lambda_R, \\ \mathbf{F}_R, & \lambda_R < 0. \end{cases} \tag{4.77}$$

The values $\mathbf{U}_L^*$ and $\mathbf{U}^*$ are determined from the relationships at the contact discontinuity:

$$\varrho_\alpha^* = \varrho_\alpha \frac{\lambda_\alpha - u_\alpha}{\lambda_\alpha - \lambda_M}, \tag{4.78}$$

$$p_\alpha^* = p_\alpha + \varrho_\alpha(\lambda_\alpha - u_\alpha)(\lambda_M - u_\alpha), \tag{4.79}$$

$$u_\alpha^* = \lambda_M, \tag{4.80}$$

$$v_\alpha^* = v_\alpha, \tag{4.81}$$

$$w_\alpha^* = w_\alpha, \tag{4.82}$$

$$e_\alpha^* = \frac{(\lambda_\alpha - u_\alpha)e_\alpha - p_\alpha u_\alpha + p_\alpha^* \lambda_M}{\lambda_\alpha - \lambda_M}, \tag{4.83}$$

where $\alpha = L, R$; u, v and w denote x, y and z components of velocity, respectively. The minimum and maximum velocity of propagation of the perturbations is determined from λ_L = min $|\lambda_1(\mathbf{U}_L)$, λ_1 $(\mathbf{U}_R)]$ and R = max $[\lambda_5(\mathbf{U}_L)$, λ_5 $(U_R)]$, respectively; $\lambda_1 = [u] + c_s$ and $\lambda_5 = |u| - c_s$ – maximum and minimum eigenvalues of the Jacobi matrix.

It should be mentioned that the results presented in this book for the numerical modelling of the system of gas dynamics equations are based on the TVD variant of the MUSCL scheme [650, 824]. The equations are approximated by the method of finite volumes which guarantees the fulfilment of the laws of conservation on a discrete level [106]. The applied numerical scheme belongs to the schemes of the Godunov type of the second order of accuracy with respect to time and the third order with respect to space. The flows of the physical quantities through the boundaries of the cells were calculated by the modified HLLC method [833] which makes it possible to model indirectly the shock waves and contact and tangential discontinuities. The scheme of the second order of accuracy with respect to time is modelled using the predictor–corrector scheme combined with the condition of no decrease of the total variation of the numerical solution (TVD condition [650]). The third order of approximation with respect to space in the regions of smooth flows achieved by the application of the MUSCL procedure for the construction of the simple quantities (density, pressure and velocity vector) at the faces of the calculation cells. This simple and economical computing method produces shock waves spread over three cells. To improve the resolution of the tangential contact discontinuities in interpolation it is recommended to use the appropriate parameter responsible for artificial compression.

4.1.4. The particle method for modelling gas disks

We describe the methods of smoothed particles hydrodynamics (SPH) for modelling the non-stationary three-dimensional hydrodynamics. The basic concept of the SPH algorithm consists of the discretisation of a continuous medium by a finite set of the Lagrangian particles which move at the speed of the flow and permit an arbitrary connection between them so that it is not necessary to use the numerical grid (see [675, 777, 943, 1089, 1175]). Each function, included in the system of equations, can be written in the form

$$f(\mathbf{r}) = \int_{\Omega} f(\mathbf{r}')\delta(|\mathbf{r}-\mathbf{r}'|)d\mathbf{r}', \tag{4.84}$$

where δ is the Dirac function, f is any parameter determined in the spatial coordinates x. Integration with respect to the volume is carried out in the region Ω occupied by matter. This integral can be modified by replacing the δ-function by the smoothing core W with the characteristic length h so that the following condition is fulfilled

$$\lim_{h\to 0} W(|\mathbf{r}-\mathbf{r}'|,h) = \delta(|\mathbf{r}-\mathbf{r}'|,h). \tag{4.85}$$

Consequently, we obtain

$$f(\mathbf{r}) = \int_{\Omega} f(\mathbf{r}')W(|\mathbf{r}-\mathbf{r}'|,h)d\mathbf{r}' + \mathcal{O}(h^2). \tag{4.86}$$

The function of the core is normalised with respect to the following rule

$$\int_{\Omega} W(|\mathbf{r}-\mathbf{r}'|,h)\, d\mathbf{r}' = 1. \tag{4.87}$$

The resultant integral (4.86) can be replaced by a finite sum with the mass element $\varrho \cdot dV$ with respect to the adjacent particles to the examined sum. Consequently, for any quantity, relating to the particle with the coordinate r, we obtain

$$f_s(\mathbf{r}) = \int_{\Omega} \frac{f(\mathbf{r}')}{\varrho(\mathbf{r}')} W(|\mathbf{r}-\mathbf{r}'|,h)\, \varrho(\mathbf{r}')d\mathbf{r}' + \mathcal{O}(h^2) \simeq \sum_{j=1}^{n} m_i \frac{f_j}{\varrho_j} W(|\mathbf{r}-\mathbf{r}_j|,h), \tag{4.88}$$

where n is the number of adjacent particles, m_i, ϱ_i, r_i is the mass, density and radius vector of the i-th particle, respectively. The transition from integration to summation is the basis of the entire SPH formalism.

For the gradient of any function we can write

$$\nabla f_s(\mathbf{r}) = \frac{\partial}{\partial \mathbf{r}} \int_{\Omega} \frac{f(\mathbf{r}')}{\varrho(\mathbf{r}')} W(|\mathbf{r}-\mathbf{r}'|,h)\, \varrho(\mathbf{r}')d\mathbf{r}' + \mathcal{O}(h^2) \simeq \sum_{j=1}^{n} m_i \frac{f_j}{\varrho_j} \nabla_i W_{ij}, \tag{4.89}$$

where the i-th particle is investigated: $(\mathbf{r} = \mathbf{r}_i)$, $\nabla_i \equiv \dfrac{\partial}{\partial \bar{r}_i}$ and $W_{ij} \equiv W(|r_i - r_j|, h)$.

Two particles i and j are termed neighbouring or interacting if the distance between them does not exceed $(h_i + h_j)$ (Fig. 4.6). Quantity h_i is referred to as the smoothing length for the i-th particle.

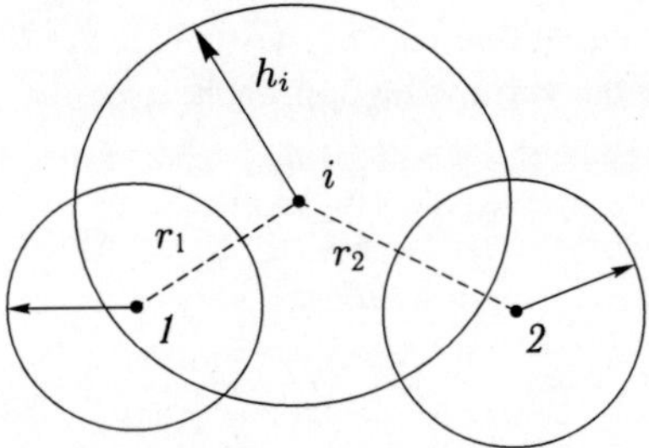

Fig. 4.6. Interaction between the particles (i, 1, 2 – numbers of particles, r_1 and r_2 are the distances from the i-th particles to the 1st and 2nd particle, respectively.,

According to the definition, the smoothing core W should yield in the limit the δ-function when the smoothing length h tends to 0 (4.85) and satisfy the normalisation condition (4.87). We may write

$$W(r,h) = \frac{1}{N(\delta)h^{\delta}} f\left(\frac{r}{h}\right), \tag{4.90}$$

here $r = |\mathbf{r} - \mathbf{r}'|$ and δ is the dimension of the problem. Consequently, the normalisation condition (4.87) can be written in the following form

$$\frac{1}{N(\delta)} \int f(q)dV = 1, \tag{4.91}$$

where $q = r/h$ and the element of the volume dV is equal to dq, $2\pi q\,dq$ or $4\pi q^2 dq$ for one-, two-and three-dimensional problems, respectively.

A suitable example of the function, satisfying these conditions, is the Gaussian core

$$W(r,h) = \frac{1}{N(\delta)h^{\delta}} e^{-q^2}, \tag{4.92}$$

where $q = r/h$ and $N(\delta) = \{\sqrt{\pi}, \pi, \pi\sqrt{\pi}\}$ for $\delta = 1, 2, 3$.

It is far more efficient to use the cores with a finite spatial length reducing the number of adjacent particles so that the computing losses can be reduced to O (n, N), where n is the number of the nearest neighbours, regardless of the additional losses for finding the adjacent particles. The core in the form of a cubic spline is used in most cases

$$f(q) = \begin{cases} 1 - \frac{3}{2}q^2 + \frac{3}{4}q^3, & 0 \le q < 1; \\ \frac{1}{4}(2-q)^3, & 1 \le q < 2; \\ 0, & 2 \le q \end{cases} \tag{4.93}$$

with the normalisation $N(\delta) = \{3/2, 7\pi/10, \pi\}$ for $\delta = 1, 2, 3$. This core satisfies the main condition (4.85) and (4.87), has a continuous first dertivative and is bounded by $2h$.

The hydrodynamics equations can be written in the SPH form which can be obtained on the basis of the variational approach.

The equation for the density

$$\varrho_i = \sum_j m_j W_{ij} \tag{4.94}$$

has the form

$$\frac{d\varrho_i}{dt} = \sum_j m_j (\vec{u}_i - \vec{u}_j) \nabla_i W_{ij}. \tag{4.95}$$

The SPH equation of motion in the absence of dissipation is:

$$\frac{d\vec{u}_i}{dt} = -\sum_j m_j \left(\frac{p_j}{\varrho_j^2} + \frac{p_i}{\varrho_i^2}\right) \nabla_i W_{ij}. \tag{4.96}$$

This form shows the conservation of the momentum because the contribution to the summation for the momentum of the *i*-th particle is equal and opposite to the momentum of the *j*-th particle. This is caused by the asymmetry of the gradient of the core.

Depending on the purpose for which the solved problem is to be used and on the nature of the flow, we can obtain the equations for the internal energy of the particles E, total energy E_{tot}, or for entropy e.

We shall discuss two forms of the equation with respect to internal energy:

$$\frac{dE_i}{dt} = \frac{p_i}{\varrho_i^2} \sum_j m_j \vec{u}_{ij} \nabla_i W_{ij} \tag{4.97}$$

or

$$\frac{dE_i}{dt} = \frac{1}{2} \sum_j m_j \left(\frac{p_j}{\varrho_j^2} + \frac{p_i}{\varrho_i^2}\right) \vec{u}_{ij} \nabla_i W_{ij}, \tag{4.98}$$

where $\mathbf{u}_{ij} = \mathbf{u}_i - \mathbf{u}_j$. The equations (4.98) and (4.96) have the same symmetry. However, the form of (4.97) is preferred in most cases.

The equation for the total energy of the *i*-th particle

$$\frac{de_i}{dt} = \sum_j m_j \left(\frac{p_i}{\varrho_i^2} \vec{u}_j + \frac{p_j}{\varrho_j^2} \vec{u}_i\right) \nabla_i W_{ij} \tag{4.99}$$

gives the equation for the total energy of the entire system

$$\frac{dE_{tot}}{dt} = \sum_i m_i \frac{de_i}{dt} = \sum_i \sum_j m_i m_j \left(\frac{p_i}{\varrho_i^2} \vec{u}_j + \frac{p_j}{\varrho_j^2} \vec{u}_i\right) \nabla_i W_{ij}. \tag{4.100}$$

The smoothing length h determines the radius of interaction of the SPH particles. In the simplest case, h = const can be used for particles. Algorithms have been proposed in which each particle is characterised by its value h_i. If it is accepted that

$$h_i \propto 1/\varrho_i^{1/\delta}, \tag{4.101}$$

it is already possible to maintain the number of neighbouring particles approximately constant, regardless of the local increase of density in some region. If the SPH system is supplemented by the evolution equation for the smoothing length

$$\frac{dh_i}{dt} = -\left(\frac{h_i}{\varrho_i \delta}\right)\frac{d\varrho_i}{dt}, \tag{4.102}$$

this may greatly improve the spatial resolution. In this case, in the equations (4.96), (4.97) it is necessary to include the terms containing gradients of h. The efficiency of this approach on the basis of the variational principle is very high [1199].

The modification of this algorithm is the Adaptive SPH algorithm (ASPH) in which the smoothing length is given in the form of a tensor which depends on the coordinates and time, and the smoothing core in this case is represented by an ellipsoid. This approach is sufficient for modelling of gas discs in the cylindrical coordinate system because the degree of inhomogeneity of the flow develops rapidly along r, φ and z.

For efficient modelling of the shockwaves, it is necessary to introduce the artificial viscosity into the numerical scheme. The selection of artificial viscosity depends on the formulation of the problem. We present a relatively general equation:

$$\Pi_{ij} = \begin{cases} \dfrac{-\mu_{ij}\alpha\bar{c}_{ij} + \beta\mu_{ij}^2}{\bar{\varrho}_{ij}}, & \vec{u}_{ij}\vec{r}_{ij} < 0, \\ 0, & \vec{u}_{ij}\vec{r}_{ij} \geq 0, \end{cases} \tag{4.103}$$

where $\mu_{ij} = h\vec{u}_{ij}\vec{r}_{ij} / (\vec{r}_{ij}^{\,2} + \epsilon^2 h^2)$, $\vec{u}_{ij} \equiv \vec{u}_i - \vec{u}_j$, $\vec{r}_{ij} \equiv \vec{r}_i - \vec{r}_j$, $\bar{c}_{ij}$, $\bar{\varrho}_{ij}$ are the mean values of the speed of sound and density for the particles i and j. The viscous term is clearly evident when the particles are compressed (i.e. $\mathbf{u}_{ij}\mathbf{r}_{ij} < 0$), and ensures the invariance in relation to the Galilean transformations, retains the total momentum and the momentum of the pulse and does not contribute in the case of solid-state rotation. The term in (4.103), containing the parameter β, becomes dominant at a large difference of the velocities (i.e. at high Mach numbers). The term with α is dominant for the small difference of the velocities. For the adequate modelling of the galactic disks it is necessary to select the values $\alpha = 1$, $\beta = 2$, $\epsilon = 0.1$, which ensures the necessary dissipation in the vicinity of the front of the shockwave and has only a slight effect on other regions. Equation (4.96) should be supplemented by the viscous term

$$\frac{d\vec{u}_i}{dt} = -\sum_j m_j \left(\frac{p_j}{\varrho_j^2} + \frac{p_i}{\varrho_i^2} + \Pi_{ij}\right)\nabla_i W_{ij}. \tag{4.104}$$

The application of the 'switch' of artificial viscosity may greatly improve the accuracy of calculations [943]. It is sufficient to write for every particle the additional equation for the parameter α:

$$\frac{d\alpha_i}{dt} = -\frac{\alpha_i - \alpha_i^*}{\tau_i} + S_i. \tag{4.105}$$

In the absence of source S, parameter α_i assumes the value α_i^* at times of the order of τ. Source S in equation (4.105) ensures the increase of α on approach to the front of the shock wave. We will discuss a simple variant of selection of the source: $S = \max(-\nabla\mathbf{u}, 0)$. The equation becomes large in the vicinity of the front of the shock wave and remains invariant in relation to the Galilean transformations. The practice of numerical calculations demonstrates the efficiency of this approach because the method is relatively simple to apply and does not require additional memory in calculations.

An important part of the SPH method is the search for the nearest particles for which it is necessary to calculate the short range pressure forces. The realisation of this type of sorting of the particles is relatively simple for identical particles (h = const) on the Euler grid, if we use the symmetry of interaction of the pair of the particles. The sorting algorithm, based on the construction of the hierarchy of grids (wood), is efficient and universal, and this is identical with the TreeCode algorithm (section 4.1.1).

We mention the main advantage of the Lagrangian method for solving astrophysical problems in comparison with the Euler grid methods (section 4.1.3) which is associated with the possibility of carrying out complicated calculations without specifying the gas–vacuum boundary, and without additional boundary conditions.

As test calculations in the construction of numerical gas-dynamic models we consider the following problems: decay of an arbitrary pressure 'jump', discharge of gas into vacuum from a vessel through a small orifice; supersonic flow of the gas through a potential well (the numerical SPH calculations, presented in this section, were performed by M.A. Butenko).

We discuss a problem of the decay of an arbitrary pressure jump with a gradient $p_1/p_2 = 10$, ϱ = const, and the adiabate exponent $\gamma = 5/3$ (Fig. 4.7). The numerical solution should be compared with the exact solution of the Riemann gas dynamics problem (Fig. 4.8) consisting of elementary solutions (shock wave, rarefaction wave, tangential discontinuity) separated from each other by the regions with constant values of the parameters. In the partial case, when there is no jump of the tangential component of velocity, but there are jumps of density and other thermodynamic

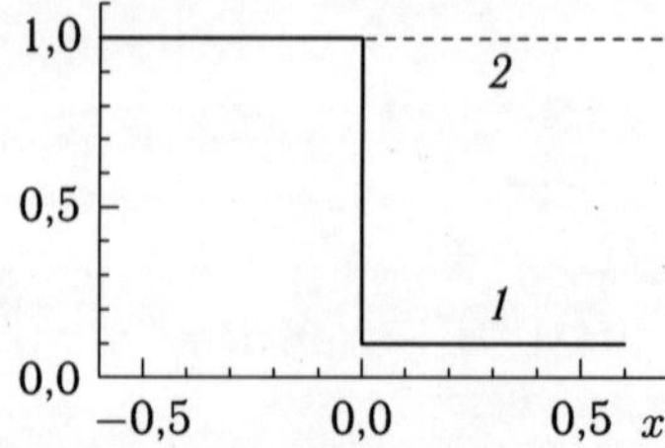

Fig. 4.7. Initial distribution of pressure (1) and density (2) in a problem of the decay of an arbitrary pressure jump.

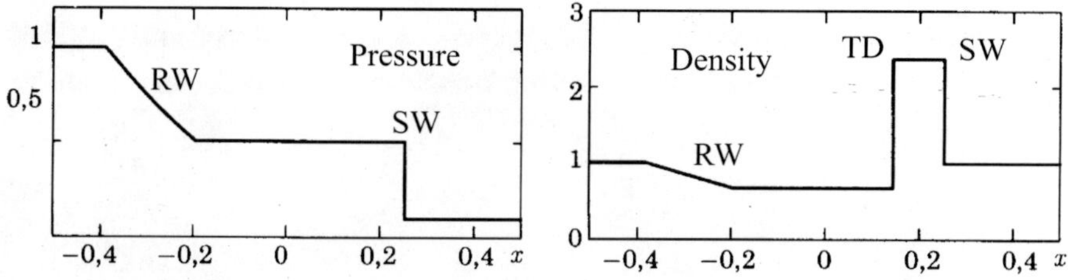

Fig. 4.8. Exact solution of the Riemann gas-dynamic problem. The distributions of pressure and density for steady flow, containing the shock wave (SW), rarefaction wave (RW) and tangential discontinuity (TD), are shown.

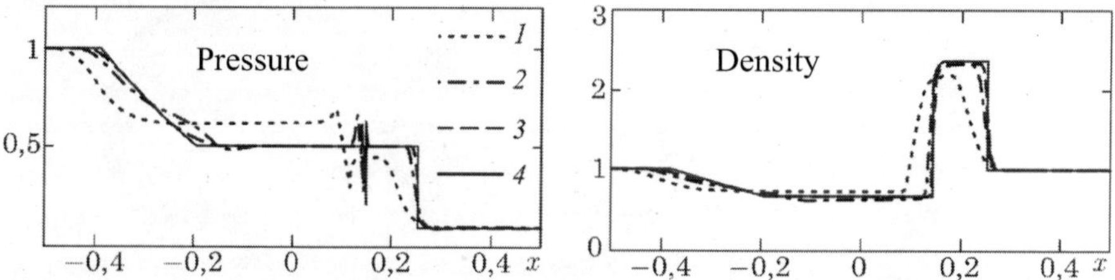

Fig. 4.9. Distribution of pressure and density, obtained by the SPH method, the calculated values of the number of particles $N = 100$ (1), $N = 400$ (2), $N = 1600$ (3), $N = 28\ 000$ (4).

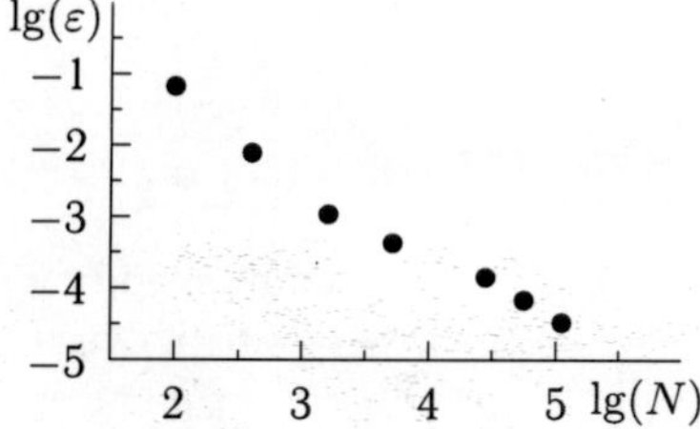

Fig. 4.10. Dependence of the relative difference between the exact solution of the solution using the SPH method on the number of particles N. The value $\varepsilon = \sqrt{\sum \frac{(\varrho - \varrho_{SPH})^2}{\varrho_{SPH}^2} \frac{1}{N(N-1)}}$ is calculated at the nodes of the mesh superposed on the SPH solution.

quantities (but not of pressure!), instead of the tangential discontinuity (TD) we should consider the presence of a contact discontinuity.

At a small number of particles N in the SPH model we obtain large differences from the exact solution, but when the number N is increased the difference becomes smaller (Fig. 4.9). By selecting N and h we can obtain good agreement with the exact solution. Figure 4.10 shows the reduction of the difference between the exact (Fig. 4.8) and the SPH solution (Fig. 4.9) with the increase of the number of particles N and a decrease of smoothing length h. However, the classic realisation of the SPH, described in this section, incorrectly calculates TD, where there is a fatal feature in pressure (Fig. 4.9). To model the Kelvin–Helmholtz instability at the tangential discontinuity, it is necessary to use a more complicated SPH algorithm.

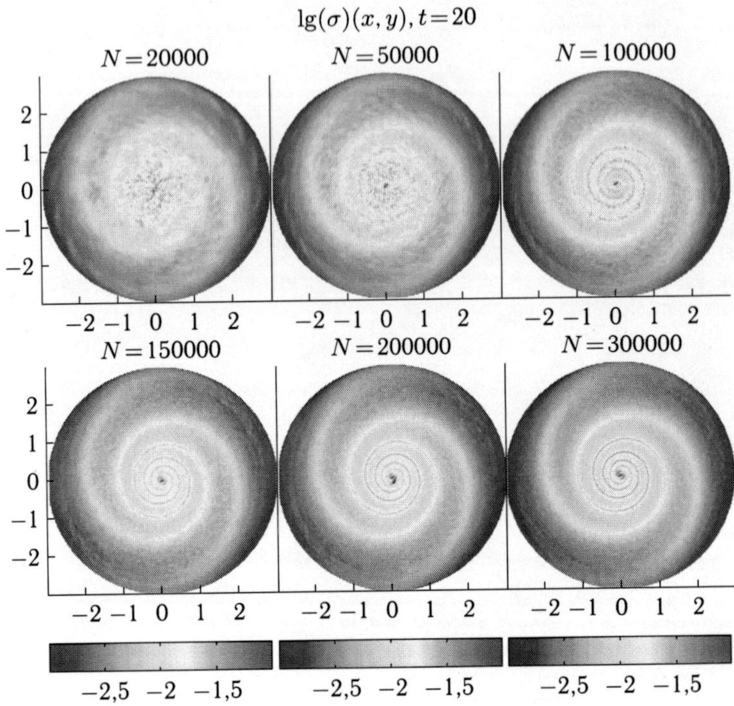

Fig. 4.11. The algorithm of the surface density in the model of a gas self-gravitating disk with the spiral structure for different numbers of particles N.

Thus, the agreement between the SPH solution and the exact solution improves with increasing N. However, to obtain high-quality results, it is necessary to use considerable resources. Even in the one-dimensional problem, it is necessary to select $N \gtrsim 3 \cdot 10^4$ to obtain a very high resolution ($\varepsilon \lesssim 10^{-4}$).

We consider a rotating gas disk with the external potential Ψ with self-gravitation taken into account. We determine the minimum number of particles at which the results of modelling cease to depend on N. As the generator of spiral waves in the disk we use the non-axisymmetry of the external field $\Psi(\varphi)$ which is determined by the distribution of matter in the stellar disk, the central bar and/or halo. The mass of the gas disk is 20% of the mass of the stellar disk. The mass of the halo within the limits of the stellar disk is $\mu = 3$.

Figure 4.11 shows the distribution of the surface density of the gas at the establishment of the quasi-stationary flow with the spiral structure for different values of the particle number N at the same moment of time with other conditions being equal. It should be seen that the models with the number of particles $N \leqslant 10^5$ show only quantitatively the structure of the spiral gap, and only starting at $N \gtrsim (2{\div}3)\cdot 10^5$ the results cease to depend on the number of particles. If self-gravitation is not taken into account, the requirements in the SPH model on the adequate description of the gas disk become less stringent.

4.2. Large-scale structure of stellar disks

4.2.1. 'Global 'criterion of stability in relation to bar excitation

We examine the results of experiments with a stellar disk without a halo ($M_h = 0$). Interesting processes is these systems take place in the central part of the disk. We investigate the evolution of the initially axisymmetric system situated mostly at the balance of gravitational and centrifugal forces ($c_r \ll V_{\text{rot}}$, see (4.21)). This system is unstable and its development is based on the growth of the dispersion of the velocities of the particles. This is accompanied by the redistribution of the density of matter, the velocity of revolution and the distortion of the axisymmetric form of the disk to bar-like.

Evidently, the first experimental attempt to characterise the stable stationary state of the collisionless gravitational disk by a single 'global' parameter was made by Ostriker and Pebbles [996]. These authors selected as the parameter the quantity equal to half the fraction of the kinetic energy of the particles of the system used for their mean (macroscopic) movement. Since the gravitating systems are described by the virial theorem in the form $2T + W = 0$, where T is the kinetic and W the potential energy of the system, the parameter mentioned above can be written in the form

$$t_m = \frac{T_{\text{m}}}{|W|}, \qquad (0 \leqslant t \leqslant 1/2) \tag{4.106}$$

where

$$T_{\text{m}} = \frac{1}{2}\int \langle \mathbf{v}(r)\rangle^2 f(\mathbf{r},\mathbf{v})d\mathbf{r}\,d\mathbf{v}, \quad \langle \mathbf{v}(\mathbf{r})\rangle = \frac{\int \mathbf{v} f(\mathbf{r},\mathbf{v})d\mathbf{v}}{\int f(\mathbf{r},\mathbf{v})d\mathbf{v}}, \tag{4.107}$$

$f(\mathbf{r}, \mathbf{v})$ is a stationary distribution function of the stars of the disk. However, instead of calculating T_{m}, Ostriker and Pebbles actually calculated the value

$$T_{\text{rot}} = \frac{1}{2}\int \langle v_\varphi(\mathbf{r})\rangle^2 f(\mathbf{r},\mathbf{v})d\mathbf{r}\,d\mathbf{v}, \tag{4.108}$$

which greatly differs from T_{m} by lower values because in the models of flat galaxies without spheroidal subsystems the excitation of large-scale non-axisymmetric excitation in the form of an oval – bar-mode – is highly efficient. In this excitation, the mean (macroscopic) movement of the particles-stars is far from circular.

In approximately one or two revolutions, the initially cold disk is rapidly heated and the energy of rotational motion is converted to the energy of random motion leading to the identity

$$t \equiv T_{\text{rot}} / |W| \simeq 0.14 \pm 0.02. \tag{4.109}$$

Equation (4.109) and the virial theory show that to ensure the stability of the disk in relation to the global bar mode $T_{\text{rot}} < 0.28\ T$. Accepting for the estimates $T_{\text{rot}} / T \sim V_{\text{rot}}^2 / (V_{\text{rot}}^2 + 3c_r^2)$, with (4.109) taken into account, we obtain $c_r \sim V_{\text{rot}}$. The last relationship results from the global criterion of stability and it should be remembered that the ratio c_r / V_{rot} depends strongly on the radial coordinate and decreases with increasing r. In the numerical experiments without a halo at the periphery of the disk $c_r/V_{rot} \simeq 0.5$ (for example [154, 155] and others).

The experimental calculation of parameter t_m, which takes into account the non-circular movement of the particles of the disk, gives [154]

$$t_m \simeq 0.20 \pm 0.02. \tag{4.110}$$

Miller and Smith [921] investigated evolution of a rotating collisionless sphere and at the end of the second rotation obtained an oblate bar in which $t_m \simeq 0.21 \div 0.24$, and $t \simeq 0.11$.

The result (4.109) require some qualification associated with the fact that the virial theory for the gravitating system ($2T + W = 0$) cannot be used directly for the experimental models. Actually, the Newton potential is a homogeneous function $k = -1 (2T = kW)$ of the degree of the coordinates [113]. However, the potential with the soft 'truncation' of the gravitational interaction at short distances (see section 4.1.2) at $r < r_c$ (r_c is the 'truncation' radius) used in the experimental models is close to the quadratic potential ($k = 2$). Taking into account that $r_c \ll R$ (R is the disk radius), it may be assumed that in the mean in the disk $k = -1 + \varepsilon$, where $0 < \varepsilon \ll 1$. Consequently, $2T + W \simeq 2\varepsilon E < 0$, $E = T + W < 0$ is the energy of the system. Comparing the experimental values of the quantities t and $t_{rot} \equiv T_{rot} / 2T$, we can estimate ε.

It is interesting to investigate the problem of the effect of the initial distribution of surface density $\sigma_{*0}(r)$ on the stability of the bar mode. Summing up the results of a large number of experiments, it may be concluded that the models for which the initial moment of time is characterised by not too high degrees of concentration of the mass in the centre of the disk, fit with respect to the parameters t the frames indicated by Ostriker and Pebbles (see (4.109)). In the models with the higher density in the central part of the disk, the quantities t, t_{rot}, t_m at the end of the experiments were lower. Evidently, this effect is associated with the high-intensity of the relaxation processes in systems with higher density. In the experiments starting from the states with the initial dispersion of the radial velocities $c_r(r) = 1.5 \cdot (2\Omega / \varkappa) \cdot c_T$ (such a disk may be gravitationally stable with respect to the small-scale perturbations (see section 2.3)), the finite value of the parameter t_m is the same as in the experiments with $c_r(r) = (2\Omega/\varkappa) \cdot c_T$.[5] However, with the increase of c_r the degree of bar-like distortion of the disk slightly decreases and t_{rot} increases by approximately 10%.

The relationships (4.109), (4.110) indicate that the flat gravitating systems without spheroidal subsystems in the stationary state should be sufficiently 'hot' – in peculiar movements of the stars there should be more than 50% of their kinetic energy. Usually, the bar-mode is excited in these systems, and the dispersion of the radial velocities of the particles even at the periphery of the disk is not smaller than 50% of the value of the circular velocity of rotation of its matter.

[5]The disk situated at the boundary of stability with respect to the small-scale perturbations ($q = c_r / c_T = 1.5 \div 2$), has $t \simeq 0.2 \div 0.25$ which, as can be seen, is not sufficient for stabilising the global bar mode).

4.2.2. Special features of the formation of bar-structures. How to avoid excitation of the bar mode?

Starting with the Hubble classification of the galaxies, it has become clear that the flat spiral galaxies can be divided into two groups. One of these groups (SB) is characterised by the presence of a bright jumper (bar) with spirals secured to its ends (in most cases almost under the right angle). In the other group of the flat galaxies (S), the spirals move away from the formation situated in the centre of the galaxy. It is clear that the investigation of the stability of the gravitating disk (analytical and numerical-experimental) should determine the factors preventing the excitation of the bar mode in many flat galaxies.

Since the bar mode is a large-scale perturbation, the shortwave approximation used in chapter 2 for the investigation of its stability is not applicable in this case. Here, it is necessary to take into account the structure of the disk as a whole and this obviously imposes relatively strict restrictions on the properties of the model. Therefore, the detailed analytical investigation of the stability of the global bar mode (GBM) was carried out only on models of solid-state rotating disks [168, 208, 738, 740]. These investigation showed that the models of the thin disk without taking into account the mass of the spherical subsystem are stable in relation to the excitation of the bar mode only at

$$t_{\text{op}} \lesssim 0.129. \tag{4.111}$$

The differential nature of disk rotation has evidently only a slight effect on the conditions of excitation of the bar mode [590]. Consequently, the result (4.111) shows that only very hot disks can be stable in relation to the bar mode – the disks in which the kinetic energy of rotation is less than a third of the total kinetic energy of the matter of the disk. It is evident that these observations do not indicate the existence of such systems.

Better results were obtained with the formulation of the problem in which the presence of the spheroidal subsystem was also taken into account in addition to the finite thickness of the disk. For example, in a model consisting of a homogeneous halo (spheroidal subsystem) and a homogeneous oblate ellipsoid of revolution (disk) with the major half-axis R and the minor half-axis $b < R$, it was found [208] that in the disk limit ($b/R \to 0$) the ellipsoid is stable in relation to the excitation of the bar mode at

$$M_h \,/\, M_d \gtrsim 1.1. \tag{4.112}$$

where M_d is the mass of the solid (disk), M_h is the mass of the halo in the sphere $r \leqslant R$. Taking into account the finite ratio b/R reduces this limit. For example, at $b/R = 0.05$, the bar mode is not excited already at $M_h/M_d \gtrsim 1.0$. The result ($M_h/M_d \gtrsim 1.18$), close to (4.112), was obtained in [21] in the model with similar macroscopic characteristics.

As already mentioned, the analytical investigation of the stability of the bar mode in the differentially rotating inhomogeneous disks of finite thickness is difficult. Evidently, this can be carried out more efficiently by numerical experiments. Early experiments (see, for example [681, 682, 996, 1049]) confirmed that in insufficiently

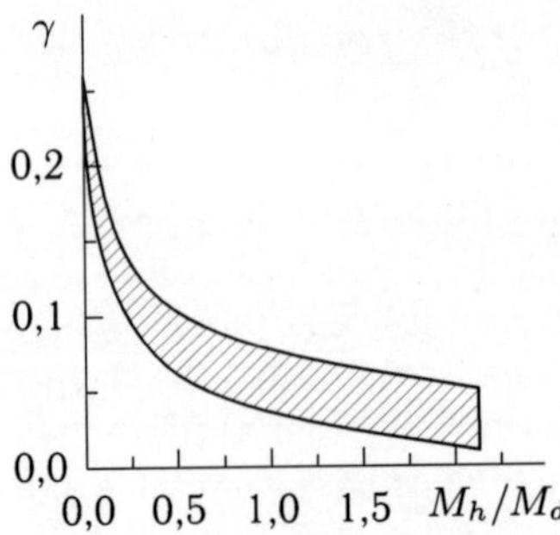

Fig. 4.12. Dependence γ (M_h/M_d) = 1 – T_{rot}/T_m, characterising the degree of development of the bar mode. In the crosshatched zone in the graph, the models at each value of M_h/M_d differ in the degree of concentration of the halo to the centre and in the initial distributuion of surface density.

hot (t_{op} $\gtrsim$ 0.14 ± 0.02) thin disks without the halo the excitation of the bar mode is unavoidable. Taking into account the finite mass of the halo shows that the bar mode is stabilised with the increase of the mass of the halo [320, 365, 473, 683, 719, 1134, 1194]. At the same time, in different models the boundaries of stability of the bar mode with respect to the parameter M_h/M_d type greatly differ but the condition $M_h/M_d \gtrsim 1$ is fulfilled.

The experimental results can be described efficiently using the parameter $\gamma = 1 - T_{rot} / T_m$, characterising the degree of bar-like distortion of the disk for different distribution of the volume density of the halo ϱ_h (r) [155]. In the model without the halo $\gamma \simeq 0.21 \div 0.26$ (Fig. 4.12). Parameter γ decreases with increasing halo mass.

Taking these results into account, it is interesting to investigate the experimental criterion [328] which confirms that to ensure the stability of the disk with respect to the excitation of the bar mode, it is important to fulfil the condition $\beta = (V_{rot})_{max}$ $(L_\sigma/GM_d)^{1/2} \gtrsim 1.1$, where $(V_{rot})_{max}$ is the maximum value of V_{rot} in the disk. According to the experimental results [155], this criterion was verified in the finite quasi-stationary states of the disk. The results show that the value β varies from $\beta \simeq 0.4$ at $M_h = 0$ to $\beta \simeq 0.9$ at $M_h = 3M_d$. Thus, the criterion [520] appears to be too rigirous. Evidently, the reason for this is that the criterion was obtained in the experiments in which the initial state of the disk was cold (c_r = 0). In this state, the excitation conditions of the bar mode were most favourable, and $(V_{rot})_{max}$ (measured at the initial moment of time) was higher than in the hot disk. In addition to this, the disks in the experiments [528] were two-dimensional which also made the excitation of the bar mode easier. Therefore, it is not surprising that this criterion results in an unusually low estimate of the ratio of the mass to the luminosity (of the order of unity) for the disks of Sc-galaxies [69]. Taking also into account the results of 3D-experiments, it should be assumed that the stellar disk is stable in relation to the excitation of the bar mode at $\beta \gtrsim 0.7$.

In [468] the authors discuss the condition $\alpha = \mathcal{L}\Omega_J/2\,|U| < 0.25$, where $\mathcal{L}$ is the total angular moment; Ω_J is the Jeans frequency, U is the potential energy. However, this criterion is difficult to use for actual galaxies. As a result of the publication of a large number of data on the determination of the spatial fields of velocity and dispersion of velocity in the galaxies, it is more efficient to re-formulate the criterion of stabilisation of the GBM in the form $c_r/V_{max} > (c_r/V_{max})c_r$, where c_r is the dispersion of the radial velocities of the disk at some fixed radius; V_{max} is the maximum velocity of rotation. Let it be that the total mass inside the sphere with radius R is M. The simple estimate

$$0.14 = E_k/|U| \simeq \frac{1}{2}\frac{c_r^2 + c_\varphi^2 + c_z^2}{GM/R} \simeq \frac{1}{2}\frac{c_r^2}{V_{max}{}^2}(1+\alpha_\varphi^2+\alpha_z^2)$$

at

$$\alpha_\varphi^2 = c_\varphi^2 / c_r^2 = 1/2,\ \alpha_z^2 = c_z^2 / c_r^2 = 1/4$$

gives $c_r/V_{max} = 0.4$, which requires clarification because both the dispersion of the velocities and the rotation curve are radially inhomogeneous, and there is no central symmetry in the distribution of mass.

In hot systems, the conditions for the development of GBM are less favourable. Therefore, if the dispersion of the velocities of the stars exceeds some critical value c_{cr}, the initially axisymmetric disk becomes stable. This condition can be written more efficiently for the dispersion of the radial velocities c_r, since the dispersion of the azimuthal velocities is, with a very high accuracy, equal to $c_\varphi = c_r \varkappa / 2\Omega$ ($\Omega = V/r$ is the angular velocity of rotation of the disk).

We consider the results of dynamic modelling without a bulge in applications to the galaxies formed at later stages. Since for the halo it should be assumed that $a \gtrsim L$, the rotation curve, with the potential (4.2) taken into account, has a relatively long region of growth. Therefore, the central part of the disk ($r \lesssim L$) is characterised by the dominance of the gravitational effect of the flat subsystem and this also determines the favourable conditions for the development of the bar mode. Even the halo with $1 \lesssim \mu \lesssim 2.5$ and dominant as regards the mass is not capable of suppressing the global bar mode if the disk is relatively cold. Starting at $M_h \lesssim 3$, at $a \lesssim L$, the gravitational instability does not lead to the formation of a long-life bar structure. In the experiments only the mass of the halo M_h was changed with the spatial scale $a = L$ and the radial profile $c_r(r)$, stabilising the GBM, was determined. Figure 4.13 shows the dependence of the ratio of the dispersion of the radial velocity of the disk c_0, averaged out over the zone $r \leqslant L$, to the maximum velocity of rotation of the disk on the relative mass of the halo. In the case of a halo with a

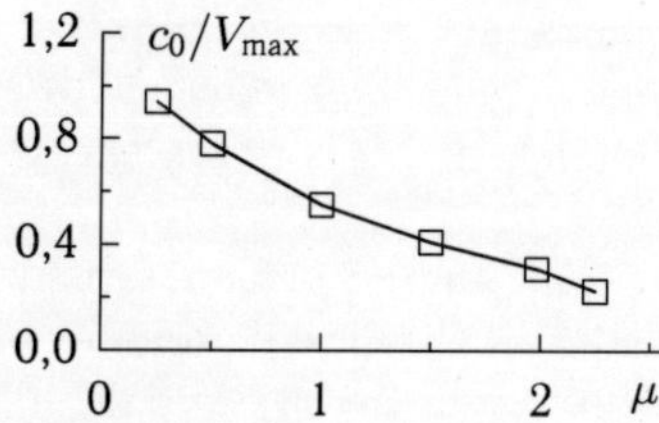

Fig. 4.13. The ratio c_0/V_{max} ensuring the stabilisation of the global bar mode at different masses of the halo in the case of equal scales of the halo and the disk $a = L$ for the thin disk at the initial moment of time of the disk ($\langle z_0\rangle/L = 0.07$).

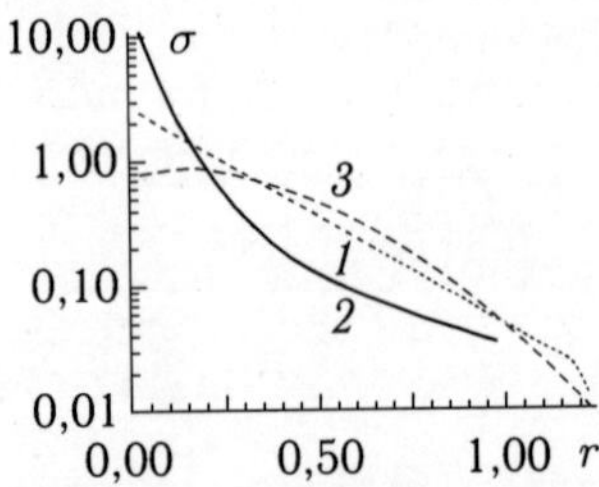

Fig. 4.14. Radial dependences of the surface density of the disk for the model 'starting' from the cold state in which extensive redistribution of density takes place. The dotted line 1 is the initial exponential profile, the solid line 2 is the final profile averaged out with respect to the azimuth, after rapid heating with the formation of a bar; the broken line 3 is the initial profile ensuring the quasi-exponential distribution at the end of the calculations.

lower density ($a \gtrsim 2L$) the formation of the bar is easier. The value μ^{crit} depends on the nature of distribution of density in the halo – in the model (4.3) on the scale a. With increasing a the stabilisation of the mode starts at high values of μ. With other conditions being equal, the value μ^{crit} decreasing with increasing disk thickness. At $\mu \simeq 1$–2.5, the formation of the bar can also take place by a complicated mechanism through the disruption of the three-sleeve mode (see paragraph 4.10).

If the profile at the initial moment of time is exponential and the disk is a relatively cold (in the final stage, the dispersion is approximately 1.5 times higher than the initial value), the formation of the bar results in extensive redistribution of matter in the disk (Fig. 4.14). Several characteristic features of these processes will now be discussed.

1. The disk spreads in the radial direction. Up to 30% of the mass can be removed from the region initially occupied by matter ($r \lesssim 4L$). As the value μ increases the amount of the mass taken away decreases.
2. The resultant profile greatly differs from the exponential one, and in the centre ($r \lesssim 2L/3$) the density increases in comparison with the initial value and then decreases. To obtain the exponential profile of the density, it is necessary to define the radial distributions of the special type (see the dotted line in Fig. 4.14).

The structure and dimensions of the bar, the ratio of the half axes and also the kinematic characteristics in the central region of the disk depend not only on the type of rotation curves but also on the laws of distribution of matter in the disk and spheroidal subsystems, and also on the initial conditions. Therefore, it is not always possible to understand the properties of a specific galaxies within the framework of general models and it is necessary to construct the model of a specific object.

It is important to mention another characteristic special feature of the development of the bar mode in the gravitationally unstable disk without the spheroidal subsystem ($\mu = 0$) when for the initial radial profile of the Toomre parameter $Q_T \simeq 0.5$–1. In this case, in addition to the central bar, the growth of even single-sleeve arms ($m = 1$) may result in the formation in the disk of additional one or two perturbations with a high amplitude with the dimensions

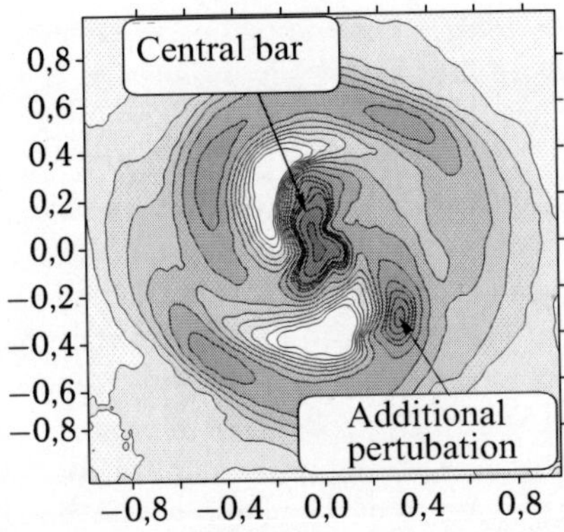

Fig. 4.15. Isolines of the logarithm of surface density in the disk without the spheroidal component according to the results of dynamic modelling. In the process of development of gravitational instability, in addition to the central jumper there is also an additional perturbation of density at the periphery whose mass is 4.5 times smaller than the mass of the central bar.

comparable with the bar (Fig. 4.15). The colder the disk, the more favourable are the conditions for the formation of these structures. These formations are long-lived, linked with the spiral branch and greatly disrupt the disk symmetry. When the spheroidal subsystem is taken into account, starting already at $\mu \gtrsim 0.5$, these structures are not detected in the models.

Summing up these considerations, it may be assumed that with other conditions in the SB galaxies being identical, the mass of the spherical subsystem on the whole should be smaller (but should exist!) than in the S-galaxies. Indirect confirmation of this result is the fact that the fraction of the SB-galaxies in the vicinity of the cluster of the galaxies in the Hair of Veronica is twice that at the periphery of the cluster. The most probable reason for this is the 'deflation' of the halo at mutual approach of the galaxies and weakening as a result of this of the stability of the disks in relation to the excitation of the bar mode [1227]. In addition, another factors causing the formation of a jumper may be the tidal interaction from the side of a massive satellite [616].

4.2.3. Effect of gas on the development of the bar mode

As in the stellar disk, the gas herself-gravitating disk may also show the development of a bar mode.[6] However, because of the collisional nature of the gas the instability (and the instability may be of the gravitational nature) can be stabilised more easily, in contrast to the collisionless stellar disk (see (4.109), it is sufficient to fulfil $t_{op} \lesssim 0.26$ [335]). Van der Voort [1257], using the linear analysis of the stability of a solid-state rotating homogeneous stellar-gas spheroid, showed that the critical

[6]In the innermost regions of some galaxies there are gas (molecular) bars. The major half-axis of these structures equals 100÷300 pc which is considerably smaller in comparison with the typical stellar bars [668, 1318]. The molecular nuclear bars are detected in both the systems with the stellar bar and in galaxies without the bar.

value t_{op} is in the range 0.15÷0.26, depending on the parameters of the model and, in particular, on the fraction of the gas mass in the system $M_g/(M_* + M_g)$.

It is well-known that the galactic gas is highly inhomogeneous on the small scale. There is an entire hierarchy of gas clouds differing in the dimensions and mass. In the previous section it was already shown that the massive gas clouds (MGC, and formations with larger scales – associates) may be characterised by the effective scattering, transferring the orbital motion of the stars and the gas to thermal one. Thus, the cloud structure of the gas can be an important factor of stabilisation of the global modes. The characteristic times of formation and life of the clouds are comparable with the duration of formation of the bar (~$5 \cdot 10^7$ years). Therefore, to explain the role of the gas in the suppression of the bar mode, it is important to investigate in detail the processes of formation of the clouds, their dynamics and star formation. A contribution to heating of the disk is also provided by dynamic friction: in movement of heavy gas clouds light stars undergo gravitational focusing, forming a higher concentration behind the cloud.

The numerical experiments, taking into account these effects, demonstrate clearly the extremely important role of the gas in the dynamics of a stellar disk [663, 1157]. It has been shown that the gas may stabilise the bar mode if it its mass amounts to only several percent of the total mass. Of course, the critical value $(M_g/M_*)_{cr}$ depends on a large number of model parameters. However, according to the authors of the above studies, for the typical galaxies it can be assumed that $(M_g/M_*)_{cr} \simeq 10\%$ (here it is important to take into account both the gas and the stars located in the region of formation of the bar).

This result will be discussed in greater detail. For the distribution of surface density in the disk $\sigma = \sigma_0/(1+ (r/a)^2)^{3/2}$, the stability criterion with respect to the order of magnitude can be written in the form [557]:

$$t_m \sim \frac{M_d / M_{\text{tot}}}{2(1+2\alpha^2)} < t_{\text{cr}} \simeq 0.1 \div 0.2,$$

here $M_{\text{tot}} = M_d + M_h$, $\alpha = c_r/\langle V_{\text{rot}}\rangle$, $\langle V_{\text{rot}}\rangle$ is the mean velocity of rotation; c_r is the dispersion of the radial velocities at the radius, where V_{rot} reaches a maximum. Taking into account the collisional nature of the gas effectively reduces the mass of the disk M_d by the value M_g. However, a more important factor is the heating of the stellar component on the massive gas clouds which results in an increase of the parameter $\alpha \propto D(t)$ with time in accordance with equation (4.134). The characteristic size of the giant gas clouds r_G is determined by gravitational instability, and the value r_G is proportional to the surface density of the gas disk σ_g. Taking into account that the mass of the cloud $M_G \sim r_G^2 \cdot \sigma_g \propto \sigma_g^3$, the following equation can be written for parameter α:

$$\alpha^2 \sim (\alpha_0^4 + \text{const} \cdot \sigma_g^4 \cdot t)^{1/2}, \quad \alpha_0 = \alpha(t=0).$$

It may be seen that the heating of the stellar subsystem depends very strongly on the density of the gas disk and, as shown by the numerical experiments, when $\sigma_g \simeq 0.1 \cdot \sigma_*$ the large-scale gravitational instability is fully stabilised. It should be remembered that this is the result of the simultaneous self-consistent effect of fragmentation of the gas, star formation and all relaxation processes.

4.2.4. Kinematics of disks with a bar

In the epicyclic approximation (a stationary axisymmetric disk in which the dispersion is small in comparison with the velocity of rotation), there is a simple relationship between the dispersion components (4.23). It would appear that in a disk with the bar the epicyclic approximation is violated. However, dynamic modelling shows that in a non-axisymmetric disk with a bar, diameter $Q_c = c_r \varkappa / (2\Omega c_\varphi) \simeq 1$ not only at the end of calculations in the conditions of the quasi-stationary state but also throughout the entire stage of formation of the bar structure (Fig. 4.16), if calculations are carried out 4 the mean values of the parameters c_r, c_φ, Ω [768] with respect to the azimuthal angle. There are small deviations ($\simeq$5%) up to the moment of formation of the bar (see Fig. 4.16b, line 1). Subsequently, the deviations increase by 10% at the periphery as a result of a large increase in the thickness of the disk in the absence of the halo (line 3) as a result of bending of the bar (section 4.5.2). In the models with a massive halo, the condition $Q_c = 1$ is fulfilled more easily (Fig. 4.16c).

Each component of the dispersion of the velocities c_r, c_φ, c_z has its own characteristic distribution which is typical for the bar mode irrespective of the experimental conditions (Fig. 4.17). The dimensions of the 'bar' as regards parameter c_r are approximately half of the dimensions with respect to surface density. The spatial distribution of the dispersion of azimuthal velocity c_φ greatly differs from c_r (x, y) because of the characteristic perpendicular structure in the centre. The

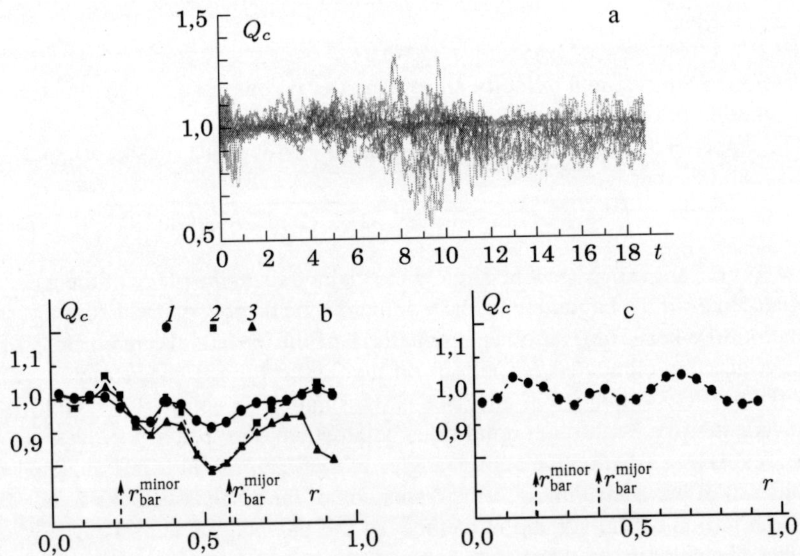

Fig. 4.16. For a model with $M_h = 0$, $M_b = 0.1\ M_d$: a) the dependence of diameter Q_c at different radii on time; b) radial distribution is $Q_c\ (r)$ averaged over time at intervals t = 0–11 (1), t = 6–11 (2), t = 11–20 (3); c) $Q_c\ (r)$ for the model $\mu = 1.5$, $a = L$. The positions of the minor and major half-axes of the bar r_{bar}^{minor}, r_{bar}^{major} are shown.

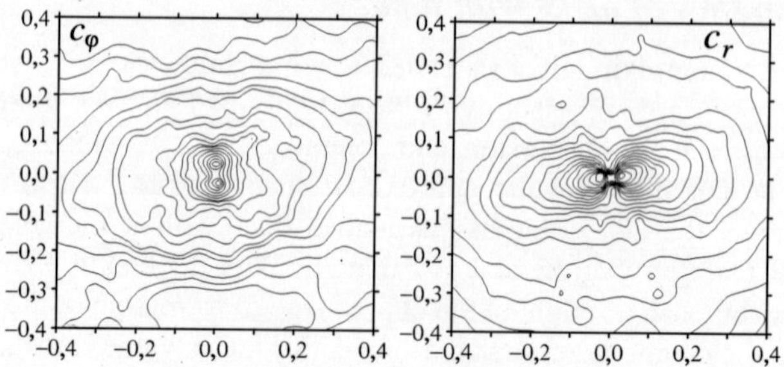

Fig. 4.17. Isolines c_σ and c_r in the central region of a powerful bar.

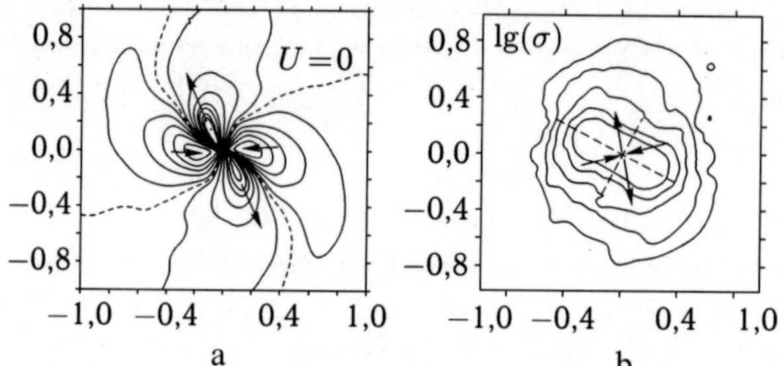

Fig. 4.18. a) the structure of the flow of the gas in the region of the bar. The images of the isolines of radial velocity U. The dotted line $U = 0$; b) the radial velocity of the stars (arrows) in the region of the bar (the isolines of the surface density are shown) according to the results of dynamic modelling.

dispersion of the vertical velocity shows smaller deviations from symmetry, with the exception of the very centre.

The dispersion of velocities along the visual ray depends on the orientation of the disk and the bar:

$$c^{\text{obs}} = \sqrt{c_z^2 \cos^2(i) + c_\varphi^2 \sin^2(i)\cos^2(\alpha) + c_r^2 \sin^2(i)\sin^2(\alpha)}, \tag{4.113}$$

where α is the angle between the slit and the major axis in the plane of the galaxy. If the major axes of the bar and the galaxy coincide, the dispersion field does not show formation of a bar [768]. At other orientations of the bar and the major axis of the Galaxy there are elongated formations on the field c^{obs}. 'The 'bar-structure' of the dispersion of the velocities in all cases is the smallest in comparison with the bar of surface density. With other conditions being equal, the 'bar of the dispersion of velocities' is more distinctive at high angles of inclination i. This is associated with a reduction of the contribution of the dispersion of the vertical velocity c_z to c^{obs}.

In the region of the bar the movement of the gas and the stars is greatly non-circular. The velocity field has a characteristic four-sector structure (Fig. 4.18). For the model of a stellar disk with the ratio of the axes at the bar of 1:4, Fig. 4.19 shows the isolines of the radial component of velocity U (Fig. 4.19a) and azimuthal velocity V (Fig. 4.19b) in the disk plane. Function $V(x, y)$ has the characteristic saddle-shaped form (Fig. 4.19a). The rotation curve $V(r)$ is often used for the determination of the

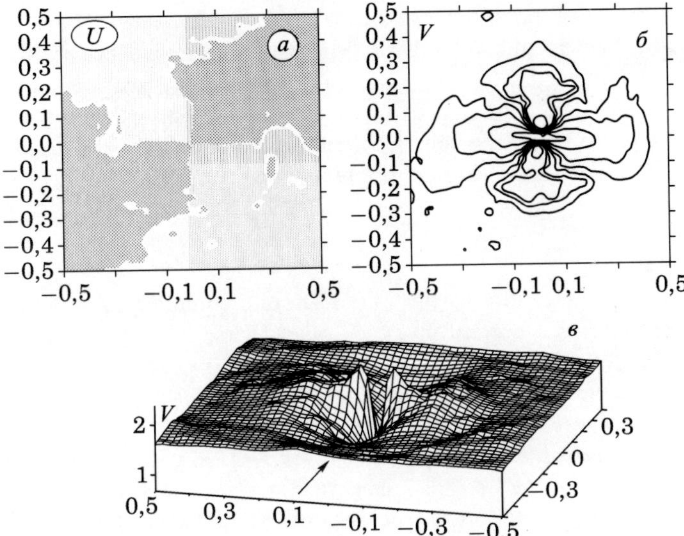

Fig. 4.19. Distribution of the components of the velocity of stars in the region of the bar according to the results of dynamic modelling: a) radial velocity U, the region where $U > 0$ and $U < 0$ are indicated by different contrast; b) the isolines of azimuthal velocity V; c) the functions $V(x, y)$ correspond to Fig. b, the arrow is in the direction of the major axis of the bar.

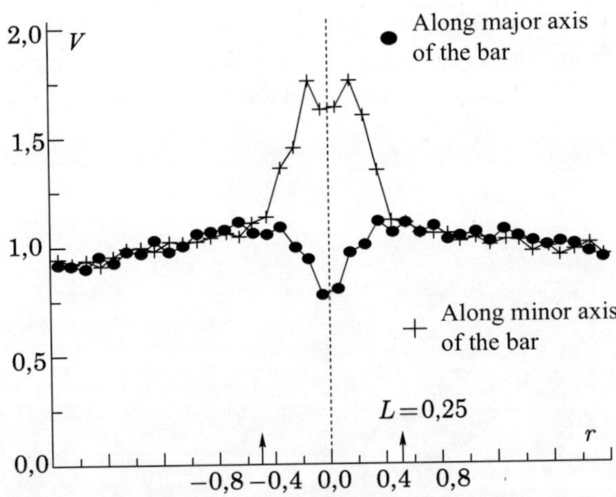

Fig. 4.20. Dependence of the azimuthal component of velocity V on distance to the centre along the major and minor axes of a very powerful bar (a model with a thin cold disk in the initial stage at $M_h = 0.25$, $a = 0.5$; $M_b = 0.3$; $b = 0.02$). The contribution to such strong azimuthal inhomogeneity V is also provided by the deviation of the radial profile of surface density σ from the exponential profile – see the increase of σ in the centre in Fig. 4.14.

law of distribution of mass in the galaxies. In the region of the bar this approach is not acceptable because the velocity field is highly inhomogeneous with respect to the azimuthal angle and there are considerable radial movements, and the estimates obtained on the basis of the balance of gravitational and centrifugal forces are highly inaccurate. The profiles of the velocity along the major and minor axes qualitatively differ (Fig. 4.19c) and the azimuthal inhomogeneity V can be quite considerable (Fig. 4.20).

Rotation of the bar

In many galaxies the central part has the form of a complicated system consisting of several components inserted into each other (bars, disks, rings, double nuclei) [287, 1168, 1170]. They also include the so-called 'mini-bars' – the major half-axis $r_{\rm bar}^{\rm major}$ of these structures is situated within the range of several hundreds of parsec. Suitable examples of the galaxies with the mini-bars are NGC 4736 with $r_{\rm bar}^{\rm major} = 225$ pc [1276], NGC 81 ($r_{\rm bar}^{\rm major} = 0.5$ kpc) [751], UGC 5600 [1150].

The dimensions of the bar are linked with the angular velocity of rotation of the bar $\Omega_{\rm bar}$. Figure 4.21 shows, on the example of the galaxy NGC 936, the radial dependences of the angular velocity of rotation of the disk $\Omega(r)$ and particles $\Omega^{(m)}(r) = \Omega - \varkappa/m$ ($m = 2, 4$, $\varkappa$ is the epicyclic frequency). There are three types of bars which differ in the angular velocity of rotation and the dimensions [1015, 1034]:

1. Fast bars with $\Omega_{\rm bar} > \Omega^{(2)}$;
2. Slow bars with $\Omega_{\rm bar} < \Omega^{(2)}$;
3. Intermediate bars with $\Omega_{\rm bar} \simeq \Omega^{(2)}$.

The fast bars, formed as a result of the instability of the global bar-mode, are well-known, starting with the classic study by Ostriker and Pebbles [996]. The model of

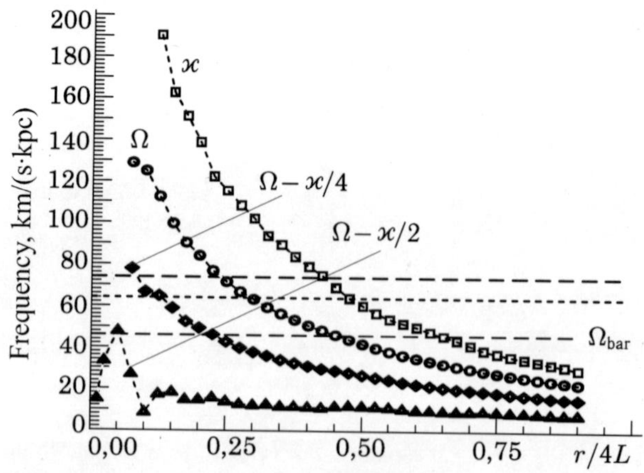

Fig. 4.21. The distribution in the disk of the angular velocity of rotation Ω, epicyclic frequency $\varkappa$ and Lindblad resonance $\Omega - \varkappa/m$ ($m = 2.4$) on the example of the NGC 936 model [255]. The angular velocity of rotation of the bar in the dynamic model $\Omega_{\rm bar} = 64$ km/(s · kpc) [255] is shown together with the restrictions from observational data $\Omega_{\rm bar}^{\rm obs} = 60 \pm 14$ km/s [903].

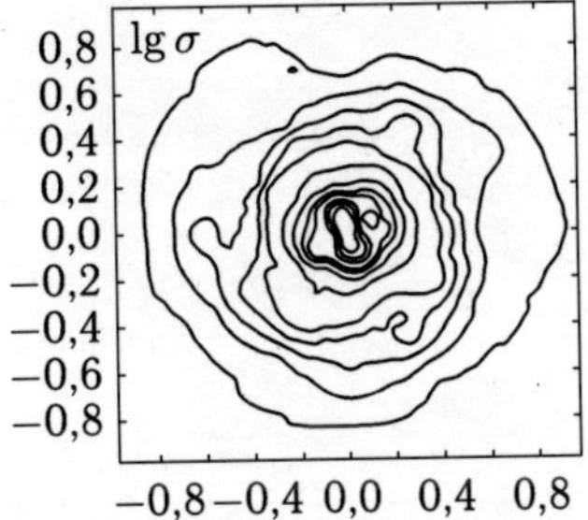

Fig. 4.22. Isolines of surface density in the dynamic model with the bar of the intermediate type with angular velocity $\Omega_{bar} = \Omega^{(2)}$.

N-bodies was used to investigate slow bars [198], and an example of an intermediate type, formed in the dynamic model, is shown in Fig. 4.22.

In the case of bulges with a small mass the rotation curve of the disk has a relatively long solid-state section in which the matter of the disk rotates with the velocity Ω_0. The massive and concentrated bulges result in a smaller solid-state section of rotation of the disk. The angular velocity of rotation of matter at the centre Ω_0 increases and, consequently, it may be that $\Omega_{bar}/\Omega_0 \ll 1$.

Dynamic models show that, with other conditions being equal, the bar rotates at a lower rate in hotter systems. It is possible that this result is determined by the additional contribution of the instability of the radial orbits which is more efficient in the hot systems [197].

It should be mentioned that the bar mode is not the only mechanism of formation of the bar. In the case of slow rotation of the system ($c_r \gg V_{rot}$) suitable conditions form for the development of the instability of the radial orbits [590]. The result of development of this instability may be jumpres (bars) in the galaxies with a massive bulge and mini-bars with the size of several hundreds of parsec in the centre of a number of flat galaxies [61, 197].

4.2.5. Central depression of stellar density in the disk

In chapter 1 we mentioned a local phenomenon such as depression ('hole') in the distribution of the surface density of the central part of the stellar disks of the galaxies. Of course, it is important to answer the question as to why the central depression of density of the stellar disk does not disappear under the effect of relaxation processes determined by collective processes. In the framework of the numerical experiment this problem was investigated in [66, 134, 136].

The dynamic evolution of the depth of the depression can be described efficiently by the following value

$$\Sigma(t) = 1 - \frac{\sigma_*(r=0,t)}{\sigma_{*\max}(r_{\max},t)}, \tag{4.114}$$

where $\sigma_{*\max}$ $(r_{\max}/t)$ is the maximum value of surface density, and $r_{\max}$ is the radius

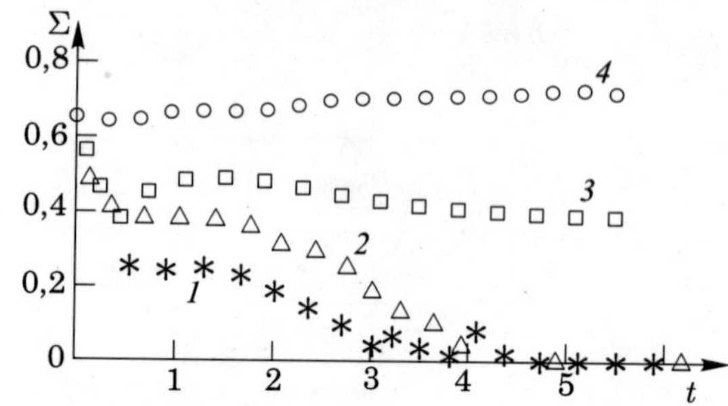

Fig. 4.23. Time evolution of the depth of depression $\Sigma(t)$ (the time normalised with respect to the period of revolution of the disk) in the case of a 'spot' nucleus of the halo $a = 0.01$ for different values of the mass of the spherical subsystem: 1) $M_h = 0.5\ M_d$; 2) $M_h = M_d$; 3) $M_h = 1.5\ M_d$; 4) $M_h = 2M_d$.

at which this maximum is reached. Modelling shows that the long-term existence of a depression in the stellar disk is possible in the systems with the maximum halo ($M_h \gtrsim M_d$) which usually has a concentrated nucleus. The maximum spherical component with $M_h \lesssim M_d$ does not delay the process of dissipation of the central depression (Fig. 4.23). At $M_h = 0.5\ M_d$ and $a = 0.01$, the central 'hole' is 'washed out' almost immediately after 0.5–1 revolutions of the stellar disk. The massive spherical component may even intensify the central depression of density. For example, at $M_h/M_d = 2$, $a = 0.01$, the value Σ approaches 0.7 (Fig. 4.23). However, this effect depends strongly not only on the total mass of the halo but also the nature of distribution of matter in it. For a 'porous' nucleus ($a = 0.2$) the depression starts to intensify only at $M_h/M_d \gtrsim 6$.

Thus, the experimental results presented here shows that only the galaxies with the massive and/or sufficiently concentrated (at the centre) spherical component can show the depression of stellar density. The duration of the depression in the disk depends on the presence of the massive bulge. This is in agreement with the observations of a number of actual galaxies (see section 1.2). It should be mentioned that the possibility of existence of the 'deficit' of surface density in the internal part of a flat stellar subsystem at specific values of the parameters of spherical subsystems was indicated in [174] where this conclusion was made on the basis of the calculations of equilibrium of the rotating disk in the halo field.

In some cases, the observations confirm the presence of the central depression of the disk density for the galaxies of late morphological types in which there is almost no bulge but there is another special feature – a jumper. Since the bar rotates at the velocity which differs from the velocity of rotation of the stellar disk, this may result in the redistribution of the angular moment and this may delay the relaxation of the central 'hole'. Evidently, in this case it is necessary to take into account the gas disk in addition to the stellar disk. The bar can not only delay relaxation but can also be responsible for the formation of a depression.

4.3. Stellar disks at the boundary of gravitational stability

In this section, the method of dynamic modelling (N-bodies) is used to investigate the condition of gravitational stability of a three-dimensional stellar disk in the gravitational field of stationary spherical componenta – bulge and halo. The main aim is to find the minimum permissible local values of the dispersion of the velocities of the stars in the disks of the galaxies with different distributions of mass at which the quasi-stationary state of the three-dimensional disks is established, with the disks starting to evolve from slightly unstable equilibrium states.

4.3.1. Problem and the criteria of gravitational stability of the disks and the stability criteria

The theoretical analysis of the problem of stability of the rotating thin disks to different types of perturbations (including bending perturbations) has been studied in detail in the monographs [203, 590]. The analytical approach to investigating the dynamics of perturbations in the gravitating disk and the determination of the condition of stability have advantages in relation to the methods of numerical modelling (mathematical rigorousness of the solution within the framework of the given task) and also a shortcoming that it can be applied only in the greatly simplified models whereas in the application to the actual objects it may result only in a rough estimate of the parameters of the disk component.

The simplification will be investigated in greater detail. Above all, the simple analytical models of contactless disks, investigated in analysis of stability, are usually based on the assumption on the small thickness of the disk – in fact, an infinitely thin disk is investigated. Regardless of the smallness of the ratios h/r and h/L (r is the radial component; h and L are the characteristic vertical and radial scales, respectively) for a large part of the stellar disk, this condition is not sufficient for neglecting the vertical movements (for this problem and also the conclusion according to which in the general case the dynamics equations of the astrophysical disk cannot be considered in the two-dimensional formulation of the problem (see chapter 3).

The analytical investigation of the dynamics of small perturbations in a stellar disk is usually based on the lower values of the radial gradient of the disk parameters (see, for example, [844, 1229, 1224]). The attempts for taking into account the gradient factors were made only within the framework of the WKB-approximation (see, for example, [151, 155, 175]). This means that the characteristic wavelength of the perturbations should be small in comparison with the local scales of the inhomogeneity of the equilibrium velocity of revolution $V(r)$, the dispersion of the radial velocities of the stars $c_r(r)$ and the surface density $\sigma(r)$. In many cases, these conditions are fulfilled with a small reserve or are even violated. The slight inhomogeneity of the angular velocity of rotation can be taken into account for slightly non-radial perturbations [151, 153]. However, highly non-axisymmetric (in the limit 'spoke-like') perturbations are more unstable. To stabilise them, with other conditions being equal, it is necessary to heat the disk to higher temperatures, i.e. high dispersion of the velocities of the stars [200] is required and this leads evidently in the non-linear stage to the increase of the dispersion of the velocities of the stars in the case of the initially cold system.

The analytical investigation of the gravitational stability of the disk is usually carried out using the epicyclic approximation $c_r \ll V$ which for the majority of galaxies is fulfilled at the periphery, where $c_r/V \simeq 0.1 \div 0.3$, but is violated on approach to the centre, where $c_r \gtrsim V$.

The second important restriction of the analytical approaches is associated with the local nature of the resultant criteria. At the same time, a number of investigations indicate the global nature of the stability condition of the disk [538, 539]. This means that if the equilibrium parameters are changed, for example, in the centre, and remain constant at the periphery of the disk, the condition of gravitational stability can change in the entire disk. In global analysis, in contrast to local analysis based on the analysis of the dispersion equations, the eigenfrequency is determined for the entire disk in solving boundary-type problems. This also determines the possibility of influencing the conditions in one of the parts of the system of the dispersion properties of the perturbations in the entire disk. Therefore, the strict approach requires investigating disk as a whole. However, global analysis has been carried out only for specific distributions of the exponential type [538, 539, 1143]. In the application to maintaining long-life spiral density waves in the galaxies, the global modes in the disks were investigated in, for example, [368, 789, 790, 792] (section 7.7.1).

Numerical experiments with collisionless systems are more flexible as regards the selection of the model. They make it possible to go outside the framework of local analysis and simple two-dimensional models and investigate directly the development of perturbations in the initially equilibrium disk. However, this approach has also its shortcomings. The most serious problems when using the method of N-bodies are, firstly, some mathematical simplifications unavoidable in any method of representing the disk by the system of N gravitating bodies, where N is considerably smaller than the number of stars in actual galaxies and, secondly, the dependence of the final state of the system after its arrival in the quasi-stationary state on the defined initial parameters which are not accurately known for the galaxies. Comparison with the actual galaxies may also result in a problem of taking into account the dissipative medium (gas) where the speed of sound is considerably lower than the dispersion of the velocities of the old stars.

In the literature there are several criteria of the gravitational stability of collisionless disks determined analytically within the framework of different models. The most important of these criteria will now be discussed [259].

1) *Toomre criterion.* For the gravitational stability of relatively axisymmetric perturbations of an infinitely thin homogeneous solid-state rotating stellar disk it is essential to fulfil the Toomre condition [1229] (2.97):

$$c_r \geq c_T = \frac{3.36 G\sigma}{\varkappa}, \text{ or } Q_T = \frac{c_r}{c_T} \geqslant 1. \tag{4.115}$$

The condition (4.150) assumes the fulfilment of the epicyclic approximation in which the difference between the velocity of rotation of the disk $V(r)$ and circular velocity $V_c(r)$ can be ignored (4.21). Although (4.115) was derived in the local analysis, the investigations of natural modes within the framework of self-modelling disks in the appropriate approximation, carried out by Evans and Read [539], confirm on the whole this relationship. Miller [970] compared the theoretical increments within the framework of the Toomre model with the results of modelling of axisymmetric modes – all other perturbations are artificially suppressed in this case. It was shown that the experimentally determined increments do not contradict the theoretical conclusions.

2) *Taking into account the finite thickness of the disk.* The finite thickness has a stabilising effect on the gravitational stability in the disk plane [590, 1254]. The generalisation of the Toomre stability criterion (4.115) for the case of a disk with finite thickness has the form (2.115) [151, 175)

$$Q_T^{(1)} = \frac{1}{1+0.974\Delta\varkappa / c_r} < 1, \tag{4.116}$$

where Δ is the half thickness of the isothermal self-gravitating disk. However, this condition proved to be insufficient for the stability of some real systems. This became clear in the initial numerical experiments carried out in the 70s and 80s which showed that at the periphery of stationary collisionless disks $c_r \simeq (1.5 \div 5) \cdot c_T$ [681, 682, 719, 914, 915, 917, 920, 996].

3) *A criterion with simplified consideration of non-axisymmetric perturbations.* One of the reasons because of which the Toomre criterion is insufficiently accurate is the stronger instability of the spiral waves in comparison with axi-symmetric ones. In studies by Polyachenko, Shukhman and Kalnajs [206, 590, 73] it was shown for the first time that non-axially symmetric modes dominate in the gravitationally unstable disk, and this problem has been examined in detail in section 2.3, 2.4. Since the dispersion of the azimuth velocities c_φ is smaller than c_r (with the exception of the very central regions), then because of the relationship (4.23) to stabilise the colder disk in the azimuthal direction it is necessary to heat it to higher temperatures. Therefore, taking into account (4.23), the condition (4.115) can be formally represented in the form:

$$Q_T^{(2)} \geqslant S, \qquad S = \frac{2\Omega}{\varkappa}. \tag{4.117}$$

Parameter S characterises the degree of differentiality of rotation of the disk. On the basis of the observed forms of the rotation curves of the galaxies, it may be assumed that $1 \leqslant S \leqslant 2$. The stability condition in this form was discussed in [151, 152].

Thus, the dispersion of the azimuthal velocities determines the elasticity of the medium for highly non-axisymmetric perturbations, and to suppress gravitational instability it is necessary to heat the disk $2\Omega/\varkappa$ more times. The criterion (4.117) is referred to as the Toomre criterion with the simplified consideration of the non-axisymmetric perturbations.

4. *The Morozov criterion.* In the framework of the WKB approximation, the dynamics of weakly non-radial perturbations in an inhomogeneous disk was described in sections 2.2, 2.4, [151, 153, 175]. The stability criterion (2.140),

(2.141) takes into account a large number of factors (the radial inhomogeneity of surface density σ and c_r, the thickness of the disk, the differentiality of rotation), and it can also be written conveniently in the form $Q_T^{(M)} = c^{(M)} / c_T$ ($c^{(M)}$ is determined from equation (2.14)). In a general case, the determination of $c^{(M)}$ is reduced to the integration of the given differential equation (see [137]). The dispersion branches of the non-axisymmetric perturbations and the reason for the increase of the increments of gravitational instability in an inhomogeneous disk were investigated in [175]. The main shortcoming of the criterion is that it was derived as a result of the dynamics of highly twisted spiral waves ($m/r \ll k_r$, k_r is the radial wavenumber), and this is followed by the formal transition to the spoke-like perturbations.

If it is assumed that the characteristic scale of the inhomogeneity $L_c \equiv \{d \ln c^{(M)} / dr\}^{-1}$ is a fixed value, the differential equation (2.144) is reduced to the algebraic relationship (2.141) [151, 153, 173]. For the typical forms of the rotation curves $D \simeq 0.6 \div 0.8$. Introducing the multiplier $D=(1+0.974\varkappa\,\Delta/(S^2 c_T))^{-1}$ we can formally take into account the thickness of the disk also in the criteria (4.117) and (4.118).

5) *The Polyachenko–Polyachenko–Strel'nikov (PPS) criterion.* In contrast to the previously examined criteria, in the studies by Polyachenko et al [200] attention was given to the maximally non-axisymmetric perturbations in the thin disk. In this case, the boundary of stability is determined exclusively by the rotation curve because it depends only on the parameter $n \equiv -rd\Omega/(\Omega\, dr)$. The dependence of the dimensionless dispersion of the radial velocities at the stability boundary on $\alpha = 2/n$ is shown in Fig. 4.24 [200]. For applications, it is convenient to use the approximating function

$$Q_T^{(P)} = \frac{c^{(P)}}{c_T} = 1.88\sqrt{1.1 + \frac{8}{\exp(\alpha - 1/4) - 1}}, \tag{4.118}$$

where $Q_T^{(P)}$ is the minimum value of the Toomre parameter for a stable disk (Fig. 4.24). This approximation has the accuracy sufficient for the applications in the range $1.2 < \alpha^2$. The criterion proposed by Polyachenko et al [200] is a one-parameter criterion because the shape of the rotation curves fully determines the boundary of

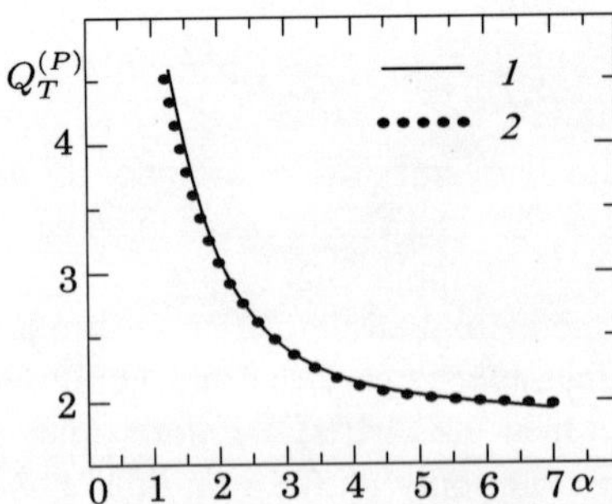

Fig. 4.24. Dependence $Q_T^{(P)}(\alpha)$. The data for the curve 1 were taken from Fig. 1 from [200]. Dependence 2 was plotted in accordance with equation (4.118). The rotation flat rotation curve corresponds to the value $\alpha = 2$ which gives $Q_T^{(P)}(\alpha) = 3.1$. At high values of α we have the case of solid state rotatiuon.

stability. In particular, for the region with the constant speed of rotation ($n = 1$) it should be expected that $Q_T^{(P)} \simeq 3$.

6. *Taking into account the gas system in the analysis of the stability of stellar disks*. The presence of a colder component result in additional destabilisation of the stellar disk. This problem was investigated in detail in a large number of studies [152, 728, 729, 845, 1052, 1280] where the stability of stellar-gas disks in relation to radial perturbations in linear approximation was analysed. A more general problem was investigated in [994], the effect on the stability of the thin disk with respect to small radial perturbations of the inhomogeneity of its composition where it consists of the particles with a specific mass spectrum, with the heavy particles having a lower dispersion of velocities.

The problem is simplified if the mass of the 'colder' component is relatively low. Taking into account that the surface density of the gas σ_{gas} is usually considerably lower than in the stellar disk σ_{star}, and $c_s \ll c_r$ (c_s is the adiabatic speed of sound in the gas), we can write the following equation for the dispersion of the velocities of the stellar disk, containing the gas, at the stability boundary [1280]:

$$\frac{c_r^{\mathrm{cr}}}{c_T} = 1 + \frac{\sigma_{\mathrm{gas}}}{\sigma_{\mathrm{gas}} + \sigma_{\mathrm{star}}} \frac{1-(c_s / c_T)^2}{1+(c_s / c_T)^2}. \tag{4. 119}$$

For example, for the parameters of the solar vicinity of the Galaxy it follows that the presence of a gas subsystem whose contribution to the surface density of the disk equals approximately 20% increases in accordance with (4.119) by ~10% the dispersion of the velocities of the stars, sufficient for the stabilisation of the radial perturbations. The gravitational stability of the stellar–gas disks was discussed in greater detail in section 8.5, see also [152, 1052].

It should be mentioned that to determine the parameters of the gas subsystem investigations were also carried out into other approaches based on the possibility of existence of different structures in the disk (spiral density waves or bar) taking into account the specific mechanism of formation and maintenance (for example, swing amplification [314, 317]), instead of considering the conditions of local stability.

7. *Analysis of the gravitational stability of the disk in numerical experiments*. The numerical experiments, describing the dynamic evolution of the disk, make it possible, defining the specific initial conditions, to investigate the instability of the disk taking into account the inhomogeneity in the distribution of mass and velocities. The formation of highly non-linear and non-axisymmetric structures, such as the bar and transient spirals, is taken into account in a natural manner. However, the complicated nature of this approach is the dependence of the results on the initial condition since the evolution of the disk, starting from the highly unstable states, may take place by a different mechanism than from the subcritical level. In addition to this, the disk can also be subjected to very slow secular instabilities which are difficult to take into account in the numerical experiments.

The evolution models of N-bodies are usually directed either to investigating the nature of development of the instability or, which is of special interest in this case, to analysis of the state of the system after many revolutions of the disk. From the large number of the published studies, it is important to mention those which are most important for the determination of the stability conditions of the disk.

Carlberg and Selwood [450, 1142] investigated the effect of small non-stationary perturbations of the potential on the evolution of the function of distribution of velocities. In particular, special attention was given to the scattering on non-stationary spiral waves; the resultant increase of the dispersion of the velocities of the stars with time is in good agreement with the results of numerical experiments. The critical value of the dispersion of the velocities of the stellar disks (parameter Q_T) has been calculated many times from the results of dynamic modelling within the framework of the problem of *N*-bodies (see, for example, [137, 320, 450, 596, 682, 917, 1127, 1136, 1140, 1141, 1144]). A stable disk with $Q_T < 1$ was not obtained in any case. Usually, at the disk periphery parameter $Q_T(r)$ increases with increase of the distance from the centre. This special feature has also been observed in the experiments with three-dimensional disks [69, 335, 415, 416, 596, 1145] and in a number of theoretical studies [1027].

In a classic study, Atanassul and Selwood [320] concluded that the two-dimensional disk is known to be stable if the Toomre parameter exceeds the value $Q_T \gtrsim 2.2 \div 2.4$. However, this conclusion was made without taking vertical movements into account. Three-dimensional disks are gravitationally more stable and, consequently, the two-dimensional models result in underestimating the mass of the marginally stable disk. It should also be mentioned that the parameter Q_T in [320] was averaged out over the radius, and modelling was carried out for the specific distribution of the surface density – the Toomre-Kuz'min model was investigated. Therefore, it cannot be conclude that the results of this study can be used for the actual three-dimensional galactic disks.

4.3.2. Parameters of stellar disks at the stability boundary

An insufficiently hot disk is gravitationally unstable. Its further evolution is characterised by heating expressed in the increase of the dispersion of the velocities c_r, c_φ particles with time. We shall discuss the results of numerical experiments for the dynamic evolution of the disks to the stable states in relation to the initial values $c_r(r)$ and $c_z(r)$, the parameters of the bulge (4.11) and the halo (4.3).

The most important special features of heating of the disk, obtained in the numerical experiments [258, 259], will be mentioned.

1. The heating time of the disk as a whole is considerably longer than the mean period of rotation of the particles (Fig. 4.25). In the initial stage ($t \lesssim 1$) the value c_r remains almost constant – until the disk remains axisymmetric. In the case of a small-mass halo, the evolution of the disk is determined by the global bar mode, and heating is determined by the process of formation of a non-axisymmetric central jumper and the two-arm spiral associated with it. In the models with the sufficiently massive halo the bar mode does not grow but a complicated transient system of small scale spiral waves, leading to the transition of the kinetic energy of rotation to the chaotic component of velocity, i.e. to heating of the disk, forms.

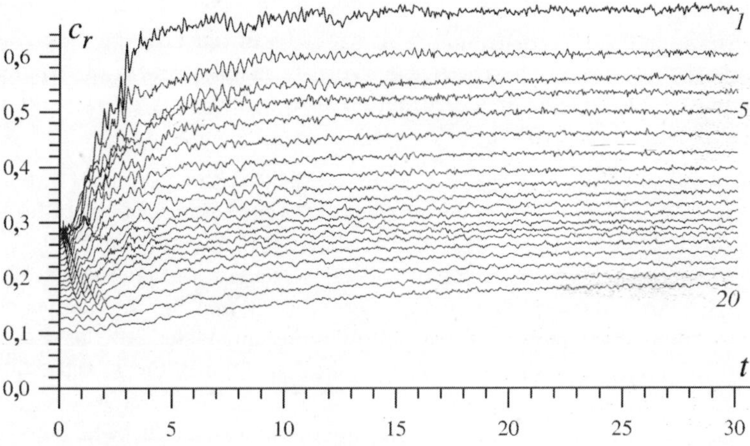

Fig. 4.25. Evolution of the initial insufficiently hot (unstable) disk. The long-term dependence of the dispersion of the radial velocities on time for 20 radial zones is shown (1 – the centre of the disk, 20 – for $r = 4L$). The period of rotation of the disk by $r = 4L$ is $\simeq 3.4$.

2. The heating of the initially cold disk ($0.5 \lesssim Q_T \lesssim 1$) starts in the central part. The process of heating of the periphery is considerably slower. The dispersion of the velocities usually starts to increase when the quasi-stationary state has already been established in the centre. At the same time, the processes in the centre and at the periphery are mutually linked: fast instability in the central part is capable of accelerating the heating of the external region of the disk, and its stability reduces the rate of this process.
3. If a bar does not form, the amplitude of the perturbations starts to decrease with increase of the dispersion of the velocities. In turn, a reduction of the amplitude of the waves also reduces the rate of increase of the dispersion of the radial velocities c_r. Heating is almost completely interrupted after decay of the transient spiral waves. The integral amplitudes of the Fourier harmonics (4.28) for all numbers of the modes m decrease with time but at a considerably lower rate than for the modes at $m = 2$. The density distribution of the disk becomes similar to axisymmetric (if the mass of the spherical component is sufficiently large to avoid the formation of a bar), and the dispersion c_r on the whole retains its level over tens of revolutions in the absence of relaxation processes in the vertical direction.
4. If the three-arm or even four-arm nodes dominate in the initial stage of evolution of the initially cold disk ($Q_T (r \leqslant 2L) \lesssim 1$), the nature of the spiral structure changes when the initial state is defined as subcritical (unstable but not too cold, for which $Q_T \gtrsim 1$). The main mode is the two-arm mode, although the harmonics $m = 3$ also plays a significant role, especially in the initial stages of evolution. A relatively characteristic special feature is the closing in the external region of the disk of the spirals into a ring-shaped structure. The spatial structure of the perturbations also depends strongly on the relative mass of the spheroidal subsystem.

5. In the case of a very cold initial state in the models with a low-mass spheroidal subsystem, extensive redistribution of matter takes place in the disk with the change of the form of the rotation curve $V(r)$ in the process of heating and formation of the bar. The final distribution of surface density $\sigma(r)$ greatly differs from exponential (Fig. 4.14).[7]

Another special feature of the models of the disks with a very cold initial state ($Q_T \lesssim 1$) is that the dispersion of the velocities at the end of calculations (after 10–15 revolutions) is slightly higher than that required for gravitational stability. This effect is caused by heating as a result of collective processes – wave-like movement with a high amplitude formed at strong instability. When the disk is heated and reaches the threshold stability, these perturbations attenuate, but the process of attenuation of the waves is of the inertia type: the dispersion of the velocities already ensures the stability of the disk, whereas the spiral waves have not yet decayed (the Fourier analysis of the perturbations of density of the disk confirms this) and cause additional heating. Therefore, to obtain the minimum dispersion of the velocities, required for the stability, it is necessary to use the rotation algorithm of successive approximations of the initial dispersion of the velocities to the stability boundary [259].

As expected, the stability boundary depends also on the initial thickness of the disk.[8] If the disk is initially thick (the vertical scale $h \gtrsim 0.2L$) and unstable only in the plane ($c_r < c_r^{\rm cr}$), then heating of the disk takes place at a lower rate and is interrupted at lower values of the dispersion of the radial velocities in comparison with the case of the initially thin disk. This effect is determined by two factors: the stabilising effect of the finite thickness of the disk and the slower rate of the relaxation processes in the vertical direction in comparison with the characteristic heating time in the disk plane.

The results of dynamic modelling can be used to determine the parameters of the disk at the boundary of gravitational stability (when the dispersion of the velocities ceases to change – after 5÷20 revolutions of the disk at the external edge).

In the case of the halo scale $a \gtrsim L$ the rotation curves have an elongated (conditions referred to as 'solid-state') section of growth of $V_c(r)$, changing in the

[7]In principle it is possible to 'select' the initial density profile which leads at the end of calculations to exponential (the dashed line in Fig. 4.14), but a similar approach appears to be artificial. Evidently, the stellar disks of the majority of galaxies do not pass through the stage of strong dynamic instability).

[8]The initial thickness of the disk of actual galaxies depends on their formation conditions. For example, if high-intensity start formation took place as a result of a collapse of the gas disk prior to the disk reaching the quasi-stationary state, the resultant stellar disk may be considerably thicker and characterised by a considerably higher dispersion of velocities c_z in comparison with the minimum value required for stability; however, this scenario is difficult to match with the existence of very thin ($h/L < 0.2$) stellar disks in the galaxies visible edge-on. The dependence of the thickness of the disks on the relative mass of the halo is also in agreement with the assumption that the dispersion of the velocities of the stars is close to that expected for the marginal stability [258].).

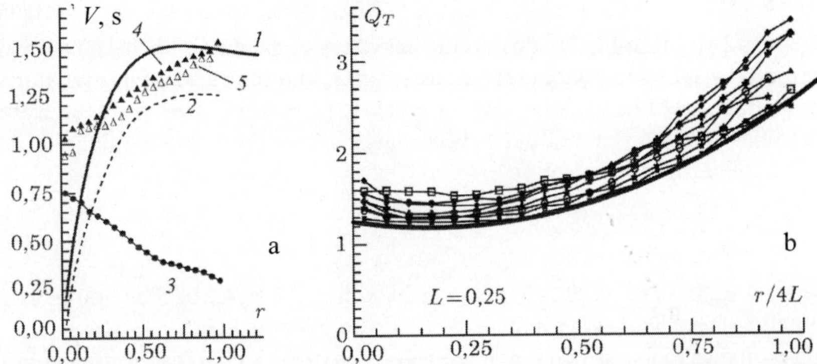

Fig. 4.26. The parameters of the disk, pre-evolving to the stable state, at the stability boundary in the models without the bulge: a) for the model in which the mass of the halo in the range $r \leqslant 1 = 4L$ is equal to the mass of the disk and the scale of the halo $a = L$, the graph shows the radial dependences of the secular velocity V_c(r) (line 1), the velocity of rotation of the disk $V(r)$ (line 2), the dispersion of the radial velocities $c_r(r)$ (line 3) and the parameters $S = 2\Omega/\varkappa$, calculated for the circular velocity of rotation (line 4) and the velocity of rotation of the disk (line 5), respectively; b) the Toomre parameter $Q_T(r)$ at the stability boundary for a series of the models without the bulge and with different halo parameters. The thick solid line was calculated from equation (4.120).

region $r \gtrsim 2L$ into a plateau $V_c \simeq$ const (line 1 in Fig. 4.26a). Figure 4.26a shows for a model of a galaxy without a bulge with $\mu = 1$ the radial distribution of the circular velocity of rotation $V_c(r)$ (line 1), the mean velocity of rotation of the particles $V(r)$ (line 2), the dispersion of the radial velocities $c_r(r)$ (line 3), and also the parameters of differentiality of rotation $S = 2\Omega/\varkappa$ (see (4.117)), calculated separately for the velocities $V_c(r)$ and $V(r)$ (symbols 4, 5), respectively.

Knowing the finite values of the dispersion of the velocities, and also the density of the disk and the velocity of rotation, we can compare, for the selected model, the values of the Toomre stability parameter (4.115) Q_T with the analytical local criteria, discussed previously in section 4.3.1, and also compare the latter with each other. It should be mentioned that the analytical criteria were obtained assuming the small value of the difference between the velocity of circular rotation $V_c(r)$ and the mean velocity of rotation $V(r)$ of the gravitating points. However, the difference between them for collisionless disks may be large. In the case of massive halo ($\mu \gtrsim 2$), the difference between the stability parameters, calculated for $V_c(r)$ and $V(r)$, is small but may become large at $\mu \leqslant 1$. This is associated with the increase of the difference $V_c - V$ with the increase of the dispersion of the velocities in accordance with (4.21). In most cases, the condition $S \simeq Q_T^{(M)} \leqslant Q_T \leqslant Q_T^{(P)}$ is fulfilled.

Although none of the criteria explains the modelling dependences $Q_T(r)$ at all r, the Polyachenko–Polyachenko–Strel'nikov (PPS) criterion, generalised for the case of finite thickness using equation (4.116), and Morozov's criterion [259] appears to be closest to them.

An important result is that the radial dependence of the Toomre parameter $Q_T(r)$, calculated for circular velocity $V_c(r)$, has the qualitatively identical form for all the investigated models without the bulge in the case $\mu = M_h/M_d \gtrsim 1$ (Fig.

4.26b). In this case, $Q_T(r)$ is approximately constant in the region $0 \leqslant r/L \leqslant 2$ on the level $Q_T \simeq 1.2 \div 1.6$ (see Fig. 4.26b). At the periphery of the disk ($r \gtrsim 2L$) an increase of the radius results in a monotonic increase of Q_T and the value of this parameter reaches at the end of the disk ($r \gtrsim 4L$) the values $Q_T \simeq 2.5 \div 3$. However, the scatter of the values Q_T for different models is not large so that the following function can be used as the restricting function

$$Q_T^{(*)} = A_0 + A_1 \cdot \left(\frac{r}{L}\right) + A_2 \cdot \left(\frac{r}{L}\right)^2, \tag{4.120}$$

where $A_0 = 1.25$, $A_1 = -0.19$, $A_2 = 0.134$ (thick line in Fig. 4.26). The minimum of this function is reached at the point $r/L = 0.7$ where it equals 1.2.

The estimates of the dispersion of the velocities of the stellar disk, obtained from the observations, usually relate to the internal region of the disk $r < 2L$. Therefore, the approximate constancy of Q_T in this region is suitable for estimating the surface density of the disk and its mass M_d (assuming its stability), if the rotation curve of the galaxy and the radial exponential scale of the disk L are known. The connecting relationship is the equation

$$c_r = Q_T^{(*)}(r)\frac{3.36G\sigma_0 \exp(-r/L)}{\varkappa}, \tag{4.121}$$

where $Q_T^{(*)}$ for the galaxies with the extended section of growth of $V_c(r)$ is determined by the relationship (4.120). In turn, the estimate of the mass of the disk $M_d = 2\pi\sigma_0 L^2$ for the given speed of revolution makes it possible to 'isolate' the mass of the dark halo within the limits of the given radius.

4.3.3. *Effect of the bulge on the boundary of gravitational stability*

Now we examine the models of the galaxies in which the rotation curve in the central region ($r \leqslant L$) is determined mainly by the bulge.

Outside the bulge ($r \gtrsim L$), the rotation curves are in most cases almost flat $V \simeq$ const. Figure 4.27 shows the typical modelling radial dependences of the disk parameters for this case. In the presence of a bulge, the parameter Q_T rapidly increases in the central region of the disk where the dynamics is determined by the bulge potential. However, outside this region in the zone $r \simeq (1/2) \cdot L$ the radial dependence Q_T retains its form (see (4.120) and Fig. 4.26b). With increase of r, the value Q_T monotonically increases to 2.5–3.5. The resultant radial distributions of the parameter Q_T are typical for the systems with a not too massive ($M_b/M_d \leqslant 0.25$) and not too long bulge ($r_b^{\max} \leqslant L$).

As the bulge becomes more massive and compacted, the value of Q_T in the centre of the disk increases. This increase of the Toomre parameter is determined mostly by the increase of epicyclic frequency $\varkappa$. With increase of the mass of the highly concentrated bulge, the dispersion of the radial velocities in the central region of the disk also slightly increases. It should be mentioned that in the model shown in Fig. 4.27a, the additional heating in the vicinity of the centre ($r \leqslant 0.5L$) of the disk subsystem is not associated directly with gravitational instability. In the initial stage,

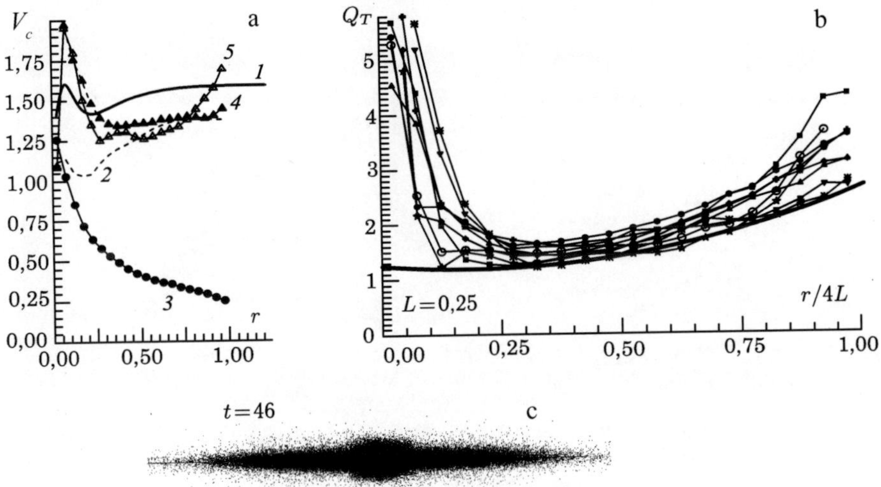

Fig. 4.27. Parameters of the disk at the boundary of stability in the model with the bulge: a the radial dependences of the disk parameters (the symbols are the same as in Fig. 4.26). The mass of the halo in the range $r \leqslant 1 = 4L$ is equal to the mass of the disk, the scale of the halo $a = 3.6L$. The parameters of the bulge $M_b = 0.24M_d$, $b = 0.04L$, $r_b^{\max} = 0.5L$; b) Toomre parameter $Q_T(r)$, calculated using $V_c(r)$ and the boundary of stability for a series of models with the bulge. The thick solid line was plotted using the equation (4.120); c) the edge-on view of the disk at the end of calculations in the model shown in Fig. 4.27a. The points indicates the positions of the particles. In the central region, the thickness of the disk greatly increases.

this model with $\mu = 1$ is characterised by the formation of a bar which fails with time as a result of scattering of the particles taking place in the vicinity of the concentrated nucleus whose scale is $b = 0.01$. The mechanism of failure of the bar is identical to the effect of a massive black hole [653] and is discussed in greater detail in chapter 9.

Consequently, the disk is additionally heated and its thickness in the central region increases (Figure 4.27c). The degree of additional heating and the circular velocity V_c in the centre of the disk are highly sensitive to the bulge parameters, mostly the radius of the nucleus b (in particular, already at $b \gtrsim 0.05L$, the rate of this failure of the bar is greatly reduced). Therefore, in the range $r \leqslant 0.5L$, the Toomre parameter Q_T can greatly change in different models (see Fig. 4.27b).

4.3.4. *Differentiality of rotation as a factor increasing the gravitational stability threshold*

Two limiting cases will be investigated – a disk whose rotation is close to solid-state rotation, and a quasi-Kepler disk where the speed of revolution decreases as $V \propto r^{-1/2}$. Evidently, these galaxies are very rare but their analysis is important from the viewpoint of methodology. Since the differentiality of the rotation is a destabilising factor, for the solid-state rotating disk we should expect lower values of Q_T with other conditions being equal.

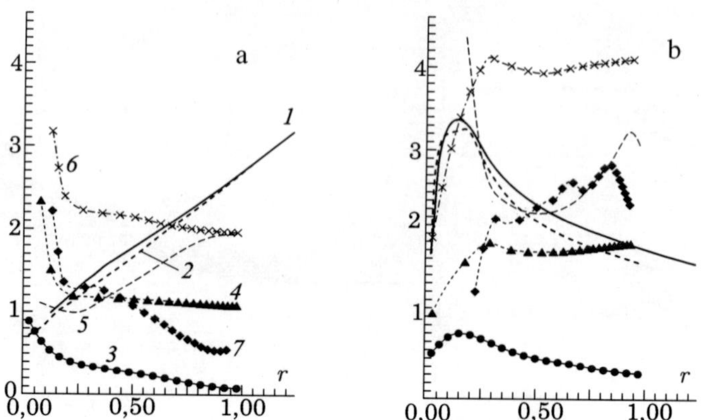

Fig. 4.28. Radial dependences of the disk parameters: a) a large part of the disk rotates almost like the solid-state disks; b) rotation in the range $r > L = 0.25$ is similar to Kepler rotation. The notations: 1) circular velocity $V_c(r)$; 2) the velocity of rotation of the disk (stars) $V(r)$; 3) the dispersion of the radial velocities $c_r(r)$; 4) parameter S calculated from equation (4.117) for the circular velocity; 5) the Toomre parameter Q_T, calculated for the circular velocity; 6) the critical Toomre parameter $Q_T^{(P)}$, calculated using the criterion (4.118); 7) the critical Toomre parameter $Q_T^{(M)}$, calculated from $V_c(r)$.

The extended section of almost solid-state rotation in a galaxy is usually detected in cases in which the mass of the disk with the limits of a large part of the disk ($r \leqslant 2L$) is dominant and, consequently, to suppress the bar-forming instability, the disk must be heated to higher temperatures than in the presence of massive spherical components. Therefore, to explain the role of differentiality of rotation we examine the model in which the formation of a bar is suppressed by a massive halo with a 'solid-state' section in the external region of the disk ($r \gtrsim L$). The differentiality parameter of rotation in this model $S(r > L) = 2\Omega/\varkappa = 1.1 \div 1.2$ is close to the value $S = 1$ which corresponds to the strictly solid-state rotation (Fig. 4.28a). The radial dependences of the parameters, characterising the stability of the system Q_T, $Q_T^{(P)}$, $Q_T^{(T)}$, S are presented in Fig. 4.28a. It may be seen that on the whole the disk is stable at lower values of the parameter Q_T (curve 6) in comparison with the cases examined previously. In the centre we have $Q_T \simeq 1$, and at the periphery of the disk $Q_T \simeq 2$.

We turn to the second limiting case which corresponds to the rotation curve close to the Kepler curve (Fig. 4.28b). In galaxies, these sections are detected only in a small number of cases [584], but we are interested in the principal problem of the effect of strong differentiality of rotation on the minimum value of the dispersion of velocities required for stability. To produce Kepler disks, massive concentrated components were introduced into the dynamic model. The constructed series of the model is a natural continuation of the models with highly massive bulges $M_b > (2 \div 4)\, M_d$. Figure 4.28b shows the results of calculations for this model. For the disks with strongly differential rotation, the value of the Toomre parameter rapidly increases so that even in the range where Q_T is minimum ($r \simeq 2L$), to ensure stability, we should have already $Q_T > 2$.

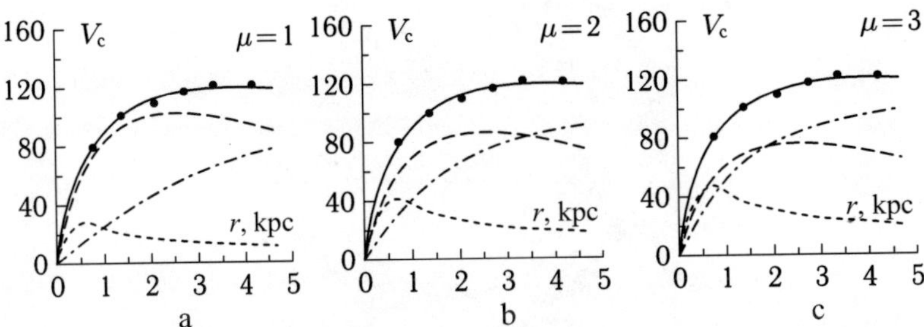

Fig. 4.29. Radial dependences of the modelling velocity of rotation (solid line) and the observed rotation curve (points) for NGC 6503 (velocity in km/s, radius in kpc) for different relative masses of the halo: a) μ = 1; b) μ = 2; c) μ = 3. The lines show the contribution of the subsystems to the modelling rotation curve: the dotted line – the bulge, the dashed line – disk; the dot-and-dash line – the halo.

4.4. Evaluation of the mass of spherical subsystems for different samples of disk galaxies

4.4.1. Dependence of the ratio of the dispersion of the velocities of the stars to the velocity of rotation in relation to the relative halo mass

Analysis of the distribution of the mass in the disk galaxies is based on investigation of the rotation curves $V(r)$ determined in most cases by the measurement of the Doppler gas velocities. If the maximum velocity of rotation of the disk can be used to obtain a rough estimate of the mass of the galaxy within the limits of the specific radius, the form of the curve reflects the distribution of mass of its main components. The ambiguity of the decomposition $V(r)$ into the components, associated with the individual components, requires us to use additional information on the mass and radial scales of the components provided by surface photometry of the disk and the bulge. This approach could provide complete information on the relative mass of the disk and the bulge, but the presence of an invisible component (dark halo) of the apriori unknown mass and with the almost unknown distribution law of density greatly complicates the problem (Fig. 4.29). The rotation curves are usually interpreted by the maximal disk method aimed at determining the upper limit of the mass of the flat component or the best fit method in which the difference between the model and observed forms of the rotational curves is minimised. However, in these cases, the result is usually ambiguous, and the shape of the rotation curve can not be explained at different relationships between the masses of the disk and the spherical component and, consequently, the mass of each of them is determined highly inaccurately.

The number of galaxies for which not only the rotation curve but also the dispersion of the velocities of the 'old' stellar disks is available increases [77, 78, 361, 413, 416, 554, 670, 962]. Analysis of the dispersion of the velocities makes it possible to reduce the range of the possible models of the galaxies explaining the observed rotation curve. The main idea is based on the application of an additional

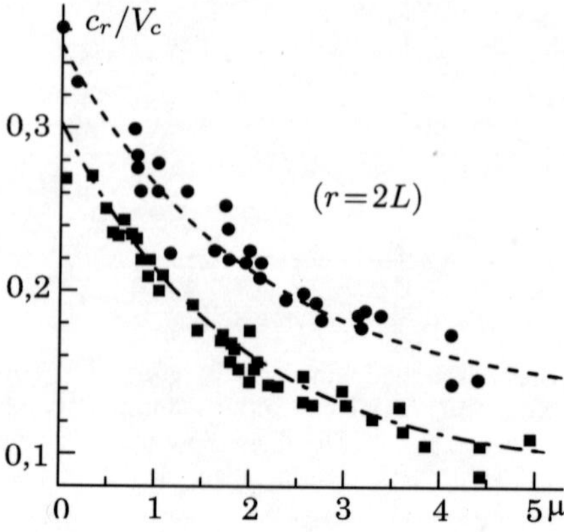

Fig. 4.30. Dependence of the ratio c_r/V_c at the radius $r = 2L$ on the relative mass of the spheroidal subsystem μ according to the dynamic modelling results. The circles refer to the models in which the initial disk was thin. The models with the initial thick disk are indicated by the squares.

condition in the modelling of the distribution of mass in the galaxies. This condition can be formulated as follows: for a stellar disk which has reached the stationary state, the modelling values of the dispersion of the velocities at different distances from the centre can be equal to or in a general case not exceed the observed values.

The numerical experiments can be used to estimate the lower limit of the mass of spherical components at which the investigated disks of the galaxies may be in the stable states. If the disk is characterised by marginal stability, these estimates should correspond to the real mass of the components.

The possibility of obtaining these estimates is associated with a reduction of the ratio of the dispersion of the velocities to the velocity of rotation c_r/V at a fixed radius with an increase of the relative mass of the spherical subsystem $\mu = M_s/M_d$ [157]. This dependence follows from the simple estimates. In the approximation of the balance of the centrifugal and gravitational forces we have $V \simeq \sqrt{r\dfrac{\partial\Phi}{\partial r}}$ and, disregarding the nature of distribution of mass in the Galaxy, on the basis of the order of magnitude we assume that $\dfrac{\partial\Phi}{\partial r} \sim G\dfrac{M_s + M_d}{r^2}$. For a disk at the boundary of gravitational stability we take into account that $c_r \propto 1/\varkappa \propto 1/\sqrt{M_s + M_d}$ (see (4.115)) and as a result we obtain the dependence: $C_r/V \propto (M_s + M_d)^{-1} \propto (1 + \mu)^{-1}$. This effect is shown in Fig. 4.30 constructed on the basis of the results of numerical modelling.

The dynamic models can be used to explain the values of the ratio of the mask of the disk and spherical component for explaining the observed velocities of rotation and the dispersion of the velocities of the old stars at different distances from the centre. The result of constructing the dynamic model is the spatial distribution of different parameters of the disk after the disk has reached the stationary state.

Evidently, the following conditions should be fulfilled throughout the disk: $c^{mod} < c^{obs}$, where c^{mod} and c^{obs} are experimental (modelling) and observed the dispersion of the velocity. In the regions where $c^{mod} < c^{obs}$, the disk is 'overheated' indicating the presence of additional factors which do not take part in the model and increase the dispersion of the radial velocities.

4.4.2. Examples of estimates of the mass of galactic components according to the observed distributions of the dispersion of velocities

On the example of NGC 6503[9] we will discuss in detail the method of matching the observed and modelling radial dependences of the velocity of rotation and dispersion [258, 1225]. The galaxy should be characterised by a small exponential scale of the disk (L = 1.16 kpc), and the estimates of the dispersion of the velocity of the stars are related to the range $r \leqslant 2L$. The rotation curve of the galaxy $V^{obs}(r)$ is taken from [354]. The observed $V^{obs}(r)$ can be explained on the basis of the models with different ratios of the spherical and disk subsystems $\mu = M_s/M_d = 0.8 \div 4.4$. The mass of the disk will be determined in the range $4L$.

The radial dependence $c_r(r)$ at the stability boundary is determined by the parameters of the spherical subsystem and by the value of the exponential scale of the disk in accordance with section 4.3. Comparison with the observed values of the dispersion of velocities requires taking into account the angle of inclination of the disk of the Galaxy i to the drawing plane. For the dispersion of the velocities along the visual ray we have

$$c^s = \sqrt{c_z^2 \cos^2(i) + c_\varphi^2 \sin^2(i)\cos^2(\alpha) + c_r^2 \sin^2(i)\sin^2(\alpha)}, \qquad (4.122)$$

where α is the angle between the direction of the slit and the major axis, measured in projection on the plane of the galaxy. According to [416], for NGC 6503 we accept $i = 74°$. The slit of the oscilloscope in observations was oriented along the major axis of the object ($\alpha = 0$). Components c_r, c_φ, c_z were taken from the numerical model.

In the case of a low-mass spherical subsystem with $\mu \leqslant 1$ at the initial distribution of the dispersion $c_r(r, t = 0)$, close to the observed distribution, the system is unstable and the disk is heated. As a result of heating, the final value of the dispersion of the velocities is considerably higher than the observed value (Fig. 4.31). The model with $\mu = 0.8$ ($a = 0.7$) corresponds with the variant of the maximum disk. At a more massive halo ($\mu = 1$, $a = 0.5$) forms inside the internal region of the disk. This is associated with the small mass and the high value of the radius of the core of the halo ('porous' halo), when the disk mass dominates in the range $r \leqslant 2L$. In these models, the value c_r and the periphery of the disk is considerably higher than the observed estimates (Fig. 4.31). In the central zone the difference is not so large but the inequality $c^{mod} > c^{obs}$ is fulfilled everywhere. In the model of the maximum disk,

[9]Examples of construction of the models of the galaxies NGC891, 936, 1167, 1169, 1566, 2179, 2273, 2712, 2775, 3198, 4150, 6340, 7331 can be found in [255, 258, 1225, 1323, 1327].

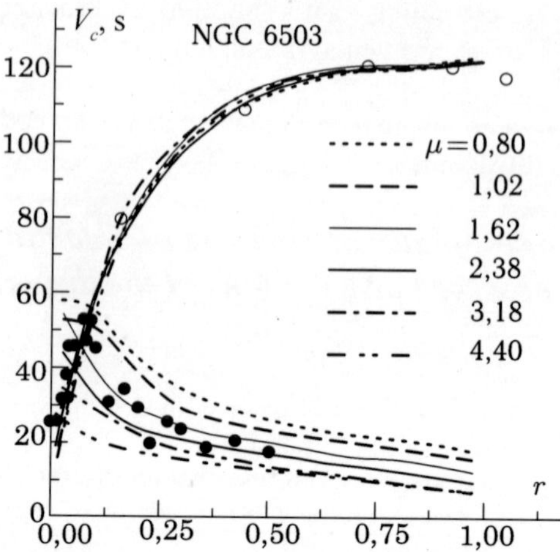

Fig. 4.31. Radial distributions of the circular velocity and the dispersion of velocities along the visual ray (in km/s) for different models of NGC 6503. The observed data are indicated by the circles [413]. On the horizontal axis L = 0.25.

the observed curve of the circular velocity of rotation is reproduced most acurately at $M_b = 1.9 \cdot 10^7\, M_\odot$, $M_d = 7.1 \cdot 10^9\, M_\odot$, $M_h = 5.6 \cdot 10^9\, M_\odot$, to which $\mu = 0.8$ corresponds. This shows that the dynamic model with these parameters results in a highly heated disk ($c_r/V \simeq 0.3$) and contradicts the observational data (Fig. 4.31). Thus, regardless of the agreement with the rotation curve, a massive spherical subsystem is required for constructing a realistic model.

A satisfactory agreement between the observed $V^{gas}(r)$ and modelling $V_c(r)$ can be obtained at very massive spherical components ($\mu \geqslant 4$). However, in these models $c^{mod} < c^{obs}$ is fulfilled everywhere. This is compatible with the observations of NGC 6503 only if it is assumed that the entire stellar disk of the galaxy is strongly superheated and has a large margin of gravitational stability. In a ring with $0.2 < r/L < 0.5$, the stability margin with respect to c_r reaches 2 or more. However, if it is taken into account that the stellar disk is close to the boundary of gravitational stability, then the models with $\mu \gtrsim 3.5$ can be rejected. The contradiction between the results of dynamic modelling and the observational data does not form at $\mu = 1.6 \div 3.5$. Of the investigated models, the models with $\mu = 1.62$ and $\mu = 2.38$ (Fig. 4.31), which gives the best agreement with the observations, are preferred. Consequently, it may be concluded that the mass of the spherical subsystem in the NGC 6503 galaxy at least 1.60 times greater than the mass of the disk. The contribution of the bulge to the mass of the spherical component remains negligible.

The conclusion on the comparatively small mass of the NGC6503 disk is in agreement with the results of study [416] concerned with modelling of this galaxy.

4.4.3. Restrictions on the mass of marginally stable stellar disks

Using the assumption on the marginal stability of the old stellar disk, we determine the mass of the disks M_d in the range $r = 4L$, using equation (4.120), which can be written in the following form

$$c_r = \frac{3.36 G\sigma}{\varkappa} Q_T^{(*)}. \tag{4.123}$$

We consider the galaxies for which the estimates of the dispersion of the velocities of the stars at a relatively large distance from the centre in the range (1.5–2)·L (L is the photometric scale of the disk) are available. In this case $Q_T^{(*)} \gtrsim 1.5$ and the effect of the bulge is completely minimised. In [77] a sample was constructed from 29 objects with the estimates of the dispersions of the velocities of the stars along the visual ray at such distances from the centre obtained in the observations. In [78], the data were collected on the kinematics of the disks (including the dispersion of the velocities of the stars outside the limits of the bulge), and the estimates of the mass of these components were obtained for 120 galaxies.

Taking into account that $\sigma(2L) = kc_rV_c/7.72LGQ_T^{(*)}$ we write the expression for the surface density of the disk and the radius $r = 2L$ in a suitable form:

$$\sigma(2L)[M_\odot/\text{pc}^2] = 6.9 \cdot 10^{-2} \frac{kc_r[\text{km/s}]V_c[\text{km/s}]}{2Q_T^{(*)} L[\text{kpc}]}, \tag{4.124}$$

where the values of the parameter $k = \varkappa/\Omega$ are distributed in the range from 2 in the case of solid-state rotation to $\sqrt{2}$ for a plateau on the rotation curve, and V_c and $Q_T^{(*)}$ related to the radial coordinate $r = 2L$. As a result, we calculate the mass of the disk with the exponential density profile $M_d = 2\pi L^2\,\sigma(2L)\,e^2$. The transition from c^{obs} to c_r is determined by (4.122).

Several results, following from the determination of the values M_d, $\sigma(2L)$, will be discussed [77].

The role of rotation in the equilibrium of the disk can be efficiently characterised by the ratio of the radial dispersion of the velocities to the velocity of circular rotation. In Fig. 4.32a, b, the ratio c_r/V_c is compared with the morphological type T and with the colour index $(B - V)_0$. Here the relationship between the compared values is clearly evident. The dependence of c_r/V_c on T (Fig. 4.32a) shows that in the galaxies of late types the rotation plays a more significant role in maintaining the equilibrium of the old stellar disk than in the Sa–SO galaxies.

Figure 4.32b shows that the dynamically 'cold' disks are characterised by more intensive star formation, as indicated by their deeper blue colour. On the other hand,

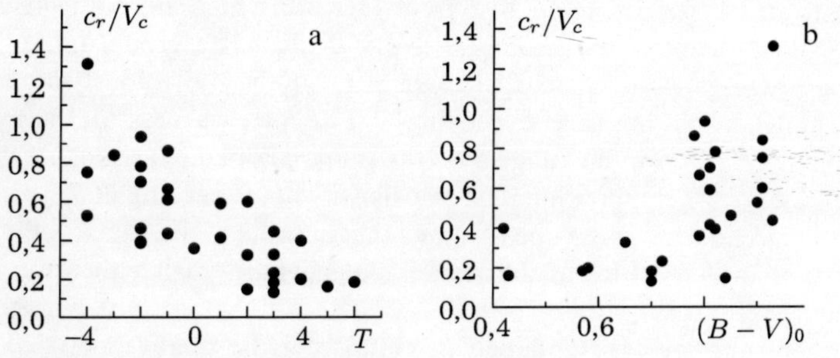

Fig. 4.32. Comparison of the ratio c_r/V_c with the morphological type T (a) and the integral colour index of the galaxy $(B–V)_0$ (b).

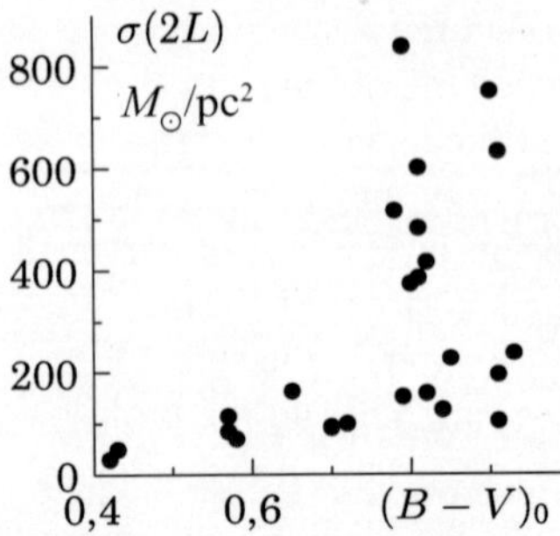

Fig. 4.33. Comparison of the local densities of the disk at $r \simeq 2L$, calculated assuming the marginal stability of the disk, with the integral colour index of the galaxies. The group of the 'red' galaxies, for which the estimates $\sigma(2L)$ are evidently too high, is indicated.

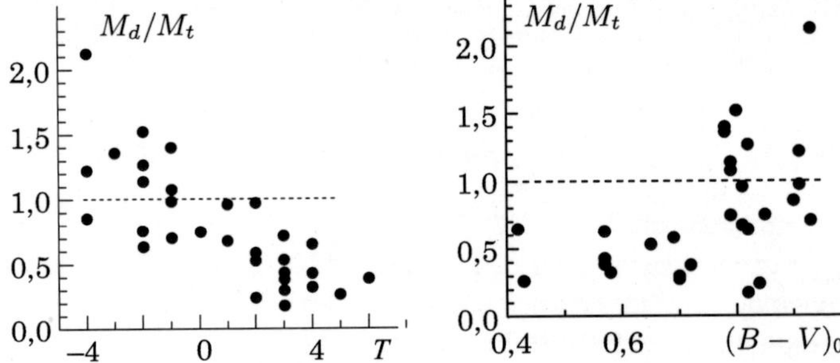

Fig. 4.34. Comparison of the ratio of the mass of the marginally stable disk to the total masts of the Galaxy M_d/M_t in the range $r = 4L$ with the morphological type T (*a*) in the integral parameter of colour of the Galaxy (b).

the 'red' galaxies (Sa–SO dominate amongst them) are characterised by greater dispersion of the kinematic properties and contain a relatively large number of strongly 'overheated' systems with a high c_r/V_c ratio. The boundary between the 'cold' and 'hot' disks corresponds to the value $(B - V)_0 \simeq 0.75$.

Figure 4.33 shows the values of the local densities of the disk at the selected distance from the centre, calculated from (4.124), which are compared with the integral colour index of the Galaxy. Clearly evident is a group of 'red' galaxies (almost all these galaxies relate to the type S0) with the anomalously high values of the critical density of the disk, exceeding 400 $M_{\odot}/\text{pc}^2$, whereas for the majority of other galaxies $\sigma(2L)$ is in the range 50–200 $M_{\odot}/\text{pc}^2$. For comparison, the disk of our Galaxy in the solar neighbourhood has the density not exceeding 60 $M_{\odot}/\text{pc}^2$ (chapter 9 [260]). Consequently, it may be assumed that in some of the galaxies the observed dispersion of the velocities of the stars is considerably higher than the minimum value required for disk stability.

Similar results are also obtained in comparing the relative mass of the disk, calculated assuming its marginal stability, with the morphological type and the colour index of the galaxies (Fig. 4.34a, b). In a general case, M_d is the upper limit of the mass of the disk because the dispersion of the velocities of the stars in the disks of the Galaxy may exceed the minimum value required for disk stability. The horizontal line separates the objects with the ratio $M_d/M_t > 1$. The latter inequality has no physical meaning and only indicates that for these galaxies the assumption on the marginal stability of the disk is not in agreement with the accepted value c_r. Either these galaxies contain strongly 'heated' (with a large stability margin) disks or the estimate of c_r is affected by the stars of the bulge which is 'hotter' than the disk. Both variants are acceptable, and the problem for each Galaxy should be solved separately. It should be mentioned however that at the radial profile $c^{obs}(r)$ the internal parts of the galaxies, where the bulge dominates, usually greatly differ from the adjacent regions of the disk where the dispersions of the velocities are lower and slowly decrease with r (see, for example [1172, 1173]). In the investigated galaxies with $M_d/M_t > 1$ (with the exception of NGC 3489), the range $r \simeq 2L$ is related to the 'disk' section of $c^{obs}(r)$ of the transition region which is an indirect argument in favour of the suggestion that the observed dissipation of the velocities, at least in some of them, is not greatly distorted by the effect of the bulge.

For a number of galaxies it may be verified whether the estimate $M_d/M_t > 1$ is the result of $c_{obs}(2L)$ estimate being too high, using the already available measurement results. For the three lenticular galaxies with $M_d/M_t > 1$ (NGC 4251, NGC 4578 and NGC 5273), the estimates of the dispersion of the velocities were obtained up to the values $r/L = 3.6$; 3.3 and 3.1, respectively, i.e. the distances at which the effect of the bulge should be considerably smaller than at $2L$. Although at high r the indeterminacy in the value of the parameter $Q_T^{(*)}$ increases, it can be excepted that for these limiting distances the condition $Q_T^{(*)} < 3$ should be fulfilled, and, consequently, we can estimate the lower limit M_d. It has been shown that in this case also the ratios M_d/M_t of the mentioned galaxies are higher than unity (1.3; 2.0 and 1.2, respectively), i.e. conclusion on 'overheating' of the disks is valid.

The situation in the case of spiral galaxies is different. The condition $M_d/M_t < 1$ is satisfied for the majority of these galaxies, but fulfilment of this condition does not yet mean that the dispersion of the velocities in the disks is close to the minimum

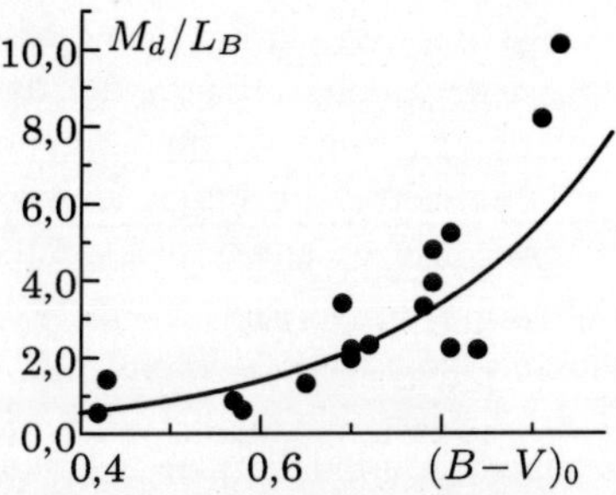

Fig. 4.35. The relationship of the ratio of the mass of the marginally stable disk to the integral luminosity of the galaxies with the integral colour index of the Galaxy. The curve is the dependence $\lg M_d/L_B = 1.937\ (B{-}V)_0{-}1.019$, obtained by the method of evolution modelling of the stellar systems (see Table A3 for the model of the 'closed box' in [358]).

value for gravitational stability. In this case, it is important to obtain independent estimates of the mass of the disks.

To compare M_d with the values, expected for the given luminosity of the galaxies, Fig. 4.35 shows the M_d/L_B diagram, where L_B is colour index for the spiral galaxies of the investigated sample. It is evident that the ratio M_d/L_B for the stellar population (usually equalling several solar units) increases with a reduction of the contribution of young stars to the integral luminosity, i.e. with increase of the colour index.

Despite the considerable dispersion of the points in Fig. 4.35 relative to the model curve, we can talk about quantitative agreement between the estimates of mass of the disks, corresponding to the condition of marginal stability, with the estimates expected for the given color and luminosity. This is a strong argument in favour of the fact that the velocity dispersion of stars in the disks of spiral galaxies is in most cases close to the minimum value required for their stability. This implies that the mechanisms of the slow dynamic 'heating' of the stationary disks in the examined spiral galaxies are not efficient enough to lead to a substantial increase in the velocity dispersion during their existence.

Thus, the stellar disks in galaxies of early types, apparently, are often strongly 'overheated'. This follows from the unacceptably high estimates of local densities and masses of integral disks, which are obtained assuming their marginal stability. At the same time the correlation of high velocity dispersion with the low rate of star formation is clearly visible. Indeed, the relative magnitude of the dispersion of velocities c_r/V increases systematically with the color index of the galaxies (Fig. 4.32b), all galaxies with $M_d/M_t > 1$ have the colour of the old stellar population $(B–V)_0 > 75$).

4.4.4. On the nature of lenticular galaxies

As noted above, overheating of the stellar disk is found most frequently amongst the galaxies of early morphological types. In connection with this we discuss the nature and current state of lenticular galaxies (S0) [68,77].

As regards the structure, the lenticular galaxies are similar to spiral galaxies of later types, having a massive stellar disk, and in many cases – also a dynamically isolated circumnuclear disk, a developed bulge and a dark halo determining the rotation velocity of their outer areas. From the mass of spiral galaxies of late types they differ mainly by higher luminosity of the bulge, lower contrast of the spiral arms or the lack thereof, a very low surface density of the gas (HI) and, consequently, extremely slow star formation.

There are several problems associated with the explanation of the observed features of the disk galaxies of early types. First of all, it is unclear whether the lenticular galaxies are a logical continuation of the morphological sequence of galaxies Sd–Sa, reflecting the conditions of their formation and subsequent character of 'quiet' evolution, or their peculiarities — this is the result of interaction of the galaxies with the environment (merging, accretion of small satellites, the loss of gas associated with the pressure of the environment). Among lenticular galaxies, there are indeed, many objects, the structure which suggests an effective external action (e.g. dynamically and chemically isolated circumnuclear disks, features of the radial profile of brightness and such unusual details of the structure as a polar ring). S0-galaxies in rich clusters apparently form due to the effect of intergalactic gas on the interstellar medium that directly or indirectly leads to a reduction of their number and 'stop' of active star formation. Obviously, the effect of the environment plays a key role, as evidenced by a lower percentage of galaxies of early types in distant clusters (Butcher–Oemler effect [435]), and also by lower rates of star formation in galaxies which are in a denser environment [497]. However, among the many galaxies of early types there are also many field galaxies, whose history can be different.

No less interesting is the problem of the gas content in lenticular galaxies. Even in cases where HI is present in detectable quantities, its gross mass is at least an order of magnitude less than one would expect as a result of a simple return to the interstellar medium of gas discharged by the pre-evolution stars of the disk [1103]. Note that the paucity of information about the thickness of stellar disks of lenticular galaxies of early types makes it difficult to compare with other galaxies on the basis of the volume of gas density or gas pressure in the plane of the disk.

The relative fraction of the dark halo in disk galaxies of early types also remains an open question. This is due primarily to complexities of assessing rotation curves at large distances from the centre of the galaxy on the baiss of the stellar absorption lines. For a small number of lenticular galaxies sufficiently long rotation curves of neutral hydrogen have been determined. According to to [972, 974] the velocity of rotation of these galaxies is often reduced to the periphery, but nevertheless the rotation curves require the presence of a sufficiently massive dark halo.

The large scatter of points on the Tully–Fisher charts (luminosity – rotational speed) for the lenticular galaxies indicates the significant heterogeneity of their properties (see, e.g. [350] and references therein) This is also evidenced by the analysis of the velocity dispersion of old stars in disks of galaxies. It shows that in parts of the disk galaxies of early types the velocity dispersion of stars significantly exceeds the minimum value required for the gravitational stability of the disk, while for spiral galaxies of later types the dispersion of the velocities of disks of the stars is apparently close to the threshold value [77]. However, the relatively low accuracy of the estimates of the velocity dispersion of stars o fthe disks outside of bright bulges and complexity of the division of the dispersion into r, φ, z, components make this conclusion only provisional.

A comparison of observational data with dynamical models, in which the velocity dispersion of the disks of the stars – both in the disk plane and in the perpendicular direction – is close to the critical values for its dynamic stability, gives an indication about the features of the dynamical evolution of the disk. When comparing the model c_ℓ and the observed values of the velocity dispersion c_{obs} of old stars of the disk along the visual ray, one of three situations can be realised.

a) $c_{obs} < c_\ell$. Based on the concept of stability of the disk, we conclude that the mass of the disk, adopted in the model, is overestimated, and to satisfy the condition of dynamic stability with respect to perturbations in the disk plane and bending perturbations it should be 'lighter'.

b) Values of c_{obs} and c_ℓ within the measurement error range are the same. In this case, we can conclude that the dynamic instabilities in the formation of the bulk of the disk 'transferred' the stellar disk to a marginally stable state, where the velocity dispersion of stars is determined by the surface density of the disk, its rotation velocity in relation to the distance r, and the internal structure. Among the non-interacting spiral galaxies, this option seems to occur most often (see for example, [77].

a) $c_{\rm obs} < c_\ell$, i.e. the observed dispersion of the velocities of stars above the model values corresponding to marginally stable disk. In this case, there is reason to assume that the stars in the disk in the process of evolution acquired a surplus (for the stability of the disk) energy of random motions, so that the disk was dynamically overheated. This could serve as evidence that the stellar population of the galaxy disk subsystem has been subjected to strong gravitational perturbations, for example, as a result of absorption of massive stellar or gas satellites, or in close interaction with close neighbours. In the inner region of the galaxy dynamical heating, in principle, can additionally be linked to the destruction of the contrast stellar bar.

To build dynamic models of galaxies, which provide estimates of the relative weight of disk/halo or the thickness of the disk, one must have estimates of the velocity of rotation and the dispersion of velocities of the stars at a greater distance from the centre, preferably on two axes of the galaxies, so it was possible to determine the velocity dispersion in radial and vertical directions .

Figure 4.36 shows the results of the construction of dynamic models for S0-galaxies NGC 1167, NGC 4150, NGC 6340, for which the observational data on the distribution of velocity dispersion in a disk was used [68]. The velocity dispersion along the line of sight according to the observational data was compared with the velocity dispersion of the maximum disk model. As one can see, all the disks of the three S0-galaxies are 'hot' even at the maximum permissible mass. This can be regarded as an argument in favour of assuming that the galaxies in the past were subjected to external gravitational perturbations, apparently, as a result of a closer interaction with nearby galaxies or absorption of sufficiently massive satellites. The effectiveness of the latter process is confirmed by numerical experiments (see, for example, [309]). The gravitational perturbation of the gas component of the disk can at the same time be a reason why the galaxies quickly lost their interstellar gas supply because of a surge in star formation, resulting the properties characteristic of lenticular systems.

For comparison, Fig. 4.36 shows the results of modelling for the SBa-galaxy NGC 2273, in which the stellar disk may be overheated but not much. We can consider the compatibility of the observed estimates with the assumption that the stellar disk is not subjected to dynamic heating and the velocity dispersion of stars in it corresponds to the minimum value that ensures its equilibrium quasi-steady state. In this NGC 2273 is similar to many spiral galaxies of later types, where the

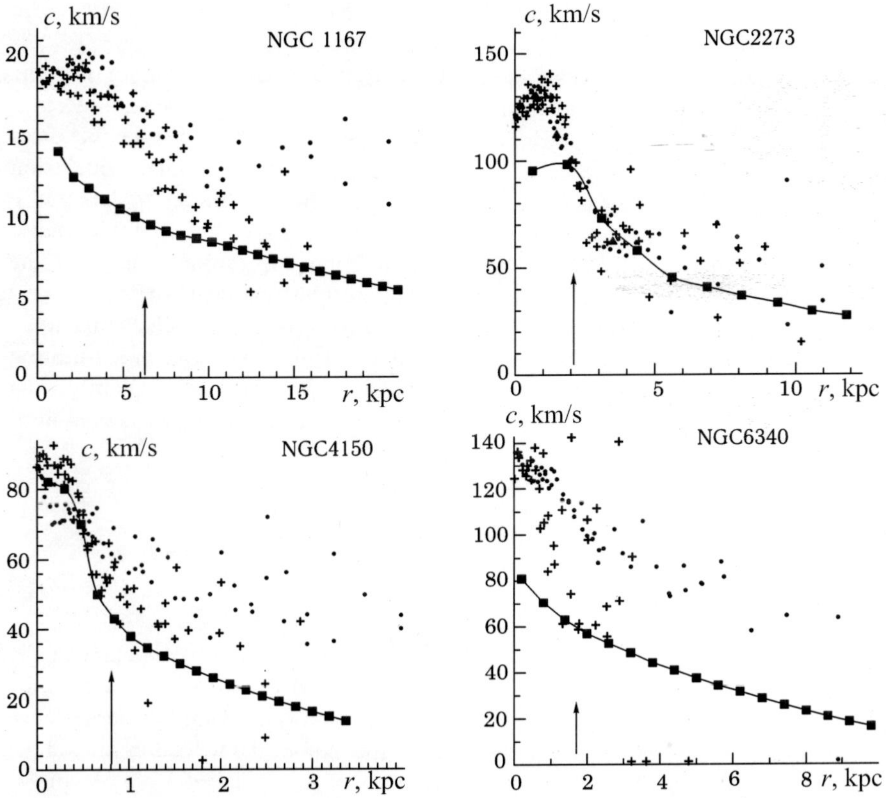

Fig. 4.36. Dispersion of the velocities for NGC 1167, NGC 2273, NGC 4150, NGC 6340. The crosses – dispersion of the velocities of the stars c_{obs} according to the observational data along the major axis, the points – along the minor axis, the lines (squares) – dispersion of the velocities of the stars of the disk along the visual ray in the model of the maximum disk along the major axis. The vertical arrow corresponds to the conventional boundary of the bulge.

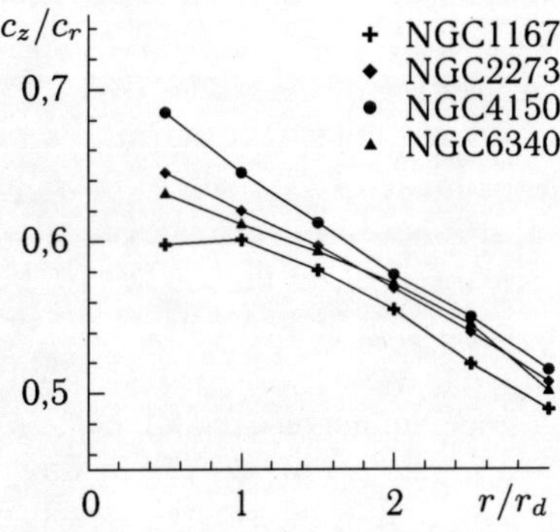

Fig. 4.37. Radial distribution of the ratio c_z/c_r in the numerical models of the maximum disk situated at the boundary of gravitational stability.

velocity dispersion of the stars over a long distance on the disk is apparently close to the threshold value.

For all four galaxies in the models of the marginally stable maximum disk, the components c_z and c_r, as well as their ratio c_z / c_r (Fig. 4.37) decrease with increase of the distance from the centre. The decrease of c_z fast enough so that the thickness of the disk remains constant: in all cases it increases with r. The same conclusion follows for the actually measured values of the velocity dispersion of stars of the considered galaxies. From the analysis of the velocity dispersion along the major and minor axis the average values of c_z / c_r were obtained. It turned out that in the range (1–2) $L\sigma$, this ratio is 0.6–0.8 and slightly decreases at the periphery.

Apparently, the low rate of star formation at present is due to such events in the history of these galaxies, which simultaneously led to strong dynamical heating of the stellar disk and to higher rates of exhaustion of gas and, consequently, high colour indices. It can be assumed that it is due to tidal influences or accretion of small stellar systems.

4.5. Bending instabilities of the stellar disk

4.5.1. The vertical structure of the stellar disks and bending instabilities

The other possibility of evaluating the mass of the halo is associated with the application of the observed thickness of the stellar disk when constructing the dynamic models (see also section 10.3). This approach is based on the dependence of the vertical scale of the disk (at the given dispersion of the velocities c_z) and the local surface density of the disk and on the mass of the spheroidal subsystem M_s. This was reported for the first time in [63] (see also [771, 911, 1243]).

The problem of the mechanism of increasing the dissipation of random velocities of the stars c_z from 6÷10 km/s in young stars up to ~ 30÷100 km/s for the old population remains one of the fundamental problems in the physics of galaxies [1035]. In addition to the mechanism of vertical heating, associated with the scattering of stars on the giant molecular clouds [1195, 1196], and on the spiral density waves [723], it is also important to consider the possibility of increase of the dispersion c_z as a result of the instability of the bending perturbations in the disk. Estimates of the ratio c_z/c_r using the observational data for different galaxies give $c_z/c_r \simeq 0.4$–0.8, but more typical values are 0.5÷0.7 [803].

Evidently, the dominant mechanism of vertical heating of the stellar disk may change in different periods. Matching the observational data with the results of dynamic modelling of the galaxies, we obtain the estimate of the mass of the halo which is the lower boundary because there are also other mechanisms of heating of the disk which are not considered in the numerical model. In the models of the N-bodies we examine different aspects of bending of the stellar galactic disks. Important results were obtained in models taking into account tidal interactions [673, 891, 1262, 1290]. The satellite generate bending and increase the thickness of the disk, as indicated by comparison of the observational data of bending in single galaxies and in groups [1061]. The formation of a disk from the edge of box-like structures in the central region as a result of bending instability was observed [1016, 1145]. The conditions of formation of the bending moments were investigated in [225, 375, 640, 1053, 1138]. The effect of the bulge was investigated in [1191]. The bending instability may develop at a considerably lower rate in comparison with the gravitational instability in the disk plane so that the characteristic times of saturation are of the order of billion of years or more [225].

Initially, we investigate the problem of the radial distributions of the ratio c_z/c_r and the vertical scale of the disk which ensure at least the stability against different bending perturbations. Since in the galaxies visible edge-on the velocity of rotation in the central region is not measured accurately (section 2.7), we examine only the galaxies of later types in which a bright bulge does not exist. Consequently, we can restore the internal part of the rotation curve if it is determined inaccurately from the observations, assuming that the shape of the curve of circular velocity of the galaxies in the central region ($r \leqslant 2L$) is determined mainly by the disk component. In addition, the presence of the bulge with other conditions being equal plays a stabilising role for the bending instability, weakening the dependence of the relative thickness of the disk h/L on the relative mass of the halo M_h/M_d [1191].

In [771, 1243], the authors constructed the dynamic models for seven galaxies (in particular, UGC 6080, UGC 9556, UGC 9422, NGC 4738], visible edge-on, with the available photometric scales of distribution of brightness in r and z directions. Of special interest is the determination of the parameters of the subsystems for the superfine Galaxy UGC 7321. In addition to the previously mentioned galaxies, the galaxies NGC 5170 and NGC 891 were also investigated. In addition to the radial and vertical skills, the radial dependences of the dispersion of the velocities of the stars are also available for these galaxies so that it is possible to estimate the mass of the components for these galaxies by two independent methods – comparing the modelling results with the observational data on the dispersion of the velocity of the stars and the thickness of the disk.

The observed ratio of the radial scale of the distribution of brightness (assuming the constancy, along the radius, of the ratio of mass to luminosity) and the density of the galactic disks to the vertical scale is in a relatively wide range $L/z_0 = 2 \div 12$ [418, 636, 639, 805–807]. This large scatter of the values may mean that the relative mass of the halo μ inside the optical radius also greatly varies from galaxy to galaxy.

The ratio c_z/c_r = const is used for the entire old stellar disk [413, 414]. This conclusion is very approximate, and dynamic modelling shows that c_z/c_r changes with the radius (section 4.5.2). The vertical scale of the stellar disks depends on the dispersion of the vertical velocities $c_z(r)$ (see section 4.1.2), and c_z itself is not independent of the dispersion of radial velocities.

Evidently, the problem of the ratio of the dispersion of the vertical velocities of the stars c_z to the dispersion of the radial velocities c_r was investigated for the first time by Toomre [1230]. The dynamics of bending (membrane) oscillations in the models of cold thin disks was investigated by Hunter and Toomre [700]. This investigation was carried out to explain the large-scale distortion of the periphery of the disks observed in a number of isolated galaxies. Another problem which required investigation of the dynamics of bending perturbations was the problem of explaining the large difference in the thicknesses and z-dispersions of the velocities of the objects of the stellar and gas disks. In [811, 877] it was shown that to solve these problems it is important to investigate the dynamics of bending perturbations in the model of the disks, hot in their plane.

Polyachenko and Shukhman [207] carried out studies using a simple model of a uniform thin self-gravitating layer and observe that to stabilise the relatively small-scale bending perturbations, it is necessary to fulfil the condition $\alpha_z \equiv c_z/c_r \geqslant 0.37$ (see section 2.6.1). Stabilisation of the bending waves is achieved at a specific value of c_z to which the minimum thickness of the stable disk corresponds. Araki [307] investigated the dynamics of the bending oscillations taking into account the non-uniform distribution of the volume density along the z-coordinate. The conclusion on the small critical value $c_z/c_r \gtrsim 0.3$ remained unchanged (see discussion in [906]].

4.5.2. Numerical modelling of bending instabilities

What is the reason for the formation of thin stellar disks? The values of the half thickness of a stellar disk are in a wide range $\Delta = 0.1 \div 1$ kpc. There is a relatively large number of edge-on galaxies with the ratio of the visible half-axes of $\chi = b/a \leqslant 0.1$. The relative thickness of the stellar disk can be naturally determined by the ratio $\chi_0 \equiv \Delta/L\sigma$. For the distribution of volume density $\varrho = \varrho \exp(-r/L\sigma)/\mathrm{ch}^2(z/\Delta_*)$, it can be accepted that $\chi/\chi_0 > 0.5$ [63]. At the same time, for the thin stellar disks ($\chi \leqslant 0.1$) it is assumed that $\chi_0 < 0.2$. The conditions in which the stationary stellar disks can have such thickness will be discussed.

The stationary system should be, firstly, gravitationally stable in relation to the perturbations situated in the plane of the disk. This imposes restrictions on the dispersion of the radial velocity $c_r \geqslant Q(r)\, c_T$. Secondly, the stability of the relatively small-scale bending perturbations requires that the condition (section 2.6.1) $\alpha^{-1} = c_z / c_r \geqslant \alpha_{\mathrm{cr}}^{-1} \simeq 0.4$ is fulfilled. In turn, the half-thickness of the disk Δ depends on the value c_z (see (2.1.4)). Assuming that the system is situated at the boundary of stability, we obtain

$$\Delta = \left(\frac{Q}{\alpha_{\mathrm{cr}}}\right)^2 \frac{3.6\, G\sigma_{*0}(r)}{\varkappa^2}. \tag{4.125}$$

The circular velocity V_{rot} in the disk is determined by the distribution of matter in planar and spherical subsystems and, consequently, we can write the following equation

$$V_{\mathrm{rot}}^2 = V_d^2 + V_h^2. \tag{4.126}$$

For the exponential density profile (1.3) the mass of the disk $M_d = 2\pi\sigma_{*0}(0) \cdot L_\sigma^2$ and in the range $r \gtrsim 2L\sigma$ the velocity of rotation is close to its maximum value $V_d^2 = 0.388\, GM_d / L_\sigma$. For the circular velocity of the halo in the vicinity of the external boundary of the disk, i.e. at $R_{\max} = 4\,L\sigma$, we have $V_h^2 = GM_h / 4L_\sigma$ (M_h is the mass of the halo inside $r \leqslant R_{\max}$). Consequently, for the velocity of rotation we have

$$V_{\mathrm{rot}}^2 = V_d^2 (1 + \mu / 1.55). \tag{4.127}$$

For the relative thickness of the disk in the case of a flat rotation curve $\varkappa = \sqrt{2}\Omega = \sqrt{2}V_{\rm rot} / r$ we have [63]

$$\chi_0(\mu) \equiv \frac{\Delta_*}{L_\sigma} = K\left(\frac{Q}{\alpha_{\rm cr}}\right)^2 \frac{\exp(-r / L_\sigma)(r / L_\sigma)^2}{1 + \mu / 1.55}, \tag{4.128}$$

where $\mu = M_h/M_d$, $K = 0.74$. Equation (4.128) shows the reduction of the value χ_0 with the increase of the relative mass of the halo μ. It is also important to note that the value α_z at the stability boundary in relation to the bending oscillations depends only slightly on μ and decreases with increasing ratio M_h/M_d. This is due to the fact that the massive halo is a stabilising factor for bending instability. To clarify these relationship, we return to the results of numerical experiments.

The large-scale axisymmetric bending instability

The non-linear dynamics of the bending oscillations with the largest scale will be considered initially. The linear analysis of the stability of the global bending perturbations was carried out in [208] (a more general case was considered in [1256], see section 2.6.3).

The evolution of the cold (in the vertical direction) systems with the low initial ratio $\alpha_z = c_z/c_r$ (cold in the vertical direction) shows clearly the increase with time of the large-scale bending oscillations which leads to vertical heating and, consequently, increase of the thickness of the disk. In particular, strong heating of the disk in the vertical direction, leading to a large increase of the disk thickness, is obtained as a result of the development of an unstable axisymmetric bending mode ($m = 0$). The axisymmetric mode is the most important phenomenon of the bending instability. Figure 4.38 shows the time dependence of the vertical coordinate of the centre of gravity of matter $\zeta\ (r, t)$, the dispersion of the vertical velocities $c_z\ (r, t)$, the vertical scale of the disk $z_0\ (r, t)$, the ratio of the dispersion of the velocities $\alpha_z \equiv c_z\ (r, t)/c_r(r, t)$, the dispersion of the radial velocities $c_r\ (r, t)$, and the velocity of rotation of particles in the disk $V\ (r, t)$.

During approximately 2.5 revolutions ($t \sim 10$) at the external edge of the disk its parameters remain almost constant (Fig. 4.38), and the intrinsic modes form in the linear stage of development of bending instabilities. After $t \gtrsim 10$ the non-linear stage of development of bending instability starts, and the amplitude of displacement of the centre of gravity in the z-direction $\zeta(r)$ increases (Fig. 4.38a). In the central zone of the disk (lines 1–5) the amplitude of the vertical displacement reaches the maximum value more rapidly and subsequently decreases to the initial value (Fig. 4.38a). The variation of ζ at the periphery of the disk (lines 11–19) is slower.

Slightly later, with a delay with respect to the value $\zeta(r)$, the dispersion of the z-velocities c_z (Fig. 4.30b) and the vertical scale of the disk z_0 (Fig. 4.38c) start to increase rapidly. The reduction in the thickness of the system, expressed in the increase of z_0, starts in this model in the central region and rapidly spreads over the entire disk. The value α_z greatly increases as a result of increase of c_z and a small reduction of c_r (Fig. 4.38d,e). The identical behaviour is shown by the dispersion

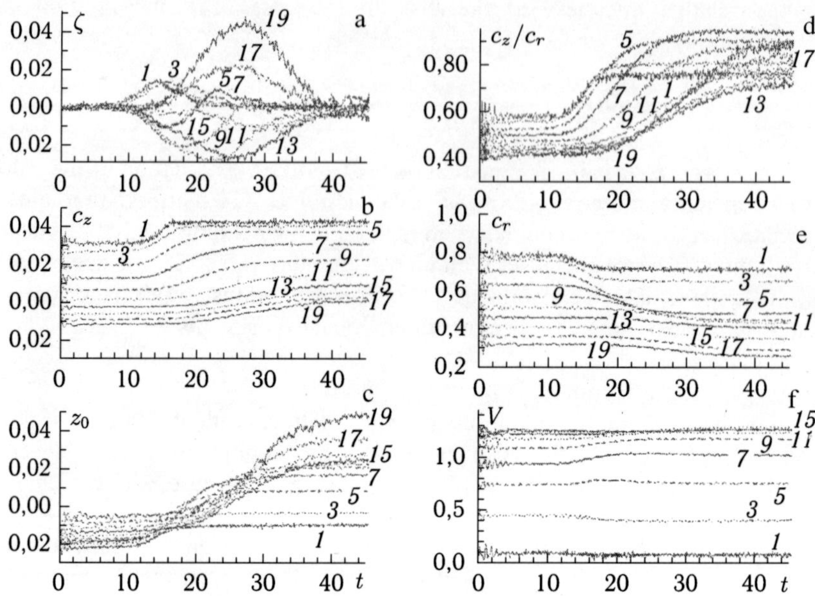

Fig. 4.38. Time dependence of the parameters of the disk in the process of development of an axisymmetric bending mode in the disk in the model with $\mu = 1$, $a = L$: a) the vertical coordinate of the centre of gravity of the matter of the disk ζ; b) the dispersion of the vertical velocities of the particles c_z; c) the vertical scale of the disk z_0; d) the ratio of c_z to the dispersion of the radial velocities c_r; e) c_r; f) the velocity of rotation of particles of the disk V. The individual curves are given for different radii r_j, the numbers indicate the value of the parameter j which determines the distance from the centre of the disk: $r_j = (0.0\ 25 + (j–1) \cdot 0.05) \cdot 4L$. Other parameters are averaged oved the the azimuthal angle.

of the azimuthal component of velocity c_φ for which the approximate equality $c_\varphi \simeq c_r \varkappa / 2\Omega$ is fulfilled[10].

The decrease of c_r/c_φ is determined by the transition of the kinetic energy of the random motion in the disk plane to the kinetic energy of vertical motion. It should be stressed that the development of this pending mode takes place at the background that is axisymmetric with respect to all parameters of the disk.

The value ζ has different signs in the centre and at the periphery of the disk (Fig. 4.38a) and, consequently, the development of instabilities is accompanied by the formation of a 'cupola' structure (Fig. 4.39). The isolines of the parameter ζ are close to circles (Fig. 4.39a). At the time $t = 25.4 = 6T_0$ the internal ring and periphery of the disk are displaced to opposite sides, and the central region ($r \leqslant 0.4$) has already returned to the initial state (Fig. 4.39b). The distribution of parameter ζ in the disk plane has the form of a 'sombrero'.

---

[10]This condition starts to be violated only in very thick disks or in regions in which $z_0/L \gtrsim 0.4$. In this case, $c_r/c_\varphi < 2\Omega/\varkappa$, i.e. the degree of anisotropy decreases. This is natural as a result of spherisation of the system).

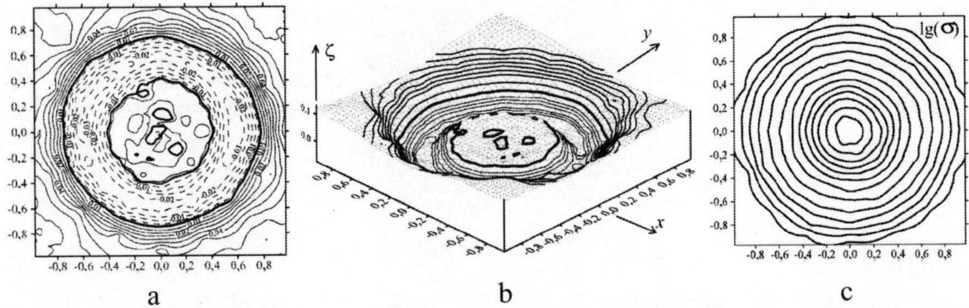

Fig. 4.39. The form of the disk in the process of development of bending instability at the time $t = 25.4$ for the calculations shown in Fig. 4.38. The graph shows the isolines of the distribution of displacement of the centre of gravity of the disk in the vertical direction $\zeta(r, \varphi)$ in the disk plane (a) and the function $\zeta(r, y)$ (b). The thicker line corresponds to the value $\zeta = 0$. For the same moment of time the graph shows the isolines of distribution of the logarithm of surface density which show the presence of axial symmetry (c).

Fig. 4.40. The edge-on image of a disk in the stage of the global axisymmetric bending mode in the model with $\mu = 4$. The scale in the vertical direction is four times greater than in the horizontal direction.

The structure of the disk in the stage of development of the bending moment $m = 0$, visible edge-on, is shown in Fig. 4.40. It has a characteristic box-like form which is typical of all the models in the stage of dominance of the axisymmetric bending mode. It should be stressed that the characteristic box-like structure is not associated with the bar, as assumed in the explanation of the observational data in a number of cases [429, 1137].

The evolution of the dispersion of the velocities c_z, c_r and the velocity of rotation of the stellar disk V starts when the amplitude of vertical displacement ζ becomes very high. The vertical heating of the disk is characterised by the increase of the ratio c_z/c_r but, starting from some values $\alpha_z = c_z/c_r$, the conditions are no longer suitable for the development of bending instability. Consequently, the system is transferred to a new steady state with a thicker disk. The characteristic duration of this process depends strongly on the parameters of the model, and taking into account the linear stage, it may be equal to tens of revolutions at the external edge of the stellar disk. If the distribution $\alpha_z(r)$ is sub-critical, then as a result of the small values of the increment of instability the linear stage increases and the intensity of heating in the linear stage is low. Thus, the amplitude of the bending moment and the very possibility of its development depend primarily in the initial radial distribution c_z/c_r.

In the models with a low mass halo ($\mu \leqslant 1$), if the disk is sufficiently hot, to ensure gravitational stability with respect to the bar mode, the main factor, heating initially the thin disk, is the axisymmetric bending moment, and the modes $m = 1, 2$ usually do not form.

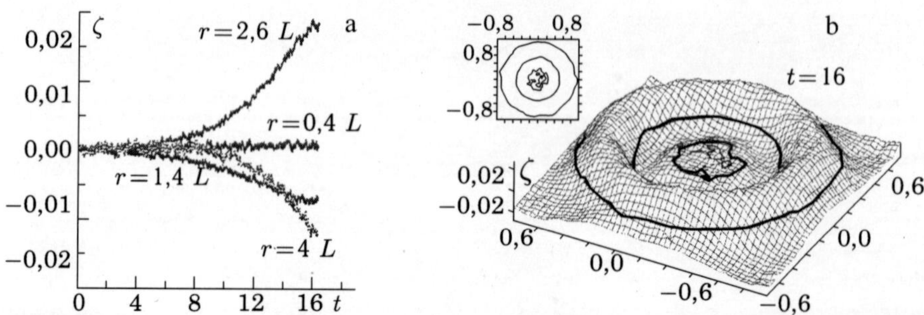

Fig. 4.41. In this model (μ = 4), the initial conditions ensure stability of the central region and the bending mode m = 0 develops in the zone $r \gtrsim L$: a) the time dependence of the vertical displacement of the disk in the initial stage for different distances from the centre; b) the form of the disk (the centre of gravity on the z-coordinate). The thick line corresponds to ζ = 0.

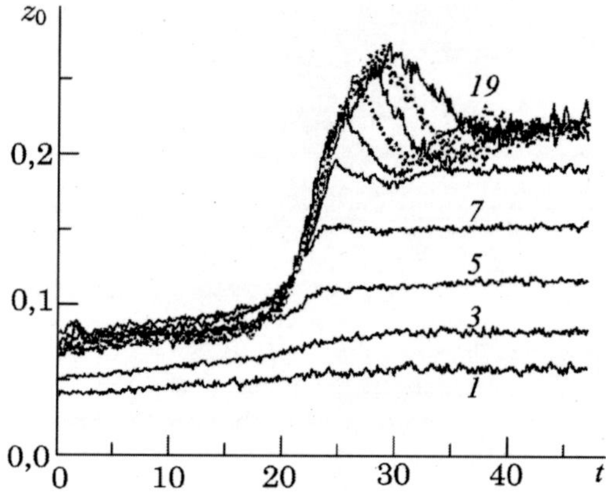

Fig. 4.42. Thickening of the disk without the halo in the presence of a bulge (M_b = 0.20 5 M_d, b = 0.2 L) as a result of the development of the mode m = 0. In the region of the bulge the vertical scale of the disk z_0 increases only slightly. The numbers in the figures – see Fig. 4.38.

It should be mentioned that the formation of the mode m = 0 does not start in the centre and starts at the periphery ($r \gtrsim L$) and does not even penetrate into the central zone of the disk. This is possible in two cases: 1) at the initial moment of time the ratio $c_z / c_r \geqslant \alpha_z^{\text{crit}}$ in the centre ensures stability, and in the external region the disk is thin and unstable (Fig. 4.41); 2) in the presence of a sufficiently massive or concentrated bulge. In these models, the bending modes in the centre have a small amplitude but the periphery thickness can greatly increase (Fig. 4.42). On the whole, the bulge plays a stabilising role, like the halo. Therefore, with other conditions being equal, the disks of the galaxies with the bulge can be thinner.

Taking into account the 'catastrophic' consequences of the development of the axisymmetric bending mode (the increase of the thickness of the disk may reach the level $z_0/L \simeq 0.3$–1), it is evident that the majority of the galaxies have not passed through this stage because this stage produces very thick disks. For example, in the model shown in Fig. 4.38 with $\mu = 1$ on average we have $\langle z_0 \rangle > /L \simeq 0.4$. Even in the presence of the bulge, the disk periphery can become thicker up to $z_0 \leqslant 0.4L$ is the result of the development of the bending instability $m = 0$ (see Fig. 4.42). In the case of a more massive halo ($\mu = 4$), the disk heated by the bending axisymmetric mode is characterised by a smaller relative thickness $\langle z_0 \rangle / L \simeq 0.16$. However, observation show the existence of galaxies with $\langle z_0 \rangle / L \leqslant 0.15$. Therefore, it should be assumed for these objects that this is caused either by a very massive halo ($\mu > 4$) or that the axisymmetric bending mode did not heat the disk. It should be mentioned that this mechanism can result in the formation of a central increase of the thickness (of the bulge type) (see Fig. 4.40) and the lens in S0 galaxies.

The bending modes $m = 1, 2$

In the presence of more massive halo ($\mu \gtrsim 2$) in the vertical heating of the disk non-axisymmetric bending modes $m = 1$ and $m = 2$ can take part in the initial stage of evolution. The axisymmetric distribution of matter does not change in this case. Figure 4 .43 shows the evolution of the initially thin disk with $\mu = 4$. In this case, in contrast to the model with $\mu = 1$ (see Fig. 4.38), in the initial stage in the central region of the disk ($r \leqslant 2L$) the bending moment $m = 2$ is the first to form (the 'saddle' type). It should be mentioned that in the framework of linear analysis, the mode $m = 2$ can be unstable in the presence of a massive halo [208]. The characteristic time of increase of the perturbations and vertical heating may greatly differ in the

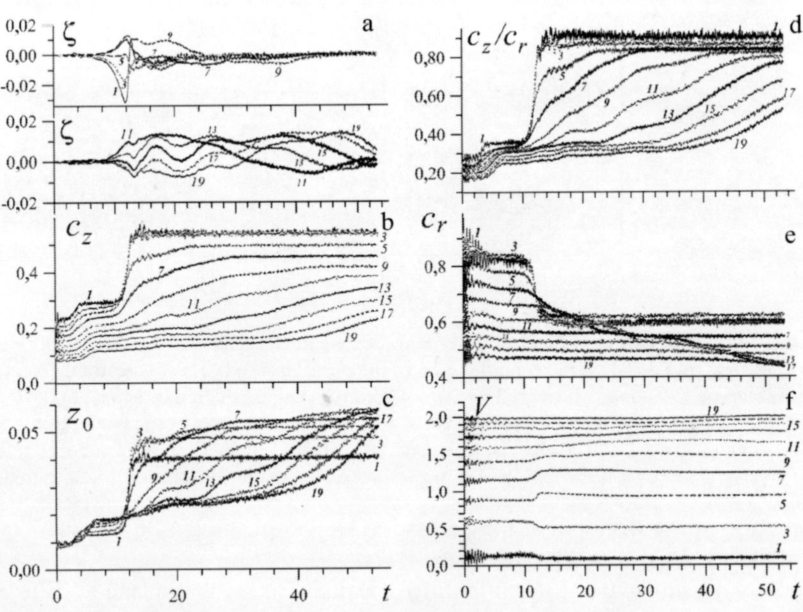

Fig. 4.43. The same as in Fig. 4.38 but for the model with $\mu = 4$ in the zone $\gamma < 8/3\ L$ (curves 1 – in the centre of the disk, 19 – at $r = 8/3\ L$). Vertical displacement ζ for $r < 4/3\ L$ and the zone $4/3\ L < r < 8/3\ L$ is shown in separate graphs (a and b, respectively).

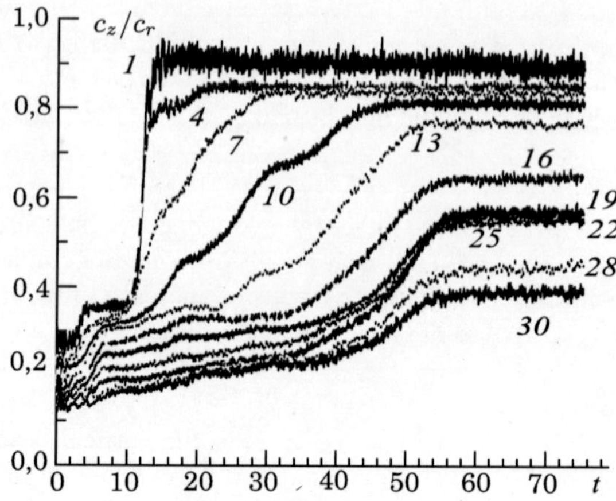

Fig. 4.44. The same as in Fig. 4.43e but for the entire disk (1 – centre, 30 – at $r = 4L$) and at long times. After $t \gtrsim 54$ the quasi-stationary state is established.

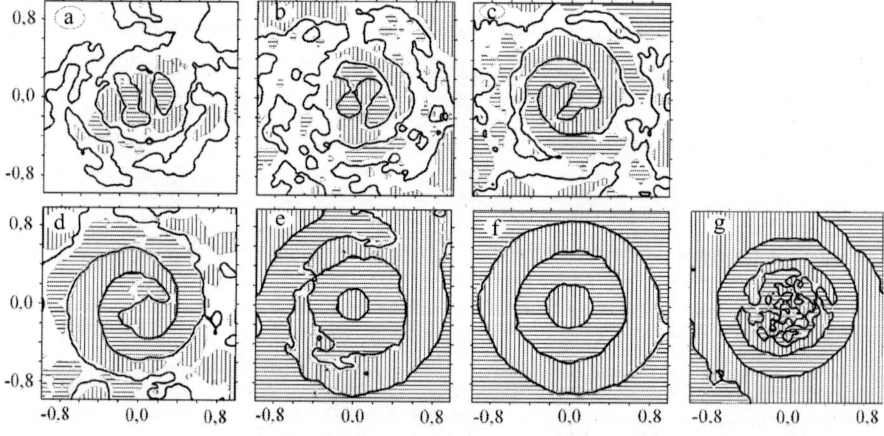

Fig. 4.45. Structure of the vertical displacement $\zeta(x, y)$ in the plane of the disk at different moments of time: a) $t = 2.7$, in the central region of the mode $m = 2$; b) $t = 3.1$, rearrangement of the modes starts $m = 2 \rightarrow m = 1$; c) $t = 3.6$; d) $t = 4.2$, the developed single-arm mode; e) $t = 10.1$, rearrangement of the mode $m = 1 \rightarrow m = 0$; f) $t = 1.4$, developed axisymmetric mode; g) $t = 51$, in the very centre at $r \leqslant L = 0.25$ the displacement has almost completely stopped and this is followed by axisymmetric perturbations, and at the periphery, the mode $m = 2$ has a low amplitude).

central zone of the disk and at the periphery. However, the final quasi-stationary state is also established in this case (Fig. 4.44).

Figure 4.45 shows the distribution of the value ζ in the disk plane at different moments of time for the model shown in Fig. 4.43. The vertical heating and, correspondingly, increase of the thickness of the disk $m = 2$, are small. However, $t \gtrsim 3$ (the period of rotation of the outer edge of the disk $T_0 = 2.7$) the nonlinear stage of the single-arm mode $m = 1$ (asymmetry) forms and, consequently, the intensity

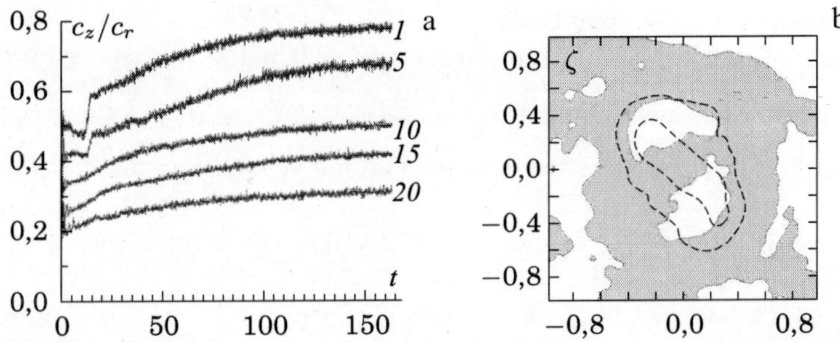

Fig. 4.46. The time dependence of the ratio of the dispersions c_z/c_r in the model with a bar at different radii; b) the structure of the bending moment in the process of formation of a bar (t = 15). The thin solid line corresponds the vertical displacement $\zeta = 0$, broken lines are the isolines of bar density. The shaded region indicates the regions $\zeta > 0$ and $\zeta < 0$. Distortion of the bar is the reason for rapid vertical heating in the central zone (lines 1, 5 in Fig. a and in the range $t = 10 \div 20$).

of vertical heating rapidly increases. Finally, the third stage of heating starts at $t \gtrsim 10$ and is associated with the buildup of the axisymmetric mode $m = 0$. In this stage, the increase of the dispersion of the velocities c_z is especially large, and the thickness of the disk increases by a factor of 2–3. Thus, the investigated model demonstrates the process of rearrangement of the bending modes and the successive transition from the wave $m = 2$ to the axisymmetric mode.

As in the case of the low-mass halo, with an increase of the initial ratio $\alpha_z = c_z/c_r$ (correspondingly, the increase of the vertical scale z_0), the conditions for global bending instability become less favourable and, starting at some distributions $\alpha_z^{\rm crit}(r)$ the distortion does not form over at least 20 periods of rotation of the disk.

It should be mentioned that in the process of development of bending instability for the modes $m = 1, 2$, the disk is not quasi-equilibrium in the vertical direction in the sense of fulfilling the balance (2.56). The distribution $c_z(r; \varphi)$ remains axisymmetric, and z_0 becomes non-axisymmetric. Thus, the condition $c_z^2 / z_0 = \text{const}$ is not fulfilled. For the axisymmetric mode, the ratio $c_z^2 \propto z_0$ is fulfilled far more efficiently, with the exception of the very central region when the disk thickness rapidly increases.

Heating in the process of bending of the bar

A model with not a very massive halo ($\mu \leqslant 1.5$) will be discussed. If the disk is gravitationally unstable at the initial moment of time, a bar can form (see section 4.2). In the formation of the bar in the initially thin cold disk, the instability of the global bar mode in the initial stage may lead to the formation of vertical bending of the bar [1053]. Figure 4.46 shows the bending of the bar whose dynamics results in an increase of the dispersion of vertical velocity with time. The amplitude of distortion of the bar decreases quite rapidly with time and with the increase of the thickness of the bar. It should be stressed that the produced bar prevents in further stages of the development of global bending modes and, above all, of the axisymmetric mode $m = 0$.

Small-scale bending perturbations

In the case of suppression of the large-scale bending perturbations, the most important mechanism of collective heating is the instability of the small-scale bending waves. The amplitude of these oscillations is small: $|\zeta| \ll z_0$ and they are not correlated. The increase of the dispersion of the velocities c_z and disk thickness is very slow: $\delta = \left(\frac{T_0\, dz_0}{z_0 dt}\right)^{-1} \sim 100$. The rate of growth depends only slightly on the radial coordinate which again confirms the local mechanism of heating. This differs from the previously examined global bending instability for which the relative time of heating δ strongly depends on r, and for the models in Fig. 4. 38, 4.43 $\delta \sim 1 \div 10$.

If the initial state permits the development of a bending global instability, the final disk after establishment of the quasi-stationary state is stable in relation to the small-scale bending perturbations. This is indicated by the absence of growth of c_z and z_0 with time in the models shown in Fig. 4.38, 4.44 (starting at some moment of time).

The c_z/c_r ratio

With the conditions being equal, the key parameter which determines the stability of the disk is the ratio c_z/c_r [771]. If the mass of the spherical subsystem is not too large in comparison with the mass of the disk, to stabilise the bending perturbations the value c_z/c_r should be considerably higher than the values 0.3–0.37 determined in linear analysis of simple models ([207, 307, 906], section 2.6).

As an example, we consider the model without the bar with a low-mass halo without the bulge and with $\mu = 1$ (Fig. 4.38). Figure 4.47a shows the initial (line 1) and final (lines 2) distributions $\alpha_z(r)$. Vertical heating as a result of the axisymmetric bending mode in this case is so strong that on average throughout the entire disk $\langle c_z/c_r \rangle = 0.82$. The non-monotonic nature of the final distribution $\alpha_z(r)$ in Fig. 4.47a is caused by the initial state (in particular, c_z/c_r), and under different conditions the

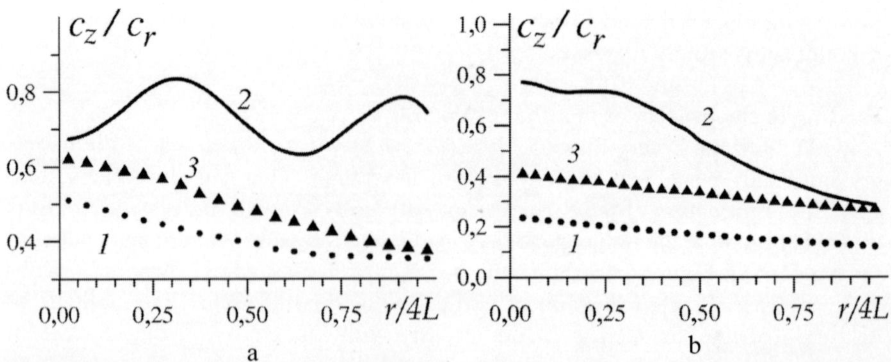

Fig. 4.47. Radial distributions of $\alpha_z = c_z/c_r$ for $\mu = 1$ (a), and $\mu = 4$ (b): 1) initial dependence, 2) final distribution, 3) critical level α_z.

dependence can be monotonic. If at the initial moment of time the parameter $\alpha_z (r)$ is defined in accordance with the line 3, the bending modes are stable in this case.

With the increase of the relative mass of the halo μ the critical value of the ratio c_z/c_r decreases. Figure 4.47b shows the initial (line 1) and final (line 2) distributions $\alpha_z (r)$ in the case $\mu = 4$. If we specified the parameter $\alpha_z (r)$ at the given moment of time in accordance with line 3, then in this case the disk is stable.

The dependence of parameter c_z/c_r at the stability boundary on the radial coordinate is important. In the case of a disk with a low-mass halo ($\mu \leqslant 1$) at the dispersion $c_r (r)$, which results in the suppression of the bar mode, to stabilise the bending instability the value c_z/c_r monotonically decreases with the radius: the typical values in the central zones are 0.7÷0.8 and at the disk periphery 0.4÷0.5 (Fig. 4.47).

We will discuss the relationship of α_z^{cr} which determines the boundary of stability, for the small-scale (SSM) in global bending modes (GBM). If the initial disk is very thin, the global bending instability is a considerably more effective mechanism of vertical heating. However, the disk is stable in relation to the GBM at lower values of the parameters $\alpha_z (r)$ than the small-scale perturbations. The disk can be stable in relation to the GBM but and stable in relation to the SSM. The latter acts as an additional factor of vertical heating. Since this mechanism and results in very slow heating, the final as in the case of development of the global axisymmetric mode.

It is important to note that the final distribution of parameters $\alpha_z (r)$, obtained as a result of the development of global bending instability, depends on the initial function $\alpha_z (r, t = 0)$. The models with the low initial vertical scale (thin disks), heated as a result of global bending instability, are characterised by a higher value of c_z/c_r than the minimum value required for stability. This is associated with the highly non-linear heating mechanism. As a result, the system 'jumps' over the boundary of instability with respect to the parameter c_z/c_r (see Fig. 4.47).

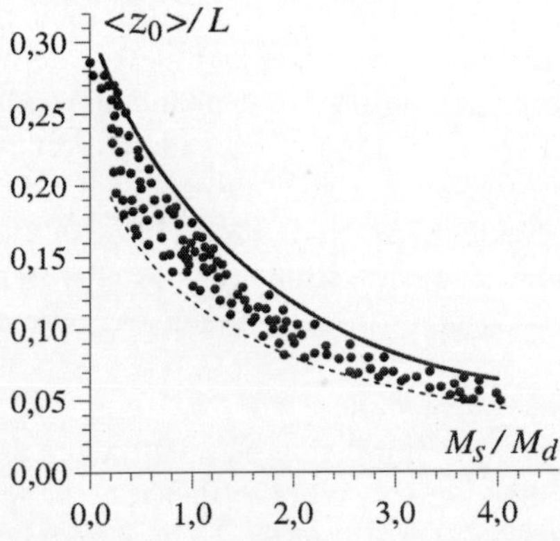

Fig. 4.48. Different symbols indicate the positions of the final states of the models on the plane of the relative vertical scale and the relative mass of the spheroidal subsystem $M_s = M_h + M_b$. The models differ in the parameters of the bulge, the halo, the bar and the dipersion of the radial velocities.

The reduction of the relative vertical scale of the disk z_0/L with increase of the relative mass of the spherical subsystem M_s/M_d is shown in Fig. 4.48. The models differ in the parameters of the spherical subsystem (bulge and halo) and also the initial distribution $c_r(r)$. The problem of the dependence of the relative vertical scale of the disk z_0/L on the relative mass M_s/M_d was investigated in a number of studies [63, 225, 911, 1191, 1243].

In the actual galaxies, there are additional factors of vertical heating and thickening of the disk. In particular, they include: density waves, giant molecular clouds, tidal interactions. Therefore, the determination of the thickness of the disk from the condition of marginal stability in respect of bending perturbations gives the lower estimate of the mass of the halo.

To construct the marginally stable disk in relation to the bending perturbations to avoid the effect of the initial conditions, it is necessary to construct a series of models at fixed values of the parameters of the spherical subsystem, the surface density of the disk σ (r), dispersion of the velocities in the disk plane c_r, c_φ with the different initial ratios $\alpha_z = c_z/c_r$. It is important to select the model ensuring the stationary vertical structure of the subcritical level, with the distribution α_z (r) minimum necessary for the bending stability. The dependence c_r (r) should ensure the gravitational stability in relation to the perturbations in the disk plane (see section 4.3). Some of the results of modelling of specific galaxies, visible edge-on, associated with the estimates of the dark mass of the halo, are presented in section 10.3.2.

4.6. Axisymmetric (lopsided) galaxies

4.6.1. Observational data

In a number of galaxies of later types with a distinctive bar the latter is displaced in relation to the centre of the disk. A classic example of this is the Large Magellanic Cloud (LMC) in which the centres of the distribution of the brightness of the disk, planetary nebulae, super giants, HI and the bar do not coincide [1260]. The kinematic centre of the disk is displaced in relation to the bar in the LMC by approximately 0.7–1.2 kpc [778, 1184]. This pattern is relatively frequently found in the galaxies of later types. Suitable examples are NGC 55 [696, 1076], NGC 1414, 4490, 4625, 4027 [472, 560, 872], NGC 1744 [529, 1029]. Of course, if a two-arm spiral structure is observed, such a structure is usually non-symmetric, one arm is developed far more extensively in comparison with the other one. For example, Pisano et al [1029] report that the NGC 921 has one powerful spiral branch, the other one is a fragmented formation.

The displaced bar and the spiral single-arm structure are the typical features of SBcd-SB galaxies (=Magellanic type) [1260]. These properties are clearly evident, for example, LMC and NGC 4618, in the NGC 925 galaxies (SAB(s)d) they are reflected in a less distinctive form, and it is usually assumed that the degree of asymmetry continuously changes in transition to galaxies of later types [1030]. Feitzinger [544] reported that the magnitude of the displacement of the centre of the bar in many cases is half of the radial scale of the disk L, although in some cases the displacement is smaller than this value. The displacement of the centre of the bar in relation to the centre of the external isophots in the NGC 4618 reached 660 pc at L = 900 pc [985]. In the NGC 925 the distance between the centre of the bar and the dynamic centre is ~950 pc at the radial scale of the external disk L = 4.2 kpc [1030]. According to the data in [1029], in these galaxies the centre of the bar r_{bar}, the centre of the galaxy according to RC3, the centre of the outer infrared isophots and the dynamic centre do not coincide. It should be mentioned that the authors attribute this asymmetry to a gas cloud $10^7 M_\odot$ which interacts with the main galaxy, but they note that the forces of these interactions are not sufficiently high.

It should be stressed that the displacement of the bar is one of the forms of disruption of the symmetry of the galactic disks. The latest investigations show the extensive propagation of lopsided galaxies, mostly in the galaxies of later types [1096] and, in particular, amongst the dwarf systems [661]. However, the individual manifestations of the asymmetry are also detected in the disks of galaxies of late and early types [476, 1134].

Modelling of the formation of the bars by the method of N-bodies showed large differences in the properties of the central bridges in the galaxies of late and early types [476, 1134]. In [293, 313, 1110, 1113, 1134], the dynamics of disk components in the gravitational field of the bar with the given displacement was studied. In construction of the models of asymmetric (lopsided) galaxies in the studies by Levine and Sparke[8 25], Noordermeer et al [973] the displacement of the centre of the halo in relation to the centre of the disk was defined initially without studying the mechanisms leading to the disruption of the axial symmetry in the distribution of matter.

Taking into account the problem of formation of the displaced bars, investigations were carried out into the mechanisms associated with the initial asymmetry of proper galaxies and special features of subsequent accretion on a disk. One of the reasons for the displacement of the bar may also be the strong tidal effect or collision of two galaxies with subsequent merger. Another possibility is the dominance of the single-arm spiral m = 1 in accordance with the swing amplification mechanism [315]. The condition for the development of the single-arm wave in the linear approximation were investigated in studies by Zang [1320], Evans and Read [538, 539], and the results of linear analysis have been confirmed in the numerical models of the N-bodies [1143].

To denote different asymmetric manifestations, the term 'lopsided galaxy' (this term is used in all previously cited studies) can be regarded as established, and in this section we will discuss primarily the displacement of the bar in relation to the centre of the disk component. Of special interest are the conditions in which displaced bars can form in the process of development of gravitational instability of the global bar mode in the isolated initially axisymmetric system [75, 767]. The physical mechanism of displacement of the bar is based on the non-linear stage of combined development of the bar mode and single-arm harmonics with the azimuthal number m = 1. With increase of the mass and/or reduction of the scale of the bulge, the conditions for the displacement of the bar become less favourable.

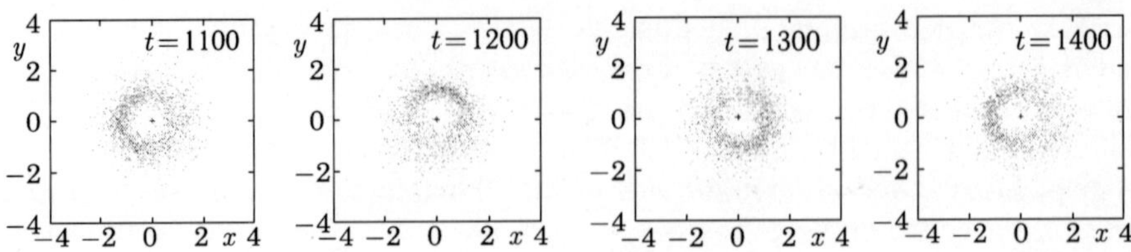

Fig. 4.49. Formation of a single-arm mode $m = 1$ in the gravitating Kepler disk [716]. The positions of the particles at different moments of time t and of the central mass (cross) are indicated.

The displacement of the bar is most effective in the case of a low-mass halo where the spatial scale greatly exceeds the radial scale of the disk.

4.6.2. Asymmetry of the galaxies as a result of the external effect

A number of numerical models which can be used to form the asymmetric structures in the central part of the galaxies will be listed.

In the nuclear, almost Kepler disk, rotating around a massive object (a massive black hole), precessing non-linear modes form in the central body-gravitating disk dynamic system [716]. This mechanism of formation of the eccentric disk is capable of explaining the formation of displaced neutral nuclei (double-peaked nucleus), for example, in M31, NGC 44 86B (Fig. 4.49).

The large-scale lopsided structures can be generated by introducing into the numerical model of the galactic disk external non-axisymmetric perturbations, for example, asymmetric distribution of matter in a dark halo or the tidal effect from the side of satellites, accreting objects, etc.

The lopsided structures forming a natural manner in the models where the centre of gravity of the disk is initially displaced in relation to the halo potential [825].

4.6.3. Formation of isolated lopsided galaxies

In this section, we examine the mechanism of formation of a lopsided system in the initially axisymmetric disk and the centrally symmetric external potential [75, 767].

One of the mechanisms of formation of the bar is the development of gravitational instability of the mode $m = 2$ in any sufficiently hot (for stability) disk when the dispersion of the radial velocities of the stars $c_r(r)$ does not exceed a critical value determined by the parameters of the disk and spherical subsystems (section 4.2.1).

The models without the spherical subsystem

We consider an isolated self-gravitating disk without taking the spherical subsystem into account. Although in the real galaxies the presence of the halo is evidently a compulsory component (see sections 4.4, 4.5, chapter 10), this example will be used to discuss several special features of bar formation.

There are two types of models differing in the initial disk thickness. These models are based on different scenarios of galaxy formation. In the first case, the collapse of a protogalactic cloud results initially in the formation of a quasi-stationary relatively cold gas disk in which a star population then forms. In the second case, a thick stellar disk forms rapidly in the course of gas compression to the plane of the galaxy, and the stars are initially characterised by a higher dispersion of vertical velocity.

The models with the thin initial disk

If the central region of the disk in the initial moment of time is relatively cold (Q_T ($r \leqslant 2L$) = 0.7) and thin $c_z/c_r \leqslant 0.4$ (dispersion of the vertical velocity is associated with the vertical half thickness and by the relationship $\Delta \simeq c_z^2 / \pi G\sigma$ in (2.57)), then in addition to the increase of the two-arm mode, leading to the formation of the bar, the single-arm perturbation with a high amplitude appears (Fig. 4.50a). The integral amplitude of the Fourier harmonics $\hat{A}$ ($m = 1$) dominates throughout the entire calculation, with the exception of the initial stage ($t \leqslant 4$). The values in (4.27), (4.28) were calculated in the system of the coordinates of the centre of gravity of the disk and the displacement of the centre at the time $t = 10$ did not exceed in the plane of the disk (x, y) the value $|\xi_{\text{disk}}| < 0.002$, which characterises the accuracy of integration of the equations of motion. The distinctive periodicity of $\hat{A}$ ($m = 1$) and also the harmonics with the numbers $m = 3, 4, 5, 6$, is the consequence of the rotation of the bar in relation to its centre which does not coincide with the centre of the disk, with the period $\tau_{\text{bar}} \simeq 2.3$.

Thus, in the process of its formation from the axisymmetric system the bar is displaced in relation to the centre of gravity of the entire disk. Figure 4.50b shows the dynamics of the centre of the bar ($x_{\text{bar}}(t)$, $y_{\text{bar}}(t)$). The formation of the quasi-stationary bar is completed at the time $t \simeq 10$ (the period of rotation of the disk $T_0 \simeq 6$). Subsequently, the movement of the centre of the bar in the inertial coordinate system is almost completely interrupted (Fig. 4.15b, 4.51). In these calculations, the displacement of the centre of the bar in relation to the centre of the mass of the

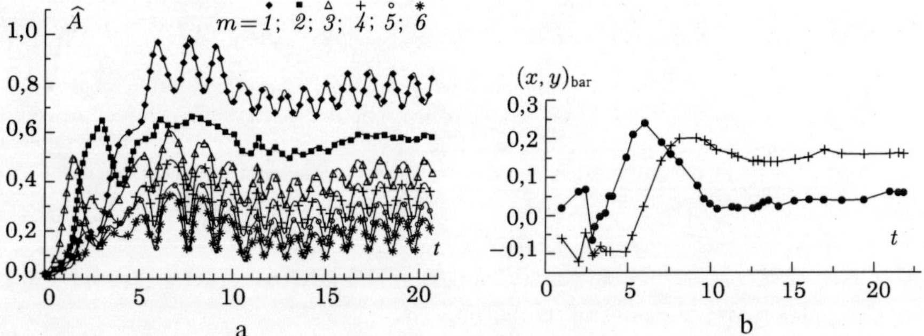

Fig. 4.50. Model of the initially cold (Q_T ($r \leqslant 2L$) = 0.7) thin (Δ/L = 0.082) disk without the spherical system with $N = 5 \cdot 10^5$; a) the time dependence of the integral amplitudes of the Fourier harmonics $\hat{A}$ for different azimuthal numbers m; b) time dependence of the coordinate of the centre of the bar (• – x-coordinate; + – y-coordinate). The centre of the coordinate system coincides with the centre of gravity of the entire disk.

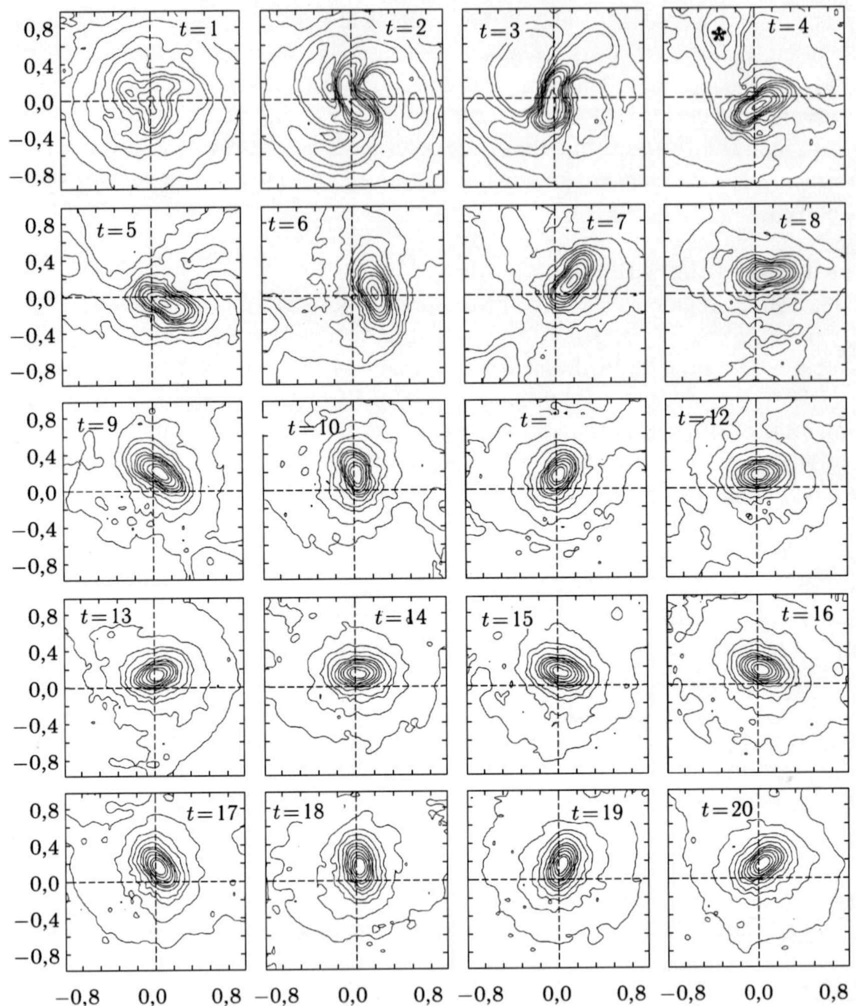

Fig. 4.51. Isolines of surface density in the model without the spheroidal subsystem, shown in Fig. 4.50, at different moments of time. The kinematic centre in the centre of gravity of the entire disk are situated in the centre of the coordinate (the coordinate axes are indicated by the dotted lines). 70% of the initial mass of the disk remains in the quasi-stationary state in the range $r < 1 \simeq$.

entire disk is $\xi_{\rm bar} = \sqrt{x_{\rm bar}^2 + y_{\rm bar}^2} = 0.17 = 0.7L$. The kinematic centre of the entire disk coincides with the centre of its inertia.

In the isolated disk, its centre of gravity $\xi_{\rm disk}$ and the kinematic centre $\xi_{\rm kc}$ should not change during evolution of the system. Therefore, the modification of the fulfilment of this conditions may be the test of quality of the numerical models. In the calculation of the gravitational force by the TREEcode method at $\Theta = 1$ at the end of calculations ($\simeq$10 periods of rotation) due to the numerical effects the centre of inertia of the disk is displaced by the value $\xi_{\text{disk}} = 0.03 < \xi_{\text{bar}}$ (ξ_{disk} decreases when selecting lower values of Θ). To verify the strength of this effect on the evolution of the system it is convenient to use the models in which the gravitational force is computed by the direct summation of the contributions from all particles (PP method, see section 4.1.1). The PP method greatly reduces the value of displacement ξ_{disk} (i.e. the error caused by numerical solution) to the values $\xi_{\text{disk}} < 10^{-5}$. All the results, associated with the displacement of the bar in the TREEcode models, remain valid.

The displacement of the bar decreases with increasing initial dispersion of the velocities c_r. In a disk hotter in the initial stage, the rate of formation of the bar decreases (in a very hot system, the bar does not form, section 4.2.1) and at the same time, the conditions for growth of the mode $m = 1$ become less favourable. Consequently, the displacement of the bar is almost insignificant already at $Q_T (r \leqslant 2L) \geqslant 1.05$.

It should be mentioned that in the case of a disk that is hotter at $t = 0$, the multi-arm waves ($m > 3$) are characterised by a considerably smaller amplitude in comparison with $m = 1, 2, 3$. In the reversed limiting case of the very cold disk ($Q_T \leqslant 0.2$), in addition to the central bar, one or more fine-scale perturbations can form at the disk periphery, because the Jeans scale $\varrho_J = c_r/\varkappa$ is very small: $\varrho_J \ll L$. These regions with high density are gravitationally bonded during several revolutions of the system (see the maximum density in Fig. 4.51, indicated by the sign * at the moment of time $t = 4$). Only the central bar-structure remains with increase of the initial dispersion of the velocities c_r.

Models with the thick initial disk

If the disk at the initial moment of time is thicker, its evolution changes. However, if the disk is relatively cold in the plane (x, y), which permits the formation of a bar mode, the displacement of the bar decreases (Fig. 4.52) and this is reflected in a reduction of the amplitude of the harmonics $m = 1$ and in slower growth in comparison with the model in which the disk is thinner and the initial moment of time (Fig. 4.52b).

The displacement of the bar disappears almost completely starting of the values of the vertical scale $\Delta/L \gtrsim 0.4$ in the absence of a halo. The thicker disks are gravitationally more stable (see (4.116)). Therefore, the characteristic time of formation of any structures, determined by self-gravitation, increases with increase of the thickness of the disk, and the amplitude of these formations decreases. The thickness of the disk is determined by the dispersion of the z-velocities c_z. Thus, there are critical initial values of the dispersion of the radial and vertical velocities which permit large displacement of the bar.

Models with the bulge

Attention will be given to the evolution of a disk taking into account the additional gravitational field of a stationary bulge, characterised by the relative mass M_b/M_d and the radial scale b. The low-mass bulge ($M_b/M_d = 0.1$, $b/L = 0.2$) in the absence of the halo in the initial stage of evolution permits the development of both the bar mode and the single-arm harmonics. In this case, the centre of the bar is close to the

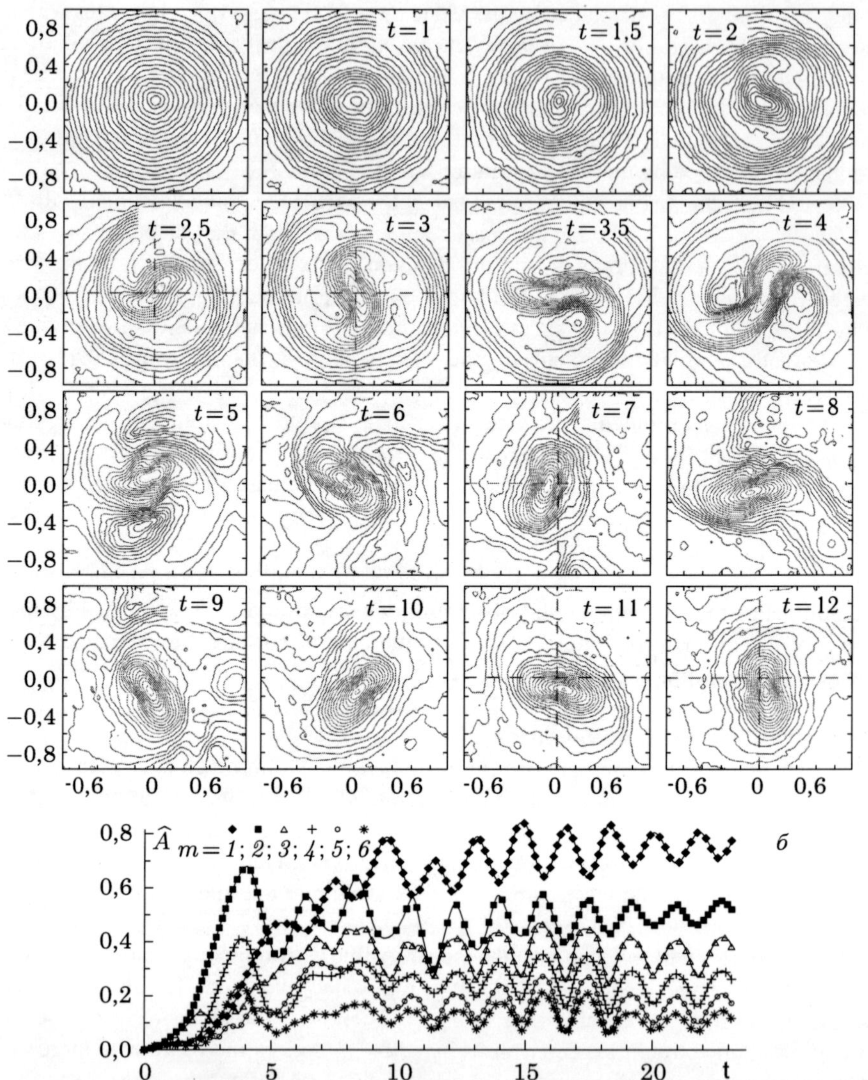

Fig. 4.52. Initial stage of evolution in the model of the initially thick disk ($\Delta/L \simeq 0.25$) without the spheroidal subsystem ($N = 10^6$): a) the isolines of the logarithm of the density surfaces at different moments of time. The displacement of the bar is approximately half the displacement in the case of the model in Fig. 4.50, 4.51; b) time dependence of the integral amplitudes of the Fourier harmonics (4.28) for different azimuthal numbers $m = 1, 2, 3, 4, 5, 6$.

centre of the bulge (this is due to the static state of the bulge), but it is displaced in relation to the kinematic centre and the centre of gravity of the disk components. The velocity field in the region of the bar has a highly non-symmetric form.

The generation of the mode with the azimuthal number $m = 1$ can also takes place in the case of a more massive bulge ($M_b/M_d > 0.1$), if the latter is not too concentrated. In the case of a low mass of the bulge $b \ll 0.1\ L$ the condition for the development of the single-arm perturbations is no longer favourable.

In the presence of a bulge, the displacement of the bar, with other conditions being equal, is smaller and with time slowly decreases as a result of the interaction of the asymmetric disk with the static bulge. This effect is identical to the effect of the halo (see below) that in the case of the bulge it is more distinctive. The reduction of the value bar with time is determined completely by the interaction of the rigid spheroidal component with ξ_{bar}. The characteristic time of the process is several periods of revolutions of the outer edge of the disk. If the 'live' bulge is taken into account, the dynamics of the process can be changed, retaining the principal conclusion on the reduction of the displacement of the bar with time.

Evidently, the stabilisation of the single-arm perturbations by the bulge and the reduction of the displacement of the bar with time taking into account the gravitational field of the bulge is the reason why the displacement of the bars is observed preferentially in the galaxies of later types, starting with the galaxies of the morphological type Scd in which usually there are no large central spherical subsystem.

The evolution of the disk with the stationary halo taken into account

The mass of the halo within the limits of the optically observed disk may exceed the mass of the disk components (see section 4.4, 4.5). Therefore, more realistic models, including the halo in accordance with (4.3), will be discussed.

The conclusion according to which the displacement of the bar ξ_{bar} in the case of the initially thin gold disk is possible is also valid when the halo is taken into account. In these with the initial Q_T ($r \leqslant 2L$) = 0.6÷0.9, as in the case without the halo, the characteristic special feature of the evolving disk is the possibility of its growth to high amplitude of the harmonics $m = 1$ which may result in the initial stage in the displacement of the bar. These effects are the strongest for a relatively porous halo whose scale is 2–4 times greater than the radial scale of the disk. In this case, the dynamics of the disks in the range $r \leqslant 2L$ is determined primarily by the distribution of mass in the disk component. Figure 4.53 shows the evolution of the disk in the model with M_h/M_d, $a = 2L$. The initial stage is characterised by the formation of an asymmetric disk in which the bar is displaced in relation to the centre of gravity and the kinematic centre of the entire disk. Subsequently, as a result of the static nature of the spheroidal subsystem, the extent of displacement of the bar decreases. In the case of smaller scales of the halo the displacement of the bar decreases and the dynamics of the system is similar to the case of the model with the concentrated bulge without the halo.

If the investigated mechanism is responsible for the formation of at least part of the lopsided galactics, then in these galactics of the late morphological types the relative mass of the halo inside the optical radius is on average smaller than in the case of the symmetric galaxies.

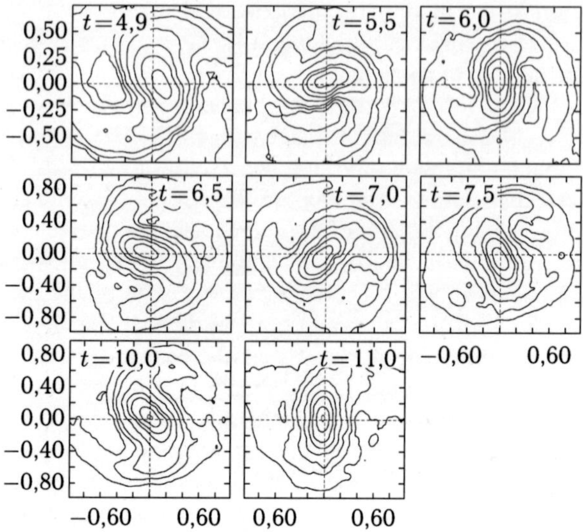

Fig. 4.53. The model with a halo: mass $M_h = M_d$, the scale of the halo $a = 2L$. The isolines of surface density at different moments of time are shown.

Taking into account the dynamics of the bulge and the halo

In the model with an active spheroidal subsystem, the centre of the mass of the entire modelling system is stationary. The initial stage of the formation of the bar in the cold disk does not differ qualitatively from the case of the stationary spheroidal subsystem (Figure 4.54a). However, the resultant displacement of the bar in relation to the centre of the disk remains unchanged. The distance between the centre of the disk in the centre of the halo is several times smaller than $\xi_{\text{bar}} = \sqrt{x_{\text{bar}}^2 + y_{\text{bar}}^2}$.

Stellar-gas models

The characteristic feature of the galaxies of later types is the larger relative fraction of gas in comparison with the early type objects. Therefore, attention will be given to the principal effect of the gas component on the effect of displacement of the bar and, on the whole, the appearance of asymmetry in the disk structure [767].

The initial state for the stellar component has the form of a disk close to the boundary of gravitational stability in the absence of gas. Thus, the initial distributions of the dispersion of the velocities are such that without taking into account the gas component, the amplitude of the single-arn harmonics $m = 1$ is very small, and the asymmetric disk does not form, although the initial stage permits the slow formation of the bar. The harmonics with $m = 1$ can form in a massive cold gas subsystem, and in the non-linear stage the harmonics results in the asymmetry of the stellar disk. Figure 4.54 shows the results of these calculations in the model with $M_{\text{gas}} = 0.40$ $7M_d$ in the range $r \leqslant 4L$ ($M_h = M_d$, $a = 1.6L$). The mechanism of formation of the lopsided galactics as a result of gravitational instability of the perturbations with the azimuthal number $m = 1$ and interaction in the non-linear stage of the single-arm

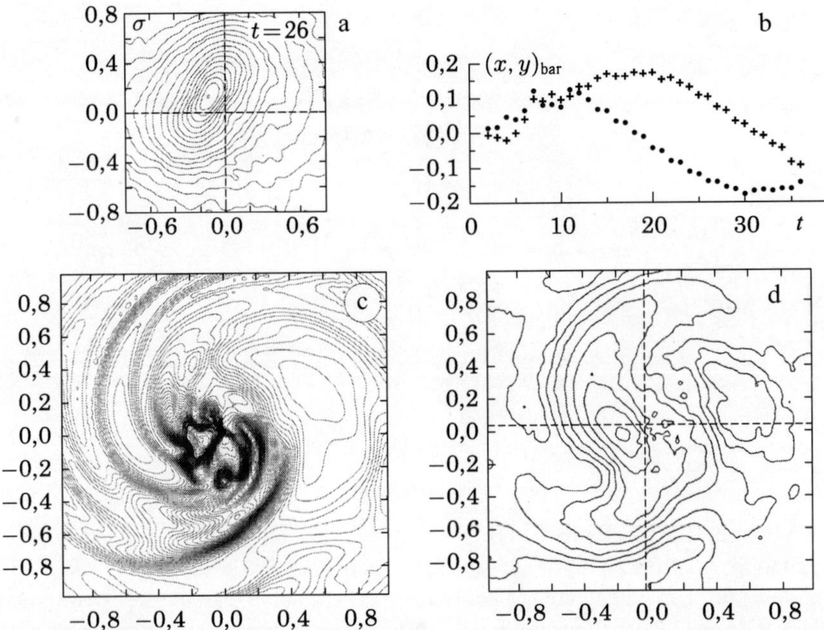

Fig. 4.54. A model with a non-stationary ('live') halo ($M_h = M_d$, $a = 2L$): a) the density isolines σ; b) the dynamics of the coordinate of the centre of the bar. The model of the stellar-gas disk: c) the isolines of the surface density of the gas, d) the stellar component. The displacement of the centre of the bar in relation to the centre of the disk and the degree of asymmetry of the spiral structure increase with increase of the relative mass of the gas.

mode with the bar mode in the case of a low-mass porous halo is more effective in the presence of the cold gas The asymmetry of the disk increases with increase of the relative mass of the gas.

4.6.4. The dynamic model of the Large Magellanic Cloud

A classic example of the galaxy of a transitional type from spirals Sc to irregular ones is the Large Magellanic Cloud (LMC). Many objects, relating not only to the type Sm but also Sd are often referred to as Magellan-type galaxies [825, 1029]. Therefore, to continue these discussions, it is important to discuss in greater detail the LMC. For this purpose, a series of the dynamic models of the LMC will be constructed and, matching these models with the observational data, we determine a number of non-observed parameters and discuss the reasons for the formation of characteristics special features of the structure of the LMC [75].

Main observational data

The investigations carried out recently evidently enable us to discuss a more complicated (than previously expected) LMC morphology. In comparison with the

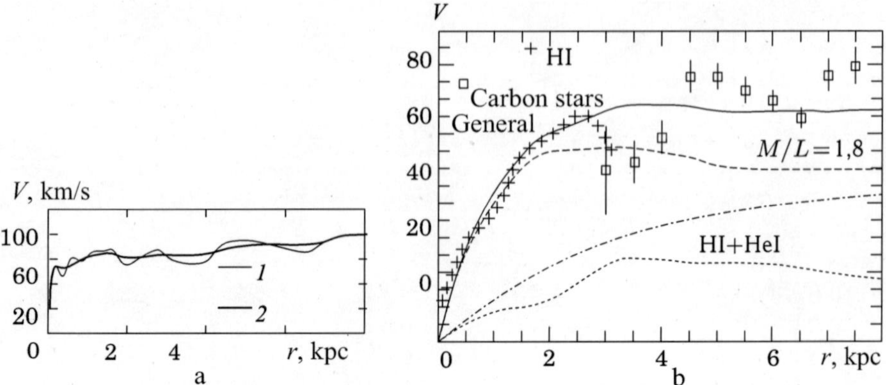

Fig. 4.55. Rotation curves for the LMC: a) from the study by Sofue [1184], (1 – observational data, 2 – the smoothed rotation curve); b) the study by Kim et al [778].

optical images, the galaxy in the HI line is more symmetric and similar to the classic disk-shaped galaxies [778]. The gas disk is observed to a radius of 7 kpc[11].The individual staff can be detected to approximately the same distances. Measurements of the rotational curve is on the basis of the observations of HI and the brightest stars·can be used to construct the models of distribution of matter in the galaxy. The data on the photometry and kinematics of the stars and the presence of the bar greatly restrict the indeterminacy of the free parameters describing the radial dependences of the density of matter in the disk and the halo.

At large distances from the centre of the LMC, the observed velocities of rotation of the disk (both with respect to the gas and the stars) are in the range 68–80 km/s without any signs of drooping of the rotation curves with the radius. Of special interest is the internal part of the disk. Sofue [1184] again analysed the data published by Kim et al [778] and constructed the rotation curve of the HI, differing in the central region ($r \leqslant 2$ kpc) from the rotation rotation curve V_{gas}^{K} plotted by Kim et al [778] by the considerably higher radial gradient of the velocity of rotation (Fig. 4.55). The reason for the difference was determined primarily by differences in the selection of the kinematic centre. In order to explain the high gradient of the velocity of the gas in the vicinity of the selected centre, Sofue was forced to introduce a dark bulge into the model which basically determines the movement in the range $r \leqslant 1$ kpc. The second special feature is the small radial scale of surface density in the Sofue model, equal to 1 kpc, which is considerably smaller than the value of the exponential scale $L = 1.6$ kpc, usually accepted for the disk of the LMC. At the same time, to explain the rotation curve presented in the study by Kim et al [778], it is not necessary to have a massive bulge. In order to investigate the possibility of existence of the proposed dark massive bulge in the LMC, the authors of the book constructed two series of dynamic models for each rotation curve.

[11]Disk-like rotation can be observed even in the Small Magellanic Cloud which gives the mass $M \sim 1.3 \cdot 10^9 M_\odot$ inside the radius $r = 2$ kpc [1206].

The distribution of matter in the disk component

The standard values of the distance to the LMC (50.1 kpc) and the angle of inclination ($i = 33°$) [814] will be used. The radial dependences of the surface brightness of the disk, long-period variables, RR Lyrae and carbon stars are described quite efficiently by the exponential law with L = 1.6 kpc (see [299] and the references therein).

The bar

An optical bar forms in the central region of the LMC. It is approximately 3.3 kpc long. A distinguishing feature of the central region of the LMC is the disagreement between the kinematic centre of the disk in the centre of the bar [608][12].

The observational data indicate that in the last $(2 \div 4) \cdot 10^9$ years there was evidently no active star formation in the central region of the galaxy [605]. It was shown in [687] that most of the stars are older than 4 billion years, and it was concluded that, regardless of the selection of the model, the region of the bar contains a considerably larger number of old stars than the external regions of the galaxy. Investigations of the age of the clusters, distributed in the bar and a disk, showed that their age differs. The history of formation of a stellar population in a bar differs from other regions, and it was concluded in [642] that the bar in the LMC formed $(2 \div 6) \cdot 10^9$ years ago. Therefore, it is assumed that the bar in the LMC is a stable structural feature according to different data, its age is estimated at $2 \div 9$ billion years [308, 642, 687]. Thus, even if the bar formed initially not as a result of the development of instability of the global bar mode but as a result of the tidal interaction or action of another mechanism, the conditions in the galaxy should be such that the duration of existence of the bar will be no less than several billion years.

Kinematics of gas and stars

The dynamic models are plotted on the basis of the rotation curves of the gas according to the studies published by Sofue [1184] ($V^S_{\rm gas}(r)$) and Kim et al [778] ($V^K_{\rm gas}(r)$) (Fig. 4.55a, b, thick lines, and Fig. 4.56, 4.57). The velocities of rotation V_* and the dispersion of the velocities $c_{\rm obs}$ of the stars are also given. The smoothed-out rotation curve of the gas was accepted as the curve of the circular velocity of rotation $V_c(r)$.

The kinematics of the stars of the disk in the LMC has been studied quite extensively. Kunkel et al [814] on the basis of 759 carbon stars constructed the rotation curve in the region from 3 to 12 kpc. This curve will be used taking into account the angle of inclination $i = 33°$ (see Fig. 4.56, 4.57). The dispersion of the velocities of the stars along the radius changes only slightly and equals $c_{\rm obs}$ = 12–16 km/s which, according to the authors of the above study, indicates that they are relatively young. In addition to these data, Figs. 4.56 and 4.57 show the radial dependences of the velocity of rotation and the dispersion of the velocities of the stars of the disk, obtained in [299].

Analysis of the kinematics [64] of more than 500 carbon stars shows, in addition to a younger population in which the dispersion of the velocity is approximately 8 km/s, also an old disk component with the dispersion of velocities $\simeq$ 22 km/s. In

[12]It should be mentioned that the problem of existence of the bar in the distribution of the stellar density of the LMC was until recently still the subject of discussion [1330].

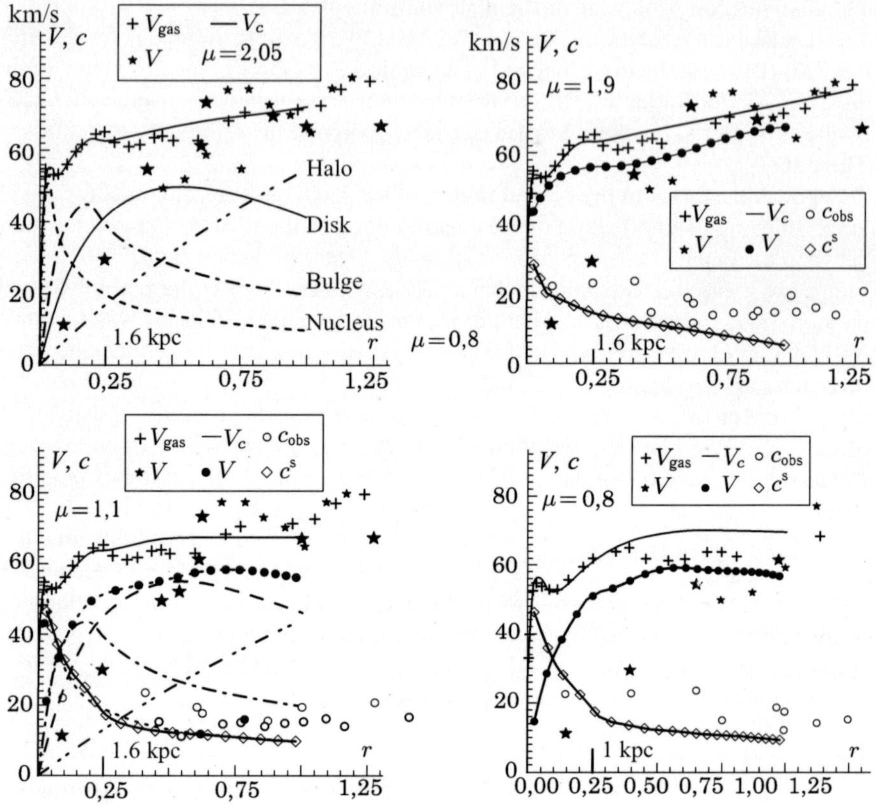

Fig. 4.56. Models with different μ and L = 1.6 kpc (*a, b, c*), L = 1 kpc (*d*); a) contribution to the circular velocity of the galaxy (continuous thick line) from the disk, the halo, the bulge and the nucleus; b) radial distribution of the observed and modelling velocities. Conventional notations: + – velocity of rotation of HI; * – the velocity of rotation of the stars (large asterisks [298], small asterisks [814]); o – the observed dispersion of the velocities of the stars; • – modelling estimates of the velocity of rotation of the stars; ◇ – modelling estimates of the dispersion of the velocities of the stars in projection on the visual ray; c) the same for the model with μ = 1.1 and L = 1.6 kpc; d) the same for a maximum disk with L = 1 kpc.

the study [1329] of 190 VRC (Vertical Red Clump) the dispersion of the velocities of the stars was 18.4±2.8 km/s. The data for the kinematics of the planetary nebulae in [899] c_{obs} = 19.1±1 km/s are in agreement with the results mentioned previously. The movement of 975 carbon starts was used to evaluate the maximum circular velocity of rotation of ≃80 km/s at a distance of 3.4 kpc from the centre and the dispersion of the radial velocities of the disk components c_r = 28 ± 14 km/s [298].

It should be mentioned that the radial dependence of the velocity of rotation and the dispersion of the velocities of the stars slightly changes if we change the position of the kinematic centre according to the study by Sofue [1184]. However, since the dispersion of the velocities c_{obs} depends only very slightly on the radial

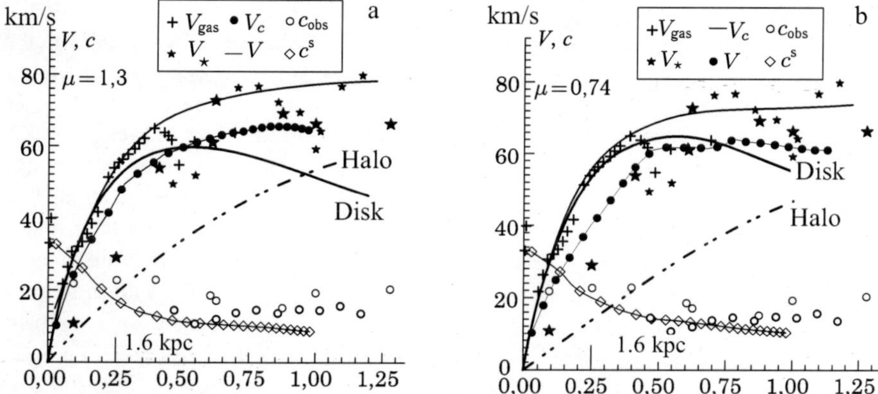

Fig. 4.57. Comparison of the observed and modelling parameters for the rotation curve $V^K_{gas}(r)$ for a model with a massive halo $\mu = 1.3$ (*a*) and with the maximum disk $\mu = 0.74$ (b). The conventional notations are the same as in Fig. 4.56.

coordinate, the course of dispersion with the radius remains almost the same, and the changes of the stellar rotation curve V_* can take place only in the central region $r \leqslant 1$ kpc. For this reason, in particular, the kinematic parameters of the stars in the vicinity of the centre of the LMC should be regarded as least reliable.

The results of dynamic modelling of the LMC

The aim of dynamic modelling is to obtain the restrictions for making selection between the rotation curve of Kim et al [778] and Sofue [1184] and, at the same time, verify the possibility of existence of a dark bulge in the LMC [75]. The main conclusion is that the rotation curve in [778] (see Fig. 4.55b) is in better agreement with the set of observational data for the LMC, and the presence of a large amount of dark mass inside the LMC is not required.

The dynamic models with different distributions of matter in the subsystems will be analysed. Since in Sofue's study the rotation curve V^S_{gas} was explained by investigating the distribution of matter, which differs from the usual distribution of brightness, it is important to analyse the dynamic models with different values of the radial scale of the disk $L = 1 \div 1.6$ kpc, determining it as a free parameter.

1) The dynamic models with the rotation curve V^S_{gas}. The case of the rotation curve $V^S_{gas}(r)$ with $L = 1.6$ kpc and a massive halo $M_h/M_d = 1.8$ will be discussed. To explain the central maximum on the rotation curve, the model includes, in addition to the massive bulge ($M_b/M_d = 0.22$), also a compact nucleus with $b_c/L = 0.036$. Figure 4.56a shows the contribution from the individual components to the circular velocity of rotation $V_c(r)$, and Fig. 4.56b shows the distribution of the observed quantities c_{obs}, V_{gas}, V_* and the appropriate modelling quantities c^s, V_c and V. The stationary state of the disk at the periphery of the galaxy is obtained at very small values of the velocity dispersion of the stars, in the range $r > 3L$, the dissipation of the radial velocities is $c_r = 10 \div 14$ km/s, which is comparable with the dispersion of the velocities of the gas. This model assumes a large reserve of stability: $c^s < c_{obs}$ in the galaxy, where c^s and c_{obs} are the dispersions along the visual ray according to

(4.1 to 2) in the model and from the observational data, respectively. The velocity of rotation of the stars V in the investigated model is comparable with the circular velocity V_c, and in the central region it greatly exceeds the observed velocities of rotation of the stars V_*. This is a shortcoming of the model. Increase of μ results in an even larger discrepancy with the observed dispersion of the velocities.

In addition, for the models with the high value of μ the long-lived bar cannot form at any initial distributions of the dispersion of the velocities of the stars. The combined effect of two factors – the presence of a relatively massive halo and a concentrated nucleus – causes that the central jumper in the process of heating of the system is short-lived, as indicated by a reduction with time of the amplitude of all Fourier harmonics, including the mode with $m = 2$ (section 4.6.3). However, if a very cold disk is considered as the initial condition, and in this disk in a wide range of the radius the Toomre parameter $Q_T < 0.5$, then after rapid heating and deformation of a bar-like structure in the initial stage the axisymmetric distribution of density is nevertheless rapidly established after ~12 periods of rotation with respect to the outer edge of the disk.

An alternative model for the selected value of L is the model of the maximum disk (MMD), for which $\mu = 1.1$ (Fig. 4.46c). The dispersion of velocities c^s in this case at the periphery of the disk is in better agreement with the observed distribution. However, in the central zone, the dispersion of the velocities along the visual ray is $c^s \simeq 40$–50 km/s, which contradicts the observations. Formation of a bar in the case of the initially cold disk is possible but this bar is also short-lived. After the final formation of the bar ($t \sim 5$–6), the bar starts to disintegrate as a result of scattering at the concentrated nucleus. Consequently, it appears that its lifetime is restricted to ~$(2$–$3) \cdot 10^9$ years which is only slightly higher than the lower estimate of the age of the bar on the basis of analysis of the stellar composition. The reason why in the previous case ($\mu = 1.9$) the lifetime of the bar was shorter than in the case with $\mu = 1.1$ is, in addition to the difference in the mass of the halo, tha the disk in the case of the MMD was thicker. Therefore, the compact nucleus, whose characteristics radius is considerably smaller than the resultant the vertical scale of the disk of 360 pc does not ensure the rapid failure of the bar mode, and the process of dissipation of the bar is extended to ~10 revolutions of the bar.

The models with the more concentrated disk ($L = 1$ kpc) and the same rotation curve V^S_{gas} are even in a less satisfactory agreement with the observed dispersion of the velocities: almost the entire disk is characterised by a large reserve of stability, whereas the reversed condition $c_{obs} < c^s$ is fulfilled in the very centre.

2) The dynamic models with the rotation curve V^K_{gas}. The results of dynamic modelling with the rotation curve $V^K_{gas}(r)$ which can be interpreted without the bulge and the nucleus, will now be discussed.

In the model with the halo $\mu = 1.3$ the condition $c^s < c_{obs}$ is fulfilled for the dispersion of the velocity of the stars, with the exception of the very centre, and there is satisfactory agreement with the velocity of rotation of the stars in the range $r > 1.5\ L$ (Fig. 4.57a). In the model of the maximum disk ($\mu = 0.7$) we obtain for $V^K_{gas}(r)$ the best agreement between the modelling results and observational data on the dispersion of the velocities of the stars and the velocity of rotation of the stars

(Fig. 4.57b). This result confirms the conclusion on the small mass of the halo at the LMC, made for the first time in [59] on the basis of the data on the kinematics of the spherical clusters, belonging to the disk, on the basis of the condition of the threshold gravitational stability of the disk.

In the model with the rotation curve $V^K_{gas}(r)$ if the initial disk is not very hot, a long-life bar always forms and exists for $\simeq 4{\cdot}10^9$ years. The angular velocity of rotation of the bar in the model is $\Omega_{bar} = 23$ km/(s·kpc).

A distinguishing special feature of the formation of the bar structure in the case of the models with $V^K_{gas}(r)$ is, under certain conditions, the high amplitude of the first mode $m = 1$ which remains high throughout the entire evolution and characterises the disruption of the symmetry of the bar in relation to the centre of the galaxy. The latter is in good agreement with the observed displacement of the kinematic centre in relation to the bar centre. The phenomenon of displacement of the bar has been studied in detail in section 4.6.3. However, it should be stressed again that the most favourable conditions form in the case in which the mass of the disk is considerably greater than that of the halo in almost the entire disk.

This is illustrated in Fig. 4.58 which shows the radial dependences of the mass of the disk and spherical subsystem inside the given radius of the LMC for different models. In the case of the rotation curve $V^S_{gas}(r)$, the mass of the disk dominates in the range 1.5 kpc $\leqslant r \leqslant$ 6 kpc (Fig. 4.58a). The model, based on the rotation curve $V^S_{gas}(r)$ allows a large part of the mass of the galaxy to be concentrated in the disk components at all radii (Fig. 4.58b). Inside the region $r < 2L = 3.2$ kpc the mass of the halo is only 20% of the mass of the disk, i.e. the dynamic role of dark matter is not significant.

Within the investigated model of the isothermal halo (4.3) it is not possible to construct a model with a less massive halo ($\mu < 0.7$) which would be in agreement with V_{gas} outside the radius $r > 4$ kpc. If the examined increase of the curve V_{gas} at the disk periphery is the result of the tidal action and not of the growth of the circular velocity, the mass of the halo may prove to be even smaller.

Conclusions

We will stress the special features of the models with the rotation curve $V^S_{gas}(r)$ (with the dark bulge) which make the model with $V^K_{gas}(r)$ (without the dark bulge) more

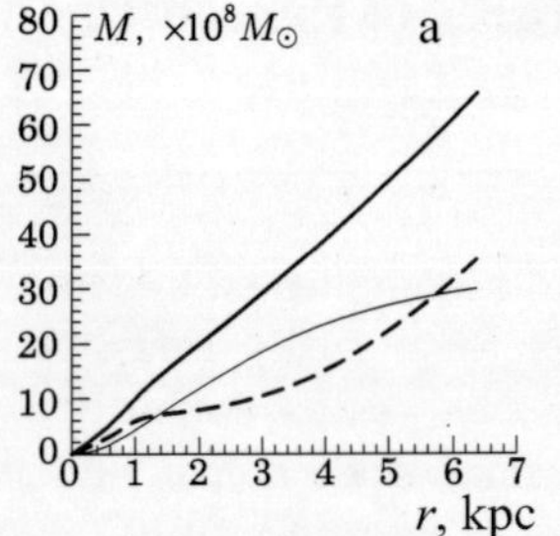

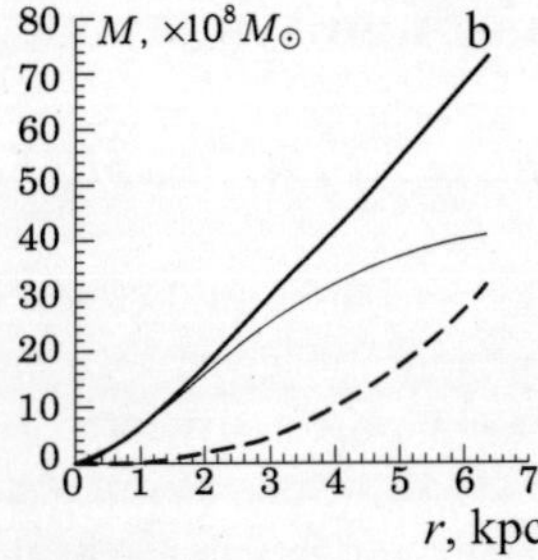

Fig. 4.58. Radial dependences of the masses enclosed inside the radius r, for a disk spherical systems in models with the rotation curves $V^S_{gas}(r)$ (a), $V^K_{gas}(r)$ (b). Thick line – total mass, thin line – disk component, dot-and-dash line – spheroidal subsystem

adequate with respect to the observations:

a. The models with $V^S_{gas}(r)$ do not explain the observed radial dependence of the velocity of the stars in the internal region of the galaxy $r < 2L$;
b. These models either do not permit the formation of the bar mode or the lifetime of the bar is not sufficiently long;
c. The amplitude of the single-arm mode $m = 1$, characterising the asymmetry in the distribution of matter, is small (on the noise level) in the models with the massive bulge throughout the entire evolution of the system, in contrast to the models with $V^K_{gas}(r)$ for which in formation of the bar the mode $m = 1$ is characterised by a high amplitude and its amplitude throughout the evolution of the system is only slightly smaller than the amplitude of the second mode. This circumstance can be used to explain naturally the observed displacement of the kinematic centre of the LMC in relation to the bar centre even without considering the tidal effects;
d. In the inner region of the disk ($r < L$) in the model with $V^S_{gas}(r)$ to ensure the stability of the disk, the dispersion of the velocities of the stars exceed 40 km/s (in the projection on the visual ray), which is considerably higher than the observed value c_{obs}. In this case, the models with the small radial scale of the disk $L \simeq 1$ kpc do not agree well with the observed distributions of the velocity of rotation and dispersion of the velocities of the stars in comparison with the case with $L = 1.6$ kpc.

Thus, the rotation curve according to Kim et al [778] is more suitable, and the assumption of the existence of the dark bulge in the LMC is not confirmed [75].

The constructed dynamic models permit the use of the model of the maximum disk for the LMC, and the mass of the halo in the range $r < 4L < 6.4$ kpc is approximately 70% of the mass of the disk components. The observed dispersion of the velocities at the periphery of the disk $c_{obs}(r \geqslant 3) \simeq 6$–$20$ km/s [298, 299, 814] indicates that this region is characterised by the stability margin even in the model of the maximum disk. It should be stressed that the constructed dynamic models do not exclude the existence of a more massive halo with $\mu \simeq 1 \div 2$ but in this case, the dispersion of the velocities of the stellar disk should be determined not by the dynamic heating as a result of gravitational instability but by other factors.

4.7. Complex non-symmetric structures in the centre of some galaxies

4.7.1. Problem of the double bars

In many galaxies, the central part appears as a complicated system consisting of several components inserted into each other (bars, disks, rings, double nuclei). Evidently, the first object suspected to contain the so-called double bar was the NGC 1291 galaxy [1258]. Special interest in the small-scale non-symmetric structures in the centre of the disks is associated with the investigation of the phenomenon of nuclear galactic activity. An important problem is the determination of the mechanisms, ensuring the inflow of the gas into the centre to the supermassive black hole ($\leqslant$ 10 pc). It is assumed that the double bars may supply gas to the active nucleus [1156].

There are more than 70 galaxies in which the presence of the double bars is possible [99]. In the majority of cases, the main argument for the presence of the second inner bar are the photometric data. The separation of the internal structures into double bars, internal disks and nuclear rings within the framework of the photometric approach [536] is evidently ambiguous. Analysis of the kinematics of the central regions of these galaxies in a number of cases does not confirm the conclusions obtained from photometric data and does not support the assumption of the existence of a large number of double bars [146, 929, 930]. We investigate the characteristic special features of the kinematics of the structures of the 'double bar' type dispersion of velocities) within the framework of the self-consistent dynamic models. An important problem of the double bars is also the problem of the dynamic stability of the systems of this type.

4.7.2. Distribution of parameters in the region of the double bar

The results of numerical modelling of a stellar disk, permitting the formation of the double bar [257, 767] will be discussed. Formation of the inner bar-like structure in the numerical models is possible in the presence of a sufficiently massive (but not concentrated!) bulge. Figure 4.59 shows the distributions lg (σ) at different moments of time in the model in which structures of the type of double bar form in certain stages.

The initial stage (1–3) is characterised by the formation of a double spiral wave, and in the centre – by a small bar-like structure with a large major axis $\simeq 0.3L$. This is followed by the formation of an outer bar (Fig. 4.59, 4, 5), normal to the inner bar. Two spiral waves from the inner bar are 'written' into the outer bar. This structure is short-lived and rearrangement takes place (Fig. 4.59, 6) resulting in the formation of a 'double bar' (7, 8, 9). In the new configuration, the dimensions of the bars are greater, especially those of the external jumper. However, the double bar fails with time (Fig. 4.59, 10, 11) and a single global bar remains (12).

It should be mentioned that the formation and failure of the inner bar is accompanied by the formation of short-lived structures resembling the 'non-perpendicular double bars'. Leading spiral waves form in the central region, and trailing spirals move away from the ends of the outer bar.

The kinematics of the central region of the disk ($r \leqslant 2L$) in the stage of the double bar will be discussed. The isolines of the dispersion of the radial velocities c_r have the oval structure (Fig. 4.60), coinciding with the major axis of the outer bar. In the region occupied by the inner bar there are two maxima of c_r. However, this characteristic feature is not the indication of the inner bar because it remains unchanged after the failure of the internal bridge. The distribution of the dispersion of the vertical velocities in the stage of the double bar differs more appreciably from the case of the single bar. Only the dispersion of the azimuthal velocities indicates more convincingly the presence of the inner bar.

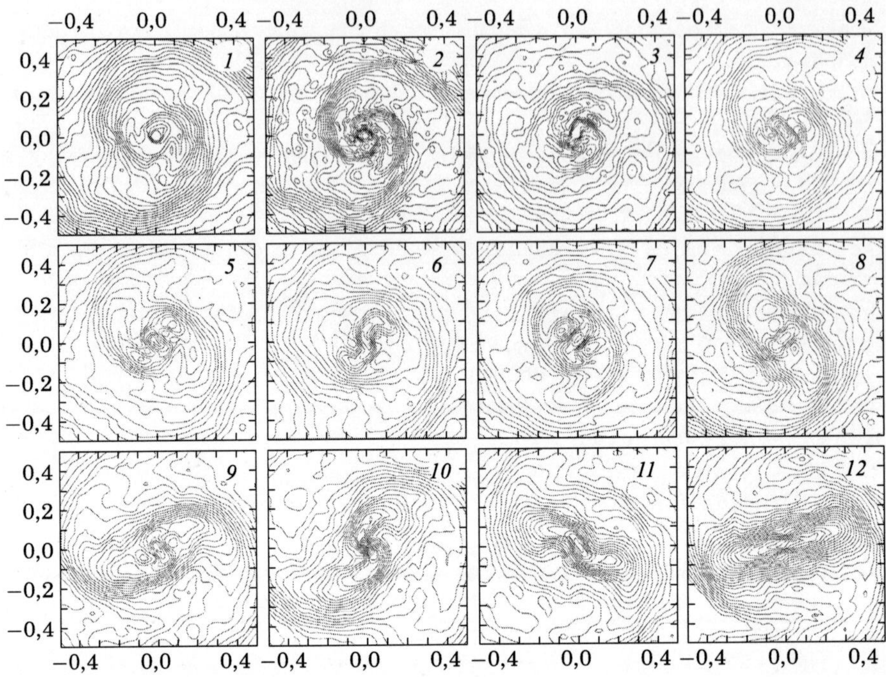

Fig. 4.59. Formation of the double bar. The isolines of the logarithm of the surface density at different moments of time in the model with $M_h = M_d$, $M_b = 0.6\ M_d$, $b = 0.16L$, $r_b^{\max} = 0.8\ L$. Only the central part of the disk $r \leqslant 2L$ is shown.

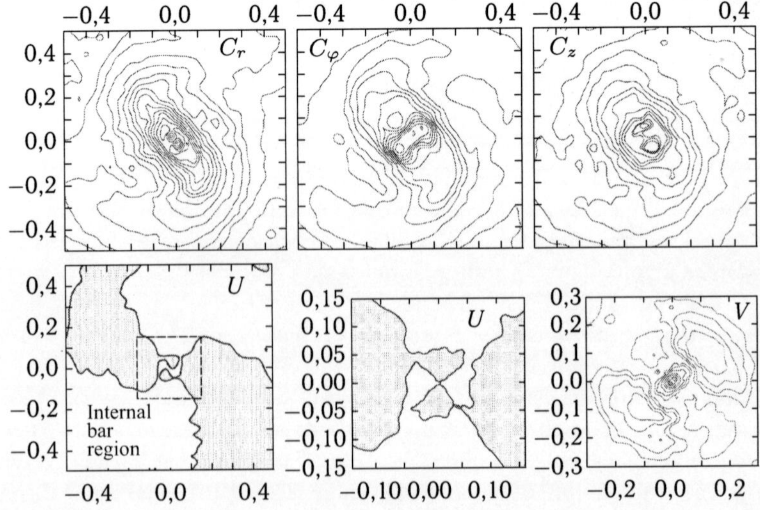

Fig. 4.60. Central part of the disk in the region of the double bar. The outer radius of the disk $R = 1$. The isolines of the dispersions of the velocities c_r, c_φ, c_z (top), radial (solid line $U = 0$) and the azimuthal (V) components of velocity in the stage of the double bar.

The presence of the inner bar is convincingly indicated by the velocity field. Figure 4.60 shows the distribution of the azimuthal component of the velocity V and radial component U. In the stage of the double bar, the isolines V are stretched along the internal jumper. The radial velocity shows the characteristic four-sector structure for both the external and inner bar. The line $U = 0$ passes along the major and minor axis of the bar.

The analysis of the vertical distribution of the matter in the disk indicates the absence of hydrostatic equilibrium (see (2.56)) in the double bar stage. This reason is associated with the phenomenon of bending of the bar during its formation. The bending instability of the bar is one of the factors leading to failure of the system of the double bar.

Thus, the formation of the structure of the type of double bar can take place in the non-linear stage of the formation of the global bar $M_b \gtrsim 0.3M_d$ and 'porous' bulge $b \simeq (0.12 \div 0.28)L$. The internal jumper forms in the models if at the initial moment: 1) the disk is thin ($z_0/L = 0.04 \div 0.1$); 2) the radial profile c_r is close to the profile of the dispersion shown in Fig. 4.27, i.e. the disk is relatively hot. In this case, the growth of the perturbations starts in the very central region of the desk (Fig. 4.59, 1, 2, 3). However, if the disk is cold ($Q_T < 1$) a system of transient spirals forms and the bar mode survives without the internal jumper stage.

The observational data on the kinematics of the central region of the double-barred galaxies indicate the 'stretching' of the distribution of the dispersion of the velocities of the stars c^{obs} along the main bar [856], but do not enable convincing assumptions to be made in the profile of c^{obs} along the inner bar.

The stability of double bars

The most important problem of the double bars is the problem of the dynamic stability of the system of this type. It was mentioned previously that the duration of the phase of the 'double bar' is not long, and one of the reasons for the disappearance of the internal jumper is the start of relaxation processes in the vertical direction – external manifestations are heating and increase of the thickness of the system. In the initial stage of the formation of the bar, the isolines of the vertical scale of the disk Δ repeat the density distribution. However, the vertical heating disrupts the orbits, stretched along the minor axis of the outer bar and this results in the failure of the internal structure.

The formation of a bar in the thin disk is accompanied by the bending instability which leads in the final analysis to thickening of the entire disk (section 4.5). Figure 4.61 demonstrates the typical bending of the bar in the non-linear stage when the single-arm bending perturbation exists in the disk (Fig. 4.61a). The wavelength of the distortion of the bar is slightly greater than its length (Fig. 4.61b). In most cases, 1–2 lengths of the bending wave are distributed along the major axis of the bar. The spiral bending modes with the azimuthal number $m = 1, 2$, bending the entire disk, are superimposed on the bending of the bar.

After stabilisation of the bending instability, when the heating of the system stops in both the disk plane and in the vertical direction, the internal jumper does not form in the hot disk.

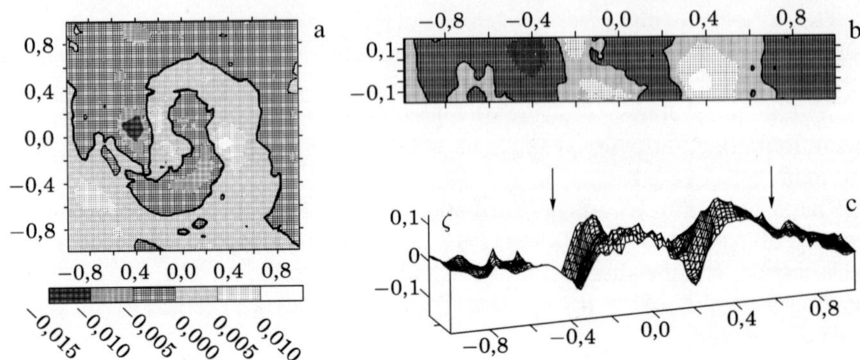

Fig. 4.61. a) the distribution of the vertical displacement of the centre of gravity ζ in the disk plane. The thick line – level $\zeta = 0$; b) only the region of the bar is shown; c) the profile of the vertical displacement along the bar. The arrows indicate the position of the ends of the bar.

If there is no bulge and the scale of the halo is $a \gtrsim L$, the formation of the global bar takes place by a qualitatively different mechanism. The structures of the type of 'double bar' do not form. As shown in section 4.6, the characteristic special features of the development of the global bar mode in the initially very cold and thin axisymmetric isolated disk is the possibility of displacement of the centre of the bar in relation to the kinematic centre in the centre of gravity of the disk component. In this case also, this factor does not allow the formation of orbits normal to the major axis of the bar.

The most favourable conditions for the formation of the double bar exist in the presence of a massive bulge. However, in the reduction of the scale of the bulge b the mechanism of failure of the bar mode, associated with the scattering of the particles in flight in the vicinity of the concentrated nucleus of the bulge (chapter 9), is important. The scattering of the particles is most efficient for the particles forming the inner bar and, consequently, the lifetime of the bar decreases with the reduction of the scale b.

Thus, the complex trajectories of the particles, providing existence of a double bar, apparently, are extremely unstable with respect to various perturbations, which, in addition to the bar-mode, occur in the initial stages (first few revolutions of the outer edge of the disk). These include transient spiral waves in the plane of the disk, bending of the bar, bending instability of the disk. A similar effect has the presence of an external asymmetric potential associated with, for example, tidal effects.

In the constructed models the structures of the type of double bars are temporary formations at a certain stage of bar formation. This finding is consistent with short lifetime of the inner bar also in [592]. In conclusion, we note a critical analysis of A.V. Moiseev 147] of the kinematic data of the observations of galaxies in which the photometric data show the presence of the double bars. It is shown that in such objects the inner bar has no effect on the kinematics of the circumnuclear gas and stars. Thus, the second bar is not the dynamically allocated galactic structure. At the same time, even in the short phase of the double bar in the constructed models the kinematics on a short phase of the 'double bar' in building a model of the kinematics of the region of the internal structure repeats the main features of the kinematics of the outer bar.

4.8. Mechanisms of stellar relaxation

The reasons which could explain the observed correlation between the age, the dispersion of the velocities of the stars and their scale of heights, have been studied in a large number of investigations (see, for example [131, 373, 585, 741, 1048] and the references therein, section 1.2).

The observations indicate the evolution of the distribution function of the velocities of the stars (see Fig. 1.13) [373, 505, 507]. At the same time, the stellar-stellar convergences in this sense are known to be ineffective. The variation of the velocity of a star as a result of scattering is proportional to the magnitude of the scattering mass. Therefore, the giant molecular clouds (GMC) can be used as an important factor influencing the dynamics of the stellar disk. The large mass and the dimensions of the GMC (section 1.3) enable stars to be scattered efficiently on them and this results in relaxation processes, increasing the effective temperature of the stellar population. The scattering on the massive non-point objects (for example, GMC) as a possible mechanism of distribution of the velocities of the stars has been studied many times [105, 741, 816, 1195, 136]. It can be referred to as mechanical, in contrast to the one described in sections 2.6, 4.5, where attention has already been given to the problem of anisotropy in the distribution of the velocity of the stars and from the viewpoint of collective processes – the stability with respect to bending modes.

4.8.1. Scattering on giant molecular clouds

Several results of the numerical modelling of the dynamics of the stellar disk taking the GMC into account and following the studies by Villumsen [1268, 1269] will be discussed. The following model will be considered. The distribution of the density of a stellar population of the disk was selected in the form (compare with (2.43))

$$\rho(r,z)=\frac{M_d}{2\pi L_\sigma^2 z_e}\exp\left\{-\frac{r}{L_\sigma}\right\}\operatorname{sech}^2\left\{-\frac{z}{2z_e}\right\}, \tag{4.129}$$

and in the halo (4.3) at $k = 1$. The distribution of the density of matter in the GMC

$$\rho_G(r)=\frac{3M_G}{4\pi r_G^3}\left[1+\frac{r^2}{r_G^2}\right]^{-5/2} \tag{4.130}$$

corresponds to the potential

$$\Phi_G(r)=-\frac{GM_G}{(r^2+r_G^2)^{1/2}}. \tag{4.131}$$

The interaction between the stars and the GMC was taken into account only if the distance between them did not exceed r_{sG} = 2 kpc. At the initial moment,

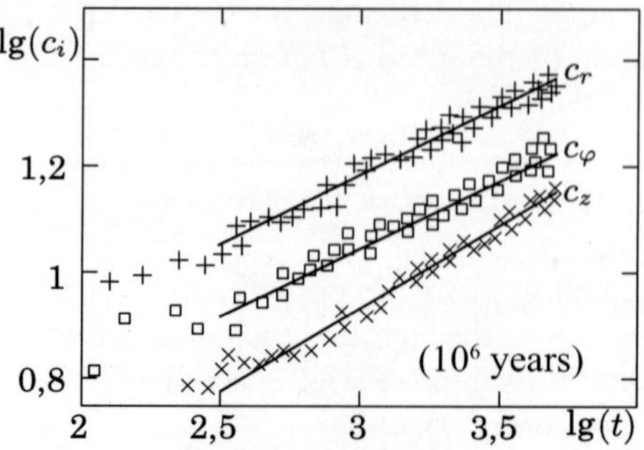

Fig. 4.62. Approximation of the time dependence of the components c_r, c_σ, c_z of the dispersion of the velocities according to the results of Villumsen's numerical modelling [1269].

the GMC formed an exponential disk with $L_G < L_\sigma$. The values of the parameters were selected in application to the Galaxy: $M_d = 0.4{\cdot}10^{11}\ M_\odot$, $L = 3.5$ kpc, z_e = 0.325 kpc, $M_h(r \leqslant 30 \text{ kpc}) = 4M_d$, $a = 0.1$ kpc, the total mass of the GMC was $4{\cdot}10^9\ M_\odot$. The initial values of the dispersion and the GMC parameters were varied in the calculations. The number of the stars N_s and the GMC N_G did not exceed 4000 for each subsystem, since $0.1 \leqslant N_G/N_s \leqslant 1$.

Figure 4.62 shows the dynamics of three components of the dispersion of the velocities of the stars. At the end of modelling, we obtain the following ratios of these components for the arbitrary radial coordinate:

$$c_r / c_\varphi / c_z = (1/0.72/0.60) \pm 0.04. \tag{4.132}$$

It is interesting to compare the data obtained by Villumsen with the data obtained by Lacey [816] who obtained the following equation, describing the evolution of the value of the dispersion $(c^2 = c_r^2 + c_\varphi^2 + c_z^2)$ as a result of collisions with the GMC:

$$\frac{dc^2}{dt} = \frac{D}{c^2}, \tag{4.133}$$

where $D = 2G^2 n_G M_G^2 \ln(\Lambda)\omega_z f(2\Omega/\varkappa)$, n_G is the surface concentration of the GMC, f is the known function. Since $2\Omega/\varkappa \simeq 1.4$, then according to Lacey, $f(1, 4) = 0.47$; ln (Λ) = ln $(r_s G/r_G)$ and for the Villumsen experiment conditions ln (Λ) = ln (2 kpc/50 pc) = 3.7. The solution of (4.133) is written in the form:

$$c(t) = \left(c_0^4 + 2Dt\right)^{1/4}, \qquad c_0 = c(t=0). \tag{4.134}$$

For the complete experimental dispersion of the velocities (Fig. 4.62) and the theoretical dependence (4.134), in the case $c_0 = 0$ we can write

$$\lg(c) = \lg(k) + n\lg(t); \qquad n \simeq 0.25. \tag{4.135}$$

In the mean, the difference between $k_{\rm exp}$ and $k_{\rm th}$ is of the order of 50% [1269].

Taking into account the dependence of the thickness of the stellar disk on the age of the stars, observed in the galaxy, it is important to mention the increase with time of quantity Δ_* in the Villumsen experiments:

$$\lg\{\Delta_*(3 \cdot 10^8 \text{ \%\%} \leqslant t \leqslant 5 \cdot 10^9 \text{ \%\%})\} = 0.91 + 0.36 \ \lg\{t / 10^6 \text{years}\}. \quad (4.136)$$

In the models proposed by Yasumoto and Fujimoto [3016] the parameters describing the GMC subsystem were varied in a wide range. In addition to these parameters, the spiral density waves were also taken into account. At the start of the experiments, the stellar disks were assumed to be cold and in the first stage the dispersion of the velocities rapidly increased and subsequently, starting at $t \simeq 10^9$ years, the following value was obtained $c \propto t^n$ with $n = 0.20$. At the time $t = 10^9$ years in different models $c \simeq 31 \div 47$ km/s. For $M_G = 10^6 M_\odot$ and the spiral wave with the twist angle of $i = 10°$, the dispersion reached the maximum value (47 km/s). The results show that the nature of the spatial distribution of the GMC has only a small effect on the rate of growth of the dispersion of the velocity of the stars[13].

It should be mentioned that already in 1961 Kuz'min [105] obtained very similar results. Assuming that the gravitational effect of the stars with the clouds of diffusion matter should increase both the radical and vertical components of the dispersion of the velocities of the stars, the following value of anisotropy was obtained at the equilibrium limit ($c_z/c_r \to$ const):

$$\alpha_b = (1 + 4\Omega^2 / \varkappa^2)^{1/2}. \quad (4.137]$$

It may easily be shown that in the non-rotating layers and in the solid-state rotating disks $(c_r / c_z)_b = \alpha_b^{-1} = 1/\sqrt{2} \simeq 0.71$, and in the differentially rotating disk, as in the solar neighbourhood ($\Omega \propto r^{-1}$), we should have $(c_z/c_r)_b \simeq 0.58$.

Thus, if the evolution of the distribution function of the velocities of the stars is determined by the process of scattering of the stars on the massive gas clouds, then for the subsystem of the older stars of the disk $c_z/c_r \simeq 0.6$. However, it should be remembered that the previously examined models take into account only approximately the actual distribution in space of the scattering objects and their dynamics at times comparable with the lifetime of the Galaxy. At the same time, the reliable observed values of c_z/c_r relate only to the very close solar neighbourhood.

However, it should be remembered that the GMC are fine structures. Their thickness h_{GMC} is considerably smaller than the thickness of the old stellar disk. Consequently, heating of the young stars in the GMC is effective only in the initial stage. With increase of the vertical component of velocity the stars stay for most of the time in the range $|z| \gg h_{\text{GMC}}$, and the heating rate rapidly decreases. Thus, the three-dimensional models of the scattering of the stars in the GMC do not explain the observational data and, evidently, only the combined effect of different mechanisms can ensure the high rate of relaxation processes.

A specific contribution to the heating of the stellar disk may be provided by the interaction with the spiral density wave [1316], spherical clusters and hypothetical massive black holes of the halo penetrating the disk subsystem. In particular, the

[13]It should be mentioned that in the discussed experiments, special attention was given to the special feature in the distribution of the GMC – the ring with increased density in the range $4 < r < 8$ kpc [1111].

possibility of heating of a stellar disk by the massive clusters of dark matter in the disk has been studied.

4.8.2. Solution of the paradox of stellar relaxation

We mention the words from the classic study by Spitzer and Schwarzschild [1195]: "All attempts for the interpretation of the observed movement of the stars in our Galaxy face the question that either any star moved on its orbit in the general gravitational field of the Galaxy without perturbation in the last 10 billion years or the pertubations should effectively change the stellar orbits for". The problem of explanation of this observed growth of the dispersion of the velocities of the stars with the different spectral class in the absence of effective heating as a result of the stellar–stellar interactions due to weak collisions of the stellar disk was referred to by K.F. Ogorodnikov in 1965 as the 'fundamental paradox of stellar dynamics'. If we use the equation (4.1) for the characteristic relaxation time as a result of the paired gravitational interactions, we obtain the times exceeding the required time by up to 1000 times.

The problem of stellar relaxation was solved basically in the study [585], and the results of this study will be mentioned briefly in the following section.

Almost all the theoretical attempts for solving this paradox were based on the proposal by A. Eddington (1943) on the efficiency of scattering on a large mass. At different times, these objects were represented by the massive gas clouds, perturbation of the density in the spiral structure, black halo holes (see discussion in [585]). It should be mentioned that the sub-halos, being the small-scale formations of the halo (chapter 10), could play this role.

The heating of a stellar subsystem as a result of the interaction of the stars and gas clouds will be discussed in greater detail. It is important to note that this interaction may lead to different relaxation mechanisms characterised by the natural heating laws.

There are two limiting scenarios of relaxation on the basis of observations. Firstly, the increase of the dispersion of the velocities of the stars in the last 10 billion years was in the range from 10 km/s to 40 km/s. The second indicates the relaxation in the first $4 \cdot 10^9$ years of the lifetime of the star and then takes place without relaxation. In the latter case, the dissipation of the velocities increases to 30 km/s.

For the evolution of the energy of the stars we can write the exponential law

$$\varepsilon(t) = \varepsilon_0 \left(1 + \frac{t}{\tau}\right)^q, \tag{4.138}$$

where τ is the characteristic time. For the first scenario we have the increase of the main energy $\langle\Delta\varepsilon\rangle \geqslant 12\varepsilon_0$, in the second case $\langle\Delta\varepsilon\rangle \geqslant 4\varepsilon_0$ is fulfilled. Let it be that the relaxation time is Δt and, consequently, the characteristic time scale is

$$\tau = \frac{\Delta t}{(\varepsilon(\Delta t)/\varepsilon_0)^{1/q} - 1}. \tag{4.139}$$

The characteristic times in accordance with (4.139) are presented in Table 4.1. The

Table 4.1. Characteristic timescales τ and the time during which the dispersion of the velocities is doubled in comparison with the initial level for different relaxation scenarios

	Characteristic scale (τ, years)		Doubling time (t_2, years)	
q	1st scenario	2nd scenario	1st scenario	2nd scenario
1	$4.2 \cdot 10^8$	$5 \cdot 10^8$	$1.3 \cdot 10^9$	$1.5 \cdot 10^9$
2/3	$8.1 \cdot 10^7$	$1.5 \cdot 10^8$	$5.6 \cdot 10^8$	$1.1 \cdot 10^9$
1/2	$1.6 \cdot 10^7$	$5 \cdot 10^7$	$2.4 \cdot 10^8$	$7.5 \cdot 10^8$
2/5	$3.2 \cdot 10^6$	$1.7 \cdot 10^7$	$1 \cdot 10^8$	$5.1 \cdot 10^8$
1/3	$6.4 \cdot 10^5$	$5.5 \cdot 10^6$	$4 \cdot 10^7$	$3.5 \cdot 10^8$

mechanisms with q = 2/5 or 1/3 are based on non-local approximation [816, 1195, 1196]. In the case with q = 1, the orbital, epicyclic and vertical movements of the stars are not taken into account in the calculations [1303].

The energy balance between the stellar and gas subsystems can be described by the equation

$$n_* \frac{d\varepsilon_*}{dt} = n_c \varepsilon_c w, \tag{4.140}$$

where n_* and n_c is the concentration of the stars and the clouds, respectively, $\varepsilon_* m_* c_*^2/2$ and $\varepsilon_c = m_c c_c^2/2$ are the mean energies of random movements, $w = \langle v n_* \sigma_E \rangle$ is the probability of the loss of energy by the clouds as a result of collisions with the stars, σ_E is the appropriate scattering cross-section, v is the relative velocity of the cloud and the star.

Equation (4.140) can be written in the form of the τ-approximation:

$$\frac{d\varepsilon_*}{dt} = \frac{\varepsilon_c}{\tau_{\text{vol}}}. \tag{4.141}$$

Plasma physics shows that

$$\tau_{\text{vol}} = \frac{\langle v^3 \rangle}{\sqrt{32\pi} G^2 n_c M_c m_* \lambda}, \tag{4.142}$$

i.e. the relaxation time is determined by the parameters of the clouds (M_c is the mass of the cloud), m_* is the mass of the star, $\lambda \sim 2$ is the Coulomb logarithm, $\langle v^3 \rangle \simeq c_*^3$ (c_* is the dispersion of the velocities of the stars). After integration of (4.141) we obtain

$$\varepsilon_* = \varepsilon_0 \left(1 + \frac{t}{\tau_*}\right)^{2/5}, \tag{4.143}$$

where $\tau_* = \frac{2}{5}\frac{m_* c_{*0}^2}{M_c c_c^2}\tau_0 = \frac{c_0^3}{10\sqrt{2\pi} G^2 n_c M_c^2 \lambda}$. For the values c_0 = 10 km/s, n_c = $2 \cdot 10^{-8}$ pc^{-3}, $M_c = 10^6 M_\odot$, $\lambda = 2$, we have $\tau_* = 5 \cdot 10^7$ years.

Similarly, we can investigate other possible models [585]: the thin layer of the clouds in the thick layer of the stars; cooling of the system of clouds; differential rotation in the collisional system of the star-cloud; friction as a result of the relative movement of the gas and the stellar disk. The results are summarised in Table 4.2. The following conclusions can be made:

Table 4.2. Properties of different mechanisms of relaxation, determined by the collision of stars with giant molecular clouds ($C_* = c_0 (1 + t/\tau)^p$, $\varepsilon_* = \varepsilon_0 (1 + t/\tau)^q$)

<table>
<tr><th colspan="2" rowspan="2">Relaxation mechanism</th><th colspan="2">Exponent</th><th rowspan="2">Time scale (τ, years)</th><th rowspan="2">Comment</th></tr>
<tr><th>p</th><th>q</th></tr>
<tr><td rowspan="2">Non-rotating collisional system of stars and clouds</td><td>Mixture of stars and clouds</td><td>1/5</td><td>2/5</td><td>$5 \cdot 10^7$</td><td rowspan="5">For all mechanisms, expect the last two, the scenario is not realised, and the second one is possible</td></tr>
<tr><td>Thin layer of clouds and thick layer of stars</td><td>1/4</td><td>1/2</td><td>$5 \cdot 10^7$</td></tr>
<tr><td rowspan="2">Differentially rotating collisional system of stars and clouds</td><td>Mixture of stars and clouds</td><td>1/3</td><td>2/3</td><td>$5 \cdot 10^7$</td></tr>
<tr><td>Thin layer of clouds and thick layer of stars</td><td>1/4</td><td>1/2</td><td>$5 \cdot 10^7$</td></tr>
<tr><td colspan="2">Viscous stress in the system of clouds in a differentially rotating stellar disk</td><td>1/4</td><td>1/2</td><td>10^8</td></tr>
<tr><td colspan="2">Friction caused by relative rotation of gas and stellar disks</td><td>1/2</td><td>1</td><td>10^{10}</td><td rowspan="2">Both mechanisms are effective mainly for hot stars</td></tr>
<tr><td colspan="2">Heating due to spatial diffusion in stellar disk</td><td>1/2</td><td>1</td><td>10^{10}</td></tr>
</table>

- the collisions of the stars with the giant molecular clouds may play a significant role in the heating of the stellar subsystem;
- various relaxation mechanisms are based on these conditions. These mechanisms may result in a wide spectrum of the dependences of the dispersion of the velocity of the stars on their age. The observed variants of the exponential laws of heating of the stellar component may be a reflection of different relaxation mechanisms;
- there are two types of scenario of heating: with and without saturation;
- heating with saturation may take place in both the non-rotating collisional system of the stars and the gas clouds and when taking into account the differential nature of rotation, irrespective of geometry (the star and the clouds are mixed in the volume or the thin layer of the clouds is in the thick layer of the stars). The investigated case also occurs when examining the viscosity in the differentially rotating disk;
- the mechanisms, associated with the differential nature of rotation, were investigated for the first time in [585]. For example, in the classic studies [816, 1195], the differential nature of rotation was not taken into account quantitatively in a correct manner;

- the heating without saturation can be determined by two mechanisms: a) friction in the presence of the relative rotation of the gas and stellar disks, b) by means of the spatial diffusion of the star;
- it should be mentioned that the last two mechanisms are characterised by the time scale $\tau \simeq 10^{10}$ years and the exponents for the age–velocity dispersion of the stars ratio p = 1/2 is in agreement with the results published in [1303];
- the relaxation mechanisms in the case of heating with saturation (second scenario) leads in similar values $\tau \simeq 5 \cdot 10^7 - 10^8$ years, although the exponents for the dependence $c_*(t)$ greatly differ: p = 1/6, 1/5, 1/4, 1/3, depending on the relaxation mechanism.

The most important result is that each group of the stars may have its own heating (cooling) law which differs from others by the exponent in the exponential law of the type (4.138).

4.9. Problem of the formation of exponential profiles of the brightness of stellar disks

The radial profile of the brightness of the stellar component of the S-galaxies is characterised, with a small number of exceptions, by the exponential law (1.2.1). It is assumed that the distribution of surface density $\sigma_*(r)$ has the same form. The universal nature of the radial dependence $\sigma(r)_* \propto \exp(-r/L_\sigma)$ requires explanation. The theory, based on considering the viscosity in the gas disk and star formation effects [223, 469, 547, 843] has been developed to the largest extent.

We use the equations of the viscous non-stationary accretion (section 11.2). In addition, it is also necessary to take into account: star formation, external accretion on the disk, the non-stationary nature of the rotation curve of the gas. Consequently, the equation for the surface density of the gas σ_g assumes the following form:

$$\frac{\partial \sigma_g}{\partial t} = -\frac{\partial}{r\partial r}\left(\frac{1}{\partial \Omega r^2 / \partial r}\frac{\partial}{\partial r}\left[\nu\sigma_g r^3\frac{\partial \Omega}{\partial r}\right]\right) + \frac{\partial}{r\partial r}\left(\frac{\dot{m}_e r}{\partial \Omega r^2 / \partial r}\left[r^2\Omega - L_e\right]\right) -$$

$$-\psi + \dot{m}_e - \frac{\partial}{r\partial r}\left(\frac{\sigma_g r^3}{\partial \Omega r^2 / \partial r}\frac{\partial \Omega}{\partial t}\right), \tag{4.144}$$

where $\dot{m}_e$ is the rate of inflow of the gas per unit area of the disk as a result of accretion, $\mathcal{L}_e$ determines the angular moments of the accreting gas, ψ is the flow rate of the gas for the formation of stars calculated per unit area, and it may be assumed that $\psi = \sigma_g/t_s$ (t_s is the characteristic star formation time). The first term on the right hand side of (4.144) is determined by the effect of viscosity ν, the second term takes into account the exchange of the momentum between the matter of the disk and the accreting gas, the last term is associated with the variation of the gravitational potential as a result of the distribution of matter in the disk. The balance of the centrifugal and gravitational forces should take into account the Poisson equation. The stars form from the gas:

$$\frac{\partial \sigma_*}{\partial t} = \frac{\sigma_g}{t_s}. \tag{4.145}$$

This model contains two main characteristic times: $t_v = r^2/\nu$ – the viscous time, and t_s. The relationship between t_v and t_s is the most important moment of the theory of evolution of the galactic disk. If $t_v \sim t_s$, the exponential distribution of the stars in the disk is established [843]. This result depends only slightly on the type of the rotation curve, the selection of the viscosity law $\nu(\sigma_g, r)$ and external accretion. This may be used to explain the universal nature of the exponential distribution of $\sigma_*(r)$ of the stellar disks of the galaxies of different types.

4.10. Three-arm spirals in a stellar disk

The two-arm mode (in particular, the bar mode) is dominant in the models of the stellar disk with the low-mass spheroidal subsystem. An increase of the mass of the halo creates suitable conditions for the formation of spirals with a high azimuthal number $m = 3, 4, \ldots$. This results is obtained, for example, in investigating the swing amplification mechanism [314, 317]. For the disks with a constant velocity of rotation V = const the swing amplification is greater for the dimensionless wavenumber $X = \lambda_t/\lambda_c$, and the value X depends on the number of arms m and the ratio of the radial forces of the halo and the disk f [372]:

$$X = \frac{2}{m}(1+f). \tag{4.146}$$

The value $f \propto \mu$, and therefore as the relative mass of the halo μ increases, the number of the arms of the spiral structure increases also.

Gravitationally unstable disks will be considered. The generation of the spiral structure as a result of the development of gravitational instability in the first approximation can be efficiently characterised by two quantities: the Toomre parameter $Q = c_r/c_T$ and the ratio of the mass of the spherical component of the disk components inside the optical radius $\mu = M_h/M_d$ (Fig. 4.63). In a very cold $Q \ll 1$ and massive disk $\mu \ll 1$, when the Jeans scale $\lambda_J = 2\pi/k_0$ (see equation (2.98)), is compared with the dimensions of the disk, suitable conditions form for the development of a single-arm mode $m = 1$ (section 4.6). The two-arm mode (bar

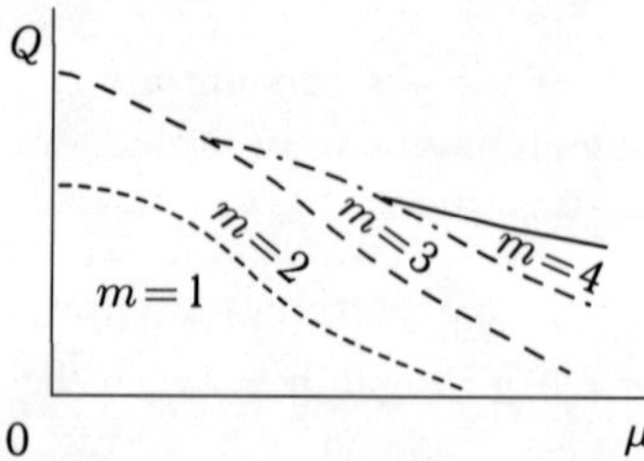

Fig. 4.63. Qualitative diagram of the position of the boundaries of the regions of the dominant development of gravitationally unstable modes in the linear stage with different azimuthal number m on the plane of the parameters (Q, μ).

mode) forms in a hotter disk already in the absence of the suitable conditions for the rapid development of the perturbation with $m = 1$.

The results of modelling of the mode $m = 3$ will be discussed within the framework of the numerical models of N-bodies. The formation of a three-arm mode requires a more massive spherical component $\mu \sim 1$ and a not very cold disk with $Q(r) \sim 1$. Figure 4.64 shows the results of non-linear development of the mode $m = 3$ from the axisymmetric state for two models differing only in the initial radial profile of the dispersion of the velocities $c_r(r)$ with the condition $c_\varphi = c_r\pi/2\Omega$ fulfilled ((equation (1.10)). It a hotter disk, the amplitude of the wave $m = 3$ increases at a lower rate (the case c, d), but the three-arm mode fails later than in the model with a smaller Q_T (a, b).

It should be mentioned that the investigation of the three-arm mode requires models of collisionless disks with a large number of particles $N \gtrsim 10^6$ and a high accuracy of integration of the equations of motion. For example, in the models with $N \lesssim 1 \cdot 10^5$ the formation of the perturbations with $m = 3$ in the non-linear stage is difficult to investigate, and in the case $N \simeq (2 \div 3) \cdot 10^5$ the three-arm spiral forms but rapidly transforms to a bar-mode. The lifetime of the mode $m = 3$ increases with increasing N, and for the convergence of the results it is necessary to use the models with the number of particles $N \gtrsim 10^6$.

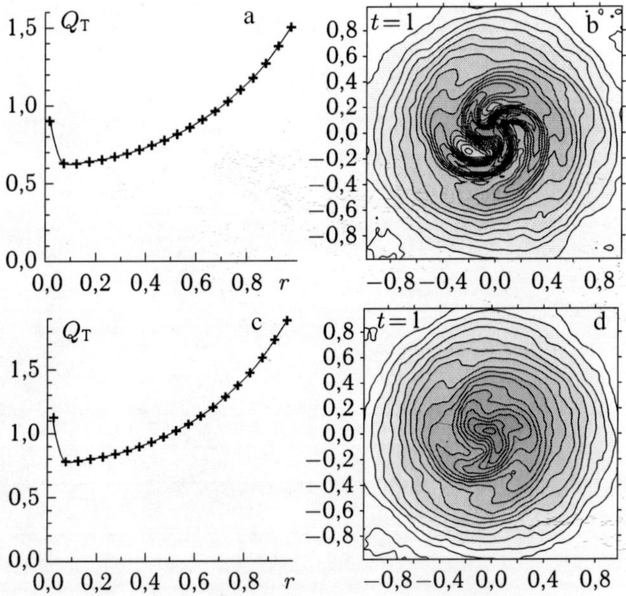

Fig. 4.64. The models of the stellar disk with a three-arm spiral structure with $\mu = 1$: a) the initial distribution of the Toomre parameter $Q_T(r)$; b) the logarithm of the surface density lg (σ) for this model; c) $Q_T(r)$ for a hotter initial disk; d) the appropriate distribution lg (σ). The period of rotation of the outer edge of the disk ($R = 1$) with respect to circular velocity is 4. The amplitude of the wave is considerably higher in the first case.

The calculation show that the three-arm modes are quite sensitive to the external effects, especially in the initial linear stage of development. The disruption of the initial axial symmetry in the distribution of the density of matter of the speed of rotation, and external effect, result in a reduction of the lifetime of the three-arm spiral or in its disappearance.

The identical effect is observed in the models situated away from the boundary of gravitational stability with respect to the parameter Q_T (Fig. 4.63, 4.26, 4.27). In this case, the perturbations $m = 3$ are accompanied by the increase of the two-arm mode, and their interaction in the non-linear stage results in the disruption of the three-arm structure in favour of the bar mode (Fig. 4.65). In this calculations, in the initial stage of evolution ($t \leq 2 \div 3$) the mode $m = 3$ increases preferentially, whereas the amplitude of the two-arm mode increases at a considerably lower rate. The three-arm mode reaches saturation earlier, but when the amplitude of the wave with $m = 2$ is compared, then its dynamics at the periphery of the disk ($r \gtrsim 0.5$) is already determined by the central bar structure.

A number of additional special features of the three-arm perturbations will be discussed. The distribution of all main kinematic parameters in the disk (the velocity of matter and the dispersion of different components of velocity) indicates the presence of the mode $m = 3$ (Fig. 4.66). In the distribution c_r c_φ the presence of the three-arm structure is more evident than in the dispersion of the vertical

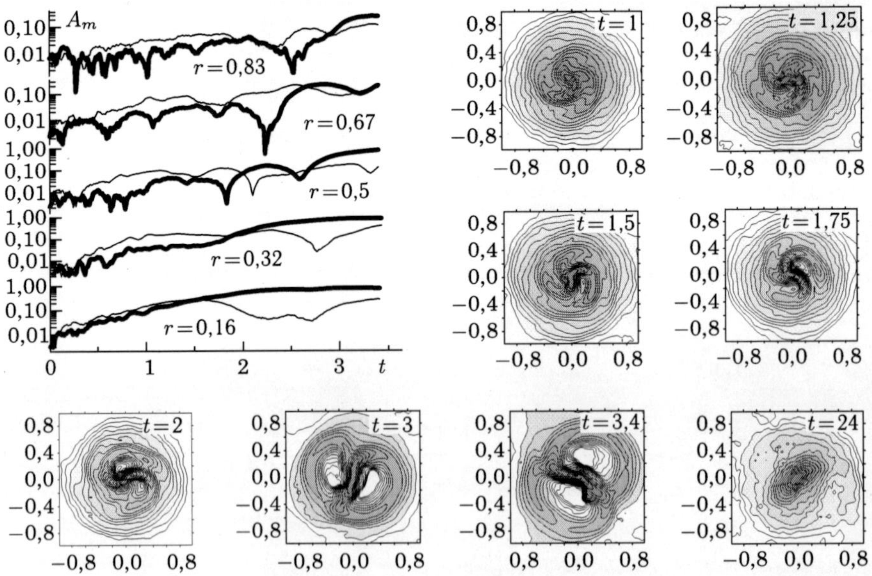

Fig. 4.65. Evolution of the disk with the initial parameters permitting the development of the modes with numbers $m = 2, 3$. On the left – the time dependence is of the Fourier harmonics A_2 (thick line), A_3 (thin line) at different distances from the centre r. On the right and below – isolines of the logarithm of the surface density at different moments of time t. The mode $m = 3$ is dominant in the initial stage and in the non-linear stage this mode is disrupted by the bar mode $m = 2$.

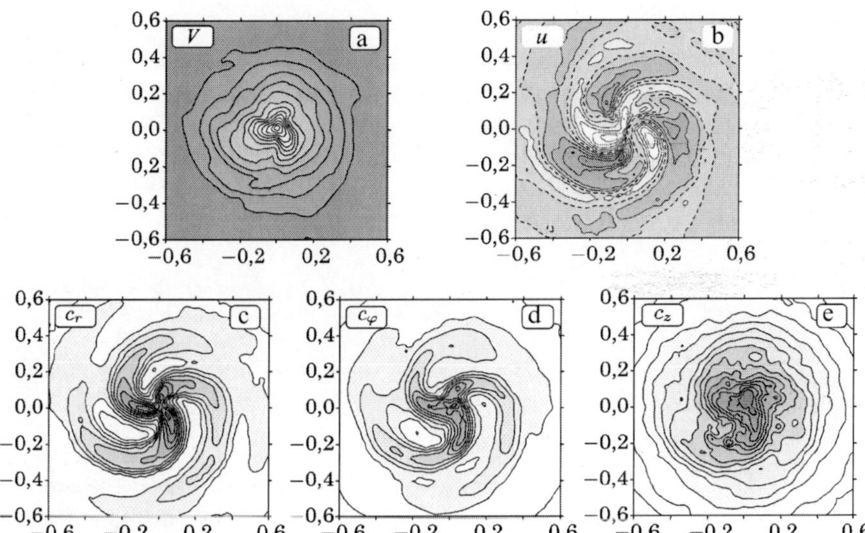

Fig. 4.66. Distribution of the kinematic parameters in the central region of the disk in the stage of dominance of the three-arm mode (dark shading indicates higher values): a) the azimuthal velocity V; b) the radial velocity u (dashed line $u = 0$); c) dispersion of radial velocities c_r; d) the dispersion of azimuthal velocities c_φ; e) the dispersion of the vertical velocities c_z.

velocities c_z. This is associated with the fact that the generation of the investigated perturbations $m = 2, 3$ is caused by the development of the gravitational instability in the disk plane, and the heterogeneity of the parameters, characterising the vertical movements, is secondary. In the region of increase density in the arm the azimuthal velocity V decreases, and the rate of rotation of the matter between the arms is greater. Radial velocity u is more complicated (Fig. 4.67), and develops in the central region and also at the periphery.

The numerical model with the gravitating collision less disks indicate the unstable nature of the three-arm structure in the non-DS stage. The phase velocities of rotation Ω_{ph} (m) for the modes $m = 2$ and $m = 3$ differ, and in the non-linear stage they interact together with the failure of the three-arm spiral in favour of the mode $m = 2$. One of the three spirals has a small amplitude, and the symmetry of the mode $m = 3$ is disrupted so that the disk has a more stable structure - a bar mode (Fig. 4.65). The identical results are obtained if there is no axial symmetry in the initial distribution of the parameters or in the process of development of the gravitational instability the disk is subjected to the external non-symmetric effect associated, for example, with the presence of a satellite or incidence of mass onto the disk.

Although the majority of the spiral galaxies of the grand design type appear to-arm on the photographs (or, in a smaller number of cases, as four-arm), Fourier analysis often indicates the presence of the harmonics $m = 3$ with a high amplitude in the field of the zero velocity of the gas (section 7.4, 7.5). In most cases, the three arms in the galaxies are not identical. One of the arms is 'weaker'. As an example, one could mention NGC 628 (Fig. 7.26).

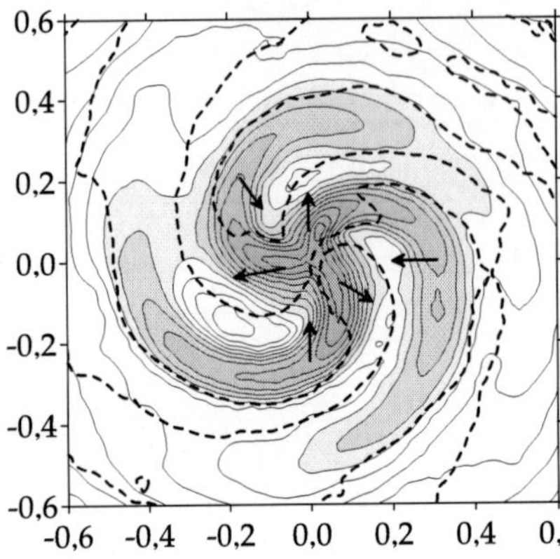

Fig. 4.67. A line at which the radial velocity $u = 0$ (the thick dashed line) is superposed on the distribution of the logarithm of surface density (the isolines are indicated by the thin lines). The arrows show the direction of radial velocity.

The calculation show that the three-arm modes are quite sensitive to the external effects, especially in the initial linear stage of development. The disruption of the initial axial symmetry in the distribution of the density of matter or the speed of rotation, or the external effect, result in a reduction of the lifetime of the three-arm spiral or in its disappearance.

The identical effect is observed in the models situated away from the boundary of gravitational stability with respect to the parameter Q_T (Fig. 4.63, 4.26, 4.27). In this case, the perturbations $m = 3$ are accompanied by the increase of the two-arm mode, and their interaction in the non-linear stage results in the disruption of the three-arm structure in favour of the bar mode (Fig. 4.65). In these calculations, in the initial stage of evolution ($t \leqslant 2 \div 3$) the mode $m = 3$ increases preferentially, whereas the amplitude of the two-arm mode increases at a considerably lower rate. The three-arm mode reaches saturation earlier, but when the amplitude of the wave with $m = 2$ is compared, then its dynamics at the periphery of the disk ($r \gtrsim 0.5$) is already determined by the central bar structure.

A number of additional special features of the three-arm perturbations will be discussed. The distribution of all main kinematic parameters in the disk (the velocity of matter and the dispersion of different velocity components) indicates the presence of the mode $m = 3$ (Fig. 4.66). In the distribution of c_r and c_φ the presence of the three-arm structure is more evident than in the dispersion of the vertical velocities c_z. This is associated with the fact that the generation of the investigated perturbations $m = 2, 3$ is caused by the development of the gravitational instability in the disk plane, and the heterogeneity of the parameters, characterising the vertical movements, is secondary. In the region of increased density in the arm the azimuthal velocity V decreases, and the velocity of rotation of matter between the arms is greater. Radial velocity u is more complicated (Fig. 4.67), and develops in the central region and also at the periphery.

The numerical models with the gravitating collisionless disks indicate the unstable nature of the three-arm structure in the non-linear stage. The phase velocities of rotation $\Omega_{\text{ph}}\,(m)$ for the modes $m = 2$ and $m = 3$ differ, and in the non-linear stage they interact together with the failure of the three-arm spiral in favour of the mode $m = 2$. One of the three spirals has a low amplitude, and the symmetry of the mode $m = 3$ is disrupted so that the disk has a more stable structure – a bar mode (Fig. 4.65). The identical results are obtained if there is no axial symmetry in the initial distribution of the parameters or in the process of development of the gravitational instability the disk is subjected to the external non-symmetric effect associated, for example, with the presence of a satellite or incidence of mass onto the disk.

Although the majority of the spiral galaxies of the grand design type appear as two-armed on the photographs (or, in a smaller number of cases, as four-armed), Fourier analysis often indicates the presence of the harmonics $m = 3$ with a high amplitude in the field of zero velocity of the gas (section 7.4, 7.5). In most cases, the three arms in the galaxies are not identical. One of the arms is 'weaker'. As an example, one could mention NGC 628 (Fig. 7.26).

4.11. Tidal effects and interaction of the galaxies

The modelling of the double galaxies is highly popular. In the simple models, the distribution of the mass of the perturbing satellite is regarded as fixed, for example, of the point type. The type of orbit of the satellite is determined by the formulation of the problem. One of the first studies in which numerical modelling of this type was carried out are those by A. Toomre and G. Toomre [1232]. The aim of these experiments is the investigation of the structures, caused by the tidal interaction [437, 616, 655] and also the processes of transfer and loss of matter [1277]. The dynamic friction in close vapours is also investigated by numerical experiments [1250]. In the study [1215] attention was given to the two-component models, including the 'cold' (gas) and 'hot' (stellar) subsystems. The formation of the spiral waves, caused by a point satellite flying on the parabolic orbit in the plane of the disk, was investigated. Barnes and Hernquist reported [337] on the modelling of the dynamics of gas at merger of two galaxies. The interesting possibility of the formation of gas rings around the galaxies as a result of collisions of the elliptical and disk galaxies was reported in [613]. It is also important to mention the possibility of generation of a bar by a flying satellite.

Numerical modelling of specific interacting galaxies is used widely. The parameters of the galaxies are adjusted (mass, including halo and the gas; their orbits; the orientation of the system in relation to the visual ray; the moment of the observation time, etc) to obtain the best agreement with the observational data; in particular, the rotation curve and the spiral pattern are mentioned. The first models already made it possible to obtain the complete image of the evolution of these systems (for example [437] – pair AM 2208-251, [655] – system M51–NGC 5195). In the latter case, the gas subsystem was also taken into account, and the authors of the model also predicted the merger of galaxies in future.

It is also important to determine the law of movement of dwarf galaxies around giant galaxies, for example, the satellite of our Galaxy – the Large Magellanic Cloud (LMC). The approximate equation of movement of the particle with mass m_i

inside the massive halo with the mass M_h with dynamic friction taken into account has the form [654, 725]:

$$\frac{d\mathbf{v}}{dt} = -\nabla\Phi(r) - 16\pi^2 G^2 M_h (m_i + M_h) \ln\Lambda \frac{\int_0^{v_{\max}} f(v) d\mathbf{v}}{|\mathbf{v}|^3} \mathbf{v}, \qquad (4.147)$$

where Φ is the potential, $\ln\Lambda = \ln\left(R_h / \varepsilon_s V_s^2\right)$ is the Coulomb logarithm; $f(\mathbf{v})$ is the distribution function, R_h is the scale of the halo, ε_s is the characteristic truncation radius. Modelling of the disk by the method of *N*-bodies with the evolution of the dark halo taken into account results in a long lifetime of these systems, in particular, the LMC is absobed as a result of dynamic friction by the Galaxy only after $1.4 \cdot 10^{10}$ years [654].

In recent years it has become possible to construct numerical models of the interacting multicomponent systems. During the passage of a dwarf galaxy through a disk galaxy modelling convincingly demonstrated the possibility of formation of bar-like, ring- and spoke-shaped structures, different asymmetric formations in the stellar and gas components [363]. Close interactions resulting in the fall of a compact satellite are capable in some cases of greatly weakening the central bar and even fracture it [362].

The viewpoint on the nature of the galaxies of the cartwheel type is generally accepted. The formation of these galaxies is associated with the fall of a satellite on a gas-rich galaxy [690]. In the evolution models, including stellar formation, radially elongated perturbations (spoke-shaped) form together with the external ring in the disk [910]. This results in the formation of characteristic gradients of the parameters of the stellar and gas subsystems [1275]. It should be mentioned that the cartwheel-type structures can also be produced in the case of simultaneous formation of multi-mode solutions ($m = 0$ and $m \neq 0$) as a result of the development of the Jeans instability.

5

Hydrodynamic instabilities in rotating gas

In previous chapters, attention was already given to a number of hydrodynamic instabilities of the gas gravitating disk evidently related to the origin of the observed structures or imposing restrictions on the values of some parameters of the galactic gas subsystem. This is mostly the gravitational (gravitational–gradient) instability, determining the minimum 'temperature' of the disk, essential for preventing the breakup of the disk to gravitationally bonded blobs of the Jeans scale. Another class of instabilities, determined by the radial heterogeneity of the density and temperature of the disk, influences the ratio of the radial gradients of these quantities and, possibly, may lead to the excitation of vortex structures of the type of Rossby solitons. The fast dissipative instabilities may play a significant role in the solution of the problem of turbulent viscosity.

However, in these instabilities there is no instability determined directly by the differential nature of rotation of the disk. The instability of this type can be excited in parts of the disk where the degree of differential rotation of matter exceeds some limit – for example, in the range of rapid decrease of $\Omega(r)$ on the outside from the internal 'hump' $V_{\text{in}}(r)$ in the galaxies with the two-hump rotation curve (Fig. 1.8).

At least, in our Galaxy in the same region ($r \leq 3$ kpc) there is a significant depression in the surface density of the gas disk. Direct evaluation on the basis of the observational data [1112] of the Jeans scale in this region of the gas subsystem of the Galaxy leads to the following results: $\lambda_J = 2\pi / k_J = 2c_s^2 / G\sigma_{g0} \simeq 10$, which is greater than the scale of the observed structure. What are the consequences of this result?

The dynamics of perturbations in the gas disk is determined by the sum of two forces $\nabla\tilde{p} / \sigma_{g0} + \nabla\tilde{\Phi}$, where $\tilde{p}, \tilde{\Phi}$, is the perturbed pressure and the gravitational potential. The relative intensity is estimated from the equation:

$$\left|\frac{\sigma_{g0}\vec{\nabla}\tilde{\Phi}}{\vec{\nabla}\tilde{p}}\right| \simeq \left|\frac{\sigma_{g0}\tilde{\Phi}}{c_s^2\tilde{\sigma}_g}\right| \simeq \left|\frac{2\pi G\sigma_{g0}(\tilde{\sigma}_g+\tilde{\sigma}_*)}{kc_s^2\tilde{\sigma}_g}\right| \simeq \frac{2\lambda}{\lambda_J}|1+\tilde{\sigma}_*/\tilde{\sigma}_g|. \tag{5.1}$$

The estimates, obtained in accordance with the results in section 3.6.2 on the basis of the observational data for equilibrium parameters of the stellar and gas flat subsystems in the investigated region of the Galaxy, show that the contribution of the stellar disk to the perturbed gravitational potential is not greater than the contribution of the gas disk. Therefore, if some hydrodynamic instability leads to buildup of the perturbations with $\lambda \ll \lambda_J$, then in the equations, describing the dynamics of the gas disk, perturbed by the gravitational force ($\nabla\tilde{\Phi}$) we can ignore in the first approximation in comparison with the perturbed hydrodynamic force $(\nabla\tilde{p}/\sigma_{g0})$. This conclusion has been confirmed in the detailed studies by Fridman and Polyachenko [590] and also in [166, 171–173]. Therefore, in this section (with the exception of 5.1.6) we will not take into account the contribution of the perturbed gravitational potential to the dynamics of perturbations of the gas disk. This problem is described in detail in [239].

The hypothesis according to which the hydrodynamic instabilities in the gas disk cause the excitation of a spiral pattern of the galaxies was formulated by Fridman in 1972. Specific considerations regarding the hydrodynamic mechanism of excitation of the spirals and the area of its operation in the galaxies were made in [566].

Later, the hydrodynamic instabilities in the three-dimensional models of the gravitating medium [166, 171–173] and in a gas disk [148, 149, 245] were investigated. Subsequent experimental investigation of the centrifugal instability in equipment with rotating 'shallow water' [164, 588] was used in the final analysis to formulate the hydrodynamic concept of the origin of the spiral pattern of the flat galaxies [237].

The term accretion disk (AD) is usually used for describing a gas disk around a massive (in comparison with the disk) compact object. The latter includes white dwarfs, neutron stars, black holes (including supermassive black holes in the centre of, evidently, the majority of galaxies). The distinguishing feature of the AD is the transition of the gravitational energy in accretion (incidence) of matter into heat followed by emission. In many cases, the concept 'accretion disk' is used in a wider sense of the word referring to the gas disk for which the process of incidence of matter on the accumulating centre takes place, for example, when we are discussing protostar, protoplanet, galactic disks.

Thus, from the viewpoint of the physical processes, many results relating to the gas disk also apply fully to the accretion disks. The difference in a number cases is very commercial and associated mainly with the type of external potential. For the disks, rotating around a compact object with mass M_1, it is of the Newton type for the point body ($\Phi \propto -M_1/r$), and self-gravitation of the gas is not significant, with the exception of the disks around the supermassive black holes. For the galactic gas disks, the external potential is determined by the distribution of stellar matter and spheroidal components (see equation (2.44)), and regions of relatively large changes of velocity and density may also exist.

5.1. Gas galactic disks in the presence of a narrow zone of rapid changes in the angular velocity of rotation. VK-galaxies

5.1.1. Dynamics of perturbations in the heterogeneous disk

We will derive the equations describing the linear waves in a heterogeneous non-self-gravitating differentially rotating disk.

In accordance with Chapter 3, it is assumed that the equilibrium state is determined by the differential equation (3.243). Thus, for the equilibrium we take into account the radial heterogeneity $V_0(r) = r\Omega(r)$, $\sigma_0(r)$, $p_0(r)$, $\Phi_0(r)$, $\Omega_z(r)$. The initial system of the equations is (3.242)–(3.280). The equations are linearised with respect to the perturbed quantities (see (3.281)) without taking into account the perturbation of the potential. The solution will be found in the form of (3.289). Thus, the waves are characterised by the frequency Ω and the azimuthal number m, and the radial structure of the perturbations is determined by the system of the equations (3.290)–(3.293) under the condition $\Phi = 0$.

We introduce the radial displacement function which is determined from the perturbed radial velocity:

$$u = d\xi / dt = \partial\xi / \partial t + \Omega \partial\xi / \partial\varphi = -i\hat{\omega}\xi . \tag{5.2}$$

As the main perturbed quantities we select $\tilde{p}$ and ξ. Excluding the perturbations of the radial and azimuthal velocity is and surface density, the system (3.290)–(3.295) can be re-written in the new form:

$$\frac{d\tilde{p}}{dr} = \left\{ \begin{array}{l} \sigma_0\left(\hat{\omega}^2 - \varkappa^2\right) + \dfrac{p_0}{\gamma_s}(\ln p_0)' \cdot \left(\ln \dfrac{p_0}{\sigma_0^{\gamma_s}}\right)' + \dfrac{p_0}{\gamma_s}\left(\ln \Omega_z^D\right)' \times \\ \times\left[\ln\left(p_0^{(D+1-\gamma_s)/D} \Big/ \sigma_0^{\gamma_s}\Omega_z^{\gamma_s - 1}\right)\right]' \end{array} \right\} \cdot \xi + \left\{ \frac{2m\Omega}{r\hat{\omega}} + \left[\ln \frac{p_0^{1/\gamma_s}}{\Omega_z^{D(\gamma_s-1)/\gamma_s}}\right]' \right\} \cdot \tilde{p}, \tag{5.3}$$

$$\frac{d\xi}{dr} = \left\{ \frac{m^2}{r^2\hat{\omega}^2\sigma_0} - \frac{1}{\gamma_s p_0} \right\} - \left\{ \frac{2m\Omega}{r\hat{\omega}} + \frac{1}{r} + \left(\ln p_0^{1/\gamma_s}\right)' - \left(\ln \Omega_z^{(\gamma_s - 1)/\gamma_s}\right)' \right\} \cdot \xi, \tag{5.4}$$

where the prime denotes the derivatives with respect to the radial coordinate. This system of the equations in the appropriate partial cases ($p_0 = \text{const}$, $\sigma_0 = \text{const}$, $\Omega_z = \text{const}$) changes to the equations investigated in [148, 149, 160, 163, 166, 171, 173, 240, 242, 588].

If the system of the equations is supplemented by the boundary conditions, or by some additional restrictions (for example, assuming that the perturbations are of the short-wave type), it is possible to determine the spatial structure of the perturbations $\xi(r)$, $\tilde{p}$ and intrinsic frequency Ω. Since, regardless of their linear form, the equations (5.3), (5.4) are quite complicated because of the large number of the radial distributions of the equilibrium functions, it is convenient to start examination at different partial cases.

5.1.2. The Kelvin–Helmholtz instability

The problem of stabilisation of the tangential discontinuity of velocity in the compressed medium

Examination will start from the classic problem of the stability of a planar tangential discontinuity (TD) [112, 665, 755, 818].

It is assumed that a liquid moves along the axis x and that there is a tangential discontinuity of the x-component of velocity at point y_0 (Fig. 5.1). Figure 5.2 shows the stationary structures, detected for the first time by Kelvin for an incompressible liquid [755]. The straight arrows indicate the field of velocities whose gradient is directed along the y axis. In the spatially narrow region $\Delta y = l$, where l is small in comparison with the width of the non-perturbed flow L:

$$\frac{l}{L} \ll 1, \tag{5.5}$$

Kelvin observed the stationary vortex structure: the chains of the vortices with the centres on the straight line $y = y_0$, where the direction of the velocity vector changes to the opposite direction. The vortices are localised in a narrow layer with a thickness $l \ll L$, and their centres are distributed on the straight line $y = y_0$ of the velocity discontinuity, parallel to the x axis. Since, according to [755], the vortex structure is distributed in a negligibly small spatial region $l \ll L$, the Kelvin–Helmholtz instability was subsequently referred to as the 'instability of the tangential discontinuity of velocity' or, simply, 'the instability of the tangential discontinuity' (Fig. 5.1).

The elongated form of the vortices, stretched along the x axis (Fig. 5.1), was the reason why Kelvin referred to these vortices as the 'cats eyes' (Fig. 5.3). Since each vortex contains the velocity component y which, like the component x, changes its sign to the opposite sign in the region of the centre of the vortex, this means that in the region of localisation of the vortices the component y of velocity, not found in the remaining part of the flow, is localised there.

Thus, the vortex structure, detected by Kelvin, is also the outstanding discovery of the first instability in the liquid.

In the natural conditions, for example, in the Earth atmosphere, the velocity can change not only along the horizontal coordinate (denoted previously by y) but

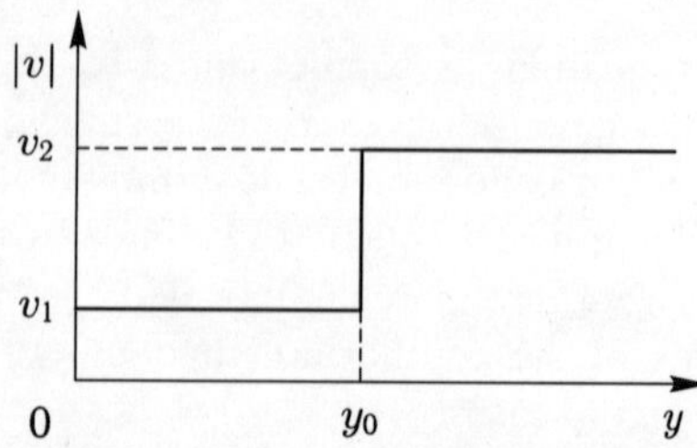

Fig. 5.1. One-dimensional dependence of the velocity vector on the coordinates $\mathbf{v} = \mathbf{v}(y)$ at which the instability of the tangential discontinuity of velocity forms.

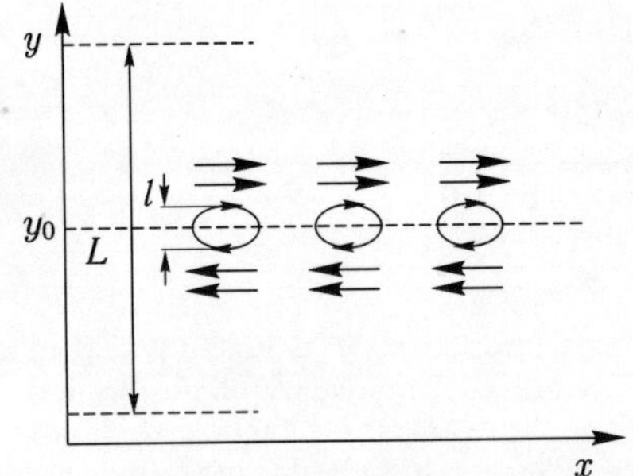

Fig. 5.2. The field of velocities in the form of stationary structures of two types. The first type – straight flow with the size L with the tangential discontinuity of velocity in the region $y = y_0$. The second type – the vortices, detected for the first time by Kelvin in 1887 and referred to by Kelvin as 'cat's eyes'.

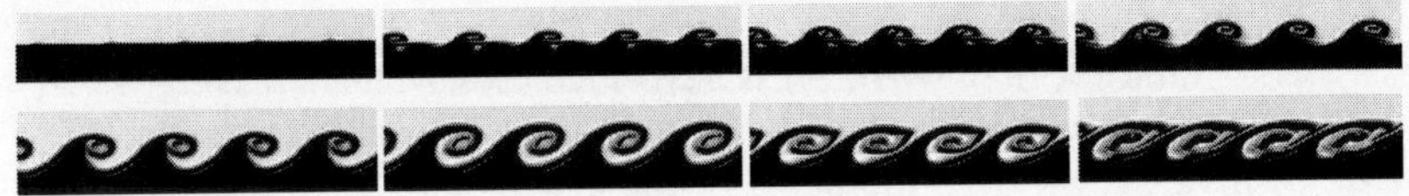

Fig. 5.3. Development of the instability of tangential discontinuity, including the highly non-linear stage, to the formation of the 'cat's eyes' (according to numerical modelling results).

also along the vertical coordinate denoted in most cases by z. Velocity inversion is detected at relatively large altitudes in the Earth atmosphere (usually above 10 km): the velocity increases with increasing z and sometimes reaches supersonic values.

In the velocity inversion range, the vector of the gradient of the modulus of velocity $\nabla|\mathbf{v}|$ is co-linear in relation to the vector of acceleration of gravitational force $\mathbf{g}$ ($\mathbf{g} \,||\nabla|\mathbf{v}|$), and the directions of these vectors differ. It should be noted that the velocity inversion region in the Earth atmosphere is detected not only at high altitudes but also at the very surface of the Earth. In fact, the boundary condition on the Earth surface (at $z = 0$) requires $\mathbf{v}(0) = 0$, whereas at any small $z \neq 0$ the inequality $|\mathbf{v}(z)| > 0$ holds. The latter after all does not exclude the possibility of the velocity vector being equal to $\mathbf{v}(x, y, z) = 0$ at a different points z_i, $i = 1, 2, 3, \ldots, n$, not only in the region of the Earth atmosphere but also in the ionosphere and magnetosphere.

Physics of instability

For the adiabatic perturbations $S = \text{const}$, discussed in [108, 231], the relationship between the thermal function W, pressure p and density ρ is determined from the relationship $dW = dp/\rho$, and for $p = A\rho^\gamma$, where A and γ are the constants (γ is the adiabate index, $\gamma = c_p/c_V$, c_p, c_V are the heat capacities at constant pressure and volume, respectively), we have

$$W = \frac{\gamma}{\gamma - 1} A^{1/\gamma} p^{(\gamma-1)/\gamma} = Bp^\alpha, \tag{5.6}$$

where

$$B \equiv \frac{\gamma}{\gamma - 1} A^{1/\gamma}, \quad \alpha \equiv \frac{\gamma - 1}{\gamma}. \tag{5.7}$$

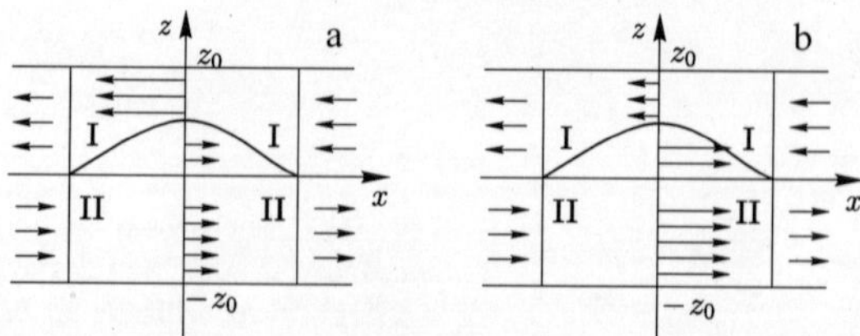

Fig. 5.4. Schematic representation of velocity distribution in different areas of the perturbed tangential subsonic and supersonic discontinuities of velocities. Perturbation of the tangential discontinuity of velocity **v**, directed along the axis x, in two opposite limiting cases: a) $M \equiv v/c \ll 1$ – the case of subsonic flow (c is the speed of sound), b) $M \gg 1$ – the case of the supersonic flow.

Thus, at any value $\gamma > 1$ ($\alpha > 0$), pressure p increases with increasing W. It was shown in [108] that the amplitude of the perturbations decreases exponentially with increase of the distance from the plane of the tangential discontinuity of velocity $z = 0$ with increasing modulus $|z|$: $\sim \exp(-|z|/z_0)$. Therefore, it is sufficient to restrict our considerations to the range $|z| < z_0$ (Fig. 5.4).

The instability of the subsonic flow will now be discussed. Region I (above the 'hump' of the perturbation) in Fig. 5.4 can be regarded as the region of the critical section of the subsonic nozzle where, as described in [123], the flow speed is maximum. Consequently, from the Bernoulli equation for the isentropic flow:

$$\frac{v^2}{2} + Bp^{\alpha} = \text{const} \tag{5.8}$$

it follows that the pressure above the hump should be minimum. This results in a further increase of the perturbation amplitude – an instability.

The effect of the Landau stabilisation of the supersonic flow will now be discussed. The region I (above the hump) in Fig. 5.4b can be regarded as the region of the narrowing channel of a supersonic diffuser ($M \gg 1$), where, as described in [123], a shock front forms which reduces the velocity of the supersonic flow to the minimum value, i.e. to the speed of sound, $v = c_s$. In this case, from the general Bernoulli equation

$$\frac{v^2}{2} + W = \text{const} \tag{5.9}$$

it follows that the thermal function increases above the hump, i.e. the pressure also increases. This results in 'penetration' of the hump back into the region II. This is therefore the effect of stabilisation of the instability of the tangential discontinuity of the velocity in the supersonic flow, detected for the first time by Landau [108][1].

We now return to the critical comment made by Syrovatskii, regarding the absence of the stabilising effect [231].

Criticism of Landau's study [108] *by Syrovatskii* [231]. We consider (in contrast to the two-dimensional flow considered in Landau's study [108]) a three-dimensional

[1]In Landau's original article [108] and in his books with E.M. Lishitz (Mechanics of solids, 1953, 1954, and Hydrodynamics, 1986) there is no qualitative explanation of stabilisation of the instability of tangential discontinuity for the supersonic flow.

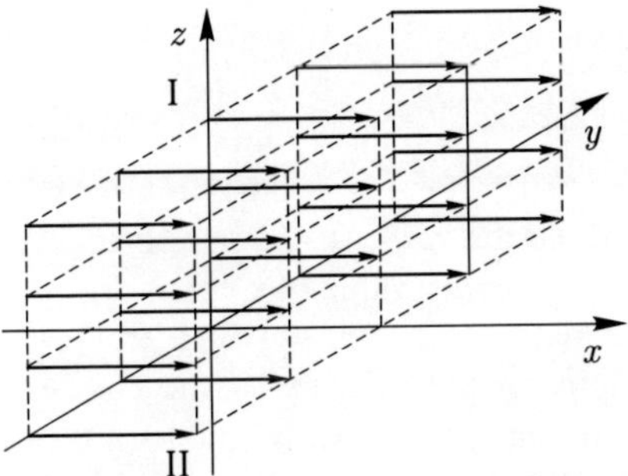

Fig. 5.5. Three-dimensional flow moving along the axis c with the tangential discontinuity of velocity. Region I corresponds to the positive values of z ($z > 0$), region II to the negative values of z ($z < 0$).

flow moving as previously along the axis with a tangential velocity discontinuity (Fig. 5.5): $v_0 = v_x\,\Theta\,(z)$, where Θ is the Heaviside function.

A dispersion equation will be derived. It is convenient to transfer from the cylindrical geometry in (5.3), (5.4) to the Cartesian geometry considered in this case. For this purpose it is sufficient to set $1/r \to 0$, $\Omega_z = 0$, and accept $\sigma_0 = \text{const} = \varrho_0$, $p_0 = \text{const}$, taking into account a velocity 'jump': $V_{01} = V_0 = r\Omega_1$, $V_{02} = r\Omega_2 = 0$. In our Cartesian coordinate system, the z axis is normal to the plane of the tangential discontinuity. As a result, the following system of equations is obtained:

$$\frac{d\xi}{dz} = \frac{\tilde{p}}{\varrho_0}\left(\frac{k_x^2 + k_y^2}{\hat{\omega}^2} - \frac{1}{c_s^2}\right), \tag{5.10}$$

$$\frac{d\tilde{p}}{dz} = \varrho_0 \hat{\omega}^2\,\xi, \tag{5.11}$$

where $c_s^2 = \gamma p_0/\varrho_0$. Assuming that displacement ξ and the perturbed pressure $\tilde{p}$ are continuous in transition through the velocity jump, a dispersion equation can be derived for the perturbations, whose amplitude decreases with increase from the TD.

The three-dimensional perturbations of the density $\tilde{\rho}$ and velocity $\tilde{v}$ are selected in the form:

$$\begin{aligned}\tilde{\rho}(x,y,z,t) &= \rho_1 \exp[i(k_x x + k_y y) - \lambda\,|\,z\,| + \gamma_\omega t],\\ \tilde{v}(x,y,z,t) &= v_1 \exp[i(k_x x + k_y y) - \lambda\,|\,z\,| + \gamma_\omega t],\end{aligned} \tag{5.12}$$

where γ_Ω is the instability increment.

Landau [108] investigated the stability of the tangential discontinuity in relation to two-dimensional perturbations, assuming that $k = k_x$, i.e. the partial case $k_y = 0$. For this partial class of the two-dimensional perturbations, Landau showed the absence of instability under the condition for the Mach number $\mathrm{M} \equiv v_0 / c_s > 2\sqrt{2}$ in the case of homogeneous density. If the values of the non-perturbed quantitie density ρ_0 and the speed of sound c_s are equal to each other on both sides of the discontinuity: $\rho_{0,\mathrm{I}} = \rho_{0,\mathrm{II}}, = \rho_0$, $c_{0,\mathrm{I}} = c_{0,\mathrm{II}} = c_s$, then in the simplest case

$$v_0^{\text{crit}} = 2\sqrt{2}c_s, \tag{5.13}$$

10 years later, Syrovatskii [231], solving a similar problem with respect to a general class of the perturbations (5.12), i.e. assuming that $k = \{k_x, k_y\} = \{k \cos\theta, k \sin\theta\}$, where θ is the angle between the vector k and its projection on the x axis, observed the presence of an instability at any value of v_0.

Assuming that the equality (5.12) is satisfied, the problem of the stability of the tangential discontinuity of the velocity of the compressed fluid with respect to arbitrary (but small) perturbations can be reduced to the following dispersion equation (here, the time dependence is selected in the form ~ exp(– $i\Omega t$):

$$k^2 c_0^2 \left[\frac{1}{(\omega - kv_0 \cos\theta)^4} - \frac{1}{\omega^4} \right] = \frac{1}{(\omega - kv_0 \cos\theta)^2} - \frac{1}{\omega^2}. \tag{5.14}$$

Reducing to the common multiplier, which has only the real root $\Omega = kv_0 \cos\theta/2$, we obtain the following equation

$$f(x) \equiv \frac{1}{(x - \mathrm{M}\cos\theta)^2} + \frac{1}{x^2} = 1, \tag{5.15}$$

where $x \equiv \Omega/(kc_s)$, $\mathrm{M} \equiv v_0/c_s$, which differs from the Landau dispersion equation [108] by the presence of $\cos\theta$ [231]. Equation (5.15) has four roots. They all are real, if the function $f(x)$ is identical with that shown in Fig. 5.6 by the thick line. However, if it is identical with the dashed line in Fig. 5.6, equation (5.15) has two real roots. Consequently, the two other roots are complexly conjugate roots, one of which describes the instability. The majorant curve is indicated in Fig. 5.6 by the dot-and-dash line. In this case, all the four roots are real: $x_1', x_4', x_2' = x_3' = (1/2)\mathrm{M}\cos\theta$, two of which are divisible. As indicated by Fig. 5.6, the critical value of the Mach number M_{cr} is determined from the equation

$$f\left(\frac{1}{2}\mathrm{M}_{\mathrm{cr}} \cos\theta\right) = 1, \tag{5.16}$$

which determines the point of contact of the majorant curve with the straight line $f(x) = 1$.

To select the roots, it is sufficient to use the result obtained for the incompressible liquid ($V_0/c_s \to 0$): Im (Ω) = $k_x V_0$. The required solution has the form

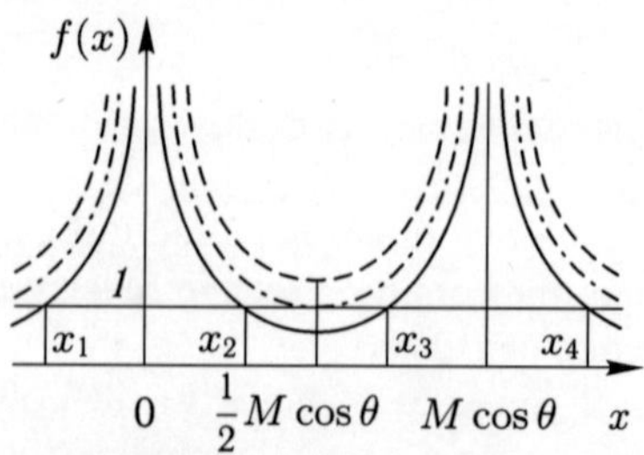

Fig. 5.6. Function $f(x)$.

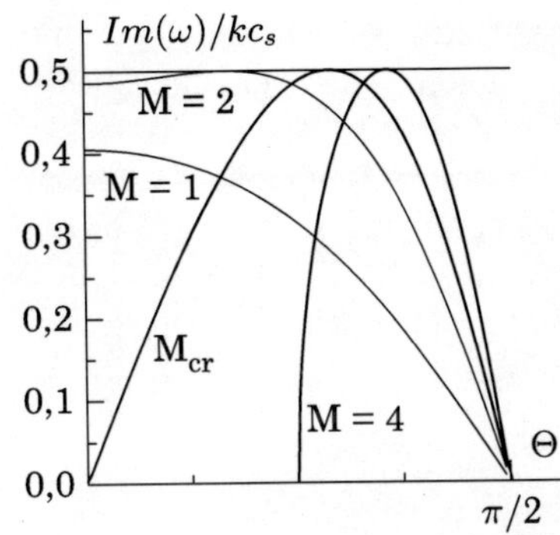

Fig. 5.7. Dimensionless increment of instability in relation to angle $\Theta = \arccos\left(k_x / \sqrt{k_x^2 + k_y^2}\right)$ for different Mach numbers.

$$\omega^2 = k^2 c_s^2 + k_x^2 V_0^2 - \sqrt{k^2 c_s^2 (k^2 c_s^2 + 4k_x^2 V_0^2)}, \tag{5.17}$$

where $k^2 = k_x^2 + k_y^2$. Figure 5.7 shows the dependences of the increment of instability of the tangential discontinuity on the angle between the wave vector and the velocity vector. It can be seen that at high Mach numbers only the 'oblique' perturbations can be unstable.

There is a critical value equal to

$$\mathrm{M_{cr}} = \frac{2\sqrt{2}}{\cos\theta}. \tag{5.18}$$

Using the expression for $\cos\theta = k_x/k_\perp$, where $\Theta = \arccos(k_x / \sqrt{k_x^2 + k_y^2})$, we obtain

$$\mathrm{M_{cr}^2} = 8\left(1 + \frac{k_y^2}{k_x^2}\right). \tag{5.19}$$

In the 'quasi-two-dimensional' systems, such as, for example, the gas disks of the galaxies, the rings around the giant planets [43, 577] and the rotating shallow water [967], only the 'longitudinal' waves are possible, $k_y/k_x \ll 1$, investigated by Landau. In this case, $\mathrm{M_{cr}}$ (5.19) transforms to M_{cr} derived by Landau [108]. The meaning of the critical comment by Syrovatskii is that the arbitrary perturbations permit examination of the opposite limiting case – 'transverse waves', $k_y/k_x \gg 1$. It is evident that, for example, at

$$\frac{k_y}{k_x} \to \infty \tag{5.20}$$

stabilisation is in principle not possible because, as indicated from (5.19), $\mathrm{M_{cr}} \to \infty$.[2]

The modified criterion of stabilisation of the instability of the tangential discontinuity of the velocity in the compressible fluid was derived in [239, 240]. In the idealised formulation of the problem [231] – tangential discontinuity of the velocity in the three-dimensional infinite space – the condition (5.20) can be fulfilled. However, the actual situation introduces two important corrections: 1) the

[2]However, it should be mentioned that at $k_y/k_x \to \infty$ the instability increment $\gamma_\Omega \to 0$ [173]. It may be shown that if the increase of the perturbation in the dissipative flows is taken into account, this has no effect on the stabilisation criterion, based on equation (5.19).

system has finite spatial dimensions according to all three measurements; 2) the tangential discontinuity of the velocity is 'spread' over some characteristic value a. The consequence of these conditions is the existence of $(k_y/k_x)_{max} \equiv (k_y)_{max}/(k_x)_{min}$. In fact, $(k_x)_{min} \sim L^{-1}$, where L is the characteristic size of the system along the x axis; the estimate $(k_y)_{max} \sim a^{-1}$ follows from the essential condition of existence of the unstable flow with the nonuniform velocity profile, $k_y a < 1$ [912].

Thus, the instability of the 'tangential discontinuity' of the velocity in the actual conditions appears to be suppressed under the condition

$$\mathrm{M} > (\mathrm{M_{cr}})_{\mathrm{mod}} \equiv 2\left[2\left(1+\frac{L^2}{a^2}\right)\right]^{1/2}, \tag{5.21}$$

where $(\mathrm{M_{cr}})_{\mathrm{mod}}$ is the modified critical Mach number, if this number is exceeded, the compressed liquid does not contain any instability of the tangential discontinuity.

Usually in practice $L^2/a^2 \gg 1$; in this case $(\mathrm{M_{cr}})_{\mathrm{mod}}$ from (5.1) is greater than $(\mathrm{M_{cr}})_{\mathrm{L}}$ (Landau [108]) L/a times:

$$(\mathrm{M_{cr}})_{\mathrm{mod}} \approx \frac{L}{a}(\mathrm{M_{cr}})_L = \frac{L}{a}2\sqrt{2}. \tag{5.22}$$

The condition of 'drift' of the perturbation has the following form

$$\frac{1}{\gamma_{\max}} \gg \frac{L}{v_0}, \tag{5.23}$$

where $\gamma \equiv \mathrm{Im}\,(\Omega)$ is the increment of instability of the tangential discontinuity of the velocity. The meaning of the criterion (5.23) is that during passage through any region of the gas along the system with length L with the velocity v_0 the perturbations do not manage to grow in this region. Actually, the characteristic instability time, described by the left part of (5.23), is considerably greater than the duration of passage by the gas of any section of the system with the size L. Thus, if the condition (5.23) is fulfilled, the instability of the tangential discontinuity can be regarded as non-existent. According to [173], $\gamma_{max} \approx 0.5\,(k_y)_{max}\,c_0 \approx 0.5\,(c_0/a)$ which, after substitution into (5.23), gives

$$\mathrm{M} \gg 0.18(\mathrm{M_{cr}})_{\mathrm{mod}}.$$

Thus, the inequality

$$v_0 > v_{\mathrm{cr}} \approx 0.4\sqrt{2}\frac{L}{a}c_0 \tag{5.24}$$

determines the criterion of stabilisation of the instability of the tangential discontinuity of the velocity in the real three-dimensional compressible gas.

5.1.3. The dispersion equation of the perturbations of discontinuity of angular velocity

Attention will be given to the dynamics of perturbations in a homogeneous gas layer rotating with a discontinuity of the angular velocity (the limiting model of the double-hump rotation curve with a narrow region of rapid decrease of $\Omega(r)$ outside of the internal hump $V_{\rm rot}(r)$, see Fig. 1.9):

$$\Omega(r) = \Omega_1 + (\Omega_2 - \Omega_1)\cdot\Theta(r - R), \tag{5.25}$$

where $\Theta(x)$ is the Heaviside theta function ($\Theta(x) = 1$ at $x > 0$ and $\Theta(x) = 0$ at $x < 0$); R is the radius of the discontinuity of angular velocity. The law of rotation (5.25) can be represented as the limit of the smooth distribution

$$\Omega(r) = \frac{\Omega_1 + \Omega_2}{2} + \frac{\Omega_2 - \Omega_1}{\pi}\,\mathrm{arctg}\left(\frac{r - R}{L}\right) \tag{5.26}$$

at $L/R \to 0$.

The differential nature of rotation $\Omega(r)$ will be taken into account but the distribution p_0 and σ_0 in (5.3), (5.4) is constant. Consequently, we have this system of equations

$$\begin{aligned}\frac{d\tilde{p}}{dr} &= \left(\frac{2m\Omega}{r\hat{\omega}} - \frac{\gamma_s - 1}{\gamma_s} D(\ln\Omega_z)'\right)\tilde{p} + \\ &+ \sigma_0\left(\hat{\omega}^2 - \varkappa^2 - D\frac{\gamma_s - 1}{\gamma_s^2}(\ln\Omega_z)'(\ln C_\ell\Omega_z)' c_s^2\right)\xi,\end{aligned} \tag{5.27}$$

$$\frac{d\xi}{dr} = \frac{\tilde{p}}{\sigma_0}\left(\frac{m^2}{r^2\hat{\omega}^2} - \frac{1}{c_s^2}\right) - \left(\frac{2m\Omega}{r\hat{\omega}} + \frac{1}{r} - \frac{\gamma_s - 1}{\gamma_s}(\ln C_\ell\Omega_z)'\right)\xi, \tag{5.28}$$

C_ℓ is linked with the right-hand part of (3.247). In the case $C_\ell\,\Omega_z = \mathrm{const}$, equations (5.27), (5.28) lead to the well-known result [148, 149]. The characteristic special feature of the system of equations is that at $\gamma_s \to 1$ none of the terms, determined by the heterogeneity of the value Ω_z, provide any contribution. Below, we restrict ourselves to the examination of the models without taking these terms into account[3] [$\propto (\gamma_s - 1)\, d\ln\Omega_z/dr$].

Transferring to the discontinuity model Ω (5.21), we solve the system (5.27), (5.28) separately on both sides of the discontinuity ($r > R$, $r < R$), assuming respectively $\Omega(r) = \Omega_{1,2}$. In this case, the system of the equations (5.27), (5.28) is reduced to one equation for $\tilde{p}(r) = p_{1,2}$:

$$\frac{d^2 p_{1,2}}{dr^2} + \frac{dp_{1,2}}{r\,dr} + \left\{\frac{\hat{\omega}_{1,2}^2 - 4\Omega_{1,2}^2}{c_s^2} - \frac{m^2}{r^2}\right\} p_{1,2} = 0, \tag{5.29}$$

and the displacement is determined from the relationship

$$\xi_{1,2} = \frac{1}{\sigma_0(\hat{\omega}_{1,2}^2 - 4\Omega_{1,2}^2)}\left\{\frac{dp_{1,2}}{dr} - \frac{2m\Omega_{1,2}}{r\hat{\omega}_{1,2}} p_{1,2}\right\}. \tag{5.30}$$

[3] If in the two-dimensional model it is formally set that $\Omega_z = \mathrm{const}$, the model will describe three-dimensional dynamics of the partial class of perturbations which have no components of the wave vector normal to the plane of the disk. Here p, σ, γ_s are not the surface but volume pressure, density and the adiabate index.

The solutions of these equations should be cross-linked at the discontinuity (at $r = R$). The corresponding cross-linking rules (the boundary conditions at the discontinuity) can be obtained as follows. We use the equations (5.27), (5.28) in which the discontinuity $\Omega(r)$ is 'spread' over a narrow transition layer with the width H and outside of this layer $\Omega = \Omega_{1,2}$. These equations are integrated with respect to the given transition layer and we transfer to the limit $H \to 0$. As a result

$$\xi\Big|_{r=R-0} = \xi\Big|_{r=R+0}, \tag{5.31}$$

$$(\tilde{p} + \sigma_0 R\Omega^2\xi)\Big|_{r=R-0} = (\tilde{p} + \sigma_0 R\Omega^2\xi)\Big|_{r=R+0}. \tag{5.32}$$

The second of these boundary conditions appears to be unusual[4]. Therefore, we shall explain its physical meaning. The radial equilibrium of the gas galactic disks is determined by the balance of the pressure gradient and centrifugal and gravitational forces: $\Omega^2 r = p_0' / \sigma_0 + \Phi_0'$, where the prime denotes the differentiation with respect to the radial coordinate. The contribution of the pressure gradient to this condition is small in comparison with the contribution of the gravitational force $| p_0' / \sigma_0 \Phi_0' | \sim c_s^2 / V_{\text{rot}}^2 \ll 1$. Therefore, the relatively sharp gradient $\Omega(r)$ in the investigated region of the disk is determined mainly by the same sharp gradient Φ_0 formed by the distribution of matter in a massive stellar subsystem. In the maximally idealised model of discontinuity Ω, the value $d\Phi_0/dr$ should be evidently discontinuous. At the same time, the total 'pressure' $\mathcal{P} = p/\sigma + \Phi$ should be continuous on the discontinuity surface distorted due to perturbations ξ, and the equilibrium pressure $\mathcal{P}_0 = p_0/\sigma_0 + \Phi_0$ is continuous at the non-perturbed discontinuity. We expand this value into a series with respect to the exponents of the amplitude of the perturbations and restrict ourselves to the linear terms and ignore in accordance with the estimate (5.1) the perturbed gravitational potential

$$\mathcal{P}(r = R + \xi) = \mathcal{P}_0(r = R + \xi) + \widetilde{\mathcal{P}}(r = R + \xi) =$$
$$= \mathcal{P}_0(R) + \mathcal{P}'_0(R)\xi + \tilde{p}(R)/\sigma_0 \equiv \mathcal{P}_0(R) + R\Omega^2\xi + \tilde{p}(R)/\sigma_0,$$

from this it follows that the combination $\tilde{p} + \sigma_0 R\Omega^2$, see (5.32), should remain continuous at the discontinuity Ω. The cross-linking conditions (5.31), (5.32) are the simplest. In a general case, the cross-linking rules may depend on the structure of the disk in the region of the 'jump' (see the detailed analysis by Fridman and Khoruzhii [245]).

The solutions of the equations (5.29), (5.30) should be restricted at $r \to 0$ and $r \to \infty$. Taking these boundary conditions into account, they have the following form

$$p_1 = A I_m(k_1 r), \quad p_2 = B K_m(k_2 r), \tag{5.33}$$

$$\xi_1 = \frac{A}{\sigma_0 c_s^2 k_1^2}\left\{\frac{2m\Omega_1}{r\hat{\omega}_1} I_m(k_1 r) - k_1 I_m'(k_1 r)\right\}, \tag{5.34}$$

[4]In the flat problem with the identical formulation, the perturbed pressure $\tilde{p}$ is continuous in transition through the surface of the velocity discontinuity [112].

$$\xi_2 = \frac{B}{\sigma_0 c_s^2 k_2^2}\left\{\frac{2m\Omega_2}{r\hat{\omega}_2}K_m(k_2 r) - k_2 K_m'(k_2 r)\right\}, \tag{5.35}$$

where

$$k_{1,2}^2 = \{4\Omega_{1,2}^2 - (\omega - m\Omega_{1,2})^2\}/c_s^2, \quad \mathrm{Re}(k_{1,2}) > 0, \tag{5.36}$$

$I_m(x)$, $K_m(x)$ are the modified Bessel functions; A, B are the arbitrary constants, and the prime indicates the differentiation of the Bessel functions with respect to their argument.

Cross-linking subsequently the solutions (5.33)÷(5.35) at the discontinuity $r = R$ according to the boundary conditions (5.31), (5.32), gives the required dispersion equation [148]:

$$k_1^2\alpha_2 - k_2^2\alpha_1 + \frac{(\Omega_1^2 - \Omega_2^2)}{c_s^2}\alpha_1\alpha_2 = 0, \tag{5.37}$$

where

$$\alpha_1 = \frac{2m\Omega_1}{\omega - m\Omega_1} - k_1 R\frac{I_m'(k_1 R)}{I_m(k_1 R)}, \tag{5.38}$$

$$\alpha_2 = \frac{2m\Omega_2}{\omega - m\Omega_2} - k_2 R\frac{K_m'(k_2 R)}{K_m(k_2 R)}, \tag{5.39}$$

To interpret the results better, the solutions of the dispersion equation (5.37) will be described by means of two dimensionless parameters $q = \Omega_2/\Omega_1$ and $M_* = R\Omega_1/c_s$, where M_* is the Mach number in our model.

5.1.4. The limiting transition to the Kelvin–Helmholtz instability

We obtain the solution of the dispersion equation (5.37) at the limit of the 'weak' discontinuity ($M_* \ll 1$). It may easily be seen that in this case according to (5.36) $|k_{1,2}R| \ll 1$. Therefore, using the representations of the Bessel functions in the form of a series with respect to the degrees of their arguments, from equation (5.37) we obtain ($m \geq 1$)

$$\omega \simeq \frac{\Omega_1}{2}[m(1+q) + (q-1)] + \frac{i\Omega_1}{2}\sqrt{(m^2-1)(1-q)^2}. \tag{5.40}$$

This shows that an instability occurs both at $q < 1$ and at $q > 1$, and the value of the increment depends on $|1 - q|$. Consequently, for buildup of such an instability it is not important which part of the disk (internal or external) rotates at a higher velocity. This means that here we are concerned with the instability of the tangential discontinuity in a slightly compressible ($M_* \ll 1$) medium. Actually, for the short-wave (in the azimuthal direction) perturbations ($m^2 >> 1$) for which the curvature effects are not important, from (5.40) we obtain

$$\omega = \frac{k_\varphi}{2}\{V_1 + V_2 + i(V_1 - V_2)\}, \tag{5.41}$$

where $k\varphi = m/R$; $V_j = R\Omega_j$ are the linear velocities of rotation. The result (5.41) accurately coincides with the increment of instability of the flat tangential discontinuity in a homogeneous incompressible medium [112].

If we ignore the gradients of the equilibrium density and pressure of the gas subsystem, the perturbed surface density $\tilde{\sigma}$ and pressure $\tilde{p}$ are linked by the relationship $\tilde{p} = c_s^2\,\tilde{\sigma}$, and it is quite easy to determine the spatial dependence $\tilde{\sigma}(r,\varphi)$. For example, in the region $r > R$ at $M_* \ll 1$, using the result (5.42) and the relationship (5.36), taking into account that the real parts of the complex amplitude have a physical meaning, we obtain

$$\tilde{\sigma}(r,\varphi) \propto \mathrm{Re}\{K_m(k_2 r)e^{im\varphi}\} \propto r^{-m}\cos\left[m\varphi + \frac{(1-q)^2\Omega_1^2\sqrt{m^2-1}}{8c_s^2}r^2\right]. \quad (5.42)$$

The condition of constancy of the perturbation phase shows that the perturbations of the density have the form of trailing spirals with the twist angle β (the angle between the tangent to the spiral and the circle), close to $\pi/2$ at $r \geqslant R$ because of the smallness of the parameter M_*. With increase of the distance from the centre of the disk at fixed $q \neq 1$, the twist angle decreases, i.e. $\beta \to 0$ at $r/R \to \infty$.

5.1.5. Centrifugal instability

In spiral galaxies, the value $V_{\rm rot}$ on the internal 'hump' of the rotation curve is considerably greater than the speed of sound in the gas subsystem and, therefore, a second limiting case of curse: $M_* \gg 1$. At this limit, using the asymptotic representations of the Bessel functions at $|k_{1,2}R| \gg m$, from equation (5.37) in the main order of these asymptotics we obtain [148]:

$$\omega = \omega^{(0)} = \frac{\Omega_1}{2}\{m(1+q) + i\,M_*(1-q^2)\}. \quad (5.43)$$

Substitution of this result into (5.36) shows that $|k_{1,2}R| \sim M_*^2$ and, consequently, the region of its applicability: $m \ll M_*^2$. In the following order with respect to the small parameter $M_*^{-2} \ll 1\left[\omega = \omega^{(0)} + \omega^{(1)}; \omega^{(1)} \propto O(M_*^{-1})\right]$, equation (5.37) gives

$$\omega^{(1)} \simeq -\frac{4m(1-q)(1-q^2)\Omega_1}{M_*^2(1-q^2)^2 + m^2(1-q)^2} - \frac{i2\Omega_1(1+q^2)}{M_*(1-q^2)}. \quad (5.44)$$

At the examined limit ($M_* \gg 1$), the instability (5.43) is no longer similar to the instability of the tangential discontinuity (5.40). Firstly, it is because this instability develops only in the case in which the internal part of the disk rotates at a higher velocity than the external part: $q^2 < 1$. Secondly, it is because its increment is almost completely independent of the wave number $k\varphi = m/R$ (compare with (5.40))[5]. Thirdly, since that the instability (5.43), in contrast to the classic instability of the tangential discontinuity, is not stabilised with $V = R(\Omega_1 - \Omega_2) > 2\sqrt{2}\,c_s$, and only does so at some higher $V/c_s \to \infty$ and, in addition to this, the instability increment (which 5.43) increases almost linearly with increasing V/c_s (stabilisation

[5]This dependence appears in the following order after (5.44): $|\,\mathrm{Im}(\omega_{m+1}) - \mathrm{Im}(\omega_m)\,|\,O(M_*^{-3})$.

of the classic instability of the tangential discontinuity in the two-dimensional gas dynamics – see (5.18)).

To explain the nature of the instability (5.43), attention will be given to the dynamics of perturbation of the discontinuity boundary having, for example, the form of a projection in the region $r > R$. The matter, found in this projection, continues to rotate at the angular velocity Ω_1 and is affected by the centrifugal force $F_1 \simeq \Omega^2_1 R$ (per unit mass). However, this projection is already found in the region where, according to the radial equilibrium condition, the gravitational force $F_2 = \partial\Phi_0 / \partial r \simeq \Omega_2^2 R < \Omega_1^2 R$ (at $\Omega_2 < \Omega_1$). The force formed in this case and oriented upwards $\Delta F = F_1 - F_2 \simeq R\left(\Omega_1^2 - \Omega_2^2\right)$ increases the amplitude of the projection and, the same time, leads to an instability. In the case $\Omega_2 > \Omega_1$, the resultant force is directed to the centre of the disk and, consequently, tries to reduce the amplitude of the projection – this explains the reason for stability in the case $q > 1$ ($\Omega_2 > \Omega_1$). These reasons are correct if the contribution of pressure to the condition of radial equilibrium of the gas disk is negligibly small, and this may occur only in the case in which $V_{\rm rot}/c_s \gg 1$ ($M_* \gg 1$). Identical considerations can be made in the case of the perturbation of the discontinuity boundary having the form of a 'depression' in the region $r < R$, and also lead to the conclusion on the instability only at $\Omega_2 < \Omega_1$ ($q <$ 1). Therefore, it is not surprising that the instability increment (5.43) is proportional to the discontinuity of the centrifugal force acting on the unit mass. Therefore, the instability (5.43) is naturally referred to as centrifugal instability.

The difference between the Kelvin–Helmholtz instability (KHI – (5.40)) and the centrifugal instability (CI – (5.43)) is clearly indicated in Fig. 5.8 which shows the dependence of the instability increment, described by the dispersion equation (5.37) at the values of the parameter $q = \Omega_2/\Omega_1 = 0.7$ and $q = 1.3$ for the mode $m = 1$. It may be seen that at $M_* \ll 1$ the increments in both cases ($q = 0.7$, $q = 1.3$) are close to each other, but at $M_* \geqslant 1$ the difference between them is quite large: at $q = 1.3$ only the KHI is formed, and at $q = 0.7$ the main contribution to the instability increment is provided by the CI mechanism. The physics of this instability is identical with the mechanism of flute instability and greatly differs from the physics of KHI.

Attention will now be given to the problem of the spatial structure of the density perturbations, excited by centrifugal instability. Taking into account that this instability occurs at $M_* \gg 1$, in (5.33) we use the asymptomatic representation of the Bessel functions. As a result, we obtain

$$\tilde{\sigma} \propto \cos\left\{ m\left[\varphi + \frac{\Omega_1(1-q)}{2c_s}(r-R) \right] \right\} \times$$

$$\times \begin{cases} \exp\left\{ M_*\Omega_1(1-q^2)\dfrac{r-R}{2c_s} \right\}, & (r < R) \\ \exp\left\{ -M_*\Omega_1(1-q^2)\dfrac{r-R}{2c_s} \right\}, & (r > R). \end{cases} \qquad (5.45)$$

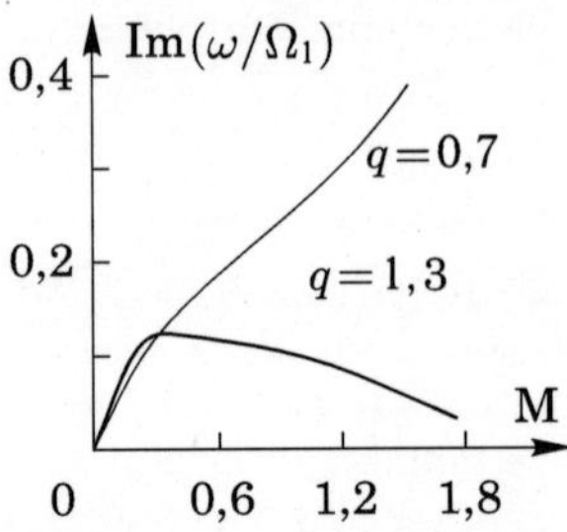

Fig. 5.8. Dependence of the instability increment, described by the dispersion parameter (5.37) on the parameter $\Delta F = F_1 - F_2 \simeq R(\Omega_1^2 - \Omega_2^2)$ at $q = 0.7$ and $q = 1.3$ for the mode $m = 1$ [233].

This shows that the density perturbations, unstable according to (5.43), have the form of trailing spirals. The pitch of such a spiral in the radial direction is determined by the relationship

$$S = m\lambda_s = \frac{2\pi m c_s}{|\,\mathrm{Re}(\omega - m\Omega)\,|} \simeq \frac{4\pi c_s}{\Omega_1(1-q)}, \tag{5.46}$$

and its angular velocity of rotation

$$\Omega_p = \frac{\mathrm{Re}(\omega)}{m} = \frac{\Omega_1}{2}(1+q). \tag{5.47}$$

At the centre, the amplitude of these perturbations decreases quite rapidly with increase of the distance from the discontinuity, and according to (5.45) the characteristic scale of the decrease of the amplitude is

$$L_s = \frac{c_s}{\mathrm{Im}(\omega)} \simeq \frac{2c_s}{\mathrm{M}_*\Omega_1(1-q^2)}, \tag{5.48}$$

and at $\mathrm{M}_* \gg 1$ value $L_s \ll S$. It is important to note that the density waves have the form of trailing spirals only in the system where the angular velocity of rotation decreases with increasing radius, i.e. at $q < 1$. The latter condition is essential for the development of CI in the system.

The results will now be summed up. The centrifugal instability is characterised by a large increment, and the density perturbations excited by this instability have the form of trailing spirals. The latter circumstance appears to be attractive from viewpoint of the possible solution of the problem of origin of the spiral pattern of the galaxies. However, within the framework of the idealised discontinuity model Ω considered here, the centrifugal instability generates too short sections of the spirals ($L_s \ll S$) and does not specify any mode with respect to the value of the increment. At the same time, it is clear that the investigation of more realistic models with the 'smeared discontinuity' Ω indicates both the most unstable (higher modes with $m \geqslant R/L$, where L is the width of smearing of the 'discontinuity' do not 'perceive' the region of rapid variation of Ω as a discontinuity and, consequently, will not be excited). On the other hand, the general decrease of the instability increment with the increase of the width of 'smearing' of the 'discontinuity' should increase the radial length of the excited structure (equation (5.48)). Therefore, it is very important to investigate in detail the centrifugal instability on more realistic model.

5.1.6. Instability of the 'jump' of the velocity of rotation of finite width

We determine the effect of smearing' of the discontinuity of angular velocity Ω ov the results discussed previously. In the first approximation, it is assumed that in the region $|r - R| < L$ with a relatively low $\Lambda = L/R \ll 1$ there is a smooth transition from the value $\Omega(r < R - L) = \Omega_1$ to Ω $(r > R + L) = \Omega_2$. We also consider the fact that in the case which is of special interest to us $M_* \gg 1$ the structure of the unstable perturbations in the radial direction is of the shortwave type $[|kR| = O(M_*^2),\ \mathrm{Re}(kR) = O(mM_*)]$. For the linear approximation $\Omega^2(r)$ in the transition region in the main order with respect to the small value

$$\Lambda^{1/2} \ll 1/M_* \tag{5.49}$$

we obtain [149]

$$\omega = \omega^{(0)} - i\frac{\Omega_1 M_* \Lambda}{4}\{m^2(1-q)^2 + M_*^2(1+q^2)^2\}, \tag{5.50}$$

where $\Omega^{(0)}$ is determined by (5.43).

This shows that in the approximation (5.49), the value Re (Ω), determined by the degree of twisting of the spirals (equation (5.46)), does not change. However, the instability increment Im (Ω) decreases quite rapidly. This leads to two important consequences. Firstly, the perturbations with a small number of the spirals are more unstable than the perturbations with $m \gg 1$. Secondly, the general (and also the main because of $M_* \gg 1$) decrease of the increment in accordance with (5.48) increases the characteristic scale L_s of the decrease of the amplitude of the perturbations in the radial direction and this expands the region of localisation of the generated spiral pattern.

In the actual spiral galactics with the double-hump rotation curves, the 'smearing' of the discontinuity Ω is considerably greater than permitted by the condition (5.49). Therefore, considering the previously obtained results in the model with slight 'smearing', from the viewpoint of determination of the tendency in the channel of Im (Ω), it is still necessary to calculate the latter tendency (as also Re (Ω)) on the models with the rotation curves similar to the actual curves. Such models are presented by the rotation curve (5.26) having the same property sso that at the limit $L/R \to 0$ the function transforms to the previously considered discontinuity Ω. Using this rotation curve, the system of equations (5.27), (5.28) can be solved numerically in a computer as the problem of the Sturm–Liouville type (determination of the eigenfunctions $\tilde{p}$ and $\tilde{\xi}$, and eigenvalues Ω) at the boundary conditions

$$\frac{\partial \ln(\tilde{p})}{\partial r}\bigg|_{r\to\infty} = i k_2^{(0)}; \quad \xi\bigg|_{r=0} = 0, \tag{5.51}$$

where

$$k_2^{(0)} = \frac{1}{c_s}\sqrt{(\omega - m\Omega_2)^2 - 4\Omega_2^2}; \quad \mathrm{Im}(k_2^{(0)}) > 0. \tag{5.52}$$

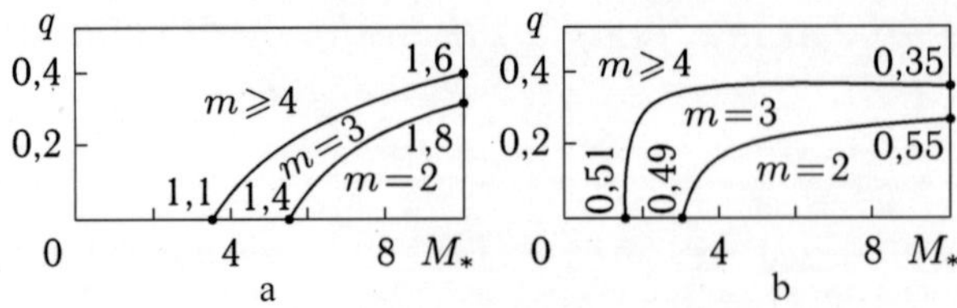

Fig. 5.9. Regions of dominance of different modes (m = 2, 3, 4,…) with respect to the increment in the plane of the parameters $M_* = R\,\Omega_1/c_s$, $q = \Omega_2/\Omega_1$: a) at $\Lambda = L/R = 0.02$; *b*) at $\Lambda = 0.1$. The numbers at the boundary points of the curves are the increment in the units of 1.

It is clear that assuming that the radial displacement in the centre of the disk is $\xi = 0$, the mode $m = 1$ is not investigated.

We describe briefly the results at the most interesting region of the parameters from the viewpoint of applications $M_* = R\,\Omega_1/c_s \gg 1$; $q = \Omega_2/\Omega_1 \ll 1$ [149, 160, 164]:

1. The higher modes ($m > 2$) are completely stabilised at $\Lambda \geqslant 1/m$. The mode $m = 2$ is stabilised at $\Lambda \geqslant 0.45$ (at the limit $q \to 0$). This result can be easily understood: in the rotating incompressible liquid, to buildup perturbations with $m \neq 0$ it is necessary to ensure that the vorticity $d(\varkappa^2/2\Omega)/dr$ changes its sign at finite $r \neq 0$ [232]. For $\Omega(r)$ (5.26), the points of the variation of the sign of this quantity exists at $\Lambda < 1/\sqrt{3} \simeq 0.58$. The deviation of the stability boundary in the present case to smaller values with respect to the parameters is caused evidently by the stabilising effect of compressibility of the medium.
2. In the region of the parameters $M_* \gg 1$, $q \ll 1$, the two-arm ($m = 2$) perturbations are most unstable (without taking into account the modes $m = 0$ and $m = 1$, see below). This result is shown in Fig. 5.9 where the plane of the parameters M_*, q shows the regions of dominance with respect to the increment of the modes at $m \geqslant 2$ at $\Lambda = 0.02$ and $\Lambda = 0.1$. It may be seen that the width of the region of dominance of the mode $m = 2$ rapidly increases with increase of parameter Λ.
3. For the two-arm perturbations, the regional instability in the plane of the parameters q, Λ is relatively large (Fig. 5.10) and its dimensions depend only slightly on the value M_* at $M_* \gg 1$. As an example, Fig. 5.11 shows three rotation curves (5.26) (in the vicinity of the inner hump), characterised by the parameters ensuring the excitation of the two-arm spiral in the model with $M_* = 10$ s with a small increment. It may be seen that to excite the spiral pattern by means of the mechanism investigated here, the flat galaxy should only be characterised by the rotation curve even with a not very distinctive double-hump form. Such rotation curves are found relatively often [9–12, 1090, 1093, 1094, 1186].
4. An increase of parameter $\Lambda = L/R$ results in a small decrease of Re(Ω). At a sufficiently small $\Lambda \leqslant M_*^{-2}$ (see the condition (5.49), the difference of Re(Ω) from the values (5.43) does not exceed 1% in accordance with the result (5.50). However, at finite Λ this effect becomes quite strong[6] and results in the appropriate

[6]For example, for two-arm perturbations at $\Lambda \geqslant 0.2$, the value Re (ω) is smaller than that obtained from (5.3) by 30–40%.

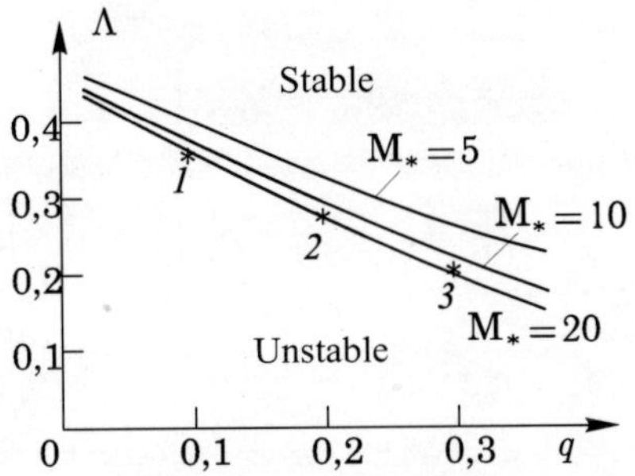

Fig. 5.10. The curves of the marginal stability of the mode $m = 2$, each point of the curves (pair of the values (q, Λ)), determines the profile of the rotation curve (5.26) for which Im $(\Omega) = +0$ at the appropriate value of the parameter M_*. The profiles $V_{rot}(r)$, corresponding to the points * –1, 2, 3 (see Figure 5.11).

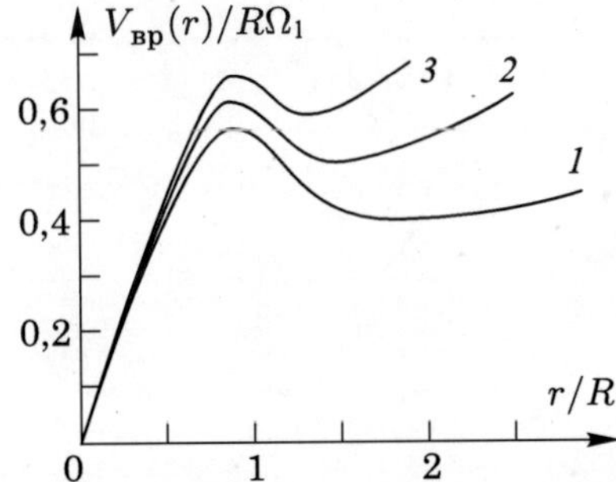

Fig. 5.11. Rotation curves (in the vicinity of the inner hump), permitting excitation of the two-arm spiral pattern with a small increment at $M_* = 10$. The set of the parameters $\{\Lambda; q; n_{min} = \min (d \ln \Omega / d \ln r)\}$ for the curves has the following values: 1 – {0.36; 0.1; –1.8}; 2 – {0.29; 0.2; –1.7}; 3) {0.24; 0.3; –1.5}.

increase of the pitch of the spiral (see equation (5.43)) and the reduction of the angular velocity of rotation $\Omega_p = \text{Re}(\Omega)/m$ (see equation (5.47)).

5. In the entire investigated range of the parameters q, M_*, Λ, the characteristic scale of the decrease of the amplitude of the perturbed density L_s with increase of the distance from the 'discontinuity' in the region $r > R$ is described satisfactorily by the equation (compared with (5.48))

$$L_s \simeq \frac{c_s}{\text{Im}(\omega)}. \tag{5.53}$$

Thus, the large decrease of the instability increment with the increase of parameter results in a very large widening of the region of localisation of the generated spiral pattern. This effect is illustrated by Fig. 5.12 (dependence $L_s(\Lambda)$) and in Fig. 5.13 which shows the examples of the eigenfunctions of the perturbed density $\tilde{\sigma}(r)$.

The problem of the dependence of the previously mentioned results on the nature of the rotation curve outside the limits of the zone of 'smearing' of the discontinuity is also important. The real rotation curves of the galaxies in the range $r \gg R$ are

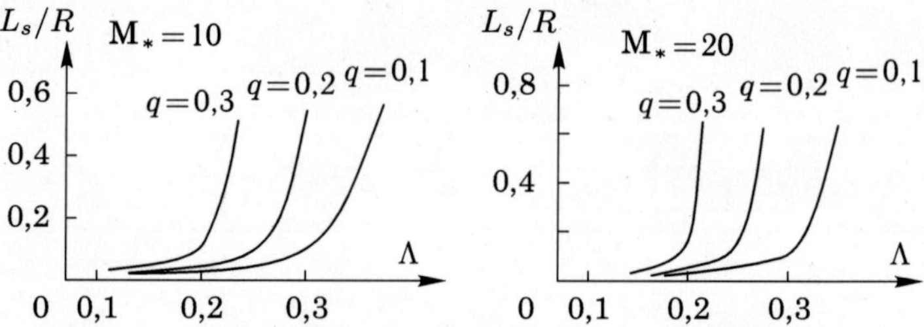

Fig. 5.12. Dependence of the scale of the radial decrease of the amplitude of the perturbed density L_s on the parameter $\Lambda = L/R$ of the rotation curve (5.26) for the mode $m = 2$ [233].

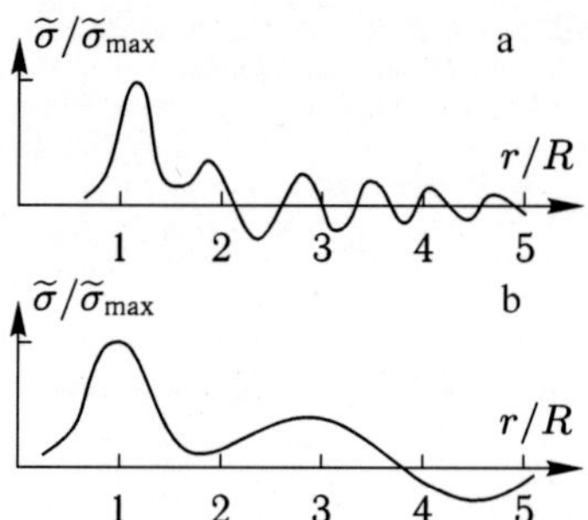

Fig. 5.13. Perturbed surface density $\tilde{\sigma}(r)$ of the unstable mode $m = 2$ in a disk with the rotation curve (5.26): $a - q = 0.1$; $\Lambda = 0.335$; $M_* = 20$; $L_s = 2R$; $b)$ $q = 0.1$; $\Lambda = 0.385$, $M_* = 5$, $L_s = 6R$.

usually characterised by the law of rotation $\Omega(r)\, r^n$ with $n = -1$, which differs greatly from the law of rotation (5.26): $\Omega \simeq \Omega_2 = \text{const}$ at $r \gg R$. Even in the range $r < R$ the rotation of the real galaxies greatly differs from the solid-state rotation. To explain the effect of these factors, comparative calculations were carried out of the stability of rotation of the gas in the galaxy M81 and in a simulation galaxy with the law of rotation (5.26) with the almost identical sections $V_{rot}(r)$ between the inner hump and the following minimum V_{rot}. The observed sections of the rotation curve in the range $r \leqslant 1.6$ kpc [626] and $r \geqslant 3.6$ kpc [1086] were cross-linked by a third degree polynomial in the range $1.6 < r < 3.6$. The distribution $\sigma_0(r)$ was also taken from the observations [1086], and $c_s(r)$ was assumed to be monotonically decreasing from $c_s(r = 0)$ 28 m/s to $c_s(r \geqslant 3) = 10$ km/s. Calculations were carried out for the mode $m = 2$ on the basis of the equations which take into account the heterogeneity of $\sigma_0(r)$ and $c_s(r)$ (compare with (5.27), (5.28)):

$$\frac{d\tilde{p}}{dr} = \frac{2m\Omega}{r\hat{\omega}}\tilde{p} + \sigma_0(\hat{\omega}^2 - \varkappa^2)\xi - \frac{g}{c_s^2}\tilde{p} + g\left(\sigma_0' + \sigma_0\frac{g}{c_s^2}\right)\xi, \tag{5.54}$$

$$\frac{d\xi}{dr} = \frac{\tilde{p}}{\sigma_0}\left(\frac{m^2}{r^2\hat{\omega}^2} - \frac{1}{c_s^2}\right) - \left(\frac{2m\Omega}{r\hat{\omega}} + \frac{1}{r} - \frac{g}{c_s^2}\right)\xi, \tag{5.55}$$

where $g = -p_0'/\sigma_0$. The modelling rotation curve is described by the following parameters: $\Lambda = 0.18$; $q = 0.9$; $R = 1.5$ kpc; $\Omega_1 = 280$ km/s/kpc ($M_* = 10$). As a result of the calculations, the following values were obtained for the curve $\Omega(m = 2) = (0.90 + i0.39)\,\Omega_1$. Thus, the calculation show that the section of 'discontinuity' $\Omega(r)$ is controlling for the parameters of instability and, consequently, of the generated spiral pattern.

Attention will now be given to the problem of excitation of the modes $m = 0$ and $m = 1$. The perturbed pressure will be rejected (5.27), (5.28) and, as a result, the following equation is obtained for $m = 0$

$$\frac{1}{r}\frac{d}{dr}\left(r\frac{d\xi}{dr}\right)+\left(\frac{\omega^2-\varkappa^2}{c_s^2}-\frac{1}{r^2}\right)\xi=0, \tag{5.56}$$

where the displacement ξ should satisfy the boundary conditions $\xi(r = 0) = 0$; $\xi(r \to \infty) = 0$. Multiplying this equation by $r\xi^*$ (the sign * denotes the complex conjugation) and integrating from $r = 0$ to $r = \infty$, we obtain

$$\omega^2=\frac{\int\limits_0^\infty\left\{\left|\frac{d\xi}{dr}\right|^2 c_s^2+|\xi|^2\,(\varkappa^2+c_s^2/r^2)\right\}r\,dr}{\int\limits_0^\infty|\xi|^2\,r\,dr}. \tag{5.57}$$

This shows that the instability of the mode $m = 0$ ($\Omega^2 < 0$) occurs only in the case if the range $\Delta r = r_2 - r_1 \neq 0$ exists and in this range ($r_1 < r < r_2$) the value $\varkappa^2$ is negative. The latter has any meaning if

$$n_{\min}=\min_r\left\{\frac{d\ln\Omega}{d\ln r}\right\}<-2. \tag{5.58}$$

For the majority of galaxies with the two-hump rotation curves $n_{min} > -2$ and the mode $m = 0$ can be excited in them (in the Galaxy $n_{\min} \simeq -1.5\pm0.2$ according to the rotation curve by Howde [250] and Clemens [470]). At the same time, the mode $m = 2$ is unstable also at $n_{\min} > -2$, as shown by the calculations in [149, 160, 233] (Fig. 5.11).

Examining the axisymmetric mechanism of the excitation of spirals in an isolated galaxy, we should exclude from this examination the mode $m = 1$ and because the buildup of these perturbations displaces the centre of the masses of the gas subsystem in relation to the centre of the mass of the stellar subsystem [$\xi(r = 0) \neq 0$]. Evidently, the latter is possible only in the presence of external effects on the investigated system.

In conclusion, we examine the problem of the effect of the perturbation of the gravitational potential on the parameters of the generated spiral structure taking into account the estimates presented in the introduction to the section: $k_J \ll k$. For this purpose, in the framework of the previously investigated discontinuity model Ω (at $M_* \gg 1$) the perturbed pressure is replaced by $(1 - 2k_J/k)\,\tilde{p}$, where $k_J = \pi G\sigma_0/c_s^2$, and the value $k^{(0)}$ (5.52) by $k = k^{(1)}$, determined from the relationship $k^2-2k_Jk-k^{(0)2}=0\ \left[k^{(1)}\simeq k^{(0)}+k_J+k_J^2/2k^{(0)2}\right]$. Therefore, solving the dispersion equation (5.37), corrected by the substitution, using the perturbation method ($\Omega = \Omega^{(0)} + \Omega^{(3)}$) we obtain [149]

$$\omega^{(3)} \simeq i\Omega_1 (k_J R)^2 (1-q^2)/\mathrm{M}_* \left\{\mathrm{M}_*^2 (1-q^2)^2 + m^2 (1-q)^2\right\}. \qquad (5.59)$$

Subsequently, the calculating the phase of the perturbed density at $r > R$, we obtain

$$\tilde{\sigma} \propto \exp\left\{ im\left[\varphi + \frac{\Omega_1(1-q) + 2k_J c_s / m}{2c_s} r\right] - \frac{\mathrm{Im}(\omega^{(0)})}{c_s} r\right\}. \qquad (5.60)$$

This shows that, irrespective of the destabilising effect (relatively weak) of the perturbations of the gravitational potential, the scale of the decrease of the amplitude of the perturbed density does not change. This is associated with the fact that in addition to the correction of Im (Ω) (5.59), the definition of Im (k) through the frequency also changes. As a result, both effects mutually compensate each other. Only the pitch of the spiral decreases (compared with (5.46)):

$$S = \frac{2\pi m c_s}{\mathrm{Re}(\omega - m\Omega_2) + k_J c_s}, \qquad (5.61)$$

but changes only slightly since at $\mathrm{M}_* \gg 1$

$$\mathrm{Re}(\omega - m\Omega_2) \simeq \mathrm{Re}(k^{(0)} c_s) \gg k_J c_s .$$

5.1.7. Taking into account the density jump in the region of development of centrifugal instability

Attention will now be given to the problem of the effect of a large change of the density of the gas disk in the vicinity of the 'discontinuity' Ω on the parameters of centrifugal instability and the excited structures. In particular, it should be mentioned that the previously investigated homogeneous models with the tangential discontinuity of angular velocity were isentropic ($dS_0/dr \equiv 0$). In the presence of a density jump, it is necessary to consider the non-isentropic models. Actually, for a discontinuous model with (5.25) and

$$\sigma_0(r) = \sigma_{01} + (\sigma_{02} - \sigma_{01})\Theta(r - R) \qquad (5.62)$$

the conditions of equilibrium (3.243) and of the non-isentropic nature ($S_0 =$ const for (3.299)) cannot be fulfilled at the same time. It should be mentioned that since the models with the pressure jump p_0 require the presence of a jump of the gravitational potential, we must restrict ourselves to the case $|g| < \infty$ ($g = -p_0'/\sigma_0$), i.e. pressure should be a continuous function of r but it should have an inflection point.

The cross-linking conditions will be written for the perturbed quantities $\tilde{p}$ and ξ on the basis of (5.54), (5.55):

$$\xi\Big|_{r=R-0} = \xi\Big|_{r=R+0},$$

$$\tilde{p}\Big|_{r=R+0} - \tilde{p}\Big|_{r=R-0} = \frac{g_1 + g_2}{2}(\sigma_{02} - \sigma_{01})\xi - \frac{\sigma_{01} + \sigma_{02}}{2} R\xi(\Omega_2^2 - \Omega_1^2) \qquad (5.63)$$

for

$$g(r) = g_1 + (g_2 - g_1)\Theta(r - R). \tag{5.64}$$

Below, we confine ourselves to the case in which we have $g \equiv 0$, i.e. $p_{01} = p_{02}$ and $\delta \equiv \frac{\sigma_{02}}{\sigma_{01}} = \frac{c_{s1}^2}{c_{s2}^2}$. Thus, acting in the sense of section 5.1.3, we obtain the dispersion equation

$$R^2 k_2^2 \alpha_1 - R^2 k_1^2 \alpha_2 + \frac{1+\delta}{2} \mathrm{M}_*^2 (q^2 - 1)\alpha_1 \alpha_2 = 0, \tag{5.65}$$

where $k_{1,2}^2 = \{4\Omega_{1,2}^2 - (\omega - m\Omega_{1,2})^2\} / c_{s_{1,2}}^2$ $\mathrm{Re}(k_1, 2) > 0$; $\mathrm{M}_* = R\Omega_1 / c_{s_1}$; for $\alpha_{1,2}$ see (5.38), (5.39).

At the limit $\mathrm{M}_* \gg m$, we obtain

$$\omega = \frac{\Omega_1}{1+\sqrt{\delta}} \left\{ m(1 + q\sqrt{\delta}) + i\mathrm{M}_* \frac{1+\delta}{2}(1 - q^2) \right\}, \tag{5.66}$$

which coincides with (5.37) for $\delta = 1$. A distinguishing special feature of the result is the independence of the increment on the magnitude of the density jump $\delta = \sigma_{02}/\sigma_{01}$ at $\delta \ll 1$ and, conversely, the proportionality $\sqrt{\delta}$ at $\delta \gg 1$.

5.1.8. Heterogeneous gas disks with two-hump rotation curves

Previously, we investigated the limiting case of the combined discontinuities in the distributions $V_{\mathrm{rot}}(r)$ and $\sigma_0(r)$. It is evident that the assumptions of the discontinuity of the equilibrium parameters that determine the instability and the superposition of these discontinuities greatly idealise the observed distributions. In addition, at least in our Galaxy, the condition p_0 = const is not fulfilled.

It is important to mention another important fact. In the numerical modelling of the process of excitation of the spiral pattern in the galaxy with the two-hump rotation curve it was shown that in a homogeneous (at the initial moment) gas disk the buildup of the instability results in the formation, in the vicinity of the internal maximum V_{rot}, of a steep density gradient resembling the one detected in the Galaxy[7] [165]. Therefore, it is very important to investigate the effect of the density jump on the parameters of the excited pattern.

We examine a group of models in which the distributions of the equilibrium thermodynamic quantities are determined by the relationship

$$\frac{dp_0}{dr} = \alpha c_s^2 \frac{d\sigma_0}{dr}, \tag{5.67}$$

where α is the numerical parameter (the value $\alpha = 1$ corresponds to the isentropic model of the disk). According to the definition of the speed of sound in the thin layer of the ideal gas $p_0 = \sigma_0 c_s^2 / \gamma$. Differentiating the latter relationship with respect to r and comparing the results with (5.67) we obtain

$$c_s(r) = \mathrm{const} \cdot [\sigma_0(r)]^{(\alpha\gamma - 1)/2}. \tag{5.68}$$

[7] It should also be noted that a dense gas ring forms in the numerical gas dynamic models with a bar [827] and in the viscous axisymmetric models of the disk [42].

Thus, on the basis of the distribution $\sigma_0(r)$ and two numerical parameters α and γ we can determine the distributions of any thermodynamic quantity in the gas disk.

We select the specific distribution $\sigma_0(r)$ in the form of a step centred at $r = R_\sigma$ and 'smeared' over the region with the width $\Delta r \simeq 2L_\sigma$:

$$\sigma_0(r) = \sigma_{01}\left[\frac{1+\Sigma}{2} + \frac{\Sigma - 1}{\pi}\operatorname{arctg}\left(\frac{r - R_\sigma}{L_\sigma}\right)\right], \tag{5.69}$$

for which $\sigma_0(r \ll R_\sigma) \simeq \sigma_{01}$ (at $L_\sigma \ll R_\sigma$) and $\sigma_0(r \gg R_\sigma) \simeq \sigma_{01}\Sigma$. At the same time, the parameter Σ characterises the magnitude of the jump $\sigma_0(r)$. The rotation curve

$$\Omega(r) = \Omega_1\left[\frac{1+q}{2} + \frac{1-q}{\pi}\operatorname{arctg}\left(\frac{r-R}{L}\right)\right]\left(1 + \nu\frac{r^2}{R^2}\right)^{-1/2} \tag{5.70}$$

ensures exit to the 'plateau' at a relatively large distance from the centre (ν = const). The dissipation of the velocities of the gas clouds in the disks of the galaxies remains almost constant along the radial coordinate (with the exception of the central part of the disk) and is equal to approximately 10 km/s [1112], and the characteristic (in the region of the plateau $r \gg R$) values are $V_{\text{rot}} \simeq 150 \div 250$ km/s [41]. Therefore, it is convenient to introduce the Mach number as follows:

$$\mathrm{M} = [V_{\text{rot}}(r) / c_s(r)]\big|_{r=5R}.$$

We start with the equations (5.54), (5.55). This system should be used for the determination of both the eigenfunctions $\tilde{p}(r)$, $\xi(r)$ and the eigenvalue – frequency Ω. Simple analysis of the equations (5.54) and (5.55) at the limit $r \to 0$ shows that $\tilde{p}(r \to 0)\ \alpha\ r^m$, $\xi(r \to 0)\ \alpha\ r^{m-1}$. The asymptotics of the solutions at $r \to \infty$ taking into account the properties of our model, has the form $\tilde{p}\,\xi \propto \exp\{i\Omega r/c_s\}$. At this asymptotics, the selection of the sign in front of the apparent unity in the exponent ensures a decrease of the amplitude of the unstable perturbations ($\mathrm{Im}(\Omega) > 0$) with increase of the distance from the region of the jump $\Omega(r)$. Thus, the natural boundary conditions for the system (5.54), (5.55) in the case of the perturbations with $m \geqslant 2$ have the following form:

$$\frac{d\ln(\tilde{p})}{dr}\bigg|_{r\to\infty} = i\frac{\omega}{c_s}; \qquad \xi\big|_{r=0} = 0. \tag{5.71}$$

The effect of the type of curve $V_{\text{rot}}(r)$ on the instability parameters was investigated in section 5.1.6. Therefore, here we present the rotation curve: $q = 0.2$; $\Lambda = L/R = 0.25$; $\nu = 0.0625$ (the form of $V_{\text{rot}}(r)$ at these values of the parameters is shown in Fig. 5.14). The number of the spirals will be assumed to be equal to ($m = 2$). We concentrate our attention on the parameters of the model of the gas disk. In particular, they include the magnitude of the jump of surface density of the gas Σ. In the Galaxy $\equiv \sigma_{02}/\sigma_{01} \simeq 3\cdot 10^{-2}$ [1112]. Parameter α (see (5.67)) which determines the magnitude of the radial gradient of pressure obviously relates to the number of difficult to observe parameters. Parameter $x_\sigma = R_\sigma/R$ can be used to study conveniently the displacement of the centres of the jumps $\sigma_0(r)$ and $\Omega(r)$. For the values of the parameters $\Lambda_\sigma = L_\sigma/R = 1/4;\ 1/16;\ \gamma = 1;\ 2;\ \mathrm{M} = 20;\ 10;\ x_\sigma = 0.9;\ 1.0;\ 1.1;$

$10^{-2} \leqslant \Sigma \leqslant 10^{1}$; $0 \leqslant \alpha \leqslant 1.25$ the numerical value of the formulated problem of the Sturm–Liouville type shows that at all values of the investigated parameters there is an instability leading to excitation of a two-arm spiral [Im ($\Omega_m = 2$) > 0] [160]. This spiral which increases in proportion to $\exp\{\mathrm{Im}\,(\Omega)t\}$ with respect to amplitude rotates with the angular velocity $\Omega_p = \mathrm{Re}(\Omega)/m$. The behaviour of the eigenfunctions $\tilde{p}(r)$ can in principle be used to determine the wavelength of the spiral pattern λ_s at the known distance from the centre of the disk. Finally, this value is local. In the case of a homogeneous disk we consider the model for $\gamma = 2$, M = 20 which gives $\Omega_p = \Omega_p^{(0)} = 0.822\,\Omega(R)$. The deviation of Ω_p from $\Omega_p^{(0)}$ is small almost on the entire plane (α, Σ). A large increase of the angular velocity of rotation of the spiral pattern Ω_p in comparison with $\Omega_p^{(0)}$ is possible only in the range of the parameters of the gas disk $\Sigma \ll 1$, $\alpha \simeq 1$. Calculations show that the value Ω_p also depends only slightly on all parameters of the models, with the exception of the parameter M: $\lambda_s \simeq (0.5 \div 0.7)\,R$ at M = 20 and $\lambda_s \simeq (1 \div 1.5)\,R$ at M = 10.

Thus, taking into account the real (finite width) 'jumps' of density in the gas disk cannot lead in the framework of the linear theory to any large change of the main dynamic parameter of the spiral pattern Ω_p.

Can any other physical parameter lead to a large decrease of Ω_p? Analogue simulation of the spiral pattern (see chapter 6) shows a difference between the value $\Omega_p^{(0)}$, predicted by linear theory, and its experimental value Ω_p ($\Omega_p^{(0)} > \Omega_p^{e}$)[8]. The width of the jumps of angular velocity $\Omega(r)$ and the thickness of the layer of 'shallow' water $H(r)$ – analogue of $\sigma_0(r)$ – in the experiments were almost identical, and the jumps themselves were almost the same. Evidently, the discussed difference between Ω_p^{e} and $\Omega_p^{(0)}$ is determined by the non-linear nature of the experiments. This is confirmed by the fact that the high-mode perturbations (the number of the spirals $m \gg 1$) are characterised by a low amplitude and for them the difference between the experimental and theoretical values of Ω_p is considerably smaller than in the case of excitation of the mode $m = 2$, characterised by a relatively high amplitude.

5.2. Instabilities of the resonant nature

5.2.1. Effect of super-reflection and development of instability

The theoretical investigations of the problem of reflection with the intensification of the monochromatic soundwave from a plane-parallel tangential discontinuity of velocity was carried out for the first time by Miles [913] and Ribner [1067]. It was shown that when the velocity of the moving medium is relatively high (wave fell on the discontinuity from a stationary medium), reflection with intensification could take place. In this process, the amplitude of the reflected waves is greater than the amplitude of the incident wave. This took place as a result of the fact that the wave with negative energy travels into the moving medium (or, more accurately, quasi-energy [896]), whereas the wave with positive energy moves into the stationary medium. This wave, irradiated into the stationary medium, supplements its energy

[8]In studies [183, 239] attention is given to possible methods and the difficulties of the observational definition of the parameter Ω_p.

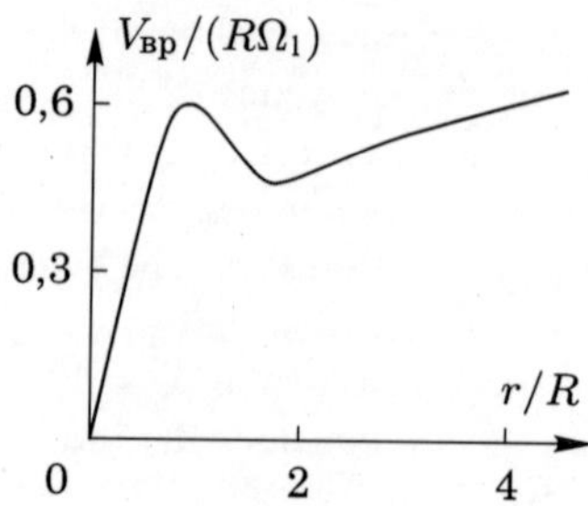

Fig. 5.14. Rotation curve of the gas disk $V_{rot} = r\,\Omega$ in the model (5.70) with the parameters $q = 0.2$; $L/R = 0.25$; $\nu = 0.0625$.

from the moving medium [226]. For the instability to develop, it is necessary to add an acoustic feedback, for example a wall, which would stop the wave reflected from the tangential discontinuity from returning to the discontinuity and again intensify.

The other source of intensification of the wave at its reflection is the resonance intensification of the sound wave, formed if the tangential discontinuity is slightly smoothed out to a thin shear layer of some finite width [401]. In this case, a thin critical layer appears inside the shear layer, and the energy of the sound wave in the layer increases. Thus, an additional branch of unstable oscillations formed due to interaction with the resonance particles can form [226].

Thus, the presence of a reflecting wall may lead to the formation of a super-reflection instability. However, the same system with the plane-parallel tangential discontinuity of the velocity may also contain the Kelvin–Helmholtz instability (tangential discontinuity) and a centrifugal instability may also exist in the system with a cylindrical tangential discontinuity of velocity (in the two-dimensional axisymmetric flow). The two latter mechanisms of hydrodynamic instabilities are evidently more powerful than the super-reflection instability. This means that the latter can be detected only in cases in which both these stronger instabilities do not form.

The hydrodynamic medium, in which the super-reflection instability was studied is rotating shallow water with a free surface. It is well-known [112] that the dynamics of this medium can be described by two-dimensional equations equivalent to the appropriate dynamic equations of the two-dimensional compressible medium. The wave perturbations of the surface density in the two-dimensional compressible medium in shallow water correspond to the wave perturbations of the depth of the layer (surface gravitational waves), where the speed of sound plays the role of the velocity of propagation of these waves.

For a two-dimensional plane-parallel compressible tangential discontinuity, the Kelvin–Helmholtz theoretical criterion of stabilisation of the instability was defined for the first time by Landau [108] (see section 5.1.2). According to this criterion, if there is only a velocity jump, the instability of the tangential breakdown does not form under the condition in which the Mach number is

$$M = |\Delta V|/c_s > 2\sqrt{2}, \tag{5.72}$$

where $|\Delta V|$ is the magnitude of the jump of the velocity of the flow.

The validity of this criterion (within the experimental error range) for the layer of rotating shallow water was confirmed by the experiments [5]. Another hydrodynamic instability should be excited in a two-dimensional compressible cylindrical tangential discontinuity of velocity when the internal part rotates at a higher velocity than the outer part and at a sufficiently large Mach number (condition (5.72) is fulfilled in this case). This instability, referred to as centrifugal instability [588], was investigated by experiments in equipment Spiral' with rotating shallow water [588, 967].

In the reversed case

$$\Omega_1 < \Omega_2, \tag{5.73}$$

where Ω_1, Ω_2 are the angular velocities of rotation of the internal and external part of shallow water, respectively, and if (5.52) is fulfilled, both instabilities (tangential discontinuity and centrifugal discontinuity) are suppressed.

Several comments should be made here. On the one hand, the theoretical and numerical investigations are restricted in this case to the linear approximation. On the other hand, we would like to apply our results to actual non-linear structures, generated in the experimental formulation. When can this be done? The amplitude of the structures (wave patterns, formed by the perturbations h (r, φ, t) of the depth of the layer of shallow water), which was predicted for the experiments, is not small in the general case, $|h| \sim \mathcal{O}(H_0)$ (where H_0 is the unperturbed depth of the layer), but their horizontal dimensions are large, i.e. $|k_\perp h| \ll 1$ (where $|k_\perp h| = |\nabla_\perp h|$). Therefore, the variation of the height of the surface is assumed to be relatively smooth. The condition $|k_\perp h| \ll 1$ is not sufficient for the application of the linearised model and, in addition to this, it is also assumed that $|h| \ll H_0$. Therefore, it is assumed that if some linear unstable mode of the perturbations is detected (for the given set of the non-perturbed parameters, corresponding to the axisymmetric stationary rotating shallow water) and its increment is considerably larger than four other possible unstable modes, then this mode in particular will be generated in the experimental equipment at the given parameters. The results of previous theoretical investigations and laboratory experiments with the Kelvin–Helmholtz instability and centrifugal instability in the investigated system confirm that for relatively unstable modes the spiral-vortex structures, calculated in the linear model, are in good agreement with the nonlinear stationary structures, observed in experimental equipment [163, 164, 588, 967].

The centrifugal instability develops at any high values of the Mach number on the condition that the angular velocity of rotation of the central part Ω_1 (internal in relation to the velocity jump) is greater than the angular velocity of rotation of the periphery Ω_2: $\Omega_1/\Omega_2 > 1$.

These velocity jumps (Fig. 5.15 a, 1.9) are observed in the disks of, evidently, a large part of the spiral galaxies (see chapter 1, [9–12]). The physics of centrifugal instability is similar to the physics of the Rayleigh–Taylor instability which describes the situation in which the liquid with the higher density is situated on the liquid in the lower density in the field of gravitational force. In the case of centrifugal instability, the centrifugal force in the central part is higher than at the periphery, as in the case of the Rayleigh–Taylor instability where the liquid with the high-

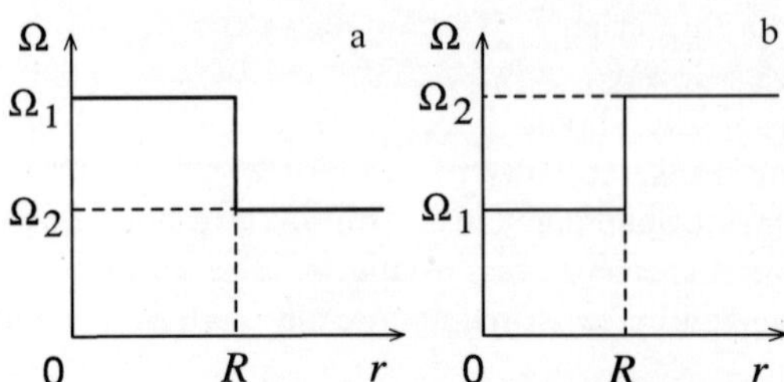

Fig. 5.15. Rotation curves: *a*) permitting the development of centrifugal instability; *b*) absence of centrifugal instability.

density is subjected to the effect of a higher gravitational force than the liquid with the lower density. Centrifugal instability generates trailing spiral waves, rotating in such a manner that the ends of the spirals are directed backwards. Therefore, these waves have a satisfactory 'aerodynamic' shape (Fig. 5.4). As the velocity jump $\Delta\Omega$ becomes greater, the number m of the trailing spiral waves, generated by centrifugal instability, decreases.

5.2.2. Super-reflection instability on the rotating 'shallow water'

The shear layer is a source of hydrodynamic instabilities. The most powerful of them are the Kelvin–Helmholtz and centrifugal instabilities. However, if the parameters of equipment are correctly selected, they can be suppressed [242].

For simplicity, we consider mainly the flow with only one discontinuity on the rotation curve $\Omega_0\,(r)$ (Fig. 5.15b). It is assumed that

$$0 \le \bar{q} = \frac{\Omega_1}{\Omega_2} < 1. \tag{5.74}$$

In this case, the centrifugal instability is suppressed. In the actual experiments, in the region of the velocity jump, viscous spreading results in the formation of a transition layer with the characteristic width $\sim 2H_0$, where H_0 is the depth of the layer. The latter is assumed to be small in comparison with all characteristic dimensions of the problem. Thus, in order to solve the situation on the qualitative level, it is sufficient to analyse the sharper discontinuity on the rotation curve (for this purpose, we will discuss the effects of the finite width of the velocity jump in the next section). The velocity jump at the discontinuity is characterised by the Mach number

$$\mathrm{M} = R\frac{\Omega_2 - \Omega_1}{c_{s0}}. \tag{5.75}$$

By analogy with a flat velocity jump for the rotating shallow water, the Kelvin-Helmholtz instability is suppressed at $\mathrm{M} \geqslant \mathrm{M}_{cr}$. Numerical analysis shows that in contrast to the flat shear layer, where $\mathrm{M}_{cr} = 2\sqrt{2}$, for the cylindrical tangential discontinuity $2 < \mathrm{M}_{cr} < 2\sqrt{2}$. The exact boundary of stability changes only slightly with the variation of the number of the azimuthal mode m. It is also assumed that the Mach number is sufficiently large in order to suppress the Kelvin–Helmholtz instability.

The dispersion equation for investigating the instability of the shear layer in shallow water with the free surface can be derived from the linearised hydrodynamic equations of the gas galactic disk [239]:

$$\frac{\partial v_r}{\partial t}+\Omega_0\frac{\partial v_r}{\partial \varphi}-2\Omega_0 v_\varphi=-\frac{\partial}{\partial r}(c_{g0}^2\eta),$$

$$\frac{\partial v_\varphi}{\partial t}+\Omega_0\frac{\partial v_\varphi}{\partial \varphi}+\frac{\kappa^2}{2\Omega_0}v_r=-\frac{1}{r}\frac{\partial}{\partial \varphi}(c_{g0}^2\eta), \qquad (5.76)$$

$$\frac{\partial \eta}{\partial t}+\Omega_0\frac{\partial \eta}{\partial \varphi}+\frac{\partial v_r}{\partial r}+\left(1+\frac{d\ln\sigma_0}{d\ln r}\right)\frac{v_r}{r}+\frac{1}{r}\frac{\partial v_\varphi}{\partial \varphi}=0,$$

where v_r, v_φ are the radial and azimuthal perturbed velocities, respectively, $c_{g0}^2\equiv c_{s0}^2-\dfrac{2\pi G\sigma_0}{|k|R_g}$, $c_{s0}^2\equiv\dfrac{dp_0}{dr}$, $\eta=\dfrac{\sigma}{\sigma_0}$ $\Omega_0(r)$ is the angular velocity of rotation, $\varkappa$ – epicyclic frequency: $\varkappa(r)=(4\Omega_0^2+r\mathrm{d}\Omega_0^2/\mathrm{d}r)^{1/2}$:

$$R_g=\begin{cases}\dfrac{1}{1+|k|h_g/2}, & \dfrac{|k|h_g}{2}\ll 1,\\[2ex] \dfrac{2}{|k|h_g}, & \dfrac{|k|h_g}{2}\gg 1.\end{cases}$$

When writing the equations of motion, we have used the linearised equation of state $p/\sigma_0=c_{s0}^2\,\eta$, and also the perturbation of the gravitational potential

$$\Psi=-\frac{2\pi G\sigma R_g}{|k|}.$$

If the following substitution is made in equations (5.76)

$$c_{g0}^2=c_{s0}^2=gH_0,\qquad \eta=H/H_0,$$

we obtain the system of linear equations of rotating shallow water. The role of the speed of sound is played by the velocity of propagation of long gravitational waves: $c_{s0}=(gH_0)^{1/2}$, where g is gravitational acceleration. We find the general solution in the form

$$f(r,\varphi,t)=\mathrm{Re}\left[\sum_m\int f(r)\exp[i(m\varphi-\omega t)]\,d\omega\right], \qquad (5.77)$$

where $\Omega=\mathrm{Re}\Omega+i\mathrm{Im}\,\omega$. In the framework of the linear approximation (because we are interested in investigating the 'linear' instability), we can use the principle of superposition, i.e. consider the separate m-th harmonics ($m\geqslant 1$).

In the case of the arbitrary continuous rotation curve and the arbitrary profile of the speed of sound, the system of differential equations (5.76) can be reduced to the following equations [239]:

$$\frac{d}{dr}(c_{s0}^2\eta)=\frac{2m\Omega_0}{r\hat{\omega}}c_{s0}^2\eta-(\chi^2-\hat{\omega}^2)\xi,$$

$$\frac{d}{dr}(rH_0\xi) = -rH_0\left[\left(1-\frac{m^2c_{s0}^2}{r^2\hat{\omega}^2}\right)\eta + \frac{2m\Omega_0}{r\hat{\omega}}\xi\right], \tag{5.78}$$

with the appropriate boundary conditions. Here $\xi(r)$ is the perturbed radial Lagrangian displacement, determined from the equality $v_r = -i\hat{\omega}(r)\xi(r)$, $\hat{\omega} = \omega - m\Omega_0$.

In the case of a sharp discontinuity on the rotation curve (Fig. 5.15b), the relative perturbations of the depth of shallow water at both sides of the discontinuity have the form of the Bessel function

$$\eta(r) = C_1H_m^{(1)}(kr) + C_2H_m^{(2)}(kr), \tag{5.79}$$

where $H_m^{(1,2)}$ is the Hankel function of the first and second kind. The analogue of the radial wave number k depends on the frequency of perturbations Ω:

$$k^2 = \frac{\hat{\omega}^2 - 4\Omega_0^2}{c_{s0}^2}. \tag{5.80}$$

Since the rotation curve is a piecewise-constant function, the wave numbers on both sides of the jump are also constant (they will be denoted by k_1 and k_2 for the internal and external region, respectively). To simplify considerations, it is assumed that the internal wall is situated at $R_a = 0$, the external wall at $R_b \to \infty$.

From the boundary conditions and the cross-linking conditions at the tangential discontinuity, we can derive the dispersion equation [239]:

$$\alpha_1\left[(x-1)^2 - \frac{4}{m^2}\right] - \alpha_2\left[(x-\overline{q})^2 - \frac{4\overline{q}}{m^2}\right] + \alpha_1\alpha_2\frac{\overline{q}^2-1}{m^2} = 0, \tag{5.81}$$

where

$$\alpha_1 = \frac{2\overline{q}}{x-\overline{q}} - k_1R\frac{J'_m(k_1R)}{J_m(k_1R)}, \qquad \alpha_2 = \frac{2}{x-1} - k_2R\frac{H_m^{\prime(1)}(k_2R)}{H_m^{(1)}(k_2R)}.$$

here the functions $J'_m(z)$, $H_m^{\prime(1)}(z)$ denote the derivatives $J_m(z)$, $H_m^{(1)}(z)$ with respect to the argument r and the dimensionless eigenfrequency $x = \Omega/(m\Omega_2)$ is introduced).

The sought eigenfrequencies are the solutions of this dispersion equation from which we can determine k_1 and k_2. If the real part of the wave number is considerably higher than the imaginary part, the eigenfunction (5.79) describes the wave solution. In the opposite case, the eigenfunction exponentially increases or decreases.

From (5.82) we obtain that k_1 is valid if

$$\mathrm{Re}\,x > \overline{q}\left(1+\frac{2}{m}\right), \tag{5.82}$$

and k_2 is valid if

$$\mathrm{Re}\,x < 1-\frac{2}{m}. \tag{5.83}$$

These considerations help to carry out the qualitative classification of the modes in accordance with their behaviour in the internal and external (in relation to the velocity jump) regions. Depending on whether k_1 and k_2 are real or imaginary, there are three different cases.

1. The modes of type I: k_1 and k_2 are real. In this case, the solutions at both sides of the Ω_0-jump have the form of a wave. This is possible if the azimuthal number is

$$m > \frac{2}{1-\mathrm{Re}\,x}. \tag{5.84}$$

From this inequality (taking into account the fact that for the unstable modes $\tilde{q} < \mathrm{Re}\,x < 1$) it follows that the minimum number of the arms in the unstable wave patterns of this type is $m = 3$.

We examine the case with $\tilde{q} = 0$. The wave in the internal part propagates between the left wall (at the origin at $r = 0$) and the Ω_0-jump. Under certain conditions, this wave reflects from the discontinuity with the amplitude higher than the amplitude of the incident wave (super-reflection) in complete analogy with the Miles–Ribner effect (Fig. 5.16).

In the presence of a feedback this leads to unstable solutions, described by Kolykhalov [97] in planar geometry. The number of unstable modes can be quite large. The real part of the eigenfrequencies can be determined from the equation

$$J'_m(k_1 R) = 0. \tag{5.85}$$

Since $k_1 R$ has the order of a high parameter Mm, the number of modes of this type has the order Mm/π.

If $x_n^{(0)}$ are the roots of the equation (5.85), the eigenfrequencies are described by the approximate formula:

$$\frac{x_n}{x_n^{(0)}} = 1 - \frac{1}{\mathrm{M}^2} + i\frac{m}{\mathrm{M}^3}\left[(1 - x_n^{(0)})^2 - \frac{4}{m^2}\right]^{1/2}. \tag{5.86}$$

2. Type II modes: k_1 – real, k_2 – imaginary. In the internal region there is a solution of the wave type, in the external region an exponential solution. This is so if Re $x > 1-2/m$. However, for these frequencies, the internal wave cannot intensify at the discontinuity, as indicated by Fig. 5.16.

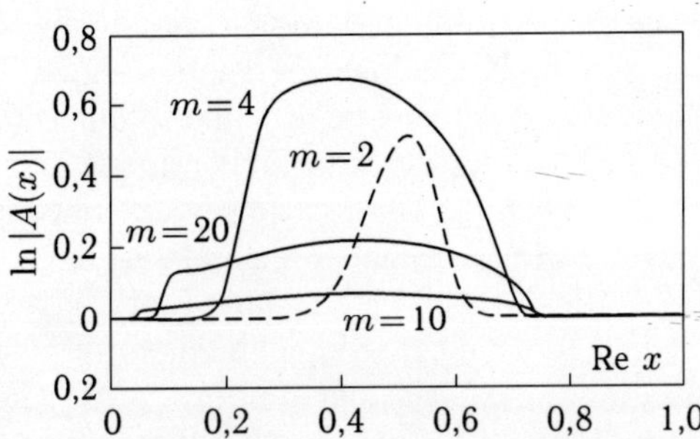

Fig. 5.16. Dependence of the logarithm of the modulus of the reflection coefficient ln $|A(x)|$ on Re (x) at $m = 8$, $q = 0$ and different Mach numbers M. At M = 4.2; the maxima of the curves becomes smaller with M at any values of the azimuthal number of the mode m. It should be mentioned that in the case of a cylindrical tangential discontinuity, intensification can already take place at M > 1 (whereas in the case of a plane-parallel at a discontinuity only at M > 2).

Nevertheless, the numerical calculations show that one mode of this type exists for any $m \geqslant 1$. With a gradual change of the Mach number this mode, alternating, becomes neutral or unstable.

The real part of frequency at $\mathrm{M} \gg m \gg 1$ is approximately equal to

$$x_0 = 1 - \frac{1}{\mathrm{M}} + \frac{2{,}25}{\mathrm{M}^2} + \ldots \tag{5.87}$$

The maximum of the imaginary frequency has the order

$$\max \operatorname{Im} x \sim \frac{1-x_0}{\mathrm{M}}\sqrt{2x_0} \sim \frac{1}{\mathrm{M}^2}\sqrt{2x_0}\,. \tag{5.88}$$

3. Type III modes: k_1 – imaginary, k_2 – real.

In the internal region, there is an exponential solution, in the external region the wave-type solution. This situation is possible only if $\tilde{q} > 0$ and $\operatorname{Re} x < \tilde{q}\,(1 + 2/m)$.

It can be shown analytically that one mode of this type exists on the condition that the azimuthal number $m \geqslant 3$. The approximate solutions for the real and imaginary parts x are relatively cumbersome and, therefore, they are not presented here in the explicit form. The real part is close to the limit: $x_0 = \tilde{q}\,(1+2/m)$. The imaginary part in the limiting case $\mathrm{M} \gg m \gg 1$ has the order

$$\operatorname{Im} x \sim \overline{q}\frac{2}{\mathrm{M}^3}\sqrt{m^2-4}\,. \tag{5.89}$$

The unstable modes of this type are of the emitting nature: they increase as a result of radiation to infinity of the wave with negative energy.

The analytical and numerical calculations for moderate values of M and m

$$2.2 < \mathrm{M} < 6, \quad m < 10, \tag{5.90}$$

show that the increments of the type I and II modes are comparable, whereas the increments of type III modes are considerably smaller. For this reason, the latter can hardly be detected in the experiments. As regards the first two types, their increments will be maximum when $\Omega_2 - \Omega_1$ is maximum, i.e. when $\tilde{q} = \Omega_1 / \Omega_2 = 0$. Therefore, we will continue to consider the case with $\Omega_1 = 0$.

If we consider the external boundary of the system at $r = R_b$, the dispersion equation (5.81) is modified and its numerical solution shows that at relatively high values of the Mach number there may be unstable solutions (corresponding to the instability of super-reflection) for every azimuthal mode $m \geqslant 1$. At any given m, these solutions correspond to one of the previously described types I or II.

The typical behaviour of the unstable roots, relating to the instability of super-reflection, with the variation of the Mach number for the given m, is shown in Fig. 5.17. The letter A in the figure indicates the solutions corresponding to the modes of type II; letter B indicates the most unstable solutions, corresponding to type I modes (with almost exponential attenuation of the amplitude in the internal region from the Ω_0-jump to the centre). The indexes at A and B correspond to the number of maxima of the radial function $\eta(r)$ in the wave zone (for A in the internal part, for B in the external part). For example, this indicates that the unstable solution A_2 is the mode of type II with two nodes along the spiral arm in the internal part.

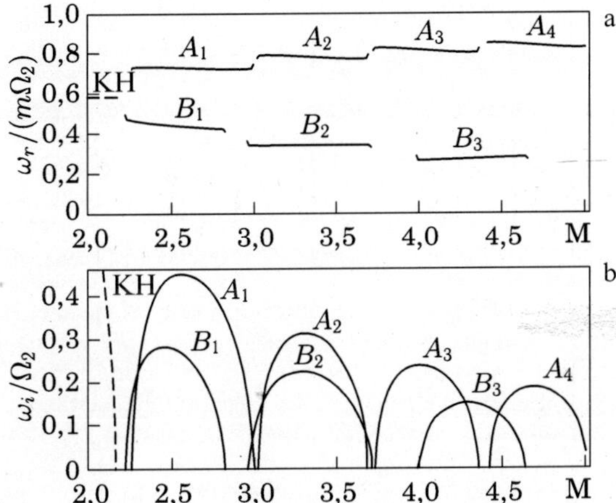

Fig. 5.17. Dependence of the dimensionless eigenfrequencies on the Mach number at $m = 5$, obtained as a result of numerical solution of the modified version of equation (5.81) for the case in which the internal and external solid walls at the radii $0.04R$ and $1.7R$, respectively, are added to the system: a) the normalised real part of the frequency, i.e. the phase angular velocity of the generated spiral structure; *b*) the normalised imaginary part of the frequency, i.e. the increment of the increase of perturbations; A_j – unstable type II mode; j – the number of nodes in the wave region (in the internal part); B_k – unstable mode of type I, k – the number of nodes in the internal region; the letters KH in the left parts of the figure – the unstable Kelvin–Helmholtz mode.

The dependence shown in Fig. 5.17 shows the resonance behaviour of the unstable eigenfrequencies, corresponding to super-reflection instability. For this reason, different modes can be detected in the experiments with the variation of the Mach number.

5.2.3. Resonance instabilities in the presence of a steep supersonic velocity gradient. Jets and disks

We continue examination of the instabilities determined by the presence in the system of regions with a steep gradient of the tangential velocity component (Fig. 5.18). In the case of a steep supersonic velocity gradient ($\Delta v > 2c_s$), this gas layer generates spontaneously sound waves with the wave length greater than or comparable with the characteristic thickness of the layer. If any reflecting surface is situated at a finite distance from this 'generator', for example, a region with a steep density gradient or a second such layer, the energy of perturbations in this waveguide layer exponentially increases with time.

The instability of the surface bending and pinch modes (fundamental harmonics) in the jet is the Kelvin–Helmholtz instability as a result of the Bernoulli effect. In addition to them, there is also a large number of higher (reflecting [548, 548]) harmonics which differ in the number of nodes of the eigenfunctions between the boundaries of the jet. Previously, it was also mentioned that the sound wave, incident on the supersonic tangential discontinuity (TD) of velocity, may be reflected with

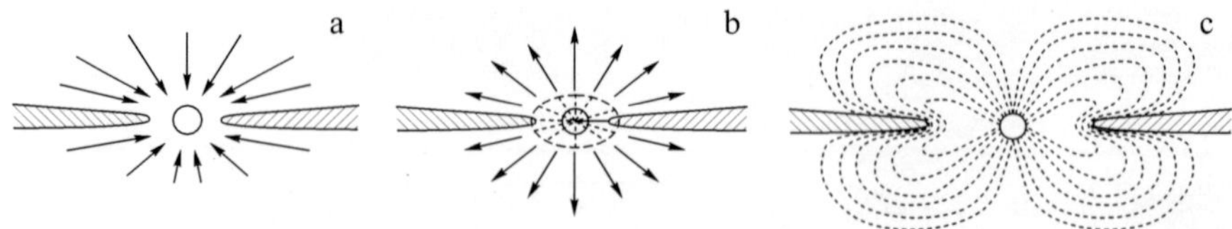

Fig. 5.18. Different schemes of the disk systems in which there are zones of rapid variation of the tangential velocity component along the vertical coordinate. The disk with a finite thickness rotates with a higher velocity $V >> c_s$; a, b) there is a corona above the disk, with the velocity of rotation of the corona considerably lower than that of the disk; c) the internal zone of the disk is compressed by the magnetic field of the star, identical configurations are possible only for galactic disks (only the presence of a marked change in the strength of the magnetic field in transition from the disk to the region of the corona along the z-coordinate is important).

intensification [112] If there are two parallel highly supersonic TDs, the amplitude of the sound wave in this gas layer increases with time as a result of multiple reflection from the TD with intensification (the super-reflection effect). This indicates the instability of the initial flow which will be referred to as the instability of the acoustic resonance type (IART)[9] [96, 98]. The instability of the deflecting harmonics is determined by the super-reflection effect of the sound waves, and there are resonance angles of incidence for which the reflection coefficient converts to infinity [112, page 454]. Smoothing of the velocity jump, retaining the possibility of intensification of the waves, results in new unstable Drazin modes [96, 517].

The Kelvin–Helmholtz instability has been studied in a large number of investigations to explain the observed structure of the jet (galactic jets, jet discharge in the systems with young stars, jets in creating closed double systems, wake behind supersonic planes [269, 648, 649, 978, 997, 1245]). However, the presence of two parallel TDs (or, more accurately, the regions with the large variation of the velocity vector, where the sound point z_s ($v/c_s = 1$) is located) may also occur in the case of disk accretion, if it is assumed that there is gas above the disk (with a considerably lower density), and the velocity of the vector of this gas greatly differs from the vector of the velocity of rotation of the disk. The substance above the disk will be conventionally referred to as the corona (different models consider corona, wind, spherical accreting flow [24, 99, 118, 273, 407, 850, 1147]). Two approximately parallel sound surfaces, situated at the distance $2z_s$, may ensure the instability of sound perturbations [182, 688, 689].

The linear analysis of the stability and numerical non-linear gas-dynamic modelling of the supersonic jets are in agreement with experimental data and, in particular, have made it possible to understand some properties of the observed structure of astrophysical jets – galactic and stellar jets [269]. The mechanisms of modification of the amplitude of the waves in the presence of the regions with large changes of the parameters of the supersonic flow have been studied extensively [32,

[9]The physical mechanism of this instability differs from the classic Kelvin–Helmholtz instability (KHI) which occurs in the case of a single TD. However, in the English literature, the general term KHI is used for denoting all instabilities, associated with the sharp velocity gradient, including the velocity discontinuity between the plasma and the magnetic field [1282].

189, 234]. The effect of super-reflection for the TD in the presence of the magnetic field has been investigated in [285, 549, 649].

In application to disk accretion, it is also important to take into account the effect of rotation, the vertical gravitational force and the z-heterogeneity of the equilibrium parameters on the frequency of reflective and fundamental harmonics. We will discuss the stability of the gas layer compressed by the magnetic field of a neutron star or a wide dwarf [119, 306, 1282].

5.2.3.1. Eigenfrequencies of oscillations in a vertically heterogeneous disk with the magnetic field taken into account

In this section, we formulate a mathematical model which determines the dynamics of linear perturbations on the background of the equilibrium vertically heterogeneous gas disk with the magnetic field taken into account. The equilibrium state is characterised by the functions $\varrho_0(z)$, $P_0(z)$, $v_0(z) = r\Omega$, $\mathbf{B}_0(z)$. The tangential component of velocity v_0 is situated in the disk plane, and at some point $z = \pm z_s$ we have $v_0 = c_s$.

The shortwave approximation in the disk plane will be investigated. No restrictions will be made in the vertical structure of the perturbations. The initial system of the MHD equations has the form:

$$\frac{\partial \mathbf{u}}{\partial t} + \left(\mathbf{u}\vec{\nabla}\right)\mathbf{u} = -\frac{1}{\varrho}\vec{\nabla}\left(P + \frac{B^2}{8\pi}\right) - \vec{\nabla}\Psi + \frac{1}{4\pi\varrho}\left(\mathbf{B}\vec{\nabla}\right)\mathbf{B}, \tag{5.91}$$

$$\frac{\partial \mathbf{B}}{\partial t} = \operatorname{rot}\left[\mathbf{u}\times\mathbf{B}\right], \tag{5.92}$$

$$\operatorname{div}\mathbf{B} = 0, \tag{5.93}$$

which should be supplemented by the continuity equation, the equation of state of the ideal gas and the law of conservation of entropy.

We apply the procedure of linearisation of the system of equations, representing each quantity in the framework of the WKB-approximation along r, in the form

$$\begin{gathered} f(r,\varphi,z,t) = f_0(r,z) + \tilde{f}(r,\varphi,z,t), \\ \tilde{f} = \hat{f}(z)\cdot\exp\left\{-i\omega t + ik_r r + im\varphi\right\}. \end{gathered} \tag{5.94}$$

The radial dependence at the equilibrium parameters must be taken into account to investigate accurately the rotation effects, determined by the Coriolis force. We confine ourselves to the case $v_0 = \{0,\, v_0 = r\Omega,\, 0\}$, $\mathbf{B}_0 = \{B_0,0,0\}$), leading to the equilibrium conditions

$$\frac{\partial}{\partial z}\left(P_0 + \frac{B_0^2}{8\pi}\right) = -\varrho_0\, g, \qquad \frac{\varrho_0\, v_0^2}{r} = \frac{\partial P_0}{\partial r} + \varrho_0\frac{\partial \Psi}{\partial r}, \tag{5.95}$$

where $g = \partial\Psi/\partial z$. The vertical displacement ξ is expressed through the z-component of velocity:

$$w = \tilde{w} = \frac{d\xi}{dt} = \frac{\partial \xi}{\partial t} + \Omega\frac{\partial \xi}{\partial \varphi} = -i\hat{\omega}\,\xi, \tag{5.96}$$

where $\hat{\omega} = \omega - m\Omega$. Excluding the perturbations of velocity $\tilde{u}, \tilde{v}$, the magnetic field $\tilde{B}_r$, $\tilde{B}_\varphi, \tilde{B}_z$ and density $\tilde{\varrho}$ we obtain the linear system of equations with respect to the vertical displacement ξ and the quantity $\hat{J} = \hat{\mathcal{P}} + \dfrac{\mathbf{B}_0\hat{\mathbf{B}}}{4\pi}$ (has the meaning of the total perturbed pressure with the magnetic pressure taken into account):

$$\frac{d\hat{J}}{dz} = \varrho_0\left\{\hat{\omega}_A^2 + \frac{g}{\varrho_0}\frac{\partial\varrho_0}{\partial z} + \frac{g^2}{c_s^2+V_A^2}\left[1 - \frac{\Omega_M^2 M_A^2}{\hat{\omega}_\Omega^2}\right]\right\}\cdot\hat{\xi} - \frac{g}{1+M_A^2}\left\{1+\frac{\Omega_M^2}{\hat{\omega}_\Omega^2}\right\}\cdot\frac{\hat{J}}{c_s^2}, \tag{5.97}$$

$$\frac{d\hat{\xi}}{dz} = \frac{g/c_s^2}{1+M_A^2}\left\{1+\frac{\Omega_M^2}{\hat{\omega}_\varkappa^2}\right\}\cdot\hat{\xi} + \frac{\hat{J}}{\varrho_0\hat{\omega}_A^2}\cdot\left\{k^2 - \frac{\hat{\omega}^4 - \hat{\omega}^2\varkappa^2 k^2 c_s^2(1+M_A^2)/\hat{\omega}_A^2}{(c_s^2+V_A^2)\cdot\hat{\omega}_\Omega^2}\right\}, \tag{5.98}$$

where the following notations were introduced: $k^2 = k_r^2 + k_\varphi^2$, $c_s^2 = \gamma\dfrac{\mathcal{P}_0}{\varrho_0}$, $V_A = \dfrac{B_0}{\sqrt{4\pi\varrho_0}}$ is Alfvén velocity;

$$\Omega_A = (\mathbf{B}_0\mathbf{k})/\sqrt{4\pi\varrho_0}, M_A = V_A/c_s,$$

$$\Omega_M^2 = \Omega_A^2/(1+M_A^2),\ \hat{\omega}_A^2 = \hat{\omega}^2 - \Omega_A^2,\ \hat{\omega}_\Omega^2 = \hat{\omega}^2 - \frac{\varkappa^2\hat{\omega}^2}{\hat{\omega}^2 - \Omega_A^2} - \Omega_M^2,$$

$\hat{\omega}_\varkappa^2 = \hat{\omega}^2 - \dfrac{\varkappa^2\hat{\omega}^2}{\hat{\omega}^2 - \varkappa^2} - \Omega_M^2$. The system of equations holds for the arbitrary dependences $v_0(z), g(z), B_0(z), \rho_0(z)$, and $v_0(z), B_0(z)$ and $u_0(z)$ can be discontinuous if (5.95) is fulfilled.

The system of equations (5.97), (5.98) will be supplemented by the boundary conditions at the point $z = 0$. Two types of oscillations will be investigated (Fig. 5.19): 1) symmetric oscillations at which ($\tilde{\xi}(z) = -\tilde{\xi}(-z)$ or $\hat{\mathcal{P}}(z) = \hat{\mathcal{P}}(-z)$) and, consequently

$$\hat{\xi}(0) = 0 \quad or \quad \frac{d\hat{P}}{dz}\Big|_{z=0} = 0 \tag{5.99}$$

– these oscillations are referred to as the pinch mode or *AS*-mode; 2) antisymmetric oscillations at which ($\tilde{\xi}(z) = -\tilde{\xi}(-z)$ or $\hat{\mathcal{P}}(z) = -\hat{\mathcal{P}}(-z)$) and, consequently

$$\hat{\mathcal{P}}(0) = 0 \quad \text{or} \quad \frac{d\hat{\xi}}{dz}\Big|_{z=0} = 0, \tag{5.100}$$

which corresponds to the bending oscillations of the disk or *AS*-mode. On the surface of the disk in the absence of a corona in the linear approximation the following condition should be fulfilled

$$\hat{\mathcal{P}}(h) + \frac{\partial P_0}{\partial z}\Big|_{z=h}\hat{\xi}(h) = 0. \tag{5.101}$$

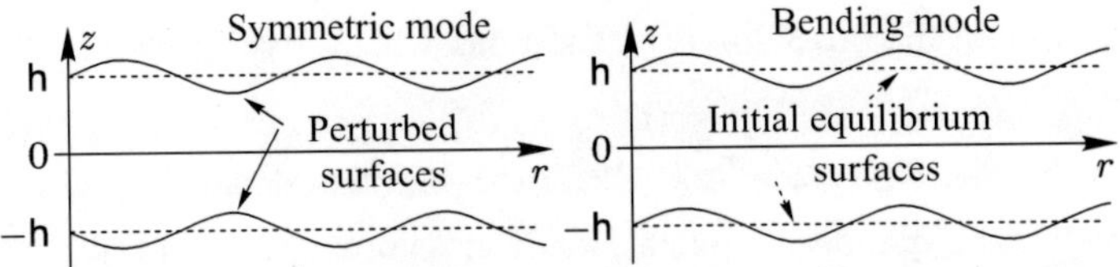

Fig. 5.19. Symmetric (*S*-mode or pinch-mode) and bending perturbations (*AS*-mode).

In the presence of a corona it is required that the perturbed quantities should not increase at infinity:

$$\xi(z=\pm\infty)=0, \quad J(z=\pm\infty)=0. \tag{5.102}$$

We obtain the problem of eigenvalues. The system of equations (5.97), (5.19) coincides with the results of [173] where attention was given to the 'flat' geometry in the case $\varkappa=0, B_0=0$. We analyse different partial models, based on the equations (5.97), (5.98).

5.2.3.2. Acoustic resonance instability (ARI) without the magnetic field

We examine the model without the magnetic field. In addition to the fundamental *S*- and *AS*-modes, the system contains additional harmonics which differ from each other by the number of the nodes of the perturbed pressure across the disk plane because the following estimate holds for the characteristic wave number in the z-direction k_z:

$$k_z h \simeq \pi j \quad (S\text{-mode}), \qquad k_z h \simeq \pi(j+1/2) \quad (AS\text{-mode}), \tag{5.103}$$

where j is the number of the harmonics. The fundamental ($j = 0$) and reflecting ($j > 0$) harmonics exist for both symmetric (*S*-) and antisymmetric (*AS*-) modes.

For two parallel TDs unstable perturbations always exist [112]. In the framework of the simple model of the planar jet (the width of the transition zone $\delta \to 0$), we investigate the effect of the Coriolis forces ($\varkappa \neq 0$) on the parameters of the ARI at the limit $g = 0$, $B_0 = 0$. Let it be that at the points $z = \pm h$ there are 'jumps' of the velocity v_0 and density ϱ_0, the indexes 'ex' and 'in' relate to respectively the external ($|z| > h$) and internal ($|z| < h$) regions. Under these assumptions, the boundary value is reduced to the dispersion equation obtained as a result of cross-linking of the solutions of the equations (5.97), (5.98) for a laminar-homogeneous medium at the points $z = \pm h$. The solutions in the internal region $-h < z < h$ are found in the form $\propto D_1\exp(k_z^{\rm in} z)+D_2\exp(-k_z^{\rm in} z)$, and in the internal region $\propto \exp(-k_z^{\rm ex}|z|)$. Condition (5.102) of the non-increasing amplitude of the perturbations at $z \to \pm\infty$ imposes the requirement Re $(k_z^{\rm ex}) > 0$. The rules of the cross-linking of the solutions at the points $z = \pm h$ are obtained from the system (5.97), (5.98). The equilibrium quantities ϱ_0, u_0 for $z > 0$ are presented in the form

$$f_0(z)=f_0^{\rm in}+(f_0^{\rm ex}-f_0^{\rm in})\cdot\Theta(z-h), \tag{5.104}$$

where Θ is the Heaviside function, and the derivatives of this function $\Theta'(z)=\delta(z)$ is the Dirac delta function. Integrating the equations (5.97), (5.98) with respect to the narrow layer with the thickness 2ε in the vicinity of the planes $z = \pm h$ and, tending

$\varepsilon \to 0$, we obtain the cross-linking conditions which give the required dispersion equation:

$$R\,\hat{\omega}_{\rm ex}^2\, k_z^{\rm in}\left[{\rm th}(kh\beta_{\rm in})\right]^{\wp} = -\hat{\omega}_{\rm in}^2\, k_z^{\rm ex}, \tag{5.105}$$

where $\wp = 1$ is the symmetric (pinch-) mode in accordance with (5.99), $\wp = -1$ is the bending mode (see equation (5.100)), $R = \dfrac{\varrho_0^{\rm ex}}{\varrho_0^{\rm in}}$, $k_z = k\sqrt{\dfrac{1}{1-\varkappa^2/\hat{\omega}^2} - \dfrac{\hat{\omega}^2}{k^2 c_s^2}}$. In the cases in which $v_0^{\rm in} = 0$, $v_0^{\rm ex} \neq 0$ or $v_0^{\rm ex} = 0$, $v_0^{\rm in} \neq 0$, the real part of the frequencies Re(ω) differ by the apparent part of the frequency naturally do not depend on whether the internal layer rotates and the external does not, or vice versa. For determinacy $v_0^{\rm ex} = v_0, v_0^{\rm in} = 0$. In the model (5.105) we have free parameters $\mathrm{M} = \dfrac{k_\varphi v_0}{kc_s^{\rm in}}$, R, $\chi = \dfrac{h\varkappa}{c_s^{\rm in}}$. We introduce the dimensionless frequency $W = \dfrac{\omega}{kc_s^{\rm in}}$. Since $\dfrac{\varkappa^2}{\hat{\omega}^2} = \dfrac{\chi^2}{(W-M)^2(kh)^2}$, then at the limit $\chi = 0$ we have a flat non-rotating jet, and the equation (5.105) changes to that investigated in the studies [548, 648].

At the limit $kh \to \infty$ (single TD), the dispersion curves of the fundamental *S*- and *AS*-modes merge, degenerating into the well-known mode of the Kelvin–Helmholtz instability. For the case $R = 1$ this mode is described analytically [112]:

$$b = \frac{\omega}{kc_s} = \frac{\mathrm{M}}{2} + \sqrt{\frac{\mathrm{M}^2}{4} + 1 - \sqrt{\mathrm{M}^2+1}}. \tag{5.106}$$

The spectrum of the eigenfrequencies, determined by (5.105), is discrete. Figure 5.20*a*, *b* shows the typical dispersion curves for the supersonic flat jet at $R = 1$. The harmonics differ in the number of nodes j of eigenfunctions between the discontinuities. The oscillations with $j = 0$ (fundamental mode) and reflecting solutions ($j > 1$) may be unstable. The reflecting harmonics at the longwave limit are stabilised, splitting into two neutral ones, for which $W = \mathrm{M} \pm 1$. Figure 5.20c, d shows the calculations for $R = 0.003$ taking into account $\varkappa$ ($\chi = 1$). The frequency dependences retain their characteristic form. Taking into account the epicyclic frequency $\varkappa$ gives the additional solution $W = \mathrm{M}$, but this has almost no effect on the dispersion properties of the unstable harmonics, with the exception of the longwave limit $kh \leqslant 0.1$ (Fig. 5.20, d, e).

The effect of the vertical heterogeneity of the equilibrium parameters will be investigated, in particular, the finite width of the transition zone δ_v and gravitational force $g(z)$ [182]. The number of the free parameters includes: $K = kh$ – dimensionless wave number; the effective Mach number $\mathrm{M} = \mathbf{k}\mathbf{v}_0/kc_s$ at the point $z = 0$; δ_v is the width of the transition zone in which the velocity changes linearly from the velocity of the disk $v^{\rm disk}$ to the velocity of the corona $v^{\rm cor}$, and also the parameters which determine the equilibrium disk. We restrict ourselves to the limit of the flow jet ($\chi = 0$) and the vertical profile of the density $\varrho_0(z) = \varrho_0(0)[(1-\epsilon)\cdot\exp(-z^2/h^2)+\epsilon]$, where the parameters $\epsilon = \varrho_0(\infty)/\rho_0(0)$ is

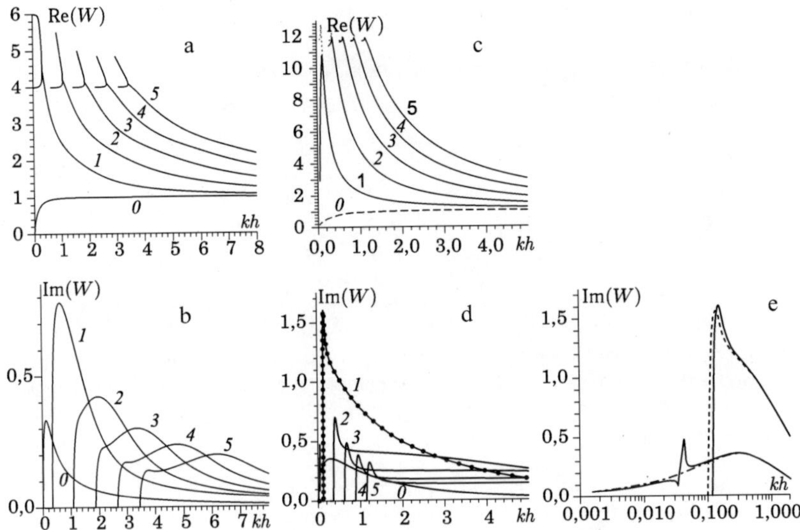

Fig. 5.20. Dependences $W(kh)$ for the S-mode: M = 5, $R = 1$, $\chi = 0$ (a, b) ; M = 30, $R = 0.003$, $\chi = 1$ (c, d, e), for comparison the broken line shows the calculations for $\chi = 0$ which indicates the absence of differences, with the exception of the longwave limit (e). The numbers of the curves are the numbers of the harmonics.

small. The numerical integration of the system of equations (5.97), (5.98), together with the conditions (5.102), (5.99) (or (5.100)) makes it possible to determine the eigenfrequency ω. Verification shows that this approach at the limit $g \to 0$, $\delta \to 0$ gives the same result as the dispersion equation (5.105).

Figure 5.21 shows the typical dependence of the increment Im (ω) on the wave number for the first harmonics j. The perturbations ($j \gg 1$) along the z-coordinate with a smaller scale are stabilised in the first instance in the case of a very wide transition zone $\delta_v \sim h$, and only the largest scale waves in the z-direction are unstable. The final width of the transition zone δ_v for the velocity is a stabilising factor: with increase of δ_v the increments decrease up to stabilisation[10]. In thc first instancc, at the values of the parameters at which the angles of propagation the perturbations greatly differ from the resonance angles, which leads to distinctive indication of the sharp peaks on the curves Im (Ω) (Fig. 5.21a)[11] in the case of the angles being close to resonance angles. The presence of the gravitational force stabilises longwave ($kh \ll 1$) perturbations (see the curves for $j = 0$ in Fig. 5.20, b,e and Fig. 5.21).

Thus, the instability of the type of acoustic resonance may develop in disk systems in which the characteristic scales of variation of the physical quantities (the velocity and density of matter, gravitational force) are comparable with the disk thickness. The fundamental modes ($j = 0$) in the region of long waves, leading in the

[10]In the case of a single 'discontinuity', for the velocity profile, linear in the vicinity of the critical point, this was shown in [98].

[11]Identical peaks were not shown on other graphs.

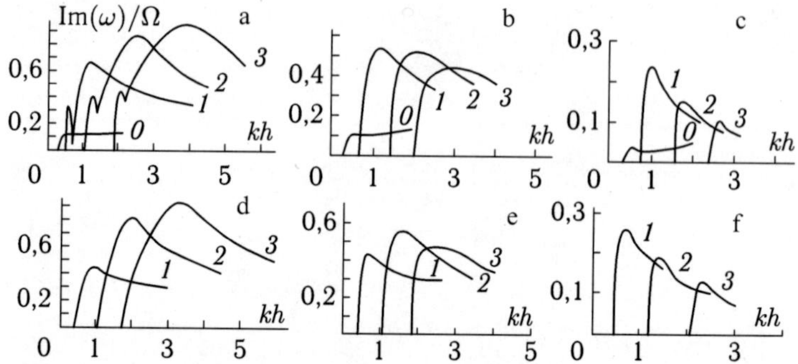

Fig. 5.21. Dependence of $\mathrm{Im}(\omega)/\Omega$ on the value kh for the pinch mode in the case of the effective Mach number M = 5 at different values of the width of the transition zone: $\delta_v = 0.01$ (a); $\delta_v = 0.1$ (*b*); $\delta_v = 0.5$ (*c*). Identical dependences for the bending oscillations (d, e, f). The increment of the fundamental mode ($j = 0$) never exceeds 0.03.

model of the symmetric flat jet to disruption of the initial flow [978], are stabilised in the case of the disk by the transverse gravitational force.

5.2.3.3. The mechanisms of instability of the acoustic resonance type

These mechanisms are based on two effects: reflection of waves with intensification with respect to the amplitude from the gas layer in which the wave is stationary ($\omega \equiv \mathbf{kV}$), and the resonance exchange of energy between the wave and the main flow taking place in the same critical layer. Although the consequence of both effects is the intensification of the sound waves, their physics differs.

The first effect (super-reflection) can be easily understood on the example of the incidence of the sound waves from a stationary medium on the supersonic ($\Delta V > c_1 + c_2$) velocity discontinuity. If the velocity of the phase of the transmitted wave along the vector of the velocity of the gas is lower than the latter, then the lines of the same phase (for example, pressure maxima) move in relation to the gas in the direction opposite to the wave vector[12]. Since the the relative movement of the perturbation and the medium in which the perturbation propagates have the physical meaning, this means that the transmitted wave transfers energy to the discontinuity and not away from it. Since the direction of the energy flux and of the incident and reflected waves is as usual, the energy, supplied to the discontinuity by the incident and transmitted wave, can be taken away only by the reflected wave. Consequently, the energy (and amplitude) of the reflected waves exceeds if needed the amplitude of the incident wave.

The physics of resonance radiation of energy from the critical layer is not so trivial and is identical with the physics of Cherenkov radiation and beam instability (the effect inverse to the Landau attenuation). This problem will be discussed again in section 5.2.4.

---

[12]In other words, the velocity of movement of the gas from the wave along the discontinuity surface is greater than the velocity at which the wave catches up with the gas.

5.2.3.4. Taking into account the magnetic field at the limit of the narrow transition zone

We take into account the magnetic field in the equations (5.97), (5.98) which greatly complicates the behaviour of the dispersion dependences (k) which is associated with the additional free parameters (Ω_A^{disk}, Ω_A^{cor}, V_A^{disk}, V_A^{cor}, δ_B) and the appearance of new types of waves as a result of the effect of the magnetic field.

In the case of a disk compressed by a magnetic field, under specific conditions the boundary of the transition zone $\delta \ll h$ [306]. More complicated stationary configurations of the magnetic field in the disk accretion regime are discussed in [21, 27]. We examine the limiting case of a homogeneous flat jet (the width of the transition zone $\delta \to 0$), which is separated from the homogeneous surrounding stationary medium (corona) by two parallel tangential discontinuities. In this model, the transverse gravitational force should be ignored. It is assumed that at the points $z = \pm h$ there are jumps of velocity v_0, density ϱ_0, magnetic field $\mathbf{B}_0 = B_{0x}\mathbf{e}_x + B_{0y}\mathbf{e}_y$ and the gas-dynamic pressure $\mathcal{P}$, the indexes 'ex' and 'in' refer respectively to the external $|z| > h$ and internal $|z| < h$ regions.

With these assumptions, the boundary-value problem is reduced to the dispersion equation derived as a result of cross-linking of the solutions of the equations (5.97), (5.98 for a layered–homogeneous medium at the points $z = \pm h$. Similarly, we determine the rules of cross-linking of the solutions at the points $z = \pm h$ for the equilibrium value ϱ_0, v_0, B_0 of the type (5.104), which leads to the continuity of $\hat{\xi}$ and $\hat{J}$ at points $z = \pm h$. As a result, we obtain the generalised equation (5.105) with the magnetic field [689] taken into account:

$$\left[\text{th}(kh\beta_{\text{in}})\right]^{\wp} = -\frac{Z_{\text{in}}}{Z_{\text{ex}}}, \tag{5.107}$$

where $\wp = 1$ is the symmetric (pinch-) mode in accordance with (5.99), $\wp = -1$ is the bending moment (see (5.100)),

$$Z_{\text{in}} = \frac{W^2 - A_{\text{in}}^2}{W\beta_{\text{in}}}, \qquad Z_{\text{ex}} = \frac{R[(W-M)^2 - A_{\text{ex}}^2]}{W\beta_{\text{ex}}}, \qquad \beta_{\text{in}} = \sqrt{1 - \frac{W^4}{W^2(a_{\text{in}}^2+1) - A_{\text{in}}^2}},$$

$$\beta_{\text{ex}} = \sqrt{1 - \frac{s(W-M)^4}{(W-M)^2(sa_{\text{ex}}^2+1) - A_{\text{ex}}^2}}, \quad W = \omega / kc_{\text{sin}}, \quad M = (\text{kv0}) / kc_{\text{sin}},$$

$$s = c_{s_{\text{in}}}^2 / c_{s_{\text{ex}}}^2, \quad R = \varrho_{\text{ex}} / \varrho_{\text{in}}, \quad a_{\text{in}} = B_{\text{in}} / \sqrt{4\pi\varrho_{\text{in}}}\, c_{s_{\text{in}}}, \quad a_{\text{ex}} = B_{\text{ex}} / \sqrt{4\pi\varrho_{\text{ex}}}\, c_{s_{\text{in}}},$$

$$A_{\text{in}} = (\mathbf{kB}_{\text{in}}) / \sqrt{4\pi\varrho_{\text{in}}}\, kc_{s_{\text{in}}}, \quad A_{\text{ex}} = (\mathbf{kB}_{\text{ex}}) / \sqrt{4\pi\varrho_{\text{ex}}}\, kc_{s_{\text{in}}}.$$

At the limit $kh \to 0$ equation (5.107) for accretion was investigated in [305]. The quantities Z_{in}, Z_{ex} are the wave impedances of the appropriate media. The conditions Re $(\beta_{ex}) \geqslant 0$ should be fulfilled. The balance of equilibrium pressures at the points $z = \pm\, h$ leads to the expression

$$s = \frac{c_{\text{in}}^2}{c_{\text{ex}}^2} = \frac{2R}{2 + \gamma(a_{\text{in}}^2 - Ra_{\text{ex}}^2)}. \tag{5.108}$$

At least two classes of instabilities may occur: magneto-hydrodynamic instability of the type of acoustic resonance and buildup of the unstable surface modes of discontinuities between the matter of the disk and the external magnetic field (investigated by Northrop [979] – it is referred to as the Kelvin–Helmholtz instability (KHI)). In the first case, the presence of matter outside the disk is not essential, in the second case, the instability develops also at $\varrho_{ex} = 0$ if the bias flow is taken into account (see for more detail [689, 1282]). As shown by analysis [689], the results of the investigated and hydrodynamic (without magnetic fields) cases coincide qualitatively for the arbitrary M, kh and $R \sim 1$. In the case $R \ll 1$, the dependence $\omega(k;\, \mathrm{M}, \mathbf{B}_0)$ is relatively complicated – the surfaces of the levels $\mathrm{Im}(\omega)$ and $\mathrm{Re}(\omega)$ are multivalent. The increments of both the *AS*- and *S*-mode reach the maximum at low values of the parameter $R \sim 1/\mathrm{M}^2$. Figure 5.22 shows an example of calculating the frequency in the absence of the magnetic field inside the disk $a_{\mathrm{in}} = 0$ but $a_{ex} \neq 0$. The fundamental mode and the first 8 reflecting harmonics are shown. The stability of the fundamental modes depends strongly on the mutual orientation of the vectors $\mathbf{k}$, $\mathbf{B}_{ex}$, $\mathbf{B}_{in}$, $\mathbf{v}_0$.

It should be mentioned that at the limit of the 'weakly compressed medium' ($\mathrm{M} \ll 1$, but $v_0 > c_s$) the instability of the fundamental harmonics is the classic Kelvin–Helmholtz instability [112]. For an incompressible MHD liquid at the limit of the single TD we obtain the well-known stability condition $[\mathbf{B}_{\mathrm{in}} \times \mathbf{B}_{\mathrm{ex}}]^2 \geq \dfrac{4\pi\varrho_{\mathrm{ex}}}{1+R}\left\{[\mathbf{B}_{\mathrm{ex}} \times \mathbf{v}_0]^2 + [\mathbf{B}_{\mathrm{in}} \times \mathbf{v}_0]^2\right\}$, $Ra_{\mathrm{ex}}^2 + a_{\mathrm{in}}^2 > \dfrac{R}{1+R}\dfrac{v_0^2}{c_s^2}$ [111]. With increase of the effective Mach number at the fixed values of other parameters, stabilisation takes place. This effect is identical to that for the case without the magnetic field, where at $v_0 < 2\sqrt{2}c_s$ the perturbations with the arbitrary orientation of the vector $\mathbf{k}$ are unstable, and at $v_0 > 2\sqrt{2}c_s$ only the perturbations with $\mathbf{k}\mathbf{v}_0 > 2\sqrt{2}c_s k$ become unstable [111]. In the case of two plane-parallel tangential discontinuities of velocity the fundamental modes remain unstable also at $\mathrm{M} \gg 1$ as a result of the super-reflection effect.

Thus, in the case of disk accretion on the magnetised object in the accretion disk, in addition to the Kelvin–Helmholtz instability [979, 1282], developed at the

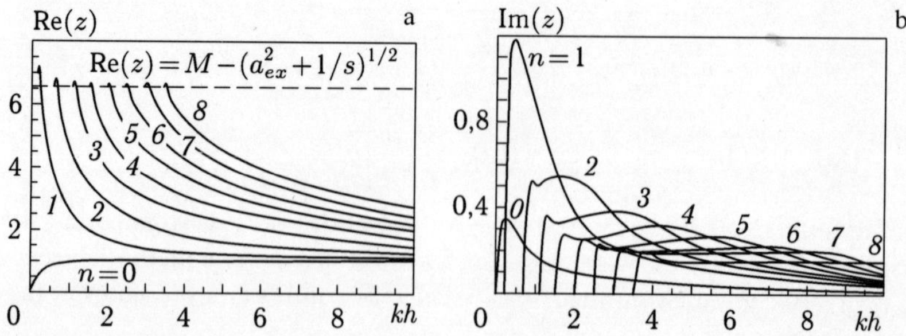

Fig. 5.22. Dependences of the dimensionless phase velocity $\mathrm{Re}(\omega)/kc_s^{\mathrm{in}}$ and of the increment on the value kh for the symmetric mode in the case $\mathrm{M} = 10$; $R = 0.1$; $s = 1$; $\mathbf{k} \| \mathbf{v}_0$; $k_\perp \mathbf{B}_{ex}$; $B_{in} = 0$ (diamagnetic disk) [689].

discontinuities between the matter of the accretion and the magnetic field in vacuum, magnetic-hydrodynamic ARI can also form and its buildup requires the presence of matter in the magnetosphere. In addition to the fundamental modes, buildup of shortwave ($kh \geqslant 1$) reflecting harmonics can also take place. In the presence of the Kelvin–Helmholtz instability, the higher harmonics can also exist, but they are neutral [688], and only the fundamental modes are unstable.

5.2.3.5. Instability of slow magnetosonic waves

Previously, we examined unstable solutions for the fast magnetosonic branches of the oscillations. The reason for the buildup of reflecting harmonics ($j \geqslant 1$) is primarily the resonance interaction of the fast magnetosonic waves (FMW) with the flux. This instability will be referred to as the FMW instability. In [254] it was shown that the instability, determined by the presence of slow magnetosonic branches of oscillations (SMW) can develop. When stabilisation of the main instability takes place on the fast magnetosonic waves, the SMW may remain unstable.

The laws of dispersion of FMW and SMW for a homogeneous medium with the parameters of the external regions give the condition $Z_{ex} = 0$ (see equation (5.107)) from which we obtain

$$W_{1,2}^2 = \frac{1 + sa_{\text{ex}}^2 \pm \sqrt{(1 + sa_{\text{ex}}^2)^2 - 4A_{\text{ex}}^2 s}}{2s}, \tag{5.109}$$

where for the frequency of FMW W_1 it is necessary to select the sign +, for the frequency of SMW W_2 – the sign –.

At the shortwave limit ($kh >>1$), the branches corresponding to the roots $W = M + W_1$ tend to the value of the frequency of the fast magnetosonic waves for the mean layer

$$W_4 = \frac{1}{2}\left\{1 + a_{\text{in}}^2 + \sqrt{(1 + a_{\text{in}}^2)^2 - 4A_{\text{in}}^2}\right\}. \tag{5.110}$$

When the condition

$$\text{M} > W_2 + W_4 \tag{5.111}$$

is fulfilled, the range of the unstable wavelengths (Fig. 5.23) always exists. The criterion (5.111) is 'softer' in comparison with the condition of instability of FMW for which it must be that $M > W_1 + W_4 > 2$. Actually, since from (5.109) we obtain $0 < W_2$, and from (5.110) $W_4 \geqslant 1$, then at the specific ratios of the parameters the instability can occur at $1 < \text{M} < 2$.

It is interesting to investigate the region of the values of the parameters at which the FMW instability is stabilised, but the SMW instability remains. In addition to the previously mentioned case of the small effective mass numbers ($1 < \text{M} < 2$), there is another possibility, associated with the effect of stabilisation of the deflecting harmonics of FMW at $a_{in} \leqslant 1$, $s \sim 1$, $R \leqslant 1/\text{M}^2 << 1$, for which in accordance with (5.108) one should have $a_{ex}^2 \geqslant 1/R$. In the presence of a sufficiently strong external magnetic field the condition $\text{M} - W_1 < W_4$ is fulfilled, and the FMW instability disappears, but in this case the condition (5.111) may remain valid and the buildup

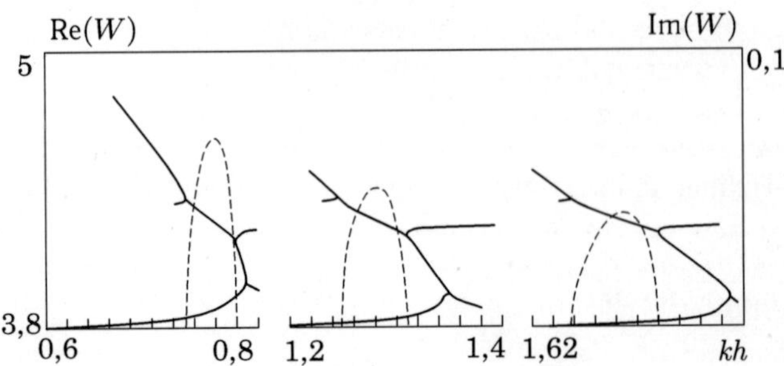

Fig. 5.23. Dependence of the real (the left axis, the solid line) and imaginary (right axis, dashed line) part of the frequency on the wave number. Only the regions of instability of the slow magnetosonic waves (Im (W) > 0) are shown.

of the perturbations, associated with SMW, continues. It should be mentioned that since the effective Mach number $\mathrm{M} = \mathbf{ku}_0/kc_{sin}$, then for the given wavelength $\lambda = 2\pi/k$ there is always a direction of the propagation of the perturbations in the strongly supersonic jet ($v_0 \gg c_s$) at which $1 < \mathrm{M} < 2$ is fulfilled.

5.2.3.6. Non-linear effects

The previously mentioned results were obtained within the framework of linear theory and, strictly speaking, are valid whilst the amplitude of the increasing perturbations is still small. Linear analysis makes it possible to evaluate only the spatial dimensions of the buildup of the perturbations and the characteristics growth time of these perturbations, relating only to the initial stage. Of course, the fate of the perturbations in the non-linear stage can be investigated only in the framework of the numerical hydrodynamic experiment [649, 978]. It is characteristic that the detailed comparison of the results of linear analysis with non-linear calculations of the instability of the type of acoustic resonance in the model of the flat strongly supersonic ($\mathrm{M} \gg 1$) jet, carried out by Norman and Hardy [648, 978], did not show any principal differences.

In the non-linear stage, the lower ($j \leqslant 3$) the reflecting harmonics evolve into a system of weak oblique shockwaves. The higher reflecting harmonics ($j > 3$) are saturated at considerably lower amplitudes but result in the formation of a hierarchy of spatial and time scales. This multimode process should result in developed acoustic turbulence.

These studies can be applied in astrophysical jets (galactic and stellar). Neglecting the role of the magnetic fields, the main difference between the jet and the AD is the presence in the latter case of a transverse component of the gravitational force, determined by the central object. However, the external force g (z) stabilises the longwave perturbations ($kh \ll 1$) and has no effect on the buildup of the waves with $kh \geqslant 1$.

The special feature of multimode instabilities in the application to the problem of turbulisation of the medium is associated with the presence of a series (hierarchy) of separated spatial scales in the disk, corresponding to the maximum values of the increments of different harmonics.

5.2.4. The Papaloizou–Pringle instability

Previously, we discussed some special features of the instabilities, based on the resonance mechanism of amplification of the sound waves. In 1984, Papaloizou and Pringle [1008] obtained unstable modes of the resonance type in the plane of a differentially rotating gas disk (see also [1009–1011, 1115]).

We will discuss thick gas disks[13] (or tori) which are often used to explain the phenomenon of active galactic nuclei – quasars, Seyfert and radio galaxies. A disk may become geometrical thick at the luminosity, exceeding the Eddington limit. If the speed of sound of the gas becomes comparable with the circular velocity, the thickness of the disk becomes comparable with the radius. Pressure starts to determine the equilibrium structure of the system in the radial direction and, consequently, the law of rotation may greatly differ from the Kepler law.

We construct a simple equilibrium configuration of the differentially rotating non-self gravitating thick disk (torus) in the external potential Ψ:

$$-\frac{1}{\rho}\vec{\nabla}P-\vec{\nabla}\Psi+\Omega^2 r\,\vec{e}_r=0.\% \tag{5.112}$$

The gravitational potential of the point mass M_1 has the form

$$\Psi=-\frac{G\,M_1}{\sqrt{r^2+z^2}}. \tag{5.113}$$

We use the model of the gas

$$\mathcal{P}=A\rho^{\gamma}, \tag{5.114}$$

$n = 1/(\gamma - 1)$ is the polytrope exponent. Using the Poincaré theorem, we confine ourselves to examine $\partial\Omega / \partial z = 0$ [1008]. We determine the value Ψ_{rot}:

$$\frac{\partial\Psi_{\text{rot}}}{\partial r}=-\Omega^2 r. \tag{5.115}$$

Taking into account (5.114) and (5.115), the equation (5.112) will be presented in the new form:

$$\nabla\left\{\Psi_{\text{rot}}+\Psi+(n+1)\frac{\mathcal{P}}{\rho}\right\}=0 \tag{5.116}$$

and, consequently

$$(n+1)\mathcal{P}/\rho+\Psi+\Psi_{\text{rot}}=C=\text{const}. \tag{5.117}$$

If on the surface of the torus $P = \rho = 0$, then this quantity is determined by the equation

$$\Psi+\Psi_{\text{rot}}=C. \tag{5.118}$$

The simplest model for analysis is the model with the constant angular moment $l \equiv r^2\Omega = \text{const}$. From equation (5.115) it follows that $\Psi_{rot} = l^2/(2r^2)$ and, taking into account (5.113), the equation (5.117) is reduced to the form

[13]See chapter 11.

$$\frac{\mathcal{P}}{\rho}=\frac{GM_1}{(n+1)r_0}\left\{\frac{r_0}{\sqrt{r^2+z^2}}-\frac{1}{2}\left(\frac{r_0}{r}\right)^2-C'\right\}, \tag{5.119}$$

where $r_0 = l^2/GM_1$, $C' = -Cr_0/GM_1$. The pressure on the surface of the torus converts to 0 and the equation (5.119) gives the radial coordinates $r_\pm$, which restrict the torus in the plane $z = 0$:

$$r_\pm = r_0/(1\mp\sqrt{1-2C'}). \tag{5.120}$$

Figure 5.24 shows the equilibrium configurations for different C'.

The disk is thick and, consequently, it is necessary to use the three-dimensional equations of gas dynamics in the cylindrical coordinate system. We apply the standard procedure of linearisation of these equations – with explanation in the sense of chapter 4, and obtain the following equation for the perturbed quantities $\propto \exp\{im\varphi - i\omega t\}$ (indicated by the sign ~):

$$\tilde{u}_r = \frac{i}{\hat{\omega}^2-\varkappa^2}\left(\hat{\omega}^2\frac{\partial W}{\partial r}-\frac{\varkappa^2\hat{\omega}mW}{2r\Omega}\right), \tag{5.121}$$

$$\tilde{u}_\varphi = \frac{1}{\hat{\omega}^2-\varkappa^2}\left\{-\frac{m\hat{\omega}^2W}{r}+\frac{\hat{\omega}\varkappa^2}{\Omega}\left(\frac{\partial W}{\partial r}-\frac{mW}{\hat{\omega}}\frac{d\Omega}{dr}\right)\right\}, \tag{5.122}$$

$$\tilde{u}_z = i\frac{\partial W}{\partial z}, \tag{5.123}$$

$$i\hat{\omega}\tilde{\rho} = \operatorname{div}(\rho_0\overline{\mathbf{u}}), \tag{5.124}$$

$$\widetilde{\mathcal{P}}/\mathcal{P}_0 = \gamma\tilde{\rho}/\rho_0, \tag{5.125}$$

where $W = -\tilde{\mathcal{P}}/(\rho_0\hat{\omega})$; $\hat{\omega} = \omega - m\Omega$. Using (5.521)–(5.123) and (5.125), we exclude from (5.124) the perturbed density and velocity and, consequently, obtain the equation for W [1008]:

$$\frac{D^2}{r}\frac{\partial}{\partial r}\left\{\frac{\rho_0 r}{D}\left(\hat{\omega}^2\frac{\partial W}{\partial r}-\varkappa^2\frac{\hat{\omega}mW}{2r\Omega}\right)\right\}+\rho_0 D\left\{-\frac{m^2W}{r^2}\left(\frac{\varkappa^2 r}{2\Omega}\frac{d\Omega}{dr}+\hat{\omega}^2\right)+\right.$$
$$\left.+\frac{\varkappa^2\hat{\omega}m}{2r\Omega}\frac{\partial W}{\partial r}\right\}+D^2\frac{\partial}{\partial z}\left(\rho_0\frac{\partial W}{\partial z}\right)+D^2\hat{\omega}^2\rho_0^2W/\gamma\mathcal{P}_0^2=0, \tag{5.126}$$

where $D \equiv \hat{\omega} - \chi^2$.

The equation (5.126) describes the dynamics of small perturbations in the torus with arbitrary equilibrium dependences ρ_0 (r, z), $\Omega(r)$. Equation (5.126) was

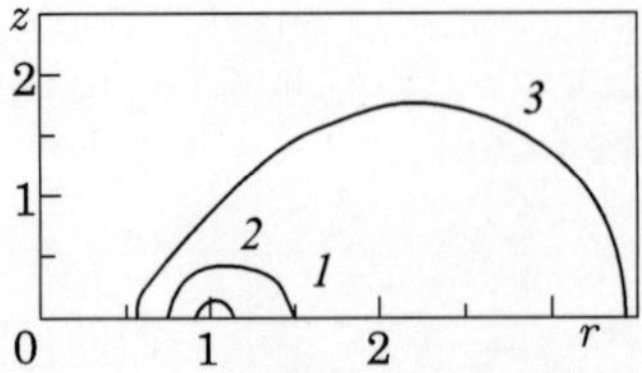

Fig. 5.24. The zero density surface of the torus in the plane (r, z) for different values of C': 1 – $C' = 0.495$; 2 – $C' = 0.444$; 3) $C' = 0.25$.

analysed in [279, 1008, 1009, 1011, 1115] using both the analytical and numerical approaches.

$$W = W_0 \exp\{i(k_r r + k_z z)\}. \tag{5.127}$$

$$\hat{\omega}^2 = \varkappa^2 k_z^2 / (k_z^2 + k_r^2), \tag{5.128}$$

Naturally, of special interest is the examination of the possibility of increase of the perturbations, i.e. formation of the conditions in which Im (ω) > 0. In particular, we discuss the role of compressibility of the gas. We confine ourselves to the model with the constant specific angular moments $l = r^2\Omega = \text{const}$. Since in this case $\varkappa = 0$, equation (5.126) has the following form:

$$\frac{\partial}{r\partial r}\left(\rho_0 r \frac{\partial W}{\partial r}\right) - \frac{m^2}{r^2}\rho_0 W + \frac{\partial}{\partial r}\left(\rho_0 \frac{\partial W}{\partial z}\right) = -\frac{\hat{\omega}^2 \rho_0}{c_s^2} W. \tag{5.129}$$

We determine the value $\omega_c \equiv \omega - m\Omega_c$, where Ω_c is some fixed value of the angular velocity. Equation (5.129) is multiplied by rW^* (W^* is the complexly conjugate W) and integrated with respect to the plane (r, z). As a result we obtain

$$A\,\omega_c^2 - B\,\omega_c - C = 0, \tag{5.130}$$

where

$$A = \int |W|^2 \rho_0 c_s^{-2} r\,dr\,dz, \tag{5.131}$$

$$B = \int |W|^2 2m(\Omega - \Omega_c)\rho_0 c_s^{-2} r\,dr\,dz, \tag{5.132}$$

$$C = \int \rho_0 \left[\left|\frac{\partial W}{\partial r}\right|^2 + \left|\frac{\partial W}{\partial z}\right|^2 + m^2 |W|^2 \left(\frac{1}{r^2} - \frac{(\Omega - \Omega_c)^2}{c_s^2}\right)\right] r\,dr\,dz. \tag{5.133}$$

The solution of the equation (5.130) is written in the form

$$\omega_c = \left(B \pm \sqrt{B^2 + 4AC}\right) / (2A). \tag{5.134}$$

The perturbations [Im (ω) > 0] can increase only in the case of high negative values of C. Equation (5.133) shows that there is always some critical value of the speed of sound at which $B^2 + 4AC = 0$. At the incompressible limit ($c_s \to \infty$) the perturbations are known to be stable. The compressible gas can be unstable, and as the value of c_s decreases, the conditions for the development of instability become more favourable. According to the second conclusion, the axisymmetric perturbations ($m = 0$) are stable.

To explain the nature of unstable global modes, it should be mentioned that the gas disk (or torus) contains three characteristic radii, determined by the equation

$$\omega - m\Omega(r) = l\varkappa(r), \tag{5.135}$$

where $l = 0$ corresponds to the corotation radius, $l = -1$ to the inner Lindblad resonance, and $l = 1$ to the external Lindblad resonance. Between the Lindblad resonances, there is a region forbidden for the neutral waves in which the solution of

(5.129) with Im (ω) = 0 cannot oscillate on the radial coordinate, and on the outside there are the so-called vibrational cavities.

The instability is based on two mechanisms:

1. The resonant interaction of the modes with the opposite signs of energy. This effect is based on the tunnelling of the waves through the forbidden band and the corresponding exchange of energy between the modes, localised in different vibrational planes. In fact, this phenomenon is the super-reflection from the vicinity of the corotation radius and does not differ greatly from that described in the previous point.
2. Investigations of the angular moment and energy from the corotation resonance. The physics of this phenomenon is based on the resonance exchange between the wave and the flow and is identical with the Landau attenuation (amplification). It should be mentioned that for the stellar disk, the exchange of energy at the corotation radius was discussed in detail in a monograph by Polyachenko and Fridman [203] but it greatly differs because of the collisionless nature of the plasma of the stars and the difference in the sign of the energy density of the acoustic waves in the gas and the waves buildup in the gravitational part of the oscillations of the stellar disk.

It was shown in [1010] that for a polytropic gas the system (5.121)–(5.125) can be used to derive the balance equation of the angular moments with the source

$$\frac{\partial h}{\partial t}+\frac{\partial (rf)}{r\,\partial r}=-\frac{\mathrm{Im}(\omega)|\tilde{u}_r|^2\sigma_0^2 r}{|\omega-m\Omega|^2}\frac{d}{dr}\left(\frac{1}{\sigma_0 r}\frac{dH}{dr}\right), \tag{5.136}$$

where $h=r\cdot\langle\tilde{\sigma}\tilde{u}_\varphi\rangle_\varphi$ is the density of the angular moment and $f=r\sigma_0\cdot\langle\tilde{u}_r\tilde{u}_\varphi\rangle_\varphi$ is the density of the radial flow of the angular moment of the perturbations, averaged out with respect to the azimuthal angle, $H=\Omega(r)r^2$ is the distribution of the angular moment of matter in the disk. If the right-hand part of (5.136) is positive, it describes radiation, and if it is negative it describes the absorption of the angular moment of the perturbations in unit time in the vicinity of the corotation radius (where the parameter $|\omega-m\Omega|$ is minimum).

Inside the corotation radius, the wave pattern rotates slower than the gas and on the outside faster; therefore, the angular moment of the wave is negative in the internal region and positive in the external region. This determines the possibility of development of instability both in the case of radiation and absorption of the angular moment of the perturbations in corotation, and differences in the localisation of the unstable perturbations in the disk. In the first case, the angular moment is emitted at corotation and is transferred to the outer 'positive' mode, and in the second case the angular moment is taken from the mode inside the corotation radius which has anyway the negative angular moment. Consequently, the absolute value of the latter increases[14].This description of the instability mechanisms shows that the rate of

[14]As shown by the numerical solution, the flow density of the angular moment of perturbations everywhere in the disk is directed to the outside. This is the reflection of the fundamental LeChatelier principle – the unstable spiral modes transfer to the outside the angular moment of matter which is accreting and at the same time releases tgravitational energy.

increase of the perturbations is determined by the rate of transfer of energy and the angular moment by the sonic wave, i.e. the increase of the perturbations takes place on the dynamic timescale:

$$\mathrm{Im}(\omega) \sim c_s / L, \tag{5.137}$$

where L is the characteristic radial scale in the disk.

In conclusion, it should be mentioned that the resonance instabilities of the gas disks and the tori are of the multimode type – for every azimuthal mode (with fixed number m we can detect up to 30 unstable harmonics differing in the number and distribution of the nodes of the function of the perturbed pressure on the radial coordinate and the speed of rotation of the wave pattern (corotation radius). On the one hand, this multimode nature and, on the other hand, the relative smallness of the increment (see equation (5.137)), leads to the possibility of simultaneous existence in the disk of a large number of such harmonics. Therefore, regardless of the fact that every spiral wave has a small amplitude, the total transfer of angular moment by these waves to the periphery of the disk may prove to be considerable.

5.2.5. Low-frequency centrifugal instability

The considerations described above showed one unstable mode caused at $\mathrm{M}(1 - q^2) \gg 1$ by the centrifugal mechanism, and at $\mathrm{M} \ll 1$ by the mechanism of the Kelvin–Helmholtz instability (section 5.1.2). At the same time, the existence, together with the main unstable modes, of its higher harmonics in the plane-parallel supersonic gas flows is a well-known fact (see [517, 548, 648, 1018], section 5.2.3). In the axisymmetric supersonic flows with differential rotation, the higher unstable harmonics were discovered only recently in systems with the exponential dependence of the velocity of rotation on the radius of the type $\Omega(r) \propto r^{-n}$, $n > 0$ (see [1009, 1010, 1115], section 5.2.4).

In [162], using the model described by the equations (5.54), (5.55), with the rotation curve (5.26), characteristic of the gas disks of the galaxies, investigations showed the presence of a second unstable mode. This mode that is new to us differs from the previously examined mode by the lower values of both $\mathrm{Re}(\omega)$ and $\mathrm{Im}(\omega)$ (Fig. 5.25).

In accordance with this, the main mode of the centrifugal instability will be referred to as the high-frequency mode, the second one asked the low-frequency one. In addition to the fact that the high-frequency mode has a higher increment in the entire investigated region of the values of the parameters (Fig. 5.25), the perturbation of this mode can increase in the ranges of the parameters in which the low-frequency mode is stabilised, namely at a lower 'jump' of the velocity of rotation and at low Mach numbers ($\mathrm{M} \leqslant 2$).

The development of the perturbations of both models results in the generation of trailing logarithmic spiral waves. In a relatively wide range of the values of the parameters, the dependences $\mathrm{Re}(\omega)$ and $\mathrm{Im}(\omega)$ on M for both modes are approximately parallel, and there are sections in which the difference does not exceed 10÷30% [162].

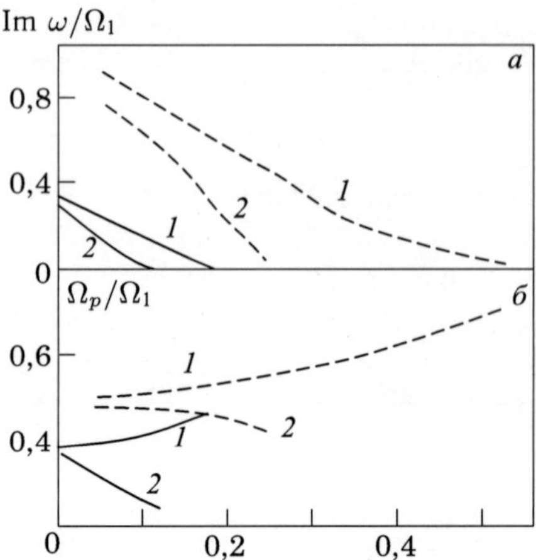

Fig.5.25. Dependence of the dimensionless increment $\mathrm{Im}(\omega/\Omega_1)$ (a) and dimensionless angular velocity of rotation of the spiral pattern Ω_p / Ω_1 (b) on the relative jump of the velocity of revolution $q = \Omega_2 /\Omega_1$ for high-frequency (1) and low-frequency modes. The solid curves correspond to the case $L/R = 0.3$, $M = 10$, dashed lines to $L/R = 0.15$, $M = 20$.

In [161] it was assumed that there are identical high-frequency and low-frequency modes for the perturbations with $m > 2$ for the galactic rotation curves of the type (5.26), (5.70) (at least in the models with the exponential dependences of the velocity of rotation on r there is a large number of unstable and deflecting harmonics for different $m \gg 1$ (see section 5.2.4)).

The physical mechanism of the buildup of the low-frequency mode is of the mixed centrifugal-resonance nature and, therefore, the investigated case is similar to the case of models with the exponential law of rotation in which the instability develops as a result of the resonance radiation of energy at the corotation radius (at which there is synchronous rotation of the wave pattern with the matter of the disk) and interaction of the waves with the opposite signs of energy (super-reflection). The importance of the centrifugal defects for maintaining the low-frequency mode is evident because it is stabilised at $\varkappa^2 > 0$. The resonance nature of this mode is indicated, in particular, by its stabilisation at $L \to 0$ [15] because as a result of the simplifications due to the resonance interaction of the wave with the flow there must be a critical layer of a finite thickness in the vicinity of the corotation radius where the velocity profile is monotonic. Another reason is its greatly supersonic nature – the low-frequency mode is stabilised with the reduction of the Mach number to $\mathrm{M} \simeq 2$, and this coincides with the threshold value for super-reflection where the resonance amplification becomes impossible (see Fig. 5.2.4 [112, 913, 1067]).

It should be mentioned that the buildup of the low-frequency mode requires either the fulfilment of the condition $mL \leqslant R$, but not $L/R \to 0$, i.e. the presence of finite 'spreading' of the velocity jump, or at $L/R \to 0$ the presence of the internal (in relation to the velocity discontinuity) reflecting surface (solid wall or a density

[15]This factor also complicates the determination of the low-frequency mode because there is no such instability in the discontinuity model.

jump) situated at the radius R_r where the condition $(R - R_r)/R \ll 1$ does not hold (the relationship $mL \le R$, used here, is identical with the condition at which there maybe an unstable flat shear layer: $kL \lesssim 1$, where k is the wave number of perturbations along the layer; L is is characteristic thickness; the last claim is evident if it is taken into account that the value m/R has the meaning of the azimuthal wave number). This shows that the spiral patterns, determined by the low-frequency mode, cannot be detected in the experiments with 'shallow water' because, generally speaking, none of the previously mentioned conditions was fulfilled in them.

5.3. Instabilities, caused by the radial heterogeneity of pressure and entropy

In the previous two sections, attention was given to the instabilities caused by the heterogeneity of the velocity of rotation of matter. In section 3.8.4, gradient–entropy instability (GEI) in the plane of the heterogeneous thin gas disk was investigated. Direct analogues of GEI are the convective and Rayleigh–Taylor instabilities. They can developed in a heterogeneous medium situated in the gravitational field[16]. Therefore, it is necessary to discuss the claims found in the literature assuming that since the matter, rotating in the Kepler orbits, is weightless, the instability of the convective type in the case of disk accretion has no significant role. Of course, this holds for the case with constant pressure but such a model is relatively artificial. We examine a stationary equilibrium heterogeneous disk without the radial movement and the viscosity for which the balance of forces in the radial direction is determined by (3.243). Although the last two terms provide a small contribution to equilibrium, they should be taken into account in the development of instabilities in the disk plane. It is convenient to introduce a specific equilibrium force for which, taking into account equation (3.243), we can write

$$g \equiv \frac{d\Phi}{dr} - \frac{v_0^2}{r} + D\frac{p_0}{\sigma_0}\frac{d\ln\Omega_z}{dr} = -\frac{dp_0}{\sigma_0\, dr}. \tag{5.138}$$

This force is small in the case of a thin disk ($g \ll d\Phi/dr \simeq r\Omega^2$) but this appears to be sufficient for the development of convective instabilities with the characteristic time of growth of the perturbations $\tau = 1/\mathrm{Im}\,(\omega) \sim (h^2k/r)\Omega$ (this estimate can be easily obtained from (3.343) or (5.145)). For the equilibrium parameters $f = \{\sigma_0, p_0, v_0 = r\Omega, \ldots\}$ we determine the scale of the radial heterogeneity $L_f = (d\ln(f)/dr)^{-1}$.

The possibility of the buildup of GEI is determined, as indicated in section 3.8.4, by the ratio of the characteristic scales of the heterogeneity of the surface density and temperature (L_σ and L_T). The radial profiles of the surface brightness in gas galactic disks are often relatively complicated and non-monotonic (see Chapter 1, Fig. 1.17). At present, there are already a large number of axisymmetric models of

[16]In plasma physics, they are referred to as diffusion and permutable instabilities. A heavy liquid above a light one, cold above hot, plasma above magnetic field – these all are examples of the development of this type of instability]

the accretion disks, differing in the final analysis in the radial distribution of the surface density and temperature and, consequently, the values of L_σ and L_T [$L_f = \{d(\ln f)/dr\}^{-1}$] [178]. Therefore, it is interesting to investigate the instabilities of the convective type due to the radial heterogeneity of the parameters in the rotating disks.

5.3.1. The Rayleigh–Taylor instability in a differentially rotating disk

In the dynamics of different solid media, the Rayleigh–Taylor instability (RTI) has its own name, associated with the history of its discovery in the given medium. In the hydrodynamics, this instability is known as the instability of a heavy liquid above a light one, in plasma physics - as flute instability, and so on. The development of this instability can be observed if a small amount of water is poured into a glass with liquid oil. There will be a large number of water flutes, descending to the bottom of the glass (i.e. flute instability). The final stage of the process is evident: water, being the heavy liquid, it is found at the bottom of the glass displacing the oil upwards.

The inversion of not only density, as in the investigated example, but also other most important parameters of the medium results in the identical instability. Letter A denotes to the function growing with increasing density ρ, and/or temperature T, and/or pressure p, i.e.

$$A = A(\rho(z), p(z), T(z)). \tag{5.139}$$

where z is the vertical coordinate since the vector of acceleration of the gravitational force $\mathbf{g}$ is directed to the side opposite to the side of the growth of the coordinate z. Consequently, the condition of development of the Rayleigh–Taylor instability can be written in the form (Fig. 5.26)

$$\frac{\partial A}{\partial z} > 0. \tag{5.140}$$

A suitable example of such an inversion is the heterogeneity of temperature. In this case, we are concerned with convective (or) thermal instability. In both cases, the Archimedes force is regarded as the physical mechanism.

There are two limiting cases of the development of the Rayleigh–Taylor instability. For the internal gravitational waves (IGW), the length of the perturbation is small in comparison with the characteristic scale of the heterogeneity ($kL \gg 1$). A

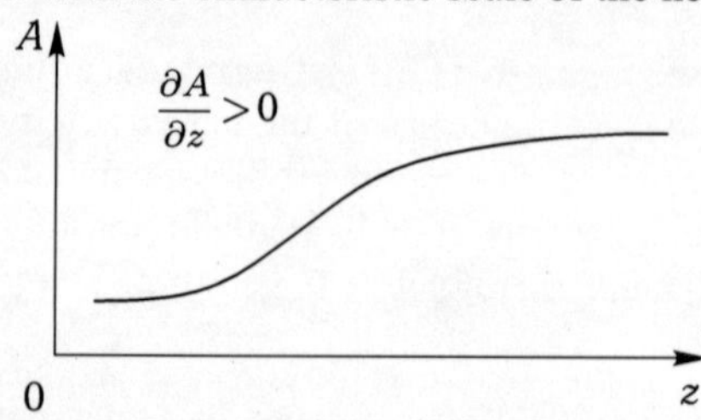

Fig. 5.26. Dependence A (z) at which the Rayleigh–Taylor instability is generated.

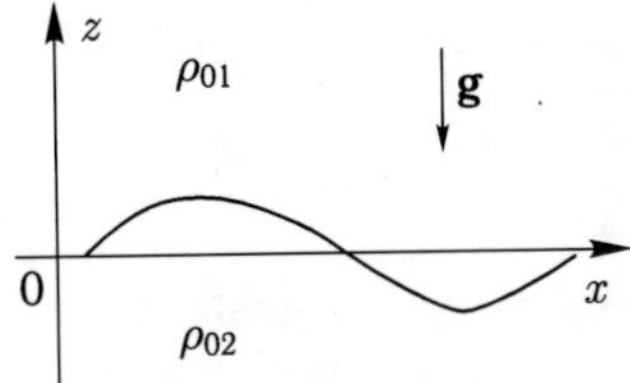

Fig. 5.27. A discontinuity of the density (contact discontinuity) in the external gravitational field.

typical example of the IGW are the waves in the atmosphere of a planet for which the vertical heterogeneity of the thermodynamic parameters is associated with the presence of the gravitational force g. For the frequency, the following equation holds

$$\omega^2 = -g\left(\frac{d\ln\varrho_0}{dz} + \frac{g}{c_s^2}\right)\frac{k_\perp^2}{k_\perp^2 + k_z^2}. \tag{5.141}$$

These waves are transverse and have low frequency and, in contrast to the sonic waves, with $\omega = kc_s$, they are characterised by a dispersion.

In the second limiting case, when the characteristic scale of the heterogeneity is small in comparison with the wavelength, the instability of the convective type can also develop. If there is an interface between the two media (Fig. 5.27), situated in the external gravitational field, then this system is characterised by the propagation of surface gravitational waves (SGW or surface gravitational-capillary waves if the capillary effects are taken into account) with the frequency

$$\omega^2 = \sqrt{k_x^2 + k_y^2}\, g\frac{\varrho_{02} - \varrho_{01}}{\varrho_{02} + \varrho_{01}}.$$

For the instability it is sufficient that $\varrho_{02} < \varrho_{01}$.

If we consider a general case in which two thermodynamic parameters, including heterogeneous entropy $s_0 \propto \ln(\mathcal{P}/\varrho^\gamma)$, change only coordinate z, then for the instability it is necessary [112] that

$$\frac{dP_0}{P_0 dz}\left(\gamma\frac{d\varrho_0}{\varrho_0 dz} - \frac{dP_0}{P_0 dz}\right) < 0. \tag{5.142}$$

The two instabilities (Kelvin–Helmholtz and the Rayleigh–Taylor), considered in this chapter, are the strongest of the subsequently discovered tens of other instabilities in different solid media. Therefore, only these two instabilities are found in both the old (classic) and the latest monographs dealing with the mechanics of solid media (hydrodynamics) in the chapter, concerned with the physics of instabilities of these media [242].

Regardless of the salience of the previously mentioned two instabilities of the solid (hydrodynamic) medium from tens of other instabilities, they are not equally important from the viewpoint of physics. Actually, the condition of generation of the Rayleigh–Taylor instability requires the existence of the inverse distribution of at least one of the parameters of the medium with height. The existence of the Kelvin–Helmholtz instability requires only the presence of some velocity gradient (regardless of its size) which usually is always found in the atmosphere or oceans.

5.3.2. Dynamics of linear waves

We discuss all functions in the form $\tilde{f}(r, \varphi, t) = f_0(r) + (r, \varphi, t)$ and linearise the equation in respect of the perturbed quantities taking equation (5.138) into account. In the framework of the WKB-approximation, we consider for the perturbations $\tilde{f} = f_1 \times \exp\{-i\omega t + ik_r r + im\varphi\}$. The condition of existence of the non-trivial solutions for the linearised system results in the dispersion equation of the fourth degree in respect of frequency ω. If we consider separately the radiation-dominated disk $p_{\text{rad}} \gg p_{\text{gas}}$ and the reversed case $p_{\text{rad}} \ll p_{\text{gas}}$, then as a result the dispersion equation has the following form [178]:

$$\hat{\omega}^4 - \hat{\omega}^2 \cdot \left[\varkappa^2 + k^2 c_s^2 \cdot \left(1 - 2\sin^2(\Theta) \frac{r\, d\Omega^2}{\varkappa^2\, dr} \right) \right] - 2s\Omega\hat{\omega}kc_T^2\Gamma \times$$

$$\times \frac{d}{dr} \ln\left(\frac{2\Omega p_0^{2/\Gamma}}{\varkappa^2 \sigma_0} \Omega_z^{2(1-\Gamma)/\Gamma} \right) - \left(\frac{2s\Omega kc_T^2}{\varkappa} \right)^2 \frac{d\ln(p_0\Omega_z)}{dr} \cdot \frac{d}{dr} \ln \frac{p_0}{\sigma_0^\Gamma \Omega_z^{\Gamma-1}} = 0, \tag{5.143}$$

where $s = \sin(\Theta) = k_\varphi / k = k_\varphi / \sqrt{k_r^2 + k_\varphi^2}$ determines the degree of non-axisymmetry; $k_\varphi = m/r$, $\Gamma = \gamma_s$ is the plane adiabate parameter. In the case $p_{\text{rad}} \gg p_{\text{gas}}$ it should be assumed that $\gamma = 4/3$ and $\Gamma = 9/7$ (see (3.248)). Strictly speaking, equation (5.143) holds at $k_r r \gg 1$, $k_r L_f \gg 1$ (the heterogeneity was taken into account in the first order), $|\sin(\Theta)| \ll 1$. The dispersion equation follows directly from (3.330) at the limit $2\pi G\sigma_0 \to 0$.

At the limit of the isentropic model $\dfrac{p_0}{\sigma_0^\Gamma \Omega_z^{\Gamma-1}} = \text{const}$ (for more details see section 3.6.1), the order of the equation (5.143) decreases because the entropy part of the oscillations is degenerated $\hat{\omega} = 0$. In a general case, equation (5.143) describes two high-frequency parts of the oscillations (acoustic) for which we can approximately write $\hat{\omega}^2 \sim \varkappa^2 + k^2 c_s^2$, and two low-frequency ones (entropy and vortex).

The formal transition to (5.143) in the non-rotating medium $\Omega = 0$ and $\Omega_z = \text{const}$ gives for the low-frequency waves $\omega^2 = \dfrac{\sin^2(\Theta)}{\gamma^2} \dfrac{1}{L_p} \left(\dfrac{\gamma}{L_\sigma} - \dfrac{1}{L_p} \right) \cdot c_s^2$, and for stability it should be

$$\frac{1}{L_p} \left(\frac{\Gamma}{L_\sigma} - \frac{1}{L_p} \right) > 0, \tag{5.144}$$

which coincides accurately with the condition of convective stability in the heterogeneous medium (5.142). It should be mentioned that in the axisymmetric model taking into account the vertical movements, the stability is determined by the Hoyland criterion for a rotating gas flow [709]; it was also investigated in [221]. The possibility of development of convective instability in a rotating gas disk in application to the problem of the spiral structure was discussed in [186].

The boundaries of convective stability. Equation (5.143) can be used to determine the boundaries of convective instability taking into account the differentiality of rotation ($1/L_\Omega \neq 0$) and the finite thickness of the disk in the main approximation ($1/L_z \equiv d \ln \Omega_z/dr \neq 0$). Equation (5.143) will be written at the limit of the low-frequency waves ($|\hat{\omega}^2| << \Omega^2$, the smallness of $|\omega|/\Omega^2$ is not the required):

$$\Lambda \cdot \nu^2 + \delta \cdot \sin(\Theta) \cdot \left[\frac{2r}{\Gamma L_p} - \frac{r}{L_\sigma} - 2\frac{\Gamma - 1}{\Gamma}\frac{r}{L_z} + \frac{r}{L_\Omega} - 2\frac{r}{L_\varkappa} \right] \cdot \nu + \\ + \sin^2(\Theta)\frac{\delta^2}{\Gamma^2} \cdot \left[\frac{r}{L_p} - \Gamma\frac{r}{L_\sigma} - (\Gamma - 1)\frac{r}{L_z} \right] \cdot \left[\frac{r}{L_p} + \frac{r}{L_z} \right] = 0, \tag{5.145}$$

where $\nu = \hat{\omega}/\Omega$, $\Lambda = \frac{\varkappa^2}{\Omega^2} + \frac{k^2 c_s^2}{\Omega^2} \cdot \left[1 + 6\sin^2(\Theta)\right]$, $\delta = 2kc_s^2/(\Omega^2 r)$. Taking into account the estimates for the half thickness of the disk $h \sim c_s/\Omega$, parameter Λ can have the values from $\Lambda \sim 1$ in the longwave limit to $\Lambda \leqslant 100$ for the perturbations in which the wavelength is comparable with the thickness of the disk. For many stationary models of the accretion disks we can except the exponential dependence of the equilibrium parameters of the disk on the radial coordinate: $r/L_f =$ const. If the condition $\text{Im}(\nu) > 0$ is satisfied, the unstable solutions, investigated in [178] in the case $r/L_z = 0$ are obtained.

Figure 5.28 shows on the plane of the parameters r/L_σ and r/L_p the boundaries of convective instability, determined from the condition $\text{Im}(\nu) = 0$, for the basic model $r/L_\Omega = r/L_z = -3/2$, $\gamma = 5/3$, $\Lambda = 10$ (the case with $r/L_\Omega = -3/2$ corresponds to Kepler rotation). Here, for comparison there are the boundaries of stability for the non-rotating medium for which the equilibrium is ensured only by the external force and the pressure gradient, and also for the model $r/L_z = 0$. It can be seen that the rotation and the finite thickness of the disk greatly change the conditions of existence of the

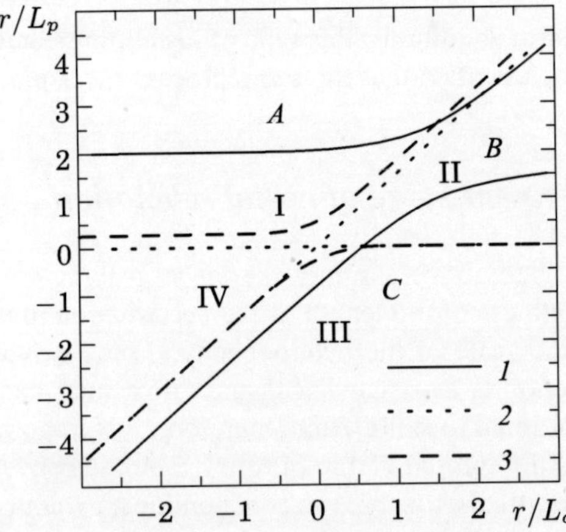

Fig. 5.28. Boundaries of convective stability for the models: 1) $r/L_\Omega = r/L_z = -3/2$, $\gamma = 5/3$, $\Lambda = 10$ (in regions *A* and *C* perturbations are unstable, *B* is the zone of stability); 2) for the non-rotating atmosphere in accordance with (5.144) (in region I, III the perturbations are unstable; II, IV are the stability zones); 3) for the case $r/L_z = 0$.

convective instability. Depending on the signs of L_σ and L_p, the regions on the plane $(r/L_\sigma, r/L_p)$, favourable for the formation of gradient instability, can both increase and decrease in size. For the equilibrium distributions with $r/L_\sigma < 0$ and $r/L_p < 0$, the differential nature of the rotation and the finite thickness of the disks are the stabilising factors. At any values r/L_σ, the critical value $|r/L_p|$ becomes smaller with a reduction of γ. This effect is in agreement with the criterion (5.144).

Parameter Λ characterises the spatial structure of the perturbations. The high values of the parameter $\Lambda \gg 1$ are obtained for the shortwave (in the radial direction) waves $k \sim \Omega/c_s$. The value Λ is higher for the perturbations with a higher azimuthal number m. It should be stressed that the perturbations most unstable from the viewpoint of the equations (5.143), (5.145) are the maximally non-axisymmetric perturbations $\sin(\Theta) = 1$ (since $\text{Im}\,\omega \propto \sin\{\Theta\}$), for which the approximations, forming the basis of the dispersion equations, are known to be disrupted. The boundaries of the convective instability in the disk plane depend on the degree of non-axisymmetry of the perturbations only through the parameter $\Lambda(\Theta)$. The parameter Λ has the strongest effect on the model with $r/L_p > 0$. In the case in which $r/L_p > 0$ only the fine-scale perturbations can be unstable.

The presence of the limiting transitions indicates that the physical mechanism, determining the growth of the perturbations with time, is identical with the classic convective instability in the presence of an entropy gradient, codirectional with the external force, or of the Rayleigh–Taylor instability, if the gradients of the equilibrium pressure and density have different signs. For the instability which basically has $r/L_p < 0$, it is necessary to differentiate between two regions. In one of the regions, the surface density decreases with the increase of the radial coordinate $(r/L_\sigma < 0)$, and the direct analogue of such an instability is the convective instability in a homogeneous field of gravity. In the second case, surface density increases with increase of the distance from the centre $(r/L_\sigma > 0)$, and this corresponds to the Rayleigh–Taylor instability.

Since in the 'Earth conditions' this type of instability results in effective mixing of matter, it may be assumed that the same process takes place in the plane of the accretion disk.

5.3.3. The non-linear stage of radial relaxation

The analysis described above assumes that at least three conditions are fulfilled: 1) linearity; 2) smallness of the length of the perturbation in the radial direction in comparison with the scale of the heterogeneity; 3) the waves should be strongly twisted (the restriction is associated with the differential nature of rotation). The increment is proportional to $\sin(\Theta)$ and, therefore, it is important to investigate the effect of the strong differential nature of rotation on the convective cells formed in the linear stage. We will discuss the results of nonlinear evolution of the convectively unstable disk by the TVD-E method (section 4.1.3) without taking viscosity into account, confining ourselves to $\Omega_z = \text{const}$, using the dimensionless coordinates and time: $t = 1$ is the period of revolution at the radius $r = 1$ [766].

At the initial moment of time we define the exponential dependences of density and pressure (r/L_p = const, r/L_σ = const). If the initial equilibrium state, determined by the functions $p_0(r)$ and $\sigma_0(r)$, ensures the stability of the disk in accordance with (5.143) (Im $(\omega) = 0$), there is no growth of perturbations with time in the numerical models.

We consider a model with $r/L_p = -3/2$ and $r/L_\sigma = -1/2$, $\gamma = 5/3$, which fits in the unstable region in accordance with (5.143). Regardless of the amplitude of the initial perturbations, the waves growing with time and having the characteristic spiral-cellular structure form. Figure 5.29 shows the isolines of the ratio of the density $\sigma(r, \varphi)$ to the equilibrium value $\sigma_0(r)$ at two moments of time t_1 = 15 000 and t_2 = 41 000. The initial stage is characterised by the formation of typical convective cells (Fig. 5.29a) with a small relative amplitude of the surface density $|\sigma - \sigma_0| / \sigma_0 \leqslant 5\%$. With time, the amplitude of the perturbations increases and the spatial structure becomes more complicated because of the differential nature of rotation of the disk (Fig. 5.29b).

The law of rotation differs only slightly from the equilibrium law ($\leqslant$3–5%). The field of velocity perturbations clearly demonstrates the vortex nature of the flow. The parameters k and m, characterising the spatial structure of the perturbations, are independent within the framework of linear analysis. Formally, with other conditions being equal, the most unstable waves are those with the angle of twist of $\Theta \to 90°$. In the numerical model, the waves with $\Theta \simeq 15°\div25°$ grow as a result of the compromise between the instability and differential nature of rotation. The structure of the unstable perturbations in the non-linear stage is determined by the parameters of the model (M, m, δ_p, δ_σ, γ), and the azimuthal number m can be changed by varying the initial perturbations along the angle φ. With increase of the azimuthal number the scale of the perturbations becomes finer also in the radial direction. The identical effect is obtained with increase of the Mach number.

Linear analysis shows that the increment of the instability is proportional to Im $\omega \propto m$ which on the whole is confirmed in the initial stage of evolution of the perturbations. Figure 5.30 shows the time dependence of the maximum values of

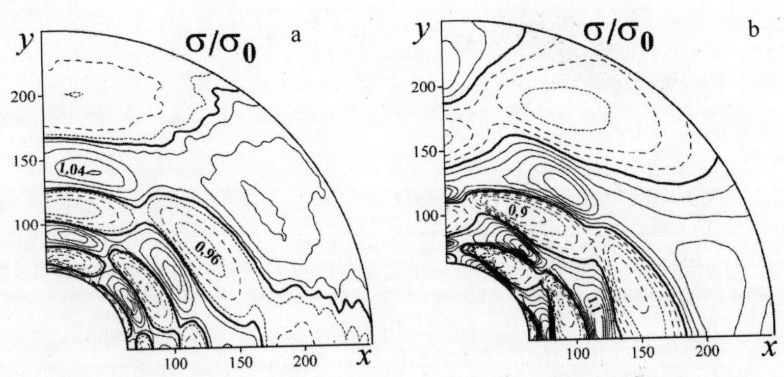

Fig. 5.29. Isolines of the surface density σ/σ_0 at two different moments of time t_1 = 15 000 (a), t_2 = 41 000 (b) [766].

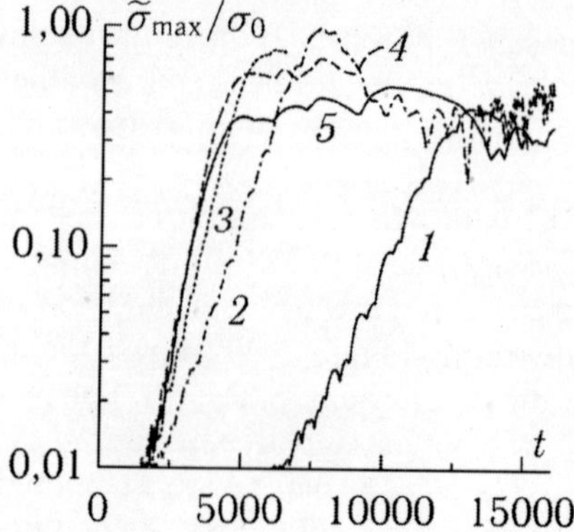

Fig. 5.30. Dependence of the maximum value of the perturbations of density $\tilde{\sigma}/\sigma_0$ for a convective cell on time for different models: M = 10, 1 – m = 2, 2 – m = 4, 3) m = 6, 4 – m = 8, 5) m = 12. As the number of harmonics increases, the rate of increase of the perturbations, which follows the exponential law to saturation, also increases.

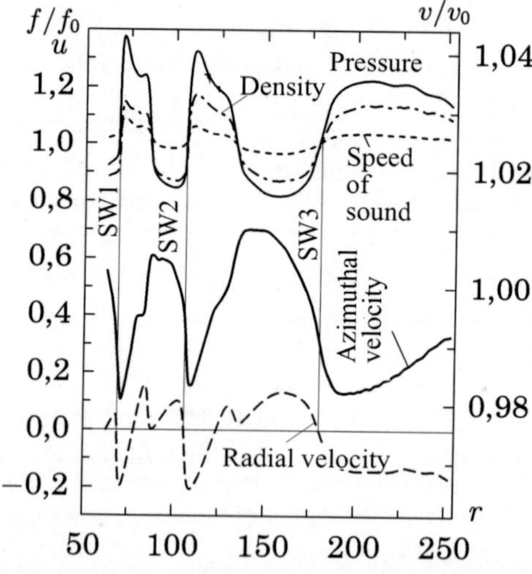

Fig. 5.31. Radial dependences σ/σ_0, p/p_0, c_s/c_{s0}, u (the left-hand axis of the ordinates) and v/v_0 (right-hand axis) for the azimuthal angle φ = 0 at the moment of time $t = 2{\cdot}10^4$, the positions of the shock waves (SW) are shown.

the relative perturbations for the selective convective cell for different values of m and M. During the characteristic time $t^{(sat)} \simeq 350$ M the perturbations grow to a highly non-linear stage, close to saturation. The amplitude of the perturbations with the small amplitude number increases at a lower rate.

The supersonic inflow of the gas to the region with increased density in the zone of the spiral wave results in the formation of weak shock waves at the rear edge. The radial profiles of the flow are identical with those described in section 11.3.3 in the examination of the dissipative instabilities because the waves are strongly twisted (Fig. 5.31).

The convection in the disk plane results in the mean in the fall of matter on the gravitational centre. The flow of matter $M = r\int_0^{2\pi} \sigma u d\varphi$ is on the whole negative. The flow of the specific angular moments is directed to the outside and the mean radial movement to the centre is associated with the displacement of the angular moments by the spiral waves resulting in accretion as a result of the redistribution of matter.

The experimental results, obtained within the framework of the quasi-Kepler disk, are basically valid in the case of the laws of rotation with $r/L_\Omega > -3/2$. In application to the gas subsystems of the disk galaxies, the values of the parameter r/L_Ω are situated in the range from 0 (an almost solid-state central zone) to – 1 – a curve of the 'plateau' type, characteristic of the majority of the galaxies in the large part of the disk.

6

Simulation of galaxies in Spiral equipment

6.1. The problem of spiral structure in galaxies

6.1.1. Background

In 1845, Lord Ross on his family estate Bir Castle in Ireland designed and built an 183 cm reflector with a focal length of 17 m, which remained the largest telescope in the world for a long time. It allowed him first to describe the spiral structure of many of the nebulae. For the next eight decades it was believed that they belong to our stellar system. Only in 1924, Edwin Hubble with the largest at that time telescope with a mirror diameter of 2.5 m at the observatory Mount Wilson (USA) firmly established that these nebulae are individual spiral galaxies.

Around the same time, the author of the book Astronomy and Cosmology (Jeans, 1929) wrote that the spiral nebulae were generated entirely by unknown forces. Indeed, at first glance, we meet with a serious paradox. Galactic disks, which contain spiral arms in their plane, rotate differentially: the angular velocity of the galaxy decreases with increasing radius. This means that if the spiral arms represent the condensation of gas and young stars, then differentially rotating with time, they must be stretched so that in 1–2 rotation periods the spiral pattern almost completely disappears. However, it has been reliably established that the stars of the Galaxy in the solar neighborhood have made about 50 rotations, while maintaining a spiral structure.

In 1938, Bertil Lindblad resolved this paradox. He suggested that the spiral arms of galaxies are density waves. The wave front rotates with constant angular velocity and the phase velocity of the density wave is constant. Therefore, any 'extension' of the wave is absent, despite the differential rotation of the disk. Later Lin and Shu, together with their counterparts, significantly developed the theory of density waves, re-rediscovering the wave nature of the spiral arms. As a result, a number of fundamental issues appeared for observational astronomy. The main one is the definition of the angular velocity of the spiral pattern. This problem was similar to the problem of detecting the corotation radius, where the constant angular velocity of spirals coincides with the angular velocity of the differentially rotating galactic disk. In the gravitational concept, this value is not found – it is a free parameter, determined from observations. Only a few years ago studies appeared with a description of a reliable method of determining this most important parameter of spiral galaxies (Lyakhovich et al, 1997; Fridman et al, 1997). Below we describe a method that allows the solution of the most important problem of detection of anticyclones, the existence of which in the gaseous disks of galaxies was previously predicted theoretically.

6.1.2. Spiral pattern as density waves

It should be regarded as a reliably established fact that global spiral galactic structures have a wave nature. It took several decades and the efforts of many outstanding researchers (see the historical review [1014]) to confirm this. It should be mentioned that initial studies in this direction were conducted by B. Lindblad in the mid 20s [116, 848].

We list the main observations that point to the wave nature of the spiral pattern.

1. Differential rotation of the galactic star–gas disk. Any non-wave structure, extended along the radius, virtually disappears after a few rotations.
2. The velocity fields of stars and gas show a spiral structure.
3. The spiral pattern is seen in virtually all galactic components: the gas in general, neutral hydrogen HI, optical HII-regions, molecular clouds and star formation regions, young stars (spectral types A and B), stellar complexes, dust, magnetic field, the old stellar population.
4. The existence of a galactic shock wave, which manifests itself in the form of a narrow strip of high density gas and dust in the inner zone of the spiral arm inside the corotation region.
5. There is a characteristic age gradient of stars across the spiral arm.
6. Observations indicate the existence of Lindblad resonances and corotation (ring structures).
7. Large-scale galactic vortex structures have been detected.
8. Observations indicate a vertical motion in the spiral arms.
9. A significant (albeit indirect) argument is a wide range of various examples of physical phenomena, when the rotating systems show the formation of stationary or quasistationary spiral structures. Spiral waves are found in planetary rings, protostellar disks, in rotating shallow water, and other natural phenomena.

6.1.3. Gravitational–hydrodynanic concept of the generation of spiral structures

The gravitational–hydrodynamic concept [149, 237, 239] includes the concept of gravity as a special case, when in addition to self-gravitating forces there are also important gradients of basic equilibrium (unperturbed) disk parameters – density,

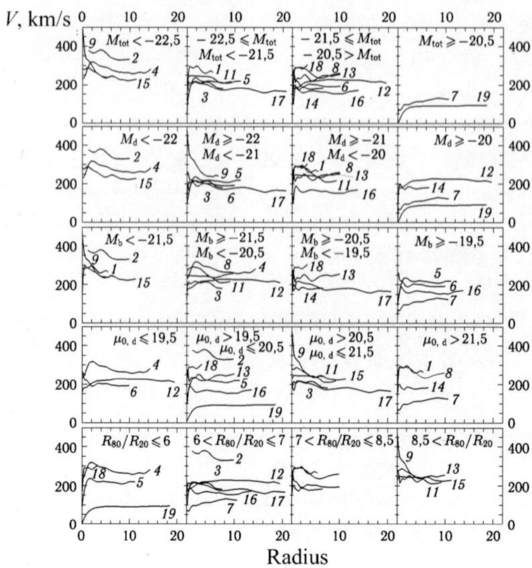

Fig. 6.1. Many of the rotation curves of galaxies with different properties show two features: 1) non-monotonic character in the centre, 2) non-decreasing form at very large distances [974]. The graph indicates the absolute values of the stellar disk, bulge, the central surface brightness of the disk and the bulge, the ratio of effective radii R_{80} and R_{20} containing 80% and 20% of the luminosity, respectively. The radius is shown in units of the scale of the disk.

rotation velocity and the speed of sound. These gradients are often neglected in the gravity concept. Figures 1.8, 1.10, 1.18, 9.1 show radial profiles of the azimuthal components of velocity and surface density. The relative magnitude of the 'jump' of the speed of rotation in the central region of the gas disks at a distance of 0.5–1.5 kpc from the centre can be $\simeq$25%, and the difference of the surface density of gas in the region of the maximum velocity gradient in some galaxies is almost two orders of magnitude. Is it possible to neglect these gradients, if in this case hydrodynamic instability is much faster than gravitational one?

At the suggestion of one of the authors (A.M.F), a group of employees of the Special Astrophysical Observatory of the Russian Academy of Sciences (SAO RAS) and the Sternberg State Astronomical Institute of the Moscow State University (SSAI) carried out a series of studies of rotation curves (dependence of the speed of rotation on the radius) of spiral galaxies using a 6 m telescope of the SAO RAS [8–12].

As it turned out, half of a random sample of spiral galaxies have a relative 'jump' of the velocity of rotation of the order of 10–15% (Fig. 1.8).

Numerous observations of the kinematics of spiral galaxies in the past decades show that the regions of the sharp change in the verocity of rotation in the central regions of galaxies are very common (Fig. 6.1, [974, 1183, 1186]).

Such observable regions of fairly rapid changes in the speed of rotation may lead to centrifugal instability [16, 149, 163, 183, 239] which can generate spiral density waves – the spiral arms of galaxies.

According to the gravitational–hydrodynamic concept, the spiral arms of some galaxies are density waves, growing up to non-linear amplitudes due to the development of hydrodynamic (non-Jeans) instabilities in the gas galactic disk. In any case, the presence of areas of sharp changes in the radial velocity of rotation and / or thermodynamic quantities requires consideration when constructing models of the spiral structure. Various aspects of this concept are the subject of several review articles and monographs [164, 184, 237, 590].

Underlying this approach is the following fact. The gas disk of the galaxies, as a rule, has a mass much smaller than that of the stellar disk (see 1.3.2) and, therefore, $\rho_{gas} \ll \rho_*$. From this it seems it should be concluded that the gas does not make a significant contribution to the gravitational potential. This is true for axisymmetric models. However, it should be taken into account that in the collisionless plasma of stars everywhere, except the central regions of the SB-galaxies, the amplitude of the density of non-axisymmetric features is much smaller than the total density of the stellar disk ($|\tilde{\rho}_*| \ll \rho_*$), whereas due to the effect of the collisional nature of the gas the density perturbations propagating in it can reach significant amplitudes $\tilde{\rho}_{gas} \sim \rho_{gas}$ (~ indicates perturbed quantities). If the values of ρ_{gas} and $|\tilde{\rho}_*|$ equal up to several percent of ρ_*, the following situation is possible $|\tilde{\rho}_{gas}| \sim |\tilde{\rho}_*|$ or even $|\tilde{\rho}_{gas}| \gg |\tilde{\rho}_*|$ (see [239, 590]).

Let a density perturbation occur in the gas disk. For estimates (for details see the paper [239]), we use the expression for the perturbed gravitational potential taking into account the finite thickness of the gas disk

$$\tilde{\Phi} = -\frac{2\pi G \tilde{\sigma}_{gas}}{k \mathrm{I}_{gas}}, \qquad \mathrm{I}_{gas} = \frac{1}{1+kh}. \tag{6.1}$$

We write the expression for the perturbed surface density of the stellar component, neglecting the inhomogeneity (see (2.86), (2.94))

$$\tilde{\sigma}_* = -\frac{\sigma_{*0}}{1+k\Delta_*}\frac{\tilde{\Phi}}{c_r^2}\left\{1 - e^{-\tilde{z}}\mathrm{I}_0(\tilde{z}) - 2\sum_{n=1}^{\infty}\mathrm{I}_n(\tilde{z})\frac{\hat{\omega}^2}{\hat{\omega}^2 - n^2\varkappa^2}e^{-\tilde{z}}\right\}. \tag{6.2}$$

Taking into account the condition $\tilde{z} \gg 1$, which holds for the Galaxy [239], $\tilde{\sigma}_* = -[\sigma_{*0}/(1+k\Delta_*)](\tilde{\Phi}/c_r^2)$. The last relation together with (6.1) yields

$$\frac{\tilde{\sigma}_*}{\tilde{\sigma}_{gas}} = \frac{G\sigma_{*0}}{c_r^2}\frac{2\pi}{k}\frac{1+kh}{1+k\Delta_*}. \tag{6.3}$$

For typical values of the parameters the ratio $\tilde{\sigma}_* / \tilde{\sigma}_{gas}$ is small. As one can see, in the study of the hydrodynamic mechanisms of generation of the spiral structure in a first approximation we can ignore the perturbation of the stellar subsystems.

The most likely candidate for a generator of the spiral structure (in the time of development and parameters of the emerging spirals in linear analysis) amongst hydrodynamic instabilities is centrifugal. This instability is excited in the region of the negative gradient of the velocity in galaxies with double-humped rotation curves (in section 5.1 in different models we analyse in detail the conditions of the centrifugal instability and the properties of the emerging spiral pattern).

The problem discussed here admits a unique opportunity for laboratory simulation ('galaxy on the kitchen table'). Nezlin and Snezhkin [184] described in detail the methodology for experiments in equipment with rotating 'shallow water', and the results. Therefore, below we briefly discuss only the highlights.

The capabilities of laboratory modelling of the spiral galactic pattern are based on the equivalence of the systems of equations, describing the dynamics of the vast central regions of gas disks of flat galaxies and a thin layer of a rotating incompressible fluid. The thickness of the liquid layer h corresponds to the surface density of the compressed gas, and the role of the speed of sound in the gas is played by the characteristic velocity of waves in shallow water $c_s \to c_0 = \sqrt{gh}$ [112]. The two-humped rotation curve ('jump' of velocity) can be modelled by two paraboloids rotating with different angular velocities.

In addition to centrifugal instability, the hydrodynamic concept also studies other instabilities. Between the edge of the central molecular disk and a special feature on the rotation curve (break or discontinuity) there appears a waveguide layer, in which instabilities of the resonant type can 'swing'. Similar instabilities were discussed in section 11.3. Other possible mechanism – centrifugal–resonance – is discussed in section 5.2. The emergence of a large-scale spiral pattern should be also caused by the development of the Papaloizou–Pringle instability with lower azimuthal numbers $m = 2 \div 6$, even in the disk without features in the rotation curve (see section 5.2.4).

The centrifugal instability belongs to a class of shear flow instabilities. As a general rule, a necessary condition for the growth of large-scale perturbances in such instability is $kL \lesssim 1$, where k is the wave vector of the perturbation along the velocity jump, L is the characteristic radial scale of the velocity gradient. Since $k \equiv 2\pi/\lambda$ (λ is the wavelength of the perturbation), then the instability condition implies that the typical width of the shear flow L is small compared with the wavelength. In the galactic disk, the coordinate along the velocity jump is azimuthal angle φ. As a result of the instability the value of L is 'smeared' so long until the border of stability is reached at which there may be only waves satisfying the condition

$$k_\varphi L \simeq 1, \tag{6.4}$$

where k_φ is the azimuthal wavenumber. By definition we have

$$k_\varphi \equiv \frac{m}{R}, \tag{6.5}$$

here m is the number of arms of the galaxy, R is radius at which the 'jump' in the speed rotation is observed.

We introduce a parameter $q \equiv \Omega_2/\Omega_1$ (Ω_1, Ω_2 are respectively the angular velocity of rotation of the central and peripheral regions of the galaxy before and after the 'jump' in velocity), then the larger the difference of the angular velocity of rotation, i.e. the smaller the parameter q, the greater the value of the characteristic scale of the velocity gradient L. In an approximate form, this dependence can be written as

$$L \simeq Aq^{-\alpha}, \quad \alpha > 0, \tag{6.6}$$

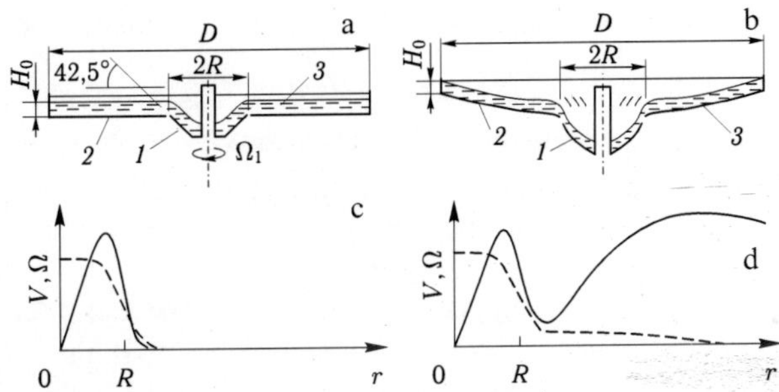

Fig. 6.2. First (a) and second (b) modifications of Spiral equipment; 1) central part rotating with the angular velocity Ω_1; 2) outer part, stationary (in model a) and rotating with angular velocity Ω_2 (b); 3) the shallow water layer; $D = 30$ cm, $R = 4$ cm (a); $D = 60$ cm, $R = 8$ cm (b). Schematic representation of the radial profiles with a large 'jump' in the velocity of rotation of shallow water (c) and a gas galactic disk (d); solid curves correspond to the linear velocity of rotation, dashed curves to angular velocity.

where A is a constant.

Substituting (6.5) and (6.6) into (6.4) we obtain a qualitative relationship between the number of arms and the dimensionless value of the velocity jump q:

$$m \simeq \frac{R}{A} q^{\alpha}, \quad \alpha > 0. \tag{6.7}$$

This dependence means that the greater the velocity difference (smaller q), the smaller is the number of arms in the galaxy.

An exact formula between m and q was obtained in the framework of the gravitational–hydrodynamic theory. The correctness of this dependence was checked for spiral galaxies with known rotation curves, and in experiments with rotating shallow water. However, before turning to the experimental results, it is necessary to obtain confirmation that the approach developed here correctly describes the generation of galactic spiral structures in the gravitational–hydrodynamic concepts.

6.2. Spiral equipment – a tool for studying galaxies

Vortices between the spirals, amazingly reminiscent of spiral arms of the galaxies, were first detected on a desktop setup with rotating 'shallow water', which was proposed by one of the authors of the book (A.M.F) and constructed at the Russian Research Center, the Kurchatov Institute, by E.N. Snezhkin, specifically for modelling the generation of spiral arms of the galaxies; the equipment was named 'Spiral' (Fig. 6.2a, b). In the first modification of this setup the central part was conical with an almost flat peripheral portion, in the second variant – both sides were in the form of paraboloids. The fluid is drawn into rotation by the bottom part, and between the central part and the periphery a jump formed in the velocity of rotation with radius R and the width close to the depth of the fluid H_0. As a result, a layer of shallow water gets a jump on the radial velocity profile of rotation (Fig. 6.2c), simulating a jump in the rotation curve of the galaxy that can cause the generation (through hydrodynamic instability) of spiral density waves. Therefore,

'Spiral' equipment has been used to model dynamic processes in gas galactic disks with a jump in the speed of rotation.

The aim of the experiments was to verify the possibility of generation of spiral density waves by gravitational–hydrodynamic instability which in the actual galaxies may be associated with the rapid local decrease (jump) of the speed of rotation of the gas observed in many cases in the galaxies and clearly visible on the rotation curve of the galaxies obtained previously at the BTA [8]. A large jump of velocity is shown schematically in Fig. 6.2d. The observations show that (taking into account the angular resolution of the telescope) approximately 50% of the spiral galaxies are characterised by the relative jumps of velocity $\Delta v/v$ on the rotation curve of no less than 10–15%. However, as shown by numerical calculations carried out by Baev and Fridman [16], the role played by the hydrodynamic effects in the mechanism of formation of the spiral arms in the gas discs of the galaxies with such velocity jumps is as important as that played by the gravitational effects.

This view point regarding the mechanism of formation of the spiral arms in the galaxies, predicted by one of the authors of the book (A.M.F) in 1972 (see [237] and the literature therein) differed from the viewpoint of the purely 'gravitational' theory of generation of spiral arms which took into account only the self-gravitation forces [847]. The gravitational theory did not take into account the observed (sometimes very large!) gradients of the main parameters of the disk.

First of all, Spiral equipment was used to verify the validity of guidelines of the gravitational–hydrodynamic theory of the spiral structure: the relationship of the relative size of the jump of the speed of rotation $\Delta v/v$ and the relative size of the spatial domain of its localization $\Delta R/R$ with the number of generated spiral arms. For this purpose, the installation was set up to produce a jump in speed, imitating the observed jumps in the rotation curves of galaxies (Fig. 6.2c). Rotating shallow water at the facility is described by equations of two-dimensional gas dynamics [112]. This is the dynamic relationship of rotating shallow water with the gas disk of spiral galaxies, which served as a starting idea for the simulation. (Strictly speaking, the theoretical basis for modelling is the identity of the dynamic equations of the self-gravitating gas disk and rotating shallow water [239, 565].

The results of modelling tests in shallow water [163] not only confirmed the correctness of the gravitational–hydrodynamic theory of generation of spiral structures, but also led to unexpected and original solutions of certain issues related to such features of the spiral structure as branching of the arms and rarely observed 'leading' spirals. Moreover, they have produced also something new: anticyclonic vortices were found between the spiral arms in shallow water [183] (Fig. 6.3). Hence, these vortices must be observed in real spiral galaxies! The history of the discovery of these structures is described in section 7.2.

6.2.1. 'Shallow water' and the model of gas galactic disks

We discuss in detail what is common between Spiral equipment and the gas galactic disk? Shallow water equations [112] and the equation for the gas galactic disk show a remarkable similarity. However, there are two significant differences.

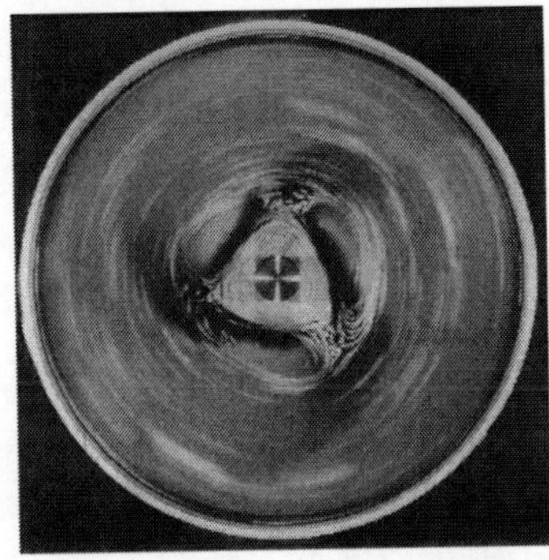

Fig. 6.3. Anticyclones between spiral density waves in shallow water of 'Spiral' equipment: three spiral arms correspond to three anticyclones.

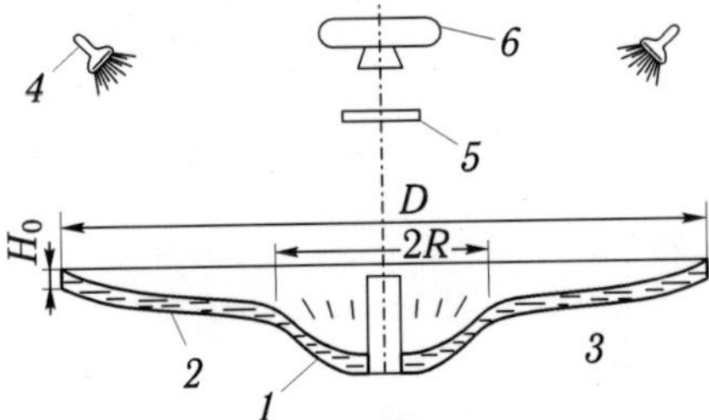

Fig. 6.4. 'Spiral' equipment: 1) the core; 2) the periphery; 3) a layer of shallow water; 4) incandescent bulbs; 5) a red filter; 6) camera. First modification: $D = 28$ cm, the radius of 'the break' $R = 4$ cm; the core of a truncated cone with an angle of inclination of the generatrix to the horizontal of 65°. The second modification: $D = 60$ cm, $R = 8$ cm. The nucleus rotates clockwise in top view.

First, the Spiral equipment has a solid foundation, which is why shallow water is characterised by a viscous Ekman boundary layer. Obviously, this does not exist in the gas in the galaxy. Second, a galactic disk is a self-gravitating subsystem while self-gravitation can not be taken take into account in the laboratory facility.

In [239] it is shown that both of these 'fundamental' differences are illusory. First, the Ekman number of Spiral equipment is small, i.e. low near-bottom viscosity. But the chief feature, perhaps, is the fact that the time of viscous damping of perturbations in shallow water in Spiral equipment is considerably longer than the development of centrifugal instability. In other words, the near-bottom viscosity does not affect the process of generation (as a result of the instability) of spiral density waves in shallow water. On the other hand, if there was not benthic viscosity, we would not have such a simple way to make a rotating fluid to form the desired velocity profile of circular rotation with a jump. In spiral galaxies, the profile of the velocity of rotation of the gas disk is produced by the gravitational potential of mainly stellar subsystems as much more massive (see, e.g. [227]). In this sense, the stellar massive subsystem (disk) together with the spheroidal component (bulge + halo) plays the same role in the formation of the rotation curve of gas in the disk as the bottom in Spiral equipment. As regards the role of viscosity in the generation of spiral arms of real galaxies in the gas disk, it is negligible both in the development of the centrifugal instability caused by the velocity gradient and in gravitational instability in the absence of appreciable velocity gradients. In the absence of the above-mentioned instabilities viscosity can lead to the formation of structures that occur, for example, in planetary rings [43]. However, we know of no case of generation due to the viscosity of large-scale spiral arms in the galaxy.

We discuss in greater detail the influence of molecular viscosity on the mechanism of the development of gradient instabilities in shallow water in Spiral equipment.

Experiments were performed on two versions of Spiral equipment (Fig. 6.4). The working fluid was a green solution of $NiSO_4$ in water, which yields contrasting pictures of spirals in red light illuminating the solution and reflected from the white bottom of the vessel. The second modification was the installation twice as large and consisted of two paraboloids, whose form corresponded to a layer of shallow water of constant thickness, H_0 = 1.5–3.5 mm both in the centre (with $\Omega_1 = 13\ s^{-1}$), and at the periphery (at $\Omega_2 = 2.6\ s^{-1}$).

Glycerin was added to the $NiSO_4$ solution which increased the viscosity of the solution two or three times (in all experiments in which the effect of viscosity was not specifically studied, the total viscosity of the solution was not more than 5–10 times greater than the viscosity of water under normal conditions); this additive significantly facilitates the formation of the spiral–vortex structure in the initial stage of its formation. The rotation mode in both versions was chosen so that the periphery of the radial dimension $D / 2$ was much greater than the Rossby–Obukhov scale r_R (similar to the characteristic Larmor radius of the plasma ions):

$$r_{\mathrm{R}} = \frac{(gH_0)^{1/2}}{2\Omega_2}, \qquad (6.8)$$

where g is gravitational acceleration. This choice of modes allowed check whether the generated spirals are stable with respect to the partition to vortices with the size $\sim r_{\mathrm{R}}$. (Note that selected ratios Ω_1/Ω_2 and $r_{\mathrm{R}}/(D/2)$ are close to the actually observed in spiral galaxies). In experiments with rotating shallow water there are some dimensionless numerical parameters whose values determine the relative importance of viscosity. First of all, the Ekman number [1019]

$$E_\nu = \frac{\nu}{\Omega_0 H^2}, \qquad (6.9)$$

where ν is kinematic molecular viscosity [108], Ω_0 – angular velocity of the bottom of the vessel, H – depth of the liquid layer. In Spiral equipment, the layer depth H varied from 0.2 to 0.4 cm, i.e. taking into account that for the wave $\nu = 0.01\ \mathrm{cm}^{-1}$ [163, 164, 588] $(E_\nu)_{\max} \approx 1/4(\Omega_0(r))_{\min}$. In experiments with the 'rotating periphery' $(\Omega_0(r))_{\min} \approx 2\ \mathrm{cm}^{-1}$, therefore, $(E_\nu)_{\max} \approx 1/8$, i.e. the effects of viscosity are small.

By definition, [1019] the Ekman number $E_\nu \equiv \delta^2 / H^2$, where $\delta = (\nu/\Omega_0)^{1/2}$ is the depth of the Ekman layer. Despite its relative smallness ($\delta_2 \ll H^2$), the Ekman layer in our experiment can effectively change the momentum of the fluid layer with depth H for some characteristic time τ_{sp}. This time is called the characteristic time of viscous damping (spin down time), or the characteristic time of viscous setting (spin up time). In experiments [163, 164, 588] for the abrupt change in speed Ω_{01} a number of modes was observed: the m_1-arm spiral was replaced by the m_2-arm spiral corresponding to the new rotation mode. The time for establishing a new steady-state is τ_{sp}. We estimate its magnitude. From the equations of motion for an order of magnitude we have

$$\frac{\partial v_r}{\partial t} \sim \nu \frac{\partial}{\partial z}\left(\frac{\partial v_r}{\partial z}\right). \tag{6.10}$$

Since v_r depends on z only in the typical interval $(0, \delta)$, we can approximately write

$$\frac{\partial v_r}{\partial z} \sim \frac{\partial v_r}{\partial z} \cdot \Theta(\delta - z),$$

where $z > 0$, $\Theta(z)$ is the Heaviside function. Integrating (6.10) with respect to z from 0 to H, we find

$$\frac{\partial}{\partial t} v_r H \sim \nu \left.\frac{\partial v_r}{\partial z}\right|_0^\delta \sim \nu \frac{v_r}{\delta}.$$

Multiplying this equality by density ρ and the bottom area S, we obtain the equation for the change of the radial pulse $\mathcal{P}_r = \rho\, SHv_r$

$$\frac{\partial \mathcal{P}_r}{\partial t} \sim \frac{\mathcal{P}_r}{\tau_{sp}}, \quad \tau_{sp} \sim \frac{H\delta}{\nu} \sim \frac{H}{(\nu\Omega_0)^{1/2}}. \tag{6.11}$$

We estimate the time of establishment of the steady-state regime, for example, mode $m = 2$, and to change the number of arms in the case where we have dramatically changed the angular velocity of rotation of the periphery by the value $\Delta\Omega$. In the first case ($m = 2$), this time will certainly be greater than $(\tau_{sp}) \approx 0.47\ s^{-1}$ at $\tau = \tau_{min}$ at $\Omega = \Omega_1 = 18\ s^{-1}$, $H \approx 0.2$ cm (in the central region of Spiral equipment $H > 0.2$ cm). In the second case for the $\Delta\Omega \approx 2\ s^{-1}$ we have $\tau_{sp} = 1.4$. The time of instability τ_{in} is much shorter than these times τ_{sp}: $\tau_{in} \ll \tau_{sp}$. The time of instability depends on the 'blurring' of velocity discontinuity. If this blurring is caused by molecular viscosity (laminar Ekman layer), then $\delta_{lam} \sim (\nu/\Omega_1)^{1/2} \approx 2.4\times10^{-2}$, i.e. $\delta_{lam} \ll \lambda$, where λ is the radial wavelength ($\lambda \approx$ 6 cm for $m = 2$) and $\gamma \approx \Omega_1$ ($\Omega_1 \approx 18\ s^{-1}$ for $m = 2$), and $\tau_{in} \sim 1/\gamma \approx 5\times10^{-2}$ s.

The arguments in favour of a negligible role of viscous friction in the process of generation of shear flow instabilities in Spiral equipment are based on the results of experiments with coloured water for which $\nu \approx 0.01\ cm^2\ s^{-1}$. In [966] it is noted that the increase of the viscosity of the working solution by a factor of 10 does not change the qualitative picture of the spiral pattern. At an even larger increase of the viscosity the spiral structure disappears.

A negative answer to the question of the influence of turbulent viscosity on the formation of structures in Spiral eqipment is given in [304] where it is shown that 'the viscous lifetime' of the structures is completely determined by laminar viscosity.

6.2.2. On the equivalence of the linearised dynamic equations of the gas of the galactic disk and rotating shallow water in Spiral equipment

We turn now to the second 'fundamental' difference between shallow water and the gas galactic disk that is associated with the presence in the latter of gravitational forces. In [239] a new quantity was introduced into the system of the dynamic equations for a self-gravitating gas disk – the speed of sound in a self-gravitating medium. This value is less than the speed of sound in the non-self-gravitating medium as the elastic properties of the gas – the ability to expand at initial compression of the gas element – worsen in the self-gravitating medium: the compressed element is retained by gravitational forces. The equations for the galactic disk by replacement of the speed of sound in the self-gravitating gas by the velocity of wave propagation in shallow water transform into equations for shallow water.

Indeed, given that $\widetilde{\Psi}$ depends mainly on the $\widetilde{\sigma}_{\text{gas}}$ [239], the system of linearised dynamic equations of the gas disk of the Galaxy can be written as follows:

$$\frac{\partial v_r}{\partial t}+\Omega_0\frac{\partial v_r}{\partial\varphi}-2\Omega_0 v_\varphi=-\frac{\partial}{\partial r}(c_{g0}^2\eta), \tag{6.12}$$

$$\frac{\partial v_\varphi}{\partial t}+\Omega_0\frac{\partial v_\varphi}{\partial\varphi}+\frac{\kappa^2}{2\Omega_0}v_r=-\frac{1}{r}\frac{\partial}{\partial\varphi}(c_{g0}^2\eta), \tag{6.13}$$

$$\frac{\partial\eta}{\partial t}+\Omega_0\frac{\partial\eta}{\partial\varphi}+\frac{\partial v_r}{\partial r}+(1+r\ln'\sigma_0)\frac{v_r}{r}+\frac{1}{r}\frac{\partial v_\varphi}{\partial\varphi}=0, \tag{6.14}$$

Here we introduce the following notation:

$$c_{g0}^2\equiv c_{s0}^2-\frac{2\pi G\sigma_0}{|k|R_{\text{gas}}},\quad c_{s0}^2\equiv\frac{p_0}{\sigma_0},\quad \eta\equiv\frac{\sigma}{\sigma_0}, \tag{6.15}$$

$$\kappa^2\equiv4\Omega_0^2\left(1+\frac{r}{2}\frac{\Omega_0'}{\Omega_0}\right),\qquad R_{\text{gas}}\approx\left(1+\frac{|k|h}{2}\right)^{-1},$$

where the prime denotes differentiation with respect to r; σ is surface density; k is the wave vector; h is the half thickness of the gas disk; stationary quantities are marked with an index value of '0', the symbol tilde over the perturbed quantities is omitted and it will be used in future only to denote the amplitudes of the perturbed quantities. When writing (6.12) and (6.13), we used the linearised equation of state

$$\frac{p}{\sigma_0}=c_{s0}^2\eta. \tag{6.16}$$

If in equations (6.12)–(6.14) we make the substitution

$$c_{g0}^2=c_{w0}^2\equiv gH_0,\quad \eta=\frac{H}{H_0}, \tag{6.17}$$

we obtain a system of linearised dynamic equations of rotating shallow water [1019]. It is this system of equations that is used to describe small disturbances in shallow water in Spiral equipment.

Therefore, we can conclude that Spiral equipment with rotating shallow water can actually be used in laboratory simulation of the generation of spiral structures of the gas galactic disks, caused by the development of hydrodynamic instability due to the presence of a velocity gradient of rotation.

Spirals are the result of a new hydrodynamic instability, called centrifugal [588]. It was first found in Spiral equipment. The centrifugal instability as well as the Kelvin–Helmholtz instability develops in the presence of the velocity jump. However, as shown above, in the case a supersonic velocity jump, which takes place in the galactic disks and Spiral equipment', the Kelvin–Helmholtz instability is stabilised and only the new instability develops – centrifugal. For the latter to take place, an outward centrifugal force is required. In this sense, the physics of centrifugal instability is similar to the physics of Rayleigh–Taylor instability, which requires the gravitational force for its appearance.

6.2.3. Mechanism of instability in the presence of the velocity jump

In section 5.1 we discussed in detail the hydrodynamic instability associated with the presence of a strong rotation velocity gradient along the disk radius. We also note several important points for understanding this.

Figure 5.4 shows the structure of the perturbation of the tangential discontinuity of velocity **v**, directed along the x-axis for the two limiting cases: when the Mach number ($M \equiv v/c$, where c is the speed of sound) is much smaller and much greater (see Figure 5.4b) than unity. The amplitudes of the perturbations on both sides of the plane of the tangential discontinuity ($z = 0$) decrease exponentially: $\sim\exp(-|z|/z_0)$ [108]. Therefore, it is sufficient to confine considerations to the region $|z| < z_0$.

Region I (for 'maximum' perturbations) in both figures can considered as an area of motion inside a nozzle (more precisely its longitudinal half). It is known that the flow pattern is in principle different in sub- and supersonic flows [112,123]. In the narrowest (critical) section of the subsonic nozzle the velocity is maximum, as is observed at the narrowest point of a river. In the critical section of a supersonic diffuser, in contrast, the velocity is minimal. Because of this, the dynamics of tangential discontinuities is not the same as in the sub- and supersonic cases. These differences are based on the constancy of the Bernoulli integral of motion: $v^2/2 + W = \text{const}$, where W is enthalpy. In subsonic flows, the speed under a convexity is less than the speed above it ($v_{II} < v_I$), therefore $W_{II} > W_I$. As in normal situations, the pressure increases ($P_{II} > P_I$), the bulge (convexity) continues to gow under the influence of pressure changes. This is the physics of the instability of a tangential velocity discontinuity or the smoothed velocity profile, i.e. Kelvin–Helmholtz instability [240].

In the case of a supersonic velocity jump (Fig. 5.4b) we have the opposite inequality: $P_{II} < P_I$, i.e. the bulge is compressed by the pressure gradient, and the instability is stabilized. Landau was the first who drew attention to the effects of instability in the case of a supersonic tangential velocity discontinuity [108]. This effect occurs in a two-dimensional medium such as a thin gas layer or a layer of shallow water. In a real three-dimensional environment, this effect is preserved, but appears at high Mach numbers [240].

All the observed gradients in velocity in the gas galactic disks are supersonic, so the Landau stabilisation criterion is fulfilled for such disks and the instability of a tangential discontinuity of velocity or the Kelvin–Helmholtz instability does not develop in them. However, the disks may have a different centrifugal instability that can develop at arbitrarily high Mach numbers if the angular velocity of the central part Ω_1 (inner portion of the disk) is greater than Ω_2: $q = \Omega_2/\Omega_1 < 1$. It is remarkable that this condition is always satisfied for the disks of spiral galaxies.

So, if in shallow water we simulate the jump of the rotation velocity, which is greater than the characteristic velocity of propagation of perturbations $\sqrt{gH_0}$ (g is gravitational acceleration), then we have the analogue of a supersonic velocity jump which is observed in a significant part of the spiral galactic disks (chapter 1, [8–12]).

6.2.4. The main results of field experiments

6.2.4.1. Spiral equipment and diagnostics

Both versions of Spiral equipment have the same concept – a round vessel, which consists of two parts: 'central' and 'periphery', and the central part can be rotated independently from the periphery. In the first modification the central part has a conical shape, the peripheral part is almost flat (see Fig. 6.2a). In the second modification (see Fig. 6.2b) both parts consist of two paraboloids with dimensions twice as large as in the first modification. The fluid is drawn into rotation, a rotation velocity jump with the initial width close to the depth of the fluid H_0 forms between the central part and the periphery of the liquid. As a result, a layer of shallow water has a jump on the radial velocity profile of rotation (see Fig. 6.2c). The diagnosis system used in Spiral equipment makes it possible to determine both the structure of density waves and the field of the perturbed velocities of shallow water.

The vessel with a white bottom is filled with a green solution; in black and white photographs through a red filter sections of 'humps' of the density waves appear darker than the areas of 'troughs'. The velocity field is determined by the direction and length of the track (during exposure) of paper circles with a diameter 1–2 mm floating on the surface of the liquid. The camera rotates with a variable angular velocity and its axis of rotation coincides with the axis of rotation of the spiral. More details on Spiral equipment, diagnostics and shallow water experiments can be found in [184].

6.2.4.2. The relationship between the number of arms and the amplitude of the velocity jump

The agreement of the results of the generation of spiral structures in the hydrodynamic theory with the results of full-scale simulation is demonstrated in Fig. 6.5, where the increase in the number of spiral arms with increasing parameter q according to (6.7) is clearly visible.

These experiments showed that the instability leads to the formation of spiral waves of surface density with different azimuthal modes (m = 0, 1, 2, ... 8). The spirals are lagging behind. The rotation of the spiral pattern occurs at a velocity intermediate between the velocity of rotation of the central part of the periphery. For some modes the angular velocity of rotation of the spiral pattern Ω_p is a monotonically increasing function of Mach number $M = \Omega_1 R / (gH_0)^{1/2}$ (Fig. 6.6). Each mode can exist only within a certain range of parameters. Thus, in the conditions of Fig. 6.6 the modes m = 6, 5, 4, 3 can be realised if the Mach number does not exceed the values respectively M = 3.3; 3.7; 4.0; 4.4. At a smooth increase in M the system goes through these boundaries and the given mode is rearranged into one with a larger scale (m decreases). The restructuring of the modes is essentially a non-linear process – this is evident from the fact that transitions between modes are abrupt and show a hysteresis.The general pattern is that the number of spirals on the perimeter of the system decreases with increasing number M. Comparison of the results of laboratory experiments with the conclusions of the linear theory (i.e. the dependence of m(M) – we are not talking, of course, about the amplitude) indicates that the differences do not exceed 30%.

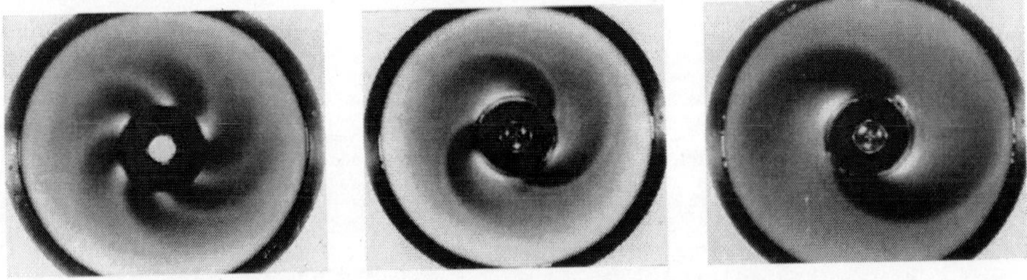

Fig. 6.5. Spiral density waves, which are formed in shallow water due to the centrifugal instability as a result of velocity gradient $q \equiv \Omega_2 / \Omega_1$. With the decrease of q the number of spiral arms decreases. The parameter q decreases in photographs from left to right.

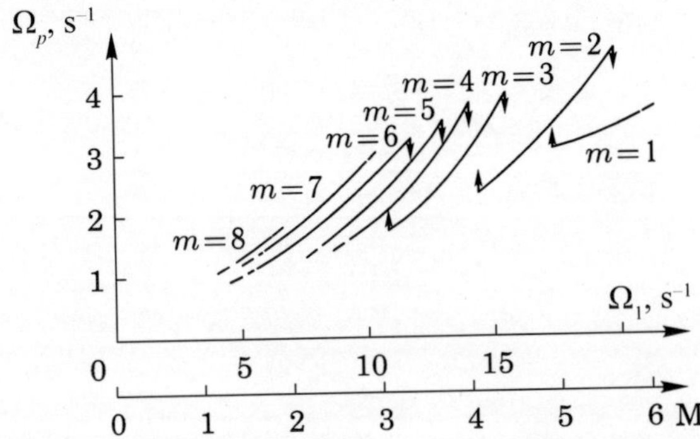

Fig. 6.6. The angular velocity of rotation of the spiral pattern Ω_p depending on the Mach number M. The numbers of the modes m are given, the arrows indicate the direction of adjustment of the modes (based on [184, 564]).

6.2.4.3. Lagging and leading spirals

Different types of the spiral pattern of galaxies are characterised not only by number of spiral arms, but also by their shape. First of all, we distinguish between 'lagging' and 'leading' spirals (Fig. 6.7). The former rotate back with their ends and, therefore, have a good 'aerodynamic' form. The vast majority of the spiral arms belongs to the lagging spirals.

Leading spirals are extremely rare. One of the authors (A.M.F) hypothesized that a sufficient condition for the generation of leading spirals is the presence of a close satellite which rotates in the opposite direction to the rotation the galaxy. The tidal bulge caused by the counter-spin of the satellite relative to its own galaxy, the periphery of the disk slows down first, and only then begins to rotate with the angular velocity of the satellite (Fig. 6.8).

The Earth–Moon system rotates synchronously and as a result we see always only one side of the Moon. Therefore, over time, a large part of the gas galactic disk begins to rotate in the opposite direction and, as a result, the lagging spirals become leading. This scenario may occur when the generation of spirals is due to the instability of the disk, which is stronger than the excitation due to tidal forces. In the

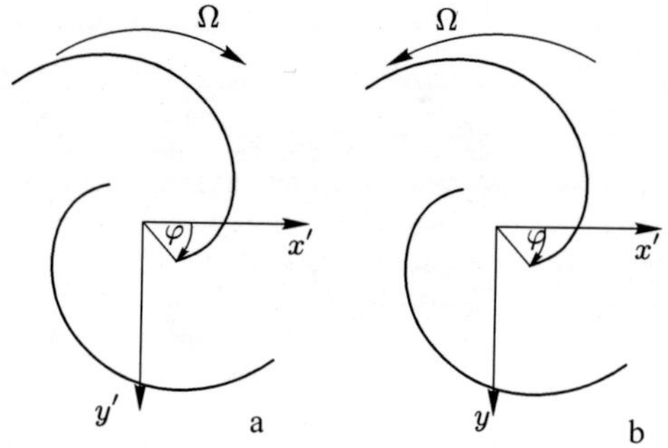

Fig. 6.7. Schematic representation of lagging (a) and leading (b) spirals. The arrows indicate the direction of rotation of the galactic disk.

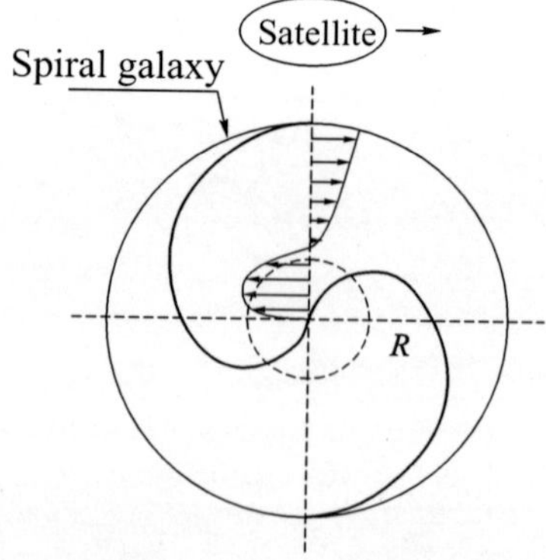

Fig. 6.8. A possible mechanism for the excitation of leading spirals in the presence of a satellite moving in the opposite direction (orbital angular momentum of the galaxy satellite is the opposite of galactic spin).

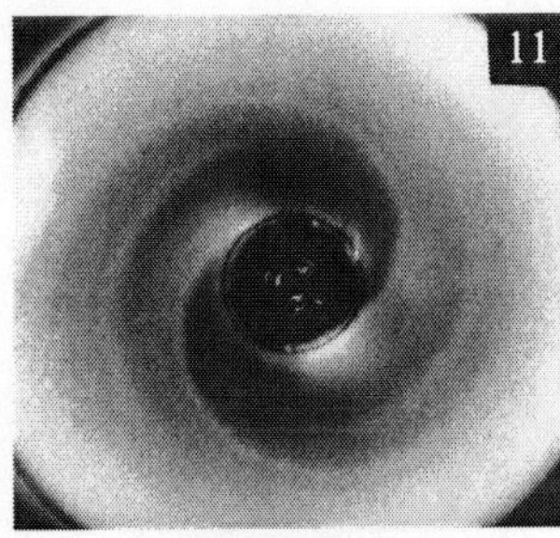

Fig. 6.9. Branching of spiral arms in shallow water in Spiral equipment in the process of changing the velocity jump.

latter case, the appearance of leading spirals is demonstrated by numerous examples of numerical simulation.

Using a random sample of 109 spiral galaxies, I. Pasha showed that only three of them have leading spirals [190]. Thus, each of these three galaxies have a close satellite, but only for one of them was it able to measure the direction of the orbital angular momentum of the satellite relative to the direction of 'spin' of the galaxy: they were opposed.

If the sign of the angular velocity of rotation of the periphery in Spiral equipment is changed, i.e. by replacing $\Omega_2 \to -\Omega_2$, leading spiral density waves can be found.

6.2.4.4. Branching spirals

This phenomenon is so widespread that its absence can be classified as an exception. Typically, the branching is 'upward': moving along the arm to the periphery, we meet area in which the arm splits. That is the picture we see on the photo of the spiral pattern in shallow water (Fig. 6.9). Recall that different but constant at the time photographing values of q correspond to different numbers of spiral arms. In contrast to the image shown in Fig. 6.2d, where the value of the jump q does not change, this image was captured when the parameter q decreased. Moreover, if at the initial time the value of q was consistent with the boundary of the instability that generates four-arm waves ($m = 4$), at the time of photographing two-arm waves were generated ($m = 2$). Generation of waves occurs at the velocity jump, where we see the mode $m = 2$. The periphery, as always, 'finds out' with a delay that the centre has switched to another mode and, therefore, at the periphery,we see the 'old' mode ($m = 4$). The result is 'branching' that reflects the evolution of the disk: an increase in the velocity of its central part compared with that of the periphery or, conversely, the relative decrease in velocity at the periphery. The first is possible, for example, at compression of the central part, the second one – when expanding the periphery.

6.2.5. Vortex structures between the spirals in shallow water

One of the results of full-scale simulation in Spiral equipment was the detection of vortices which appear to be a part of the overall spiral–vortex structure in the differentially rotating medium.

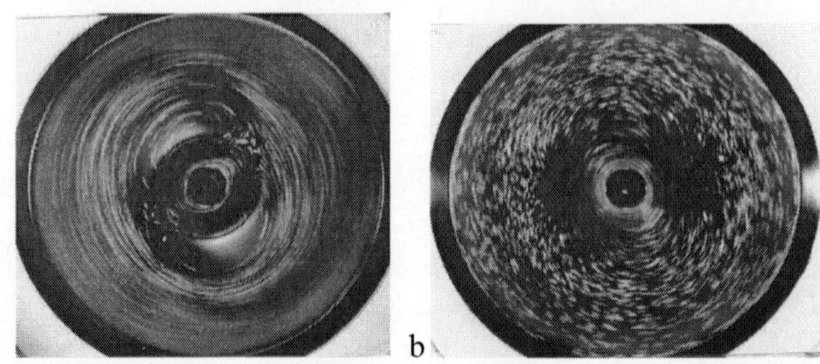

Fig. 6.10. Flow lines in the case of the generation of a two-arm density wave in Spiral equipment: the camera is fixed relative to the rotating shallow water (left), the camera rotates with the angular velocity of spirals (right).

With a fixed camera one can capture the current lines in rotating shallow water. This kind of image is often seen in astronomical studies devoted to determining the velocity field in galaxies (Fig. 6.10a). Figure 6.10b shows a picture of anticyclones which was obtained using a camera which rotates at the same speed as the spiral pattern.

6.2.6. Numerical simulation of instability super-reflection

Let us turn to full-scale simulation of the instability of super-reflection (section 5.2.2).

We consider a realistic model of a rotating shear layer. The parameters of this model are carefully chosen so that they satisfy certain physical conditions and maximize the instability growth rate of super-reflection which should prevail in magnitude over viscous damping and dispersive spreading. Then these optimal parameters may be used in construction of the experimental setup for studying super-reflection instability [242].

The proposed schematic geometry of the experimental setup is shown in Fig. 6.11. It consists of two parts: one inside the other. Inside there is a fixed horizontal disk with a radius of R_{in}. The outer part is a conical surface bounded by inner radius R_{in} and outer radius R_{out}. In the present experiment, the length of the waves that are generated on the surface of shallow water must be several times the wavelength of the capillary water (1.73 cm) in order to be able to neglect the effects of surface tension.

A small amount of water is poured into the resulting cavity so that a thin layer of liquid covers the surface of both sides of the bottom (during rotation of the outer part). The thickness of the water layer is a few millimeters. It should be several times greater than the thickness of the Ekman layer [1019], but small enough to remain in the shallow water approximation [108].

The profile of the bottom of the outer part should be chosen so that the shallow water depth was uniform if possible. In a real situation, the unperturbed free surface for $r > R$ has the form close to a paraboloid. However, the shape of a paraboloid

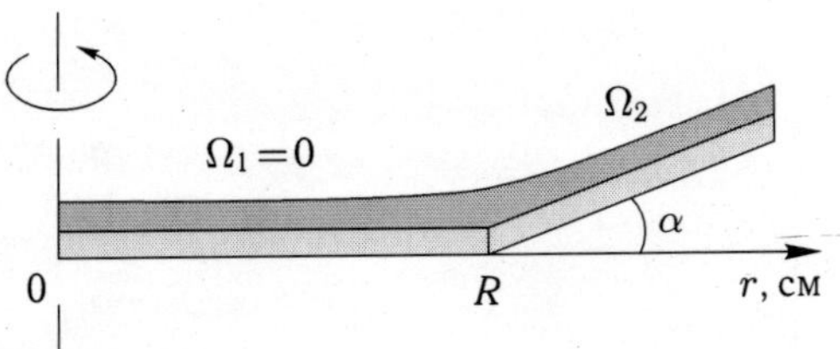

Fig. 6.11. The experimental geometry for simulation of instability of super-reflection.

depends on the angular velocity of the outer part. Consequently, the condition of constant depth shallow water can be strictly satisfied only for certain values of angular velocity. When the jump in velocity is fixed, the only way to change the Mach number is the change in the depth of shallow water. However, this method is less convenient than the change in the velocity of rotation. Therefore, we propose a constant slope of the bottom of the outer part ($\alpha_0 \sim 15°$), which allows us to assume that the depth is approximately constant in the range of Mach numbers $2.2 < \mathrm{M} < 6.0$.

The velocity shear layer in the proposed pilot plant is not infinitely narrow. For this reason, it is necessary to consider the effect of smoothing the velocity jump on the mechanism of generation of unstable modes. Therefore, in addition to a sharp break a small blurring of the jump as a result of viscosity was also studied:

$$\Omega(t) = \frac{\Omega}{2}\left(1 + \mathrm{th}\frac{r-R}{L}\right), \tag{6.18}$$

where L is the width of the smoothing of the tangential discontinuity, $L \leqslant H_0$.

Such a dynamical system can be viewed using numerical analysis. Rather than solving the dispersion equation, it is more convenient to numerically solve the eigenvalue problem for the system (5.78) with appropriate boundary terms and conditions.

The main result of these calculations is that for small smoothing of the tangential discontinuity, discussed earlier, increments of unstable solutions decrease – instability of super-reflection is weakened. However, the increments still remain large enough to be expected that the instability of super-reflection will show itself in the experiment. For a given Mach number, the modes characterised by the type and the azimuthal number, may have close increments. Nevertheless, with increasing Mach number we can observe a monotonic decrease in the number of arms m of the most unstable mode (if the increase in the Mach number is achieved by increasing the angular velocity Ω of rotation of the periphery).

Figures 6.12–6.14 show the wave patterns obtained by numerical simulation for the rotation curve (6.18). In all cases, $L = H_0 / 2$. The direction of rotation of the patterns and of the periphery – counter-clockwise.

Figure 6.12 shows two patterns with five arms, which may occur during the development of the instability of super-reflection at a Mach number $\mathrm{M} = 2.7$. Figure 6.13 shows the pattern of unstable modes with three and four arms, which can occur with super-reflection instability in the case of $\mathrm{M} = 2.8$. Figure 6.14 shows

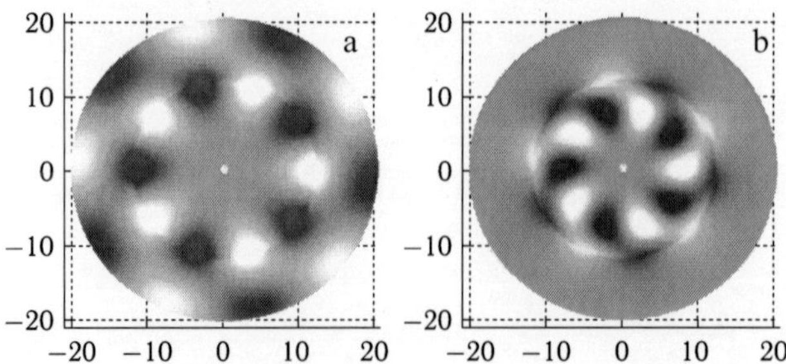

Fig. 6.12. Unstable modes B_1 with five arms of type I (a) and A_1 of type II (b), obtained from the numerical solution of (5.78) for the eigenvalues. The Mach number M = 2.7.

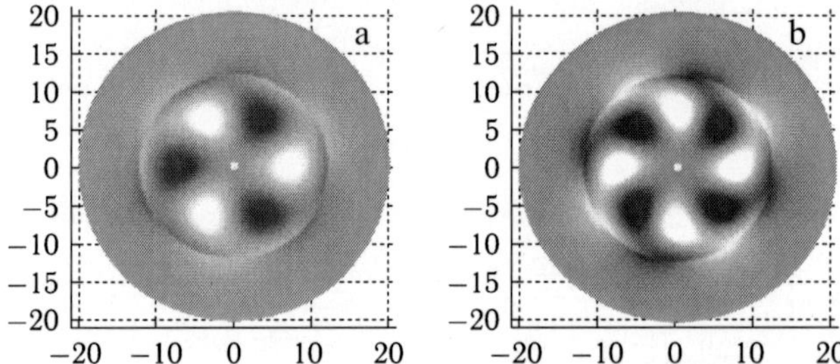

Fig. 6.13. Three-arm (a) and four-arm (b) unstable modes (A_1) of type II. The Mach number M = 2.8.

for comparison the patterns of three-arm unstable Kelvin–Helmholtz modes and of super-reflection.

Thus, the super-reflection instability generates structures with wave patterns that are shown in Figs. 6.12b, 6.13b and 6.14b. These large-scale structures are leading spirals (Fig. 6.7b) generated by the field of a smoothed jump of the speed of rotation.

For a fixed number m of the azimuthal harmonic the dependence of the increment on the Mach number for the two main families (I and II) of unstable roots of super-reflection is resonant (see Fig. 5.17).

Since with increasing Mach number the maximum growth rate of the super-reflection instability decreases rapidly, then the recommended range of the experimental study of the super-reflection instability corresponds to $2.2 < M < 4.5$. Taking into account the smoothing of the velocity jump to the half-width values of $L \sim H_0 / 2$ the small-scale structures ($m \sim\geqslant 8$) are stabilised. Increase of the Mach number in this range may lead to a rearrangement of the modes (in accordance with their maximum growth rate) from $m \sim 6.7$ to $(m-1)$, $(m-2)$,..., up to $m = 3.4$.

The wave patterns that occur during the development of super-reflection instability differ from the patterns generated by the Kelvin–Helmholtz instability (Fig. 6.14a). The latter have almost radial structures, monotonically decreasing on both sides of the discontinuity of angular velocity.

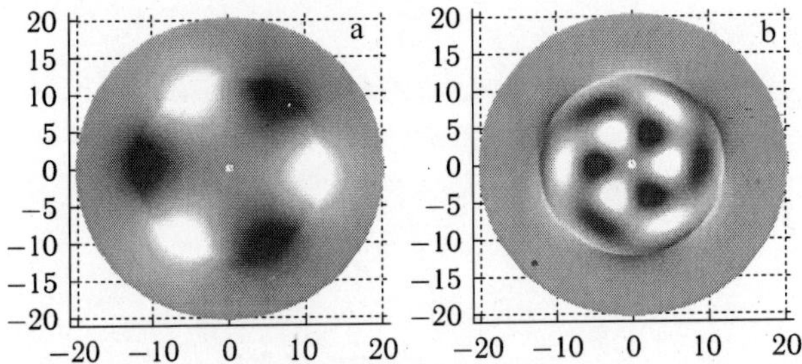

Fig. 6.14. Unstable Kelvin–Helmholtz three-arm modes for Mach number M = 2.0 (a) and super-reflection (A_2) of type II for the Mach number M = 4.2 (b).

In addition, the wave patterns excited by super-reflection instability differ from the patterns generated by centrifugal instability. The latter instability generates lagging (Fig. 6.7a) tightly coiled spirals, localized near the velocity jump. The centrifugal mechanism can be excited only at a sufficiently rapid decrease of the angular velocity with radius (faster than $\sim r^{-2}$, in the jump region).

6.2.7. The experimental discovery of super-reflection instability

In accordance with the above, a smooth layer with a free surface is located at the differentially rotating bottom. The original two-dimensional hydrodynamics equations describing the evolution of the liquid surface are published in the monograph [1019]. The speed of sound in these equations corresponds to $(gH)^{1/2}$, where g is gravitational acceleration, and H is the thickness of the liquid layer. The angular rotation velocity increases with the radius, thereby suppressing the strong centrifugal instability [588]. Another mechanism capable of preventing the experiment is the instability of the tangential discontinuity. The latter practically does not develop at supersonic speed at which super-reflection instability may form. If the condition $\Omega_2 > \Omega_1$ is fulfilled in the equipment (Ω_1 and Ω_2 are the angular velocities of rotation in the central and peripheral parts, respectively), then the competing centrifugal instability does not develop [588]. At the same time, this condition is not a hindrance for the development of super-reflection instability.

The described experiments were performed on the apparatus shown in Fig. 6.15. A thin layer of liquid covers the bottom level, reminiscent of 'a pan'. The outer part of the bottom, rotating with angular velocity Ω_0, is tapered (angle was 15°). The inside of the bottom is fixed, $\Omega_0 = 0$. The mechanical interaction of the liquid with the bottom is provided by the rotation forces with a sharp jump in the zone between the appropriate parts of the layer, the outer 'periphery' with an outer diameter D = 41 cm and the inner 'core'.

The apparatus was equipped with a black and white TV camera. The fields of the perturbation were obtained by optical densitometry, which works well in the presence of the white bottom [967, 1099]. The liquid is common water, tinted green, so that through a red filter layer thickening appeared darker than the field with lower surface density. Relative perturbations in the experiments were less than 20%.

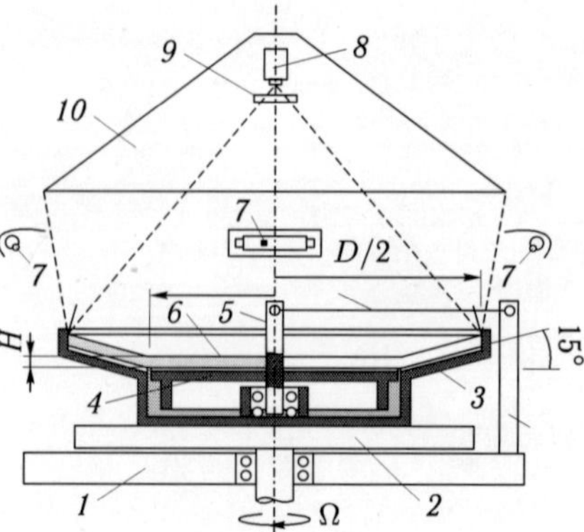

Fig. 6.15. The experimental setup: 1) the fixed plate, 2) swing-rotating table with angular velocity Ω_0, 3) rotating vessel with a conical bottom, 4) fixed flat part of the bottom, 5) axial column 8 mm in diameter, immersed in the liquid layer, 6) a layer of greenish liquid (with a free surface), 7) white halogen lamps, 8) black and white TV camera, 9) red filter, 10) black-matte screen.

The experiment began with the steady state in which the liquid completely covered the bottom layer which was as thin as possible. The core remained motionless, while the periphery rotated as a rigid body. The Mach number $M = \Omega_0 R_0/(gH)^{1/2}$ was very slowly reduced by lowering the speed Ω_0 or by increasing the thickness of the liquid H, adding the liquid to the vessel. (The value of H was measured near the centre of the core). For some value $M_0 \equiv M(R_0)$ the layer thickness became perturbed in a circular area of the core, bordering on the rotating periphery. Perturbations appeared at the periphery in the form of nearly radial 'spokes' changing the length and width along the azimuth at different speeds of rotation. Over time, even if the value M_0 remained unchanged, there was a disorder in a previously stable single structure (wave pattern) rotating in the same direction with a constant angular velocity Ω_{ph}. With the rotational symmetry of order m, the structure clearly shows the development of a hydrodynamic instability with the azimuthal wave number m (Fig. 6.16 shows a typical structure with $m = 6$, there are also structures with different number m).

With a further decrease of M_0 the system comes back into order and related structures with higher and higher m appear. Increasing the number M_0, we observed the reverse transitions to structures with progressively diminishing number m. Both transitions – decrease and increase of M_0 – show signs of hysteresis, which indicates the apparent non-linearity of the observed structures. The wave numbers of the modes observed ranged from 3 to 10. The modes, appearing at their stability-instability threshold, can be efficiently controlled. The number of modes depends on many factors: background, interim regime of change of M_0, method of adding the fluid, viscosity, etc. Note that the most resistant to changes in the conditions were the modes 5 and 6. Other characteristics have changed in the same way as it was in the case of centrifugal instability [588]: Ω_{ph} decreases with a decrease of Ω_0 and increase of m. If M_0 is sufficiently small the system is in the zone of predominance

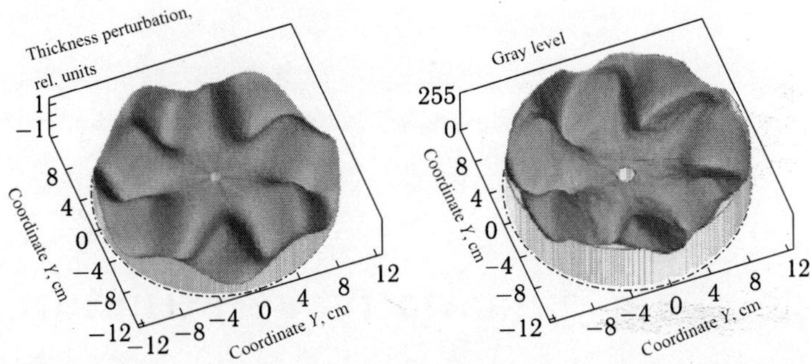

Fig. 6.16. Typical structures with mode $m = 6$, generated by the instability of super-reflection. (a) Theoretically calculated eigenfunction in the process of development of super-reflection instability [573, 574] ($H = 5.0$ mm, $\Omega_0 = 4.83$ rad s^{-1} ($M_0 = 2.62$), $\Omega_{ph} = 4.25$ rad s^{-1} ($M_{ph0} = 2.3$), the characteristic width of the region close to H). The direction of rotation – clockwise. (b) the perturbation of the thickness of the liquid layer in the experiment [576] in arbitrary units (H = 4.0 mm, $\Omega_0 = 3.98$ rad s^{-1} ($M_0 = 2.4$), $\Omega_{ph} = 3.13$ rad s^{-1}, ($M_{ph0} = 1.90$). There is good agreement between the theoretical and experimental data.

of the instability of tangential discontinuity (Kelvin–Helmholtz), characterized by the development of the azimuthal modes $m = 1$ and 2.

To identify the observed instability, it will be attempted to compare the structure shown in Fig. 6.16b with the eigenfunction of super-reflection instability, calculated in the theory [573, 574] for conditions that are physically close to the experimental conditions (Fig. 6.16a). The eigenfunction appears as a superposition of two-coherent systems. In turn, each system consists of six leading spiral waves moving outward toward the velocity shear At some distance from the width the waves moving inward are intense than those moving out. This is exactly the same as what was expected from theory [573, 574]. The similarity between pairs of figures is quite remarkable, given that we compare the *apriori* results of non-linear structures in the generation of which viscosity exerts an effect, with the results of *de facto* linear inviscid theory.

7

Large-scale galactic wave structures

7.1. Dynamic factors and the spiral structure

The effectiveness and direction of dynamic processes, and the observed spiral patterns undoubtedly caused by them, depend on the ratio of three physical factors:

1. The relative and absolute content of the gas. This factor completely determines the possibility of star formation and the existence of the spiral structure. This applies to the flocculent spirals, which very weakly perturb the disk subsystem of old stars and are the product of complex processes in a dynamically cold subsystem of gas, dust and young stars. Apparently, the role of the gaseous component for global spiral arms (grand design) is also a decisive factor. A good example of this are lens-shaped galaxies with a bar (SB0). In them outside of the central jumper the stellar disk, as a rule, does not reveal any large amplitudes of spiral perturbations: a bar does exist – spirals no.
2. The ability of the stellar disk to respond to an external impact is determined by its 'temperature, which in the first approximation is characterized by the variance of the speeds of the three velocity components (c_r, c_φ, c_z). This is associated with the possibility of development of gravitational instability in the stellar disk in its plane and with respect to flexural perturbations. For hot stellar disks the spiral pattern is a less common phenomenon. In a cold stellar disk ($c_r / V \lesssim 0.3$ at $L_\sigma \lesssim r \gtrsim 2L_\sigma$) in the presence of gas the spiral pattern is the rule. For galaxies that are visible from edge-on, the temperature of the disk is reflected in the relative thickness of the stellar disk.
3. Spheroidal subsystems (halo + bulge) are to some extent always present in S-galaxies. The relative mass of the spheroidal subsystem in relation to the disk system can vary in a wide range $\mu = M_s / M_d \simeq 0.5–10$. The disk component is dynamically more active and parameter μ significantly affects all processes in the stellat–gas disk. Since the gravitational influence of the bulge is limited to the central region of the disk and the halo effect spreads over a large distance of the galaxy, it is important to consider separately the relative mass of the bulge M_b / M_d and halo M_h / M_d separately. This is already apparent in the Hubble sequence based on the correlation between the massive bulge and the geometry of the spiral pattern.

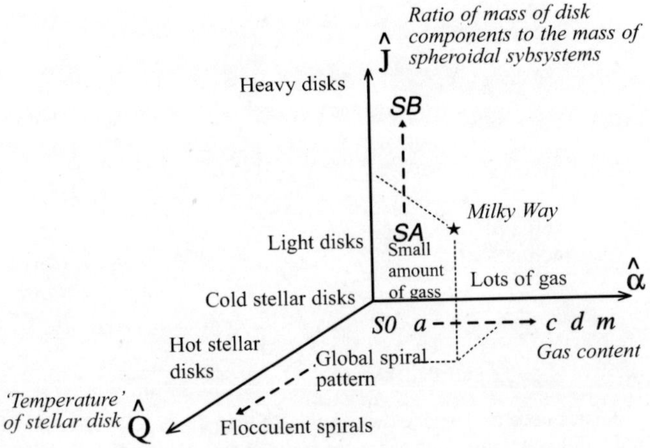

Fig. 7.1. Dynamic classification and spiral structure of galaxies [367] using the notation of the Hubble (Fig. 1.2) and de Vaucouleurs classification (Fig. 1.2)

It should be emphasized that these factors are interrelated. For example, a very cold / thin disk can not be imagined without a massive bulge / halo.

These factors can, for clarity, be presented in the form of a three-dimensional diagram, where the axes are (Fig. 7.1): $\hat{\alpha}$ – the fraction of the gas; $\hat{Q}$ – the parameter of gravitational stability, suitable for the characteristics of the dynamic temperature of the disk $\hat{J}(M_b / M_d, M_h / M_d)$ characterizes the dynamic activity of the disk component through the masses of spheroidal subsystems to the disk.

7.2. The spiral–vortex galactic structures

We proceed to the presentation of the results that led to the discovery of spiral-vortex structures in real galaxies [125, 241, 569, 571, 579, 581, 586, 587].

7.2.1. Anticyclones in the galaxy Mrk 1040

The initial discovery of the vortex structures in the differentially rotating disk was made in field experiments with shallow water (section 6.2.5).

The identity of the equations for shallow water and the gas galactic disk [239], the quantitative agreement of the experimental results on the dependence of the number of spirals on the Mach number[163] at the velocity jump, experimental confirmation of the purely 'astronomical' hypotheses about the origin of branching and leading spirals – all point in favour that the rotating shallow water in Spiral equipment perfectly simulates the dynamic processes and structures in the gas disks of spiral galaxies. Therefore, the appearance of anticyclones on the surface of shallow water between the spiral arms only one or two times smaller than the arms, did not leave doubt that similar vortices exist in spiral galaxies. If so, then the challenge is finding them.

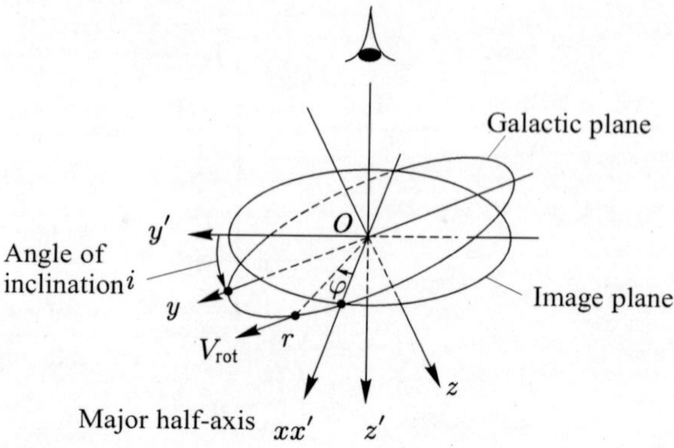

Fig. 7.2. Coordinates used to describe the position of the galaxy relative to the observer and the positions of individual sections of the gas in the plane of the galactic disk. The angle of inclination of the galaxy i – the angle between the image plane (the plane perpendicular to the line of sight) and the plane of the galaxy; x', y' – Cartesian coordinates in the image plane; x, y – Cartesian coordinates in the plane of the galaxy; r, φ – polar coordinates in the plane of the galaxy, r – galactocentric radius; φ – galactocentric azimuthal angle measured from the major line of the galaxy (also known as 'line of nodes') – lines of intersection of the galactic plane and the image plane. The direction of the line of sight coincides with the axis z.

The question arises: what conditions must be satisfied by the spiral galaxy as the first candidate for the detection of vortices?

As can be seen from the results of the experiment, the centres of anticyclones are located at the velocity jump, and structures – spirals and vortices – are caused by hydrodynamic instability. We should therefore find a galaxy with a large jump in speed which obviously would cause this instability.

But this is not the only condition that must be satisfied by the spiral galaxy as a candidate for the detection of vortices. It must be 'suitably oriented' in space relative to the observer (Fig. 7.2). We discuss this in more detail.

Anticyclones which we want to find in the coordinate system relatively fixed in relation to the spiral arms are characterized by the presence of closed streamlines in the plane of the galactic disk with centres at the corotation radius. At this radius, the disk rotation speed relative to the spiral arms is equal to zero. Different from zero is only the perturbed velocity of gas clouds caused by the same instability, which formed spiral density waves and vortices.

If we could measure, for example, only the perturbed azimuthal velocity, then for the two-arm galaxy, we would see two hypothetical points symmetrically located on both sides of the centre at the corotation circumference, in the internal and external neighborhood of each of which the perturbed azimuthal velocities have opposite directions. This property can be used as the criterion of observation of the anticyclone in the galactic disk.

However, for each cloud, we measure using the Doppler shift only the radial velocity component, which lies on the line connecting the cloud to the telescope. In general, except for the azimuthal velocity, two other velocity components of the

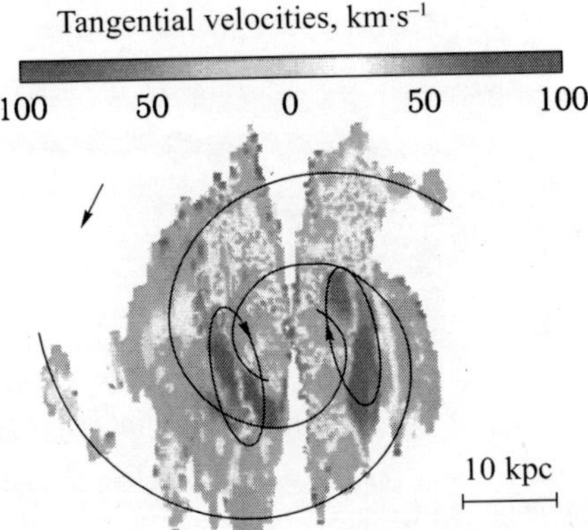

Fig. 7.3. Map of tangential velocities of the galaxy Mrk 1040 after 'de-projection' (i.e. turning in such a way as though we are seeing the galaxy, perpendicular to the disk). We have: $v \sim v_{obs}/\cos(\varphi)\sin(i)$, where v_{obs} is the velocity along the line of sight, φ is the azimuthal angle measured from the line of nodes. Anticyclones are marked by a closed solid line in the form of ellipses with arrows, and their centres are located relatively close to the line of nodes (in this case – the horizontal line).

cloud contribute the radial velocity: the radial and z-component (directed along the axis of rotation). However, if the gas galactic disk was oriented so that the azimuthal velocity perturbation in the vicinity of the centres of the vortices would have been directed along the visual ray, and the radial and z-components – perpendicular to it, then the contribution of the latter to the radial velocity would be equal to zero. We would measure only the azimuthal velocity. There is no contribution of the z-component velocity, if we observe the disk almost edge-on ($i = \pi / 2$). The contribution of the radial velocity in the region of the centres of the vortices is absent if centres of the vortices lie almost on the line of nodes (major dynamic axis) of the galaxy (Fig. 7.2).

These conditions of orientation with respect to a ground observer are satisfied by the galaxy Mrk 1040, which has an abnormally large and sharp jump in the speed of rotation. Indeed, the slope of the galaxy is $i = 73^{\circ}$, which differs by only 17° from the edge-on galaxy. As we see from Fig. 7.3, the centres of the anticyclones (indicated by the closed curves with arrows) located near the line of nodes [14].

The jump in the radial velocity along the line of nodes, i.e. where the vortices are localised, as shown in Fig. 7.4, is huge – more than 150 km /s, i.e. $\Delta v / v > 0.6$. Averaged over the azimuth, the rotation velocity has a jump $\Delta v/v \approx 0.3$. The radial velocity gradient is also abnormally large:

$$d(\ln v) / d(\ln r) \simeq -3{,}1, \tag{7.1}$$

which corresponds to a strong hydrodynamic instability [149].

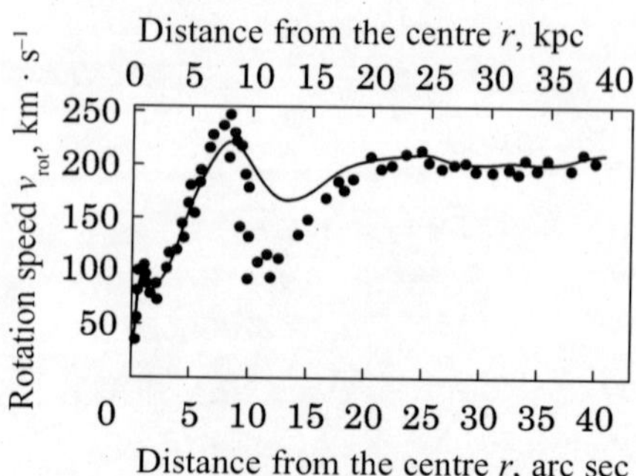

Fig. 7.4. Comparison of the rotation curve averaged over the azimuth with the distribution of velocities along the line of nodes (circles) in the galaxy Mrk 1040.

7.2.2. On the possibility of the existence of the vortex structure in the solar neighborhood

As already mentioned, to identify galactic anticyclones it is necessary to find the velocity field in a coordinate system rotating together with a spiral pattern. In experiments on the Spiral equipment, the rotation speed of the spiral arms could be measured directly. However, for real galaxies the definition of this rate is a separate and very difficult problem. The fact that the characteristic time of rotation of galactic disks is about a hundred million years, it is clear that in any reasonable observation time it is not possible to record the amount of displacement of the spiral arms due to their rotation.

The problem of determining the speed of rotation of the spiral pattern is equivalent to determination of the corotation radius, i.e. radius at which a galactic disk rotates with the same speed as the spiral. If a spiral wave is formed as a result of the instability caused by the velocity jump, its corotation radius coincides with the jump [149, 239]. The presence in the galaxy Mrk 1040 of an anomalously sharp jump in the rotation curve and the satisfied conditions of hydrodynamic instability immediately removed the question of the position of the corotation radius of the galaxy and facilitated the identification of giant anticyclones there [14]. Nevertheless, for most external galaxies the problem of the position of the corotation radius remains open.

Not surprisingly, the next on the list of candidates for detecting vortex structures is our own Galaxy, for which the position of the corotation radius has been studied most extensively.

To determine the corotation radius in this case, the data on the so-called gradient of stellar ages are used. Stars form mainly in the spiral arms, so they there is an increased concentration of young, newly born stars. With time, stars travel further and further from the spiral arms. Since the angular velocity of rotation of the gas disk around centre of the Galaxy decreases with decreasing distance from the centre, a spiral pattern rotates as a unit with a constant angular rate and the region of the disk inside the corotation circle travels ahead of the spiral arms, and the area outside the disk lags behind. Stars have a 'memory' of the gas disk from which they

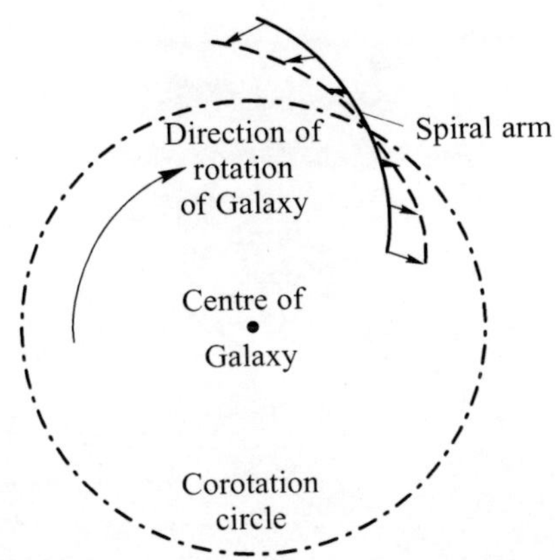

Fig. 7.5. Age gradient of stars near the corotation circle. The area of high concentration of young stars (shown by the solid line) coincides with the observed localization of the spiral arm. Older star, born earlier in the arm have moved fromthe arm to the region shown by the dashed line. Their tracks are marked by small arrows, the length and direction of which are determined by the rotation speed of the disk in the reference frame connected with the arm; the large arrow shows the direction of rotation of the Galaxy in the laboratory frame. Inside the corotation circle (dotted circle), the disk rotates faster than the spiral arms, so the direction of the small arrows is the same as the direction of the large arrow. Outside the corotation circle the direction of the small arrows is the opposite. The age gradient of the stars is oriented in the direction of the small arrows, so the radius at which it changes direction corresponds to the corotation radiu (dotted circle).

were formed, so that at distances from the galactic centre smaller than corotation radius, the stars are moving faster than the spiral arms, and at distances larger than the corotation radius – slowly. As a result, the direction of the gradients of the age of stars in both sides of the corotation circle should be the opposite (Fig. 7.5, see the explanation in the caption).

The above-described feature of the age gradient was found in the solar neighbourhood [3, 51], indicating that the position of the corotation radius close enough to the solar circle (orbit of the Sun around the centre of the Galaxy). Research of the field of radiation velocity of Cepheids in the solar neighborhood has also shown that in this region the azimuthal velocity of the spiral pattern in the Galaxy coincides with the rotaton speed of the disk [927]. Thus, we see that different researchers have come to the conclusion that the Sun is near the corotation radius[1]. In addition, the observed rotation curve of the gaseous disk of our Galaxy has a depression just in the vicinity of the solar orbit. Based on these facts, it was suggested [566] that in the vicinity of the galactic orbit of the Sun there are centres of giant anticyclones.

This hypothesis was tested using data on the radial speed of 316 molecular clouds, 256 classical Cepheids and 106 young open clusters [581].

Figure 7.6 shows the model velocity field in a reference frame rotating with the angular velocity equal to the average angular velocity of rotation of stars near the

[1]A different conclusion regarding the position of the corotation radius in the vicinity of the end of the bar of the Galaxy has also been made ($r_c \simeq 3$ kpc).

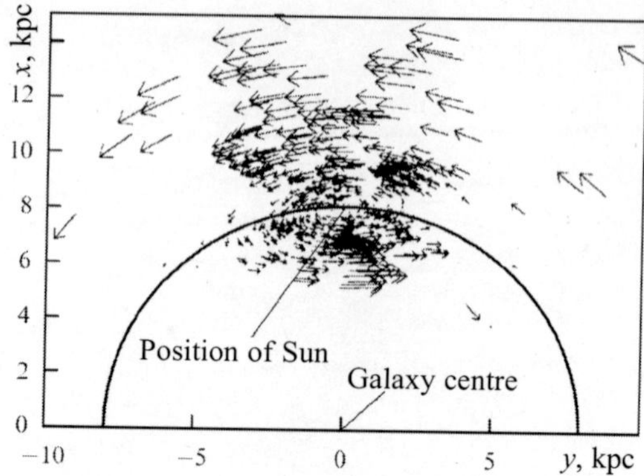

Fig. 7.6. Modelling the velocity field obtained by the movement of stars and gas [581].

Sun [581]. This velocity field has the form of an anticyclone centred in the solar neighborhood. It should be noted that in our Galaxy study of the velocity field is prevented by interstellar dust which absorbs the light. This leads to the fact that we can investigate only a small region of the Galaxy and almost do not see the parts of the Galaxy on the opposite side of its centre.

Observations of external galaxies allow us to see the entire galaxy simultaneously, which makes it possible to more accurately determine its characteristics. Studies have therefore been continuing in this direction.

7.2.3. The galactic vortices – as a property of a spiral density wave

Detection of predicted anticyclones in the gaseous disk Mrk 1040 and the solar vicinity of our Galaxy was quite natural. In both cases, the observed rotation curve of the gas disk has a region of a sharp fall (depression) at some radius. The presence of such a region seems to indicate the existence of significant local deviations from circular motion. Vortex structures were first discovered as a result of experimental simulation in Spiral equipment.

However, as mentioned in section 7.2.1, the galaxy Mrk 1040 shows an abnormally large velocity jump, and in the presence of a close satellite. In the case of the solar neighborhood the lack of statistical data leads to errors which are so large that it cannot be concluded that the discovery of anticyclones has been reliably proven.

In this case, the found anticyclones could be considered as some special structures inherent in only peculiar galaxies with sharp jumps in speed and not as a constant attribute of any spiral galaxy with the wave nature of the spirals.

A simple qualitative argument shows that it is not so and gas vortices are a fixture of the spiral density waves, regardless of the nature of forces acting in the disk: hydrodynamic, gravitation or tidal forces, caused by, for example, a bar or a satellite of the galaxy (for a more rigorous, mathematical proof of this fact, see [125]).

Suppose that as a result of a collective mechanism (instability) in the galactic disk there are perturbations of velocity and density. While the perturbation amplitudes are small, their growth is described by the linear theory of instability. Then all the perturbed quantities can be easily expressed in terms of one, i.e. they are proportionate to each other. The field of perturbed velocities of the disk, transferring a part of mater from some areas of the disk to the other, creates a certain local thickening in the form of spiral density arms, while in others – spiral regions of reduced density.

As a resut of a further increase in the amplitude of the perturbed quantities the density of the gas in the spiral arms can grow several times to form a non-linear density wave, which we observe as a spiral arm. However, it follows from the observations that the values of perturbed velocities in both disks, gas and stellar, are much smaller than the speed of rotation. Increase of the perturbations causes stabilization of the instability and the spiral wave becomes quasi-stationary. Features of the velocity field, caused by such a wave, can be studied effectively in the reference frame rotating with the angular velocity of the spiral pattern. This rotating coordinate system is remarkable primarily because the spiral arms in it are fixed, i.e. the perturbed surface density of the disk and the components of the perturbed velocity associated by linear relations with the perturbed linear density can be written as functions independent of time:

$$\begin{aligned} \tilde{\sigma}(r,\varphi) &= C_\sigma(r)\cos(2\varphi - F_\sigma(r)) \\ \tilde{v}_r(r,\varphi) &= C_r(r)\cos(2\varphi - F_r(r)), \end{aligned} \tag{7.2}$$

where r is the galactocentric radius, φ is the azimuthal angle measured from the major dynamic axis of the galaxy (scc Fig. 7.4), C_i is the amplitude and F_i the phase of the i-th parameter. For definiteness, we restrict ourselves to the case of a two-arm galaxy.

Since the disk of the galaxy rotates differentially, the internal part of the disk (up to the corotation radius) will rotate faster than the spiral pattern and external, on the contrary, slowly. The observer, located at the corotation radius r_c, will see that the angular speeds of the disk on both sides of the corotation circle are directed in opposite directions. In accordance with (7.2), for each value of the radius of the perturbed surface density $\tilde{\sigma}(\varphi)$ four times (twice the number of spiral arms) changes its sign with a change in the azimuthal angle. Accordingly, the radial velocity also changes its sign four times. As a result, near the corotation region, where the value of the total azimuthal velocity is low, there are closed streamlines – vortices (see Fig. 7.7).

As the gas moves in them in the opposite direction to that of rotation of the disk, they are anticyclones (Fig. 7.8b).

It should be emphasized that these anticyclones can not be considered as a mathematical feature of the adopted model of rotation, in which the position of the closed trajectories is completely determined by the definition of the angular velocity of the reference system. Only in the region of actual corotation the interstellar gas in its movement does not cross the spiral branches, so the gas flows (assuming a low velocity dispersion compared to the speed of rotation) are destined to move there permanently or in the vicinity of a particular spiral arm (cyclones, in Fig. 7.8a), or between the branches (anticyclones, Fig. 7.8b). The presence of vortex motion in

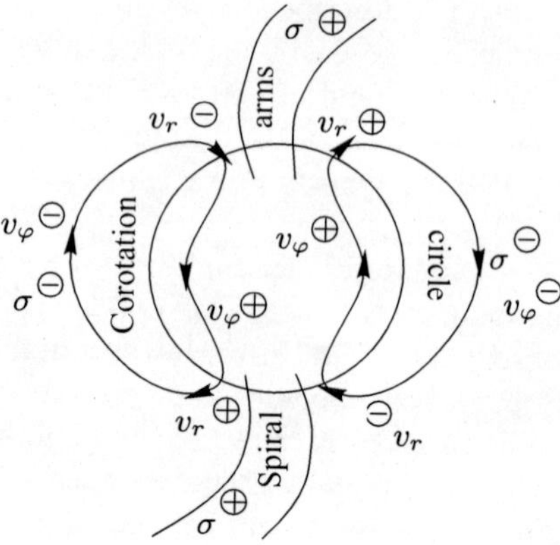

Fig. 7.7. Position of two anticyclones near the corotation radius in a coordinate system rotating with the angular velocity of the two-arm spiral pattern. The sign of the perturbed density varies on the circle along the azimuthal angle four times (on the arms – a plus sign, between them – minus). Schematic trajectories of the particles, and the arrows show the direction their motion in anticyclones.

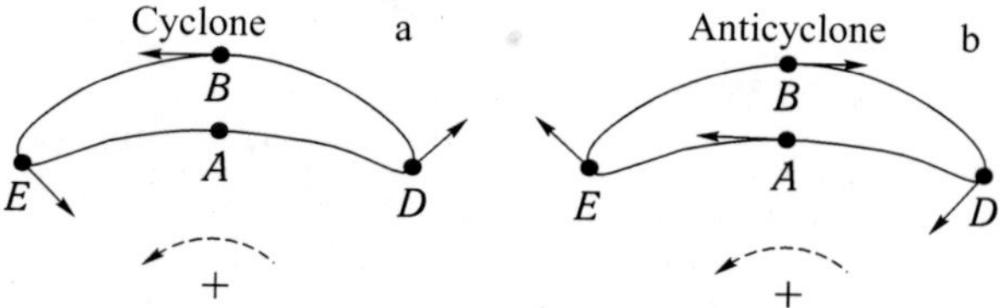

Fig. 7.8. Schemes of cyclonic (a) and anticyclonic vortices (b). The solid lines show the trajectories of material points, arrows indicate the direction of their movement; dotted line with an arrow shows the direction of rotation in the galaxy. + symbol is placed in the centre of rotation.

the corotation region is a physical feature of galaxies. Therefore, the evolution of star formation and gas content in the corotation region are of particular interest.

Thus, throughout the disk, small pertubed velocities do 'their work', collecting material in glowing arms, but the arms somewhat do not appear, being 'crushed' by the large circular velocity of the disk. Spiral perturbed velocities appear in the form of giant anticyclones only at the corotation radius where the circular velocity of the gas is small in the coordinate system of the rotating spiral. It demonstrates that the spirals and vortices are different manifestations of a single collective mechanism, whose name is – density waves. The above-described qualitative picture of the formation of anticyclones in the disks of the galaxies is based on only one assumption – the assumption on the wave nature of the spiral structure. Therefore, the vortex structure should be a universal attribute of spiral galaxies, regardless of the composition of the disk matter (gas or stars) and the mechanism of generation of density waves.

7.2.4. Position of the centres of stationary vortices

Thus, the above analysis indicates that giant anticyclonic vortices are a full-fledged component of the large-scale structure of spiral galaxies. However, interest in the galactic vortices is caused not only the fact that there are new structures in spiral galaxies but also ny the unique information that can be obtained from their analysis. A remarkable feature is the giant anticyclones dependence of the position of their centres with respect to the spiral arms (Fig. 7.9a–c) on the mechanism of formation of wave patterns.

The reason for this dependence is easily understood from general considerations. Indeed, the existence of stationary non-circular motions requires a certain balance of forces. The balance of forces in turn is determined by the density distribution – spiral pattern – and by the forces which dominate – self-gravitational forces or pressure.

To understand the general laws defining the azimuthal position of the vortices relative to the spiral arms, we consider the areas in which the gas moves strictly radially (Fig. 7.10). The distribution of the specific angular momentum, defined in the zeroth approximation by the rotation speed of the disk, $M = \Omega r^2$, inhomogeneous along the radius. As a result, the transition of the gas particle to another radius requires a change in its angular momentum, i.e. azimuthal forces acting on the particle should 'twist' it or, conversely, inhibit, depending on whether it is moving in the direction of increasing or decreasing torque of the disk. Figure 7.9 shows the direction of the forces acting on the particle for the case when the angular momentum increases with the radius.

The azimuthal force acting on a particle of gas is determined by the azimuthal behaviour of the perturbation pressure $\tilde{P}$ and gravitational potential $\tilde{\Phi}$ caused by the density wave and, possibly, by the external tidal force F_{ext} acting on the gas in the presence in the galaxy of a satellite or a large bar:

$$F_{\varphi} = -\frac{1}{\rho r}\frac{\partial \tilde{P}}{\partial \varphi} - \frac{1}{r}\frac{\partial \tilde{\Phi}}{\partial \varphi} + F_{ext}\,. \tag{7.3}$$

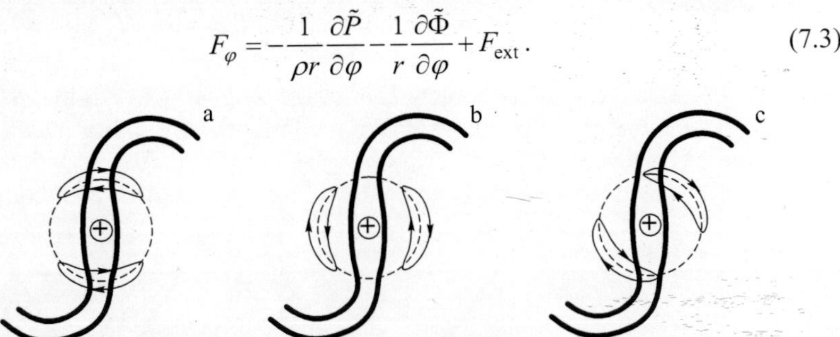

Fig. 7.9. Schematic representation of the locations of anticyclones relative to the spiral arms in the two-arm density wave: a – vortices in spirals; b – between the spirals; c – an intermediate position of the vortices. If the angular momentum of the disk in the corotation region increases with radius, then: the case a corresponds to the pressure forces being dominant; b – self-gravitating force, and the case c occurs when the tidal forces play a significant role in the corotation region. Solid lines with arrows show the motion of a material particle in a vortex; bold lines are the lines of equal surface density in the spiral arms.

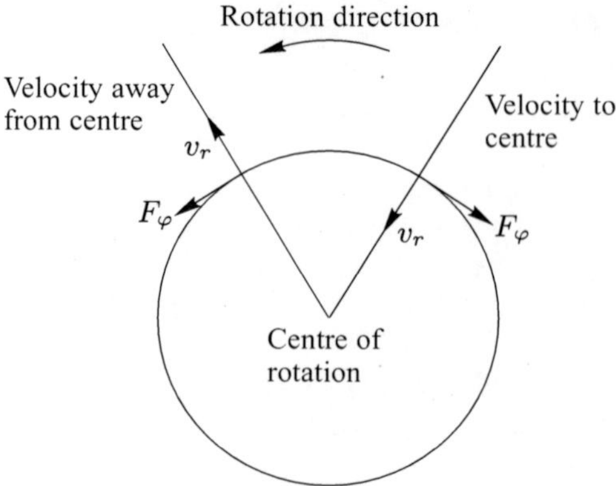

Fig. 7.10. When the angular momentum increases with radius, and if the radial velocity v_r is directed towards the centre, the azimuthal force F_φ is oriented against the direction of rotation, and if v_r is directed from the centre, then F_φ is oriented in the direction of rotation.

Therefore, if the pressure forces are predominant, then the balance of the forces necessary for the existence of a stationary anticyclonic vortex will be achieved by placing the centre of the vortex at the maximum of the perturbed pressure (Fig. a). Since the position of the maximum pressure in the spiral wave coincides with the maximum density, this means that the vortex is located on the spiral arm (Figure 7.7a). The same pattern is well known in meteorology: anticyclones in the Earth atmosphere are related to the field of higher pressure, and low pressure regions are associated with cyclones.

If the gas density in a certain section of the disk increases, the test particles in the disk will have an additional gravitational attraction to this volume, i.e. the perturbed gravitational force is directed toward increasing density (Fig. 7.11b). Consequently, if the spiral wave is of the gravitational nature and the corotation region is dominated by self-gravitational forces, as usual in spiral galaxies, the centres of the anticyclones should be located at the minimum density, i.e. between the spiral arms (Fig. 7.9b).

If the corotation region is characterized by a strong influence of the external tidal forces, the centres of anticyclones will not coincide with with a maximum nor a minimum of density (Fig. 7.9c).

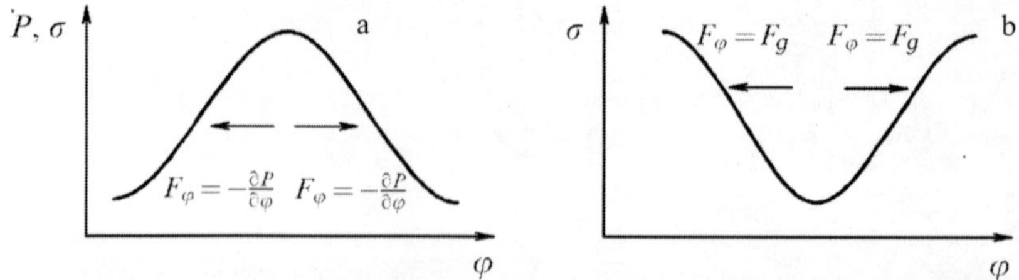

Fig. 7.11. If the azimuthal force is directed from the centre of the vortex and hydrodynamic pressure forces prevail in the disk (pressure gradient), in the centre of the vortex there should be the maximum pressure and surface density (a), and if gravitational force F_g dominate, then the centre of the vortex should contain the minimum of surface density (b).

Table 7.1. The location of the vortex centre relative to the spiral branches depending on the direction of growth of the relative angular moment M of the disk and on which of the forces dominates: hydrodynamic pressure force or gravitational force

Dominant force	Position of the centre of the vortex	
	$dM/dr > 0$	$dM/dr < 0$
Pressure gradient	On spiral	Between spirals
Self-gravitation	Between spirals	On spiral
External gravitational force	The centre of the vortex can be displaced in relation to the extremum of density	

If the specific angular momentum in the corotation region decreases with radius, which is possible in the case of a very sharp drop in the angular velocity of rotation, the azimuthal direction of the force necessary for stable existence of an anticyclone is reversed compared with that shown in Fig. 7.10. Accordingly, the location of the centres of the vortices relative to the spiral arms will also be reversed (Table 7.1).

The available information on the laws of manifestation of galactic anticyclones is fully consistent with the above analysis. Laboratory simulation of shallow water [183] was carried out for the case when the corotation radius shows a sharp drop in the rotation speed, which corresponds to the angular momentum decreasing wth radius, $dM/dr < 0$. Therefore, the vortex structures were observed between the spirals. Numerical simulation (see, e.g. [16]) showed that in the gas galactic disks with rotation curves that have a sharp jump of angular velocity, the vortex structures are located between the spiral density arms. It is between the spirals where the anticyclones were found in the galaxy Mrk 1040 (see Section 7.2.1). [126] describes the numerical simulation of hydrodynamic instability, which leads the generation of spiral waves in disks with sufficiently smooth rotation curves, $dM/dr > 0$. The vortex structure were excited simultaneously with the spiral arms and placed on them.

The problem of closed banana-like trajectories of the stars in the gravitational field of a spiral wave, corresponding to the vortex structures in the stellar disk, was considered by Kontopulos [478]. For small amplitudes of the spiral gravitational potential in the disk whose momentum increases with radius, he found that the vortex structures are located between the spirals. Thus, the position of the vortices in the stellar and gaseous self-gravitating disks is the same (see Table 7.1, the row 'self-gravitation').

Naturally, the question may arise: if the vortex structures should exist in all galaxies where there are spiral density waves, and their detection gives much valuable information the dynamic properties of the galaxy, why we can refer to only one or two objects for which the indication of the existence of such structures follows from the observations?

The reason lies in the incommensurability of the efforts of detection of vortex structures in comparison with the discovery of the spirals. In the first case it is necessary to solve complex tasks both purely observational and related with the methods of data processing. In particular, it is necessary to determine such an important fundamental parameter as the corotation radius of the spiral pattern.

As mentioned in chapter 6, if the spiral structure formation is due to the hydrodynamic instability at the speed jump, the position of the corotation radius coincides with that of the jump. Note that the observed jump on the rotation curve does in some cases reflect the actual feature of the changes of the equilibrium circular velocity with variation of r, and is the consequence of perturbations of the velocity field by the density wave[2]. However, if the rotation curve of the galaxy is smooth, then the corotation radius can not be found independently. In [126] it was shown that in the latter case, to determine the corotation radius and detect giant anticyclones, it is necessary to restore the field of the total (three-component) velocity vector in the gas disk of the galaxy.

7.3. Construction of the three-dimensional velocity field of gas from observations

7.3.1. Field of radial velocities from the observations

The problem of determining the total velocity vector is caused by the fact that in astronomy for distant objects we can directly determine the velocity only by the Doppler shift of the observed spectral lines. Thus, only one component of the velocity is known – along the line of sight. To restore the other components of velocity we need a model of the object.

As we shall see below, this turns out to be only one additional conditions, implying that for the considered galaxies the perturbed surface density and velocity can be presented in the form of trigonometric functions (see also equation 7.2)):

$$C(r)\cos[m\varphi - F(r)], \tag{7.4}$$

where C and F are the the amplitude and phase, m is the azimuthal number equal to the number of arms of the galaxy.

Turning to the question of the validity of this assumption, first we note that the method of restoring the full vector field of velocity from the observed radial velocity field, described below, is valid only for galaxies with a regular spiral structure (grand design spirals). These galaxies clearly have a number of spiral arms, i.e. their structure is almost unchanged when turning through the angle $2\pi/m$.

To test the validity of the assumption (7.4), several independent observational tests were developed which were applied to a number of galaxies with a regular spiral structure, and in all cases they confirmed the possibility of describing the perturbed parameters of the gaseous disk of the galaxy in the form (7.4).

Let us explain why such a simplified representation for $\tilde{\sigma}$ and $\tilde{v}$ in real galaxies may be valid with satisfactory accuracy. The dynamics of gas galactic disks is described by a system of hydrodynamic equations. From this point of view approximation (7.4) means that, firstly, the system of equations describing the behaviour of galaxies with the regular spiral structure is nearly linear and, secondly, the general solution of this system can be represented as a Fourier series in azimuthal harmon-

[2]Or circular motions are superposed on the inhomogeneity of the circular velocity. For our Galaxy this problem is discussed in chapter 9.

ics, dominated by the term corresponding to a single harmonic with the azimuthal number *m*, which determines the number of spiral arms.

With regard to the correctness of the linear approximation in hydrodynamic description of gas galactic disks, it should be noted in these galaxies perturbed velocities are small (20–30 km/s) compared to the unperturbed velocity of rotation (typically > 200 km/s). With respect to the perturbed surface density it should be noted that the Fourier spectra of the non-axisymmetric perturbation of the surface brightness of the studied galaxies indicate the absolute predominance of the second Fourier harmonics above the rest.

According to the theoretical research [140, 141], nonlinear effects stabilize the instability which forms the observed structures, even at low amplitudes of the perturbed quantities, when the linear approximation still retains its validity for describing the relationships between individual parameters of the perturbation. This final state is dominated by the harmonic which initially had a maximum growth rate.

Although all physical theories require experimental validation, as a rule there is no complete synchronicity in the development of the theory and experiment. In our case, this inconsistency is manifested by the fact that the task of detection of giant anticyclones in the galaxies was formulated before any technical capabilities of its solution were developed.

The first attempts to find vortices in the velocity fields of gas disks of galaxies were carried out using the data available at the time – one-dimensional distribution of radiation velocities along the slit of the spectrograph with its different orientations. But this approach did not provide the required level of accuracy – vortex detection requires the knowledge of the radial velocities at several thousands of points of the disk to within a few kilometers per second. The situation changed only recently.

Significant progress in the observations of the radial velocity fields of gas disks of external galaxies has been made in the last 5–10 years. Today it is possible to determine the velocity along the line of sight at the same time at 10000 (or more) points of the disk. In addition, for regions of ionized gas the accuracy of a single measurement of the radial velocity is ~5 km/ s, and for the neutral gas is 2 km/s (in radio observations).

Typical fields of the radial velocity of gas of spiral galaxies and the appropriate brightness maps are shown in Fig. 7.12 (galaxy NGC 157) and Fig. 7.13 (Galaxy NGC 1365).

Studies of velocity fields of spiral galaxies have shown that velocity perturbations in the spiral arms are much less than in the circular ones. This allows us to apply successfully the linear or quasilinear theory for analysis.

According to the linear theory, the disks of the galaxies show have density waves with a definite azimuthal number. At the same time, it can be shown that all three components of velocity in a flat disk should contribute to the various Fourier harmonics of the observed radial velocity. The main idea of restoring the full velocity vector field is very simple – to be determined from observations of the Fourier coefficients of the radial velocity field and, using simple algebraic relations, find all three components of velocity of the gas in the galaxy. Within the same theory it

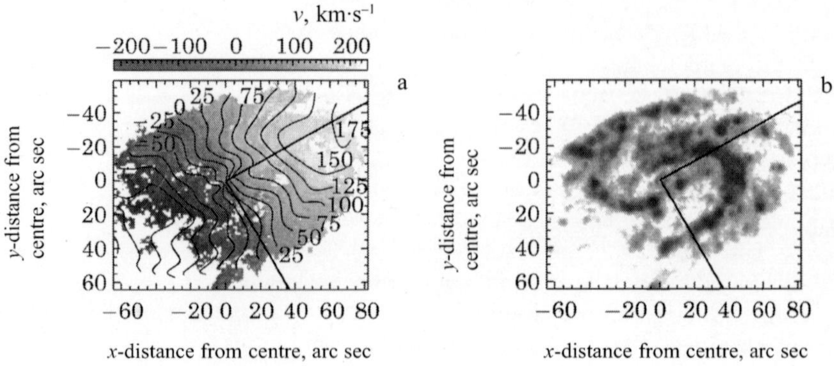

Fig. 7.12. Radial velocity field of NGC 157 galaxy, obtained from the Doppler shift in the lines of H_α, with the values of the isolines (a). Maps of surface brightness of NGC 157 in the line of H_α (b).

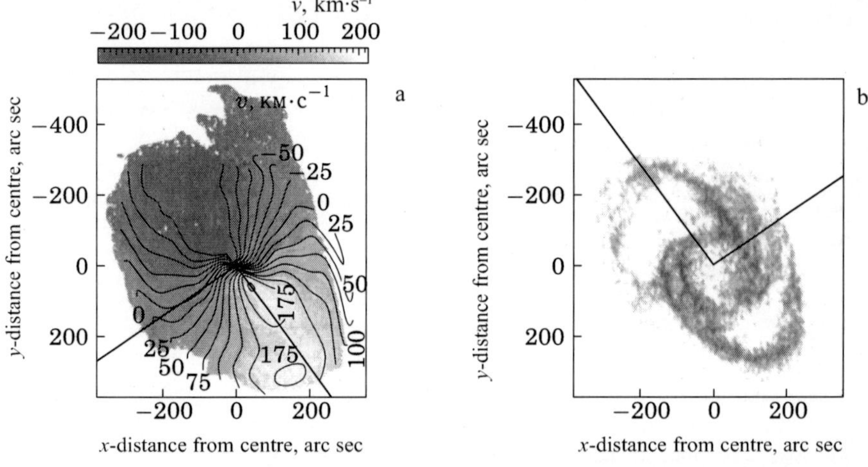

Fig. 7.13. Radial velocity field of NGC 1365 galaxy, obtained from the Doppler shift in the 21 cm line (HI), with the values of the isolines (a). Maps of surface brightness in the 21 cm line (b).

appears that the phase relations between different components of the velocity jump change their character in corotation. Therefore, the restoration of all three velocity components automatically solves the problem of determining the corotation radius of the spiral pattern.

7.3.2. How to determine the three velocity components of a single observable component?

Consider the problem of determining the velocity components in more detail. First of all, we find a relation between the observed radial velocity of the gas cloud C, moving in a galactic disk, with three components of its velocity.

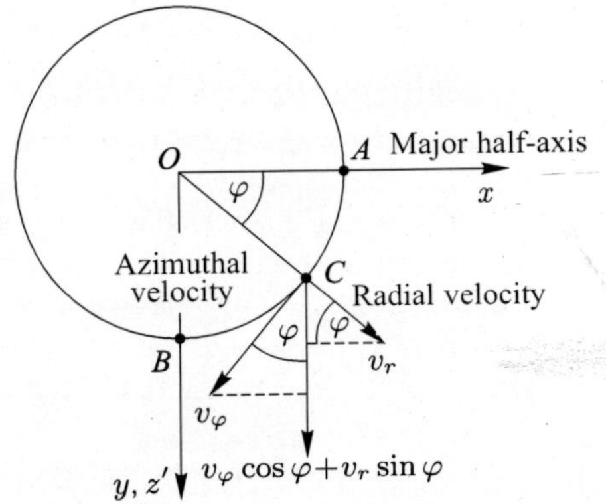

Fig. 7.14. The case where the galaxy is seen edge-on (inclination of the galaxy $\pi/2$. The velocity of the point C $v_{obs} = v_{sys} + v_\varphi \cos \varphi + v_r \sin \varphi$, where v_{sys} is the systemic velocity of the galaxy along the line of sight.

We first consider the simplest case, when the angle of inclination of the galaxy is $i = \pi/2$ (Fig. 7.14). The observer measures only radial velocity of point C, i.e. only the velocity component along the axis z'. The contribution to the radial velocity of the azimuthal velocity component $(v_\varphi)_{z'} = v_\varphi \cos \varphi$ and that of the radial velocity component is equal to $(v_r)_{z'} = v_r \sin \varphi$. The contribution to the radial velocity must also be provided by the systemic velocity of the galaxy along the line of sight v_{sys}. The vertical (perpendicular to the plane of the galaxy) component of the velocity v_z for $i = \pi/2$ does not give any contribution to the measured velocity. As a result, the radial velocity at $i = \pi/2$ is expressed through the velocity components as follows:

$$v_{obs} = v_{sys} + v_\varphi \cos\varphi + v_r \sin\varphi .$$

In the case of an arbitrary angle of inclination of the galaxy

$$v_{obs} = v_{sys} + (v_r \sin\varphi + v_\varphi \cos\varphi)\sin i + v_z \cos i .$$

For the density wave, which has a definite value azimuthal number, the perturbed density $\tilde{\sigma}$ and the perturbed velocities $\tilde{v}_r, \tilde{v}_\varphi, \tilde{v}_z$ can be written as

$$\tilde{\sigma}(r,\varphi) = C_\sigma(r)\cos[m\varphi - F_\sigma(r)], \tag{7.5}$$

$$\tilde{v}_r(r,\varphi) = C_r(r)\cos[m\varphi - F_r(r)], \tag{7.6}$$

$$\tilde{v}_\varphi(r,\varphi) = C_\varphi(r)\cos[m\varphi - F_\varphi(r)], \tag{7.7}$$

$$\tilde{v}_z(r,\varphi) = C_z(r)\cos[m\varphi - F_z(r)]. \tag{7.8}$$

Now, considering the speed of the disk of the galaxy as the sum of the perturbed and the unperturbed velocity:

$$\begin{aligned} v_r(r,\varphi) &= \tilde{v}_r(r,\varphi), \\ v_\varphi(r,\varphi) &= v_{\text{rot}}(r) + \tilde{v}_\varphi(r,\varphi), \\ v_z(r,\varphi) &= \tilde{v}_z(r,\varphi), \end{aligned} \tag{7.9}$$

and substituting the full velocity into the formula for v_{obs} we obtain the expansion of the observed velocity into the Fourier series in the azimuthal angle φ. A series will consist of eight terms, describing the zero harmonic proportional to v_{sys}, the first even harmonics proportional to v_{rot}, as well as the even and odd $(m-1)$-th, m-th and $(m+1)$-th harmonic. The corresponding Fourier coefficients are given by:

$$a_{m-1} = \frac{C_\varphi \cos F_\varphi + C_r \sin F_r}{2} \sin i, \tag{7.10}$$

$$b_{m-1} = \frac{C_\varphi \sin F_\varphi - C_r \cos F_r}{2} \sin i, \tag{7.11}$$

$$a_{m+1} = \frac{C_\varphi \cos F_\varphi - C_r \sin F_r}{2} \sin i, \tag{7.12}$$

$$b_{m+1} = \frac{C_\varphi \sin F_\varphi + C_r \cos F_r}{2} \sin i, \tag{7.13}$$

$$a_m = C_z \cos F_z \cos i, \tag{7.14}$$

$$b_m = C_z \sin F_z \cos i. \tag{7.15}$$

The rest of the expansion coefficients v_{obs} in a Fourier series are equal to zero.

From these relations it is clear that the contributions of various components of the velocity of the gas in the galaxy to the azimuthal Fourier harmonics of the observed radial velocity are as follows:

- the systemic velocity of the galaxy contributes to the zeroth harmonic of the observed velocity;
- the velocity of a purely circular motion contributes to the coefficient at the cosine of the first harmonic of the observed velocity;
- the radial and azimuthal velocity components contribute to the $(m-1)$-th and $(m+1)$-th harmonic of the observed velocity;
- vertical velocity contributes to the m-th harmonic of the observed velocity.

Thus, if the galaxy has m spiral arms, then the velocity field observed in the line of sight should contain, in addition to zeroth and first harmonics, two of each $(m-1)$-th, m-th and $(m+1)$-th harmonics.

Defining the Fourier coefficients from the observations of the velocity field, frm the system of equations (7.9)–(7.15) we find the unknown parameters of the vector velocity field in the galaxy.

Indeed, in this case it is easy to see eight non-zero coefficients of the Fourier series which is just about sufficient to determine the eight unknown functions:

$v_{sys}, v_{rot}, C_r, C_\varphi, C_z, F_r, F_\varphi, F_z$. The task of restoring the three velocity components from the observed radial velocity field can be considered solved for galaxies with the number of arms $m > 3$.

For a two-arm galaxy for the same eight unknown functions it is possible to determine only seven coefficients of the Fourier series: a_0, a_1, a_2, a_3, b_1, b_2, b_3. We require one more equation which links the unknown functions.

It was suggested that there are three independent ways to restore the three velocity components with different types of additional input relationships between the parameters of the density wave. The first two methods are based on the ratio between the phases of unknown functions, obtained from the hydrodynamic equations for a single assumption: the perturbation has the form of a wave with the azimuthal number $m = 2$.

The first of these methods is based on the fact that for relatively smooth (such that $\varkappa^2 \equiv 2\Omega\ (2\Omega + rd\Omega/dr) > 0$) rotation curves between phases radial and azimuthal velocity the following relations must be satisfied (before and after the corotation radius r_c)

$$F_\varphi - F_r = -\frac{\pi}{2} \quad \text{at} \quad r < r_c,$$
$$F_\varphi - F_r = \frac{\pi}{2} \quad \text{at} \quad r > r_c. \tag{7.16}$$

For a dramatically changing rotation curve ($\varkappa^2 < 0$) the following relations are satisfied

$$F_\varphi - F_r = \frac{\pi}{2} \quad \text{at} \quad r < r_c,$$
$$F_\varphi - F_r = -\frac{\pi}{2} \quad \text{at} \quad r > r_c. \tag{7.17}$$

The second method is based on the fact that the tightly wound spiral phase radial and azimuthal velocity must satisfy the relations

$$F_\varphi - F_\sigma = \pi \quad \text{at} \quad r < r_c,$$
$$F_\varphi - F_\sigma = 0 \quad \text{at} \quad r > r_c. \tag{7.18}$$

In particular, the above 'jumps' of the phase differences allow us to determine the corotation radius of the spiral structure directly from observations of the radial velocity field.

The third method is to determine the equilibrium rotation curve, obtained from the mass distribution in the galaxy. The method presented here is based on a study of the photometric brightness distribution in galaxies.

Here we consider the features of different methods of recovery and verification for real galaxies.

7.4. Anticyclones in spiral galaxies

7.4.1. Spiral perturbations in real galaxies and monochromatic waves of small amplitude

In line with the strategy proposed above, the first step in the recovery of the velocity field of the gaseous disk of the galaxies is verification of the applicability to real galaxies with a regular structure of the simplified description of a monochromatic perturbation (7.5)–(7.8). The majority of the spiral galaxies are two-arm. Therefore, verification consists of two parts:

1. The shape of functions (7.5)–(7.8) implies the dominance of the second Furier harmonic and the Fourier transform for each perturbed function.

2. As it follows from (7.12)–(7.15), all higher harmonics, starting with the fourth, are absent in the Fourier expansion of velocity along the line of sight V_{obs}. This means that in the Fourier spectrum of real galaxies, for which the velocity is tobe restored by the proposed method, all the harmonics, starting with the fourth, are negligibly small values as compared with the first three harmonics.

Figure 7.15 shows the spectra of the surface brightness maps of the NGC 157 galaxy in the near infrared (K) range (Fig. 7.15a) and NGC 1365 in the 21 cm line of neutral hydrogen (Fig. 7.15b). As can be seen, both galaxies show a distinct predominance of the second Fourier harmonic of surface brightness, which indicates the prevalence of the same harmonic in the spectrum of surface density $\tilde{\sigma}$.

The appearance in the spectrum of the radial velocity of the first three harmonics is a consequence of the projection effect. From a mathematical point of view this is due to the multiplication at projection on the line of sight of monochromatic (dependent on 2φ) components of gas velocity on also monochromatic functions of the argument φ (cos φ and sin φ). Thus, the verification of dominance in the

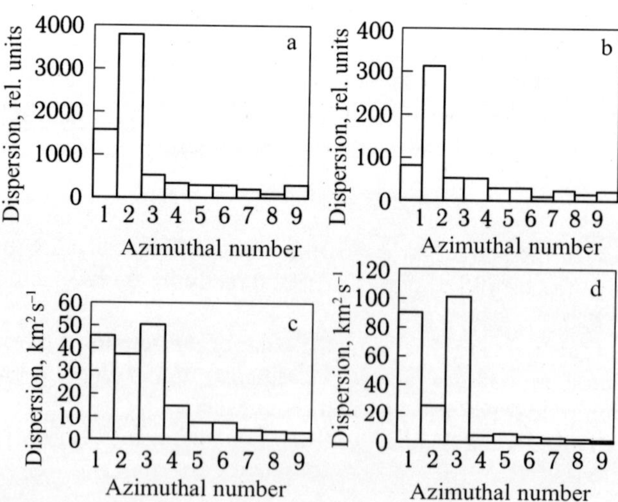

Fig. 7.15. Spectra of the axial–non-symmetric maps of surface brightness of the galaxies NGC 157 in the K-band (A) and NGC 1365 in the 21 cm line (b). The Fourier spectra of radial velocity fields of the galaxies NGC 157 (c) and NGC 1365 (d), are presented at the in Figs. 7.12 and 7.13.

spectrum of the radial velocity of the first three harmonics is, in essence, a test of closure in the amplitude of the second harmonic in the components of the velocity vector v_r, v_φ and v_z.

Figure 7.15c, d show the Fourier spectra of radial velocity fields of the galaxies NGC 157 and NGC 1365, less the contributions of the systematic velocity and purely circular motion of gas (more precisely, the contribution of the first odd harmonics). The first of the fields was obtained in the line of ionized hydrogen H_α, and the second – in the 21 cm line of neutral hydrogen. In both cases we can see the predominance of the first three harmonics, which confirms the good accuracy of the description of real spiral perturbations in the galaxies as monochromatic.

Data on reduced rates provide a direct proof of the wave nature of the observed spiral arms. This method also gives a method to obtain the solutions of some new problems, most important among them is the discovery of new galactic structures - giant anticyclones.

For a number of spiral galaxies it was possible, by processing the velocity along the line of sight, to recover the full velocit vector (NGC 157, NGC 6148, NGC 1365, NGC 3893, etc). In all cases, the results obtained by independent methods are in good agreement with each other. Thus, the theory leaves a 'hypothetical' level crossing into the area of proven astronomical facts.

7.4.2. Observational test: are spiral arms density waves?

The test presented in section 7.4.1 shows the possibility of describing real spiral perturbations of a single Fourier harmonic. This feature allows us to restore the full velocity vector field from the one-componentradial velocity field (see 7.3). However, the velocity fields reconstructed in such a way could be nothing more than a 'snapshot', not showing the complex processes taking place in the galactic disk. However, if the spiral perturbation is a density wave, the situation dramatically changes – the restored field after the transition to the coordinate system corotating with the spiral pattern, is stationary and fully characterizes the dynamic state o fthe disk.

Thus, it is necessary to verify by observations whether the spiral structure of galaxies are density waves. Earlier we wrote that the Lindblad hypothesis on the wave nature of the spiral arms solved the paradox according to which the spiral structure is not 'blurred' by differential rotation. An observational test was carried out to test the hypothesis about the wave nature of the spiral arm of our galaxy – a change of the sign of the age gradient of stars inside the corotation circle in the spiral arm (see Fig. 7.5). It was noted that this test is applicable only in the solar neighborhood of our galaxy. And what about the observational test of the wave nature of spiral arms in other galaxies?

In this section we describe one of the proposed observational tests to determine the wave nature of spiral galaxies [126, 581]. In the proposed method, besides the theory of natural oscillations of the disk, we also use the ratio between the observed radial velocity of the gas disk and the perturbed velocity components of the gas. The relations (7.12) and (7.13) show the relationship between the third harmonic of the radial velocity v_{obs} with wave motions only in the disk plane, i.e. only with two components of the perturbed velocity of the gas $\tilde{v}_r$ and $\tilde{v}_\varphi$. This means that there is some functional dependence of the phase of the third harmonic of the radial velocity of the phase F_3 on the phases F_r and F_φ, which can be written as

$$F_3 = f_1(F_r, F_\varphi). \tag{7.19}$$

However, it follows from the theory of natural vibrations of the disk that the phases F_r and F_φ are not independent, but linked by (7.16) or (7.17) depending on the sign of $\varkappa^2$. Therefore, the functional dependence (22) simplifies to:

$$F_3 = f_2(F_r). \tag{7.20}$$

(7.18) implies a unique relationship between the phases F_r and F_σ. Hence, we finally have

$$F_3 = f(F_\sigma). \tag{7.21}$$

Because of the dominance in the Fourier spectrum of surface brightness of the second harmonic (see Fig. 7.15a, b) phase F_σ actually coincides with the phase of the second harmonic of brightness. The latter is easily determined from observations. Phase F_3 in (7.21) is also easily computed as the phase of the third Fourier harmonic of the observed radial velocity v_{obs}.

Thus, the relation (7.21) can be easily tested, which is the observational test of the wave nature of the spiral structure.

According to [126, 581], ratio (7.21) is explicitly written as follows:

$$\Phi \equiv F_3 - \frac{\pi}{2} = F_\sigma \quad \text{for} \quad r > r_{cr}, \tag{7.22}$$

$$\Phi \equiv F_3 - \frac{\pi}{2} = F_\sigma \quad \text{for} \quad r < r_{cr},\ C_\varphi > C_r, \tag{7.23}$$

$$\Phi + \pi \equiv F_3 + \frac{\pi}{2} = F_\sigma \quad \text{for} \quad r < r_{cr}, C_\varphi < C_r. \tag{7.24}$$

Fulfillment of (7.22)–(7.24) for an actual galaxy indicates, as already noted, that it observed spiral arms are density waves.

The verification of the described observational test can be made very convincing when one considers that for galaxies for which the condition $C_\varphi > C_r$, two curves defined by equations

$$\cos(2\varphi - F_\sigma) = 1, \tag{7.25}$$

$$\cos(2\varphi - \Phi) = 1, \tag{7.26}$$

according to (7.22) (7.23) should coincide over the entire disk. Equation (7.25) defines the curve of the maximum surface brightness of the second harmonic, and (7.26) – a test curve, which describes a two-arm spiral with the radiation velocity phase of the third Fourier harmonic F_3, shifted by $\pi/2$.

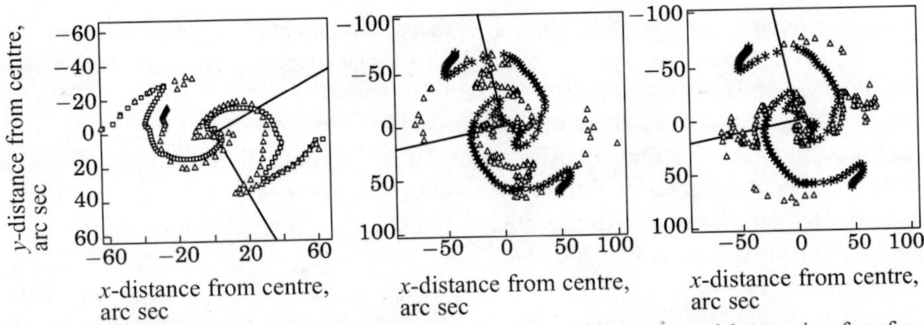

Fig. 7.16. a – superposition of the maxima of the curves of the second harmonic of surface brightness in the *K*-band (squares), described by equation (7.25), and the test curve (triangles), described by the equation (7.26) for the galaxy NGC 157, characterized by the condition $C_\varphi > C_r$. b, c, – superposition of these curves for the galaxy NGC 3631, for which $C_\varphi < C_r$ at $10'' < r < 50''$ [587]: the position of the maxima of the second harmonic of surface brightness in the red (*R*) band (asterisk), the position of the maxima of the test curve (triangles in Fig. b) and position of the minimum test curve (triangles in Fig. c).

For galaxies, characterized by the inverse ratio, $C_\varphi < C_r$, both of the above curves should coincide in the region outside the corotation circle, $r > r_{c2}$. In the region $r < r_{c2}$ the test curve must coincide with the curve of the minimum of the second harmonic of surface brightness. In other words, if $C_\varphi < C_r$ the test curve should have a jump in the corotation region, passing, roughly speaking, between the spiral arms within the corotation circle and, jump changing its location, be combined with the spiral arms – with the curve of maximum of their brightness – outside the corotation circle.

Such a correlation of the curves is seen in Fig. 7.16, where the triangles show the curve (7.26). Figure 7.16a shows the curve (squares) of the maximum of the second harmonic of surface brightness in the *K*-band of galaxy NGC 157, for which $C_\varphi > C_r$. Good agreement of the curves can be seen immediately. Figure 7.16b, shows a comparison of the lines of the maxima of the second harmonic in the *K*-band with maxima and minima of the test curve for the galaxy NGC 3631. For this galaxy at $10'' < r < 50''$ the amplitude of the perturbed azimuthal velocity is less than the amplitude of the perturbed radial velocity, $C_\varphi < C_r$ [587]. It is seen that the maxima of the test curves fall on the surface brightness maximum in the very centre of the disk also behind the corotation circle. In the interval, they almost coincide with the minima of surface brightness as it should be for the galaxies with $C_\varphi < C_r$. Thus, it is possible to show the wave nature of the spiral structure also in the galaxy NGC 3631. Therefore, it follows that the spiral arms of galaxies are in fact density waves.

7.5. Vertical motion and velocity field in the plane of the gaseous disk

7.5.1. Two types of vertical motions in the gas disk

The test discussed in section 7.4.1 shows the connection with two-arm density waves of the first and third Fourier harmonics of the radial velocity field, determined by movements in the plane of the disk. It is these movements that interest us in terms of identifying giant anticyclones, similar to those observed in experiments with shallow water. However, in order to be able to restrict ourselves to considering only these harmonics we mus show that the second Fourier harmonic is caused by vertical movements in the same density wave. This paragraph is an example of use for this purpose of observational tests which we developed [247].

To clarify the basic idea of this observational test, we present some facts from the theory of oscillations of galactic disks.

We know two types of vertical motions in the disks of galaxies excited by completely different mechanisms [590]. The first of these is associated with density waves, and together with the already discussed movements in the plane of the disk forms a single three-component velocity vector field in a spiral density wave. This type of movement does not deform the central plane of the disk $z = 0$, since the component of velocity v_z in the density wave is an odd function of z: $v_z(-z) = -v_z(z)$, so that $v_z = 0$ in the plane $z = 0$.

The second type of vertical motion leads to bending of the disk simililar to the surface of a vibrating membrane. The vertical velocity in this movement is an even function of the coordinate across the disk z: $v_z(-z) = v_z(z)$, $v_z \neq 0$ in the $z = 0$ plane. This behaviour of the gas disk can be caused either by the tidal influence of the galaxy-satellite, or by the motion of the stellar disk in which a gas disk is immersed.

In the stellar disk, membrane (bending) vibrations are excited by the so-called hose-pipe instability, which arises when the velocity dispersion of stars in the plane of the disk is much larger than in the vertical direction. A small initial bend of the disk in this case grows as a result of the centrifugal force acting on the stars, moving along curved paths. Exactly in the same way the centrifugal forces bend the hose with a fast current flow in it, from which the name of this instability originated.

Thus, we need to figure out what type of vertical movements, differing in the class of symmetry of the function $v(z)$ with respect to the plane $z = 0$ is observed in the gas disk of spiral galaxies and gives rise to the second Fourier harmonic in the spectrum of the radial velocity field. To solve this problem, it is naturally first to choose a galaxy oriented toward the observer so that the vertical velocity $v(z)$ provides the main contribution to the observed velocity along the line of sight. Then we have to measure the radial velocity so that we obtain different results with different symmetry $v(z)$.

The first requirement is satisfied if the galaxy has a small angle of inclination i (the so-called galaxies seen 'flat') because the ratio of the contributions of vertical velocity and velocity in the plane of the disk is proportional to $\cos i$.

As for the method sensitive to the symmetry of v_z, then it may consist in comparing the results of measurements of the same vertical velocity field at different wavelengths and varying degrees of transparency in the gas disk. For one of the wave bands the optical thickness of the disk should be less than unity, and for another – more. These requirements are satisfied by, for example, measurements of the velocity of the Doppler shift of the 21 cm line and the line H_α.

We explain what kind of differences in measurement results can be expected using different wavelengths in the disks with different optical thickness for a particular type of vertical motion of gas in the disk of the galaxy. In the H_α line, because of the large optical thickness of the disk for this line, we see mostly its part closest to us. Therefore, for both types of vertical movement measurements in the H_α line give the maximum vertical velocity. Indeed, in the case of the odd behaviour v_z is zero in the plane $z = 0$ and increases with distance from this plane, reaching a maximum at the edge of the disk, i.e. where we measure it with the line H_α. The value of v_z in the bending vibrations practically does not depend on z (with an accuracy of $2\pi/\lambda \ll 1$, where λ is the wavelength[590].

When measurements are made using the 21 cm line for which the disk is almost transparent, the contributions of sections of the disk on different sides from the central plane $z = 0$ are approximately equal in magnitude. In the case of odd behaviour of $v_z(z)$ these contributions mutually cancel each other out – areas with opposite vertical velocities will cause a Doppler shift of the line in opposite directions, leaving its maximum unmoved. Consequently, for the vertical movements in the density wave the radial velocity (less the systemic velocity of the galaxy), measured in the 21 cm line, will be significantly less than the measured in the H_α line. In the case of membrane vibrations and even the behaviour of $v_z(z)$ measurements in both lines should be close to each other, i.e. almost the same.

7.5.2. Observations of vertical motions in the gas disk of NGC 3631 galaxy

Described above in section 7.5.1 the observational test to determine the nature of vertical motions in the galatic disk has been tested in the study of the velocity field of the spiral galaxy NGC 3631, which has a distinct two-arm structure (Fig. 7.17) and is oriemted almost 'flat' ($i = 17°$). This galaxy is isolated, which eliminates the possibility of distortion by tidal forces.

Observations in the H_α line, used in this study, were obtained at the 6-m telescope of SAO RAS [247, 587]. Observations with the 21 cm line were made by J.H. Knapen on the Westerbork Synthesis Radio Telescope [783]. Due to some technical reasons, the observations in the optical range (H_α) and radio range (21 cm) had significantly different angular resolutions (2″ and 15″2″ × 11). Therefore, for a more adequate comparison, the optical field of radial velocities was smoother up to 13″ resolution, comparable with that of the radio data. The initial optical field, the smoothed optical field and the rf field of the radiation velocity of the galaxy NGC 3631 are shown in Fig. 7.18a–c.

The Fourier spectra of all the fields, shown in Fig. 7.18a–c, demonstrate the dominance of the first three harmonics (Fig. 7.18e, f), which is a consequence of the two-arm spiral structure in this galaxy (see the ratios (7.9)–(7.15)).

The results presented in Fig. 7.18d, f agree with the assumption that the second Fourier harmonic of the radial velocity field owes its presence to vertical movements of the gas in the disk of NGC 3631. Indeed, in the data in the H_α line, shown in Fig. 7.18d, e, only the second harmonic has a significant amplitude, which should be if the harmonics is caused by vertical movements, while the first and third harmonics

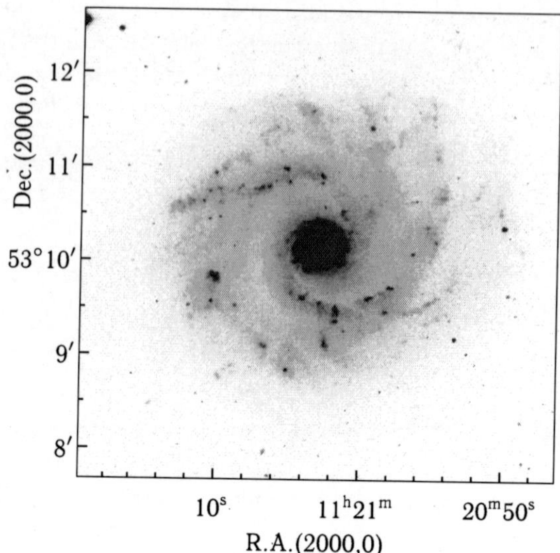

Fig. 7.17. The image of the spiral galaxy NGC 3631 in the *K*-band (from the archives of the Isaac Newton Telescope, Canary Islands, ING).

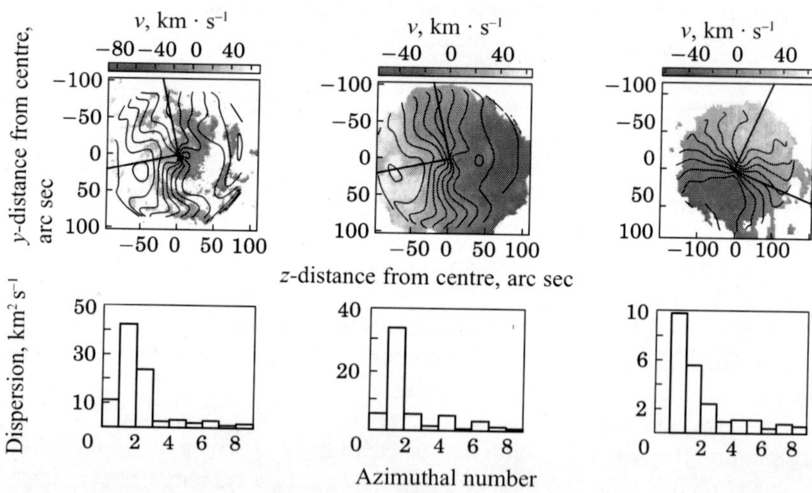

Fig. 7.18. The fields of the radial velocities of gas in the galaxy NGC 3631 (see Fig. 7.17), obtained from the following measurements: a – in the H_α line with high angular resolution, ~2″; b – from the same measurements, but smoothed to the resolution of ~2″; c – from measurements in the 21 cm line with a resolution of 15.2 × ~11″. The contribution of different Fourier harmonics to the deflection of the field of the radial velocities of the galaxies from purely circular motion, averaged over the area of the disk ($r <$ ~80″); d – velocity field in the H_α line with high angular resolution, e – smoothed velocity field in the line of H_α, f – the velocity field in the 21 cm line.

are caused by movements in the plane of the disk, whose contribution is suppressed due to the effect projection (the projection of the velocity is proportional to the line of sight sin i). For NGC 3631 $\sin i \simeq 0.3$. (Dispersion, measured in units of [km^2/s^2], presented at the in Fig. 7.18d–f, is suppressed even more strongly: it is proportional to $\sin^2 i \simeq 0.1$. At the same time, the amplitude of the second harmonic of the radial velocity field derived from radio data, is much lower (Fig. 7.18f).

Comparing Fig. 7,18 d, e, and f, we conclude that the smaller amplitude of the first and third harmonics of the velocity field in the 21 cm line compared with the amplitudes of the initial field in the H_α line is connected with the influence of resolution – these harmonics have similar amplitudes in the field of radial velocities in the 21 cm line (Fig. 7.18f) and the smoothed velocity field in the H_α line. The second harmonic in the field of radial velocities in the 21 cm line is much smaller than in the smoothed velocity field in the H_α line (Fig. 7.18e). This supports the fact that in NGC 3631 we observe vertical motions in the density wave, rather than bending vibrations [247, 587].

7.5.3. Gas velocity field in the plane of the galactic disk

As follows from section 7.5.2, all independent observational tests favour the applicability of our method to recover the full vector velocity field of gas in spiral galaxies with the regular spiral structure of the radial velocity field.

The literature describes the results of treatment of the radial velocity fields and restoration of the full velocity vector of about ten spiral galaxies – among them NGC 157, NGC 6148, NGC 1365, NGC 3893 and others [569–571, 581, 584, 587]. In all cases, independent methods gave consistent results with each other. This allowed us to go to a completely different level of conclusiveness of the results. In astronomy, we can not 'touch' the objects under study, or to formulate experiments with them, as is done in physics (with the exception of the nearest solar system objects). The comparison between the results of a number of independent methods of analysis of observational data may serve as proof of the correctness of the model used.

At the error of determination of velocity of the element of the image of the order of 15 km/s and using more than 10^5 velocity determinations iy was possible to detect the amplitudes of the velocity in the spiral arms – 20–30 km/s with an accuracy of 2–5 km/s.

The restoration of velocities can immediately solve two classical objectives: to determine the corotation radius, and give a direct proof of the wave structure of the observed spiral arms. In addition, this method made it possible to solve a number of new challenges. One of the most important task is to discover new structures in galaxies – giant anticyclones.

Figure 7.19 show for the two galaxies NGC 157 and NGC 1365 the velocity field in the plane of a gas disk in a coordinate system rotating with the angular velocity of the spiral pattern superimposed on the de-projected image of the appropriate galaxies. If the spiral pattern is stationary or quasi-stationary, then this velocity field will also be stationary or sufficiently slowly varying with time. As can be seen from Fig. 7.19, in both cases there are two well-marked anticyclones near the corotation radius. Such anticyclonic vortices were predicted based on laboratory experiments on shallow water (chapter 6). Recall that the centres of anticyclones are at the corotation radius.

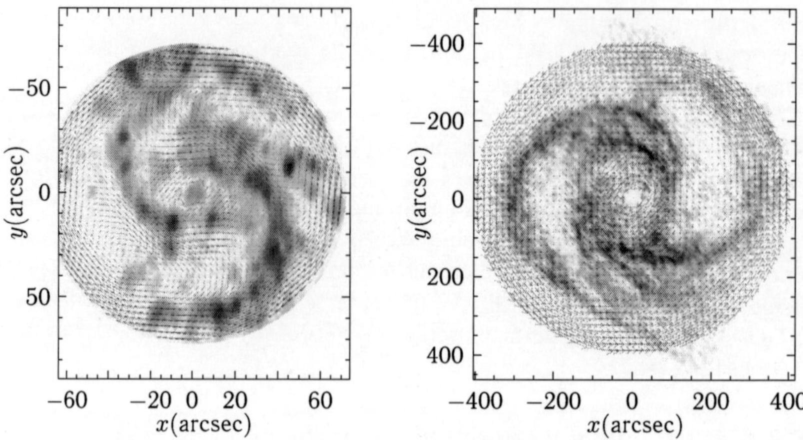

Fig. 7.19. Gas velocity field in the plane of the galactic disk: a – NGC 157, superimposed on the map of its surface brightness in the H_α line; b – NGC 1365, superimposed on the map of its surface brightness in the 21 cm line.

We note that the perturbed velocity of the gas in the spiral branches at radii smaller than the corotation radius are directed toward the centre, and at radii larger than the corotation radius – from the centre of galaxies. This behaviour of velocity is predicted by the theory of density waves.

Comparison of the position of the spiral arms with the position of anticyclones shows that the centres of anticyclones are located between the spiral arms. The latter means that for these galaxies self-gravitation forces in the gas exceed the hydrodynamic pressure forces (see Table 7.1).

Unfortunately, the analysis presented here can still be carried out only for a very small number of galaxies. But the analysis clearly demonstrated the great potential offered by method for studying the Fourier components of the azimuthal distribution of the radial velocity, when there is sufficient observational data for this.

Despite the unity of the spiral–vortex structure, 50 years passed between the discovery of the spirals and vortices. The spirals 'shine like the tip of the iceberg', surprising astronomers by the variety and dynamism of their forms. Vortices were the 'underwater' part of the same 'iceberg', which could not be found by many telescopes at various wavelengths. And the reason for this is not a lack of power of telescopes, but the need for making predictions at the level of laboratory and numerical simulation, analytical studies, which revealed the hidden nature of these structures, the place of their location and the method of detecting them from the obersvational data. Long-term efforts to search for these structures have now resulted in a dynamic picture the spiral galaxies.

7.6. Giant cyclones in galaxies

As noted above (see Figs. 7.5 and 7.6), in the reference frame rotating at the velocity of spiral arms, the direction of rotation of the disk inside and outside the corotation circle is opposite (solid lines with arrows in Fig. 7.20). For the two-arm galaxy

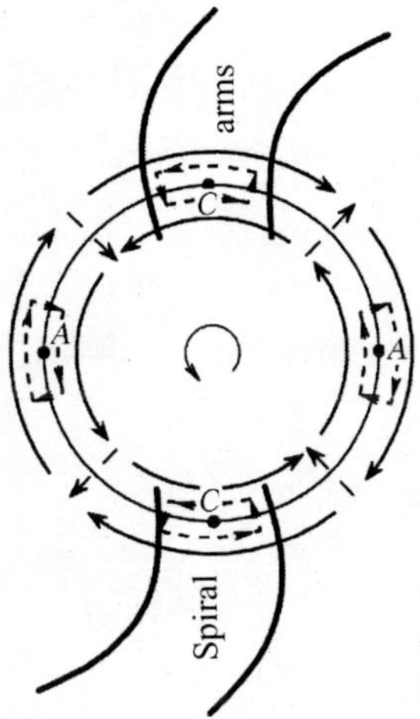

Fig. 7.20. Schematic representation of the stationary (solid lines with arrows) and perturbed (dashed lines with arrows) velocity fields in the vicinity of the corotation circle in the coordinate system rotating with the spiral pattern.

shown in Fig. 7.20, the perturbed velocity (dashed lines with arrows) changes the sign four times during the change in azimuth angle by 2π. By themselves, the perturbed velocities form four vortex-like structures – two cyclones and two anticyclones (Fig. 7.20). However, due to the fact that the directions of the perturbed (in the density wave) azimuthal velocities coincide with the direction of rotation between the spiral arms and are opposite to the direction on the arms, the rotation of the unperturbed rotation supports the anticyclonic vortices and suppresses cyclones. For this reason anticyclonic vortices always form between the arms.

Cyclones on the arms may be formed only when at the prevalence of perturbed velocity, rotation only suppresses them. Therefore, the condition for the existence of cyclonic vortices in the full velocity field of the galaxy is the predominance of the gradient of the perturbed azimuthal velocity over the velocity gradient in the coordinate system connected with a spiral pattern:

$$\left|\frac{\partial \tilde{v}_\varphi}{\partial r}\right| > \left|\frac{\mathrm{dv}_{\mathrm{circ}}}{\mathrm{dr}}\right| \equiv \left|\frac{\mathrm{d}(\mathrm{v}_{\mathrm{rot}} - \Omega_{\mathrm{ph}} \mathrm{r}_{\mathrm{c}})}{\mathrm{dr}}\right|. \tag{7.27}$$

It should be noted that in the spiral galaxy in the coordinate system of the galactic centre the perturbed velocity is always less than the rotational velocity, and in most cases even much smaller $(|\tilde{v}_\varphi| \ll v_{\mathrm{rot}})$. However, the gradient of the perturbed velocity can, in some regions of the disk, be steeper than the gradient of the rotational velocity due to the difference in the characteristic scale of variation of these quantities. The presence of such regions in the corotation circle indicates the possibility of appearance in the given galaxy of cyclones in the velocity field. The above representation led to the prediction of the existence of cyclones in the galaxies with a highly developed spiral structure [584].

One can show that there are four different versions of the location of the cyclonic vortices.Three of them consist in the formation of a pair of cyclones with centres at the corotation circle, inside or outside of it. The fourth option combines the second

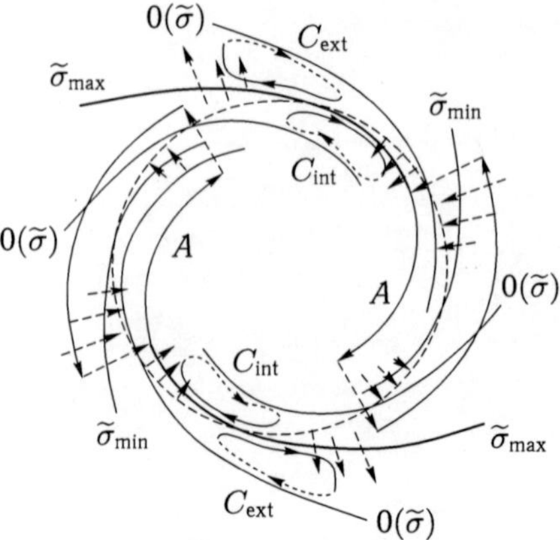

Fig. 7.21. The velocity field in the coordinate system rotating with the spiral pattern in the plane of the disk of the galaxy with a strong density wave: in addition to the two anticyclones, located between the spiral arms, the vortex structure includes two pairs of cyclone centres on opposite sides of the corotation circle (dashed circle) and the line of maxima of surface density (thick lines).

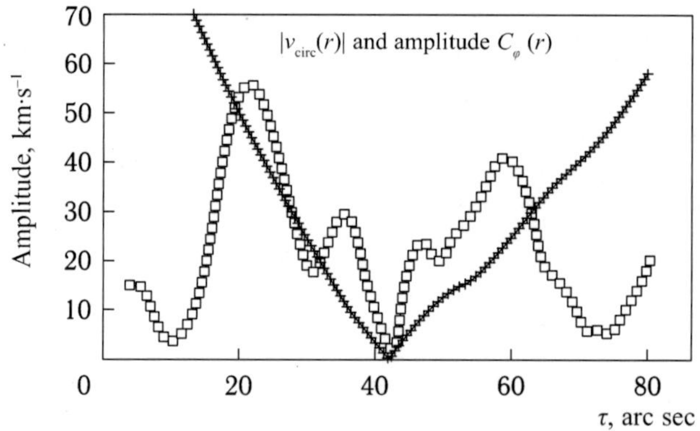

Fig. 7.22. The amplitude of the perturbed azimuthal velocity C_φ (squares), calculated as a function of radius, superimposed on the profile of speed of rotation (crosses) in the accompanying spiral system reference $|v_{circ}| = |\, v_{circ} \,|=|\, v_{rot} - \Omega_{ph} r_c \,|, r_c \sim 42''$.

and third – four cyclonic vortices formed on both sides of the corotation circle. At the same time, the cyclones are slightly shifted relative to the line of the maxima of surface density (Fig. 7.21) [586].

The theoretical prediction of the possible existence of cyclonic vortices in the velocity field of galaxies with a strong spiral structure [584] was confirmed when the velocity field in the plane of the disk galaxy NGC 3631 was restored [586, 587]. Figure 7.22 compares the radial dependence of the perturbed azimuthal velocity $|\tilde{v}_\varphi|$, and velocity of rotation of the disk in NGC 3631 in the appropriate coordinate system v_{circ}. The corotation radius in this galaxy is about 42″. As can be seen from Fig. 7.22, the condition (7.27) holds on either side of the corotation circle (the slope

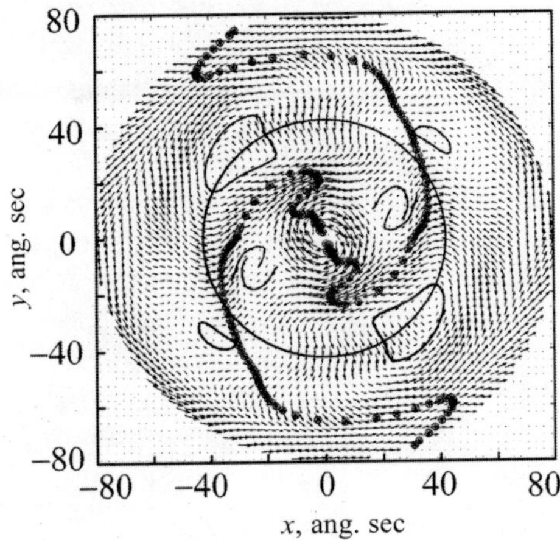

Fig. 7.23. Restored velocity field of gas in a coordinate system rotating with the spiral pattern in the disk of the galaxy NGC 3631, having a strong density wave. Asterisks show the position of maxima of surface brightness of the galaxies in the *K*-band (see Fig. 7.17), the location of the cyclonic vortices discovered in this galaxy and the location as predicted by the theory of density waves (see Fig. 7.21).

of the curve describing the behaviour of $\tilde{v}_\varphi$ is greater than that of the curve v_{circ}). Consequently, according to the prediction, we expect the detection of cyclones on both sides of the corotation circle, as shown in Fig. 7.21.

The restored velocity field of gas in the disk of NGC 3631 in the coordinates, rotating with the spiral pattern, is shown in Fig. 7.23. Here we see the structure of the velocity field which includes two anticyclones between the spiral arms and four cyclone, the location of which is in complete agreement with the theoretical predictions (see Fig. 7.21) [586, 587].

7.7. Spiral patterns as global eigenmodes in an unstable disk

Despite the generally accepted view that spiral phenomenon in the galaxies is of a wave nature, the question of the mechanisms of its generation is actively discussed. In particular, we consider two possibilities. In the first, it is thought that the spiral pattern is a long-lived structure [366–368], which can change its characteristics under the influence of external (tidal interactions, accretion) and internal (e.g. a general decrease in the gas content due to star formation), but the observed pattern is a manifestation of unstable global modes associated with the development of gravitational (see sections 2, 3, 4) or hydrodynamic instabilities (see chapters 5 and 6).

An alternative view is based on the short-lived, recurrent nature of the spiral structure, which is updated by the swing amplification mechanism in the shear flow, or under the influence of external forces, for example, by the satellites

7.7.1. Global density modes in a gravitationally unstable disk

In this section we mention the possibility of describing the observed spiral structure of concrete galaxies, based on numerical nonlinear models of the dynamics of global eigenmodes in a gravitationally unstable disk [789, 790, 792, 993].

The simplest model to describe the multicomponent gravitationally interacting (stellar–gas) subsystem is the approximation based on hydrodynamic equations in the traditional manner (chapter 3) when the gas and stellar components are described in the hydrodynamic approximation[3] for a polytropic gas

$$p_{s,g} = K_{s,g}\,\sigma_{s,g}^{\gamma_{s,g}}, \tag{7.28}$$

where s, g denote the stars and gas, respectively. In these works, as the dynamical equations we use (3.277), (3.278) and (3.279) separately for the parameters of the stellar disk σ_s, u_s, v_s, and the gas component σ_g, u_g, v_g.

The total potential is determined by the contribution of gas, stars and spheroidal subsystem (halo + bulge):

$$\Phi(r,\varphi) = -\int_{R_{\rm in}}^{R_{\rm out}}\int_0^{2\pi}\frac{\left[\sigma_s(r',\varphi')+\sigma_g(r',\varphi')\right]r'\,dr'd\varphi'}{\sqrt{r^2+r'^2-2rr'\cos(\varphi-\varphi')}} + \Phi_h(r,z), \tag{7.29}$$

providing a link between the gas and stars.

More complicated are the models with the interaction between components in the presence of star formation. We write equations for the three-components disk [789, 790]: gas disk (index 'g'), 'active' stellar disk (index '*'), 'inactive' stellar disk (index 'n'). The system of three equations

$$\frac{D_j\sigma_j}{Dt} = \begin{cases} -C_*\,\sigma_g^2 + \eta\,\sigma_s/\tau, & j=g,\\ \zeta\,C_*\,\sigma_g^2 - \sigma_s/\tau, & j=*,\\ (1-\zeta)C_*\,\sigma_g^2 + (1-\eta)\sigma_s/\tau, & j=n, \end{cases} \tag{7.30}$$

ensures the conservation of the total mass in the disk, taking into account that: C_* is a constant of the star formation rate in accordance with the Schmidt law, η defines the mass of the stars which is returned to the interstellar medium, τ is the average lifetime of massive stars, ζ, $(1-\zeta)$ is the efficiency of transfer of gas to the *- and n-components, respectively. We introdyce the notation D_j/dt for the full (continuous) time derivative for the j-th subsystem (see the left side of equation (3.277).

Then the system of equations (3.278) (3.279) already for the three components can be written in the form [790]

$$\sigma_g\frac{D\mathbf{u}_g}{Dt} + \nabla p_g + \sigma_g\nabla\Psi = -C_*\,\sigma_g^2\mathbf{u}_g + \eta\frac{\sigma_*}{\tau}\mathbf{u}_*, \tag{7.31}$$

---

[3]For stellar components instead of the kinetic equation we used the so-called the model of stellar hydrodynamics.

$$\sigma_* \frac{D\mathbf{u}_*}{Dt} + \nabla p_* + \sigma_* \nabla \Psi = \zeta\, C_*\, \sigma_g^2 \mathbf{u}_g - \eta \frac{\sigma_*}{\tau} \mathbf{u}_*, \tag{7.32}$$

$$\sigma_n \frac{D\mathbf{u}_n}{Dt} + \nabla p_n + \sigma_n \nabla \Psi = (1-\zeta) C_*\, \sigma_g^2 \mathbf{u}_g + (1-\eta) \frac{\sigma_*}{\tau} \mathbf{u}_*. \tag{7.33}$$

To model the spiral structure of a specific galaxy, the initial state should be an axisymmetric equilibrium disk with the rotation curve $V(r)$, velocity dispersion of the stars $c_*(r)$ (for gas – the speed of sound $c_s(r)$), surface density $\sigma_{*,g}(r)$ in accordance with these observations. The possibility of gravitational instability is determined by the choice of the parameter $Q(r)$ (see section 4.3). Note that the presence of a massive cold gas component has a significant gravitational destabilizing effect on a hotter stellar disk. Figure 7.24 shows, for the galaxy NGC 1566, the dependence $Q(r)$, which was used in [790] for selecting the initial state of the numerical model. Figure 7.25 shows an example of producing a global pattern for NGC 1566 [790].

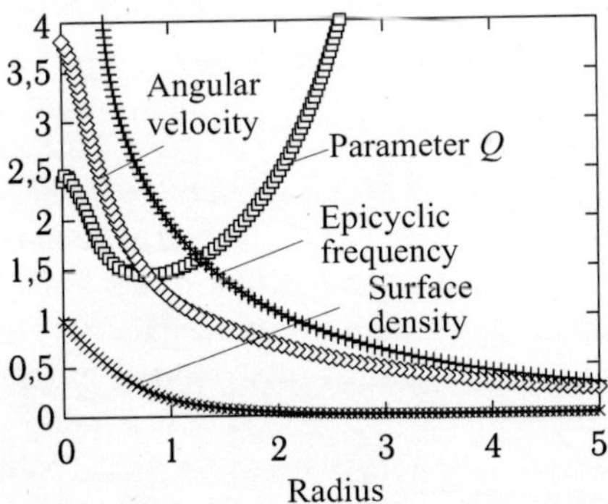

Fig. 7.24. The radial distribution of the equilibrium parameters for NGC 1566 in the model which allows the generation of a two-arm spiral [790].

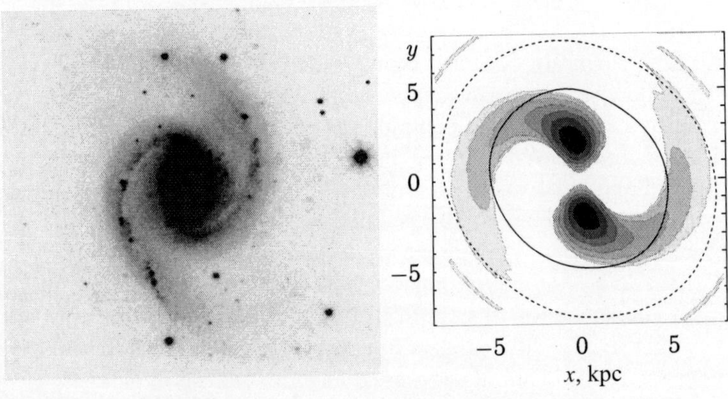

Fig. 7.25. Image of NGC 1566 in the band I (a). The density distribution in the numerical model by non-linear evolution of a gravitationally unstable global mode $m = 2$, (b). The corotation radius (solid line) and the outer Lindblad resonance (dotted line, see section 1.4.1) are shown.

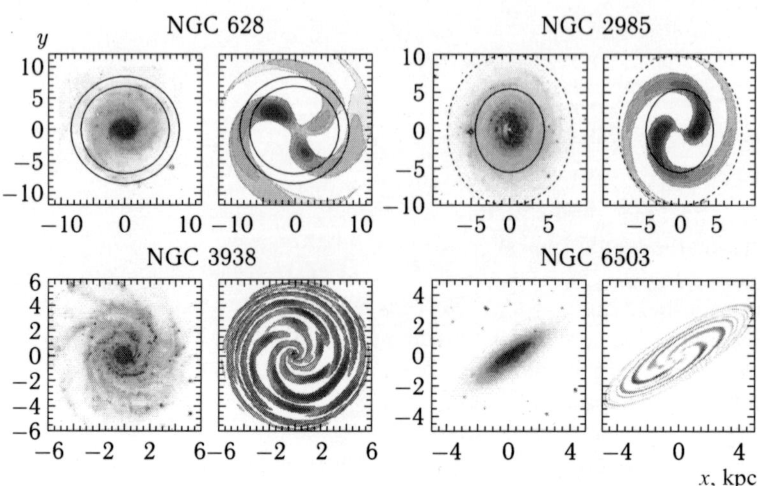

Fig. 7.26. The observed images of galaxies and the results of simulation in [792].

These models have been used to explain the spiral galaxies with different geometry of the pattern and number of arms: NGC 488, NGC 628, NGC 2985, NGC 3938, NGC 6503 (Fig. 7.26). The data used in these models included the observed radial profiles of the rotational speed of the disk, dispersion of velocity components and the mass distribution in accordance with the photometric data, which provided the agreement between the spiral patterns of the galaxies and their models.

A similar approach to the construction of models of the spiral structure for the specific galaxies, based on models of collisionless *N*-bodies, was also shown to be effective. The initial stage is the formation of the spiral pattern, the morphology of which is close to the observed one.

7.7.2. Complex spiral patterns of galaxies

Beautiful correct spirals, obtained by the theorists in the Lindblad wave approach (regardless of the mechanism of generation), are in the nature the exception rather than the rule. Many have often have complex structures [37]. There are γ- and Θ-shapes, stacked form, branches (including branches beginning, as it were 'nowhere'); interacting galaxies appear more complex structures ('tails', 'jumper','whiskers', 'antennae') [85]. We will not address the question of the origin of flocculent spirals and interacting systems. Despite all the difficulties of the wave theory, excitation and self-maintenance of gravitational or hydrodynamic modes, generally speaking, can explain the generation of a spiral structure with a given number of arms in one way or another in a specific galaxy. To understand the reasons for the formation of observed more complex spiral patterns (branching of arms in galaxies such as NGC 151, 1288, 1297, 2997, 4725, 4736, 5033, 6946, 7412, etc.) we need additional assumptions. For example, laboratory simulation in 'shallow water' revealed the effect of restructuring of azimuthal modes (changing the number of arms of the generated spirals) with a smooth variation of the rotational velocity of the internal part of the apparatus (the 'core' of the galaxy) in the course of the experiment (see Fig. 6.5). In this regard, it was hypothesized that the galaxies observed with branching arms are in an essentially non-stationary state and, therefore, the branching of the arms is a transient process of restructuring the azimuthal modes [164, 184]. At the same time the prevalence of galaxies with branching arms indicates the long-term existence of these formations, which does not quite fit into the framework of the assumption of the highly non-stationary nature of such systems.

The general approach proposed by A.M. Fridman [237] explains the multilevel nature of the spiral structure – different number of arms in various regions of the disk with respect to r – by the fact that in these areas there are different conditions for the preferred development of any instability. Each of these instabilities is characterized by its dependence of the maximum growth rate ofn the azimuthal mode number m, which leads to various kinds of spiral structure in different areas of the system.

We discuss a similar idea, based on the simultaneous existence (superposition) in the disk of the galaxy of two or more vibrational modes with the same or different azimuthal numbers and with the same or different mechanisms of excitation (gravity and / or hydrodynamic) [161, 162]. We emphasize that here we discuss quasi-stationary modes determined by the simultaneous development of various unstable modes in the same area of the disk. Under this assumption, in addition to the branching of the arms (NGC 151, 1288, 2997, 4725, 4736, 5033, 6946, 7412), we can explain the presence of a dense gas ring superimposed on the spiral structure (NGC 1024, 2223, 3124, 3344, 7329, etc.), existence in a spiral pattern of sleeves with different twist angles, the presence of the spiral structure in the central area of the bar (NGC 1512), etc. The analysis shows the presence of different harmonics in the distribution of brightness for many galaxies with the global pattern of [480].

An example of the existence in the system of the primary unstable mode and its higher harmonics is a supersonic jet – the simplest model has the form of two plane-parallel tangential discontinuities of the velocity (chapter 5, [648, 649, 978, 997, 1018, 1245]). A similar situation is possible in an axisymmetric supersonic flow of a differentially rotating gas ($\Omega \propto r^{-\delta}$), which also may contain higher unstable harmonics (section 5.2.4 and [279, 1009–1011, 1115).

In [162] the authors considered complex spiral patterns of galaxies as a result of the superposition of two concrete centrifugal modes – high- and low-frequency (see section 5.2.5) for the rotation curve (5.70), which is characterized by a jump of angular velocity q and the width of the transition region L.

We compare the properties of high-and low-frequency centrifugal instabilities. Figures 7.27 and 7.28 show the dependence of the complex frequency and the dimensionless corotation radius on the relative rotational speed of the central region of the disk $M = R\Omega_1/c_s$ and L for both modes. Figure 7.27 shows the dependence of the width L, and Fig. 5.25 the dependence on the parameter q. It is significant that in a sufficiently wide range of the parameters the values of frequency and growth rates of both modes are comparable, and there are areas where the difference does

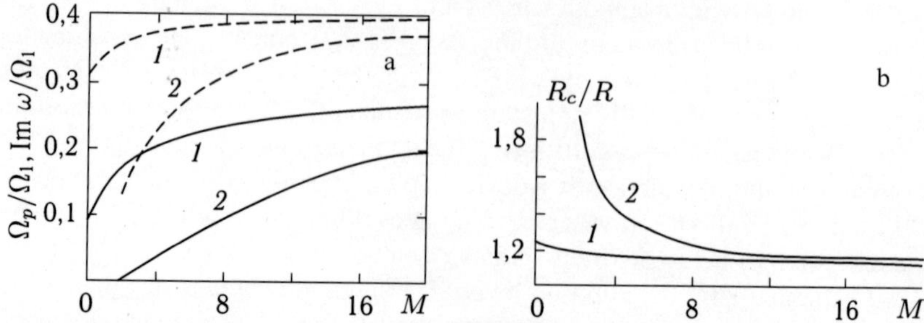

Fig. 7.27. Depending on the increment Im (ω) (solid lines) and the angular velocity of rotation of the spiral pattern Ω_p (dashed lines) on M (a) and dependence of R_c/R on $M = R\Omega_1/c_s$, (b). 1 – high-frequency mode, 2 – low-frequency, $L = 0.3$, $q = 0.05$.

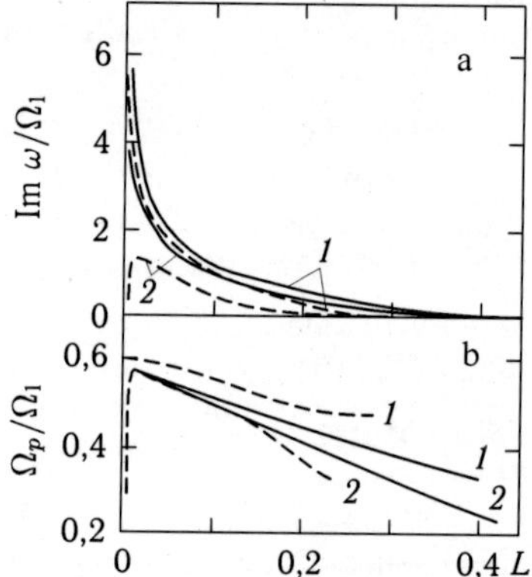

Fig. 7.28. Dependence of the increment Im (ω) (top) and the angular velocity of rotation of the spiral pattern Ω_p (bottom) on L: 1 – high-frequency mode, 2 – low-frequency mode, solid lines: $M = 10$, $q = 0.05$; dashed lines: $M = 20$, $q = 0.2$.

not exceed 10–30%. Thus, there is the possibility of simultaneous excitation of two unstable modes.

Determination of the wave pattern for each of the modes $[\tilde{\sigma}(r,\varphi),\Omega_p]$ and overlay them on each other makes it possible to obtain a spiral pattern with the following characteristics (Fig. 7.29):

1. the presence of branching of spiral arms, with different values of the parameters q, L and M can give a combination of branching spirals with different geometries;
2. the presence of bar-like jumpers between the spiral arms;
3. the presence of a 'ring' of higher density, which connects the spiral arms at a radius $r = 2R$;
4. the presence of four regular spiral arms, although the azimuthal mode number is $m = 2$

Cases 1) and 2) may occur in the same system, but this depends on the time of observation. Note also the interesting possibility of generation of a gas bar in the disk by the hydrodynamic mechanism.

Fig. 7.29. The geometry of the eigenfunctions of the perturbed density for high- (a, d) and low-frequency (b, e) modes, as well as the results of their superposition (c, f) for two different sets of the values of the parameters. Shaded areas indicate positive perturbations of surface density, the circle is drawn at the radius $r = R$.

These results refer to the mode $m = 2$, but the existence of similar modes with the azimuthal number $m \geqslant 3$ must also be considered. This is indicated by the results of calculations for the reflection harmonics $m \gg 1$ in the study of the Papaloizou–Pringle instability (section 5.2.4).

The physical mechanism of excitation of low-frequency modes is similar to that investigated by Papaloizou and Pringle [1010] and Glatzel in models with the power-law rotation $\Omega \propto r^{\delta}$, where the instability develops due to the resonance at the corotation radius and the interaction between waves with opposite signs of energy. The resonant nature of this mode is indicated by its stabilization at $L \to 0$, since due to increase the resonant interaction of waves with the flow there must be a critical layer of finite thickness near the corotation radius, where the velocity profile is monotonic (see Fig. 7.27). The second reason is the need to generate a substantially supersonic flow. The low-frequency mode is stabilized at $M \simeq 2$, which coincides with the boundary of super-reflection, and at $M \lesssim 2$ resonant amplification is no longer possible.

Of course, the results based on linear analysis, are tentative. However, if the perturbations of the considered modes can reach the stage of saturation of significant amplitudes, other interesting effects are likely to occur: non-linear interaction of disturbances during the passage of the spiral arm of one mode through the arm of the other mode; the interaction of the arms of the low-frequency mode with banana-shaped anticyclonic vortices, observed in the experiments between the arms of the high-frequence mode in the corotation circle [184], etc. At the same time, there are indications that the distinction between the linear theory and the non-linear simulation results will not be too large. First, experiments performed in Spiral equipment [184] have shown that the high-frequency mode reaches saturation at the stage of formation of shock waves, and the parameters of the excited spirals differ from those predicted by the linear theory by no more than 30%. Second, the low-frequency mode has a distinct reflective in nature, and in non-linear numerical simulations of Norman and Hardy [978] and in more recent works on modelling of a planar supersonic jet it is shown that perturbations of the reflection type are saturated at the stage of formation of shock waves, the spatial structure of which is perfectly consistent with the predictions of linear analysis (see section 5.2.4).

7.8. Slow galactic bars in spiral galaxies

7.8.1. Some features of the formation of bars

From the classical theory of liquid figures of rotation it is well known that if we start rotating a gravitating sphere from an incompressible fluid, it turns first to a biaxial ellipsoid, and at faster rotation – to a three-axial one [459]. From this we infer a relationship between the angular momentum of a gravitating liquid body and its form.

Such a link is missing in collisionless stellar systems. It means that, on the one hand, we can point to a spherical system with a non-zero angular momentum and, on the other – to virtually non-rotating ellipsoids, in which the ratio of rotational energy to gravitational (potential) energy is a small parameter. For elliptical galaxies this ratio is approximately 13%. The shape of elliptical galaxies is mainly determined by the anisotropy of their 'temperature' – the ratio of velocity dispersions along and across the axis of rotation is less than unity. As for the possibility of rotation of a spherically symmetric stellar system, then as shown by Lynden-Bell [861] this situation is in principle feasible.

This is not difficult to understand when one considers that, beginning with a non-rotating spherical system with the distribution function $f_0 = f_0\,(E,L)$, where E is energy, and L is the momentum of a star, one can always get an infinite set of distribution functions describing the rotating spherical system:

$$F_0(E,L,L_z) = \mu f_0(E,L)\Theta(L_z) + (1-\mu) f_0(E,L)\Theta(-L_z), \qquad (7.34)$$

where μ is a parameter, $0 \leqslant \mu \leqslant 1$ is a step function. The transformation described by equation (7.34) means that we convert the velocity of a group of stars so that the number of particles with $L_z > 0$ was equal to μ, and with $L_z < 0$ it becomes $1-\mu$, $\mu \neq 1/2$. In this case, it is easy to see that there is no change either in the density of the system $\rho_0\,(r)$ or created by its gravitational potential $\Phi_0\,(r)$.

As an example, we consider a system of particles (stars), rotating in circular orbits (Fig. 7.30a, b). The distribution function [143]

$$f_0 = \frac{\rho_0(r)}{2\pi v_0(r)} \delta(v_r)\delta[v_\perp - v_0(r)],$$

$$v_\pm^2 \equiv v_\Theta^2 + v_\varphi^2, \quad v_0^2(r) = r\Phi_0'(r), \qquad (7.35)$$

where $\Phi_0\,(r)$ is the gravitational potential, describes a situation where in the plane tangent to an arbitrary field at an arbitrary point the sum of the velocities of all particles is zero, i.e.system as a whole does not rotate.

We choose for convenience a certain direction – the axis z, and create in the equatorial plane the situation with the predominance of particles rotating in one direction around the axis in relation to the particle rotating in the opposite direction. Such a system will possess a non-zero total angular momentum.

In the stationary and axially symmetric case, when $\partial / \partial t = \partial / \partial \varphi = 0$, and $\Phi_0 = \Phi_0(r)$, we obtain the following kinetic equation [143.590]:

$$\frac{v_\perp}{r}\left(\cos\alpha\frac{\partial f_0}{\partial\theta} - \sin\alpha\cos\theta\frac{\partial f_0}{\partial\alpha}\right) + v_r\left(\frac{\partial f_0}{\partial r} - \frac{v_\perp}{r}\frac{\partial f_0}{\partial v_\perp}\right) + \left(\frac{v_\perp^2}{r} - \frac{\mathrm{d}\Phi_0}{\mathrm{dr}}\right)\frac{\partial f_0}{\partial v_r} = 0. \quad (7.36)$$

One can see that equation (7.36) satisfies, for example, the function of the form [230.248]

$$f_{0\mu} = \frac{\rho_0}{2\pi v_0}\delta(v_r)\delta(v_r - v_0)(1 + \mu\sin\theta\sin\alpha), \quad |\mu| \le 1. \quad (7.37)$$

Now we can calculate the angular velocity of the homogeneous ($\rho 0$ = const) sphere [230, 248]. By definition, the linear velocity of rotation at the point **r**

$$\langle v_\varphi(\mathbf{r})\rangle = \frac{1}{\rho_0}\int v_\varphi F_{0\mu}\,\mathrm{d}\alpha\, v_\perp\,\mathrm{d}v_\perp\,\mathrm{d}v_r = \frac{1}{2}\mu\Omega_0 r\sin\theta. \quad (7.38)$$

Here we used the formula $v_\varphi = v_\perp \sin \alpha$. The expression for $\langle v_\varphi(\mathbf{r})\rangle$ shows that the angular velocity of the system Ω_{rot} is directly associated with the parameter μ:

$$\Omega_{\mathrm{rot}} = \frac{\mu\Omega_0}{2}, \quad (7.39)$$

where Ω_0 is the rotational speed of the particles in circular orbits. As follows from this result, the homogeneous sphere rotates as a solid body.

As we can describe qualitatively the formation of bars in galaxies? According to the observations, the distribution of galaxies in terms of specific moments is relatively wide. Protogalaxies, which received a relatively large momentum, formed massive disks, in which a large 'fast' bar formed are a result of the so-called bar-instability. At the ends of the bar, where the lagging spiral arms start, their linear velocity and the linear velocity of the differentially rotating disk are the same (if the bar is rigidly connected with the spirals). This area of the disk at the ends of the bar is the place for the main – corotation – resonance of the galactic disk. Such a bar is called fast in contrast to the slow bar whose ends rotate much more slowly than the adjacent area.

The slow bar could form in a protogalaxy with a specific momentum smaller than the one in which the fast bar. Formation of the slow bar could be the result of instability of radial orbits. The central 'seal', formed by radial collapse and having as a result radially elongated orbits of stars, turned out to be 'colder' in the transverse direction than in the radialone. The main gravitational energy was used in the radial movement at collapse. It is not therefore surprising that the radial velocity dispersion is considerably greater than the transversal one. Small perturbations in the transverse direction begin to grow and this cannot be stoped by the small transverse dispersion (Fig. 7.30c). In fact, the anisotropic Jeans instability in the transverse direction and leads to an increase in the dispersion in this direction and thus to an increase in yje 'eccentricity' of stellar orbits (the orbits in general are not closed).

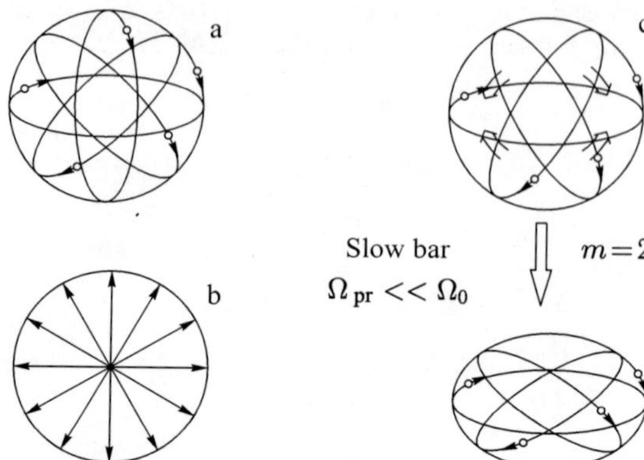

Fig. 7.30. Einstein's model of a spherically symmetric star cluster: a – all particles move in circular orbits around the centre of mass in the direction indicated by arrows; b – in the plane tangent to an arbitrary sphere at any of its points, the sum of the velocities of all particles is equal to zero; c – scheme of transformation of a spherical collisionless ellipsoidal system as a result of the development of the instability of radial orbits.

This instability of the radial (or elongated) orbits was predicted by Zel'dovich et al [80] in 1972. The instability of radial orbits belongs to a class of Jeans instability. This means that in this instability the amplitude of the most large-scale collective oscillation mode increases most rapidly and converts the spherical system to the biaxial ellipsoid (Fig. 7.30c) [196]. The latter slowly rotates with the precession frequency much lower than the characteristic frequency of rotation of stars ($\Omega_{pr} \ll \Omega_0$), hence the name – 'the slowest bar'. A necessary condition for instability of radial orbits: $d\Omega_{pr}/dL > 0$, where L – the momentum of the star, coincides exactly with the condition of formation of a slow bar as a result of the capture of stars by the gravitational bar-potential [864].

7.8.2. Discovery of slow bars

According to V.L. Polyachenko, the slow bar must have the following features:

1) its radius is limited by the inner–inner Lindblad resonance;

2) In the case of a slow bar the major spiral arms of galaxies, being lagging spirals, are limited on the side of the central field, by the outer–inner Lindblad resonance;

3) the ends of the slow bar are connected to the main spiral arms bythe leading spirals wit the azimuthal extent of $\approx 180°$.

Figure 7.31 shows the response of the gas disk to the gravitational potential of a slow bar, first calculated in [1034]. As we see, the response consists of two spirals. The first is part of the lagging spiral (external), which is a continuation of the main lagging spiral arms of the galaxy. The second spiral is leading. Occupying a narrow section in the radial direction the disk, the latter is very tightly twisted and makes half a turn between the bar and the outer spiral.The leading spiral is localized between the inner–inner (IILR) (bar ends), and outer–inner (OILR) (start of the

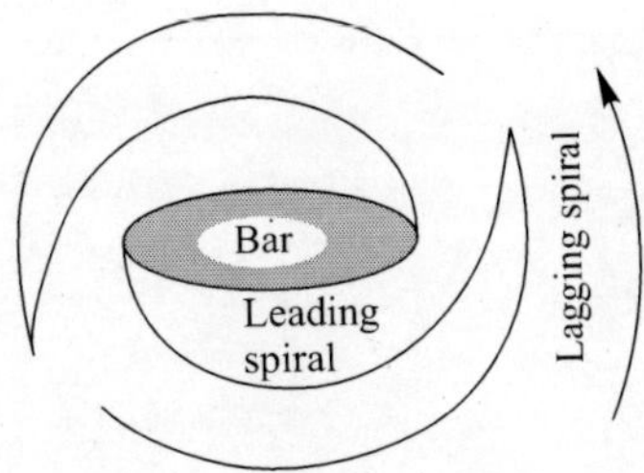

Fig. 7.31. The response of the gaseous disk in the gravitational potential of a slow bar [1034].

lagging spiral) Lindblad resonances. This form of spiral arms in the vicinity of the inner Lindblad resonance of the galaxies NGC 157 was found [578] in the analysis of photometric data on this galaxy, obtained with the Hubble Space Telescope (data available through the existence of the archive Hubble Space Telescope). Figure 7.32 shows the position of maximum of the second harmonic of the maps of the surface brightness of the central part of the galaxy NGC 157, obtained in the visible range (central wavelength ≈5852 A, filter width ≈1873 A), depending on the radius. A bar with length 10″ is clearly visible when the second harmonic of the brightness maps obtained in the near Infrared (IR) band H is superposed on the image. Resonance IILR is found at 5″, OILR at 8″, leading spirals connect ends of the bar at 5″ with the start of the lagging spiral, located at 10″, 2″ from OILR.

Thus, the agreement with the theoretical prediction was very good, which is a serious argument in favour of the existence of a slow bar in NGC 157.

7.8.3. The dynamics of gas in the vicinity of a slow bar

Let us consider the peculiarities of the gas in the centre of the disk, which rotates in a non-axisymmetric gravitational potential:

$$\Psi(r,\varphi,z,t) = \Psi_0(r,z)(1+\varepsilon(r,\varphi,z,t)), \tag{7.40}$$

where r, φ, z are cylindrical coordinates. The main part of the centrally symmetric part of the potential Ψ_0 determines the rotation of the disk. We believe that at the initial time $t = 0$, there is no small non-axisymmetric addition. Slowly (for the period of rotation of the outer edge of the disk) we include a non-axisymmetric perturbation. What is the response of gas to this external effect in the case of the initially axi-symmetric density distribution in the gas disk?

The perturbation of potential ε. Ψ_0 is due, of course, to the non-axisymmetric distribution of gravitating mass in the stellar disk, for example, in the form of a central 'jumper', which can rotate rigidly with angular velocity Ω_b, – we call this structure a bar. As a simple model of a bar in the plane of the disk, we select [1273, 1274]

$$\varepsilon(r,\varphi,z=0) \propto \frac{r^2}{(r^2+b^2)^{5/2}}\cos(2(\varphi-\Omega_b t)), \tag{7.41}$$

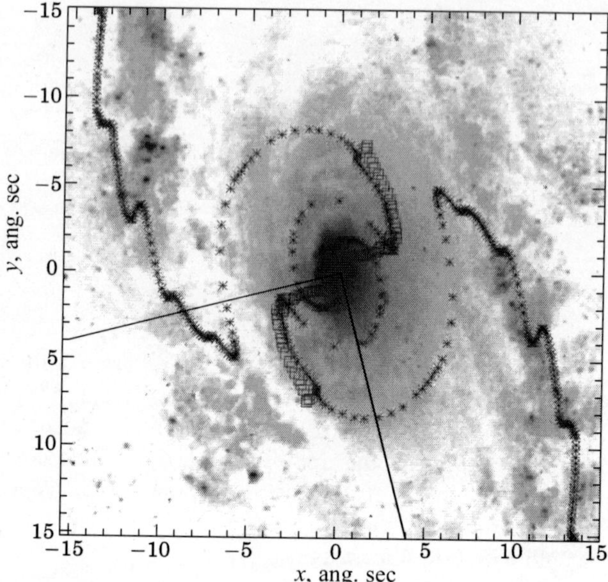

Fig. 7.32. Position of the maxima of the second Fourier harmonics of the broadband optical image (*) and images in near-infrared band H (squares), superimposed on the optical image of the central part of galaxy NGC 157 [578]. The galaxy rotates counter-clockwise, which means that the leading spiral starts from the ends of the bar and makes a half turn and then goes backward into the lagging spiral forming the main component of the spiral pattern of this galaxy. We can clearly see a qualitative agreement with the structure shown in Fig. 7.31.

where b is the characteristic spatial scale of the bar, for the vertical scale $c \ll b$. We are interested in a situation of slow rotation compared with rotation of the gas (the limiting case $\Omega_b = 0$).

Figure 7.33 shows the results of our calculations of the dynamics of the initially axisymmetric disk as a result of appearance of the non-axisymmetric component at the potential during time $t = 3$ [264]. Only the most central part of the calculation area of the disk ($r \leqslant 1$) is shown, and modelling was carried out for the gas disk extending up to $r \leqslant 15$. This allowed us to exclude the influence of boundaries in the numerical model. The shape of the spiral structure for the surface density of gas in the 2D- and 3D-models without self-gravity taked into account differs slightly.

A leading spiral forms in the centre already at the stage of the slow bar ($t = 2$). At $t \geqslant 3$ the bar parameters do not change. The amplitude of the leading spiral increases and the formation of a lagging two-arm spiral starts with a certain delay (after $t \gtrsim 6$). Thus, after a typical pattern already formed after three rotations of disk when in the very centre of the disk there is a two-arm leading spiral and then, almost away from it all, a two-arm lagging spiral moves way from its ends. This structure is long-lived – in the numerical model it was observed for tens of revolutions.

The amplitude of the waves depends on the parameters of the model, in particular, the degree of non-axisymmetry of the potential ε_0, which in different models varies widely. The above picture is typical and does not depend on the specific

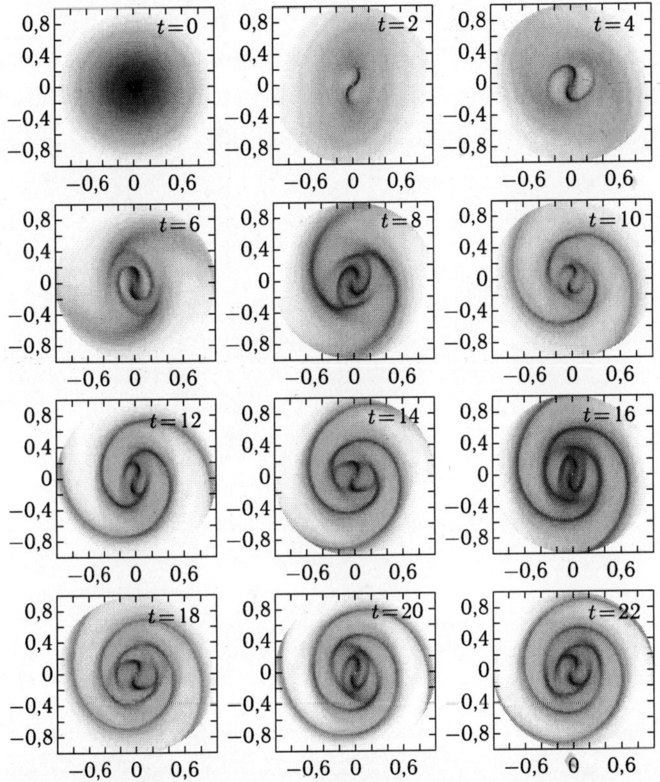

Fig. 7.33. The distribution of the logarithm of the surface density of gas lg $[\sigma(r,\varphi)]$ in the central region of the disk ($r < 1$) at different times in the presence of the non-axisymmetric potential ('low bar'). The disk rotates counter-clockwise. Period of complete revolutions of the disk at radius $r = 1$ is $t \simeq 3$. Evolution of the disk from the axisymmetric to quasi-stationary state is shown.

choice (7.41) – structures retain their qualitative form in the models with other types of functional dependences (7.41).

The main parameters that significantly alter the geometry and other properties of the flow (but retaining the general form) should in the first approximation include the following:

– Adiabatic index γ (in more complex models cooling due to radiation is taken into account). If γ markedly differs from 1 and cooling is absent, strong heating of the gas takes place under the effect of shock waves and this has a depressive effect on the obtained structure. Starting with $\gamma \lesssim 5.1$ there is almost no heating and the model with $\gamma = 1.005$–1.05 give the same result.

– The important thing is the ratio of the speed of rotation gas to the speed of sound. It is convenient to choose the initial value at a radius $r = 1{:}M = V/c_s$, because at large distances c_s decreases slightly and V reaches a plateau. Figure 7.34 shows the simulation results for different values of M. The most distinct structures are obtained at high Mach numbers. In the hotter disks the wave amplitude decreases,

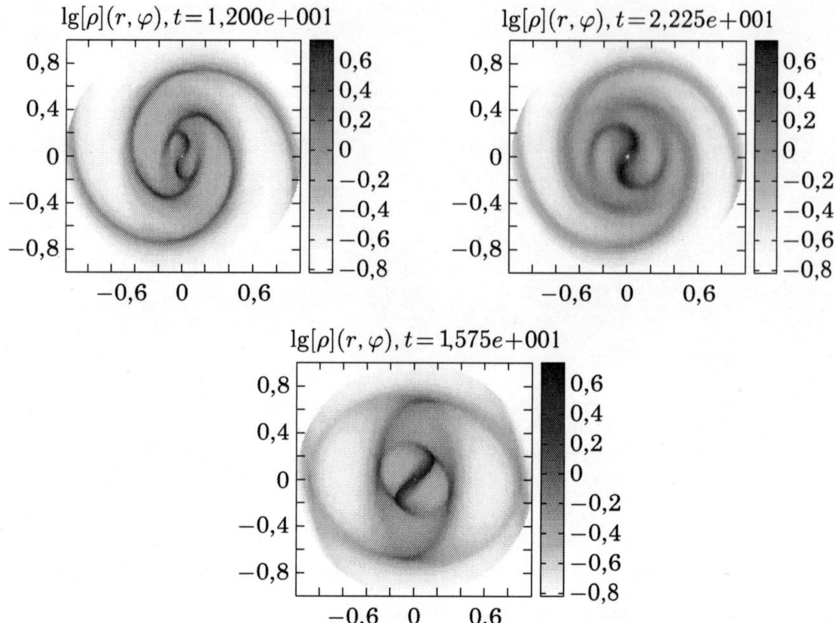

Fig. 7.34. The distribution of the logarithm of the surface density of gas at establishment of a quasi-stationary pattern for different values of characteristic Mach number M = 20, 14 (top), 9 (bottom).

and their geometry significantly changes. For small values of M the wave structure is more complicated – there are characteristics Θ-structure, which can be nested within each other;

- The relative amplitude of the non-axisymmetric potential significantly affects the characteristic times of formation of the quasi-stationary pattern and differential parameters of the gas in the wave. For small values of the perturbation the birth of shock waves does not occur – there is stabilisation in the stage of non-linear waves;

– An important parameter is the angular velocity of the bar Ω_b with respect to the rotational speed of the gas in the disk. Or, in other words, the position of the corotation radius, which is described by the calculations far beyond the area shown.

Thus, the presence of the inner two-arm leading spiral which changes with the distance from the centre to a lagging spiral, should be regarded as the typical manifestation of a slowly rotating non-axisymmetric external gravitational perturbation.

7.9. Comparison of flocculent galaxies and grand design spirals

In this chapter, we have so far considered galaxies in which the gas disk contains two clearly visible spiral arms (grand design). Elmegreen and Elmegreen (1982, 1987) proposed a classification of spiral galaxies based on 12 types of spiral structure. In their table, the galaxies with a regular extended spiral pattern and flocculent galaxies occupy the opposite sides, acting as the antipodes.

Figure 7.35a shows a typical flocculent galaxy NGC 4414 in the H_α line observed using the 6 m telescope of the Special Astrophysical Observatory (SAO) of the Russian Academy of Sciences. We do not see any regular structure. But this only applies to gas disks of flocculent galaxies. Since the mid-1990s, regular structures have been detected in the stellar disks of flocculent galaxies in observations in the *K*′-band (2.1 μm) (Fig. 7.35b) (Thornley and Mundy, 1996, 1997; Block, Elmegreen, and Wainscoat, 1996; Grosbol and Patsis, 1998; Elmegreen et al., 1999).

Let's try to clarify the nature of the differences between these types of galaxies, based on the results of Fourier analysis of brightness and a velocity field map along the line of sight for the flocculent and regular spiral galaxies.

Let us compare the flocculent galaxy NGC 4414 (Fig. 7.35) with the galaxy NGC 157 (see Fig. 7.12) which, according to the work by the Elmegreens (1987), belongs to the highest 12-second type galaxies – grand design. Generally speaking, there is no clear evidence of significant differences between the flocculent and

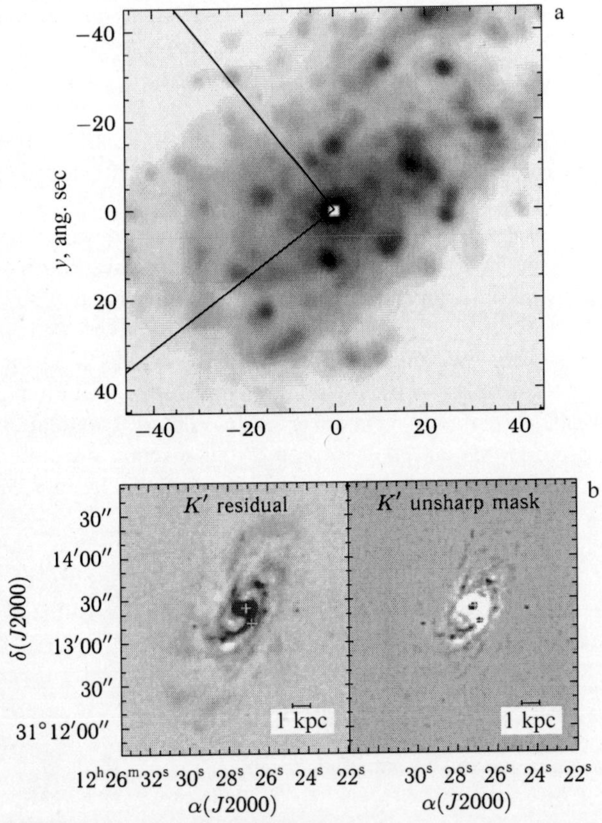

Fig. 7.35. a – an image of NGC 4414 flocculent galaxy in the emission lines H_α (6-m telescope of SAO RAS), corresponding to an ionized component in the gas disk. b – stellar disk in the band *K*′ 2.1 mm [Thornley and Mundy, 1997]. A notable feature is clearly the correct spiral structure in the stellar component and the absence of a global pattern in the gas disk.

regular galaxis of Sa-Sc-types when consideri0ng their photometric properties, star formation rate and gas content in the range of large dispersion values of their properties (Elmegreen and Elmegreen, 1986; Romanishin, 1985). However, one should take into account two circumstances. First, there are virtually no galaxies of the grand design type amongst later types than Sc (Elmegreen and Elmegreen, 1986), considering that the late-type galaxies that have the largest relative mass HI relative to the total mass (see, for example, Broeils and Rhee, 1997). Second, the grand design galaxies have systematically lower radial gradients of the circular velocity in the outer regions of disks in relation to flocculent galaxies, which is a proof of the fact that in the first type of disk galaxies, the ratio of the disk mass to the total mass exceeds this value in the objects of the second type (Biviano et al., 1991). This indicates a trend that galaxies with regular spirals are rarer than gas-rich light disk relative to the total mass.

The reason for the observed morphological differences in gas disks of the flocculent and grand design galaxies is clearly seen from Fig. 7.36a and 7.37a. The figures show the Fourier spectra of the deviations from azimuthal symmetry in various galactocentric rings for brightness maps H_α (6-m telescope of SAO RAS) of the flocculent galaxy NGC 4414 (Fig. 7.36a) and the regular spiral galaxy NGC 157 (Fig. 7.37a). In the first case (NGC 4414) we can not find a dominant harmonics throughout the disk. In the case of NGC 157 the second harmonics dominates everywhere the except zone 11″–19″, where resonance 4:1 is localized[4]. Some rings shows the predominance of higher even harmonics.

Thus, the spectra of the Fourier brightness maps of flocculent are grand design galaxies are qualitatively different, and the corresponding Fourier spectra of the stellar disks are similar in appearance. Indeed, the Fourier analysis of the brightness maps in the band *R* in Fig. 7.36b and 7.37b (HST archives, and Jacobus Captein Telescope) reveals the predominance of the second harmonic in the stellar disk of the grand design galaxy in 8 out of 9 annular zones shown in the figure, and in the case of the flocculent galaxy this is observed in 6 out of 9 annular zones[5]. This fact explains the different visual manifestation of the gaseous and stellar disks in flocculent galaxies, given the close correlation between spiral structures in the stellar and gas disks in the galaxies with a regular spiral pattern.

More detailed analysis showed that the gas disks of the flocculent galaxies are dynamically linked to the stellar subsystem. In addition, in some respects there is a similarity between flocculent and regular spirals. Note almost the exact match second phase Fourier harmonics of the brightness of the gas (squares) and star (asterisk) disks in both cases [6](Fig. 7.38a, b).

[4]The relatively high amplitudes of the first harmonic in NGC 157 are related to unequal brightness of the two spiral arms due to the asymmetry of the brightest areas of star formation. If we hide these regions, the amplitude the first harmonic is halved.

[5]The strong first harmonics in the region $r < 40''$ in the case of NGC 4414 and in the region $r < 10''$ in the case of NGC 157 is caused by decrease of the bright spherical component and does not correspond to any disk structure, see also Fig. 4 in Fridman and Khoruzhii (2000).

[6]Coincidence is broken only in a narrow range from 25″ to 35″ (Fig. 7.38a). Only in this region (Fig. 7.39) one can see the predominance of the seventh harmonic, which is 2.8 times greater than the second and, therefore, could distort the latter, interacting with it.

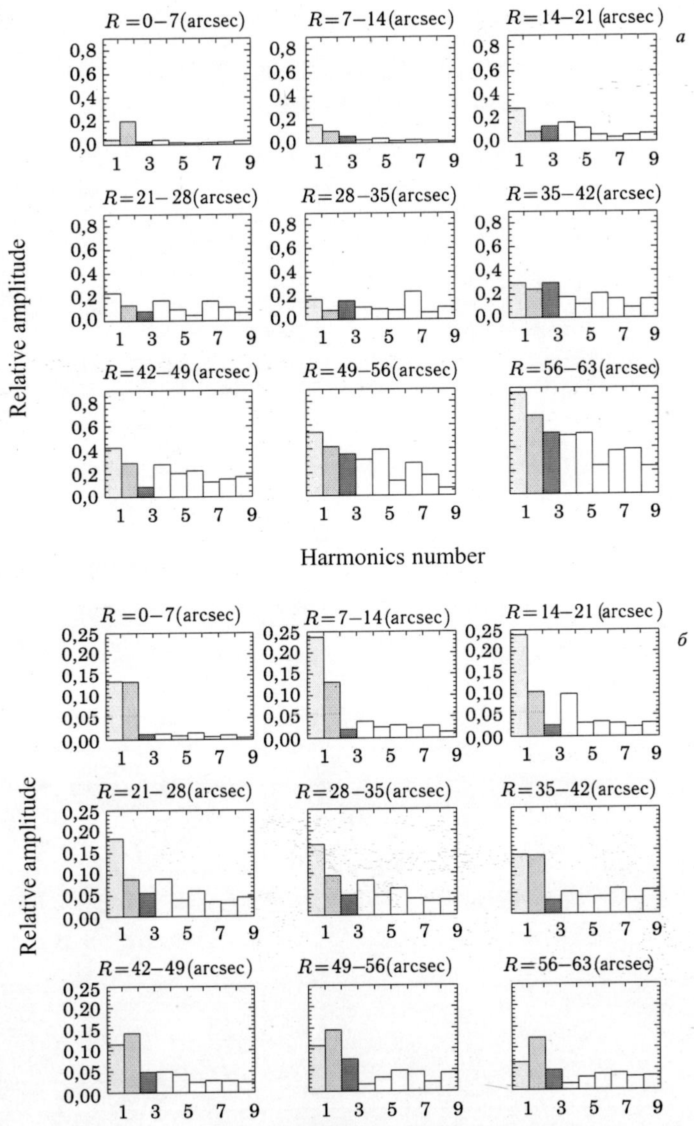

Fig. 7.36. Fourier spectrum of the brightness maps of the gas disk (H_α) (a) and the stellar disk (b) in the band *R* for flocculent galaxy NGC 4414.

In both cases the spectra of the velocity of the gas disks along the line of sight after subtracting the disk rotation speed show the dominance of the first three Fourier harmonics (Fig. 7.38c, d). This fact should be considered as evidence of the dominance of the second harmonic of the perturbed velocity components $\tilde{V}_r, \tilde{V}_\varphi, \tilde{V}_z$ (i.e. in (7.5)–(7.8) we can set $m = 2$) in both galaxies (Lyakhovich et al., 1997; Fridman et al., 1997, 2001).

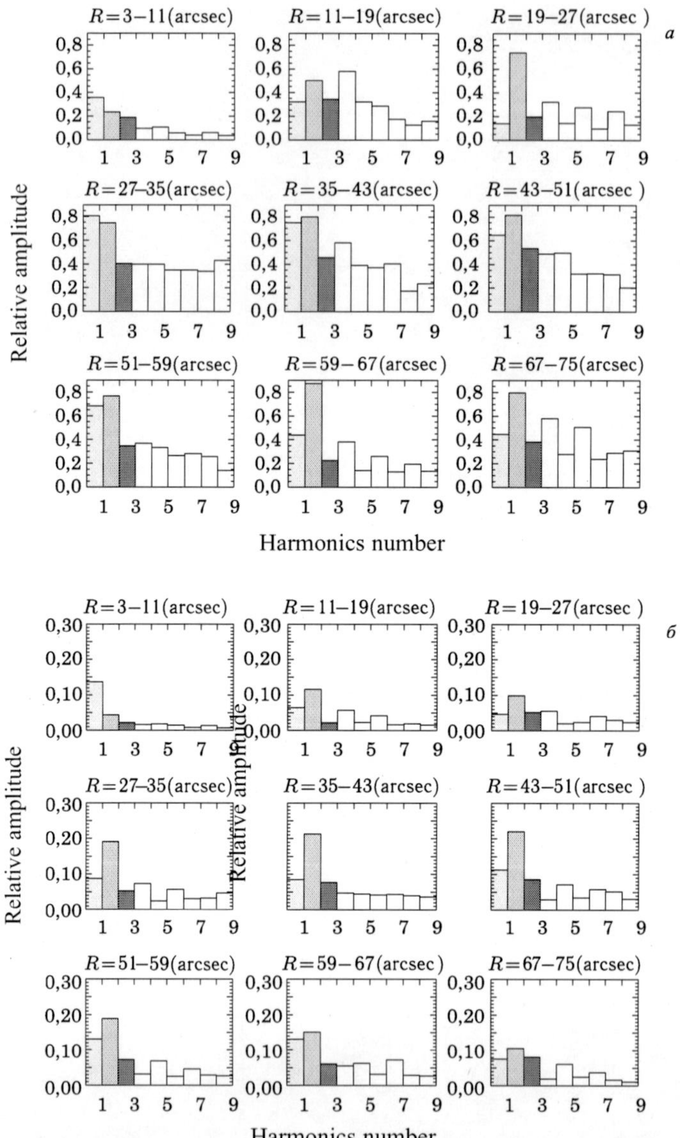

Fig. 7.37. Fourier spectrum of the brightness maps of the gas disk (H_{α}) (a) and Fourier spectrum of the brightness of the stellar disk (b) in the band R for the galaxy NGC 157.

Thus, in both the regular spiral patterns and at least in some flocculent galaxies, the observed behaviour of the stellar and gas discs is caused by the same dynamic mechanism.

The simplest model to explain the above observation facts is shown in Fig. 7.40a, b. The value of k_r is the radial component of the eigenvector, σ_i, and c_i are respectively,

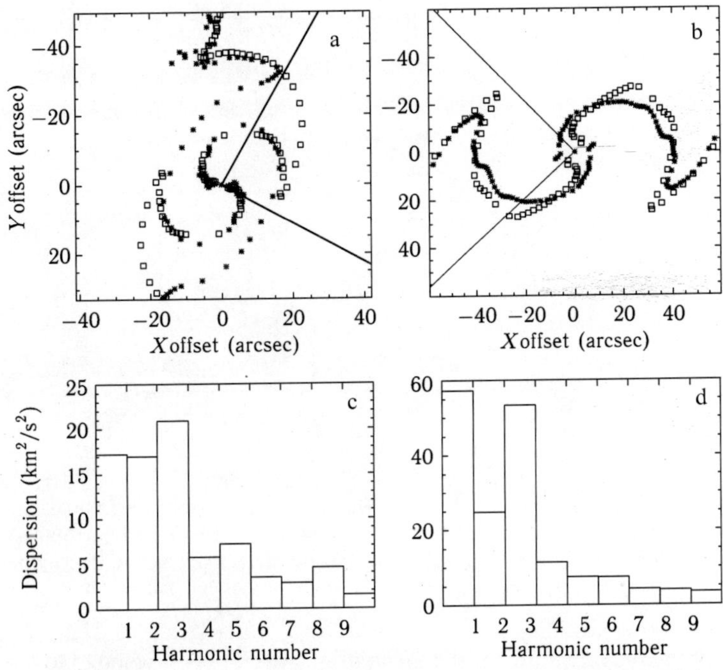

Fig. 7.38. Some similarities between the properties of flocculent galaxy NGC 4414 (a and c) and the grand design galaxy NGC 157 (b and d).

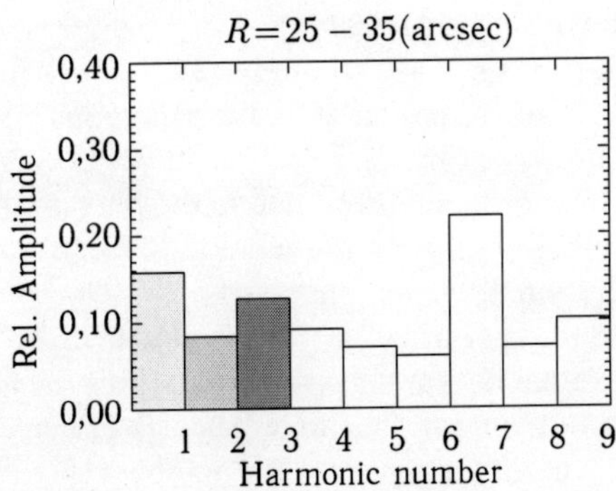

Fig. 7.39. Fourier spectrum of the brightness map in the H_α for NGC 4414 in the radial ring $25'' < r < 35''$.

the surface density and velocity dispersion, the index g denotes parameters of the gas disk, the index * allocates stellar disk parameters, the perturbations are indicated by ~, $\varkappa$ is the epicyclic frequency. In the case of a flocculent galaxy (Fig. 7.40a), we used the following parameters: $Q_*^2 = (\varkappa c_* / \pi G \sigma_*)^2 = 1.7$, $\sigma_g / \sigma_* = 0.2$, $c_g / c_* = 0.2$, which gives $Q_g^2 = 1.7$. In the case of spiral galaxies with a global pattern (Fig. 7.40b) $Q_*^2 = 1.25$, $\sigma_g / \sigma_* = 0.1$, $c_g / c_* = 0.2$, which leads to $Q_g^2 = 5$. These parameters are

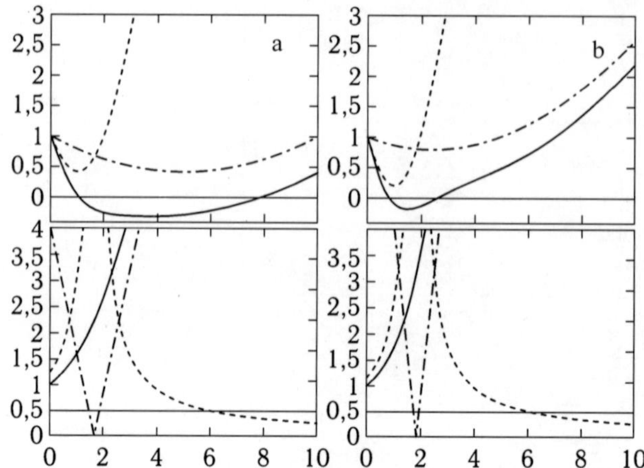

Fig. 7.40. Spectral characteristics of intrinsic perturbations in the gas disk (dashed line), stellar disk (dotted line) and for a system of these disks. The dependences on the dimensionless radial wave number $k_r/(\pi G\sigma_*/c_*^2$ are shown. The upper graphs show the squares of the dimensionless frequency $\omega^2/\varkappa^2$, at the bottom – the relationship of the dimensionless amplitude of density perturbations $(\tilde{\sigma}_g/\sigma_g)/(\tilde{\sigma}_*/\sigma_*)$. a) flocculent galaxy, b) spiral galaxy with a global pattern.

chosen as fairly typical for the demonstration and do not correspond to any particular galaxy. Note however, that they are consistent with the properties described at the beginning: the flocculent galaxies have a higher ratio of the mass of the gaseous disk to the stellar disk, and a lower relative weight of the disk to the total mass compared with the grand design galaxies.

In the flocculent galaxy (Fig. 7.40a), both the stellar disk in the absence of the gas disk and the gas disk without the stellar disk are marginally stable. As a result of gravitational interaction between the disks an instability occurs in a wide range of wave numbers (Fig. 7.40a, top). However, due to the higher velocity dispersion of the stellar disk there is a high relative amplitude of the perturbations of the surface density only for the very small wavenumbers (Fig. 7.40a, bottom). The potential is mainly determined by the perturbations of the stellar disk. Therefore, the field velocity has the form similar to the density perturbations in the stellar disk. In a galaxy with a regular spiral pattern the single gas disk is much more stable, since the instability occurs in a narrow range of wave numbers (Fig. 7.40b).

In conclusion, we make two observations. First, the dispersion curves in Fig. 7.40 correspond to the local dispersion relation, while for a more accurate analysis the calculations should be performed for the global eigenmodes (see Bertin and Lin, 1995; Bertin, 2000). Second, in analysis we examined the conditions of gravitational instability of only radial perturbations. However, it is well known that the non-axisymmetric waves k_φ are more unstable (see chapter 2), but this will only lead to quantitative changes. The qualitative nature of the differences in the spectral characteristics, as shown in Fig. 7.40, does not change for more accurate calculations.

8

Non-linear effects in astrophysical disks

8.1. Non-linear perturbations in a gravitating disk near the boundary of stability. Soliton solutions

8.1.1. Formulation of the problem

Consider the problem of gravitational stability of waves of small but finite amplitude, focusing on the role of this non-linear effects [41]. Therefore, when choosing a model, we restrict ourselves to the simplest system of equations (3.239), (3.242) with Ω_z = const, and the Poisson equation can be written as

$$\frac{\partial}{r\,\partial r}\left(r\frac{\partial \Psi}{\partial r}\right)+\frac{1}{r^2}\frac{\partial^2 \Psi}{\partial \varphi^2}+\frac{\partial^2 \Psi}{\partial z^2}=4\pi G\sigma\delta(z), \tag{8.1}$$

where $\delta(z)$ is the delta-function.

Confining ourselves to the adiabatic case without external gravitational fields, we write:

$$p=\frac{c_s^2}{\gamma_s}\sigma_0\left(\frac{\sigma}{\sigma_0}\right)^{\gamma_s},\quad c_s^2=\gamma_s\frac{p_0}{\sigma_0},\quad \gamma_s=3-\frac{2}{\gamma}. \tag{8.2}$$

All are traditional symbols (γ, γ_s are three-dimensional and two-dimensional adiabatic indices, respectively).

We represent all the physical quantities as the sum of the equilibrium quantities f_0 and disturbance $\tilde{f}$. Then for the perturbation of the radial velocity and surface density we have:

$$\tilde{u}=-\frac{1}{\eta_0}\hat{L}\tilde{v}, \tag{8.3}$$

$$\tilde{\sigma}=\frac{\sigma_0}{\eta_0}\frac{\partial \tilde{v}}{\partial r}, \tag{8.4}$$

where we have introduced the notations

$$\hat{L} = \frac{\partial}{\partial t} + \Omega_0 \frac{\partial}{\partial \varphi}, \quad \eta_0 = \frac{1}{r}\frac{\partial r v_0}{\partial r}.$$

Substituting (8.2)–(8.4) in equation (3.15), we obtain

$$\hat{L}^2 \tilde{v}_\varphi = \left(c_s^2 \frac{\partial^2}{\partial r^2} - 2\Omega_0 \eta_0 \right) \tilde{v}_\varphi + \eta_0 \frac{\partial \tilde{\Psi}}{\partial r} + \frac{(\varkappa - 2)}{2\eta_0} c_s^2 \frac{\partial}{\partial r} \times$$
$$\times \left[\left(\frac{\partial \tilde{v}_\varphi}{\partial r} \right)^2 + \frac{\varkappa - 3}{3\eta_0} \left(\frac{\partial \tilde{v}_\varphi}{\partial r} \right)^3 \right] - \eta_0 \frac{\tilde{v}_\varphi^2}{r}. \tag{8.5}$$

The approach is based on the assumption of slow variation with time of $\frac{\partial \tilde{f}}{\tilde{f} \partial t} \ll \Omega$.

The non-linearity of the above equation is due to the terms in square brackets and the last term.

We seek solutions in the form:

$$\tilde{v}_\varphi(r,t) = \tilde{v}(t) \exp\left[i \int_0^r k(r) dr \right], \quad \tilde{\Psi}(r,t) = \tilde{\Psi}(t) \exp\left[i \int_0^r k(r) dr \right], \tag{8.6}$$

considering $k_\varphi = m/r \ll k_r = k$. Due to the linearity of (8.1) we obtain the relationship between velocity and potential:

$$\tilde{\Psi}(t) = -i \frac{2\pi G \sigma_0}{\eta_0} \tilde{v}(t) \operatorname{sign} k, \quad \operatorname{sign} k = \frac{k}{|k|}. \tag{8.7}$$

Obviously, in the linear approximation we have the dispersion relation (2.111).

Confining ourselves to the case when Ω_0 = const, we start from the equilibrium density profile

$$\sigma_0(r) = \sigma(0)\sqrt{1 - r^2/R^2} \tag{8.8}$$

and the speed of sound

$$c_s^2(r) = c_s^2(0)(1 - r^2/R^2), \tag{8.9}$$

thereby aligning with the dispersion equation, we obtain

$$|k|_{1,2} = (1 - r^2/R^2)^{-1/2} [k_0(0) \pm (k_0^2(0) + a^2)^{1/2}], \tag{8.10}$$

where

$$k_0(0) = \pi G \sigma(0) / c_s^2(0), \quad a^2 = (\omega^2 - 4\Omega_0^2)/c_s^2(0). \tag{8.11}$$

Considering the entire disk ($r \leqslant R$), we rewrite (8.6) in the form of the Bohr-Sommerfeld quantization rule

$$\int_0^R k(r, \omega) dr = \pi \left(N + \frac{1}{2} \right). \tag{8.12}$$

After integration we obtain for the stability boundary $\omega = 0$

$$\left(N+\frac{1}{2}\right)_{1,-1} = [k_0(0)R/2] \pm \sqrt{[k_0(0)R^2/4] - \Omega_0^2 R^2 / c_s^2(0)}. \tag{8.13}$$

Considering

$$k_0(0)R \geq 2\, v_0(R)/c_s(0), \tag{8.14}$$

for the stability boundary

$$N_1 = [k_0(0)R/2] + \sqrt{[k_0(0)R^2/4] - v_0^2(R)/c_s^2(0)},$$
$$N_{-1} = [k_0(0)R/2] - \sqrt{[k_0(0)R^2/4] - v_0^2(R)/c_s^2(0)}. \tag{8.15}$$

Assuming that only the quantum number N_0 corresponds to the unstable region, we have $N_1 – N_{-1} = 2$, or

$$k_0^2(0)R^2/4 - v_0^2(R)/c_s^2(0) = 1,$$

where

$$k_0(0)R = 2\sqrt{1 + v_0^2(R)/c_s^2(0)}. \tag{8.16}$$

We now turn to the integration of nonlinear effects. We believe

$$\int_0^r k_N(r)dr = \bar{k}_N\, r = nk_0 r,$$

where $k_0 = k_{N_0}$. We write for the perturbed quantities the Fourier series:

$$\tilde{v} = \sum_n v_n \exp(ink_0 r), \quad \tilde{\Psi} = \sum_n \Psi_n \exp(ink_0 r). \tag{8.17}$$

Substituting (8.17) into (8.5) with (8.7), and assuming

$$\omega_l^2 = 2\Omega_0\eta_0 + l^2 k_0^2 c_s^2 - 2\pi G\sigma_0 l\,|\,k_0\,|, \tag{8.18}$$

we obtain the equation

$$\sum_l \hat{L}^2 v_l(t,\varphi)\exp(ilk_0 r) = -\sum_l \omega_l^2 v_l(t,\varphi)\exp(ilk_0 r) -$$
$$-\frac{\varkappa-2}{2\eta_0}\sum_l lk_0 c_s^2\Big[i\sum_n nk_0^2(l+n)v_l(t,\varphi)v_n(t,\varphi)\exp(i(l+n)k_0 r) -$$
$$-\frac{\varkappa-3}{3\eta_0}\sum_{n,m} k_0^3 nm(l+n+m)v_l(t,\varphi)v_n(t,\varphi)v_m(t,\varphi)\times$$
$$\times\exp(i(l+n+m)k_0 r)\Big] + \frac{\eta_0}{r}\sum_{l,n} v_l(t,\varphi)v_n(t,\varphi)\exp(i(l+n)k_0 r). \tag{8.19}$$

Equation (8.18) can be rewritten in the form of $\sum_\ell A_\ell \exp\{i\ell k_0 r\}$. The case

A_0 corresponds to the Toomre linear approximation. For equations $A_1 = 0$ and $A_2 = 0$, we have:

$$\hat{L}^2 v_1(t,\varphi) = -\omega_{k_0}^2 v_1(t,\varphi) + \frac{\varkappa - 2}{2\eta_0} k_0^3 c_s^2 \Big[4i v_2(t,\varphi) v_{-1}(t,\varphi) - \\ -\frac{\varkappa - 3}{\eta_0} k_0 \,|\, v_1(t,\varphi) \,|^2\, v_1(t,\varphi) \Big] + 2\frac{\eta_0}{r} v_2(t,\varphi) v_{-1}(t,\varphi), \tag{8.20}$$

$$\hat{L}^2 v_2(t,\varphi) = -\omega_{2k_0}^2 v_2(t,\varphi) - \frac{\varkappa - 2}{2\eta_0} k_0^3 c_s^2 2i v_1^2(t,\varphi) + \frac{\eta_0}{r} v_1^2(t,\varphi), \tag{8.21}$$

where

$$\omega_{2k_0} = \pm\sqrt{2\Omega_0 \eta_0} \quad \eta_0 > 0. \tag{8.22}$$

$$v_2(t,\varphi) = \frac{1}{\hat{L}_{k_0}^2 + \omega_{2k_0}^2} \left(i\frac{2-\varkappa}{\eta_0} k_0^3 c_s^2 + \frac{\eta_0}{r} \right) v_1^2(t,\varphi), \tag{8.23}$$

here $\varkappa$ is the epicyclic frequency.

The resulting equation is easy to reduce to a non-linear equation:

$$\hat{L}^2 v_1(t,\varphi) = \gamma_{k_0}^2 v_1(t,\varphi) + \frac{(2-\varkappa) k_0^2 \Omega_0}{\eta_0} \cdot \left[\frac{8(2-\varkappa)\Omega_0 \eta_0}{\omega_{2k_0}^2} - \\ -(3-\varkappa) \right] \cdot |\, v_1(t,\varphi) \,|^2\, v_1(t,\varphi). \tag{8.24}$$

where γ_{k0} is the maximum growth rate corresponding to the Jeans instability.

8.1.2. Explosive gravitational instability in the disk

When the condition

$$\varkappa < 5/3 \tag{8.25}$$

(equivalent to γ <3/2) is satisfied, the amplitude of the perturbation can be arbitrarily large in finite time. We rewrite (8.24) in its simplest form:

$$\frac{\partial^2 v_1(t)}{\partial t^2} = \left[\gamma_{k_0}^2 + 3\frac{\Omega_0}{\eta_0}(2-\varkappa)\left(\frac{5}{3} - \varkappa\right) k_0^2 \,|\, v_1(t) \,|^2 \right] v_1(t). \tag{8.26}$$

Equation (8.26) can be written in quadratures (index 1 is omitted):

$$t - t_0 = \int_{v_0}^{v} \frac{dv}{\sqrt{w_0(r) + \gamma_{k_0}^2 + A v^4}}, \tag{8.27}$$

where

$$A = \frac{3}{2}\frac{\Omega_0}{\eta_0}(2-\varkappa)\left(\frac{5}{3} - \varkappa\right) k_0^2,$$

v_0 is the initial velocity perturbation at time t_0, and $w_0\,(r)$ is an arbitrary function of the radial coordinate. In the limit $\gamma k_0^2 v_0^2 \ll A v_0^4$

$$v / v_0 = 1 / [1 - \sqrt{A} v_0 (t - t_0)]. \tag{8.28}$$

As we see, in a finite time perturbations grow to infinite values, indicating that an explosive instability forms.

8.1.3. Solitons in the disk

In the parameter region

$$5/3 < \varkappa < 2 \tag{8.29}$$

equation (8.24) has stationary solutions of the soliton type. The amplitudes of the perturbations grow to values:

$$|v_1|^2 = \frac{\gamma_{k_0}^2 \eta_0}{3(2-\varkappa) k_0^2 \Omega_0 (\varkappa - 5/3)}, \tag{8.30}$$

$$\frac{|\sigma_1|^2}{\sigma_0^2} = \frac{k_r^2}{k_0^2} \frac{\gamma_{k_0}^2}{3(2-\varkappa)\Omega_0 \, \eta_0 (\varkappa - 5/3)}. \tag{8.31}$$

Consider a sufficiently narrow wave packet $\Delta k/k_0 \ll 1$ in a neighborhood of the wave number k_0. We use the decomposition

$$\gamma_k^2 = \gamma_{k_0}^2 + \frac{1}{2} \frac{\partial^2 \gamma_k}{\partial k^2}\bigg|_{k=k_0} (k - k_0)^2 + \ldots \tag{8.32}$$

Given $\dfrac{\partial^2 \gamma_k^2}{2\partial k^2} = -c_s^2, k - k_0 = kr - k_{0r}, k_r >> k\varphi,$ we rewrite (8.32) as

$$\gamma_k^2 = \gamma_{k_0}^2 - k_{1r}^2 c_s^2 . \tag{8.33}$$

Referring to equation (8.32) taking into account the conversion gives

$$\int v_1(k,t) e^{ik_1 r} dk_1 = v_1(r,t).$$

As a result, we finally obtain the equation

$$\hat{L}^2 v_1(\vec{r},t) = (\gamma_{k_0}^2 + c_s^2 \Delta_r) v_1(\vec{r},t) + 3\frac{(2-\varkappa)k_0^2 \Omega_0}{\eta_0} \times \left(\frac{5}{3} - \varkappa\right) |v_1(\vec{r},t)|^2 v_1(\vec{r},t), \tag{8.34}$$

where $\Delta_r = \dfrac{\partial^2}{\partial r^2}$.

Turning to the local coordinate system rotating with the angular velocity of the disk, we assume that $v_1(\mathbf{r}, t) = V(\xi) = V(\mathbf{r} - \mathbf{u}t)$. Regarding 'the travelling coordinate' ξ we get instead of the partial differential equation the ordinary differential equation

$$\alpha^2 \frac{d^2 V}{d\xi^2} = \gamma_0^2 V - \beta^2 V^3 = -\frac{\partial W}{\partial V}, \tag{8.35}$$

in which

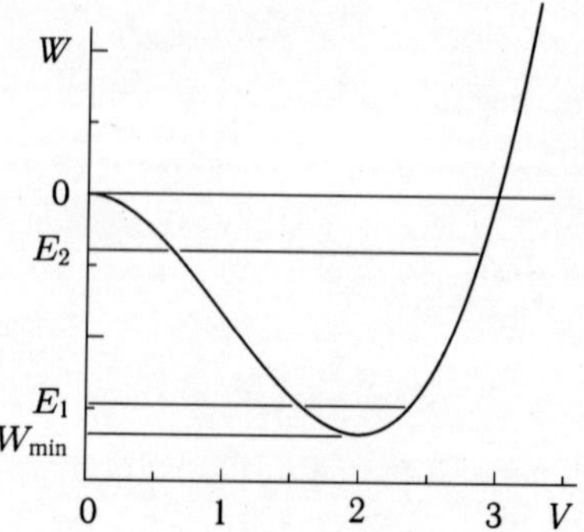

Fig. 8.1. The energy level in the 'potential well' $W(V)$ determines the type of wave. At E_1 vibrations are slightly different from harmonic (linear). At E_2 the non-linearity becomes important and cnoidal waves are formed. The level $E = 0$ corresponds to motion along the separatrix in the phase plane.

$$\alpha^2 = u^2 - c_s^2, \quad \beta^2 = 3(2-\varkappa)\left(\varkappa - \frac{5}{3}\right)\frac{\Omega_0}{\eta_0}k_0^2. \tag{8.36}$$

Equation (8.35) is similar to the equation of motion of a particle in a potential $W = W = -\gamma_0^2 V^2/2 - \beta^2 V^4/4$ (V plays the role of 'the coordinate', and ξ of 'time' (Fig. 8.1).

The solution (8.35) is a soliton

$$V(\xi) = \sqrt{2}\frac{\gamma_0}{\beta}\,\mathrm{sch}\frac{\gamma_0 \xi}{\sqrt{u^2 - c_s^2}}. \tag{8.37}$$

8.2. The observed structure of the oscillations of the soliton envelope in the spiral arms of NGC 1365 (grand design)

The bulk of the interstellar gas in galaxies accounts for neutral hydrogen, so the most direct way of measuring the field velocity of the gas disk are observations in the line of HI 21 cm. However, the angular resolution of radio observations, which may be achieved on a large field, is greatly inferior to the resolution of the observations in the optical range. It is for this reason that almost all studies in detail of the velocity field of spiral galaxies (except for the closest ones) were obtained by measuring radial velocities in the line of H_α.

The special position amongst the radiodata is occupied by the observation of the Galaxy NGC 1365. This giant galaxy, close to our Galaxy, has a strong spiral structure. Therefore, the resolution in the 21 cm line is sufficient to study the detailed internal structure of spiral arms in the gas disk of the galaxy.

Observations show that the neutral hydrogen (HI) is concentrated in the spiral arms. The difference density is determined by the pressure on both sides of the shock wave front [112]:

$$\frac{\varrho_1}{\varrho_2} = \frac{(\gamma+1)p_1 + (\gamma-1)p_2}{(\gamma-1)p_1 + (\gamma+1)p_2}, \tag{8.38}$$

where in the area 1 the flow is supersonic, and in 2 subsonic. Formula (8.38) is valid for an ideal gas. Having the volume adiabatic index $\gamma = 5/3$, the atomic gas can have a density at the front of the 'classical' strong shock wave four times exceeding the density before the front, which follows from (8.3) in the limit $p_2/p_1 \to \infty$.

However, observations show that this ratio can be almost an order of magnitude larger. Additional compression of gas in comparison with the compression expected in the 'classical' shock wave is caused by strong cooling of gas due to radiation (free–free transitions). However, we are interested in structures whose scale is substantially greater than the width of the shock front of the spiral wave density in the atomic gas.

The question of the structure of spiral arms in the atomic gas is closely linked to another issue (almost 50 years 'old') that the gas disks of galaxies are (with the set of their basic parameters) near the boundary of dynamic instability. This situation is natural: as the instability increases the velocity dispersion also increases, the disk is 'heated' and is becoming more and more stable until the very reason for heating the disk disappears – the disk reaches the stability boundary. Studies [74], as well as more recent ones, confirmed this assumption by the example of the different samples of spiral galaxies.

Mikhailovskii, Petviashvili, and Fridman [140, 141] have shown that the non-linear dynamics of the gas of the galactic disk at the boundary of gravitational instability is described by the non-linear Schrödinger equation (NSE):

$$i\left(\frac{\partial A}{\partial t}+V_0\frac{\partial A}{\partial x}\right)-\frac{V_0}{4k_0}\frac{\partial^2 A}{\partial x^2}-V_0k_0^3|A|^2A=0, \tag{8.39}$$

where ω_0 – frequency, k_0 – wave number of the fundamental wave, $V_0=\left(\frac{\partial\omega}{\partial k}\right)_0=\frac{\omega_0}{2k_0}$ is the group velocity.

In the canonical form of the cubic NSE can be written as:

$$i\frac{\partial u}{\partial t}+\frac{\partial^2 u}{\partial x^2}+\beta|u|^2u=0, \tag{8.40}$$

which formally coincides with the quantum mechanical Schrödinger equation for a particle in the potential, $\beta|u|^2$. With the help of this equation we can describe the distribution of the envelopes of wave packets in a dispersive medium in the weakly non-linear limit (Fig. 8.2).

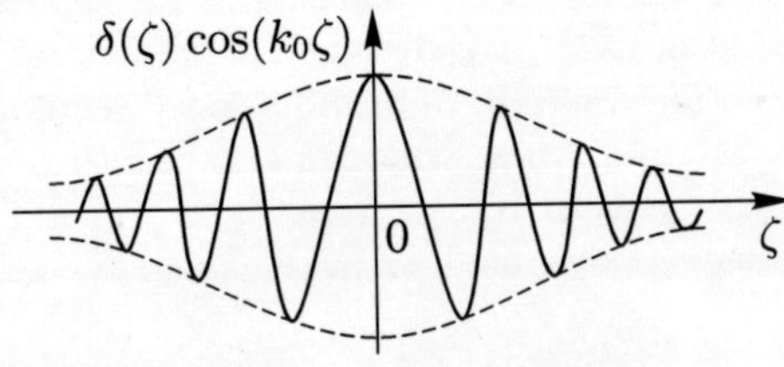

Fig. 8.2. The theoretically predicted structure of spiral arms as a soliton of the envelope [140, 141], ζ is the wave variable characterizing the incident wave, δ is the function defining the shape of the soliton envelope.

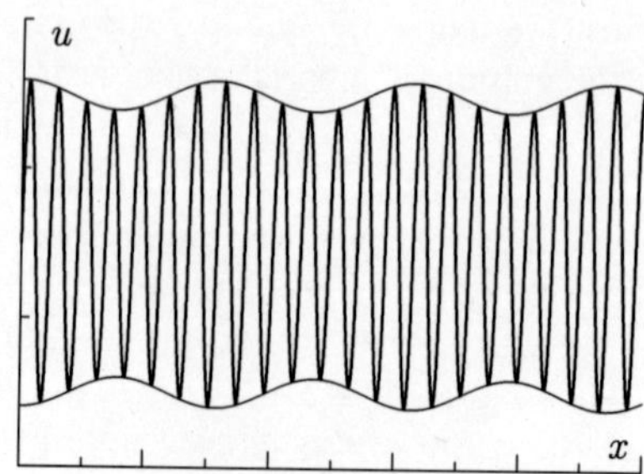

Fig. 8.3. Harmonic oscillations with the slowly changing amplitude.

Consider the harmonic solution

$$u = a\exp\{i(kx - \omega(k)t)\}, \tag{8.41}$$

assuming the amplitude a to be small but finite. Changing the amplitude of a harmonic wave may be due to the impact of the higher harmonics produced by the non-linear term in the NSE. As a result, the wave envelope varies slowly in the spatial and temporal coordinates (Fig. 8.3)

NSE-type equations describe the dynamics of waves in plasma physics, hydrodynamics, non-linear optics and other branches of science:

– propagation of electromagnetic waves in a medium where the refractive index depends on the electric field;

– amplitude-modulated radio waves;

– optical self-focusing in propagation of the laser beam in an environment in which dielectric polarization depends in a non-linear manner on the electric field;

– dynamics of the amplitude of the wave packets on the surface of the reservoir of finite depth;

– instability of Stokes waves in deep water;

– oscillatory processes in α-helices (peptide groups in the presence of hydrogen bonds) of protein molecules;

– when describing the two-layer Kelvin–Helmholtz instability;

– Langmuir waves in electron plasma.

This is not a complete list and the dynamics of astrophysical disks should be added.

The theoretical results and observational data make it possible to propose a hypothesis that the spiral arms of galaxies are the solitons of the envelope (the latter described by one of the solutions of the non-linear Schrödinger equation, see Fig. 8.2). The idea of the spiral arms, which retain their shape in a differentially rotating galactic disk, as a soliton which has the property of invariance in propagation in inhomogeneous flows and media and, by definition, look very natural and attractive.

Figure 8.4 shows that the real spiral arms in the gas disk of NGC 1365 contain an oscillating structure – the large scale envelope describes two spiral arms, within each of which there is a filling of small-scale quasi-periodic variations in density. Such a structure is in good agreement with that predicted by the theory

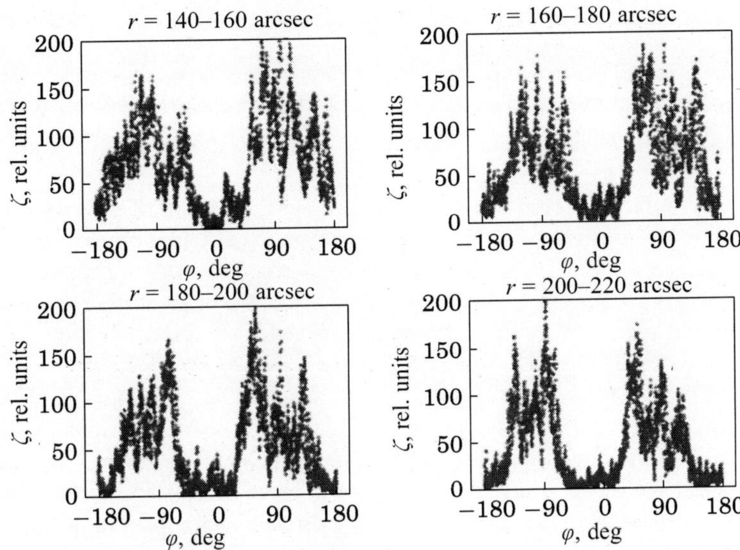

Fig. 8.4. The azimuthal distribution of radio radiation intensity I in the 21 cm line of the spiral galaxy NGC 1365 at different distances r from the centre of the galaxy.

of the structure of the soliton envelope (Fig. 8.2). Note that in principle similar observational manifestations could also have a secondary gravitational instability, which generates more short-wave perturbations in the secondary gas, 'compressed' as a result of the primary instability.

8.3. Observable manifestations of chaos in spiral galaxies

8.3.1. Statement of the problem

In chapter 7 it was shown that the resulting three-dimensional reconstruction of the velocity field is a very rich source of information on the dynamics of the galactic disk. For example, a steep drop in the velocity of the rotation curve, as in the case of Mrk 1040, testifies in favour of the hydrodynamic instability as a generator of spiral density waves. The position of the 'jump' of velocity indicates the place where the major resonance in the disk – corotation – is situated.

The centres of giant anticyclones are located on the corotation radius. These structures are an integral part of the spiral density waves, regardless of the nature of the mechanism of its generation. In the case of a smooth rotation curve, when the only mechanism of excitation of the spiral-vortex structure is the gravitational instability, in the process of recovery of the complete velocity field of the gas disk there are several independent ways to determine the corotation radius: with a jump of the phase shift (see section 7.3.2) and also based on an assessment of the positions of the centres of giant anticyclones visible in the reconstructed velocity field in the plane of the disk (see in Fig. 7.19). The location of these centres in relation to the spiral arms (taking into account the sign of $\varkappa^2$) gives one additional independent criterion for determining the nature of the generator of the spiral vortex structure of galaxies (see Table in section 7.2.4).

It was shown above how the knowledge of some parameters of the velocity field makes it possible to unambiguously determine the type of vertical movements of the gas and to answer the question whether the observed spiral structure in this galaxy is really a density wave.

Here we point out some additional features of movement of the gas in galactic disks. These features were first observed in real galaxies, based on the reconstructed three-dimensional velocity field – a very informative tool for research of the dynamics of galaxies. In this section, we discuss the theoretical results and the observational data, indicating the presence of chaos in spiral galaxies [244, 575, 579].

As follows from the observations, the gas disks of the spiral galaxies (the set of main parameters which characterise them) lie near the stability boundary. This fact is quite natural: in development of instability the velocity dispersion increases and the disk reaches the stability boundary [259]. Since the instability that generates spiral arms and vortices is saturated, the three-dimensional motion of the gas should be stationary in a reference frame rotating with the density wave. As shown by Sagdeev et al [56], the equations of motion of the three-dimensional steady-state velocity field can be reduced to dynamic equations in the two-dimensional phase space. The latter may be used as the basis for the study of stochastic dynamics of the stationary velocity fields of galactic gas disks. In this way, an important task is to demonstrate the fact that analysis of the observed velocity field of the galactic disk may be the source of our understanding of the stochastic dynamics of the galaxies.

This problem has more to do with the gas than with the stellar disk. Despite the obvious progress in the fields of radial velocity measurements of stellar disks, our understanding of stochastic star dynamics of external galaxies is composed almost entirely from the analysis of theoretical models. The reason that the information extracted from observational data in the fields of radial velocities of stars is more than modest is as follows.

Generally, the galaxies are not resolved into individual stars. Using the analysis of the absorption lines we can measure the field of radial velocities of the stars, averaged over a small spatial region. The trajectory of this small region of the stellar disk can significantly differ from the trajectories of the stars themselves in the field. Thus, for example, the stellar bar, whose ends are connected with the spiral sleeves, rotates with the sleeve as a rigid body, while the trajectories of stars in the bar may repeatedly intersect each other, be extremely complicated and very different from trajectories of a simple rotation. This behaviour is typical for collisionless self-gravitating systems.

Using observational data, we can construct the gravitational potential as a function of the coordinates. However, the change in the potential within the measurement error often leads to the conversion of regular stellar orbits into chaotic and vice versa.

The field of radial velocities of the gas disk, at first sight, also cannot be directly used to study the stochastic dynamics of the disk. However, using the method of recovery of the three-dimensional velocity field from the observed radial velocity field (chapter 7, [127, 579, 581, 584, 587] allows (by calculating the Lyapunov characteristic exponents) to identify regular and chaotic trajectories [567].

As is well known (see, e.g. [835]), the Lyapunov exponents are calculated for the trajectories in the phase space, whereas the recovery method (see chapter 7) gives a three-dimensional velocity field in the coordinate space. This difficulty, however, can be overcome. As shown in [56], the hydrodynamic equations describing the three-dimensional steady incompressible flow are reduced to the time-dependent dynamical equations in the two-dimensional phase space. The same is true for the three-dimensional compressible steady flow [631]. The idea of transformation is as follows.

The stationary three-dimensional flow is described by a set of equations

$$\frac{dx}{dt} = v_x(x,y,z), \quad \frac{dy}{dt} = v_y(x,y,z), \quad \frac{dz}{dt} = v_z(x,y,z), \tag{8.42}$$

which can be rewritten as follows:

$$\frac{dx}{v_x} = \frac{dy}{v_y} = \frac{dz}{v_z} = dt, \tag{8.43}$$

and then

$$\frac{dx}{dz} = \frac{v_x}{v_z} \equiv f_1(x,y,z), \quad \frac{dy}{dz} = \frac{v_y}{v_z} \equiv f_2(x,y,z). \tag{8.44}$$

The last equation shows that we are dealing with a 'transient' task for the two-dimensional dynamic system in the phase space (x, y), where z plays the role of time τ. Then we have

$$\frac{dx}{d\tau} = f_1(x,y,\tau), \quad \frac{dy}{d\tau} = f_2(x,y,\tau). \tag{8.45}$$

These equations describe the dynamic system in which chaotic and regular trajectories can co-exist.

8.3.2. Restored velocity field of the NGC 3631 spiral galaxy

Figure 8.5 shows the reconstructed velocity field of NGC 3631 in the plane of the galaxy with superimposed lines of the constant phase of the vertical (perpendicular to the plane of the disk) speeds. Maxima are marked by squares of absolute values of the vertical gas velocity for each radius. Asterisks show the positions of zeros of the vertical velocity (the amplitude of the vertical velocity is not shown), thin lines – the position of the vortices – anticyclones (upper left and lower right) and cyclones.

As noted in Section 8.3.1, the velocity field in the gas disk of a spiral galaxy is stationary.This means that the trajectories coincide with the streamlines in the three-dimensional coordinate space. In areas of the zero vertical movements each fluid 'particle' is involved only in two-dimensional motion and, therefore, the same trajectories must coincide with their two-dimensional streamlines. Perhaps this is why the separatrix of two anticyclones and two cyclones whose centres are

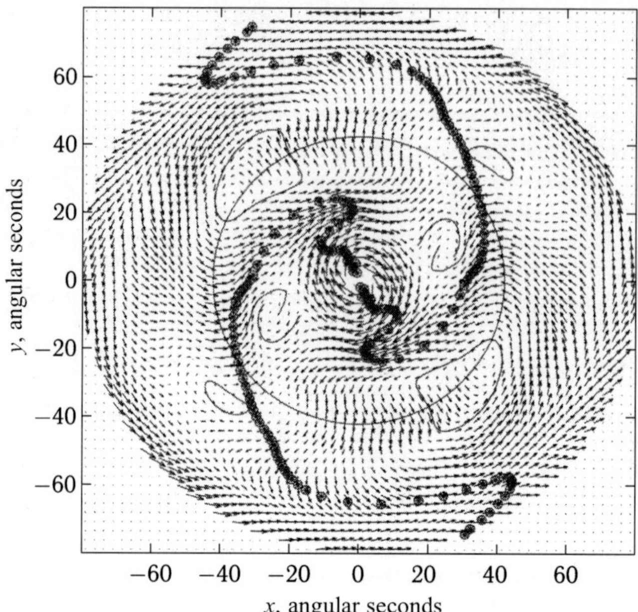

Fig. 8.5. Restored velocity field of NGC 3631 in a galaxy plane and superimposed lines of constant phases of vertical movements. Squares mark the azimuthal maxima of the absolute values of vertical gas velocity for each radius. Asterisks show zeros of vertical velocity, thin lines – the position of vortices – anticyclones (upper left and lower right) and cyclones.

near zero vertical velocities are closed. At the same time, the separatrices of two other cyclones whose centres lie between the lines of maximum and zero vertical velocities are not closed.

The 'unsteady' nature of the motion of fluid particles in the disk plane within these cyclones may be caused by vertical movement, which, unlike previous cases, can not be considered negligible.

In addition to the specific lines – the separatrices surrounding the vortices – the two-dimensional velocity field in the plane of the galaxy contains singular points – saddle (marked with crosses in Fig. 8.6). Choosing the start of the streamlines near the separatrix or near the saddle point, it can be seen that the streamlines diverge. It is interesting to ask about the velocity of divergence of the trajectories: is this velocity exponential or not, that is, are the streamlines chaotic or regular. To do this, we must calculate the Lyapunov exponents.

8.3.3. Computation of characteristic Lyapunov exponents

Under the chaos we imply the rapid mixing that occurs due to the rapid (exponential) divergence of the close trajectories. Consider two initially close points at a distance d_0 from each other. After some time t, they will be at distance $d(t)$. The exponential divergence of trajectories means that

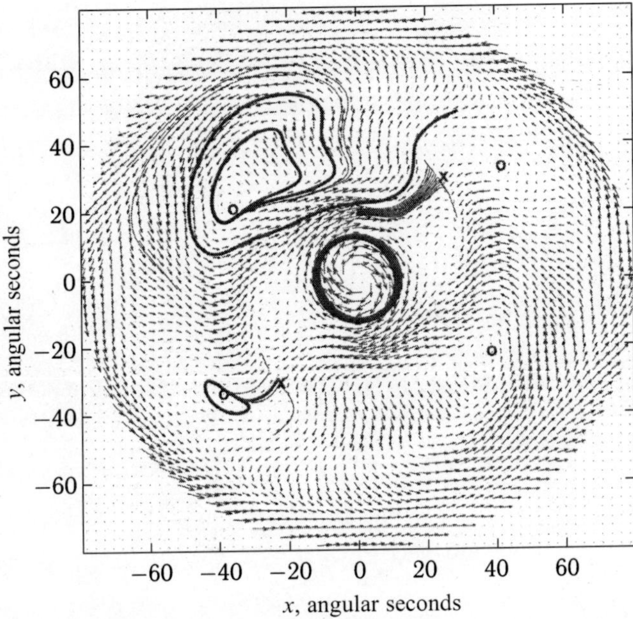

Fig. 8.6. The divergent trajectories near separatrices of vortices and the beam trajectories near the saddle point, superimposed on the reconstructed two-dimensional velocity field. Bold solid lines show trajectories which are used to calculate the stochastic characteristics near the vortex, 'o'-symbols indicate the position of centres of the vortices 'x' – the position of the saddle points. In the centre of the disk there is a family of regular orbits.

$$d(t) \sim d_0 \exp\{\lambda t\}, \tag{8.46}$$

where the exponent averaged over a large interval of time $\lambda > 0$ is the Lyapunov exponent (the maximum), and it characterizes the exponential divergence in the limit $t \to \infty$ and $d_0 \to 0$. In other words, the Lyapunov exponent is determined by the double limit [835]:

$$\lambda = \lim_{\substack{t\to\infty \\ d_0\to 0}} \frac{1}{t} \ln \frac{d(t)}{d_0}. \tag{8.47}$$

You can see that this limit exists and is the same for almost all mutual orientations of the originally selected points. Only for some mutual orientations the limit (8.47) is less than the maximum value [835]. Here we consider the situation of general position and drop the term 'maximum' for the Lyapunov exponent (see [835]).

The mathematically convenient formulation of the Lyapunov exponent is uncomfortable in the computer calculations, even in the study of idealised (in terms of real systems) theoretical models. A significant assumptions made in these calculations is the assumption that the distance $d(t)$ remains sufficiently small so that evolution of the vector[1] $\mathbf{w}\,(t)$, which connects points on two close streamlines

[1]The vectors and matrices will be denoted in bold.

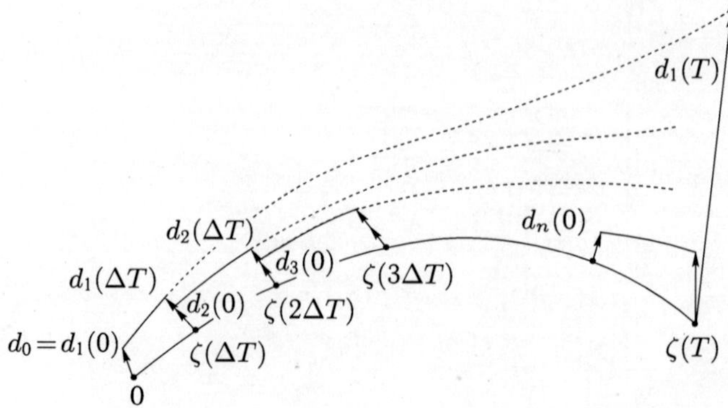

Fig. 8.7. The method of calculating the Lyapunov exponent for two trajectories. The main path – the lower solid curve, a few subsidiary trajectories are shown by short solid lines. The subsequent course of the subsidiary curves is shown by dashed lines.

(Fig. 8.7), is described by some linearised equation [835]. Indeed, by definition of the velocity **V** (ζ), we obtain the following equation:

$$\frac{d\zeta}{dt} = \mathbf{V}(\zeta),$$

$$\zeta = \zeta_0 + \mathbf{w},$$

$$\mathbf{V}(\zeta) = \mathbf{V}_0(\zeta_0) + \frac{d\mathbf{V}}{d\zeta}\bigg|_{\zeta=\zeta_0} \mathbf{w} + \frac{1}{2}\frac{d^2\mathbf{V}}{d\zeta^2}\bigg|_{\zeta=\zeta_0} \mathbf{w}^2 + \frac{1}{6}\frac{d^3\mathbf{V}}{d\zeta^3}\bigg|_{\zeta=\zeta_0} \mathbf{w}^3 + \ldots.$$

Using these relations, as well as the expression of $\mathbf{V}_0(\zeta_0) = d\zeta_0 / dt$, we find the time derivative of the small vector **w** as the sum of:

$$\frac{d\mathbf{w}}{dt} = \frac{d}{dt}(\zeta - \zeta_0) = \mathbf{V}(\zeta) - \mathbf{V}_0(\zeta_0) =$$

$$= \frac{d\mathbf{V}}{d\zeta}\bigg|_{\zeta=\zeta_0} \mathbf{w} + \frac{1}{2}\frac{d^2\mathbf{V}}{d\zeta^2}\bigg|_{\zeta=\zeta_0} \mathbf{w}^2 + \frac{1}{6}\frac{d^3\mathbf{V}}{d\zeta^3}\bigg|_{\zeta=\zeta_0} \mathbf{w}^3 + \ldots.$$

The linear term of the linear operator **w** contains $dbfV(\zeta) / d\zeta\,|_{\zeta=\zeta_0} \equiv \mathbf{M}(\zeta)$ which is used to determine the Lyapunov exponent. Hence, their computation can be easily implemented in the case of neglecting all non-linear terms in the expansion in comparison with the linear term. This is possible under the condition that the value of $d \equiv |\mathbf{w}|$ is much smaller than the characteristic value R_{ch} of changes of the velocity field. Thus, when $d/R_{ch} \ll 1$ we come to a linear equation

$$\frac{d\mathbf{w}}{dt} = \mathbf{M}(\zeta_0)\mathbf{w}. \tag{8.48}$$

This equation can be written in the equivalent form as

$$\frac{dw_i}{dt} = M_{ik} w_k; \qquad i,k = 1,2,3.$$

The matrix **M** is a second order tensor.

In addition to the theoretical restriction – use of linear approximation for finding the Lyapunov exponent – there are a number of other limitations resulting from the analysis of observational data.

1. The duration of observation is much shorter than the characteristic time ($\sim 1/\lambda$) of the exponential divergence of two points moving along neighboring trajectories.
2. The minimum distance between two neighbouring trajectories $d_{\min}$ may be less than the spatial resolution of the telescope δ, $d_{\min} \geq \delta$. For the 6 m telescope of the SAO the mean δ is equal to about 1.5 arc sec, which, even for the considered close galaxy NGC 3631 is about 0.02 of the radius of the 'working' part of its gas disk.
3. The ratio of spatial resolution δ to the characteristic scale of changes in the velocity field R_{ch}, δ/R_{ch} is not always a small parameter, as required by the condition of the linear approximation. For some areas the value δ/R_{ch} is of the order of unity.

We have already noted that, as follows from the analysis of observations, the field of the gas velocities in the galactic disks should be stationary in the coordinate system rotating with the spiral arms. As is well known (see, for example, [123]), the stationary trajectory can be regarded as the streamline. Two other features of the real observed systems limit the allowable maximum length of the trajectory (and, consequently, the maximum time for computing the Lyapunov exponent) in the case when we consider only two neighboring trajectories. Indeed, since the initial distance between the neighbouring trajectories cannot be chosen arbitrarily small, the exponential divergence of trajectories the distance between them is relatively large (so that the linear approximation is longer valid) already in a relatively short period of time.

To overcome this difficulty, we use the method first described in [454, 1271] and consisting of the following. We divide the so-called 'core' trajectory (lower solid curve in Fig. 8.7) into equal time intervals ΔT. At the end of each segment $\zeta(\Delta)$, $\zeta(2\Delta T)$, etc. we construct vectors $\mathbf{w}(\Delta T)$, $\mathbf{w}(2\Delta T)$, etc., which are then shortened to the value $|\mathbf{w}_1(0)| = d_0$. From the end of each shortened vector we form 'auxiliary' trajectory (a short segment of the solid line corresponding to time interval ΔT, and then – dashed curve). Thus, at each time step, we use a new auxiliary path, located closer than the previous one, to the main path. At the end of each time step n, we estimate the corresponding Lyapunov exponent according to the following definition:

$$\lambda^{(n)} = \frac{1}{n\,\Delta T}\sum_{i=1}^{n} \ln\frac{|\mathbf{w}_i(\Delta T)|}{d_0}. \tag{8.49}$$

It can be shown that the expression (8.49) coincides with the physical definition of the Lyapunov exponent (see Fig. 8.7):

$$\lambda = \frac{1}{T}\ln\frac{|\mathbf{w}_1(T)|}{d_0} \tag{8.50}$$

for sufficiently large T.

Standing in the definition of the Lyapunov exponent (8.47) the limit $t \to \infty$ (in a real situation $T \to \infty$) means that we are interested in the result of dynamical evolution observed in a time much greater than the characteristic time of the process. The characteristic time of the dynamic process is the time of local instability $\tau \sim \lambda^{-1}(T)$. Consequently, the limit $T \to \infty$ for real systems means

$$\zeta \equiv T\lambda(T) \gg 1. \tag{8.51}$$

Obviously, during this time (a lot longer than the local instability time), the chaos could fully develop.

In some situations, the trajectories are very short, so that (8.51) is not satisfied. In this case, we consider more than one path, the whole family, consisting of N close trajectories [479]. The Lyapunov exponent for the family of trajectories is given by

$$\lambda^{(n)} = \frac{1}{N}\sum_{k=1}^{N}\lambda_k^{(n)}, \tag{8.52}$$

where $\lambda^{(n)}$ is calculated using (8.49), while the k-th path is considered as main, N is the total number of basic trajectories. In this case, condition (8.51) acquires an extra large factor $N \gg 1$, so that the following can be satisfied:

$$\eta \equiv N\xi = NT\lambda(T) \gg 1. \tag{8.53}$$

The above-described method for determining the Lyapunov exponent with linearised equations requires the calculation of the Jacobian matrix M_{ik} (8.48), whose elements are the components of the velocity gradients. We used the method for calculating the Lyapunov exponent for two trajectories illustrated in Fig. 8.7. The only different between this method and that described above is the method of computing the evolution of the vector $\mathbf{w}(t)$. By specifying the number of time steps n, we compute the trajectories starting at points $\zeta(0)$ and $\zeta(0) + \mathbf{w}_1(0)$ over time $\Delta T = T/n$. Further, we renormalise the vector $\mathbf{w}_1(\Delta T)$ as shown in Fig. 8.7 and find a new starting point for the auxiliary trajectory. After performing n steps, the Lyapunov exponent is evaluated by the same formula (8.49).

To calculate the Lyapunov exponent, is necessary to define the Riemann metric d. We use the metric

$$d(\mathbf{w}) \equiv \sqrt{g_1 x^2 + g_2 y^2}, \tag{8.54}$$

which depends on two constant positive coefficients g_1 and g_2 (x and y are components of the vector $\mathbf{w}$). For brevity, we denote these different metrics indicating factors: (g_1, g_2). For example, the commonly used metric $d \equiv \sqrt{x^2 + y^2}$, has the form (1.1.) Figure 8.8 shows the behavior of $\lambda^{(i)}$ depending on $i\Delta T/T_0$ ($T_0 = 7.5 \cdot 10^7$ years) for various n, calculated for the main trajectory shown by the thick line near the separatrix in the anticyclone in Fig. 8.6. For this trajectory $\lambda^{(n)} = 0.87T_0^{-1}$ and $\xi = 9.57$ (see (8.51)). As can be seen from Fig. 8.8, the behavior of $\lambda^{(i)}$ weakly depends on the number n of time steps along the path. The initial distance d_0 is taken equal to 1.4 arc sec, i.e. of the order of the angular resolution of the field of the velocities of the galaxies NGC 3631.

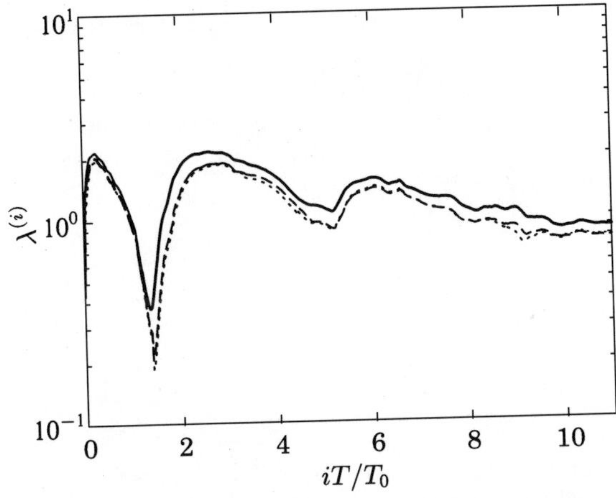

Fig. 8.8. Dependence of $\lambda^{(i)}$ from $i\Delta T/T_0$ for different number of steps n, calculated for the main trajectory in the vicinity of the anticyclone separatrix: n = 10000 (solid line), n = 1000 (dotted line); n = 100 (dashed line).

Oseledets' theorem (1968) argues that the calculation result of the Lyapunov exponent is independent of the Riemann metric. It is interesting to note that for the considered real system the Lyapunov exponents, computed using different metrics, also slightly differ (Fig. 8.9). Near the separatrix of the anticyclone for the metric (1,1) $\lambda^{(n)} = 0.8768\ T_0^{-1}$, for the metric (1,0) $\lambda^{(n)} = 0.8931T_0^{-1}$, for the metric (0,1) $\lambda^{(n)} = 0.8513T_0^{-1}$. For the trajectory of the cyclone near the separatrix, using the metric (1.1), we get $\lambda^{(n)} = 1.04\ T_0^{-1}$ (Figure 8.10) and $\xi = 6.24$ (for the metric (1,0) $\lambda^{(n)}= 1.0884T_0^{-1}$, for the metric (0,1) $\lambda^{(n)} = 0.8892T_0^{-1}$).

Because of the rather short trajectories, in the neighborhood of the saddle point we should not consider only a single trajectory, but a family of close trajectories. We took N = 10 initial points located equidistantly on the y-axis between y = 18.5 and y = 20.5 angular second (see Fig. 8.6). Dependence of $\lambda^{(i)}$, calculated by the formula (8.52) is shown in Fig. 6.52. Evaluation of the Lyapunov exponent $\lambda = 0.79T_0^{-1}$. Taking into account that $T = 2T_0$ and $N = 10$, we can obtain $\eta = 15.8$ (see 8.53), i.e. result is quite reliable.[2]

As can be seen from Figs. 8.8–8.11, the Lyapunov exponents for trajectories near an anticyclone, a cyclone and a saddle point are positive, i.e. they are chaotic trajectories. In Fig. 8.12 we can see an example of $\lambda \simeq 0$ for the same trajectory of the family of the same type of stream lines near the centre of the galaxy. This trajectory is regular.

In addition to the method used to calculate the Lyapunov exponent from two trajectories, the Lyapunov exponents were recalculated by the method using the

[2]According to the definition of the metric – non-negative function $d\ (x, y)$, $d = 0$ only if $x = y = 0$, so the metric coefficients must be strictly positive. In the case of metrics (1,0) and (0,1) this means that the corresponding coefficients are close to zero, remaining strictly greater than zero.

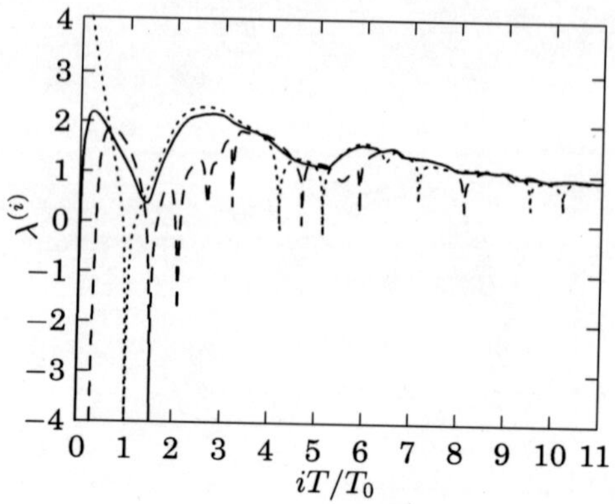

Fig. 8.9. Dependence of $\lambda^{(i)}$ from $i\Delta T/T_0$ for the trajectory near the separatrix of the anticyclone that is calculated for different metrics (1,1) – solid line, (1,0) – dashed line, (0,1) – dotted line. The number of time steps $n = 1000$.

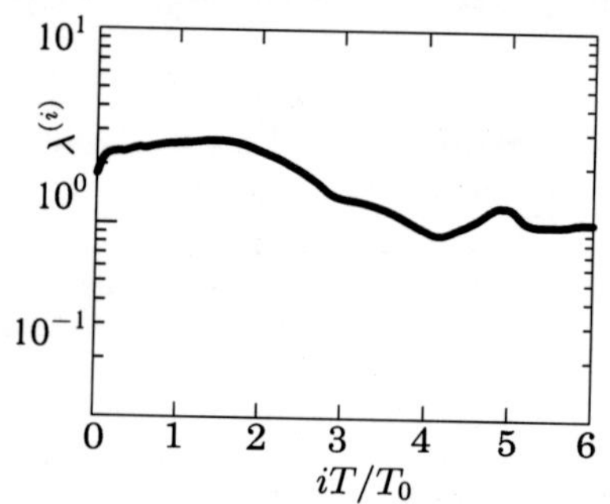

Fig. 8.10. The behavior of $\lambda^{(i)}$, calculated for the main path near the separatrix of the cyclone.

Fig. 8.11. Dependence of $\lambda^{(i)}$ for a family of close trajectories near the saddle point.

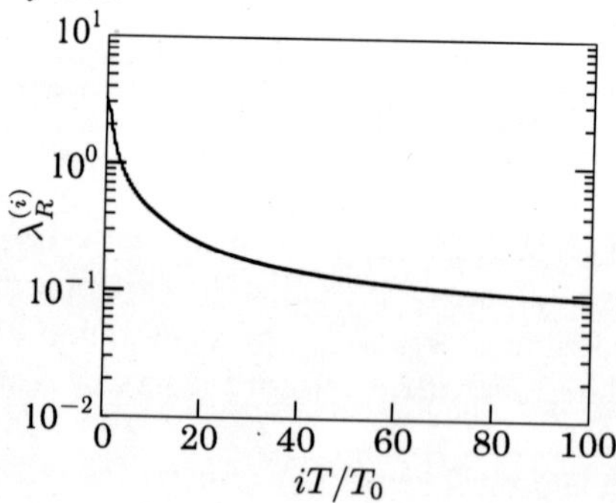

Fig. 8.12. Dependence of $\lambda^{(i)}$ for a regular trajectory near the the galactic centre.

linearised equation (8.48). It was found that for sufficiently smooth interpolation of the vector field, both methods give identical results.

8.3.4. The range of expansion coefficients

Using the definition

$$a_i = \frac{1}{\Delta T} \ln \frac{|\mathbf{w}_i(\Delta T)|}{d_0} \tag{8.55}$$

of the expansion coefficients[3], we find the range of this quantity along the trajectory. By definition, the spectrum of the expansion coefficient is equal to [479]

$$S(a, x_0, y_0) = \lim_{n\to\infty} \frac{1}{n} \frac{dn(a)}{da}, \tag{8.56}$$

where $dn\,(a)$ is the number of expansion coefficients a_i, falling in the interval $(a, a + da)$. The spectra of chaotic and regular trajectories, computed using different metrics, are shown in Fig. 8.13–8.16. As can be seen from the data in these figures, the spectra are not invariant with respect to changes of the metric. An analytical proof of this fact, apparently, was first given in [530]

The range of expansion coefficients plays an important role in the stochastic dynamics, since the Lyapunov exponent is its first moment:

$$\lambda = \int a\, S(a)\, da. \tag{8.57}$$

The invariance of the Lyapunov exponent with respect to changes in the metric has an analytic proof and has been confirmed many times numerically for various models of dynamic systems. The study [244] presents, apparently for the first time, the numerical proof of invariance of λ and the invariance of $S(a)$ with respect to the metric for the real existing system. Let's try to explain at a qualitative level different behaviours of the spectrum and its first moment with respect to the change of the metric.

The difference in the spectra at different metric is easy to understand. Suppose that in step i the vector $\mathbf{w}_i(\Delta T)$ was almost parallel to the y-axis. Then the expansion coefficient a_i, calculated using the metric (1,1), may be greater than zero, while using the metric (1,0) we obtain $d_i \sim 0$ and the corresponding large negative coefficient a_i. The alignment of the vectors $\mathbf{w}_i$ parallel to the coordinate axes is explained by the presence of broad wings with $a \to \pm\infty$ in the shape of the spectra found by using the metrics (1,0) and (0,1).

In order to understand the reason for the independence of the Lyapunov exponent on the change of the metric, we consider two arbitrary metrics: $d\,(g_1, g_2)$ and $\bar{d}\,(\bar{g}_1, \bar{g}_2)$. First of all, note that there exist such positive constants C_1 and C_2 that for the arbitrary vector $\mathbf{w}$

$$C_1 \bar{d}(\mathbf{w}) \le d(\mathbf{w}) \le C_2 \bar{d}(\mathbf{w}). \tag{8.58}$$

---

[3]Here we follow the terminology first introduced by Oseledets. The same quantity is called the local Lyapunov exponent and the expansion coefficient.

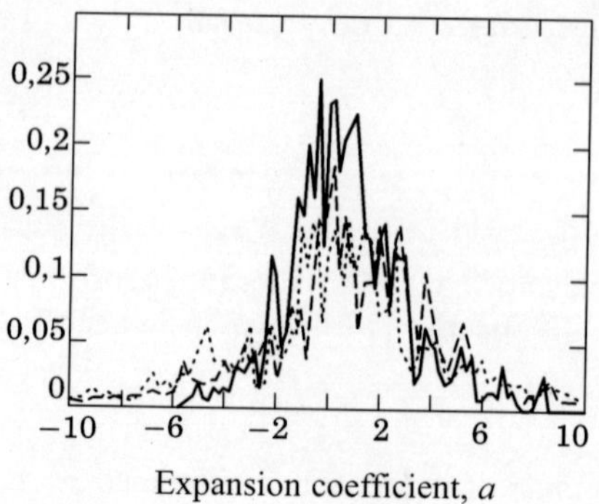

Fig. 8.13. The spectrum for the chaotic trajectories in the vicinity of the anticyclone (see 8.6), calculated for various metrics: (1,1) – solid line, (1,0) – dashed line, (0,1) – the dotted line. The number of time steps $n = 1000$.

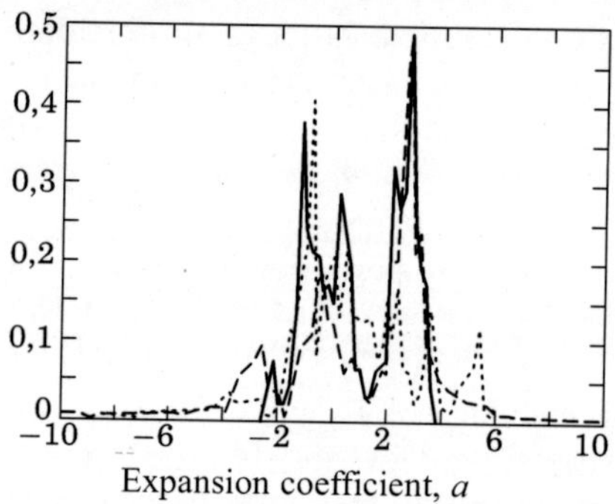

Fig. 8.14. The spectrum for the chaotic trajectories in the vicinity of the cyclone (see in Fig. 8.6), calculated for different metrics (1,1) – solid line, (1,0) – dashed line, (0,1) – the dotted line. The number of time steps $n = 1000$.

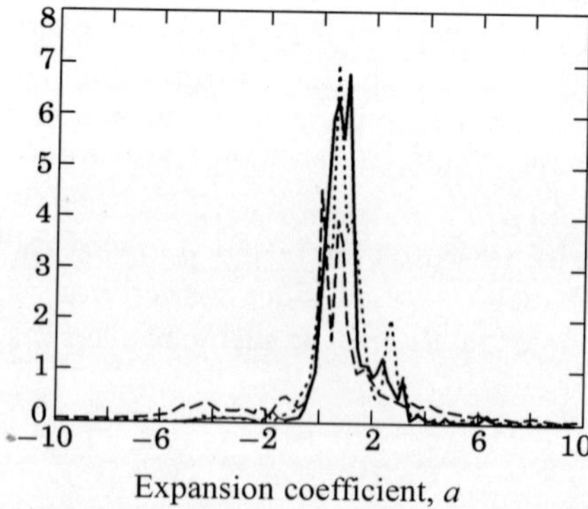

Fig. 8.15. The spectrum for a family of chaotic trajectories in the neighborhood of the saddle point (see Fig. 8.6), calculated for different metrics: (1,1) – solid line, (1,0) – dashed line, (0,1) – dotted line. The number of steps time $n = 1000$.

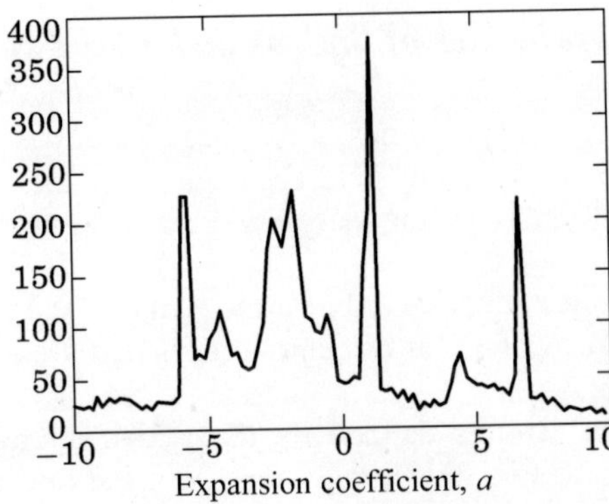

Fig. 8.16. The range for a regular trajectory around the centre of the disk (see Fig. 8.6), calculated for the metric (1,1).

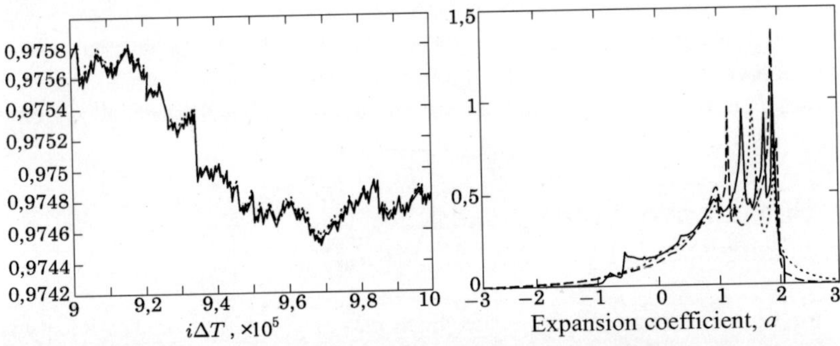

Fig. 8.17. Lyapunov exponents (left) and spectra of the Chirikov standard model (right) calculated for various metrics: (1,1) – solid line, (0,1) – dashed line, (1,0) – dotted line.

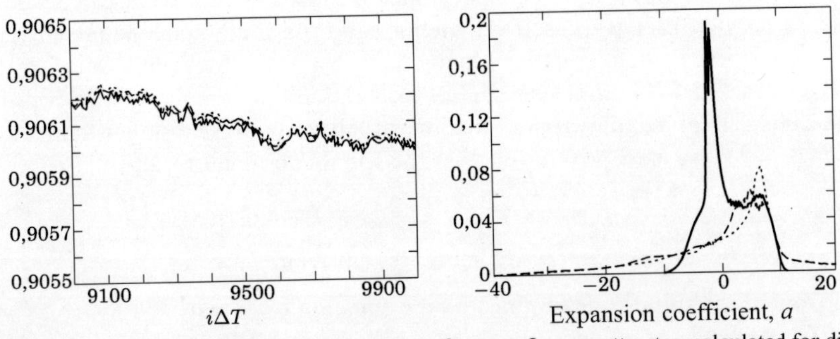

Fig. 8.18. Lyapunov exponents (left) and spectra of strange Lorenz attractor, calculated for different metrics: (1,1) – solid line, (0,1) – dashed line, (1,0) – dotted line.

Figures 8.17 and 8.18 show the Lyapunov exponents and spectra calculated using various metrics for the well-known dynamic models – the Chirikov standard model and the strange Lorenz attractor. We can to see that in both cases the Lyapunov exponent, calculated using different metrics, is the same, while the shape of the spectra is different.

8.5.3. Dynamic properties of the gas disk of the galaxy NGC 3631

Here are the main dynamic properties of the gase disk of the galaxy NGC 3631.

– The three-component velocity field, recovered from the observed velocity field of the gas disk of NGC 3631, is stationary and there coexist both regular and chaotic trajectories of the gas.

– Regular trajectories are observed near the centre of the disk, stochastic – near the saddle points and separatrices of giant vortices that are located in the velocity field.

– The type of trajectories defined by the two methods (calculation of the Lyapunov exponent for two trajectories, and using the linearised equation of the vector of the deviation), gave similar results.

– Lyapunov exponents obtained for the real galaxies were invariant with respect to a change in the metric, in full compliance with the Oseledets mathematical theorem.

– In [244] it was numerically shown that the shape of the spectrum of the expansion coefficients depends on the metric used.

Note that the last two conclusions are also valid for classical systems: the Chirikov standard model and the strange Lorenz attractor.

8.4. Shock waves in the gas disk

8.4.1. Large-scale shocks in the gas disk

The possible existence of large-scale shock waves (SW) in the spiral density waves was predicted by Fujimoto [249]. In subsequent papers [194, 327, 1070, 1072, 1163) the supersonic flows of gas passing through the density wave (that is, in fact, a potential well) with the appearance of shock waves were calculated. The role of thermal processes, self-gravitation, and the multiphase nature of the gas was described in detail in the monographs by Marochnik and Suchkov [131] and Gorbatskii [41]. In these studies contexts either the flow was considered one-dimensional or the approximation of slightly twisted logarithmic spirals ($\sin(i) \ll 1$) was used. At transition to the spiral coordinates η and ξ:

$$\eta = \cos(i)\ln(r/R) + (\varphi - \Omega_B t)\sin(i), \tag{8.59}$$

$$\xi = (\varphi - \Omega_B t)\cos(i) - \sin(i)\ln(r/R), \tag{8.60}$$

the front of the spiral is determined by the equation η = const.[4]

The existence of galactic shock waves in the gas disk has been confirmed by several observational facts (see chapter 1). First of these include an increased concentration of young stars and dust in a narrow strip along the spiral wave (in the shock wave front region), that is caused by compression of matter in the shock wave improving the conditions for star formation.

[4]This approach is similar to that considered in section 11.4.

8.4.2. Supersonic flow of gas through the one-dimensional potential well

Before proceeding to consideration of the gas disk, consider the problem of a supersonic flow of gas through a potential well in one-dimensional approximation, following the results of [797], where the analytical and numerical approaches were used to study the evolution of the flow from the start of the formation of a shock wave to the establishment of the stationary picture.

Here we confine ourselves only to the description of the numerical modelling results based on the solution of non-stationary equations of hydrodynamics. Let the potential energy depend only on one coordinate $\psi(x)$, and the condition

$$\Psi(-x) = \Psi(x), \quad \Psi(-\infty) = \Psi(\infty) = 0.$$

is fulfilled.

For definiteness, we choose the potential in the form

$$\Psi = \Psi_0 \exp\left\{-x^2 / a^2\right\}, \tag{8.61}$$

which is characterised by the width a of the potential well and depth ψ_0.

Let a uniform at the initial moment supersonic gas flow moves from left to right and is characterised by Mach's number $M = u_0 / c_{s0}$. The presence of an external force $\propto -d\psi/dx$ leads to the formation of non-linear perturbations, which originally appear on leaving the potential well in the region $x > 0$ in Fig. 8.19. The perturbation transforms to a shock wave quite rapidly without changing its position. However, this situation is unstable and the shock wave moves to the other side of the well. After that, a steady flow with the shock wave at the entrance of the potential well ($x < 0$)

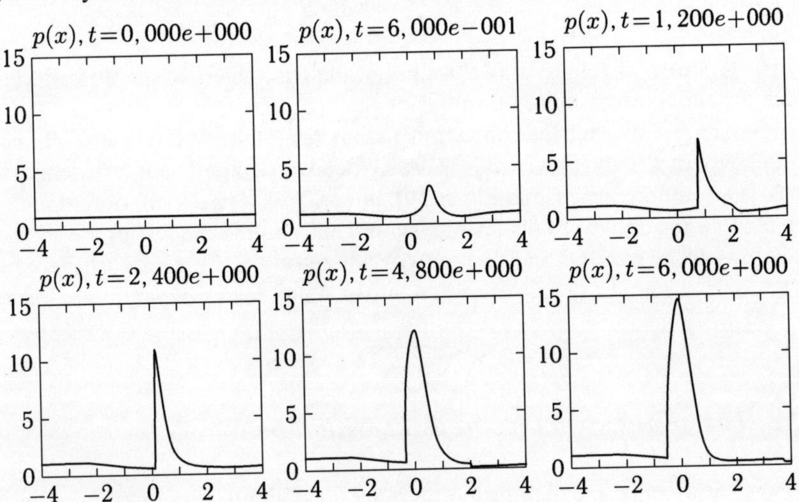

Fig. 8.19. The evolution of the supersonic flow in the flow through the potential well. The pressure distribution at different points in time, t, is shown. At the initial time p = const. The x coordinate is normalised to the value of the parameter a.

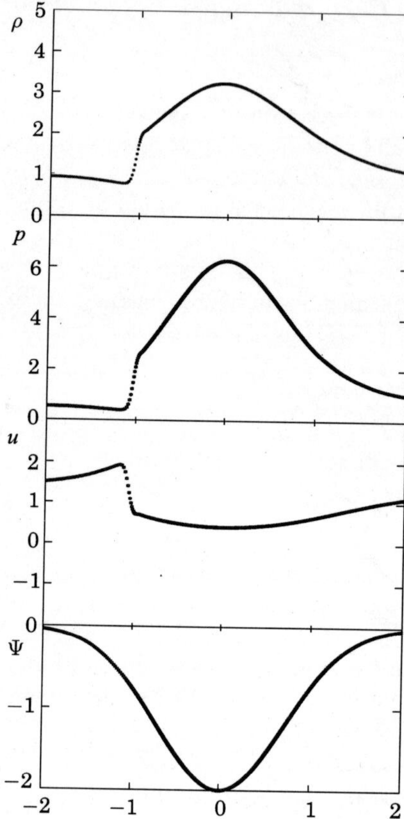

Fig. 8.20. The distributions of density ρ, pressure p, and velocity u along the dimensionless coordinate x/a, after the establishment of a steady supersonic viscous gas flow through the potential well Ψ (shown below). The dots show the values of hydrodynamic parameters at the nodes of the mesh. Because of the viscosity, we have a finite width the shock front, in contrast to the flow in Fig. 8.19.

forms. Of course, there are critical values of the parameters of the flow a, M, ψ_0, at which the shock wave cannot form inside the potential well.

Figure 8.20 shows the hydrodynamic parameters of the flow in a strongly viscous gas. As we can see, the presence of viscosity leads to a significant broadening of the shock front (unlike the case in Fig. 8.19), but otherwise the result does not change.

Note that the above results change significantly in the models of the interstellar medium with heating Q^+ and cooling Q^-, where on the right side of the equation for the energy sources Q^+, Q^- are taken into account, for example, in the form [30]

$$Q^- = k\,\Lambda(T)\varrho^2\,, \quad Q^+(T) \simeq \text{const}\,. \tag{8.62}$$

8.4.3. The spiral structure in disks with a bar

We now turn to the gas disk in the presence of a non-axisymmetric gravitational potential.

The gravitational potential well, leading to the emergence of shock waves, may be due to the presence of a spiral wave of the stellar density, bars, a triaxial bulge (or massive satellite). The influence of the central bar-like perturbation on the excitation process and the evolution of the spiral pattern of galaxies has been studied in many gas-dynamic numerical experiments [103, 104, 295, 313, 887].

At the heart of the work under discussion is the numerical solution of non-stationary equations of gas dynamics in an external potential

$$\Phi = \Phi_0(r) + \Phi_1(r, \varphi, t), \tag{8.63}$$

where $\Phi_0(r)$ is the axisymmetric part of the potential, and the non-axisymmetric perturbation is usually chosen in the form

$$\Phi_1 = \Phi_0 \, \varepsilon(r, t) \cos[2(\varphi - \Omega_B t)]. \tag{8.64}$$

For the dimensionless amplitude $\varepsilon(r, t)$ we can be write

$$\varepsilon = \varepsilon_0 \frac{(r/a)^2}{[1 + (r/a)^2]^{3/2}}, \tag{8.65}$$

is a is the characteristic scale of the bar.

In regions of the disk rotating more slowly than the bar a two-arm spiral pattern is generated. In the vicinity of corotation $[\Omega_B = \Omega(r_c)]$ ring structures can form. The generation of the spiral pattern is strongly dependent on the ratio of the angular velocities of the bar and and the disk. With other conditions being equal, the higher the rotation velocity of the bar, the greater the twist angle of spirals [103, 104]. With the increase of the parameter ε_0 the angle i decreases. In the experiments, depending on the model parameters, both the logarithmic and Archimedean spirals are obtained. According to the results of [887], formation of shock waves is observed even for $\varepsilon_0 = 0.01$.

8.4.4. Numerical experiments with the evolving jumper

Consider the dynamics of gas in the central part of the disk ($r \lesssim a$) in the gravitational field of the growing bar, following [13, 827]. Interest in this problem is related, in particular, to the problem of explanation of active galactic nuclei (section 1.5). First of all, there is a correlation between the activity of the nucleus and the presence of a bar (or a satellite) [438, 1174, 1211]; in nearby Seyfert galaxies significant non-circular motion of the gas emitting in the forbidden lines was detected [15, 849], which indicates the possible presence of shock waves.

There is a view that galaxies with active nuclei are a special, quite rare type of galaxy [50]. According to the opposing concept, most of the galaxies passes through a rather short active phase. In favour of the second concept is the fact that the overall structure of the Seyfert galaxies (excluding the active nucleus) is practically no different from the conventional planar structure of galaxies [7] and most of the close Seyfert galaxies have at least a small bar [1211].

Let us discuss the results of the simulation of the gas disk in the potential of the form (8.64), (8.65), but with the amplitude increasing with time [13, 827]:

$$\varepsilon_0 = \varepsilon_{00}[1 - \exp(-t/\tau_B)] \tag{8.66}$$

with $\tau_B \simeq 2\pi/\Omega_B$. In the region of the bar with the growing amplitude a shock wave front forms, and the bisymmetric sectorial structure of the flow appears: the gas flows from the central part of the disk in narrow sectors and flows into the much broader sectors (Fig. 8.21). The maximum value of u_r in the outgoing gas is slightly greater than the half speed of the gas $r \simeq a$ (Fig. 8.22) which for the Seyfert galaxies amounts to $\sim 100 \div 150$ km/s. The velocity of the flows directed to the centre of the disk is not higher than $30 \div 50$ km/s.

Streams of the gas flowing in ($\dot{M}_{in}$) and flowing out ($\dot{M}_{out}$) only partially compensate each other (Fig. 8.23). The total radial flow of gas depends on the value of ε_{00}. In the case of $\varepsilon_{00} = 0.05$ the total gas flow is directed almost always outside. In the case of $\varepsilon_{00} = 0.1$ the value of $\dot{M}$ changes its quasiperiodic direction.

Note the similarity of the shock wave, obtained in the experiments, with the shock waves that are directly observable in the nearest galaxies NGC 3786 and 1365 [15, 849]. In central parts of active galaxies the observed velocity of escaping gas exceeding the velocity of influent gas.

Analysis of the balance of flows of of the bisectorial gas flows is important for understanding the nature of the activity of galactic nuclei. The most significant are two results: the quasiperiodicity of the value of the total flow through the circle with a small diameter in the centre ('pumping' of gas through the core) and the dependence on the contrast of the picture of the gravitational potential of the bar. The high luminosity of active galactic nuclei (10^{46} erg/s) is usually associated with accretion of gas on a massive central source at a rate of $\dot{M} \sim 1\ M_\odot$ / year on the scale $r \simeq 10^{16} \div 10^{17}$. However, this scheme assumes that the reason for the activity is related to the dynamic structure of the galaxy, and in this case the lifetime of the active nuclei is of the order of $10^9 \div 10^{10}$ years – 'long' scale of activity. Estimation of the time of the active phase on the basis of the frequency of occurrence of active

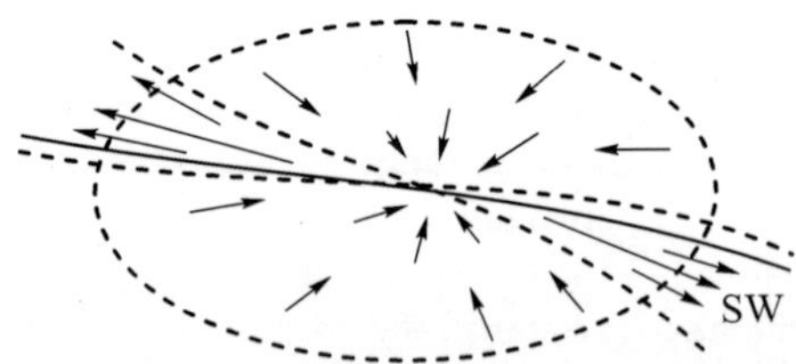

Fig. 8.21. Schematic representation of the structure flow in the bar. Shows the location of the shock wave (SW).

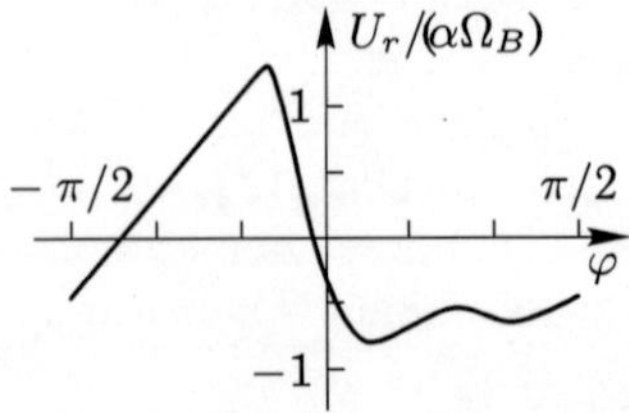

Fig. 8.22. The azimuthal dependence of the radial velocity component $u_r(\varphi)$ at $r = a$ for $\varepsilon = 0.1$.

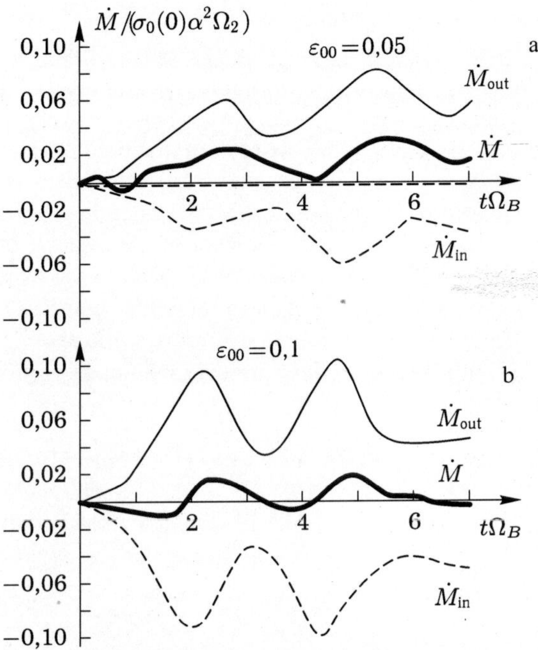

Fig. 8.23. The velocity of gas flowing out $\dot{M}_{out}$, flowing in $\dot{M}_{in}$ and the total velocity $\dot{M} = \dot{M}_{in} + \dot{M}_{out}$ of gas flow through a circle of radius $r = 0.1$, $a = \varepsilon_{00}$ for 0.05 (a), 0.1 (b). The dashed line (at $\simeq$−0.002) is the velocity at which gas enters the central region in the absence of the bar due to the numerical (circuit) viscosity.

galactic nuclei gives $10^7 \div 10^8$ years. The recurrent nature of the activity is associated only with the non-stationary properties of the system.

Models with the evolving bar provide an opportunity to withdraw the contradiction between the 'long' and 'short' scales of activity. The fact that the majority of Seyfert galaxies have a bar indicates that the activity of of some of the galactic nuclei can be linked with the work of the 'quasiperiodic' pumping mechanism of the central 'accretion machines' of the gas of the disk. Excitation of the bar-mode in the central parts of stellar disks of galaxies leads to a restructuring of the gas in which the nucleus of the galaxies may be periodically receive a rather significant mass of gas. Thus, for typical values of a = 3 kpc, Ω_B = 100 km·s/ kpc and σ_g $(r \lesssim 0.3) \simeq 2\cdot 10^3\ M_\odot/\mathrm{pc}^2$ the value of $\dot{M} \simeq 1\ M_\odot$ / year (see Fig. 8.23). 'Processing' of this gas by the accreting central object can, in principle, provide the observed activity of galactic nuclei with a characteristic period of the manifestation of the order of tens of millions of years.

Note that the reason for the formation of a bar in addition to the excitation of the bar-mode in the stellar disk due to gravitational instability may be a flying satellite of a galaxy which may also lead to the quasiperiodic flow of gas in the centre of the disk [438].

8.4.5. Self-similar shock waves

The assumption that in the gas disks orbiting compact objects there may be spiral shock waves has been expressed repeatedly [863, 1161]. The appeal of these studies is due to the fact that the spiral shock waves can transfer angular momentum from the interior of the disk to external parts. In thin accretion disks ($h/r \ll 1$) the flow is supersonic, which allows for the possibility of formation of shock waves. Causes of shock waves may be the second component in the system, or the asymmetric magnetosphere around the compact object [909]. Due to the dissipative processes on the wave front, the matter may fall in a spiral to the centre.

Consider the steady flow containing two or more spiral shock waves in the self-similar approach [1200]. We write the stationary equations of gas dynamics in the following form:

$$\frac{\partial}{r\partial r}(h\rho u r)+\frac{\partial}{r\partial\varphi}(h\rho v)=0, \tag{8.67}$$

$$\rho\left(u\frac{\partial u}{\partial r}+\frac{v\partial u}{r\partial\varphi}-\frac{v^2}{r}\right)=-\frac{\partial p}{\partial r}-\rho\frac{GM_1}{r^2}, \tag{8.68}$$

$$\rho\left(u\frac{\partial v}{\partial r}+\frac{v\partial v}{r\partial\varphi}+\frac{uv}{r}\right)=-\frac{\partial p}{r\partial\varphi}. \tag{8.69}$$

We assume that the half thickness h can depend only on the radial coordinate. In particular, we study two cases:

$$h=h_0=\text{const}, \tag{8.70}$$

$$h(r)=\sqrt{R\hat{T}/(\mu\Omega^2)}. \tag{8.71}$$

The first case corresponds to a strictly two-dimensional flow in the disk plane. In the second, it is assumed that the gas is in hydrostatic equilibrium in the z-direction [cf. with (11.24)]. In (8.71) the value of $\hat{T}(r)$ is the characteristic temperature at distance r.

For definiteness, we assume that the gas rotates around a relatively massive body, then $\Omega(r)=\sqrt{GM_1/r^3}$. The equation for the entropy of the radiation from the surface of the disk taken into account can be written in the form of

$$h\rho T\left(u\frac{\partial S}{\partial r}+\frac{v\partial S}{r\partial\varphi}\right)=-\sigma_{SB}T_s^4, \tag{8.72}$$

where T_s is temperature of the surface of the disk, σ_{SB} is the Stefan–Boltzmann constant. For the entropy of an ideal gas we have:

$$S=c_V\ln\{p/\rho^\gamma\}. \tag{8.73}$$

In the case of thermodynamic equilibrium for the large optical thickness $\tau \gg 1$ we can assume [224]:

$$T^4=\frac{3}{4}\tau T_s^4, \qquad \tau=h\hat{\rho}k. \tag{8.74}$$

We believe that the opacity k and the characteristic density $\hat{\rho}$ are functions of the radial coordinate. The azimuthal velocity has the form

$$v = r\Omega + v_1. \tag{8.75}$$

We use the dimensionless quantities marked at the top by ~:

$$(u, v_1) = \Omega_0 r_0 (\tilde{u}, \tilde{v}); \quad h = r_0 \tilde{h}; \quad p = r_0^2 \Omega_0^2 \rho_0 \tilde{p}; \quad \rho = \rho_0 \tilde{\rho}, \tag{8.76}$$

where r_0 is a characteristic radius, ρ_0 is the arbitrary scale of density. We introduce new 'spiral' coordinates [1200]:

$$x = r / r_0, \quad \Psi = \varphi + \beta(x), \tag{8.77}$$

Function $\beta(x)$ is defined below, but it is clear that the case Ψ = const determines the spiral. We believe that the opacity is a function of the radial coordinate:

$$k = k_0 x^{\nu}. \tag{8.78}$$

In view of (8.75)–(8.77), (8.67)–(8.69) take the form:

$$\tilde{\rho}\left[\left(x^{-3/2} + \frac{\tilde{v}}{x}\right)\frac{\partial \tilde{u}}{\partial \Psi} + \tilde{u}\frac{\partial \tilde{u}}{\partial x} + \frac{d\beta}{dx}\tilde{u}\frac{\partial \tilde{u}}{\partial \Psi} - \frac{\tilde{v}^2}{x} - \frac{2\tilde{v}}{x^{3/2}}\right] = -\frac{\partial \tilde{p}}{\partial x} - \frac{d\beta}{dx}\frac{\partial \tilde{p}}{\partial \Psi}, \tag{8.79}$$

$$\tilde{\rho}\left[\left(x^{-3/2} + \frac{\tilde{v}}{x}\right)\frac{\partial \tilde{v}}{\partial \Psi} + \tilde{u}\frac{\partial \tilde{v}}{\partial x} + \frac{d\beta}{dx}\tilde{u}\frac{\partial \tilde{v}}{\partial \Psi} + \frac{\tilde{u}}{2x^{3/2}} + \frac{\tilde{u}\tilde{v}}{x}\right] = -\frac{\partial \tilde{p}}{x\partial \Psi}, \tag{8.80}$$

$$\frac{\partial}{\partial x}(\tilde{h}\tilde{\rho}\tilde{u}x) + \frac{\partial}{\partial \Psi}\left[\tilde{h}\tilde{\rho}\left(\frac{d\beta}{dx}\tilde{u}x + \tilde{v} + x^{-1/2}\right)\right] = 0. \tag{8.81}$$

Looking for solutions to the self-similar form: $q_i = x^n Q_i(\Psi)$, $n = n_i$. The values of n_i are determined from the condition of the independence of the equations on the variable x. To do this, it was assumed

$$\tilde{u} = x^{1/2} U(\Psi), \quad \tilde{v} = x^{-1/2} V(\Psi),$$

$$\tilde{\rho} = x^{-n} R(\Psi), \tilde{p} = x^{-n-1} P(\Psi), \quad \frac{d\beta}{dx} = x^{-1} B, \quad \tilde{h} = x^m C, \tag{8.82}$$

where B, C, m, n are constants. If Θ is the angle between the tangent to the spiral (Ψ = const) and the radial direction (Figure 8.24), then

$$B = \text{tg}(\Theta). \tag{8.83}$$

It is easy to see that the case B = const corresponds to the logarithmic spiral. The index m is equal to zero for a disk with constant thickness (8.70). In the case of the law (8.71) $m = 1$ and is valid for the temperature $T \propto \tilde{p} / \tilde{\rho} \propto x^{-1}$. If we choose as the characteristic value the temperature at $\Psi = 0$, then

$$C = \sqrt{P(0) / R(0)}. \tag{8.84}$$

As you can see, the parameter C is the isothermal speed of sound in units of the Keplerian velocity $\Omega_0 r_0$ or the reverse Mach number. In view of (8.82) the equations (8.79)–(8.81) take the form of ordinary differential equations with respect to Ψ:

$$W\,U' - \frac{1}{2}U^2 - V^2 - 2V = (n+1)\frac{P}{R} - B\frac{P'}{R}, \tag{8.85}$$

$$W\,V' + \frac{1}{2}U\,V + \frac{1}{2}U = -\frac{P'}{R}, \tag{8.86}$$

$$(R\,W)' + \left(\frac{1}{2} + m - n\right)R\,U = 0, \tag{8.87}$$

Here $W = 1 + V + B$, U. We determine the value of n. We integrate (8.87), as a result we obtain

$$(RW)\Big|_0^{2\pi} + \left(\frac{1}{2} + m - n\right)\int_0^{2\pi} R\,U\,d\Psi = 0. \tag{8.88}$$

The first term in (8.88) is zero. The second term determines the accretion rate, which is different from zero in the case of shock waves. Consequently,

$$\frac{1}{2} + m - n = 0. \tag{8.89}$$

So, in the case h = const $n = 1/2$, and for $h \propto xn = 3/2$. In view of (8.89), the continuity equation (8.87) can be rewritten as

$$(R\,W)' = 0. \tag{8.90}$$

To preserve the self-similarity of (8.72), using (8.73), (8.74) and (8.78), we must put $\nu = -1/2$, then the law of variation of energy takes the form:

$$W\left(\frac{P'}{P} - \gamma\frac{R'}{R}\right) + U(\gamma\,n - n - 1) + F\frac{P^3}{R^4} = 0, \tag{8.91}$$

where

$$F = \frac{4}{3}\frac{\sigma_{SB}T_0^3}{\rho_0^2 k_0 c_V \Omega_0 r_0^2 C^2}. \tag{8.92}$$

Here $T_0\mathcal{R}/\mu = p_0/\rho_0 = \Omega_0^2 r_0^2$.

The system of equations (8.85)–(8.92) for the unknowns U, V, P, R must be supplemented by boundary conditions. Consider N identical shock waves, separated by a fixed angle $\Delta\Psi$, then the solutions must be periodic with period $\Delta\Psi = 2\pi/N$. We write the expression for the normal and tangent to the line Ψ = const of the velocity components (see Fig. 8.24):

$$V_{\parallel} = \sin\Theta\left\{\frac{U}{B} - 1 - V\right\} = (1+B^2)^{-1/2}\{U - B(1+V)\}, \tag{8.93}$$

$$V_{\perp} = \cos\Theta\{1 + V + BU\} = (1+B^2)^{-1/2}\,W. \tag{8.94}$$

Thus, the normal component of the flow of matter is proportional to W. In passing through the shock wave front there must be the continuous normal component of the flow of matter, the tangential component of velocity, momentum flow, the energy flux [112]. As we use the notation of these conditions can be written in the following form:

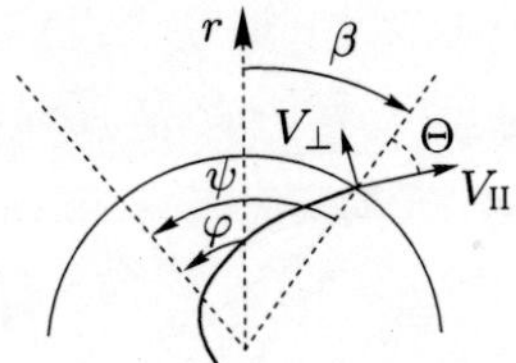

Fig. 8.24. Problem of self-similar accretion.

$$\{RW\} = 0, \tag{8.95}$$

$$\{V_{\|}\} = 0, \tag{8.96}$$

$$\{P + RW^2 / (1 + B^2)\} = 0, \tag{8.97}$$

$$\left\{ \frac{\gamma}{\gamma - 1} \frac{P}{R} + \frac{1}{2} \frac{W^2}{1 + B^2} \right\} = 0. \tag{8.98}$$

Here, $\{A\}$ denotes a difference in the values of the quantity A on the opposite sides of the shock wave front. Condition (8.95) because of (8.90) is satisfied automatically. The conditions (8.96), (8.97) and (8.98) define only three constants of integration. For the fourth condition we reduce system (8.85), (8.86), (8.90), (8.91) to the equation

$$\left[(1 + B^2)c^2 - W^2\right]V' = \frac{1}{2}UW(V+1) - BQc^2 - Uc^2\left(n - \frac{n+1}{\gamma}\right) + F\frac{P^4}{R^5}, \tag{8.99}$$

where $Q = \dfrac{1}{W}\left[\dfrac{1}{2}U^2 + V^2 + 2V + \dfrac{1}{2}BU(V+1) + (n+1)\dfrac{c^2}{\gamma}\right], c^2 = \gamma P / R$

is the dimensionless adiabatic speed of sound. Equation (8.99) with $(1 + B^2)$, $c^2 = W^2$ fulfilled (or, taking into account (8.94) $V_{\perp}^2 = c^2$) has a singularity, which corresponds to the presence of a sound point at $\Psi = \Psi_s$. The equality to zero of the right-hand side of (8.99) at the sonic point Ψ_s gives the fourth condition

$$\left[\frac{1}{2}UW(V+1) - BQc^2 - Uc^2\left(n - \frac{n+1}{\gamma}\right) + \frac{FP^4}{R^5}\right]_{\Psi=\Psi_s} = 0. \tag{8.100}$$

Conditions (8.86–8.98), (8.100) determine the constants of integration of the equations (8.85), (8.86), (8.90), (8.91).

In the limiting case of a large number of shock waves the problem can be solved analytically [1200]. We consider only the adiabatic flow [$F \equiv 0$ in (8.91)]. If we proceed from the smallness of the parameter $\epsilon = 1/N \ll 1$ and the assumption that the functions U, V, R, P are linear between the shock waves, we can write the relation between γ and the angle of the spiral Θ [1200]:

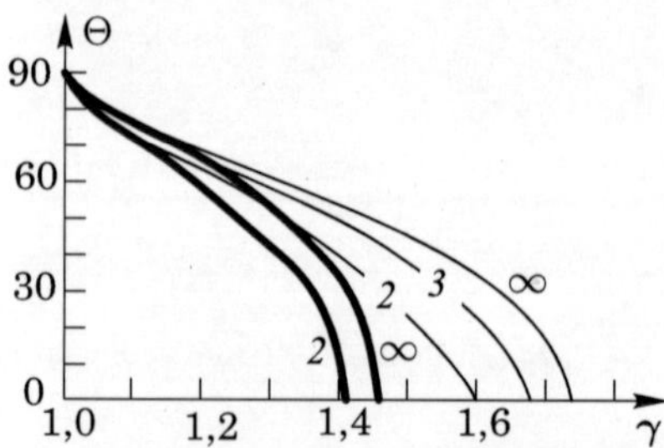

Fig. 8.25. The dependence of angle Θ on γ for different numbers of shock waves [symbol ∞ corresponds to the analytical solution of (11.77)]. The thin lines belong to the disk model with a constant thickness, thick lines – to the disk model with (8.71).

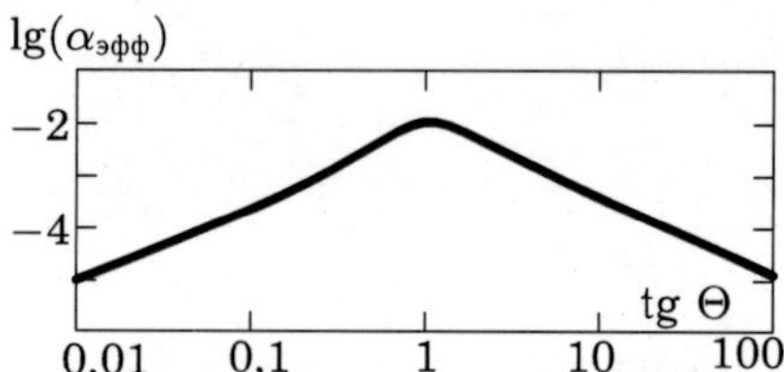

Fig. 8.26. The dependence of α_{eff} on angle Θ for two shock waves.

$$B^2 = \mathrm{tg}^2\Theta = \frac{2}{\gamma - 1} - 2\frac{n+1}{\gamma} - 1. \tag{8.101}$$

A numerical approach to solving the above problem makes it possible to consider an arbitrary number of shock waves, including taking into account the radiative losses. The results of such calculations are shown in Fig. 8.25. The inclusion of radiation losses allows us to estimate the effective α-parameter appearing in the 'viscous axisymmetric models' (see section 11.2). If we define the average radial flow

$$\langle u \rangle = \mathrm{R}\, u\rho\, d\varphi \,/\, \mathrm{R}\, \rho\, d\varphi \tag{8.102}$$

and the accretion rate

$$\dot{M} = -2\pi r \sigma \langle u \rangle, \tag{8.103}$$

then, comparing with the results derived from the standard model of the AD:

$$\dot{M} = 3\pi\sigma\nu_{\mathrm{eff}} = 3\pi\sigma\alpha_{\mathrm{eff}}\Omega h^2, \tag{8.104}$$

we get

$$\alpha_{\mathrm{eff}} = \frac{2}{3} r \Omega^{-1} h^{-2} \,|\langle u \rangle|. \tag{8.105}$$

Figure 8.26 shows the coefficient α_{eff} as a function of the angle Θ. In the case of two shock waves ($N = 2$) there is a maximum at $\Theta \simeq \pi/4$ and max $\{\alpha_{\mathrm{eff}}\} = 1.02 \cdot 10^{-2}$.

8.4.6. The position of the front of the galactic shock wave

In section 8.4.2 it is shown that during the one-dimensional (flat) supersonic flow of gas through the potential gravitational well the shock wave (SW) forms at the entrance to the well [797]. Let us turn to the observational data. It is believed that the dust bands define the position of the shock wave – see images of nearby galaxies with high resolution. We consider as typical galaxy NGC 1300 with a bar, visible almost flat. We assume that the corotation radius r_c is at the ends of the bar, from which strong spiral arms run (Fig. 8.27). Let us turn to the reference frame of the spiral pattern, rotating with frequency Ω_p. Inside the jumper (inside corotation), the gas accumulates, as shown by the arrows. The angular velocity of rotation of the outer spirals is higher than that of gas and the latter accumulates at the leading edge of the arm. This picture is typical and well traced in the closest SB-galaxies, and objects without the bar, for example, M51, M74, M83, NGC 1365, etc.

Thus, the shock wave in galaxies appears to be located at the exit from the potential well – at the trailing edge of the spiral arm and the front edge of the bar. This conclusion contradicts the solution for a plane flow of gas (see section 8.4.2). This result is related to the inapplicability of the univariate analysis for galactic gas disks.

Consider the formation of shock waves in a differentially rotating radially inhomogeneous gas disk in the gravitational potential $\Psi(r, \varphi, z)$, taking into account non-axisymmetric helical components in the form [1273.1274]

$$\Psi(r,\varphi,z) = \Psi_0(r,z)\cdot[1+\varepsilon_0\Psi_1(\xi)\cos\Theta_p], \quad \Psi_1 = \frac{\xi^2}{(1+\xi^2)^{3/2}}, \tag{8.106}$$

where

$$\xi = \sqrt{(r/b)^2 + (z/h_*)^2}, \quad \Theta_p = m\left[\varphi - \varphi_p(r_0) + \Omega_p t - \frac{\ln(r/r_0)}{\operatorname{tg} i}\right], \tag{8.107}$$

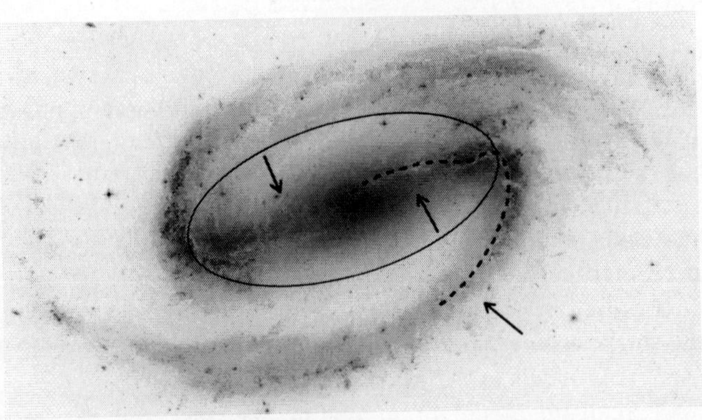

Fig. 8.27. Image of NGC 1300 (HST). The thin solid line localizes the bar. The dotted line shows the main dust band – presumably the position of the shock wave. Arrows indicate the direction of gas flow in a rotating system of the coordinates.

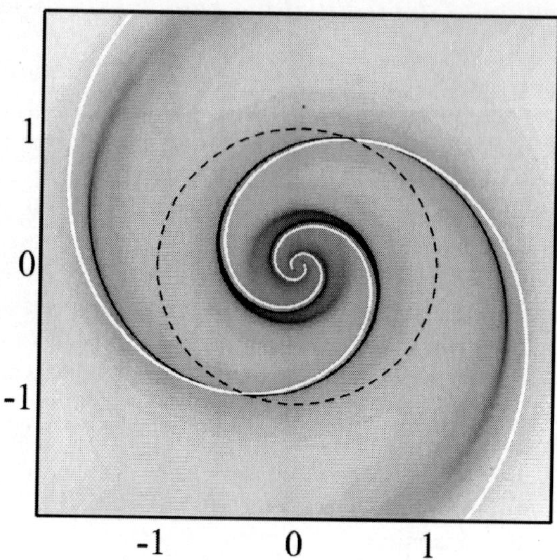

Fig.8.28. The distribution of the logarithm of the surface density of the gas disk (shades of gray). The darkest colour indicates the position near the shock wave front. The white line – the minimum of the potential energy of the stellar density wave according to (8.106). A circle with a radius of corotation $r_c = 1$ is shows by the dotted line.

b, h_* are, respectively, radial and vertical scales, m is the number of arms, i is the angle of twist of spirals, Ω_p is the angular velocity of rotation of the spiral pattern, ε_0 characterises the depth of the potential well of the spiral wave, Ψ_0 is due to the halo and the axisymmetric distribution of matter in the stellar disk, $\varphi_p(r_0)$ determines the initial position of the pattern, the parameter r_0 affects the geometry of the spiral, Ψ_1 describes a decrease in the amplitude of the spiral wave with distance from the centre of the disk. We restrict ourselves to a model without a bar with a rotation curve that goes on the 'plateau' at a radius $r \gtrsim 0.3$.

Figure 8.28 shows, according to our results of numerical simulation of gas dynamics of the disk, the position of the fronts of shock waves in relation to the potential and the spiral density wave of the stellar components($\propto\Theta_p(\varphi)$ in (8.106)). In this case, the corotation radius $r_c = 1$. In the region $r < r_c$ the shock wave is located on the front edge of the potential well (for gas – at the exit of well). At the periphery $r > r_c$ the situation is reversed: the shock wave is located on the trailing edge of the sleeves, but also the flowing gas at the outlet of the potential well. The position of corotation $r_c = 1$ is well distinguished in Fig. 8.28, as the intersection of the white (minimum potential energy of the spiral density wave) and dark lines (front of the shock wave). Note that this result is obtained in two-dimensional and three-dimensional models in the gas disk.

Thus, in a realistic model of a rotating gas disk the discrepancy with the observational data disappears. Of course, there remains the question of how much the dust band determines the spatial position of the shock wave; this requires studies in the multiphase models.

8.5. Some problems of the excitation of gravitational density waves

8.5.1. The mechanism of gravitational instability in the stellar–gas disk

The population of disk galaxies is inhomogeneous in its dynamic characteristics – the partial contributions to the total surface density of the disk and the velocity dispersion of its constituent objects. Thus, the subsystems of quite young early-type stars (O, B, A) are characterised by low surface density and small velocity dispersion, while the subsystems of old stars are much more massive and dispersions of their radial velocities are high. The relative contribution of the gas as a subsystem to the total density of the disk increases with distance from the centre with the nearly constant dispersion of the velocity of gas clouds (much less than the velocity dispersion of the old stars).

We investigate the dynamics of small perturbations of the gravity type in the plane of the multicomponent disk, ignoring in the first approximation the radial heterogeneity of its equilibrium parameters ($kL \gg 1$, where k is the wave number; L is the scale of the radial inhomogeneity of the disk). This approximation allows to exclude from consideration the branches of the gradient type, and thus explore independently the dispersion properties of the gravitational spiral density waves important to the theory of Lin and Shu [844, 845]. This problem was first posed by Lin and Shu [845] (they investigated a two-component stellar–gas disk with a $c_s / c_r = \sqrt{0.1} \simeq 0.316$), as set out below were obtained in [152].

The equilibrium distribution functions of stars are assumed to be of the Schwarzschild type (see section 2.1.2) with partial surface densities of stellar subsystems σ_{*_i} ($i = 1, 2, ..., N$) and the corresponding radial velocity dispersions c_{ri}. The gas subsystem is characterised by similar parameters σ_{g0}, c_g. In the approximation the perturbed density of the i-th star subsystem $\tilde{\sigma}_i$ is associated with a perturbed gravitational potential $\tilde{\Phi}$ by the ratio (see section 2.2.2)

$$\tilde{\sigma}_i = -\frac{\sigma_{*i}}{c_{ri}^2}\tilde{\Phi}\cdot\left\{1 - \mathrm{I}_0(z_i)\exp(-z_i) - 2\sum_{n=1}^{\infty}\frac{\hat{\omega}^2 \mathrm{I}_n(z_i)\exp(-z_i)}{\hat{\omega}^2 - n^2\varkappa^2}\right\}, \tag{8.108}$$

where the $z_i = k^2 c_{ri}^2 / \chi^2$, and the remaining notations are given in section 2.2. In the same approximation, the perturbed density of the gas subsystem has the form (see section 3.7)

$$\tilde{\sigma}_g = \frac{\sigma_{g0} k^2}{\hat{\omega}^2 - \varkappa^2 - k^2 c_g^2}\tilde{\Phi}. \tag{8.109}$$

Substituting the expressions (8.108) and (8.109) in the Poisson equation for a multicomponent disk

$$\tilde{\sigma}_g + \sum_{i=1}^{N}\tilde{\sigma}_i = -k\tilde{\Phi} / 2\pi G, \tag{8.110}$$

we obtain the desired dispersion equation:

$$\frac{k}{2\pi G} = -\frac{\sigma_{g0}k^2}{\hat{\omega}^2 - \varkappa^2 - k^2 c_g^2} +$$

$$+\sum_{i=1}^{N} \frac{\sigma_{*i}}{c_{ri}^2}\left\{1 - \mathrm{I}_0(z_i)\exp(-z_i) - 2\sum_{n=1}^{\infty} \frac{\hat{\omega}^2\, \mathrm{I}_n(z_i)\exp(-z_i)}{\hat{\omega}^2 - n^2\varkappa^2}\right\}. \quad (8.111)$$

Further analysis was carried out in the framework of two-component stellar disk ($N = 2$), one of the components of which ($i = 1$) is a massive subsystem of old stars, and the other one ($i = 2$) is less massive ($\sigma_{*2} < \sigma_{*1}$) and colder ($c_{r2} < c_{r1}$) subsystem of young stars. For the gas subsystem we assume that the inequalities $\sigma_{g0} < \sigma_{*1}$, $c_g < c_{r1}$. In the study of the dispersion properties of the perturbations with $\hat{\omega}^2 < \pi^2$ in (8.111) we can neglect terms with $n \geqslant 2$ in the sums of the modified Bessel functions. In this approximation (8.111) can be written as

$$1 = \frac{1}{k k_{T_1} \rho_1^2}\left\{1 - a_1 + \frac{\alpha}{\beta^2}(1 - a_2) - 2\frac{\hat{\omega}^2}{\hat{\omega}^2 - \varkappa^2}\left(b_1 + \frac{\alpha}{\beta^2} b_2\right) - \frac{z_1 \varkappa^2 \delta}{\hat{\omega}^2 - \varkappa^2 - k^2 c_g^2}\right\}, \quad (8.112)$$

where

$$\alpha = \sigma_{*2}/\sigma_{*1};\ \delta = \sigma_{g0}/\sigma_{*1};\ \beta = c_{r2}/c_{r1};\ \rho_i = c_{ri}/\chi;$$
$$k_{Ti} = \chi^2/2\pi G\sigma_{*i};\ a_i = I_0(z_i)\exp(-z_i);\ b_i = I_1(z_i)\exp(-z_i).$$

The boundary of gravitational stability of the model disk determined from the condition $\hat{\omega}^2 = 0$ at the minimum dispersion curve $\hat{\omega}^2 = \hat{\omega}^2(k)$. Condition $\hat{\omega}^2 = 0$, is obtained directly from (8.112)

$$\sqrt{z_1} k_{T_1} \rho_1 + (a_1 - 1) + \frac{\alpha}{\beta^2}(a_2 - 1) - \frac{z_1 \delta}{1 + z_g} = 0, \quad (8.113)$$

where $z_g = k^2 c_g^2/\varkappa^2$, and the minimum condition for dispersion curve – differentiating (8.112) with respect to k and setting $\hat{\omega}^2 = d\hat{\omega}^2/dk = 0$:

$$\left[1 - a_1 - 2z_1(b_1 - a_1)\right] + \frac{\alpha}{\beta^2}\left[1 - a_2 + 2z_2(b_2 - a_2)\right] - \frac{z_1(1 - z_g)\delta}{(1 + z_g)^2} = 0. \quad (8.114)$$

The joint solution of (8.113) and (8.114) defines the Jeans scale $\lambda_J = 2\pi/k_J$ (the scale of marginally stable perturbations) and the condition of the marginal stability of the disk. In particular, for one-component stellar disk ($\alpha = \delta = 0$) from this system we obtain the well-known Toomre result [1229]: $k_{j\rho 1} = 0.974$; $k_{T\rho 1} \approx 0.535$ or $c_{r1} = c_T = 3.36\ G_{\sigma *1}/\chi$; $\lambda_J = 21.7\ G\sigma_{*1}/\varkappa^2$.

As a first example, consider a model in which the disk I is a massive and 'hot' subsystem of old stars, the disk II – a low-mass and 'cold' subsystem of young stars ($\alpha < 1, \beta < 1$), and the gas subsystem is not ($\delta = 0$). The results of the solution of (8.113), (8.114) in the framework of this model are shown in Fig. 8.29a and 8.30a. Figure 8.29a shows that when $\beta = c_{r2}/c_{r1} \gtrsim 0.2$ the dependence $\lambda_J(\sigma_{*2}/\sigma_{*1})$ is unambiguous, and with increasing $\alpha = \sigma_{*2}/\sigma_{*1}$ and a decrease in β the stability boundary of the disk is shifted to shorter wavelengths, but only slightly. At transition to smaller values $\beta \lesssim 0.2$ there is a qualitative change – the dependence $\lambda_J(\alpha)$ is ambiguous: in some areas the parameter α [$\alpha_{\min}(\beta) \le \alpha \le \alpha_{\max}(\beta)$], equation (8.114)

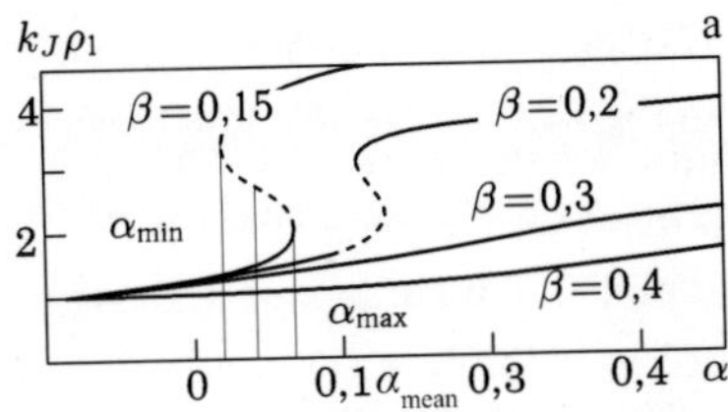

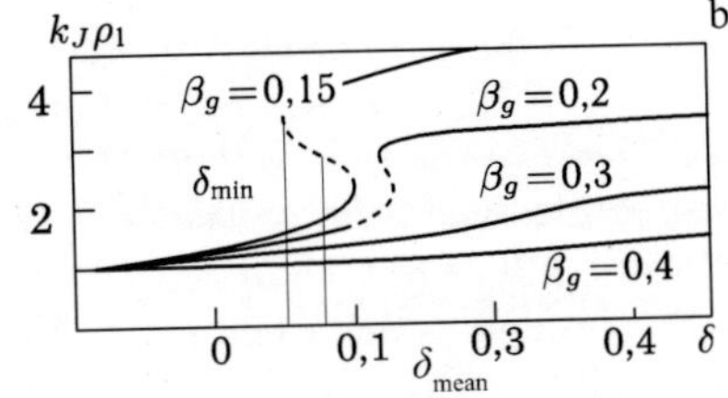

Fig. 8.29. The dependence of the inverse Jeans scale $k_J\rho_1$ on model parameters: a – two-component stellar disk, and b – stellar–gas disk.

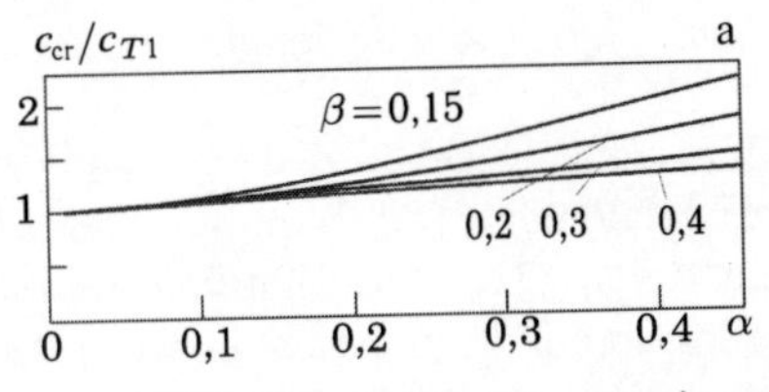

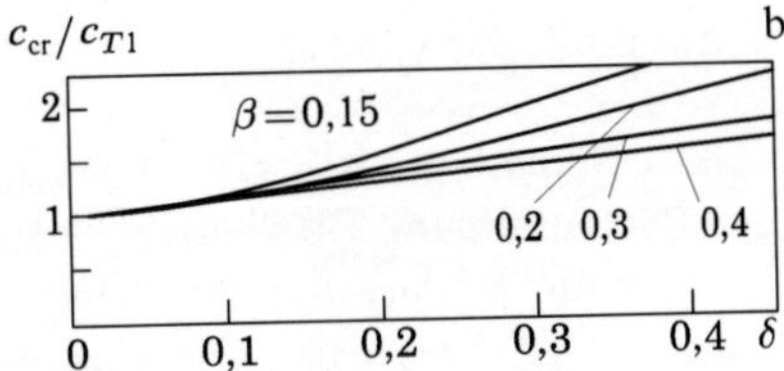

Fig. 8.30. The dependence of the margin required for the stability of the disk as a whole, the dispersion of radial velocities in massi4ve stars of the old subsystem ($c_{cr} = c_{r1}$) of the model parameters: a – two-component stellar disk; b – a stellar–gas disk.

has three real roots. Two of them correspond to the two minima on the dispersion curve $\hat{\omega}^2 = \hat{\omega}^2(k)$, and the third to the maximum between them (dotted line in Fig. 8.29a). The long-wavelength minimum is due to a massive subsystem of old stars, and it defines the marginal stability of the disk $\alpha < \alpha_{mean}$ (see Fig. 8.29a). The short-wave minimum is created by the 'cold' subsystem of young stars. When $\alpha > \alpha_{max}$ (β) dispersion curve $\hat{\omega}^2 = \hat{\omega}^2(k)$ has again a minimum, which lies already in the short-wave area and is due to the low-mass 'cold' subsystem of the stars.

Figure 8.29a clearly shows the interesting feature of the composite (two-component) model of a gravitating disk, if one of the subsystems is sufficiently 'cold' ($\beta \lesssim 0.2$), then, even as a very low-mass subsystme ($\beta \lesssim \alpha \ll 1$), it determines the Jeans scale λ_J and the stability of the whole disk. In this connection it should be noted that if both subsystems are sufficiently 'cool', then the two minima of the dispersion curve can be at $\hat{\omega}^2 < 0$. Thus, there may be two isolated k-regions (two rings on the k-plane) of unstable perturbations. This means that in such a disk structures with two very different spatial scales ($\lambda_{J1}/\lambda_{J2} \sim c_{r1}/c_{r2}$) can be excited. In general, in the N-component disk under certain restrictions on the parameters of subsystems there can be N isolated k-regions of gravitationally unstable perturbations.

The value of $c_{r1} = c_{cr}(\alpha)$, required for marginal stability of the whole disk with a fixed $\beta = c_{r2}/c_{r1}$, is obviously a function of the parameter α that c_{cr} ($\alpha = 0$) $\equiv c_{T1} = 3.36\, G\sigma_{*1}/\varkappa$. This functional dependence is shown in Fig. 8.30a. It is seen that in the range $\alpha \lesssim \beta$ this dependence is rather weak: $c_{cr}(\alpha) \simeq c_{T1}(1 + \alpha)$, i.e. in accordance with the results of Toomre [1229], $c_{cr} \propto (\sigma_{*1} + \sigma_{*2})$. However, if one of the subsystems is rather 'cold' ($\beta \lesssim 0.2$), then in the range $\alpha \lesssim \beta$ the c_{cr} value increases much more rapidly with the increase of the parameter α. Nevertheless, it is important to note that the stability of the disk with a rather massive cold subsystem II ($\alpha > \beta$) can be maintained by a sufficiently hot subsystem of old stars.

Let us now consider another composite model – a two-component stellar–gas disk consisting of a subsystem of old stars (σ_{*1}, c_{r1}) and the gas subsystem ($\sigma_{g0} = \delta\sigma_{*1}$; $c_g = \beta_g c_{r1}$). At the first glance it might seem that the results of this model will be identical to the results obtained for the two-component stellar disk, up to relabelling $\alpha \to \delta$, $\beta \to \beta_G$. However, it is not, since the dynamics of perturbations in the gas disk is described by the hydrodynamic equations, and in the stellar disk by a collisionless kinetic equation, and in the wavelength region $kc_r/\varkappa \gtrsim 1$ (or if we have $kc_g/\varkappa \gtrsim 1$) the descriptions of these methods are not equivalent. The last circumstance is important, because even for the 'coldest' subsystems of young stars and gas $\min(c_g, c_r)$, $k_s/\varkappa \sim 1$ (where $k_s = 2\pi/\lambda_s$; λ_s is the wavelength of the spiral pattern).

The calculation results in a model of a stellar-gaseous disk are shown in Fig. 8.29b and 8.30b. Qualitatively, these results do not differ from those described above for the model of the two-component stellar disk. Quantitative differences are also as follows. First, the short-wave minimum occurs at a higher relative proportion of 'cold' (gas) subsystem: $\delta_{\min}(\beta_g = \beta) > \alpha_{\min}(\beta)$ (see in Fig. 8.29a and 8.29b). Secondly, the value required for the marginal stability of the disk as a whole $c_{r1} = c_{cr}(\delta)$, with the increase of parameter δ increases faster than with increase of parameter $\alpha = \sigma_{*2}/\sigma_{*1}$ in the two-component model of the stellar disk (see Fig. 8.30a and 8.30b).

Let us dwell on the three-component model of the disk (the old stars + young stars + gas) and the role of the finite thickness of the subsystems. We assume that the light subsystem of young stars (spectral types O, B) has the same value of dispersion of the radial velocity as the gas, i.e. $\beta = c_{r2}/c_{r1} = c_g/c_{r1} \equiv \beta_g$ [1112, 1301]. Then this model is characterised by three parameters: α, δ, β. The results of studies of this model at the most interesting limit of $\alpha \ll \delta$ can be summarised as follows. The dependence of the inverse (normalised) Jeans scale k_{Jp1} on the parameters of the model almost matches that shown in Fig. 8.29b, if the horizontal axis gives not δ but the sum $\hat{\delta} = \delta + \alpha$. The value of $c_{cr}(\hat{\delta}, \beta)$ is close to that shown in Fig. 8.30b, but somewhat lower.

However, the most significant impact on the above results is provided by taking into account the finite thickness of the disk subsystem. The method for accounting for these variables is known (see 2.3.2, 3.7.2), and the corresponding calculations [589] showed that for reasonable values of $h/\rho_1 \simeq 0.05 \div 0.1$ and $\Delta_{*1}/\rho_1 \simeq 0.2 \div 0.4$, $\hat{\delta}_{\text{mean}} \simeq 0.3 \div 0.4$ (short-wave minimum becomes 'deeper' than the long-wave one, and it determines the boundary of gravitational stability of a disk with $\hat{\delta} > \hat{\delta}_{\text{mean}}$). Similar results were also obtained by Jog and Solomon [729] who investigated the gravitational instability of a two-component disk model, in which both the stellar and gase subsystems are described in the hydrodynamic approximation.

In conclusion, we estimate the typical value λ_J of the parameters. We assume that $\sigma_{*1} \simeq 83\ M_\odot/\text{pc}^2$; $R = 8$ kpc, $\varkappa = 38$ (km/s)/kpc, and therefore, only for the old stars $\lambda_J = 21.7\ G\sigma_{*1}/\varkappa^2 \simeq 5.5$ kpc. Let the gas density be $\sigma_{g0\odot} \simeq 12\ M_\odot/\text{pc}^2$. Hence, $\delta \simeq 0.14$, and $\hat{\delta} = \delta + \alpha \simeq 0.15$ (here parameter α takes into account only the O- and B-stars with $c_r \lesssim 10$ km/s). According to the old stars with the thick disk taken into account $c_{r1} \simeq 48$ km/s, and for the gas $c_g \simeq 7 \div 8$ km/s [600]. Thus, the lower bound is $\beta \simeq 0.15 \div 0.16$, and according to Fig. 8.30b, $c_{r1} \simeq 1.3\ c_{T1}$, and from Fig. 8.29b, it follows that the minimum of the dispersion curve $\hat{\omega}^2 = \hat{\omega}^2(k)$ closest to the boundary of the stability is a long-wavelength minimum with $k_J\ c_{r1}/\hat{\omega}^2 \simeq 1.35$. Then we obtain the initial value of $\lambda_J \simeq 5.5$ kpc.

8.5.2. The role of star formation

For sufficiently gas-rich flat galaxies, the above-discussed short-wavelength gravitational instability can excite the structure whose characteristic scales are of the order of the wavelength of the spiral pattern. Unclear, however, remains the excitation mechanism of instability, because the gravitational instability tends to have a tendency for quite fast self-suppression [1159]. We discuss the possibility of excitation of the gravitational instability of the process of star formation. For the first time this possibility was reported by Kato [742]. In particular, Kato showed that in the model of the two-component stellar–gas disk with a $\beta = \sqrt{0.1} \simeq 0.316$ and $c_{r1} = c_{T1} = 3.36\ G\sigma_{*1}/\varkappa$ the birth of stars from the gas leads to a buildup of gravitational instability, which the maximum growth rate for perturbations of Jeans scale ($k_{\rho 1} \simeq 1$).

We investigate this issue within the considered three-component models of stellar–gas disk. Assume that at the initial time the disk was marginally stable, that is $c_{r1} \equiv c_{\rm cr}(\alpha, \beta, \delta, k_J)$. Then, during the time interval dt part of the gas transforms into young stars with $c_{r2} = c_g$, i.e. α and δ received increments related to the ratio

$$d\alpha + d\delta = 0. \tag{8.115}$$

In a system with the modified α and δ the stability boundary should be shifted along the wavelength by dk_J from k_J and $c_{\rm cr}$ should change by the value $dc_{\rm cr}$. If for $d\alpha > 0$ we have $dc_{\rm cr} > 0$, then the whole system becomes unstable, since $c_{r1} = c_{\rm cr}(\alpha, \beta, \delta, k_J) < c_{\rm cr}(\alpha + d\alpha, \beta, \delta + d\delta, k_J + dk_J) = c_{\rm cr}(\alpha, \beta, \delta, k_J) + dc_{\rm cr}$. Thus, the instability due to star formation occurs

$$\frac{dc_{\rm cr}}{d\alpha} = \left\{\frac{\partial}{\partial\alpha} - \frac{\partial}{\partial\delta} + \frac{dk_J}{d\alpha}\frac{\partial}{\partial k_J}\right\} c_{\rm cr} > 0, \tag{8.116}$$

where $dk_J/d\alpha$ can be calculated directly from (8.114). Calculations show that in the entire range of the disk parameters studied above ($0.15 \leqslant \beta \leqslant 0.4$; $0 \leqslant \alpha \leqslant 0.5$; $0 \leqslant \delta \leqslant 0.5$) the value $dc_{\rm cr}/d\alpha > 0$.

Thus, the process of star formation leads to a buildup of gravitational instability in an initially stable multicomponent galactic disk. In this case, apparently, in the disk of the galaxy there would be relaxation processes taking place and increasing the velocity dispersion of the objects that make up the disk subsystems, and, therefore, seek to recover the system to a stable state. However, the ongoing star formation and the delay of the relaxation processes will support such 'smoldering' instability until the disk reaches the steady state in which the processes of star formation and destruction of stars will be mutually compensated. Thus, the process of star formation is always a constantly acting generator of the Jeans scale structures in a multicomponent disk. The value of the Jeans scale depends on the proportion of 'cold' subsystems in the disk and on their 'temperature'.

8.5.3. Resonance of the radial flow of gas with the gravitational spiral wave

An interesting mechanism for the excitation of gravitational spiral density waves may be a mechanism due to instabilities in a system with the relative motion of its subsystems. Instabilities of this type are usually called beam instabilities. For the first time, apparently, such a mechanism was studied [202] in the hydrodynamic models of a multicomponent disk with subsystems rotating with significantly different angular velocities. Later, Kato [743] considered a more realistic model of the stellar–gas disk, where the gas, in addition to participation in the general rotation of the disk, is experiencing further macroscopic motion in the radial direction. The dispersion equation of the model has the form (cf. (8.111))

$$\frac{k}{2\pi G}=\frac{\sigma_{g0}k^2}{(\omega-m\Omega_g-ku_g)^2-\varkappa_g^2-k^2c_g^2}+$$

$$+\frac{\sigma_*}{c_r^2}\left\{1-\mathrm{I}_0(z)\exp(-z)-2\sum_{n=1}^{\infty}\frac{(\omega-m\Omega)^2\,\mathrm{I}_n(z)\exp(-z)}{(\omega-m\Omega)^2-n^2\varkappa^2}\right\}, \qquad (8.117)$$

where u_g is the radial velocity of the gas flow; Ω, Ω_g are angular velocities; $\varkappa$, $\varkappa_g$ is the epicyclic frequency of the stellar and gas disks, respectively; $z = (kc_r/\varkappa)^2$.

The dispersion equation (8.117) was solved by Kato [743] only numerically for several values of the parameters in it, so that the received results cannot be used to determine the role of a beam mechanism to initiate the spiral pattern of different galaxies. We present an analytical study of equation (8.117) and obtain corresponding estimates for the Galaxy [150]. We assume that $|u_g|\lesssim c_r \ll r\Omega$. Then

$$|\Omega_g/\Omega-1|\lesssim c_r^2/r^2\Omega^2 \ll 1,$$

and thus we can set $\Omega_g \simeq \Omega$. For use in the theory of Lin and Shu [845] of perturbations with frequencies $|\omega-m\Omega|<\varkappa$ the dispersion equation (8.117) takes the form (cf. (8.112))

$$1-\sqrt{z}\,k_T\rho-a-\frac{2b\nu^2}{\nu^2-1}-\frac{z\delta}{(\nu-\eta)^2-(1+z_g)}=0, \qquad (8.118)$$

where the $\nu=(\omega-m\Omega)/\varkappa$, $\eta=ku_g/\varkappa$, $z_g=k^2c_g^2/\varkappa^2$. If the gas subsystem does not form ($\delta = 0$), then the solution of (8.118) are the gravitational oscillation modes of the stellar disk (see (2.96)):

$$\nu^2=\nu_0^2=\frac{1-\sqrt{z}k_T\rho-a}{1-\sqrt{z}k_T\rho-a-2b}. \qquad (8.119)$$

In accordance with the observational data we assume $\delta=\sigma_{g0}/\sigma_*\ll 1$. Then, if the value of ν_0 is not too close to $\nu_1=\eta+\hat{s}\sqrt{1+z_g}$ (here $\hat{s}=\pm1$), the dispersion equation (8.18) leads to the appearance of real corrections of the order of the frequency

ν_0. However, in the resonant case[5] when $\nu_0 \simeq \nu_0$, equation (8.118) gives an unstable solution $\nu = \nu_0 + \varepsilon$ with Im (ε) $\gg \nu_0 - \nu_1$. Indeed, from (8.118) in this case,

$$\varepsilon^2 = \frac{\hat{s}\, z(\nu_0^2 - 1)\delta}{4\nu_0\sqrt{1+z_g}(1-\sqrt{z}k_T\rho - a - 2b)}. \tag{8.120}$$

Since $(\nu_0^2 - 1) < 0$ and in a stable disk according to Toomre [1229] $(1-\sqrt{z}k_T\rho - a - 2b) < 0$, the desired resonant beam instability ($\varepsilon^2 < 0$) occurs at $\hat{s} = -\text{sign}(\nu_0)$.The condition for excitation of this instability (resonance condition) has the form

$$u_g \simeq \frac{\varkappa}{k}\left(|\nu_0| + \sqrt{1+z_g}\right)\text{sign}(\nu_0), \tag{8.121}$$

and the maximum instability growth rate, according to (8.120)

$$\text{Im}(\omega) \simeq \varkappa\sqrt{\delta}\left\{\frac{z(1-\nu_0^2)}{4|\nu_0|\sqrt{1+z_g}(a+2b+\sqrt{z}k_T\rho - 1)}\right\}^{1/2}. \tag{8.122}$$

To assess the need for the excitation of the observed spiral structure in the Galaxy with the radial velocity of the gas flow u_g we used the observational data listed chapter 1 and set $\lambda_s \simeq 1.7$ kpc [1051]. Assuming that the spiral is trailing and two-arm ($m = 2 > 0$, $k > 0$), from (8.119) we obtaine $\nu_0 = -0.97$ (the sign of ν_0 is selected in accordance with the theory of Lin and Shu [846], and from (8.121) the value required for the excitation of spirals $u_g \simeq -22$ km/s. Assuming then $\delta = 0.14$ (see chapter 1), from (8.122) we obtain $\text{Im}(\omega) = 0.3\varkappa\sqrt{\delta} \simeq 0.16\,\Omega$. In this way, even in the presence of the instability u_g required for the buildup, the characteristic time of excitation of the spiral pattern τ, estimated from the maximum growth rate, is equal to 6 revolutions of the disk, which is almost twice the time of drift of the wave spiral packet to the centre of the Galaxy [1231].

There are apparently no radial gas flows in the solar neighborhood with $|u_g| \geqslant 3$ km/s. This implies that the mechanism of resonant beam instability of the radial gas flow in the vicinity of $r = R_\odot$ can not be a generator of gravitational spiral density waves. Note also that in the theory of Lin and Shu, according to (8.121) the resonance beam instability can occur only if the radial gas flow is directed towards the galactic centre.

It is easy to see that the minimum value $|u_g / c_g| = \sqrt{1/z_g + z_g}$ required for the buildup of gravitational spiral density waves by the resonant beam instability can be reached at the corotation radius, where $\nu_0 = 0$. The absolute minimum $|u_g| \simeq \sqrt{2}c_g$ is achieved if in this area $z_g = (2\pi c_g / \varkappa\lambda_s)^2 = 1$. This assessment indicates that the beam mechanism is unlikely to be the mechanism of excitation of the spiral pattern in the flat galaxies.

[5]The width of the resonance area is not computed since, as will be seen from the estimates of the maximum growth rate of the studied instability is of the order of or less than the inverse of the radial drift time of the package of spiral gravitational waves [1231].

8.5.4. Non-linear saturation of the resonance amplification of spiral waves at the corotation radius

Due to differential rotation of the spiral disk of the galaxy $\Omega = \Omega\ (r)$ at some distance from its centre $r = r_c$, the angular velocity of the spiral pattern is $\Omega_p = \Omega\ (r_c)$ (referred to as the value of the corotation radius r_c). Therefore, the disk stars, which are localized in the vicinity of the corotation radius, are in resonance with the spiral wave. Lynden–Bell and Kalnajs [866] showed that in the linear theory the resonant interaction of waves with the stars of the disk leads to amplification of the wave (physics of this process was described in detail by Fridman and Polyachenko [203, 590].

The growing wave must therefore have an impact on the dynamics of the resonant stars, changing their distribution function and also the conditions of wave amplification. It is therefore clear that the study of the effectiveness of the mechanism of resonance amplification of the spiral gravitational waves at the corotation radius in the non-linear regime is very important. We shall examine this question, following [180].

In the reference frame rotating with the angular velocity of the spiral pattern Ω_p, the value of the energy of an individual star can be expressed as

$$H = \frac{v_r^2}{2} + \frac{[r_0\Omega(r_0) + v_\varphi]^2}{2} - r_0\Omega_p[r_0\Omega(r_0) + v_\varphi] + \Phi_0 + \tilde{\Phi}, \qquad (8.123)$$

where r_0 is the radius of the orbit of the centre of the epicycle of the star (see 1.2.3). The Hamiltonian (8.123) in the epicyclic approximation is conveniently written in action–angle variables

$$H = \varkappa(r_0)\mathrm{I}_1 + \frac{2\Omega_p\Omega'_p}{r_c\varkappa_c}\mathrm{I}_2^2 + \tilde{\Phi}(w_1, w_2, \mathrm{I}_1, \mathrm{I}_2), \qquad (8.124)$$

where the $\mathrm{I}_1 = v_\perp^2 / 2\varkappa(r_0)$; $\mathrm{I}_2 = r_0^2\Omega(r_0) - r_c^2\Omega_p$ are the action variables (epicyclic constants of motion); w_1, w_2 are the corresponding angular variables (w_1 is the phase of rapid motion of the star along the epicyclic trajectory, w_2 is the phase of slow motion of the centre of the epicycle in the reference frame rotating with angular velocity Ω_p). The index 'c' for a number of variables means that they are evaluated at $r = r_c$.

The equations of motion of individual stars derived from the Hamilton equations: $d\mathrm{I}_i / dt = -\partial H / \partial w_i$; $dw_i / dt = \partial H / \partial \mathrm{I}_i$. Confining ourselves to a small neighborhood of the corotation radius

$$|x| = |r_0 - r_c| = |(2\Omega_p / r_c\varkappa_c^2)\mathrm{I}_2| \ll r_c, \qquad (8.125)$$

these equations can be written as:

$$\frac{d\mathrm{I}_1}{dt} = -\frac{\partial\tilde{\Phi}}{\partial w_1}, \qquad (8.126)$$

$$\frac{d\mathrm{I}_2}{dt} = \frac{r_c\varkappa_c^2}{2\Omega_p}\frac{dx}{dt} = -\frac{\partial\tilde{\Phi}}{\partial w_2}, \qquad (8.127)$$

$$\frac{dw_1}{dt} = \varkappa_c + x\varkappa_c' + \frac{\partial \tilde{\Phi}}{\partial \mathrm{I}_1}, \tag{8.128}$$

$$\frac{dw_2}{dt} = \frac{2\Omega_p}{r_c \varkappa_c^2}(\varkappa_c' \mathrm{I}_1 + \Omega_c' \mathrm{I}_2) + \frac{\partial \tilde{\Phi}}{\partial \mathrm{I}_2}. \tag{8.129}$$

From (8.127) and (8.129) it can be seen that the resonance condition of the star in the plane (I_1, I_2) has the form $(\varkappa_c' \mathrm{I}_1 + \Omega_c' \mathrm{I}_2) = 0$. In particular, the among the stars with circular orbits ($\mathrm{I}_1 = 0$) resonant are only those whose orbital radius is equal to r_c; stars at which $\mathrm{I}_1 \neq 0$ enter into resonance with the wave, if the radius of the orbit of the guiding centre of the epicycle lies in some vicinity of the corotation radius.

From equations (8.128) and (8.129) it follows that $w_1 \gg w_2$. This allows us to simplify the task by transferring from the system (8.126)–(8.129), which describes the motion of the individual star, to the equations describing the motion of the centre of its epicycle. Such a transition can be made using the method of averaging over the 'fast' phase w_1, developed by Bogolyubov and Mitropol'skii [29][6]. For this we represent variables 'action–angle' in the form of $\mathrm{I}_i = \langle \mathrm{I}_i \rangle + \tilde{\mathrm{I}}_i; w_i = \langle w_i \rangle + \tilde{w}_i$ where $\tilde{\mathrm{I}}_i$ and $\tilde{w}_i$ describe the 'trembling' motion with the fast phase w_1, and $\langle \mathrm{I}_i \rangle$ and $\langle w_1 \rangle$ – smooth movement with the slow phase w_2. Leaving the previous notations (without the averaging sign) in further calculations for the smoothly varying values, from the system (8.126)–(8.129) we obtain

$$\dot{\mathrm{I}}_2 = \left(\frac{r_c \varkappa_c^2}{2\Omega_p}\right)\dot{x} = -\frac{\partial \langle \tilde{\Phi} \rangle}{\partial w_2}, \tag{8.130}$$

$$\dot{w}_2 = \frac{2\Omega_p}{r_c \varkappa_c^2}(\varkappa_c' \mathrm{I}_1 + \Omega_c' \mathrm{I}_2) + \frac{\partial \langle \tilde{\Phi} \rangle}{\partial \mathrm{I}_2}, \tag{8.131}$$

and, moreover,

$$\mathrm{I}_1 = \mathrm{const}. \tag{8.132}$$

We represent the gravitational potential of spiral waves in the reference frame rotating with angular velocity Ω_p in the form

$$\tilde{\Phi} = S(r)\cos\{m[\varphi + \chi(r)]\}, \tag{8.133}$$

where
$$r = r_0 + \hat{\rho}\sin(w_1);\ \varphi = w_2 + (2\Omega / r_0 \varkappa)\hat{\rho}\cos(w_1);$$

$$w_1 = \varkappa_c t;\ \hat{\rho} = \sqrt{2\mathrm{I}_1 / \varkappa}\ \text{(see (2.70))}.$$

Then, averaging $\tilde{\Phi}$ over the quick phase w_1, we obtain

$$\langle \tilde{\Phi} \rangle = S(r_0) J_0(\tilde{k}\hat{\rho})\cos\{m[w_2 + \chi(r_0)]\}, \tag{8.134}$$

where the value $\tilde{k}$ value is defined in (2.76), $k_r = m\left(\frac{\partial \chi}{\partial r}\right)_{r=r_0}$; $J_0(\tilde{k}\hat{\rho})$ is the Bessel

[6]A similar approach allowed to develop a theory of the drift motion of charged particles in a magnetic field by averaging their actual movement over the 'fast' phase of Larmor rotation.

function of zeroth order. In further calculations we assume a spiral pattern to be tightly twisted $[\mathrm{tg}(i) = m / k_r r_0 \ll 1]$ and two-arm ($m = 2$). With this in mind, in the neighborhood of corotation resonance we can consider

$$\tilde{k} = k_r(r_c) = k;\ \hat{\rho} = \sqrt{2\mathrm{I}_1 / \varkappa_c};\ S(r_0) = S(r_c) = S_c;\ m\chi(r_0) = m\chi(r_c) + kx.$$

Assuming then, without loss of generality $\chi(r_c) = 0$, and denoting the phase of the spiral wave $\Theta = w_2 + kx/2$, we reduce the system (8.130), (8.131) to the equation

$$\ddot{\Theta} + \frac{4\Omega_p |\Omega_c'|}{r_c \varkappa_c^2} S_c J_0(k\hat{\rho}) \sin(2\Theta) = 0, \tag{8.135}$$

where, in accordance with the observations, it was considered $\Omega_c' < 0$, and the relationship of the coordinates of the centre of the epicycle x, φ with the phase Θ is determined by the relations

$$x = \frac{1}{|\Omega_c'|}\left(\frac{2\Omega_p \varkappa_c'}{r_c \varkappa_c^2}\mathrm{I}_1 - \dot{\Theta}\right); \qquad \varphi = \Theta - \frac{kx}{2}. \tag{8.136}$$

Equation (8.135), which is the solution of the problem of the motion of the centre of the epicycle of the star in the spiral wave field, is the energy integral

$$\hat{E} = \frac{\dot{\Theta}^2}{2} + \frac{\omega_0^2}{2}\begin{cases}\sin^2\Theta & \text{at } J_0(k\hat{\rho}) > 0,\\ \cos^2\Theta & \text{at } J_0(k\hat{\rho}) < 0,\end{cases} \tag{8.137}$$

where

$$\omega_0^2 = \frac{8\Omega_p |\Omega_c'|}{r_c \varkappa_c^2} |J_0(k\hat{\rho})| S_c. \tag{8.138}$$

The solution (8.135) can be obtained in terms of elliptic functions, and in a visual form – asymptotic. However, even without solving this equation, we can obtain a qualitative picture of the nature of the centres of epicycles (for brevity – particles). The effective potential (see 8.137), in which such particles move, is shown in Fig. 8.31. It is seen that all particles can be divided into trapped in a wave for which $q^2 = \omega_0^2 / 2\hat{E} > 1$, and passing particles for which $q^2 < 1$. The trajectories of these particles in the phase plane (Θ, Θ) are shown in Fig. 8.31b.

The trapped particles, due to the fact that the effective potential (see Fig. 8.31a) is not quadratic, depending on the value of the parameter q^2, have different periods of motion along the coordinate Θ. Thus, in the vicinity of the corotation radius occurs mixing of particles in the phase space so that the function of the distribution of these particles contains a plateau over the time of the order $\tau_0 = 2\pi/\omega_0$ (similar nonlinear effects in plasma physics of this phenomenon were described by Kadomtsev [83].

To describe the intensity of the wave by the dimensionless amplitude of the gravitational force $F = kS_c / r_c\Omega_p^2$, the characteristic time for $\tau_0 \sim T_c / \sqrt{F\mathrm{tg}i}$ where $T_c = 2\pi/\Omega_p$ (with respect to parameters of spiral waves in the Galaxy $\tau_0 \sim 10T_c$). After the formation of the plateau the growth of the amplitude wave due to the gradient of the distribution function in the phase space, must cease. Therefore,

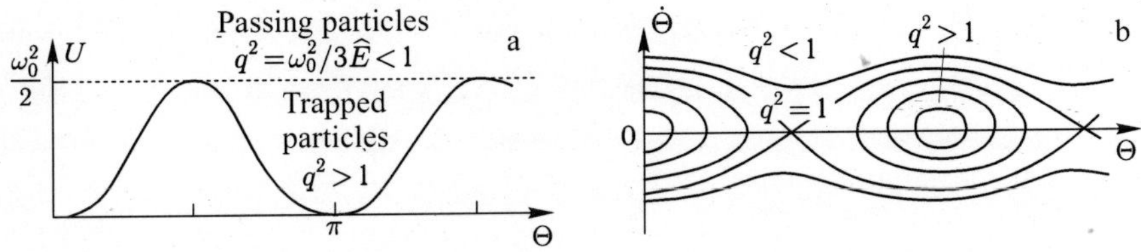

Fig. 8.31. a – the effective potential $U = \omega_0^2 \sin^2 \Theta / 2$ for movement to the centre of the epicycle (the case of $J_0^2(k\hat{\rho}) > 0$), b – trajectory of the centres of the epicycles in the phase plane $(\Theta, \dot{\Theta})$.

to test the effectiveness of the mechanism of amplification of spiral waves proposed Lynden–Bell and Kalnajs [866], it is necessary to calculate the saturation amplitude of the wave to the point of 'switching off' this mechanism and the magnitude of the time interval of its operation.

The spiral wave as a wave of negative energy in the range $r < r_c$, should be resonantly amplified in the neighbourhood of the corotation radius due to the transfer of the momentum $\mathcal{L}$ and energy $\mathcal{E}$ to the resonant stars. The rate of such transfer under the linear theory is defined by the following relations:

$$\dot{\mathcal{L}} = -\frac{\pi S_c^2 r_c \varkappa_c^2}{2\Omega_c \,|\,\Omega_c'\,|}\frac{\partial \hat{f}}{\partial \mathrm{I}_2}\bigg|_{\mathrm{I}_2=0}, \qquad \dot{\mathcal{E}} = \Omega_p \dot{\mathcal{L}}, \tag{8.139}$$

where

$$\hat{f} = \hat{f}(\mathrm{I}_2) = \int f_0(\mathrm{I}_1, \mathrm{I}_2)\, dw_1\, dw_2\, d\mathrm{I}_1 = 4\pi^2 \int f_0(\mathrm{I}_1, \mathrm{I}_2)\, d\mathrm{I}_1, \tag{8.140}$$

and $f_0(\mathrm{I}_1, \mathrm{I}_2(r_0))$ is the initial function distribution of the particles. In the vicinity of the resonance $\varkappa_c' \mathrm{I}_1 + \Omega_c' \mathrm{I}_2 = 0$ represent this function as a series

$$f_0(\mathrm{I}_1, r_0) \simeq f_0(\mathrm{I}_1, r_c - b\mathrm{I}_1) + \frac{\partial f_0}{\partial r_0}\bigg|_{r_0=r_c-b\mathrm{I}_1} \cdot (x + b\mathrm{I}_1) =$$

$$= f_0(\mathrm{I}_1, r_c - b\mathrm{I}_1) - \frac{\omega_0 \mathrm{sign}(x + b\mathrm{I}_1)}{q\,|\,\Omega_c'\,|}\sqrt{1 - q^2 \sin^2 \Theta}\, \frac{\partial f_0}{\partial r_0}\bigg|_{r_0=r_c-b\mathrm{I}_1}, \tag{8.141}$$

where $b = 2\Omega_p \varkappa_c' / r_c \varkappa_c^2 \Omega_c'$. The phase mixing will lead to the distribution function of the trapped particles becoming constant [83]:

$$f_{\mathrm{fin}}^{(\mathrm{tr})} = f_0(\mathrm{I}_1, r_c - b\mathrm{I}_1), \tag{8.142}$$

and the distribution function of circulating particles becomes

$$f_{\mathrm{fin}}^{(\mathrm{circ})} = f_0(\mathrm{I}_1, r_c - b\mathrm{I}_1) + \frac{\pi\, \omega_0 \mathrm{sign}(x + b\mathrm{I}_1)}{2\,q K(q)\,|\,\Omega_c'\,|}\frac{\partial f_0}{\partial r_0}\bigg|_{r_0=r_c-b\mathrm{I}_1}, \tag{8.143}$$

where $K(q)$ is the complete elliptic integral of the first kind.

A visual representation of the distribution function (8.142) (8.143) can be obtained from in Fig. 8.32. As the initial distribution function f_0 (I_1, r_0) in the construction of this figure was the Schwarzschild function (2.42) with $\Omega(r_0) = \Omega_p r_c / r_0$; $c_r / r_c \Omega_p = 0.17$; $\sigma_*(r_0) = \sigma_0 \exp(-r_0/L)$; $L = 0.5 r_c$ and also the parameters of

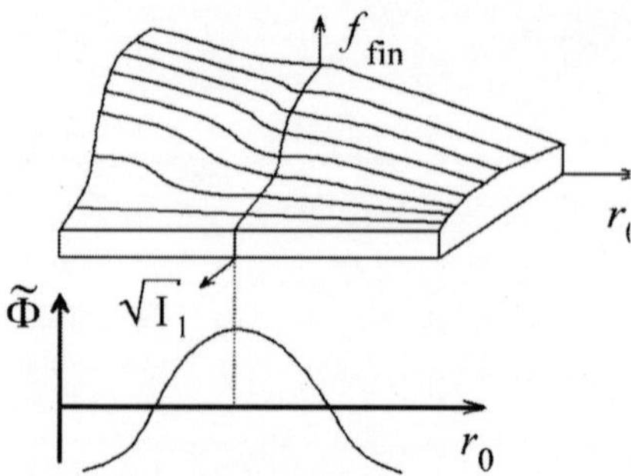

Fig. 8.32. The ultimate function of the distribution of stars in the direction $\varphi = 0$. The origin of coordinates $(r_0, \sqrt{I_1})$ located at the corotation radius.

the wave $F = kS_c / r_c\Omega_p^2 = 0.05$; $\text{tg}\, i = 2/kr_c = 0.14$ [1319]. Figure 8.32 shows $f_{\rm fin}$ in the direction $\varphi = 0$ in the spiral gravitational potential (8.133) with $\chi(r_0) = k\,(r_0 - r_c)$.

The value of the momentum transferred from resonant particles to the wave

$$\delta L = \text{R}\left[f_{\rm fin} - f_0\right]\cdot\left(\text{I}_2 + r_c^2\Omega_p\right)dw_1 dw_2 dI_1 dI_2 \tag{8.144}$$

is calculated, assuming a low velocity dispersion of stars and thus neglecting the deviation of the resonance line $r = r_c$. As a result, we obtain

$$\delta L = -\frac{32\, S_c^2 r_c \varkappa_c^2}{\omega_0\, |\Omega_c'|\, \Omega_p}\frac{\partial \hat{f}}{\partial \text{I}_2}\bigg|_{\text{I}_2=0}\cdot\{Q^{(tr)} + Q^{(circ)}\}, \tag{8.145}$$

where the quantities

$$Q^{(\text{tr})} = \int_0^1 \frac{q}{\pi}\left[E(q) + (q^2 - 1)K(q)\right]dq \simeq 0.071, \tag{8.146}$$

$$Q^{(circ)} = \int_0^1 \left[\frac{E(q)}{\pi} - \frac{\pi}{4K(q)}\right]\frac{dq}{q^4} \simeq 0.028 \tag{8.147}$$

characterise the contributions to the transfer of momentum and energy to the spiral wave from the trapped and untrapped particles, respectively ($E\,(q)$ is the complete full elliptic integral of the second kind).

Comparison of (8.139) with (8.145) gives

$$\delta\mathcal{L} \simeq 0.32\,\tau_0\dot{\mathcal{L}}. \tag{8.148}$$

Hence it is clear that the characteristic time of operation of the mechanism of amplification of the spiral wave is about three revolutions of the Galaxy at the corotation radius (see the estimate of τ_0 made above for $F = 0.05$; tg $(i) = 0.14$).

We now assume that the Galaxy has spiral waves of the infinitesimal amplitude, and estimate the amplitude of their saturation due to the mechanism of resonant amplification of such waves at the corotation radius. To do this, we equate the momentum that can be transmitted to the wave by the resonant stars (8.145), to the momentum of the wave [737]

$$\mathcal{L}_w = \int \frac{kmr_0}{4\varkappa G} \tilde{\Phi}^2 \frac{\partial D}{\partial \nu} dr_0 d\varphi, \tag{8.149}$$

where $\nu = m(\Omega_p - \Omega)/\varkappa$; $D = D(k,\nu)$ is the 'dielectric permittivity' of the stellar disk. The latter in the approximation used to compute $\delta\mathcal{L}$ (8.145) can be written as follows

$$D \simeq \frac{2\pi G\sigma_* k}{\varkappa^2(1-\nu^2)}. \tag{8.150}$$

Then the order of magnitude

$$\mathcal{L}_w \simeq 2\pi^2 \frac{S_c^2 \sigma_*}{\varkappa^3 \mathrm{tg}^2 i}. \tag{8.151}$$

Assuming in (8.145) the distribution function of stars to be of the Schwarzschild type, we obtain the saturation amplitude of the wave

$$F_s \sim \eta^2 \mathrm{tg}^3 i \sim \mathrm{tg}^3 i, \tag{8.152}$$

where the $\eta = d\ln(\mu(r_0))/d\ln(r_0); \mu(r_0) = 2\pi\sigma_* r_0 dr_0 / dI_2^2$ is the mass per unit of the square of the angular momentum.

For the value tg (i) = 0.14 [144, 1319], normally accepted for the Galaxy, this gives $F_s \sim (3 \div 4)\cdot 10^{-3}$, which is at least an order of magnitude less than the values obtained by comparing the calculated and observed kinematic effects associated with the dynamics of spiral waves in the Galaxy.

The problem solved here is close in meaning to the problem of non-linear interaction of charged particles with the wave of the electric potential [83]. At the same time, unlike plasma, where the resonance with the wave is a small fraction of particles in the velocity space, in this case the wave is in resonance with almost all particles of of the velocity space, localized, however, in a small neighborhood of the corotation radius. This factor 'spoils' the spatial distribution of the gravitational potential and to ignore its influence, the width of the plateau formed Δx must be small compared with the wavelength, $\lambda_s = 2\pi/k$. From (8.136)–(8.138) we obtain the estimate $\Delta x = 2\sqrt{S_c}/\Omega_p$, and therefore $\Delta x/\lambda_s \simeq \sqrt{2F/\pi^2 \mathrm{tg} i} \lesssim 0.3$ for the parameters of the spiral structure of the Galaxy.

It should also be noted that, according to Galeev and Sagdeev [38], the results of the problem of the growth of the wave to the finite amplitude, in contrast to the results of the problem of damping of waves of finite amplitude, should be considered only as an estimate because in the calculations the growth of the potential of particles was neglected. Nevertheless, the resulting conclusion on the ineffectiveness of resonant amplification of spiral waves at the corotation radius is correct, because the estimate of the saturation amplitude of the wave (8.152) shows that $F_s \ll F_{\mathrm{obs}}$.

We also note that similar to the plasma formulation in [142], we solved the problem of non-linear interaction of the beam of the stars moving along the axis of the self-gravitating cylinder with a wave propagating in the same direction.

9

The model of the Galaxy (Milky Way)

9.1. Kinematics and the distribution of matter

9.1.1. The law of rotation of the Galaxy in the solar neighborhood

Our understanding of the physics of and structure of the Galaxy changes as the photometric and kinematic data from obtained from the observations become more accurate. The rotation curve belongs to the most important characteristics. There are a large number of works which determine the radial dependence of the rotational velocity of a dynamically cold component [213, 420, 470, 831, 1186] et al). Different galactic subsystems have different rotational velocities [351]. A stellar disk rotates more slowly than a gas component ($V < V_{gas}$, where $V_{gas}(r)$ is the rotational velocity of the gas and young stars, and $V(r)$ characterizes the old stellar population). In constructing the model of the Galaxy the major parameter is the angular rotational velocity $V_c(r)$, which determines the spatial distribution of the gravitational potential and, hence, mass. Along with the stellar and gas discs, an essential role is played by the spheroidal subsystem, as evidenced by the models of the Galaxy constructed so far [441, 500, 991, 1081, 1121]. The disk and spheroidal subsystems have a complex structure, which is reflected in the allocation of individual components that differ in their physical properties. In the stellar disk we can distinguish small and large components [351], and in the spheroidal subsystem – the halo, the bulge and the nucleus.

The same rotation curve can be constructed with very different parameters of the disk and spheroidal components (see chapter 1). If we consider the results of determination of the surface density of matter in the solar neighborhood $\sigma_{\odot}$, radial L and vertical h scales of the disk from these observations, then this ambiguity is reduced. Additional restrictions are imposed on the use of the data model of the Galaxy of the velocity dispersion of stars. At the heart of this capability is a natural assumption that the disk, formed by the old stellar population in which the bulk of mass is concentrated, is gravitationally stable. Additional restrictions on the distribution of the galactic material are caused by the demand for stability against bending perturbations because this imposes conditions on the variance of vertical velocities c_z and the vertical structure of t he disk [63, 207].

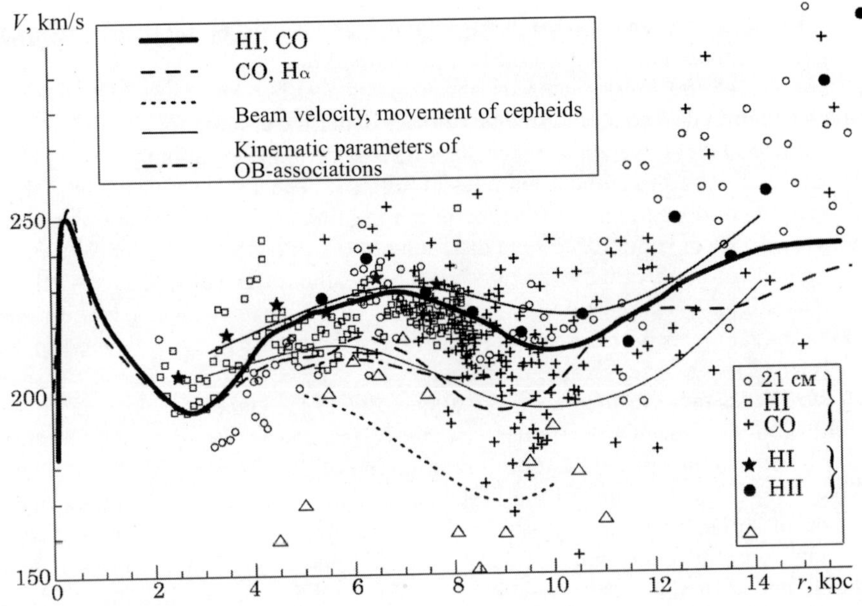

Fig. 9.1. Rotation curves of the Galaxy (taken from Ref. 260), obtained by different authors: [470] – the solid thick line, according to HI, CO data, $R_{\odot}$= 8.5 kpc; [1186] – dashed line, CO, H_{α}, $R_{\odot}$= 8.5 kpc; [48] – the dotted line, according to the beam velocities and natural movement of classic cepheids, $R_{\odot}$= 7.1 kpc; [214] – solid thin line, for the cases $R_{\odot}$= 7.5 kpc and $R_{\odot}$= 8.5 kpc; [49] – dot-and-dash line, according to the kinematic parameters of the system of OB-associations, $R_{\odot}$= 7.1 kpc; [968] – (21 cm line), □(HI), + (CO), $R_{\odot}$= 8.2 kpc; [420] – rotation velocity according to HI (*), for regions HII (•), $R_{\odot}$= 8.5 kpc; [300] – rotation velocity obtained in analysis of the kinematics of planetary nebulae and AGB-stars (Δ), $R_{\odot}$= 7.9 kpc.

In constructing models of the rotation curve, as a rule, we assume that the condition $V_c = V_{gas}$ is fulfilled, i.e. in the presence of only circular motions the pressure gradient in the radial balance of forces is neglected. One of the first models of the Galaxy, constructed by Schmidt [1121], was developed and significantly refined in models [155, 227, 326, 441, 500, 991, 1081, 1102]. These axisymmetric models in some cases significantly vary in the mass of its main components. The seven-components model of the Galaxy proposed in [2289] has revealed the role of pressure gradients for the rotation curve in some areas. Construction of non-axisymmetric models, in particular, those including the bar, is only possible by means of dynamic modeling [601, 602].

The rotation curve for the young population of the galaxy was determined in many studies (Fig. 9.1 shows the observational data by different authors). The galactic rotation curve is plotted using used kinematic data, not only on the gas component, but also other objects (OB-stars, planetary nebulae, Cepheids, regions HI and HII, carbon stars). All these rotation curves, which are based on observational data of the young population, will be denoted by V_{gas}.

We note two features in the observed galactic rotation curve (see Fig. 9.1):

– A local minimum (trough) in the solar neighbourhood's orbit ($r \simeq 6 \div 11$ kpc);

– A local inner maximum in the central kiloparsec.

Following [260, 769], we discuss the consequences in which these features result in constructing the model of circular velocity under the assumption $V_c = V_{gas}$.

For definiteness, we choose the rotation curves from the works by Sofue [1183, 1186] (see thick solid and dashed lines in Fig. 9.1). The conclusions made in these studies do not depend on a difference in the rotation curves of the various authors who use different values of $R_\odot$ and different objects of observation. For example, if the values of $R_\odot$ are varied the range 7.5÷9.5 kpc, the rotational speed in [420] varies only slightly. By varying the parameters $R_\odot$ and V_{gas} we change absolute values of the spatial scales of different components and their mass in dynamic models, but the relative magnitudes change only slightly..

We discuss these features of the rotation curve of the Galaxy.

Local deflection of the circular velocity of rotation in the solar neighborhood may be due to the special feature of distribution of density in the disk component [968, 969] or in the halo. In the first case, we can assume that the radial distribution of surface density is described, for example, by the sum of two exponential laws with different radial scales $L_{\sigma 1}$ and $L_{\sigma 2}$, and the disk with the short scale should end abruptly in the region $r \simeq 6 \div 7$ kpc, and the second disk at a radius of $r \simeq 12 \div 15$ kpc. Examples of radial profiles of brightness in other galaxies are well known[1] (see chapter 1). The circular velocity for a model with a sufficiently massive disk (total mass $M_d = 7 \cdot 10^{10}$ $M_\odot$) in the case of an infinitely thin disk ($z_0 = 0$) is shown in Fig. 9.2a. For the amplitude of deflection of $\Delta V = 15$ km/s the total surface density in a narrow zone of $r = 6 \div 6.4$ kpc should be approximately halved (Fig. 9.2b). The gravitational contribution of the disk material is dominant in region 3 kpc $\lesssim r \lesssim$ 15 kpc. This model is a model of the maximum disk, which, as will be shown in section 9.2.1, does not allow us to obtain agreement with the observed values of the velocity dispersion of stars.

Taking into account the finite disk thickness reduces the magnitude of ΔV to 6 km/s at $z_0 = 500$ pc (Fig. 9.2c, d). With increasing mass of the halo and, accordingly, weight reduction of the disk, the value of ΔV decreases as the relative contributions of disk components to the circular rotation speed according to (4.22) becomes smaller. In the case of an infinitely thin disk to obtain the amplitude of deflection of $\Delta V = 9$ km/s in models with more massive halos $\mu = M_h\,(4L)\,/\,M_d = 1.9$ (Fig. 9.2e) we require the surface density gradient $\sigma(r = 5.8\ \text{kpc})\,/\,\sigma\,(r = \text{kpc}) \gtrsim 3$ (!). For a fixed gradient of the surface density increase of the vertical scale to the observed values of $z_0 \simeq 500$ pc reduces the deflection to $\Delta V = 3$ km/s.

If we turn to other S-galaxies, the visible deviations from the exponential photometric profile are observed either at the edge of the galaxy [802, 806] or in the centre as a central depression of the stellar disk [17]. Another example are objects consisting of several exponential disks (see Fig. 1.4, 1.5). One explanation for the universality of the exponential distribution of the brightness of the disk components can be the features of the evolution of a viscous gas in conjunction with the process

[1]However, as a rule, such features occur further away from the center – about 3–5 radial scales.

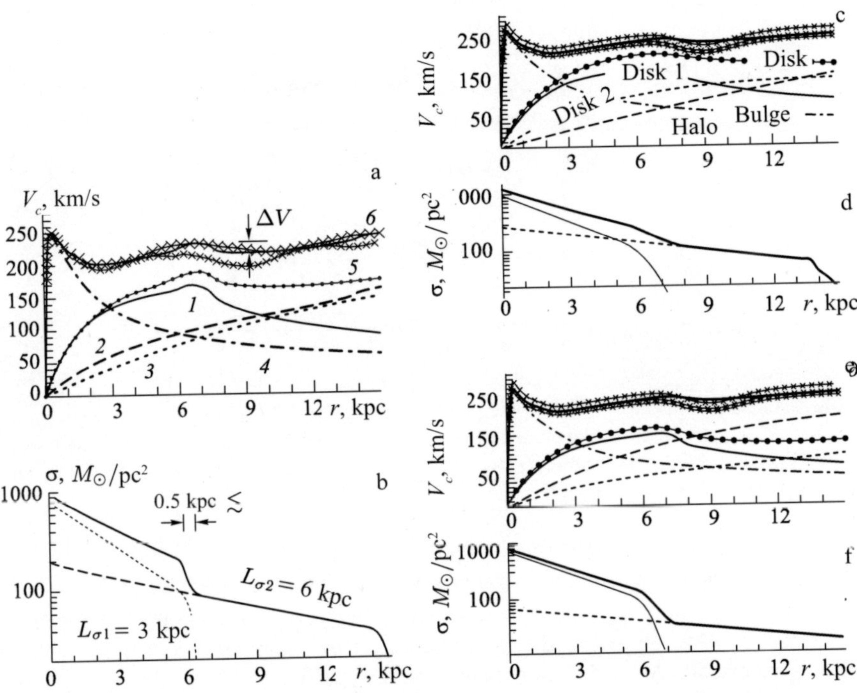

Fig. 9.2. a) circular velocity V_c in the model of the infinite thin disk (solid line 6). Crosses – the rotation velocity according to [1183, 1186]. Contributions to the circular velocity are shown from: 1st disk (1) ($L_{\sigma 1}$ = 3 kpc, the disk breaks up at 6 kpc); 2nd disk (2) ($L_{\sigma 2}$ = 6 kpc, the disk breaks up at 14 kpc); halo (3); bulge (4); the sums of the two disks (2). b) profiles of the surface density σ [$M_\odot$/pc^2] of the disk components for the model shown in Fig. a. c) radial dependence of the circular velocity V_c and the appropriate contributions of the disk components (disk 1 and disk 2), the bulge, the halo for the disk of finite thickness with the vertical scale z_0 = 0.5 kpc. The relative mass of the halo within 12 kpc μ = 0.51 kpc. As in Fig.a. d) radial distribution of surface density from two disks and the sum of these distributions in the model shown in Fig. c; e) circular velocity in the model with a more massive halo μ = 1.9 (inside 12 kpc) and the infinitely thin disk. f) radial profiles of the surface density for the model shown in Fig. e.

of star formation [547]. Calculations show that these factors form over a few billion years an exponential profile of the stellar disk, regardless of the initial distribution of gas. The presence of a region of a sharp decrease in surface density at a distance of 6÷7 kpc from the centre of the Galaxy is not directly confirmed by the observational data.

Deflection of V_c (r), in principle, can also be obtained as a result of the features in the density distribution in the halo; however, this requires a larger decrease in the volume density near $r \simeq 6$ kpc than in the case of the disk (see Fig. 9.2). If in the case of the disk the diameter of epicyclic movement in the solar neighbourhood's orbit is $\Delta r = 2c_r/\varkappa \simeq 2$ kpc, which is slightly smaller than the area of deflection, and in principle, the construction of such a jump in the density distribution is possible,

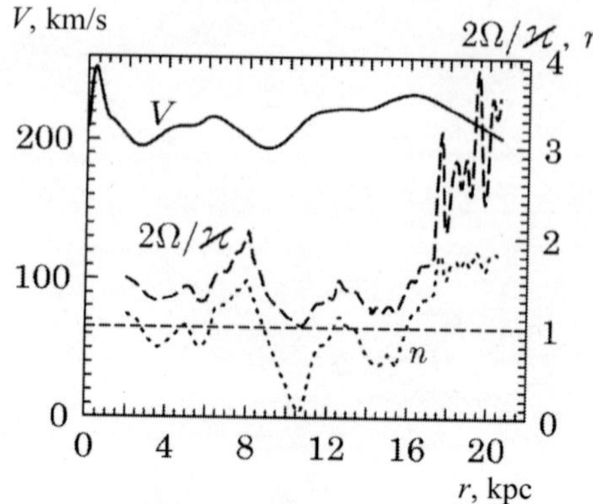

Fig. 9.3. For the rotation curve $V = r\,\Omega$ ([1196]), see the curve 2 in Fig. 9.1), the graph shows the radial dependence of the parameters $2\,\Omega/\varkappa$ and $n = -\mathrm{d}\ln\Omega/d\ln r$ (the right axis). The maxima at the values n and $2\,\Omega/\varkappa$ in the range $r \simeq 8$ kpc are determined by the local deflection of the rotation curve $V(r)$.

in the halo, where the typical velocity dispersion is of the order 100 km/s, a narrow gradient in the radial direction of distribution of matter can not be long lived.

There is an additional argument against the reality of a strong local deflection of the circular rotation speed $V_c(r)$, resulting from kinematic data. For a collisionless system particle motion occurs on epicyclic orbits, which leads to the condition (4.23) $c_r / c_\varphi = 2\Omega / \varkappa$.

Figure 9.3 shows the radial dependence of the parameters $S \equiv \Omega / \varkappa$ and $n \equiv -d\ln\Omega / d\ln r$ for the rotation curve published for Sofue [1186]. In the area $r = 8 \div 10$ kpc there is a significant change of the parameter $S \simeq 1 \div 2.1$. Moreover, in the solar neighbourhood $2\,\Omega/\,\varkappa \simeq 1.7$ which, in accordance with (4.23) gives $c_r / c_\varphi \simeq 1.7$. Let us turn to the observational data. For the old stellar population in the solar neighborhood the ratio $c_r / c_\varphi \simeq 1.55$ [507] which gives $n = 1.15$. This results is an almost flat rotation curve $V \propto r^{-0.15}$. For young stars, according to [49], $c_r / c_\varphi = 8.2/5.8 = 1.41$ which also leads to $V \simeq$ const. Analysis of the velocity distribution of the nearest stars using the Hipparcos database shows that the angular velocity of rotation in the solar neighborhood remains almost constant with radius $\Omega \propto r^{-1/9}$ [505].

In [213], the authors determined the dispersion of the velocities of young stars (cepheids, young stellar clusters) $(c_r, c_\varphi, c_z) = (15.0, 10.3, 8.5)$ km/s. The age of these objects is on average 10 million years. Of special interest is the proximity of the parameter $c_r / c_\varphi = 1.46$ to the ratio $2\,\Omega/\varkappa$, if it is assumed that $n = 1.06$ ($V \propto r^{-0.06}$). This means that the characteristic time of formation of epiccyclic motion is known to be smaller than the single period of galactic rotation.

Thus, there are arguments supporting the view that the observed large deflection of the rotation curve of the gas component in the young stars is evidently caused not as much by the special features of the radial distribution of gravitating matter as by the dynamic phenomena and is associated, for example, with the non-circular motions because of the passage of the spiral density wave. It is quite possible that a very strong deflection of the rotational curve, obtained in the previously cited studies, is determined by the systematic effects of the local system of the coordinates.

It should be mentioned that investigation of the kinematics of the OB stars and classic cepheids in the range $r = 6 \div 12$ kpc does not contradict the conclusion on the approximate constancy of the velocity of rotation of the Galaxy in this region [594]. At the same time, the kinematics of the planetary nebulae, stars with dusty shells, and carbon stars detects the minimum of the velocity of rotation at $r = 8.5$ kpc in the solar vicinity [30] (triangles in Fig. 9.1). If it is assumed that the local deflection is caused by the spiral structure, then in this case the amplitude of these perturbations should be high [300].

9.1.2. An important comment on the data on the kinematics of our Galaxy

The field of velocities and, in particular, the rotation curve in our Galaxy are still not known with sufficient accuracy. The key values are being re-examined: the distance from the Sun to the galactic centre $R_\odot$ and the angular velocity of rotation $\Omega_\odot$. In this book, we have used in most cases the values $R_\odot = 8$ kpc, $\Omega_\odot = 28$ km/s/kpc, but the scatter of these parameters in the current scientific literature is large: $R_\odot = 7.5 \div 8.4$ kpc, $\Omega_\odot = 25 \div 30.7$ km/s/kpc. The tendency of the recent years indicates in favour of higher values of $R_\odot$ and $\Omega_\odot$.

Thus, for example, using the trigonometric parallax of the regions of star formation (according to masers), in [1059] the authors obtained values for the distance $R_\odot = 8.4 \pm 0.6$ kpc, and for the linear velocity of rotation $V_\odot = 254 \pm 16$ km/s, which is considerably higher than the values considered in this book, $V_\odot = 220$ km/s. Analysis of the data for 28 masers with the known trigonometric parallaxes leads to the result: $\Omega_\odot = 31 \pm 1.2$ km/s/kpc, $V_\odot = 248 \pm 14$ km/s [402]. The results of observations of the masers are regarded as more accurate and this may require re-examination of other parameters of our Galaxy.

9.1.3. The kinematics of the central region, concentrated bulge and the bar

It has been assumed that the velocity of rotation of a cold subsystem in the central region $r \lesssim 1$ kpc has a distinctive maximum at the radius $r \simeq 0.3$ kpc [470, 1183, 1186]. If the following equality is valid $V_c = V_{gas}$, then to explain the central maximum at the circular velocity it is necessary to take into account the massive bulge with a concentrated nucleus. For the parameters of this bulge in the King model of the distribution of matter it is assumed that $a_n \simeq 100$ pc, $M_b \simeq 0.6 \cdot 10^{10}\, M_\odot$ and $r_b^{max} \simeq 0.8$ kpc. The central density of the spheroidal subsystem (bulge + halo)[2] in this case exceeds 400 $M_\odot$/parsec^3.

However, the concentrated and massive bulge does not make it possible to form a long-life bar in dynamic models [260, 769]. The bar fails with time and the failure mechanism is identical to the effect of a massive central black hole [653]. The level of non-axisymmetry in the distribution of matter in the disk at every moment of

[2]Of course, we are not discussing here the central massive black hole.

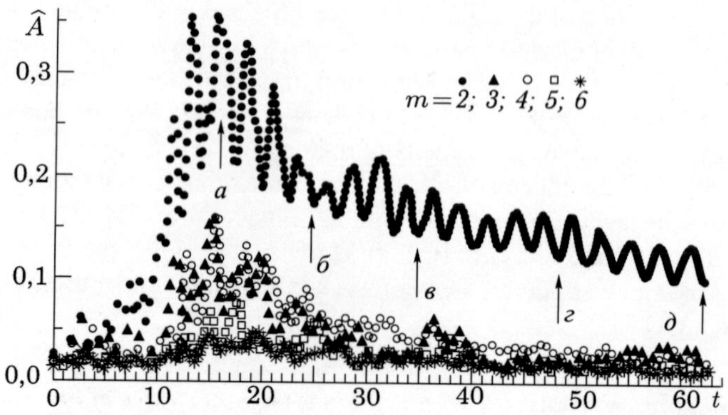

Fig. 9.4. Time dependence of the integral amplitudes of the Fourier harmonics for the azimuthal numbers $m = 2, 3,\ldots, 6$ in the model of N-bodies with a massive concentrated nucleus. A bar forms and subsequently is destroyed, as indicated by the reduction of amplitudes for the harmonics $m = 2$. The time $t = 30$ corresponds to 2.8 billion years.

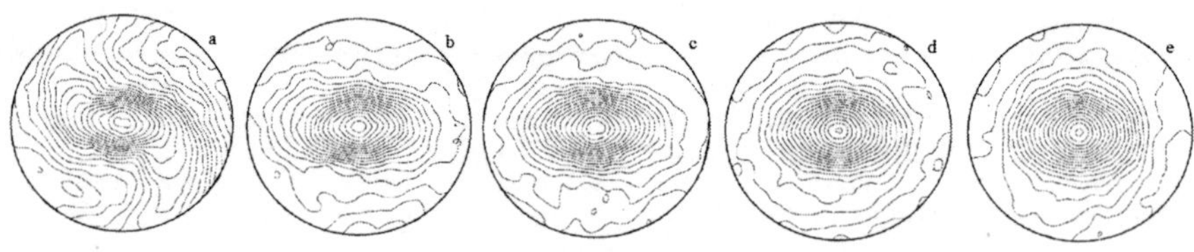

Fig. 9.5. Isolines of the surface density in the model, shown in Fig. 9.4, at different moments of time (see the position of the arrows in Fig. 9.4). Only the central part of the disk in the bar region is shown. Secular failure of the bar takes place.

time is characterised by the amplitude of the Fourier harmonics (4.28). Figure 9.4 shows the time dependence of the integral amplitudes $\hat{A}$ (4.28). The value $\hat{A}$ (2, t) determines the amplitude of the two-arm mode and, in particular, of the bar (Fig. 9.4). The dissipation of the bar takes place, and symmetrisation takes place from the central regions to the periphery of the disk. This is determined by the failure mechanism of the bar. It should be mentioned that the quasi-periodic form of the dependence $\hat{A}$ ($m = 2$, t) in Fig. 9.4 is associated with the exchange of energy between the bar-mode ($A(p = 0, m = 2)$) and spiral waves (Fig. 9.6). With time, evolution of the dynamic model is accompanied by a reduction of this parameter, indicating the dissipation of the waves with $m = 2$ (in particular, the bar) and the establishment of the initial axisymmetric distribution of matter in the disk.

As shown by the dynamic models with different pulse parameters, the lifetime of the bar increases with a reduction of the mass of the central spherical subsystem (bulge + nucleus) and/or increase of the scale of the nucleus of the bulge a_n. Thus, the concentrated massive bulge ($a_n \lesssim 100$ pc, $\rho_{b0} \gtrsim 400\ M_\odot/\mathrm{pc}^3$) cannot be matched with the presence of the long-life central bridge.

It is important to mention another consideration against the highly concentrated distribution of mass in the bulge. It is associated with the observation, along the line

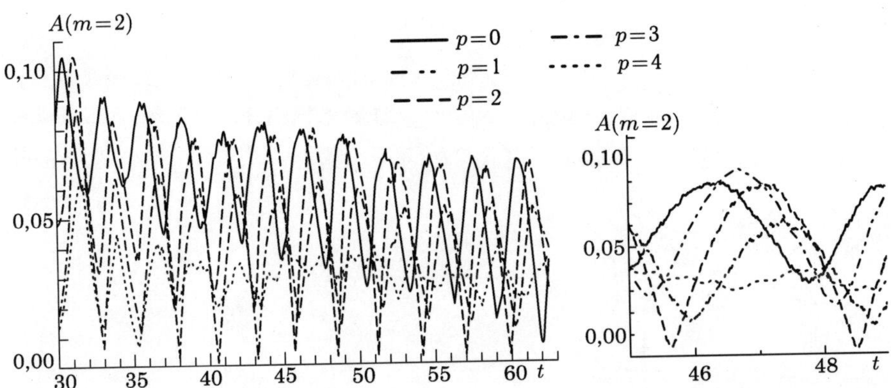

Fig. 9.6. Dependence of the value A ($m = 2, p, t$) on time for $p = 0, 2, 3, 4$. The amplitudes of the Fourier harmonics with different degrees of twisting of the waves show a phase shift. The interval $t = 45 \div 49$ for $p = 0, 1, 2, 3, 4$, is shown separately.

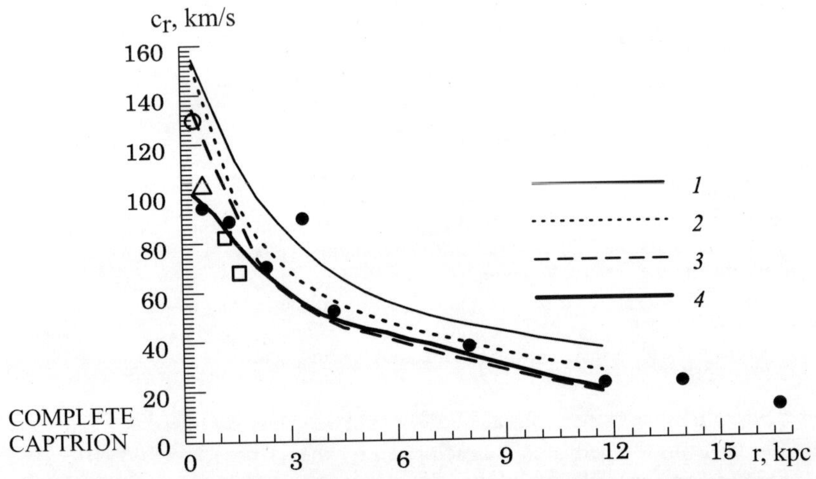

Fig. 9.7. Dispersion of the radial velocities of the stars $c_r(r)$. According to the observational data: ●– [507, 831]; O – dispersion of the velocities of the giants of the later types inside 0.3 kpc [399, 400]; Δ – [1068]; □– [926]. According to the results of dynamic modelling (the models differ in the parameters of the halo and the bulge, $\mu = M_h/M_d$ is the relative mass of the halo inside $4L = 12$ kpc) the dependences were obtained for: $\mu = 0.95$ (1); $\mu = 1.42$ (2); $\mu = 1.5$ (3); $\mu = 1.6$ (4). The first three models include the concentrated nucleus with the scale $a_n \leqslant 100$ pc, and in the fourth model there is no nucleus and the scale of the bulge is $a_b = 200$ pc. It may be seen that with increasing halo mass the value c_r decreases. In the central region $r \lesssim 2$ kpc, to reduce the dispersion of the radial velocities to the values of the order of 100 120 km/s it is necessary to consider the bulge with the scale $a_b \gtrsim 200$ kpc.

of sight, of the values of the dispersion of the velocities of the giants of later types inside the range 0.3 kpc $c^{obs} = 128 \pm 14$ km/s [399, 400]. The authors show that this value characterises the stars of the bulge and, consequently, the dispersion of the

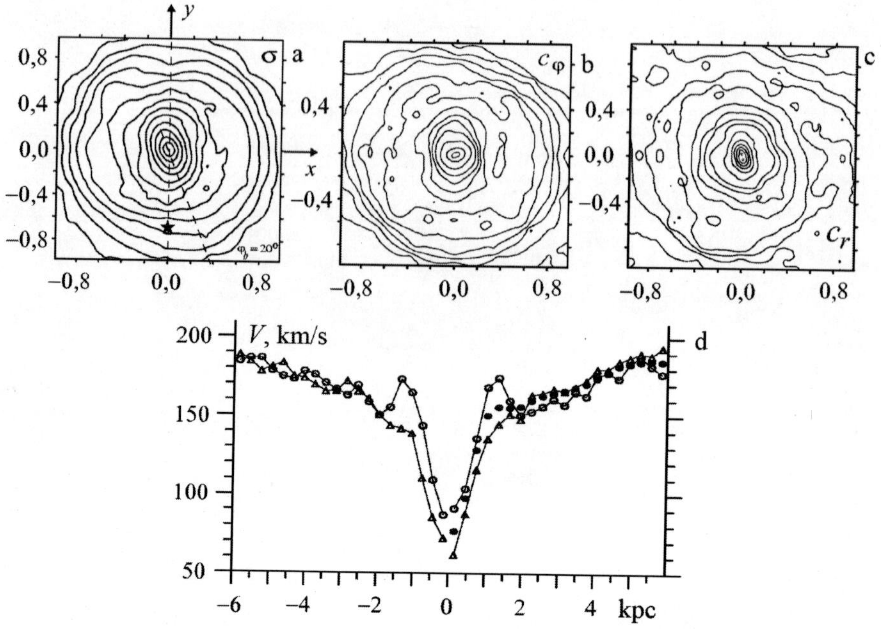

Fig. 9.8. Properties of the disk with the bar in the Galaxy according to the results of dynamic simulation of the stellar disk [260, 769]; a – the distribution in the disk plane of the surface density $\sigma(x, y)$. * indicates the position of the Sun. The angle between the major axis of the bar and the straight line passing through the centre of the disk and the solar neighbourhood is $\varphi_{\text{bar}} = 20°$; b) the isolines of the distribution of the dispersion of the azimuthal velocity is in the disk $c_\varphi\,(x, y)$; c) isolines of the distribution of the dispersion of the radial velocities in the disk $c_r\,(x, y)$; d) the velocity of rotation of matter of the disk V as a function of the distance from the centre along different directions: ○ – $V(x)$; Δ – $V(y)$. The coordinates x and y correspond to Figure a. The velocity of rotation averaged out with respect to the azimuthal angle is also shown (•).

velocities of the disk component should be lower. The selection of the K-giants in the range 500 pc from the centre is characterised by the value $c_r^{\text{obs}} = 104 \pm 10$ km/s [1068]. Luis and Freeman [831] obtain the value $c_r^{\text{obs}} = 106$ km/s for the central dispersion of the velocities of the stars of the disk. Figure 9.7 shows the radial distribution of the dispersion of the velocities c_r for the stars of the disk obtained on the basis of the observational data and the results of dynamic modelling for different relative masses of the disk. Thin lines indicate the models with the massive concentrated ($a_n \simeq 100$ pc) bulge which ensures the circular velocity of rotation in the centre in accordance with Fig. 9.1. If in the range $r \gtrsim 3$ kpc at a specific disk mass it is possible to match the observed and modelling of values of c_r, then in the centre of the disk the dispersion of the velocities, obtained in the dynamic models, exceeds 130 km/s (curves 1,2, Fig. 9.7). The dispersion of the velocities in the centre of the disk can be matched only in the models with a large bulge scale $a_n \gtrsim 200$ kpc (thick line in Fig. 9.7). In this case, the volume central density of the bulge does not exceed 100 $M_\odot/\text{pc}^3$, and the mass of the bulge is $\lesssim 10^{10}\, M_\odot$.

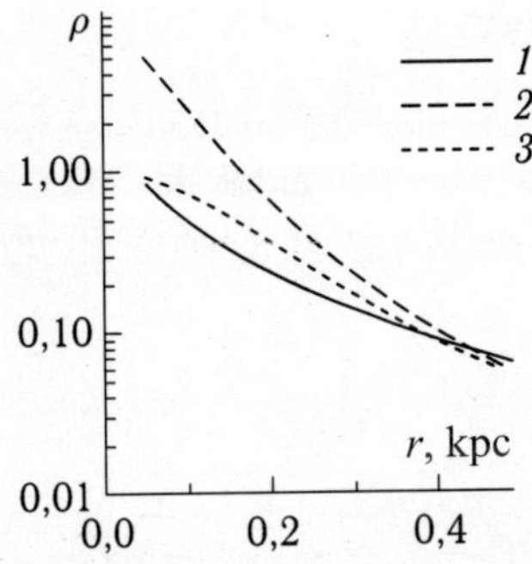

Fig. 9.9. Profiles of the volume density of the centrally symmetric bulge in the plane $z = 0$: 1) according to the results of photometric observations the dependence $\propto \exp(-(r/1.9\ \text{kpc})^2)/(1 + r/0.1\ \text{kpc})^{1.8})$ was fitted [374]; 2) in accordance with the King law (4.11) for $b = 0.1$ kpc; 3) for $b = 0.2$ kpc; normalisation of ϱ is arbitrary.

The radius of the bar, according to the data obtained by different authors, is in the range $r_{\text{bar}} = 2 \div 4.5$ kpc, the ratio of the half-axis of the bar in the plane of the disk is estimated to be 1.6 3, the angle between the large axis of the bar in the direction to the Sun is $\varphi_{\text{bar}} = 15° \div 35°$ [614, 1203, 1292]. When comparing the results of modelling with the observed values it is assumed that $\varphi_{\text{bar}} = 20°$, $r_{\text{bar}} = 3.2$ kpc. Therefore, the dynamic model should allow the existence of the long-life bar with the ratio of the half-axes of 1.6. From this viewpoint, the models with the massive or concentrated bulge with $a_b \simeq 100$ pc do not explain the observed bar. However, if a bar starts to form in the initial stage of evolution of a cold disk, then in subsequent stages the bar will be destroyed as a result of scattering of the nucleus of the bulge on the central potential.

Within the framework of the models with $a_n \gtrsim 200$ pc and $M_b \simeq 10^{10}\ M_\odot$ it is not possible to explain the distinctive central maximum at the circular velocity V_c but it is possible to match the observed and modelling values of the dispersion of the velocities in the centre and the formation of the long life bar. Figure 9.8 a,b,c shows the isolines of the surface density and dispersion of the velocities in the presence of the central bridge. As a result of the special features of movement of matter in the region of the bar, the velocity of rotation of the stars in the model in the centre greatly differs along the major and minor axes of the bar (Fig. 9.8d, chapter 4). Along the x-coordinate, the velocity of rotation of the stars V in the model has a characteristic maximum in the vicinity of the centre (which does not form in the direction 'Sun–galactic centre' along the y-coordinate).

Thus, the local maximum of the velocity in the range $200 \div 500$ kpc in Fig. 9.1 cannot be associated only with the concentrated massive bulge. By rejecting the assumption $V_c = V$gas in the region of the central peak on the rotation curves it is possible, firstly, to match the dispersion of the velocities of the stars in the centre with the observations and, secondly, it is possible to explain the existence of the long-life bar [260, 769]. The local maximum of the velocity can be determined by the movement of matter in the non-symmetric potential in the region of the bar or by the presence of the internal disk or ring [566]. It should also be mentioned that the model used for describing the density of matter in the bulge in the case of the bulge scale $b \simeq 0.2$ kpc is in better agreement with the infrared photometric profile of the bulge (according to the COBE/DIRBE data) in the range 100 pc $\lesssim r \lesssim 0.5$ kpc [374] than at $b \simeq 0.1$ kpc (Figure 9.9).

9.2. Results of dynamic modelling of the Galaxy

We obtain estimates for the surface density of a stellar disk by matching the values c_r, V_{gas}, V^* and Δ from the observations with the results of modelling of the disks at the stability boundary of the disk for both the perturbations in the plane of the system and for vertical waves [260, 769].

9.2.1. The law of distribution of matter in the disk

In accordance with paragraph 9.11, it is assumed that the circular velocity has no significant deflection at the periphery of the Galaxy. Therefore, we examine equipotential profiles of the surface density $\sigma(r) = \sigma_0 \exp(-r/L)$ with L = const. The observations do not make it possible to determine reliably the radial scale L with the estimates of the scale being in a wide range $L = 2 \div 5$ kpc (see [374, 560, 566, 604, 614, 646, 731, 760] and references in these studies). More reliable data for the determination of the radial scale of the disk are those based on observations in the near infrared range ($L = 2{\div}3.5$ kpc). A disk with $L = 3$ kpc will be regarded as the basic model. The King model (4.3) with $a_b = 0.2$ kpc and $\varrho_{b0} = 72\ M_\odot/\text{pc}^3$ is used for the bulge.

We investigate the restrictions on the surface density in the solar region $\sigma_\odot$, following [260, 769]. Evidently, the parameters in the solar vicinity of the disk determined with the highest accuracy include the dispersion of the velocities of the stars. For the oldest stellar population it is assumed that $(c_r, c_\varphi, c_z) = (38, 24, 20)$ km/s [507].

On the basis of the results of dynamic modelling, Fig. 9.10a shows the dependence of the dispersion of the radial velocities of the stars ensuring, at the given mass of the halo, the gravitational stability of the disk, on the surface density in the solar neighbourhood $\sigma_\odot$. The results of modelling for $L = 3$ kpc are shown by the sign '*'. At higher values of the surface density $\sigma_\odot \gtrsim 70\ M_\odot/\text{pc}^2$ the value c_r is higher than the observed value $c_r = 38$ km/s. In the case of a low surface density 40 $\sigma_\odot \lesssim\ M_\odot/\text{pc}^2$ there is a large reserve of stability. The best agreement between the model and observations assuming that the stellar disk is situated at the boundary of gravitational stability, is obtained at $\sigma_\odot \simeq 58\ M_\odot/\text{pc}^2$ (it should be stressed that it is the top estimate of the surface density $\sigma_\odot \leqslant 58\ M_\odot/\text{pc}^2$). In this case, the scale of the halo is approximately twice the radial scale of the disk $a_h/L = 6.3$ kpc/3 kpc (Fig. 9.10b), and inside the sphere with a radius $R_\odot = 8$ kpc for the halo mass we have $M_h = 3 \cdot 10^{10}\ M_\odot$, which equals 87% of the mass of the disk (Fig. 9.10c). The total mass of the disk in such a model is $4.4 \cdot 10^{10}\ M_\odot$ (within the range 12 kpc), and the central surface density is $\sigma_\odot = 840\ M_\odot/\text{pc}^2$. The contribution of the disk to the circumferential velocity at the radius $r = 2.2L$, where the maximum V_c^{disk} is reached for the exponential profice of surface density, equals $(V_c^{\text{disk}}/V_c)|_{r=2.2L} = 0.73$. The contribution of the bulge to the circumferential velocity in this model is at the radius $r = 2.2\ L(V_c^{\text{disk}}/V_c)|_{r=2.2L} = 0.43$. It should be mentioned that a similar result $(V_c^{\text{disk}}/V_c)|_{r=2.2L} = 0.76$ was obtained in the Fux model [601] for $L = 3$ kpc.

Within the range $r = 4L = 12$ kpc for the relative mass of the halo we have $\mu = M_h/M_d = 1.6$ which is in agreement with the results of analysis of the vertical structure of the disk ($\mu = 1.8$) [324]. Our Galaxy can be regarded as a typical system in which the dark matter of the halo not only determines the kinetics of the outer regions $r > 12$ kpc but also plays a significant dynamic role in the region of the stellar disk ($r < 4L = 12$ kpc). The resultant estimate of the surface density of the disk $\sigma_\odot \leqslant 58\ M_\odot/\mathrm{pc}^2$ indicates that the contribution of the disk to the vertical balance of the forces is controlling.

The density of the galactic disk in the solar neighbourhood according to the observational data

The observational data on the vertical distribution of the stars are characterised by a relatively large scatter of the surface density in the solar vicinity $\sigma_\odot = 39 \div 71\ M_\odot/\mathrm{pc}^2$, although the values $45 \div 60\ M_\odot/\mathrm{pc}^2$ [449, 508, 809, 1102] appear to be more reliable. At low values of the relative contribution of the halo the mass of the Galaxy should increase. One should note the results of evaluation of the surface density on the basis of the observational data of red giants in the region $|z| \leqslant 350$ pc, inside of which $\sigma_\odot = 42 \pm 6\ M_\odot/\mathrm{pc}^2$ [791]. Taking into account the contribution of the region $|z| > 350$ pc to the surface density of the disk with the vertical scale ~300 pc, we obtain for the total density $\sigma_\odot \gtrsim 50\ M_\odot/\mathrm{pc}^2$.

For the bulk density of gravitating matter in the plane $z = 0$ it can be excepted that $0.1 \div 0.2\ M_\odot/\mathrm{pc}^3$.

Measurements of the radial scale of the disk give values in the range $L = 3 \pm 1$ kpc [1102], both for the thin and thick stellar disks. Attention will be given to the model with a larger scale of the disk $L = 4$ kpc (Fig. 9.10a). In this case, for the disk to be situated at the stability boundary, the density $\sigma_\odot$ should be equal $\simeq 80\ M_\odot/\mathrm{pc}^2$, which contradicts the observational data [614, 632, 809]. The surface density can be matched only assuming that the disk is characterised by a large reserve of gravitational stability $\Delta c_r = c_r^{obs} - c_r = (88\text{-}26)$ km/s = 12 km/s. If the entire disk is taken into account, and not only the solar vicinity, it appears that the agreement between the radial distributions of c_r and c_r^{obs} in the models with $L = 40$ kpc is far worse than in the case with $L = 3$ kpc[3].

In the case of a short scale $L = 2$ kpc at the fixed value of $R_\odot = 8$ kpc even in the model of the maximum disk for which $(V_c^{disc}/V_c)|_{r=2.2L} = 0.89$, at the stability boundary we have $c_r = 32$ km/s and the surface density is low: $\sigma_\odot = 32 M_\odot/\mathrm{pc}^2$ (Figure 9.10a). The mass of the spherical subsystem in the region $r < 4L = 8$ kpc is only 63% of the mass of the disk. Thus, there is a stability margin, because the value of $\sigma_\odot$ cannot be increased to $50\ M_\odot/\mathrm{pc}^2$ (correspondingly, the dispersion of the radial velocities can be increased to the observed value $c_r^{obs} = 38$ km/s) within the framework of the given rotation curve. Thus, the model with $L = 3$ kpc is in better agreement with the set of the observational data than the disk with $L = 2$ kpc or $L = 4$ kpc.

[3]If the scale of the disk L is changed in the dynamic models, with $L/R_\odot$ remaining constant, all the conclusions also hold for the ratio $\mu = M_h/M_d$, although the absolute values of the masses of the components change.

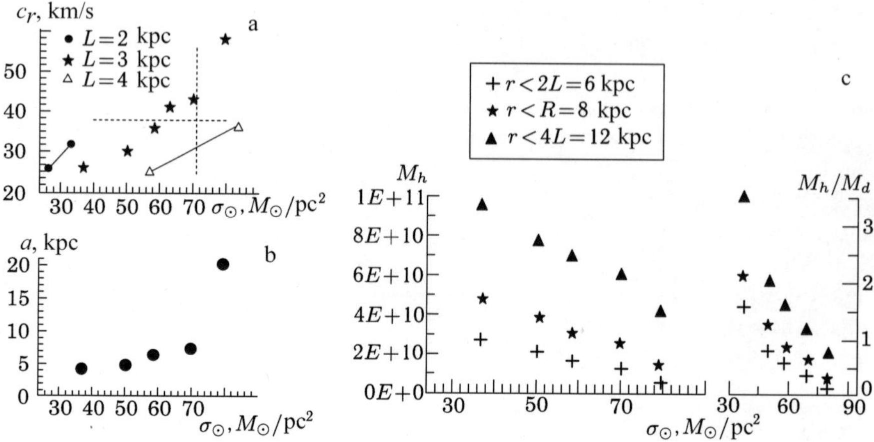

Fig. 9.10. Dependence of the parameters of the models on the surface density of the disk at the radius $R_\odot$= 8 kpc) dispersion of the radial velocities at different radial scales of the disk L = 2, 3, 4 kpc. The horizontal line indicates the observed value c_r = 38 km/s. The vertical line corresponds to the surface density of matter $\sigma_\odot$= 71±6 $M_\odot$/pc^2 included in the range $|z|$ <1.1 kpc according to the data in [809]. b) the scale of the halo a in accordance with the equation (4.3) for a series of models with the radial scale of the disk L = 3 kpc; c) the mass of the halo M_h and the ratio of the mass of the halo to the mass of the disk inside different radii (r = 6 kpc = 2L, 8 kpc, 12 kpc = 4L).

In the dynamic model which is matched with the observed rotation curve of the cold component, the velocity of rotation of the stars and the dispersions of the velocities of the disk, we obtain for the total volume density at the radius $R_\odot$ = 8 kpc $\varrho_\odot$= 0.068 $M_\odot$/pc^3 (the fraction of the disk component is $\varrho_\odot^{\rm disk}$ = 0.058 $M_\odot$/pc^3).The difference $\varrho_\odot^s$= $\varrho_\odot$– $\varrho_\odot^{\rm disk}$ = 0.01 $M_\odot$ /pc^3 [494]. If the surface density of the disk in the solar vicinity is only $\varrho_{\odot,\rm obs}$ = 40 $M_\odot$/pc^2 [487], this indicates a margin of gravitational stability of the stellar disk and a larger mass of the halo in comparison with the disk inside r = 8 kpc.

9.2.2. Ellipsoid of the velocities of stars (c_r, c_φ, c_z)

Figure 9.11a shows the radial dependences of the parameters of the dynamic model from [260] which is in best agreement with the observed distribution of the dispersion of the velocities of the stars in the solar neighbourhood (c_r, c_φ, c_z) = (38, 24, 20) km/s [507] and, consequently, c_r/c_φ= 1.58, c_z/c_r = 0.53.

Figure 9.11b shows for this model the profile of the surface density $\sigma(r)$, $Q_c \equiv c_r \varkappa / 2\,\Omega\, c_\varphi$, c_z/c_r, and the broken line shows the exponential profile with the scale L = 3 kpc. For the entire disk $Q_c \simeq 1$ (see the discussion in chapter 4, section 4.2.4). The ratio of the dispersions $\alpha_z = c_z/c_r$ monotonically decreases with increase of the distance from the centre, where α_z (r = 0) $\simeq$ 0.7. At the periphery of the Galaxy outside the radius of the Sun orbit we have only α_z (r >10 kpc) $\simeq$ 0.4.

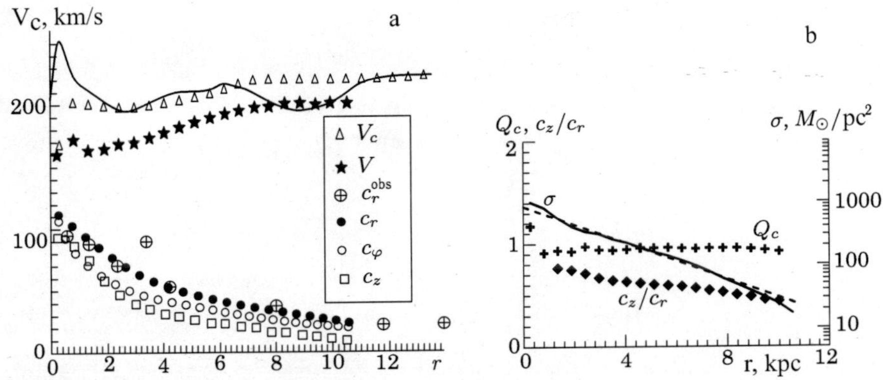

Fig. 9.11. The radial dependences of the parameters of the disk component for the optimum dynamic model of the Galaxy in comparison with the observational data; a) the kinematic parameters: the rotation curve (solid line), circumferential velocity V_c, the velocity of rotation of the disk in the model V, the observed dispersion of the velocities c_r^{obs}, modelling dispersion of the velocities c_r, c_φ, c_z; b) the radial dependences of the ratio c_z/c_r (rhomb) and the parameter Q_c(+) (the left axis), surface density σ (solid and dotted lines, right axis).

9.2.3. Kinematics of the old stellar disk and surface density in the solar neighbourhood

We continue discussing the results of constructing the numerical dynamic models of the Galaxy in which the entire stellar disk is situated at the boundary of gravitational stability [260, 769]. As a result, we obtain additional restrictions on some dynamic and kinematic parameters of the main subsystems (the disk, the bulge, the halo).

If it is assumed that the stellar disk is situated at the boundary of gravitational stability then there are at least three parameters which depend on the distribution of the surface density of the disk σ. They include the dispersion of the radial velocities of the stars c_r, the velocity of rotation of the stars V_*, and the characteristic vertical thickness of the stellar disk Δ:

1. As the value of σ increases, the value c_r, ensuring the stability of the disk (chapter 2) also increases, with other conditions being equal;
2. As the value of σ increases, the velocity of rotation of the stars V_*decreases in comparison with the circumferential velocity V_c for which in the absence of the non-axisymmetric movements of the gas it may be assumed that $V_c = V_g$.
3. With increase of the surface density for the constant distribution of the circular velocity the thickness of the disk should increase (the dispersion of the vertical velocities c_z also increases) in order to stabilise the bending instability.

The dynamic models show clearly that the reduction of the mass of the halo is accompanied by the increase of the difference between the circumferential velocity of rotation V_c and the velocity of rotation of matter in the disk (stars) V. This effect follows from (4.21) and is determined by the increase of the contribution of thermal motion of the stars to the radial balance of the forces. Thus, the difference between

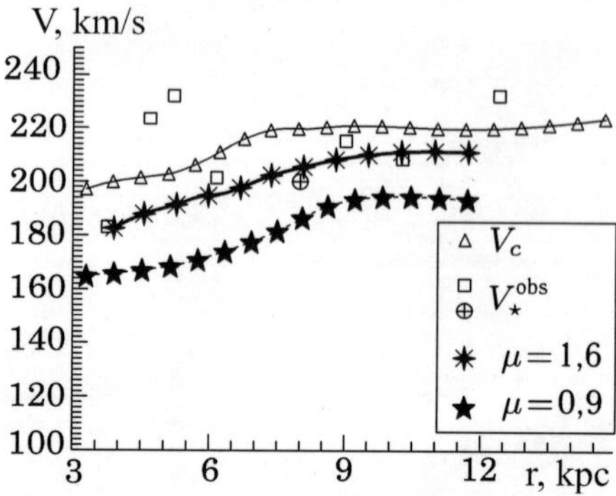

Fig. 9.12. Velocity of rotation of the stars of the disk V^{obs} according to the data in [831] (squares) [351] (circle with cross). The thicker line describes a model with the relative mass of the halo $\mu = M_h/M_d = 1.6$, and the broken line $\mu = 0.9$. The circumferential velocity is indicated by the sign Δ.

the circular velocity and the velocity of rotation of the old population allows the independent evaluation of the disk mass. According to the observational data [831], the velocity of rotation of the old stellar population in the range $r > 5$ kpc is greater than 200 km/s. In [351], in the solar neighbourhood the velocity of rotation of the stars in the plane $z = 0$ was 200 km/s. Thus, the difference in the velocity of rotation between the youngest and older stars in the vicinity of the plane $z = 0$ does not exceed 20 km/s. The latter circumstance is an additional factor in favour of the not very massive stellar disk (Fig. 9.12). If the fraction of the disk inside the Sun orbit would be represented by the mass greater than 70%, the difference between the circular velocity V_c and the velocity of rotation of the old stellar disk V would not be smaller than 35 km/s (model with $\mu = 9$ in Fig. 9.12).

9.2.4. Vertical scale of the stellar disk

Often, instead of (2.57), the vertical profile of the volume density of the mass of the disk is approximated by the exponential law:

$$\varrho(z) = \varrho_0 \exp(-z / h_{\exp}), \tag{9.1}$$

where $h_{\exp}$ is the vertical exponential scale of the disk. The estimates $(h_{\exp})_\odot$ for the old disk are in the range from 200 pc to 350 kpc [660, 791, 1058, 1248]. A positive gradient of the vertical scale can form with the increase of the radial component [604, 763]. According to the data in [374], in the central range $h_{\exp} < 210$ pc. In [351], after deducting the contribution from the stars of the thick disk, the vertical scale of the disk was 330 pc. The authors of [791] reported the value $h_{\exp} \simeq 280$ pc. The vertical distribution of the M-dwarfs is approximated more efficiently by the combined law $\varrho \propto 0.8 \cdot \mathrm{ch}^{-2}(z/323 \text{ pc}) + 0.2 \cdot \exp(-|z|/656 \text{ pc})$ [487], which corresponds approximately to the values $z_0 = 400$ pc for the law (1.4) in the range $|z| < 1$ kpc. However, the surface density of the M-dwarfs does not exceed the value $\sigma = 14\ M_\odot/\mathrm{pc}^2$ and represents a small part of the total density of the disk [487].

Figure 9.13 shows the radial dependence of the parameter

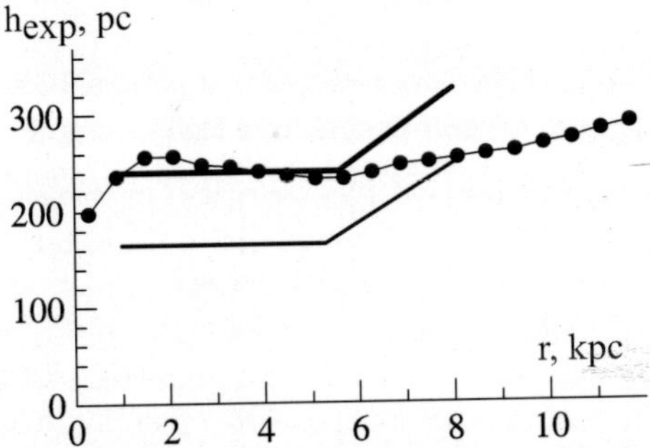

Fig. 9.13. Radial dependence of the vertical scale of the disk in the model with $\mu = M_h/M_d = 1.6$ (●). Solid lines restrict observed distribution of the exponential vertical scale $h_{exp} = 0.5\,\Delta$.

$$\Delta^2 = \frac{\int_{-\infty}^{\infty} (z - \langle z\rangle)^2 \varrho(z)dz}{\int_{-\infty}^{\infty} \varrho(z)dz}, \tag{9.2}$$

obtained in the dynamic model which explain satisfactorily the distribution of the kinematic parameters of the disk ($\langle ... \rangle$ – averaging over z). Determination of the parameter Δ in accordance with (9.2) differs only slightly quantitatively from the half thickness of the disk z_0 which is included in the law of vertical density $\varrho \propto \mathrm{ch}^{-2}(z/z_0)$ [323]. Because at large distances from the plane $z = 0$ both laws of the z-distribution of density have the same asymptotics, then in this case $z_0 \simeq 2h_{exp}$. The increase of the vertical scale with the radius at the periphery is a characteristic special feature of many galaxies, including ours [292].

9.2.5. The dark halo in our Galaxy. General conclusions

In contrast to other galaxies for which the observations provide only the dispersion of the velocities along the beam of sight, in our Galaxy in the solar neighbourhood there are three main components c_r, c_φ, c_z [26], resulting in additional restrictions on the model of distribution of matter in the Galaxy. In this paragraph, we discuss the model of the Galaxy taking into account the latest data on the distribution of the components of the dispersion of the velocities of the stars. For this purpose, for every dynamic model satisfying the rotation curve, we determine the radial distributions of the dispersion of the velocities in a stellar disk. This resulted in the construction of the model which is in best agreement with the observational data simultaneously for a wide range of the parameters, which take into account the distribution of surface density in the disk, its thickness, the velocity of rotation of the gas and stellar components, the dispersion of the velocities in the stellar disk, the presence of the bar, and the data on the bulge.

Taking into account the dispersion of the velocities of the stars together with the rotation curve of the Galaxy reduces the size of the area of permissible values of the disk mass. The assumption according to which the disk component of the Galaxy is close to the boundary of gravitational stability imposes restrictions on the distribution of matter in various subsystems and on the kinetic parameters of the disk. The analysis results yielded the following conclusions:

1. To explain local deflection of the velocity in the Galaxy in the range 6 kpc $\lesssim r \lesssim$ 12 kpc, it should be assumed that at the radius $r \simeq$ 6 kpc in a narrow zone ~0.5 kpc wide there is more than a double gradient of the surface density. The observed relationship between the dispersions of the velocities in the radial and azimuthal directions indicates that the velocity of rotation in the solar neighbourhood is almost completely independent of the radius. Thus, the observed local deflection on the rotation curve of the gas subsystem in the young stars does not reflect the behaviour of circular velocity.
2. The assumption on the presence of a distinctive maximum of the circular velocity V_c in the central region ($r \lesssim$ 1 kpc) imposes restrictions on the lifetime of the bar and the values of the dispersion of the velocities of the stars in the central zone of the disk. The constructed dynamic models restrict the values of the scale of the bulger nucleus at $a_n \gtrsim$ 200 pc. The total mass of the bulge in this case is $M_b \lesssim 1.2 \cdot 10^{10}\, M_\odot$. The results indicate that the central maximum on the rotation curve in the range $r \simeq$ 200÷300 pc is determined not only by the distribution of matter in the bulge but the presence of the bar is also very important. Thus, the condition of equality of the velocity of rotation V_{gas} and the circular velocity V_c in this case does not make it possible to determine accurately the mass of the subsystems in the central region of the Galaxy – the non-circular motion in the region of the bar is also important. Another explanation of the sharp maximum of the circular velocity of rotation in the central region is to take into account the internal disk with a sharp boundary [228, 566].
3. The model with the radial exponential scale of the disk L = 3 kpc and the vertical scale z_0 = 0.5 kpc better explains the entire set of the observed data than the models with L = 2 kpc or L = 4 kpc.
4. The conclusion on the possible destruction of the bar in the presence of a concentrated nucleus at the bulge, obtained in the construction of the models of the Galaxy, also holds in a general case. Thus, the lifetime of the bar can depend on the scale of the nucleus and the mass of the bouts.

9.3. Global spiral pattern

Up to the 90s of the 20th century, the views regarding the spiral structure were based on the spatial distribution of OB-stars, the young scattered clusters, the regions HII and neutral hydrogen according to the data for the 21 cm line. There were two fragments of the global pattern: the Sagittarius–Carina arm (internal arm) and the Perseus arm (external arm), which are part of the global two-arm structure. Optical data detects the so-called Orion arm which does not belong to the global structure (chapter 1). This is confirmed by the absence of gigantic HII regions, synchrotron radio emission, and appropriate perturbations in the field of velocities of the stars.

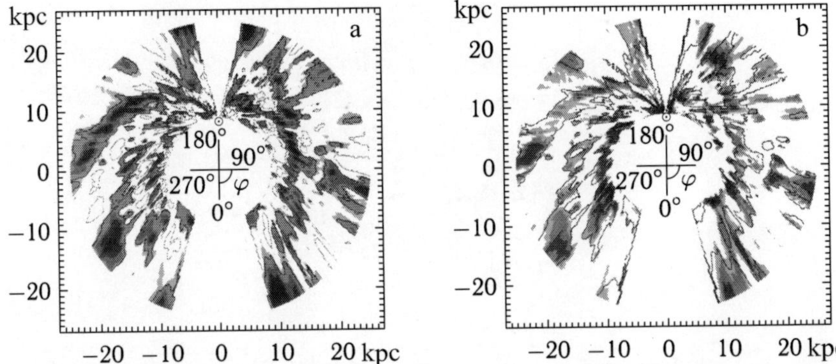

Fig. 9.14. Relative perturbations of the surface density HI (a) and of the thickness of the gas layer (b) [826], indicating the non-symmetric complex spiral structure in the outer part of the Galaxy disk (see also Fig. 1.22).

The gradient of the age of the stars across the Sagittarius–Carina arm [131] and the spiral structure of the field of velocities have been detected.

In the radio range we can observe almost the entire Galaxy and the two-arm spiral structure. The pulsars, super giants, Cepheids, neutral and molecular hydrogen, synchrotron and gamma radiation, the fields of velocities and the spatial distributions of the stars and gas play the role of dynamic indicators of the spiral structure (Figs, 9.14, Fig. 1.23, 1.22).

There are convincing indications of the existence of the vortex structure in the solar neighbourhood (see section 7.2.2).

After discovering of the bar in the centre of our Galaxy and in subsequent latest observational data on the distribution of matter in the disk (chapter 1, section 1.4), it is justified to regard our Galaxy as belonging in the group of relatively typical SB-galaxies with a global two-arm pattern in the presence of additional arms in the central zone (3 kpc, Saggitarius arm,), and the outer region (Orion arm, Outer arm).

9.4. Mini-spiral in the centre of the Galaxy – result of instability of super-reflection

9.4.1. What do we know about the mini-spiral?

In this section attention will be given to the interesting problems of the dynamics of the centre of our Galaxy associated with the mysterious structure, the so-called mini-spiral. Regardless of a large number of investigations, the nature and origin of the structure have not as yet been explained. It is completely clear that the key importance in the investigation of the kinematics and dynamics of the 'mini-spiral' belongs to the analysis of the field of velocities, in particular, to the rotation curve of the disk in the vicinity of the core disk in the galaxy centre.

The 'mini-spiral' is the gaseous fibrous structure in the central cavity with the size of 1 pc which is the most inner part of the core disk. The mini-spiral consists

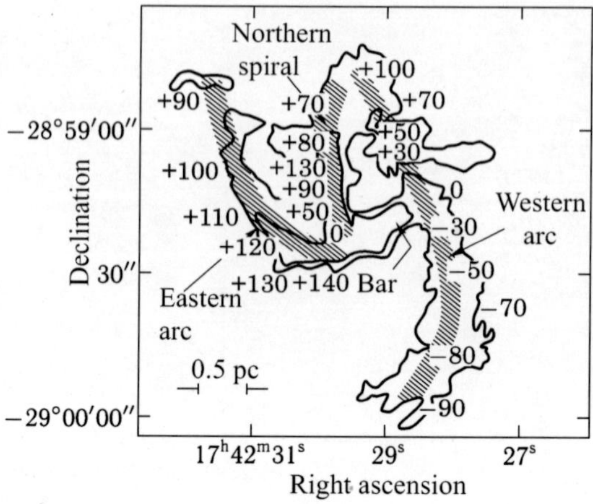

Fig. 9.15. The field of velocities of the ionised gas along the beam of view in the region of the 'mini-spiral' [612].

of several ionised filaments (Fig. 9.15): Western Arc, Bar, Northern Spiral, Eastern Spiral (Arm) which are obviously the fronts of ionisation separating the regions with low and high density [1221]. Several different approaches have been proposed to determining the nature of the 'mini-spirals' which include: 1) the inflow of matter; 2) the outflow of matter; 3) disk instability (for more details see [582, 583]). We will discuss the mechanism of the instability in order to enable the quantitative description of the most important special features of the 'mini-spiral'.

There are two mechanisms for the generation of the spiral structure in a rotating gas disk: 1) gravitational [628, 844, 847] and 2) gravitational–dynamic [237–239]. Since the self-gravitation of the gas inside the central cavity is negligible, the hydrodynamic approach is of special interest. We will show that the special mechanism of instability, based on super-reflection, results in the generation of a single-arm spiral density wave, which is compatible with the observational data for the 'mini-spiral'.

Observations of different components (dust, molecular gas, atomic gas and ionised gas) indicate the presence of flat subsystems in the central region which form a single differentially rotating disk. The innermost part of the dust ring is superposed with the atomic and ionised gas which in turn in the external section mixes with the molecular ring (Fig. 9.16a).

Table 9.1. Object and the geometry of the centre of the Galaxy (references to the original studies see in [582] in Table 1)

Object	Position angle	Angle of inclination
Near-nucleus (molecular) ring	$15° \pm 10°$	$65° \pm 5°$
Dust ring	$\simeq 22°$	$60° \pm 5°$
Atomic ring	$\simeq 25°$	$70° \pm 5°$
Ionised ring	$\simeq 19°$	$61° \pm 2°$
Western Arc	$\simeq 22°$	$\simeq 56°$
Northern arm	$\simeq 20°$	$\simeq 65° \pm 5°$

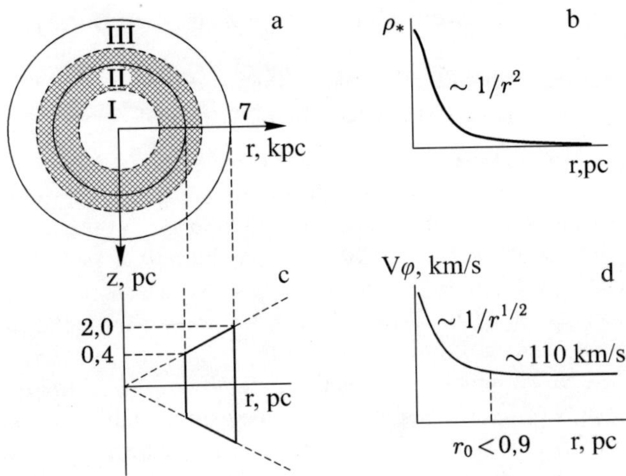

Fig. 9.16. a) the diagram of the flat subsystems in the innermost central part of our Galaxy ([583] and references therein): I) the circumnuclear molecular ring (inner radius $r_{in} \simeq 1.5$ pc, outer radius $r_{out} = 7$–12 pc); II) the dust ring ($r_{in} = 0.9 \pm 0.2$ pc, $r_{out} = 6$ pc); III) atomic gas (the total mass is $M \simeq 300\ M_\odot$, density 10^4 cm^{-2}; ionised gas ($\langle \varrho \rangle > 2.5 \cdot 10^3$ cm^{-3}, $\varrho_{\text{mini-spiral}} \simeq 10^4$ cm^{-3}); b) the thickness of the molecular ring; c) the profile of density in the stellar cluster; d) the velocity of rotation of the disk in the presence of a massive black hole in the centre of a stellar cluster.

Table 9.1 shows that for every flat subsystem (including the 'mini-spiral'), the position angles and the angle of inclination are similar to each other. All this indicates that all the flat subsystems form a single, although highly heterogeneous disk (ring). Figure 9.16b shows that the thickness of the disk around the nucleus increases with increasing radius as observed in the gas disk in the Galaxy and other spiral galaxies. This effect can be easily understood by considering the hydrostatic equilibrium of the gas.

The velocity of rotation of the circumnuclear disk and its inner part is almost constant along the radius everywhere, including possibly the innermost ($\leqslant$1pc) central zone (Genzel et al, 1985). This follows from a simple estimate based on the mass distribution. Actually, the density of the central stellar cluster gives the asymptotic deppendence $\varrho^* \propto r^{-2}$(Eckart et al, 1993, Fig. 9.16c), which follows from the equilibrium condition $V_\varphi^2 / r = GM(r)/r^2$, ensuring the constant velocity of rotation in the external region $V_\varphi^2 \propto M(r)/r \propto \varrho_* r^2 = \text{const}$. The presence of the central massive black hole or a high-density stellar nucleus should produce the Kepler rotation curve in the vicinity of the centre (Fig. 9.16d). As a result, the rotation curve of the disk has a deflection region in the vicinity of the centre of the Galaxy.

Although the total dynamics of the central region of the Galaxy may be morphologically complicated, the problem of the nature of the 'mini-spirals' is the cornerstone for many dynamic problems.

9.4.2. The nature of the 'mini-spiral' in the centre of our Galaxy

The hypotheses on the nature of the 'mini-spirals' can be divided into three categories ([580, 582, 583] and the references therein):

1. Inflow. It is assumed that the gas clouds in the central region are decelerated as a result of elastic collisions and the redistribution of the momentum of the amount of motion. Consequently, they penetrate into the zone of the central parsec, forming the 'mini-spiral'. Subsequently, the strong ultraviolet radiation from the centre dissociates and ionises the molecular gas. It is assumed that the mass of the atomic gas is $\simeq 300\ M_{\odot}$ inside the central 1.5 pc, and for the characteristic time of $\sim 10^4$ years, the rate of inflow is $(2–3)\cdot 10^{-2}\ M_{\odot}$ / year.

   This flow cannot reach the immediate vicinity of the central black hole (BH) and the 'supply of food' is suppressed by the gamma radiation from the hot stars in the central parsec [998]. However, the wind may be a direct source of 'food' for the black hole by means of Bondi–Hoyle accretion, especially if the mass of the black hole M_{BH} is sufficiently large. If M_{BH} reaches $\sim 10^6\ M_{\odot}$, the x-ray radiation from Sgr A* should greatly exceed the value observed from the galactic centre [885][4]. Although the authors of [954] presented arguments according to which the advection of radiation should be capable of greatly weakening the accretion luminosity, it became clear that this is not sufficient for resolving this contradiction [998].
2. Outflow from the centre. The assumption according to which the 'mini-spiral' is a two-sided precessing jet contradicts the absence of symmetry because Sgr A* as a natural source of outflow is displaced in relation to the centre of the 'mini-spiral'.
3. 'The mini-spiral' = wave: the Northern Arc and Western Arc are the only structure rotating in the circumnuclear disk in the presence of additional gas flows [580]. Analysis of the field of velocities along the beam of sight of the 'mini-spiral' shows that the Northern Arm and Western Arc are a coherent structure in the form of the Archimedes spiral rotating in the plane of the circumnuclear disk [819]. At the same time, the presence of a steep gradient of velocity in the region in which the Eastern Arm merges with the Northern Arm (Fig. 9.15) indicates, according to [819], the absence of any strong dynamic relation between these structures.

We will discuss the results of investigations of the kinematics of the ionised gas inside 1.5 kpc of the Galaxy, using the two-dimensional field of velocity of the ionised gas along the beam of sight, obtained from the data in [1073].

The distribution of velocity along the beam of sight on the whole has the form typical of a regularly rotating disk. Some areas are characterised by strong deviations from the general regular rotation. However, the positions of the largest devia-

[4]It should be mentioned that Lynden–Bell and Rees [1047] predicted the existence of a black hole in the Galactic centre prior to the discovery of the actual candidate (Sgr A*).

tions from the regular rotation are in agreement with the existence of the individual filaments and flows, mentioned in a number of studies (for example, [819]).

It is important to mention the Eastern Arm which is situated evidently outside the main gas disk and is characterised by a large positive difference of the velocities. In addition, this special feature is continued almost directly in the western direction, and the positive difference of the velocities can be seen to the west from the Western Arc. Possibly, we are concerned here with the gas large-scale flow, flowing on the plane of the gas disk.

The Northern Arm and Western Arc are distinguished by the emission maximum of brightness and may belong to the rotating flat gas disk.

In the investigations of the dynamics of the inner gas disk no account was made of the elements related to the regions of strongly non-circular motions. Subsequent analysis was carried out with this 'smoothing'.

Assuming that the angle of inclination of the central gas disk is $63 \pm 3°$, P.A = $32 \pm 4°$, and that the dynamic centre is connected with Sgr A*, single-element rings were designed for the field of velocities and the azimuthal dependence of the velocity along the beam of sight was plotted on a grid along the radial coordinate from 14″ to 42″ from Sgr A*.

The cosine-like form of these relationships (Fig. 9.17) indicates the dominance of the flat circular rotation of the gas inside $R \simeq 1.5$ pc (42″).

We now discuss the possibility of generating a single-arm spiral in a central gas disk whose velocity profile of rotation has a deflection region, as mentioned previously.

It is assumed that the deflection is situated at the point $r = R_0$ on the rotation curve. This means that at this point there is a 'jump' of the derivative of the velocity of rotation: $V'_{0\varphi}(R_0) \propto \Theta(r - R_0)$ and, correspondingly, $V''_{0\varphi}(R_0) \propto \delta(r - R_0)$, where Θ is the Heaviside function, δ is the Dirac delta function. In this case, the hydrodynamic equations, describing the dynamics of the rotating disk, can be reduced to the Schrödinger equation with the potential in the form of the δ-function. The latter has always at least one level, i.e. in the region $r = R_0$ there is a spiral wave. The mechanism of generation for this wave is the corotation resonance. Actually, the energy of the sound wave in the flow is determined by the equation [112]:

$$E = \frac{E_0 \omega}{\omega - \mathbf{kv}} = \frac{E_0 \omega}{\omega - m\Omega_0(r)}, \quad E_0 > 0,$$

where ω is the intrinsic frequency, $\mathbf{k}$ is the wave vector, and it is taken into account that $\mathbf{kv}_{0\varphi} = (m/r) \times (r\,\Omega_0)$, m is the azimuthal wavenumber. Evidently, energy E is positive, if $\omega/m > \Omega_0(r)$.

The form of the potential well in Fig. 9.18 corresponds to the profile of the velocity of rotation, shown in Fig. 9.16. The wave function has the exponential truncation at both sides of the potential well. Therefore, if there were no corotation resonance, the reflection coefficients on both sides of the potential barrier would be equal to 1. In reality, the wave, reflected from the corotation resonance, receives as a result additional positive energy. Therefore, the reflection coefficient in the left side potential barrier is greater than unity, i.e. instability of super-reflection becomes

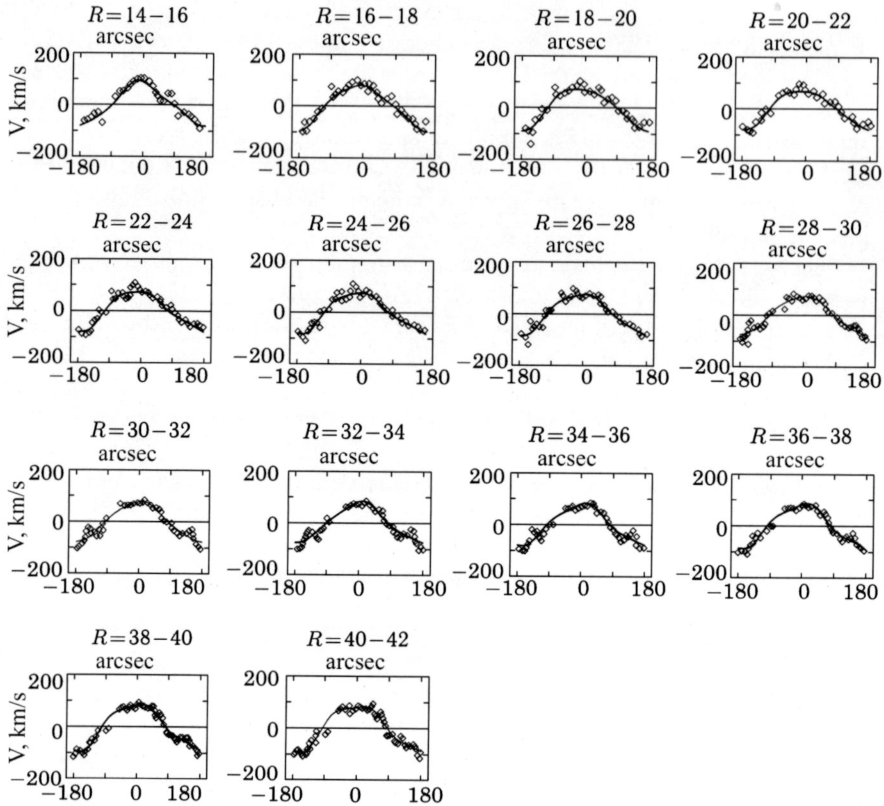

Fig. 9.17. The azimuthal dependence of velocity along the beam of sight for different distances from the centre in the disk plane. The velocities are averaged out over the ring with the width $2''$. The characteristic form of these dependences ($\propto$ cos) indicates the dominance of the circular rotation of the disk.

evident. It should be mentioned that Orr (Orr, 1907) was the first who detected the phenomenon of 'super-reflection' in hydrodynamics, and in the physics of galaxies, the super-reflection was investigated for the first time in [628] and [794] (for more details see section 5.2).

The inflection on the rotation curve, shown in Fig. 9.19a, results in the generation of a spiral single-arm wave which coincides with the Northern Arm and the Western Arc (Fig. 9.19b).

Possibly, the spiral density wave is not restricted by the central zone but continues to the circumnuclear (molecular) ring. The presence of the density wave in the ring in the circumnuclear ring may explain why the ring parameters, such as the position angle and the angle of inclination, depend on the radius. A similar position is also observed in other galaxies every time when the gravitational potential has axial symmetry [1322].

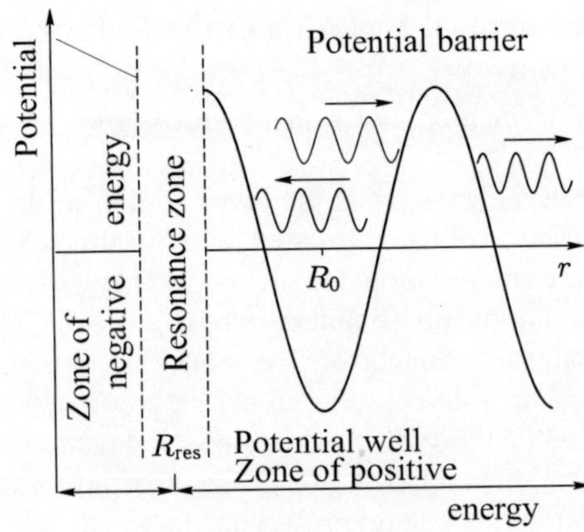

Fig. 9.18. The diagram explaining the super-reflection effect.

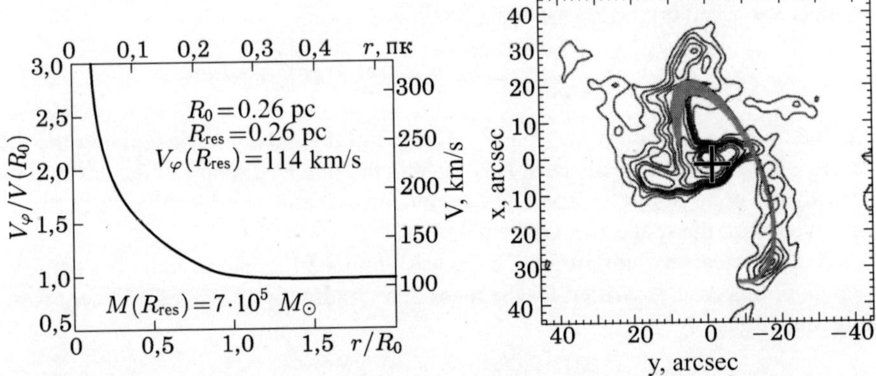

Fig. 9.19. The accepted rotation curve (a) and the chart of isolines according to the observational data [396] (b). The solid region indicates the results of numerical calculations [582]. The cross indicates the centre of rotation. The geometrical parameters are presented in Table 9.1.

The presence of the first harmonics (sin) in the observed field of velocities along the visual ray requires special attention. The most natural explanation of the observed special feature is the presence of the azimuthally symmetric radial flows, accretion or discharge in the disk. The observational data showed that the accretion takes place from the outside of 1.6 pc, and the outflow on the inside.

Accretion may be explained by the regular viscous transfer of the momentum of the pulse in the outer part of the disk. The reason for the slow outflow of the matter inside 1.6 pc to the end has not as yet been determined.

Since the region of outflow corresponds to the region with the relatively strong density wave, it is natural to connect these phenomena together. Thus, the observed outflow of matter may prove to be the manifestation of the 'acoustic drift' – the quasi-stationary radial flow of matter of the disk, caused by a non-linear density wave [580, 582, 583].

9.5. Modelling of chemical evolution of the Galaxy (D.Z. Vibe, B.M. Shustov)

9.5.1. Construction of the model of galactic evolution

The aim of modelling the galaxies is the investigation of the evolution processes, i.e. the history of variation of their structural and dynamic characteristics, chemical composition, and other important parameters – the mass of the gas, the rate of star formation (RSF), the luminosity (bolometric or in a specific photometric band), the rate of supernova explosion, as a terror. The modelling of the evolution of chemical composition has a special role because it makes it possible to compare to the maximum degree the theoretical predictions and observed restrictions.

The simplest type of the model of galaxies is the single-zone model in which the entire galaxy is regarded as a single reservoir, in which two main processes take place: transfer of gas from the interstellar medium (ISM) into young stars, and the reversed process, return of the gas into the ISM by pre-evolution stars [122]. These models are based on the law of conservation of mass:

$$\frac{dM_g}{dt} = -\psi + R + M^{\text{in}} - M^{\text{out}}, \tag{9.3}$$

i.e. the rate of variation of the mass of the gas in the galaxy M_g is equal to the difference between the rate of star formation and the rate R of return of the gas into the ISM by the pre-evolution stars. The components M^{in} and M^{out} describe the exchange of matter with the space around the galaxies.

The chemical evolution of the galaxy is simulated using an identical equation which, in this case, is written for the mass of the individual chemical element in the interstellar gas M_i:

$$\frac{dM_i}{dt} = -Z_i\psi + Y_i + M_i^{\text{in}} - M_i^{\text{out}}. \tag{9.4}$$

here $Z_i = M_i / M_g$ is the mass fraction of the *i*-th element in the ISM, Y_i is the mass of this element, injected into the ISM by the pre-evolution stars.

In most cases, when solving the problems of evolution of the galaxies it is necessary to restrict ourselves only to these equations are assuming that only the mass of the gas changes with time and the content of different chemical elements and the dimensions of the galaxies, remain constant. However, as shown in [553], the system (9.3)–(9.4) can also be supplemented by the law of conservation of energy. Assuming that the thickness of the gas disk H is determined by the energy of turbulent motion of the gas (as a gas becomes 'hot', the thickness of the disk increases), we can write the following formal equation for it

$$\frac{dH}{dt} = H_+ - H_-, \tag{9.5}$$

in which the term H_+ describes the flow of energy into the gas from the supernova explosions, and the term H_- are the losses of energy through radiation into the intergalactic space.

To solve the system (9.3)–(9.4), it must be supplemented by the relationships which determine the parameters of star formation, evolution of stars and synthesis of heavy elements.

Although the single-zone models are capable of reproducing with acceptable accuracy the 'global' parameters of the galaxies, the internal structure of the galaxies is not described here. To obtain more complete information on the structure of the galaxies, it is necessary to use multi-zone models, i.e. the models in which the equations (9.3)–(9.4) are solved separately for different subsystems of the galaxies, for example, for the halo and for the disk or for several ring-shaped zones of the disk with different radii. The multi-zone models should be supplemented by the relationships describing the interaction of different zones, for example, exchange of gas between the halo and the disk.

The first term in equation (9.3) is the rate of star formation in the galaxy, i.e. the mass of the gas transformed into stars in unit time. The birth of stars is a complicated process and its rate depends on a large number of factors. However, the assumption on the self-regulation of star formation enables us to assume that the mean rate of star formation in a galaxy depends only on its 'global' parameters. In the modelling of the evolution of the galaxies, it is usually assumed that the mean rate of star formation is directly proportional to some degree of the volume density of the gas ρ or surface density of the gas Σ:

$$\psi = f\rho^n V_{\mathrm{G}}[\mathrm{g\cdot s^{-1}}], \tag{9.6}$$

where V_G is the volume of the galaxy. For our Galaxy, the value of the coefficient f is approximately equal to 2–5·10^7 at $n = 2$ [491].

The mass distribution of the resultant stars is described by the initial mass function (IMF), usually denoted by $\phi(M)$. It is quite easy to understand the meaning of the IMF through the expression

$$\phi(M)\psi(t)dM, \tag{9.7}$$

which determines the number of stars with the mass in the range from M to $M + dM$, formed at the moment of time t. Consequently, the mass of the gas, returning into the ISM at the moment of time t is

$$\int_{M_{\min}}^{M_{\max}} \phi(M)\psi(t-\tau_M)[M-R_M]dM, \tag{9.8}$$

where $M_{\min}$ and $M_{\max}$ is the minimum and maximum mass of the stars, R_M is the mass of the residue (white dwarf, neutron star or black hole) of the star with mass M.

The values of the minimum and maximum mass, and also the type of function R_M, can be obtained by calculating the evolution of stars with different masses. It is now assumed that no thermonuclear reactions start in a collapsing protostellar clot if its mass is less than $0.08M_\odot$. The maximum possible mass of the star at which the star is still capable of 'withstanding' the pressure of introducing radiation, is 100 $M_\odot$.

In 1955, E. Saltpeter [1104], proposed to use the following simple exponential law for the initial mass function

$$\phi(M) \propto M^{-2,35}, \tag{9.9}$$

in which the coefficient of proportionality is determined from the normalisation condition

$$\int_{M_{\min}}^{M_{\max}} \phi(M)dM = 1. \tag{9.10}$$

In the last couple of decades, several more complicated formulations were proposed for the IMF, and it was also assumed that the form of the IMF changes from galaxy to galaxy and even within the limits of the same stellar system. There is no united view regarding whether the parameters of IMF change with time. Possibly, massive stars formed in the past with a higher probability, but it is usually assumed that the angle of inclination of the IMF did not change during galactic evolution, and that the values of $M_{\max}$ and, in particular, $M_{\min}$ could have changed.

The equation for Y_i is similar to equation (9.8):

$$Y_i = \int_{M_{\min}}^{M_{\max}} \phi(M)\psi(t-\tau_M)[(M-R_M)Z_i(t-\tau_M)+P_i(M)]dM. \tag{9.11}$$

The first term in the square brackets describes the return to the ISM of the *i*-th element which penetrated into the star at the moment of its formation and 'became conserved' there. The second term $P_i(M)$ is the so-called yield – the mass of the heavy element, synthesised in the star with mass *M*. Depending on the synthesis special features, the stars can be divided into three groups.

The low-mass stars ($M < 1\ M_\odot$) do not eject any mass into the interstellar medium, with the exception of the final stages of their evolution. At the same time, the lifetime of these stars is so long that even the very first stars, formed in the galaxies, are still in the main sequence of processes.

The stars with intermediate mass ($1M_\odot < M < 8 \div 10\ M_\odot$) synthesise a certain amount of carbon, nitrogen and other light elements during combustion of the layer source. These elements penetrate into the ISM when the star ejects the shell and transforms into the planetary nebula. There ar only few theoretical models of nuclear synthesis in the stars with intermediate mass. Until recently, the requirements of the models of chemical evolution of galaxies were satisfied only by the calculations [1060]. Recently, several similar studies were performed [876,1252].

The main source of heavy elements are massive stars ($M > 8$–$10M_\odot$), more precisely, the final stage of their evolution – supernova explosions ejecting into the interstellar space products of nucleosynthesis. Until recently, the models of chemical evolution of galaxies used 'classical' calculations [1310]. In recent years, new studies appeared, with particular emphasis on nucleosynthesis in very massive stars and stars of zero or very low metallicity [466, 971].

Everything said about the synthesis of heavy elements refers only to the normal evolution of single stars. Of course, there are other sources both associated with stellar evolution (e.g. new stars) or not related to it (spallation reactions in interaction with cosmic rays). From these sources we usually consider only the Ia supernovae, the origin of which is now most often associated with the final stage of the evolution of some close binary stars [676]. These supernovae are the main suppliers of one of the most important chemical elements – iron. It is more difficult to take into account their contribution than that of the type II supernovae, as the rate of Ia supernovae explosions is determined by the nature of evolution of a binary system and not of a single star, and therefore depends not only on the IMF, but also on the distribution of binary stars on the major semi-axes and the mass ratio of components.

9.5.2. Chemical evolution of the Galaxy and mass motion

In the 'closed' model of the galaxy, or in a certain volume within it (such as the solar neighborhood), i.e. in the absence of mass exchange with the surrounding space, the metallicity is only a function of time. However, observations show that, in reality, the chemical composition depends not only on time. Inside the Milky Way, it is also a function of the galactocentric distance (radial gradient). If we turn to the other galaxies of the same age, it turns out that the mean metallicity correlates with luminosity (mass) of the galaxy.

These features of the chemical composition of our galaxy and other galaxies are an indicator of dynamic processes taking place in them [1038]. In general, the relevant dynamic factors are divided into two groups: 1) the accretion of intergalacticmatter at a rate that depends on the galactocentric distance, and 2) radial flow of gas in the disk. The first explanation is used most often, as an introduction of accretion to the model can simultaneously solve the 'problem of G-dwarfs', i.e. the problem of the observed lack in the solar neighborhood of a substantial population of low-mass stars with low metallicity. To reproduce the observed gradient, one should assume that the temporal scale of accretion depends on the distance R from the centre of the galaxy in such a manner that the rate of gas accumulation in the centre of the galaxy and, consequently, production of heavy elements is faster [465, 1041].

The source of the accreting gas may be both the halo of the galaxy (by accretion in this case we understand the continuing formation of disk), and the matter accumulated by the galaxy as it moves in the intergalactic medium (IGM). Molla et al [933], who investigated theoretically the metallicity gradients in several spiral galaxies, concluded that their results are not consistent with the existence of a constant flow of gas in thedisk and therefore felt that the first option is preferable. On the other hand, in [465] it is stated that the hypothesis of the formation of the disk from the halo gas contradicts the widely divergent distributions of angular momentum in these subsystems.

Arguments against the existence of a constant flow of gas in the galactic disk are summarised by Kennikat [758]. As indicated in this review, the current observational data do not contradict the existence of accretion only if its velocity is substantially inferior to the present rate of star formation. Direct observations of accretion of gas in the galactic disk relate to separate high-velocity clouds and do not answer the question of the role played by the periodic absorption of similar clouds by the disk in its evolution. Even if they do provide a flow of intergalactic gas in a disk with an average speed of 0.5 $M_\odot$ per year [397], the sporadic fallout of the clouds is unlikely to lead to the formation of such a regular and continuous pattern in the distribution of heavy elements in the disk.

Another argument against the radial dependence of the accretion rate lies in the fact that the increased accretion rate in the central region of the galaxy should lead to more active star formation, and hence a more intense release of energy by young stars. This energy (for example, in the form of galactic fountains and winds) will stop falling out of gas on the disk.

An alternative to accretion which depends on the radius may be the radial flow of gas in the disk of the galaxy. In [817] three main causes of the radial displacement of gas in the disk were identified: 1) the fall on the disk of material with low angular momentum; 2) the viscosity of the gas disk; 3) the gravitational interaction between the gas and the spiral density waves. A number of studies have shown that the inclusion of these mechanisms, along with a radially varying accretion, explains the occurrence of a gradient of chemical composition (see e.g. [1039] and references therein). However, the theoretical rationale for the nature of the radial flow is not yet available, so in the models they have to be defined by artificial means. In addition, to align the radial distribution of gas in the radius and solve 'the problem of G-dwarfs' even in the presence of currents, it is still necessary to introduce a radially changing accretion into the model [1039].

However, in addition to accretion, there is also a reverse process: active galaxies eject matter into intergalactic space through 'galactic fountains', which merge at high altitudes, forming a powerful galactic wind. With this wind, the Galaxy loses a few tenths of $M_\odot$ in the year. It is logical to assume that the efficiency of ejection of heavy elements in galactic fountains depends on the galactocentric distance. The matter, enriched with heavy elements, leaves the galaxy at the periphery more easily due to the fact that the density of the gas disk here is lower than in the central regions of the Galaxy. If we break the model galaxy into concentric annular zones and take into account the dependence of the efficiency of emission on the mass distribution in the Galaxy, we can reproduce not only the gradient but also its constancy in time [34].

Openness of the disk galaxies to accretion of gas and to loss of heavy elements does not of course exhaust possible explanations for the negative gradient of the chemical composition. Its appearance is also caused by the introduction to the model of radius-dependent star formation parameters. For example, the distance from the centre of the galaxy may determine the minimum mass of produced stars – as a consequence of decrease of the average black-body temperature of the interstellar gas from the centre to the periphery. Increasing $M_{\min}$ increases the number of supernovae per unit mass of gas transformed into stars, i.e. improves the efficiency of processing gas into heavy elements. The proportionality factor in equation (9.6) may depend on R if we enter into the model the catalytic role of the spiral arms [1126]. Interaction of spiral arms with the gaseous disk as the cause of the appearance of a radial gradient is studied in [284].

Modelling the release of matter from the protogalaxy fragments (protospherical clusters) showed that it could lead to the formation of a slight metallicity gradient in the halo of the Galaxy – due to the pressure gradient in the protogalactic cloud [1013]. This same process can be linked with the unusual chemical composition of low-metallicity stars. Ejection of matter from protospherical clusters stopped star formation in them until the gas managed to significantly enrich itself with the products of type Ia supernovae explosions, so that the contents of chemical elements of the old halo stars is determined solely by emissions of type II supernovae whose composition depends on the initial metallicity of the star [715].

Finally, on the global scale the emission of heavy elements in galaxies may lead to the appearance of the 'mass-metallicity' correlation. As already mentioned, the main channel for the loss of heavy elements in disk galaxies is the hot galactic wind. The main source of energy for it are groups of young massive stars (OB-associations). Supernova explosions sweep the interstellar gas from the association and form the so-called supershell. The hot gas, which supports the expansion of supershells, is enriched with the products of stellar nucleosynthesis. As shown by calculations [710], if the total energy of explosion is high enough (40-50 explosions), the shell can overcome the gravitational attraction of the galactic disk and rise to a sufficiently high altitude (a few hundred parsecs). Because of the Rayleigh–Taylor instability, the shell starts to break up, opening an exit for the hot enriched gas to the halo and the galactic space. The gas flows from the individual ruptured shells merge at the height of several kiloparsecs and form a quasi-steady hot axisymmetric galactic wind. Since mostly type II supernovae explode in the groups, the wind is enriched primarily by their products. Confirmation of the existence of such winds are observations of the mild X-ray emission around spiral galaxies. Wang et al [1281] found such emission at distances up to 8 kpc from the disk of the galaxy NGC4631 Sc/SBd, known for its extensive radio halo. According to the authors, the hot gas is ejected with the wind at a rate of 1 $M_{\odot}$ per year.

Only a fraction of this wind travels into the intergalactic space. For example, according to [510] in a low-mass galaxy ($M = 1.4 \cdot 10^9\ M_{\odot}$) the fraction of supernova products released into the IGM, is significant, but does not exceed $f_{esc} \leqslant 0.6$. In massive galaxies, ejection of matter is difficult ($f_{esc} \ll 1$). The minimum mass of the galaxy at which the supershells which can not overcome the gravitational attraction of the disk is equal to $10^{12}\ M_{\odot}$ [710]. As an observational estimate of f_{esc}, one can use the surface factor of filling (i.e. the ratio of the total area of 'holes' in the distribution of hydrogen to the total area of the disk). The filling factor for the gas disk of the Andromeda Nebula (lg $M \approx 11.5$) is equal to 0.01 [425]. For M33 (lg $M \approx 10.1$) the value of the filling factor is equal to 0.4 [509]. For our Galaxy the calculations [1004] predict the value of the filling factor between 0.1 and 0.23 (f_{esc} = 0.15). On average, we can assume that the efficiency of ejection of heavy elements into the intergalactic medium is linked by the linear dependence with the logarithm of the mass of the galaxy.The inclusion of this dependence in the chemical evolution model can explain not only the 'mass–metallicity' correlation for disk galaxies, but also a considerable amount of heavy elements in the intergalactic medium [1287].

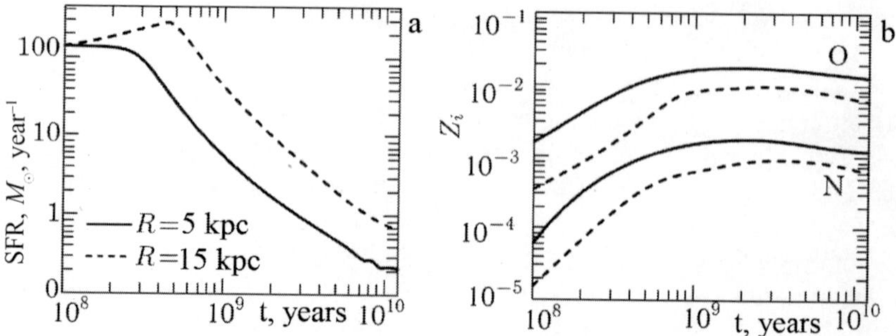

Fig. 9.20. Left graph – the estimated star formation rate (SFR) in the central region of the galaxy and on its periphery. Right graph – the evolution of oxygen and nitrogen for different galactocentric distances. The difference in the content of O and N in the central region of the Galaxy and at the periphery during the evolution of the Galaxy is almost unchanged. According to calculations [34].

9.5.3. The Milky Way – a typical galaxy

Models of chemical evolution of disk galaxies are constructed primarily for the Milky Way, and even more specifically – for the solar neighborhood, – because it is for this star system that we have accumulated the largest number of observational material. The question naturally arises of how the conclusions made on the basis of modeling the chemical evolution of the Milky Way can be applied to other galaxies of comparable mass.

In general, the observations show that the basic laws of the distribution of heavy elements in our galaxy – various metallicity values in the disk and halo, the radial gradient of the chemical composition in the disk, even two peaks in the distribution of the metallicity of globular clusters – are typical of other similar galaxies, in particular, the Andromeda Nebula. In addition, the luminosity function galaxies in clusters is such that the main contribution to the production of (optical) radiation in the Universe is made by galaxies similar to ours.

However, a more detailed analysis shows that the 'typicality' of the Milky Way is limited. In particular, in [645] comparison was made for our Galaxy and M31 with other nearby galaxies of the speed rotation of the peripheral stars V_{flat}, their metallicity and total angular momentum of the disk. It turns out that the Milky Way lags by more than 1σ from the relationship between these parameters found in other galaxies. In particular, our Galaxy is characterized by deficiency of stellar mass and metallicity fo the given speed of rotation. In general, studying the parameters of the local population of galaxies, the authors of [645] found that galaxies like the Milky Way, are only about 7% $\pm$1% of the total population of spiral galaxies. This is due to the fact that our galaxy, in contrast to most of its neighbours, has not experienced over the last 10 billion years a single major merger and therefore has a relatively peaceful history of star formation. At the moment, the Andromeda nebula is typical, so in fact the models of chemical evolution of disk galaxies should be tested using this system.

10

Dark matter inside the galaxies

The aim of this chapter is to discuss briefly the most important observational data and theoretical results, indicating the existence of dark matter within the limits of several optical radii from the centre of the galaxy, and describe the appropriate evaluations of this dark matter (see also sections 1.4.5, 9.1 and 9.2).

A dark halo is an important component of the galaxy. We shall mentioned briefly the direct and indirect results showing that inside the Galaxy within the limits of its optical radius ($R_{opt} \simeq$ (4–30) kpc for different objects) and in the nearest vicinity ($\lesssim$(2–3) R_{opt}) there is dark matter with the mass greater than the total mass of the stars and the gas. Here we shall not restrict ourselves to only S-galaxies and will also consider the elliptical galaxies.

1. The presence of massive halos explains the elongated flat rotation curves where the velocity of the gas does not decrease to the distances (23)R_{opt}. In NGC 3741 the rotation curve with respect to HI is visible up to 38 radial scales (= 8 optical radii)! In other words, the form of the rotation curve greatly differs from the expected one assuming the approximate constancy of $M/\mathcal{L}$ along the radius. The models of the maximum disk give the values $M_h/M_d \gtrsim 0.5$ already in the range R_{opt}.
2. The ratio of the mass to the luminosity $M/\mathcal{L}$ is considerably higher than that expected for the normal stellar population, if the mass is evaluated on the basis of the dynamics of the galactic disk (however, the problem of what can be regarded as a normal stellar population remains).
3. The results of measurements of the beam velocities and distances of the adjacent stellar systems are used to determine the global parameters of the galaxies and of their groups [86, 749], in particular, in the range up to 8 Mpc the fraction of baryon matter in this volume is equal to only 2.3% of the critical density of the Universe [746]. The total mass of the Local group of the galaxies inside the volume with a radius of 1 Mpc in relation to the centre of the mass (approximately in the middle between the Galaxy and the Andromeda Nebula) equals $1.9 \pm 0.2 \cdot 10^{12} M_\odot$ [747]. Nevertheless, the ratio of the baryonic matter of the Galaxy and the Andromeda Nebula pair to the total mass can be estimated as ~1/10.

4. The observed effects of the gravitational lensing indicates the presence of dark matter. The strong lensing of the galaxies and clusters of the galaxies, detected in the form of rings and arches, can be used to estimate the mass of dark matter and its form [1238]. We will mention the results of weak lensing on the basis of the statistical estimate of the directions of elongation of the form of the images of the far galaxies of the background which indicate the profile of the density of the total mass $\rho \propto r^{-2}$ faraway outside the limits of the stellar subsystem [610].
5. The existence of the cold stellar disks without bars indicates the large mass of the halo. Without the latter, the disk should either have a bar or should be very hot ($c_r/V \gtrsim 0.5$) at the periphery (chapters 4 and 10).
6. At the periphery of the E-galaxies, the local values of the ratio of the mass to luminosity $M/\mathcal{L}_B$ increase and greatly exceed the values of 5–10 solar units, typical of the central regions. This growth indicates the presence of dark matter.
7. The detection of the hot (x-ray) gas coronas with the temperature of $\sim 10^6$ K indicates the presence of a large amount of dark matter. The observed parameters of hot gas can be used to estimate the mass required for maintaining them, and investigate the variation of the mass with the distance from the centre of the Galaxy.
8. The galaxies with low surface brightness (LSB) have the velocities of rotation up to 100–200 km/s, which is difficult to explain without considering the large amount of dark matter. If it is not assumes that the dark matter is situated in the disk, the mass of baryonic matter in the form of conventional stars and the gas at dwarf LSB evidently does not exceed 20% of the total mass [523].
9. The large part of the spiral galaxies at the periphery of the disk ($r \gtrsim 2L_\sigma$) is dynamically cold: the dispersion of the radial velocities c_r is small in comparison with the velocity of rotation: $c_r/V_{\max} \simeq 0.3 \div 0.1$. This indicates the presence of a massive halo within the limits of the stellar disk (section 4.4, 10.2).
10. In the majority of the galaxies, visible edge-on, the ratio of the characteristic vertical scale of the stellar disk to the radial scale is in the range $\Delta_*/L \simeq 0.1 \div 0.3$. Without considering the dark halo it is not possible to explain the comparatively small thickness of these stellar disks (section 4.4, 10.2).
11. The direct estimate of the mass of dark matter inside the galaxy and in its immediate vicinity is provided by the kinematics of polar rings [421, 713, 1065, 1298]. Comparison of the kinematics in two orthogonal directions (the plane of the dis and the polar ring) offers the principal possibility of investigating the form of the dark halo – its difference from the spherical symmetry [92].
12. It is important to note that without considering the dark matter it is not possible to explain, within the framework of the accepted cosmological models, the formation of the galaxies from the initially expanding matter.
13. The kinematics of the so-called hypervelocity stars (HVS), with approximately 20 stars of this type already detected [423], depends strongly not only on the mass of the dark matter but also on the anisotropy in the distribution of density.

Thus, the results of observations of the HVS with collection of the statistical data in principle provide restrictions for the halo parameters.

14. The dynamics of the gas and stellar flows, observed in the halo of our Galaxy (high-velocity clouds, immediate velocity clouds, Magellan stream, tidal streams of the stars and the gas – Sagittarius, Orphan) also indicates the presence of dark matter with a very large mass [359].

It has been assumed that the conclusion on the presence in the limits of the optical radius of the stellar disk of the dark matter comparable or greater than the total mass of the halo and the stars is unavoidable, at least for the majority of galaxies. The main mass of dark matter should be located outside the disk, forming a spheroidal subsystem. The estimates of the ratio of dark matter to baryonic matter inside the optical radius using different methods and different observational material are in agreement together with the accuracy to a factor of two.

Below, we examine in detail some observational data which can be used for estimating the parameters of dark matter.

10.1. Flat rotation curves of the galactic gas

Approximately 80% of galaxies with high luminosity belong to the group of spiral galaxies. Investigations of the kinematics of the S-galaxies in the 70s presented the first confirmation of the existence of dark matter inside them [1093, 1094].

To explain the flat rotation curves (V = const) far outside the limits of the outer boundary of the stellar disk, a hypothesis was proposed for a relatively massive dark halo – a spheroidal subsystem with the stellar disk inside it (Fig. 1.32). The halo also contains stars (including neutron stars, white and brown dwarfs), gas and spherical clusters, i.e. 'visible' (baryonic) matter. However, its stellar mass does not exceed 10%. In this chapter, the mass of the halo M_h describes the mass inside the so-called optical radius R_{max} determined by the outer boundary of the stellar (but not gas) disk. In most cases, we have $R_{max} \simeq (3\text{-}5) \cdot L_\sigma$ (L_σ is the radial exponential scale of the disk).

The circular velocity of rotation in the disk plane V_c is determined from the balance of the radial gravitational forces, acting from the side of all subsystems

$$\frac{V_c^2}{r} = f_{\rm disc} + f_{\rm bulge} + f_{\rm halo}, \tag{10.1}$$

where the specific forces from the side of the visible matter f_{disc}, f_{bulge} determine the velocity $V_c^{(vm)}$. Within these or other assumptions of the nature of the distribution of the dark matter in the halo $\rho_h(r)$ we determine the specific force f_{halo} and the appropriate circular velocity $V_c^{(h)} = \sqrt{r \cdot f_{\rm halo}}$.

Figure 10.1 shows the rotation curves of the gas extending far outside the limits of the optical radius of the galaxies R_{max} where it makes contact with the stellar disks in which the main mass of visible matter is concentrated (in particular stars). Additionally, we have constructed $V_c(r)$ which coincides with the rotation curve

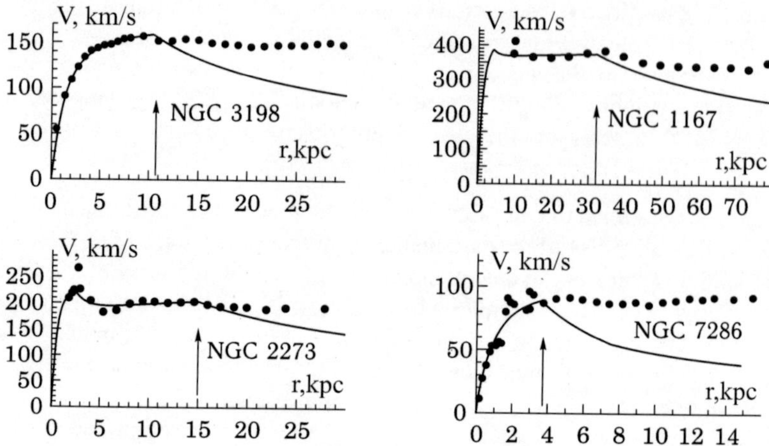

Fig. 10.1. Examples of several nearest galaxies with the known rotation curves of the plateau type which extend far outside the limits of the boundaries of the stellar disk: NGC 3198 [353], NGC 1167, NGC 2273, NGC 7286 [972]. The graphs show the radial dependences of the velocity of rotation of the gas on the basis of the observations V_{obs} (full symbols) and the circular velocities $V_c^{(vm)}$, constructed assuming that outside the limits of the stellar disk there is no mass (the solid line). The vertical arrow indicates the boundary of the stellar disk of the galaxy (the optical radius R_{max}).

of the gas inside R_{max}, but in the region $r > R_{max}$ the matter, including the dark component, does not exist. In this case, we ignore the assumption on the constancy of the 'mass/luminosity' ratio along the radius ($M/\mathcal{L}\neq$const), ensuring the constancy $V_c = V_{obs}$ inside R_{max}. At the distance $2R_{max}$ the relative difference of the velocities $(V_{obs} - V_c^{(vm)})/V_{obs}$ is approximately 30%. In a number of galaxies, the observed velocity of rotation of the gas remains constant up to $3R_{max}$ and even further. In NGC 3741 the rotation curve with respect to *HI* is visible up to 38 radial scales, i.e. approximately to 8 optical radii.

The shape of the rotation curve greatly differs from the expected shape assuming the constancy along the radius ($M/\mathcal{L}$), i.e. if the surface density of the disk is proportional to its surface brightness.

Considering a relatively massive halo (comparable with the mass of the stellar disk and greater) we can easily explain the observed flat rotation curves of the gas (Fig. 10.2). Here, it will be assumed that the ratio of the surface density of the disk to its brightness remains unchanged along the radius of the galaxy, if not mentioned otherwise.

This approach to the decomposition of the rotation curve does not make it possible to separate unambiguously the contribution of the individual components, especially the disk and the halo, to the rotation curve of the galaxy (Fig. 10.2). We can explain the observed law of rotation using two models which greatly differ in the ratio of the mass of the halo inside the optical radius $M_h = M_h\,(r \leqslant R_{max})$ to the mass of the disk M_d: $\mu = M_h/M_d$ (Fig. 10.3). In principle, the radial profile $V(r)$ can

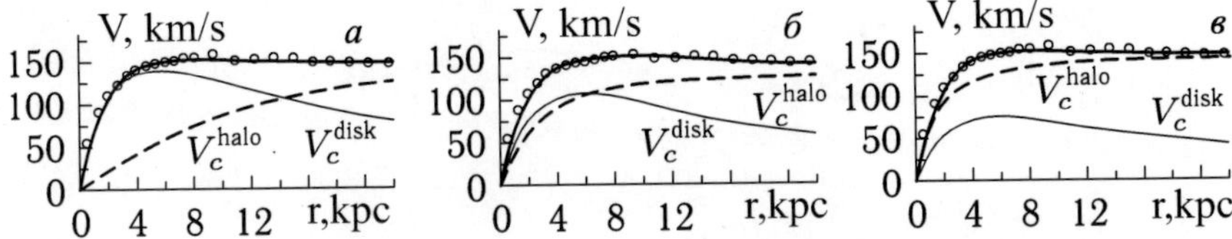

Fig. 10.2. The results of decomposition of the rotation curve of the NGC 3198 galaxy in the models which differ in the ratio of the mass of the halo the mass of the disk within the range $r \leqslant R = 4L = 10.4$ kpc: a) $\mu = 0.65$; b) $\mu = 2$; c) $\mu = 5.5$. The rotation curve was taken from [353], $L = 2.6$ kpc, $z_0 = 0.4$ kpc. The distribution of density in the halo (4.3) and the disk (2.57) is used. When introducing additional components (for example, the bulge) into the model, the degree of ambiguity of the decomposition is even greater. The model *a* is the model of the maximum disk in the case of the laws (4.3), (2.57), for NGC 3198, o indicates observational data,

Fig. 10.3. The kinematics of the galaxies can be used to obtain the estimate of only the total mass within the limits of the given radius. For example, the velocity of rotation of the Sun around the centre of our Galaxy is ≃200 km/s. Using this value for the periphery of the disk, we obtain the estimate of the mass of the order of $10^{11}\, M_{\odot}$ within the sphere with a radius of 12 kpc. The rotation curve indicates only the nature of distribution of mass as a whole. The same curve $V(r)$ can be explained by the massive halo and by the small mass disk or by the massive disk a low mass halo. The following question should be answered: what is the ratio of the visible (baryonic) matter and the dark matter, and where is the latter situated.

be described by restricting ourselves to the contribution of only the halo, i.e. it is assumed that $\mu \gg 1$ – this limiting case of a light disk is shown in Fig. 10.2c.

In the reversed limit, at the given law of distribution of the density $\rho_h(r)$ there is a minimum value of the mass of the halo inside the specific radius at which the condition $V \simeq V_{obs}$ is fulfilled. Since in this case the mass of the disk component is maximum possible, this model is referred to as the model of the maximum disk. This approach gives the lowest estimate of the mass of the halo. Inside the optical radius of the galaxy $r \leqslant R_{max}$ the relative mass of the halo within the framework of such a model remains approximately $\mu = M_h/M_d \simeq 0.4$–0.7. However, in the range $r \lesssim 2R_{max}$ to which the gas disk often extends, we may obtain $\mu = M_h/M_d \simeq 2$–2.5.

It is important to note that in the majority of cases it is not possible to explain the rotation curve without considering the dark matter even inside the optical radius, if it is assumed that the mass/luminosity $M/\mathcal{L}$ ratio is constant along the radius of the galaxy. Figure 10.4 shows the results of construction of the rotation curve

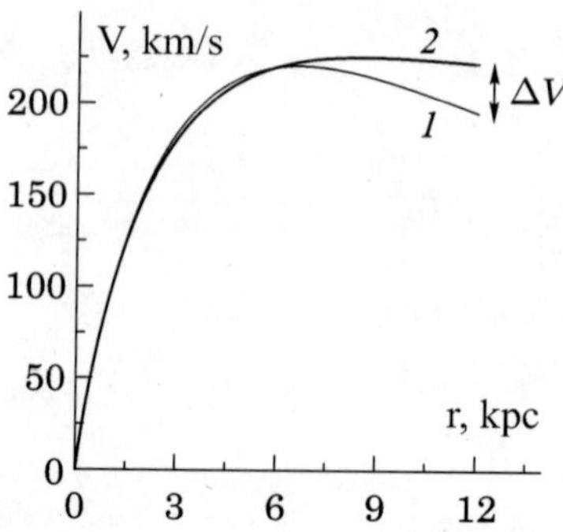

Fig. 10.4. Example of the decomposition of the rotation curve without taking the dark halo into account. Curve 1 – the exponential disk with $L_\sigma = 3$ kpc, $z_0 = 0.25$ kpc, assuming constant ratio $M/\mathcal{L}$. The maximum velocity is observed at the radius $r = 2.2L_\sigma$. 2 – the ratio $M/\mathcal{L}$ changes along the radial coordinate ensuring the increase of the velocity of rotation at the external boundary of the disk by $\Delta V \simeq 35$ km/s so that the observed rotation curve coincides with the modelling rotation curve.

of the plateau type within the limits of the optical radius without considering the spheroidal subsystem. If it is assumed that the ratio $M/\mathcal{L}$ is constant in the entire disk, then on its outer boundary we consider the difference in the velocity of rotation of approximately 15% of V_{max} which equals 35 km/s in the given example.

Disregarding the assumption on the constancy of $M/\mathcal{L}$, we obtain in principle any rotation curve in the range $r < R_{max}$ (Fig. 10.4). It should be stressed that outside the limits of the stellar disk it is also necessary in this case to introduce the 'hidden' matter into the model (Fig. 10.1).

The concept of the massive dark halo, considered in this section and arising from the need to explain the non-decreasing nature of the rotation curve outside the limits of the radius $r > 2.2L_\sigma$ (see the curve 1 in Fig. 10.4) is not the only possible. It may be assumed that the hidden mass is situated in the disk component. The curve 2 in Fig. 10.4 can also be interpreted as the result of presence, in addition to the stellar disk with $M/\mathcal{L}$ = constant, also of a dark disk component. However, the assumption on the dark mass of this component, explaining the observed rotation curve, is in contradiction with a number of other observational results. In particular, it does not make it possible to explain the cold stellar disk, when in the range $r \gtrsim 2L_\sigma$ we have the ratio $c_r/V \lesssim 0.2$, and the thin stellar disks with $z_0/L_\sigma \lesssim 0.2$ (chapter 4). Finally, the cosmological scenarios stand firmly on the position of existence of the massive spheroidal (triaxial in a general case) halo and the various non-contradicting observational confirmations favouring this viewpoint. These problems are discussed in the subsequent sections.

It is important to note the hypothesis which makes it possible in principle to explain the flat rotation curves without considering the dark matter. In this case, we are concerned with the so-called MOND (modification of Newton dynamics) based on the revision of the classic Newton law at large distances. Instead of the gravitational potential for the point body $\Psi = -GM/r$ at small accelerations, which

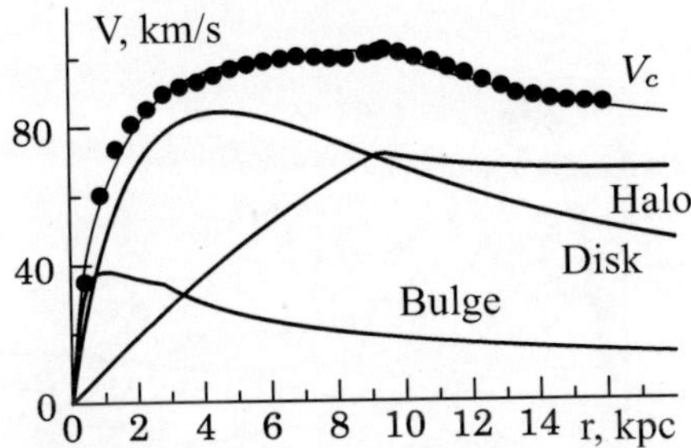

Fig. 10.5. Rotation curve of the gas of NGC 4244 (● – the data from [990]) and the results of decomposition. The parameters of the stellar disk: $L = 2$ kpc, $z_0 = 0.42$ kpc [990].

take place at large r, in this case, we use other radial dependences ensuring the constant velocity of rotation V = const. This approach, which formally solves the problem of flat rotation curves, results in additional physical and astronomical problems. Therefore, the MOND hypothesis will not be discussed further. It is not assumed that this radical step is at present efficiently substantiated.

We shall consider a number of galaxies with peculiar rotation curves.

NGC 4244. The nearby galaxy NGC 4244 belongs to the type SA (s) cd and has the maximum velocity of rotation V_{max} = 100 km/s (Fig. 10.5). The stellar disk breaks up at the radius $R_{max} \simeq 10$ kpc, the gas disk extends to 15 kpc. The characteristic special feature is the large reduction of the velocity of rotation of the gas, starting with the radius R_{max}. Figure 10.5 shows the results of construction of the model of the galaxy assuming $V_{gas} = V_c$. The presence in the stellar disk with the radial scale $L = 2$ kpc of a sharp boundary at the distance of 10 kpc from the centre is not capable of explaining the drop of the rotation curve. The observed velocity of rotation of the disk can be described only by the special feature of the distribution of matter in the halo.

We shall mention other examples of the galaxies in which the velocity of rotation decreases at the periphery: in the NGC 891 at $r \simeq 7'$, at NGC 3709 at $r \sim 4'$, at M51 at $r \sim 4'$, and also at NGC 1808 and NGC 3031.

NGC 157 is an even more surprising galaxy (SABb) and has been investigated in a large number of studies, for example [584, 1098].

Within the limits of the stellar disk ($r \lesssim 8$ kpc) the rotation curve of the NGC 157 can be explained without considering the dark matter (Fig. 10.6). The gravitation of the dark halo should prevail only starting at a distance of 25–30 kpc from the centre, in order to explain the constancy of V to 45 kpc. In the zone 8 kpc $\lesssim r \lesssim$ 20 kpc, the rotation curve is similar to the Kepler law, indicating that the contribution of the halo is small.

Figure 10.8 shows the results of the composition of the rotation curve in the model of the maximum disk where within the limits of the optical radius R_{max} = 7 kpc the fraction of the disk component is equal to only 35% of the total mass ($1.84 \times 10^{10} M_\odot$, parameter $\mu = 1.86$). In a sphere with a radius of 20 kpc the mass of the halo is absolutely dominant and its fraction is equal to 92% of the entire mass. A

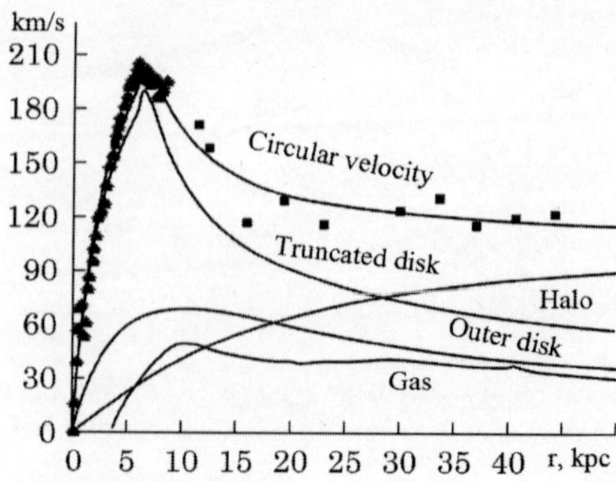

Fig. 10.6. The rotation curve of NGC 157 and the results of decomposition [1098].

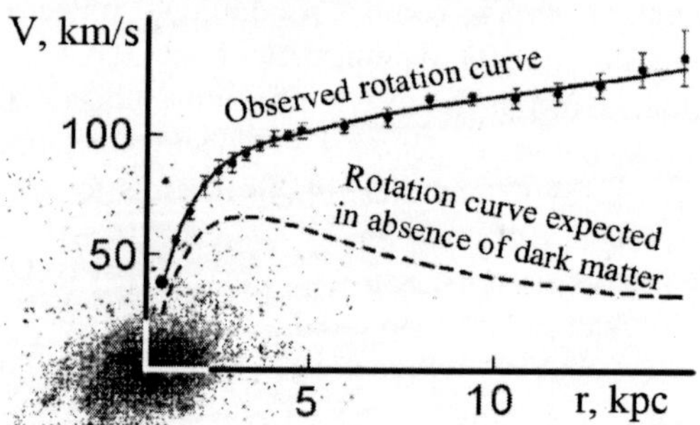

Fig. 10.7. The rotation curve of M33 according to the observational data in [485]. The image of the galaxy which extends to ~7 kpc is also shown.

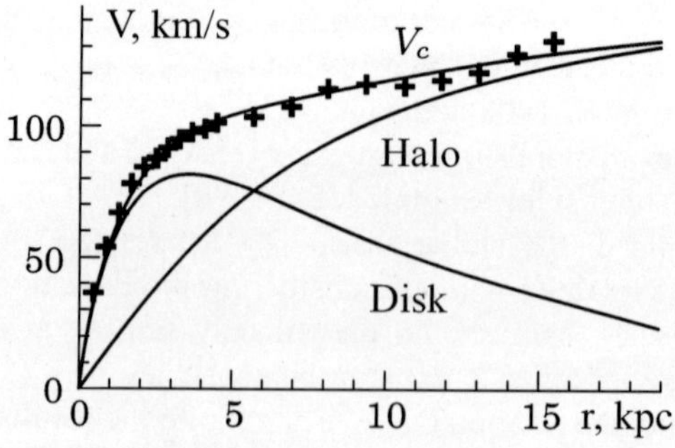

Fig. 10.8. The results of the composition of the rotation curve of M33 (crosses) in the model of the maximum disk. The circular velocity V_c is determined by the contribution from the disk and the halo.

general property of such a monotonically growing rotation curve is the fact that the scale of the halo a (see (4.3)) should be considerably greater than the exponential scale of the disk L_σ (C (1.3)). In the case of M33 we have $L_\sigma = 1.7$ kpc and $a = 6$ kpc.

A large number of examples of this behaviour of the rotation curve can be found in the sample of 967 galaxies with the rotation curves [1025]. It is important to mention the galaxies in which the growing form of the rotation curve extends far beyond the limits of the stellar disk and the maximum velocity is obtained at the periphery of the stellar disk ($r \gtrsim 3 \div 4L_\sigma$) or even far beyond the limits of the optical radius: NGC 1560, DD0 170, NGC 3109, IC 2559, IC4315. This situation is most typical of dwarf galaxies where the velocity of revolution is ~50–100 km/s: UGC 2023, UGC2455, UGC4278, UGC5829, UGC7323, and also low-luminosity LSB galaxies.

10.2. Radial profiles of the dispersions of the velocities in the stellar disk and the mass of the halo

10.2.1. The hypothesis on marginal stability of stellar disks and the halo mass

The flat rotation curves are the reason for the existence of dark matter inside the galaxies and in the immediate vicinity, but they also permit with similar success the presence of dark matter in the disk and may be used as a source of only the lower estimate of its mass within the framework of the model of the maximum disk. The determination from the observations with the high accuracy of the radial profiles of the dispersion of the velocities of the stars $c^{obs}(r)$ may provide additional information on the parameters of the halo and, consequently, on the mass and surface density $\sigma(r)$ of the visible disk component.

The main concept of this approach is associated with the assumption that the stellar disk should be situated in the vicinity of the boundary of gravitational stability and, consequently, $c^{obs}(r) \simeq c^{crit}(r)$. In principle, the disk may be superheated: $c^{obs} > c^{crit}$ which is actually detected for a number of galaxies S0–Sa, but the reversed condition cannot be fulfilled for any long periods of time because the unstable disk is heated quite rapidly in the non-linear stage of development of gravitational instability.

On the other hand, the value $c^{crit}(r)$ depends on the ratio $\mu = M_h/M_d$. A more massive disk requires for its stabilisation high values of the dispersion of velocities. This is evident because in the other limiting case of a very light disk the disk is always gravitationally stable. Thus, this condition, imposed on the dispersion of the velocities of the stars in the disk, can provide an additional estimate of the mass of the halo (or the disk), since the total mass of the galaxy is determined from the rotation curve.

Figure 10.9 shows the effect of the dependence of the relative degree of heating of the disk c_r/V_c on the relative mass of the halo μ.

Some specific galaxies will now be investigated.

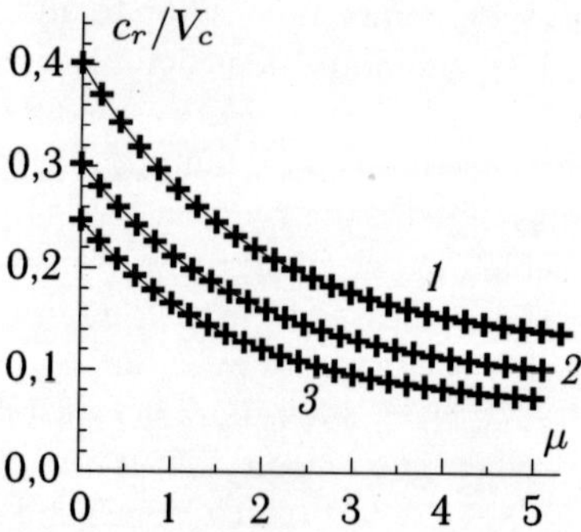

Fig. 10.9. Dependence of the ratio c_r/V_c at different radii: 1) $r = L_\sigma$, 2) $2L_\sigma$, 3) $3L_\sigma$, on the relative mass of the spheroidal subsystem μ according to the results of dynamic modelling of collisionless disks at the boundary of gravitational stability by the method of N-bodies without bulges.

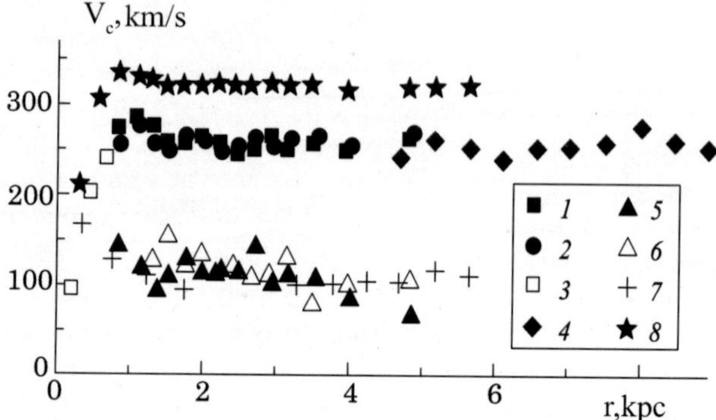

Fig. 10.10. The data on the kinematics of NGC 3115 galaxy along the major axis: the velocity of rotation of the stars (1 and 2 – to different sides from the centre [712], 3 – mean for both sides, 4 – according to the studies [448, 531]; the dispersion of the velocities of the stars (5, 6 – according to [712], 7 – according to [531]); 8) the circumferential velocity V_c taking into account the projection effects (for $i = 85°$ [448, 531]).

10.2.2. Lenticular galaxies NGC 3115

The NGC 3115 galaxy is a classic example of the galaxy of the type S0 without the bar and almost without any gas. Figure 10.10 shows the observed radial dependences of the velocity of rotation and the dispersion of velocities of the stars along the line of sight. The data on the very central region $r \lesssim 5'' = 0.24$ kpc, where the kinematics is determined by a massive black hole [531, 796], are not shown in Fig. 10.7 because it is not our task to describe this region. It is assumed that the distance to the galaxy is $D = 9.7$ kpc [531, 712], for the exponential scale of the disk $L = 50'' = 2.35$ kpc [712] and for the outer boundary of the disk $R \simeq 200''$ [796].

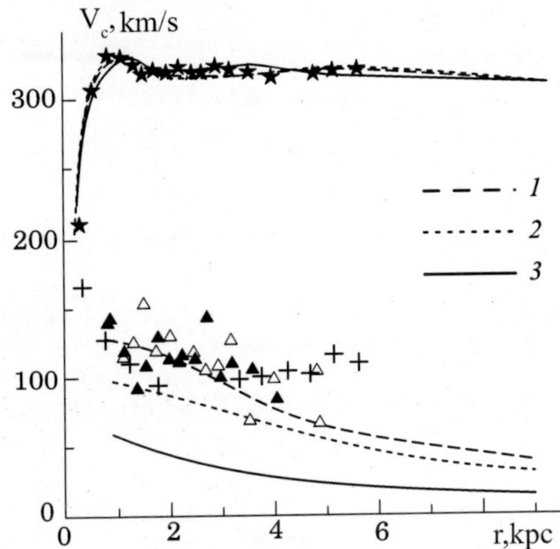

Fig. 10.11. NGC 3115. The radial dependences of the circular velocity V_c and the dispersion of the velocities along the line of sight c_ℓ for three models with different μ: 1) $\mu = 0.67$; 2) $\mu = 1.33$; 3) $\mu = 4.1$. The distribution c^{obs} is also shown (for the notations see Fig. 10.10).

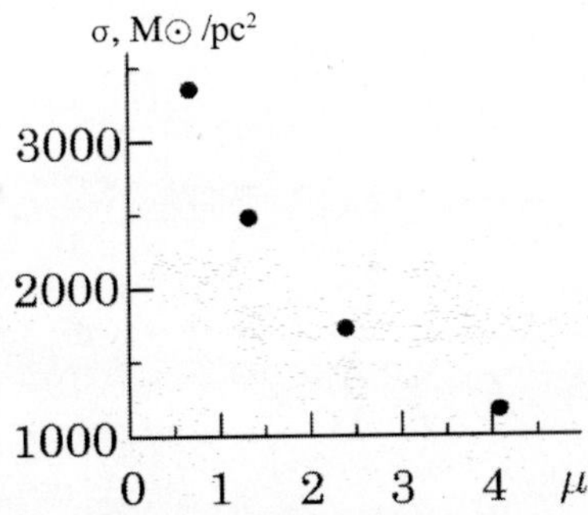

Fig. 10.12. Dependence of the central surface density of the disk σ_0 on μ for different models of NGC 3115.

A series of models was constructed for this galaxy which differ by the relative contribution of the disk and spheroidal components (Figure 10.11). Figure 10.12 shows the reduction of the central surface density of the disk (within the framework of the distribution (1.3)) with μ. Model 1 is the model of the maximum disk (Fig. 10.11, 10.13a). The radial distribution of the dispersion of the velocities of the stars is in favour of the massive spheroidal subsystem with $\mu \gtrsim 1.3$, and the mass of the bulge is $M_b \gtrsim 0.3\, M_d$. However, when $\mu = 1.3$ (Fig. 10.12), the central surface disk density is very high ($\sigma_0 \simeq 2500\, M_\odot/\text{pc}^2$) and this may act as an indirect indication of the superheating of the disk (the problem of 'superheating' of S0-galaxies, based on a large general sample, is discussed in section 4.4.4 [77, 1327]).

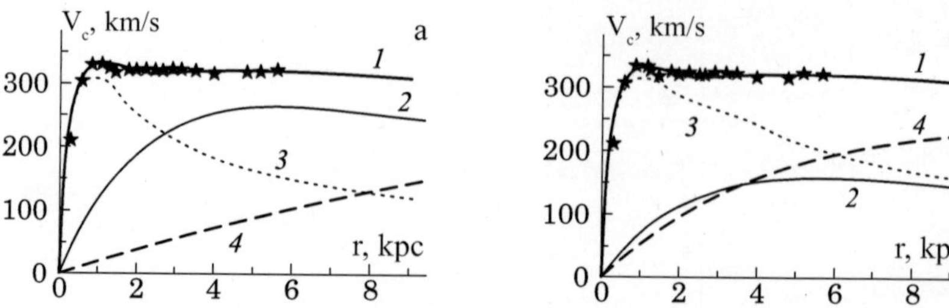

Fig. 10.13. For the NGC 3115 galaxy: a) the results of decomposition of the rotation curve in the model of the maximum disk; b) for the model 2 (Fig. 10.11). The individual curves show the contribution to the circular velocity from different subsystems: 1) the total circular velocity; 2) the contribution from the disk; 3) from the bulge; 4) from the halo.

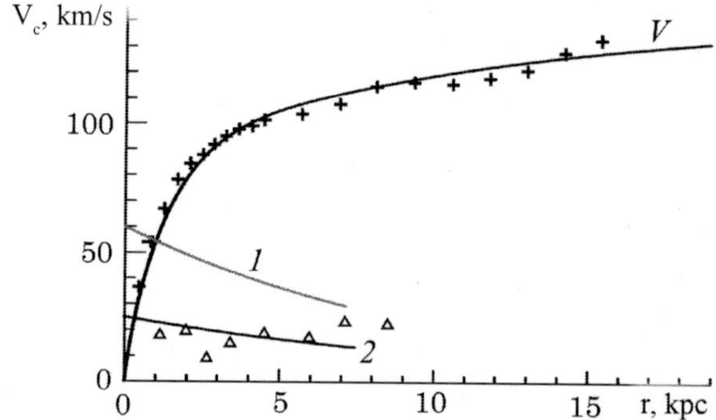

Fig. 10.14. M33 galaxy: the symbols '+' and the solid line are the velocity of rotation of the gas according to the observations and in the model, respectively; Δ is the observed distribution of the dispersion of the radial velocities of the planetary nebulas, curve 1 is the dispersion of the velocities in the model of the maximum disk, 2) in the model with the massive halo $\mu = 3$.

10.2.3. A galaxy of later type M33

The dynamic model of the M44 galaxy (Fig. 10.1) will be constructed using the radial distribution of the dispersion of velocities of planetary nebulas belonging to the disk of the galaxy (Fig. 10.14). In the model of the maximum disk at the boundary of gravitational stability the dispersion of the radial velocities is considerably higher than the observed value (curve 1). Agreement is obtained in the case of the massive halo and, correspondingly, a light disk for which $\mu \simeq 3$.

10.2.4. The conclusions

The identical approach to the evaluation of the halo mass 'from the bottom' was also developed in a number of studies [77, 255, 258, 260, 416, 1327]. Examples of the application of dynamic modelling for the decomposition of the galaxies into disk and spheroidal components for different types of objects have also been published in section 4.4.

The following general conclusions can be made:

1. Taking into account the dispersion of the velocities of the stars in the disk in the evaluation of the relative halo mass results in a higher value of parameter μ in comparison with the value yielded by the restrictions on the side of the flat rotation curves. The condition of gravitational stability requires more massive halos, approximately by a factor of 1.5–2.
2. The model of the maximum disk cannot be used in the majority of cases, especially for the galaxies of later morphological types.
3. It may be assumed that the disks of the galaxies of the late morphological types are situated in most cases in the vicinity of the stability boundary. The parameter μ for these galaxies is on average higher than for the objects of earlier types. Here we are not talking about the external disks far away from the centre of the disk regions which are often dynamically superheated.
4. The disks of the galaxies of early types, especially lenticular ones, are often superheated. Their observed dispersions of the velocities of the stellar disk are highly than the value required by gravitational stability even in the model of the maximum disk. In the group of the galaxies of the later type thc strongly superheated disks are evidently found only in a small number of cases.

10.2.5. Evaluation of the halo mass for the galaxies of later type

The extended radial dependence of the dispersion of the velocities of the stars in the disk is known to exist for a small number of objects [68, 77, 78]. It is therefore interesting to estimate the halo mass only on the basis of the observed central dispersion of velocities c^{obs}. This is in principle possible if the galaxy has no bulge and then c^{obs} can be related to the disk component. In these cases, it is convenient to consider the galaxies of the later type (Sc, Sd, Sm). In contrast to the approach based on the construction of the dynamic model for the specific galaxy taking into account its kinematic and photometric special features along the radius (see section 4.4.3), here we discuss the possibility of decomposition of a disk-halo within the framework of a simple model which can be used only for the galaxies in the absence of a large bulge. The value c^{obs} should not be related to the very centre of the galaxy where the presence of a supermassive black hole may result in a large increase of the dispersion of the velocities in the immediate vicinity $r \lesssim 10$ pc.

The investigations carried out in [256] are based on the dependence of the ratio of the dispersion of the radial velocities c_r ($r = 0$) to the maximum velocity of rotation $\beta \equiv c_r(0)/V^{(m)}$ on the relative mass of the halo $\mu = M_h/M_d$, obtained using dynamic modelling. Since the number of galaxies with the known distribution c^{obs} is small in comparison with the object for which the central dispersion of the velocity of the stars has been determined[1], the method described here can be used for a large sample.

[1]It should be stressed that the velocity dispersion should related to the region in which the effect of the central black hole or central core is not longer significant.

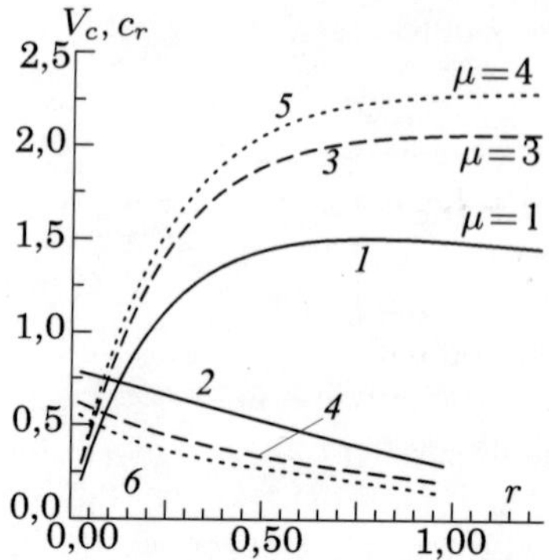

Fig. 10.15. Radial dependences of the circumferential velocity of rotation V_c (curves 1, 3, 5) and the dispersion of the radial velocities c_r (curves 2, 5 6, respectively) at different values of μ and $a = L = 0.25$ according to the results of dynamic simulation.

The rotation curves of the galaxies without large bulges usually contain a long section of the monotonic increase of the velocity of rotation in the range $r \lesssim 2L$, which is explained quite efficiently by the exponential law of the distribution of matter in the disk if the condition $a \gtrsim L$ is fulfilled for the halo scale. Another special feature of the rotational curves is the elongated plateau in the range $r \gtrsim 2L$ which is often found to be considerably greater than the optical radius.

With the increase of the contribution of the halo to the mass of the galaxies the ratio of the dispersion of the velocities of the stars to the velocity of their rotation V decreases (Fig. 4.30). The identical effect is observed for the central region (Fig. 10.15). Figure 10.16 shows the dependences $\beta(\mu)$ obtained on the basis of the results of dynamic modelling [256]. The velocity of rotation of the stars V is smaller than the circumferential velocity V_c, and this difference rapidly increases in the case of a small-mass halo (Fig. 10.16). Therefore, for comparison with the observational data, it is preferred to use the dependence $\beta(\mu)$, computed using the circumferential velocity of rotation.

Function $\beta(\mu)$ depends on the scale of the halo a (Fig. 10.16). At the same μ, the increase of a increases the ratio $c_r(0)/V_c^{(m)}$. Therefore, if the parameter μ is estimated using the known value of β using the curve 1 in Fig. 10.16 (the case $a = L$), the result is the lower estimate of the mass of the halo. The dependence $\beta(\mu)$ in the case $a > L$ gives high values of μ thus reducing the disk mass. In comparison with the observational data, since the dispersion of the velocities of the stars in the bulge is greater than in the disk, we obtain the lower estimate of the relative mass of the halo μ.

It is assumed that the maximum circumferential velocity of rotation $V_c^{(m)}$ is equal to the maximum velocity of rotation of the gas, measured either in the optical range $V_{\rm opt}^{(m)}$ (H_α) or on the basis of radio observations $V_{\rm rad}^{(m)}$ (HI) – these values are provided by the electronic catalogue LEDA. The catalogue also contains the maximum velocity of rotation $V_{\rm rot}^{(m)}$ related to a single system, and the double velocity of rotation without taking into account the angle of inclination of the galaxies $\Delta V^{(m)}$.

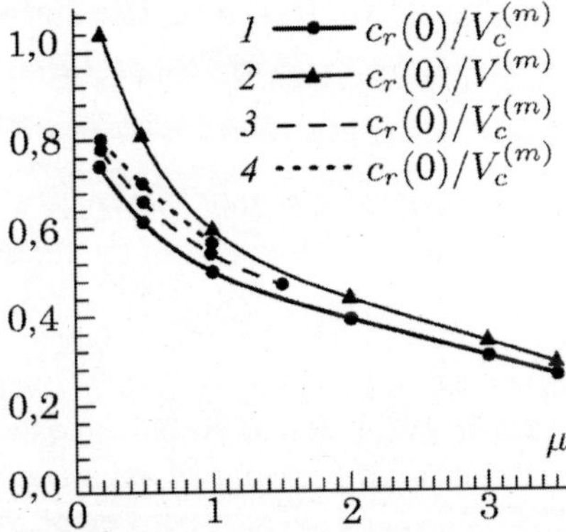

Fig. 10.16. The ratio of the dispersion of the radial velocities of the stars in the centre of the disk $c_r(0)$ to the maximum circumferential velocity $V_c^{(m)}$ (curves 1) and the maximum velocity of rotation of the stars $V^{(m)}$ (curve 2) for different relative halo masses $\mu = M_h/M_d$ with $a = L$ using the results of dynamic modelling of the disk in the absence of the bulge. Curves 3 and 4 were plotted for the models with $a = 2L$ and $a = 3L$, respectively.

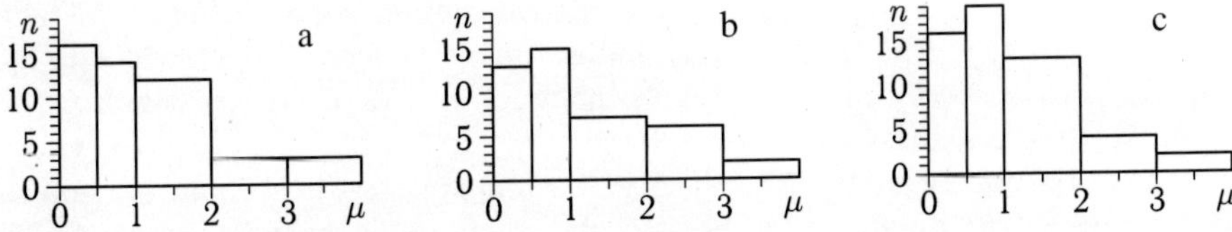

Fig. 10.17. Distribution of the number of galaxies of the later type ($t \geqslant 4$) with respect to the relative halo mass μ, calculated using: $V_{\text{rad}}^{(m)}$ (a); $V_{\text{opt}}^{(m)}$ (b); $V_{\text{rot}}^{(m)}$ (c).

The value of the parameter β, determined from the observational data, can be used with the help of the dependence 1 in Fig. 10.16 to estimate the relative mass of the halo [256]. In this case, the galaxies with $\beta > 0.9$ should not be considered because the dependence shown in Fig. 10.16 cannot be used for these objects. The reason for the high values of β may be associated with a large bulge, as for example in NGC 628, 4254, 4536, 5480 [322] so that for these objects it is not possible to estimate the mass of the halo only on the basis of the central dispersion of the velocities. Figure 10.17 shows the distribution of the number of galaxies with respect to the relative mass of the halo μ [256]. The calculation of the mean values gives: $\left\langle c_r(0)/V_{\text{rad}}^{(m)}\right\rangle = 0.54; \left\langle c_r(0)/V_{\text{opt}}^{(m)}\right\rangle = 0.55; \left\langle c_r(0)/V_{\text{rot}}^{(m)}\right\rangle = 0.56$,which results in the mean relative mass of the halo $\langle\mu\rangle \simeq 0.8$.

This approach is used to separate the galaxies with small β (for example, NGC 598, 1353, 2998, 3198, 3938, 4565, 7171, 7217, 7541) which can be characterised by a massive halo, larger than the mass of the disk by a factor of two or more, and these objects require special examination.

10.3. The vertical scale of the disk and the halo mass

10.3.1. Dependence of the signals of the stellar disk on the halo mass

The second independent method of dividing the mass into disk and spheroidal components may be based on the measurement of the thickness of the disk for the edge-on galaxies.

The existence of a relationship between the relative thickness of the stellar disk z_0/L_σ and the relative mass of the halo $\mu = M_h / M_d$ was suggested in [63] and subsequently this approach for evaluation of the halo mass was developed further in the studies [771, 911, 1192, 1243] (the results of dynamic modelling are presented in section 4.5). The problem of the effect of a compact massive bulge on the development of bending instability and on the disk thickness was studied in [225, 1191].

It will be assumed that the stellar disk is situated in the vicinity of the boundary of gravitational stability with respect to the perturbations in the disk plane. This leads to restrictions on the dispersion of the radial velocities. On the other hand, we have a relationship between the vertical scale of the disk z_0 and the dispersion of the vertical velocities, resulting from the vertical equilibrium of the disk. The ratio $\alpha_z = c_z/c_r$ is always smaller than unity and is a function decreasing slowly with the radius. As indicated by equation (4.128), the ratio $\chi_0 = z_0/L_\sigma$ decreases with increase of the relative halo mass μ (section 4.5.2).

Thus, with other conditions being equal, the more massive halo contains a thinner disk. If we ignore the dependences of Q and α_z on the parameter μ, from (4.128) we have $\chi_0 \propto (1+ \mu/1.55)^{-1}$. The approach based on the construction of the stellar system simultaneously at the boundary of gravitational stability in the disk plane and the boundary of the relative bending instability may provide only the lower limit of the halo mass (parameter μ) because there are various mechanisms which additionally increase the disk thickness. They include in particular external factors associated with the tidal interaction, accretion of matter, and absorption of satellites.

10.3.2. Results of modelling of the edge-on galaxies

When constructing the model of an edge-on galaxy the free parameters include the mass of the halo M_h inside the radius R_{max} (which corresponds to $r = 4L$ or the radius of break of the disk in the observations), the radial scale of the halo and the central surface density of the disk σ_0 [1243]. The parameter $\mu = M_s/M_d$ (ratio of the mass of the spheroidal subsystem $M_s = M_h + M_b$ to the disk mass M_d) is determined from the condition of agreement of the observed and modelling vertical scale and the velocity of rotation of the disk $V(r)$. In the models with the low-mass halo ($\mu \gtrsim 0.5$), the thickness of the disks is usually greater than the observed value, and in the models with the very massive spheroidal subsystem ($\mu \gtrsim 4$) the modelling disk is thinner than that detected in the investigated galaxies. The optimum model of the galaxy is the one in which the vertical scale of the disk is close to the observed

Table 10.1. The parameters of the edge-on galaxies [1243]

Galaxy	*D*, mpc	*L*, kpc	$\langle z_0 \rangle$, kpc	$\langle h_z \rangle$, kpc	R_{max}, kpc	μ
UGC 6080	32.3	2.9	0.69	0.48	9.99	0.6
NGC 4738	63.6	4.7	1.30	0.70	19.20	0.5
UGC 9556	30.6	1.5 (3.6)	0.51	–	9.00	1.1
UGC 9422	45.6	3.5	0.80	0.51	14.60	0.6
NGC 5170	20.0	6.8	0.82	–	26.20	1.6
UGC 8286	4.8	2.0	0.26	0.13	8.00	1.6
UGC 7321	10.0	2.1	0.17[a]	0.14[b]	8.15	3.4
NGC 891	9.5	4.9	0.98	0.49	21.00	0.9 (1.7)

Comment: *D* is the distance to the galaxy; *L* is the exponential scale of the disk; z_0 is the vertical scale averaged according to the law (2.570); h_z is the vertical scale averaged with respect to the radius for the exponential law; R_{max} is the break radius of the disk; $\mu = M_s/M_d$ is the relative mass of the spheroidal subsystem. For UGC 7321: a – the reduced vertical scale for the disk periphery; b – for the disk centre.

one. This approach makes it possible to obtain the upper estimates of the disk mass (since the disks during evolution can also be heated as a result of the effect of the factors which are not taken into account in these models) and, correspondingly, the lower limit of the mass of the dark halo of the galaxies[2].

We examine the edge-on galaxies using the results from [390, 391, 771, 1243]. The disk parameters are presented in Table 10.1. The values of the vertical scales are shown for the two laws of variation of brightness and are averaged with respect to radius. In selecting the distances to the galaxies UGC 6080, NGC 4738, UGC 9556, UGC 9422, the Hubble constant was assumed to be equal to $H_0 = 75$ km/(s·Mpc). In addition to the vertical scale, for the NGC 5170 and NGC 891, we also know the radial dependence of the dispersion of the velocities of the stars so that in constructing the mathematical model the results can be matched using these two parameters.

As an example, we examine in detail the model of UGC 9422 [1243]. The distance to this Galaxy is $D = 45.6$ Mpc. For the radial scale of the disk it was assumed that $L = 16'' = 3.5$ kpc, for the vertical scale: $h_z = 2.3'' = 0.5$ kpc, $z_0 = 3.6'' = 0.8$ kpc. A sharp discontinuity in the disk is detected at the radius $R_{max} = 14.6$ kpc, and the bulge is not detected. The rotation curve (obtained from the line H_α) was taken from [87]. Figure 10.18a shows the velocities of rotation for this galaxy V^{obs}, circumferential velocities V_c and the velocities along the line of sight V^s without taking the internal absorption into account (section 2.7).

A series of models was constructed for this galaxy. The models are characterised by similar circular velocities but differ in the mass of the disk M_d and spheroidal M_h components. Consequently, the models are characterised by different vertical scales of the disks z_0. Figure 10.18b shows the radial distribution z_0 according to the observational data and the results of dynamic modelling. In the case of a low-mass halo ($\mu = 0.34$), the disk is known to be thicker than the observed disk. This result is the reason against considering the model of the maximum disk in which the central

[2]In addition to the bending instabilities, the factors of vertical heating include the non-stationary density waves, scattering of stars on the LMC, tidal interactions.

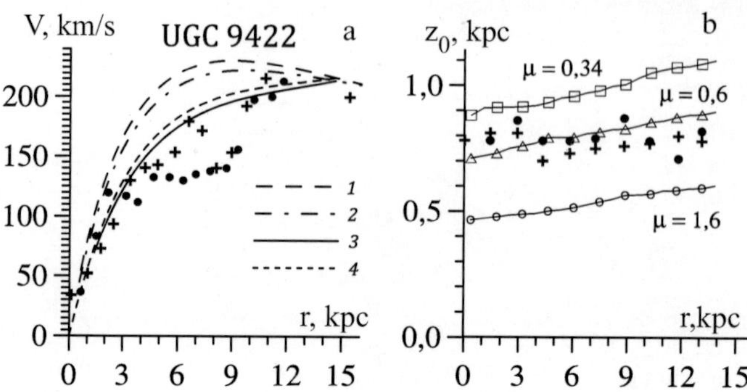

Fig. 10.18. Characteristics of the galaxy UGC 9422: a) the radial dependences V_c (lines 1, 2 – respectively for $\mu = 0.3$ and 0.6), V^{obs} (+ and ●), V_s (lines 3, 4 – respectively for $\mu = 0.6$ and $\mu = 0.3$); b) the vertical scale according to the observational data z_0^{obs} (+ and ●); the modelling vertical scales z_0 for: $\mu = 0.34$ (□), $\mu = 0.6$ (Δ), $\mu = 1.6$ (○). The symbols + and ● relate to the velocities on different sides from the centre of the disk [890].

surface density $\sigma_0 = 1440\ M_\odot/pc^2$, and the contribution of the disk to the circular velocity is maximum and equals $V_c^{disk}/V_{c|r=2.2L} = 0.94$.

With increase of the mass of the halo the value z_0 decreases and already at $\mu = 0.6$ we obtain satisfactory agreement with the observational data. In this model we have $M_d = 8.6 \cdot 10^{10}\ M_\odot$, $M_h = 5.2 \cdot 10^{10}\ M_\odot$, $\mu = 1200\ M_\odot/pc^2$ and $V_c^{disk}/V_{c|r=2.2L} = 0.87$. For a disk with a smaller mass ($\mu = 1.6$) the condition $z_0 < z_0^{obs}$ (Fig. 10.18) is fulfilled.

The UGC 7321 galaxy can be included in the group of superthin galaxies [627]. According to the data in [880] there is no bulge, $D = 10$ kpc, $L = 2.1$ kpc, $R_{25} = 8.15$ kpc (the radius at the 25th isophote). The latter value is used as R_{max} for constructing the dynamic model. The special features of the galaxy include the absence of the thin equatorial dust band. In [889] the authors published the radial distribution of the vertical scale. In the central region of the disk it is more efficient to use the exponential law for the vertical profile with the scale h_z, and at the periphery the function sech (z/h_{ch}).

The rotation curve from [890] will be used. For the galaxies observed edge-on, the effects of protection and absorption of radiation by the dust are least intensive at the disk periphery (see section 2.7 [76]). We select the value of the systematic velocity $V_{sys} = 394$ km/s which gives similar values of the velocity $V \simeq 100$ km/s in the external region for both sides from the centre of the galaxy.

The thin disk assumes the presence of a relatively massive halo with $\mu = 3.4$ (see Table 10.1, [771]). In this model $M_d = 0.53 \cdot 10^{10}\ M_\odot$, $M_h = 1.8 \cdot 10^{10}\ M_\odot$ and the contribution of the stellar disk to the circumferential velocity at the radius $r = 2.2L$ is only 60%. In this case, for the central surface density we have $\sigma_0 = 220\ M_\odot/pc^2$. The latter result is in agreement with the conclusions made in [890] according to which UGC 7321 is a galaxy with a low surface luminosity.

The results presented here and previously in section 4.5 show that the model of the maximum disk, used frequently in the literature, cannot be used for the majority of galaxies because it cannot be used to explain the observational data for the dispersion of the velocity of the stars and the vertical scales in them.

10.3.3. The thickness of the disk and the dispersion of the velocities of the stars in the edge-on galaxies (NGC 891 and NGC 5170)

It should be stressed that the previously described approach to evaluating the mass of the halo on the basis of information on the thickness of the disk is independent in respect of the method of determination of μ on the basis of the velocity dispersion in the disk plane (section 10.2). At the same time, both methods provide the lower estimate of the mass of the halo within the limits of the optical radius. Unfortunately, the number of the edge-on galaxies with the known radial profile of the dispersion of the velocities is small. Therefore, the estimates of μ for such galaxies on the basis of the observed dispersion of the velocities and the thickness of the stellar disk are of special interest [1243].

NGC 5170

The radial and vertical scales, the dispersion of the velocities and the circular velocity of rotation were published in [417] where special attention was given to the indeterminacy in the evaluation of the disk thickness associated with the ambiguity of the estimate of the angle of inclination of the galaxy: at $i = 87°$ for the vertical scale $z_0 = 0.82$ kpc, and at $i = 85°$ $z_0 = 0.57$ kpc. In this case, for determinacy $z_0 = 0.82$ kpc. It should be mentioned that when selecting $i = 87°$, the mass of the disk determined on the basis of the numerical experiments, is higher than at $i = 85°$. The NGC 5170 stellar disk extends to $R_{max} = 26.2$ kpc on the radial scale $L = 6.8$ kpc. The photometric data indicate the presence of the bulge $R^{(b)}_{max} \simeq 2$ kpc (see equation (1.19)) [417].

Figure 10.19 shows the velocities of rotation and the dispersion of the velocities in the model of the maximum disk (a) and for a disk with the smaller mass with $\mu = 1.6$ (b). In the first case, the dispersion of the velocities in the disk plane is considerably greater than the observed values. In the model of the maximum disk, the mean vertical scale $\langle z_0 \rangle = 1.1$ kpc is higher than the observational value of 0.82 kpc. The agreement with the observational data is obtained in the model with the smaller disk mass. In the model with $\mu = 1.6$, the mean vertical scale is $\langle z_0 \rangle = 0.82$ kpc. It is noteworthy that for the NGC 5170 galaxy the agreement of the model with the observational data for the two independent parameters (the dispersion of the velocities of the stars along the line of sight and the vertical scale of the disk) gives a similar estimate of the relative mass of the halo $\mu \simeq 1.6$ [771].

NGC 891

The NGC 891 galaxy contains a bulge [418, 942] and, therefore, in constructing the dynamic model the bulge was taken into account with the parameters $M_b = 1.7 \cdot 10^{10}\, M_\odot$, $b = 1$ kpc, $R^{(b)}_{max} = 3$ kpc. Photometric decomposition using the data published in [1036] gives $M_b / M_d \simeq 0.1$. The definition of the law of

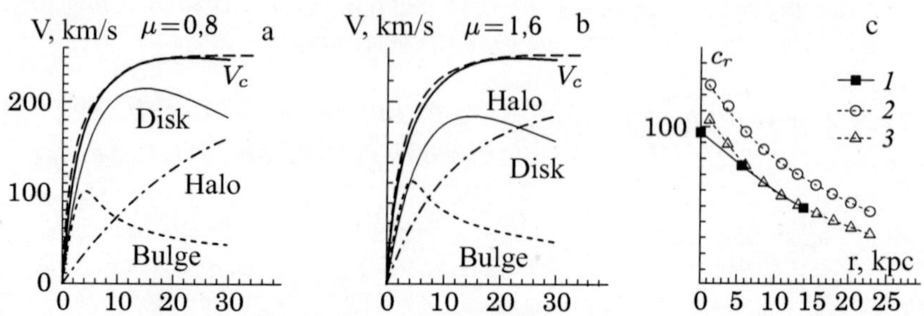

Fig. 10.19. Two models for the NGC 5170 galaxy: $\mu = 0.83$ (a), $\mu = 1.6$ (b). The radial dependences of the circumferential velocities and the contribution of different components to V_c are presented; c) the dispersions of the radial velocities: 1) according to the observational data [117], 2) a model with $\mu = 0.83$, 3) $\mu = 1.6$.

vertical distribution of the luminosity of the disk has been studied in a number of investigations. The vertical exponential scale h_z of the old stellar disk varies with different authors in the range from 400 to 650 pc [325, 806, 942, 1152, 1315). Van der Kruit and Searle [805, 806] obtained for NGC 891: $D = 9.5$ mpc, $i = 89°$, $L = 107'' = 4.9$ kpc, $R_{max} = 21$ kpc, $z_0^{obs} = 0.99$ kpc. In the outer region the thickness of the stellar disk was greater [1107].

We will use the rotation curves obtained according to the observations HI [354, 418, 1107] (Fig. 10.20a). For NGC 891 we have observational estimates of the dispersion of the velocities of the stars in the disk plane [418] and the velocity of rotation of both the gas and stellar component. The molecular ring at the radius $r \simeq 3.5$ kpc allows the presence of the bar to be assumed [1185][3]. Therefore, the dynamic model, including the bar in the centre of the disk, was also constructed.

The mass of the components of the NGC 891 will be estimated, using the observed values of the dispersion of the velocities of the stars and taking into account the assumption of the marginal gravitational stability of the stellar disk with respect of the perturbations in the plane. According to the data in [418], in the centre of the galaxy the dispersion of the velocities of the stars in the bulge is 160 km/s, and the dispersion of the radial velocities of the stars in the centre of the disk is $c_r(0) = 120 \pm 20$ km/s. To explain the observed radial distribution of the dispersion of the velocities we should assume the presence of the halo with $M_h/M_d \gtrsim 1.5$ (see the curve 3 in Fig. 10.20a). It should be mentioned that for a slightly different rotation curve of NGC 891, the same method gave a similar estimate of $\mu = 1.77$ [258].

Figure 10.20b shows that to explain the observed vertical scale of the stellar disk of the NGC 91 it is sufficient to assume that the mass of the halo inside 20 kpc is approximately equal to the disk mass. In may be seen that the resultant estimate of μ on the vertical scale is slightly lower than the value obtained using the dispersion of the velocities. The latter result may be associated with the fact that the observed thickness of the disk may be associated with other factors of vertical heating which

[3]The presence of the bar and of the molecular ring is linked by a dynamic relationship [949].

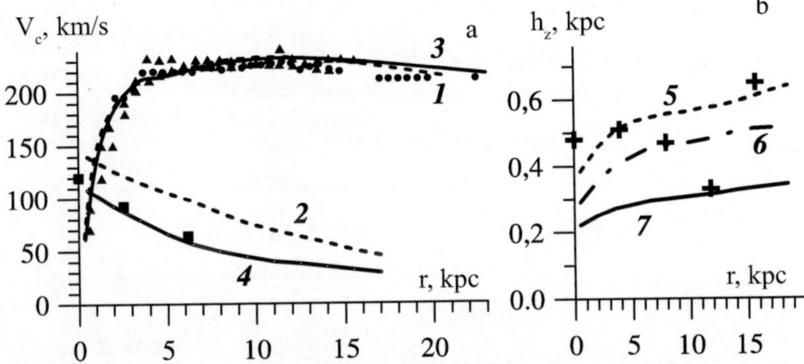

Fig. 10.20. Characteristics of the NGC 891 galaxy: a) the data on the velocity of rotation: ● – HI [418], ▲ – HI for two sides of the disk according to [1107], ■ – dispersion of the radial velocities according to study [418]; b) the vertical scale h_z: + – observational data, lines – modelling dependences for three values of μ. The modelling results are indicated by the lines for $\mu = 0.46$: V_c (1), c_r (2); for $\mu = 1.5$ the lines 3 and 4, respectively. For the height scale: $\mu = 0.4$ (line 5), $\mu = 0.7$ (line 6), $\mu = 1.7$ (line 7).

are not considered here. A more rarefied 'thick' disk with the thickness larger than the thickness of the main disk should also be considered in this case.

According to the results published in [807], inside $R_{max} = 21$ kpc the mass of the disk $M_d = 6{\cdot}10^{10}\ M_\odot$ is only 34% of the total mass of the galaxy. It should be mentioned that the estimate of the mass of the disk on the basis of the form of the rotation curve ($15{\cdot}10^{10}\ M_\odot$ [418]) leads to the mass/luminosity ratio $(M/\mathcal{L}) = 13.5{\pm}4$ which is higher than the identical ratio obtained using other methods (5÷10). This also contradicts the model of the maximum disk for NGC 891.

In conclusion, it should be mentioned that in [942] in the determination of the vertical scale the authors use a more complicated law for the vertical profile in the form of a sum of two exponential distributions which takes into account the presence of the thick disk. However, taking this fact into account has almost no effect on the previously mentioned restrictions for the parameter μ.

10.4. The form and structure of the dark halo

10.4.1. What do we know about the triaxial halo and is structure?

There is a large number of theoretical and observational data indicating the absence of central symmetry in the distribution of dark matter with the limits of the disk component of the S-galaxies. The density of the dark matter of the halo at large distances (~20–200 kpc) evidently follows the law $\rho_h \propto 1/r^3$ (see section 4.1.2, (4.4)). On the background of the monotonic decrease of density from the centre there are more fine-scale heterogeneities. Modelling within the framework of the cosmological CDM model (Cold Dark Matter) indicate not only the triaxial distribution of the mass in the CDM halo but also the existence of fine-scale structures – subhalos – and the latter are situated anisotropically with respect to the halo with preferential

Table 10.2 The data on the form of the halo

Objects/models/ methods	$s = c/a$	$q = b/a$	Reference	Comment
Cosmological ΛCDM models	$\langle s\rangle \simeq 0.7$	$\langle q\rangle \simeq 0.8$	[428]	WDM-halo 10% more spherical than CDM models
Cosmological CDM models	$\langle s\rangle \simeq 0.6$	$\langle q\rangle \simeq 0.75$	[347]	
Cosmological CDM models. Formation of stellar halos due to tidal intereference of satellites	$c/a \sim 0.3$		[483]	Inside 10 kpc
Cosmological LCDM and SCDM models	0.6	0.75	[726]	
Cosmological ΛCDM models	0.5	0.8	[291]	
Via Lactea project with number of particles $N = (1–2.3)\cdot 10^8$	0.64		[512]	Estimate of density of dark matter along the line of sight
Via Lactea model	0.45	0.5	[810]	In central region for density
Form of gravitational potential of halo according to results of CMD modelling	$s_0 = 0.72$	$q_0 = 0.78$	[659]	s_0, q_0 – in centre, at periphery more symmetrically
Radial dependence of vertical scale HI in M 31	0.4		[332]	
Microlensing	0.6		[1106]	Massive compact halo objects (MACHO)
Lyra RR, red giants, horizontal branch stars	$c/a \sim 0.2$		[941]	Inner stellar halo
Magellan flow modelling	0.8 – 0.9		[643]	
Stellar halo in MW	0.63	0.75	[963]	Major axis forms the angle of ~60° with the line connecting Sun and Galaxy centre
Grillmair and Dionatos stellar streams	$c/a \simeq 1$		[1288]	At fixed values of disk and halo masses

orientation along the major axis of the halo [810, 1329]. The number of the subhalos may exceed 10^5.

The triaxial form of the halo can be characterised by three spatial scales: a, b, c. The following can be used as the triaxiality parameter:

$$T = \frac{1-(b/a)^2}{1-(c/a)^2}, \tag{10.2}$$

where for the half-axes $a \geqslant b \geqslant$ c. There are oblate halos (at the poles) with $T < 1/3$, prolate halos with $T > 2/3$, and triaxial halos with $1/3 < T < 2/3$. The results of modelling of the formation of the galaxies on the basis of the cosmological models of *N*-bodies (CDM) indicate the triaxial distribution of the mass in the CDM halo with the ratios of the half-axes $s = c/a \sim 0.5$–0.7, $q = b/a \sim 0.6$–0.9 [(810, 1329], Table 10.2). The triaxiality parameters *s*, *q* depend on the halo mass, red shift and the distance to the halo centre. In the models of hot dark matter the distributions are more spherically symmetric than in the CDM models.

The observed spatial distribution of the dwarf satellites (subhalo?) in the Galaxy and other large galaxies may be a unique marker of the shape of the dark halo. However, the number of the observed dwarf satellites around MW and M31 is considerably smaller than predicted. Evidently, there is a large difference in their spatial distributions: the radial profiles of the subhalo are longer than in the observed dwarf satellites. In addition to this, there are indications of the anisotropic nature of the distribution of the dwarf novae in relation to the Galaxy and Andromeda Nebula [1329].

It is important to mention the Holmberg effect, associated with the preferential position of the dwarf satellites along the minor axis of projection of the optical galaxies not belonging to the Local group. The latest investigations indicate the existence of such an effect at the distances of ~200 kpc for a large number of objects. On the other hand, the satellites may be located preferentially along the major axis of the optical galaxy, and the degree of anisotropy decreases with distance from the optical centre.

Another indirect source of information are the observational data on the stellar halo for which the volume density decreases in accordance with the law $\propto r^{-3}$. Here we are concerned with the spherical clusters, variable of the type RR Lyra, blue stars of the horizontal branch, giants with a low content of metals, red giants, Virgin type cepheids, F-dwarfs [754]. Direct kinematic observations of different objects of the halo can be used to determine the radial profile of the dispersion of the velocities which indicates, in turn, within the framework of the specific models of the halo the nature of distribution of the gravitating mass. The non-symmetric nature of distribution of matter in the halo may be indicated by the anisotropy of the dispersion of the velocities of different objects in the halo of our Galaxy. The stellar halo shows large differences from the centrally symmetric distribution $c/a \sim 0.3$ [483] and even $c/a \sim 0.2$ for the inner halo [941]. Analysis of the databases for the SDSS and SEGUE reviews indicates the asymmetry of the galactic spheroid

Table 10.3. Data of the shape of the halo (continuation)

Objects/models/ methods	$s = c/a$	$q = b/a$	Reference	Comment
Tidal dwarf galaxy Sagittarius	0.72 (0.44)	0.99 (0.97)	[822]	For density in the range 20 kpc $< r <$ 60 kpc (in the brackets - for the potential
Sagittarius, M-giants 2MASS	0.66		[667]	Dark halo
Stellar flow GD-1 (photometry and kinematics)	0.83–0.94		[788]	Model of the dark halo potential close to $r \approx 15$ kpc
Branches A and B of the Sgr flow	0.92–0.97		[545]	For the potential
Tidal flows in Galaxy	< 0.7		[701]	Comparison with models of N-bodies
Distribution of HI along radius in Galaxy	0.5–1.25		[991]	Most probable value q = 0.8 [991]
Radial distribution of HI in NGC 891, NGC 4244	0.3		[991]	
Radial distribution of HI in UGC 7321	1		[982]	
Weak gravitational lensing	0.67		[680]	For projection on the drawing plane
Stellar halo for 1047 edge-on galaxies	0.6		[1331]	SDSS review
Polar ring of NGC 4650A	0.4		[1100]	
Kinematics of polar rings	~ 0.6		[713]	
Distribution and kinematics of atomic hydrogen	0.2–0.8	> 0.9	[904]	Estimates for the potential
Stellar disks visible face-on (in IR range)		$\gtrsim 0.95$	[1069]	For the potential
X-ray isophots around E-galaxies	0.5		[904, 991]	NGC 720, NGC 1332, NGC 3923

[964], and the stellar halo is triaxial and the minor axis forms the angle of 13° with the z-direction [963].

The stellar tidal flows in the form of arches, detected in the project Sloane Digital Sky Survey (SPSS), are evidently the result of absorption of the dwarf galaxies representing highly promising sources of information on the dark halo. The special features of this type in the distribution of the observed matter in the halo include the structure Virgo Over Density. It is assumed that these flows are associated with the destruction of the dwarf galaxies in the massive halo. The dynamics of the destructed satellites depends on the nature of distribution of the main mass of the halo, nevertheless, the spatial and kinematic parameters of greatly differing objects can be regarded as the most important markers of the properties of the dark spheroidal subsystem and, in particular, its form and orientation. We should mention:

- High velocity close, the Sagittarius dwarf tidal stream. Hercules-Aquila [360];
- Object Complex H loses gas and an impact front is detected;
- Stellar tidal stream in the NGC 5907 Galaxy, an estimate was obtained for the halo $c/a = 0.5$ [1101], a similar stream is detected in the case of NGC 5055;
- Magellan stream [643];p
- Giant tidal stream of stars with a low metallicity from M32 and NGC 205 to the halo of the Andromeda Nebula ($c/a = 3/5$ [702]);
  Cetus polar stream [965];
  Retro gradually returning stellar flows Grillmair and Dionatos [1288];
- Dwarf satellites in constellations Coma Berenices, Canes Venatici (II), Leo (IV), Hercules, Segue 1 spherical cluster [359].

The triaxial form of the halo is also indicated by the observations of the Sagittarius tital stream (Sagittarius dSph) which give the ratio of the axes c/a 0.8 [727] but if this stream is dynamically young than possibly $c/a \simeq 0.6$ [666].

In addition to cosmological modelling and the application of the observational data of the tidal streams there are other approaches which can be used to estimate the form of the dark halo within the limits of the optical radius of the galaxy and in its immediate vicinity on the basis of the characteristic special features of the visible baryonic matter – the stars and the gas (Table 10.2). There are several methods for determining the form of the halo (Fig. 10.21): 1) on the basis of the expansion of the gas disk HI with movement away from the centre (flaring gas layer) – with other conditions being equal, the thickness of the gas disk decreases with increase of the oblateness of the halo; 2) on the basis of the warping of the gas layer; 3) on the basis of the isophots of x-ray radiation; 4) on the basis of the external polar rings – a dynamic model of the polar ring is constructed taking into account the non-symmetric halo and this is matched with the observational data; 5) on the basis of the parameters of the precessing gas disk.

Tables 10.2 and 10.3 show that the quantitative estimates vary in a relatively wide range but the difference from the central symmetry is a general rule.

10.4.2. The fine-scale structure of the halo

A special feature of the halo (host halo) is the presence of different substructures (subhalo) and local heterogeneities [810], and the identical images are obtained for star halos which have also a complicated structure consisting of efficiently mixed components, tidal streams, cavities, arcs [483, 882]. The possibility of existence of objects with an even smaller scale has also been discussed [47].

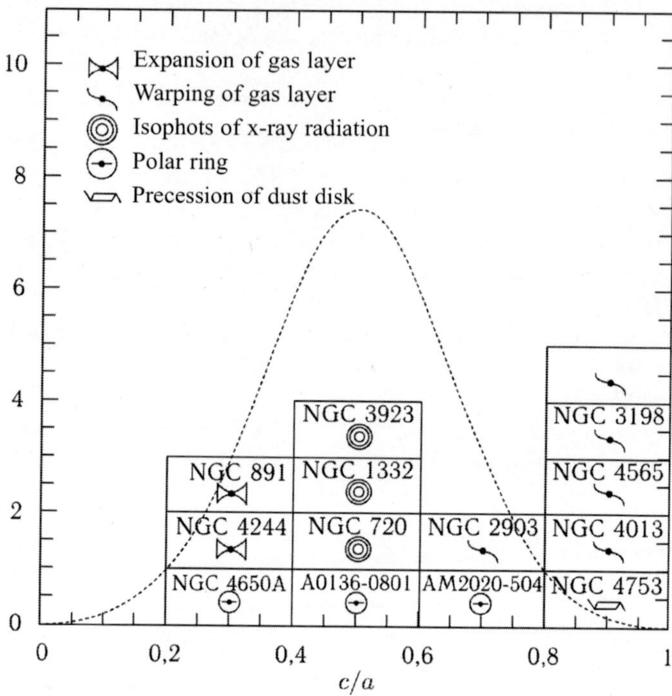

Fig. 10.21. Estimate of the parameter c/a for the dark halo for different galaxies using different methods according to [904, 991].

The total number of the subhalos in the numerical models is estimated at $\gtrsim 10^5$, and their fraction represents approximately 6% of the mass of the halo [810] and, with the increase of the spatial resolution of the cosmological models, this fraction increases. It should also be mentioned that the forms of the dark halo and the subhalo can greatly differ. This is indicated by the results of modelling CDM in which the subhalo appears to be more symmetric with the ratio of the half-axes $c/a \simeq 0.68$ and $b/a \simeq 0.83$ [810]. The halo and the subhalo in the central region greatly differ from the spherically symmetric case in comparison with the periphery. The ratios $a/b/c$ depend only slightly on the mass of the subhalo, but the degree of triaxiality of the subhalo increases at the periphery of the halo. The effect of the tidal forces in the halo–subhalo system should result in spheroidisation of the subhalo. The subhalos are also triaxial structures but in them this effect is less pronounced in comparison with the host halo. The orientation of the subhalos in the space is complicated – the major axis of the subhalos its on average directed to the galactic centre. Evidently, the spatial distribution of the concentration of the subhalo has the form $\propto 1/r^2$.

More than 200 dwarf galaxies–satellites have been found in our Galaxy. Although the luminosity of these galaxies varies in a wide range $(10^3–10^9) \cdot \mathcal{L}_\odot$, the mass of 18 objects, measured on the basis of the kinematics of the stars of these systems, is approximately $10^7\ M_\odot$ within the range of 300 pc from their centre

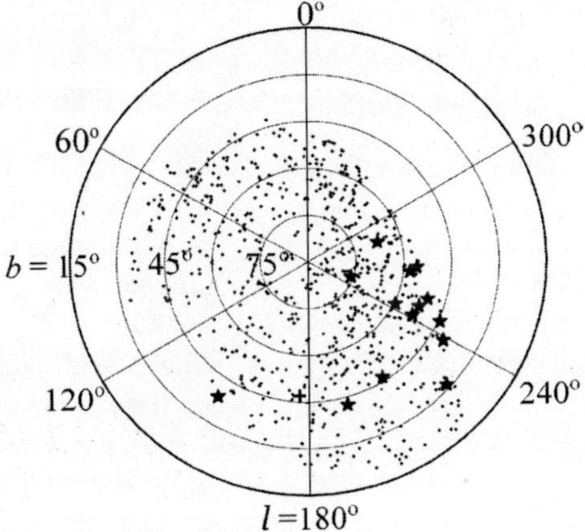

Fig. 10.22. Position of the hypervelocity stars on the plane of the galactic coordinate (l, b) (*, +) on the background of other stars from the sample [424].

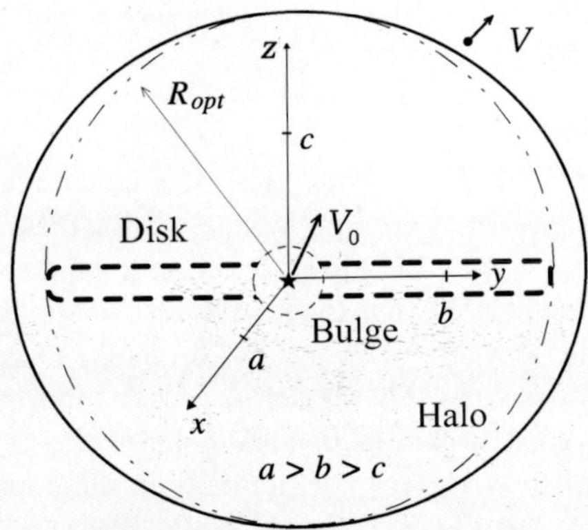

Fig. 10.23. A stellar disk and a bulge immersed in a triaxial halo. A star leaves the centre of the galaxy (the * symbol) with the velocity of $V_0 \gtrsim 1000$ km/s. Outside the limits of the optical radius R_{opt}, the velocity of the star is $V \sim 500$–800 km/s.

[1210]. Thus, the 'weakest' objects evidently consist mainly of dark matter with the mass of up to $10^7\ M_\odot$ which is probably the lower limit of the subhalo mass. Even in the immediate vicinity of 1–2 kpc from the Sun, the kinematics of the fast stars with a low metallicity indicates the possibility of the existence of the substructures of the halos in accordance with the CDM scenario [1056].

One can mention another possible representation of the fine-scale heterogeneities of the dark halo, resulting in the warping of the galactic disk as a result of interaction of the Magellan Cloud with the matter of the dark halo, where the perturbations in the clouds leading to the bending of the periphery of the gas disk [1291].

10.4.3. Hypervelocity stars

Recently, a new class of stars has been detected, the so-called hypervelocity stars (HVS) with the characteristic feature of the stars being the high velocity of motion $v \gtrsim 500$ km/s, indicating departure from the Galaxy [423]. The first HVS detected had the velocity of 853 km/s [423]. The number of the detected HVS objects is now close to 30. The typical velocities equal 600–800 km/s. One of the fastest objects is the neutron star B1508+55, whose velocity is greater than 1000 km/s [462]. The spatial position of the HVS is anisotropic (Fig. 10.22).

HVS have also been detected in other galaxies. The star HE0437-5439, genetically connected with the Large Magellan Cloud (LMC), has the velocity of 740 km/s and is situated at a distance of 61 kpc from the Sun and 18 kpc from the LMC [641]. The HVS properties have also been discussed for the Andromeda Nebula [1153].

Various mechanisms of the generation of high velocity in the HVS, associated with the dynamics of the star in the vicinity of a black hole in the centre of the Galaxy, have been considered: [677, 980, 1153]:

- disruption of the binary stellar system by a massive black hole in the centre of the galaxy;
- ejection of a star by a black hole with intermediate mass;
- scattering of the stars by a cluster of black holes with stellar mass, rotating around a supermassive black hole.

However, a star can acquire high velocity also as a result of the gravitational interaction in the system of 3–4 stars in the core of a stellar cluster.

The special interest in the HVS objects is associated with the potential possibility of investigating the properties of the dark halo, restoring the trajectory of movement of the hypervelocity star. It should be mentioned that the galactic bulges often also represent triaxial systems [902], and together with the stellar bar in the disk they form a complex single structure. The disk additionally disrupts the symmetry of the potential of the system and this results in a deflection from the strictly radial position. With the buildup of the statistical material the HVS-type stars can be regarded as markers of the triaxiality of the halo on the condition that they travel from the centre of the Galaxy [423, 623, 641] (Fig. 10.23).

10.5. Dynamics of galactic disks in the external non-axisymmetric potential

The galactic gas and stellar disks are axisymmetric only in the very first approximation. The spiral density waves, the central bridge, the non-axisymmetric (triaxial) nature of the dark massive halo, galaxies–satellites, and the triaxial bulge disrupt the axial symmetry of the 'outer' gravitational potential in which the galactic disk components are situated.

We will investigate the dynamics of a rotating gas and stars in the gravitational non-axisymmetric field. Of special interest is the effect of the global asymmetry of the potential, associated with the distribution of the matter of the dark massive halo.

It is natural to expect that the disk responds to this effect by the formation of spiral wave structures [264, 265, 356, 1242].

10.5.1. The problem of the spiral structure and the triaxial halo

We examine the hypothesis according to which the non-axisymmetric nature of the spheroidal components (mainly the dark halo) may be responsible for the formation of the spiral structures in the galactic disks, at least in some galaxies.

In the first approximation, the distribution of the volume density in a halo can be described by the following equation

$$\varrho_h(x,y,z)=\varrho_h(\xi),\quad \xi\equiv\sqrt{\frac{x^2}{a^2}+\frac{y^2}{b^2}+\frac{z^2}{c^2}},\tag{10.3}$$

where the case $a = b = c$ ensures the centrally symmetric distribution, and the model $a = b$ c determines the axisymmetric halo. The value of the scale c (or, more accurately, its difference from the values a, b) has only a small effect on the dynamics of the disk which is assumed to be situated in the plane (x, y). The nature of distribution of the mass in the dark halo is not determined convincingly by the observations. The simple model of the potential of the halo will be represented by the distribution (see other radial profiles in section 4.1.2)

$$\Psi_h(x,y,z)=4\pi G\varrho_{h0}a^2\cdot\left\{\ln(\xi)+\frac{\operatorname{arctg}(\xi)}{\xi}+\frac{1}{2}\ln\frac{1+\xi^2}{\xi^2}\right\},\tag{10.4}$$

and the Poisson equation gives the appropriate density distribution

$$\begin{aligned}\varrho_h(x,y,z)=\frac{\varrho_{h0}a^2}{\xi^2}\left\{\left(\frac{1}{a^2}+\frac{1}{b^2}+\frac{1}{c^2}\right)\left[1-\frac{\arctan(\xi)}{\xi}\right]-\right.\\ \left.-\frac{1}{\xi^2}\left(\frac{x^2}{a^4}+\frac{y^2}{b^4}+\frac{z^2}{c^4}\right)\left[2+\frac{1}{1+\xi^2}-3\frac{\arctan(\xi)}{\xi}\right]\right\}.\end{aligned}\tag{10.5}$$

In the limiting case of the symmetric halo $a = b = c$ the equations (10.4) and (10.5) yield the model of the quasi-isothermal halo (4.3).

It is sufficient to confine our considerations to the examination of the model with $a \neq b = c$. If the halo is oblate in the vertical direction ($c < b$), it is efficient to simply increase the mass of the halo because in the plane $z = 0$ the gravitational force from the halo is greater. At $a/b \neq 1$ in the plane of the disk we have a non-axisymmetric halo which in a general case may rotate with a low angular velocity Ω_h. Cosmological models permit the values $\varepsilon_h = |q-1| \simeq 0 \div 0.2$ (section 10.4.1). In a general case, we can consider the models in which the quantities a, b, c in (10.4) depend on the spatial coordinates x, y, z.

It is assumed that the angular velocity of rotation of the halo Ω_h is lower than the angular velocity of rotation of the outer edge of the disk $\Omega(R) = V(R)/R$. If the typical values of the linear velocity $V(R)$ for the galactic disks in the region of the halo are situated in the range $V_{max} \simeq 100 \div 300$ km at the distance $R \simeq 10 \div 15$ kpc ($\Omega(R) \sim 5 \div 30$ km/s/kpc), then for the angular velocity of rotation of the halo we have the estimate $\Omega_h \lesssim 5$ km/s/kpc.

Further, when examining the dynamics of the stellar and gas disks, we will use the potential of the triaxial halo in the form (10.4). Different models of the halo differ in the circular velocities V_c at large distances ($r > R_{max}$, R_{max} is the boundary of the stellar disk) (section 4.1.2). In the inner region of the galaxy we have $V_c^{(iso)} < V_c^{Bur} < V_c^{NWF}$. Therefore, the selection of the model (10.4) with other conditions being equal should be regarded as the lower estimate of the effect of the non-axisymmetric halo. Within the framework of the models of N-bodies and the gas disk we consider the principal problem of the possibility of generation of different structures in the non-axisymmetric halo [264, 265].

Various aspects of the effect of the non-axisymmetric nature of the gravitational potential on the dynamics of the galactic disk have been studied in [356, 658, 659, 884, 1242]. In particular, the triaxial halo is capable of explaining the rotation curves of LSB-galaxies [658], for example, F568-3 [1286]. The problem of the generation of the central bar and special features of its dynamics in the presence of a non-axisymmetric halo has been studied in [869]. Modelling of the disk in the triaxial halo makes it possible to produce warps of the galaxies [520, 1088]. Below, we examine in detail the morphology and kinematics of the spiral structure in the galactic disks in a non-axisymmetric massive dark halo using the results of numerical modelling in which the disk is situated in the plane (x, y) of the halo and the axis of the Galaxy is identical with the z-axis of the halo.

10.5.2. The stellar disk consisting of N-bodies in a non-axisymmetric halo

In order to exclude the effect of the gravitational instability of the disk which is capable of generating spiral waves in a symmetric halo, the initial state for the problem studied here should be represented by the distribution of the dispersions of the radial, azimuthal and vertical velocities in the disk similar to those which ensure the gravitational stability with respect to the perturbations in the disk plane and bending waves (chapter 4). We can use the iteration approach to constructing a collisionless stellar disk with a small reserve of stability (section 4.3.2 [259]). A direct confirmation of the gravitational stability of the initial state of the disk are test calculations in the symmetric halo in which the radial distributions of the main parameters of the disk (dispersions of the velocities $c_r(r)$, $c_\varphi(r)$, $c_z(z)$, the velocity of rotation V, surface density σ, the vertical scale of the disk h) retain their initial values throughout many periods of rotation of the disk.

The initial distribution of the halo mass can be represented conveniently by the centrally symmetric distribution and, subsequently, after some characteristic time τ_h, it is convenient to transfer to a non-axisymmetric halo. Since the disk is relatively thin, the special feature of the variation of the density of the halo along the z-axis does not have any significant role. Figure 10.24 shows the typical results

Fig. 10.24. Isolines of the surface density of a collisionless stellar self-gravitating disk in the gravitational field of a non-axisymmetric massive dark halo at different moments of time (logarithmic scale). The spiral structure forms rapidly in the disk which is initially axisymmetric with the exponential profile (L = 0.25 cm (1.3)). Under the same conditions in the case of the centro-symmetric halo, the disk remains axisymmetric with the duration of several tens of periods of rotation of the external edge of the disk component (of the order of 10 billion years).

of evolution of a stellar self-gravitating collisionless disk in the model of N-bodies in a non-axisymmetric halo [264].

The main results of modelling the dynamics of the stellar disk in the non-axisymmetric halo can be formulated as follows:

1. Regardless of the marked non-axisymmetry of the halo and the associated very powerful (for the stellar disk) non-axisymmetric perturbations in the form of a two-arm spiral wave, on the whole the disk in the later stages of evolution is symmetric as in the case of modelling the gravitationally unstable disk in the symmetric halo.
2. In the process of formation of the spiral structure and later during many revolutions of the disk, the parameters of the disk averaged out with respect to the azimuthal angle (surface density, three components of the local dispersion of the velocity of the stars c_r, c_φ, c_z, the velocity of rotation V, disk thickness h_z) do not show any significant systematic changes (heating of the disk or redistribution of matter along the radius). Consequently, the initial source of generation of the spiral structure is retained, ensuring the long-life spiral pattern.

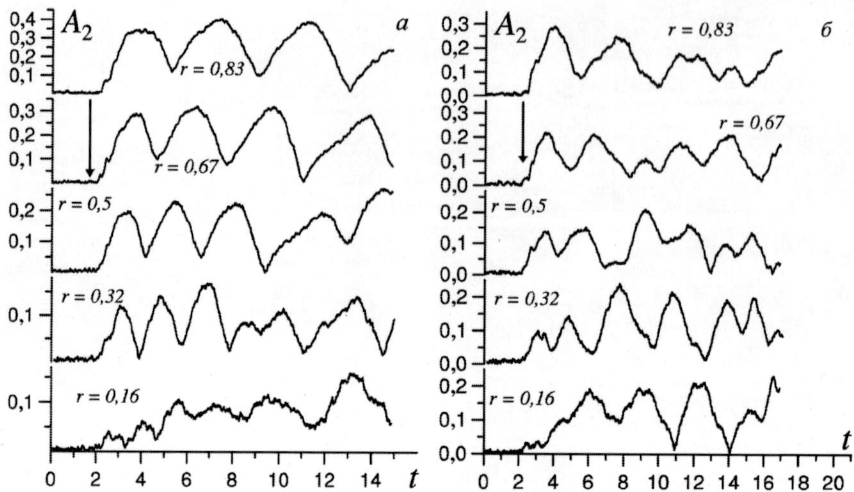

Fig. 10.25. The time dependence of the amplitude of the Fourier harmonics $m = 2$ at different distances from the centre of the disk r: a) $\mu = 4$, $a = 0.3$, b) $\mu = 1.8$, $a = 0.3$. The rotation period of the outer edge of the disk is equal to 3.

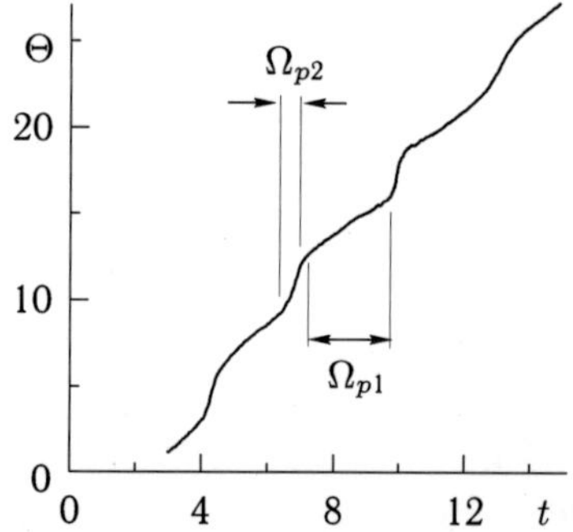

Fig. 10.26. Dependence of the wave phase on time $\Theta(t)$ at a fixed radius $r = 0.67$ (see Fig. 10.25a). There are two stages, characterised by different angular velocity of rotation $\Omega_p = d\Theta/dt$: slow rotation with Ω_{p1} for a long period of time, and a short phase with rapid rotation $\Omega_{p2} > \Omega_{p1}$.

3. The spiral pattern in the stellar disk is non-stationary. The evolution is quasi-periodic. The non-linear spiral wave rotates with the characteristic frequency $\Omega_p = \mathrm{Re}(\omega/m)$ which depends on time. At the fixed radius r in determining the azimuthal motion, the wave amplitude changes in the quasi-periodic fashion (Fig. 10.25). The phase shift at different radii is observed.
4. There are two characteristic velocities of rotation of the pattern Ω_{p1} and Ω_{p2} at each fixed radius (Fig. 10.26). For a long period of time the wave moves at a lower frequency Ω_{p1} for which the corotation radius r_{c1} is situated at the periphery of the disk or outside its limits. The duration of the phase with the rotational frequency $\Omega_{p2} > \Omega_{p1}$ is considerably shorter than with the frequency Ω_{p1}. This

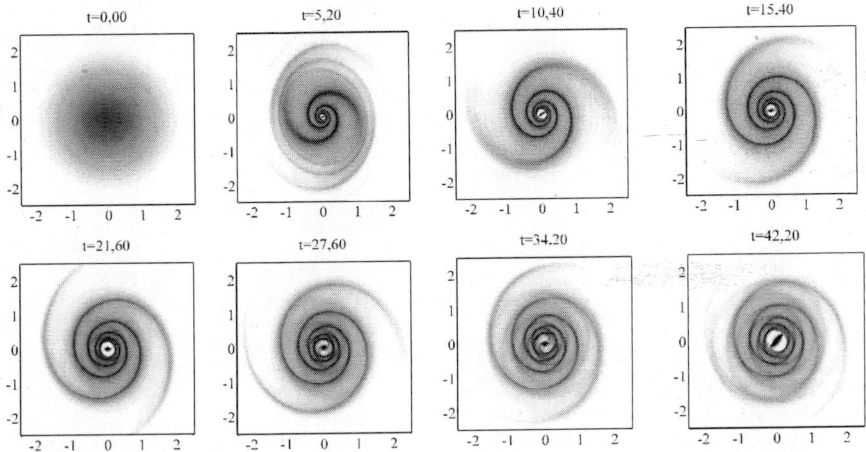

Fig. 10.27. The distribution of the logarithm of surface density in the disk plane at different moments of time in the model: $d = 0.02$, $\varepsilon_h = 0.1$, $\mathcal{M}_0 = 10$, $L_3 = 0.7$.

is determined by passage through the potential well of the halo. This nature of movement of the density wave is typical in the presence of a non-axisymmetric gravitational field of a non-rotating or a very slowly rotating halo ($\Omega_h \ll \Omega_p$). Frequency Ω_{p2} corresponds to the corotation radius r_{c2}, which is situated within the limits of the disk in the range $r \simeq (2\text{–}3)\cdot L_\sigma$.

5. The non-axisymmetric halo may support the formation of a central bridge. In the case of the formation of a powerful global bar in the disk centre (in the model shown in Fig. 10.24 its size is small $r_{\text{bar}} \lesssim L_\sigma$) the nature of evolution is greatly complicated. In particular, the perturbation with the azimuthal number $m = 3$ forms together with the mode $m = 2$.

10.5.3. A gas disk in a non-axisymmetric halo

The results of modelling of a gas disk, based on the numerical approach described iun section 4.1.3, will be discussed. Calculations were carried out using the cylindrical (r, φ, z) and Cartesian (x, y, z) high-resolution grids [264, 265]. The dimensions of the numerical grades for 2-D models were: 5000 × 5000 for the Cartesian and ($N_r \times N_\varphi = (2400 \times 720)$ for the polar grades. In the 3-D model, the grid parameters were: $N_r = 600$, $N_\varphi = 360$, $N_z = 200$. To evaluate the effect of the selected numerical methods on the disk dynamics, hydrodynamic calculations were carried out using the smoothed-particle method (section 4.1.4). SPH calculations were conducted with the number of particles $N = 10^5 \div 10^6$ and showed good agreement with the solutions obtained on the basis of the TVD grid method.

At the initial moment of time $t = 0$ the axisymmetric equilibrium disk is in the axisymmetric potential $\Psi(r, t = 0)$ with $\varepsilon_h = 0$ ($a = b$). Subsequently, during the time period τ_h the value of the scale a increased in accordance with the linear law to $\varepsilon_h = 0.001 \div 0.2$ for different models, and after 'switching on' the non-axisymmetry at the time $t > \tau_h$ the halo potential no longer changed.

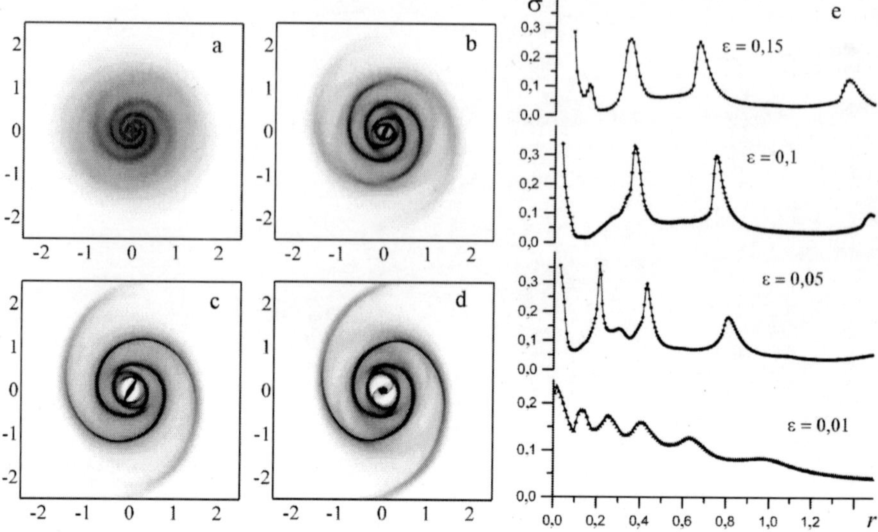

Fig. 10.28. Models with different values of $\varepsilon_h = 0.01$ (a); 0.05 (b); 0.1 (c); 0.15 (d) at $\mathcal{M}_0 = 10$, $d = 0.1$ at the same moment of time: a–d) the distribution of the logarithm of surface density in the disk plane; e) the radial profiles of the surface density for the fixed value of the azimuthal angle $\varphi = 0$ for the given values of the parameter ε_h.

The main free parameters of the model and their typical dimensionless values will be listed:

1. The degree of non-axisymmetry of the halo $\varepsilon_h = 0 \div 0.2$ ($q = b/a = 1 - \varepsilon_h$).
2. The width of the quasi-solid-state section of the velocity of rotation in the central region of the disk $d = 0.015 \div 0.5$.
3. The characteristic scales of the radial distribution of the surface density $L = 0.25 \div 1.5$.
4. The speed of sound of the gas or disk temperature, characterised by the effective Mach number $\mathcal{M}_0 = V_{max}/c_s\ (r = 0) = 5 \div 30$.
5. The relative mass of the halo $\mu_g = M_g / M_0$ (M_0 is the mass of the galaxy inside the radius $r \leqslant R = 2$).
6. The angular velocity of rotation of the halo $\Omega_h = 0\ 0.2$, which corresponds to the radius of corotation at the periphery of the gas disk.
7. The duration of transition from the axisymmetric to non-axisymmetric halo h = 0 30.

The nature of the flow in the gas disk depends greatly on the set of the values of these model parameters. However, in all cases, the generation of the spiral structure is observed, and in the main region of the disk $0.5 \lesssim r \lesssim 2$ there is a long-life quasi-stationary two-arm wave. Figure 10.27 shows the typical long-term evolution of the disk ($t = 40$ corresponds to 6 rotations of the periphery of the disk $r \sim 2$). The two-arm perturbation includes almost the entire disk and is slow with the exception of the very centre where more complex formations are detected.

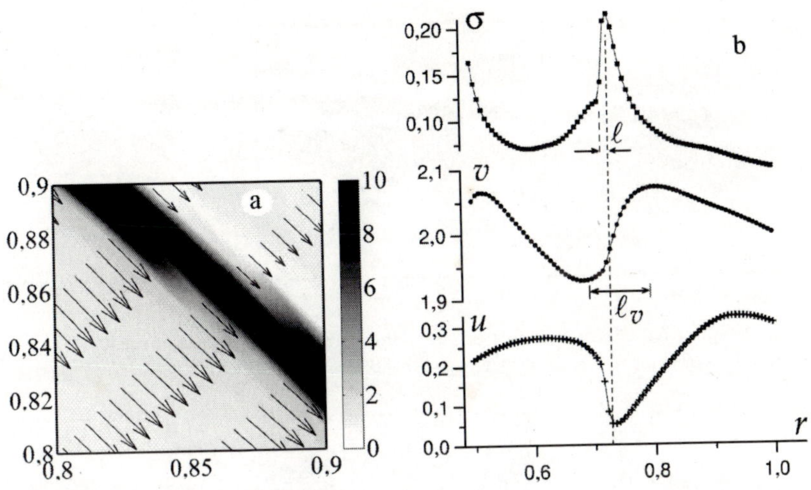

Fig. 10.29. Structure of the flow in the vicinity of the spiral wave. The model with $\varepsilon_h = 0.1$, $d = 0.02$, $L_3 = 0.7$, $\mathcal{M}_0 = 0$, $\Omega_h = 0$. a) Distribution lg((div u)²) (in grey colour shades) together with the velocity field u. b) Radial profiles of the radial velocity u, azimuthal component v, surface density σ in the vicinity of the spiral density wave. The symbols indicate the values of the functions in the nodes of the numerical grid.

Several characteristic special features of the morphology and kinematics of the disk obtained on the basis of the numerical experiments will be mentioned.

1. The presence of the non-axisymmetric halo leads the formation of the spiral structures in the gas disk (see the distribution of the logarithm of surface density in Fig. 10.28a–d). Under the typical conditions the intensity of perturbations increases during 1–3 rotations of the disk to a highly non-linear stage forming a system of shock waves (Fig. 10.8e). The time of formation of the quasi-stationary spiral structure increases with decrease of ε_h. With other conditions being equal, with decrease of ε_h the spirals become more tightly twisted with a smaller amplitude (Fig. 10.28e). The massive non-axisymmetric halo is capable of generating non-linear waves in the gas component even at small $\varepsilon_h < 0.01$, but the duration of increase in this case exceeds eight periods of rotation of the disk periphery.
2. Analysis of the structure of the gas flow indicates that it is complicated and includes a system of shock waves (SW) and regions of strong shear flows (Fig. 10.29). The non-linear (shock) waves have the large-scale structure 'trapping' the disk down to low density values $\varrho/\varrho_{max} \sim 10^{-3}$. The form of the spiral depends strongly on the model parameters but it is a large-scale formation rotating by the angle ~2π and in some cases exceeding 4π. The amplitude of the spiral waves depends on the degree of non-axisymmetry of the halo ε_h, but at $\varepsilon_h \gtrsim 0.28$ a shock wave forms. The zone of shear flow forms in the region of the spiral density wave in the vicinity of the front of the shock wave (Fig. 10.29a). The region of variation of the tangential component of velocity is larger

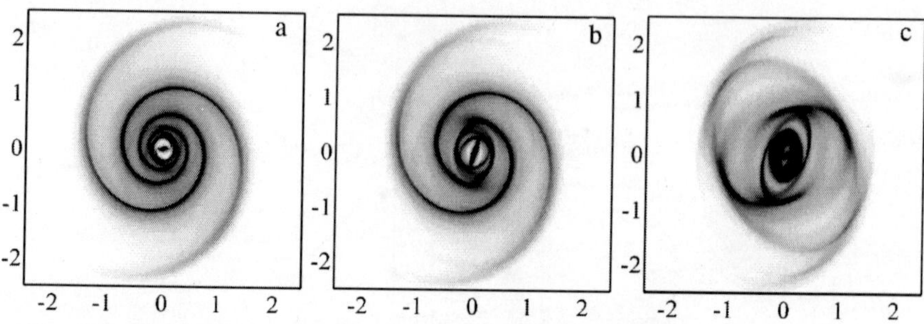

Fig. 10.30. Spiral patterns with different geometry in the central region at the same moment of time for $\mathcal{M}_0 = 10$ (a), $\varepsilon = 0.1$, in models with: $d = 0.2$ (a), 0.1 (b) and 0.3 (c).

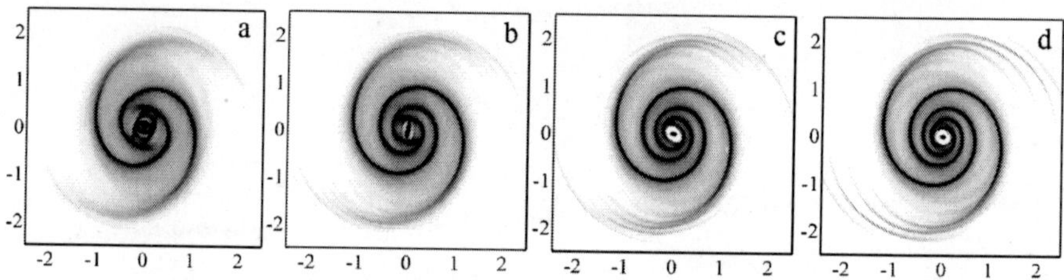

Fig. 10.31. Models with different Mach numbers $\mathcal{M}_0 = 10$ (a); 15 (b), 20 (c), 30 (d) for $\varepsilon = 0.1$, $d = 0.1$ at the same moment of time.

in comparison with the front of the shock wave whose width ℓ is determined by the properties of the numerical model (Fig. 10.29b). The values σ and u at the front of the shock wave change almost 'in a jump' (within the range of ℓ), in contrast to the component v whose range of variation is considerably larger than ℓ. The width of the shock wave front ℓ decreases with the increase of resolution of the numerical scheme (with increase of the number of cells N_r, N_φ), since the front of the shock wave is described by only 2–4 numerical cells. The width of the region of the shear flow ℓ_v is finite ($\ell_v \gg \ell$) and does not depend on the selection of the parameters of the numerical grid.

3. The geometry of the spiral pattern depends on the rotation curve of the gas $V(r)$, the nature of distribution of density in the dark halo $\varrho_h(r, \varphi)$, the radial profile of the surface density of the disk $\sigma(r)$ and the speed of sound $c_s(r)$. In particular, leading spirals can form in the central part of the disk and they can always change at the periphery to trailing spirals, forming complicated Θ-structures (nested Θ-structures can also form) in the centre (Fig. 10.30). In the figures, the gas disk rotates in the anticlockwise direction. Thus, the trailing spirals are untwisted in the clockwise direction, and the trailing ones in the opposite direction. The presence of the leading spirals in the very central zone of the disk is indicated in Fig. 10.28d, 10.30c and 10.30 4a, b.
4. The generation of the distinctive Θ-structures, formed by the leading spirals in the central region, takes place in the models of the gas disk without taking into account the perturbations of the density of the stellar subsystem. Since the Θ-structures do not form in the models of the stellar collisionless disks, the

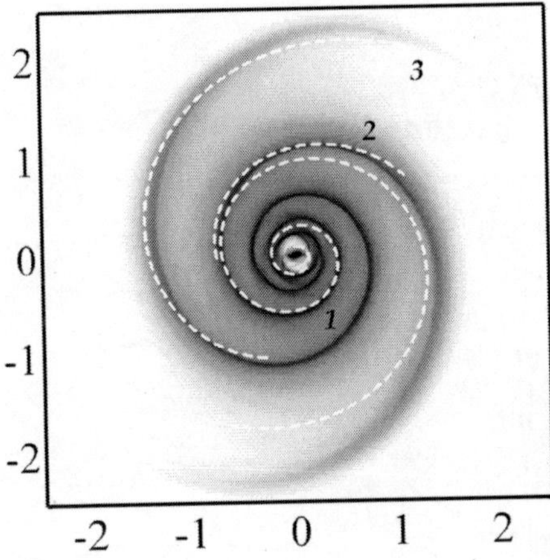

Fig. 10.32. Geometry of the spirals in the model $\mathcal{M}_0 = 10$, $\varepsilon_h = 0.1$, $d = 0.02$. On the background of the distribution of the logarithm of surface density (grey colour) the lines show the sections of the logarithmic spirals with different twist angles: 1) $i = 10^{\circ}$, 2) $i = 12^{\circ}$, 3) $i = 14^{\circ}$.

central Θ-structures in the self-consistent stellar–gas systems are less distinctive and the problem of the dynamics of the central part requires special examination. It should be mentioned that the complicated structures of the gas, reflected in the form of, for example, dust streaks inside the bar, are clearly visible on the images of the SB-galaxies (for example, NGC 1672, NGC 1300, NGC 1365, NGC 3627, etc). In addition to the gas–dust spirals from the centre there are structures formed along the major axis repeating the form of the bar (see, for example, NGC 1097, NGC 1288, NGC 2903) or the outer spirals do not begin from the end of the bar and include only part of the leading edge of the bar (NGC 4548, VCC 1615). The so-called 3-kpc arms in our Galaxy have the same form.

5. The reduction of the size of the zone of quasi-solid-state rotation in the centre of the gas disk (like the reduction of parameter ε_h) results in the generation of more heavily twisted spirals at the centre. The conditions for the appearance of the Θ-structures in the centre become less favourable. The trailing spirals already form in the very centre, affecting the entire disk (Fig. 10.30).
6. In colder disks (at higher Mach numbers) more heavily twisted and thin spirals form in the non-linear stage (Fig. 10.31).
7. The typical values of the twist angles of the spirals i are in the range $i \simeq 5° \div 20°$. Widely open patterns do not form in the investigated models. Figure 10.32 shows the example of the approximation of the resultant structure by sections of the logarithmic spiral $r = r_a \exp(\chi\varphi)$ with the twist angle $i = \pi/2 - \arccos\left(\chi/\sqrt{1+\chi^2}\right)$. The angle i for the given case increases from 10° in the central zone to 14° at the periphery. The small increase of angle i with the radius is a typical behaviour for the constructed models.

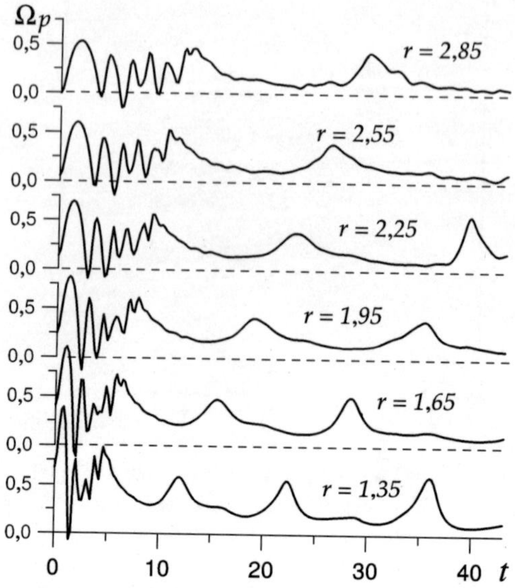

Fig. 10.33. The dependences $\Omega_p(t)$ of the two-arm spiral for the rings of a disk situated at different distances from the centre r.

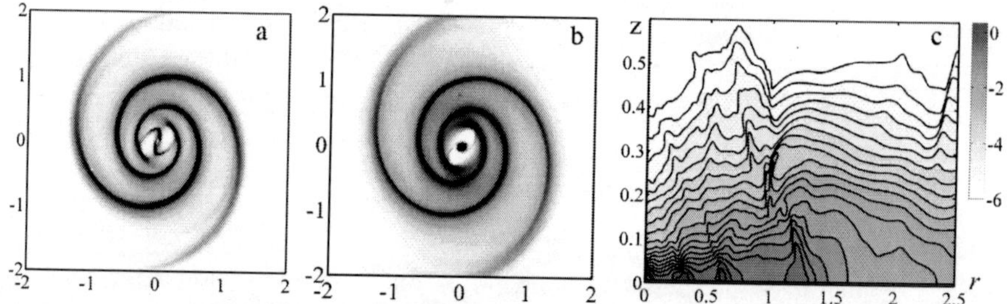

Fig. 10.34. Distribution of the surface density (in the plane (x, y)) in 3D- (a) and 2D-models (b) for: $\varepsilon_h = 0.1$, $\mathcal{M}_0 = 15$, $d = 0.125$. The vertical structure of the volume density in the plane $(x, 0, z)$ for the azimuthal angle $\varphi = 0$ (c).

8. The problem of the angular velocity of rotation of the spiral pattern Ω_p will be discussed. Figure 10.33 shows the dependences of the angular velocity of rotation of the two-harm spiral Ω_p on time for different radii r. There is the initial stage of formation of the pattern ($t \lesssim 10$, closer to the centre the rate of these processes is higher than at the periphery) and this is followed by the formation of the quasi-periodic mode of rotation of the non-linear perturbations. The angular velocity of rotation Ω_p is low and the appropriate corotation radius is situated at the periphery of the disk or outside its limits. It should be stressed that the spiral pattern also rotates in the case of the stationary halo $\Omega_h = 0$. Calculations with a slowly rotating halo $\Omega_h \lesssim 0.2$ show that the geometry of the spiral patterns

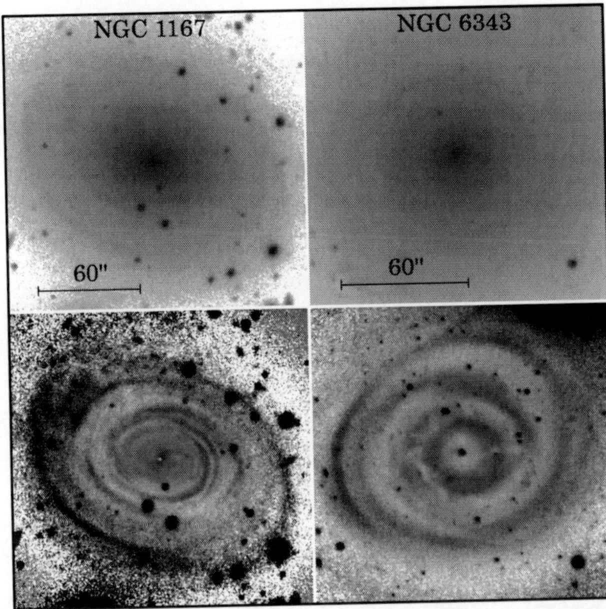

Fig. 10.35. The results of observation of lenticular galaxies NGC 1167 and NGC 6340 on the BTA [68]. The images in the filter *V* on the logarithmic scale of brightness are shown: initial images (at the top) and the difference between the observed and modelling distribution of surface brightness (bottom) are indicated.

and their kinematics depend only slightly on Ω_h, and in this case $\Omega_p > \Omega_h$. The angular velocity of the wave in the disk slowly increases with increasing value of Ω_h. For the mechanism of formation of the spiral structure described here, it is interesting to study galaxies in which the spiral structure rotates at a relatively low velocity and the corotation radius is situated at the periphery or even outside the external limits of the spiral structure as, for example, in NGC 3359 [1017] (see the examples in [434]).

9. There is a qualitative agreement between the results of two-dimensional (2D) and three-dimensional (3D) gas-dynamic calculations (Fig. 10.34). In the initial stages (several rotations) of generation of the spirals the formation of the front of strongly non-linear waves is detected with the wave propagating in the vertical direction under the angle to the plane of the gas disk. The geometry of the trailing spiral pattern in the main part of the disk has the same form in 2D- and 3D-models (see Fig. 10.34a, b). Differences are found in the central region where complicated fine-scale structures can form. In this case, the presence of the vertical measurement should greatly change the dynamics of the gas because the speed of sound in the centre is known to be not low in comparison with the velocity of rotation of the system, and the significance of the vertical motions is very high (see discussion in 3.1 and 3.2).
10. Taking self-gravitation into account as a whole increases the efficiency of the formation of the spiral pattern of the non-axisymmetric halo.

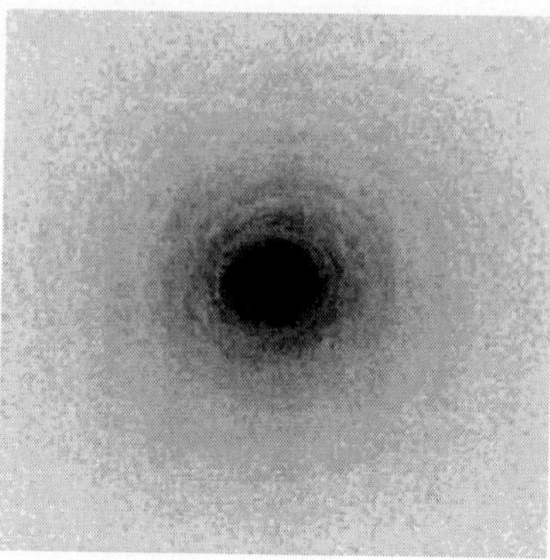

Fig. 10.36. The image of NGC 524 (HST according to [1166]) shows the thin almost axisymmetric spirals, associated with young stars.

10.5.4. Thin gas spirals in lenticular galaxies

The gas content in the galaxies of different types S0–Sa is low but in some cases there are heavily twisted very narrow spiral patterns almost without any star formation.

A distinctive example is NGC 1167. The images, obtained in BTA [68], show a system of ring-shaped arcs or spirals (Fig. 10.35). The spirals are surprisingly thin and smooth without any characteristic unevenness and distinctive nodules which usually make the spiral arms looking like dust and local regions of star formation. This indicates that they do not contain O-stars forming the bright regions HII. The image of the galaxy NGC 6340 (type S0/a) shows a bright internal part of the disk (lens) and low-contrast blurred external spiral arms (Fig. 10.35). Tidal veins in the central part of the disk are clearly visible. However, special attention should be given to the thin and low-contrast odds and ends of the rings or spiral structure similar to those detected in NGC 1167 but enclosed in a smaller zone of the disk. Another suitable example is NGC 524.

The nature of these thin and smooth arms has not as yet been explained. They cannot be caused by gravitational instability mainly due to the small surface density of the gas. Also, the small radial wavelength excludes perturbation in the collisionless old stellar disk. At all distances from the centre r the gas density remains considerably lower than the critical value required for the gravitational instability of the gas layer and this also explains the absence of the observed regions of star formation in the galaxy. In a number of studies, the similar structures in the central region are interpreted as 'acoustic spirals without star formation' (for example [303]) whose existence is associated with the bar. It should be mentioned that no bar was found in the NGC 1167 and NGC 6340.

The slightly non-axisymmetric halo may ensure the generation and maintenance of heavily twisted very narrow spiral patterns in the gas in the absence of spiral perturbations of the density with a high amplitude in a hot stellar disk formed from old stars. The high Mach number $\mathrm{M} = V/c_s$ supports the formation of heavy twisting of the waves. It should be mentioned that the NGC 1167 galaxy is one of the most rapidly rotating galaxies, its maximum velocity of rotation V_{max} = 400 km/s [68], which gives M $\simeq$ 40 for the speed of sound c_s = 10 km/s. The maximum velocity of rotation of the NGC 6340 is $V_{max} \simeq$ 220 km/s [68]. The identical fine structure was detected in the lenticular galaxy NGC 524 (Fig. 10.36), and the observations also indicate the high velocity of rotation $V \gtrsim$ 340 km/s [1166].

Figure 10.37 shows the results of modelling of the spiral waves of the gas disk in a non-axisymmetric halo. The small difference of the parameter q in the model (10.4) in comparison with unity ($q = 0.999 \div 0.97$) appears to be sufficient for the formation at long times of heavily twisted thin spirals in the gas, but the massive stellar disk remains axisymmetric. At a high non-axisymmetric form of the halo ($q \lesssim 0.95$) the spirals (more open ones) start to appear in the stellar disk and this has a strong effect on the dynamics of the gas subsystem – the geometry of the waves in the gas is adjusted under the spiral pattern in the stellar component. At high Mach numbers and low non-axisymmetric form of the halo ε_h it is possible to obtain almost axisymmetric waves with a small radial wavelength (Fig. 10.38). The rotation curves in these models are regarded as typical for the galaxies of different types – with the azimuthal velocity rapidly reaching a plateau.

10.5.5. The spiral structure outside the limits of the optical radius

We discuss interesting observations of powerful spiral structures at the periphery of the galaxies outside the limits of the optical radius R_{opt} where the external spiral pattern is detected with respect to HI and/or in the ultraviolet range. According to the GALEX data for the NGC 5012, the outer spirals extend over distances several times greater than R_{opt} and the same situation is also detected for the spiral pattern with a spectral HI in NGC 2915, NGC 5055, NGC 6744, NGC 6946 (Fig. 10.39, [405]).

These remote outer spirals in the gas component can be explained by the spiral wave of the stellar density or bar. Because of the low density of the gas at the periphery, it is evident that the gravitational instability of the gas also cannot ensure the formation of a global spiral pattern. A realistic hypothesis is the one according to which the non-axisymmetric halo may be responsible for the formation of the spiral arms in the gas far away outside the limits of the stellar disk.

10.5.6. Spheroization of the halo

The numerical calculation results, presented above, showed that the triaxial halo (non-axisymmetric in the disk plane) is capable of leading, in a galactic disk, to the generation of a spiral pattern whose properties depend strongly on the disk parameters and the halo. The problem of the responsibility of the non-axisymmetric halo for the formation of the real spiral pattern in the galaxies grand design is of course far from being solved. The actual pattern is greatly complicated by the formation of transient spirals, the central stellar bridge, or global modes as a result of the development of

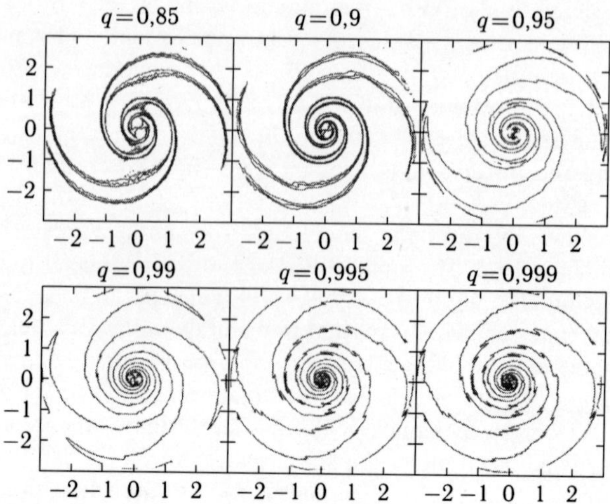

Fig. 10.37. Perturbations of the surface density in the gas disk in the presence of a non-axisymmetric halo at a different values of the parameter q.

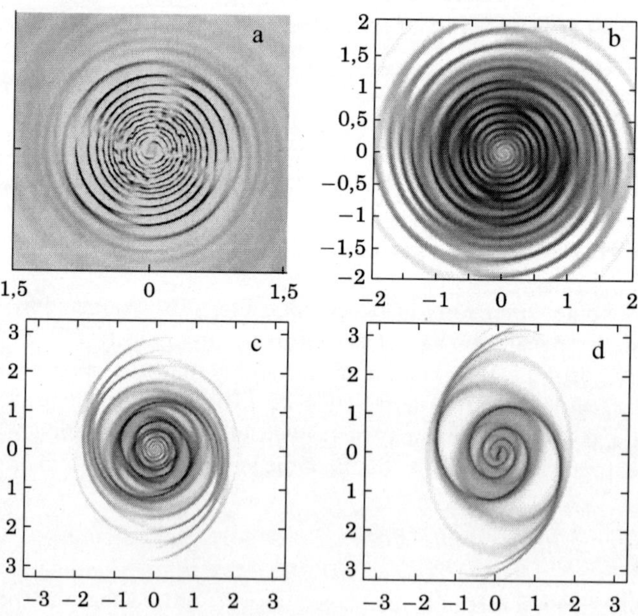

Fig. 10.38. Structure of a gas disk without self-gravitation at different values of the parameters of the model: a) $\varepsilon_h = 0.01$, M = 25; b) $\varepsilon_h = 0.01$, M = 30; c) $\varepsilon_h = 0.05$, M = 30; d) $\varepsilon_h = 0.1$, M = 30. The surface density logarithm is shown. The darker colour corresponds to higher values of the surface density of the gas. The initial radial profile of surface density: $\varrho(r)$ = const (a); $\varrho(r) = (1+ (r/1.5)^2)^{-5/2}$ (b–d).

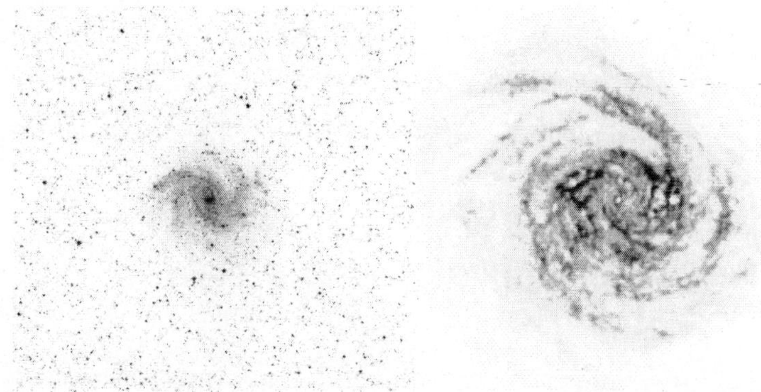

Fig. 10.39. Images of NGC 6946 on a single scale: according to the DSS data $R_{\text{opt}} = 10$ kpc (left), the distribution HI (right) [405]. The spirals in the gas reach the distance $\gtrsim$25 kpc from the centre.

gravitational instability [790]. In any case, the models considered here can be used to produce some restrictions on the form of the dark halo in the plane of the galactic disks. If the properties of the observed spiral pattern in the galaxy are incompatible with the modelling results, this may indicate the high degree of axisymmetry of the dark halo in the disk plane.

For example, for lenticular galaxies with a low gas content where the spiral perturbations in the stellar disk from old stars in the bands *I*, *K*, *R* are detected on the noise level, the numerical experiments give the limiting values of the magnitude of the deviation of the distribution of the mass in the dark halo from the axial symmetry $\lesssim$2–3%.

Almost all cosmological models yield a triaxial dark halo with the parameters $s \sim q \sim 0.5\text{–}0.8$ cm (section 10.4.1). The results of modelling of the stellar and gas disks with a highly non-axisymmetric potential indicate the need for the generation of powerful global two-arm spiral structures (sections 10.5.2, 10.5.3). Evidently, this mechanism cannot be regarded as universal because of the very large number of the morphological properties of the observed spiral pattern. In particular, the spiral structure of the S0-galaxies is almost impossible to detect. In addition to this, the spirals, generated by the non-axisymmetric halo have characteristic kinematic special features: rotation of the pattern is nonstationary $\Omega_p(t)$ (Fig. 10.26, 10.33), and during a long period of time the rotation is slow because the corotation radius is situated at the disk periphery.

A number of observations permits the presence of a slight non-axisymmetry of the dark halo ($q \simeq 0.9\text{–}1$, section 10.4.1) but contradict the results of cosmological modelling indicating the more extensive non-axisymmetric nature. Evidently, the formation of the triaxial halo with the typical values $s \sim q \sim 0.5\text{–}0.8$ and the formation of a galactic disk are followed by further evolution of the dark halo in the direction of spheroidisation, at least in the disk plane. This may be caused by the strong interaction of the non-axisymmetric halo with the disk components in early stages of the dynamics of the system.

11

Dynamics of accretion disks

The term accretion disk (AD) is used normally for describing a gas disk around a massive (in comparison with the disk) compact object. The latter includes white dwarfs, neutron stars, black holes. The main feature of the accretion disk is the transition of the gravitational energy in accretion (fall) of matter into heat followed by radiation. The term 'accretion disk' is often used in a wider meaning of the word taking into account the gas disk for which the process of fall of matter on the accumulating centre is important, for example, when discussing the protostar, protoplanet and galactic disks.

Thus, from the viewpoint of the physical processes, many results relating to the gas disks, fully relate to the accretion disk. Therefore, the difference between this chapter and the previous chapters is highly conditional and associated mainly with the type of external potential. For the disks, rotating around compact objects with mass M_1, it is Newtonian for a point body ($\Phi \propto -M_1/r$), and self-gravitation of the gas is insignificant, with the exception of the disks around the supermassive black holes. In the case of the galactic gas disks, the external potential is determined by the distribution of stellar matter (see (2.44)) and the self-gravitation often appears to be a very important factor.

One of the key problems of the physics of accretion disks is the problem of the mechanisms of removal of the angular momentum referring to the fall of matter on the gravitating centre. The advances in this area are based in most cases on viscous models of accretion disks (AD) [311, 312, 344, 393, 1002, 1147, 1179, 1219). Evidently, the first study of the mechanism of transfer of the pulse momentum as a result of viscosity was carried out by von Weizsacker in 1948 [1293]. A second possibility is the removal of the angular momentum by means of large-scale spiral shockwaves which can form during the flow of the gas in closely packed binary systems taking into account the gravitational effect of the donor star [23, 385, 1200, 1201].

The viscous models of the accretion disks, which are based on the hypothesis on turbulent viscosity (α-model proposed by Shakura and Sunyaev [272, 1147] and its numerous modifications) can be used to explain many observations in the closely packed binary systems, active galactic nuclei, protostar disks, but the satisfactory non-phenomenological model of turbulent viscosity is not yet available. The viscosity can be very important in the gas galactic disks [42, 222]. It should be mentioned that regardless of the achievements obtained in recent years [330, 393, 944, 1012, 1222, 1241], the active examination of turbulence in the accreting systems is only starting and the main results have not as yet been obtained.

There is a large number of studies concerned with the linear analysis of the stability of viscous accretion disks [43, 169, 261, 262, 283, 290, 329, 330, 389, 442, 765, 838, 895, 923, 1148, 1219, 1278, 1279]. However, the problem of consequences of the development of instabilities in highly non-linear stages has been studied only recently with the framework of numerical hydrodynamic modelling [263, 282, 288, 331, 705, 934, 1241].

The problem of convection in the gas disks is stressed in the studies by von Weizsacker [1293] and Safronov [221]. The development of ADAF-models (advectively dominant accretion flows) for explaining the low luminosity of x-ray binary and active galactic nuclei with black holes [281, 707, 957] has resulted in the construction of CDAF-models (convectively dominant accretion flows) in which flows form at a low viscosity with strong turbulence as a result of the development of convective instability [331, 709, 955, 1209]. Another example of vertical convection in the gas disk is the instability in the radiation-dominant region [288, 389]. When taking into account the ionisation of hydrogen, we obtain unstable vertical distribution of temperature, resulting in convective heat transfer [720, 908, 1179].

The characteristic feature of the accreting systems is the wide spectrum of non-stationary manifestations (dwarf novae of different types: SS Swan; SU Big Bear, Z Giraffe; X pulsars, x-ray bursters of I and II types; quasi-periodical oscillations of x-ray sources; variability of radiation in the candidates for black holes: Cyg X-1, GX 339-4, Cir X-1). To explain some variabilities of radiation, it is important to consider the radiation of non-stationary flows determined by the development of magnetic hydrodynamic (MHD) flows.

The presence of a magnetic field greatly complicates the hydrodynamic flows in the gravitational field of the compact object where the variety of instabilities of plasma is evidently not smaller than for the thermonuclear plasma [138]. The magnetic fields play a significant role in the dynamics of a number of accretion-jet systems in the vicinity of compact stars. In particular, they influence the structure of jet and the nature and rate of accretion [27, 393, 944, 1012].

Investigation of the physical processes, taking place in the accretion disks, is a very complicated task, and the description of these processes has been the subject of a large number of monographs and original articles (see, for example [342, 419, 556, 577, 590, 924, 1046, 1147, 1146, 1147, 1283]). Without mentioning the problem of detailed description of the theory of accretion disks, in this chapter we will discuss only several problems of their stability and turbulence.

The active investigation of turbulence in the accretion systems should lead, in particular, to the construction of non-phenomenological models of turbulent viscosity [20, 169, 393, 484, 572]. The important fraction of the gas in the galaxies, protoplanets and protostar disks and the major fraction of the gas in the accreting systems represent completely or partially ionised plasma. Plasma can contain a large number of unstable modes whose development effectively turbulises matter. It can be said without exaggeration that turbulence is the natural state for plasma. Many other mechanisms can also operate at the same time and lead to turbulence through the non-linear stage of instabilities.

The energy of radiation of accretion disks is in the range from 6 to 30% of Mc^2, where M is the mass of the accretion disk, c is the speed of light (compare with 0.3% Mc^2 for nuclear reactions in the interior of the stars). At the same time, the accretion disks are objects with maximum efficiency in the group of all other objects in the universe. The accretion disks are characterised by the anomalously high coefficient of turbulent viscosity (its value is 9–11 orders of magnitude greater than the coefficient of molecular viscosity), which transforms the kinetic energy of the plasma, rotating in the disk, to radiation energy. The problem of existence of developed turbulence in the accretion disks is of considerable importance.

In section 11.2 of this chapter we examine the models of axisymmetric accretion disks based on the assumption of the transfer of the angular momentum on the disk as a result of viscosity. Section 11.3 examines some magneto-gas-dynamic instabilities which can play, in specific conditions, a significant role in the accretion disks. The interest in this type of investigation is associated with the fact that the instabilities may lead to effective stabilisation of matter in the accretion disk, determine the mechanism of penetration of plasma to magnetised compact objects and result in different types of non-stationary processes. Alternative results, associated with the possibility of formation of spiral waves (including shockwaves) in gas disks in a binary system, are also presented.

Finally, section 11.7 describes the developed turbulence on the basis of the instability of super-reflection on the example of close-packed binary stars [243, 379–382, 386, 419].

11.1. Problems of accretion on compact objects

11.1.1. Close binary systems (CBS)

The process of accretion may take place most effectively in binary systems[1] where one of the stars transforms to a compact relativistic object (white dwarfs,, neutron stars, black hole) and the other one fils its Roche region [211]. At the present time, it is believed that these systems formed inevitably in specific stages of evolution of relatively close binary systems[2] [211, 1047]. Through the internal Euler point (E_1, see [572], and also below), the matter leaves the normal optical star and is situated in the Roche region of the compact object. As a result of the presence of the angular momentum, the gas does not fall down immediately and forms a ring rotating around the relativistic star which in the presence of effective mechanisms

[1]This was discovered by Ya.B. Zel'dovich [79].

[2]The term 'close binary system' usually refers to the presence of mass exchange between components.

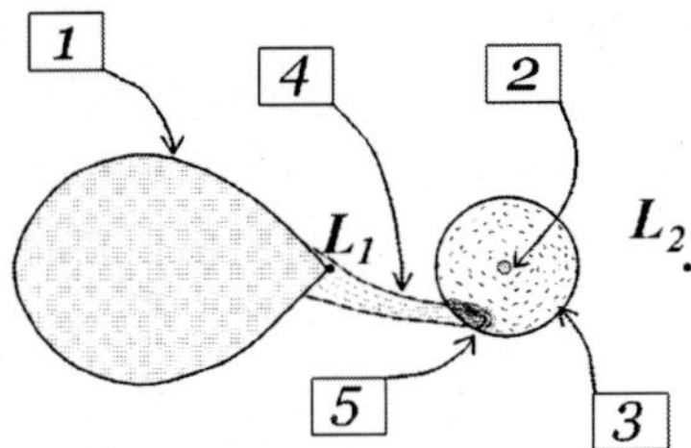

Fig. 11.1. Schematic representation of the structure of the close binary system in the disk accretion mode: 1) optical star; 2) compact star; 3) gas disk; 4) the jet; 5) the region of the impact front, manifested in a bright (hot) line.

of momentum transfer may spread into an accretion disk. The gas jet, leaving the star through the internal Euler point (or libration) impacts on the disk with the generation of a large amount of energy. This region of the accretion disk is observed as a bright line. The structure of this type of accreting systems is shown schematically in Fig. 11.1.

The accretion disks in the close binary systems are not rare objects in the universe. In fact, they are found quite frequently in our Galaxy. The reason for this is that approximately 80% of all stars are formed in pairs (see, for example [132]). Having a large mass, the star travels faster (in comparison with the star with a lower mass) through its 'life path' transforming to a compact object. Depending on the mass, the final product of evolution can be a wide dwarf, a neutron star or a black hole. Its companion – the second star – increases in size in the same period of time. When the radius of the star reaches the libration point at which all the forces with the exception of the pressure gradient are balanced, the process of flow and fall (accretion) of matter on the compact object starts. Therefore, the latter is regarded as the accretor (we will also refer it to as star 1), and the star losing the mass is a donor or star 2 (Fig. 11.2).

It is well-known that the compact objects have the following radii [273]:

1) for a black hole $R_{\mathrm{BH}} = R_g = 2GM_1/c^2$;
2) for neutron stars $R_{\mathrm{NS}} \simeq (1 \div 2) \cdot 10^6$ cm;
3) the radii of white dwarfs are approximately 100 times smaller than the radius of the Sun $R_{\mathrm{WD}} \simeq 7 \cdot 10^8$ cm, i.e. the radius is approximately equal to the Earth radius.

If the relativistic star does not have a magnetic field, it may be assumed that the internal boundary of the accretion disk r_1 extends to the surface of the star. The magnetic field may disrupt the accretion disk at distances greatly exceeding the radius of the accreting object. According to the data published in [1042, 1239, 1294], the magnetic fields in the neutron stars may reach $(2 \div 6) \cdot 10^{12}$ G. Pringle and Rees [1047] noted that the disk may fail under the effect of the magnetic field at distances of the order of the Alvén radius $r_1 \sim 10^8 \div 10^9$ cm.

The values of the radius of the external boundary of the accretion disks R are determined by the parameters of the binary system, in particular by the orbital period P and the mass of the optical star M_1, the rate of its mass loss, and also by the

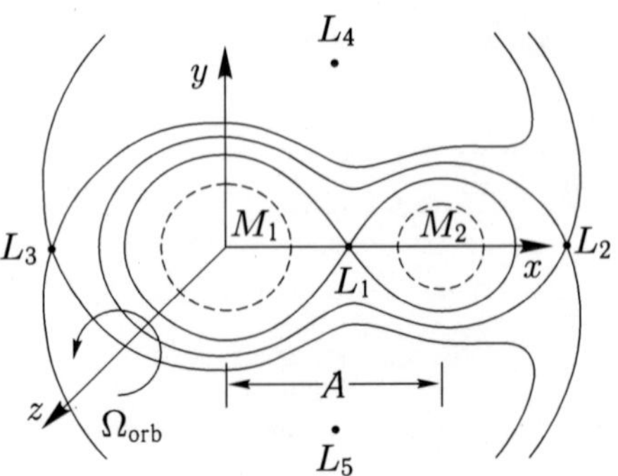

Fig. 11.2. Equipotential Roche surfaces in the equatorial plane of the binary system with the mass ratio $q = M_2/M_1 = 0.5$. The libration points are denoted as L_1–L_5.

processes taking place in the accretion disk [44]. As a result of the rapid development of physical conditions in the binary close systems, there are many situations which can be encountered, and the accretion disk may occupy almost the entire Roche cavity of the compact objects. The distances between the components can vary greatly, $0.2 \lesssim a/R_\odot \lesssim 2000$. An important parameter is the ratio of the mass of the components[3]:

$$q \equiv \frac{M_2}{M_1}. \tag{11.1}$$

Usually, the system of binary stars is investigated using the Roche approximation, based on the following assumptions:

1) the gravitational field of each star is strictly spherically symmetric;
2) the orbits of each star are circular around their mass centre;
3) intrinsic rotation of the stars is synchronised by tidal forces with their orbital rotation is similar to synchronisation of rotation in the Earth–Moon system (consequently, only one side of the Moon surface can be observed).

We will consider a system of binary stars with the distance between the star centres A, the angular velocity of orbital rotation Ω_{orb} and the period of orbital rotation $P_{orb} = 2\pi/\Omega_{orb}$. The components of the system – stars 1 and 2 – have the mass M_1 and M_2, respectively. We use the right coordinate system rotating synchronously with a system of the binary stars. The origin of the coordinate system is placed in the centre of the star 1, the axis x is directed along the line connecting the centres of the stars, and the axis z along the axis of rotation (see Fig. 11.2).

In accordance with classic mechanics [113], the effect of rotation can be described using an additional term in the expression for the energy of the rotating body with mass M. This term, which depends only on the coordinate of the body, is proportional to the square of the angular velocity of rotation: $-(1/2)\, M\, [\Omega_{orb} \times (\mathbf{r} - \mathbf{r}_c)]^2$, where $\mathbf{r}_c$ is the radius vector of the centre of the mass of the binary system.

[3]Different notation $q = M_1/M_2$ is used sometimes in the literature.

This additional term is referred to as centrifugal energy [113]. The gravitational potential of the field of forces in the previously mentioned coordinate system has the following form:

$$\Phi = -\frac{GM_1}{\sqrt{x^2+y^2+z^2}} - \frac{GM_2}{\sqrt{(x-A)^2+y^2+z^2}} - \frac{\Omega_{orb}^2}{2}\left[\left(x - \frac{AM_2}{M_1+M_2}\right)^2 + y^2\right], \tag{11.2}$$

where G is the gravitational constant.

The examples of the equipotential surfaces, described by the equation (11.2), are shown in Fig. 11.2. For high values of $|\Phi|$ the surfaces have the form of two spheroidal cavities surrounding the gravitating masses M_1 and M_2 (the dashed circles in Fig. 11.2). With the reduction of $|\Phi|$ the surfaces expand and touch each other at the point referred to as the first libration point.

It should be mentioned that in the astrophysical literature, the first libration point is referred to as the internal Lagrange point L_1, although this contradicts the historical facts because the first three libration points (L_1, L_2, L_3) where discovered by Euler in 1767. Five years later, in 1772, Lagrange discovered the final two libration points L_1, L_5 which in stellar mechanics are referred to as the triangular libration points or Lagrange points.

The typical feature of the libration points is that the sum of all three forces in them – two gravitational attraction forces to the masses M_1 and M_2, and the centrifugal force – is equal to $\nabla\Phi = 0$. The first three libration points are unstable: a test particle, placed at these points, leaves them at any small perturbation. At the same time, since the Lagrange points L_4 and L_5 are stable at the ratio of the masses of the components of the close-packed binary system $q < 0.04$ [1217] and therefore the particles can accumulate in them. This phenomenon has been observed in the solar system since 1906 when a family of Trojan asteroids was discovered in the vicinity of the triangular libration points of the Sun–Jupiter system. However, it should be mentioned that in a limited problem of three bodies, associated with the dynamics of the solar system, the condition $q < 0.04$ is always fulfilled and the appropriate points L_4 and L_5 are stable. The reversed inequality $q > 0.04$ is fulfilled for the close binary systems. This means that in the close binary systems all five libration points are unstable

The region bounded by an equipotential surface passing through libration point L_1 is called the Roche lobe. This area plays an important role in the physics of binary stars as the size of the star can not can significantly exceed its radius. When one of the stars fills its Roche lobe, the matter of the star under the action of an unbalanced pressure force begins to flow through the neighborhood of L_1 in the area of the Roche star–accretor. If the radius of the accretor is small, a significant proportion of inflowing material, having an excess of angular momentum, forms the so-called accretion disk around the star–accretor.

Observations of close binary stars (especially in a short-wave spectral range) indicate a significant flow of matter on the accretor. This is possible only if the the outer parts of the disk receive a greater share of the angular momentum of the accreting plasma. This requires huge quantities of viscosity – 9–11 orders of magnitude greater than molecular viscosity, which is possible only in the case of highly developed turbulence in the accretion disk. As a mechanism for removal angular momentum we have considered not only developed turbulence, which may cause severe instability, but other physical processes. Here are the main ones: tidal interaction [511, 842, 1007], spiral shock waves [909, 1118, 1202, 1216], convection [388, 841, 999, 1149], wind from the accretion disk and jets [394, 447, 858, 856], transfer of angular momentum to the propagating waves [840, 857, 1250], as well as various instabilities of the accretion disk – magnetohydrodynamic [35, 329, 460, 1307], parametric [629, 859], baroclinic [440, 779, 832]. Analysis of different methods of transferring angular momentum (see, for example, reviews [328, 1006] shows that all the mechanisms encounter some difficulties in explaining the properties of accretion disks.

The most famous and widely used mechanism is magnetorotational instability (MRI) [35]. However, its implementation requires a magnetic field. From the observations we know that there are many systems in which the field is so small that it does not manifest itself. To implement the MRI it is necessary that the field had a significant value not only on the surface of the star–accretor but also at a significantly greater distance of the order of the radius of the accretion disk. There are no CBS and, hence, the mechanism of MRU cannot be applied here.

We note one important fact – the mechanism of transfer angular momentum must work in disks with high temperature. The point is that at the time of formation of the disk the density of disk matter is small and, therefore, the process of radiative cooling is ineffective. At the same time, the star–accretor typically has a high temperature and, hence, the equilibrium temperature of the disk material at the initial time will also be high.

Section 11.5 describes the proposed [572] effective mechanism for the transfer of angular momentum, which can operate in hot disks of close binary systems in the absence of a magnetic field.

Observable manifestations of CBS are very diverse, primarily because different objects are very different from each other as regards their physical conditions and, of course, it is nor possible to build any universal model of AD. Therefore, we briefly describe, on the basis of observational data, some types of CBS in which the presence of gas disks is highly likely, and they can be responsible for some of the observed features.

11.1.2. The stars of the U Gemini type, or dwarf novae

The distinguishing special features of these objects is the quasi-periodic flash activity with the characteristic times T_U greatly exceeding the orbital period P. At the present time, more than 300 objects of this type are known [44]. The dwarf novae (U Gemini type stars) are usually characterised by their photometric characteristics.

- The objects of the SS Swan type (typical representatives U Gem, SS Cyg, BV Cen, RU Peg, VY Aqr, SS Aur, T Leo, VZ Aquarii, etc)). They are characterised

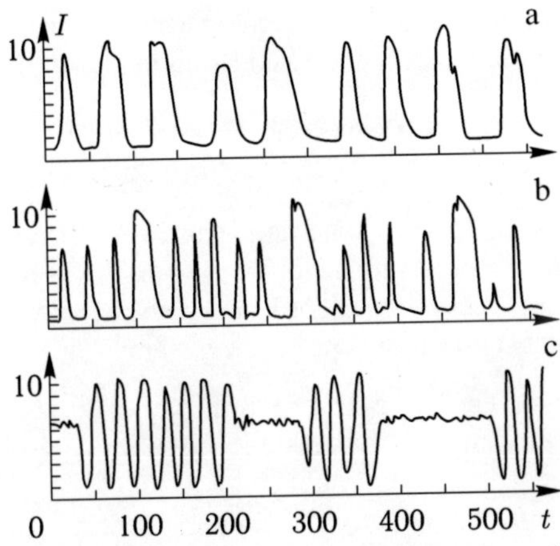

Fig. 11.3. Visual curves of luminosity of dwarf novae different types (1 – intensity in arbitrary units): a) the stars of the SS Swan; b) the stars of the type SU Great Bear; c) the stars of the type Z Giraffe.

by a relatively long period of rest between flaress (Fig. 11.3a). The duration of the cycle in the type SS Aur objects can differ: approximately 43 for VZ Aquarii, 57 days for SS Aur, 103 days for U Gem, 558 days for EX Hya..

- The distinguishing special feature of the flares in the objects of the type SU Great Bear (TU Men, YZ Cnc, SU UMa, VW Hyi, HT Cas, etc.) is that in addition to the usual shine maxima after 3–10 cycles we observe the so-called super maxima distinguished by stronger shine and duration (Fig. 11.3b). The stars of the type SU UMa have been reviewed by Charles [461].
- In the stars of the Z Giraffe type the rest period is comparable with the flare period. The mean length of the cycle is 9–40 days. However, there are also long periods of time during which there is no activity (Fig. 11.3c).

A subgroup of dwarf novae, characterised by a high amplitude of flares (6–9 stellar units) is often mentioned, and they are referred to as the stars of the type WZ Sge. Regardless of quantitative differences, there are no principal differences between the stars of the type WZ Sagittae and SU Big Bear, and the dwarf novae of the type WZ Sge are the limiting case of the stars of the type SU UMa [987].

Outside large surges, the photometric behaviour of the stars of the type U Gemini is very complicated. There are both irregular fluctuations of shine and periodic ones, determined by orbital motion. It is also important to note that there is no strict correlation between the duration of the quasi-periodic cycles T_U, the amplitude of variability of shine and the orbital period P.

In addition to the observed non-stationary behaviour of luminosity, some dwarf novae showed the dynamics of the external radius of the disk R during a flare. For example, in the case of the U-Gem in the non-active state (between flares), the value R is $0.3a$, increasing to $R \simeq 0.4a$ during a flare and subsequently returning to the initial value. The behaviour of Z Cha and OY Car is very similar [481, 482, 986, 1265]. An interesting detail in the behaviour of $R(t)$ was reported by O'Donoghue

[986] – at the very beginning of a flare there is a short fast reduction of the value R (the disk is comprised) and this is followed by an increase of the size of the disk.

11.1.3. New stars

The new stars are characterised by the rapid increase of the shine (by 6–13 stellar units), and the bright state exists for several days and this is followed by a reduction lasting usually several weeks[4]. In accordance with the characteristic times of the shine curve, there are usually four types of new stars:

1) Na – fast new stars. The increase of luminosity is very rapid (within $5{\cdot}10^5$ s). After reaching a maximum, the reduction of luminosity by 3^m does not last longer than 10^7 s.
2) Nb – slow new stars. The reduction of shine by 3^m lasts more than 100 days.
3) Nc – very slow new stars. This group includes, for example, the previously mentioned RT Ser.
4) Nr – recurrent novae characterised by recurrent flares, separated by decades.

It is believed that all the novae are in the final analysis recurrent.

Usually, several years after the maximum shine, a gas envelope forms around a new star propagating at a velocity of $\sim 100 \div 1000$ km/s. The masses of the envelopes are estimated at $10^{28} \div 10^{29}$ g. All this indicates that a strong explosion took place on the star as a result of which the system loses $0.001 \div 0.01\%$ of its mass and radiates the energy of $\sim 10^{45}$ erg.

11.1.4. Systems with white and red dwarfs

In most cases, it is not possible to establish directly from the observations the dual nature of cataclysmic variables (which also include dwarf novae and novae), and the complicated shine curves were considered on the basis of the following modelling representations. The new stars and the dwarf novae are low-mass systems consisting of a white dwarf (primary component) and a red star with $M_2 \lesssim M_\odot$, which fills its critical Roche cavity and loses mass through the internal vibration point L_1 (see Fig. 1.11). An accretion disk appears as a result of conservation of the momentum of the amount of motion around a compact object. The contribution of the primary component of the spectrum is associated mostly with the absorption lines. The main contribution to the continuous spectrum and the emission lines is provided by the disk and the bright line. The velocity of movement of matter in the jet v may reach several hundreds of kilometres/s, and the rate of flow of matter $\dot{M} \lesssim 10^{-8} \div 10^{-7} M_\odot/$year, at the same time for $v = 300$ km/s it is quite easy to estimate the luminosity of the bright line $L_{bs} = \dot{M}v^2/2 \simeq 3{\cdot}10^{33}$ erg/s $L \lesssim L_\odot$. As a result of accretion in the stationary stage, half of the potential energy is transformed into thermal energy and is then radiated. Naturally, for the cataclysmic variables the stationary model gives

4 An example with RT Ser is mentioned in [44], where the reduction by 4.5^m lasted 27 years.

the estimate of the upper value of luminosity of the accretion disk. In the inactive state $L_{AD} \ll \dot{M}GM_1/2r_1 \simeq 100\, L_{\odot}$, but it may be assumed that $L_{AD} \sim L_{\odot}$, whereas the luminosity of the dwarfs is considerably smaller than the value $L_{\odot}$.

One of the observed special features of the new stars and dwarf new stars is the oscillation of the luminosity at the minimum (inactive state). Four components of such variability have been found. Irregular rapid flickering with the characteristic time $\sim 10 \div 10^4$ s is evidently associated with the disk and/or the bright line. Flickering completely disappears during darkening of the line and the internal parts of the accretion disk by the optical star [44, 711, 893]. In the objects with a low intensity of the flow of matter the bright line is clearly dominant, as indicated in U Gem [959]. On the other hand, in the objects with a strong flow of matter, which is typical of novae (UX UMa), the disk is dominant [1075].

An interesting special feature of cataclysmic variables is that in the objects the halo amplitude of the flare is greater and the orbital periods are shorter than in the galactic disk [692, 693].

11.1.5 Disks around neutron stars

We will briefly discuss the accretors – magnetised neutron stars situated in the accretion stage. In this case, we are concerned with x-ray pulsars (sources of pulsating x-ray radiation) and x-ray bursters (sources of variable non-periodic radiation). The x-ray sources are characterised by the complicated temporal behaviour of the shine: there are periodic, quasi-periodic and random changes. The reasons for them greatly vary and special analysis must be performed for every object [118].

X-ray pulsars

There are several tens of x-ray pulsars, including systems with both massive OB-stars gas and red dwarfs and giants. The primary component is the rotating neutron star with a strong magnetic field $\sim 10^8 \div 10^{12}$ G (see the review in [947]), in Her X-1 the strength of the magnetic field reaches $\simeq 3 \cdot 10^{12}$ G [1284]. Convective motion and rapid rotation during the formation of a neutron star may result in the formation of fields of up to 10^{15} G. These objects are referred to as magnetars and the results of observation of sources SGR 1806-20, SGR 1900+14, XTE J1819-197 indicate the existence of such a strong magnetic field [703].

A special feature of the x-ray pulsars is the variable non-periodic x-ray radiation with the characteristic times of $0.7 \div 10^3$ s (the values $\gtrsim 100$ s are typical). It is generally accepted that the variability of radiation is associated with the powerful magnetic fields. Long-period variability is also detected which, for example, in Centaur X-3 indicates that the optical star fills its critical Roche cavity. The luminosity of the x-ray pulsars equals $10^{33} \div 10^{39}$ erg/s, with the accretion rate for these being $10^{-13} \div\ 10^{-7}\, M_{\odot}$/year. And if in the case of super-giants the outflow in the form of quasi-spherical stellar wind ensures the observed luminosity, then in the systems with *Be*- and low-mass stars the outflowing matter forms an accretion disk.

The second observed fact, indicating the existence of the accretion disk, is the acceleration of rotation of the neutron star with time. This effect is observed in the majority of x-ray pulsars and may be explained by the fact that in the disk accretion mode the matter, falling into the magnetosphere, transfers to the magnetosphere its momentum of the amount of motion.

X-ray bursters

The x-ray bursters are characterised by the flare activity in the x-ray range ($\lesssim$1 keV). There are around 100 x-ray bursters. The observed flares of the type I are repeated with the interval of several hours, and faster flares of the type II ('fast burster' MXB 1730-335 with the characteristic recurrence time of ~6÷1000 s). The luminosity of the bursters during a flare reaches ~1038 erg/s, the total energy, irradiated during this period in the x-ray range is $\sim 10^{39}$ erg. Between these flares, the x-ray bursters are slowly changing x-ray sources with the mean luminosity of $\sim 10^{36} \div 10^{37}$ erg/s. The entire set of the data indicates that the reason for the formation of flares is associated with the nuclear explosions of matter built up between the flares on the surface of the neutron star as a result of accretion. In the majority of x-ray bursters there is no eclipse and bright optical twins which indicates that the optical stars are low-mass red dwarfs ($M_1 >> M_2$), filling the Roche cavity, since accretion takes place in the disk mode. Thus, the radiation of the bursters between the flares is determined by the existence of a 'stationary' accretion disk. The observed radiation of the x-ray bursters does not show any strictly periodic pulsations and this may be explained by the low strength of the magnetic field of the neutron stars ($B_0 \ll 10^8$ G). In this situation, the accretion disk may reach the star surface [851].

The flares of type II may be explained by the generation of energy during accretion of matter on the neutron star with the magnetic field in the vicinity of surface of $B_0 \sim 10^8$ G, when the matter penetrates in portions into the magnetosphere as a result of the development of some instability at the boundary of the magnetic field and the accreting plasma [647].

Quasi-periodic oscillations

The galactic bulge contains x-ray sources without flare activity but showing quasi-periodic oscillations (QPO) of the x-ray flux with the characteristic frequencies of $\nu = 1 \div 40$ Hz. The QPO, discovered by the x-ray satellite EXOSAT [781], were found in large quantities on the Ginga satellite [1024]. The QPO were also detected in x-ray spectrum of some x-ray pulsars, at low frequencies, for example, in X 1627-673 which is a low-mass binary system, $\nu = 0.04$ Hz [1155], and also in sources with an ultrasoft x-ray spectrum, in the candidates for black holes (according to [522], in LMC X-1 $\nu = 0.0$ Hz). The presence of the quasi-periodic oscillations in the candidates for black holes indicated that the QPO are associated with the processes taking place in the inner parts of the accretion disk and not depend on the solid surface of the star or its magnetic field. However, this holds if the nature of the QPO for all objects is the same.

The current classification of the QPO is based on the two-colour x-ray diagram where the so-called horizontal branch, the normal and flashing parts, forming the characteristic Z-shaped figure, are clearly visible. In many cases, the same source has several harmonics. The power, the frequency of the centroid and the width of the peak in the power spectrum depend strongly on the intensity of the source.

11.1.6. Disks around black holes

A list of candidates for black holes is constantly supplemented. At the same time, new observations sometime lead to the removal of suspicions regarding certain objects. The following candidates for black holes are most reliable: A 0620-00, Cyg X-1, LMC X-3, SS 433 [187, 267, 268]; there are also other candidates. A distinguishing feature is that the mass of the optical star in A 0620-00 does not exceed 0.7 $M_{\odot}$. Therefore, this confirms the presence of the accretion disk around the relativistic component because the power of the stellar wind at the low-mass stars is not high. According to the results [730], the emission lines have two-peak profiles, characteristic of the accretion disk. The observed asymmetry may be associated with the bright line at the edge of the disk. However, in this case, the dimensions of the disk should be close to the critical Roche cavity. This is not typical of the low- mass binary systems.

The object SS 433 is best known because of its relativistic bipolar jets. The discovery of the optical eclipse in the SS 433 system has been used to conclude that this object is a close binary system consisting of a massive ($M_2 \simeq 5 \div 10\ M_{\odot}$) OB-star and a relativistic object ($M_1 > 4\ M_{\odot}$), surrounded by a bright, optically thick (for X-ray radiation) precessing accretion disk [463, 874, 875, 1208].

The Cyg X-1 object is the 'oldest' candidate for black holes (see the reviews in [268, 874, 984]). The range of temporal variability is very large: from 10^{-2} s to 10^7 s. In the x-ray range, the Cyg X-1 source shows transitions from the 'still' to the 'excited' state. In the first regime, the hard x-ray component is radiated (~100 keV) . The holding time in the 'excited' state is approximately one month. This is accompanied by the formation of a high-intensity soft spectral component in the 3–6 keV band, and the total luminosity is approximately doubled. The observed x-ray spectrum is modelled quite efficiently by the radiation from the accretion disk [1155, 1228]. At the same time, there is still no acceptable explanation of the reason for the transition from the 'still' to 'excited' state. The duration of the processes evidently indicates that they are associated with the transfer of matter on the accretion disk.

The variability of radiation is a common property of the compact stellar x-ray sources. However, the random constant variation in the subsecond range is detected only in a small number of cases: Cyg X-1, GX 339-4, Cir X-1. Possibly, these sources are the accretion disks around black holes. This temporal behaviour is usually explained by the heterogeneities [1148] rotating in the innermost regions of the accretion disk ($r \sim R_g$). The detected QPO in the LMC X-1 also indicates the presence of the accretion disk in the system (see section 11.1.5).

11.1.7 Protostars

The stars form from gas-dust clouds. If the medium initially has the angular momentum, a disk may form during compression. Without it it is difficult to imagine the mechanism of formation of sufficiently close binary stars and planetary systems. The observation of the protostars (the stars of the type T Tau) confirms the presence of gas disks around the central object [46, 652, 854, 945]. However, insufficient information is available on the physical conditions in the structure of the observed disks.

The statistical processing of the observations confirms the presence of eclipses and illumination of the stars by the disks around the stars [764]. Direct images of the disks at the young stars were obtained using CCD receivers [341]. However, the main amount of information on the dusty disks around the stars is provided by polarisation and infrared measurements because the main source of infrared radiation is the stellar dust with a temperature of $(3\div15)\cdot10^3$ K [471]. According to the data in [63], in β Pic the disk extends to 650 atomic units and its thickness is 48÷65 AU. The dust disk around the HL Tau has the diameter of approximately 270 AU [349]. The observation of the object SS V13, including emission, indicates the presence of a relatively massive disk ($\lesssim 12M_\odot$) [1108]. In addition to the massive disks with the diameter of 0.001÷0.1pc, there are also considerably smaller massive disks. Thre is a large number of indications of the existence of accretion disks in the stars of the type T Tau.

An indirect confirmation of the presence of the disks are the bipolar jets: evidently, the accretion disks play a significant role in the formation of the jets in the case of young stars. The objects with the jets are the stars of the type T Tau with the disks or protostars with a smaller mass with powerful dust envelopes and the developed accretion process. The observed properties of the bipolar discharge of matter from the stars have been discussed in [46, 945, 946]. The massive disks are sources of matter accreting the stellar nucleus. According to Sato [1114], many observations of the young stars from infrared sources invisible in optics to the stars of the type T Tau are explained satisfactorily on the basis of the evolution of the stellar gas-dust disk.

Although the process of formation of the planetary system from the protoplanetary disk has not as yet been completely explained, the very fact of existence of the protoplanetary disks is not doubted. At the present time, all the existing models of the formation of the solar system start with the concept of the collapse of a gas-dust cloud with the formation of the Sun and a protoplanetary disk [427].

Planetary rings (Saturn, Jupiter, Uranus, Neptune) are the objects with the finer scale in the hierarchy of the astrophysical disks. The observational data and detailed examination of the Phoenix of planetary rings can be found in the monographs by Gor'kavyi and Fridman [43].

11.1.8 Overflow through the internal libration point and formation of the disk

We will now discuss several special features of the overflow of the gas from a normal star to a close binary star following [243]. In the classic formulation, the close binary system is studied assuming that the system has a critical surface and when the surface is filled by one of the stars mass exchange starts to take place in the system. This surface is regarded as anonymous with the internal critical surface (the Roche surface) in a limited problem of three bodies. Matter exchange takes place through the vicinity of the libration point L_1 in which the pressure gradient is not balanced by the gravitational force. The gas dynamics of the outflow of matter from the vicinity of the point L_1 has been studied many times by a large number of authors. For the first time, analysis of this problem, carried out in semi-analytical approximation, was published in the studies of Paczynski and Sienkiewicz [1003] and Lubow and Shu [860], and its current state, together with the list of the most important tasks, has been published in, for example, reviews [419, 1158, 1283].

We will consider what takes place with the gas after it passed the libration point L_1. With increase of the distance from L_1, the gas velocity increases under the effect of the attraction force of the star–accretor (star 2) and all types of flow become supersonic (see, for example [860, 1046]). In addition to this, under the effect of the Coriolis force, the jet of matter is deflected from the line connecting the centres of the stars. In accordance with the analysis performed by Lubow and Shu [840], the equation for the angle of deflection of the jet ϑ_s has the form

$$\cos(2\vartheta_s) = -\frac{4}{3}g_\vartheta^{-1} + \sqrt{1-\frac{8}{9}g_\vartheta^{-1}}, \tag{11.3}$$

where

$$g_\vartheta(q) = \left[\frac{q}{x_{L_1}/A} + \frac{1}{(1-x_{L_1}/A)^3}\right](q+1)^{-1}. \tag{11.4}$$

The angle of deflection of the jet ϑ_s depends only on the ratio of the masses of the components. Figure 11.4a shows that in the examined simplest case (without considering the interaction of the jet with the non-accreted gas), the angle of rotation of the jet changes only slightly and is in the range from –28.4° to –19.5°.

In a number of cases, the trajectory of the jet in the system is regarded in the ballistic approximation. The formal basis for the admissibility of this approximation is the supersonic nature of the flow which makes it possible to ignore the pressure effects. The equation of motion, describing the trajectory of the test particle in the absence of the magnetic field and pressure, can be written in the form:

$$\frac{d^2\mathbf{r}}{dt^2} = -\nabla\Phi - 2\Omega\times\frac{d\mathbf{r}}{dt} \tag{11.5}$$

The results of calculations of the trajectories in the orbital plane of the system with equal masses of the components ($q = 1$) are shown in Fig. 11.4b. As indicated by the figure, at low initial velocities[5] all the trajectories of the particles are more or less similar and do not depend on the angle of entry into the system.

Analysis of the trajectories shows that the jet travels quite close to the point mass representing star 2 in the Roche approximation. The minimum obtained distance

[5]The velocity of the particle is compared here with the characteristic orbital velocity of the binary system $\sqrt{G(M_1+M_2)/A}$

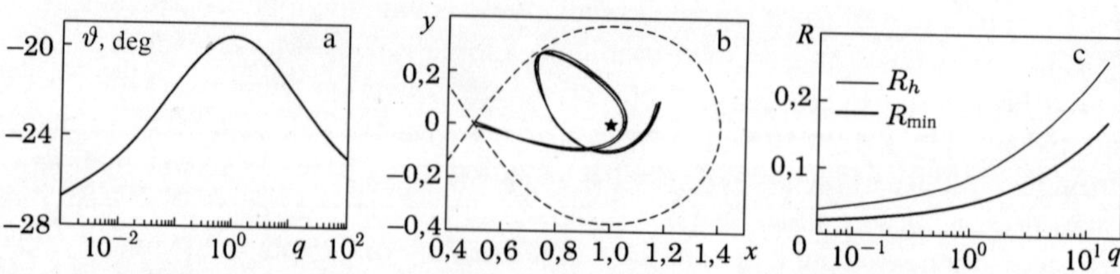

Fig. 11.4. a) function $\vartheta_s(q)$, determining the angle of deflection of the jet [860]; b) the trajectories of the particles in the orbital plane of the binary system with the ratio of the masses of the components $q = 1$; in the calculations it was assumed that the particles fly out from L_1 with the flow velocity $u = 0{,}03\sqrt{G(M_1+M_2)/A}$ in the directions 0° and ±45° in relation to the line, connecting the centres of the components; c) the minimum distance to the star 2 to which the particle, emitted from L_1, arrives during the initial flight, R_{min} and the radius of the initial ring R_h as a function of the ratio of the masses of the components q.

R_{min} between the trajectory of a separate article and the centre of the star for the systems with different mass ratios of the components is shown in Fig. 11.4c [860]. In the range $0.05 < q < 1$, the value R_{min} with the accuracy to 1% can be approximated by the equation [1283]

$$\frac{R_{min}}{A} = 0.0488\,q^{-0.464}. \tag{11.6}$$

For the binary systems, in which the radius of the second component $R_2 > R_{min}$, the jet of matter falls directly on the surface of the star. However, in the case in which $R_2 < R_{min}$, the flow of matter bends around the star and finally intersects itself at some point (see the trajectory of the particles in Fig. 11.4b). It should be mentioned that the presence of the intersecting trajectories in the system indicates the formation in the system of powerful interaction of the gas flows and it is not correct to take this interaction into account within the framework of the ballistic approximation. The complicated nature of the interaction of the jet with matter of the inter-component shell makes it necessary to solve the total system of gas-dynamic equations for describing the process of transfer of matter in the semidetached binaries. Examples of this description will be presented in section 3, here, following the studies [342, 860] we examine on the qualitative level the pattern of the flow on the basis of the analysis of the trajectories of the test particles.

We examine the angular momentum of the particle after passage through a point at which the trajectory intersects itself. Taking into account the fact that the point of intersection of the trajectories of the test particles is quite close to the star 2 and, consequently, the effect of star 1 on the flow in the vicinity of the collision point can be ignored, it is assumed that the angular momentum of the element of the gas is retained with respect to star 2. Since for the particle with the fixed angular momentum, rotating around a point mass, the minimum energy is reached in the circular orbit, it should be expected that the gas after collision forms a ring with a radius R_h around the star 2. The gas in the ring rotates around the star 2 with the azimuthal velocity v_φ which, assuming the absence of pressure forces, is determined from the balance between the centrifugal force v_φ^2/r and the gravitational attraction of the star GM_2/r^2. Equating them, we obtain a gas ring with the Kepler rotation law $v_K = (GM_2/r)^{1/2}$. The angular momentum in this ring increases in the direction from the centre of the ring as $\sim \sqrt{r}$.

The radius of the initial ring R_h can be estimated using the procedure proposed in [860], according to which it is determined as the distance to the accretor in which the jet of matter has the same tangential velocity as the particles on the closed periodic orbit, passing through the given point. However, this approach is very complicated for algorithmic transformation. The R_h value can be also determined with sufficient accuracy as the point at which the tangential velocity of the jet is equal to the tangential velocity of the ring. If we take into account the presence of the Coriolis forces in the system, we obtain an even more accurate expression. Since in this case, the distribution of the velocity in the ring is described by the law

$$v_\phi^{\text{ring}} = \sqrt{\Omega^2 r^2 + v_K^2} - \Omega r, \tag{11.7}$$

from the condition $v_\phi^{\text{ring}}(R_h) = v_\phi^{\text{stream}}(R_h)$ we obtain the dependence $R_h(q)$, which coincide with the dependence calculated in [860]. The dependence of R_h on q is shown in Fig. 11.4c.

The resultant dependence of the angular velocity of matter $\Omega_K = v_K/r$ on the radius shows that the gas in the ring rotates in a differential fashion, i.e. shear stresses, different from zero, operate in the gas flow. The presence of the dissipative processes leads to situation in which the rotation of the gas particles in the inner orbits will slow down and, on the other hand, in the outer orbits it will accelerate. Consequently, this will result in the redistribution of the angular momentum in the ring which will spread in the accretion disk.

11.2. Axisymmetric disk accretion

11.2.1. Diffusion approximation

We derive the equations of non-stationary axisymmetric disk accretion and gradually determine the applicability conditions.

The very concept of the mechanism, leading to the transfer of the angular momentum in the axisymmetric disk and, consequently, accretion of matter, is quite simple. According to this concept, the medium rotating in a differential fashion is characterised by the interaction between the adjacent layers associated, for example, with the existence of the magnetic field, turbulence, molecular or radiation viscosity, etc[6]. In the typical conditions, molecular viscosity cannot ensure the accretion rate found in the observations, and the radiation itself is the consequence of accretion. The magnetic field falls onto the disk, for example, together with the matter discharged from the normal star. Under specific conditions, the strength of the magnetic field may reach $H^2/8 \sim \rho GM_1/r$, which is sufficient for accretion.

[6]In this approach, the analogy with the falling orbital satellite, travelling to the upper layers of the Earth atmosphere, is fully applicable.

The axisymmetric problem, examined below, is based on the assumption according to which the disk thickness $2h$ is everywhere small in comparison with the radial coordinate r. Effective cooling results in a geometrically thin disk. If the disk is thin, it can be described by the surface density

$$\sigma = 2\int_0^h \rho\,\mathrm{d}z, \tag{11.8}$$

where h is the half thickness of the disk. The variation of surface density with time also determines the dynamics of the accretion disk [342, 577, 590, 1045, 1283]. The presence of the dissipative processes in the disk leads to the radial movement of matter in the direction to the accretor with the velocity v_{rad}.

We write the gas dynamics equations taking into account dissipation for the axisymmetric disk around the central body with the mass M_1. The law of conservation of the mass is written taking into account the source of matter:

$$\frac{\partial\sigma}{\partial t}+\frac{\partial}{r\partial r}\left(r\sigma u\right)=\frac{1}{2\pi r}\frac{\partial \dot{M}_e}{\partial r}. \tag{11.9}$$

Here $\sigma = 2h\bar{\rho}$ is surface density, u is radial velocity, $\dot{M}_e(r, t)$ is the given function describing the rate of inflow of matter into the accretion disk. The rate of accretion is:

$$\dot{M} = -2\pi\int r\rho u\,dz = \text{const}, \tag{11.10}$$

which does not depend on the radial coordinate in the stationary case.

In the law of conservation of the momentum

$$\frac{\partial}{\partial t}\left(\sigma l\right)+\frac{\partial}{r\partial r}\left(r\sigma l u\right)=\frac{\partial(r^2W_{r\varphi})}{r\partial r}+\frac{l_e\partial\dot{M}_e}{2\pi r\partial r}-\sigma B \tag{11.11}$$

the value $l = r\,v$ is the specific momentum of matter situated at distance r, $l_e(r, t)$ is the specific momentum of matter which is determined by the source M_e, the component of the tensor of viscous stresses, integrated with respect to the z-coordinate, has the form

$$W_{r\varphi} = \sigma\nu\left(\frac{\partial v}{\partial r}-\frac{v}{r}\right), \tag{11.12}$$

where ν is the first (shear) kinematic viscosity. The last term in (11.11) describes the effective tidal interaction determined by the second component of mass M_2, and B is the specific rate of loss of the angular momentum by the matter.

When describing the large-scale motion, the turbulent medium can be regarded as a liquid characterised by turbulent viscosity ν_t, which differs from the true (molecular) kinematic viscosity [112]. The developed turbulence can be regarded as the hierarchy of turbulence pulsations, differing in the spatial scales. If ν_t and l_t

denote respectively the velocity and scale of the main (mostly large-scale) turbulent motion, then as regards the order of magnitude we can write

$$\nu_t \sim v_t\, l_t. \tag{11.13}$$

Assuming isotropic turbulence, it is natural to assume that $l_t \lesssim h$. Since the supersonic turbulence pulsations rapidly dissociate, $v_t \lesssim c_s$. Consequently, we have

$$\nu_t \lesssim c_s\, h. \tag{11.14}$$

In 1973, Shakura and Sunyaev [1147] noted: 'now, in the absence of the complete theory of turbulence, on the one hand, and the observational data for the existence of turbulence in the disks, on the other hand, we can only assume its presence'. This comment is still valid after more than 30 years. Regardless of this, the models of the accretion disks with turbulent viscosity have been developed quite extensively and are very popular. The problem of viscosity in the theory of accretion disks is unfortunately most unclear. As already mentioned, it is not doubted that the molecular viscosity $\nu_{\rm mol} \simeq c_s l f_t$ (l_{ft} is the free path length) cannot ensure the required accretion rate. Therefore, when we discuss viscosity in the theory of disk accretion, we normally mean the turbulent viscosity $\nu_{\rm t} \gg \nu_{\rm mol}$.

In the absence of a complete theory of turbulence, the study by Shakura [272 has become extremely important. In this study, 'all our ignorance of turbulence has been reduced to a single dimensionless parameter α'[118]. For the tensor of viscous stresses, it is assumed that

$$W_{r\varphi} = -\alpha\, 2\, h\, P, \tag{11.15}$$

where P is the pressure averaged out with respect to the z-coordinate; α is a free parameter. The models, based on the relationship of the type (11.15), are referred to as the α-models or standard models of the accretion disks.

In the equation of state of matter we take into account the gas, radiation and magnetic pressures:

$$P = P_{\rm gas} + P_{\rm rad} + P_{\rm m}, \tag{11.16}$$

where $P_{\rm gas} = \mathcal{R}\sigma T/(2h\mu)$, $P_{\rm rad} = aT^4/3$, T is temperature, $\mathcal{R} = 8.31\cdot 10^7$ erg (mole · K), radiation constant $a = 7.56\cdot 10^{-15}$ erg/(cm^3· K^4), μ is the molar mass (in the case of completely ionised hydrogen plasma $\mu = 0.5$). To determine the magnetic pressure P_{m}, it is necessary to make additional assumptions regarding the structure of the magnetic field.

We examine the structure of the disk in the z-direction. It is assumed that in the z-direction matter is in hydrostatic equilibrium (see (3.234)):

$$\frac{\partial \mathcal{P}(z)}{\partial z} = \rho(z)\frac{\partial}{\partial z}\frac{G M_1}{\sqrt{r^2 + z^2}}. \tag{11.17}$$

For a thick disk, for which $z \ll r$, we have

$$\frac{\partial P}{\partial z} = \frac{GM_2 \rho z}{r^3}, \tag{11.18}$$

or as regards the order of magnitude

$$\frac{\partial p}{\partial z} \approx \frac{P_0}{h}, \quad z \approx h, \quad c_s \approx \sqrt{\frac{P_0}{\rho_0}}, \tag{11.19}$$

where the index 0 corresponds to the values taken on the equatorial plane. As a result, we can write for the disk half thickness h

$$h \simeq \sqrt{\frac{c_s^2}{GM_2 / r^3}} = \frac{c_s}{v_K} r, \tag{11.20}$$

where c_s is the speed of sound.

The energy balance equation in the z-direction is

$$\frac{1}{\rho} \frac{dF}{dz} = \frac{F_v}{\sigma}, \tag{11.21}$$

where $F = -\frac{c}{3k\rho} \frac{d}{dz}(aT^4)$ is the flux of radiant energy; F_v is the source of energy, $k(\rho, T)$ is opacity; c is the speed of light. The combined solution of the equations (11.17, 11.21) determines the z-structure of the thin disk. Taking into account that $h \ll r$ is fulfilled, the hydrostatic equilibrium condition (11.17) can be written in the form[7]

$$p = 2hP = C^2 \sigma h^2 \frac{G M_1}{r^3}, \tag{11.22}$$

where $C^2 \sim 1$ (see section 3.6.1).

We turn to the radial component of the equation of motion (3.242)

$$\frac{\partial(\sigma u)}{\partial t} + \frac{\partial(r \sigma u^2)}{r \partial r} - \frac{\sigma v^2}{r} = -\frac{\partial p}{\partial r} + \frac{3Dp}{2r} - \frac{\sigma G M_1}{r^2} + \frac{\partial(r W_{rr})}{r \partial r} - \frac{W_{\varphi\varphi}}{r}, \tag{11.23}$$

where $W_{rr} = 2\sigma v(\partial u / \partial r) + \sigma(\mu - v)\mathrm{div}(\mathbf{u})$; $\mu(\sim v)$ is the second kinematic viscosity; $W_{\varphi\varphi} = 2\sigma v u / r + \sigma(\mu - v)\mathrm{div}(\mathbf{u})$. Using the evident estimate, resulting from (11.22):

$$c_s \simeq \sqrt{\frac{p}{\sigma}} \simeq \frac{h}{r} v, \tag{11.24}$$

7In the literature equations of the type (11.15), (11.22), (11.28) are used quite often, $\sigma = 2\bar{\rho} h$ and so on, which differ in the coefficients of the order of unity. This is associated with differences in averaging over to the z-coordinate.

from (11.9), (11.11), with (11.14) and (11.24) taken into account, we obtain

$$|u| \sim v/r \sim c_s h/r. \tag{11.25}$$

Thus, with the accuracy to the small value $(h/r)^2$ in the radial direction (see (11.23)) there is a balance of only gravitational and centrifugal forces and we can write

$$v = r\Omega = \sqrt{\frac{GM_1}{r}}. \tag{11.26}$$

It is important to stress the approximate nature of the relationship (11.26); this is especially important when examining the dynamics of sound perturbations in the disk plane. The differences between the velocity of rotation and the Kepler velocity is determined practically only by pressure since the dissipative terms in (11.23) are small $[\propto(h/r)^4]$. The deviation of the disk from the strictly Kepler disk is also associated with the propagation of the soundwave.

Using (11.26), we exclude the radial velocity from (11.9) and (11.11) and, as a result, we obtain the evolution equation for surface density

$$\frac{\partial\sigma}{\partial t} = \frac{3\partial}{r\partial r}\left[\frac{\sqrt{r}\partial}{\partial r}(\nu\sigma\sqrt{r})\right] + \frac{\partial\dot{M}_e}{2\pi r\partial r} + \frac{\partial}{\pi r\partial r}\left\{r\left[1-\sqrt{\frac{r_e}{r}}\right]\frac{\partial\dot{M}_e}{\partial r}\right\} + \frac{2\partial}{r\partial r}\frac{B\sigma}{\Omega}, \tag{11.27}$$

where $r_e = l_e^2/GM_1$ is the Kepler radius, corresponding to the specific angular momentum l_e, and the latter is determined by the value $q = M_2/M_1$. The main term, which determines the structure of the disk, is the first time in the right-hand side of (11.27), determined by viscosity. The other terms of the equation may play any role only in the external region of the disk. Thus, we have the typical equation of the diffusion type (especially clear from (11.74)). Taking into account (11.12), (11.15), and (11.22) for the viscosity we can write the following equation

$$\nu = \frac{2}{3}\alpha h^2 \Omega. \tag{11.28}$$

Taking into account (11.9) and (11.11), we easily obtain the expression for the radial velocity:

$$u = \frac{1}{\sigma}\left\{-\frac{3}{\sqrt{r}}\frac{\partial}{\partial r}(\sqrt{r}\sigma\nu) + \frac{1}{\pi}\left(\sqrt{\frac{r_e}{r}}-1\right)\frac{\partial\dot{M}_e}{\partial r} - \frac{2\sigma B}{r\Omega}\right\}. \tag{11.29}$$

In a stationary disk with no account made for the last two terms, the expression for the velocity has the simple form: $u = -3\nu/(2r)$.

Viscosity redistributes the angular momentum in the disk taking it away from the inner part of the disk to the external one. This results in the accretion of matter - the movement of the gas in the direction to the centre. We investigate the temporal revolution of the surface density using equation (11.29) in the simple case ν = const and without taking the last two terms into account. At the initial moment of time, the matter is in a thin ring (δ-function) at the radius $r = R_d/2$. Temporal evolution for the four moments of time: $t = 1\cdot10^{-3}t_0$, $2\cdot10^{-3}t_0$, $4\cdot10^{-3}t_0$, $9\cdot10^{-3}t_0$ is shown in Fig. 11.5

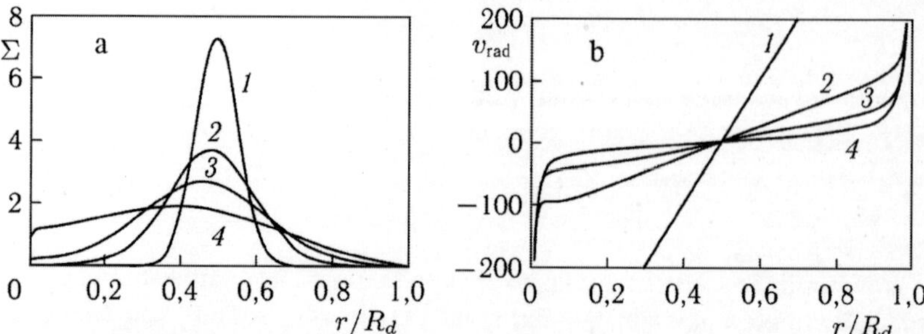

Fig. 11.5. Temporal evolution of a gas ring under the effect of viscosity. The graph shows the distribution of surface density (a) and the radial velocity of spreading $v_{\rm rad}$ (b) on the normalised radius r/R_d for four moments of time [243] (see the explanation in the text).

(curves 1–4, respectively). The timescale is determined on the basis of the accepted values of viscosity and the disk radius R_d as $t_0 = R_h^2/\nu$. Figure 11.5 also shows the value $v_{\rm rad}$ (in the units of ν/R_d) for the same moments. As indicated by the results of the calculations, a large part of the matter of the ring moves inwards, losing energy and angular momentum. A small pad of matter moves outwards, acquiring the angular momentum. Consequently, the initial ring, as expected, spreads into a disk.

In the stationary state, the outer regions of the disk should get rid themselves constantly of the excess angular momentum. The loss of the angular momentum by the periphery of the disk may be associated with the effect of different mechanisms: as a result of tidal forces, acting on the disk from the side of the donor star, and the interaction with the gas of the inter-component shell which subsequently leaves the binary system, magnetic wind, etc.

The equation of energy balance for the matter of the disk has the following form:

$$\sigma T\left(\frac{\partial S}{\partial t}+u\frac{\partial S}{\partial r}\right)=F_\nu - F_{\rm rad}+F_{\rm conv}+F_\varkappa+Y, \qquad (11.30)$$

where S is the specific entropy of matter. On the right there are the terms describing the energy supplied to the unit area (or removed from this area) per unit time. Energy generates as a result of viscous participation

$$F_\nu = \frac{W_{r\varphi}^2}{\sigma\nu} = \nu\sigma\left(r\frac{d\Omega}{dr}\right)^2 = \frac{9}{4}\nu\sigma\Omega^2. \qquad (11.31)$$

The radiation from the surface is described by the term $F_{\rm rad}$. Within the framework of the model of the optically thick accretion disk, taking two of its sides into account, we can write [273]:

$$F_{\rm rad} = \frac{4}{3}\frac{acT^4}{\sigma k}. \qquad (11.32)$$

Energy can be transferred by convection in the vertical direction $F_{\rm conv}$ (see, for example [1179]. The value $F_\varkappa$ is determined by the diffusion transfer of energy along the radial component. The matter, flowing into the disk, brings in not only the pulsed momentum but also the energy and this is taken into account by the term Y. This energy is associated primarily with the transformation to heat of the kinetic energy of the gas jet flowing into the external region of the accretion disk with the formation of a shockwave.

The main sources of opeacity of matter in the accretion disk are usually [273]: scattering on free electrons

$$k_{es} = 0.4 f_e \text{ cm}^2 / \text{g}, \qquad k_T = 0.4 \text{ cm}^2 / \text{g}, \tag{11.33}$$

free–free absorption (reversed bremsstrahlung)

$$k_{ff} = 0.645{\cdot}10^{23} \frac{f_e f_i Z}{A} g_{ff}\, \rho T^{-7/2} \text{ cm}^2 / \text{g}, \tag{11.34}$$

where $g_{ff} \simeq 1$; f_e is the number of free electrons per baryon; f_i is the fraction of the atoms ionised to the given state, Z is the charge, A is atomic weight.

Thus, the task of determination of the evolution of the accretion disk within the framework of the constructed model is reduced to the combined solution of the system of equations (11.27)–(11.34), (11.16), (11.22). Consequently, for a specific initial state we calculate the dependence of the disk parameters (σ, T, h, ν, k, u,…) nn time and the radial coordinate. In this approach, the functions $M_e(r, t)$, $Y(r, t)$ should be determined from some additional considerations. The procedure of introduction of matter and energy into the disk was described in [343–345]. Consideration of the tidal interaction within the framework of the axisymmetric models was the subject of studies [853, 1007, 1044, 1179].

The characteristic times of the disk accretion are: dynamic time $t_\varphi \simeq 1/\Omega$, the time of establishment of hydrostatic equilibrium in the z-direction $t_z \simeq h/c_s$, the thermal timescale $t_{th} \simeq \sigma c_s^2/F_{rad}$ and determined by viscosity $t_\nu \simeq r/u \simeq (r/h)^2 t_\varphi$. The relationship between these quantities depends on the radial coordinate but on the whole we may write

$$t_\varphi \simeq t_z \lesssim t_{th} \ll t_\nu . \tag{11.35}$$

Because of these estimates, since $TdS/dt = C_q dT/dt$, heat capacity C_q is often ignored and considerations are restricted to the following approximation (11.30): $F_\nu - F_{rad} + F_{conv} + F_\varkappa + Y = 0$.

We shall discuss in somewhat greater detail the tidal interaction. Under the effect of viscous forces the disk in the absence of tidal interaction spreads without limits in the radial direction. The disk radius tends to infinity. The angular momentum as a result of friction is taken away from the internal regions of the disk to the external ones and in the presence of an object/donor, the disk can spread only to the radius at which the tidal interaction is capable of removing the entire angular momentum from the disk. In the stationary case, the total angular momentum of matter is a constant quantity and, consequently, the following balance should be fulfilled

$$\dot{M}_e \sqrt{G M_1 r_e} = \int_{r_1}^{R} \sigma B 2\pi r\, dr, \tag{11.36}$$

where r_1, R is the inner and outer radius of the disk, respectively. Taking into account the tidal force, the accretion disk is finite in the radial direction.

The majority of studies in which the tidal interaction was taken into account in (11.27), based on the results obtained by Pringle and Papaloizou [1007]. The quantity B can be conveniently described by the equation

$$B = \frac{r^2\Omega^2}{\pi q^2} f(r), \tag{11.37}$$

where $q = M_2/M_1$ is the mass of the object–donor, and $f(r)$ is the known dimensionless function [1007]. Smak [1179] used the simple approximation

$$f = \text{const} \cdot r^5; \tag{11.38}$$

this equation describe the results of calculations obtained byPapaloizou and Pringle [1007] for the case $q \simeq 1$. For the ratio of the masses $q \gg 1$ we can write

$$f \sim \frac{1}{[(a/r)-1]^2[1-(\omega/\Omega)]^2}, \tag{11.39}$$

where $\omega^2 = G(M_1 + M_2)/a^3$ is the distance between the objects.

11.2.2. Stationary models

When examining the stationary models it is sufficient in all evolution equations in section 11.2.1 to set $\frac{\partial}{\partial t} \equiv 0$. In the framework of some assumptions on the type of laws of opacity, viscosity and energy transfer, it is possible to obtain the analytical dependences of the parameters of the accretion disk on the radial coordinate. Usually, according to Shakura and Syunyaev [1147], the disk is divided into several regions in which the main role is played by the specific processes of transfer of radiation (k_{es} or k_{ff}) and pressure (P_{gas} or P_{rad}), the tidal interaction is not taken into account, and it is assumed that $P = F_{\text{rad}}$.

1. In the innermost radiation-dominated zone ($p_{\text{rad}} \gg p_{\text{gas}}$) of the disk 'a', taking into account the fulfilment of $k_T \gg k_{ff}$ we have

$$\begin{aligned} &\sigma = 2.4 R^{3/2} (\alpha \dot{m} \mathcal{L})^{-1} \left[\text{g/cm}^2\right], \quad T = 2.3 \cdot 10^7 R^{-3/8} (\alpha m)^{-1/4} \left[\text{K}\right], \\ &h = 3 \cdot 10^6 \dot{m} m \mathcal{L}\ [\text{cm}], \quad u_r = 14 \cdot 10^{10} \alpha \dot{m}^2 R^{-5/2} \mathcal{L}\ [\text{cm/s}], \end{aligned} \tag{11.40}$$

here $m = M_1 / M_\odot$, $\dot{m} = \dot{M} / (m \cdot 3 \cdot 10^{-8}) M_\odot [\text{year}]$, $R = r/(3R_g) = r/(9m)[\text{km}]$, $\mathcal{L} = R_1^{-1/2} - R^{-1/2}$ [year]. The formation of the function $\mathcal{L}(r)$ is determined by the boundary condition $W_{r\varphi}(R = R_1) = 0$.

2. For the region 'b', where $p_{\text{gas}} \gg p_{\text{rad}}$ and $k_T \gg k_{ff}$, we can write

$$\sigma = 0.81 \cdot 10^5 \dot{m}^{3/5} m^{1/5} \alpha^{-4/5} R^{-3/5} \mathcal{L}^{3/5} \left[\text{g/cm}^2\right],$$
$$T = 3.1 \cdot 10^8 (\dot{m}\mathcal{L})^{2/5} (\alpha m)^{-1/5} R^{-9/10} \text{ [K]}, \tag{11.41}$$
$$h = 1.64 \cdot 10^4 (\dot{m}\mathcal{L})^{1/5} m^{9/10} \alpha^{-1/10} R^{21/20} \text{ [cm]},$$
$$u_r = 4.1 \cdot 10^6 \alpha^{4/5} \dot{m}^{2/5} m^{-1/5} R^{-2/5} \mathcal{L}^{-3/5} \text{ [cm/s]}.$$

3. In the outer region of the disk 'c' ($p_{gas} \gg p_{rad}$, $k_t \ll k_{ff}$):

$$\sigma = 3.3 \cdot 10^5 \alpha^{-4/5} \dot{m}^{7/10} m^{1/5} R^{-3/4} \mathcal{L}^{7/10} \left[\text{g/cm}^2\right],$$
$$T = 7.6 \cdot 10^7 \alpha^{-1/5} \dot{m}^{3/10} m^{-1/5} R^{-3/4} \mathcal{L}^{3/10} \text{ [K]}, \tag{11.42}$$
$$h = 8.1 \cdot 10^3 \alpha^{-1/10} \dot{m}^{3/20} m^{9/10} R^{9/8} \mathcal{L}^{3/20} \text{ [cm]},$$
$$u_r = 10^6 \alpha^{4/5} \dot{m}^{3/10} m^{-1/5} R^{-1/4} \mathcal{L}^{-7/10} \text{ [cm/s]}.$$

It should be mentioned that, depending on the specific values of the parameters of the system, some region may not be found in the disk.

At the present time, there is a large number of studies concerned with the determination of the radial structure of the accretion disk under specific conditions. Some of these models are described in [178, 262, 393].

11.2.3. Temperature in the accretion disks

We examine in greater detail the problem of the radial distribution of temperature in the accretion disks and the dependence of T on the system parameters, following [243].

In (11.30) we confine ourselves only to the energy balance as a result of viscous heating and radiation cooling:

$$Q^+ - Q^- = 0. \tag{11.43}$$

For an optically fine case, in which Q^+ is determined by volume radiation heating and viscosity heating, Q^- by volume radiation cooling, we obtain

$$Q^+_{\text{visc}}(\rho, T) + n^2[\Gamma(T, T_{\text{wd}}) - \Lambda(T)] = 0. \tag{11.44}$$

Here $\Gamma(T, T_{wd})$ is the function of radiation heating, which depends on the gas temperature T and the temperature of the central object (white dwarf) T_{wd}, $\Lambda(T)$ is the radiation cooling function, Q^+_{visc} (φ, T) is viscous heating.

For am optically thick case Q^+ is determined by viscous heating, Q^- by radiant heat conductivity[8] and convection in the vertical direction

$$Q^+_{\text{visc}}(\rho, T) - \frac{\partial F_{\text{rad}}}{\partial z} - \frac{\partial F_{\text{conv}}}{\partial z} = 0. \tag{11.45}$$

[8]Molecular heat conductivity is ignored since it is small in comparison with radiant heat conductivity.

Here F_{rad} and F_{conv} are the radiation and convective energy fluxes.

To determine the functions included in the equations (11.44) and 11.45), we use the continuity equation in the form (11.8), and the equation of balance of angular momentum (11.11), assuming that $\partial/\partial t = 0$, $t_e = B = 0$, and the expression for viscous heating Q^{+}_{visc} (11.31). We use the condition of hydrostatic equilibrium in the vertical direction (11.18) in the equation of state of the ideal gas with radiation (11.16).

To cause the system of equations, it is necessary to specify the viscosity coefficient ν = $\alpha c_s h$ (see (11.28)), where $c_s = \sqrt{\mathcal{R}T + (1/3)aT^4/\rho}$ (here a is the radiation constant). The value c_s can be determined on the basis of temperature in the equatorial plane of the disk $z = 0$. Consequently, we obtain the equation linking $\dot{M}, T|_{z=0}$ and ρ at the given values of r and α [243]:

$$\dot{M} = 2\pi\,\alpha\,\Omega_K^{-2}\rho c_s^3 = 2\pi\,\alpha\,\Omega_K^{-2}\left(\mathcal{R}T\rho^{2/3} + \frac{1}{3}aT^4\rho^{-1/3}\right)^{3/2}. \tag{11.46}$$

Considering (11.46) with respect to the variable $\rho^{1/3}$, for the positive real roots in the case of radiation pressure we have the condition

$$\dot{M} > \frac{\sqrt{3}\pi\sqrt{\mathcal{R}}a\alpha T^{9/2}}{\Omega_K^2}, \tag{11.47}$$

which gives the minimum value of the rate of accretion at the given values of T, r, α. The estimate (11.47) can be re-written in the new form

$$T < 7\times 10^5\left(\frac{r}{R_{\text{wd}}}\right)^{-2/3}\left(\frac{\dot{M}}{10^{-9}\,M_{\odot}\text{year}^{-1}}\right)^{2/9}\left(\frac{\alpha}{0.1}\right)^{-2/9}\ [\text{K}], \tag{11.48}$$

where $R_{\text{wd}} = 10^9$ cm is the radius of the accretor – the white dwarf.

We examine the condition (11.47) for the outer parts of the accretion disk. It is assumed that $r = A/5$, where A is the distance between the component of the binary system, $A = 1.42R_{\odot}$, which corresponds to IP Peg and, consequently

$$\dot{M} > 10^{-9}\left(\frac{T}{10^5 K}\right)\left(\frac{\alpha}{0.1}\right)[M_{\odot}\text{year}^{-1}]. \tag{11.49}$$

If the condition (11.49) is fulfilled, the roots of the equation (11.46), linking ρ, T and $\dot{M}$ at the given values r and α, can be presented in the form

$$\rho = (\mathcal{R}T)^{-3/4}\left(\frac{\dot{M}\Omega_K^2}{2\pi\alpha}\right)\sin^3\left[\frac{1}{3}\arcsin\left(\sqrt{\mathcal{R}}aT^{9/2}\frac{2\pi\alpha}{\dot{M}\Omega_K^2}\right)\right], \tag{11.50}$$

$$\rho = (\mathcal{R}T)^{-3/4}\left(\frac{\dot{M}\Omega_K^2}{2\pi\alpha}\right)\times\cos^3\left[\frac{1}{3}\arcsin\left(\sqrt{\mathcal{R}}aT^{9/2}\frac{2\pi\alpha}{\dot{M}\Omega_K^2}\right)+\frac{\pi}{6}\right], \tag{11.51}$$

where for simplicity we ignore the numerical multipliers $\sqrt{3}/2 \approx 1$. The first of the roots of the equation (11.46) corresponds to the so-called disks with dominance of radiation pressure: $\beta = (1/3)\, aT^4/\rho\mathcal{R}T > 1$, and the second one – to the disks with dominance of gas pressure: $\beta < 1$.

These equations describe two parts of the two-parameter dependence $\rho(\dot{M},T)$. To fulfil the task, i.e. to calculate the dependence $T(\dot{M})$, it is necessary to use an additional equation – the thermal balance equation (11.43). As indicated in section 2.2.1, the form of equation (11.43) differs depending on the optical thickness of the disk and, consequently, for further analysis it is necessary to calculate the optical thickness.

Optical thickness τ is determined by the product of the absorption coefficient $\varkappa_{\text{abs}}$, density and the geometrical thickness of the layer: $\tau = \varkappa_{\text{abs}}\,\rho l$. The controlling parameter for disk accretion is the ratio of the geometrical thickness of the layer at which the value $\tau = 1$ is obtained, to the half thickness of the disk: $l^{\tau=1}/H$. After several transformations:

$$\frac{l^{\tau=1}}{H} = \frac{2\pi\alpha}{\varkappa_{\text{abs}}\,\dot{M}} c_s^2 \Omega_K^{-1}, \tag{11.52}$$

The absorption coefficient $\varkappa_{\text{abs}}$ depends in a complicated manner on T and ρ (and also on the degree of ionisation, chemical composition, etc). In this review, we use the simple approximation for $\varkappa_{\text{abs}}(T,\rho)$ [296, 841, 55]:

$$\varkappa_{\text{abs}}(T,\rho) = \begin{cases} \varkappa_1 \rho^{2/3} T^3, & \varkappa_1 = 10^{-8}, \\ \varkappa_2 \rho^{1/3} T^{10}, & \varkappa_2 = 10^{-36}, \\ \varkappa_3 \rho T^{-5/2}, & \varkappa_3 = 1.5 \cdot 10^{20}, \\ \varkappa_4, & \varkappa_4 = 0.348. \end{cases}$$

According to [357], these four sub-regions correspond to scattering on molecular hydrogen, atomic hydrogen, free-free and free-bounded transitions, and also Thompson scattering. The boundaries of the sub-regions at which transitions from one expression to the other one take place, are determined by the condition of equality of the values of $\varkappa_{\text{abs}}$, calculated from these expressions. The graphs of the dependence of $\varkappa_{\text{abs}}$ on T and ρ are presented in Fig. 11.6. The regions with $d\varkappa_{\text{abs}}/dT > 0$ are clearly visible. In these regions, thermal instability can develop in which the dependence between surface density in the disk temperature is expressed by the S-curve on the plane (Σ, T_{eff}). The thermal instability is often used for explaining flares of the dwarf novae (see, for example [444, 894]), but it is clear that this is valid only for relatively cold discs.

We returned to the expression (11.46). It is assumed that $\alpha = 0.1$, $r = A/5$ and we consider the disks in which gas pressure dominates for which $\beta = (1/3)\, aT^4/\rho\mathcal{R}T < 1$. Figure 11.7a in the plane $(T, \dot{M})$ shows by crosshatching the region corresponding to all possible solutions for the investigated disks. The dashed line corresponds to the condition (11.49) of existence of the solution of equation (11.46) – in the region situated below this line there are no solutions. The solid line shows the boundary between the optically thick and optically thin disks; horizontal crosshatching shows the region of optically thick disks, and the vertical crosshatching – optically thin disks. Identical interpretation for the disks with the dominance of radiation pressure ($\beta > 1$) is shown in Fig. 11.7b.

Figure 11.7a shows that at the realistic values $\dot{M} \in [10^{-12}, 10^{-7}]\, M_{\odot}\text{year}^{-1}$ the disks with the dominance of gas pressure are mostly optically thick, although at low $\dot{M}$ solutions corresponding to the optically scene cold disks are also possible. Figure 11.7b shows that the disks with the dominance of radiation pressure are mostly optically thin and only at high values of $\dot{M}$ optically thick hot disks can exist.

Equation (11.46), linking $\dot{M}$, T and at ρ the given values of r and α, with the additional thermal balance equation (11.46) taken into account, makes it possible to reduce the number of unknown quantities and obtain the required relationship between $\dot{M}$ and T. The vertical distribution of temperature in the optically thick disk is described by the equation of radiant heat conductivity with the source determined by viscous heating (11.46), which can be presented in the following form

$$\frac{\partial e}{\partial t} = \frac{\partial}{\partial z}\left[\frac{1}{\varkappa_{\text{abs}}\rho}\frac{\partial}{\partial z}\left(\frac{1}{3}acT^4\right)\right] + \rho\alpha c_s^2\Omega_K, \qquad (11.53)$$

where e is the specific internal energy, c is the speed of light. To solve the equation (11.53), it is necessary to specify the boundary conditions. As a result of symmetry of the problem, the derivative of temperature at the equatorial plane should be equal to 0, i.e. $T'|_{z=0} = 0$. The temperature at the upper boundary of the disk is determined from the condition $\Gamma(T_*, T_{\text{wd}}) = \Lambda(T_*)$. The functions $\Gamma(T, T_{\text{wd}})$ and $\Lambda(T)$ are complicated but they are well-known (see, for example, [493, 1197]). The temperature, determined from the equality of these functions (for the temperature of a central object – the white dwarf, – equal to $T_{\text{wd}} = 70000$ K), has the value $T(H) = T_*) = 13\,600$ K.

$$t_{\text{diff}} \simeq \frac{R\varkappa_{\text{abs}}\rho^2 H^2}{acT^3}, \qquad (11.54)$$

The solution of the equation (1153) reaches the stationary mode when the characteristic time of heat conductivity becomes comparable with the time of viscous heating:

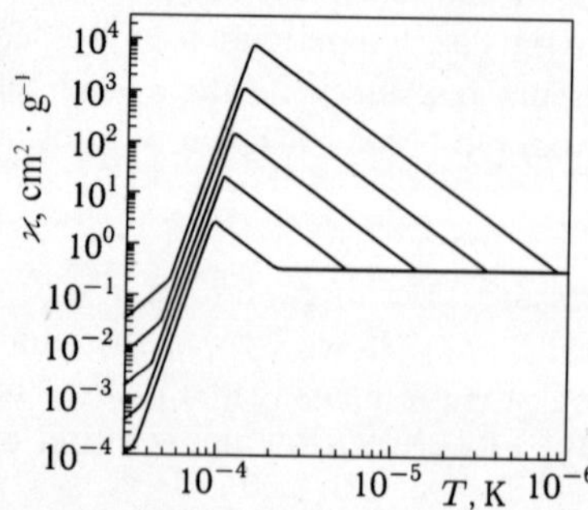

Fig. 11.6. Dependence of the absorption coefficient $\varkappa_{\text{abs}}(T)$ for values $n = 10^{18}$ cm^{-3}, 10^{16} cm^{-3}, 10^{15} cm^{-3}, 10^{14} cm^{-3} (from top to bottom) [357].

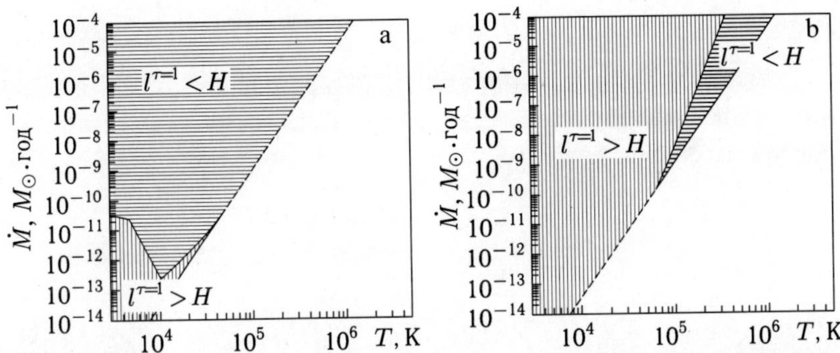

Fig. 11.7. Solution of the equation (11.46) for the disks with the dominance of gas pressure (a) and disks with dominance of radiation pressure (b) in the plane (T, $\dot{M}$) for $\alpha = 0.1$ and $r = A/5$. The horizontal crosshatching shows the optically thick disks, the vertical crosshatching – optically thin disks; the boundary between these regions is indicated by the solid line. The dashed line reflects the condition (11.49) of existence of the solution of the equation (11.46) – in the region situated below this line there are no solutions.

$$t_{\text{heat}} \simeq \frac{RT}{\alpha c_s^2 \Omega_K} \simeq \alpha^{-1}\Omega_K^{-1}. \tag{11.55}$$

It should be mentioned that for the stationary case equation (11.53) can be integrated analytically. We denote $U = T^4$, $U_* = T_*^4$, $U_0 = U|_{z=0}$ and again assume that is not dependent on z and, consequently

$$\frac{d}{dz}\left[\frac{1}{\varkappa_{\text{abs}}\rho}\frac{d}{dz}\left(\frac{ac}{3}U\right)\right] = -\rho\alpha c_s^2\Omega_K. \tag{11.56}$$

After integration with respect to z we obtain

$$\frac{1}{\varkappa_{\text{abs}}\rho}\frac{d}{dz}\left(\frac{ac}{3}U\right) = -\rho\alpha c_s^2\Omega_K z, \tag{11.57}$$

the integration constants is equal to 0 because of $U'|_{z=0} = 0$. To facilitate analysis, the last equation is transformed as follows

$$\frac{1}{\varkappa_{\text{abs}}}\frac{dU}{dz} \equiv \frac{dB}{dz} = -\frac{3}{ac}\rho^2\alpha c_s^2\Omega_K z, \tag{11.58}$$

where the function $B(U)$ which is determined from the differential equation

$$\frac{dB}{dU} = \frac{1}{\varkappa_{\text{abs}}(U,\rho)}, \tag{11.59}$$

for the fixed value can be written in the analytical form. Integrating the last equation with respect to z, we obtain

$$B(U) = B(U_*) + \frac{3}{2ac}\rho^2\alpha c_s^2\Omega_K(H^2 - z^2), \tag{11.60}$$

or for $z = 0$

$$B(U_0) = B(U_*) + \frac{3}{2ac}\rho^2 \alpha c_s^2 \Omega_K H^2. \tag{11.61}$$

Taking into account the expressions

$$c_s^2 = \left(\mathcal{R}U_0^{1/4} + \frac{1}{3}\frac{aU_0}{\rho}\right), \tag{11.62}$$

$$H^2 = \left(\mathcal{R}U_0^{1/4} + \frac{1}{3}\frac{aU_0}{\rho}\right)\Omega_K^{-2}, \tag{11.63}$$

we obtain the algebraic equation for U_0

$$B(U_0) = B(U_*) + \frac{3}{2ac}\rho^2 \alpha \Omega_K^{-1}\left(\mathcal{R}U_0^{1/4} + \frac{1}{3}\frac{aU_0}{\rho}\right)^2. \tag{11.64}$$

The last equation determines in the implicit form of the dependence $U_0(\rho)$, i.e. $T(\rho)$. Expressing $\dot{M}$ through ρ and T, we obtain the dependence $\dot{M}(\rho) = \dot{M}(T(\rho), \rho)$ which also gives the required dependence $T(\dot{M})$ in the parametric form. Formally, the solution was obtained also in the optically thin regions but because of the assumptions made these points were rejected.

We examine the graphical representation of the results of solution of this equation. The asterisks in Fig. 11.8a show the dependence $T(\dot{M})$ for $\alpha = 1$ and $r = A/5$. As in Fig. 11.7, the dashed line restricts at the bottom the region of existence of solution of the equation (11.46), the solid line separates the regions of the optically thin and optically thick disks. Figures 11.8b,c show the solutions for $\alpha = 0.1$ and $\alpha = 0.01$, respectively. Figure 11.8d shows the dependence of the accretion time on disk thickness. It may be seen that all the resultant disks are geometrically thin, i.e. $H \ll r$.

The radiant heat conductivity is not the only mechanism of transfer of the generated heat into the optically thin regions. Under specific conditions, a significant role may also be played by convection. Without taking radiation pressure into account, the convective flow can the written in the form [387, 1128]:

$$F_{\text{conv}} = c_P \rho \left(\frac{|g|}{T}\right)^{1/2} \frac{l^2}{4}[\Delta(\nabla T)]^{3/2}, \qquad \Delta(\nabla T) = -\frac{T}{c_P}\frac{\partial S}{\partial z}.$$

Here c_p is the heat capacity at constant pressure, $S = \mathcal{R}\ln(T^{3/2}/\rho)$ is specific entropy, $g = -2\Omega_K^2 z$ is gravitational acceleration, l is the mixing length which is assumed to be equal to $l = \alpha H$. To determine the vertical distribution of temperature with convection taken into account, we should now solve the equation

$$\frac{\partial e}{\partial t} = \frac{\partial}{\partial z}\left[\frac{1}{\varkappa_{\text{abs}}\rho}\frac{\partial}{\partial z}\left(\frac{1}{3}acT^4\right)\right] - \frac{\partial F_{\text{conv}}}{\partial z} + \rho\alpha c_s^2 \Omega_K \tag{11.65}$$

with the same boundary conditions as for equation (11.53). Equation (11.65) no

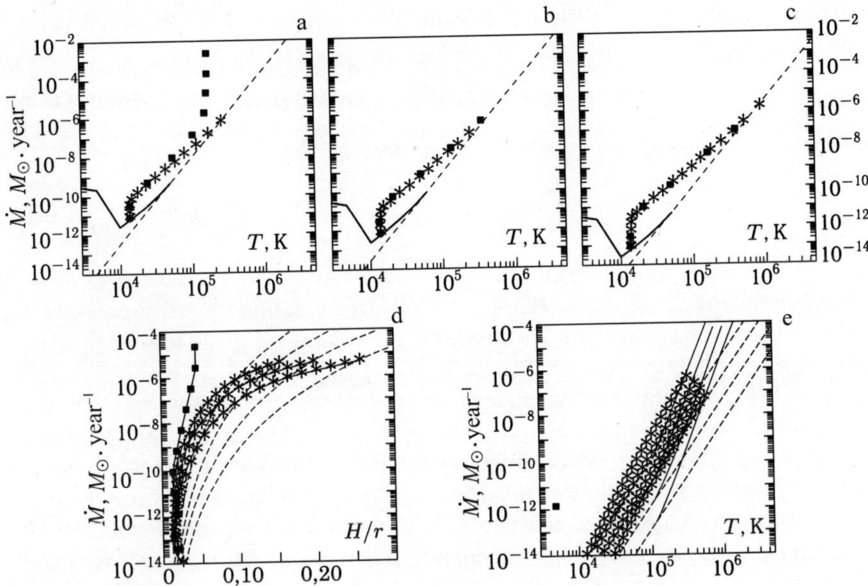

Fig. 11.8. Solution of equation (11.53) at $r = A/5$ for the optically thick disk (asterisks) and the solution of equation (11.65) with convection taken into account (squares), the dashed line restricts at the bottom the region of existence of the solution of equation (11.46), the solid line separates the regions of the optically thin and optically thick disks: a) at α = 1; b) at α = 0.1; c) at α = 0.01; d) the solutions for the disk with the dominance of gas pressure. The solid curves show the possible states of the disk on the plane (H/r, $\dot{M}$) for $r = A/5$. The curves with the asterisks are the solutions of equation (11.53), taking into account the radiant heat conductivity, convection and viscous heating for $\alpha = 1$. The dashed curves restrict at the bottom the region of existence of the solution (11.46) for $\alpha = 1$, $\alpha = 0.1$, $\alpha = 10^{-2}$, $\alpha = 10^{-3}$ (from top to bottom); d) solutions for the optically thin disk (asterisks) for α = 1, = 10^{-3}) from top to bottom); e) the solutions for the optically thin disk (*) for $\alpha = 1$, $\alpha = 0.1$, $\alpha = 10^{-2}$, $\alpha = 10^{-3}$ (from top to bottom) and $r = A/20$. The dashed lines restrict at the bottom the region of existence of solution of the equation (11.56), the solid lines separate the regions of the optically thick and optically thin disks.

longer permits a simple analytical solution and it must be solved numerically by the adjustment method. The solution of the equation is indicated in Fig. 11.8a–d by the squares. It may be seen that convection is important only for the case $\alpha \simeq 1$.

Summing up the results of analysis, it may be accepted that in the optically thick case at low values of $\dot{M}$ the temperature of the disk is constant $T = T_* =$ 13 600 K, and at high $\dot{M}$ the disk temperature increases as $T \propto \dot{M}^{1/3}$. Thus, for the realistic parameters of accretion disks in the close binary systems $\dot{M} \simeq (10^{-12}–10^{-7})$ $M_\odot$year^{-1}, α $\simeq 10^{-1}–10^{-2}$, the temperature of the gas of the disk in the outer regions ($r \simeq A/5–A/10$) is in the range from ~10^4 to ~10^6 K.

Solving equation (11.53) for different r, we can also calculate the dependences $T(r)$ and $\rho(r)$. Analysis of the results shows that $T \propto r^{-0.8}$, $\rho \propto$ r$^{-1.8}$, which is in good agreement with the results $T \propto r^{-3/4}$ from the study [1147].

The temperature of the optically thin disks is determined by the balance between the radiation heating $\Gamma(T, T_{wd})$ and viscous heating (11.31), on the one hand, and radiation cooling $\Lambda(T)$ on the other hand. The thermal balance equation (11.44) which can be written in the form

$$\rho\alpha c_s^2\Omega_K + \rho^2 m_{\rm p}^2(\Gamma(T,T_{\rm wd}) - \Lambda(T)) = 0, \tag{11.66}$$

is reduced to the quadratic equation with respect to ρ:

$$\alpha(\rho\mathcal{R}T + \frac{1}{3}aT^4)\Omega_K + \rho^2 m_{\rm p}^{-2}(\Gamma(T,T_{\rm wd}) - \Lambda(T)) = 0. \tag{11.67}$$

Solving this equation at fixed r and α, we obtain the dependence $\rho(T)$ and, subsequently, the dependence $T(\dot{M})$. Formally, the solution was also obtained for the optically thick regions but because of the assumptions made in this case these points are rejected.

Thus, the disks with the dominance of gas pressure are mostly optically thick and only at low $\dot{M}$ solutions are possible corresponding to the optically thin disks. However, the disks with the dominance of radiation pressure are mostly optically thin. The dominance of radiation pressure is possible only in the inner regions of the disk and, therefore, further analyses will be carried out for the case $r = A/20$ which for a typical dwarf Nova IP Peg corresponds to five radii of the accretor – the white dwarf.

The calculation results are presented in Fig. 11e where the asterisks show the dependences $T(\dot{M})$ for $\alpha = 1$, 0.1, 10^{-2}, 10^{-3} (from top to bottom) and $r = A/20$. The disks obtained in the solutions are geometrically thick, i.e. $H \simeq r$. It should be mentioned that the initial assumptions of the model restrict its applicability only to the case of geometrically thin disks and, therefore, the solutions for the geometrically thick disks are formal.

Analysis of the main processes of heating and cooling of matter in the accretion disks of the binary system shows that for the realistic parameters of the accretion disks in the close binary ($\dot{M} \simeq [10^{12}$–$10^{-7} M_{\odot}$ year^{-1} and $\alpha \simeq 10^{-1}$–10^{-2}), the gas temperature in the outer part of the disk is in the range from $\sim10^4$ K to $\sim10^6$ K.

11.2.4. Models with vertical convection

In addition to the radiant energy transfer from the inner layers of the disk, energy can also be transferred outside by convective motion (see (11.30)). However, radiation transfer always takes place at high temperature and convection may not form. The role and the conditions of formation of convection in the application to the internal structure of the stars have been examined in detail (see, for example [22]). There is a number of factors which do not permit formal transfer of the results of theoretical studies of the stars to the accretion disks. In particular, there are differences in the dependence of the gravitational force on the coordinate and viscosity is strong. The importance of taking into account convection in the gas disks was stressed by Paczynskii [999]; Vila [1266, 1267] investigated convective models of cold disks in cataclysmic binary and massive hot disks, and Lin and Papaloizou [841] studied a protoplanetary disk with convection. The construction of non-stationary models, based on detailed calculations of the z-structure, was described in [543, 893, 1178, 1179], etc. In application to the galactic cores, the convective instability was investigated by Mineshige and Osaki [923].

For the convective heat flow we can write

$$F_{\text{conv}} = C_p \rho v l \left\{ \left| \frac{dT}{dz} \right| - \left| \frac{dT}{dz} \right|_{\text{ad}} \right\}, \tag{11.68}$$

where v, l is respectively the velocity of climbing and the linear dimensions of the convective gas cells during mixing; C_p is heat capacity. As indicated by (11.68), the controlling role is played by the fact that the true temperature gradient is greater than the adiabatic one.

Meyer and Meyer–Hofmeister [892] detected the unstable distribution of temperature determined by ionisation of hydrogen. In ionisation of hydrogen ($T \simeq 10^4$ K) the opacity k (ρ, T) greatly changes in magnitude so that the radiation mechanism of energy transfer cannot longer ensure the required rate. This results in the formation of a steep temperature gradient between the internal layers where the main amount of heat is generated, and the outer layers from which flashing takes place. Consequently, suitable conditions form for convective motion. In the detailed calculations of transfer of radiation it is naturally not possible to use the simple relationships for opacity of the type (11.33), (11.34), taking into account also the fact that the approximation of the optically thick disk may be violated [893, 1179]. The opacity tables are used in most cases, see, for example [22].

The typical dependence of the effective surface temperature of the disk T_{ef} ($z = \pm h$) on the surface density of matter is shown in Fig. 11.9. Each curve has the characteristic S-shape – consists of three sections: cold section AB, hot section CD, transition section BC. The section BC corresponds to instability, whereas the hot and cold solutions are thermally stable. Since matter is completely ionised at $T \gtrsim 10^4$ K, the hot section is determined quite accurately. This cannot be said of the section AB. In the range of lower temperatures, the opacity law is not so accurate, the cold section can be optically thin and, therefore, the section AB is determined less accurately and differences were observed in the data obtained by different authors [1265].

11.2.5. Models of dwarf novae

The characteristic feature of the dwarf novae (the stars of the type U Gem, etc.) is their non-stationary behaviour (section 11.1.1). At the present time, there are quite a number of mechanisms proposed for explaining the phenomenon of dwarf novae and, on the whole, cataclysmic variables [44]. The explosive nature of the behaviour of many systems is evidently determined by the non-stationary mode of disk accretion. The problem is to determine the position of the 'valve' which 'opens' for a certain period of time which leads, in the final analysis, to a surge. There are three types of mechanisms:

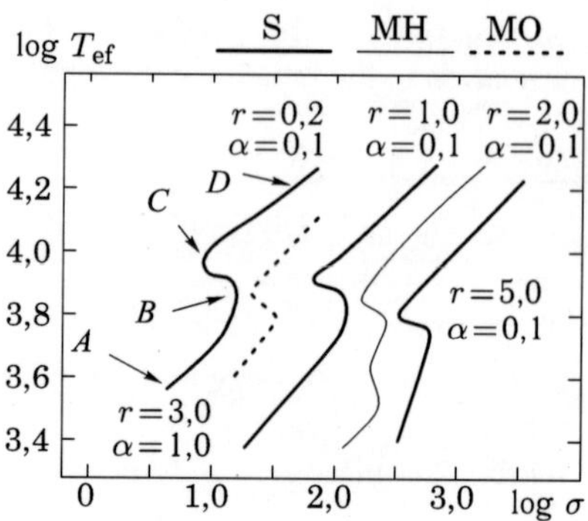

Fig. 11.9. Dependence of the effective surface temperature of the accretion disk T_{ef} on the surface density σ at different values of α and r (in the units of 10^{10} cm) in the case of the central object with the mass 1 $M_{\odot}$. The curves were obtained by: S – Smak [1179]; MH – Meyer and Meyer–Hoffmeister [892]; MO – Mineshige and Osaki [922].

1. The reason for the non-stationary nature is associated with a normal star. The instability of discharge of matter from a red dwarf leads to quasi-periodic oscillations in the buildup of matter by the disk and, at the same time, in the luminosity of the gas disk [342]. Since the gas in the accretion disk is supplied in individual portions, the non-stationary accretion (and, consequently, the flash) is simply the response of the disk to changes in the external conditions. Within the framework of the approach proposed in section 11.2.1, the rate of supply of matter into the accretion disk $\dot{M}_e(t)$ is a function of time which should be specified.
2. The mechanism of quasi-periodic activity may operate in the accretion disk itself [1178, 1179]. If the viscosity in the disk is relatively small, matter builds up in the outer region of the disk. When the density in the disk reaches a specific critical value, the rate of accretion increases as a result of some reasons (development of instability, turbulisation of the medium) and this results in a flash. Thus, the 'valve' is situated in the accretion disk itself.
3. The absence of accretion between the flashes can be explained by the 'propeller' effect [711]. The magnetic field of a rapidly rotating white dwarf prevents the fall of matter on its surface. This is accompanied by the buildup of matter in the vicinity of the boundary of the magnetosphere which leads to slow approach of the inner boundary of the disk of the white dwarf. Since this results in a reduction of the velocity of motion of the lines of force as a result of the solid-state nature of rotation, the 'propeller effect' disappears at a specific moment resulting in powerful accretion (flash). Subsequently, the system is in the initial state. Thus, within the framework of the described scenario, the valve is represented by the magnetic field of the accreting object.

Unstable accretion disks

The results of modelling the non-stationary accretion disks, based on taking into account convective instability, investigated in section 11.2.4, will be discussed

[543, 893, 1179, 1265]. A detailed study was carried out by Smak [1179]. The equations of disk accretion (section 11.2.1) were solved numerically, taking into account convective transfer in the z-direction (11.2.4). The rate of supply of matter to the accretion disk $\dot{M}_e$ and the parameter α remained constant throughout the entire calculation period. Of special interest are the temporal dependences of the luminosity of the disk $L_d(t)$ and the outer radius of the disk $R(t)$ because these quantities are observable (section 11.1.1). The quasi-periodic accretion modes were obtained in almost all cases. There were two types of solution: type A – initially the instability forms in the outer zone of the disk and, affecting all the inner regions, propagate through the centre of the accretion disk (Fig. 11.10a). The shape of the luminosity curve during a flash is non-symmetric, and the repeatability is almost strictly periodic. Type B is characterised by the formation of an instability initially in the inner region [$r \simeq (0.1 \div 0.2)R$] and subsequent propagation into and outside the accretion disk[9]. In this case, the instability sometimes does not reach the outer region of the disk (see Fig. 11.10b). On the whole, the flashes of the type B are less regular. With other conditions being equal, the flashes of type A are typical of higher rates of inflow of matter. Combined flashes AB can also take place, when instabilities form simultaneously in the outer and inner regions of the accretion disk.

The nature of flash activity of both types becomes apparent in the examination of evolution of the surface density between the active phases. In this period, the surface density is lower than the stationary value – matter builds up. When the surface density reaches a specific value, corresponding to the critical value – the point of rotation on the curve $T_{ef}(\sigma)$ (see Fig. 11.9, point B), and instability develops - a flash starts. The flash energy is gravitational energy, i.e. the radial flow of matter becomes more intensive. The region of instability propagates into both sides from the initial area. The rate of accretion increases, the reserve of matter in the disk decreases and, in the final analysis, the conditions for convection disappear and this results in an interruption of the flash.

It is important to mention the distinguishing special feature in the behaviour of $R(t)$: in the case of a type A flash, the radius of the disk increases by $\lesssim 20\%$, and the flash of type B is characterised by a very small variation of the value R ($\lesssim 7\%$) and, in some cases, the 'instability wave' does not even extend from the inner to the outer region.

Convective instability across the disk plane is not the only possible instability resulting in the quasi-periodic accretion mode. Identical consequences result from the gradient–entropy (GE) instability in the plane of the disk (see section 3.8.4 and 5.3), investigated by the authors of [179]. The development of a GE instability requires fulfilment of the specific relationships between the characteristic scales of the inhomogeneities of surface density L_σ and temperature L_T. Let it be that at the initial moment of time the gradients of the values $\sigma(r)$ and $T(r)$ are such that the accretion disk is GE-stable. However, the matter arriving in the stationary manner at the outer edge of the accretion disk ($\dot{M}_e$ = const) in heating as a result of the 'bright line' increase the gradients of surface density and temperature and this in

[9] The solution of this type was obtained for the first time in [542].

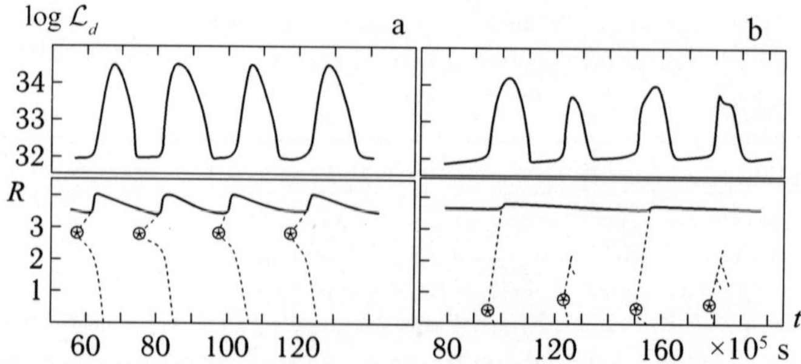

Fig. 11.10. Temporal evolution of luminosity $\mathcal{L}_d$ and the outer radius R of the accretion disk [1149]. The circle indicates the position of nucleation of the instability, the dashed line shows the propagation of the instability through the accretion disk: a) solution of type A; b) solution of type B.

the final analysis creates the conditions required for GE instability. The increase of the amplitude of perturbations in the accretion disk increases the level of turbulent viscosity. The matter, built up in the outer part of the accretion disk, effectively accretes on the compact object (flash), and the gradients $\sigma(r)$ and $T(r)$ evolve to the values at which the accretion disk becomes GE-stable. The turbulent viscosity, not sustained by the instability, and the accretion processes which determining this viscosity attenuate but the matter arriving in the stationary manner at the outer edge of the accretion disk prepares the system for a new accretion cycle (flash) (the form of the curve $\mathcal{L}_d(t)$ is identical with that shown in Fig. 11.10).

Taking these results into account, the following conclusions can be made. In both cases of the development of both convective and GE-instability the problem is mathematically reduced to solving the non-linear equation of diffusion with a source. It is well-known that many phenomena of self-organisation (establishment in the dissipative non-equilibrium medium of special structures, evolving with time) are described on the basis of single models mathematically expressed by non-linear equations of the diffusion type [212].

The models with the non-stationary inflow of mass

We discuss the response of an accretion disk to the increase of the rate of inflow of matter $\dot{M}_e(t)$ (discussion of the reasons for the non-stationary form of the value M_e is not the subject of the present monograph)[10]). Bath and Pringle [344, 345] carried out detailed calculations of a non-stationary accretion disk without taking into account the tidal interaction with the function $\dot{M}_e(t)$, the typical form of which is shown in Fig. 11.11a. Figure 11.11b shows the variation of the luminosity of the accretion disk in the form of the response to the external effect $[\dot{M}_e(t)]$ at a constant value of parameter α. The movement of matter in the radial direction is determined by the effect of viscosity (equation (11.29) gives the estimate $u \sim -\nu/r$), i.e. by the value α. Analysis of the dynamics of the processes and comparison with

[10]The variation of the rate of outflow of matter from the optical component into the CBS may be caused by dynamic instability in the star [211, 342, 369, 525, 1005].

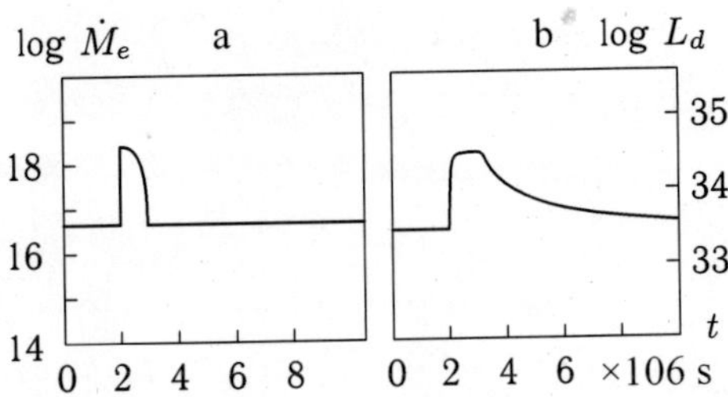

Fig. 11.11. Dependence of the rate of inflow of matter into the accretion disk $\dot{M}_e$ on time and the appropriate response of luminosity $\mathcal{L}_d$ [344].

the observations can be used to evaluate the value of parameter α. A satisfactory agreement is obtained at $0.01 \lesssim \alpha \lesssim 0.5$.

Inclusion of the tidal interaction in the form of (11.38a) in the calculations enabled Livio and Verbunt [853] to investigate the dynamics of the radius of the disk $R(t)$, caused by changes in the rate of overflow of matter. Let the evolution of the disk be governed by equation (11.36) which, as we understand, is the consequence of the law of conservation of the momentum of motion. Equation (11.36) is transformed taking into account (11.37):

$$\dot{M}_e \sqrt{G M_1 r_e} = \frac{2 G M_1}{q^2} \int_{r_1}^{R} \sigma f \, dr. \tag{11.69}$$

The left-hand part of the relationship (11.69) is linear in relation to $\dot{M}_e$. Under the integral sign only the quantity σ depends on $\dot{M}_e$; is is assumed that $\sigma \propto \dot{M}_e^n$. Equations (11.41), (11.42) show that $n = 0.6$; 0.7. These values are quite typical [986]. If $n < 1$, then the increase of $\dot{M}_e$ will be accompanied by an increase of the disk radius R. Taking this into account, we can describe quantitatively the reaction of the disk to the increase of $\dot{M}_e$. Let us assume that the equilibrium radius R_0 corresponds to the value $\dot{M}_{e0}$, and $R1$ corresponds to $\dot{M}_{e1}$. If $\dot{M}_{e0} < \dot{M}_{e1}$, then $R_0 < R_1$. The matter flowing into the disk has the specific angular momentum l_e to which the radius $r_e = l_e^2 / GM_1 < R_0$ corresponds. Consequently, when the value of $\dot{M}_e$ increases from $\dot{M}_{e0}$ to $\dot{M}_{e1}$at the initial moment the disk is compressed to a radius R_2 ($r_e < R_2 < R_0$). Consequently, with the increase of the density of matter the viscosity in the outer region of the disk also increases ($\nu \sim \alpha \, \Omega \, h^2 \propto \sigma$). At the same time, the outer boundary of the disk moves to a new equilibrium radius R_1. During this time, the mass and luminosity of the accretion disk increase. If the rate of inflow of matter returns to the previous value $\dot{M}_{e0}$, the disk has the initial dimensions. The characteristic times are determined by the viscous processes. Figure 11.12 shows the results of numerical solution of the equations of non-stationary accretion.

The most important difference between the two investigated models (unstable disk and a variable rate of supply of matter) is the presence in the second case of a short period of time when the size of the disk rapidly decreases and only then starts to increase. There are data indicating such a pecularity for Z Cha (see section 11.1.2).

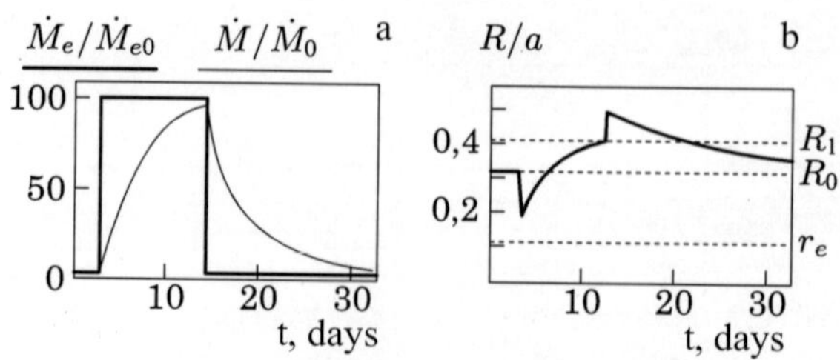

Fig. 11.12. Evolution of the accretion disk as a result of changes in the inflow of matter $\dot{M}_e(t)$ [853]. The following functions are indicated: a) $\dot{M}_e(t)$ and the rate of accretion onto the central object $\dot{M}_e(t)$; b) $R(t)$.

The previously investigated models are very simple primarily as a result of of the phenomenological approach to constructing the viscous models of the accretion disk based on the equations of the type of (11.28). Of course, when the disk state changes (density, temperature, etc.), the viscosity may evolve with a considerable delay and the presence and growth of the fine-scale magnetic fields can also be very important [445].

11.2.6 Self-modelling non-stationary solutions

Taking into account the previously investigated non-stationary solutions, obtained in numerical analysis, it is very interesting to consider the self-modelling approach developed in [550, 551]. We introduce new notations which will be used only in this section:

$$F = -W_{r\varphi} r^2, \qquad \dot{M} = -2\pi r \sigma u. \tag{11.70}$$

The value $\dot{M}$ is equal to the mass passing through the radius r in the unit time. If the effect of the second component of the system is ignored, equation (11.9) at transition to the variable $l = \sqrt{GM_1 r}$ has the form

$$\frac{\partial \sigma}{\partial t} = \frac{1}{2\pi r}\frac{\partial \dot{M}}{\partial r} = \frac{1}{4\pi}\frac{(GM_1)^2}{l^3}\frac{\partial \dot{M}}{\partial l}. \tag{11.71}$$

Similarly, equation (11.11): $\dot{M}(r,t) = 2\pi \dfrac{\partial F}{\partial l}$. The latter is substituted into (11.71) and we obtain

$$\frac{\partial \sigma}{\partial t} = \frac{1}{2}\frac{(GM_1)^2}{l^3}\frac{\partial^2 F}{\partial l^2}.s \tag{11.72}$$

To determine another relationship between σ and F, we can use the energy balance equation (11.30) in the approximation

$$F_v = F_{\text{rad}}. \tag{11.73}$$

As a result, the evolution equation has the following form

$$\frac{\partial F}{\partial t} = A \frac{F^m}{l^n} \frac{\partial^2 F}{\partial l^2}, \tag{11.74}$$

Here A is a constant, and the values of the parameters m and n depend on the selection of the specific model (laws of viscosity and opacity, the equations of the state of matter). Thus, for example, if the gas pressure is greater than the radiation pressure, then $m = 2/5$, $n = 6/5$ for the case $k_T \gg k_{ff}$ and $m = 0.3$, $n = 0.8$ in the reverse limit $k_{ff} \gg kT$.

Let it be that at the initial moment of time the matter in the outer region of the disk at the radius r_0 is present in the form of a ring which then accretes onto a compact object. In the self-modelling approach, we obtain three stages of accretion. In the first stage, the matter reaches the accreting object. In the second stage, the matter accretes, the rate of accretion $\dot{M}(t)$ and luminosity $\mathcal{L}(t)$ increase with time:

$$\dot{M} \propto t^{2,47} \qquad (k_{ff} \gg k_T); \quad \dot{M} \propto t^{1,66} \qquad (k_{ff} \ll k_T). \tag{11.75}$$

The third stage is characterised by the decrease of $\dot{M}(t)$ and $\mathcal{L}(t)$ with time:

$$\dot{M} \propto t^{-1,44} \qquad (k_{ff} \gg k_T), \dot{M} \propto t^{-1,25} \qquad (k_{ff} \ll k_T). \tag{11.76}$$

During stage I the energy is mostly emitted from the range $r \sim r_0$ and it is low because the value r_1/r_0 is small. In the stages II and III, a large part of energy leaves the region r_1 and it is proportional to the rate of accretion.

11.2.7. 'Thick' accretion disks

At accretion of matter, part of gravitational energy[11] is used for heating the gas, with the inner regions of the disk being the hottest. In the final analysis, the thermal energy is taken away by radiation which encounters along its path the matter moving in the opposite direction and prevents accretion. The main source of opacity for the completely ionised plasma is the scattering of the photons on the free electrons which is determined by the Thomson equation:

$$\sigma_T = \frac{8\pi}{3}\left(\frac{e^2}{m_e c^2}\right)^2 = 0.665 \cdot 10^{-24}\ \text{cm}^2, \tag{11.77}$$

where σ_T is the Thomson scattering cross-section. In the mean, the photon during collision transfers its entire momentum to the electron and subsequently to the proton as a result of electrostatic connection. Let the luminosity be equal to $\mathcal{L}$(erg/s). Since the photon energy $\sim pc$ (p is the photon pulse), the number of the photons intersecting the unit area in unit time, is equal to $N = \mathcal{L}/(4\pi r^2 pc)$. During unit time, the electron undergoes $N\sigma_T$ collisions. Let it be that on average the photon transfers to the electron the momentum p in a single collision. Since the force, acting on the electron, is the rate of transfer of the momentum, then

[11]In the stationary model, the total luminosity of the disk is determined by half of the gravitational energy of incident matter.

$$f_{\text{rad}} = \frac{\mathcal{L}\sigma_T}{4\pi r^2 c}. \tag{11.78}$$

The gravitational force, acting on the proton and, consequently, also on the electron is also proportional to $1/r^2$. Therefore, in the approximation of spherical symmetry there is critical luminosity (referred to as the Eddington limit) determined by the balance of the gradient of radiation pressure and gravitational force

$$\mathcal{L}_{\text{cr}} = \frac{4\pi c G M_1 m_p}{\sigma_T} = 1.3 \cdot 10^{38} (M_1 / M_\odot) \text{ erg/s}. \tag{11.79}$$

The critical value of luminosity corresponds to the critical value of the rate of accretion $\dot{M}_{\text{cr}}$.

In the case of supercritical disk accretion on approach to the central body the disk ceases to be thin ($r \sim 2h$). The radius at which the luminosity reaches $\mathcal{L}_{\text{cr}}$, is referred to as the spherisation radius

$$r_{\text{sph}} = \frac{G M_1 \dot{M}_{\text{cr}}}{L_{\text{cr}}} = \frac{\dot{M}_{\text{cr}}\sigma_T}{4\pi m_p c} = \frac{2}{3} 10^{14} \frac{\dot{M}_{\text{cr}}}{M_\odot / \text{year}} (\text{cm}). \tag{11.80}$$

The identical estimate is obtained from the equality $r_{\text{sph}} = 2h$ for the value h determined by (11.40). A proportion of matter starts to be discharged from the range $r < r_{\text{sph}}$ in the form of a quasi-spherical shell. As assumed by Shakura and Syunyaev, in the case $\dot{M} > \dot{M}_{\text{cr}}$ the self-consistent accretion mode starts to operate which results in the luminosity only slightly exceeding the Eddington limit. In this case, in the range $r > r_{\text{sph}}$ the disk remains thin and its structure is identical to the subcritical mode. For a black hole of stellar mass it is assumed that $r_{\text{sph}} = 10$ km and, consequently, from (11.80)

$$\dot{M}_{\text{cr}} = \frac{4\pi r_{\text{sph}} m_p c}{\sigma_T} = \frac{3}{2} 10^{-8} M_\odot / \text{year}. \tag{11.81}$$

The structure of the 'thick disk', close to hydrostatic equilibrium and situated at the balance of the gradient of pressure and centrifugal and gravitational forces [280, 301, 352, 619, 721, 834, 1001] will be discussed. In particular, it should be mentioned that in the presence of a strong pressure gradient the velocity of rotation of matter can greatly differ from the Kepler velocity and, at the same time, the general rate of dissociating energy can be considerably smaller than the value $\sim G M_1 \dot{M} / r_1$, characteristic of thin accretion disks.

In the theory of accretion disks, the problem of viscosity is one of the unexplained problems and only as a result of the assumption on the small thickness of the accretion disk ($h/r \ll 1$) it is possible to construct relatively accurate and simple models. However, in the case of a thick disk, the problem of the value of viscosity is far more complicated, and in the studies [280, 721, 1001] models of thick disks without viscosity were constructed. Later, in a number of investigations [176, 177, 301, 352, 619, 834] attention was given to the range $r_1 \ll r \ll r_{\text{sph}}$ taking into account dissipative processes. Without detailed explanation, we shall mention only the principal approach to the construction of the models of this type in the spherical coordinate system (r,φ,Θ). This approach is based on constructing stationary axisymmetric ($\partial/\partial\varphi \equiv 0$) solutions, characterised by the self-consistent distribution of the parameters of the flow of the gas in the central gravitational field along the radial and meridional coordinates Θ[12]. This is possible if the ratio between the radial components of the forces (gravitational, centrifugal and the force determined by the pressure gradient) acting on the rotating matter remains constant along the radial coordinate. In this case, the Θ-distributions of all quantities at different distances from the centre will be similar to each other and the flow of matter will be purely radial. The solution of the equations of gas dynamics is presented in the form $f(r,\Theta) = f_*(\Theta)\, r^{-a_f}$, where $f = \{\rho, \mathbf{u}, T, s, p\ldots\}$, a_f = const. In this case, the parameters a_f are selected to ensure that in each equation all the terms are proportional to the same degree of the radial coordinate. These solutions are naturally referred to as self-modelling. As a result of applying this procedure, the equations in the partial derivatives are reduced to the conventional differential equations with respect to the variable Θ (or $t \equiv z/r$, as in the case of Liang [834], in the cylindrical coordinate system). This is followed by definition of specific boundary conditions at $\Theta = \pi/2$ and $\Theta = \Theta_0$ ($0 \leqslant \Theta_0 < \pi/2$); the resultant boundary-value problem is solved numerically. As a result, we determine the distributions of all physical quantities on the plane (r, Θ).

One of the main advantages of the self-modelling solutions is that they are simple to construct so that we can investigate the effect of different factors on the structure of the accreting flow. However, it is evident that within the framework of the self-modelling approximation we can only consider the formation of the quantitative pattern and it is difficult to expect any reliable quantitative results. In addition to this, there are problems associated with the stability of the models of thick accretion disks [970]. It should be mentioned that the range of application of the models of dissipative thick disks may be represented by the zone $r_1 \ll r \ll r_{\mathrm{sph}}$ in 'thin' accretion disks.

Taking into account natural difficulties in the construction of analytical (not self-modelling) two-dimensional models, it should be mentioned that the numerical modelling of axisymmetric flows in the field of a black hole, including the relativistic effects, is now very popular. Initial calculations already show that in the process of fall of cold matter with a non-zero pulse momentum the gas is heated by a standing shock wave resulting in the formation of a thick ($h \sim r$) disk, sustained by the pressure gradient [1305]. In fact, a torus, rotating with a non-Kepler velocity, forms in this case. In [657] the authors modelled the regions closest to the black hole for the case of a relatively small momentum of the pulse of accreting matter. The gas was pulled quite strongly along the spiral trajectory into the black hole without undergoing, as assumed by the authors, viscous interaction.

ADAF- and CDAF-models

Within the framework of the standard α-model of the accretion disks [1147] there

[12]Liang [834] constructed models in a cylindrical coordinate system (r, φ, z).

is a number of problems both theoretical and with explanation of observational data. The latter include the explanation of the low luminosity of x-ray binary and active galactic nuclei with black holes ([953], see review in [951] and references therein). One of the most extensively investigated objects of this type is a source in the centre of the galaxy Sgr A^* [900]. The luminosity is reduced in the so-called ADAF-models (advection-dominant accretion flows) [281]. They are hotter and the Eddington luminosity is achieved at a lower accretion rate. Observations of a number of objects, for example A0620-00, GRO J6055-40, and a central source in the active galaxy M 106 can be used to interprete the spectral data on the basis of the ADAF-model [644, 820, 954].

The starting point of the investigations of ADAF-models was the study [957] in which the authors constructed self-modelling solutions of stationary accretion whose characteristic special feature is accretion at weak radiation or even in the absence of cooling as a result of radiation losses. The main mechanism of energy transfer is the radial advective flow, and the transfer of the angular momentum is ensured by α-viscosity. The structure of the flow is very similar to spherical and the rotation of the gas greatly differs from the Kepler law. In the presence of strong viscosity ($\alpha \gtrsim 0.3$) numerical modelling gives the results identical with the ADAF-models [707]. However, in the calculations with a low value of parameter α we obtain flows with strong turbulence as a result of the development of convective instability which greatly changes the spatial structure of the flow [707, 1209]. These models are referred to as 'convection-dominant accretion flows' (CDAF) [331].

The most important properties of these models are the generation of convective turbulence, transfer of the angular momentum to the centre, the compensating flow of the angular momentum to the outside as a result of viscosity (or, for example, magnetorotating instability [290, 329, 956] and references therein), together with the flow of thermal energy along the radius. Numerical modelling demonstrates a lower rate of accretion and a strong energy flux to the outside as a result of strong radial convection. The fundamental problem of the CDAF models is the presence of the flow of the angular momentum to the inside [331].

For the CDFA it was possible to construct self-modelling solutions within the framework of the approximation of the mixing path [955], and numerical modelling was carried out to investigate convection in a gas torus in the plane r–z [709]. The main results for ADAF and CDAF were using axisymmetric models [706, 707, 709, 957, 958, 1209]. Advective cooling starts to play a significant role when the disk ceases to be thin, i.e. when $h \sim r$ [1002]. The ratio of the speed of sound to the Kepler velocity of rotation in the previously cited numerical models reaches $c_s/r\,\Omega_K \simeq 0.2 \div 0.5$.

Evidently, the transition to the three-dimensional ADAF-models does not change quantitatively the conclusions of axisymmetric calculations [708]. It should be mentioned that important results on convective transfer of the angular momentum inside were obtained already when constructing the models of rapidly rotating stars with a convective core [26].

The models of the convectively-dominating accretion flows are attractive for explaining the low luminosity of the accreting black holes and x-ray binary and galactic cores in the presence of a supermassive black hole.

The presence of the magnetic field (MHD CDAF) can greatly change the properties of the flow, in particular, convection may result in the flow of the angular momentum both into the side and on the periphery along the radial coordinate. Evidently, this is associated with the effect of magneto-rotational instability [331, 705]. In the model of the gas axisymmetric thick disk in the presence of the magnetic field it was shown that perturbations with the wavelengths greater than the vertical scale of the disk remain convectively unstable [956].

It should be mentioned that the internal radiation-dominating regions of the accretion disk can be unstable in relation to vertical convection also in a relatively thin disk [389]. The non-linear stage of this convection in the thin axisymmetric accretion disk was investigated for the radiation-dominant zone of the standard model of the accretion disk for r–z perturbations in study [288].

In section 5.3 we considered the consequences of development of convective instability in the plane of a non-axisymmetric disk. In this case, non-axisymmetric perturbations are unstable and the reason for convection is associated with the radial heterogeneities of the quasi-Kepler disk. Convective mixing in the plane of the disk may result in the mean in the radial movement of matter (accretion) without taking into account the action of viscous forces. Thus, the convective instability may play a significant role in the physics of accreting systems and the possibility of formation of convective flows has been demonstrated within the framework of various accretion models.

11.3. Large-scale instabilities in accretion disks

The name of this section indicates that the number of pages dealing with this subject may be considerably greater than the size of this book. Since the accretion disks are gas disks, they can be described fully by the results relating to the possibility of development in gas disks of the instabilities investigated in chapter 5 and section 11.2.4.

The matter in accretion disks is fully or partially ionised plasma. Plasma may contain a very large number of unstable modes [138, 139] whose development effectively turbulises the matter. It may be said that the turbulent state is natural for plasma. As shown by the estimates, these models are mostly small-scale in comparison with the thickness of the disk. However, for successive solution of the problem of stability our considerations regarding the physical conditions on the scales are not sufficiently accurate. For this reason, the automatic transfer of the achievements of plasma physics to the systems studied here does not give the realistic picture.

In this section, in addition to the already investigated factors, we will discuss large-scale instabilities. For these instabilities it is not essential to know in greater detail the vertical structure of the disks.

11.3.1. Instability of a radiation-dominant region

We examine the standard model of the accretion disk (α-model, section 11.2). The gas temperature in the accretion disk increases on approach to the accreting object. In fact, taking into account the balance of energy generated as a result of the effect of viscosity (11.31) and carried away by radiation (11.32), in the approximation of black body radiation we obtain

$$\frac{1}{2}W_{r\varphi}r\frac{d\Omega}{dr}=\frac{3}{8\pi}\dot{M}\frac{GM_1}{r^3}\left[1-\sqrt{\frac{r_1}{r}}\right]=\sigma_{SB}T^4, \qquad (11.82)$$

where SB is the Stephan–Boltzmann constant; r_1 is the inner boundary of the disk. From (11.82) for $r \gg r_1$ we determine the radial distribution of temperature

$$T=\left(\frac{3}{8\pi\sigma_{SB}}\frac{\dot{M}GM_1}{r^3}\right)^{1/4}\propto r^{-3/4}. \qquad (11.83)$$

Firstly, an increase of temperature is accompanied by the increase of the contribution of radiation pressure $P_{rad} \propto T^4$ in comparison with the gas pressure P_{gas} and, secondly, the role of Thompson scattering on free electrons becomes controlling. Section 11.2.2 presents the solution for the three regions: a) $p_{rad} \gg p_{gas}$, $k_T \gg k_{ff}$; b) $p_{gas} \gg p_{rad}$, $k_T \gg k_{ff}$; c) $p_{gas} \gg p_{rad}$, $k_{ff} \gg k_T$. It is quite easy to obtain equations for the boundaries between the zones [1147]

$$\frac{r_{ab}}{3R_g}\simeq 10^2\left(\alpha m\right)^{2/21}\left(\dot{m}\right)^{16/21}, \qquad (11.84)$$

$$\frac{r_{bc}}{3R_g}\simeq 10^4\left(\dot{m}\right)^{2/3}. \qquad (11.85)$$

It can be seen that in the case of black holes and neutron stars with a weak magnetic field the presence of the radiation-dominated region is quite realistic ($r_1 \simeq 3R_g \lesssim r \lesssim r_{ab}$).

The problem of the stability of disk accretion of stellar mass on the black hole was formulated for the first time in [837, 838]. Detailed analysis of the stability in relation to the axisymmetric perturbations taking into account radiation pressure was carried out by Shakura and Syunyaev [1148]. In subsequent studies, they investigated: the stabilising effect of the effects of the general theory of relativity [278], general polytrope models [442], non-axisymmetric perturbations [895], taking into account sonic and epicyclic oscillations [261].

Prior to examining the dispersion properties of a viscous disk with radiation, it should be mentioned that in the case $p_{rad} \gg p_{gas}$ the dynamic viscosity in the stationary state is expressed through the combination of constants. Actually, equating (1131) and (11.32) and using (11.22), (11.33) for $p = p_{rad}$, we obtain

$$\frac{\nu\sigma}{2h}=3.5\cdot 10^{10}\,C^2\quad \text{g/s}\cdot\text{cm}\simeq(1\div 3.5\cdot 10^{10})\quad \text{g/s}\cdot\text{cm}. \qquad (11.86)$$

It should not be expected that the value of turbulent viscosity in any disk will be determined only by the constant value (11.86). Consequently, this cast doubts on the possibility of development of a thermal instability, determined by the inequality $F_\nu \neq F_{\text{rad}}$.

We confine ourselves to examination of the axisymmetric perturbations $\propto \exp\{-i\omega t + ikr\}$. For the perturbations $h \ll 1/k \ll r$ we will retain in the linearised equations only the terms $\propto (kh)^2$ and ignore the terms of the order $(h/r)^2$ and kh^2/r. For axisymmetric perturbations, the disk remains a Kepler disk with the accuracy to $(h/r)^2$, kh^2/r and, consequently, taking into account the viscosity law (11.28) we can use the equation (11.27) $\dot{M}_e \equiv D \equiv 0$:

$$\frac{\partial \sigma}{\partial t} = 2\alpha \frac{\partial}{r\partial r}\left(\sqrt{r}\frac{\partial}{\partial r}\left(\Omega h^2 \sigma \sqrt{r}\right)\right). \tag{11.37}$$

The functions $\sigma(r, t)$ and $h(r, t)$ will be presented in the form

$$\sigma = \sigma_0(r)[1 + \Sigma(r,t)], \quad h = h_0(r)[1 + H(r,t)], \tag{11.88}$$

where Σ and H are small. After linearisation we obtain

$$\frac{\partial \Sigma}{\partial t} = 2\alpha\Omega h_0^2 \frac{\partial^2}{\partial r^2}(\Sigma + 2H). \tag{11.89}$$

We use the equation of the energy balance in the form (compare with (3.244))

$$\frac{d}{dt}\left(\frac{\varepsilon}{\rho}\right) = \frac{\mathcal{P}}{\rho^2}\frac{d\rho}{dt} + q^+ - q^-, \tag{11.90}$$

This equation takes into account the work of the pressure forces, dissipation q^+ and radiation q^-, and $d/dt = \partial/\partial t + u\partial/\partial r + w\partial/\partial z$. Multiplying (11.90) by ρdz and integrating over the entire disk, assuming the homogeneity of expansion of compression of the disk along the axis z:

$$\frac{\partial}{\partial t}(Eh) + P\frac{dh}{dt} + \frac{\partial}{r\partial r}(E+P)uhr - u\frac{\partial}{\partial r}(Ph) = F_\nu - F_{\text{rad}}, \tag{11.91}$$

where $P(r, t)$ and $E(r, t)$ are the mean values with respect to the z-coordinate of respectively pressure and internal energy ($p(r, t) = R\mathcal{P}dz = 2hP$ is surface pressure). Equation (11.91) is the generalisation of (3.245) taking into account the dissipative effects for the axisymmetric models. In addition to the gas-dynamic pressure, we will also take into account the radiation pressure [110]

$$E = \frac{P_{\text{gas}}}{\gamma - 1} + 3P_{\text{rad}} = \frac{1 + \beta(3\gamma - 4)}{\gamma - 1}P, \tag{11.92}$$

where the coefficient $\beta = P_{\text{rad}} / P (P = P_{\text{rad}} + P_{\text{gas}})$ characterises the fraction of radiation in the total pressure; γ is the volume adiabate index. Using the continuity equation (3.239), (11.92) and the relationship $p \propto \Omega^2 \sigma h^2$ (see (3.238)), we can write the equation (11.91) in relation to surface pressure (see (3.247), (3.252)):

$$\frac{\partial p_{\text{gas}}}{\partial t}+\gamma_s p_{\text{gas}}\frac{\partial ru}{r\partial r}+u\frac{\partial p_{\text{gas}}}{\partial r}+7\frac{\gamma-1}{\gamma+1}\left\{\frac{\partial p_{\text{rad}}}{\partial t}+\frac{9}{7}p_{\text{rad}}\frac{\partial ru}{r\partial r}+u\frac{\partial p_{\text{rad}}}{\partial r}\right\}=$$
$$=4\frac{\gamma-1}{\gamma+1}\left[F_\nu-F_{\text{rad}}+\frac{pu}{2\Omega}\frac{d\Omega}{dr}\right], \quad (11.93)$$

where γ_s is determined by (3.248). Excluding the pressure using (11.22) and taking into account the expressions for F_ν, F_{rad}, after linearisation the equation (11.93) is reduced for the case $\gamma = 5/3$ to the form

$$(8+51\beta_0-3\beta_0^2)\frac{\partial H}{\partial t}+3(4\beta_0^2+3\beta_0+1)\frac{\partial \Sigma}{\partial t}=$$
$$=\frac{2}{3}(9\beta_0^2+18\beta_0+5)\alpha\Omega h_0^2\frac{\partial^2}{\partial r^2}(\Sigma+2H)+6\alpha\Omega[(1+\beta_0)\Sigma+(5\beta_0-3)H], \quad (11.94)$$

where the parameter β_0 corresponds to the non-perturbed state. In the framework of the approximation $h \ll 1/k \ll r$ we find the solution of the system (11.89), (11.94) in the form $\Sigma, H \propto \exp\{-i\omega t+ikr\}$, which results in the following dispersion equation[13]:

$$(8+51\beta_0-3\beta_0^2)\left(\frac{\omega}{\alpha\Omega}\right)^2+i[2(4+23\beta_0-3\beta_0^2)(kh_0)^2+$$
$$+6(3-5\beta_0)]\left(\frac{\omega}{\alpha\Omega}\right)-4(5-3\beta_0)(kh_0)^2=0. \quad (11.95)$$

Solving the equation (11.95) is not difficult, and Fig. 11.13 shows the dependences $\omega(k)$ for different values of β_0. For the instability [Im $(\omega) > 0$] it is necessary to fulfil

$$\frac{1}{3}(4+23\beta_0-3\beta_0^2)(kh_0)^2<5\beta_0-3.$$

At $\beta_0 < 3/5$ perturbations with any kh_0 are stable. In the case $\beta_0 > 3/5$ the waves with $k < k_{cr}$ are unstable. In the shortwave range $k_{cr} < k < k_0$ (β_0) only one unstable mode for which Re$(\omega)\neq 0$ changes, i.e. the perturbations have the form of concentric waves running on the disk. In the case $k < k_0$ there are two unstable branches with Re$(\omega) \equiv 0$, which corresponds to the standing waves. In the limit of long waves for the increments from (11.95) we obtain the following asymptotics

$$\text{Im}(\omega_T)\simeq\frac{6(5\beta_0-3)}{8+51\beta_0-3\beta_0^2}\alpha\Omega,$$
$$\text{Im}(\omega_D)\simeq\frac{2(5-3\beta_0)}{3(5\beta_0-3)}(kh_0)^2\alpha\Omega+O\left(\frac{kh^2}{r}\right). \quad (11.96)$$

[13]Small differences in the numerical coefficients for various authors are associated with the method of averaging the initial equations with respect to the z-coordinate.

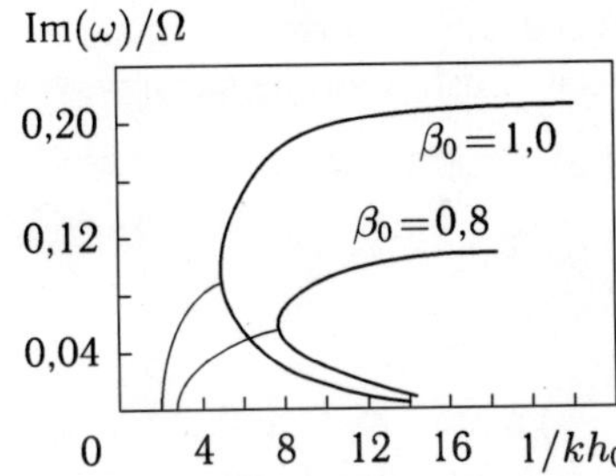

Fig. 11.13. Dependence of the increment on the wave number at different $\beta_0 = p_{rad}/p$. At $\beta_0 < 3/5$, the disk-model is stable (Imω < 0).

The physics of these instabilities differs. Using the relationship between the perturbation Σ and H, resulting from (11.94), and the dispersion equation (11.95), we obtain that on the lower branch for long wavelength values $\Sigma = -H$ is fulfilled, i.e. viscosity is not perturbed $[\tilde{\eta} = (\Sigma + H)\eta_0 \simeq 0]$. The instability is dynamic or viscous.

For the other unstable branch the increment $\mathrm{Im}(\omega_T)$ increases with increase of the wavelength and parameter β_0. At $\beta_0 = 1$ we obtain $\mathrm{Im}(\omega_T) \simeq 0.2\alpha\,\Omega$. In the limit $kh_0 \ll 1$ easily obtain the estimate $\Sigma/H \ll 1$. Thus, the upper branch describes the thermal instability.

Investigations of the modified α-models, in which the viscosity is not proportional to the total pressure and it is proportional to gas pressure[14] $[W_{r\varphi} = -\alpha(1-\beta)2hP$, compare with (11.15)], indicate the are stable with respect to the axisymmetric perturbations even in the radiation-dominated range. The analysis, carried out in [262], confirms that the stability criterion for thermal and viscous modes $\beta < \beta^{cr} = 3/5$ [1148], is valid only at $\Delta h = 0$, $\Delta\sigma = 0$, $\delta_h = 2$, $\delta_\sigma = 0$ (see section 11.3.3). The critical value β^{cr} in a general case depends on the values of these parameters. The thermal mode may be unstable also in the case of dominance of gas pressure ($\beta = 0$), in particular, in the model of the optically thin accretion disk.

In the framework of the homogeneous equilibrium model the dispersion properties of the non-axisymmetric ($\propto \exp\{-it\omega + ikr + im\varphi\}$) perturbations are described by the equation (11.97) after substituting ω by $\omega - m\Omega$. Thus, the increments of the investigated instabilities do not change.

The flat adiabate parameter

In conclusion, we derive the expression for the effective flat parameter of the adiabate γ_s taking into account the radiation pressure (section 3.6.1) [253]. In the limiting case $p_{rad} \ll p_{gas}$, equation (11.93) leads to (3.248). In the reverse limit $p_{rad} \gg p_{gas}$ from the equality to zero of the value in the braces in (11.93) we obtain $\gamma_s = 9/7$. Since for the radiation $\gamma = /3$, the equation (3.248) remains valid. At an arbitrary value of the parameter β the expression for the flat parameter of the adiabate can be derived by investigating the dynamics of small axisymmetric perturbations at the background of the equilibrium homogeneous state without taking dissipation

[14] This viscosity is referred to as magnetic viscosity. This is associated with the fact that the turbulent dynamo generates the magnetic field $\mathbf{B} = (B_r, B_\varphi)$, and $B_r / B_\varphi \simeq \sqrt{h/r}$ and the magnetic pressure $P_m = B_\varphi^2/8$ is comparable with gas pressure.

and self-gravitation into account. In the shortwave approximation, equations (11.93), (3.239) and (3.242) lead to the dispersion relationship for the sonic waves $\omega^2 = \gamma_s p_0 k^2/\sigma_0$ with

$$\gamma_s = 1 + 2\frac{a_3 - a_1}{a_2}, \qquad (11.97)$$

where $a_1 = 1 + 3\beta + 4(3\gamma - 4)\beta^2$; $a_2 = (1 + 3\beta)(1+ \gamma) + \beta(3\gamma - 4)(9 - \beta)$; $a_3 = (1 + 3\beta)(\beta(3\gamma - 4) + \gamma)$ [261]. The relationship (11.97) corresponds to the standard definition $\gamma_s = (\partial \ln p/\partial \ln \sigma)_s$ taking into account the last term in the right-hand side of equation (11.93). At $\gamma > /3$ taking the radiation pressure into account results in a monotonic reduction of the value γ_s. At $\gamma_s \leqslant /3$, the condition $\gamma \leqslant 9/7$ is fulfilled but the dependence $\gamma_s(\beta)$ is not monotonic – it has a minimum. This results is similar to the well-known effect for a bubble liquid in which the speed of sound is lower than in the gas and liquid respectively. It should be mentioned that at specific values of the quantities γ and β, the relationship (11.97) permits $\gamma_s < 0$, which indicates the formation of unstable solutions in which an increase of the surface density reduces the value of surface pressure.

11.3.2. Tidal instability

In a gas disk rotating around a compact object with the mass M_1 an instability may develop which is associated with the tidal effect from the side of the second component with the mass M_2 of a close binary system. The physics of this instability can be easily understood by considering the movement of a test particle in the gravitational field of the binary system.

Let us assume that in the case $M_2 = 0$ the test particle moves along a periodic orbit. It is well-known that small oscillations in the vicinity of the trajectory takes place with the epicyclic frequency $\varkappa = \varkappa_0$, the deviations are described by the conventional equation of the harmonic oscillator (see the section 1.2.3)

$$\frac{d^2\xi}{dt^2} + \varkappa^2\xi = 0. \qquad (11.98)$$

Taking into account the normal star ($M_2 \neq 0$), rotating with the period $T_0 = 2\pi/\Omega_0$, the frequency of oscillations in (11.19) will be a periodic function of time and it can be assumed that

$$\varkappa^2 = \varkappa_0^2\left[1 + \sum_{n=1}^{\infty} \varepsilon_n \cos(n\Omega t)\right], \qquad (11.99)$$

where Ω is the mean angular velocity at movement of the particle in the periodic orbit, the coefficients ε_n are determined by the potential of the second star [678]. In this case, equation (11.98) describes the parametric resonance [112]. At specific ratios between the natural frequency and the frequency of the driving force, the deviation ξ stars to increase rapidly with time – the tidal instability develops. Since the period of the binary system T_0 depends on the relative mass of the perturbing body $\mu = M_2/(M_1 + M_2)$, the parameters ε_n are also determined by the value μ, and only odd terms differ from zero (because of the symmetry of the potential). The resonance with $\varepsilon_3 = 1.85\mu$ is most intensive and, consequently, a tidal instability forms when the period of the test particle equals $\simeq 1/3$ of the period of the second star in the inertial system[15] [678, 679, 1297].

In the case of a Kepler disk, the period of rotation of the particles increases with the radius and there is the value r_0 at which resonance starts. As shown in sections 11.2.1, 11.2.5, as a result of the tidal interaction in the CBS (close binary system), the disk has a finite size. The outer radius of the disk R is determined by the relationship $q = M_2/M_1$ – as q increases, the value of R decreases. For $q \ll 1$ the disk extends outside the radius r_0. Consequently, a tidal instability develops in the outer region of the disk and results in the formation of a slowly precessing elliptical disk. The period of rotation of such an elliptical disk is 3–6% greater than the orbital period of the binary system. At $q \simeq 1/4$, the equality $R \simeq r_0$ is fulfilled. Therefore, no tidal instability develops in the systems with $q \geqslant 1/4$.

The formation of a tidal instability is determined by the parametric resonance 3:1 between the orbital motion of the gas in the disk and the orbital rotation of the binary system, investigated in detail within the framework of numerical modelling of the gas disk in the CBS [678, 679, 1297]. Possibly, the so-called superhumps in the systems of the type SU UMa (periodic increase of shine by 20–30% with the period slightly greater than the orbital period during the superflash period (section 11.1.2)) are caused by the elliptical disk.

11.3.3. Dissipative–acoustic instability. The dynamics of longwave perturbations. Qualitative examination

The characteristic feature of the standard model of the accretion disk is the presence of the dependence of turbulenct dissipative coefficients on the disk parameters. In section 3.3 the dynamics of small perturbations with no account made for perturbations of the value of viscosity was investigated. The dissipative instability was studied in detail in section 3.3.1. Using a very simple model, we examine the effects of the dependence of dynamic viscosity on density ($\eta = \sigma \nu \propto \sigma^{1+\delta}$) on the nature of sonic oscillations in the disk plane in order to understand better the physics of dissipative instabilities (exact analysis in section 3.3.1).

Taking into account the perturbations of dynamic viscosity, equalling $\eta_1/\eta_0 = (1+\delta)\,\sigma_1/\sigma_0$, for the Kepler disk the system of equations for relatively longwave perturbations (3.353)–(3.355) has the following form

$$-i\omega\sigma_1 + i\sigma_0 k u_1 = 0, \tag{11.100}$$

$$-i\omega u_1 - 2\Omega v_1 = -ikc_T^2 p_1 / p_0 - \lambda_0 k^2 u_1, \tag{11.101}$$

[15]This resonance is well known in the stellar mechanics in a limited problem of three bodies. Examples include special features of the motion of asteroids in the Sun–Jupiter system and some special features of the structure of the Saturn rings taking planet satellites into account [43].

$$-i\omega v_1 + \frac{1}{2}\Omega u_1 = -k^2 \nu_0 v_1 - ik\frac{3}{2}\Omega \nu_0 \eta_1 / \eta_0 . \qquad (11.102)$$

To a first approximation it will be assumed that $p_1/p_0 = \gamma_s\, \sigma_1/\sigma_0$ (in [261, 262], the thermal balance equation took into account the dissipative factors). As a result, we obtain the dispersion equation

$$\omega^3 + ik^2(\lambda_0 + \nu_0)\omega^2 - (\Omega^2 + k^2 c_s^2 + k^4 \nu_0 \lambda_0)\omega - ik^2 \nu_0 \Omega^2 [3(1+\delta) - c_s^2 k^2 / \Omega^2] = 0 . \qquad (11.103)$$

We will derive an equation for the frequency of the longwave sonic oscillations in the form $\omega/\Omega = 1 + \varepsilon k^2$ ($|\varepsilon k^2| \ll 1$). Thus, for the apparent part of frequency we have

$$\mathrm{Im}(\omega) \simeq \frac{k^2 \nu_0}{2}\left(2 + 3\delta - \lambda_0 / \nu_0\right). \qquad (11.104)$$

If the condition $(2+3\delta)\nu_0 > \lambda_0$ is fulfilled, we have an instability of acoustic oscillations. It should be mentioned that for the α-model in the region of dominance of the gas pressure and opacity k_T we have $\delta = 2/3$ (see section 11.2). The increase of the intensity of perturbations is determined completely by the dependence of dynamic viscosity η on the disk parameters. Taking into account the thermal processes, radiation, radiation pressure, dependence of viscosity on temperature, the vertical structure of the disk and the vertical oscillations changes the values of the increments and retains the possibility of formation of an instability, Im $(\omega) > 0$ [261, 262, 765].

The dependences of viscosity and opacity on the surface density and the thickness of the disk $\left(\nu \propto \sigma^{\delta_\sigma} h^{\delta_h}, \bar{\varkappa} \propto \sigma^{\Delta_h} h^{\Delta_h}\right)$ have a controlling effect on the dissipative instabilities of acoustic and also viscous and thermal (section 11.3.1) part of the oscillations. The large number of models of the accretion disks available at the present time appear to be in both stable and unstable regions in respect of the parameters Δ_h, Δ_σ, δ_h, δ_σ for different branches of the oscillations [262].

In section 3.3.1 we discussed in detail the dynamics of linear perturbations of the gas non-self-gravitating disk taking into account the vertical structure of both equilibrium parameters and the perturbations.

The non-linear modelling of viscous accretion disks

In [263] the authors investigated the non-linear dynamics of the previously discussed dissipative–acoustic instability using different numerical methods of solving the hydrodynamic equations.

The results show that in the presence of initial perturbations with the amplitude 0.1% of the equilibrium values, the acoustic oscillations can increase to low-intensity shock waves during the passage of a sound wave through the disk along the radial coordinate (Fig. 11.14). Consequently, a non-stationary system of fine-scale shock waves forms in the disk and the relative amplitude of these shockwaves can exceed 50% when $\alpha \gtrsim 0.2$ (Fig. 11.15). The consequence of these wave movements in the accretion disk is the non-stationary component of the luminosity of the disk with the amplitude of several percent (Fig. 11.16).

The efficiency of the investigated processes (formation of shock waves and quasi-periodic oscillations of luminosity) depends strongly on the level of turbulent viscosity, and the formation of a system of waves with a high amplitude without taking into account the effects of reflection from the boundaries in these models is possible only at 0.05.

11.4. Non-axisymmetric disk accretion

11.4.1. Gas-dynamic modelling of overflow of matter in the close binary system. Disk formation conditions

As shown in section 11.2, within the framework of the axisymmetric models it is quite easy to understand many observed manifestations of the accretion disks. At the same time, in close binary systems, accretion disks are principally non-axisymmetric because of the gravitational effect from the side of the normal star and the fact that the matter arrives in the accretion disk in the form of a jet through the inner libration point. If relatively long time periods which greatly exceed the rotation period are investigated, then it would appear that the standard models of the accretion disks are a good approximation. Since the matter carries out many revolutions during its movement to the accreting object, the matter is stirred in the angular direction during this period. The gravitational force of the normal star on approach to the compact object becomes very small in comparison with the force determined by the mass M_1.

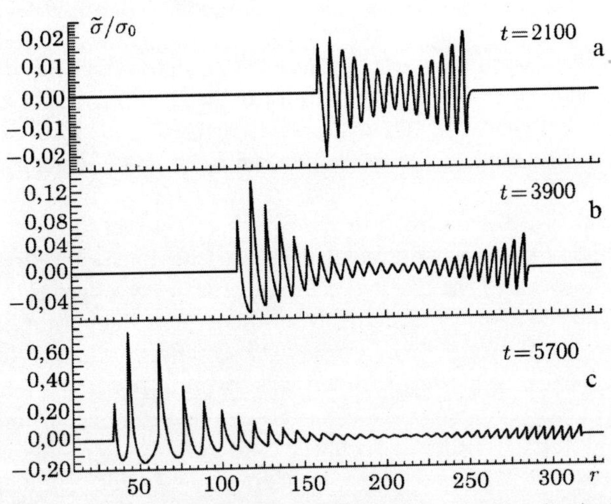

Fig. 11.14. Radial dependences of the relative perturbations of surface density $\tilde{\sigma}/\sigma_0$ at different moments of time for the basic model. At the moment $t = 2100$ (a) the relative amplitude of surface density does not exceed 2% and the shock waves do not form, but the amplitude of perturbations increases with increase of the distance from the generator ($r_g = 208$). Starting at the moment of time $t \gtrsim 3500$ suitable conditions form for the formation of shock waves in the perturbed regions furthest away from the source $r \gtrsim 270$ and $r \gtrsim 130$ (b). At the time $t \gtrsim 6000$ the perturbations reach the inner (r_{in}) and outer (r_{out}) boundaries (c).

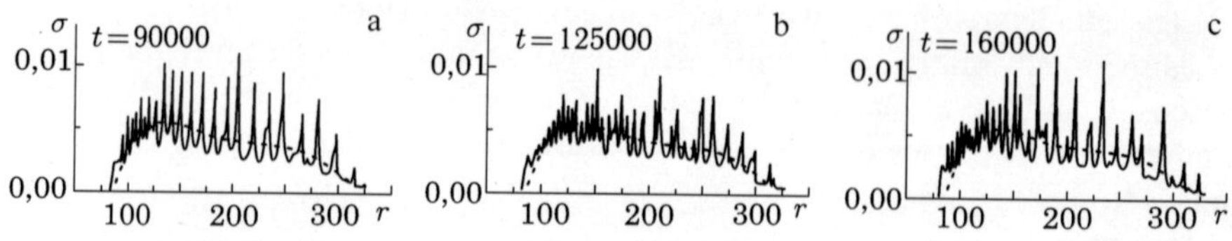

Fig. 11.15. Radial distributions of the surface density at different moments of time in the accreting ring at $\alpha = 0.3$, $M = 200$. The dotted line indicates the initial profile.

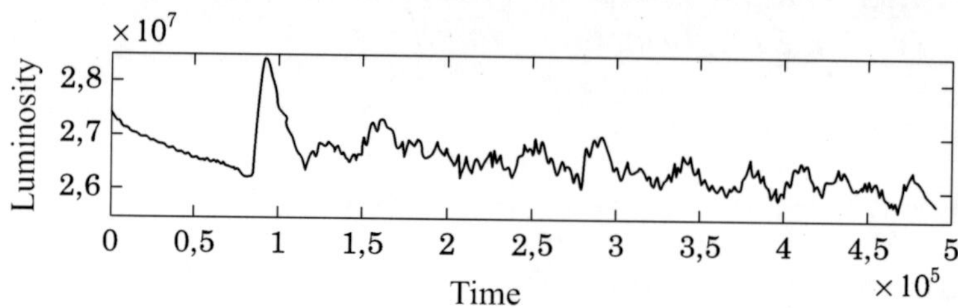

Fig. 11.16. Long-term dependence of the luminosity (in the conventional units) of an unstable ring situated in the region $70 \gtrsim r \gtrsim 350$ in the model with $M = 200$.

There is a number of interesting questions: what are the conditions for the formation of the accretion disk in the close binary system? Which part of matter is lost by the system? Will the matter accrete without viscosity? The problems of this type appear completely insoluble for the supporters of analytical solutions because of the non-stationary and non-uniform nature. The only solution is numerical modelling.

Matter can leave an optical star in the form of solar wind, i.e. from the entire surface of the star. The second mode can form during filling of the critical Roche region of the normal star when the matter is discharged in the form of a jet through a relatively small vicinity of the inner libration point. If the velocity of the gas is sufficiently high, it is quite difficult to expect the formation of a disk.

More general considerations show that shock waves can form during accretion in the close binary systems. Biermann [371] was obviously the first investigator who in the framework of the hydrodynamic approach investigated the flow of gas in a close binary system in the stellar wind mode. The method of characteristics was used to examine only the supersonic flow. Obviously, such a solution cannot contain shock waves. In [1190] attention was given to the formation of a conical shock wave behind the accreting object. However, the finite difference method characterised by the first order of accuracy used in this case results in a very high numerical viscosity. In addition to this, the Cartesian grid does not make it possible to specify accurately the boundary conditions on the surface of both stars.

In [886, 888, 1116–1119, 1201] the authors used the numerical schemes of the second order on a curvilinear grid whose coordinate lines were close to the isolines of the effective potential of the system consisting of two bodies ($q = M_2/M_1 = 1$), situated at the distance a from each other and rotating with the angular velocity Ω_0. One of the stars fills its critical Roche region, and the radius of the other star

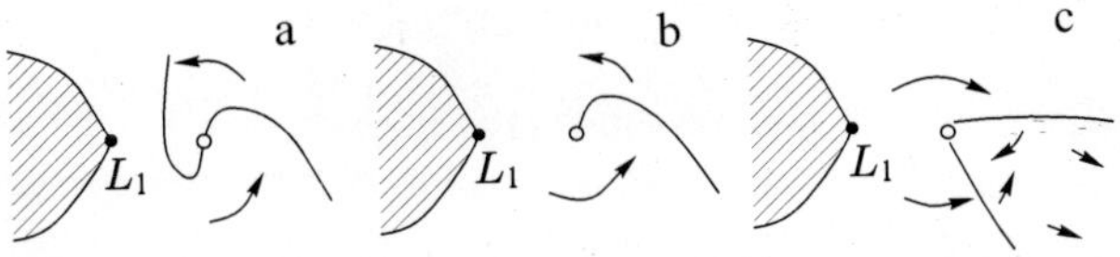

Fig. 11.17. Characteristic distribution of shock waves for different values of $c_0/a_0\Omega$: a) $c_0 \ll a_0\Omega$; b) $c_0 \simeq a_0\Omega$; c) $c_0 \gtrsim a_0\Omega$.

does not exceed $0.03a$. The effects, associated with cooling, heating, viscosity[16] and magnetic fields were not taken into account. The values of density ρ_0 and the speed of sound c_0 were defined on the surface of the normal star.

The flows at different $\beta = c_0/a\Omega_0$ were investigated in [886, 1118] for $q = 1$. If the speed of sound is low ($\beta \lesssim 0.75$), then a disk with two spiral shock waves[17] (Fig. 11.17a) forms around the compact object. The maximum Mach number $M = |\mathbf{u}|/c_s$ does not exceed 3. At $0.75 \lesssim \beta \lesssim 0.9$ flow rearrangement takes place: the disk becomes less distinctive and only one spiral shock wave remains (Fig. 11.17b). At the values β, situated in thc range $0.9 \div 1.0$, a conical shock wave forms (Fig. 11.17c), and the flow inside the flow becomes distinctively subsonic. With a further increase of the speed of sound ($\beta \gtrsim 1$) a distinctive mode of stellar wind forms. With increase of the speed of sound the angle between the shock waves on the surface of the star-donor becomes smaller. Passing through the conical shock wave, the speed of gas greatly decreases and part of the gas accretes on the compact object. A large part of matter leaves the system. Similar results wcre obtained in [1154].

Thus, the type of accreting flow (discharge with the formation of a disk or in the form of stellar wind with the formation of a conical shock wave) in the system in which its Roche cavity was filled by the star-donor is determined by the value of the parameter $c_0/a\Omega_0$. The typical estimate of the investigated systems is $a\Omega_0 \sim 10^7$ cm/s, which corresponds to a temperature of 10^6 K. In the absence of a stellar corona the temperature of the matter discharged from the stars is considerably lower than 10^6 K. Consequently, the discharge regime through the inner libration point with the formation of the accreting disk around the compact object is most likely to form. During the overflow of matter through the internal libration point the specific angular momentum of matter is a very large and this results in the formation of a disk. In the case of stellar wind the specific angular momentum is relatively small and the disk does not form [711].

In [23, 385, 936] investigations were carried out into the structure of overflow of the gas in the close binary system within the framework of three-dimensional models. Special attention was given to the effects of interaction of the gas flow through the inner libration point L_1 (see Fig. 11.17). The results of the calculations show the more complicated nature of the flow in comparison with the two-dimensional one (section 11.5). Gas flows with a small specific angular momentum

---

[16]In this case we are concerned with physical viscosity but the numerical schemes used in most cases also have numerical viscosity.

[17]For more details see section 11.4.2.

form above the the plane of the system and this results in the formation of a hot corona. Three-dimensional numerical modelling was carried out to investigate the bending instabilities of the accretion disks [94].

11.4.2. Spiral shockwaves in the CBS. Gas-dynamic modelling

In section 11.4.1 we already mentioned several results of the numerical gas-dynamic modelling of the overflow of matter in a close binary system. Here, we will discuss in greater detail the problem of spiral shock waves in a gas disk initiated by the gravitational potential of the satellite – the normal star.

In particular, in [1117, 1118] it was shown that:

1. The gas is lost by the normal star through the vicinity of the point L_1 in the form of a supersonic jet (see Fig. 11.1);
2. The main part of matter rotates around the compact object in the form of the accretion ring/disk;
3. The tidal interaction results in the formation of one, two or three spiral shock waves (SW);
4. The gas is heated in the shock wave, loses its angular momentum in relation to the accreting star. The amount of the angular momentum, lost in the shockwave, is greater than that lost as a result of numerical (scheme) viscosity;
5. The system may lose a large part of matter through the point L_2;
6. The rate of loss of matter by the optical star (t) may oscillate quite considerably, whereas the rate of accretion $\dot{M}_e(t)$ is a 'smoother function (Fig. 11.18);
7. The ratio $\dot{M} / M_e$ depends strongly on the parameters of the system and equals ~20÷90%.

The formation of the shock waves is caused by the second component, i.e. the generator is situated at the periphery of the accretion disk. At the same time, it is necessary to determine the distance to which the shock waves can come in relation to the accreting object. To solve this problem, a series of experiments was carried out in [1119] in which the size of the compact object was equal to $r_* = 0.01a$. Since the value of a for the close binary systems with the period from several hours to several days is $\sim 10^{10} \div 10^{11}$ cm, then $r_* \simeq 10^8 \div 10^9$ cm, which corresponds to the radius of the white dwarf. If the compact object is represented by a neutron star with the magnetic field $\sim 10^{12}$ G, the disk fails at a distance of $\sim 5 \cdot 10^8$ cm [556]. Calculations show convincingly that the shock waves extend up to $r = r^*$.

The effect of numerical viscosity will be discussed. We use the numerical schemes for solving the equations of hydrodynamics which have the second order of accuracy and give the scheme Reynolds number Re $\sim (r/\delta r)^2$ (δr is the size of the cell). In the vicinity of the compact object Re ~100. Thus, the angular momentum is transferred to the outside and the gas falls on the centre even in the presence of the axisymmetric potential (without shock waves). The effect of numerical viscosity can be reduced by reducing the value δr. To answer the following question: which part of the angular momentum is lost in the shock wave, experiments were carried

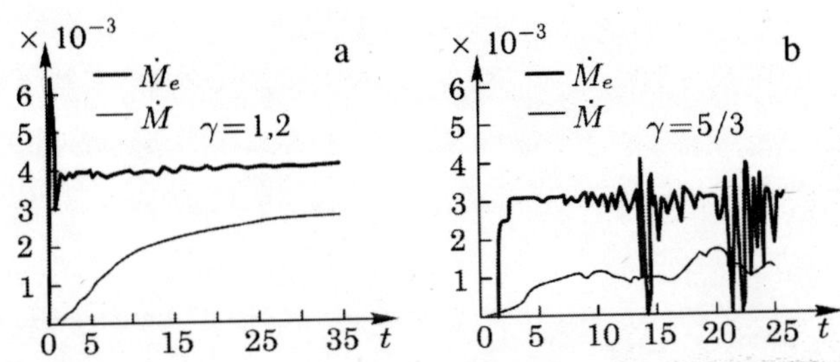

Fig. 11.18. Characteristic behaviour of $\dot{M}_e(t)$ and $\dot{M}(t)$ determined on the basis of the experimental results.

out in [1117] in which at time $t = 22/\Omega_0$ (the disk is in the quasi-stationary state) each spatial cell in the radial direction was divided into halves and the calculations were continued up to $t = 27/\Omega_0$. On the whole, the global structure of the flow did not change, and the average value $\dot{M} / M_e$ decreased from 0.9 to 0.7. Thus, according to the estimates of the authors, approximately 60÷70% of the total losses of the angular momentum are associated with the shock waves.

The accretion process can be efficiently characterised by the accretion time $\tau = M/\dot{M}$ (M is the mass of the disk). Figure 11.19 shows the experimental dependence of τ on the ratio of the masses of the components $q = M_2/M_1$ [888]. The horizontal line corresponds to the axisymmetric model ($q = 0$) in which accretion is determined completely by numerical viscosity. Within the framework of the viscous standard model of the accretion disk, the value $\tau = r^2/\nu$ (section 11.2.1). For the viscosity $\nu = \alpha\, c_s^2/\Omega$ we have

$$\tau = \frac{1}{\alpha\Omega}\left(\frac{\Omega r}{c_s}\right)^2 = \frac{\mathrm{M}^2}{\alpha\Omega}, \qquad (11.105)$$

where M is the Mach number. Assuming for the outer regions M ~3 and returning to Fig. 11.19, for $q = 10^{-3} \div 1$ we obtain $\alpha_{\mathrm{ref}} \simeq 5 \cdot 10^{-4} \div 4 \cdot 10^{-1}$.

In the studies cited previously the calculations did not include the radiation losses and this resulted in a high temperature, close to virial temperature. Taking into account the cooling processes should, on the one hand, increase the characteristic Mach number. On the other hand, within the framework of the self-modelling approach a reduction of temperature reduces the amplitude of the shock waves. Explanation of the role of these factors requires detailed analysis.

We will discuss the results obtained in the previously mentioned gas-dynamic modelling, in comparison with self-modelling solutions (section 8.4.5). Figure 8.25 shows that the stationary self-modelling solutions, containing two spiral shock waves in a constant thickness disk, are not possible for $\gamma > 1.6$. Numerical modelling at $\gamma = 5/3$ results in strongly oscillating flows (Fig. 11.18), and direct comparison of the angle of the spiral Θ of the self-modelling wave with the experimental results gives an acceptable agreement. Comparison of the results of numerical modelling of the shock waves, self-modelling solutions and the standard theory of disk accretion was carried out in [888]. The dependence of the twist angle Θ of the shockwave

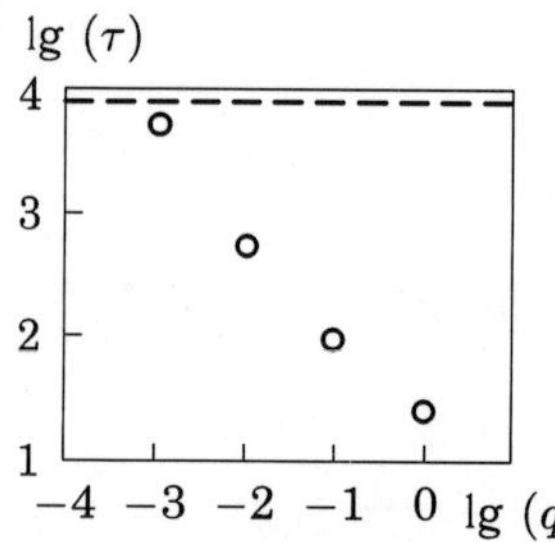

Fig. 11.19. Dependence of τ on q according to the experimental results [888].

on the adiabate parameter[18] is shown in Fig. 11.20. Measurements of Θ relate to the inner zone of the accretion disk where the effect of the second component is minimal. In the region $\gamma \lesssim 1.4$ there is a good agreement. In numerical experiments at $\gamma \gtrsim 1.55$ the two stationary shock waves did not form and the self-modelling approach also forbids their existence at $\gamma > 1.595$ (see Fig. 8.25 and section 11.20). In the range $1.595 < \gamma < \sqrt{3}$ stationary solutions are possible for the number of shock waves greater than two. Figure 11.21 shows radial dependences of the Mach number of the shock wave M. Large differences can be seen in comparison with the self-modelling solutions in the outer region of the accretion disk; the differences decrease on approach to the centre. Evidently, this behaviour is caused by the fact that the tidal interaction was not taken into account when constructing the self-modelling solutions.

A surprising at first sight aspect of these results is the possibility of accretion without radiation losses. In contrast to this, within the framework of the standard disk accretion all dissipative energy is radiated out. Therefore, we will examine self-modelling waves with the radiation (section 8.4.5). The dependence of temperature on the rate of accretion in the case of hydrostatic equilibrium in the z-direction is shown in Fig. 11.22. Here, the dimensionless rate of accretion is determined from the following equation

$$\dot{m} = \dot{M}/K, \quad K = 4\sqrt{\pi a c(\gamma - 1)}\left(\frac{\mu}{\mathcal{R}}\right)^2 (G M_1)^{7/4} k^{-1/2} r^{-1/4}, \qquad (11.106)$$

where a is the radiation constant; μ is the mean molecular mass; k is the opacity. There is the critical value $\gamma = \gamma_{cr} \simeq \sqrt{6} - 1 \simeq 1.45$. At $\gamma < \gamma_{cr}$ the temperature reaches the finite value, whereas the rate of accretion is not limited. In the case of the high accretion rate, the flow becomes optically opaque so that the radiation losses do not play any role and the solution tends asymptotically to the adiabatic solutions (horizontal dashed line). According to [888] it will be shown that the α-models are characterised by the same behaviour.

Equation (11.30), expressing the energy balance, is written for the stationary state in the case

[18]As regards the selection of the values of the adiabate parameter it should be mentioned that the value $\gamma = 5/3$ in the three-dimensional case corresponds to $\gamma = 3/2$ in two-dimensional models (section 3.6.1).

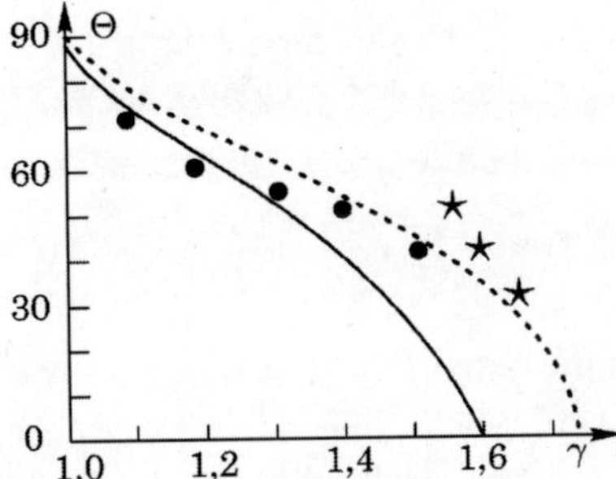

Fig. 11.20. The dependence of angle Θ on γ. The solid line shows the self-modelling solutions for two shockwaves ($N = 2$), the dashed line for $N = \infty$ (see Fig. 8.25). The circles indicate the results obtained in the numerical experiments for the stationary state, and the asterisks – for the models far away from the stationary state [888].

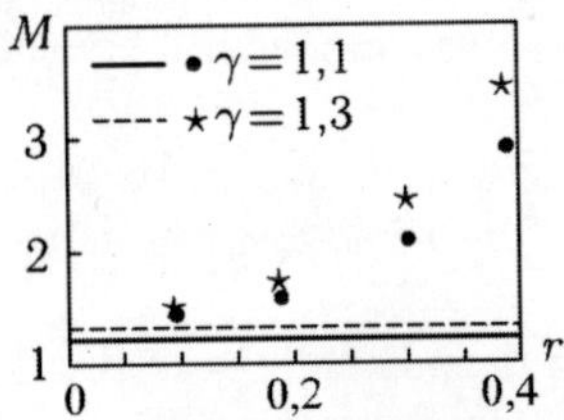

Fig. 11.21. Dependence of the Mach number of the shockwave on the radial coordinate. The lines show the self-modelling solutions.

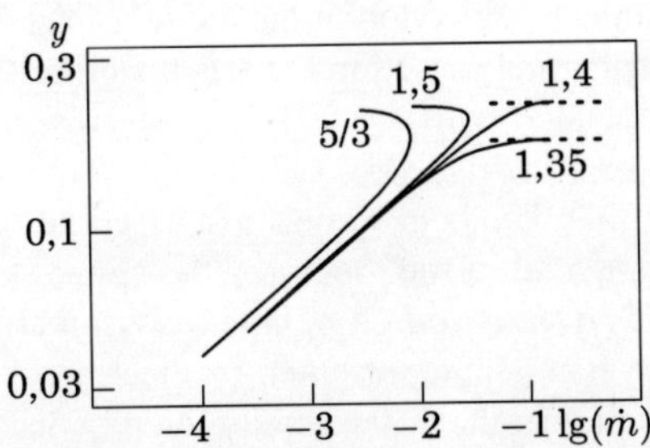

Fig. 11.22. Dependence of dimensionless temperature $y = T/T_{\rm vir}$ $[T_{\rm vir} = \mu GM_1/(\mathcal{R}r)]$ on the rate of accretion for the disks with the self-modelling shockwaves for different adiabate parameters. The solid line – radiation losses taken into account, the dashed line – adiabatic accretion.

$$\sigma T u \frac{dS}{dr} = \frac{9}{4}\sigma \nu \Omega^2 - \frac{4acT^4}{3\sigma k}. \tag{11.107}$$

Using the relationships $\dot{M} = -2\pi r\sigma u$, $u = -3\nu/(2r)$ the value σ is excluded from (11.107):

$$\dot{M}^2\left(\frac{2}{3}Tr\frac{dS}{dr} + \Omega^2 r^2\right) = \frac{16\pi^2 acT^4}{3k}\nu r^2. \tag{11.108}$$

If it is assumed that $k = k_0 r^{-1/2}$, the self-modelling solutions with $h \propto r$, $p \propto r^{-5/2}$, $\rho \propto r^{-3/2}$ are possible. For the ideal gas, taking into account (11.28), we can write the equation for $\dot{M}$:

$$\dot{M}^2 \left[y(\gamma - 5/3) + \gamma - 1 \right] = \frac{32\pi^2}{9} \frac{ac}{k_0} \left(\frac{\mu}{\mathcal{R}} \right)^4 (G M_1)^{7/2} \alpha y^5, \qquad (11.109)$$

where $\mathcal{R} = Tr/(\mu G M_1)$ is the ratio of the disk temperature to the virial temperature. Taking into account (11.106) relationship (11.109) can be written in the new form

$$\dot{m} = \frac{\sqrt{2\pi\alpha}}{3} \frac{y^{5/2}}{\sqrt{(\gamma - 5/3)y + \gamma - 1}}. \qquad (11.110)$$

The dependences $y(\dot{m})$ are identical with the case with the shockwaves below the critical value γ the disk temperature tends to its asymptotic value. In the limiting case of adiabatic accretion ($a \to 0$) from (11.109) we obtain that the dimensionless disk temperature is

$$y = \frac{\gamma - 1}{5/3 - \gamma}. \qquad (11.111)$$

Equation (11.111) shows that the adiabatic accretion is possible only for $\gamma < 5/3$. For the values close to unity, the disk temperature is low $y \ll 1$. It may be seen that there is the maximum value of the adiabate parameter below which the rate of accretion may be arbitrary. When $\gamma > \gamma_{cr}$ the rate of accretion is maximum possible. The value γ_{cr} is determined by selection of the model of the accretion disk. It should be mentioned that for spherical accretion there is the critical value of γ and $\gamma_{cr} = 5/3$ [404]. The latter is associated with the fact that the gravitational field is the same in both cases.

It is clearly evident that the shock waves discussed here are very similar to the non-linear waves investigated in the theory of the spiral structure of the galaxies. We examine a (possibly weak) source of non-axisymmetric perturbations in the outer region of the disk. It can be represented by the second component or by some instability. This perturbation in the differentially rotating disk generates spiral waves which transfer the angular momentum, interacting with the matter of the disk. This problem has been studied extensively in applications to galaxies [952, 1160, 1231].

11.5. Waves in accretion disks (theory and numerical modelling)

The correct description of gas dynamics of matter in semi-detached binary systems is necessary to solve the complete three-dimensional system of equations. Unfortunately, the implementation of this problem has long been limited to the development of computer technology and, therefore, to analyze the flow structure two-dimensional models were used. In the last fifteen years, starting with the works [678, 935, 948] an opportunity to numerically study the transport of gas dynamics of matter in more realistic three-dimensional models appeared. In this section we provide an overview of the results of three-dimensional numerical studies of the flow pattern in semi-detached binary systems obtained in [377, 379–383, 386, 936]. The TVD method used (Total Variation Diminishing) for solving the system of gas dynamic equations allows us to consider the morphology of gas flows in a system, despite the presence of significant density gradients. The gas dynamics of mass transfer in a semi-detached system was investigated at wide temporal intervals, which allowed us to consider the main features of the picture of the flow in the steady (stationary) mode.

11.5.1. Morphology of flows and shock waves

Analysis of the basic processes of heating and cooling of matter in accretion disks in binary systems, presented in 11.2.3, shows that for realistic parameters of accretion disks the gas temperature in the outer parts of the disk is in the range from 10^4 to $\sim 10^6$ K. This means that both hot and cold accretion disks can form in these binary systems.

Let us consider the gas dynamics of matter in binary systems with hot disks. These solutions were obtained for cases in which the temperature of the outer parts of the accretion disk was $\sim(100\text{–}200)\cdot 10^3$ K.

Information on the morphology of the flow in these systems can be obtained from Fig. 11.23a and b. The distribution of density and velocity vectors in the equatorial plane of the system is shown in Fig. 11.23a, the velocity field with the tracks of the set of particles is shown in Fig. 11.23b. Figure 11.23a also shows the current line (light line with circles) limiting the accretion disk and the current line passing through the shock wave at the edge of the jet (dark line with squares). According to the monograph [419], the gas dynamics of flow of matter in a semi-detached binary system is determined by the presence of a jet of matter from L_1 by the quasi-elliptic accretion disk, the disk halo and the intercomponent envelope. The introduced classification of the main elements of the flow pattern is based on their physical properties: 1) if the gas motion is not determined by the gravitational field of the accretor, then it forms an intercomponent envelope, filling the space between the components; 2) if the gas is rotates around the accretor and then in mixed with the matter of the jet, then it does not belong to the disk and the halo forms around the disk; 3) the accretion disk is formed from a jet of matter which, once in the gravitational field of the accretor, does not interact further with the jet, and moves toward the star, losing angular momentum.

Analysis of the results revealed that the self-consistent solution of the problem does not consider the interaction of the jet outflowing from libration point L_1 with the resultant accretion disk ('hot spot'). Interaction of gas near the disk halo and the intercomponent envelope with the jet leads to the formation of a shock wave located along the edge of the jet. This shock wave, the so-called hot line, is clearly visible on Fig. 11.23 a, b.

The tidal effect of the donor star leads to a spiral shock wave, denoted by the dashed curve in Fig. 11.23a. The solution with a high-temperature gas contains only one arm of the spiral shock wave, while in the field, where the second arm is placed, the flow structure is determined by the jet of matter from L_1. Note also that in this case the tidal spiral wave goes deep into the inner parts of the disk.

Let us consider the morphology of the flow of matter in semi-detached binary system with a stationary cold (T~1.4·10⁴ K) disk. Figure 11.23c shows the density distribution and velocity vectors in the equatorial plane of the system. Shock waves are formed in the disk, seen in Fig. 11.23c as a thickening of the isoline density. Densening of the isolines at the edge of the accretion disk corresponds to a sharp drop in density – from the characteristic values for the disk to the background density. The region of interaction of the jet and the disk halo in the equatorial plane of the system is shown in Fig. 11.23d, e: density isolines and velocity vectors (Fig. 11.23d) are shown together with the velocity field (Fig. 11.23e). As can be seen from the presented results, the interaction of the disk halo and the jet has all the typical features of an oblique collision of two streams.The resulting structure of the two shock waves and a tangential discontinuity between them is clearly visible on Fig. 11.23d, e. The region of shock interaction of the gas of the disk halo with the jet is outside the disk and has a complicated shape. Parts of the halo away from the disk have low density and the shock wave due to their interaction with the jet

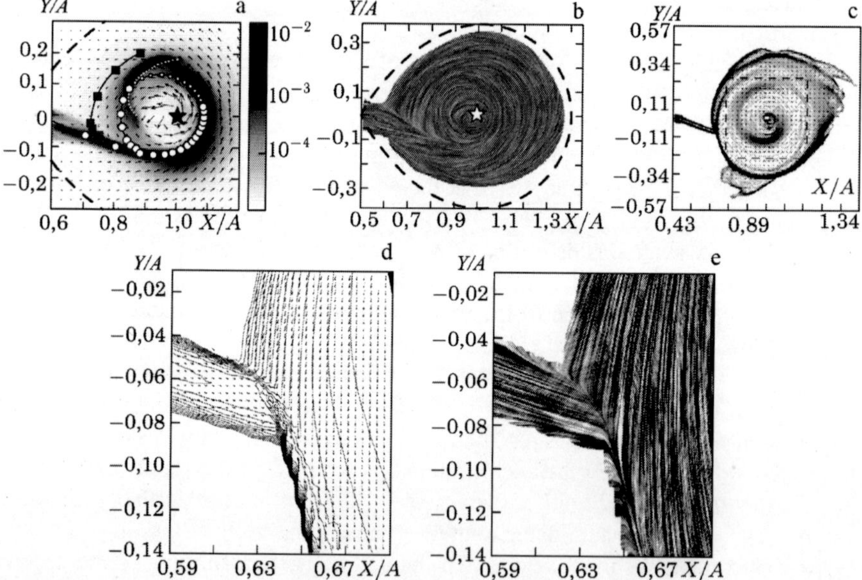

Fig. 11.23. The distribution of density and velocity vectors in the equatorial plane of the system for the solution with the hot (a) and cold (c) accretion disks. X and Y coordinates are in units of distance A between the components. The accretor is marked with an asterisk. The dashed curves – the boundary of the Roche lobe. The current line limiting the accretion disk is shown by the light curve with circles. The current line passing through the shock wave at the edge of the jet is indicated by the dark curve with squares. The larger icons (circles and squares) correspond to the location of the shock wave. The dotted curve denotes the arm of the tidal spiral shock wave. b – visualization of the velocity field in the equatorial plane of the system. Isolines of the density and velocity vector (d) and visualization of the velocity field (f) in the interaction of the jet and the disk halo in the equatorial plane of the system for the solution with the cold accretion disk.

is located along the edge of the jet. With increase of the halo gas density the shock wave bends and at the end all is in the position along the edge of the disk. The resulting shockwave is relatively long and can be called the hot line. As follows from the above general characteristics of the flow pattern, in the interaction region the halo gas and the jet gas pass through the shock waves that correspond to their flow, mixing, and their mixture moves along the tangential discontinuity between the two shock waves. Subsequently, this matter forms their own disk, the halo and intercomponent envelope.

The revealed features of the flow for the cases of hot and cold accretion disks, respectively, are shown schematically in Fig. 11.24a, b, showing of the part of the donor filling its Roche lobe, the position of the libration point L_1, the jet of material flowing out from L_1, as well as the position of the accretor. Three-dimensional gas-dynamic calculations show that the solution for the case of the cold accretion disk has the same qualitative features as the solution for the case of high temperatures: interaction of the jet and the disk takes place without impacts, the area of higher energy generation – shock wave (HL) – is caused by interaction of the gas around the halo and the jet and is located outside the disk. At the same time, for the cold case the accretion disk (zones A and B in Fig. 11.24) is essentially denser (compared to the density of the substance of the jet), its height decreases, and the shape changes from quasi-elliptical to almost circular. The circumdisk halo (zone C in Fig. 11.24) is also significantly reduced in size. The second arm of the tidal spiral shock waves forms in the disk and both arms do not move closer to the accretor and are located in the outer parts of the disk.

11.5.2 The precessional density wave in cold disks

Taking into account the effect of a weak jet of matter on the dense internal parts of the disk, and the fact that the shock waves (hot line and the two arms of the tidal waves) are located in the outer parts of the disk, we can separate in the low-temperature solution another element of the flow structure, namely the inner region of the disk (zone A in Fig. 11.24), in which the influence of gas dynamic perturbations can be neglected. We consider the behaviour of matter in the inner parts of the disk not subject to gas-dynamic perturbations. In the absence of external influences the

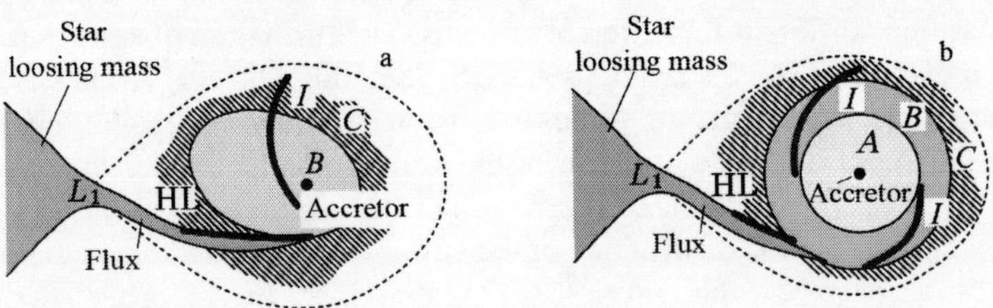

Fig. 11.24. Schematic representation of the main features of the flows in semi-detached binary systems at high (a) and low (b) gas temperature. Zones A and B – accretion disks, C – the halo around the disj, HL – hot line, I – arms of the tidal spiral shock wave, L_1 – libration point. The dashed curve shows the Roche equipotential, passing through libration point L_1.

particle will rotate around the gravitating centre – accretor – in an elliptical orbit. It is known (see, e.g.[1283]) that in a dual star the effect of the second component will lead to a retrograde precession of the orbit of a particle, rotating around the star–accretor (i.e. to rotation of the major axis of the orbit in the direction toward the orbital motion). The rate of precession decreases when approaching the accretor.

The accretion disk is formed from a variety of particles, each of which moves in its elliptical orbit. Since the particles interact with each other and form a gas, the disk should be considered in the gas dynamic approximation and, thus, from orbits we should go to the current lines, which will also be elliptical. It is known that the gas can not contain mutually intersecting flows and, therefore, the current lines can only touch each other.

From geometrical considerations it is clear that a disk can be made from non-intersecting ellipses can only be inserting ellipses into each other. If the eccentricity of elliptical current lines is zero, we obtain a circular disk. If the eccentricities of the current lines are not equal to zero, we obtain an 'equilibrium' solution when the major semi-axes of all current lines lie on the same straight line. If the system has an external action (which is inevitable in a double star) and the rate of precession of the orbit increases with increase of the semi-major axis, then the current lines remote from the accretor will overtake the current lines with smaller semi-axes. In this case, since in the gas-dynamic disk there cannot be intersecting current lines, equilibrium is established with time and all the current lines start precessing with the same angular velocity, i.e. rigidly. Due to the precession velocity depending on the characteristic size of the orbit, the current line more distant from the accretor must be rotated at a greater angle, and in the direction opposite to that of the rotation of matter, because the precession is retrograde.

The precession velocity is in the range between the velocities of precession of external ('fast') and internal ('slow') orbits. The position of the inner orbit is determined by the fact that in this region of the disk the influence of the second component compared with the influence of the gravitational field of the accretor can be neglected. The position of the outer orbit is determined by the size of the region with no gas-dynamic perturbations, since the presence of disturbances disrupt the regularity of the precession of current lines. It is obvious that the position of the inner and outer orbits depends actually on both the system parameters and the parameters of mass transfer. Therefore, we should expect that the average velocity of precession is different for different systems. The formation of spiral structures in accretion disks was considered in [380, 868, 886, 888, 988, 1117, 1118].

Analysis of the results of three-dimensional numerical modelling (see Fig. 11.23c) fully confirms the hypothesis of the formation of spiral density waves in the inner parts of the cold disk. The nature of the wave motion can be judged from Fig. 11.25a, which shows the distribution of density and velocity vectors in the equatorial plane of the system for the two time points that differ by 1.5 orbital periods. The two-arm spiral wave is at rest in the non-inertial coordinate system connected with a double star, while the precessional wave is moving at low velocity in the laboratory coordinate system (the coordinate system of the stationary observer); in the non-inertial system its rotational period slightly exceeds the orbital period. Figure 11.25b shows the two kinds of distribution of the radial flow of matter in the accretor to the disk in the equatorial plane of the system. The flux of matter to the accretor is normalized to its maximum value. It is clearly seen that the line connecting the peaks of the flux of matter to the accretor has a spiral shape. As a result of increase of the radial flux of matter behind the precessional wave the accretion rate increases by approximately an order of magnitude as compared with the velocity obtained in the solution without a wave.

The coincidence of the qualitative analysis of the results of the calculations enables us to discuss confidently the mechanism of formation of the precessional wave.

11.5.3. Density wave in hot disks

In [383, 398], it was found that even a small change in the rate of mass transfer in a binary system leads to a perturbation of the equilibrium state of a hot accretion disk and, consequently, the formation of a bunch of matter moving across the disk at a velocity about 6 times larger than Ω_{orb} (Fig. 11.26).

The results of calculations show that the bunch is a long-lived formation, not significantly altering its mean characteristics at times of the order of several tens of

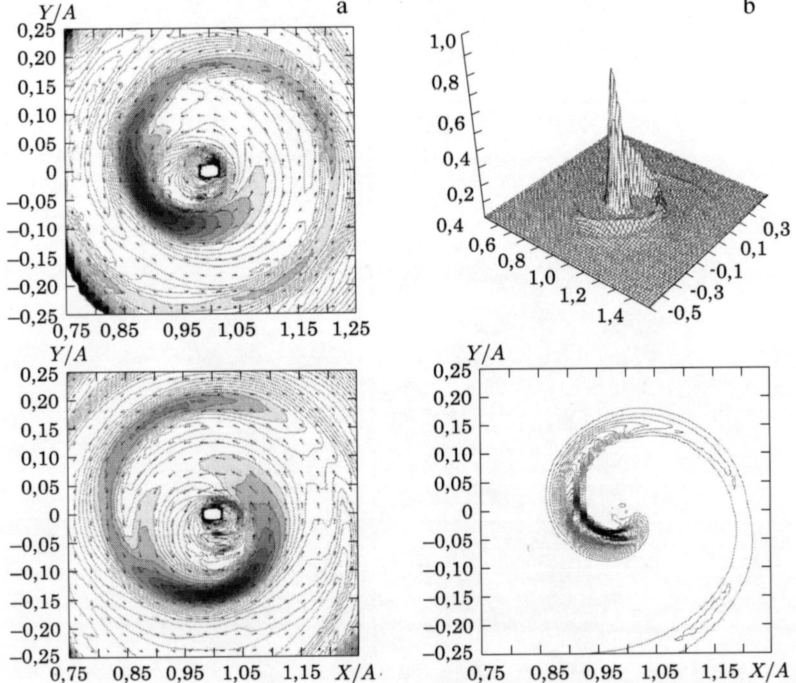

Fig. 11.25. a – density isolines and velocity vectors in the equatorial plane of the system near the accretor (in the region, indicated in Fig. 11.23c by the dashed squares) for the two time points that differ by 1.5 orbital period; b – the distribution in the disk of the radial flow of matter, directed to the accretor, the value of the flow is normalized to its maximum value.

the orbital periods of the system. An illustration of this is in Fig. 11.27 which shows the time dependence of the average density of matter passing through the half-plane, perpendicular to the plane of the system and cutting the disk along the line connecting the components in the direction from the accretor to the outer Lagrange point (plane XZ, $Y = 0 / X > A$). Figure 11.27 shows three curves corresponding to the decrease (halving) in the rate of mass transfer by factor of 2 (upper curve), 10 (middle curve) and 10^5 (lower curve) times. It is evident that, along with the general change of density its quasi-periodic oscillations are caused by passage of the bunch. Analysis of the curves shows that after the change (decrease) in the rate of mass transfer a densening appears in the disk moving with a period of ~(0.15–0.18) P_{orb}. As follows from the data presented, the bunch does not spread under the influence of dissipative processes (the density varies only in the initial transition period), and the frequency of its rotation remains unchanged almost until the disappearance of the disk.

Fridman et al [241, 572] published arguments in favour of the wave nature of the bunch, which is nothing more than a single-arm spiral density wave. The latter, as is known from a well-developed theory of galactic spiral density waves rotating rigidly and not stretched by differential rotation.

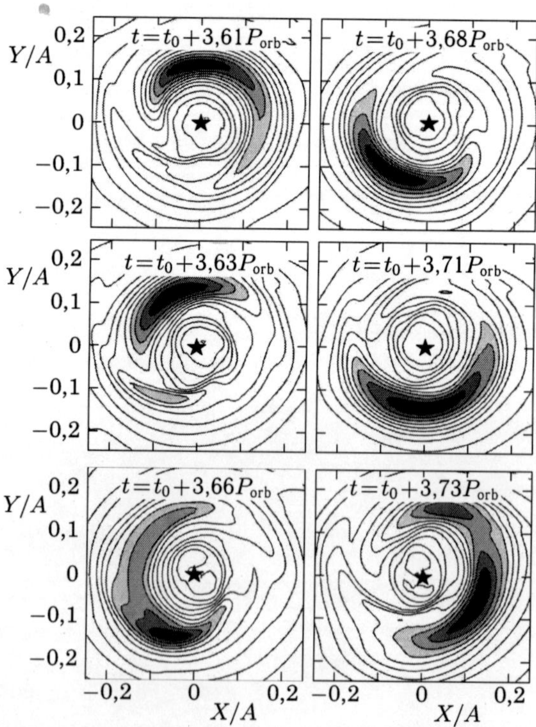

Fig. 11.26. Density distribution in the equatorial plane of the system near the accretor (black asterisk) for the six time moments (indicated in the figure), covering the period rotation of the bunch. For all the distributions the density maximum $\rho_{max} \simeq 0.035\rho(L_1)$.

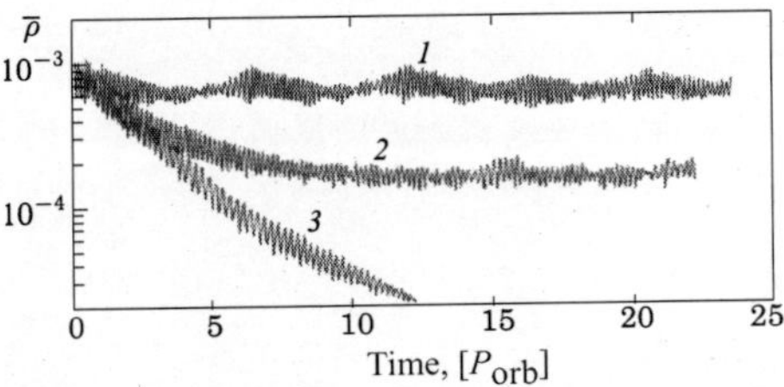

Fig. 11.27. Curves of changes in time mean density of matter, passing through the half-plane, perpendicular to the plane of the system and cutting the disk along the line connecting the components in the direction from the accretor to the outer libration point. Curve 1 corresponds to a two-fold decrease in the rate of mass transfer, curve 2 – to 10-fold, and curve 3 – 10^5 times.

As noted in the introduction of this chapter, the flow of matter in a double system determined by the strength of the field given by the Roche potential (11.2). In order to clarify the nature of the bunch, first consider how variations in the plane of the disk can excite the gravitational potential. Subtract from the total potential the spherically symmetric potential accretor $\Phi_A : \tilde{\Phi} = \Phi - \Phi_A$, and consider the Fourier spectra of the azimuthal harmonics of the potential $\tilde{\Phi}$ in different radial regions of the disk. Figure 11.28a shows the amplitudes of the components of the Fourier spectrum of the potential Φ for different radial regions of the accretion disk. It is seen that in the considered area is dominated by the harmonic $m = 2$. The dominance of the second harmonic demonstrates the nature of the tidal effect of the donor, which is also confirmed by the monotonic increase of the absolute value potential with increasing disk radius. In the peripheral region of the disk closest to the donor the absolute magnitude of the non-spherical part of the potential is maximum.

We can see that in the presented calculation results the gravitational force is not decisive for the distribution of disk material. Consider the range of 'purified'[19] surface density $\tilde{\sigma}$ after a perturbation, i.e. after the mass transfer rate changed and a bunch formed in the accretion disk. The corresponding harmonic spectra of the surface density are shown in Fig. 11.28b. It is seen that the mode $m = 1$ dominates in the Fourier spectrum of surface density which is consistent with the presence of the investigated densening in the disk. A similar pattern is observed in the distribution of harmonics of the radial velocity (Fig. 11.28c).

Analyzing the phase curve of the maximum density of the first azimuthal harmonics of the density profile at different times, we note that the position of the phase curve of maximum density of the mode $m = 1$ at every moment of time is rigidly associated with the bunch of matter. Moreover, if we consider all the phase curves of the maximum density of the first azimuthal harmonics of the density profile for the period of rotation of the bunch (Fig. 11.29a), we see that, having made a complete turn, the phase curve of maximum density has not practically changed its shape, which indicates that the differential rotation of the disk is not 'eroding' this harmonic.

[19]'Purified' function is determined by subtracting the distrbution average over the period.

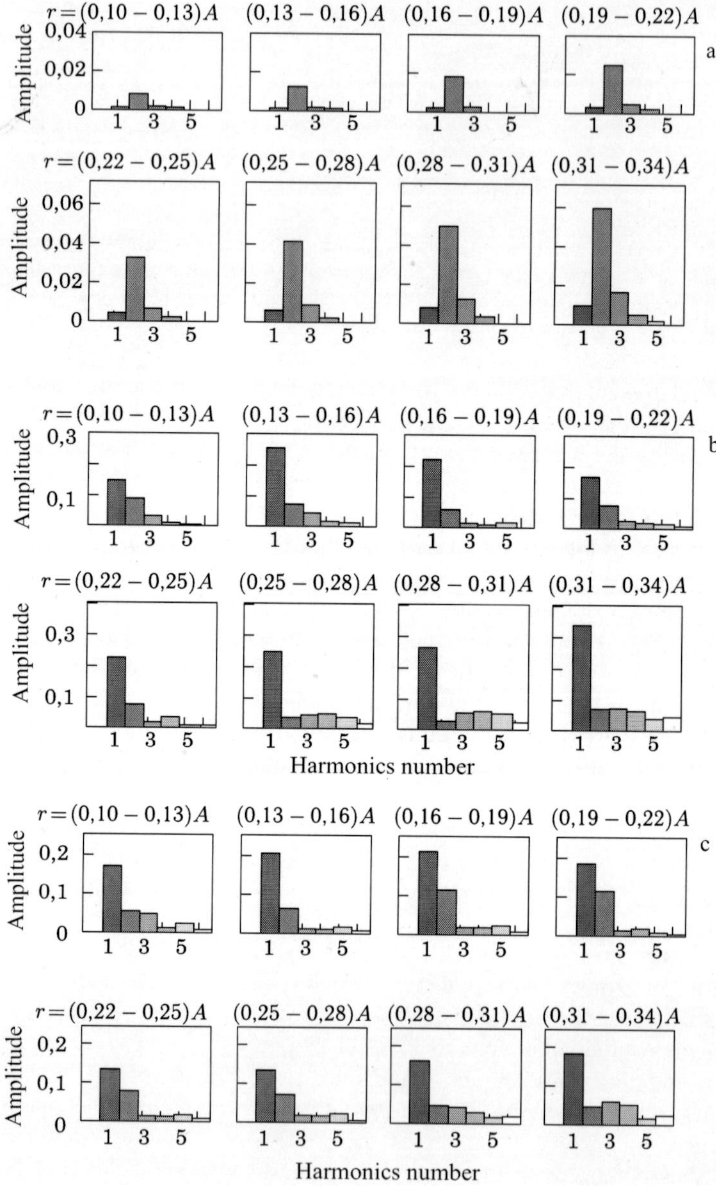

Fig. 11.28. The relative amplitudes of the first five components of the Fourier spectrum for different radial equatorial regions of the accretion disk: a – non-axisymmetric part of the potential, b – non-axisymmetric part of the 'purified' surface density, c – non-axisymmetric part of 'purified' radial velocity.

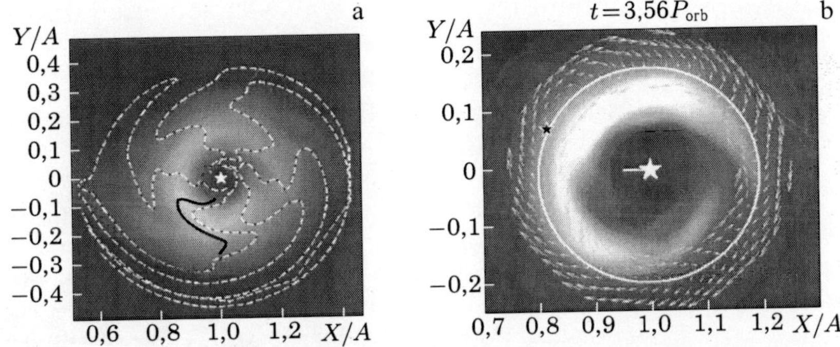

Fig. 11.29. a – phase curves of the maximum density of the first azimuthal harmonic of the density profile for the period of rotation of the bunch in the orbital plane; the solid dark line shows the phase curve of maximum density at the time of one revolution of the bunch, the asterisk denotes the position of the accretor; b – the distribution of density and velocity in the equatorial plane in the coordinate system associated with the bunch, for the moment time $t = 3.56P_{orb}$. The bright star represents the accretor and dark one the centre of the anticyclone. The segment emanating from the accretor shows the direction to the donor star. The light solid line shows the corotation circle, where the phase velocity of the single-arm spiral wave coincides with the velocity of rotation of the disk.

The form of the phase velocity of rotation of the mode $m = 1$, which differs from the tidal mode, and the solid-state nature of its rotation suggest that the bunch is a spiral density wave. According to the theory of spiral density waves in galactic disks, the presence of a single-arm spiral wave must be accompanied by an anticyclonic vortex centred on the corotation radius (see, for example, [241]). Anticyclone centre should be located in the vicinity of the maximum surface density. Analysis of the velocity field in the plane of the disk in the system of coordinates associated with the bunch shows the presence of the anticyclonic vortex together with the densening (Fig. 11.29b). The anticyclone centre, shown by the asterisk, is in full agreement with the theory in the vicinity of the maximum surface density – in the vicinity of the bunch. Moreover, as predicted by the theory of spiral density waves, the centre of the anticyclone is located exactly at the corotation circle marked by the light solid line in Fig. 11.29b.

Summing up the above, we present the evidence that the mode $m = 1$, is a single-arm spiral–vortex structure.

1. Mode $m = 1$ rotates with a phase velocity different from the phase velocity of the tidal mode m = 2;

2. The deformation of the phase of the maximum values of the surface density of the disk by differential rotation in the $m = 1$ mode is almost negligible.

3. Since the calculated radius of the corotation mode $m = 1$, was in the area ofstrong dominance of this mode, we can expect to analogy with the galactic disk, the presence of a single anticyclone centred on the corotation circle and in the corresponding area – near the maximum surface density and the corotation circle. The anticyclone was located precisely in this area of the disk.

11.6. Observational manifestations of waves in accretion disks

11.6.1. Hot line

One of the most informative methods of observational studies of close binary systems is the analysis of their shine curves. A short period (several hours), inherent in these stars, allows in a relatively short time of observation to determine the properties and characteristic parameters of the processes occurring in the system. In the shine curves of eclipsing CBSs there are additional features that can not be explained by the simple model of the cold star–white dwarf–accretion disk system. In particular, the absolute majority of the shine curves of the eclipsing CBS has the so-called hump. To explain this, Gorbatskii [630] and Smak [1177] suggested that the hump is the glow of a hot spot occurring at the border of the accretion disk, in the area of collision of the jet of matter from L_1 with the disk. In the last 30–35 years, the model of the hot spot has been widely used to interpret the shine curves of cataclysmic binaries.

In the gas-dynamic studies of matter exchange in the CBS it was shown (see section 11.4) that the jet and the accretion disk are morphologically a single entity, and their interaction takes place without impact. Naturally, in such a solution in the region of contact of the jet and the disk there is no increase of temperature and, therefore, we are forced to abandon the hot spot hypothesis to explain the hump in the shine curves of the eclipsing close binaries. Three-dimensional calculations of the flow structure of matter in the interacting TDS showed that in the steady state the impact interaction, leading to an increase in temperature, occurs a at a collision of the matter of the halo around the disk and of the jet travelling from L_1. This interaction forms an extended shock wave (hot line), the radiation of which can explain the observed effects on the shine curves of cataclysmic variables, and, in particular, appearance of normal and abnormal bumps that accompany the eclipse of the accretion disk by the star–donor.

The solution of the inverse problem for the interpretation of eclipsing shine curves of various types of cataclysmic variables in the two alternative models showed that using the hot line model we can better describe the observed shine curves both during a superburst and in the inactive state of the system than in a model of hot spots (see, e.g. [773, 775, 776]).

As an example of such a comparison Fig. 11.30a, b shows the the observed and synthetic shine curves of IP Reg, located in a quiet state, in a V-filter. In the hot line model we can reproduce more efficiently the width of the humps on the shine curves, the form of the eclipse and details of shine changes away from the eclipse. Comparison of the hot spot model with the hot line model proves advantages of the latter in the interpretation of shine curves of the close binaries so than we can talk about the calculations resulting in a shock wave (hot line) as a really existing structure in the accretion disks of close binary stars.

Another proof of the existence of the hotline may be the study of the possible nature of dips in the shine curves of the semi-detached binaries with stationary disks [386]. Observations of low-mass X-ray binaries (LMXBs) have revealed the presence of dips in the X-ray shine curves of several systems. These dips were explained using the hypothesis of the presence of thickening at the outer edge of the accretion disk in the phase of ~8 corresponding to the position of this feature on the shine curve. The presence of matter, surrounding the X-ray source, at a considerable height above the orbital plane of the system, as well as the fact that matter is distributed unevenly in the azimuth, may be explained by the gravitational influence of the companion on the accretion disk and by the interaction of the jet of matter from the point L_1 with the disk. The coincidence of the phase of the observed dips with the expected position of the region in which the jet from L_1 approaches the outer edge of the disk has focused the attention of researchers on this particular area. Today, the most promising idea for explaining the appearance of the substance at altitudes well above the thickness of the disk, is the idea of ricocheting of the jet from the outer edge of the disk, which allows us to explain the existence of the observed dips in the shine curves of the LMXBs.

The presence of dips on the shine curves caused by the appearance of matter in certain places high above the disk, can be explained most easily in the LMXBs, as the LMXBs contain a compact source at the centre of the disk. Nevertheless, similar features of the shine curves in different wavelength ranges were observed in a number of cataclysmic binaries during a burst (U Gem, OY Car, Z Cha). Further observations have shown that for systems that are in the steady state, it may cause dips to appear on the shine curves. For example, studies of ultraviolet shine curves of eclipsing nova-like cataclysmic binaries UX UMa and RW Tri [883] confirmed this hypothesis and give grounds to speak about the universal nature of this phenomenon in semi-detached binary systems with accretion disks. It is interesting to note that in contrast to the explosive systems, the stars with stationary disks show the presence of pre-eclipse dips at much earlier phases[20] 0.6–0.7 [595, 883].

The question arises: what will lead to the appearance of matter at a considerable height above the accretion disk in the case where the interaction of the jet with the disk is stationary? Gas dynamic studies of the established flow pattern in semi-detached binary systems show that in this case the interaction of the jet and the drive is shock-free. In contrast to the non-stationary case (model with a hot spot), in which the jet is assumed to impact on the edge of the accretion disk, in the stationary

[20]The phase angle ϕ is counted (as in the analysis of the observational data) from the line connecting the centres of the stars in the direction opposite to the direction of rotation of the system.

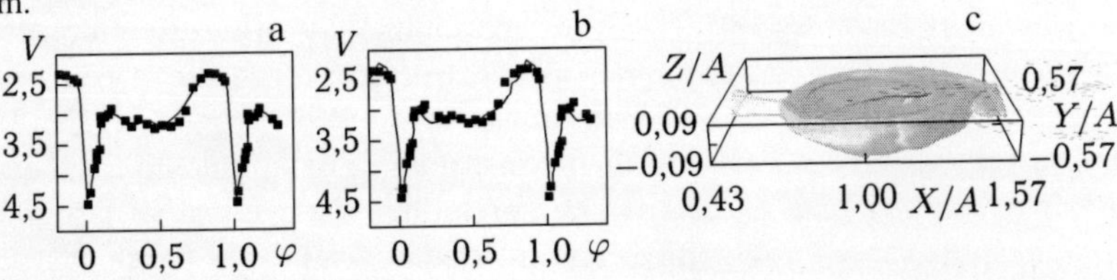

Fig. 11.30. The synthetic shine curves calculated in the models of the hot line (a) and hot spots (b) are shown by solid curves. Squares – the average shine curve of the system IP Peg, located in a quiet state in the V-filter. c) a three-dimensional density isosurface $\rho = 5\ 10^{-11}$ g cm^{-3}; a fragment of the current lines, starting in the neighborhood of L_1, is also shown.

case the jet interacts with the gas of the halo around the disk, forming a long area of high energy – a hot line. This means that that in the steady state the jet will not rebound and therefore, this effect can not be used to explain dips in the shine curves of binary systems with stationary disks.

In [386] attention was given to studies of the thickness of the disk halo in a CBS with stationary disks and it was shown that in the absence of a direct collision between the jet and the disk may cause thickening of the halo above the disk. In the gas-dynamic flow pattern described by a model with a hot line, in the interaction of jets with the disk halo a significant part of matter receives vertical acceleration. The vertical movement of gas caused by the presence of the z-component of velocity, together with its movement along the outer edge of the disk leads to a gradual increase in the thickness of the disk halo. The field of vertical acceleration is limited by the area of the hotline the angular dimensions of which do not exceed ~65°. However, after the passage of this region the gas acquires a sufficiently large vertical velocity component that causes the gas to rise higher and higher until the reserves of kinetic energy are depleted. The point at which upward movement stops is at the maximum height and corresponds to a phase ~0.7, i.e. it goes far beyond the hot line. The height of thickening is several times greater than the typical height of the disk and reaches values ~0.04 A (this corresponds to the ratio of the height of thickening to the distance from the accretor >0.1), and its angular dimensions are greater than ~130°. The maximum calculated thickening of the halo above the outer edge of the disk falls on the phase ~0.7 (see Fig. 11.30c), which coincides with the observed values for cataclysmic variables with stationary disks. This confirms the hot line model previously proposed for describing the pattern of flow in semi-detached binary stars and offers additional possibilities to interpret the shine curves of such systems.

11.6.2. Tidal spiral wave

The use of gas-dynamic modelling techniques in conjunction with Doppler tomography can identify the main elements of the flow on Doppler maps without solution of the inaccurately defined inverse problem.

In recent years, the results of observations of binary systems in the form of consecutive spectrograms of a certain emission line $I\ (\lambda,\ t)$ (or, in other words, $I\ (V_R,\phi)$ have been used widely. The best method for the analysis of spectrograms is Doppler tomography [881]. This method converts the orbital variability of the intensity of emission lines in the map of the luminosity in the two-dimensional velocity space. The Doppler tomogram is constructed as a result of converting a series of spectrograms for consecutive time points $I\ (V_R,\ t)$ (or, equivalently, for successive orbital phases $I\ (V_R,\phi)$) in the distribution of luminosity in the plane $(V_x,\ V_y)$. The Doppler map is easier to interprete than the original spectrogram, in addition, in some cases tomography may show some features of the flow structure (or at least give a hint). In particular, the double-humped profile lines corresponding to the circular motion (e.g. in the accretion disk) change to a blurred ring on the Doppler map. In other words, the components of the binary system are resolved in the velocity space, although they may not be spatially resolved through direct observation, and, therefore, the method of Doppler tomography is a powerful tool for studying binary systems.

Unfortunately, the problem of reconstruction of the spatial distribution of the intensity of emission lines on the basis of the Doppler maps in general has no solution, because the points situated at large distances may have the same velocity, and contribute to the same region of the Doppler map. Thus, the transformation $I(V_x, V_y) \rightarrow I(x, y)$ is impossible without some apriori assumptions about the structure of the velocity field.

The situation changes dramatically when, along with the Doppler tomography technique, we use gas-dynamic calculations [815]. In this case, there is no need to solve the inverse problem, since the Doppler tomogram can be built directly: $\rho(x,y) - T(x,y) \rightarrow I(x,y)$ and $I(x,y) - V_x(x,y) \; - V_y(x,y) \rightarrow I(V_x, V_y) \rightarrow I(V_R, \phi)$.

In recent years a large number of publications has been devoted to monitoring the binary system IP Peg, and in particular, the Doppler tomograms of the system. One of the common findings of these studies is the lack of recognition of the traditional hot spots in the IP Peg system, which is consistent with our model of the hot line. The observed Doppler tomograms for the double IP Peg system can be divided into two subgroups: for quiescent and active states of the system. The dwarf nova IP Peg shows a flash with a period of approximately 3 months, where about 10 day falls on the flash and 80 days – on the quiescent state (respectively ~60 and ~480 orbital periods).

The tomograms obtained for the quiescent state of the system correspond to a solution with the hot accretion disk. The distribution of the intensity in the equatorial plane of IP Peg in the quiescent state and the synthetic Doppler map are shown in Fig.11.31a, b.

Summarizing the results of the analysis presented by synthetic Doppler tomograms of IP Peg in the quiescent state, it can be concluded that the main contribution to the luminosity should come from four elements of the flow: the hot line, the brightest part of the jet from L_1, the densening near the disk apogee and the densening behind the arm of the spiral shock wave. Obviously, depending on the specific characteristics of the systems the contribution of these elements to the total luminosity may be different. It is also clear that the results of the modelling calculations can not be used to determine which of the elements will dominate the system. However, a comparison of synthetic tomograms with observations will not only answer this question but also give the opportunity to determine which of the parameters of the gas-dynamic model needs adjusting, and, consequently, to clarify the details of the flow pattern in the system.

A typical example of the observed Doppler tomograms for the quiescent state of the system is shown in Fig. 11.31e [939]. Characteristic features of these tomograms are a bright spot in the field A and increase of brightness in region C (Fig. 11.31a). Comparison with synthetic tomograms (Fig. 11.31b) shows that the main contribution to the luminosity in the observational tomograms of IP Peg in the quiescent state is provided by a shock wave at the edge of the jet – hot line – and the densening near the apogee of the drive. Contributions to the luminosity from the spiral wave and the matter jet are relatively small.

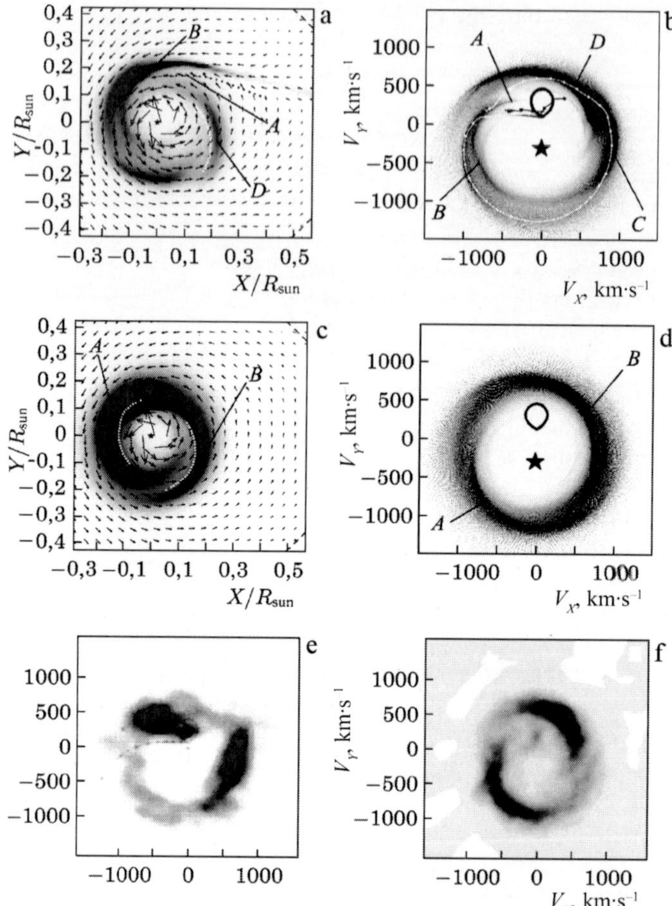

Fig. 11.31. The intensity distribution in the equatorial plane of the IP Peg system in the quiescent state (a) and active state (c), as well as synthetic Doppler maps for this system at quiescent (b) and active (d) states. In Figures b and d A, B, C and D are the the zone of maximum intensity: A – hot line, B – jet, C – the apogee of the quasi-elliptical accretion disk, D – the arm of the tidal spiral waves, the Roche lobe of the donor star (thick solid curve) and accretor (star) are shown. e and f – Doppler tomograms of IP Peg in the quiescent and active states, respectively [939, 1308].

In the active state the flow structure can be estimated from the solution with a cold accretion disk. The intensity distribution in the equatorial plane of the IP Peg system in the active state and the synthetic Doppler map are shown in Fig. 11.31c, d. In this case the main contribution to the luminosity comes from the two branches of the tidal spiral wave. A typical example of the observed Doppler tomograms for the active state of the system is shown in Fig. 11.31f [1308]. Characteristic features of these tomograms are bright spiral arms in the two quadrants of the Doppler map. A comparison with Fig. 11.31c, d, shows that in the active state the two bright spots are caused by emission of radiation from regions of higher density, located behind the two arms of the spiral tidal shock wave.

In addition to clarifying the gas-dynamic model, a comparison of the observed tomograms with synthetic ones identifies the shock waves in accretion disks of close binaries. Depending on the system state, the tomograms show both the hot line and the arms of the tidal spiral wave.

11.6.3. The precessional spiral density wave in cold disks

From the gas-dynamic calculations we know that after the occurrence of the precessional density wave in the disk the accretion rate rises sharply (up to an order of magnitude). The resulting increase in the accretion rate caused by the wave should lead to the development of a flash in the system. Matter is brought to the surface of the accretor along the precessional wave and, therefore, the area in which matter accretes is localized in the azimuth and a luminous spot hence forms on the surface of the accretor. This should lead to a hump in the shine curve. The wave slowly precesses in the coordinates associated with a stationary observer, causing the displacement of the region of enhanced accretion (i.e. hump) with every new rotation of the system. Beats between the orbital period of the system and the precession wave period is determined during this hump.

Indeed, similar phenomena were observed in some systems. 'Superhumps' – modulations on the shine curve of a binary system the period of which differs by a few percent different from the orbital period, are observed mainly during the superflashes in systems such as SU UMa. The basic observational features of superhumps were described in [1283]. To explain the phenomenon of superhumps various researchers proposed a number of models (a brief description and criticism are, for example, in [1283]). Today, the most popular model is the one which explains these variations in luminosity by the precession of the outer parts of the accretion disk. The presence of the 3:1 Lindblad resonance in the disk leads to the development of instability, which results in the outer part of the disk starting to precess with the period several times greater than the orbital period. Beats between the two periods – orbital and precession - causing periodic shine fluctuations associated with the superhump. It should be noted that this model has a number of shortcomings, the most important of which is to limit the maximum mass ratio of the components of the system. In order to ensure that the 3:1 Lindblad resonance was inside the accretion disk, the mass ratio of components q (the ratio of the mass of the donor to the mass of the accretor) should not exceed ~0.33 [1000], while there is a number of known systems in which the superhumps are observed at much larger q. An example of such a system is TV Col, where q is in the range of ~0.6–0.9.

A new mechanism for the formation of superhumps in systems such as SU UMa, based on the assumption that the occurrence in the accretion disk of a precessional density wave was proposed in 2004 [381]. The superflash model, based on the assumption that the presence of the precessional spiral wave in the accretion disk was first used to explain all important observational properties of the superflashes and superhumps in systems such as SU UMa, including the period, duration, energy release and anticorrelation of luminosity and colour temperature at the maximum of the normal superhumps, the phenomenon of 'late superhumps', etc. In addition, the proposed model does not impose severe restrictions on the ratio of the mass of the binary system and can be used to explain the mechanism of formation of superhumps in systems with a large mass ratio – up to $q = 0.93$.

11.6.4. Spiral density wave in hot disks

The presence of a bunch in the rotating accretion disk can be manifested in changes in the shine of TDS. As objects for finding the bunch we should consider eclipsing cataclysmic variables with large angles of inclination of the orbit. We can assume that the presence of a bunch, rotating in the accretion disk, can lead to fluctuations in the luminosity of the shine curves away from the eclipse. For search of the observational evidence of single-arm spiral density waves in the accretion disks, we chose the IP Peg and WX Ari systems. The IP Peg system is an eclipsing cataclysmic variable with the orbital period of 3.8 h. The mass ratio of the components of the systems $q = 1.7$, the angle of inclination of the orbit is ~68°. The WX Ari II system is also an eclipsing cataclysmic variable with the inclination of the orbit of ~72° and an orbital period of 3.34 h. Even a cursory view allows us to identify variations in the luminosity of the non-eclipsing parts of the shine curves of these systems; however, more accurate results can be obtained by Fourier analysis.

Figure 11.32 shows the shine curves of IP Peg [939], obtained 5 and 6 days after the flash in the continuum, the lines H_β, Hγ, HeI 4474 *A* and HeII 4686 *A*.

Fourier analysis shows that the curves are distinguished by strong harmonics with a frequency of 5.5 Ω_{orb}. The data on the spectra of light curves can serve as evidence of the presence in the accretion disk at the cataclysmic variable IP Peg of a spiral density wave, rotating with a frequency 5.5 times higher than the frequency of the orbital rotation system. The presence of a bunch is most clearly demonstrated by the shine curve for the HeII line.This can be explained by the fact that the density wave, which is optically dense, moving through the accretion disk, periodically 'hides' from the observer the hot region (emitting in this line) between the wave and the second component. The manifestation of this phenomenon is also partly visible in the shine curve for the continuum.

Further study of the shine curves of cataclysmic variables showed the presence of similar harmonics and other objects of this class, e.g. WX Ari. The shine curves of WX Ari are characterized by the dominance of the harmonic with a frequency of approximately 6.25 Ω_{orb}. Thus, the analysis of the shine curves of the actually observed cataclysmic variables with the characteristics that are sufficiently relevant with respect to characteristics of the numerical model, allows to say that there are changes in the luminosity of the non-eclipsing parts of these curves with a frequency of about one fifth of the orbital period, which coincides with the orbital period of the bunch, obtained in the calculations.

11.7. Developed turbulence in accretion disks

11.7.1. Super-reflection instability in accretion disks

Above it was shown that periodic variations of the quasi-stationary mode of the

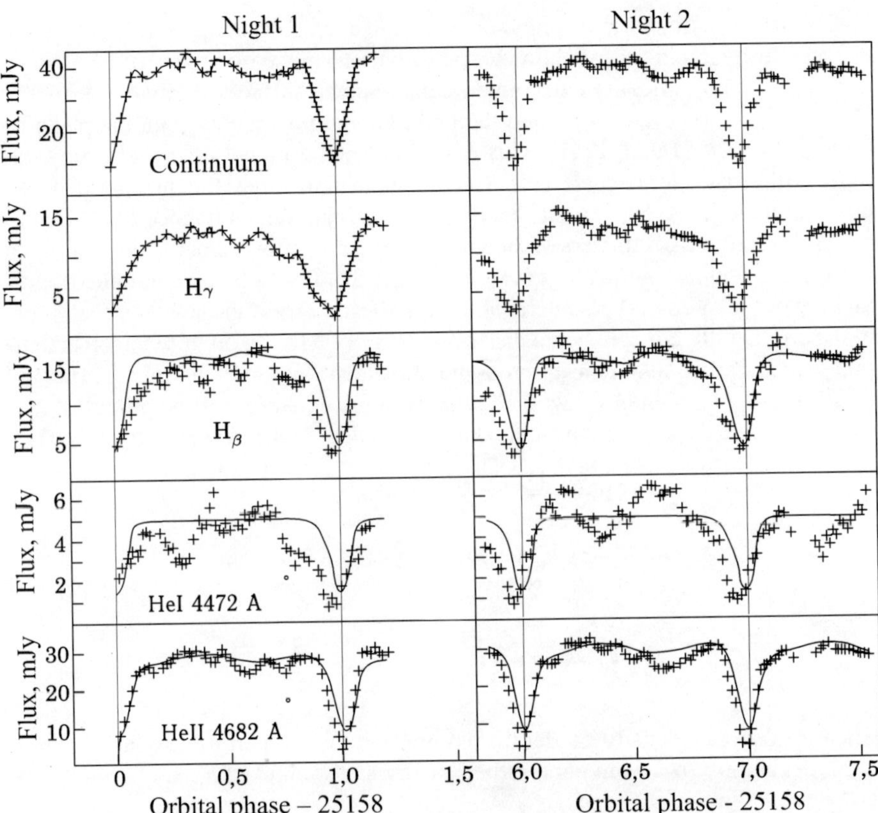

Fig. 11.32. Shine curves of IP Peg in continuum, lines H_β, H_γ, HeI 4472 A, and HeII 4686 A [939].

accretion disk in three-dimensional numerical calculations are caused by the single-arm spiral–vortex structure with a period (in the coordinates, rotating with the angular velocity of the system) $\tilde{T} \simeq (1/6)P_{\mathrm{orb}}$. This period is not related to the period, caused by the force of the tidal interaction of the donor. In this section, we consider the above-mentioned single-arm spiral–vortex structure, which is generated by super-reflection instability in the accretion disk.

To clarify the reasons for the origin of this mode, it is necessary to reduced the influence of tidal forces. Here are the steps that allowed to solve this problem [243].

1. Results of three-dimensional numerical calculations have been transferred to the new coordinate system $\{x',y',z'\}$ with centre at the accretor. This system revolves around the centre of mass of the double star, in accordance with the movement of the accretor and its axes are parallel to the axes of the inertial coordinate system[21]. The transition was then made to the system $\{r,\phi,z\}$ ($z = 0$ plane in the new and old coordinate systems is unchanged).

[21]Note that in the new system of the coordinates the variation period $T_1 \simeq (1/5)\, P_{\mathrm{orb}}$.

2. All distributions were integrated over z and averaged over period P_{orb}. Thus, we analyze the dynamics of small (linear) perturbations in the isolated two-dimensional disk model with stationary axisymmetric parameters: surface density $\sigma_0(r)$, angular velocity $\Omega_0(r)$, pressure $P_0(r)$ and the gravitational potential $\Phi_0(r)$in the plane $z = 0$ (see, e.g. Fig. 11.33a) [241]. This two-dimensional model of the accretion disk inherits the basic large-scale properties of the initial three-dimensional disk, but does not include the tidal effect caused by the gravitational influence of the donor.

Analysis of the parameters of the model of the disk shows that the distribution of the specific angular momentum is an important feature – a break (smoothed) in the vicinity of the radius, where the region of manifestations of the single-arm structure in the original three-dimensional disk starts (Fig. 11.33b). In order to understand the importance of the kink in the graph of the radial distribution of the specific (per unit mass) angular momentum[22], we write the dynamic equations of the perturbed state two-dimensional model of an isolated accretion disk in the new coordinate system:

$$\begin{aligned}
&\frac{\partial \sigma}{\partial t}+\partial 1 r\frac{\partial(r\sigma V_r)}{\partial r}+\frac{1}{r}\frac{\partial(\sigma V_\varphi)}{\partial \varphi}=0,\\
&\frac{\partial V_r}{\partial t}+V_r\frac{\partial V_r}{\partial r}+\frac{V_\varphi}{r}\frac{\partial V_r}{\partial \varphi}-\frac{V_\varphi^2}{r}=-\frac{1}{\sigma}\frac{\partial P}{\partial r}-\frac{\mathrm{d}\Phi_0}{\mathrm{d}r},\\
&\frac{\partial V_\varphi}{\partial t}+V_r\frac{\partial V_\varphi}{\partial r}+\frac{V_\varphi}{r}\frac{\partial V_\varphi}{\partial \varphi}+\frac{V_r V_\varphi}{r}=-\frac{1}{\sigma r}\frac{\partial P}{\partial \varphi},
\end{aligned} \qquad (11.112)$$

where σ is surface density. Under the assumption that all perturbed quantities have small deviations $\delta\tilde{F}$ from unperturbed steady-state values of F_0

$$F(r,\varphi,t)=F_0(r)+\delta\tilde{F}(r,\varphi,t),$$

where

$$\delta\tilde{F}(r,\varphi,t)=\delta F(r)\exp[i(m\varphi-\omega t)],$$

system of equations (11.112) after linearization can be reduced to an ordinary differential second-order equation for the perturbed enthalpy δW [126]:

$$\frac{\chi}{r\sigma_0}\frac{\mathrm{d}}{\mathrm{d}r}\left[\frac{r\sigma_0}{\chi}\frac{\mathrm{d}}{\mathrm{d}r}(\delta W)\right]-\frac{\chi}{r\sigma_0\hat{\omega}}\left[\frac{\mathrm{d}}{\mathrm{d}r}\left(\frac{2m\Omega_0\sigma_0}{\chi}\right)\right]\delta W-\left[\frac{m^2}{r^2}+\frac{\chi}{c_0^2}\right]\delta W=0. \qquad (11.113)$$

The following notations are used here

$$\chi(r,\omega)\equiv[\hat{\omega}^2(r)-\varkappa_0^2(r)],\quad \varkappa_0^2\equiv 2\Omega_0\left(2\Omega_0+r\frac{\mathrm{d}\Omega_0}{\mathrm{d}r}\right),$$

$$\hat{\omega}(r,\omega)\equiv\omega-m\Omega_0(r),\quad c_0^2\equiv\frac{\partial P_0}{\partial\sigma_0}.$$

---

[22]It is obvious that kink in the graph of the specific angular momentum means a break in the graph $\Omega_0(r)$.

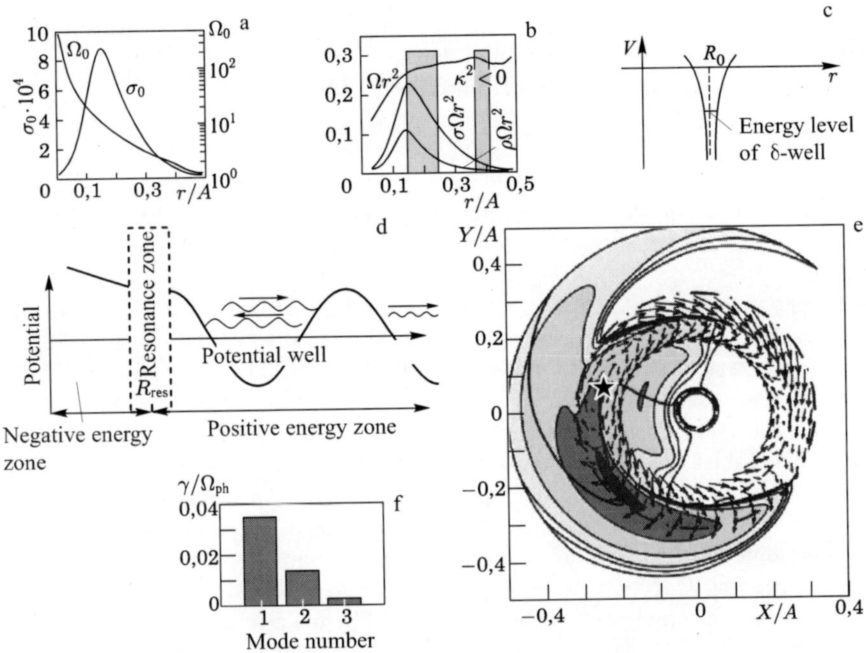

Fig. 11.33. a) the radial distribution of surface densities σ_0 (as a fraction of the density of matter in the interior libration point) and the angular velocity Ω_0 of the isolated model of of a two-dimensional accretion disk in the new coordinate system (see text). b) the distribution of specific angular momentum per unit mass (upper curve), per unit area (middle curve) and per unit volume (lower curve) is shown as the area (left shaded box) of localization the first (maximum amplitude) harmonics; it can be seen that the position of the kink in the curves of specific angular momentum coincides with the boundary of the localization of the first harmonic. c) the energy level in the δ-well. d) the usual well, in which there are incident and reflected waves. e, f) the results of a numerical study of linear modes, generated by super-reflection instability in the two-dimensional model of an isolated accretion disk: e) increments the first three large-scale modes ($m = 1,2,3$); f) the eigenfunctions of the perturbed surface density (the value of which exceeds the background density $\sigma_0(r)$ and the perturbed velocity (indicated by arrows) for the most unstable single-arm spiral mode ($m = 1$).

In the limiting case of a sharp kink $\Omega_0(r)$ at radius $r = R_0$ its first derivative $d\Omega_0(r)/dr$ is proportional to $\theta(r-R_0)$, and the second derivative $d^2\Omega_0(r)/dr^2$ is proportional to $\delta(r - R_0)$, hence

$$\frac{d\chi}{dr} \propto \frac{d\varkappa_0^2}{dr} \propto \frac{d^2\Omega_0}{dr^2} \propto \delta(r - R_0).$$

Eliminating the first derivative of (11.113), we reduce this equation to the Schrödinger equation with the 'δ-well' [126], which always contains the energy level (Fig. 11.33c). In a real situation of the accretion disk the δ-well transforms into an ordinary hole (Fig. 11.33d). In this case, in the potential well there are two waves: incident and reflected. The latter is reflected from the resonance zone (in the corotation circle) with increasing amplitude. The additional energy is drawn from the kinetic energy of rotation of the disk, which increases when approaching the centre. This leads to super-reflection instability, which generates a single-arm density wave [126].

To investigate spiral density waves, caused in the accretion disk by the supra-reflection instability, the boundary-value problem for the equation (11.113) was solved numerically. In particular, the eigenvalues of the problem were calculated: the angular phase velocity of rotation $\Omega_{ph} \equiv \mathrm{Re}\,(\omega)/m$ and the instability increment $\gamma \equiv \mathrm{Im}\,(\omega)$.

Comparison of the main results, obtained in the two-dimensional model of the disk with the appropriate results of numerical modelling of the accretion disk in the binary system shows that they are in almost complete agreement [241, 572]. The results, obtained in the linear approximation, are shown in Fig. 11.33e, f. Figure 11.33e shows the values of the increments of the increase of the first, second and third mode in the units of phase velocity Ω_{ph}. Figure 11.33f shows the main eigenfunctions for the most unstable single-arm spiral ($m = 1$). The perturbed surface density (only the value greater than the value of the stationary background) is represented by the dark area, the thick curve is the phase curve of maximum surface density. The vector field of the perturbed velocity (in the coordinate system rotating with the single-arm mode) is indicated by the arrows, the asterisk denotes the centre of the anticyclone vortex. The dashed circle corresponds to the corotation radius r_c, the dot-and-dash line corresponds to the outer Lindblad resonance (OLR). It may be seen that the corotation circle passes through the centre of the anticyclone in accordance with the theory of density waves [241].

Note that:

1) the general spiral–vortex structure rotates with a period of

$$T^{(m=1)} \equiv 2\pi / \Omega_{ph}^{(m-1)} \simeq T_1 \simeq (1/5)P_{orb}$$

in the same direction (anticlockwise) as the accretion disk, thus the single-arm spiral density wave is lagging;
2) the phase curve of the density wave does turn on $\pi/2$ in the positive direction in the range of (r_c, r_{OLR});
3) The centre of an anticyclonic vortex in the perturbed vector velocity field, is localized, as already mentioned, on the corotation circle, rotates with the spiral density wave with the same angular velocity $\Omega_{ph}^{(m-1)}$, and plays an important role in the redistribution of the angular momentum of the disk.

All the investigated large-scale unstable modes exhibit similar properties: they are localized in the vicinity of their corotation radius. These radii are close to each other, lie in the inner part of the disk, and can be approximated as as follows:

$$r_c^{(m)} \simeq r_c^{(1)}(1+\beta)^{m-1}, \quad r_c^{(1)} \simeq 0.25A, \quad \beta \simeq 0.024, \quad m = 1, 2, 3.$$

These values are close to the value of $r_c^{(1)} \simeq 0.21$ A, estimated in three-dimensional numerical calculations. As for the rotational frequencies, they can be approximated as[23]:

$$\omega_r^{(m)} \equiv 2\pi f^{(m)} \equiv m\Omega_{\text{ph}}^{(m)} \simeq m\Omega_{\text{ph}}^{(1)}(1+\beta_1)^{-(m-1)}, \quad \beta_1 \simeq 0.045.$$

Thus, we have some frequency shifts:

$$\Delta f^{(m)} \equiv \frac{f^{(m)} - mf^{(1)}}{f^{(m)}} \simeq -\beta_1(m-1); \quad m = 2,3.$$

The above results in two-dimensional model of an isolated accretion disk with real parameters, coinciding with the parameters of the three-dimensional model of the accretion disk in a close system of binary stars, are not unexpected. The observed rate of rotation of the 'nuclear' ionized disk in the centre of the galaxy has a smooth kink in the neighborhood of the radius 0.9 pc (see chapter 9, Fig. 9.19), where the shape of the rotation velocity graph is caused by the presence in the disk centre of a massive black hole surrounded by an 'isothermal' globular cluster. In such a system, as we saw above, a structure similar to that observed forms due to the super-reflection instability.

11.7.2. Estimates of the coefficient of turbulent viscosity and the magnitude of α in accretion disks

In the monographs and reviews on plasma physics the formula for estimating the turbulent viscosity coefficient is written in the form (see, for example [736])

$$\nu_{\text{turb}} \approx \frac{(\gamma_L)_{\max}}{(k_\perp^2)_{\min}}, \tag{11.114}$$

where $(\gamma_L)_{\max}$ is the maximum increment of linear instability, which causes the corresponding turbulent viscosity, $(k_\perp)_{\min}$ is the the minimum wave number corresponding to the maximum length of the wave generated by a given instability.

Equation (11.114) can be easily obtained from dimensional considerations. Indeed, the dimension of the turbulent viscosity ν_{turb}, as well as molecular ν_{mol} viscosity is

$$[\nu_{\text{turb}}] = [\nu_{\text{mol}}] = \text{cm}^2 \cdot \text{s}^{-1}.$$

Substituting into the right-hand side of this expression the values of the appropriate dimension: $[\text{cm}^2] = [k^{-2}]$ $[\text{s}^{-1}] = [\gamma]$, we arrive at the formula (11.114).

The developed three-dimensional Kolmogorov turbulence in the accretion disks means that the maximum size of turbulent fluctuations is much smaller than the disk half-thickness h, i.e. $k_\perp h \gg 1$, or in the limiting case $(k_\perp)_{\min} h \geqslant 1$. Then, considering the maximum increment of super-reflection instability for the mode ($m = 1$) (Fig.

[23]In the case of pure Keplerian rotation, we have: $\omega_T^{(m)} \simeq m\Omega_{\text{ph}}^{(1)}(1+B)^{-3(m-1)/2}$:

11.33e) and the equilibrium condition of the disk in the z-coordinate, $h \approx c_0/\Omega_0$,

$$(\gamma_L)_{\max} \approx 0.035\Omega_{\rm ph} = 0.035\Omega_{\rm cor} = 3.5\times10^{-2}\left(\frac{c_0}{h}\right), \quad (k_\perp^2)_{\min} \approx h^{-1}.$$

Substituting these expressions into (11.114), we obtain the coefficient of turbulent viscosity in the form

$$\nu_{\rm turb} \approx 3.5\times10^{-2}\, c_0 h.$$

Following [1146, 1147], we obtain

$$\alpha \approx 3.5\times10^{-2}$$

which is consistent both with the scope of the numerical viscosity of our calculations, $10^{-2} \ll 10^{-1}$, and with the intervals defined in the relevant articles of the observers and theorists (see [867, 894, 1162, 1180].

11.7.3. Turbulence spectrum

It is well known that, in accordance with Kolmogorov's scenario of the developed stationary turbulence the energy flux is directed from small wave numbers $k_\perp$ to higher values. The power of the energy flux per unit mass is constant:

$$\varepsilon \approx (\delta V)^2 t^{-1} \approx (\delta V)^3 \lambda^{-1} = \text{const.} \tag{11.115}$$

Here we assume that turbulent fluctuations with a spatial size $\sim\lambda$ have characteristic velocity pulsations[24] δV. From (11.115) we find the well-known Kolmogorov-Obukhov range of turbulence [787, 983]:

$$\delta V \approx (\lambda\varepsilon)^{1/3}, \quad \varepsilon = \text{const.}$$

Note that condition (11.115) can be satisfied only when the turbulent medium is homogeneous and isotropic, which is not true for our study of heterogeneous and differentially rotating accretion disks.

If we immerse both hands in shallow water in a bath and start to rotate them in the same direction, then we will see two vortices, seeking to merge into one (Fig. 11.34), i.e. the characteristic scale in this case will be increased, in contrast to the scale of the Kolmogorov cascade, which is reduced. This case corresponds to a two-dimensional (Batchelor) turbulence, for which the inequality converse to the Kolmogorov inequality is fulfilled: $k_\perp h \ll 1$ or $(k_\perp)_{\max} h \leqslant 1$.

As we saw above, on the non-eclipsing side of the shine curve the visible amplitude of the variations varies with a period $T_1 \simeq (1/5)T_{\rm orb}$ (where $T_{\rm orb}$ is the orbital period of the binary system)[25] . These variations manifest themselves most

[24]A detailed description of turbulent fluid motion as a superposition of the averaged and pulsation components is contained, for example, in [938],

[25]It is clear that the tidal variations of large amplitude are also present. Their period $T_{\rm tdl} \simeq T_{\rm orb}/2$.

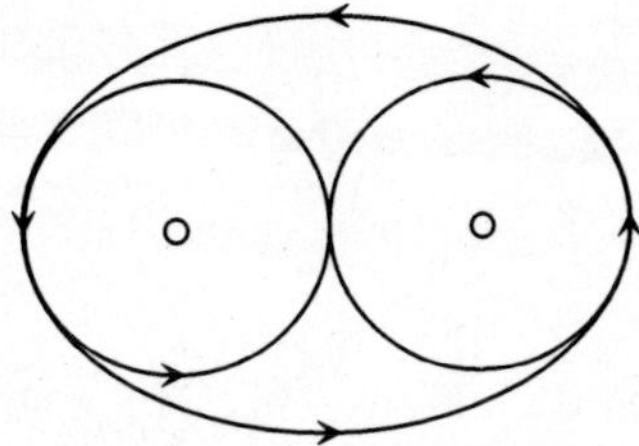

Fig. 11.34. Schematic representation of vortices in shallow water. The characteristic feature – tendency of the two vortices to merge into a single one.

clearly in the emission lines of doubly ionized helium, HeII 4686 A (see Fig. 11.32). In the IP Reg system the angle of inclination of the orbit is ~ 68°, so that the colder and denser bunch – single-arm density wave – will hide periodically the inner hot area from the observer. Hence the visible variations in the non-eclipsing parts of the shine curves. The HeII 4686 A line is clearly out of the inner parts of the disk with the temperature ≃51 eV higher than the ionization threshold of He II. By analyzing the luminosity variations in this line, we can get idea of the structures screening the the radiation in the outer parts of the accretion disk.

In accordance with the results described in 11.5, the single-arm spiral density wave in the accretion disk may be caused by super-reflection instability. This dense structure is localized in the radial region $r \sim (0.15\text{–}0.30)\,A$ and has a period of close to T_1.

We represent the density in the equatorial plane of the accretion disk ($z = 0$) at some radius r at a fixed azimuthal angle φ as

$$\rho(t) = \langle\rho\rangle + \Delta\rho(t), \quad \Delta\rho(t) = \delta\rho_{\text{tdl}}(t) + \sum_{m=1}^{m_{\max}} \delta\rho^{(m)}(t).$$

Here we have assumed that density variations around the average level $\langle\rho\rangle$ caused by the superposition of tidal harmonics $\delta\rho_{\text{tdl}}$ and its principal azimuthal modes $m(m_{\max} \leqslant 3\text{–}4)$, associated with the super-reflection instability. Then we can approximate in the same way the observed intensity of radiation in the HeII 4686 A line as a superposition of the average level $\langle I\rangle$ and some fluctuations $\Delta I(t)$, i.e.

$$I(t) = \langle I\rangle + \Delta I(t), \quad \Delta I(t) = \delta I_{\text{tdl}}(t) + \sum_{m=1}^{m_{\max}} \delta I^{(m)}(t).$$

Here $\delta I_{\text{tdl}}(t)$ is the tidal harmonic with frequency $f_{\text{tdl}} = 2 f_{\text{orb}}, f_{\text{orb}} \equiv 1/T_{\text{orb}}$. Other fluctuations correspond to the principal eigenmodes and have a simple harmonic relationship:

$$\delta I^{(m)} \cong A_m \cos[2\pi f_m (t - t_{0m})].$$

Therefore, the amplitude of the m-th harmonic A_m should be proportional to the amplitude $|\delta\rho^{(m)}|$ of the appropriate eigenmode, $A_m \equiv A(k_\perp^{(m)})\ \alpha\ |\delta\rho^{(m)}|$. The

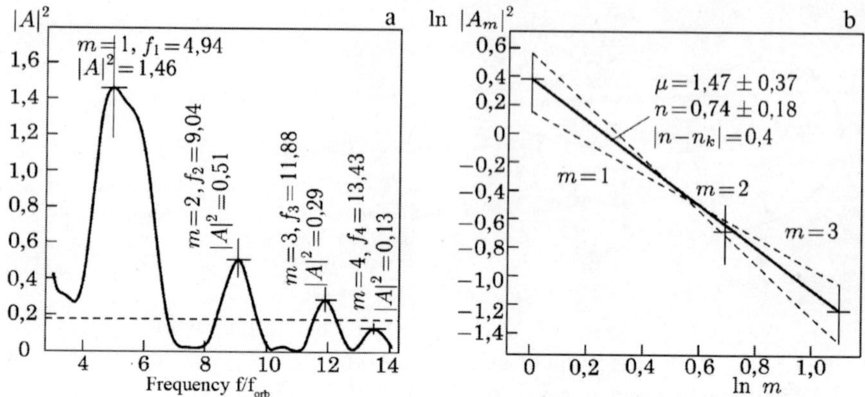

Fig. 11.35. a – Fourier spectrum of the shine variability of the IP Reg system in HeII 4686 A line (1st night in Fig. 11.32). The horizontal dashed line indicates the level corresponding to the observed signal/noise ratio equal to unity; b – a linear representation of the power spectrum. The dashed lines indicate the limits corresponding to the maximum errors; $n_k = 1/3$ is the Kolmogorov spectrum.

calculated (in [575]) Fourier spectrum (square modulus of the amplitude $|A_m|^2$ of the normalized orbital frequency $f_{\rm orb}$) of the fluctuations $\Delta I(t)$ shows four major peaks in addition to the tidal harmonics (Fig. 11.35a). The first peak, having a frequency $f_1 \simeq 4.94$, presumably corresponds to the mode $m = 1$. Therefore, according to a numerical study of the linear approximation of an isolated two-dimensional accretion disk with the main parameters of the three-dimensional disk in a binary system (see section 11.5), two subsequent peaks correspond qualitatively to the modes $m = 2$ and $m = 3$ of the investigated the model[26] .

The specific heat (elastic) energy of the m-th harmonic of the density wave can be estimated as (see, e.g. [112])

The specific kinetic energy in the same wave is expressed as [112]

$$\delta E_T^{(m)} \cong \frac{c_0^2}{2\langle\rho\rangle}(\delta\rho^{(m)})^2 \propto (\delta\rho^{(m)})^2.$$

where $\delta V^{(m)}$ is the amplitude of the perturbed velocity in the m-th harmonic wave. As shown in [112], $\delta E_T^{(m)} \simeq \delta E_K^{(m)}$ from which

$$\delta E_K^{(m)} \cong \frac{\langle\rho\rangle}{2}(\delta V^{(m)})^2 \propto (\delta V^{(m)})^2,$$

$$\delta E_T^{(m)}\,\mathrm{d}k \propto |\delta\rho^{(m)}|^2 \propto |A(k_\perp^{(m)})|^2 .$$

Assuming a power law distribution of spectral amplitudes of the peaks, we can write

[26]Shifts of the frequencies in the spectrum, shown in Fig. 11.35a, are estimated as $\Delta f_1^{(2)} \simeq -0.09$ and $\Delta f_1^{(3)} \simeq -0.25$. They agree qualitatively with the results obtained in this model.

$$(\delta V^{(m)})^2 \propto \ |A(k_\perp^{(m)})|^2 \simeq \frac{B}{(k_\perp^{(m)})^\mu} \simeq \frac{\tilde{B}}{m^\mu}, \tag{11.116}$$

where

$$k_\perp^{(m)} \simeq k_\varphi^{(m)} \simeq \frac{m}{r_c^{(m)}} \simeq \frac{m}{r_c^{(1)}}(1+\beta)^{-(m-1)}, \quad m=1,2,3,$$

$$\tilde{B} = B r_c^{(m)}$$

and the corotation radii $r_c^{(m)}$ of the large-scale modes are assumed to be close to each other (according to the results presented in section 11.7.1, we are able to assume here $\beta \sim 0.05 \ll 1$).

Taking the logarithm of (11.116), we obtain

$$\ln\{|A_m|^2\} \simeq \ln \tilde{B} - \mu \ln m.$$

As we see from Fig. 11.35b, three peaks corresponding to the modes $m = 1, 2, 3$, are linked by a linear (with a fairly high degree of accuracy) dependence. The slope of this linear dependence gives the following exponent n:

$$\delta V^{(m)} \propto \left(k_\perp^{(m)}\right)^{-n}, \quad n \equiv \frac{\mu}{2} \simeq 0.74 \pm 0.18.$$

Here, the estimated value of the error corresponds to uncertainties of the calculated (using the discrete Fourier transform) power spectrum (indicated by crosses in Fig. 11.35).

The value of the exponent $n \simeq 0.74 \simeq 3/4$, calculated for an accretion disk in IP Reg is significantly different from the exponent of fully developed turbulence $n_K = 1/3$, calculated by Kolmogorov and Obukhov – by the value $\Delta n \simeq 0.4$. Such a large difference arises from the fact that, in any accretion disk, the two conditions required to implement the Kolmogorov–Obukhov range – homogeneity and isotropy – are not fulfilled.

Thus, a comprehensive analysis of large-scale processes in accretion disks of binary stars, including three-dimensional numerical simulation, the numerical solution of linear boundary-value problems to find the eigenfunctions and eigenvalues of the two-dimensional isolated accretion disk with the main parameters obtained in the corresponding three-dimensional calculations and Fourier analysis fluctuations in the power spectrum of the shine curves for the non-eclipsing part of the accretion disk of IP Reg, yielded the following results (see [243] and references therein).

1. The existence of global tidal mode $m = 2$, caused by the gravitational influence of the donor star on the accretion disk, has been demonstrated.

2. The presence of increasing amplitude of the collective oscillation modes $m = 1, 2, 3$, caused by super-reflection instability, has been confirmed.

3. It has been shown that the maximum increment $\gamma \simeq 0.035\Omega_{ph}$ has the first mode, $m = 1$.

4. The calculated maximum increment of instability and evaluation of the minimum value of the wave vector $(k_\perp)_{min} \simeq h^{-1}$ for the first time allowed to obtain from the first principles the turbulent viscosity coefficient and the parameter value $\alpha = 0.035$. The value of α obtained is consistent with the observational and theoretical estimates of many authors.

5. The spectrum of the fully developed turbulence of the accretion disk in the IP Peg, which is a power law with exponent $n \simeq 3/4$, has been determined for the first time.

References

1. Abramyan M.G., Density waves in inserted discks, Astrofizika, 1978, v. 14, 579–590.
2. Avedisova B.C., Pis'ma v AZh, 1985, v. 11, 448.
3. Avedisova B.C., Astrofizika, 1989, v. 30, 140.
4. Antonov V.A., Uch. zap. LGU, Ser. Mat. Nauk, 1971, v. 47, No. 350, 64–85.
5. Antipov S.V., Pis'ma ZhETF, 1983, v. 37, No. 3, 319.
6. Antonov V.A., et al., in: LOVAGO Dynamics and evolution of stellar systems, Moscow and Leningrad, 1975, 289–301.
7. Afanas'ev V.L., Usp. Fiz. Nauk, 1986, volo. 150, 313-315.
8. Afanas'ev V.L., et al., Astrofizika, vol. 28, 243; 1988, v. 29, 155.
9. Afanas'ev V.L., et al., Astrofizika, 1988, v. 28, 142–150.
10. Afanas'ev V.L., et al., Astrofizika, 1988, v. 29, 497-504.
11. Afanas'ev V.L., et al., Astr. Zh., 1991, v. 68, 1134-1149.
12. Afanas'ev V.L., et al., Astr. Zh., 1992, v. 69, 19-37.
13. Afanas'ev V.L., et al., Kinetics of gas in the field of a rotating evolving jumper, Preprint VolGU 6-89, Volgograd, 1989.
14. Afanas'ev V.L., et al., Pis'ma AZh., 1993, v. 19, 787.
15. Afanas'ev V.L. and Shapovalova A.I., Astrofizika, 1981, v. 17, 403.
16. Baev P.V., et al., Pis'ma AZh., 1987, v. 13, 964.
17. Barabanov A.V. and Zasov A.V., AZh., 1979, v. 56, 252-257.
18. Bartaya R.A., Byul. Abastum, AO, 1979, v. 50, 238.
19. Bektasova N.K. and Petrovskaya I.V., Uch. zap. LGU, Ser. Matem. Nauk, 1983, v. 60, No. 38, 127-141.
20. Belotserkovskii O.M., Turbulence - New approaches, Nauka, Moscow, 2003.
21. Bisnovatyi-Kogan G.S., Astrofizika, 1984, v. 20, 547-563.
22. Bisnovatyi-Kogan G.S., Physical fundamentals of the theory of star evolution, Nauka, Moscow, 1989.
23. Bisikalo D.V., et al., AZh., 2001, v. 78, 707-716.
24. Bisnovatyi-Kogan G.S. and Blinnikov S.I., Pis'ma AZh., 1976, v. 2, 489-493.
25. Bisnovatyi-Kogan G.S. and Blinnikov S.I., Sov. Astr., 1981, v. 25, 175.
26. Dynamics of Accretion Disks Bisnovatyi-Kogan G.S., et al., AZh., 1979, v. 56, 770-780.
27. Bisnovatyi-Kogan G.S. and Ruzmaikin A.A., Astroph. Space Sci., 1976, v. 42, 375-399.
28. Blandford R., in: Numerical modelling in astrophysics, Mir, Moscow, 1988.
29. Bogolyubov N.N. and Mitropol'skii Yu.A., Asymptotic methods in the theory of non-linear oscillations, Nauka, Moscow, 1974.
30. Bochkarev N.G., Fundamentals of the physics of the interstellar medium, Mos cow State University, Moscow, 1999.
31. Brekhovskikh L.M. and Goncharov V.V., Introduction into the mechanics of solid media, Nauka, Moscow, 1982.

32. Brekhovskikh L.M. and Godin O.A., Acoustics of laminated media, Nauka, Moscow, 1989.
33. Badsell Ch. and Langdon A., Physics of plasma and numerical modelling, Energoatomizdat, Moscow, 1989.
34. Vibe D.Z. and Skustov B.M., Astr. Zh., 2001, v. 78, 977.
35. Velikhov E.P., Zh. Eks. Teor. Fiz., 1959, v. 36, 1398-1404.
36. Vil'kovskii E.Ya., Quasars, Nauka, Moscow, 1985.
37. Vorontsov-Vel'yaminov B.A., Extragalactic astronomy, Nauka, Moscow, 1978.
38. Galeev A.A. and Sagdeev R.Z., The Non-linear theory of plasma, Vopr. Teor. Plazmy, No. 7, 1973, 3-145.
39. Genkin I.L. and Miroshkin A.D., Tr. AFI AN KazSSR, 1983, v. 40, 87-90.
40. Gerasimov A.G. and Petrovskaya I.V., KFNT, 1990, v. 6, 17-24.
41. Gorbatskii V.G., Introduction into physics of galaxies and galactic clusters, Nauka, Moscow, 1986.
42. Gorbatskii V.G. and Serbin V.M., Astrofizika, 1983, v. 19, 79-89.
43. Gor'kavyi N.N. and Fridman A.M., Physics of planetary rings, Nauka, Moscow, 1994.
44. Hoffmeister K., et al., Variable stars, Nauka, Moscow, 1990.
45. Gradshtein I.S. and Ryzhik I.M., Tables of integrals, sums, series and products, GIFML, Moscow, 1963.
46. Grivnev E.M. and Fridman A.M., AZh, 1990, v. 67, 21-32.
47. Gurevich A.V., et al., Usp. Fiz. Nauk, 1997, v. 167, 913-943.
48. Dambis A.K., et al., Pis'ma v AZh, 1995, v. 21, 331-347.
49. Dambis A.K., et al., Pis'ma v AZh, 2001, v. 27, 68-75.
50. Dibai E.A., ATs, 1971, v. 660, 1-6.
51. Efremov Yu.N., Pis'ma v AZh, 1983, v. 9, 94.
52. Efremov Yu.N., Regions of star formation in galaxies: stellar complexes and spiral arms, Nauka, Moscow, 1989.
53. Efremov Yu.N., Inside the universe, Editorial URSS, Moscow, 2002.
54. Efremov Yu.N., Stellar islands, Fryazino: Vek2, 2005.
55. Efremov Yu.N., et al., Usp. Fiz. Nauk, 1989, v. 157, 599-629.
56. Zaslavskii G.M., et al., Weak chaos and quasi-regular structures, Nauka, Moscow, 1991. 57.
58. Zasov A.V., INT, 1981, v. 18, 3-47.
59. Zasov A.V., Pis'ma v AZh, 1985, v. 11, 730-736.
60. Zasov A.V., ATs, 1986, v. 1443, 730-736.
61. Zasov A.V., The physics of galaxies, Moscow State University, Moscow, 1988.
62. Zasov A.V., Pis'ma v AZh, 1989, v. 15, 210-215.
63. Zasov A.V., et al., Pis'ma v AZh, 1991, v. 17, 884-892.
64. Zasov A.V., et al., Pis'ma v AZh, Effect of the massive halo on the stability of the bending modes of the stellar disc. The dynamics of the gravitating systems and methods of analytical stellar mechanics, Nauka, Alma-Ata, 1987, 23-24.
65. Zasov A.V., The effect of the massive Halo on the stability of the large-scale bending modes of the stellar disc. Problems of stellar mechanics an stellar dynamics, Nauka, Alma-Ata, 1990, 87-91.
66. Zasov A.V., et al., AZh, 1990, v. 67, 222-229.
67. Zasov A.V. and Sil'chenko O.K., Usp. Fiz. Nauk, 2010, v. 180, 434-439.
68. Zasov A.V., et al., AZh, 2008, v. 52, 79-93.
69. Zasov A.V. and Morozov A.G., AZh, 1985, v. 62, 475-481.
70. Zasov A.V. and Morozov A.G., ATz, 1985, v. 1356, 1-3.
71. Zasov A.V. and Neizvestnyi S.I., Pis'ma v AZh, 1989, v. 15, 963-970.
72. Zasov A.V. and Postnov K.A., General astrophysics, Fryazino, 2006.

73. Zasov A.V. and Sil'chenko O.K., Pis'ma v AZh, 1987, v. 13, 455-463.
74. Zasov A.V. and Simakov S.G., Astrofizika, 1988, v. 29, 1909-198.
75. Zasov A.V. and Khoperskov A.V., AZh, 2002, v. 79, 195-204.
76. Zasov A.V. and Khoperskov A.V., Pis'ma v AZh, 2003, v. 29, 497-507.
77. Zasov A.V., et al., Pis'ma v AZh, 2004, v. 30, 653-662.
78. Zasov A.V., et al., Pis'ma v AZh, 2011.
79. Zel'dovich Ya.B., DAN, 1964, v. 155, 67-69.
80. Zel'dovich Ya.B., et al., Preprint SibIZMIR SO RAN SSSR, No. 7-72, Irkutsk, 1972.
81. Zel'dovich Ya.B., Relativistic astrophysics, Nauka, Moscow, 1967.
82. Kadomtsev B.B., Plasma turbulence, in: Problems of plasma theory, Atomizdat, Moscow, 1964, v. 4, 188-339.
83. Kadomtsev B.B., Collective phenomena in plasma, Nauka, Moscow, 1988.
84. Kaplan S.A. and Pikel'ner S.B., Physics of the interstellar medium, Nauka, Moscow, 1978.
85. Karachentsev I.D., Binary galaxies, Nauka, Moscow, 1987.
86. Karachentsev I.D., Usp. Fiz. Nauk, 2001, v. 171, 860-863.
87. Karachentsev I.D. and Chzhou Shu, Pis'ma AZh., 1991, v. 17, 321-326.
88. Karachentseva V.E. and Karachentsev I.D., Soobshch. Byurokanskoi AO, 1967, v. 38, 49-57.
89. Karimova D.K. and Pavlovskaya E.D., AZh., 1974, v. 51, 597-605.
90. Karimova D.K. andf Pavlovskaya E.D., AZh., 1976, v. 53, 495-500.
91. Karimova D.K. andf Pavlovskaya E.D., AZh., 1981, v. 58, 984-995.
92. Karimova D.K. andf Pavlovskaya E.D., AZh., 1984, v. 61, 66-74.
93. Kasparova A.V. and Zasov A.V., PAZh, 2008, v. 34, 174-184.
94. King A.R., Introduction to classic stellar dynamics, Editorial URSS, Moscow, 2002.
95. Kovalenko I.G. and Lukin D.V., Pis'ma AZh., 1999, v. 25, 260-269.
96. Kolykhalov P.I., Preprint No. 825, IKI AN SSSR, Moscow, 1983.
97. Kolykhalov P.I., Izv. AN SSSR, Ser. Mekhan. Zhidk. Gaza, 1984 v. (3) 1, 45.
98. Kolykhalov P.I., DAN, 1985, v. 280, 95-98.
99. Kolykhalov P.I. and Syunyaev R.A., Pis'ma AZh., 1979, v. 5, 338-344.
100. Kolykhalov P.I. and Syunyaev R.A., Pis'ma AZh., 1980, v. 6, 680-686.
101. Korchagin V.I. and Marochnik L.S., Azh., 1976, v. 52, 15-25.
102. Korchagin V.I. and Petviashvili V.I., Pis'ma AZh., 1985, v. 11, 121-122.
103. Korchagin V.I. and Shevelev Yu.G., Astrofizika, 1980, v. 16, 757-767.
104. Korchagin V.I. and Shevelev Yu.G., Astrofizika, 1981, v. 17, 455-468.
105. Kuzmin G.G., Publ. Tartuskoi AO, 1961, v. 33, 351-370.
106. Kulikovskii A.G., et al., Mathematical methods of numerical solution of hyperbolic systems of equations, Fizmatlit, Moscow, 2001.
107. Kyazumov G.A., Pis'ma AZh., 1980, v. 6, 398-401.
108. Landau L.D., DAN, 1944, v. 44, 151.
109. Landau L.D. and Lifshits E.M., Quantum mechanics, Nauka, Moscow, 1974.
110. Landau L.D. and Lifshits E.M., Statistical physics, Part I, Nauka, Moscow, 1976.
111. Landau L.D. and Lifshits E.M., Electrodynamics of solid media, Nauka, Moscow, 1976.
112. Landau L.D. and Lifshits E.M., Hydrodynamics, Nauka, Moscow, 1986.
113. Landau L.D. and Lifshits E.M., Mechanics, Nauka, Moscow, 1988.
114. Levi V.V. and Morozov A.G., AZh., 1987, v. 64, 919-928.
115. Levi V.V. and Morozov A.G., ATs., 1989, v. 1535, 3-4.
116. Lindblad B., Structure of stellar systems, IL, Moscow, 1962.
117. Lipunov V.M., AZh., 1980, v. 57, 1253-1265.

118. Lipunov V.M., Astrophysics of neutron stars Nauka, Moscow, 1987.
119. Lipunov V.M. and Makura N.I., Pis'ma AZh., 1980, v. 6, 28-33.
120. Lipunova G.V. and Shakura N.I., AZh., 2002, v. 79, 407-421.
121. Lifshits E.M. and Pitaevskii L.P., Statistical physics, part 2, Nauka, Moscow, 1978.
122. Lifshits E.M. and Pitaevskii L.P., Physical kinetics, Nauka, Moscow, 1979.
123. Loitsyanskii L.G., Mechanics of liquids and gas, Nauka, Moscow, 1973.
124. Lominadze G.H., et al., Byul.. Abastumanskoi AO, 1985, No. 58, 211-226.
125. Lyakhovich V.V., et al., AZh., 1996, v. 73, 24.
126. Lyakhovich V.V., et al., in: Unstable processes in the Universe, Kosmosinform, Moscow, 1997, 194.
127. Lyakhovich V.V., et al., AZh., 1997, v. 74, 509.
128. Makarov D.I., et al., Pis'ma AZh., 1999, v. 25, 813-819.
129. Makarov D.I., et al., Pis'ma AZh., 1997, v. 23, 734-739.
130. Makarov D.I., et al., Pis'ma AZh., 1997, v. 23, 509-514.
131. Marochnik L.S. and Suchkov A.A., Galaxy, Nauka, Moscow, 1984.
132. Masevich L.S. and Tutukov A.V., Evolution of stars: theory and observations, Nauka, Moscow, 1988.
133. Mashonkina L.I., et al., AZh., 2004, v. 81, 209-233.
134. Mikhailova E.A. and Morozov A.G., ATs., 1988, v. 1528, 3-4.
135. Mikhailova E.A. and Morozov A.G., in: Classical graviphysics, Proc. 2nd National Conference, Volgograd State University, Volgograd, 1989, 18-19.
136. Mikhailova E.A. and Morozov A.G., in: Problems of stellar mechanics and dynamics, Nauka, Alma Ata, 1990, 124-128.
137. Mikhailova E.A. and Khoperskov A.V., AZh., 1992, v. 69, 1112-1116.
138. Mikhailovskii A.B., Theory of plasma instabilities, v. 1, Atomizdat, 1975.
139. Mikhailovskii A.B., Electromagnetic instabilities in heterogenous plasma, Energoatomizdat, Moscow, 1991.
140. Mikhailovskii A.B., et al., Pis'ma ZhETF, 1977, v. 26, 129.
141. Mikhailovskii A.B., et al., AZh., 1979, v. 56, 279.
142. Mikhailovskii A.B., et al., ZhETF, 1977, v. 73, 20-30.
143. Mikhailovskii A.B., et al., ZhETF, 1970, v. 59, 1609.
144. Mishurov Yu.N., et al., AZh., 1979, v. 56, 268-278.
145. Mishurov Yu.N., et al., AZh., 1976, v. 53, 268-272.
146. Moiseev A.V., Morphology and kinematics of gas and stars in galaxies with jumpers, Dissertation, Nizhnii Arkhyz, 2002.
147. Moiseev A.V., Pis'ma AZh., 2002, v. 28, 840-854.
148. Morozov A.G., Pis'ma AZh., 1977, v. 3, 195-199.
149. Morozov A.G., AZh., 1979, v. 56, 498-504.
150. Morozov A.G., ATs., 1980, v. 1137, 1-3.
151. Morozov A.G., AZh., 1980, v. 57, 681-686.
152. Morozov A.G., Pis'ma AZh., 1981, v. 7, 9-13.
153. Morozov A.G., Pis'ma AZh., 1981, v. 7, 197-200.
154. Morozov A.G., AZh., 1981, v. 58, 34-40.
155. Morozov A.G., AZh., 1981, v. 58, 734-742.
156. Morozov A.G., AZh., 1982, v. 59, 864-860.
157. Morozov A.G., Pis'ma AZh., 1983, v. 9, 716-720.
158. Morozov A.G., AZh., 1983. v. 60, 466-470.
159. Morozov A.G., AZh., 1985, v. 62, 209-217.
160. Morozov A.G., KFNT, 1989, v. 5, 75-80.
161. Morozov A.G. and Mustsevoi V.V., ATs., 1991, No. 1550, 1-2.
162. Morozov A.G. and Mustsevoi V.V., Pis'ma AZh., 1992, v. 1, 46-55.
163. Morozov A.G., et al., Pis'ma AZh., 1984, v. 39, 504.
164. Morozov A.G., et al., Usp. Fiz. Nauk, 1985, v. 145, 161.

165. Morozov A.G., et al., ATs., 1985, v. 1414, 7-8.
166. Morozov A.G., et al., AZh., 1976, v. 53, 946-949.
167. Morozov A.G., et al., Preprint SibIZMIR SO AN SSSR, Irkutsk, 1974, 2-74.
168. Morozov A.G., et al., Preprint SibIZMIR SO AN SSSR, Irkutsk, 1973, 1-73.
169. Morozov A.G., et al., Pis'ma AZh., 1985, v. 11, 231-238.
170. Morozov A.G., et al., Nauch. Inform. Astrosoveta AN SSSR, 1986, No. 61, 110.
171. Morozov A.G., et al., in: LOVAGO: Dynamics and evolution of stellar systems, Nauka, Moscow and Leningrad, 1975, 238.
172. Morozov A.G., et al., ZhETF, 1976, v. 71, 1249-1262.
173. Morozov A.G., et al., DAN, 1976, v. 231, 588-591.
174. Morozov A.G. and Fridman A.M., AZh., 1973, v. 50, 1028-1033.
175. Morozov A.G. and Khoperskov A.V., Astrofizika, 1986, v. 24, 467-476.
176. Morozov A.G. and Khoperskov A.V., KFNT, 1989, v. 5, 37-42.
177. Morozov A.G. and Khoperskov A.V., KFNT, 1990, v. 6, 36-40.
178. Morozov A.G. and Khoperskov A.V., Pis'ma AZh., 1990, v. 16, 567-573.
179. Morozov A.G. and Khoperskov A.V., ATs., 1991, v. 154, 11-12.
180. Morozov A.G. and Shukhman I.G., Pis'ma AZh., 1980, v. 6, 87-91.
181. Morozov A.G. and Khoperskov A.V., Physics of disks, Volgograd State University, Volgograd, 2005.
182. Mustsevoi V.V. and Khoperskov A.V., Pis'ma AZh., 1991, v. 17, 281-288.
183. Nezlin M.V., et al., Pis'ma AZh., 1986, v. 12, 504-521.
184. Nezlin M.V., et al., in: Astrophysics and plasma physics in shallow water experiments, Nauka, Moscow, 1990.
185. Nigmatulin R.I. Dynamics of multiphase media, part II, Nauka, Moscow, 1987.
186. Nikonov S.V. and Solov'ev L.S., Pis'ma ZhETF, 1986, v. 44, 163-166.
187. Novikov I.D. and Frolov V.P., Usp. Fiz. Nauk, 2001, v. 171, 307-324.
188. Ogorodnikov K.F. and Osipkov L.P., Byul. Abastumanskoi AO, 1980, v. 52, 37-42.
189. Ostrovskii L.A., et al., Usp. Fiz. Nauk, 1986, v. 150, 417-437.
190. Pasha I.I., Pis'ma AZh., 1985, v. 11, 3.
191. Pasha I.I., ATs., 1985, v. 1387, 4-6.
192. Pasha I.I. and Tsitsin F.A., ATs., 1979, v. 1071, 1-3.
193. Petrovskaya I.V., INT, 1985, v. 26, 113-154.
194. Pikel'ner S.B., AZh., 1970, v. 47, 752-759.
195. Polyachenko V.L., Pis'ma AZh., 1977, v. 3, 99-103.
196. Polyachenko V.L., Pis'ma Astr. Zh., 1992, v. 69, 10-18.
197. Polyachenko V.L., AZh., 1992, v. 69, 10-18.
198. Polyachenko V.L. and Polyachenko E.V., Pis'ma AZh., 1996, v. 22, 337-344.
199. Polyachenko V.L. and Polyachenko E.V., Pis'ma AZh., 1997, v. 23, 551-560.
200. Polyachenko V.L. and Polyachenko E.V., Pis'ma AZh., 1997, v. 23, 598-605.
201. Polyachenko V.L. and Strel'nikov A.V., ATs., 1988, v. 1529, 1-2.
202. Polyachenko V.L., et al., AZh., 1971, v. 48, 1175-1182.
203. Polyachenko V.L. and Fridman A.M., Equilibrium and stability of gravitating systems, Nauka, Moscow, 1976.
204. Polyachenko V.L. and Fridman A.M., Pism'ma AZh., 1981, v. 7, 136-141.
205. Polyachenko V.L., et al., AZh., 1980, v. 57, 497-504.
206. Polyachenko V.L. and Shukhman I.G., Preprint, SibIZMIR SO AN SSSR, Irkutsk, 1972, No. 1-2.
207. Polyachenko V.L. and Shukhman I.G., Pis'ma AZh., 1977, v. 3, 254-257.
208. Polyachenko V.L. and Shukhman I.G., AZh., 1979, v. 56, 724-731.
209. Polyachenko V.L., et al., Astr. Lett., 2008, v. 34, 86-103.
210. Polyachenko V.L., et al., Astr. Lett., 2008, v. 34, 163-178.
211. Pringle G.E. and Wade R.A. (eds), Interacting binary stars, Fizmatlit, Moscow, 1993.

212. Rabinovich M.I. and Trubetskov D.I., Introduction to theory of oscillations and waves, Nauka, Moscow, 1984.
213. Rastorguev A.S., et al., Pis'ma AZh., 1999, v. 25, 689-703.
214. Rastorguev A.S., et al., Pis'ma AZh., 2002, v. 28, 454-464.
215. Reshetnikov V.P., Surface photometry of galaxies,
216. Reshetnikov V.P., Usp. Fiz. Nauk, 1960, v. 130, 53-56.
217. Rohlfs K., Lectures in theory of density waves, Mir, Moscow, 1980.
218. Saslaw W.C., Gravitational physics of stellar and galactic systems, Mir, Moscow, 1989.
219. Safronov V.S., DAN, 1960, v. 130, 53-56.
220. Sakhikov F.K. and Smirnov M.A., Sov. Astr., 1990, v. 34, 347-348.
221. Safronov V.S., Evolution of the protoplanetary cloud and formation of the Earth and planets, Nauka, Moscow, 1969.
222. Sil'chenko O.K. and Lipunov V.M., Astrofizika, 1987, v. 26, 443-456.
223. Simakov S.G., Pis'ma AZh., 1990, v. 16, 679-689.
224. Sobolev V.V. Lectures in theoretical astrophysics, Nauka, Moscow, 1985.
225. Sotnikova N.Ya. and Rodionov S.A., Pis'ma AZh., 2002, 2003, v. 29, 367.
226. Stepanyants Yu.A. and Fabrikant A.L., Usp. Fiz. Nauk, 1989, v. 159, 83.
227. Sumin A.A., et al., Pis'ma AZh., 1991, v. 17, 698-701.
228. Sumin A.A., et al., Pis'ma AZh., 1991, v. 17, 779-786.
229. Suchkov A.A., Pis'ma AZh., 1983, v. 9, 44-47.
230. Synakh V.S., et al., DAN, 1971, v. 201, 827-830.
231. Syrovatskii S.I., ZhETF, 1954, v. 27, 121.
232. Timofeev A.V., Usp. Fiz. Nauk, 1970, v. 102, 185-210.
233. Torgashin Yu.M., Disseration, Moscow, 1986.
234. Fabrikant A.L., Izv. VUZ, Radiofizika, 1987, v. 30, 275.
235. Fadeeva V.N. and Terent'ev A.M., Tables of values of the probability integral from the complex argument, Gostekhizdat, Moscow, 1954.
236. Fridman A.M., Itogi Nauki Tekhniki: Astronomiya, 1975, v. 10, 61-60.
237. Fridman A.M., Usp. Fiz. Nauk, 1978, v. 125, 352-255,
238. Fridman A.M., AZh., 1986, v. 63, 884-895.
239. Fridman A.M., ZhETF, 1990, v. 98, 1121-1137.
240. Fridman A.M., Usp. Fiz. Nauk, 1990, v. 160, 179-183.
241. Fridman A.M., Usp. Fiz. Nauk, 2007, v. 177, No. 2, 121-148.
242. Fridman A.M., Usp. Fiz. Nauk, 2008, v. 178, No. 3, 225-242.
243. Fridman A.M. and Bisikalo D.V., Usp. Fiz. Nauk, 2008, v. 178, No. 6, 577-604.
244. Fridman A.M. et al., in: Galaxies and chaos, eds. G. Contopoulos and N. Voglis, Springer-Verlag, 2003, 223-234.
245. Fridman A.M. and Khoruzhii O.V., Usp. Fiz. Nauk, 1993, v. 163, 79-85.
246. Fridman A.M. and Khoruzhii O.V., Appendix to Physics of planetary rings, Nauka, Moscow, 1994.
247. Fridman A.M., et al., Pis'ma AZh., 1998, v. 24, 883-893.
248. Fridman A.M. and Shukhman I.G., DAN SSSR, 1972, v. 202, 67.
249. Fujimoto M., in Symposium MAC 29, Byurokan, Erevan, 1968, 453-463.
250. Howde W.A., Pis'ma AZh., 1979, v. 5, 124-127.
251. Hodge P., Galaxies, Nauka, Moscow, 1992.
252. Hockney R. and Eastwood G., Numerical modelling by the particle method, Mir, Moscow, 1987.
253. Khoperskov A.V., Akust. Zh., 1995, v. 41, 1-3.
254. Khoperskov A.V., Izv. VUZ, Radiofizika, 1996, v. 39, 891-900.
255. Khoperskov A.V., AZh., 2001, v. 78, 985-989.
256. Khoperskov A.V., Pis'ma AZh., 2002, v. 28, 723-727.
257. Khoperskov A.V., Vestnik VolGU, 2002, ser 1: Matematika. Fizika, No. 7, 66-71.
258. Khoperskov A.V., et al., AZh., 2001, v. 78, 213-228.

259. Khoperskov A.V., et al., AZh., 2003, v. 80, 387-408.
260. Khoperskov A.V. and Tyurina N.V., AZh., 2003, v. 80, 483-489.
261. Khoperskov A.V. and Khrapov S.S., Pis'ma AZh., 1995, v. 21, 388-393.
262. Khoperskov A.V. and Khrapov S.S., AZh., 1999, v. 76, 256-269.
263. Khoperskov A.V., et al., Pis'ma AZh., 2003, v. 29, 288-299.
264. Khoperskov A.V., et al., Vestnik VolGU, 2007, ser 1: Matematika. Fizika, No. 11, 105-114.
265. Khoperskov A.V., et al., AZh., 2010, (2010arXiv2007.2298).
266. Chandrasekkar S., Principles of stellar dynamics, IL, Moscow, 1948.
267. Cherepashchuk A.M., in: Current problems of physics and evolution of stars, Nauka, Moscow, 1989, 133-168.
268. Cherepashchuk A.M., Usp. Fiz. Nauk, 2003, v. 173, No. 4, 345-384.
269. Sentrell G., et al. (eds), Numerical modeling in astrophysics, Mir, Moscow, 1988, 95.
270. Churilov S.M. and Shukhman I.G., Issled. Geomagn. Aeron. Fiz. Solntsa, 1981, v. 57, 75-79.
271. Churilov S.M. and Shukhman I.G., ATs., 1981, v. 1157, 1-2.
272. Shakura N.I., AZh., 1972, v. 49, 921-929.
273. Shapiro S.A. and T'yukolski S.A., Black holes, white dwarfs and neutron stars, Mir, Moscow, 1985.
274. Shatsova R.B., Planck distribution of velocities of stars in the solar neighbour hood, Rostov-on-Don University, 1965.
275. Shatsova R.B., Construction and application of the function of the phase density of stars, PhD Dissertation, 1975.
276. Einasto J., et al., ATs., 1974, v. 811, 3-4.
277. Aarseth S.J., Direct integration methods of the N-body problem, Astroph. Space Sci., 1971, vol. 14, 118–132.
278. Abramowicz M.A., Innermost parts of accretion discs are thermally and secularly stable, Nature, 1981, vol. 294, 235.
279. Abramowicz M.A., Blaes O.M., Ghosh P., A new example of the Papaloizou–Pringle instability, Astrophys. J., 1987, vol. 323, 629–633.
280. Abramowicz M.A., Calvani M., Nobili L., Thick accretion disks with super-Eddington luminosities, Astrophys. J., 1980, vol. 242, 772–788.
281. Abramowicz M.A., Chen X., Kato S., Lasota J.-P., Regev O., Thermal equilibria of accretion disks, Astrophys. J., 1995, vol. 438, L37–L39.
282. Abramowicz M.A., Igumenshchev I.V., Quataert E., Narayan R., On the radial structure of radiatively inefficient accretion flows with convection, Astrophys. J., 2002, vol. 565, 1101–1106.
283. Abramowicz M.A., Kato S., Constraints for transonic black hole accretion, Astrophys. J., 1989, vol. 336, 304–312.
284. I.A. Acharova, J.R.D. Lépine, Yu.N. Mishurov, MNRAS, 2005, vol. 359, 819.
285. Acharova I.A., Lepine J.R.D., Mishurov Yu.N., et al. A mechanism for the formation of oxygen and iron bimodal radial distribution in the disc of our Galaxy, MNRAS, 2010, vol. 402, 1149–1155.
286. Acheson D.J., On over-reflexion, J. Fluid Mech., 1976, vol. 77, 433–472.
287. Afanasiev V.L., Sil'chenko O.K., Complex Formation History of the Lenticular Galaxies with Stellar Counterrotation: NGC 4138 and NGC 4550, Astron. J., 2002, vol. 124, 706–721.
288. Afanasiev V.L., Sil'chenko O.K., A lopsided chemically distinct nucleus in NGC 5055, Astron. Astrophys., 2002, vol. 388, 461–469.
289. Agol E., Krolik J., Turner N.J., Stone J.M., Two-dimensional hydrodynamic simulations of convection in radiation-dominated accretion disks, Astrophys. J., 2001, vol. 558, 543–552.

290. Aguerri J., Elias-Rosa N., et al. Photometric properties and origin of bulges in SB0 galaxies, Astron. Astrophys., 2005, vol. 434, 109.
291. Akiyama S., Wheeler J.C., Meier D.L., Lichtenstadt I., The Magnetorotational Instability in Core-Collapse Supernova Explosions, Astrophys. J., 2003, vol. 584, 954–970.
292. Allgood B., Flores R.A., Primack J. R., et al. The shape of dark matter haloes: dependence on mass, redshift, radius and formation, MNRAS, 2006, vol. 367, 1781–1796.
293. Alard C., Flaring and warping of the Milky Way disk: not only in the gas, astro-ph/0007013.
294. Albada van G.D., Roberts W.W., A high-resolution study of the gas flow in barred spirals, Astrophys. J., 1981, vol. 246, 740–750.
295. Albada van T.S., Bahcall J.N., Begeman K., Sancisi R., Distribution of dark matter in the spiral galaxy NGC 3198, Astrophys. J., 1985, vol. 295, 305–313.
296. Albada van T.S., Sanders R.H., Periodic orbits and gas flow in barred spirals, MNRAS, 1982, vol. 201, 303–316.
297. Alexander D.R., Augason G.C., Johnson H.R., An opacity-sampled treatment of water vapor, Astrophys. J., 1989, vol. 345, 1014.
298. Allen C., Santillan A., An improved model of the galactic mass distribution for orbit computations, ReV. Mexicana Astron. Astrof., 1991, vol. 22, 255–263.
299. Alves D.R., Rotation of the Disk of the Large Magellanic Cloud. In: Galaxy Disks and Disk Galaxies /Eds. J.G. Funes, S.J. and E.M. Corsini – ASP Conf. Series, vol. 3x108, 2000, 537–540.
300. Alves D.R., Nelson C.A., The rotation curve of the Large Magellanic Cloud and the implications for microlensing, Astrophys. J., 2000, vol. 542, 789–803.
301. Amaral L.H., Ortiz R., Lepine J.R.D., Maciel W.J., The rotation curve of the Galaxy obtained from planetary nebulae and AGB stars, MNRAS, 1996, vol. 281, 339–347.
302. Anderson M., Thermal conduction and self-similar accretion flows, MNRAS, 1987, vol. 227, 623–638.
303. Andrievsky S.M., Bersier D., Kovtyukh V.V., et al. Using Cepheids to determine the galactic abundance gradient. II. Towards the galactic center, Astron. Astrophys., 2002, vol. 384, 140.
304. Ann H.B., Thakur P., Formation of Nuclear Spirals in Barred Galaxies, Astrophys. J., 2005, vol. 620, 197–209.
305. Antipov S.V., et al., Sov. J. Plasma Phys., 1988, vol. 14, 648.
306. Anzer U., Börner G., Accretion onto rotating, magnetic neutron stars – The inner edge of the disk, Astron. Astrophys., 1983, vol. 122, 73–78.
307. Anzer U., Borner G., Meyer-Hofmeister E., The influence of external magnetic fields on the structure of thin accretion disks, Astron. Astrophys., 1987, vol. 188, 85–88.
308. Araki S. A, Theoretical Study of the Stability of Disk Galaxies and Planetary Rings: Thesis PH.D. / Massachusetts Institute of Technology. Boston, 1986.
309. Ardeberg A., Gustafsson B., Linde P., Nissen P.-E., On the history of star formation in the bar of the Large Magellanic Cloud/Ardeberg A., Gustafsson B., Linde P., Nissen P.-E., Astron. Astrophys., 1997, vol. 322, 13–16L.
310. Ardi E., Tsuchiya T., Burkert A., Constraints of the Clumpiness of Dark Matter Halos through Heating of the Disk Galaxies, Astrophys. J., 2003, vol. 596, 204.
311. Arnaud M., Evrard A.E., MNRAS, 1999, vol. 305, 631.

312. Artemova I.V., Bisnovatyi-Kogan G.S., Bjoernsson G., Novikov I.D., Structure of accretion disks with optically thick–optically thin transitions, Astrophys. J., 1996, vol. 456, 119–123.

313. Artemova I.V., Bisnovatyi-Kogan G.S., Igumenshchev I.V., Novikov I.D., On the structure of advective accretion disks at high luminosity, Astrophys. J., 2001, vol. 549, 1050–1061.

314. Athanassoula E., Bar-driven spiral structure, Astron. Astrophys., 1980, vol. 88, 184–193.

315. Athanassoula E., The spiral structure of galaxies – Amsterdam: North-Holland. 1984. (Athanassoula E. The spiral structure of galaxies, Physics Reports, 1984, vol. 114, 321–403.)

316. Athanassoula E., Some recent results on the dynamical evolution of galactic structures. Towards understanding galaxies at large redshifts / Eds. R. Kron, A. Renzini – Kluwer Acad., 1988, 111–116.

317. Athanasoula E., Bar-halo interaction and bar growth, Astrophys. J., 2002, vol. 569, 83–86L.

318. Athanasoula E., Bosma A., Papaioannou S., Halo parameters of spiral galaxiesl, Astron. Astrophys., 1987, vol. 179, 23–40.

319. Athanassoula E., Fady E., Lambert J.C., Bosma A., Optimal softening for force calculations in collisionless N-body simulations, MNRAS, 2000, vol. 314, 475–488.

320. Athanassoula E., Misiriotis A., Morphology, photometry and kinematics of N-body bars. I Three models with different halo central concentrations, MNRAS, 2002, vol. 330, 35–52.

321. Athanasoula E., Sellwood J.A., Bi-symmetric instabilities of the Kuz'min – Toomre disc, MNRAS, 1986, vol. 221, 213–232.

322. Badhwar G.D., Stephens S.A., Hydrostatic equilibrium of gas, extent of cosmic ray confinement, and radio emission in the Galaxy, Astrophys. J., 1977, vol. 212, 494–506.

323. Baggett W.E., Baggett S.M., Anderson K.S.J., Bulge-Disk decomposition of 659 spiral and lenticular galaxy brightness profiles, Astron. J., 1998, vol. 116, 1626–1642.

324. Bahcall J.N., The distribution of stars perpendicular to galactic disk, Astrophys. J., 1984, vol. 276, 156–168.

325. Bahcall J.N., Casertano S., Velocity dispersions of disk stars in spiral galaxies, Astrophys. J., 1984, vol. 284, 35–38L.

326. Bahcall J.N., Kylafis N.D., On the spatial distribution of Population II stars in SB and later type galaxies, Astrophys. J., 1985, vol. 288, 252–258.

327. Bahcall J.N., Soneira R.M., The Universe at faint magnitudes. I. Models for the Galaxy and the predicted star counts, Astrophys. J. Suppl. Ser., 1980, vol. 44, 73–110.

328. Baker P.L., Barker P.K., The interaction of interstellar gas with stellar density wave packets, Astron. Astrophys., 1974, vol. 36, 179–189.

329. Balbus S., Local interstellar gasdynamical stability and substructure in spiral arms, ApJ, 1988, vol. 324, 60–74.

330. Balbus S.A., Enhanced Angular Momentum Transport in Accretion Disks, Annu. Rev. of Astron. Astrophys., 2003, vol. 41, 555–597.

331. Balbus S.A., Hawley J.F., A powerful local shear instability in weakly magnetized disks. Linear analysis, Astrophys. J., 1991, vol. 376, 214–222.

332. Balbus S.A., Hawley J.F., Instability, turbulence, and enhanced transport in accretion disks, Rev. Modern Phys., 1998, vol. 70, 1–53.

333. Balbus S.A., Hawley J.F., On the Nature of Angular Momentum Transport in Nonradiative Accretion Flows, Astrophys. J., 2002, vol. 573, 749–753.

334. Banerjee A., Jog C.J., The Flattened Dark Matter Halo of M31 as Deduced from the Observed H I Scale Heights, Astrophys. J., 2008, vol. 685, 254–260.

335. Barbuy. B., Oxygen in 20 halo giants, Astron. Astrophys., 1988, vol. 191, 121.

336. Barbuy B., Erdelyi-Mendes M., Oxygen in old and thick disk stars, Astron. Astrophys., 1989, vol. 214, 239.

337. Bardeen J.M., Global instabilities of disk. Dynamics of stellar systems – IAU Symp., 1975, vol. 69, 297–320.

338. Barnes J., Dynamical instabilities in spherical stellar systems / Dynamics of Star Clusters, 1985, IAU SymP. No. 114, 297.

339. Barnes J., Hernquist L.E., Fueling starburst galaxies with gas-rich mergers, Astrophys. J., 1991, vol. 370, 65–68.

340. Barnes J., Hut P., A Hierarchical O(NlogN) Force-Calculation Algorithm, Nature, 1986, vol. 324, 446–449.

341. Barnes J.E., Hut P., Error analysis of a tree code, Astrophys. J. Suppl. Ser., 1989, vol. 70, 389–417.

342. Barton W.B., Gordon M.A., Carbon monoxide in the Galaxy. III: The overall nature of its distribution the equatorial plane, Astron. Astrophys., 1978, vol. 63, 7–27.

343. Bastien P., Menard F., Asselin L., Turbide L., The circumstellar environment of two young stars in Cassiopeia. Model Stellar Environ.: How and Why? Proc. 4th IAP Astrophys. Meet. Hon. Jean–Claude Pecker. – Paris, 1988, 185–188.

344. Bath G.T., Dynamical instabilities and mass exchange in binary systems, MNRAS, 1975, vol. 171, 311–328.

345. Bath G.T., Edwards A.C., Manthle V.J., The evolution of viscous discs. IV: Stream penetration effects, MNRAS, 1983, vol. 205, 171–185.

346. Bath G.T., Pringle J.E., The evolution of viscous discs. I: Mass transfer variations, MNRAS, 1981, vol. 194, 967–986.

347. Bath G.T., Pringle J.E., The evolution of viscous discs. II: Viscous variations, MNRAS, 1982, vol. 199, 267–280.

348. Bath G.T., Pringle J.E., "Cataclysmic variables: An theoretical review", in Interacting Binary Stars / Eds. Pringle J.E., Wade R.A. – Cambridge: Cambridge UniV. Press, 1985, Cambridge Astrophys. Ser., vol. 6.

349. Bailin J., Steinmetz M., Internal and External Alignment of the Shapes and Angular Momenta of lCDM Halos, Astrophys. J., 2005, vol. 627, 647–665.

350. Beck R., Magnetic field structure from synchrotron polarization, EAS Publ. Series, 2007, vol. 23, 19–36.

351. Beckwith S., Zuckerman B., Skrutskie M.F., Dyck H.M., Discovery of solar system–size halos around youg stars, Astrophys. J., 1984, vol. 287, 793–800.

352. Bedregal A.G., Aragon-Salamanca A., Merrifield M.R., The Tully-Fisher relation for S0 galaxies, MNRAS, 2006, vol. 373, 1125.

353. Beers T.C., Sommer-Larsen J., Kinematics of metal-poor stars in the Galaxy, Astrophys. J. Suppl. Ser., 1995, vol. 96, 175–221.

354. Begelman M.C., Meier D.L., Thick accretion disks: self–similar, supercritical models, Astrophys. J., 1982, vol. 253, 873–896.

355. Begeman K.G., H I rotation curves of spiral galaxies. I – NGC 3198, Astron. Astrophys., 1989, vol. 223, 47–60.

356. Begeman K.G., Broels A.H., Sanders R.H., Extended rotation curves of spiral galaxies: dark haloes and modified dynamics, MNRAS, 1991, vol. 249, 523–537.

357. Begum A., Chengalur J.N., Karachentsev I.D., Sharina M.E., Baryonic Tully-Fisher relation for extremely low mass Galaxies, MNRAS, 2008, vol. 386, 138–144.

358. Bekki K., Freeman K.C., Extended H I Spiral Structure and the Figure Rotation of Triaxial Dark Halos, Astrophys. J. Lett., 2002, vol. 574, 21–24.

359. Bell, K.R., Lin, D.N.C., Using FU Orionis outbursts to constrain self-regulated protostellar disk models, Astrophys. J., 1994, vol. 427, 987.

360. Bell E., de Jong R.S., Stellar Mass-to-Light Ratios and the Tully-Fisher Relation, Astrophys. J., 2001, vol. 550, 212–229.

361. Belokurov V., et al. Cats and Dogs, Hair and a Hero: A Quintet of New Milky Way Companions, Astrophys. J., 2007, vol. 654, 897–906.

362. Belokurov V., Evans N.W., et al. The Hercules-Aquila Cloud, ApJL, 2007, vol. 657, 89–92.

363. Beltran J.C.V., Pizzella A., Corsini E.M., et al. Kinematic properties of gas and stars in 20 disc galaxies, Astron. Astrophys., 2001, vol. 374, 394–411.

364. Bensby T., Feltzing S., Lundström I., Ilyin I., α-, r-, and s-process element trends in the Galactic thin and thick disks, Astron. Astrophys., 2005, vol. 433, 185.

365. Berentzen I., Athanassoula E., Heller C.H., Fricke K.J., Numerical simulations of interacting gas-rich barred galaxies: vertical impact of small companions, MNRAS, 2003, vol. 341, 343–360.

366. Berentzen I., Heller C.H., Fricke K.J., Athanassoula E., Numerical Simulations of Interacting Gas-Rich Barred Galaxies, Astroph. Space Sci., 2001, vol. 276, 699–706.

367. Van Den Bergh S., The galaxies of the local group. CUP, 2000.

368. Berman R.H., Brownrigg D.R.K., Hockney R.W., Numerical models of galaxies I. The variability of spiral structure, MNRAS, 1978, vol. 185, 861–875.

369. Berman R.H., Mark J.W.-K., Are massive galactic haloes necessary to prevent rapid, global bar formation?, Astron. Astrophys., 1979, vol. 77, 31–36.

370. Bertin G., Dynamics of galaxies – Cambridge: University Press, 2000, 414.

371. Bertin G., Lin C.C., Spiral structure in galaxies: a density wave theory – Cambridge, MA MIT Press, 1996, 271.

372. Bertin G., Lin C.C., Lowe S.A., Thurstans R.P., Modal approach to the morphology of spiral galaxies. I - Basic structure and astrophysical viability, Astrophys. J., 1989, vol. 338, 78–103.

373. Bianchini A., Solar-type cycles in close binary systems, AsJ, 1990, vol. 99, 1941–1952.

374. Bien R., Just A., Berczik P., Berentzen I., High resolution in z-direction: The simulation of disc-bulge-halo galaxies using the particle-mesh code SUPERBOX, Astr. Nachr., 2008, vol. 329, 1029–1032.

375. Biermann P., A simple gasdynamical model of mass exchange in close binary systems, Astron. Astrophys., 1971, vol. 10, 205–212.

376. Binney J., Tremaine S. Galactic Dynamics – Princeton: Princeton University Press, New Jersey, 2008.

377. Binney J., Dehnen W., Bertelli G., The age of the Solar neighbourhood, MNRAS, 2000, vol. 318, 658–664.

378. Binney J., Gerhard O., Spergel D., The photometric structure of the inner Galaxy, MNRAS, 1997, vol. 288, 365–374.

379. Binney J., Jiang I.-G., Dutta S., The persistence of warps in spiral galaxies with massive haloes, MNRAS, 1998, vol. 297, 1237–1244.

380. Birdsall C.K., Fuss D., Clouds-in-clouds, clouds-in-cells physics for many–body plasma simulation, J. Coniput. Phys., 1969, vol. 3, 494–511.

381. Bisikalo D.V., Numerical Modeling of Mass Transfer in Close Binaries, Astrophys. Space Sci., 2005, vol. 296, 391–401.

382. Bisikalo D.V., Boyarchuk A.A., Kuznetsov O.A., Chechetkin V.M., Observational manifestations of matter flow features in symbiotic stars, Astr. Rep., 1996, vol. 40, 653–661.
383. Bisikalo, D. V., Boyarchuk, A. A., Kaigorodov, P. V., Kuznetsov, O. A., Morphology of the Interaction Between the Stream and Cool Accretion Disk in a Semidetached Binary System, Astron. Rep., 2003, vol. 47, 809.
384. Bisikalo D.V., Boyarchuk A.A., Kaigorodov P.V., et al. The Structure of Cool Accretion Disks in Semidetached Binaries, Astron. Rep., 2004, vol. 48, 449.
385. Bisikalo D.V., Boyarchuk A.A., Kaigorodov P.V., et al. A Model for Superoutbursts in SU UMa-type Binaries, Astron. Rep., 2004, vol. 48, 588.
386. Bisikalo D.V., Boyarchuk A.A., Kaigorodov P.V., et al. The Formation of Precessional Spiral Density Wave in Accretion Disks and a New Model for Superoutbursts in SU UMa-type Binaries, Chin. J. Astron. Astrophys., 2006, vol. 6, 159.
387. Bisikalo D.V., Boyarchuk A.A., Kil'Pio A.A., et al. The Structure of Matter Flows in Semi-Detached Binaries after the Termination of Mass Transfer, Astron. Rep. 2001, vol. 45, 611.
388. Bisikalo D. V., Boyarchuk A.A., Kilpio A.A., Kuznetsov O.A., A possible manifestation of spiral shock waves in the accretion disks of cataclysmic variables. Astron. Rep., 2001, vol. 45, 676.
389. Bisikalo D.V., Boyarchuk A.A., Chechetkin V.M., et al. Three-dimensional numerical simulation of gaseous flow structure in semidetached binaries, MNRAS, 1998, vol. 300, 39–48.
390. Bisikalo D.V., Kaigorodov P.V., Boyarchuk A.A., Kuznetsov O.A., The Possible Nature of Dips in the Light Curves of Semidetached Binaries with Stationary Disks, Astron. Rep., 2005, vol. 49, 701.
391. Bisnovatyi-Kogan G.S., Book Review: Stellar physics I Fundamental concepts and stellar equilibrium – Berlin: Springer-Verlag, 2001.
392. Bisnovatyi-Kogan G.S., Blinnikov S.I., A hot corona around a black-hole accretion disk as a model for Cygnus X-1SoV, Astron. Lett., 1976, vol. 2, 191.
393. Bisnovatyi-Kogan G.S., Blinnikov S.I., Disk accretion onto a black hole at subcritical luminosity, Astron. Astrophys., 1977, vol. 59, 111–125.
394. Bizyaev D., Mitronova S., Photometric parameters of edge-on galaxies from 2MASS observations, Astron. Astrophys., 2002, vol. 389, 795–801.
395. Bizyaev D., Kajsin S., The Stellar Disk Thickness of Low Surface Brightness Galaxies, Astrophys. J., 2004, vol. 61, 886–897.
396. Bizyaev, D.V., Zasov A.V., Disk Boundaries in Spiral Galaxies, Astron. Reports, 2002, vol. 46, 721–730.
397. Bisnovatyi-Kogan G.S., Lovelace R.V.E., Advective Accretion Disks and Related Problems Including Magnetic Fields, New Astron. Reviews, 2001, vol. 45, 663–742.
398. Blandford R.D., Payne D.G., Hydromagnetic flows from accretion discs and the production of radio jets, Mon. Not. R. Astron. Soc., 1982, vol. 199, 883.
399. Blitz L., The rotation curve of the Galaxy to R=16 kiloparsecs, Astrophys. J., 1979, vol. 231, 115–119L.
400. Blitz L., Binney J., Lo K.Y., et al. The centre of the Milky Way, Letters to Nature, 1993, vol. 361, 417–424.
401. Blitz L., Spergel D.N., Teuben P.J., et al. High-Velocity Clouds: Building Blocks of the Local GrouP, Astrophys. J., 1999, vol. 514, 818.
402. Block D.L., Stockton A.S., Structure in host galaxies of steep-spectrum radio QSOs, AsJ, 1991, vol. 102, 1928–1932.
403. Blum R.D., Carr J.S., DePoy D.L., et al. Radial velocities of M giants at 300 parsec projected radius from the galactic center, Astrophys. J., 1994, vol. 422, 111–117.

404. Blum R.D., Carr J.S., Sellgren K., Terndrup D.M., M giant kinematics in off-axis fields between 150 and 300 parsecs from the galactic center, Astrophys. J., 1995, vol. 449, 623–634.

405. Blumen W., Drazin P.G., Billings D.F.J., Shear layer instability of an inviscid compressible fluid, Fluid Mech., 1975, vol. 71, 305.

406. Bobylev V.V., Bajkova A.T., Galactic parameters from masers with trigonometric parallaxes, MNRAS, 2010, vol. 408, 1788–1795.

407. Bode P., Ostriker J.P., Tree Particle-Mesh: An adaptive, efficient, and parallel code for collisionless cosmological simulation .. Astrophys. J. Suppl. Ser., 2003, vol. 145, 1–13.

408. Bondi H., On spherical symmetrical accretion, MNRAS, 1952, vol. 112, 195.

409. Boomsma R., Oosterloo T.A., et al. HI holes and high-velocity clouds in the spiral galaxy NGC 6946, Astron. Astrophys., 2008, vol. 490, 555–570.

410. Borkova T.V., Marsakov V.A., Globular Cluster Subsystems in the Galaxy, Astr. Rep., 2000, vol. 44, 665–684.

411. Börner G., Hayakawa S., Nagase F., Anzer U., Disk formation at the magnetosphere of wind-fed pulsars – Application to VELA X-1, Astron. Astrophys., 1987, vol. 182, 63–70.

412. Boroson T.A., Oke J.B., Spectroscopy of the galaxy components of N and Seyfert galaxies, PASP, 1987, vol. 99, 809–815.

413. Bottema R., The stellar kinematics of the spiral galaxies NGC 3198 and NGC 3938, Astron. Astrophys., 1988, vol. 197, 105–122.

414. Bottema R., The stellar velosity dispersion of the spiral galaxies NGC 6503 and NGC 6340, Astron. Astrophys., 1989, vol. 221, 236–249.

415. Bottema R., The intricate kinematics of the Sb spiral galaxy NGC 2613, Astron. Astrophys., 1989, vol. 225, 358–368.

416. Bottema R., The stellar velocity dispersion of the spiral galaxies NGC 1566 and NGC 2815, Astron. Astrophys., 1992, vol. 257, 69.

417. Bottema R., The stellar kinematics of galactic disks, Astron. Astrophys., 1993, vol. 275, 16–36.

418. Bottema R., The maximum rotation of a galactic disc, Astron. Astrophys., 1997, vol. 328, 517–525.

419. Bottema R., Simulations of normal spiral galaxies, MNRAS, 2003, vol. 344, 358–384.

420. Bottema R., Gerritsen J.P.E., An investigation of the kinematics of the spiral galaxy NGC 6503, MNRAS, 1997, vol. 290, 585–597.

421. Bottema R., van der Kruit P.C., Freeman K.C., The stellar velocity dispersion in the disk of the spiral galaxy NGC 5170, Astron. Astrophys., 1987, vol. 178, 77–90.

422. Bottema R., van der Kruit P.C., Valenijn E.A., The stellar velocity dispersion of the edge-on spiral galaxy NGC 891, Astron. Astrophys., 1991, vol. 247, 357–367.

423. Boyarchuk A.A., Bisikalo D.V., Kuznetsov O.A., Chechetkin V.M., Mass Transfer in Close Binary Stars – London: Taylor & Francis, 2002. Adv. in Astron. and Astro phys., vol. 6.

424. Brand J., Blitz L., The velocity field of the outer Galaxy, Astron. Astrophys., 1993, vol. 275, 67–90.

425. Brosch N., Kniazev A., Moiseev A., Pustilnik S., On the nature of the apparent ring galaxy SDSS J075234.33+292049.8, MNRAS, 2010, vol. 401, 2067–2080.

426. Brown W.R., Geller M.J., et al. Discovery of an Unbound Hypervelocity Star in the Milky Way Halo, Astrophys. J., 2005, vol. 622L, 33–36.

427. Brown W.R., Geller M.J., Kenyon S.J., MMT Hypervelocity Star Survey, Astrophys. J., 2009, vol. 690, 1639–1647.

428. Brown W.R., Geller M.J., et al. The Anisotropic Spatial Distribution of Hypervelocity Stars Astrophys. J., 2009, vol. 690L, 69–71.
429. Brinks E., Bajaja E., A high resolution hydrogen-line survey of Messier 31. III - H I holes in the interstellar medium, Astron. Astrophys., 1986, vol. 169, 14.
430. Brun A.S., Toomre T., Turbulent convection under the influence of rotation: sustaining a strong differential rotation, Astrophys. J., 2002, vol. 570, 865–885.
431. Brush S.G., Theories of the origin of the solar system 1956–1985, Rev. Mod. Phys., 1990, vol. 62, 43–50.
432. Bullock J.S., Shapes of dark matter halos. In: The Shapes of Galaxies and Their Dark Halos, ed. P. Natarajan (Singapore: World Scientific), 2002, 109.
433. Bureau M., Freeman K.C., The nature of Boxy/Peanut-Shaped bulges in spiral galaxies, Astron. J., 1999, vol. 118, 126–138.
434. Burkert A., The Structure of Dark Matter Halos in Dwarf Galaxies, ApJL, 1995, vol. 447L, 25–28.
435. Burstein D., Structure and origin of SO galaxies. III. The luminosity distribution perpendicular to the plane of the disks in SO's, Astrophys. J., 1979, vol. 23, 829–836.
436. Burton W.B., Galactic structure derived from neutral hydrogen observations using kinematic models based on the density wave theory, Astron. Astrophys., 1971, vol. 10, 76–96.
437. Buta R., Combes F., Galactic Rings, Fund. Cosmic Physics, 1996, vol. 17, 95–281.
438. Buta R., Mitra S., de Vaucouleurs G., Mean morphological types of bright galaxies, Astron. J., 1994, vol. 107, 118–134.
439. Buta R., Purcell G.B., NGC 3081 - Surface photometry and kinematics of a classic resonance ring barred galaxy, Astron. J. 1998, vol. 115, 484–501.
440. Buta R., Zhang X., Pattern corotation radii from potential-density phase-shifts for 153 osubgs sample galaxies, ApJSS, 2009, vol. 182, 559–583.
441. Butcher H., Oemler A., The evolution of galaxies in clusters. V - A study of populations since Z approximately equal to 0.5, Astrophys. J., 1984, vol. 285, 426.
442. Bylly J., Observations of bipolar outflows, Publ. Astr. Inst. Czechosl. Acad. Sci., 1987, vol. 69, 9–16.
443. Byrd G.G., Klaric M., The spiral compact-galaxy pair AM 2208–251: computer simulations versus observations, Astron. J., 1990, vol. 99, 1461–1469.
444. Byrd G.G., Sundellus B., Valtonen M., Tidal triggering of syfert galaxies and quasars: occurence in multiple systems, Astron. Astrophys., 1987, vol. 171, 16–24.
445. Byrd G., Freeman T., Buta R., The Inner Resonance Ring of NGC 3081. II. Star Formation, Bar Strength, Disk Surface Mass Density, and Mass-to-Light Ratio, AJ, 2006, vol. 131, 1377–1393.
446. Byun Y.I., Freeman K.C., Kylafis N.D., Diagnostics of dust content in spiral galaxies: Numerical simulations of radiative transfer, Astron. J., 1994, vol. 432, 114.
447. Cabot W., The nonaxisymmetric baroclinic instability in thin accretion disks. Astrophys. J., 1984, vol. 277, 806.
448. Caldwell J.A.R., Ostriker J.P., The mass distribution within our Galaxy: a three component model, Astrophys. J., 1981, vol. 251, 61–87.
449. Camenzind M., Demole F., Straumann N., The stability of radiation–pressure–dominated accretion discs, Astron. Astrophys., 1986, vol. 158, 212–216.
450. Canizares C.R., Donahue M.N., Trinchieri G, Astrophys. J., 1986, vol. 304, 312.
451. Cannizzo J.K., Kenyon S.J., The accretion disk limit cycle mechanism in GK Persei, Astrophys. J. Lett., 1986, vol. 309, L43.
452. Cannizzo J.K., Kenyon S.J., The secondary outburst maximum of T Coronae Borealis: implications for the physics of accretion disks, Astrophys. J., 1992, vol. 386, 17–21.

453. Canzian B., A new way to locate corotation resonances in spiral galaxies, Astrophys. J., 1993, vol. 414, 487–492.
454. Cao X., Spruit H.C., Instability of an accretion disk with a magnetically driven wind, Astron. Astrophys., 2002, vol. 385, 289.
455. Capaccioli M., Cappellaro E., Held E.V., Vietri M., Deep kinematics and dynamics of edge-on S0 galaxies. I. NGC 3115, Astron. Astrophys., 1993, vol. 274, 69–78.
456. Cardone V., Sereno M., Modelling the Milky Way through adiabatic compression of cold dark matter haloes, Astron. Astrophys., 2010, vol. 438, 545–556.
457. Carlberg R.G., Sellwood J.A., Dynamical evolution in galactic discs, Astrophys. J., 1985, vol. 292, 79–89.
458. Carollo D., Beers T.C., Lee Y.S., et al. Two stellar components in the halo of the Milky Way, Nature, 2007, vol. 450, 1020.
459. Carraro G., Ng Y.K., Portinari L., On the Galactic disc age-metallicity relation, MNRAS, 1998, vol. 296, 1045.
460. Carter D., Weak disks in rapidly rotating elliptical galaxies, Astrophys. J., 1987, vol. 312, 514–517.
461. Casartelli M., Diana E., Galgani L., Scotti A., Phys. Rev., 1976, vol. A13, 1921.
462. Cayatte V., Kotanyi C., Balkowski C., van Gorkom J.H, Astron. J., 1994, vol. 107, 1003.
463. Casasola V., Combes F., Bettoni D., et al. Are truncated stellar disks linked to the molecular gas density?, 2007, astro-ph/0702018.
464. Casoli F., Combes F., Stark A.A., Mapping of a molecular complex in a northern spiral arm of M31, Astron. Astrophys., 1987, vol. 173, 43–48.
465. Ciotti L., Bertin G., Analytical properties of the $R^{1/m}$ law, Astron. Astrophys., 1999, vol. 352, 447.
466. Chandrasckhar S., Ellipsoidal Figures of Equilibrium – New Haven: YU Press, 1969.
467. Chandrasekhar S., Hydrodynamic and Hydromagnetic Stability – New York: Dover Publ., 1981.
468. Charles P.A., The mysterious SU UMa stars, Sky & Telesc., 1990, vol. 79, 607–610.
469. Chatterjee S., Vlemmings W.H.T., Brisken W.F., et al. Getting Its Kicks A VLBA Parallax for the Hyperfast Pulsar B1508+55, Astrophys. J., 2005, 630L, 61.
470. Cherepashchuk A.M., SS 433 as an eclipsing binary, MNRAS, 1981, vol. 194, 761–769.
471. Cherepashchuk A.M., Afanas'ev V.L., Zasov A.V., Katkov I.Yu., Kinematics of disk galaxies with known masses of their supermassive black holes. Observations, Astron. Reports, 2010, vol. 54, 578–589.
472. Chernin A.D., Vorontsov-Vel'yaminov's rows in giant spiral galaxies: geometrical properties and physical interpretation, Astrophys., 1998, vol. 41, 399–407.
473. Chernin A.D., Spiral patterns with straight arm segments, MNRAS, 1999, vol. 308, 321–332.
474. Chernin A.D., Zasov A.V., Arkhipova V.P., et al. ASP Confer. Ser., 2001, vol. 230, 147.
475. Chernin A.D., Zasov A.V., Arkhipova V.P., et al. Polygonal arms and hexagonal rings: Morphology and physical interpretation, Astron. Astroph. Trans., 2001, vol. 20, 139–146.
476. Chernin A.D., Korolev V.V., Kovalenko V.V., Global Irregularities of Spiral Patterns in Galaxies: Manifestation of hydrodynamic Instabilities?, Astroph. and Space Science Library. 2006, vol. 337, 321.
477. Chiappini C., Matteucci F., Gratton R., The Chemical Evolution of the Galaxy: The Two-Infall Model, Astrophys. J., 1997, vol. 477, 765.

478. Chieffi A., Limongi M., The Explosive Yields Produced by the First Generation of Core Collapse Supernovae, Astrophys. J., 2002, vol. 577, 281–294.
479. Chilingarian I., Evolution of dwarf early-type galaxies, MNRAS, 2009, vol. 394, 1229.
480. Christodoulou D.M., Shlosman I., Tohline J.E., A new criterion for bar-forming instability in rapidly rotating gaseous and stellar systems. II. Nonaxisymmetric form, Astrophys. J., 1995, vol. 443, 563–569.
481. Clarke C.J., Chemical evolution of viscously evolving galactic disks, MNRAS, 1989, vol. 238, 283–292.
482. Clemens D.P., Massachusetts-stony brook galactic plane CO survey: the galactic disk rotation curve, Astrophys. J., 1985, vol. 295, 422–436.
483. Cohen M., The Tauri stars. Phys. Rep., 1984, vol. 116, 173–249.
484. Colin J., Athanassoula E., Asymmetrical barred galaxies, Astron. Astrophys., 1989, vol. 214, 99–105.
485. Combes F., Sanders R.H., Formation and properties of persisting stellar bars, Astron. Astrophys., 1981, vol. 96, 164–173.
486. Combes F., Distribution of CO in the Milky Way, Annu. Rev., Astron. Astrophys., 1991, vol. 29, 195.
487. Combes F., Properties of dark matter haloes, New Astron. Reviews., 2002, vol. 46, 755–766.
488. Combes F., Elmegreen B.G., Bars in early- and late-type galaxies, Astron. Astrophys., 1993, vol. 271, 391–401.
489. Combes F., Gerin M., Spiral structure of molecular clouds in response to bar forcing: a particle simulation, Astron. Astrophys., 1985, vol. 150, 327–338.
490. Contopoulos G., Periodic orbits near the particle resonance in galaxies, Astron. Astrophys., 1978, vol. 64, 323.
491. Contopoulous G., Grosbol P., Stellar dynamics of spiral galaxies – Nonlinear effects at the 4/1 resonance, Astron. Astrophys., 1986, vol. 155, 11–23.
492. Contopoulos G., Voglis N., A fast method for distinguishing between ordered and chaotic orbits, Astron. Astrophys., 1997, vol. 317, 73.
493. Considere S., Athanassoula E., Analysis of spiral components in 16 galaxies, Astron. Astropliys. Suppl. Ser., 1988, vol. 76, 365–404.
494. Cook M.C., High–speed photometry of Z Chamaeleontis covering the 1983 March normal outburst, MNRAS, 1985, vol. 216, 219–239.
495. Cook M.C., High-speed photometry of the eclipsing dwarf nova OY Carinae, MNRAS, 1985, vol. 215, 211–232.
496. Cooper A.P., Cole S., Frenk C.S., et al. Galactic stellar haloes in the CDM model, MNRAS, 2010, vol. 406, 744–766.
497. Coppi B., Coppi P., Angular momentum transport in thin accretion disks and intermittent accretion, Phys. Rev. Lett., 2001, vol. 87, 1101–1103.
498. Corbelli E., Salucci P., The extended rotation curve and the dark matter halo of M33, MNRAS, 2000, vol. 311, 441–447.
499. Corder S., Sheth K., Scoville N., Koda J., Vogel S., Ostriker E., Detection of Dense Molecular Gas in Interarm Spurs in M51, ApJ, 2008, vol. 689, 148–152.
500. Cote S., Carignan C., Sancisi R., A dark–halo–dominated galaxy: NGC 5585, Astron. J., 1991, vol. 102, 904–913.
501. Could A., Bahcall J.N., Flynn C., Disk M dwarf luminosity function from Hubble Space Telscope star counts, Astrophys. J., 1996, vol. 465, 759–768.
502. Courant R., Isaacson E., Rees M., On the solution of nonlinear hyperbolic differential equations by finite differences, Comm. Pure Appl. Math., 1952, vol. 5, 243–255.

503. Courteau S., Rex H., Maximal disks and the Tully-Fisher relation, Astrophys. J., 1999, vol. 513, 561–571.
504. Courvoisier T.J.-L., Clavel J., Observational constraints on disc models for quasars and Seyfert galaxies, Astron. Astrophys., 1991, vol. 248, 389–394.
505. Cox D.P., Self-regulating star formation – The rate limit set by ionizing photons, ApJL. 1983, vol. 265, L61.
506. Cox D.P., The three-phase interstellarmediim revisited, Annu. Rev. Astron. Astroph., 2005, vol. 43, 337–385.
507. Cox D.P., Daltabuit E., Radiative Cooling of a Low-Density Plasma, Astrophys. J., 1971, vol. 167, 113.
508. M. Creze, E. Chereul, O. Bienayme, C., Pichon / The distribution of nearby stars in phase space mapped by Hipparcos? I. The potential well and local dynamical mass, Astron. Astrophys., 1998, vol. 329, 920–936.
509. Cruz-Gonzales C., Local density gradient and galactic parameters determined from nearly stars, MNRAS, 1974, vol. 168, 41–52.
510. Culsrud R.M., Mark J.W.-K., Caruso A., The hose-pipe instability in stellar systems. Astroph. Space Sci., 1971, vol. 14, 52–55.
511. Daniel C., Zabludoff A.J, Astrophys. J., 2005, vol. 621, 201.
512. Danver C.-G., A morphological investigation of some near galaxies, Annals of the observatory of Lund. 1942, vol. 10, 7.
513. Dather M., Biermann P.L., Accretion in disk galaxies: the hydrodynamical evolution of molecular rings, Astron. Astrophys., 1990, vol. 235, 55–59.
514. Dauphole B., Colin J., Globular clusters as a new constraint for the potential of our Galaxy, Astron. Astrophys., 1995, vol. 300, 117–125.
515. Davies J.I., Burstein D., eds., "The Opacity of Spiral Disks", Proceedings of the NATO Advanced Research Workshop, Kluwer, Dordrecht, 1995.
516. Debattista V.P., Sellwood J.A., Dynamical friction and the distribution of dark matter in barred galaxies, Astrophys. J., 1998, vol. 493, 5–8L.
517. Debattista V.P., Sellwood J.A., Constraints from dynamical friction on the dark matter content of barred galaxies, Astrophys. J., 2000, vol. 543, 704–721.
518. L. Deharveng, M. Pena, J. Caplan, R. Costero., Oxygen and helium abundances in Galactic Hii regions - II. Abundance gradients, MNRAS, 2000, vol. 311, 329.
519. Dehnen W., The distribution of nearby stars in velocity space inferred from HIPPARCOS data, Astron. J., 1998, vol. 115, 2384–2396.
520. Dehnen W., Towards optimal softening in three-dimensional N-body codes. I. Minimizing the force error, MNRAS, 2001, vol. 324, 273–291
521. Dehnen W., Binney J.J., Local stellar kinematics from *Hipparcos* data, MNRAS, 1998, vol. 298, 387–394.
522. Dehnen W., Binney J.J., Mass models of the Milky Way, MNRAS, 1998, vol. 294, 429–438.
523. Deul E.R., den Hartog R.H., Small-scale structures in the H I distribution of M33, Astron. Astrophys., 1990, vol. 229, 362.
524. De Young D.S., Gallagher J.S., Selective loss of metals from low-mass galaxies, Astrophys. J. Lett., 1990, vol. 356, L15.
525. Dgani R., Livio M., Regev O., On the effects of tidal interaction on thin accretion disks: an analytic study, Astrophys. J., 1994, vol. 436, 270.
526. Dias W.S., Lépine J.R.D., Direct Determination of the Spiral Pattern Rotation Speed of the Galaxy, Astrophys. J., 2005, vol. 629, 825–831.
527. Diemand J., Kuhlen M., Madau P., Dark matter substructure and gamma-ray annihilation in the Milky Way halo, Astrophys. J., 2007, vol. 657, 262–270.

528. Dicaire I., Carignan C., et al. Hα kinematics of the Spitzer Infrared Nearby Galaxies Survey – II, MNRAS, 2008, vol. 385, 553–605.

529. Digel S., Bally J., Thaddeus P., Giant molecular clouds in the outer arm of the Galaxy, Astrophys. J., 1990, vol. 357, 29–33.

530. Ding S.-X., Yang L.-T., Wu X.-B., Lu Ye., The radial-azimuthal instability of a hot two-temperature accretion disc with advection, MNRAS, 2000, vol. 317, 737–742.

531. Dobbs C.L., Bonnell I.A., Spurs and feathering in spiral galaxies, MNRAS, 2006, vol. 367, 873–878.

532. Domingue D.L., Kee W.C., White R.E., Seeing galaxies through thick and thin. II. Direct measures of extinction in spiral disks through spectroscopy of overlapping galaxies, Astrophys. J., 2000, vol. 545, 171–180.

533. Drazin P.G, Reid W.H., Hydrodynamic Stability – Cambridge; New York: Cambridge University Press, 1981 (Drazin P.G., Davey A., J. Fluid Mech., 1977, vol. 82, 255).

534. Driel van W., Woerden van H., Distribution and motions of H I in the SA galaxies NGC 1169 and NGC 3898, Astron. Astrophys., 1994, vol. 286, 395–407.

535. Drimmel R., Spergel D.N., Three-dimensional Structure of the Milky Way Disk: The Distribution of Stars and Dust beyond 0.35 *R??*, Astrophys. J., 2001, vol. 556, 181–202.

536. Dubinski J, Chakrabarty D., Warps and Bars from the External Tidal Torques of Tumbling Dark Halos, Astrophys. J., 2009, vol. 703, 2068–2081.

537. Dutta S.N., Substructure in clusters of galaxies and the value of Omega, MNRAS, 1995, vol. 276, 1109–1115.

538. K. Ebisawa, K. Mitsuda, H. Inoue, T., Dotani/Discovery of the 0.08 Hz quasi periodic oscillation from the black hole candidate LMC X–1. Phys. Neutron stars and black holes – Tokyo, 1988, 149–153.

539. Eder J., Schombert J., Gas-rich Dwarfs from the PSS-II. III. H I Profiles and Dynamical Masses, ApJS, 2000, vol. 131, 47–70.

540. Edvardsson B., Andersen J., Gustafsson B., et al. The Chemical Evolution of the Galactic Disk – Part One – Analysis and Results, Astron. Astrophys., 1993, vol. 275, 101.

541. Edwards D.A., Instability in the red star of semi–detached binary systems. III, MNRAS, 1988, vol. 231, 25–36.

542. Einasto J., Tenjes P., Barabanov A.V., Zasov A.V., Central holes in disks of spiral galaxies, Astroph. Space Sci., 1980, vol. 67, 31–43.

543. Efremov Yu.N., On the chains of star complexes and superclouds in spiral arms, MNRAS, 2010, vol. 405, 1531–1543.

544. Efstathion G., Lake G., Negroponte J., The stability and masses of disk galaxies, MNRAS, 1982, vol. 199, 1069–1088.

545. Elmegreen B.G., Properties of spurs in spiral galaxies, Astrophys. J., 1980, vol. {242}, 528–532.

546. Elmegreen B.G., Wilcots E.M., Pisano D.J., HI observations of the spiral arm pattern speed in the late-type barred galaxy NGC 925, Astrophys. J., 1998, vol. 494, 37–39L.

547. Elskens Y., Physica D, 1997, vol. 100, 142.

548. Emsellem E., Dejonghe H., Bacon R., Dynamical models of NGC 3115, MNRAS, 1999, vol. 303, 495–514.

549. England M.N., Gottesman S.T., Hunter J.H., High–resolution observations, kinematics, and dynamics of the barred spiral NGC 1073, Astrophys. J., 1990, vol. 348, 456–466.

550. Englmaier P., Pohl M., Bissantz N., The Milky Way Spiral Arm Pattern, Tumbling, Twisting, and Winding Galaxies: Pattern Speeds along the Hubble Sequence / Edt. E.M. Corsini and V.P. Debattista, 2008
551. Erickson R.R., The third and fourth moments of the local stellar velocity distribution, Astrophys. J., 1975, vol. 195, 343–358.
552. Erwin P., Sparke L.S., Triple bars and complex central structures in disk galaxies, Astrophys. J., 1999, vol. 521, 37–40L.
553. Erwin P., Sparke L., Double bars, inner disks, and nuclear rings in early-type disk galaxies, Astron. J., 2002, vol. 124, 65–77.
554. Eskridge P.B., Frogel J.A., What is the true fraction of barred spiral galaxies? Astroph. Space Sci., 1999, vol. 269-270, 427–430.
555. Evans N.W., Read J.C.A., Stability of Power-Law Disks I. The Fredholm integral equation, MNRAS, 1998, vol. 300, 83–105.
556. Evans N.W., Read J.C.A., Stability of power-law discs – II. The global spiral modes, MNRAS, 1998, vol. 300, 106–130.
557. Evstegneev E.A., Reshetnikov V.P., Parameters of bulges and disks of interacting galaxies, AstL, 1999, vol. 25, 582–590.
558. Faulkner J., Lin D.N.C., Papaloizou J., On the evolution of accretion disc flow in cataclysmic variables. I. The prospect of a limit cycle in dwarf novae systems, MNRAS, 1983, vol. 205, 359–375.
559. Faulkner J., Lin D.N.C., Papaloizou J., Workshop on cataclysmic variables and low-mass X-ray binaries – Cambridge: Mass, 1983.
560. Faulkner J., Lin D.N.C., Papaloizou J., On the evolution of accretion disc flow in cataclysmic variables, MNRAS, 1985, vol. 212, 105–149.
561. Feitzinger J.V., Magellanic-type galaxies, Space Sci Review, 1980, vol. 27, 35–105.
562. Fellhauer M., Belokurov V., Evans N.W., et al. The Origin of the Bifurcation in the Sagittarius Stream, Astrophys. J., 2006, vol. 651, 167–173.
563. Fellhauer M., Kroupa P., et al. SUPERBOX – an efficient code for collisionless galactic dynamics, New Astr., 2000, vol. 5, 305–326.
564. Ferguson A.M.N., Clarke C.J., The evolution of stellar exponential discs, MNRAS, 2001, vol. 325, 781–791.
565. Ferrari A., Massaglia S., Trussoni E., Magnetohydrodynamic Kelvin-Helmholtz instabilities in astrophysics. III - Hydrodynamic flows with shear layers, MNRAS, 1982, vol. 198, 1065–1079.
566. Ferrari A., Trussoni E., Zaninetti L., Magnetohydrodynamic Kelvin-Helmholtz instabilities in astrophysics. II Cylindrical boundary layer in vortex sheet approximation, MNRAS, 1981, vol. 196, 1051–1066.
567. Filipov L.G., Self–similar problems of the time–dependant discs accretion and the nature of the temporary X–ray sources, Adv. Space Res., 1984, vol. 3, 305–313.
568. Filipov L., Shakura N., Ljubarskii Yu., Self–similar processes in accretion discs, Adv. Space Res., 1988, vol. 8, 163–169.
569. Fillmore J.A., Boroson T.A., Dressler A., Internal kinematics of spiral galaxies: gas and stellar rotation curves and dispersion profiles, Astrophys. J., 1986, vol. 302, 208–233.
570. Firmani C., Tutukov A., Evolutionary models for disk galaxies, Astron. Astrophys., 1992, vol. 264, 37.
571. Fisher D., Kinematic profiles of S0 galaxies, Astron. J., 1997, vol. 113, 950–974.
572. Finoguenov A., Reiprich T.H., Boehringer H, Astron., Astrophys., 2001, vol. 368, 749.

573. Frank J., King A.R., Raine D.J., Accretion power in astrophysics – Cambridge: Cambridge University Press, 1985.

574. Frank J., Shlosman I., On the stability of Kuzmin/Toomre disks, Astrophys. J., 1989, vol. 346, 118–121.

575. Freeman K.C., On the disks of spiral and SO galaxies, Astrophys. J., 1970, vol. 160, 811–830.

576. Freeman K.C., The distribution of light in galaxies. Struct. and ProP. Nearby Galaxies: SymP. IAU. N 77., Bad Münstereifel., 1978, 3–14.

577. Freeman K.C., Kinematics and dynamics of the Magellanic Clouds. Structure and evolution of the Magellanic Clouds, IAU SymP. N 108.

578. 1984, 107–114.

579. Freeman K.C., The old population. The Milky Way galaxy: SymP. IAU N 106. The Netherlands, 1985, 113–122.

580. Freeman K.C., The galactic spheroid and old disk, Ann. Rev. Astron. Astrophys., 1987, vol. 25, 603–632.

581. Freudenreich H.T., COBE's Galactic bar and disk, Astrophys. J., 1998, vol. 492, 495–510.

582. Fridman A.M., Morozov A.G., et al. On the Hydrodynamical Conception of the Spiral Structure in Galaxies with a Velocity "kink" on the Rotation Curve / Observational Evidence of Activity in Galaxies, Ed. E.E. Khachikian et al., 1987, vol. 121, 147–157.

583. Fridman A.M., "Theory of gradient instabilities of a gaseous galactic disc and rotating shallow water", in Dynamics of Astrophy steal Discs / Ed. J A Sellwood – Cambridge: Cambridge Univ. Press, 1989, 185.

584. Fridman A.M., Dynamics of disks in the Milky Way: some solved problems and some puzzles. Physics of the gaseous and stellar disks of the Galaxy / King I.R., ed. ASP Conference Series., 1994, vol. 66, 15–28.

585. Fridman A.M., Khoruzhii O.V., Polyachenko E.V., Observational Manifestation of Chaos in the Gaseous Disk of the Grand-Design Spiral Galaxy NGC 3631, Space Science Rev., 2002, vol. 102, 51.

586. Fridman A.M., Koruzhii O.V., Zasov A.V., et al. Vertical motions in the gaseous disk of the spiral galaxy NGC 3631, Astr. Lett., 1998, vol. 24, 764–773.

587. Fridman A.M., et al. "Discovery of new structures - giant antycyclones – in disks of spiral galaxies", in Searching for Absolute Values and Unity in the Sciences. Science for the Benefit of Humanity. Commemorative Volume of the Twenty-first Intern. / Conf. on the Unit of the Sciences, 1997 – Lexington, Ky.: ICCU, 1997.

588. Fridman A.M., et al. "Investigation of the dynamics of spiral galaxies on the base of 3D vector velocity field of their gaseous disks reconstructed from observed line-of-sight velocity field", in The Combination of Theory, Observations, and Simulation for the Dynamics of Stars and Star Clusters in the Galaxy, 23rd Meeting of the I A U, Joint Discussion 15, 25 August 1997, Kyoto, Japan, Meeting Abstracts / Ed. J Andersen – Dordrecht: Kluwer Acad. Publ., 1998, 39.

589. Fridman A.M., et al. New structures in galactic disks: predictions and discoveries, Astron. Soc. of the Pacific Conf. Ser., 2001, vol. 230, 187.

590. Fridman A.M., Boyarchuk A.A., Bisikalo D.V., et al. The collective mode and turbulent viscosity in accretion discs, Phys. Lett. A., 2003, vol. 317, 181–198.

591. Fridman A.M., et al. The Over-Reflection Instability: Myth or Reality?, Astroph. and Space Science Library, 2006, vol. 337, 3–22.

592. Fridman A.M., et al. On the possibility of experimental detection of the over-reflection instability, Phys. Lett. A., 2006, vol. 349, 198.

593. Fridman A.M., et al. Fully developed turbulence in accretion discs of binary stars: turbulent viscosity coefficient and power spectrum. In: Chaos in Astronomy, 2007 / Eds P. Patsis, G. Contopoulos – Berlin: Springer-Verlag, 2008.

594. Fridman A.M., et al. Over-reflection of waves and over-reflection instability of flows revealed in experiments with rotating shallow water, Phys. Lett. A, 2008.

595. Fridman A.M., Gor'kavyi N.N., Physics of Planetary Rings – New York: Springer, 1999.

596. Fridman A.M., Khoruzhii O.V., Slow bars in spiral galaxies, Phys. Lett. A., 2000, vol. 276, 199.

597. Fridman A.M., Khoruzhii O.V., Progress in the Study of Galaxies: Structures, Collective Phenomena and Methods, Space Sci. Rev., 2003, vol. 105, 1.

598. Fridman A.M., Khoruzhii Î.V., Lyakhovich V.V., Is the mini-spiral in the innermost center of the galaxy caused by a new hydrodynamical instability? Physics of the Gaseous and Stellar Disks of the Galaxy. ASP Conf. Series, 1994, vol. 66, 285–303.

599. Fridman A.M., Khoruzhii O.V., Lyakhovich V.V., et al. Spiral-Vortex structure in the gaseous disks of galaxies, Astroph. Space Sci., 1997, vol. 252, 115–131.

600. Fridman A.M., Khoruzhii Î.V., et al. Mini-spiral at the galactic center: A link between its structure and the value of a central point mass. Unsolved problems of the Milky Way/ eds. Blitz L., Teuben P. – Dordrecht Kluwer, 1996, 241–246.

601. Fridman A.M., Khoruzhii Î.V., Lyakhovich V.V., et al. The dynamics of the Galactic center: origin of the mini-spiral. The Galactic Center / Edit. by R. Gredel, ASP, 1996, vol. 102, 335–344.

602. Fridman A.M., Khoruzhii O.V., Lyakhovich V.V., Silchenko O.K., Zasov A.V., Afanasiev V.L., et al. Restoring the full velocity field in the gaseous disk of the spiral galaxy NGC 157, Astron. Astrophys., 2001, vol. 371, 538–559.

603. Fridman A.M., Khoruzhii O.V., Piskunov A.E., On the solution of the main paradox of the stellar dynamics, ASP Conference Series / King I.R., ed. 1994, vol. 66, 215–234.

604. Fridman A.M., Khoruzhii O.V., Polyachenko E.V., Zasov A.V., Sil'chenko O.K., Afanas'ev V.L., Dodonov S.N., Moiseev A.V., Giant cyclones in gaseous discs of spiral galaxies, Phys. Lett. A., 1999, vol. 264, 85–93.

605. Fridman A.M., Khoruzhii O.V., Polyachenko E.V., Zasov A.V., Sil'chenko O.K., et al. Gas motions in the plane of the spiral galaxy NGC 3631, MNRAS, 2001, vol. 323, 651–662

606. Fridman A.M., Morozov A.G., Nezlin M.V., Snezhkin E.N., Centrifugal instability in rotating shallow water and the problem of the spiral structure in galaxies, Phys. Lett. A., 1985, vol. 109, 228–231.

607. Fridman A.M., Morozov A.G., Palous J., Piskunov A.F., On the gravitational instability of three–component model of Galaxy / Symp. Stellar clusters and associations, Prague, 1983, 98.

608. Fridman A.M., Polyachenko V.L., Physics of gravitating systems – N.Y.: Springer-Verlag, 1984, vol. 1, 468 vol. 2, 358.

609. Fridman A.M., Polyachenko V.L., Instabilities of gravitating systems. vol. 1. Layer, Cylinder, and Sphere – M: Izdatelskii centr "Akademiya", 1998.

610. Friedli D., Martinet L., Bars within bars in lenticular and spiral galaxies: a step in secular evolution?, Astron. Astrophys., 1993, vol. 277, 27–41.

611. Frick P., Beck R., Shukurov A., et al. Magnetic and optical spiral arms in the galaxy NGC 6946, MNRAS, 2000, vol. 318, 925–937.

612. Frink S., Fuchs B., Roser S., Wiellen R., Derivation of the galactic rotation curve using space velocities, Astron. Astrophys., 1996, vol. 314, 430–437.

613. Froning C.S., Long K.S., Knigge C., Accretion and Outflow in Interacting Binary Systems: Far Ultraviolet Spectroscopic Explorer Observations of the Nova-like Cataclysmic Variable UX Ursae Majoris, Astrophys. J., 2003, vol. 58, 433.

614. Fuchs B., von Linden S., Dynamical stability and evolution of the discs of Sc galaxies, MNRAS, 1998, vol. 294, 513–522.

615. Fuchs B., NGC 2613, 3198, 6503, 7184: Case studies against "Maximum" disks. Galaxy Dynamics / ASP Conf. Ser. / Eds. Merrit D.R. et. al. 1999, vol. 182, 365–368.

616. Fuchs B., Constraints on the decomposition of the rotation curves of spiral galaxies. Identification of Dark Matter/ Eds. Spooner N.J.C., Kudryavtsev V. – Singapore: WS, 2003, 72–77

617. Fuchs B., Möllenhoff C., Heidt H., Decomposition of the rotation curves of distant field galaxies, Astron. Astrophys., 1998, vol. 336, 878–882.

618. Fuchs B., Thielheim K.O., On the hydrostatic equilibrium of the interstellar gas, Astrophys. J., 1979, vol. 227, 801–807.

619. Fux R., 3D self-consistent N-body barred models of the Milky Way. I. Stellar dynamics, Astron. Astrophys., 1997, vol. 327, 983–1003.

620. Fux R., 3D self-consistent N-body barred models of the Milky Way. II. Gas dynamics, Astron. Astrophys., 1999, vol. 345, 787–812.

621. Fux R., Gas dynamics in the galactic bar region from N-Body and SPH simulations. Astrophysics and Space Science, 2001, vol. 276, 367–374.

622. Fux R., Martinet L., The disc scale length of the Galaxy, Astron. Astrophys., 1994, vol. 287, 21–24L.

623. Gallagher J.S., Cole A.A., Holtzman J., Smecker-Hane T., Star Formation History of the Disk of the Large Magellanic Cloud. New Views of the Magellanic Clouds. IAU Symposium #190, 1999, 306–312.

624. Galliano E., Alloin D., Near-IR 2D-spectroscopy of the $4"$x$4"$ region around the Active Galactic Nucleus of NGC 1068 with ISAAC/VLT. Astron. Astrophys., 2002, vol. 393, 43–56.

625. Gomez C.G., Athanassoula E., Analysis of the distribution of HII regions in external galaxies. II. Analysis of the spiral structure, AAS, 1993, vol. 100, 431–464.

626. Gardiner L.T., Turfus C., Putman M.E., Numerical simulation of asymmetric spiral structure in the Large Magellanic Cloud, Astrophys. J., 1998, 507, L35–38.

627. Gavazzi G., Pierini D., Boselli A., The phenomenology of disk galaxies, Astron. Astrophys., 1996, vol. 312, 397.

628. Gavazzi R., et al. The sloan lens acs survey. iV. the mass density profile of early-type galaxies out to 100 effective radii, Astrophys. J., 2007, vol. 667, 176.

629. Gelato S., Chernoff D.F., Wasserman I., An adaptive hierarchical particle-mesh code with isolated boundary conditions, Astrophys. J., 1997, vol. 480, 115–131.

630. Genzel R., Townes C.H., Physical conditions, dynamics, and mass distribution in the center of the Galaxy, Ann. Rev. A&A, 1987, vol. 25, 377–423.

631. Gerber R.A., Lamb S.A., Balsara D.S., Combined hydrodynamical and N–body studies of colliding galaxies: the formation of ring galaxies, Bull. Amer. Astron. Soc. 1990, vol. 22, 1243.

632. Gerhard O.E., Structure and mass distribution of the Milky Way bulge and disk. Galaxy Disks and Disk Galaxies / Ed. by Jose G. Funes, S.J. and Enrico Maria Corsini – San Francisco: Astron. Soc. of the Pacific, 2001, 21–30.

633. Gerhard O., Jeske G., Saglia R.P., Bender R., Breaking the degeneracy between anisotropy and mass. The dark halo of the E0 galaxy NGC 6703, MNRAS, 1998, vol. 295, 197–215.

634. Gerhard O., Pattern speeds in the Milky Way, MmSAI, 2011, vol. 18, 185–188.

635. Gerin M., Combes F., Athanassoula E., The influence of galaxy interactions on stellar bars, Astron. Astrophys., 1990, vol. 230, 37–54.

636. Ghez A.M., Salim S., Hornstein S.D., et al. Stellar Orbits around the Galactic Center Black Hole, Astrophys. J., 2005, vol. 620, 744–757.

637. Gildreich P., Lynden–Bell D., Gravitational instability of uniformly rotating disks, MNRAS, 1965, vol. 130, 97–124.

638. Gilham S., Scale–free axisymmetric accretion with weak viscosity, MNRAS, 1981, vol. 195, 755–763.

639. Gilmore G., Reid N., New light on faint stars. III - Galactic structure towards the South Pole and the Galactic thick disc, MNRAS, 1983, vol. 202, 1025-1047.

640. Giovanelli R., Avera E., Karachentsev I.D., Spectroscopy of edge-on spirals, Astron. J., 1997, vol. 114, 122–137.

641. Giovanelli R., Haynes M.P., The inner scale length of spiral Galaxy rotation curves, Astrophys. J., 2002, vol. 571, 107–111L.

642. Gnedin O.Y., Gould A., Miralda-Escude J., Zentner A.R., Probing the Shape of the Galactic Halo with Hypervelocity Stars, Astrophys. J., 2005, vol. 634, 344.

643. Graham A.W., An Investigation into the Prominence of Spiral Galaxy Bulges, Astron. J., 2001, vol. 121, 820.

644. Graham A.W., A K-band central disc surface brightness correlation with scale-length for early-type disc galaxies, and the inclination correction, MNRAS, 2001, vol. 326, 543–552.

645. Goad J.M., Kinematics of the gas in the nuclear disk of M81. Astrophys. J. Suppl. Ser., 1976, vol. 32, 89–113.

646. Goad J.W., Roberts M.S., Spectroscopic observations of superthin galaxies, Astrophys. J., 1981, vol. 250, 79–86.

647. Goldreich P., Lynden-Bell D. I., Gravatational stability of uniformly rotating disks, MNRAS, 1965, vol. 130, 97–124.

648. Goodman J., A local instability of tidally distorted accretion disks, Astrophys. J., 1993, vol. 406, 596.

649. Gorbatskii V.G., Effects of gas streams in close binary systemsn, Astrophys. 1967, vol. 3, 116.

650. Govorukhin V.N., Morgulis A., Yudovich V.I., Zaslavsky G.M., Phys. Rev. E, 1999, vol. 60, 2788.

651. Gould A., Galactic disc column density by maximum likelihood, MNRAS, 1990, vol. 244, 25–28.

652. Gradie J., Hauashi J., Physical properties of the Beta Pictoris circumstellar disk, Lunar and Planet Sci., 1987, vol. 18, 351–352.

653. Graff D.S., Gould A.P., et al. The velocity structure of Large Magellanic Cloud carbon stars: young disk, old disk, and perhaps a separate population, Astrophys. J., 2000, vol. 540, 211–216.

654. Greengard L., The numerical solution of the N-body problem, Computers in physics, 1990, No. 3–4, 142–152.

655. de Grijs R., Edge-on disk galaxies. A structure analysis in the optical and near-infrared. PhD thesis, UniV. Groningen, The Netherlands, 1997.

656. de Grijs R., The global structure of galactic discs, MNRAS, 1998, vol. 299, 595.

657. de Grijs R., Peletier R.F., van der Kruit P.C., The z-structure of disk galaxies towards the galaxy planes, Astron. Astrophys., 1997, vol. 327, 966–982.

658. de Grijs R., van der Kruit., Structure analysis of edge-on spiral galaxies, Astron. Astropliys. Suppl. Ser., 1996, vol. 117, 19–37.

659. Griv E., Yuan C., Gedalin M., The first detection of warping of outer stellar disk in N-body simulations of isolated and rapidly rotating disk-shaped galaxies, Astrophys. J., 2002, vol. 580, 27–30L.

660. Grosbol P., Dottori H., Pattern speed of main spiral arms in NGC 2997. Estimate based on very young stellar complexes, Astron. Astrophys., 2009, vol. 499, 21L–24L.

661. Gualandris A., Zwart S.P., A hypervelocity star from the Large Magellanic Cloud, MNRAS, 2007, 376L, 29.

662. Gusev A.S., Peculiarities of the stellar population of the LMC bar, Astron. Rep., 1998, vol. 42, 446–452.

663. Haghi H., Zonoozi A.H., Rahvar S., Magellanic Stream A possible tool for studying dark halo model, NewA, 2009, vol. 14, 692–699.

664. Hameury J.-M., Lasota J.-P., McClintock J.E., Narayan R., Advection-dominated flows and the X-ray Delay in the outburst of GRO J1655–40, Astrophys. J., 1997, vol. 489, 234–243.

665. Hammer F., Puech M., Chemin L., et al. The Milky Way, an Exceptionally Quiet Galaxy: Implications for the Formation of Spiral Galaxies, Astrophys. J., 2007, vol. 662, 322.

666. Hammersley P.L., Cohen M., Garzon F., et al. Structure in the First Quadrant of the Galaxy: an Analysis of "TMGS" Star Counts using the "SKY" Model, MNRAS, 1999, vol. 308, 333–363.

667. Hanawa T., Hirotani H., Kawai N., A theory of the rapid burster (MXB 1730-335). X–Ray Binaries – Paris, 1989, 423–426.

668. Hardee P.E., Norman M.L., Spatial stability of the slab jet. I. Linearized stability analysis, Astrophys. J., 1988, vol. 334, 70–79.

669. Hardee P.E., Rosen A., Dynamics and Structure of Three-dimensional Trans-Alfvenic Jets. II. The Effect of Density and Winds, Astrophys. J., 2002, vol. 576, 204–221.

670. Harten A., Hight resolution schemes for hyperbolic conservation laws, J. Coniput. Phys., 1983, vol. 49, 357–393.

671. Harten A., Lax P.D., van Leer B., On upstream differensing and Godunov-type chemes for hyperbolic conservation laws, SIAM Review, 1983, vol. 25, 35–61.

672. Hartmann L., Kenyon S.J., Further evidence for disk accretion in FU Orionis objects, Astrophys. J., 1987, vol. 312, 243–253.

673. Hasan H., Norman C., Chaotic orbits in barred galaxies with central mass concentrations, Astrophys. J., 1990, vol. 361, 69–77.

674. Hashimoto Y., Funato Y., Makino J., To circularize or not to circularize? Orbital evolution of satellite galaxies, Astrophys. J., 2003, vol. 582, 196–201.

675. Haward S., Byrd G.G., A self–gravitating simulation of the M51 system, Astron. J., 1990, vol. 99, 1798–1812.

676. Hawley J.F., Gammie C.F., Balbus S.A., Local three-dimensional magnetohydrodynamic simulations of accretion disks, Astrophys. J., 1995, vol. 440, 742–763.

677. Hawley J.F., Smarr L.L., Wilson J.R., A numerical study of nonspherical black hole accretion. I. Equations and test problems, Astrophys. J., 1984, vol. 277, 296–311.

678. Hayashi E., Navarro J.F., Hiding cusps in cores: kinematics of disc galaxies in triaxial dark matter haloes, MNRAS, 2006, vol. 373, 1117–1124.

679. Hayashi E., Navarro J.F., Springel V., The shape of the gravitational potential in cold dark matter haloes, MNRAS, 2007, vol. 377, 50–62.

680. Haywood M., Robin A.C., Creze M., The evolution of the Milky Way disc. II. Constraints from star counts at the galactic poles, Astron. Astrophys., 1997, vol. 320, 440–459.

681. Heller A.B., Brosch N., Almoznino E., et al. Lopsidedness in dwarf irregular galaxies, MNRAS, 2000, vol. 316, 569–587.

682. Heller C.H., Shlosman I., Fueling nuclear activity in disk galaxies: starbursts and monsters: Preprint. 1993.
683. Heller C.H., Shlosman I., Fueling nuclear activity in disk galaxies: starbursts and monsters, Astrophys. J., 1994, vol. 424, 84–105.
684. Hellsten U., Sommer-Larsen J., Dynamical and chemical evolution of viscous, star-forming galactic discs, MNRAS, 1992, vol. 255, 650–654.
685. Helmholtz H., Philos. Mag., 1868, vol. 36, 337.
686. Helmi A., Is the dark halo of our Galaxy spherical?, MNRAS, 2004, vol. 351, 643–648.
687. Helmi A., Velocity Trends in the Debris of Sagittarius and the Shape of the Dark Matter Halo of Our Galaxy, Astrophys. J., 2004, vol. 610L, 97–100.
688. Henkel C., Baan W.A., Mauersberger R., Dense molecular gas in galactic nuclei, Astron. Astrophys.Rev., 1991, vol. 3, 47.
689. Heraudeau Ph., Simien F., Stellar kinematical data for the central region of spiral galaxies. I., Astroph. Space Sci., 1998, vol. 133, 317–323.
690. Heraudeau Ph., Simien F., Maubon G., Prugniel Ph., Stellar kinematic data for the central region of spiral galaxies. II. Astroph. Space Sci., 1999, vol. 136, 509–514.
691. Hernquist L., An analytical model for spherical galaxies and bulges, Astrophys. J., 1990, vol. 356, 359–364.
692. Hernquist L., N-body realizations of compound galaxies. Astrophys. J. Suppl. Ser., 1993, vol. 86, 389–400.
693. Hernquist L., Heyl J.S., Spergel D.N., Bending instabilities in galaxy merger remnants, Astrophys. J., 1993, vol. 416, 9–12L.
694. Hernquist L., Hut P., Makino J., Discreteness noise versus force errors in N-body simulations, Astrophys. J., 1993, vol. 402, 85–88.
695. Herquist L., Katz N., TreeSPH: a unification of SPH with the hierarchical TREE method. Astrophys. J. Suppl. Ser., 1989, vol. 70, 419–446.
696. Hillebrandt W., Niemeyer J.C., Type IA Supernova Explosion Models, ARA&A, 2000, vol. 38, 191.
697. Hills J.G., Hyper-velocity and tidal stars from binaries disrupted by a massive Galactic black hole, Nature, 1988, vol. 331, 687.
698. Hirose M., Osaki Y., Hydrodynamic simulations of accretion diks in cataclismic variables, Publ. Astron. Soc. Japan, 1990, vol. 42, 135–163.
699. Hirose M., Osaki Y., Three–dimensional structure of accretion disks in close binary systems, Publ. Astron. Soc. Japan, 1991, vol. 43, 809–821.
700. Hoekstra H., Yee H.K.C., Gladders M.D., Properties of Galaxy Dark Matter Halos from Weak Lensing, Astrophys. J., 2004, vol. 606, 67–77.
701. Hohl F., Dynamics of plane stellar systems, Astron. and Space Sci., 1971, vol. 14, 91–109.
702. Hohl F., Numerical experiments with a disk of stars, Astrophys. J., 1971, vol. 168, 343–359.
703. Hohl F., Effect on halo component on bar formation in disk galaxies, Colloq. Int CNRS, 1975. ¹ 241, 55–63.
704. Hohl F., Three–dimensional galaxy simulations, Astrophys. J., 1978, vol. 83, 768–778.
705. Hohl F., Hochney R.W., A computer model of disks of stars, J. Comput. Phys., 1969, vol. 4, 306–323
706. Holmberg J., Nordström B., Andersen J., The Geneva-Copenhagen survey of the Solar neighbourhood II., Astron. Astrophys., 2007, vol. 475, 519.
707. Holtzman J.A., Gallagher J.S., Cole A.A., et. al. Observations and implications of the star formation history of the Large Magellanic Cloud, Astron. J., 1999, vol. 118, 2262–2279

708. Hoperskov A.V., Mustsevaya Ju.V., Mustsevoj V.V., Shear–layer instabilities in accretion disks around magnetized compact objects, Astroph. Space Sci., 1992, vol. 195, 373–387.

709. Hoperskov A.V., Mustsevaya Ju.V., Mustsevoj V.V., Disc accretion onto magnetized compact objects. Astron. Astrophys. Trans., 1993, vol. 4, 65–80

710. Horellou C., Combes F., A Model for the Cartwheel Ring Galaxy. Astroph. Space Sci., 2001, vol. 276, 1141–1149.

711. Horiuchi T., Kato S., A model of hydromagnetic turbulent viscosity in radiation-pressure-dominated disks. Publ. Astron. Soc. Japan, 1990, vol. 42, 661–674.

712. Howell S.B., Szhody P., High galactic latifude cataclismic variables, Astrophys. J., 1990, vol. 356, 623–629.

713. Howell S.B., Szhody P., Cataclismic variables at high galactic latifude: photometric survey results after three years of observations, Bull. Amer. AS, 1991, vol. 23, 943.

714. Huang K., Statistical Mechanics. 4-nd edition. Wiley, 1987.

715. Hubble E., Extragalactic nebulae, Astrophys. J., 1926, vol. 64, 321

716. Hummel E., Dettmar R.J., Wielebinski R., Neutral hydrogen and radio continuum observations of NGC 55, Astron. Astrophys., 1986, vol. 166, 97–106

717. Hunter C., The structure and stability of self–gravitating disks, MNRAS, 1963, vol. 126, 299–315

718. Hunter C., Self-gravitating gaseous disks, Ann. Rev. Fluid Mech., 1972, vol. 4, 219-242.

719. Hunter C., Patterns of waves in the galactic disks, Astrophys. J., 1973, vol. 181, 685–705.

720. Hunter C., Toomre A., Dynamics of the bending of the galaxy, Astrophys. J., 1969, vol. 155, 747–776.

721. Ibata R., Lewis G.F., et al. Great circle tidal streams: evidence for a nearly spherical massive dark halo around the Milky Way, ApJ, 2001, vol. 551, 294–311.

722. Ibata R., Irwin M., et al. A giant stream of metal-rich stars in the halo of the galaxy M31, Nature, 2001, vol. 412, 49–52.

723. Ibrahim A.I., Swank J.H., Parke W., New Evidence of Proton-Cyclotron Resonance in a Magnetar Strength Field from SGR 1806-20, Astrophys. J., 2003, vol. 584, 17–21L

724. Ichikawa S., Osaki Y., Time evolution of the accretion disk radius in a dwarf nova. Publ. Astron. Soc. Japan, 1992, vol. 44, 15–26

725. Igumenshchev I.V., On angular momentum transport in convection-dominated accretion flows, Astrophys. J., 2002, vol. 577, 31–34L

726. Igumenshchev I.V., Abramowicz M.A., Rotating accretion flows around black holes: convection and variability, MNRAS, 1999, vol. 303, 309–320

727. Igumenshchev I.V., Abramowicz M.A., Two-dimensional models of hydrodynamical accretion flows into black holes, Astrophys. J. Suppl. Ser., 2000, vol. 130, 463–484

728. Igumenshchev I.V., Abramowicz M.A., Narayan R., Numerical simulations of convective accretion flows in three dimensions, Astrophys. J., 2000, vol. 537, 27–30L

729. Igumenshchev I.V., Chen X., Abramowicz M.A., Accretion discs around black holes: two-dimensional, advection-cooled flows, MNRAS, 1996, vol. 278, 236–250

730. Igumentschev I.V., Shustov B.M., Tutukov A.V., Astron. Astrophys., 1990, vol. 234, 396

731. Illarionov A.F., Sunyaev R.A., Why the number of galactic X-ray stars is so small? Astron. Astrophys., 1975, vol. 39, 185–195

732. Illingworth G., Schechter P.L., Velocity and velocity dispersion profiles in NGC 3115, Astrophys. J., 1982, vol. 256, 481–496

733. Iodice E., Arnaboldi M., Bournaud F., et al. Polar ring galaxies and the Tully Fisher relation: implications for the dark halo shape, Astrophys. J., 2003, vol. 585, 730–738

734. Ivezić Z., Sesar B., Jurić M., Bond N., et al. The Milky Way Tomography with SDSS. II. Stellar Metallicity, Astrophys. J., 2008, vol. 684, 287

735. Iwamoto N., Umeda H., Tominaga N., et al. The First Chemical Enrichment in the Universe and the Formation of Hyper Metal-Poor Stars. Science, 2005, vol. 309, 451

736. Jacobs V., Sellwood J.A., Long-lived lopsided modes of annular disks orbiting a central mass, Astrophys. J., 2001, vol. 555, L25–L28

737. Jahnke E, Emde F, Losch F, Tables of Higher Functions. New York: McGraw-Hill, 1960

738. Jalali M.A., Hunter C., Unstable bar and spiral modes of disk galaxies, Astrophys. J., 2005, vol. 630, 804–829

739. James R.A., Sellwood J.A., Galactic models with variable spiral structure, MNRAS, 1978, vol. 182, 331–344

740. Janiuk A., Czerny B., Siemiginowska A., Szczerba R., On the turbulent α-disks and the intermittent activity in active galactic nuclei, Astrophys. J., 2004, vol. 602, 595–602

741. Jaroszynski M., Abramowicz M.A., Paczynski B., Supercritical accretion disks around black holes. Acta Astron.., 1980, vol. 30, 1–34

742. Jeans J.H., Astronomy and Cosmogony 2nd ed. Cambridge: The UniV. Press, 1929

743. Jenkins A., Binney J., Spiral heating of galactic discs, MNRAS, 1990, vol. 245, 305–317

744. Jernigan J.C., Porter D.H., A tree code with logarithmic reduction of force terms, hierarchical regularization of all variales, and explicit accuracy controls, Astrophys. J., 1989, vol. 71, 871–893

745. Jiang I.-G., Binney J., The orbit and mass of the Sagittarius dwarf galaxy, MNRAS, 2000, vol. 314, 468–474

746. Jing Y.P., Suto Y., Triaxial Modeling of Halo Density Profiles with High-Resolution N-Body Simulations, Astrophys. J., 2002, vol. 574, 538–553.

747. Johnston K.V., Law D.R., Majewski S.R., A Two Micron All Sky Survey View of the Sagittarius Dwarf Galaxy. III. Constraints on the Flattening of the Galactic Halo, Astrophys. J., 2005, vol. 619, 800–806

748. Jog C.J., Solomon P.M., Two-fluid gravitational instabilities in a galactic disk, Astrophys. J., 1984, vol. 276, 114–126

749. Jog C.J., Solomon P.M., A Galactic disk as a two-fluid system: consequences for the critical stellar velocity dispersion and the formation of condensations in the gas, Astrophys. J., 1984, vol. 276, 127–134

750. Johnston H.M., Kulkarni S.R., Oke J.B., The black hole A 0620–00 and its accretion disk, Astrophys. J., 1989, vol. 345, 492–497

751. Jones T.J., Ashley M., Hyland A.R., Ruelas-Mayorga A., A search for the infrared counterpart of type II OH masers I., MNRAS, 1981, vol. 197, 413–428

752. de Jong R.S., Near-infrared and optical broadband surface photometry of 86 face-on disk dominated galaxies. II., Astron. Astropliys. Suppl. Ser., 1996, vol. 118, 557–573,

753. de Jong R.S., Near-infrared and optical broadband surface photometry of 86 face-on disk dominated galaxies. III., Astron. Astrophys., 1996, vol. 313, 45

754. Julian W.H., Toomre A., Non-Axisymmetric Responses of Differentially Rotating Disks of Stars, Astrophys. J., 1966, vol. 146, 810–830
755. Just A., Penarrubia J., Large scale inhomogeneity and local dynamical friction, Astron. Astrophys., 2005, vol. 431, 861–877
756. Kadomtsev B B, in Reviews in Plasma Physics Vol. 4 /Ed. M.A. Leontovich. New York: Consultants Bureau, 1971
757. Kalnajs A.J., Dynamics of flat galaxies. I., Astrophys. J., 1971, vol. 166, 275–293
758. Kalnajs A.J., The equilibria and oscillations of a family of uniformly rotating stellar disks, Astrophys. J., 1972, vol. 175, 63–76
759. Kalnajs A.J., Dynamics of flat galaxies. IV - The integral equation for normal modes in matrix form, Astrophys. J., vol. 212, 637–644
760. Kalnajs A.J., Athanassoula G.E., The bar modes of uniformly rotating stellar disks, MNRAS, 1974, vol. 168, 287–290
761. Kamahori H., Fujimoto M., Velocity dispersion of stars and giant molecular clouds, Publ. Astron. Soc. Japan, 1986, vol. 38, 77–94.
762. Kato S., Oscillation and overstability of density waves in a rotating disk-like star-gas system, Publ. Astron. Soc. Japan, 1972, vol. 24, 61–86.
763. Kato S., Excitation of spiral density waves by gas flow in a star–gas disk, Publ. Astron. Soc. Japan, 1973, vol. 25, 231–242.
764. Kato T., Nelson P., Stockdale Ch., et al. Photometric study of new southern SU UMa-type dwarf novae and candidates - III. NSV 10934, MM Sco, AB Nor and CAL 86, MNRAS, 2004, vol. 347, 861–872.
765. Karachentsev I.D., Karachentseva V.E., Parnovsky S.L., Flat Galaxy Catalogue, Astron. Nachr., 1993, vol. 314, 97–222.
766. Karachentsev I.D., Karachentseva V.E., Huchtmeier W.K., Makarov D.I., A Catalog of Neighboring Galaxies, Astron. J., 2004, vol. 127, 2031–2068.
767. Karachentsev I.D., Kashibadze O.G., Makarov D.I., Tully R.B., The Hubble flow around the Local GrouP, MNRAS, 2009, vol. 393, 1265–1274.
768. Karachentsev I.D., Mitronova S.N., et al. The 2MASS Tully-Fisher relation for flat edge-on galaxies, Astron. Astrophys., 2002, vol. 396, 431–438.
769. Karachentsev I.D., Makarov D.I., Binary galaxies in the local supercluster and its neighborhood, Astroph. Bull., 2008, vol. 63, 299–345.
770. Kaufer A., Szeifert Th., Krenzin R., et al. The galactic abundance gradients traced by B-type stars, Astron. Astrophys., 1994, vol. 289, 740.
771. Kaufman M., Bash F.N., Crane P.C., Jacoby G.H., The arc and other structures in the center of M81, Astron. J., 1996, vol. 112, 1021–1034.
772. Kawai A., et al. GRAPE-5: a special-purpose computer for N-body simulations, Publ. Astron. Soc. Japan, 2000, vol. 52, 659–676.
773. Keel W.C., White R.E., Dust Structures as Revealed by HST Imaging of Overlapping Galaxies, AAS, 1999, vol. 195, 104.
774. Keel W.C., The road to galaxy formation. Springer, 2007.
775. Keller S.C., Murphy S., et al. Revealing Substructure in the Galactic Halo The SEKBO RR Lyrae Survey, Astrophys. J., 2008, vol. 678, 851.
776. Kelvin W., Philos. Mag., 1887, 5, vol. 24, 188.
777. Kenney J.D.P., Young J.S, Astrophys. J., 1989, vol. 344, 171.
778. Kennicutt R.C., The star formation law in galactic disks, Astrophys. J., 1989, vol. 344, 685
779. Kennicut R.C., Proc. IAU Symp. N 171, 1996 / Eds Bender R., Davies R.L. – Dordrecht: Kluwer, 11.
780. Kent S.M., CCD surface photometry of field galaxies. I. Observations, ApJS, 1984, vol. 56. 105–141.

781. Kent S.M., CCD surface photometry of field galaxies. II – Bulge/disk decompositions, ApJS, 1985, vol. 59, 115–159.
782. Kent S.M., Dark matter in spiral galaxies. I. Galaxies with optical rotation curves, Astron. J., 1986, vol. 91, 1301–1327.
783. Kent S.M., Dark matter in spiral galaxies. II. Galaxies with H I rotation curves, Astron. J., 1987, vol. 93, 816–832.
784. Kent S.M., Dame T.M., Fazio G., Galactic structure from the *Spacelab* infrared telescope. II. Luminosity models of the Milky Way, Astrophys. J., 1991, vol. 378, 131–138.
785. Kenyon S.J., Hartmann L.W., On the apparent positions of T Tauri stars in the H–R diagram, Astrophys. J., 1990, vol. 349, 197–207.
786. Khoperskov A.V., Khrapov S.S., Instability of high-frequency acoustic waves in accretion disks with turbulent viscosity, Astron. Astrophys., 1999, vol. 345, 307–314.
787. Khoperskov A.V., Khrapov S.S., Computer modeling of non-stationary gas quasi-Keplerian disk / Ed.: A.M.Fridman et al., Astroph. and Space Science Libr., Springer, 2006, vol. 337, 217–229.
788. Khoperskov A.V., Khrapov S.S., Self-consistent gas and stellar dynamics of disk galaxies: a problem of dark mass / Ed.: A.M.Fridman et al., Astroph. and Space Science Libr., Springer, 2006, vol. 337, 337–343 (arXiv1007.4332).
789. Khoperskov A.V., Moiseev A.V., Chulanova E.A., Dynamical modeling of SB galaxies, Bull. Spec. Astrophys. Obs. of the Russian AS, 2001, vol. 52, 135–145.
790. Khoperskov A.V., Tiurina N.V., Milky Way parameters by the results of N-body simulation / Commun. from the Konkoly Obs. Hungarian AS. Budapest, 2003, vol. 13, Part 3, [1] 103, 163–166.
791. Khoperskov A.V., Just A., Korchagin V.I., Jalali M.A., High resolution simulations of unstable modes in a collisionless disc, Astron. Astrophys., 2007, vol. 473, 31–40.
792. Khoperskov A.V., Bizyaev D., Tyurina N.V., Butenko M., Numerical modelling of the vertical structure and dark halo parameters in disc galaxies, Astron. Nachr., 2010, vol. 331, 731–745.
793. Khoperskov S.A., Khoperskov A.V., Eremin M.A., Butenko M.A., Polygonal structures in a gaseous disk: Numerical simulations, Astronomy Letters, 2011, vol. 37, 563–575.
794. Khoperskov A.V., Eremin M.A., Khoperskov S.A., Butenko M.A., Morozov A.G., Dynamics of gaseous disks in the nonaxisymmetric dark halo, Àñòðîí. æóðí., 2012, vol. 89, 1.
795. Khosroshahi H.G., Wadadekar Y., Kembhavi A., Correlations among Global Photometric Properties of Disk Galaxies, Astrophys. J., 2000, vol. 533, 162.
796. Khruzina T.S., Cherepashchuk A.M., Bisikalo D.V., et al. Interpretation of Light Curves of IP Peg in a Model with Shockless Interaction between the Gas Stream and Disk, Astron. Rep., 2001, vol. 45, 538.
797. Khruzina T.S., Cherepashchuk A.M., Bisikalo D.V., et al. Light curve interpretation for “Quiescent” X-ray novae in a model with noncollisional interaction between the flow and disk. The system GU Mus = GRS 1124-68, Astron. Rep., 2003, vol. 47, 621.
798. Khruzina T.S., Cherepashchuk A.M., Bisikalo D.V., et al. Interpretation of Orbital Light Curves of U Geminorum, Astron. ReP., 2003, vol. 47, 848.
799. Khruzina T.S., et al. Interpretation of the light curves of “quiescent” X-ray novae in a model with a noncollisional interaction between the flow and disk. The XTE J1118+480 system, Astron. Rep., 2005, vol. 49, 79.

800. Kim W.-T., Hong S.S., Yun H.S., The smoothed particle hydrodynamics and the binary tree combined into BTSH, J. Kor. Astron. Soc., 1994, vol.27, 13–29.

801. S. Kim, L. Staveley-Smith, M.A. Dopita, et al./An HI aperture synthesis mosaic of the Large Magellanic Cloud, Astrophys. J., 1998, vol.503, 674–688.

802. Kim W.-T., Ostriker E., Formation and Fragmentation of Gaseous Spurs in Spiral Galaxies, ApJ, 2002, vol. 570, 132–151.

803. Kim C.-G., Kim W.-T., Ostriker E., Galactic Spiral Shocks with Thermal Instability, ApJ, 2008, vol. 681, 1148–1162.

804. Kim Ch.-G., Kim W.-T., Ostriker E.C., Galactic Spiral Shocks with Thermal Instability, Astrophys. J., 2008, vol. 681, 1148–1162.

805. Klahr H.H., Bodenheimer P., Turbulence in accretion disks: vorticity generation and angular momentum transport via the global baroclinic instability, Astrophys. J., 2003, vol.582, 869–892.

806. Klessen R.S., Kroupa P., Dwarf Spheroidal Satellite Galaxies without Dark Matter: Results from Two Different Numerical Techniques, Astrophys. J., 1998, vol.498, 143–155.

807. van der Klis M., Jansen F., Paradijs van J., et. al. Intensity–dependent quasiperiodic oscillations in X–ray flux of GX5–1, Nature, 1985, vol.316, 225–230.

808. van der Klis M., Millisecond oscillations in X-ray binaries, Ann. Rev. Astron. Astrophys., 2000, vol.38, 717–160.

809. Knapen J.H., “Atomic hydrogen in the spiral galaxy NGC 3631 41”, Mon. Not. R. Astron. Soc., 1997, vol.286, 403.

810. Knapen J.H., Shlosman I., Peletier R.F., A subarcsecond resolution near-infrared study of Seyfert and “normal” galaxies: II. Morphology, Astrophys. J., 2000, vol.529, 93–100.

811. Knobloch E., Spruit H.C., Baroclinic waves in a vertically stratified thin accretion disk, Astron. Astrophys., 1986, vol.66, 359–365.

812. Koda J., Sofue Y., Wada K., A unified scaling law in spiral galaxies, Astrophys. J., 2000, vol.531, L17—L20.

813. Kolmogorov A.N.C.R., The Local Structure of Turbulence in Incompressible Viscous Fluid for Very Large Reynolds’ Numbers, Acad. Sci. URSS, 1941, vol.30, 301.

814. Koposov S.E., Rix H-W., Hogg D.W., Constraining the Milky Way Potential with a Six-Dimensional Phase-Space Map of the GD-1 Stellar Stream, Astrophys. J., 2010, vol.712, 260–273.

815. Korchagin V., Theis Ch., Global spiral modes in star-forming gravitating disks, Astron. Astrophys., 1999, vol.347, 442–454.

816. Korchagin V., Kikuchi N., Miyama S.M., Orlova N., Peterson B.A., Global spiral modes in NGC 1566: observations and theory, Astrophys. J., vol.541, 565–578.

817. Korchagin V.I., Girard T.M., et al. Local surface density of the galactic disk from a three-dimensional stellar velocity sample, Astron. J., 2003, vol.126, 2896–2909.

818. Korchagin V., Orlova N., Kikuchi N., Miyama S.M., Moiseev A., Application of the global modal approach to spiral galaxies, Astro-ph/0509708

819. Kormendy J., The velocity dispersion in the disk of the SO galaxy NGC 1553, Astrophys. J., 1984, vol.286, 116–131.

820. Kormendy J., The velocity dispersion in the disk of the SBO galaxy NGC 936, Astrophys. J., 1984, vol.286, 132–143.

821. Kormendy J., McClure R.D., The nucleus of M33, Astron. J., 1993, vol.105, 1793–1812.

822. Kormendy J., Richstone D., Evidence for a supermassive black hole in NGC 3115, Astrophys. J., 1992, vol.393, 559–578.

823. Kovalenko I.G., Levy V.V., Steady gas flow with a shock wave in a potential well, Astron. Astrophys., 1992, vol. 264, 406–414.

824. Kregel M., van der Kruit P.C., Radial truncations in stellar discs in galaxies, MNRAS, 2004, vol. 355, 143–146.

825. Kruit van der P.C., The thickness of the hydrogen layer and the three-dimensional pass distribution in NGC 891, Astron. Astrophys., 1981, vol. 99, 298–304.

826. Kruit van der P.C., The radial distribution of surface brightness in galactic disks, Astron. Astrophys., 1987, vol. 173, 59–80.

827. Kruit van der P.C., The three-dimensional distribution of light and mass in disks of spiral galaxies, Astron. Astrophys., 1988, vol. 192, 117.

828. Kruit van der P.C., Truncations in stellar disks/ Edited by J.G. Funes and E.M. Corsini, Galaxy Disks and Disk Galaxies. San Francisco: Astron. Soc. of the Pacific, 2001, vol. 230, 119–126.

829. Kruit van der P.C., de Grijs R., On the axis ratio of the stellar velocity ellipsoid in disks of spiral galaxies, Astron. Astrophys., 1999, vol. 352, 129–137.

830. Kruit van der P.C., Freeman K.C., Stellar kinematics and the stability of disks in spiral galaxies, Astrophys. J., 1986, vol. 303, 556–572.

831. Kruit van der P.C., Searle L., Surface photometry of edge-on spiral galaxies. I. A model of the three-dimensional distribution of light in galactic disks, Astron. Astrophys., 1981, vol. 95, 105–115.

832. Kruit van der P.C., Searle L., Surface photometry of edge-on spiral galaxies. II. The distribution of light and colour in the disk and spheroid of NGC 891, Astron. Astrophys., 1981, vol. 95, 116–126.

833. Kruit van der P.C., Searle L., Surface photometry of edge-on spiral galaxies. III. Properties of the three-dimensional distribution of light and mass in disks of spiral galaxies, Astron. Astrophys., 1982, vol. 110, 61–78.

834. Kruit van der P. C., Shostak G. S., Vertical motion and the thickness of HI disks: mplications for galactic mass models. Internal kinematics and dynamics of galaxics, Symp. IAU, 100, 1983, 69–76.

835. Kuijken K., Gilmore G., The Galactic disk surface mass density and the galactic force K_z at z=1,1 riloparsecs, Astrophys. J., 1991, vol. 367, 9–13L.

836. Kuhlen M., Diemand J., Madau P., The Shapes, Orientation, and Alignment of Galactic Dark Matter Subhalos, Astrophys. J., 2007, vol. 671, 1135.

837. Kulsrud R.M., Mark J.W.-K., Caruso A., The hose-pipe instability in stellar systems, Astroph. Space Sci., 1971, vol. 14, 52–55.

838. Kumar S.S., On gravitational instability, Publ. Astron. Soc. Japan, 1960, vol. 12, 552–555.

839. Kundt W., Robnik M., Dipole confined by a disk, Astron. Astrophys., 1980, vol. 91, 305–310.

840. Kunkel W.E., Demers S., Irwin M. J., Albert L., The dynamics of the Large Magellanic Cloud periphery: mass limit and polar ring, Astrophys. J., 1997, vol. 488, L129–132.

841. Kuznetsov O.A., Bisikalo D.V., Boyarchuk A.A., et al. Synthetic Doppler Tomograms of Gas Flows in the Binary System IP Peg, Astron. Rep., 2001, vol. 45, 872–881.

842. Lacey C.G., The influence of massive gas clouds on stellar velocity dispersions in galactic discs, MNRAS, 1984, vol. 208, 687–707.

843. Lacey C.G., Fall S.M., Chemical evolution of the galactic disk with radial gas flows, Astrophys. J., 1985, vol. 290, 154.

844. Lamb H. Hydrodynamics 6th / ed. Cambridge: The Univ. Press, 1932.

845. Lacy J.H., Achtermann J.M., Serabyn E., Galactic center gasdynamics – A one-armed spiral in a Keplerian disk, Astrophys. J., 1991, vol. 380L, 71–74.

846. Lasota J.-P., Abramowicz M.A., Chen X., et al. Is the accretion flow in NGC 4258 advection-dominated?, Astrophys. J., 1996, vol. 462, 142–146L.
847. Laurikainen E., Salo H., Buta R., Knapen J.H., Properties of Bars and Bulges in the Hubble Sequence, MNRAS, 2007, vol. 381, 401.
848. La Vigne M.A., Vogel S.N., Ostriker E.C., A Hubble Space Telescope Archival Survey of Feathers in Spiral Galaxies, ApJ, 2006, vol. 650, 818–834.
849. Law D.R., Majewski S.R., The Sagittarius Dwarf Galaxy: A Model for Evolution in a Triaxial Milky Way Halo, Astrophys. J., 2010, vol. 714, 229–254.
850. La Vigne M.A., Vogel S.N., Ostriker E.C., A Hubble Space Telescope Archival Survey of Feathers in Spiral Galaxies, Astrophys. J., 2006, vol. 650, 818–834.
851. Lee, Kim, Franco and Hong, JKOA, 2004, vol. 37, 249.
852. van Leer B., Towards thå ultimate conservative difference scheme. V. A second order sequel to Godunov's method, J. Comput. Phys., 1979, vol. 32, \No 1, 101–136.
853. Levine S.E., Sparke L.S., A Model for lopsided galactic disks, Astrophys. J., 1998, vol. 496, 13–16L.
854. Levine E.S., Blitz L., et al. The Warp and Spiral Arms of the Milky Way, Mapping the Galaxy and Nearby Galaxies. Springer New York, 2008, 85–90.
855. Levy V.V., Mustsevoj V.V., Sergienko V.A., The quasi-periodic regime in barred galaxies, PASP, 1994, vol. 66.
856. Levy L., Rose J.A., et al, Astron. J., 2007, vol. 133, 1104.
857. Lewis B.M., Face-on galaxies, Astrophys. J., 1984, vol. 285, 453–457.
858. Lewis G.F., Babul A., Katz N., et al. The effects of gasdynamics, cooling, star formation, and numerical resolution in simulations of cluster formation, Astrophys. J., 2000, vol. 536, 623–644.
859. Lewis J.R., Freeman K.C., Kinematics and chemical properties of the old disk of the Galaxy, Astron. J., 1989, vol. 97, 139–162
860. Li H., et al, Astrophys. J., 2000, vol. 533, 1023.
861. Li S., An HLLC Riemann solver for magneto-hydrodynamics, J. Comput. Phys., 2005, vol. 203, 344.
862. Liang E.P., Two–dimensional self–similar accretion flows. I. Generalized α-viscosity models, Astrophys. J., 1988, vol. 334, 339–348.
863. Lichtenberg A.J., Lieberman M.A., Regular and Stochastic Motion – Springer-Verlag, New York, Heidelberg, Berlin, 1983.
864. Lighthill J., Waves in fluids – Cambridge: Univ. press. 1978.
865. Lightman A.P., Time–dependent accretion disks around compact objects. I. Theory and basic equations, Astrophys. J., 1974, vol. 194, 419–427.
866. Lightman A.P., Eardley D.M., Black holes in binary systems: instability of disk accretion, Astrophys. J., 1974, vol. 187, 1–3L.
867. Lin D.N.C., Convective accretion disk model for the primordial solar nebula, Astrophys. J., 1981, vol. 246, 972–984.
868. Lin D.N.C., Papaloizou J., Tidal torques on accretion discs in binary systems with extreme mass ratios, MNRAS, 1979, vol. 186, 799.
869. Lin D.N.C., Papaloizou J., On the structure and evolution of the primordial solar nebula, MNRAS, 1980, vol. 191, 37–48.
870. Lin D.N.C., Papaloizou J.C.B., Savonije G.J., Propagation of tidal disturbance in gaseous accretion disks, Astrophys. J., 1990, vol. 365, 748.
871. Lin D.N., Pringle J.E., The formation of the exponential disk in spiral galaxies, Astrophys. J., 1987, vol. 320, 87–91L.
872. Lin C.C., Shu F.H., On the spiral structure of disk galaxies, Astrophys. J., 1964, vol. 140, 646–655.

873. Lin C.C., Shu F.H., On the spiral structure of disk galaxies. II, Proc. Nat. Acad. Sci. USA, 1966, vol. 55, 229–234.

874. Lin C.C., Yuan C., Shu F.H., On the spiral structure of disk galaxies. III. Comparison with observations, Astrophys. J., 1969, vol. 155, 721–746.

875. Lindblad B., Stockholms Observat. Ann., 1941, vol. 29, 155.

876. Lindbland B., On the nature of the spiral nebulae, MNRAS, 1927, vol. 87, 420–426.

877. Lindblad P.O., Jorsater S., Large scale galactic shocks in barred galaxies, Proc. 10th Europ. Reg. Astr. Meeting IAU. Praha, 1987, 289–298.

878. Lipunov V.M., A model of two-stream non-radial accretion for binary X-ray pulsars, Astroph. Space Sci., 1982, vol. 82, 343–361.

879. Lipunov V.M., Postnov K.A., Accretion spin-up of low magnetic neutron stars, Astroph. Space Sci., 1984, vol. 106, 103–115.

880. Little B., Carlberg R.G., The long-term evolution of barred galaxies, MNRAS, 1991, vol. 250, 161–170.

881. Livio M., Verbunt F., The response of the accretion disc radius to a temporary enhancement of mass transfer, MNRAS, 1988, vol. 232, 1–5.

882. Liu C.M., Najita J., Tokunaga A.T., A survey for circumstellar disks around young substellar objects, Astrophys. J., 2003, vol. 585, 372–391.

883. Longair M.S., S.L. Galaxy Formation. Springer, 2008.

884. Lopez-Corredoira M., Cabrera-Lavers A., Garzon F., Hammersley P.L., Old stellar Galactic disc in near-plane regions according to 2MASS: Scales, cut-off, flare and warp, Astron. Astrophys., 2002, vol. 394, 883–900.

885. de Lorenzo-Caceres A., Falcon-Barroso J., Vazdekis A., Martinez-Valpuesta I., Stellar Kinematics in Double-Barred Galaxies: The σ-Hollows, Astrophys. J., 2008, vol. 684, 83–86L.

886. Lubow S.H., Vertically driven resonances in accretion disks, Astrophys. J., 1981, vol. 245, 274.

887. Lubow S.H., Papaloizou J.C.B., Pringle J.E., On the Stability of Magnetic Wind-Driven Accretion Discs, MNRAS, 1994, vol. 268, 1010.

888. Lubow S.H., Pringle J.E., Kerswell R.R., Tidal Instability of Accretion Disks, Astrophys. J., 1993, vol. 419, 758.

889. Lubow S.H., Shu F.H., Gas dynamics of semidetached binaries, Astrophys. J., 1975, vol. 198, 383.

890. Lynden-Bell D., Can spherical clusters rotate?, MNRAS, 1960, vol. 120, 204.

891. Lynden-Bell D., Statistical mechanics of violent relaxation in stellar systems, MNRAS, 1967, vol. 136, 101–121.

892. Lynden-Bell D., On spiral generating, Galaxies and relativistic astrophisics /Eds. Barbanis B., Hadjidemetriou J.D. – Berlin: Springer, 1974, 114–119.

893. Lynden-Bell D., On a mechanism that structures galaxies, MNRAS, 1979, vol. 187, 101–107.

894. Lynden-Bell D., On why discs generate magnetic towers and collimate jets, MNRAS, 2003, vol. 341, 1360.

895. Lynden-Bell D., Kalnajs A.J., On the generating mechanism of spiral structure, MNRAS, 1972, vol. 157, 1–30.

896. Lynden-Bell D., Pringle J.E., The evolution of viscous disks and the origin of nebular variables, MNRAS, 1974, vol. 168, 603–637.

897. Lyubarskij Yu.E., Postnov K.A., Prokhorov M.E., Eccentric Accretion Discs, MNRAS, 1994, vol. 266, 583.

898. Machado R.E.G., Athanassoula E., Loss of halo triaxiality due to bar formation, MNRAS, 2010 (arXiv:1004.3874)

899. Makarov D.I., Karachentsev I.D., Burenkov A.N., Survey of rotation curves for northern spiral edge-on galaxies, 2000, Astro-ph/0006158.
900. Gaburov E., Harfst S., Zwart S., SAPPORO: A way to turn your graphics cards into a GRAPE-6, New Astr., 2009, vol. 14, 630–637.
901. Marcelin M., Athanassoula E., Kinematics and dynamics of the barred spiral galaxy NGC 1313, Astron. Astrophys., 1982, vol. 105, 76–84.
902. Marconi A., Hunt L.K., The relation between black hole mass, bulge mass, and near-infrared luminosity, Astrophys. J., 2003, vol. 589, 21–24L.
903. Margon B., Observations of SS 433, Ann. Rev. Astron. Astrophys., 1984, vol. 22, 507–536.
904. Margon B., SS 433 continues to perlex, Nature, 1987, vol. 328, 293–294.
905. Marigo P., Chemical yields from low- and intermediate-mass stars: Model predictions and basic observational constraints, Astron. Astrophys., 2001, vol. 370, 194.
906. Mark J.W.-K., Collective instabilities and waves for inhomogneous stellar systems. II., Astrophys. J., 1971, vol. 169, 455–475.
907. Mark J.W.-K., On density waves in galaxies. IV. Wave amplification through processes that remove angular momentum from galactic disks, Astrophys. J., 1976, vol. 206, 418–434.
908. Marochnik L.S., Mishurov Yu.N., Suchkov A.A., On the spiral structure of our Galaxy, ASS, 1972, Ò. 19. Ñ. 285–292.
909. Márquez I., Moles M., Effects of the interaction on the properties of spiral galaxies. I. The data, Astron. Astropliys. Suppl. Ser., 1996, vol. 120, 1–34.
910. Marsh T.R., Home K., Images of accretion discs. II - Doppler tomography, MNRAS, 1988, vol. 235, 269.
911. Martinez-Delgado D., Penarrubia J., Gabany R.J., et al. The Ghost of a Dwarf Galaxy: Fossils of the Hierarchical Formation of the Nearby Spiral Galaxy NGC 5907, ApJ, 2008, vol. 689, 184–193.
912. Mason K.O., Drew J.E., Knigge C., Pre-eclipse dips in the ultraviolet light curves of two eclipsing nova-like cataclysmic variables, MNRAS, 1997, vol. 290, 23L.
913. Masset F.S., Bureau M., On the Spiral Structure of NGC 2915 and Dark Matter, Astrophys. J., 2003, vol. 586, 152–169.
914. Mastichiadis A., Ozernoy L.M., X-ray and gamma-ray emission of Sagittarius A* as a wind-accreting black hole, Astrophys. J., 1994, vol. 426, 599–603.
915. Matsuda T., Inoue M., Sawada K., Spin-up and spin-down of an accreting compact object, MNRAS, 1987, vol. 226, 785–811.
916. Matsuda T., Inoue M., Sawada K., et al. A reinvestigation of gas response to an ovaly deformed gravitational potential, MNRAS, 1987, vol. 229, 295–314.
917. Matsuda T., Sekino N., Shima E., et al. Mass transfer by tidally induced spiral shocks in an accretion disk, Astron. Astrophys., 1990, vol. 235, 211–218.
918. Matthews L.D., The extraordinary superthin spiral galaxy UGC 7321. II. The vertical disk structure, Astron. J., 2000, vol. 120, 1764–1778.
919. Matthews L.D., Gallagher J.S., van Driel W., The extraordinary superthin spiral galaxy UGC 7321. I. Disk color gardients and global properties from multinavelength observations, Astron. J., 1999, vol. 118, 2751–2766.
920. Mayer L., Governato F., Colpi M., et al./The metamorphosis of tidally stirred dwarw galaxies, Astrophys. J., 2001, vol. 559, 754–784.
921. Meyer F., Meyer–Hofmeister E., On the elusive cause of cataclysmic variable outbursts, Astron. Astrophys., 1981, vol. 104, 10–12.
922. Meyer-Hofmeister E., Meyer F., The pre-outburst state of dwarf novae, Astron. Astrophys., 1988, vol. 194, 135–142.

923. Meyer-Hofmeister E., Ritter H., Accretion disks in close binaries Astrophys. and Space Sci. Library, 1993, vol. 177, 143.

924. McKee M.R., The radial-azimuthal stability of accretion disks around black holes, Astron. Astrophys., 1990, vol. 235, 521–525.

925. McIntyre M.E., On the "wave-momentum" myth, J. Fluid Mech., 1981, vol. 106, 331-347.

926. McMillan S.L.W., Aarseth S.J., An $O(N \log N)$ integration scheme for collisional stellar Systems, Astrophys. J., 1993, vol. 414, 200–212.

927. McWilliam A., Abundance Ratios and Galactic Chemical Evolution, ARA&A, 1997, vol. 35, 503.

928. Meatheringham S.J., Dopita M.A., Ford H.C., Webster B.L., The kinematics of the planetary nebulae in the Large Magellanic Cloud, Astrophys. J., 1988, vol. 327, 651–663.

929. Melia F., Falcke H., The supermassive black hole at the Galactic center, Ann. Rev. Astron. Astrophys., 2001, vol. 39, 309–352.

930. Melott A.L., Shandarin S.F., Splinter R.J., Suto Y., Demonstrating discreteness and collision error in cosmological N-body simulations of dark matter gravitational clustering, Astrophys. J., 1997, vol. 479, 79–83L.

931. Mendez-Abreu J., Aguerri J.A.L., Corsini E.M., Simonneau E., Structural properties of disk galaxies I., Astron. Astrophys., 2008, vol. 487, 555.

932. Merrifield M.R., Kuijken K., The pattern speed of the bar in NGC 936, MNRAS, 1995, vol. 274, 933–938.

933. Merrifield M.R., Halo tracing with atomic hydrogen, The shapes of galaxies and their dark halos. Ed. P. Natarajan, Singapore, 2002, 170.

934. Merritt D., Ferrarese L., The M - σ relation for supermassive black holes, Astrophys. J., 2001, vol. 547, 140–145.

935. Merritt D., Sellwood J.A., Bending instabilities in stellar systems, Astrophys. J., 1994, vol. 425, 551–567.

936. Mestel L., On galactic low rotation, MNRAS, 1963, vol. 126, 553–575.

937. Meyer F., Meyer-Hofmeister E., On the elusive cause of cataclysmic variable outbursts, Astron. Astrophys., 1981, vol. 104, 10–12L.

938. Michel F.C., Hydraulic jumps in "viscous" accretion disks, Astrophys. J., 1984, vol. 279, 807–813.

939. Mihos J.C., Hernquist L., Star-forming galaxy models: blending star formation into TREESPH, Astrophys. J., 1994, vol. 437, 611–624.

940. Mikhailova E.A., Khoperskov A.V., Sharpak S.S., The Z-structure of Disk Galaxies. Numerical Experiments . Stellar dynamics: from classic to modern / Edited by Ossipkov L.P., Nikiforov I.I. – Saint Petersburg, 2001, 147–153.

941. Mikhailovskii A.B., Theory of Plasma Instabilities Vol. 2 --New York: Consultants Bureau, 1974.

942. Miles J.W., On the reflection of sound at an interface of relative motion, J. Acoustic Soc. Amer., 1957, vol. 29, 226–228.

943. Miller R.H., Numerical experiments in collisionless systems, Astroph. Space Sci., 1971, vol. 14, 73–90.

944. Miller R.H., On the stability of a disk galaxy, Astrophys. J., 1974, vol. 190, 539–542.

945. Miller R.H., Validity of disc galaxy simulations, J. Coniput. Phys., 1976, vol. 21, 400–437.

946. Miller R.H., Numerical experiments on the stability of disklike galaxies, Astrophys. J., 1978, vol. 223, 811–823.

947. Miller R.H., On the stability of disklike galaxies in massive haloes, Astrophys. J., 1978, vol. 224, 32–38.
948. Miller R.H., Recent investigations on disk galaxies in massive halos, Celest. Mech., 1985, vol. 37, 307–324.
949. Miller R.H., Prendergast K.H., Quirk W.J., Numerical experiments on spiral structure, Astrophys. J., 1970, vol. 161, 903–916.
950. Miller R.H., Smith B.F., Dynamics of stellar bar, Astrophys. J., 1979, vol. 227, 785–797.
951. Mineshige S., Osaki Y., Disk-instability model for outbursts of dwarf novae: time-dependent formation and one-zone model, Publ. Astron. Soc. Japan, 1983, vol. 35, 377–396.
952. Mineshige S., Shields G.A., Accretion disk thermal instability in galactic nuclei, Astrophys. J., 1990, vol. 351, 47–61.
953. Mineshige S., Wheeler J.C., (Eds) Disk Instabilities in Close Binary Systems: 25 Years of the Disk-Instability Model / Frontiers Sci. Ser. 79. No. 26 – Tokyo: Universal Academy Press, 1999.
954. Minniti D., Velocities of supergiants in the bulge of M 33, Astron. Astrophys., 1996, vol. 306, 715–720.
955. Minniti D., White S.D.M., Olszewski E.W., Hill J.M., Rotation of the galactic bulge, Astrophys. J., 1992, vol. 393, 47–50L.
956. Mishurov Yu. N. Zenina I.A., Dambis A.K., et al. Is the Sun located near the corotation circle?, Astron. Astrophys., 1997, vol. 323, 775.
957. Mishurov Yu.N., Lepine J.R.D., Acharova I.A., Corotation: Its Influence on the Chemical Abundance Pattern of the Galaxy, Astrophys. J. Lett., 2002, vol. 571, 113–115.
958. Miyamoto M., Nagai R., Three-dimensional models for the distribution of mass in galaxies, Publ. Astron. Soc. Japan, 1975, vol. 27, 533–543.
959. Moiseev A.V., Velocity dispersion of stars and gas motion in double-barred galaxies, Bull. SAO., 2001, vol. 51, 140–150.
960. Moiseev A.V., 2D spectroscopy of double-barred galaxies. “Galaxies: the third dimension” / Eds: Rossado et. al., ASP Conf. Ser., 2002, vol. 264, 232–239.
961. Moiseev A.V., 3D spectroscopy of candidate double-barred lenticular galaxies, arXiv:0811.3527.
962. Moiseev A., Bizyaev D., 3D spectroscopic study of galactic rings: Formation and kinematics, New Astr. Rev., 2009, vol. 53, 169–174.
963. Mollá M., Ferrini F., DíA.I. az., Evolution of Spiral Galaxies. VII. Time Evolution of the Radial Distributions of Abundances, Astrophys. J., 1997, vol. 475, 519.
964. Molteni D., Acharya K., Kuznetsov O., et al./Bending instability of an accretion disc around a black hole, Astrophys. J., 2001, vol. 563, 57–60L.
965. Molteni D., Belvedere G., Lanzafame G., Three-dimensional simulation of polytropic accretion discs, MNRAS, 1991, vol. 249, 748.
966. Molteni D., Kuznetsov O.A., Bisikalo D.V., et al. On the angular momentum transfer on to compact stars in binary systems, MNRAS, 2001, vol. 327, 1103–1110.
967. Monet D.G., Richstone D.O., Schechter P.L., The effect of massive disks on bulge isophotes, Astrophys. J., 1981, vol. 245, 454–458.
968. Monin A.S., Yaglom A.M., Statistical Fluid Mechanics vol. 1, 2 – Cambridge, MA: MIT Press, 1971, 1975.
969. Morales-Rueda L., Marsh T.R., Billington I., Spiral structure in IP Pegasi: how persistent is it?, MNRAS, 2000, vol. 313, 454.
970. Morrison H.L., Flynn C., Freeman K.C., Where does the disk stop and the halo begin?, Astron. J., 1990, vol. 100, 1191–1222.

971. Morrison H.L., Helmi A., Sun J., et al. Fashionably Late? Building Up The Milky Way'S Inner Halo, Astrophys. J., 2009, vol. 694, 130.

972. Morrison H.L., Miller E.D., Harding P., et al./Stellar populations in edge-on galaxies drom deep CCD surface photometry. II. One-dimensional fits of NGC 891, Astron. J., 1997, vol. 113, 2061–2074.

973. Morris J.P., Monaghan J.J., A switch to reduce SPH viscosity, J. Comp. Phys., 1997, vol. 136, 41–50.

974. Moss D., Shukurov A., Accretion disc dynamos opened up by external magnetic fields, Astron. Astrophys., 2004, vol. 413, 403–414.

975. Mundt R. Jets from young stars, Mitt. Astr. Ges., 1987, vol. 70, 100–115.

976. Mundt R., Brugel E.W., Buhrke T., Jets from young stars: CCD imaging, long-slit spectroscopy, and interpretation of existing data, Astrophys. J., 1987, vol. 319, 275–303.

977. Murakami T., Progress in gamma–ray burst astronomy with Ginga, Adv. Space Res., 1990, vol. 10, 63–67.

978. Muraoka K., Kohno K., Tosaki T., et al. ASTE CO (3-2) Mapping Toward the Whole Optical Disk of M 83: Properties of Inter-arm Giant Molecular-Cloud Associations, Astrophys. J., 2009, vol. 706, 1213–1225.

979. Nagasawa M., Matsuda T., Kuwahara K., Smoothed Particle Rendering for Fluid Visualization "Three–Dimensional Accretion Disk and Jet Formation", Numer. Astrophys. Jpn., 1991, vol. 2, 27.

980. Nakai N., Radial distribution of CO in barred spirals and the bar structure of our Galaxy, Publ. Astron. Soc. Japan, 1992, vol. 44, 27–30L.

981. Nakanishi H., Kuno N., Sofue Y., et al, Astrophys. J., 2006, vol. 651, 804.

982. Narayan R., Why Do AGN lighthouses switch off? lighthouses of the Universe: The most luminous celestial objects and their use for cosmology. Proc. of the MPA/ESO/. 2002, 405 (astrp-ph/0201260).

983. Narayan R., Goldreich P., Goodman J., Physics of model in a differentially rotating system–analysis of the shearing sheet, MNRAS, 1987, vol. 228, 1–41.

984. Narayan R., Mahadevan R., Quataert E., Advection-dominated accretion around black holess. Theory of black hole accretion discs / Ed. Abramowicz M.A., Bjornsson G., Pringle J.E. – Cambridge: Cambridge Univ. Press, 1998, 148–157.

985. Narayan R., McClintock J.E., Yi I., A new model for black hole soft X-Ray transients in quiescence, Astrophys. J., 1996, vol. 457, 821–833.

986. Narayan R., Igumenshchev I.V., Abramowicz M.A., Self-similar accretion flows with convection, Astrophys. J., 2000, vol. 539, 798–808.

987. Narayan R., Quataert E., Igumenshchev I.V., Abramowicz M.A., The magnetohydrodynamics of convection-dominated accretion flows, Astrophys. J., 2002, vol. 577, 295–301.

988. Narayan R., Yi I., Advection-dominated accretion: a self-similar solution, Astrophys. J., 1994, vol. 428, 13–16L.

989. Narayan R., Yi I., Advection-dominated accretion: Self-similarity and bipolar outflows, Astrophys. J., 1995, vol. 444, 231–243.

990. Nather R.E., High-speed photometry, Vistat Astron., 1973, vol. 15, 91–112.

991. Navarro J.F., Frenk C.S., White S., A Universal Density Profile from Hierarchical Clustering, Astrophys. J., 1997, vol. 490, 493–508.

992. Nees C.L., Yoss K.M., Kinematic and abundance gradients in the galactic disk, Astron. J., 1988, vol. 95, 463–495.

993. E. Neistein, Maoz D., Rix H.-W., Tonry J.L., A Tully-Fisher relation for S0 galaxies, Astron. J., 1999, vol. 117, 2666–2675.

994. Newberg H.J., Yanny B., The Milky Way's stellar halo – lumpy or triaxial?, J. Physics: Conf. Ser., 2006, vol. 47, 195–204.

995. Newberg H.J., Yanny B., Cole N., The Overdensity in Virgo, Sagittarius Debris, and the Asymmetric Spheroid, Astrophys. J., 2007, vol. 668, 221–235.

996. Newberg H.J., Yanny B., Willett B.A., Discovery of a New, Polar-Orbiting Debris Stream in the Milky Way Stellar Halo, ApJL, 2009, vol. 700, 61–64.

997. Nezlin M.V. Polyachenko V.L., Snezhkin E.N., et al. nterarm vortices predicted by laboratory simulation of spiral structure, Sov. Phys. JETP, 1987, vol. 65, 1.

998. Nezlin M.V., Snezhkin E.N., Rossby Vortices, Spiral Structures, Solitons: Astrophysics and Plasma Physics in Shallow Water Experiments – Berlin: Springer-Verlag, 1993.

999. Nikiforov I.I., Small Galaxy Groups: IAU Colloquium 174 / Edited by Mauri J. Valtonen and Chris Flynn. 2000, 403.

1000. Nikiforov I.I., Stellar dynamics: from classic to modern/Ed. Ossipkov L.P., Nikiforov I.I. – Saint Petersburg, 2001, 28–31.

1001. Nityananda R., Narayan R., The relevance of the Eddington limit to thick accretion disks, Adv. Space Res., 1984, vol. 3, 29–34.

1002. Nomoto K., Tominaga N., Umeda H., et al. Nucleosynthesis yields of core-collapse supernovae and hypernovae, and galactic chemical evolution, Nuclear Physics A., 2006, vol. 777, 424.

1003. Noordermeer E., The distribution of gas, stars and dark matter in early-type disk galaxies / Rijkuniversitet Groningen, Thesis, 2004.

1004. Noordermeer E., Sparke L.S., Levine S.E., The kinematics of lopsided galaxies, MNRAS, 2001, vol. 328, 1064–1080.

1005. Noordermeer E., van der Hulst J.M., Sancisi R., Swaters R.S., van Albada T.S., The mass distribution in early-type disc galaxies: declining rotation curves and correlations with optical properties, MNRAS, 2007, vol. 376, 1513–1546.

1006. Noordermeer E., van der Hulst J.M., Sancisi R., et.al. The Westerbork HI survey of spiral and irregular galaxies. III., Astron. Astrophys., 2005, vol. 442, 137.

1007. Nordström B., Mayor M., Andersen J., et al. The Geneva-Copenhagen survey of the Solar neighbourhood. Ages, metallicities, and kinematic properties of 14 000 F and G dwarfs, Astron. Astrophys., 2004, vol. 418, 989.

1008. Norman C.A., Ikeuchi S., The disk-halo interaction: superbubbles and the structure of the interstellar medium, Astrophys. J., 1989, vol. 345, 372–383.

1009. Norman M.L., Hardee P.E., Spatial stability of the slab jet. II. Numerical simulations, Astrophys. J., 1988, vol. 334, 80–94.

1010. Northop T., Helmholtz instability of a plasma, Phys. Rev., 1956, vol. 103, 1150–1155.

1011. O'Leary R.M., Loeb A., Production of Hypervelocity Stars through Encounters with Stellar-Mass Black Holes in the Galactic, MNRAS, vol. 383, 86.

1012. Oblak E., Mayor M., The galactic radial gradient of velocity dispercion / 10th Eorop. Reg. Astron. Meet. – IAU. Praha, 1987, 263–265.

1013. O'Brien J.C., Freeman K.C., van der Kruit P.C., The dark matter halo shape of edge-on disk galaxies. IV. UGC 7321, Astron. Astrophys., 2010, vol. 515, 63

1014. Obukhov A.M., C.R., Acad. Sci. URSS, 1941, vol. 32, 22.

1015. Oda M., Cyg X-1 – a candidate of the black holes, Space Sci. Rev., 1977, vol. 20, 757.

1016. Odewahn S.C., Properties of the Magellanic-type spirals. I - Surface photometry of NGC 4618 and NGC 4625, Astron. J., 1991, vol. 101, 829–844.

1017. O'Donoghue D., The radius of the accretion disc in Z Cha between outbursts, MNRAS, 1986, vol. 220, 23–26L.

1018. O'Donoghue D., Chen A., Marang F., et al. WX Cet and the WZ Sge stars, MNRAS, 1991, vol. 250, 363–372.

1019. Ogilvie G.I., Non-linear fluid dynamics of eccentric discs, MNRAS, 2001, vol. 325, 231.

1020. Okumura S.K., Ebisuzaki T., Kinematic structures of merger remnants, Publ. Astron. Soc. Japan, 1991, vol. 43, 781–793.

1021. Olling R.P., NGC 4244: A Low Mass Galaxy With a Falling Rotation Curve and a Flaring Gas Layer, Astron. J., 1996, vol. 112, 457–480.

1022. Olling R.P., Merrifield M.R., Two measures of the shape of the dark halo of the Milky Way, MNRAS, 2000, vol. 311, 361–369.

1023. O'Neil K., Bothun G., The Space Density of Galaxies through $\mu B(0)$=25.0 Magnitudes per Inverse Arcsecond Squared, Astrophys. J., 2000, vol. 529, 811.

1024. Orlova N., Korchagin V., Theis Ch., Global spiral modes in multi-component disks, Astron. Astrophys., 2002, vol. 384, 872–878.

1025. Ortega V.G., Volkov E., Monte-Lima L., Axisymmetric instabilities in gravitating discs with mass spectrum, Astron. Astrophys., 2001, vol. 366, 276–280.

1026. Osher S., Riemann solvers, the entropy condition, and difference approximations, J. Numer. Anal., 1984, vol. 21, 217–235.

1027. Ostriker J.P., Peebles P.J.E., A numerical study of the stability of flattened galaxies: or can cold galaxies suvive?, Astrophys. J., 1973, vol. 186, 467–480.

1028. Ouyed R., Clarke D.A., Pudritz R.E., Three-dimensional simulations of jets from Keplerian disks: self-regulatory stability, Astrophys. J., 2003, vol. 582, 292–319.

1029. Ozernoy L.M., Genzel R., Is Sgr A^* Underfed, Underefficient, or Underdone? / Proc. 169th Symp. Intern. Astr. Union, 1996, 181–187.

1030. Paczynski B., Close binaries, Coments on Astrophysics. 1976, vol. 6, 95–98.

1031. Paczynski B., A model of accretion disks in close binaries, Astrophys. J., 1977, vol. 216, 822.

1032. Paczyński B., Biita P.J., Thick accretion disks and supercritical luminosities, Astron. Astrophys., 1980, vol. 88, 23–31.

1033. Paczyński B., Bisnovatyi-Kogan G., A model of a thin accretion disk around a black hole, Acta Astron., 1981, vol. 31, 283–291.

1034. Paczynski B., Sienkiewicz R, Acta Astron., 1972, vol. 22, 73.

1035. Palla F., Zinnecker H., Maeder A., Meynet G., Physics of star formation in galaxies. Springer, 2002.

1036. Palouš J., Tenorio-Tagle G., Franco J., Star formation in differentially rotating galactic disks: The physics of self-propagation, MNRAS, 1994, vol. 270, 75.

1037. Papaloizou J.C.B., Bath G.T., Stellar stability in close binary systems, MNRAS, 1975, vol. 172, 339–357.

1038. Papaloizou J.C.B., Lin D.N.C., Theory Of Accretion Disks I: Angular Momentum Transport Processes, Annu. Rev. Astron. Astrophys., 1995, vol. 33, 505.

1039. Papaloizou J., Pringle J.E., Tidal torques on accretion discs in close binary systems, MNRAS, 1977, vol. 181, 441–454.

1040. Papaloizou J.C.B., Pringle J.E., The dynamical stability of differentially rotating discs with constant specific angular momentum, MNRAS, 1984, vol. 208, 721–750.

1041. Papaloizou J.C.B., Pringle J.E., The dynamical stability of differentially rotating discs. II, MNRAS, 1985, vol. 213, 799–820.

1042. Papaloizou J.C.B., Pringle J.E., The dynamical stability of differentially rotating discs. III, MNRAS, 1987, vol. 225, 267–283.

1043. Papaloizou J.C., Savonije G.J., Instabilities in self-gravitating gaseous discs, MNRAS, 1991, vol. 248, 353–369.

1044. Pariev V.I., Blackman E.G., Boldyrev S.A., Extending the Shakura-Sunyaev approach to a strongly magnetized accretion disc model, Astron. Astrophys., 2003, vol. 407, 403–421.

1045. Parmentier G., Jehin E., Magain P., et al. The Old Halo metallicity gradient: the trace of a self-enrichment process, Astron. Astrophys., 2000, vol. 363, 526.

1046. Pasha I.I., Density-wave spiral theories in the 1960s. I. II., Astro-ph/0406142, astro-ph/0406143.

1047. Pasha I.I., Polyachenko V.L., The Lynden-Bell slow bar as a determinant of the spiral-ring structure in barred galaxies, MNRAS, 1994, vol. 266, 92–96.

1048. Patsis P.A., Grosbol P., Hiotelis N., Interarm features in gaseous models of spiral galaxies, Astron. Astrophys., 1997, vol. 323, 762–774.

1049. Patsis P.A., Athanassoula E., Grosbol P., Skokos Ch., Edge-on boxy profiles in non-barred disc galaxies, MNRAS, 2002, vol. 335, 1049–1053.

1050. Patsis P.A., Kaufmann D.E., Gottesman S.T., Boonyasait V., Stellar and gas dynamics of late-type barred-spiral galaxies NGC 3359, MNRAS, 2009, vol. 394, 142–156.

1051. Payne D.G., Gohn H., The stability of confined radio jets: the role of reflection modes, Astrophys. J., 1985, vol. 291, 655–667.

1052. Pedlosky J., Geophysical Fluid Dynamics – N.Y: Springer-Verlag, 1986.

1053. Peimbert M., Rayo J.F., Torres-Peimbert S., Abundance gradients in the Galaxy derived from H II regions, Astrophys. J., 1978, vol. 220, 516.

1054. Peletier R.F., Falcon-Barroso J., Bacon R., et al. The SAURON project – XI. Stellar populations from absorption-line strength maps of 24 early-type spirals, MNRAS, 2007, vol. 379, 445–468.

1055. Peletier R.F., Knapen J.H., Shlosman I., et al./A subarcsecond-resolution near-infrared study of Seyfert and "Normal" galaxies. I. Imaging Data, Astrophys. J. Suppl. Ser., 1999, vol. 125, 363–407.

1056. Pence W.D., Blackman C.P., Gas dynamics in barred spiral galaxies. II. NGC 7496 and 289, MNRAS, 1984, vol. 210, 547–563.

1057. Penninx W., Lewin W.H.G., Mitsuda K., et al./ Quasiperiodic oscillations in GX 17+2, MNRAS, 1990, vol. 243, 114–125.

1058. Persic M., Salucci P., Rotation Curves of 967 Spiral Galaxies, ApJS, 1995, vol. 99, 501–541.

1059. Petrosian A.R., Movsessian T., Comte G., et al. Multi-wavelength study of a new sample of blue compact dwarf galaxies. IV., Astron. Astrophys., 2002, vol. 391, 487–508.

1060. Pichon C., Lynden-Bell D., Equilibria of flat and round galactic disks, MNRAS, 1996, vol. 282, 1143–1158.

1061. Pinkney J., Gebhardt K., Bender R., et al. Kinematics of 10 early-type galaxies from Hubble Space Telescope and Ground-based spectroscopy, Astrophys. J., 2003, vol. 596, 903–929.

1062. Pisano D.J., Wilcots E.M., Elmegreen B.G., The HI distribution and dynamics in two late-type barred spiral galaxies: NGC 925 and NGC 1744, Astron. J., 1998, vol. 115, 975–999.

1063. Pisano D.J., Wilcots E.M., Elmegreen B.G., Structure and star formation in NGC 925, Astron. J., 2000, vol. 120, 763–776.

1064. Pfenniger D., Combes F., Martinet L., Is dark matter in spiral galaxies cold gas? I. Observational constraints and dynamical clues about galaxy evolution, Astrophys. J., 1994, vol. 285, 79–93.

1065. Pohlen M., Balcells M., Lütticke R., Dettmar R.-J., Evidence for a large stellar bar in the low surface brightness galaxy UGC 7321, Astron. Astrophys., 2003, vol. 409, 485–490.

1066. Pohlen M., Trujillo I., The structure of galactic disks. Studying late-type spiral galaxies using SDSS, Astron. Astrophys., 2006, vol. 454, 759–772.

1067. Polychenko V.L., Galactic bars and associated structures. Physics of the gaseous and stellar disks of the Galaxy/ASP Conference Series / Ed. King I.R. 1994, vol. 66, 103–115.

1068. Polyachenko V.L., Rastorguev A.S., Fridman A.M., Eichhorn H.K.,/On the possibility of reconstructing the phase density for star clusters, Astron. Lett. 1996, vol. 22, 459–465.

1069. Popescu C.C., Misiriotis A., Kylafis N.D., et al./Modelling the spectral energy distribution of galaxies I. Radiation fields and grain heating in the edge-on spiral NGC 891, Astron. Astrophys., 2000, vol. 362, 138–150.

1070. Popham R., Narayan R., Does accretion cease when a star approaches breakup?, Astrophys. J., 1991, vol. 370, 604.

1071. Portinari L., Chiosi C., On star formation and chemical evolution in the Galactic disc, Astron. Astrophys., 1999, vol. 350, 827.

1072. Portinari L., Chiosi C., On radial gas flows, the Galactic Bar and chemical evolution in the Galactic Disc, Astron. Astrophys., 2000, vol. 355, 929.

1073. Pounds K.A., Stanger V.J., Turner T.J., et al. Discovery of a strong soft X-ray excess in Mkn 335 – evidence for an accretion disk?, MNRAS, 1987, vol. 224, 443–452.

1074. Prantzos N., Boissier S., Chemo-spectrophotometric evolution of spiral galaxies - III. Abundance and colour gradients in discs, MNRAS, 2000, vol. 313, 338.

1075. Pravdo S.H., Boldt E.A., Holt S.S., Serlemitsos P.J.,/ X-ray spectra of Hercules X-1. II. The pulse, Astrophys. J. Lett., 1977, vol. 216, 23–26.

1076. Press W.H., Teukolsky S.A., Vetterling W.T., Flannery B.P., Numerical recipes in FORTRAN – Cambridge: Cambridge Univ. Press, 1992.

1077. Priedhorsky W.C., Verbunt F., Tidal forces and mass transfer instabilities in low–mass X-ray binaries, Astrophys. J., 1988, vol. 333, 895–905.

1078. Pringle J.E., Accretion discs in astrophysics, Annu. Rev. Astron. Astrophys., 1981, vol. 19, 137.

1079. Pringle J.E., in Interacting Binary Stars Ch. 1, Cambridge Astrophys. Ser., Vol. 6. / Eds J E Pringle, R A Wade – Cambridge: Cambridge Univ. Press, 1985.

1080. Pringle J.E., Rees M.J., Accretion disc models for compact X-ray sources, Astron. Astrophys., 1972, vol. 21, 1–9.

1081. Quillen A.C., Garnett D.R., The saturation of disk heating in the solar neighborhood and evidence for a merger 9 Gyrs ago, ASP Conference Series, 2001, vol. 230, 87–88.

1082. Quirk W.J., Numerical experiments in spiral structure II., Astrophys. J., 1971, vol. 167, 7–19.

1083. Quirk W.J., On the gas content of galaxies, Astrophys. J., 1972, vol. 176, 9–14L.

1084. Quiroga R.J., The spiral structure in the inner parts of the Galaxy, Astrophys. J. Suppl. Ser., 1977, vol. 50, 281–300.

1085. Rafikov R.R., The local axisymmetric instability criterion in a thin, rotating, multicomponent disc, MNRAS, 2001, vol. 323, 445–452.

1086. Raha N., Sellwood J.A., James R.A., Kahn F.D., A dynamical instability of bars in disk galaxies, Nature, 1991, vol. 352, 411–412.

1087. Rayleigh Lord, Proc. Math. Soc. London Ser. A., 1916, vol. 93, 148.

1088. Raymond J.C., Cox D.P., Smith B.W., Radiative cooling of a low-density plasma, Astrophys. J., 1976, vol. 204, 290.

1089. Rautiainen P., Salo H., Laurikainen E., Model-based pattern speed estimates for 38 barred galaxies., MNRAS, 2008, vol. 388, 1803–1818.

1090. Re Fiorentin P., Helmi A., Lattanzi M.G., Spagna A., Structure in the motions of the fastest halo stars, Astron. Astrophys., 2005, vol. 439, 551–558.
1091. Rees M.J., Accretion and quasar phenomena, Physica Scripta., 1978/ vol. 17, 193.
1092. Reid N.I., Majewski S.R., Star counts redivivus. I. A new look at the Galaxy at faint magnitudes, Astrophys. J., 1993, vol. 409, 635–662.
1093. Reid M.J., Menten K.M., Zheng X.W., et al. Trigonometric Parallaxes of Massive Star-Forming Regions. VI, Astrophys. J., 2009, vol. 700, 137–148.
1094. Renzini A., Voli M., Advanced evolutionary stages of intermediate-mass stars. I - Evolution of surface compositions, Astron. Astrophys., 1981, vol. 94, 175.
1095. Reshetnikov V., Battaner E., Combes F., Jiménez-Vicente J., Statistics of galaxy warps in the HDF North and South, Astron. Astrophys., 2002, vol. 382, 513–521.
1096. Reshetnikov V.P., Combes F., Tidally-triggered disk thickening. II. Results and interpretations, Astron. Astrophys., 1997, vol. 324, 80.
1097. Reshetnikov V., Combes F., Statistics of optical WARPS in spiral disks, Astron. Astrophys., 1998, vol. 337, 9.
1098. Reshetnikov V., Combes F., Spiral galaxies with large optical warps, Astron. Astropliys. Suppl. Ser., 1999, vol. 138, 101–107.
1099. Reshetnikov V.P., Sotnikova N.Ya., The polar stellar ring and dark halo of NGC 5907, Astron. Lett., 2000, vol. 26, 277–284.
1100. Richer H.B., Harris W.E., Fahlman G.G., Bell R.A., et al. Age Gradient and the Second Parameter Problem in the Galactic Halo, Astrophys. J., 1996, vol. 463, 602.
1101. Ribner H.S., Reflection, transmission and amplification of sound by a moving medium, J. Acoustic. Soc. Amer., 1957, vol. 29, 435–441.
1102. Rich R.M., Kinematics and abundances of K giants in the nuclear bulge of the Galaxy, Astrophys. J., 1990, vol. 362, 604–619.
1103. Rix H.-W., Zaritsky D., Nonaxisymmetric Structures in the Stellar Disks of Galaxies, Astrophys. J., 1995, vol. 447, 82–102.
1104. Roberts W.W., Large-scale shock formation in spiral galaxies and its implications on star formation, Astrophys. J., 1969, vol. 158, 123–143.
1105. Roberts W.W.Jr., Roberts M.S., Shu F.H., Density wave theory and the classification of spiral galaxies, Astrophys. J., 1975, vol. 196, 381–405.
1106. Roberts W.W., Yuan C., Application of the density-wave theory to the spiral structure of the Milky Way system. III. Magnetic field: large–scale hydromagnetic shock formation, Astrophys. J., 1970, vol. 161, 877–902.
1107. Roberts D.A., Goss W.M., Multiconfiguration VLA H92-alpha observations of Sagittarius A West at 1 arcsecond resolution, Astrophys. J. Suppl. Ser., 1993, vol. 86, 133–152.
1108. Roberts W., Hausman M., Spiral structure and star formation. I - Formation mechanisms and mean free paths, ApJ, 1984, vol. 277, 744–767.
1109. Roberts W., Stewart G., The role of orbital dynamics and cloud-cloud collisions in the formation of giant molecular clouds in global spiral structures, ApJ, 1987, vol. 314, 10–32.
1110. Robin A.C., Creze M., Mohau V., The radial structure of the galactic disc, Astron. Astrophys., 1992, vol. 265, 32–40.
1111. Robinson E.L., The structure of cataclysmic variables, Ann. Rev. Astron. Astrophys., 1976, vol. 14, 119–14.
1112. Robinson B.J., van Damme K.J., 21 cm observations of NGC 55, Aust. J. Phys., 1966, vol. 19, 111–127.
1113. Roe P.L., The use of the Riemann problem in finite difference schemes, Lect. Notes Phys., 1981, vol. 141, 354–359.

1114. Roe P.L., Approximate Riemann problem solvers, parameter vectors, and difference schemes, J. Comput. Phys., 1981, vol. 43, 357–372.

1115. Rohlfs K., Lectures on density wave theory – N.Y.: Springer–Verlag, 1977.

1116. Rohlfs K., Chini R., Wink J.E., Bohme R., The rotation curve of the Galaxy, Astron. Astrophys., 1986, vol. 158, 181–190.

1117. Rohlfs K., Kreitsehmann J., A realistic model of the Galaxy, Astrophys. J. Suppl. Ser., 1981, vol. 79, 289–319.

1118. Romeo A.B., How faithful are N-body simulations of disc galaxies?, Astron. Astrophys., 1994, vol. 286, 799–806.

1119. Romeo A.B., Modelling gravity in N-body simulations of disc galaxies. Optimal types of softening for given dynamical requirements, Astron. Astrophys., 1998, vol. 335, 922–928.

1120. Romeo A.B., Horellou C., Bergh J., N-body simulations with two-orders-of-magnitude higher performance using wavelets, MNRAS, 2003, vol. 342, 337–344.

1121. Rothberg B., Joseph R.D., A Deep K-Band Photometric Survey of Merger Remnants, Astron. J., 2004, vol. 128, 2098.

1122. Rots A.H., Distribution and kinematics ofneutral hydrogen in the spiral galaxy M 81. II. Analysis, Astron. Astrophys., 1975, vol. 45, 43–55.

1123. Rosenbluth M.N., Post R.F., Phys. Fluids, 1965, vol. 8, 547.

1124. Roskar R., Debattista V.P., et al. Misaligned Angular Momentum in Hydrodynamic Cosmological Simulations: Warps, Outer Discs, and Thick Discs, MNRAS, 2010 (1006.1659)

1125. Rosswog S., Astrophysical Smooth Particle Hydrodynamics, New Astr. ReV., 2009, vol. 53, 78–104.

1126. Rubin V.C., Burstein D., Ford W.K., Thonnard N., Rotation velocities of 16 Sa galaxies and a comparison of Sa, Sb and Sc rotation properties, Astrophys. J., 1985, vol. 289, 81–104.

1127. Rubin V.C., Ford W.K., Rotation of the Andromeda nebula from a spectroscopic survey of emission regions, Astrophys. J., 1970, vol. 159, 379–403.

1128. Rubin V.C., Ford W.K., Kumar C.K., Stellar motion near the nuclens of M 31, Astrophys. J., 1973, vol. 181, 61.

1129. Rubin V.C., Ford W.K., Thonnard N., Rotational properties of 21 Sc galaxies with a large range of luminosities and radii, from NGC 4605 (R=4 kps) to UGC 2885 (R=122 kps), Astrophys. J., 1980, vol. 238, 71–487.

1130. Rubin V.C., Ford W.K., Thonnard N., Burstein D., Rotational properties of 23 Sb galaxies, Astrophys. J., 1982, vol. 261, 439–456.

1131. Rudnick G., Rix H.-W., Lopsidedness in early type disk galaxies, Astron. J., 1998, vol. 116, 1163–1168.

1132. Rudnick G., Rix H.-W., Kennicutt R.C.Jr., Lopsided galaxies, weak interactions, and boosting the star formation rate, Astrophys. J., 2000, vol. 538, 569–580.

1133. Rudolph A.L., Fich M., Bell G.R., et al. Abundance Gradients in the Galaxy, ApJS, 2006, vol. 162, 346.

1134. Ryder S.D., Zasov A.V., Sil'chenko O.K., et al. The peculiar rotation curve of NGC 157, MNRAS, 1998, vol. 293, 411–428.

1135. Rylov A.Yu., Snezhkin E.N., Titishov K.B., Velocity Field in the Spiral-Wave Pattern Observed in Rotating Shallow-Water Experiments, Astron. Rep., 2004, vol. 48, 275.

1136. Sackett P.D., Rix H.-W., Jarvis B.J., Freeman K.C., The flattened dark halo of polar ring galaxy NGC 4650A: A conspiracy of shapes? Astrophys. J., 1994, vol. 436, 629–641.

1137. Sackett P.D., Morrisoni H.L., Harding P., Boroson T.A., A faint luminous halo that may trace the dark matter around spiral galaxy NGC 5907, Nature, 1994, vol. 370, 441–443.

1138. Sackett P.D., Does the Milky Way have a maximal disk?, Astrophys. J., 1997, vol. 483, 103–110.

1139. Sage L.J., Welch G.A., The Cool ISM in S0 Galaxies. II. A Survey of Atomic Gas, Astrophys. J., 2006, vol. 644, 850.

1140. Salpeter E.E., The Luminosity Function and Stellar Evolution, Astrophys. J., 1955, vol. 121, 161.

1141. Salucci P., Persic M., Maximal halos in high-luminosity spiral galaxies, Astron. Astrophys., 1999, vol. 351, 442–446.

1142. Samurovic S., Cirkovic M.M., Milosevic-Zdjelar V., Flattened galactic haloes and baryonic dark matter, MNRAS, 1999, vol. 309, 63–79.

1143. Sancisi R., Allen R.J., Neutral hydrogen observations of the edge-on disk galaxy NGC 891, Astron. Astrophys., 1979, vol. 74, 73–84.

1144. Sandell G., Aspin C., Duncan W.D., et al. The structure of dust discs, Astron. Astrophys., 1990, vol. 232, 347–355.

1145. Sanders D.B., The distribution of molecular clouds in the Galaxy. Kinematics, dynamics and structure of the Milky Way – Vancouver, 1982, 115–125.

1146. Sanders R.H., Huntley J.M., Gas response to oval distortions in disk galaxies, Astrophys. J., 1976, vol. 209, 53–65.

1147. Sanders D.B., Scoville N.Z., Solomon P.M., Giant molecular clouds in the Galaxy. II. Characteristics of discrete features, Astrophys. J., 1985, vol. 289, 373–387.

1148. Sanders D.B., Solomon P.M., Scoville N.Z., Giant molecular clouds in the Galaxy. I. The axisymmetrical distribution of H_2, Astrophys. J., 1984, vol. 276, 182–203.

1149. Sanders R.H., Tubbs A.D., Gas as a tracer of barred spiral dynamics, Astrophys. J., 1980, vol. 235, 803–820.

1150. Sato S., Circumstellar materials of young stellar objects, Progr. Theor. Phys. Suppl., 1988, vol. 96, 37–49.

1151. Savonije G.J., Heemskerk M.H.M., Non-axisymmetric unstable modes in a thin differentially rotating gaseous disc, Astron. Astrophys., 1990, vol. 240, 191–202.

1152. Sawada K., Matsuda T., Anzer U., et al. Inhomogeneous wind accretion: comparison between 3D and 2D computations, Astron. Astrophys., 1989, vol. 221, 263–272.

1153. Sawada K., Matsuda T., Hachisu I., Spiral shocks on a Roche lobe overflow in a semidetached binary system, MNRAS, 1986, vol. 219, 75–88.

1154. Sawada K., Matsuda T., Hachisu I., Accretion shocks in close binary system, MNRAS, 1986, vol. 221, 679–686.

1155. Sawada K., Matsuda T., Inoue M., Hachisu I., Is the standart accretion disc model invulnerable?, MNRAS, 1987, vol. 224, 307–322.

1156. Scarano S., Lepine J., Marcon-Uchida M., Breaks in the radial oxygen abundance and corotation radius of three spiral galaxies, MNRAS, 2011, MNRAS. vol. 412, 1741–1754.

1157. Schaye J., Star formation thresholds and galaxy edges: why and where, Astrophys. J., 2004, vol. 609, 667—682.

1158. Schmidt M., Stars and stellar systems. 1965, vol. 4, 513–527

1159. Schmidt M., The mass of the galactic halo derived from the luminosity function of high-velocity stars, Astrophys. J., 1975, vol. 202, 22–29.

1160. Schmitt H.R., The Frequency of Active and Quiescent Galaxies with Companions: Implications for the Feeding of the Nucleus, Astron. J., 2001, vol. 122, 2243–2256.

1161. Schoenmakers R.H.M., Franx M., de Zeeuw P.T., Measuring non-axisymmetry in spiral galaxies, MNRAS, 1997, vol. 292, 349–364.
1162. Schommer R.A., Christian C.A., et al. Kinematics of star clusters in M 33, Astron. J., 1991, vol. 101, 873–883.
1163. Schönrich R., Binney J., Chemical evolution with radial mixing, MNRAS, 2008, vol. 396, 203–222.
1164. Schroeder H.C., Comins N.F., Relaxation times in two-dimensional galaxy simulations, Astrophys. J., 1989, vol. 346, 108–117.
1165. Schwarzschild M., Structure and Evolution of the Stars – Princeton: Princeton Univ. Press, 1958.
1166. Scodeggio M., Gavazzi G., Astrophys. J., 1993, vol. 409, 110.
1167. Scorza C., Bender R., A disk in the elliptical galaxy NGC 3610, Astron. Astrophys., 1990, vol. 235, 49–54.
1168. Scoville N.Z., Young J.S., The molecular gas distribution in M 51, Astrophys. J., 1983, vol. 265, 148–165.
1169. Sedov L.I.,A Course in Continuum Mechanics – Groningen: Wolters-Noordhoff, 1971.
1170. Sellwood J.A., Galaxy models with live halos, Astron. Astrophys., 1980, vol. 89, 296–307.
1171. Sellwood J.A., Bar instability and rotation curves, Astron. Astrophys., 1981, vol. 99, 362–374.
1172. Sellwood J.A., The art of N–body building, Ann. Rev. Astron. Astrophys., 1987, vol. 25, 151–186.
1173. Sellwood J.A., Meta-stability in galactic discs, MNRAS, 1989, vol. 238, 115–131.
1174. Sellwood J.A., Peanut shaped bars. Galactic bulges / Eds. Dejonghe H., Habing H.J. 1993, vol. 153, 391–392.
1175. Sellwood J.A., Axisymmetric bending oscillations of stellar disks, Astrophys. J., 1996, vol. 473, 733–745.
1176. Sellwood J.A., Astrophys. J. Suppl. Ser., 2000, vol. 272, 31.
1177. Sellwood J.A., Athanassoula E., Instabilities of hot stellar discs. Instabilities of hot stellar disks, Internal Kinematics and dynamics of galaxies, 1983, vol. 100, 203–204.
1178. Sellwood J.A., Athanassoula E., Unstable modes from galaxy simulations, MNRAS, 1986, vol. 221, 195–212.
1179. Sellwood J.A., Carlberg R.G., Spiral instabilities provoked by accretion and star formation, Astrophys. J., 1984, vol. 282, 61–74.
1180. Sellwood J.A., Evans N.W., The stability of disks in cusped potentials, Astrophys. J., 2001, vol. 546, 176–188.
1181. Sellwood J.A., Lin D.N.C., A reccurent spiral instability cycle in self-graviting particle discs, MNRAS, 1989, vol. 240, 991–1007.
1182. Sellwood J.A., Merritt D., Instabilities of counterrotating stellar disks, Astrophys. J., 1994, vol. 425, 530–550.
1183. Shakura N.I., Disk Model of Gas Accretion on a Relativistic Star in a Close Binary System, Sov. Astron., 1973, vol. 16, 756.
1184. Shakura N.I., Sunyaev R.A., Black holes in binary systems. Observational appearance, Astron. Astrophys., 1973, vol. 24, 337–355.
1185. Shakura N.I., Sunyaev R.A., A theory of the instability of disk accretion onto black holes and the variability of binary X-ray sources, galactic nuclei and quasars, MNRAS, 1976, vol. 175, 613–632.
1186. Shakura N.I., Sunyaev R.A., Zilitinkevich S.S, Astron. Astrophys., 1978, vol. 62, 179.

1187. Shalyapina L.V., Moiseev A.V., Yakovleva V.A., A spectroscopic study of the peculiar galaxy UGC 5600, \litPAZ, 2002, vol. 28, 505–515.

1188. Shapiro S.L., Lightman A.P., Eardley D.M., A two-temperature accretion disk model for Cygnus X-1: structure and spectrum, Astrophys. J., 1976, vol. 204, 187.

1189. Shaw M.A., Gilmore G., The luminosity distributions of edge-on spiral galaxies. I. A two-dimensional model and its application to NGC 891 and 4565, MNRAS, 1989, vol. 237, 903–927.

1190. Sherwin B.D., Loeb A., O'Leary R.M., Hypervelocity Stars from the Andromeda Galaxy, MNRAS, 2008, vol. 386, 1179–1191.

1191. Shetty R., Ostriker E.C., Global Modeling of Spur Formation in Spiral Galaxies, Astrophys. J., 2006. vol. 647, 997–1017.

1192. Shima E., Matsuda T., Takeda H., Sawada K., Numerical calculations are performed for axisymmetric flow, MNRAS, 1985, vol. 217, 367–386.

1193. Shinoda K., Kii T., Mitsuda K., et al./Discovery of the quasi–periodic oscillations from the X–ray pulsar X1627–673, Publ. Astron. Soc. Japan, 1990, vol. 42, 27–32.

1194. Shlosman I., Frank J., Begelman M.C., Bars within bars – A mechanism for fuelling active galactic nuclei, Nature, 1989, vol. 338, 45–47.

1195. Shlosman I., Noguchi M., The effects of gas on global stability galactic discs: radial flows, Astrophys. J., 1993, vol. 414, 474–486.

1196. Shore S., Livio M., van den Heuvel E., Interacting Binaries (Saas- Fee Adv. Course, Lect. Notes. 1992. N. 22 / Eds H. Nussbaumer, A. Orr – Berlin: Springer-Verlag, 1994

1197. Shu F.H., Models of partially relaxed stellar disks, Astrophys. J., 1969, vol. 158, 505–518.

1198. Shu F.H., On the density wave theory of galactic spirals. II. The propagation of the density wave action, Astrophys. J., 1970, vol. 160, 99–112.

1199. Shu F.H., Mass transfer in semi-detached binaries, IAU SymP. ' 73.

1200. 1976, 253–264.

1201. Shu F.H., Galli D., Lizano S., et al. Mean Field Magnetohydrodynamics of Accretion Disks, Astrophys. J., 2007, vol. 665, 535.

1202. Shu F., Milione V., Roberts W.W., Nonlinear gaseous density waves and galactic shocks, Astrophys. J., 1973, vol. 183, 819–841.

1203. Shu C.W., Osher S.J., Efficient implimenation of essentially non-oscillatory shock capturing scheme II, J. Comput. Phys., 1989, vol. 83, 32–78.

1204. Shu F.H., Stachnik R.W., Yost J.C., On the density-wave theory of galactic spirals. III. Comparison with external galaxies, Astrophys. J., 1971, vol. 166, 465–479.

1205. Shu F.H., Physics of Astrophysics II. Gas dynamics. USB, 1992.

1206. Sil'chenko O.K., Face-on galaxies ngc 524 and ngc 6340: chemically decoupled nuclei and inclined circumnuclear disks, Astron. J., 2000, vol. 120, 741–751.

1207. Sil'chenko O.K., A Chemically Decoupled Nucleus and Inner Polar Ring of the SBb Galaxy NGC 4548, Astr. Letter., 2002, vol. 28, 207–216.

1208. Sil'chenko O.K., Afanasiev V.L., Inner Polar Rings in Regular Lenticular Galaxies, Astron. J., 2004, vol. 127, 2641–2658.

1209. Sil'chenko O.K., Moiseev A.V., Nature of nuclear rings in unbarred galaxies: ngc 7742 and ngc 7217, Astron. J., 2006, vol. 131, 1336—1346.

1210. Sil'chenko O.K., Moiseev A.V., Afanasiev V.L., et al. The Leo I cloud: secular nuclear evolution of NGC 3379, NGC 3384, and NGC 3368?, Astrophys. J., 2003, vol. 591, 185–203,

1211. Sil'chenko O., Young circumnuclear disks in elliptical galaxies, ApSS, 2009, vol. 320, 51

1212. Simien F., Prugniel Ph., Kinematical data on early-type galaxies. V., Astron. Astropliys. Suppl. Ser., 2000, vol. 145, 263–267.

1213. Simien F., Prugniel Ph., Kinematical data on early-type galaxies. VI, Astron. Astrophys., 2002, vol. 384, 371–382

1214. Simkin S.M., Su H.J., Schwartz M.P., Nearby seyfert galaxies, Astrophys. J., 1980, vol. 237, 404–413.

1215. Simpson J.S., Numerical techniques for tree–dimensional smoothed particle hydrodynamics simulation: applications to accretion disks, Astrophys. J., 1995, vol. 448, 822–831.

1216. Sinha R.P., A reexamination of the rotation curve for the Galaxy, Astron. Astrophys., 1978, vol. 69, 227–233.

1217. Smak J., Eruptive Binaries. I. Hot Spots and Distorsions of the Radial Velocity Curves, Acta Astron., 1970, vol. 20, 311.

1218. Smak J., Accretion in cataclismic binaries. I., Acta Astron., 1982, vol. 32, 199.

1219. Smak J., Accretion in cataclysmic binaries. IV., Acta Astron., 1984, vol. 34, 161–189.

1220. Smak J., Dwarf nova outbursts. VI., Acta Astron., 1999, vol. 49, 391–401.

1221. Smirnova A.A., Moiseev A.V., Afanasiev V.L., Seyfert galaxies that are undergoing merging but appear non-interacting, MNRAS, 2010, vol. 408, 400–406.

1222. Smith M.C., Evans N.W., et al. Kinematics of SDSS subdwarfs: Structure and substructure of the Milky Way halo, MNRAS, 2009. vol. 399, 1223–1237.

1223. Sofue Y., The most comletely sampled rotation curves for galaxies, Astrophys. J., 1996, vol. 458, 120–131L.

1224. Sofue Y., Dark bulge, exponential disk, and massive halo in the Large Magellanic Cloud, Publ. Astron. Soc. Japan, 1999, vol. 51, 445–448.

1225. Sofue Y., Nakai N., CO observations of edge-on galaxies. III. NGC 891: Threshold radius for a star-formation disk, Publ. Astron. Soc. Japan, 1993, vol. 45, 139–151.

1226. Sofue Y., Tutui Y., Honma M., et al. Central rotation curves of spiral galaxies, Astrophys. J., 1999, vol. 523, 136–146.

1227. Solomon P.M., Rivolo A.R., Mooney T.J., et al. Masses, luminosities and dynamics of galactic molecular clouds. Star formation in Galaxies / Ed. C. Persson. 1986.

1228. Solomon P.M., Sunders D.B., Giant molecular clouds as a dominant component of interstellar matter in the Galaxy. Giant Mol. Clouds Galaxy, 3rd Gregynog. / Astrophys. Workshop. Oxford e.a., 1980, 41–73.

1229. Sommer-Larsen J., Vedel H., Hellsten U., On the global structure of self-gravitating discs for softened gravity, MNRAS, 1988, vol. 294, 485–488.

1230. Sorensen S.A., Matsuda T., Sakurai T., Computer simulations of gas flow around close binary systems, Astroph. Space Sci., 1975, vol. 33, 465–480.

1231. Sotnikova N.Ya., Rodionov S.A., Bending instability of stellar disks: the stabilizing effect of a compact bulge, Astr. Letter., 2005, vol. 31, 15–29.

1232. Sotnikova N.Ya., Rodionov S.A., Estimating the dark halo mass from the relative thickness of stellar disks, Astr. Letters, 2006, vol. 32, 649–660.

1233. Soubiran C., Bienaymé O., Mishenina T.V., Kovtyukh V.V., Vertical distribution of Galactic disk stars., Astron. Astrophys., 2008, vol. 480, 91.

1234. Sparke L.S., Sellwood J.A., Dissection of an N-body bar, MNRAS, 1987, vol. 225, 653–675.

1235. Spitzer L., Schwarzschild M., The possible influence of interstellar clouds on stellar velocities, Astrophys. J., 1951, vol. 114, 385.

1236. Spitzer L., Schwarzschild M., The possible influence of interstellar clouds on stellar velocities. II, Astrophys. J., 1953, vol. 118, 106–112.

1237. Spitzer L (Jr.), Physical Processes in the Interstellar Medium – New York: Wiley, 1978.
1238. Splinter R.J., A nested-grid particle-mesh code for high-resolution simulations of gravitational instability in cosmology, MNRAS, 1996, vol. 281, 281–293.
1239. Springel V., Hernquist L., Cosmological smoothed particle hydrodynamics simulations: the entropy equation, MNRAS, 2002, vol. 333, 649–664.
1240. Spruit H.C., Stationary shocks in accretion disks, Astron. Astrophys., 1987, vol. 184, 173–184.
1241. Spruit H.C., Matsuda T., Inoue M., Sawasa K., Spiral shocks and accretion discs, MNRAS, 1987, vol. 229, 517–527.
1242. Spruit H.C., Stationary shocks in accretion disks, Astron. Astrophys., 1987, vol. 184, 173–184.
1243. Stanek K.Z., Udalski A., Szymaski M., at al. Modeling the Galactic bar using Red Clump Giants, Astrophys. J., 1997, vol. 477, 163–175.
1244. Stanford L.M., Da Costa G.S., Norris J.E., Cannon R.D., The Age and Metallicity Relation of ω Centauri, Astrophys. J., 2006, vol. 647, 1075.
1245. Stark A.A., Kinematics of molecular clouds. I.Velocity dispersion in the Solar neighborhood, Astrophys. J., 1984, vol. 281, 624–633.
1246. Staveley-Smith L., Kim S., Stanimirovic S., Neutral hydrogen in the Magellanic Clouds. New views of the Magellanic Clouds. / IAU Symposium #190 / Edited by Chu Y.-H., Suntzeff N., Hesser J., Bohlender D. 1999, 37.
1247. Steinicke W., Jakiel R., Galaxies and How to Observe Them – Springer, 2007.
1248. Stewart G.C., Watson M.G., Matsuoka M., et. al. Simultaneous observations of the X-ray and optical eclipses of SS 433 and their implication, MNRAS, 1987, vol. 228, 293–303.
1249. Stone J.M., Pringle J.E., Begelman M.C., Hydrodynamical non-radiative accretion flows in two-dimensions, MNRAS, 1999, vol. 310, 1002–1016.
1250. Strigari L.E., Bullock J.S., A common mass scale for satellite galaxies of the Milky Way, Nature, 2008, vol. 454, 1096–1097.
1251. Su H.J., Simkin S.M., Disk structure and nuclear activity in Seyfert galaxies, Astrophys. J., 1981, vol. 238, 1–5.
1252. Sugimoto D., Chikada Y., Makino J., et al. A special purpose computer for gravitational many-body problems, Nature, 1990, vol. 345, No. 6270, 33–35.
1253. Sumin A.A., Fridman A.M., Haud U., A Three-Component Dynamical Model of the Galaxy, Sov. Astron. Lett., 1991, vol. 17, 295.
1254. Sun M., Jones C., Forman W., et al., Astrophys. J., 2007, vol. 657, 197.
1255. Sundelius B., Thomasson M., Valtonen M.J., Byrd G.G., Tidal spiral arms in two-component galaxies. Density waves and swing amplification, Astron. Astrophys., 1987, vol. 174, 67–77.
1256. Syer D., Narayan R., Steady flow on to a conveyor belt - Causal viscosity and shear shocks, MNRAS, 1993, vol. 262, 749.
1257. Szebehely V., Theory of Orbits: the Restricted Problem of Three Bodies – New York: Academic Press, 1967.
1258. Szuszkiewicz E., Slim accretion discs with different viscosity prescriptions, MNRAS, 1990, vol. 244, 377–383.
1259. Szuszkiewicz E., Miller J.C., On the thermal stability of transonic accretion discs, MNRAS, 1997, vol. 287, 165–179.
1260. Taam R.E., Lin D.N.C., The evolution of the inner regions of viscous accretion disks surrounding neutron stars, Astrophys. J., 1984, vol. 287, 761–768.
1261. Telesco C.M., Davidson J.A., Werner M.W., 10–30 Micron Maps of the Central 5 Parsecs of the Galaxy, Astrophys. J., 1996, vol. 456, 541–556.

1262. Tevzadze A.G., Chagelishvili G.D., Zahn J.-P., et al. On hydrodynamic shear turbulence in stratified Keplerian disks: Transient growth of small-scale 3D vortex mode perturbations, Astron. Astrophys., 2003, vol. 407, 779–786.

1263. Tikhonov N., Galazutdinova O., Drozdovsky I., Thick disks and halos of spiral galaxies M 81, NGC 55 and NGC 300, Astron. Astrophys., 2005, vol. 431, 127-142,

1264. Tinsley B., Evolution of the Stars and Gas in Galaxies, Fundamentals of Cosmic Physics, 1980, vol. 5, 287.

1265. Tiurina A.V., Khoperskov A.V., Zasov A.V., Estimation of masses of components of galaxies by the simulations N-body, Astron. Astrophys. Trans., 2001, vol. 20, 155–159.

1266. Thomasson M., Donner K.J., Sundelius B., et al. Formation of leading spiral arms in retrograde galaxy encounters, Astron. Astrophys., 1989, vol. 211, 25–36.

1267. Thompson L.A., Bar instabilities in Coma cluster galaxies, Astrophys. J., 1981, vol. 244, 43–45L.

1268. Thorne K.S., Price R.H., Cygnus X-1: An interpretation of the spectrum and its variability, Astrophys. J., 1975, vol. 195, 101.

1269. Toomre A., On the gravitational stability of a disk of stars, Astrophys. J., 1964, vol. 139, 1217–1238.

1270. Toomre A., Notes on the Summer Study Program in Geophysical Fluid Dynamics at the Woods Hole Oceanographic Institution. Geophys, Fluid Dyn., 1966, vol. 66-46, 111.

1271. Toomre A., Group velocity of spiral waves in galactic discs, Astrophys. J., 1969, vol. 158, 899–913.

1272. Toomre A., Toomre J., Galactic bridges and tails, Astrophys. J., 1972, vol. 178, 623–666.

1273. Toro E.F., Riemann Solvers and Numerical Methods for Fluid Dynamic. A Practical Introduction – Berlin: Springer, 1997.

1274. Tout C., Accretion disk viscosity, Astrophys. and Space Sci. Library, 1996, vol. 208, 97.

1275. Tremaine S., Slow Modes in Keplerian Disks, Astron. J., 2001, vol. 121, 1776–1789.

1276. Tremaine S., Gerhardt K., et al. The slope of the blak hole mass versus velocity dispersion correlation, Astrophys. J., 2002, vol. 574, 740–753.

1277. Tremaine S., Secular Stability and Instability in Stellar Systems Surrounding Massive Objects, Astrophys. J., 2005, vol. 625, 143–155.

1278. Treu T., Koopmans L.V.E., et al. The dark matter halos of spheroidal galaxies and clusters of galaxies, IAUS, 2004, vol. 220, 159–164

1279. Trumper J., Peitsch W., Reppin C., et al. Evidence for strong cyclotron line emission in the hard X-ray spectrum of Hercules X-1, Astrophys. J., 1978, vol. 219, 105–110.

1280. Tully R.B., Fisher J.R., A new method of determining distances to galaxies, Astron. Astrophys., 1977, vol. 54, 661.

1281. Turner N.J., Stone J.M., Krolik J.H., Sano T., Local three-dimensional simulations of magnetorotational instability in radiation-dominated accretion disks, Astrophys. J., 2003, vol. 593, 992–1006.

1282. Tutukov A.V., Fedorova A.V., The role of close passages of galaxies and the asymmetry of their dark haloes in the formation of their spiral patterns, Astr. Rep., 2006, vol. 50, 785–801.

1283. Tyurina N.V., Khoperskov A.V., Bizyaev D., Bending instability in galaxies: the stellar disk thickness and the mass of spheroidal component. / Ed.: A.M.Fridman

et al., Astroph. and Space Science Libr., Springer, 2006, vol. 337, 291–306. (astro-ph/0405142, AAS, 2003, 202.40.13 (astro-ph/0306193))
1284. Unwin S.C., Neutral hydrogen in the Andromeda nebula. III. The velocity field, MNRAS, 1983, vol. 205, 773–786.
1285. Urpin V., Instability of relativistic sheared jets and distinction between FRI and FRII sources, Astron. Astrophys., 2002, vol. 385, 14–20.
1286. Valentijn E.A., Opaque spiral galaxies, Nature, 1990, vol. 346, 153.
1287. Valenzuela O., Klypin A., Secular bar formation in galaxies with a significant amount of dark matter, MNRAS, 2003, vol. 345, 406–422.
1288. Vallenari A., Bertelli G., Schmidtobreick L., The Galactic disk: study of four low latitude Galactic fields, Astron. Astrophys., 2000, vol. 361, 73–84.
1289. Valluri M., A model for the declining rotation curves of cluster spiral galaxies, Astrophys. J., 1994, vol. 430, 101–106.
1290. Valtonen M.J., Valtaoja L., Sundelius B., et al.Dynamical friction on a satellite of a disk galaxy: the curcular orbit, Celest. Mech., 1990, vol. 48, 95–113.
1291. van den Bosch F.C., Robertson B.E., Dalcanton J.J., Constraints on the structure of dark matter halos from the rotation curves of Low Surface Brightness Galaxies, Astron. J., 2000, vol. 119, 1579–1591.
1292. van den Hoek L.B., Groenewegen M.A.T.., New theoretical yields of intermediate mass stars, A&AS, 1997, vol. 123, 305.
1293. Vandervoort P.O., The equilibria of highly flattened, rapidly rotating galaxy, Astrophys. J., 1970, vol. 161, 67–86.
1294. Vandervoort O.P., Density waves in a highly flattened, rapidly rotating galaxy, Astrophys. J., 1970, vol. 161, 87–102.
1295. Vandervoort P.O., New applications of the equations of stellar hydrodynamics, Astrophys. J., 1975, vol. 195, 333–341.
1296. Vandervoort P.O., Modes of oscillation of a uniformly rotating, homogeneous spheroid of stars, Astrophys. J., 1991, vol. 377, 49–71.
1297. Vandervoort P.O., On the effects of interstellar gas on the oscillations and the stability of galaxies, Astrophys. J., 1991, vol. 383, 498–510.
1298. Vaucouleurs de G., Southern galaxies. VII. The remarkable lenticular barred galaxy NGC 1291, Astrophys. J. Suppl. Ser., 1975, vol. 29, 193–218.
1299. Vaukouleurs de G., On the scale length of the exponential disk of the Galaxy, Observatory, 1979, vol. 99, 128–129.
1300. Vaucouleurs de G., Freeman K., Structure and dynamics of barred spiral galaxies, in particular of the Magellanic type, Vistas Astron., 1972, vol. 14, 163–294.
1301. Vaucouleurs de G., Vaucouleurs de A., Corwin H., et al. Third Reference Catalogue of Galaxies. N.Y.: Springer-Verlag, 1991
1302. Velazquez H., White S.D.M., Sinking satellites and the heating of galaxy discs, MNRAS, 1999, vol. 304, 254–270.
1303. 2Veltz L., Bienayme O., Freeman K.C., et al. Galactic kinematics with RAVE data. I. The distribution of stars towards the Galactic poles, Astron. Astrophys., 2008, vol. 480, 753–765.
1304. Verbunt F., Accretion disks in stellar X-ray sources - A review of the basic theory of accretion disks and its problems. Space Sci. Rev., 1982, 32379.
1305. Verbunt F., Theory and observations of time-dependent accretion disks. The phisics of accretion onto compact objects / Ed. Mason K.O., Watson M.G., White N.E. – Berlin: Springer, 1986, 59–75.
1306. Vila S.C., Convection as a source of viscosity for accretion disks of cataclysmic binaries, Astrophys. J., 1978, vol. 223, 979–981.

1307. Vila S.C., Disk models with convective viscosity, Astrophys. J., 1981, vol. 247, 499–502.
1308. Villumsen J.V., The vertical growth and structure of galactic disks, Astrophys. J., 1983, vol. 274, 632–645.
1309. Villumsen J.V., Evolution of the velocity distribution in galactic disks, Astrophys. J., 1985, vol. 290, 75–85.
1310. Vishniac E.T., Diamond P., A self-consistent model of mass and angular momentum transport in accretion disks, Astrophys. J., 1989, vol. 347, 435.
1311. Voglis N., Contopoulos G., J. Phys. 1994, vol. A27, 4899.
1312. Vorobyov E.I., Bizyaev D., Radial B–V/V–K color gradients, extinction-free Q_{BVK} combined color indices, and the history of star formation of the Cartwheel ring galaxy, Astron. Astrophys., 2001, vol. 377, 835–844.
1313. Wada K., Koda J., Multi-Phase Gas Dynamics in a Weak Barred Potential, Publ. Astron. Society of Japan., 2001, vol. 53, 1163–1170.
1314. Wada K., Koda J., Instabilities of Spiral Shocks I: Onset of Wiggle Instability and its Mechanism, MNRAS, 2004, vol. 349, 270–280.
1315. Wada K., Instabilities of Spiral Shocks. II. A Quasi-Steady State in the Multiphase Inhomogeneous ISM, Astrophys. J., 2008, vol. 675, 188–193.
1316. Wainscoat R.J., Freeman K.C., Hyland A.R., Astrophys. J., 1989, vol. 337, 163.
1317. Waller W.H., et. al. Ultraviolet signposts of resonant dynamics in the starburst-ringed Sab Galaxy, M94 (NGC 4736), Astron. J., 2001, vol. 121, 1395–1412.
1318. Waller W.H., Bohlin R.C., Cornett R.H., et al. Ultraviolet Signatures of Tidal Interaction in the Giant Spiral Galaxy M101, Astrophys. J., 1997, vol. 481, 169–173.
1319. Wallin J.F., Stuart B.V., Mass trunsfer and loss in interacting disk galaxies, Astrophys. J., 1992, vol. 399, 29–37.
1320. Wallinder F.H., Stability properties of an isothermal accretion disk, Astron. Astrophys., 1990, vol. 237, 270–274.
1321. Wallinder F.H., The stability of slim accretion disks, Astron. Astrophys., 1991, vol. 249, 107–117.
1322. Wang B., Silk J., Gravitational instability and disk star formation, Astrophys. J., 1994, vol. 427, 759–769.
1323. Wang Q.D., Walterbos R.A.M., Steakley M.F., et al. ROSAT detection of diffuse hot gas in the edge-on galaxy NGC 4631, Astrophys. J., 1995, vol. 439, 176.
1324. Wang Y.M., Welter G.L., Plasma-magnetospheric interaction in X-ray sources – an analysis of the linear Kelvin-Helmholtz instability, Astron. Astrophys., 1982, vol. 113, 113–117.
1325. Warner B., Cataclysmic Variable Stars – Cambridge: Cambridge UniV. Press, 1995.
1326. Wasserman I., Leading between the lines, Nature, 1990, vol. 346, 6281, 222–223.
1327. Westpfahl D.J., The Speed and Origin of the H i Spiral Pattern in M81, Astrophys. J., 1998, vol. 115, 203–224.
1328. Widrow L.M., Dynamical Models for Disk Galaxies with Triaxial Halos, Astrophys. J., 2008, vol. 679, 1232–1238.
1329. Wiebe D.S., Shustov B.M., Tutukov A.V., Role of disk galaxies in the chemical enrichment of the intracluster medium, \liAA, 1999, vol. 345, 93.
1330. Willett B.A., Newberg H.J., Zhang H., et al. An Orbit Fit for the Grillmair Dionatos Cold Stellar Stream, Astrophys. J., 2009, vol. 697, vol. 207–223.
1331. Weinberg M.D., Detection of a large-scale stellar bar in the Milky Way, Astrophys. J., 1992, vol. 384, 81–94.
1332. Weinberg M.D., Dynamics of an interacting luminous disc, dark halo and satellite companion, MNRAS, 1998, vol. 299, 499–514.

1333. Weinberg M.D., Blitz L., A Magellanic Origin for the Warp of the Galaxy, Astrophys. J., 2006, vol. 641L, 33–36.

1334. Weiner B.J., Sellwood J.A., The properties of the galactic bar implied by gas kinematics in the inner Milky Way, Astrophys. J., 1999, vol. 524, 112–128.

1335. von Weizsacker C. Rotation kosmischer Gasmassen, Z. Naturforsch., 1948, vol. 3*a*, 524

1336. Wheaton W.A., et. al. An absorption feature in the spectrum of the pulsed hard X–ray flux from 4U 0115+63, Nature, 1979, vol. 282, 240.

1337. Wheeler J.C., Sneden Ch., Truran J.W., Abundance ratios as a function of metallicity. ARA&A, 1989, vol. 27, 279.

1338. White R.E., Keel W.C., Conselice C.J., Seeing galaxies through thick and thin. I. Optical opacity measures in overlapping galaxies, Astrophys. J., 2000, vol. 542, 761.

1339. Whitehurst R., Numerical simulations of accretion discs. I. Superhumps: a tidal phenomenon of accretion discs, MNRAS, 1988, vol. 232, 35–51.

1340. Whitemore B.C., McElroy D.B., Schweizer F., The shape of the dark halo in polar-ring galaxies, Astrophys. J., 1987, vol. 314, 439–456.

1341. Whitmore B.C., Rubin V.C., Ford W.K., Stellar and gas kinematics in disk galaxies, Astrophys. J., 1984, vol. 287, 66–79.

1342. Whittet D.C.B., Dust in the galactic environment – Bristol: IOP, 1992.

1343. Wielen R., The kinematics and ages of stars in Gliese's catalogue, Highlights of astronomy. 1974, vol. 3, 395–408.

1344. Wielen R., The diffusion of stellar orbits derived from the observed age-dependent of the velocity dispersion, Astron. Astrophys., 1977, vol. 60, 263–275.

1345. Wielen R., Dettbarn C., Fuchs B., Jahrei H., Radons G., Dynamics of Stellar Populations in Galactic Disks. In: The Stellar Populations of Galaxies, Proc. IAU Symp., 1992, vol. 149, 81–92.

1346. Wilkerson M.S., Neutral-hydrogen observations of smooth-arm spiral galaxies, Astrophys. J., 1980, vol. 240, 115–119L.

1347. Wilson J.R., Numerical study of fluid flow in a Kerr Space, Astrophys. J., 1972, vol. 173, 431–438.

1348. Winter L., Sion E.M., Composite accretion disk and white dwarf model analyses of the quiescence of dwarf novae: Cygni E.M., Orionis C.Z., Ceti W.W, Astrophys. J., 2003, vol. 582, 352–357.

1349. Winters W.F., Balbus S.A., Hawley J.F., Chaos in turbulence driven by the magnetorotational instability, MNRAS, 2003, vol. 340, 519.

1350. Wolf S., Barwig H., Bobinger A., et al. A comprehensive study of multi-emission sites in IP Peg, Astron. Astrophys., 1998, vol. 332, 984.

1351. Wood J.H., Marsh T.R., The ephemeris and variations of the accretion disc radius in IP Pegasi, MNRAS, 1989, vol. 239, 809–824.

1352. S.E. Woosley, T. Weaver., The Evolution and Explosion of Massive Stars. II. Explosive Hydrodynamics and Nucleosynthesis, ApJS, 1995, vol. 101, 181.

1353. Wu X.-B., Li Q.-B., Zhao Y.-H., Yang L.-T., The radial-azimuthal instability of accretion disk. I. radiation pressure domination, Astrophys. J., 1995, vol. 442, 736–742.

1354. Wu H., et al. Intermediate-Band Surface Photometry of the Edge-on Galaxy NGC 4565, Astron. J., 2002, vol. 123, 1364–1380.

1355. Wyse R.F.G., Gilmore G., Franx M., Galactic Bulges, ARA&A, 1997, vol. 35, 637.

1356. Xilouris E.M., Alton P.B., Davies J.I, et al., Optical and NIR modelling of NGC 891, Astron. Astrophys., 1998, vol. 331, 894.

1357. E.M. Xilouris, Y.I. Byun, N.D. Kylafis, et al./Are spiral galaxies optically thin or thick?, Astron. Astrophys., 1999, vol. 344, 868–878.

1358. Yanez M.A., Norman M.L., Martos M.A., Hayes J.C., Resonance-related Spiral Substructure in a Galactic Gaseous Disk, ApJ, 2008, vol. 672, 207–213.

1359. Yasumomu M., Fujimoto M., Dynamical evolution of galactic disk stars due to giant moolecular clouds and spiral density waves, Publ. Astron. Soc. Japan, 1989, vol. 41, 823–840.

1360. Yee H.C., Construction of explicit and implicit symmetric TVD schemes and their applications, J. Comput. Phys., 1987, vol. 68, 151–179.

1361. Young J.S., Scoville N.Z., Molecular gas in galaxies. Ann. Rev., Astron. Astrophys., 1991, vol. 29, 581–625.

1362. Yuan C., Application of the density–wave theory to the spiral structure of the Milky–Way system. II. Mygration of stars, Astrophys. J., 1969, vol. 158, 889–898.

1363. Zang T.A., The stabilityof a model Galaxy / Ph.D. Thesis, 1976 – Massachussetts Institute of Technology, Cambrigde, MA.

1364. Zaritsky D., Shectman S.A., Thompson I., et.al Constraints on intervening stellar populations toward the LMO, Astrophys. J., 1999, vol. 117, 2268–2285.

1365. Zasov A.V., Sil'chenko O.K., Minibars in the Centers of Normal Galaxies, Astr. Soc. Pac. Conf. Ser., 1996, vol. 91, 207–211.

1366. Zasov A.V., Khoperskov A.V., Tiurina A.V., The determination of mass of stellar disks of galaxies from the kinematic data / Stellar dynamics: from classic to modern / Edited by Ossipkov L.P., Nikiforov I.I. – Saint Petersburg, 2001, 95–102.

1367. Zasov A.V., Abramova O.V., The star-formation efficiency and density of the disks of spiral galaxies, Astr. Rep., 2006, vol. 50, 874–886.

1368. Zasov A.V., Abramova O.V., Midplane Gas Density and the Schmidt Law, arXiv:0712.1149 (2008)

1369. Zasov A.V., Khoperskov A.V., Static Pressure of Hot Gas: Its Effect on the Gas Disks of Galaxies. Astr. Lett., 2008, vol. 34, 739–744.

1370. Zasov A.V., Moiseev A.V., Khoperskov A.V., Sidorova E.A., Early-type disk galaxies: Structure and kinematics, Astron. Rep., 2008, vol. 52, 79-93.

1371. Zasov A.V., Smirnova A.A., The Gas Content in Galactic Disks: Correlation with Kinematics, Astron. Lett., 2005, vol. 31, 160–170.

1372. Zentner A.R., Kravtsov A.V., Gnedin O.Y. Klypin A.A., The anisotropic distribution of galactic satellites, Astrophys. J., 2005, vol. 629, 219.

1373. Zhang X., Buta R.J., The Potential-Density Phase-Shift Method for Determining the Corotation Radii in Spiral and Barred Galaxies, Astron. J., 2007, vol. 133, 2584–2606.

1374. Buta R., Zhang X., Pattern speeds and galaxy morphology, MmSAI, 2011, vol. 18, 13–21.

1375. Zhao H.S., Evans N.W., The so-called "bar" in the Large Magellanic Cloud, Astrophys. J., 2000, vol. 4 545, 35–38L.

1376. Zibetti S., White S., Brinkmann J., Haloes around edge-on disc galaxies in the Sloan Digital Sky Survey, MNRAS, 2004, vol. 347, 556–568.

1377. Zweibel E.G., Heiles C., Magnetic fields in galaxies and beyond, Nature, 1997, vol. 385, 131–136.

1378. Abramowicz M.A., Kato S., Constraints for transonic black hole accretion, Astrophys. J., 1989, vol. 336, 304–312.

1379. I.A. Acharova, J.R.D. Lépine, Yu.N. Mishurov, MNRAS, 2005, vol. 359, 819.

1380. Acharova I.A., Lepine J.R.D., Mishurov Yu.N., et al. A mechanism for the formation of oxygen and iron bimodal radial distribution in the disc of our Galaxy, MNRAS, 2010, vol. 402, 1149–1155.

1381. Acheson D.J., On over-reflexion, J. Fluid Mech., 1976, vol. 77, 433–472.

1382. Afanasiev V.L., Sil'chenko O.K., Complex Formation History of the Lenticular Galaxies with Stellar Counterrotation: NGC 4138 and NGC 4550, Astron. J., 2002, vol. 124, 706–721.

1383. Afanasiev V.L., Sil'chenko O.K., A lopsided chemically distinct nucleus in NGC 5055, Astron. Astrophys., 2002, vol. 388, 461–469.

1384. Agol E., Krolik J., Turner N.J., Stone J.M., Two-dimensional hydrodynamic simulations of convection in radiation-dominated accretion disks, Astrophys. J., 2001, vol. 558, 543–552.

Index

H

I